INTRODUCTORY ALGEBRA

ALAN S. TUSSY
CITRUS COLLEGE

DIANE R. KOENIG
ROCK VALLEY COLLEGE

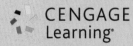

Australia • Brazil • Mexico • Singapore • United Kingdom • United States

*To the memory of my parents, **Jeanene** and **Bill Tussy**.*

AST

*In memory of my parents, **Darryl** and **Joan Vaupel**, who taught me to set my goals high and work hard to achieve them.*

DRK

Introductory Algebra, **Fifth Edition**
Alan S. Tussy, Diane R. Koenig

Product Manager: Marc Bove
Content Developer: Cynthia Ashton
Content Coordinator: Cynthia Ashton
Product Assistant: Kathryn Clark
Media Developer: Guanglei Zhang, Bryon Spencer
Brand Manager: Mark Linton
Market Development Manager: Danae April
Content Project Manager: Jennifer Risden
Art Director: Vernon Boes
Manufacturing Planner: Becky Cross
Rights Acquisitions Specialist: Don Schlotman
Production Service: Scratchgravel Publishing Services
Photo Researcher: PreMedia Global
Text Researcher: PreMedia Global
Copy Editor: Peggy Tropp
Illustrator: Lori Heckelman Illustration; Scratchgravel Publishing Services
Text Designer: Diane Beasley
Cover Designer: Terri Wright
Cover Image: Jill Fromer/Getty Images; Dana Hursey/Masterfile
Compositor: Graphic World, Inc.

© 2015, 2011 Cengage Learning

WCN: 01-100-101

ALL RIGHTS RESERVED. No part of this work covered by the copyright herein may be reproduced, transmitted, stored, or used in any form or by any means graphic, electronic, or mechanical, including but not limited to photocopying, recording, scanning, digitizing, taping, Web distribution, information networks, or information storage and retrieval systems, except as permitted under Section 107 or 108 of the 1976 United States Copyright Act, without the prior written permission of the publisher.

> For product information and technology assistance, contact us at
> **Cengage Learning Customer & Sales Support, 1-800-354-9706.**
> For permission to use material from this text or product,
> submit all requests online at **www.cengage.com/permissions**.
> Further permissions questions can be e-mailed to
> **permissionrequest@cengage.com**.

Library of Congress Control Number: 2013944901

ISBN-13: 978-1-285-42957-1

ISBN-10: 1-285-42957-5

Cengage Learning
200 First Stamford Place, 4th Floor
Stamford, CT 06902
USA

Cengage Learning is a leading provider of customized learning solutions with office locations around the globe, including Singapore, the United Kingdom, Australia, Mexico, Brazil, and Japan. Locate your local office at **www.cengage.com/global**.

Cengage Learning products are represented in Canada by Nelson Education, Ltd.

To learn more about Cengage Learning Solutions, visit **www.cengage.com**. Purchase any of our products at your local college store or at our preferred online store **www.cengagebrain.com**.

Printed in the United States of America
1 2 3 4 5 6 7 18 17 16 15 14

CONTENTS

Study Skills Workshop   S-1

CHAPTER 1

An Introduction to Algebra   1

- **1.1** The Language of Algebra   2
- **1.2** Fractions   13
- **THINK IT THROUGH** *Budgeting*   24
- **1.3** The Real Numbers   29
- **1.4** Adding Real Numbers; Properties of Addition   41
- **THINK IT THROUGH** *Calculating Sleep Debt*   47
- **1.5** Subtracting Real Numbers   51
- **1.6** Multiplying and Dividing Real Numbers; Multiplication and Division Properties   58
- **1.7** Exponents and Order of Operations   69
- **1.8** Algebraic Expressions   82
- **1.9** Simplifying Algebraic Expressions Using Properties of Real Numbers   96

    Chapter Summary and Review   106

    Chapter Test   116

CHAPTER 2

Equations, Inequalities, and Problem Solving   119

- **2.1** Solving Equations Using Properties of Equality   120
- **2.2** More about Solving Equations   130
- **2.3** Applications of Percent   140
- **THINK IT THROUGH** *Percent of Increases in College Costs*   145
- **2.4** Formulas   149
- **2.5** Problem Solving   163
- **2.6** More about Problem Solving   172
- **2.7** Solving Inequalities   183

    Chapter Summary and Review   197

    Chapter Test   206

    Cumulative Review   208

CHAPTER 3

Graphs, Linear Equations, and Inequalities in Two Variables; Functions   211

- **3.1** Graphing Using the Rectangular Coordinate System   212
- **3.2** Equations Containing Two Variables   222
- **3.3** Graphing Linear Equations   236
- **3.4** Rate of Change and the Slope of a Line   249
- **THINK IT THROUGH** *Average Rate of Tuition Increase*   254
- **3.5** Slope–Intercept Form   263
- **THINK IT THROUGH** *Prospects for a Teaching Career*   264
- **3.6** Point–Slope Form; Writing Linear Equations   271
- **3.7** Graphing Linear Inequalities   280
- **3.8** Functions   291
  - Chapter Summary and Review   305
  - Chapter Test   315
  - Cumulative Review   317

CHAPTER 4

Solving Systems of Equations and Inequalities   319

- **4.1** Solving Systems of Equations by Graphing   320
- **THINK IT THROUGH** *Bridging the Gender Gap in Education*   325
- **4.2** Solving Systems of Equations by Substitution   334
- **4.3** Solving Systems of Equations by Addition (Elimination)   343
- **4.4** Problem Solving Using Systems of Equations   352
- **THINK IT THROUGH** *College Students and Television Viewing*   355
- **4.5** Solving Systems of Linear Inequalities   364
  - Chapter Summary and Review   374
  - Chapter Test   381
  - Cumulative Review   383

CHAPTER 5

Exponents and Polynomials   385

5.1 Natural-Number Exponents; Rules for Exponents   386

THINK IT THROUGH   *PIN Code Choices*   388

5.2 Zero and Negative Integer Exponents   397

5.3 Scientific Notation   407

THINK IT THROUGH   *STEM Majors and Space Travel*   411

5.4 Polynomials   415

5.5 Adding and Subtracting Polynomials   422

5.6 Multiplying Polynomials   431

5.7 Dividing Polynomials by Monomials   442

5.8 Dividing Polynomials by Polynomials   448

Chapter Summary and Review   455

Chapter Test   461

Cumulative Review   463

CHAPTER 6

Factoring and Quadratic Equations   465

6.1 The Greatest Common Factor; Factoring by Grouping   466

6.2 Factoring Trinomials of the Form $x^2 + bx + c$ 474

6.3 Factoring Trinomials of the Form $ax^2 + bx + c$ 485

6.4 Factoring Perfect-Square Trinomials and the Difference of Two Squares   494

6.5 Factoring the Sum and Difference of Two Cubes   502

6.6 A Factoring Strategy   506

6.7 Solving Quadratic Equations by Factoring   511

6.8 Applications of Quadratic Equations   517

THINK IT THROUGH   *Pythagorean Triples*   520

Chapter Summary and Review   527

Chapter Test   534

Cumulative Review   536

CHAPTER 7

Rational Expressions and Equations 539

- **7.1** Simplifying Rational Expressions 540
- **7.2** Multiplying and Dividing Rational Expressions 550
- **7.3** Adding and Subtracting with Like Denominators; Least Common Denominators 559
- **7.4** Adding and Subtracting with Unlike Denominators 567
- **7.5** Simplifying Complex Fractions 575
- **7.6** Solving Rational Equations 584
- **7.7** Problem Solving Using Rational Equations 593
- **7.8** Proportions and Similar Triangles 599
- **THINK IT THROUGH** *Student Loan Calculations* 601
- **7.9** Variation 612
- **THINK IT THROUGH** *Study Time vs. Effectiveness* 618

  Chapter Summary and Review 622

  Chapter Test 632

  Cumulative Review 634

CHAPTER 8

Roots and Radicals 637

- **8.1** Square Roots 638
- **THINK IT THROUGH** *Traffic Accidents* 640
- **8.2** Higher-Order Roots; Radicands That Contain Variables 650
- **8.3** Simplifying Radical Expressions 657
- **8.4** Adding and Subtracting Radical Expressions 668
- **8.5** Multiplying and Dividing Radical Expressions 675
- **8.6** Solving Radical Equations; the Distance Formula 685
- **8.7** Rational Exponents 696

  Chapter Summary and Review 704

  Chapter Test 712

  Cumulative Review 714

CHAPTER 9

Quadratic Equations 719

9.1 Solving Quadratic Equations: The Square Root Property 720

9.2 Solving Quadratic Equations: Completing the Square 727

9.3 Solving Quadratic Equations: The Quadratic Formula 735

9.4 Graphing Quadratic Equations 746

9.5 Complex Numbers 759

Chapter Summary and Review 769

Chapter Test 777

Cumulative Review 779

APPENDIXES

Appendix 1   Statistics   A-1
Appendix 2   Roots and Powers   A-6
Appendix 3   Answers to Selected Exercises   A-7
(appears in Student Edition only)

Index   I-1

PREFACE

We are excited to present the Fifth Edition of *Introductory Algebra*. We believe the revision process has produced an even stronger instructional experience for both students and teachers. First, we have fine-tuned several of the popular features of our series, including the *Strategy* and *Why* example structure, the problem-solving strategy, and the real-life applications.

Second, we have expanded the quantity and types of problems available in Enhanced WebAssign® to include more vocabulary and application problems. In addition we have added a wider variety of *Guided Practice* and *Try It Yourself* problems.

We want to thank all of you across the country who provided suggestions and input about the previous edition. Your insights have proven invaluable. Throughout the revision process, our fundamental belief has remained the same: Algebra is a language in its own right. As always, the prime objective of this textbook (and its accompanying ancillaries) is to teach students how to read, write, speak, and think using the language of algebra.

NEW TO THIS EDITION

Six-Step Problem-Solving Strategy

In an effort to better describe the problem-solving strategy used in this book, we have inserted a new second step in what was formerly a five-step problem process. This additional step (*Assign a variable to represent an unknown value in the problem*) better delineates the thought process students should use as they solve application problems. The six steps of the problem-solving strategy are now: *Analyze the problem*, *Assign a variable*, *Form an equation*, *Solve the equation*, *State the conclusion*, and *Check the result*.

Extensively Updated Application Problems

Many real-life application problems have been updated to keep them current and relevant. Students can see how math relates to their everyday lives through these timely topics.

Additional *Try It Yourself* Problems

In response to reviewers' suggestions, we have inserted additional problems in the *Try It Yourself* sections of selected Study Sets.

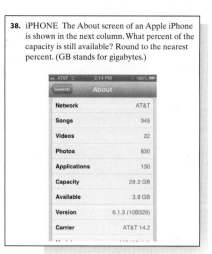

38. iPHONE The About screen of an Apple iPhone is shown in the next column. What percent of the capacity is still available? Round to the nearest percent. (GB stands for gigabytes.)

TRUSTED FEATURES

Emphasis on Study Skills

Introductory Algebra begins with a *Study Skills Workshop* module. Instead of simple, unrelated suggestions printed in the margins, this module contains one-page discussions of study skills topics, followed by a *Now Try This* section offering students actionable skills, assignments, and projects that will impact their study habits throughout the course.

Chapter Openers That Answer the Question: When Will I Use This?

Instructors are asked this question time and again by students. In response, we have written chapter openers called *From Campus to Careers*. This feature highlights vocations that require various algebraic skills. Designed to inspire career exploration, each includes job outlook, educational requirements, and annual earnings information. Careers presented in the openers are tied to an exercise found later in the *Study Sets*.

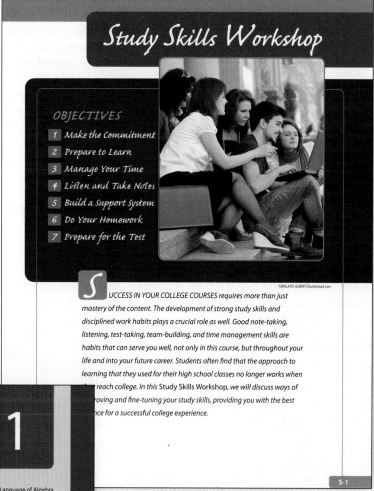

Examples That Tell Students Not Just How, But WHY

Why? That question is often asked by students as they watch their instructor solve problems in class and as they are working on problems at home. It's not enough to know *how* a problem is solved. Students gain a deeper understanding of the algebraic concepts if they know *why* a particular approach was taken. This instructional truth was the motivation for including a *Strategy* and a *Why* explanation in each worked example.

Examples That Offer Immediate Feedback

Each worked example includes a *Self Check*. These can be completed by students on their own or as classroom lecture examples, which is how Alan Tussy uses them. Alan asks selected students to read aloud the *Self Check* problems as he writes what the student says on the board. The other students, with their books open to that page, can quickly copy the *Self Check* problem to their notes. This speeds up the

Examples That Ask Students to Work Independently

Each worked example ends with a *Now Try* problem. These are the final steps in the learning process. Each one is linked to a similar problem found within the *Guided Practice* section of the *Study Sets*.

Emphasis on Problem Solving

New to *Introductory Algebra*, the six-step problem-solving strategy guides students through applied worked examples using the *Analyze, Assign, Form, Solve, State,* and *Check* process. This approach clarifies the thought process and mathematical skills necessary to solve a wide variety of problems. As a result, students' confidence is increased, and their problem-solving abilities are strengthened.

Detailed Author Notes

These notes (shown in red) appear to the right of the solution to every worked example and guide students step by step.

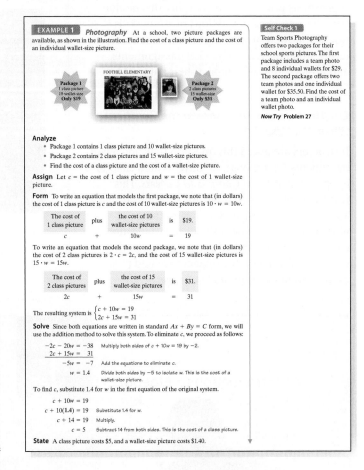

Integrated Focus on the Language of Algebra

Language of Algebra boxes draw connections between mathematical terms and everyday references to reinforce the language of algebra philosophy that runs throughout the text.

> **The Language of Algebra** The preposition *per* means for each, or for every. When we say the rate of change is 3 permits *per* month, we mean 3 permits for each month.

Guidance When Students Need It Most

Appearing at key teaching moments, *Success Tips* and *Caution* boxes improve students' problem-solving abilities, warn students of potential pitfalls, and increase clarity.

> **Success Tip** By the commutative property of multiplication, we can change the *order* of factors. By the associative property of multiplication, we can change the *grouping* of factors.

> **Caution!** Be careful when translating division. As with subtraction, order is important. For example, *s* divided by *d* is *not* written $\frac{d}{s}$.

Useful Objectives Help Keep Students Focused

Each section begins with a set of numbered *Objectives* that focus students' attention on the skills that they will learn. As each objective is discussed in the section, the number and heading reappear to remind readers of the objective at hand.

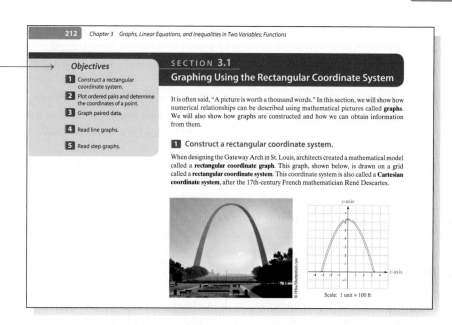

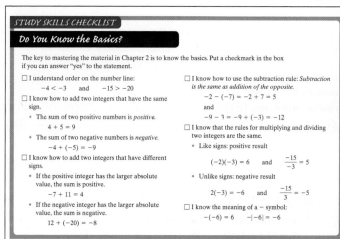

Study Skills That Point Out Common Student Mistakes

In Chapter 1 we have included four *Study Skills Checklists* designed to actively show students how to effectively use the key features in this text. Subsequent chapters include one checklist just before the *Chapter Summary and Review* that provides another layer of preparation to promote student success. These *Study Skills Checklists* warn students of common errors, giving them time to consider these pitfalls before taking their exam.

Study Sets

Study Sets are found in each section and offer a multifaceted approach to practicing and reinforcing the concepts taught in each section. They are designed for students to methodically build their knowledge of the section concepts, from basic recall to increasingly complex problem solving, through reading, writing, and thinking mathematically.

Vocabulary—Each *Study Set* begins with the important *Vocabulary* discussed in that section. The fill-in-the-blank vocabulary problems emphasize the main concepts taught in the chapter and provide the foundation for learning and communicating the language of algebra.

Concepts—In *Concepts*, students are asked about the specific subskills and procedures necessary to successfully complete the *Guided Practice* and *Try It Yourself* problems that follow. The *Concepts* problems build layers of understanding, using an instructional technique known as scaffolding.

Notation—In *Notation*, the students review the new symbols introduced in a section. Often they are asked to fill in steps of a sample solution. This strengthens their ability to read and write mathematics and prepares them for *Guided Practice* problems by modeling solution formats.

Guided Practice—The problems in *Guided Practice* are linked to a corresponding worked example or objective from that section. This feature promotes student success by referring them to the proper examples if they encounter difficulties solving homework problems.

Try It Yourself—To promote problem recognition, the *Try It Yourself* problems are thoroughly mixed and are *not* linked to worked examples, giving students an opportunity to practice decision making and strategy selection as they would when taking a test or quiz.

Applications—The *Applications* provide students the opportunity to apply their newly acquired algebraic skills to relevant and interesting real-life situations.

Writing—The *Writing* problems help students build mathematical communication skills.

Review—The *Review* problems consist of randomly selected problems from previous chapters. These problems are designed to keep students' successfully mastered skills up to date before they move on to the next section.

Using Your CALCULATOR

This is an optional feature that is designed for instructors who wish to use calculators as part of the instruction in this course. This feature introduces keystrokes and shows how scientific and graphing calculators can be used to solve problems. In the *Study Sets*, icons are used to denote problems that may be solved using a calculator.

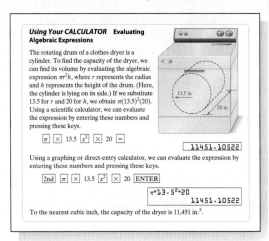

Think It Through

The *Think It Through* feature makes the connection between mathematics and student life. These relevant topics often require algebra skills from the chapter to be applied to a real-life situation. Topics include tuition costs, student enrollment, job opportunities, credit cards, and many more.

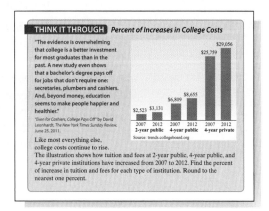

Comprehensive End-of-Chapter Summary with Integrated Chapter Review

The end-of-chapter material has been designed to function as a complete study guide for students. Chapter summaries include definitions, concepts, and examples—organized by section. Review problems for each section immediately follow the summary for that section. Students will find the detailed summaries a very valuable study aid when preparing for exams.

Chapter Tests

Appearing at the end of every chapter, Chapter Tests can be used as preparation for the class exam.

Cumulative Reviews

The *Cumulative Review* follows the end-of-chapter material and keeps students' skills current before they move on to the next chapter. For reference, each problem is linked to the associated section from which the problem came. The final *Cumulative Review* is often used by instructors as a Final Exam Review.

ANCILLARIES

For the Student	For the Instructor
	Annotated Instructor's Edition (ISBN: 978-1-285-42958-8) This special version of the complete student text contains all answers printed next to all respective exercises.
Student Solutions Manual (ISBN: 978-1-285-86065-7) Author: Rhoda Oden, Gadsden State Community College The Student Solutions Manual provides worked-out solutions to all of the odd-numbered exercises in the text.	**Complete Solutions Manual** (ISBN: 978-1-285-86066-4) Author: Rhoda Oden, Gadsden State Community College The Complete Solutions Manual provides worked-out solutions to all of the problems in the text.
Student Workbook (ISBN: 978-1-285-86068-8) Author: Maria H. Andersen, former math faculty at Muskegon Community College and now working in the learning software industry The Student Workbook contains the entire student Assessments, Activities, and Worksheets from the Instructor's Resource Binder for classroom discussions, in-class activities, and group work.	**Instructor's Resource Binder** (ISBN: 978-0-538-73675-6) Author: Maria H. Andersen, former math faculty at Muskegon Community College and now working in the learning software industry Each section of the main text is discussed in uniquely designed Teaching Guides containing instruction tips, examples, activities, worksheets, overheads, assessments, and solutions to all worksheets and activities.
	Instructor's Companion Website Everything you need for your course in one place! This collection of book-specific lecture and class tools is available online via www.cengage.com/login. Formerly found on the PowerLecture, access and download PowerPoint presentations, images, and more.
Enhanced WebAssign Homework with LOE Access (Printed Access Card ISBN: 978-1-285-85770-1, Online Access Code ISBN: 978-1-285-85773-2) Enhanced WebAssign (assigned by the instructor) provides you with instant feedback on homework assignments. This online homework system is easy to use and includes helpful links to textbook sections, video examples, and problem-specific tutorials.	**Enhanced WebAssign Homework with LOE Access** (Printed Access Card ISBN: 978-1-285-85770-1, Online Access Code ISBN: 978-1-285-85773-2) Exclusively from Cengage Learning, Enhanced WebAssign® combines the exceptional mathematics content that you know and love with immediate feedback, rich tutorial content, and interactive, fully customizable eBooks (YouBook), helping students to develop a deeper conceptual understanding of their subject matter. Online assignments can be built by selecting from thousands of text-specific problems or can be supplemented with problems from any Cengage Learning textbook.
	Solution Builder This online database offers complete worked solutions to all exercises in the text, allowing you to create customized, secure solutions printouts (in PDF format) matched exactly to the problems you assign in class. For more information, visit www.cengage.com/solutionbuilder.

For the Student	For the Instructor
Math Study Skills Workbook, 4e (ISBN: 978-0-8400-5309-1) Author: Paul D. Nolting This best-selling workbook helps traditionally unsuccessful students learn to effectively study mathematics. Typically used for a Math Study Skills course, Freshman Seminar, or as a supplement to class lectures, the Nolting workbook helps students identify their strengths, weaknesses, and personal learning styles in math. Nolting offers proven study tips, test-taking strategies, a homework system, and recommendations for reducing anxiety and improving grades.	**Math Study Skills Workbook, 4e** (ISBN: 978-0-8400-5309-1) Author: Paul D. Nolting This best-selling workbook helps traditionally unsuccessful students learn to effectively study mathematics. Typically used for a Math Study Skills course, Freshman Seminar, or as a supplement to class lectures, the Nolting workbook helps students identify their strengths, weaknesses, and personal learning styles in math. Nolting offers proven study tips, test-taking strategies, a homework system, and recommendations for reducing anxiety and improving grades.

ACKNOWLEDGMENTS

We want to express our gratitude to our accuracy checker, Brenda Keller, as well as to all those who helped with this project: Pat Foard, Rhoda Oden, Steve Odrich, Mary Lou Wogan, Paul McCombs, Maria H. Andersen, Sheila Pisa, Laurie McManus, Alexander Lee, Ed Kavanaugh, Karl Hunsicker, Cathy Gong, Dave Ryba, Terry Damron, Marion Hammond, Lin Humphrey, Doug Keebaugh, Robin Carter, Tanja Rinkel, Jeff Cleveland, Jo Morrison, Sheila White, Jim McClain, Paul Swatzel, and the Citrus College library staff (including Barbara Rugeley) for their help with this project. Your encouragement, suggestions, and insight have been invaluable to us.

We would also like to express our thanks to the Cengage Learning editorial, marketing, production, and design staff for helping us craft this new edition: Marc Bove, Danielle Derbenti, Mark Linton, Rita Lombard, Stefanie Beeck, Cynthia Ashton, Kathryn Clark, Heleny Wong, Jennifer Risden, Vernon Boes, Diane Beasley, Anne Draus, and Graphic World.

Additionally, we would like to say that authoring a textbook is a tremendous undertaking. A revision of this scale would not have been possible without the thoughtful feedback and support from the following colleagues listed below. Their contributions to this edition have shaped this revision in countless ways.

Alan S. Tussy
Diane R. Koenig

Reviewers

Tonya Adkins, *Johnson & Wales University*
Darla Aguilar, *Pima Community College*
Catherine Albright, *Brookdale Community College*
Patricia Barrientos, *El Paso Community College*
Gail Burkett, *Palm Beach State College*
Dorothy Carver, *Portland Community College*
Kristin Chatas, *Washtenaw Community College*
Seth Daugherty, *St. Louis Community College at Forest Park*
Mary Deas, *Johnson County Community College*
Steve Dostal, *College of the Desert*
Joe Estephan, *Victor Valley College*
Suzie Goss, *Lone Star College Kingwood*

Sheyleah Harris-Plant, *South Plains College*
Shelle Hartzel, *Lake Land College*
Steven Jackson, *Ivy Tech Community College*
Cynthia Johnson, *Arizona Western College*
Greta Johnson, *Amarillo College*
Jonathan Lee, *University of Oklahoma, Oklahoma City Community College*
Edith Lester, *Volunteer State Community College*
Susan McLoughlin, *Union County College*
Stacey Nichols, *Anne Arundel Community College*
Heather Niehoff, *Ivy Tech Community College*
Edward Pierce, *Ivy Tech Community College*
Jim Pierce, *Lincoln Land Community College*
Glenn Sandifer, *San Jacinto College, Central Campus*

Larry Small, *Pierce College*
Alexis Thurman, *County College of Morris*
Robyn Toman, *Anne Arundel Community College*
Diane Valade, *Piedmont Virginia Community College*
Betty Vix Weinberger, *Delgado Community College*
Lynn White, *Jones County Junior College*
Mary Wolyniak, *Broome Community College*

Reviewers of Previous Editions

Sheila Anderson, *Housatonic Community College*
Cedric E. Atkins, *Mott Community College*
William D. Barcus, *SUNY, Stony Brook*
David Behrman, *Somerset Community College*
Kathy Bernunzio, *Portland Community College*
Linda Bettie, *Western New Mexico University*
Michael Branstetter, *Hartnell College*
Joseph A. Bruno, Jr., *Community College of Allegheny County*
Girish Budhwar, *United Tribes Technical College*
Sharon Camner, *Pierce College–Fort Steilacoom*
Robin Carter, *Citrus College*
John Coburn, *Saint Louis Community College–Florissant Valley*
Eric Compton, *Brookdale Community College*
Joy Conner, *Tidewater Community College*
Sally Copeland, *Johnson County Community College*
Ann Corbeil, *Massasoit Community College*
Ben Cornelius, *Oregon Institute of Technology*
Ruth Dalrymple, *Saint Philip's College*
Nathalie Darden, *Brookdale Community College*
Carolyn Detmer, *Seminole Community College*
John D. Driscoll, *Middlesex Community College*
James Edmondson, *Santa Barbara Community College*
LaTonya Ellis, *Bishop State Community College*
David L. Fama, *Germanna Community College*
Steven Felzer, *Lenoir Community College*
Rhoderick Fleming, *Wake Technical Community College*
Maggie Flint, *Northeast State Technical Community College*
Charles Ford, *Shasta College*
J. Donato Fortin, *Johnson and Wales University*
Heather Gallacher, *Cleveland State University*
Barbara Gentry, *Parkland College*
Kathirave Giritharan, *John A. Logan College*
Marilyn Green, *Merritt College and Diablo Valley College*
Joseph Guiciardi, *Community College of Allegheny County*
Geoff Hagopian, *College of the Desert*
Deborah Hanus, *Brookhaven College*
A.T. Hayashi, *Oxnard College*
Michael Heeren, *Hamilton College*
Cheryl Hobneck, *Illinois Valley Community College*
Laurie Hoecherl, *Kishwaukee College*
Todd J. Hoff, *Wisconsin Indianhead Technical College*
Judith Jones, *Valencia Community College*
Therese Jones, *Amarillo College*
Joanne Juedes, *University of Wisconsin–Marathon County*
Susan Kautz, *Cy-Fair College*
Jack Keating, *Massasoit Community College*

Russ Alan Killingsworth, *Seattle Pacific University*
Dennis Kimzey, *Rogue Community College*
Monica C. Kurth, *Scott Community College*
Sally Leski, *Holyoke Community College*
Sandra Lofstock, *St. Petersberg College–Tarpon Springs Center*
Lynn Marecek, *Santa Ana College*
Lois Martin, *Massasoit Community College*
Mikal McDowell, *Cedar Valley College*
Chris Mirbaha, *The Community College of Baltimore County*
Elizabeth Morrison, *Valencia Community College*
Jan Alicia Nettler, *Holyoke Community College*
Marge Palaniuk, *United Tribes Technical College*
K. Maggie Pasqua, *Brookdale Community College*
Gregory Perkins, *Hartnell College*
Scott Perkins, *Lake-Sumter Community College*
Angela Peterson, *Portland Community College*
Euguenia Peterson, *City Colleges of Chicago–Richard Daley*
Jane Pinnow, *University of Wisconsin–Parkside*
Carol Ann Poore, *Hinds Community College*
Christopher Quarles, *Shoreline Community College*
George Reed, *Angelina College*
J. Doug Richey, *Northeast Texas Community College*
Patricia C. Rome, *Delgado Community College*
Patricia B. Roux, *Delgado Community College*
Rebecca Rozario, *Brookdale Community College*
Angelo Segalla, *Orange Coast College*
Eric Sims, *Art Institute of Dallas*
Lee Ann Spahr, *Durham Technical Community College*
Annette Squires, *Palomar College*
John Squires, *Cleveland State Community College*
John Strasser, *Scottsdale Community College*
June Strohm, *Pennsylvania State Community College–Dubois*
Rita Sturgeon, *San Bernardino Valley College*
Stuart Swain, *University of Maine at Machias*
Celeste M. Teluk, *D'Youville College*
Jo Anne Temple, *Texas Technical University*
Sharon Testone, *Onondaga Community College*
Bill Thompson, *Red Rocks Community College*
Barbara Tozzi, *Brookdale Community College*
Marilyn Treder, *Rochester Community College*
Sven Trenholm, *Herkeimer County Community College*
Donna Tupper, *Community College of Baltimore County–Essex*
Thomas Vanden Eynden, *Thomas More College*
Andreana Walker, *Calhoun Community College*
Jane Wampler, *Housatonic Community College*
Arminda Wey, *Brookdale Community College*
Stephen Whittle, *Augusta State University*
Mary Lou Wogan, *Klamath Community College*
Valerie Wright, *Central Piedmont Community College*
Kevin Yokoyama, *College of the Redwoods*
Mary Young, *Brookdale Community College*

ABOUT THE AUTHORS

Alan S. Tussy

Alan Tussy has taught all levels of developmental mathematics at Citrus College in Glendora, California. He has written nine math books—a paperback series and a hardcover series. A creative and visionary teacher who maintains a keen focus on his students' greatest challenges, Alan Tussy is an extraordinary author, dedicated to his students' success. Alan received his Bachelor of Science degree in Mathematics from the University of Redlands and his Master of Science degree in Applied Mathematics from California State University, Los Angeles. He has taught up and down the curriculum from Prealgebra to Differential Equations. He is currently focusing on the developmental math courses. Professor Tussy is a member of the American Mathematical Association of Two-Year Colleges.

Diane R. Koenig

Diane Koenig received a Bachelor of Science degree in Secondary Math Education from Illinois State University in 1980. She began her career at Rock Valley College in 1981, when she became the Math Supervisor for the newly formed Personalized Learning Center. Earning her Master's Degree in Applied Mathematics from Northern Illinois University, Ms. Koenig in 1984 had the distinction of becoming the first full-time woman mathematics faculty member at Rock Valley College. In addition to being nominated for AMATYC's Excellence in Teaching Award, Diane Koenig was chosen as the Rock Valley College Faculty of the Year by her peers in 2005, and, in 2006, she was awarded the NISOD Teaching Excellence Award as well as the Illinois Mathematics Association of Community Colleges Award for Teaching Excellence. In addition to her teaching, Ms. Koenig has been an active member of the Illinois Mathematics Association of Community Colleges (IMACC). As a member, she has served on the board of directors, on a state-level task force rewriting the course outlines for the developmental mathematics courses, and as the association's newsletter editor.

APPLICATIONS INDEX

Examples that are applications are shown with **boldface** page numbers.
Exercises that are applications are shown with lightface page numbers.

Agriculture. *See* **Farming**

Ancient history
archaeologist, **637**
archaeology, 667–668
classical Greek columns, 429
Greek architecture, 429
Maya civilization, 40
pyramids, 162

Animals
birds in flight, 598
birds of prey, 372
dog kennels, 598
dog shows, 81
dolphins, 421, 522
hamster habitats, 161
horses, 161
livestock auction, 463
pesticides, 373
pets, 206, 485
snails, 201
whales, **151**
zoology, 460

Architecture
architecture, 262, 621
classical Greek columns, 429
drafting, 39
Greek architecture, 429
igloos, 162
pyramids, 162

Arts and entertainment. *See* **Entertainment**

Business and industry
accounting, 68, 117, 169
advertising, **165**, 453, 610, 778
agents, 148
auctions, **165–166**
Avon products, 160
banking, 39, 158
bankruptcy, 610
business loans, 779
candy stores, 379
cattle auctions, 170
commemorative coins, 363
commissions, **144**, 206
computing a paycheck, 610
computing interest, 636
consignment, 148
copiers, 310
corporate investments, 180
counter space, 105, 196
credit, 160
currency exchange rates, 633
customer service, 523
data conversion, 170
depreciation, 278–279
discount coupons, 342
discounts, 200
Disney theme parks, **87**
down payments, 206
eBay, 146
endorsements, 379
entrepreneurs, 158
exports, 148
extra income, 180
Facebook, 11
furnace repair, 599
grand openings, 206
high-risk companies, 180
Hollywood, 158
home sales, 207
income, 9
income property, 270
insurance costs, 148
inventories, 183, 290
investing, 209
investments, 207
listing price, 170
loans, 180
making sporting goods, 290
manufacturing, 234–235, **356–357**, 745
market research, **275–276**
market value of a house, 235
maximizing revenue, 421
metal fabrication, 745
mixing candy, 383
office furniture, 494
online holiday sales, 715
online sales, 717
packaging, 420
PayPal, 146
Post Office, 68
price guarantees, 147
produce departments, 421
production planning, 289
profits and losses, 51
property values, **426–427**
quality control, 610
real estate, 67, 148
rentals, 183, 303
restaurants, 201
retailing, 745
room taxes, 147
salaries, **A-4**
sales, 599, **A-4**
savings accounts, 170, 657
savings bonds, 460
selling a home, 779
service charges, 195
shopping, 116, 201
signs, 205
stock exchange, 50
stock market, 463
student enrollments, **86–87**
supermarket displays, **418–419**
sweeps week, 80
television, 117
trade, 39
TV shopping, 148
U.S. jobs, 57
vacations, 170
Valentine's Day, 158
Wal-Mart, 262
warehousing, 183

Careers
air traffic controller, **319**
archaeologist, **637**
automotive service technician, **119**
dental assistant, 291
DOD, 414
earning money, **285–286**
elementary school teacher, **465**
employment service, 271
graphic designer, **719**, 744
job testing, **A-4**
physical therapist, **211**
physician assistant, **1**
police officer, **385**
prospects for a teaching career, **264**
recreation director, **539**
retirement years, 407
work schedules, 312

Clothing and footwear
clothes shopping, 382
clotheslines, 171
costume designs, 685
inventories, 290
men's shoe sizes, **274–275**
quality control, 610
sewing, 712
sewing costs, 270–271
shopping, 612
shopping for clothes, 610
sunglasses, 440

Collectibles
collectibles, **86**, 200
commemorative coins, 363
stamps, 440

Communication
advertising, 778
autographs, 204
chain letters, 80
children's stickers, 716
comics, 744
commercials, 147
communications, 453
data conversion, 170
dictionaries, 170
Facebook, 11
flags, **152–153**, 160, 524, 744
freeway signs, **25**
graphic design, 545
graphic designer, **719**, 744
hello/goodbye, 181
Internet usage, 200
iPhones, 117, 147
lighthouses, 726
mailing breakables, 505
naval operations, 430
phone bills, 448
road signs, 705
satellite orbits, 694–695
signal flags, 501
signing petitions, 202
signs, 205, 441
social networks, 129
sound systems, 202
stop signs, 129
talk radio, 148
telephones, 196
television, 117
TV coverage, 333
TV news, 780
TV programming, 361
TV towers, 171, 633
word processors, 545

Community life
awards, 207
child care, 147
cleaning highways, 633
community gardens, 744
crash investigations, 379
crime scenes, **167**
genealogy, 147
identity theft, **143**
New York City, 777
parenting, 181
pesticides, 373
redevelopment, 373
security gates, 772
sewage treatment, 598
sidewalks, 745
social networks, 129
St. Louis, 778
white-collar crime, **357–358**

Community service
fund-raising, 538
service clubs, 158
walk-a-thons, 113

xviii

Applications Index   xix

Construction. *See also* **Engineering**
accessibility, 261
architecture, 262
blueprints, 611
bridge construction, 220
carpentry, 169, 309, 713
carpeting, **555–556**
chair production, 11
classical Greek columns, 429
construction, 524, 533
construction costs, 453
counter space, 105, 196
deck design, 695
decking, 744
drafting, 39, 649
drainage, 261
engineering, 171
grade of a road, 261
height of a bridge, 691–692
height of a building, 611
highway designs, 695
house construction, 525
insulation, 523
kitchen floor plans, 744
making gutters, 745
pool design, 260
reading blueprints, 429
roof design, **645**
roofing, 318, 545, 782
roofing houses, 597
scale models, **606**
staircase production, 11
suspension bridges, 421
swimming pool borders, 525
swimming pools, 170
value of a house, 429
value of two houses, 429
waste disposal, 169
windows, 196

Demographics
aging populations, **141–142**
demographics, 333
genealogy, 147
U.S. life expectancy, 148
world population, 457

Education
average rate of tuition increase, **254**
bachelor's degrees, 376
college enrollments, 306
college fees, 270
college students and television viewing, **355**
credit, 160
DMV written test, 147
education, 351
education costs, 247
elementary school teacher, **465**
field trips, 169
gender gap, **325**
grades, **192,** 195
grading, **A-2**
grading papers, 598
graduations, 196
high school sports, 342
loans, 158
occupational testing, 195
paying tuition, **172–173**
percent of increase in college costs, **145**
prospects for a teaching career, **264**
STEM majors and space travel, **411**
student enrollments, **86–87**
student loan calculations, **601**
student loans, 362
study time *vs.* effectiveness, **618**
test scores, 147
tutoring, 170

Electricity
electric bills, 448
electricity, 162, 461, 620, 767

Electronics and computers. *See also* **Technology**
computer drafting, 271
computer memory, 147
computers, 397
electronics, 407, 583, 767
HDTVs, 160
keyboards, 12
supercomputers, 414
value of a computer, 429
value of two computers, 430
word processors, 545

Energy
automation, 280
CFL bulbs, 351
choosing a furnace, 362
electric bills, 448
electricity, 461, 620, 767
electronics, 407
energy conservation, 95
energy savings, 159
energy usage, 81
filling an oil tank, **593–594**
fuel efficiency, 12, **A-5**
gas consumption, 610
gas mileage, 195, 221, **A-4**
irrigation, 261
light, 80
mixing fuel, 611
New York City, 777
oil reserves, 414
power output, 783
power usage, 694
solar heating, 169
speed of light, 461
ticket sales, 290
water management, **55**
windmills, 656, 695

Engineering. *See also* **Construction**
blueprints, 674
bridge construction, 220
carpentry, 309
computer drafting, 271
drafting, 649
drainage, 261
engineering, 171
freeway design, 279
height of a bridge, 691–692
highway designs, 695
model railroads, 611
pool design, 260
reading blueprints, 429
scale models, **606**
suspension bridges, 234, 421
tunnels, 421

Entertainment. *See also* **Sports**
amusement park rides, 667
art, 170, 362
art history, 396
ATVs, 535
autographs, 204
awards platforms, 684
ballooning, 533
billiards, 105, 234, 524
birthday parties, **243**
boating, 222, **358–359,** 524, 598, 633, 715
bouncing balls, 396
building a high ropes adventure course, **644**
buying CDs and DVDs, **369**
buying compact discs, 373
buying tickets, 361
camping, 201
card games, 50–51
carousels, 734
celebrity birthdays, 170
choreography, 523
cliff divers, 770
collectibles, **86**
college students and television viewing, **355**
comics, 744
concert tours, **175**
concerts, 169
crafts, 523
crossword puzzles, 758
cycling, 598
daredevils, 726
decorations, 169
designing a tent, 524
dimensions of a painting, **437–438**
dining area improvements, **178–179**
dining out, 146
Disney theme parks, **87**
diving, 407, 458
dog shows, 81
endorsements, 379
entertaining, 530
exhibition diving, 522
Ferris wheels, 710
field trips, 169
filling a pool, 597, 717
films, **149–150**
five-card poker, 781
flutes, 115
game shows, **46**
games, 220
gift shopping, 380
graduations, 196
Grand Canyon, 772
Halloween, 202
HDTVs, 160
hip-hop, 610
holiday decorating, 702
Hollywood, 158
hot-air balloons, 158, 656
hotel reservations, **77**
hula hoops, 40
investment clubs, 384
jewelry, 201
keyboards, 12
kites, 160
limo service, **266–267**
lottery winnings, 202
markup, 363
mixing perfume, 610
mountain bicycles, 171
movie losses, 50
movie stunts, 501, **724**
museum tours, 183
musical instruments, 464
NASCAR, 57, 202, 523
national parks, 169
New York City, 170
number puzzles, 196
online holiday sales, 715
organ pipes, 545, 649
paper airplanes, 648
parades, 158, 525
photo enlargement, 611
photographic chemicals, 537
photography, 40, 182, 353–354, 485
picture frames, **643**
piñatas, 430
Ping-Pong, 105
playground equipment, 674
pogo sticks, 633
production costs, 270
projector screens, 685
racing, 57, 383
recreation director, **539**
restaurants, 201
river tours, 598–599
rocketry, **90–91**
roller coasters, 726
rolling dice, 290
Rubik's cube, 770
sailing, **740**
sales, 599
scale models, **606**
Scrabble, 81
search and rescue, 714
seesaws, 621
set designs, 685
sewing, 105
sharing the winning ticket, 129
skateboarding, 162
ski runs, 612
sound systems, 202

speakers, 701
sport fishing, **A-4**
sweeps week, 80
swimming pools, 170
synthesizers, 129
talk radio, 148
television, 117, 207, 648, **741**
tents, 674
theater screens, 362
theater seating, 705
thrill rides, 525
tightrope walkers, 533
touring, 169
toy designs, 702
trampolines, 280, 758
trumpet mutes, 95
TV coverage, 333
TV history, 170
TV news, 780
TV programming, 361
TV sales, 148
TV shopping, 148
TV shows, **216**
TV towers, 171, 633
vacations, 170, 382, 464
video cameras, 626
walk-a-thons, 113
walking and bicycling, 204
water balloons, 421
water skiing, 430
wedding pictures, 361
wishing wells, 183
Wizard of Oz, 649
wrapping gifts, 81
yo-yos, 161
YouTube video contest, 81

Farming
cattle auctions, 170
farming, 620
growing sod, 94
milk production, 262

Finance
accounting, 68, 117, 169
advertising, **165,** 610
agents, 148
art, 362
auctions, 148, **165–166**
auto insurance, 80
average rate of tuition increase, **254**
Avon products, 160
banking, 39, 50, 158, **A-1–A-2**
bankruptcy, 610
boating, 222
bronze, 182
budgeting, **24**
business loans, 779
buying boats, 373
buying CDs and DVDs, **369**
buying compact discs, 373
buying furniture, 373
buying paint supplies, 361
buying tickets, 361
candy stores, 379

car registration, 311
cattle auctions, 170
certificates of deposit, 201
CFL bulbs, 351
choosing a furnace, 362
clothes shopping, 382
coffee sales, 363
college fees, 270
commemorative coins, 363
commissions, **144,** 206
comparing interest rates, 599
comparing investments, **595–596,** 599
computing a paycheck, 610
computing interest, 636
concerts, 169
consignment, 148
construction costs, 453
corporate investments, 180
cost analysis, 758
cost of living, 200
credit, 160
credit cards, 50
currency, 414
currency exchange rates, 633
data conversion, 170
depreciation, 9, 261, 278–279, 316, 621, 657
dining, 342
discount coupons, 342
discounts, 148, 200
down payments, 206
earning money, **285–286**
eBay, 146
economics, 333
education costs, 247
employment service, 271
endorsements, 379
energy savings, 159
entrepreneurs, 158
Executive branch, 361
extra income, 180
family budgets, 200
field trips, 169
films, **149–150**
financial planning, 362, 382, 715
fund-raising, 538
gift shopping, 380
graduations, 196
grand openings, 206
grocery shopping, **605,** 610
group rates, 247–248
health club discounts, **144**
high-risk companies, 180
hiring babysitters, 290
Hollywood, 158
home sales, 207
house cleaning, 629
income, 9
income property, 270
income taxes, 202
insurance costs, 148
insured deposits, 414
inventories, 290
investing, 209, 290, 379, 397, 745
investment, 318

investment clubs, 384
investment income, 204
investment plans, 181
investments, 181, 207, 629
iPads, 270
landscaping, **369**
listing price, 170
livestock auction, 463
loans, 158, 180
loose change, 779
lottery winnings, 202
making sporting goods, 290
making tires, 362
manufacturing, **356–357**
markdown, 363
market research, **275–276**
market value of a house, 235
markup, 363
maximizing revenue, 421
mixing candy, 318, 383, 781
mixing nuts, 363
mixtures, 204
movie losses, 50
museum tours, 183
new cars, 158
old coins, 180
online holiday sales, 715
online sales, 717
ordering furnace equipment, 373
paying tuition, **172–173**
PayPal, 146
pennies, 317
percent of increase in college costs, **145**
personal loans, 181
photography, 353–354
piggy banks, 183
PIN code choices, **388**
Post Office, 68
price guarantees, 147
printing presses, 271
production costs, 270
production planning, 289
profits and losses, 51
property values, **426–427**
prospects for a teaching career, **264**
Queen Mary, 68
raising a family, 280
real estate, 148
rentals, 183
restaurants, 201
retailing, 745
retirement, 180
retirement income, **150**
retirement years, 407
room taxes, 147
salad bar, 270
salaries, **A-4**
sales, 148, 599, **A-4**
salvage value, 311
savings, 158
savings accounts, 170, 657
savings bonds, 460
selling a home, 779
selling ice cream, 362

service charges, 195
service clubs, 158
sewing costs, 270–271
shopping, 116, 201, 379, 612
shopping for clothes, 610
snack foods, 182
stock exchange, 50
stock market, 463
stockbrokers, 148
student loan calculations, **601**
student loans, 362
sunscreen, 382
supermarket displays, **418–419**
tax tables, 146
taxes, **142**
1099 forms, 180
ticket sales, 290
trade, 39
truck mechanics, 169
Tupperware, 200
TV sales, 148
TV shopping, 148
U.S. budget, 39, 146
vacations, 382, 464
Valentine's Day, 158
value of a computer, 429
value of a house, 429
value of two computers, 430
value of two houses, 429
Wal-Mart, 262
walk-a-thons, 113
warehousing, 183
wedding pictures, 361
white-collar crime, **357–358**
wishing wells, 183

Food and beverages. *See also* **Health and nutrition**
baking, **606–607**
beverages, 333
blending coffees, 714
blending tea, 634
cider, 620
coffee blends, 182
coffee sales, 363
cooking, 28, 162, 474, 597, 610, 636
counting calories, 169
dining, 342
energy drinks, 714
frying foods, 159
got milk, 279
making cheese, 182
milk, 204
milk production, 262
mixing candy, 182, 781
mixing nuts, 363
mixtures, 204
nutrition, 384, 611, **A-5**
packaged salad, 182
packaging fruit, 598
pickles, 207
produce departments, 421
production planning, 289
raisins, 182
salad bar, 270

selling ice cream, 362
snack foods, **177–178,** 182
stone-ground flour, 441
tea, 207
watermelons, 536
world's coldest ice cream, 57

Games and toys. *See also*
Entertainment
air hockey, 684
ballistics, 303
billiards, 105, 234, 524
card games, 50–51
checkerboards, 447
checkers, 501, 534
chess, 726
crossword puzzles, 758
darts, 501
five-card poker, 781
games, 220
lottery winnings, 202
number puzzles, 196
paper airplanes, 648
piñatas, 430
Ping-Pong, 105
playground equipment, 674
playpens, 538
pogo sticks, 633
poker, 635
pool, 447
rocketry, 535
rolling dice, 290
Rubik's cube, 770
Scrabble, 81
sharing the winning ticket, 129
table tennis, 234
toy designs, 702
toys, 440

Gardening and lawn care
community gardens, 744
environmental protection, 668
fertilizer, 182
finding the height of a tree,
  607–608
gardening, **355–356,** 440, 533, 648
gardening tools, 524, 583
geometry, 362
groundskeeping, 597
height of a tree, 611, 636
herb gardens, 667
irrigation, 261
landscaping, 95, **369,** 534
lawn care, 354
lawn seed, 182
lawn sprinklers, 303
lawnmowers, 684
mixing fuel, 611
pesticides, 373
tree trimming, 361

Geography
Alaska, 361
astronomy, 111
California coastline, **163–165**
Colorado, 207

elevations, 379
finding location, 648
flags, **152–153,** 160, 524, 744
geography, 57, 111, 161, 221, 413
Grand Canyon, 772
Gulf Stream, 363
Hawaii, 305
jet stream, 363
land elevations, 56
landmarks, 220
latitude and longitude, 332
national parks, 169
New York City, 170
Niagara Falls, 694
Sahara Desert, 50
Sunshine State, 57
Washington Monument, 694

Geometry
complementary angles, 171
concentric circles, 636
construction, 524
dimensions of a triangle, 524
geometry, 39, 196, 203, 207, 342,
  362, 620, 649, 695, 716, 772
geometry tools, 536
height of a triangle, 649, 743, 782
isosceles triangles, **167–168**
perimeter of a rectangle, **519–520**
perimeter of a square, 649
right triangles, **521**
supplementary angles, 171
triangles, 171

Government
awards, 207
daily tracking polls, 332
DOD, 414
Executive branch, 361
flags, **152–153,** 160
foreign policy, 57
health department, 758
Marine Corps, 362
military science, 49
NATO, 289
political surveys, 235
politics, 50, 51
redevelopment, 373
signing petitions, 202
taxes, **142**
U.S. budget, 39

Hardware
auto mechanics, 429
blueprints, 674
car radiators, 68
cutting steel, 635
depreciation, 9, 621
drafting, 39
engineering, 171
exercise equipment, 702
frames, 28
gardening tools, 524, 583
gears, 782
geometry tools, 536
hard drives, 621

hardware, 28, 524, 674
lawnmowers, 684
locks, 170
making gutters, 745
manufacturing, 234–235
metal fabrication, 745
organ pipes, 545
parts, 303
pendulums, 694
playground equipment, 674
power tools, 106
pulley designs, 441
robots, 169
speakers, 701
tools, 95

Health and nutrition. *See also*
Medicine and health
accessibility, 261
blending coffees, 714
calculating sleep debt, **47**
calories, 28, 169
childbirth, 396
commercials, 147
counseling, 279
CPR, 610
dentistry, 147
energy drinks, 714
exercise, 196, 629
exercise equipment, 702
food labels, 148
graphic design, 545
growth rates, 259
gymnastics, 383
health club discounts, **144**
health department, 758
health risks, 633
high school sports, 342
losing weight, 782
medical questionnaires, 49
nutrition, 147, 384, 611, **A-5**
physical fitness, 310, 598
physical therapist, **211**
sleep debt, **47**
sunscreen, 382
treadmills, 261

History
American Red Cross, 105
archaeologist, **637**
archaeology, 667–668
art history, 203, 396
bachelor's degrees, 376
commemorative coins, 363
costume designs, 685
daredevils, 726
foreign policy, 57
history, 57, 523, 537
Marine Corps, 362
Maya civilization, 40
memorials, 160
NATO, 289
old coins, 180
presidents, 523
pyramids, 162
quilts, 783

science history, 726
supercomputers, 414
1099 forms, 180
TV history, 170
U.S. history, **166,** 289
world history, 161
world population, 457

Home management
adjusting a ladder, 648
air conditioning, 447, 463
antifreeze, 95
antiseptics, 146, 182
appliances, 459
baking, **606–607**
birthday parties, **243**
blueprints, 611
buying furniture, 373
buying paint supplies, 361
carpeting, **555–556**
CFL bulbs, 351
chair production, 11
child care, 147, 382
childbirth, 396
choosing a furnace, 362
clotheslines, 171
cooking, 28, 162, 474, 597, 610
curling irons, 430
deck design, 695
decking, 744
decorating, 28
decorations, 169
dining area improvements,
  178–179
dog kennels, 598
draining a tank, 636
electric bills, 448
entertaining, 530
fencing, 674
filling pools, 597
firewood, 161
flood damage, 182
floor mats, 118
food preparation, **85**
frames, 28
furnace filters, 453
furnace repair, 599
furniture, 524
grand king size beds, 634
grocery shopping, **605,** 610
groundskeeping, 597
hiring babysitters, 290
holiday decorating, 597, 702
house cleaning, 629
ice chests, 462
igloos, 162
insulation, 523
interior decorating, 182, 649
irons, 206
kitchen floor plans, 744
landscaping, 95, **369**
locks, 170
mailing breakables, 505
making gutters, 745
miniblinds, 448
ordering furnace equipment, 373

painting, **86,** 441
painting equipment, 379
painting houses, 629
phone bills, 448
picture frames, 473, **643**
playpens, 538
quilts, 783
raising a family, 280
reach of a ladder, **644–645**
roofing houses, 597
room temperatures, 196
rugs, 159
sewage treatment, 598
sewing, 105, 712
sewing costs, 270–271
staircase production, 11
teepees, 161
Tupperware, 200
unit comparisons, 406
waste disposal, 169
wedding pictures, 361
wrapping gifts, 81

Law enforcement
crash investigations, 379
crime scenes, **167**
identity theft, **143**
job testing, **A-4**
lie detectors, 56
locks, 170
police officer, **385**
white-collar crime, **357–358**

Mathematics
abacus, 745
concentric circles, 636
confirming formulas, 448
decimal numeration system, 406
dimensions of a triangle, 524
finding the median, **A-3**
finding the mode, **A-3**
geometry, 39, 203, 342, 362, 620, 649, 695, 716, 772
geometry tools, 536
height of a triangle, 649, 743, 782
integer problem, 441
isosceles triangles, **167–168**
mathematical models, 302
number problems, **593,** 629
perimeter of a rectangle, **519–520**
perimeter of a square, 649
plotting points, 523
Pythagorean triples, **520**
right triangles, **521**
surface area, 714
volume of a pyramid, 525

Measurement. *See* **Weights and measures**

Medicine and health. *See also* **Health and nutrition**
anatomy, 674
causes of death, 361
childbirth, 396
CPR, 610

dental assistant, 291
dentistry, 147, 221, 630, 711
dosages, 620
exercise, 196, 629
eyesight, 57
first aid, 171
forensic medicine, 522
growth rates, 259
health department, 758
health risks, 633
medical technology, **359**
medical tests, 701
medication, 318, 620
medicine, 221, 592
nutrition, 611
physical fitness, 598
physical therapist, **211**
physical therapy, 362
physiology, 248, **A-1**
sunscreen, 382
U.S. life expectancy, 148
viruses, **A-4**

Miscellaneous
aluminum cans, 170
angle of elevation, 463
backpacks, 714
bulletin boards, 524
childbirth, 396
children's stickers, 716
copiers, 310
designing a tent, 524
dictionaries, 170
draining a tank, 636
filling an oil tank, **593–594**
finding the height of a tree, **607–608**
food preparation, **85**
gender gap in education, **325**
genealogy, 147
holiday decorating, 597
house cleaning, 629
ice chests, 462
identity theft, **143**
IQ tests, 407
lie detectors, 56
luggage, 440
mailing breakables, 505
maps, 220
melting ice, 505
mixing perfume, 610
office furniture, 494
packaging, 396, 420, 656
packaging fruit, 598
parenting, 181
photo enlargement, 611
picture framing, 473
playpens, 538
projector screens, 685
quality control, 610
reach of a ladder, **644–645**
rubber bands, 159
sewage treatment, 598
storage, 494
storage cubes, 783
straws, 159

tape measures, 597
tubing, 525
visibility, 702

Natural resources
aluminum cans, 170
bronze, 182
environmental protection, 668
gas mileage, 195
geography, 161
geology, **614–615**
goldsmithing, 209
national parks, 169
Niagara Falls, 694
oil reserves, 414
properties of water, 160
solar heating, 169
tides, 303
tornadoes, 181
toxic cleanup, 278
water heaters, 316
water management, **55**
water pressure, 220
width of a river, 611
windmills, 656, 695

Reading and language
autographs, 204
bookbinding, 441
dictionaries, 170
hello/goodbye, 181
international alphabet, 558
reading, 207

Recordkeeping
accounting, 68, 117, 169
computer drafting, 271
computer spreadsheets, 68
data analysis, 583
drafting, 649
earned run average, 583
maps, 220
spreadsheets, 81

Safety
bicycle safety, 734
CPR, 610
driving safety, 535, 779
earthquakes, 744
fireboats, 234
first aid, 171
lighthouses, 726
locks, 170
rescues at sea, **174**
road safety, 694
search and rescue, 714
security gates, 772
traffic accidents, **640**
traffic safety, 12

Science
antifreeze, 182
antiseptics, 146, 182
archaeologist, **637**
archaeology, 667–668
astronomy, 67, 111, **410–411,** 413, 657, 716

atoms, **411,** 413, 457
bacteria, 407
bacterial growth, 67
ballistics, 303
biology, 159, 407, 461, 758
botany, 28
chemistry, 592, **617**
chemistry experiment, 270
computing pressures, 621
converting temperatures, 279
data analysis, 583
Earth science, 372
earthquakes, 744
Earth's atmosphere, 314
electricity, 162, 461, 620, 767
electronics, 583, 767
escape velocity, 734
falling objects, 744
flood damage, 182
fluid temperature, 67
forensic medicine, 522
free fall, 462
freezing points, 159
genetics, 501
geography, 57, 111, 161
geology, **614–615**
goldsmithing, 209
gravity, 620
Hooke's law, 621
light, 68, 80
light years, 414
lunar gravity, 620
magnification, 112
making cheese, 182
marine biology, 363
Mars exploration, 414
medicine, 592
melting ice, 505
metallurgy, 159
military science, 49
mixing solutions, **176–177**
oil reserves, 414
optics, 592
pendulums, 694
photographic chemicals, 537
physics, 57, 68, 501, 558
power usage, 694
properties of water, 160
protons, 414
reflections, 302
research experiments, 248
room temperatures, 196
salt solutions, 182
satellite orbits, 694–695
science history, 726
speed of light, 461, **556**
speed of sound, 413, 414
STEM majors and space travel, **411**
studying microgravity, 726
temperature change, 67
temperature conversion, **89**
temperature on Mars, 95
thermodynamics, 162
tides, 303
time of flight, 522

water pressure, 220
wavelengths, 413
wind speed, 629
work, 209
world's coldest ice cream, 57
zoology, 460

Sports
agents, 148
air hockey, 684
archery, 40, 161, 777
awards platforms, 684
badminton, 734
baseball, 441, 523, 648
basketball, 183
bulls-eye, 161
cliff divers, 770
competitive swimming, **85**
cross-training, 181
cycling, 181, 598
diving, 407, 458
earned run average, 583
exercise equipment, 702
exhibition diving, 522
filling a pool, 717
golf, 49–50, 220
gymnastics, 383
high school sports, 342
ice skating, 209
lightning bolt, 207
markdown, 363
mathematical models, 302
mountain bicycles, 171
NASCAR, 57, 202, 523
officiating, 522
Olympics, 50
packaging, 396
platform diving, 303
pole vaulting, 279
pool design, 260
pro wrestling, 726
racing, 57, 383
racing programs, 147
sailing, **740**
skateboarding, 162
soccer, 170
softball, 161, 235, **517–518**
sport fishing, **A-4**
sports equipment, 205
swimming, 158
swimming pool borders, 525
table tennis, 234
tennis, 170
ticket sales, 290
track and field, **594–595**, 758
treadmills, 261
triathlons, 118
walk-a-thons, 113
walking and bicycling, 204
water skiing, 430
women's tennis, **518–519**
wrestling, 649

Technology
automation, 280
computer memory, 147

computers, 397
cutting steel, 635
electronics, 407, 583, 767
Facebook, 11
hard drives, 621
HDTVs, 160
improving horsepower, 206
iPads, 270
iPhones, 117, 147
magnification, 112
medical technology, **359**
model railroads, 611
New York City, 777
parts, 303
power output, 783
power tools, 106
printing, 182
pulleys, 162
quality control, 116
robots, 169
rocketry, **90–91**
satellite orbits, 694–695
skyscrapers, 726
software, 183
sound systems, 202
speakers, 701
St. Louis, 778
STEM majors and space travel, **411**
supercomputers, 414
synthesizers, 129
telephones, 196
television, 207, 648, **741**
TV coverage, 333
TV history, 170
TV sales, 148
TV towers, 171, 633
value of a computer, 429
value of two computers, 430
video cameras, 626
wheelchairs, 161
word processors, 545
YouTube video contest, 81

Travel and transportation
air traffic control, 181
air traffic controller, **319**
air travel, 158
aircraft carriers, 474
airplanes, 204
Alaska, 361
angle of elevation, 463
antifreeze, 95, 182, 379, 382
ATVs, 535
auto insurance, 80
auto mechanics, 429
automotive service technician, **119**
aviation, 363, 780
ballooning, 533
bicycle safety, 734
boat travel, 599
boating, 222, **358–359**, 363, 379, 524, 598, 633, 715
buying boats, 373
California coastline, **163–165**

campers, 160
camping, 201
car radiators, 68
car registration, 311
cargo space, 717
child care, 382
cleaning highways, 633
commercial jets, 262
comparing travel, 598
concert tours, **175**
crash investigations, 379
delivery trucks, 108
DMV written test, 147
driver's licenses, 611
driving, 620
driving safety, 535, 779
engine output, 262
escape velocity, 734
Ferris wheels, 710
filling an oil tank, **593–594**
fireboats, 234
flight paths, 612, 633
floor mats, 118
freeway design, 279
freeway signs, **25**
fuel efficiency, 12, **A-5**
gas consumption, 610
gas mileage, 195, 221, **A-4**
gears, 782
grade of a road, 261
group rates, 247–248
Gulf Stream, 363
highway designs, 695
hot-air balloons, 158, 656
hotel reservations, 77
improving horsepower, 206
jet stream, 363
jets, 430
license plates, 636
lighthouses, 726
limo service, **266–267**
luggage, 440
making tires, 362
maps, 220
Mars exploration, 414
mixing fuel, 611
model railroads, 611
muscle cars, 598
NASCAR, 57, 202, 523
national parks, 169
naval operations, 430
navigation, 695, 745
new cars, 158
off roading, 343
Queen Mary, 68
racing, 57
racing programs, 147
rearview mirrors, 473
rentals, 303
rescues at sea, **174**
river tours, 598–599
road safety, 694
road signs, 705
road trips, 181, 318
room taxes, 147
sailing, **740**

salvage value, 311
shipping pallets, 524
shortcuts, 648
signal flags, 501
signs, 441
ski runs, 612
speed limits, **230**
speed of a plane, 634
speed of light, **556**
speed of trains, 182
STEM majors and space travel, **411**
stopping distance, 421
submarines, 50
tailwinds/headwinds, 382
time of flight, 522
tires, 40, 161
touring, 169
tourism, 309
traffic accidents, **640**
traffic safety, 12
travel times, 207, 598
traveling, 620
truck mechanics, 169
trucking, 203
unmanned aircraft, 181
value of a car, 221
visibility, 702
wheelchairs, 161
wind speed, 598, 629
winter driving, 181

Urban life
New York City, 777
security gates, 772
sidewalks, 745
skyscrapers, 726
St. Louis, 778

U.S. history
aging populations, **141–142**
American Red Cross, 105
bachelor's degrees, 376
currency, 414
DOD, 414
exports, 148
flags, **152–153**
foreign policy, 57
history, 523, 537
insured deposits, 414
Marine Corps, 362
national parks, 169
presidents, 523
St. Louis, 778
tax tables, 146
U.S. federal budget, 146
U.S. history, **166,** 289
U.S. life expectancy, 148
Washington Monument, 694

Weather
converting temperatures, 279
flood damage, 182
snowfall, 305
temperature records, 56
tornadoes, 181, 332

U.S. temperatures, **54–55**
water heaters, 316

Weights and measures
angstrom, 413
awards platforms, 684
backpacks, 714
draining a tank, 636
finding location, 648
finding the height of a tree, **607–608**
freezing points, 159
frying foods, 159
gauges, 58
height of a bridge, 691–692
height of a building, 611
height of a tree, 611, 636
height of a triangle, 649, 743, 782
keyboards, 12
length of a meter, 413
light years, 414
losing weight, 782
perimeter of a square, 649
period of a pendulum, **641**
pulleys, 162
reach of a ladder, **644–645**
speed of sound, 414
stop signs, 129
surface area, 714
tape measures, 597
temperature conversion, **89**
tires, 40
wavelengths, 413
width of a river, 611

Study Skills Workshop

OBJECTIVES

1. Make the Commitment
2. Prepare to Learn
3. Manage Your Time
4. Listen and Take Notes
5. Build a Support System
6. Do Your Homework
7. Prepare for the Test

YURALAITS ALBERT/Shutterstock.com

SUCCESS IN YOUR COLLEGE COURSES requires more than just mastery of the content. The development of strong study skills and disciplined work habits plays a crucial role as well. Good note-taking, listening, test-taking, team-building, and time management skills are habits that can serve you well, not only in this course, but throughout your life and into your future career. Students often find that the approach to learning that they used for their high school classes no longer works when they reach college. In this Study Skills Workshop, we will discuss ways of improving and fine-tuning your study skills, providing you with the best chance for a successful college experience.

1 Make the Commitment

YURALAITS ALBERT/Shutterstock.com

Starting a new course is exciting, but it also may be a little frightening. Like any new opportunity, in order to be successful, it will require a commitment of both time and resources. You can decrease the anxiety of this commitment by having a plan to deal with these added responsibilities.

Set Your Goals for the Course. Explore the reasons why you are taking this course. What do you hope to gain upon completion? Is this course a prerequisite for further study in mathematics? Maybe you need to complete this course in order to begin taking coursework related to your field of study. No matter what your reasons, setting goals for yourself will increase your chances of success. Establish your ultimate goal and then break it down into a series of smaller goals; it is easier to achieve a series of short-term goals rather than focusing on one larger goal.

Keep a Positive Attitude. Since your level of effort is significantly influenced by your attitude, strive to maintain a positive mental outlook throughout the class. From time to time, remind yourself of the ways in which you will benefit from passing the course. Overcome feelings of stress or math anxiety with extra preparation, campus support services, and activities you enjoy. When you accomplish short-term goals such as studying for a specific period of time, learning a difficult concept, or completing a homework assignment, reward yourself by spending time with friends, listening to music, reading a novel, or playing a sport.

Attend Each Class. Many students don't realize that missing even one class can have a great effect on their grade. Arriving late takes its toll as well. If you are just a few minutes late, or miss an entire class, you risk getting behind. So, keep these tips in mind.

- Arrive on time, or a little early.
- If you must miss a class, get a set of notes, the homework assignments, and any handouts that the instructor may have provided for the day that you missed.
- Study the material you missed. Take advantage of the help that comes with this textbook, such as the video examples and problem-specific tutorials.

Now Try This

1. List six ways in which you will benefit from passing this course.
2. List six short-term goals that will help you achieve your larger goal of passing this course. For example, you could set a goal to read through the entire *Study Skills Workshop* within the first 2 weeks of class or attend class regularly and on time. (***Success Tip:*** Revisit this action item once you have read through all seven *Study Skills Workshop* learning objectives.)
3. List some simple ways you can reward yourself when you complete one of your short-term class goals.
4. Plan ahead! List five possible situations that could cause you to be late for class or miss a class. (Some examples are parking/traffic delays, lack of a babysitter, oversleeping, or job responsibilities.) What can you do ahead of time so that these situations won't cause you to be late or absent?

2 Prepare to Learn

Many students believe that there are two types of people—those who are good at math and those who are not—and that this cannot be changed. This is not true! You can increase your chances for success in mathematics by taking time to prepare and taking inventory of your skills and resources.

Discover Your Learning Style. Are you a visual, verbal, or auditory learner? The answer to this question will help you determine how to study, how to complete your homework, and even where to sit in class. For example, visual-verbal learners learn best by reading and writing; a good study strategy for them is to rewrite notes and examples. However, auditory learners learn best by listening, so listening to the video examples of important concepts may be their best study strategy.

Get to Know Your Textbook and Its Resources. You have made a significant investment in your education by purchasing this book and the resources that accompany it. It has been designed with you in mind. Use as many of the features and resources as possible in ways that best fit your learning style.

Know What Is Expected. Your course syllabus maps out your instructor's expectations for the course. Read the syllabus completely and make sure you understand all that is required. If something is not clear, contact your instructor for clarification.

Organize Your Notebook. You will definitely appreciate a well-organized notebook when it comes time to study for the final exam. So let's start now! Refer to your syllabus and create a separate section in the notebook for each chapter (or unit of study) that your class will cover this term. Now, set a standard order within each section. One recommended order is to begin with your class notes, followed by your completed homework assignments, then any study sheets or handouts, and, finally, all graded quizzes and tests.

Now Try This

1. To determine what type of learner you are, take the *Index of Learning Styles Questionnaire* at http://www.engr.ncsu.edu/learningstyles/ilsweb.html, which will help you determine your learning type and offer study suggestions by type. List what you learned from taking this survey. How will you use this information to help you succeed in class?

2. Complete the *Study Skills Checklists* found at the end of sections 1–4 of Chapter 1 in order to become familiar with the many features that can enhance your learning experience using this book.

3. Read through the list of Student Resources found in the Preface of this book. Which ones will you use in this class?

4. Read through your syllabus and write down any questions that you would like to ask your instructor.

5. Organize your notebook using the guidelines given above. Place your syllabus at the very front of your notebook so that you can see the dates over which the material will be covered and for easy reference throughout the course.

3 Manage Your Time

Now that you understand the importance of attending class, how will you make time to study what you have learned while attending? Much like learning to play the piano, math skills are best learned by practicing a little every day.

Make the Time. In general, 2 hours of independent study time is recommended for every hour in the classroom. If you are in class 3 hours per week, plan on 6 hours per week for reviewing your notes and completing your homework. It is best to schedule this time over the length of a week rather than to try to cram everything into one or two marathon study days.

Prioritize and Make a Calendar. Because daily practice is so important in learning math, it is a good idea to set up a calendar that lists all of your time commitments, as well as the time you will need to set aside for studying and doing your homework. Consider how you spend your time each week and prioritize your tasks by importance. During the school term, you may need to reduce or even eliminate certain nonessential tasks in order to meet your goals for the term.

Maximize Your Study Efforts. Using the information you learned from determining your learning style, set up your blocks of study time so that you get the most out of these sessions. Do you study best in groups or do you need to study alone to get anything done? Do you learn best when you schedule your study time in 30-minute time blocks or do you need at least an hour before the information kicks in? Consider your learning style to set up a schedule that truly suits your needs.

Avoid Distractions. Between texting and social networking, we have so many opportunities for distraction and procrastination. On top of these, there are the distractions of TV, video games, and friends stopping by to hang out. Once you have set your schedule, honor your study times by turning off any electronic devices and letting your voicemail take messages for you. After this time, you can reward yourself by returning phone calls and messages or spending time with friends after the pressure of studying has been lifted.

Now Try This

1. Keep track of how you spend your time for a week. Rate each activity on a scale from 1 (not important) to 5 (very important). Are there any activities that you need to reduce or eliminate in order to have enough time to study this term?

2. List three ways that you learn best according to your learning style. How can you use this information when setting up your study schedule?

3. Download the *Weekly Planner Form* from www.cengagebrain.com/shop/search/tussy and complete your schedule. If you prefer, you may set up a schedule in Google Calendar (calendar.google.com), www.rememberthemilk.com, your cell, or your email system. Many of these have the ability to set up useful reminders and to-do lists in addition to a weekly schedule.

4. List three ways in which you are most often distracted. What can you do to avoid these distractions during your scheduled study times?

4 Listen and Take Notes

Make good use of your class time by listening and taking notes. Because your instructor will be giving explanations and examples that may not be found in your textbook, as well as other information about your course (test dates, homework assignments, and so on), it is important that you keep a written record of what was said in class.

Listen Actively. Listening in class is different from listening in social situations because it requires that you be an *active* listener. Since it is impossible to write down everything that is said in class, you need to exercise your active listening skills to learn to write down what is *important*. You can spot important material by listening for cues from your instructor. For instance, pauses in lectures or statements from your instructor such as "This is really important" or "This is a question that shows up frequently on tests" are indications that you should be paying special attention. Listen with a pencil (or highlighter) in hand, ready to record or highlight (in your textbook) any examples, definitions, or concepts that your instructor discusses.

Take Notes You Can Use. Don't worry about making your notes really neat. After class you can rework them into a format that is more useful to you. However, you should organize your notes as much as possible as you write them. Copy the examples your instructor uses in class. Circle or star any key concepts or definitions that your instructor mentions while explaining the example. Later, your homework problems will look a lot like the examples given in class, so be sure to copy each of the steps in detail.

Listen with an Open Mind. Even if there are concepts presented that you feel you already know, keep tuned in to the presentation of the material and look for a deeper understanding of the material. If the material being presented is something that has been difficult for you in the past, listen with an open mind; your new instructor may have a fresh presentation that works for you.

Avoid Classroom Distractions. Some of the same things that can distract you from your study time can distract you, and others, during class. Because of this, be sure to turn off your cell phone during class. If you take notes on a laptop, log out of your email and social networking sites during class. In addition to these distractions, avoid getting into side conversations with other students. Even if you feel you were only distracted for a few moments, you may have missed important verbal or body language cues about an upcoming exam or hints that will aid in your understanding of a concept.

Now Try This

1. Before your next class, refer to your syllabus and read the section(s) that will be covered. Make a list of the terms that you predict your instructor will think are most important.

2. During your next class, bring your textbook and keep it open to the sections being covered. If your instructor mentions a definition, concept, or example that is found in your text, highlight it.

3. Find at least one classmate with whom you can review notes. Make an appointment to compare your class notes as soon as possible after the class. Did you find differences in your notes?

4. Go to www.cengagebrain.com/shop/search/tussy and read the *Reworking Your Notes* handout. Complete the action items given in this document.

5 Build a Support System

Have you ever had the experience where you understand everything that your instructor is saying in class, only to go home and try a homework problem and be completely stumped? This is a common complaint among math students. The key to being a successful math student is to take care of these problems before you go on to tackle new material. That is why you should know what resources are available outside of class.

Make Good Use of Your Instructor's Office Hours. The purpose of your instructor's office hours is to be available to help students with questions. Usually these hours are listed in your syllabus and no appointment is needed. When you visit your instructor, have a list of questions and try to pinpoint exactly where in the process you are getting stuck. This will help your instructor answer your questions efficiently.

Use Your Campus Tutoring Services. Many colleges offer tutorial services for free. Sometimes tutorial assistance is available in a lab setting where you are able to drop in at your convenience. In some cases, you need to make an appointment to see a tutor in advance. Make sure to seek help as soon as you recognize the need, and come to see your tutor with a list of identified problems.

Form a Study Group. Study groups are groups of classmates who meet outside of class to discuss homework problems or study for tests. Get the most out of your study group by following these guidelines:

- Keep the group small—a maximum of four committed students. Set a regularly scheduled meeting day, time, and place.
- Find a place to meet where you can talk and spread out your work.
- Members should attempt all homework problems before meeting.
- All members should contribute to the discussion.
- When you meet, practice verbalizing and explaining problems and concepts to each other. The best way to really learn a topic is by teaching it to someone else.

Now Try This

1. Refer to your syllabus. Highlight your instructor's office hours and location. Next, pay a visit to your instructor during office hours this week and introduce yourself. (*Success Tip:* Program your instructor's office phone number and email address into your cell phone or email contact list.)

2. Locate your campus tutoring center or math lab. Write down the office hours, phone number, and location on your syllabus. Drop by or give them a call and find out how to go about making an appointment with a tutor.

3. Find two to three classmates who are available to meet at a time that fits your schedule. Plan to meet 2 days before your next homework assignment is due and follow the guidelines given above. After your group has met, evaluate how well it worked. Is there anything that the group can do to make it better next time you meet?

4. Download the *Support System Worksheet* at www.cengagebrain.com/shop/search/tussy. Complete the information and keep it at the front of your notebook following your syllabus.

6 Do Your Homework

Attending class and taking notes are important, but the only way that you are really going to learn mathematics is by completing your homework. Sitting in class and listening to lectures will help you to place concepts in short-term memory, but in order to do well on tests and in future math classes, you want to put these concepts in long-term memory. When completed regularly, homework assignments will help with this.

Give Yourself Enough Time. In Objective 3, you made a study schedule, setting aside 2 hours for study and homework for every hour that you spend in class. If you are not keeping this schedule, make changes to ensure that you can spend enough time outside of class to learn new material.

Review Your Notes and the Worked Examples from Your Text. In Objective 4, you learned how to take useful notes. Before you begin your homework, review or rework your notes. Then, read the sections in your textbook that relate to your homework problems, paying special attention to the worked examples. With a pencil in hand, work the *Self Check* and *Now Try* problems that are listed next to the examples in your text. Using the worked example as a guide, solve these problems and try to understand each step. As you read through your notes and your text, keep a list of anything that you don't understand.

Now Try Your Homework Problems. Once you have reviewed your notes and the textbook worked examples, you should be able to successfully manage the bulk of your homework assignment easily. When working on your homework, keep your textbook and notes close by for reference. If you have trouble with a homework question, look through your textbook and notes to see if you can identify an example that is similar to the homework question. See if you can apply the same steps to your homework problem. If there are places where you get stuck, add these to your list of questions.

Get Answers to Your Questions. At least one day before your assignment is due, seek help with the questions you have been listing. You can contact a classmate for assistance, make an appointment with a tutor, or visit your instructor during office hours.

Now Try This

1. Review your study schedule. Are you following it? If not, what changes can you make to adhere to the rule of 2 hours of homework and study for every hour of class?

2. Find five homework problems that are similar to the worked examples in your textbook. Were there any homework problems in your assignment that didn't have a worked example that was similar? (**Success Tip:** Look for the *Now Try* and *Guided Practice* features for help linking problems to worked examples.)

3. As suggested in this Objective, make a list of questions while completing your homework. Visit your tutor or your instructor with your list of questions and ask one of them to work through these problems with you.

4. Go to www.cengagebrain.com/shop/search/tussy and read the *Study and Memory Techniques* handout. List the techniques that will be most helpful to you in your math course.

7 Prepare for the Test

Taking a test does not need to be an unpleasant experience. Use your time management, organization, and these test-taking strategies to make this a learning experience and improve your score.

Make Time to Prepare. Schedule at least four daily 1-hour sessions to prepare specifically for your test.

Four days before the test: Create your own study sheet using your reworked notes. Imagine you could bring one $8\frac{1}{2} \times 11$ sheet of paper to your test. What would you write on that sheet? Include all the key definitions, rules, steps, and formulas that were discussed in class or covered in your reading. Whenever you have the opportunity, pull out your study sheet and review your test material.

Three days before the test: Create a sample test using the in-class examples from your notes and reading material. As you review and work these examples, make sure you understand how each example relates to the rules or definitions on your study sheet. While working through these examples, you may find that you forgot a concept that should be on your study sheet. Update your study sheet and continue to review it.

Two days before the test: Use the *Chapter Test* from your textbook or create one by matching problems from your text to the example types from your sample test. Now, with your book closed, take a timed trial test. When you are done, check your answers. Make a list of the topics that were difficult for you and review or add these to your study sheet.

One day before the test: Review your study sheet once more, paying special attention to the material that was difficult for you when you took your practice test the day before. Be sure you have all the materials that you will need for your test laid out ahead of time (two sharpened pencils, a good eraser, possibly a calculator or protractor, and so on). The most important thing you can do today is get a good night's rest.

Test day: Review your study sheet, if you have time. Focus on how well you have prepared and take a moment to relax. When taking your test, complete the problems that you are sure of first. Skip the problems that you don't understand right away, and return to them later. Bring a watch or make sure there will be some kind of time-keeping device in your test room so that you can keep track of your time. Try not to spend too much time on any one problem.

Now Try This

1. Create a study schedule using the guidelines given above.
2. Read the *Preparing for a Test* handout at www.cengagebrain.com/shop/search/tussy.
3. Read the *Taking the Test* handout at www.cengagebrain.com/shop/search/tussy.
4. After your test has been returned and scored, read the *Analyzing Your Test Results* handout at www.cengagebrain.com/shop/search/tussy.
5. Take time to reflect on your homework and study habits after you have received your test score. What actions are working well for you? What do you need to improve?
6. To prepare for your final exam, read the *Preparing for Your Final Exam* handout at www.cengagebrain.com/shop/search/tussy. Complete the action items given in this document.

1
An Introduction to Algebra

1.1 The Language of Algebra
1.2 Fractions
1.3 The Real Numbers
1.4 Adding Real Numbers; Properties of Addition
1.5 Subtracting Real Numbers
1.6 Multiplying and Dividing Real Numbers; Multiplication and Division Properties
1.7 Exponents and Order of Operations
1.8 Algebraic Expressions
1.9 Simplifying Algebraic Expressions Using Properties of Real Numbers
Chapter Summary and Review
Chapter Test

from Campus to Careers
Physician Assistant

Physician assistants practice medicine under the direction of doctors and surgeons. They are trained to diagnose, treat, and suggest preventative health care measures as directed by their supervising physician. Working as members of a health care team, they take medical histories, examine and treat patients, order and read laboratory tests and X-rays, and make diagnoses.

In **Problem 78** of **Study Set 1.4**, you will see how physician assistants use signed numbers to help assess a patient's risk of contracting heart disease.

JOB TITLE: Physician Assistant
EDUCATION: Most programs require 2 or more years of study after college, leading to a master's degree.
JOB OUTLOOK: Excellent. Jobs are expected to grow much faster than average for all occupations through the year 2020.
ANNUAL EARNINGS: Median salary: $94,870
FOR MORE INFORMATION:
www.bls.gov/ooh/healthcare/physician-assistants.htm

Objectives

1. Read tables and graphs.
2. Use the basic vocabulary and notation of algebra.
3. Identify algebraic expressions and equations.
4. Use equations to construct tables of data.

SECTION 1.1
The Language of Algebra

Algebra is the result of contributions from many cultures over thousands of years. The word *algebra* comes from the title of the book *Ihm Al-jabr wa'l muqābalah*, written by the Arabian mathematician al-Khwarizmi around A.D. 800. Using the vocabulary and notation of algebra, we can mathematically **model** many situations in the real world. In this section, we begin to explore the language of algebra by introducing some of its basic components.

1 Read tables and graphs.

Two-column **tables** are often used to describe numerical relationships. For example, the table below lists the number of bicycle tires a production planner must order when a given number of bicycles are to be manufactured. For a production run of, say, 300 bikes, we locate 300 in the left-hand column and then scan across the table to see that the company must order 600 tires.

The information in the table can be presented in a **bar graph** as shown below. The **horizontal axis** labeled "Number of bicycles to be manufactured" has been scaled in units of 100 bicycles. The **vertical axis**, labeled "Number of tires to order," is scaled in units of 100 tires. The bars directly over each of the production amounts extend to a height indicating the corresponding number of tires to order. For example, if 200 bikes are to be manufactured, the height of the bar indicates that 400 tires should be ordered.

Bicycles to be manufactured	Tires to order
100	200
200	400
300	600
400	800

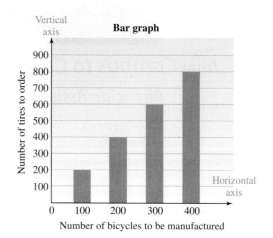

> **The Language of Algebra** *Horizontal* is a form of the word *horizon*. Think of the sun setting over the *horizon*. For *vertical*, think of vertical blinds. They are usually used as window coverings for large window areas and run up and down.

Another way to present the information is with a **line graph**. Instead of using a bar to denote the number of tires to order for a production run of a given size, we use a dot drawn at the correct height. After drawing the four data points for 100, 200, 300, and 400 bicycles, we connect them with line segments to create the following line graph.

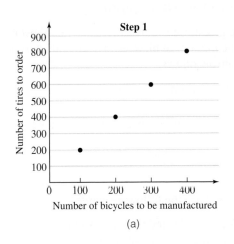

(a)

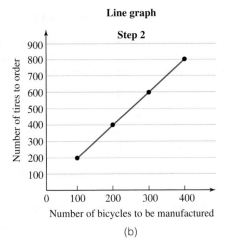

Line graph

(b)

The line graph not only presents all the information contained in the table and the bar graph, it also provides additional information that they do not. We can use the line graph to find the number of tires to order for a production run of a size not shown in the table or the bar graph.

EXAMPLE 1 Use the line graph to find the number of tires needed when 250 bicycles are to be manufactured.

Strategy We will start on the horizontal axis at 250. Then we will scan up to the line, and over, to read the number of tires to order on the vertical axis.

WHY We start on the horizontal axis because that scale gives the number of bicycles to be manufactured. We scan up and over to the vertical axis because that scale gives the number of tires to order.

Solution
First, locate 250 (between 200 and 300) on the horizontal axis. Then draw a line straight up to intersect the graph. (See the figure below.) From the point of intersection, draw a horizontal line to the left that intersects the vertical axis. We see that the number of tires to order is 500.

Self Check 1
Use the graph to find the number of tires needed when 350 bicycles are to be manufactured.

Now Try Problem 43

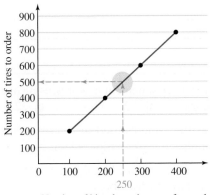

❷ Use the basic vocabulary and notation of algebra.

From the table, the bar graph, and the line graph, we see that there is a relationship between the number of tires to order and the number of bicycles to be manufactured. Using words, we can express this relationship as follows:

"The number of tires to order is two times the number of bicycles to be manufactured."

Since the word **product** is used to indicate the answer to a multiplication, we can restate the relationship this way:

"The number of tires to order is the *product* of 2 and the number of bicycles to be manufactured."

To indicate other arithmetic operations, we will use the following words.

- A **sum** is the result of an addition: The sum of 5 and 6 is equal to 11.
- A **difference** is the result of a subtraction: The difference of 3 and 2 is equal to 1.
- A **quotient** is the result of a division: The quotient of 6 and 3 is equal to 2.

Many symbols used in arithmetic are also used in algebra. For example, a + symbol is used to indicate addition, and a − symbol is used to indicate subtraction, and an = symbol means *is equal to*. Because the letter x is often used in algebra and could be confused with the multiplication symbol ×, we usually write multiplication using a **raised dot** or **parentheses**.

Symbols Used for Multiplication

×	times symbol	$6 \times 4 = 24$
·	raised dot	$6 \cdot 4 = 24$
( )	parentheses	$(6)4 = 24$ or $6(4) = 24$ or $(6)(4) = 24$

There are a few ways to indicate division. In algebra, the symbol most often used to indicate division is the *fraction bar*.

Symbols Used for Division

÷	division symbol	$24 \div 4 = 6$
)	long division	$4\overline{)24}$ with quotient 6
—	fraction bar	$\dfrac{24}{4} = 6$

Self Check 2
Express the following statement in words.

$22 \cdot 11 = 242$

Now Try Problem 47

EXAMPLE 2
Express each statement in words, using one of the words *sum, product, difference,* or *quotient*.   **a.** $22 \div 11 = 2$ **b.** $22 + 11 = 33$

Strategy We will look at the symbols used in each statement to see whether addition, subtraction, multiplication, or division is indicated.

WHY The word we should use (*sum, product, difference,* or *quotient*) depends on the arithmetic operation we have to describe.

Solution

a. The quotient of 22 and 11 is equal to 2.
b. The sum of 22 and 11 is equal to 33.

3 Identify algebraic expressions and equations.

Another way to describe the tires-to-bicycles relationship uses *variables*. **Variables** are letters (or symbols) that stand for numbers. If we let the letter b represent the number of bicycles to be manufactured, then the number of tires to order is two times b, written $2b$. In the notation, the number 2 is an example of a **constant** because it does not change value.

> **The Language of Algebra** Since the number of bicycles to be manufactured can *vary*, or change, it is represented using a *variable*.

When multiplying a variable by a number, or a variable by another variable, we can omit the symbol for multiplication. For example,

$2b$ means $2 \cdot b$ xy means $x \cdot y$ $8abc$ means $8 \cdot a \cdot b \cdot c$

We call $2b$, xy, and $8abc$ algebraic expressions.

> **Algebraic Expressions**
>
> Variables and/or numbers can be combined with the operations of addition, subtraction, multiplication, and division to create **algebraic expressions**.

Here are some other examples of algebraic expressions.

$4a + 7$ — This expression is a combination of the numbers 4 and 7, the variable a, and the operations of multiplication and addition.

$\dfrac{10 - y}{3}$ — This expression is a combination of the numbers 10 and 3, the variable y, and the operations of subtraction and division.

$15mn(2m)$ — This expression is a combination of the numbers 15 and 2, the variables m and n, and the operation of multiplication.

> **The Language of Algebra** We often refer to *algebraic expressions* as simply *expressions*.

In the bicycle-manufacturing example, if we let the letter t stand for the number of tires to order, we can translate the **verbal model** to mathematical symbols.

The number of tires to order	is	two	times	the number of bicycles to be manufactured
t	$=$	2	$\cdot$	b

The statement $t = 2 \cdot b$, or more simply, $t = 2b$, is called an *equation*. An **equation** is a mathematical sentence that contains an $=$ symbol. The $=$ symbol indicates that the expressions on either side of it have the same value. Other examples of equations are

$3 + 5 = 8$ $x + 5 = 20$ $17 - 2r = 14 + 3r$ $p = 100 - d$

> **The Language of Algebra** The equal symbol = can be represented by verbs such as:
>
>   is   are   gives   yields
>
> The symbol ≠ is read as *"is not equal to."*

Self Check 3

Translate into an equation: The number of unsold tickets is the difference of 500 and the number of tickets that have been purchased.

Now Try Problems 55 and 57

EXAMPLE 3 Translate the verbal model into an equation.

| The number of decades | is | the number of years | divided by | 10. |

Strategy We will represent the unknown quantities using variables, and we will use symbols to represent the words *is* and *divided by*.

WHY To translate a verbal (word) model into an equation means to write it using mathematical symbols.

Solution
We can represent the two unknown quantities using variables: Let d = the number of decades and y = the number of years. Then we have:

| The number of decades | is | the number of years | divided by | 10. |
| d | $=$ | y | $\div$ | 10 |

If we write the division using a fraction bar, then the verbal model translates to the equation $d = \dfrac{y}{10}$.

In the bicycle-manufacturing example, using the equation $t = 2b$ to describe the relationship has one major advantage over the other methods we have discussed. It can be used to determine the number of tires needed for a production run of *any* size.

Self Check 4

Use the equation $t = 2b$ to find the number of tires needed if 604 bicycles are to be manufactured.

Now Try Problem 59

EXAMPLE 4 Use the equation $t = 2b$ to find the number of tires needed for a production run of 178 bicycles.

Strategy In $t = 2b$, we will replace b with 178. Then we will multiply 178 by 2 to obtain the value of t.

WHY The equation $t = 2b$ indicates that the number of tires is found by multiplying the number of bicycles by 2.

Solution

$t = 2b$ This is the describing equation.

$t = 2(\mathbf{178})$ Replace b, which stands for the number of bicycles, with 178. Use parentheses to show the multiplication.

$t = 356$ Perform the multiplication: $2(178) = 356$.

To manufacture 178 bicycles, 356 tires will be needed.

1.1 The Language of Algebra

> **The Language of Algebra** To *substitute* means to put or use in place of another, as with a *substitute* teacher. In Example 4, we substitute 178 for *b*.

4 Use equations to construct tables of data.

Equations such as $t = 2b$, which express a relationship between two or more variables, are called **formulas**. Formulas are used in many fields, such as economics, biology, nursing, and construction. In the next example, we will see that the results found using the formula $t = 2b$ can be presented in table form.

EXAMPLE 5 Find the number of tires to order for production runs of 233 and 852 bicycles. Present the results in a table.

Strategy In the formula $t = 2b$, we will replace *b* with 233 and then with 852.

WHY We need to find the number of tires to order for two different-sized production runs.

Solution

Step 1 We begin by constructing a two-column table with the appropriate column headings. The size of each production run (233 and 852) is entered in the left-hand column of the table.

Step 2 We use the formula $t = 2b$ to find the number of tires needed if 233 and 852 bicycles are to be manufactured.

Bicycles to be manufactured b	Tires to order t
233	
852	

$t = 2b$
$t = 2(233)$ Replace *b* with 233.
$t = 466$

$t = 2b$
$t = 2(852)$ Replace *b* with 852.
$t = 1,704$

Step 3 We enter these results in the right-hand column of the table: 466 tires for 233 bicycles to be manufactured and 1,704 tires for 852 bicycles to be manufactured.

Bicycles to be manufactured b	Tires to order t
233	466
852	1,704

Self Check 5
Find the number of tires needed for production runs of 87 and 487 bicycles. Present the results in a table.

Now Try Problem 62

ANSWERS TO SELF CHECKS

1. 700   2. The product of 22 and 11 is equal to 242.   3. $u = 500 - p$ 4. 1,208

5.

Bicycles to be manufactured b	Tires to order t
87	174
487	974

STUDY SKILLS CHECKLIST

Get to Know Your Textbook

Congratulations. You now own a state-of-the-art textbook that has been written especially for you. The following checklist will help you become familiar with the organization of the book. Place a checkmark ☑ in each box after you answer the question.

☐ Turn to the **Table of Contents** on page iii. How many chapters does the book have?

☐ Each chapter of the book is divided into **sections**. How many sections are there in Chapter 1, which begins on page 1?

☐ **Learning Objectives** are listed at the start of each section. How many objectives are there for Section 1.6, which begins on page 58?

☐ Each section ends with a **Study Set**. How many problems are there in Study Set 1.5, which begins on page 56?

☐ Each chapter has a **Chapter Summary & Review**. Which column of the Chapter Summary found on page 106 contains examples?

☐ How many review problems are there for Section 1.1 in the **Chapter Summary & Review**, which begins on page 106?

☐ Each chapter has a **Chapter Test**. How many problems are there in the Chapter 1 Test, which begins on page 116?

☐ Each chapter beginning with Chapter 2 ends with a **Cumulative Review**. What chapters are covered by the Cumulative Review that begins on page 383?

Answers: 9, 9, 4, 86, the right column, 12, 44, 1–4

SECTION 1.1 STUDY SET

VOCABULARY

Fill in the blanks.

1. The answer to an addition problem is called a _____. The answer to a subtraction problem is called a _____.

2. The answer to a multiplication problem is called a _____. The answer to a division problem is called a _____.

3. _____ are letters or symbols that stand for numbers.

4. Variables and numbers can be combined with the operations of addition, subtraction, multiplication, and division to create algebraic _____.

5. An _____ is a mathematical sentence that contains an = symbol.

6. An equation, such as $t = 2b$, which expresses a relationship between two or more variables, is called a _____.

7. The illustration in Exercise 8 shows a _____ graph.

8. In the illustration below, the _____ axis of the graph has been scaled in units of 1 second. The _____ axis of the graph has been scaled in units of 50 feet.

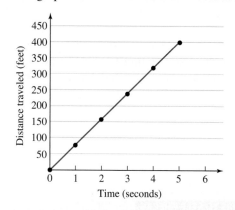

CONCEPTS

Classify each item as either an algebraic expression or an equation.

9. $18 + m = 23$

10. $18 + m$

11. $y - 1$

12. $y - 1 = 2$

1.1 The Language of Algebra

13. $30x$

14. $t = 16b$

15. $r = \dfrac{2}{3}$

16. $\dfrac{c-7}{5}$

17. a. What operations does the expression $5x - 16$ contain?
  b. What variable does the expression contain?

18. a. What operations does the expression $\dfrac{12 + t}{25}$ contain?
  b. What variable does the expression contain?

19. a. What operations does the equation $m + 1 = 20 - m$ contain?
  b. What variable does it contain?

20. a. What operations does the equation $y + 14 = 5(6)$ contain?
  b. What variable does it contain?

21. DEPRECIATION The graph below shows how a piece of machinery loses value over its lifetime.
  a. Explain what information the dashed lines help us find.
  b. As the machinery ages, what happens to its value?

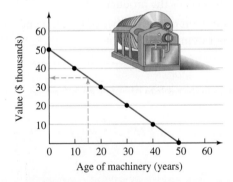

22. INCOME The graph below shows how a company's income increases as the number of customers increases.
  a. Use the graph to find the income received from 30, 50, and 70 customers.
  b. As the number of customers increases, what happens to the income?

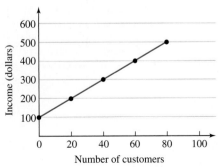

NOTATION

Write each multiplication in two other ways. First, using a raised dot · and then using parentheses ( ).

23. 5×6

24. 4×7

25. 34×75

26. 90×12

Write each expression without using a multiplication symbol.

27. $4 \cdot x$

28. $5 \cdot y$

29. $3 \cdot r \cdot t$

30. $22 \cdot q \cdot s$

31. $l \cdot w$

32. $b \cdot h$

33. $P \cdot r \cdot t$

34. $l \cdot w \cdot h$

35. $2(w)$

36. $2(l)$

37. $(x)(y)$

38. $(r)(t)$

Write each division using a fraction bar.

39. $32 \div x$

40. $y \div 15$

41. $30\overline{)90}$

42. $20\overline{)80}$

GUIDED PRACTICE

Use the line graph in Example 1 to find the number of tires needed to make the following number of bicycles. See Example 1.

43. 125 bicycles

44. 400 bicycles

45. 100 bicycles

46. 200 bicycles

Express each statement using one of the words sum, difference, product, *or* quotient. *See Example 2.*

47. $18(24) = 432$

48. $45 \cdot 12 = 540$

49. $11 - 9 = 2$

50. $65 + 89 = 154$

51. $2x = 1$

52. $16t = 9$

53. $\dfrac{66}{11} = 6$

54. $12 \div 3 = 4$

Translate each verbal model into an equation. (Hint: You will need to use variables. Answers may vary, depending on the variables chosen.) See Example 3.

55. The sale price is $100 minus the discount.

56. The cost of dining out | equals | the cost of the meal | plus | $7 for parking.

57. 7 | times | the age of a dog in years | gives | the dog's equivalent human age.

58. The number of centuries | is | the number of years | divided by | 100.

Use the given equation (formula) to complete each table. See Examples 4 and 5.

59. $d = 360 + L$

Lunch time (L) minutes	School day (d) minutes
30	
40	
45	

60. $b = 1{,}024k$

Kilobytes (k)	Bytes (b)
1	
5	
10	

61. $t = 1{,}500 - d$

Deductions (d)	Take-home pay (t)
200	
300	
400	

62. $w = \dfrac{s}{12}$

Inches of snow (s)	Inches of water (w)
12	
24	
72	

TRY IT YOURSELF

Translate each verbal model into an equation. Answers may vary, depending on the variables chosen.

63. The amount of sand that should be used is the product of 3 and the amount of cement used.

64. The number of waiters needed is the quotient of the number of customers and 10.

65. The weight of the truck is the sum of the weight of the engine and 1,200.

66. The number of classes that are still open is the difference of 150 and the number of classes that are closed.

67. The profit is the difference of the revenue and 600.

68. The distance is the product of the rate and 3.

69. The quotient of the number of laps run and 4 is the number of miles run.

70. The sum of the tax and 35 is the total cost.

Use the data in the table to complete each formula.

71. $d = \dfrac{e}{\boxed{}}$

Eggs e	Dozen d
24	2
36	3
48	4

72. $p = \boxed{}\,c$

Canoes c	Paddles p
6	12
7	14
8	16

73. $I = \boxed{}\, c$

Couples c	Individuals I
20	40
100	200
200	400

74. $t = \dfrac{p}{\boxed{}}$

Players p	Teams t
5	1
10	2
15	3

APPLICATIONS

75. CHAIR PRODUCTION Use the diagram to complete the six formulas that planners could use to order the necessary number of legs l, arms a, seats S, backs b, arm pads p, and screws s for a production run of c chairs.

$l = \boxed{}\, c \qquad b = \boxed{}$
$a = \boxed{}\, c \qquad p = \boxed{}\, c$
$S = \boxed{} \qquad s = \boxed{}\, c$

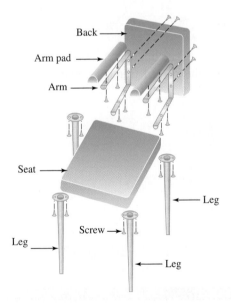

76. STAIRCASE PRODUCTION Complete the four formulas that could be used by the job superintendent to order the necessary number of staircase parts for a tract of h homes, each of which will have a staircase as shown.

$b = \boxed{}\, h \qquad p = \boxed{}\, h \qquad r = \boxed{}\, h \qquad t = \boxed{}\, h$

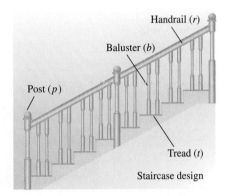

Staircase design

77. FACEBOOK The following table shows the approximate number of Facebook subscribers (in millions) by selected countries as of May 2012. Graph the data using a bar graph.

Country	Number of subscribers (in millions)
Brazil	72
India	64
Indonesia	48
Mexico	42
U.K.	31
U.S.A.	159

Source: checkfacebook.com

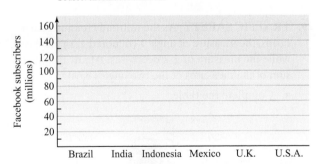

78. **FUEL EFFICIENCY** A *hybrid* car (such as a Toyota Prius) is a vehicle that is powered by two types of energy sources: an electric motor and a small combustion engine. Use the data in the following table to draw a line graph of the U.S. hybrid vehicle sales for 2002 through 2012. (Source: DOE Alternative Fuels Data Center, HybridCars.com)

Year	Hybrids sold
2002	36,035
2003	47,600
2004	84,199
2005	209,711
2006	252,636
2007	352,274
2008	312,386
2009	290,271
2010	274,210
2011	268,755
2012	382,704

Source: Bureau of Justice Statistics

79. **TRAFFIC SAFETY** As the railroad crossing guard drops, the measure of angle 1 (denoted ∠1) increases, while the measure of ∠2 decreases. At any instant, the sum of the measures of the two angles is 90°.

    a. Complete the table.

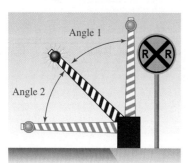

∠1	∠2
0°	
30°	
45°	
60°	
90°	

b. Use the data in the table to construct a line graph for values of ∠1 from 0° to 90°.

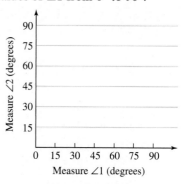

80. **KEYBOARDS** As the legs of the keyboard stand are widened, the measure of angle 1 (denoted ∠1) will increase, and, in turn, the measure of ∠2 will decrease. For any position, the sum of the measures of the two angles is 180°.

    a. Complete the table.

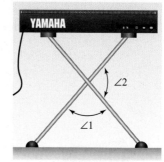

∠1	∠2
50°	
60°	
70°	
80°	
90°	

b. Use the data in the table to construct a line graph for values of ∠1 from 50° to 90°.

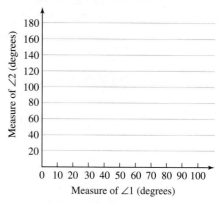

WRITING

81. Many people misuse the word *equation* when discussing mathematics. What is an equation? Give an example.

82. Explain the difference between an algebraic expression and an equation. Give an example of each.

83. Which do you think is more informative, a bar graph or a line graph? Explain your reasoning.

84. Create a bar graph that shows, on average, how many hours of television you watch each day of the week. Let Sunday be day 1, Monday day 2, and so on.

SECTION 1.2
Fractions

Objectives
1. Factor and prime factor natural numbers.
2. Recognize special fraction forms.
3. Multiply and divide fractions.
4. Build equivalent fractions.
5. Simplify fractions.
6. Add and subtract fractions.
7. Simplify answers.
8. Compute with mixed numbers.

In arithmetic, we add, subtract, multiply, and divide **natural numbers**: 1, 2, 3, 4, 5, and so on. Assuming that you have mastered those skills, we will now review the arithmetic of fractions.

1 Factor and prime factor natural numbers.

To compute with fractions, we need to know how to *factor* natural numbers. To **factor** a number means to express it as a product of two or more numbers. For example, some ways to factor 12 are

$$1 \cdot 12, \quad 6 \cdot 2, \quad 4 \cdot 3, \quad \text{and} \quad 2 \cdot 2 \cdot 3$$

The numbers 1, 2, 3, 4, 6, and 12 that were used to write the products are called *factors* of 12. In general, a **factor** is a number being multiplied.

> **The Language of Algebra** When we say "factor 8," we are using the word *factor* as a verb. When we say "2 is a *factor* of 12," we are using the word *factor* as a noun.

Sometimes a number has only two factors, itself and 1. We call such numbers *prime numbers*.

> **Prime Numbers and Composite Numbers**
>
> A **prime number** is a natural number greater than 1 that has only itself and 1 as factors. The first ten prime numbers are 2, 3, 5, 7, 11, 13, 17, 19, 23, and 29.
>
> A **composite number** is a natural number, greater than 1, that is not prime. The first ten composite numbers are 4, 6, 8, 9, 10, 12, 14, 15, 16, and 18.

Every composite number can be factored into the product of two or more prime numbers. This product of these prime numbers is called its **prime factorization**.

EXAMPLE 1 Find the prime factorization of 210.

Strategy We will use a series of steps to express 210 as a product of only prime numbers.

WHY To *prime factor* a number means to write it as a product of prime numbers.

Solution
First, write 210 as the product of two natural numbers other than 1.

$$210 = 10 \cdot 21 \quad \text{The resulting prime factorization will be the same no matter which two factors of 210 you begin with.}$$

Neither 10 nor 21 is a prime number, so we factor each of them.

$$210 = 2 \cdot 5 \cdot 3 \cdot 7 \quad \text{Factor 10 as } 2 \cdot 5 \text{ and factor 21 as } 3 \cdot 7.$$

Writing the factors in ascending order, the **prime-factored form** of 210 is $2 \cdot 3 \cdot 5 \cdot 7$.
Two other methods for prime factoring 210 are shown below.

Self Check 1
Find the prime factorization of 189.

Now Try Problem 15

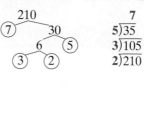

Factor tree   *Division ladder*

Work downward. Factor each number as a product of two numbers (other than 1 and itself) until all factors are prime. Circle prime numbers as they appear at the end of a branch.

Work upward. Perform repeated division until the final quotient is a prime number. It is helpful to start with the smallest prime, 2, as a trial divisor. Then, in order, try larger primes as divisors: 3, 5, 7, 11, and so on.

Either way, the factorization is $2 \cdot 3 \cdot 5 \cdot 7$. To check it, multiply the prime factors. The product should be 210.

The Language of Algebra Prime factors are often written in *ascending* order. To *ascend* means to move upward.

Success Tip A whole number is divisible by

- 2 if it ends in 0, 2, 4, 6, or 8
- 3 if the sum of the digits is divisible by 3
- 5 if it ends in 0 or 5
- 10 if it ends in 0

2 Recognize special fraction forms.

In a fraction, the number above the **fraction bar** is called the **numerator**, and the number below is called the **denominator**.

Fraction bar → $\dfrac{5}{6}$ ← Numerator
← Denominator

Fractions can describe the number of equal parts of a whole. For example, consider the circle with 5 of 6 equal parts colored red. We say that $\frac{5}{6}$ (five-sixths) of the circle is shaded.

The Language of Algebra The word *fraction* comes from the Latin word *fractio* meaning "breaking in pieces."

Fractions are also used to indicate division. For example, $\frac{8}{2}$ indicates that the numerator, 8, is to be divided by the denominator, 2:

$$\frac{8}{2} = 8 \div 2 = 4$$ We know that $\frac{8}{2} = 4$ because of its related multiplication statement: $2 \cdot 4 = 8$.

If the numerator and denominator of a fraction are the same nonzero number, the fraction indicates division of a number by itself, and the result is 1. Each of the following fractions is, therefore, a **form of 1**.

$$1 = \frac{1}{1} = \frac{2}{2} = \frac{3}{3} = \frac{4}{4} = \frac{5}{5} = \frac{6}{6} = \frac{7}{7} = \frac{8}{8} = \frac{9}{9} = \ldots$$

If a denominator is 1, the fraction indicates division by 1, and the result is simply the numerator. For example, $\frac{5}{1} = 5$ and $\frac{24}{1} = 24$.

> **Special Fraction Forms**
>
> For any nonzero number a,
>
> $$\frac{a}{a} = 1 \quad \text{and} \quad \frac{a}{1} = a$$

3 Multiply and divide fractions.

The rule for multiplying fractions can be expressed in words and in symbols as follows.

> **Multiplying Fractions**
>
> To multiply two fractions, multiply the numerators and multiply the denominators.
>
> For any two fractions $\frac{a}{b}$ and $\frac{c}{d}$,
>
> $$\frac{a}{b} \cdot \frac{c}{d} = \frac{a \cdot c}{b \cdot d}$$

EXAMPLE 2 Multiply: $\frac{7}{8} \cdot \frac{3}{5}$

Strategy To find the product, we will multiply the numerators, 7 and 3, and multiply the denominators, 8 and 5.

WHY This is the rule for multiplying two fractions.

Solution

$$\frac{7}{8} \cdot \frac{3}{5} = \frac{7 \cdot 3}{8 \cdot 5}$$ Multiply the numerators.
Multiply the denominators.

$$= \frac{21}{40}$$

Self Check 2

Multiply: $\frac{5}{9} \cdot \frac{2}{3}$

Now Try Problem 27

One number is called the **reciprocal** of another if their product is 1. To find the reciprocal of a fraction, we invert its numerator and denominator.

- $\frac{3}{4}$ is the reciprocal of $\frac{4}{3}$, because $\frac{3}{4} \cdot \frac{4}{3} = \frac{12}{12} = 1$.
- $\frac{1}{10}$ is the reciprocal of 10, because $\frac{1}{10} \cdot 10 = \frac{10}{10} = 1$.

> **Success Tip** Every number, except **0**, has a reciprocal. Zero has no reciprocal, because the product of 0 and a number cannot be 1.

We use reciprocals to divide fractions.

Dividing Fractions

To divide two fractions, multiply the first fraction by the reciprocal of the second.

For any two fractions $\frac{a}{b}$ and $\frac{c}{d}$, where $c \neq 0$,

$$\frac{a}{b} \div \frac{c}{d} = \frac{a}{b} \cdot \frac{d}{c}$$

Self Check 3

Divide: $\frac{6}{25} \div \frac{1}{2}$

Now Try Problem 31

EXAMPLE 3 Divide: $\frac{1}{3} \div \frac{4}{5}$

Strategy We will multiply the first fraction, $\frac{1}{3}$, by the reciprocal of the second fraction, $\frac{4}{5}$.

WHY This is the rule for dividing two fractions.

Solution

$$\frac{1}{3} \div \frac{4}{5} = \frac{1}{3} \cdot \frac{5}{4}$$ Multiply $\frac{1}{3}$ by the reciprocal of $\frac{4}{5}$. The reciprocal of $\frac{4}{5}$ is $\frac{5}{4}$.

$$= \frac{1 \cdot 5}{3 \cdot 4}$$ Multiply the numerators. Multiply the denominators.

$$= \frac{5}{12}$$

4 Build equivalent fractions.

The two rectangles on the right are the same size. The first rectangle is divided into 10 equal parts. Since 6 of those parts are red, $\frac{6}{10}$ of the figure is shaded.

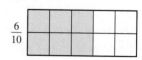

The second rectangle is divided into 5 equal parts. Since 3 of those parts are red, $\frac{3}{5}$ of the figure is shaded. We can conclude that $\frac{6}{10} = \frac{3}{5}$ because $\frac{6}{10}$ and $\frac{3}{5}$ represent the same shaded part of the rectangle. We say that $\frac{6}{10}$ and $\frac{3}{5}$ are *equivalent fractions*.

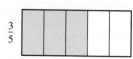

Equivalent Fractions

Two fractions are **equivalent** if they represent the same number.

Writing a fraction as an equivalent fraction with a larger denominator is called **building** the fraction. To build a fraction, we multiply it by a form of 1. Since any number multiplied by 1 remains the same (identical), 1 is called the **multiplicative identity element**.

Multiplication Property of 1

The product of 1 and any number is that number.

For any number a,

$$1 \cdot a = a \quad \text{and} \quad a \cdot 1 = a$$

EXAMPLE 4 Write $\frac{3}{5}$ as an equivalent fraction with a denominator of 35.

Strategy We will compare the given denominator to the required denominator and ask, "By what must we multiply 5 to get 35?"

WHY The answer to that question helps us determine the form of 1 to be used to build an equivalent fraction.

Solution

We need to multiply the denominator of $\frac{3}{5}$ by 7 to obtain a denominator of 35. It follows that $\frac{7}{7}$ should be the form of 1 that is used to build $\frac{3}{5}$. Multiplying $\frac{3}{5}$ by $\frac{7}{7}$ changes its appearance but does not change its value, because we are multiplying it by 1.

$$\frac{3}{5} = \frac{3}{5} \cdot \frac{7}{7} \qquad \frac{7}{7} = 1$$

$$= \frac{3 \cdot 7}{5 \cdot 7} \qquad \text{Multiply the numerators.}$$
$$\qquad\qquad \text{Multiply the denominators.}$$

$$= \frac{21}{35}$$

Self Check 4
Write $\frac{5}{8}$ as an equivalent fraction with a denominator of 24.
Now Try Problem 35

To build an equivalent fraction in Example 4, we multiplied $\frac{3}{5}$ by 1 in the form of $\frac{7}{7}$. As a result of that step, the numerator and the denominator of $\frac{3}{5}$ were multiplied by 7.

Step 1 $\frac{3}{5} = \frac{3}{5} \cdot \frac{7}{7}$

Step 2 $= \frac{3 \cdot 7}{5 \cdot 7}$ The numerator is multiplied by 7.
The denominator is multiplied by 7.

Step 3 $= \frac{21}{35}$ The result is a fraction equivalent to $\frac{3}{5}$.

This process illustrates an important property of fractions.

The Fundamental Property of Fractions

If the numerator and denominator of a fraction are multiplied by the same nonzero number, the resulting fraction is equivalent to the original fraction.

If $\frac{a}{b}$ is a fraction and c is a nonzero number,

$$\frac{a}{b} = \frac{ac}{bc}$$

Since multiplying the numerator and denominator of a fraction by the same nonzero number produces an equivalent fraction, your instructor may allow you to begin your solution to Example 4 at step 2.

Building Fractions

To build a fraction, multiply it by 1 in the form of $\frac{c}{c}$, where c is any nonzero number.

5 Simplify fractions.

Every fraction can be written in infinitely many equivalent forms. For example, some equivalent forms of $\frac{10}{15}$ are:

$$\frac{2}{3} = \frac{4}{6} = \frac{6}{9} = \frac{8}{12} = \mathbf{\frac{10}{15}} = \frac{12}{18} = \frac{14}{21} = \frac{16}{24} = \frac{18}{27} = \frac{20}{30} = \cdots$$

> **The Language of Algebra** The word *infinitely* is a form of the word *infinite*, which means endless.

Of all of the equivalent forms in which we can write a fraction, we often need to determine the one that is in *simplest form*.

Simplest Form of a Fraction

A fraction is in **simplest form**, or **lowest terms**, when the numerator and denominator have no common factors other than 1.

To **simplify a fraction**, we write it in simplest form by removing a factor equal to 1. For example, to simplify $\frac{10}{15}$, we note that the greatest factor common to the numerator and denominator is 5 and proceed as follows:

$$\frac{10}{15} = \frac{2 \cdot 5}{3 \cdot 5} \quad \text{Factor 10 and 15.}$$

$$= \frac{2}{3} \cdot \frac{5}{5} \quad \text{Use the rule for multiplying fractions in reverse: write } \frac{2 \cdot 5}{3 \cdot 5} \text{ as the product of two fractions, } \frac{2}{3} \text{ and } \frac{5}{5}.$$

$$= \frac{2}{3} \cdot 1 \quad \text{A nonzero number divided by itself is equal to 1: } \frac{5}{5} = 1.$$

$$= \frac{2}{3} \quad \text{Use the multiplication property of 1: any number multiplied by 1 remains the same.}$$

To simplify $\frac{10}{15}$, we removed a factor equal to 1 in the form of $\frac{5}{5}$. The result, $\frac{2}{3}$, is equivalent to $\frac{10}{15}$.

We can easily identify the greatest common factor of the numerator and the denominator of a fraction if we write them in prime-factored form.

Self Check 5

Simplify each fraction, if possible:

a. $\frac{24}{56}$ b. $\frac{16}{125}$

Now Try Problem 45

EXAMPLE 5

Simplify each fraction, if possible: a. $\frac{63}{42}$ b. $\frac{33}{40}$

Strategy We will begin by prime factoring the numerator and denominator of the fraction. Then, to simplify it, we will remove a factor equal to 1.

WHY We need to make sure that the numerator and denominator have no common factors other than 1. If that is the case, then the fraction is in *simplest form*.

Solution

a. After prime factoring 63 and 42, we see that the greatest common factor of the numerator and the denominator is $3 \cdot 7 = 21$.

$$\frac{63}{42} = \frac{3 \cdot 3 \cdot 7}{2 \cdot 3 \cdot 7} \quad \text{Write 63 and 42 in prime-factored form.}$$

$$= \frac{3}{2} \cdot \frac{3 \cdot 7}{3 \cdot 7} \quad \text{Write } \frac{3 \cdot 3 \cdot 7}{2 \cdot 3 \cdot 7} \text{ as the product of two fractions, } \frac{3}{2} \text{ and } \frac{3 \cdot 7}{3 \cdot 7}.$$

$$= \frac{3}{2} \cdot 1 \qquad \text{A nonzero number divided by itself is equal to 1: } \frac{3 \cdot 7}{3 \cdot 7} = 1.$$

$$= \frac{3}{2} \qquad \text{Any number multiplied by 1 remains the same.}$$

b. Prime factor 33 and 40.

$$\frac{33}{40} = \frac{3 \cdot 11}{2 \cdot 2 \cdot 2 \cdot 5}$$

Since the numerator and the denominator have no common factors other than 1, the fraction $\frac{33}{40}$ is in simplest form (lowest terms).

> **The Language of Algebra** What do Eddie Murphy, Billy Crystal, Queen Latifah, and Tom Hanks have in common? They all attended a community college. The word *common* means shared by two or more. In this section, we will work with *common* factors and *common* denominators.

To streamline the simplifying process, we can replace pairs of factors common to the numerator and denominator with the equivalent fraction $\frac{1}{1}$.

EXAMPLE 6 Simplify: $\frac{90}{105}$

Strategy We will begin by prime factoring the numerator, 90, and denominator, 105. Then we will look for any factors common to the numerator and denominator and remove them.

WHY When the numerator and/or denominator of a fraction are large numbers, such as 90 and 105, writing their prime factorizations is helpful in identifying any common factors.

Solution

$$\frac{90}{105} = \frac{2 \cdot 3 \cdot 3 \cdot 5}{3 \cdot 5 \cdot 7} \qquad \text{Write 90 and 105 in prime-factored form.}$$

$$= \frac{2 \cdot \overset{1}{\cancel{3}} \cdot 3 \cdot \overset{1}{\cancel{5}}}{\underset{1}{\cancel{3}} \cdot \underset{1}{\cancel{5}} \cdot 7} \qquad \text{Slashes and 1's are used to show that } \frac{3}{3} \text{ and } \frac{5}{5} \text{ are replaced by the equivalent fraction } \frac{1}{1}. \text{ A factor equal to 1 in the form of } \frac{3 \cdot 5}{3 \cdot 5} = \frac{15}{15} \text{ was removed.}$$

$$= \frac{6}{7} \qquad \text{Multiply the remaining factors in the numerator: } 2 \cdot 1 \cdot 3 \cdot 1 = 6. \text{ Multiply the remaining factors in the denominator: } 1 \cdot 1 \cdot 7 = 7.$$

Self Check 6

Simplify: $\frac{126}{70}$

Now Try Problem 53

We can use the following steps to simplify a fraction.

> **Simplifying Fractions**
>
> 1. Factor (or prime factor) the numerator and denominator to determine their common factors.
> 2. Remove factors equal to 1 by replacing each pair of factors common to the numerator and denominator with the equivalent fraction $\frac{1}{1}$.
> 3. Multiply the remaining factors in the numerator and in the denominator.

We have seen that the Fundamental Property of Fractions can be used to build equivalent fractions. If the sides of the equation $\frac{a}{b} = \frac{ac}{bc}$ are reversed, it can also be used to simplify fractions.

> **The Fundamental Property of Fractions**
>
> If the numerator and denominator of a fraction are divided by the same nonzero number, the resulting fraction is equivalent to the original fraction.
>
> If $\frac{a}{b}$ is a fraction and c is a nonzero number,
>
> $$\frac{ac}{bc} = \frac{a}{b}$$

> **Caution!** When all common factors of the numerator and/or the denominator of a fraction are removed, forgetting to write 1's above the slashes can lead to a common mistake.
>
> *Correct*
>
> $$\frac{15}{45} = \frac{\overset{1}{\cancel{3}} \cdot \overset{1}{\cancel{5}}}{3 \cdot \underset{1}{\cancel{3}} \cdot \underset{1}{\cancel{5}}} = \frac{1}{3}$$
>
> *Incorrect*
>
> $$\frac{15}{45} = \frac{\cancel{3} \cdot \cancel{5}}{3 \cdot \cancel{3} \cdot \cancel{5}} = \frac{0}{3} = 0$$

6 Add and subtract fractions.

In algebra as in everyday life, we can only add or subtract objects that are similar. For example, we can add dollars to dollars, but we cannot add dollars to oranges. This concept is important when adding fractions.

Consider the problem $\frac{2}{5} + \frac{1}{5}$. When we write it in words, it is apparent we are adding similar objects.

two- **fifths** + one- **fifth**
      ↑— Similar objects —↑

Because the denominators of $\frac{2}{5}$ and $\frac{1}{5}$ are the same, we say that they have a **common denominator**.

> **Adding and Subtracting Fractions That Have the Same Denominator**
>
> To add (or subtract) fractions that have the same denominator, add (or subtract) their numerators and write the sum (or difference) over the common denominator.
>
> For any fractions $\frac{a}{d}$ and $\frac{b}{d}$,
>
> $$\frac{a}{d} + \frac{b}{d} = \frac{a+b}{d} \quad \text{and} \quad \frac{a}{d} - \frac{b}{d} = \frac{a-b}{d}$$

For example,

$$\frac{2}{5} + \frac{1}{5} = \frac{2+1}{5} = \frac{3}{5} \quad \text{and} \quad \frac{18}{23} - \frac{9}{23} = \frac{18-9}{23} = \frac{9}{23}$$

> **Caution!** We **do not** add fractions by adding the numerators and adding the denominators!
>
> $$\cancel{\frac{2}{5} + \frac{1}{5} = \frac{2+1}{5+5} = \frac{3}{10}}$$
>
> The same caution applies when subtracting fractions.

Now we consider the problem $\frac{2}{5} + \frac{1}{3}$. Since the denominators are not the same, we cannot add these fractions in their present form.

two-**fifths** + one-**third**
$\quad\quad\quad$ ↳ Not similar objects ↲

To add (or subtract) fractions with different denominators, we express them as equivalent fractions that have a common denominator. The smallest common denominator, called the **least** or **lowest common denominator**, is usually the easiest common denominator to use.

Least Common Denominator (LCD)

The **least** or **lowest common denominator (LCD)** for a set of fractions is the smallest number each denominator will divide exactly (divide with no remainder).

> **Success Tip** To determine the LCD of two fractions, list the multiples of one of the denominators. The first number in the list that is exactly divisible by the other denominator is their LCD. For $\frac{2}{5}$ and $\frac{1}{3}$, the multiples of the first denominator, 5, are
>
> 5, 10, ⑮, 20, 25, …
>
> Since 15 is the first number in the list that is exactly divisible by the second denominator, 3, the LCD is 15.

The denominators of $\frac{2}{5}$ and $\frac{1}{3}$ are 5 and 3. The numbers 5 and 3 divide many numbers exactly (30, 45, and 60, to name a few), but the smallest number that they divide exactly is 15. Thus, 15 is the LCD for $\frac{2}{5}$ and $\frac{1}{3}$.

To find $\frac{2}{5} + \frac{1}{3}$, we find equivalent fractions that have denominators of 15 and we use the rule for adding fractions.

$$\frac{2}{5} + \frac{1}{3} = \frac{2}{5} \cdot \frac{3}{3} + \frac{1}{3} \cdot \frac{5}{5} \quad \text{Multiply } \tfrac{2}{5} \text{ by 1 in the form of } \tfrac{3}{3}. \text{ Multiply } \tfrac{1}{3} \text{ by 1 in the form of } \tfrac{5}{5}.$$

$$= \frac{2 \cdot 3}{5 \cdot 3} + \frac{1 \cdot 5}{3 \cdot 5} \quad \begin{array}{l}\text{Multiply the numerators.} \\ \text{Multiply the denominators.}\end{array}$$

$$= \frac{6}{15} + \frac{5}{15} \quad \text{Note that the denominators are now the same.}$$

$$= \frac{6+5}{15} \quad \text{Add the numerators.}$$
$$\quad\quad \text{Write the sum over the common denominator.}$$
$$= \frac{11}{15}$$

When adding (or subtracting) fractions with unlike denominators, the least common denominator is not always obvious. Prime factorization is helpful in determining the LCD.

> **Finding the LCD Using Prime Factorization**
>
> 1. Prime factor each denominator.
> 2. The LCD is a product of prime factors, where each factor is used the greatest number of times it appears in any one factorization found in step 1.

Self Check 7

Subtract: $\dfrac{11}{48} - \dfrac{7}{40}$

Now Try Problem 63

EXAMPLE 7 Subtract: $\dfrac{3}{10} - \dfrac{5}{28}$

Strategy We will begin by expressing each fraction as an equivalent fraction that has the LCD for its denominator. Then we will use the rule for subtracting fractions with *like* denominators.

WHY To add or subtract fractions, the fractions must have like denominators.

Solution
To find the LCD, we find the prime factorization of both denominators and use each prime factor the *greatest* number of times it appears in any one factorization:

$$\left.\begin{array}{l}10 = 2 \cdot 5 \\ 28 = 2 \cdot 2 \cdot 7\end{array}\right\} \text{LCD} = 2 \cdot 2 \cdot 5 \cdot 7 = 140 \quad \begin{array}{l}\text{2 appears twice in the factorization of 28.}\\ \text{5 appears once in the factorization of 10.}\\ \text{7 appears once in the factorization of 28.}\end{array}$$

Since 140 is the smallest number that 10 and 28 divide exactly, we write $\frac{3}{10}$ and $\frac{5}{28}$ as fractions with the LCD 140.

$$\frac{3}{10} - \frac{5}{28} = \frac{3}{10} \cdot \frac{14}{14} - \frac{5}{28} \cdot \frac{5}{5} \quad \begin{array}{l}\text{We must multiply 10 by 14 to obtain 140.}\\ \text{We must multiply 28 by 5 to obtain 140.}\end{array}$$

$$= \frac{3 \cdot 14}{10 \cdot 14} - \frac{5 \cdot 5}{28 \cdot 5} \quad \begin{array}{l}\text{Multiply the numerators.}\\ \text{Multiply the denominators.}\end{array}$$

$$= \frac{42}{140} - \frac{25}{140} \quad \text{Note that the denominators are now the same.}$$

$$= \frac{42 - 25}{140} \quad \begin{array}{l}\text{Subtract the numerators.}\\ \text{Write the difference over the common denominator.}\end{array}$$

$$= \frac{17}{140}$$

We can use the following steps to add or subtract fractions with different denominators.

1.2 Fractions 23

Adding and Subtracting Fractions That Have Different Denominators

1. Find the LCD.
2. Rewrite each fraction as an equivalent fraction with the LCD as the denominator. To do so, build each fraction using a form of 1 that involves any factors needed to obtain the LCD.
3. Add or subtract the numerators and write the sum or difference over the LCD.
4. Simplify the result, if possible.

7 Simplify answers.

When adding, subtracting, multiplying, or dividing fractions, remember to express the answer in simplest form.

EXAMPLE 8 Perform the operations and simplify:

a. $45\left(\dfrac{4}{9}\right)$ b. $\dfrac{5}{12} + \dfrac{3}{2} - \dfrac{1}{4}$

Strategy We will perform the indicated operations and then make sure that the answer is in simplest form (lowest terms).

WHY Fractional answers should always be given in simplest form.

Caution! Remember that an LCD is **not needed** when multiplying or dividing fractions.

Solution

a. $45\left(\dfrac{4}{9}\right) = \dfrac{45}{1}\left(\dfrac{4}{9}\right)$ Write 45 as a fraction: $45 = \dfrac{45}{1}$.

$= \dfrac{45 \cdot 4}{1 \cdot 9}$ Multiply the numerators. Multiply the denominators.

$= \dfrac{5 \cdot \overset{1}{\cancel{9}} \cdot 4}{1 \cdot \underset{1}{\cancel{9}}}$ To simplify the result, factor 45 as $5 \cdot 9$. Then remove the common factor 9 from the numerator and denominator.

$= 20$ Multiply the remaining factors in the numerator. Multiply the remaining factors in the denominator. Simplify: $\dfrac{20}{1} = 20$.

b. Since the smallest number that 12, 2, and 4 divide exactly is 12, the LCD is 12.

$\dfrac{5}{12} + \dfrac{3}{2} - \dfrac{1}{4} = \dfrac{5}{12} + \dfrac{3}{2} \cdot \dfrac{6}{6} - \dfrac{1}{4} \cdot \dfrac{3}{3}$ $\dfrac{5}{12}$ already has a denominator of 12. Build $\dfrac{3}{2}$ and $\dfrac{1}{4}$ so that their denominators are 12.

$= \dfrac{5}{12} + \dfrac{3 \cdot 6}{2 \cdot 6} - \dfrac{1 \cdot 3}{4 \cdot 3}$ Multiply the numerators. Multiply the denominators.

$= \dfrac{5}{12} + \dfrac{18}{12} - \dfrac{3}{12}$ The denominators are now the same.

$= \dfrac{20}{12}$ Add the numerators, 5 and 18, to get 23. From that sum, subtract 3. Write that result, 20, over the common denominator.

Self Check 8

Perform the operations and simplify: a. $24\left(\dfrac{7}{6}\right)$

b. $\dfrac{1}{15} + \dfrac{31}{30} - \dfrac{3}{10}$

Now Try Problems 67 and 71

$$= \frac{\overset{1}{\cancel{4}} \cdot 5}{3 \cdot \underset{1}{\cancel{4}}} \quad \text{To simplify } \tfrac{20}{12}, \text{ factor 20 and 12, using their greatest common factor, 4. Then remove } \tfrac{4}{4} = 1.$$

$$= \frac{5}{3}$$

THINK IT THROUGH Budgeting

"Working with a personal budget puts you in control of your money. It can also help you achieve your savings goals."

Leigh Roberts in Business Times

In the article "Budget Basics for College Graduates," Janet Bodnar offers financial guidance to new college grads beginning a job. The circle graph below shows her spending guidelines presented as a fraction of one's monthly take-home pay. How much does she recommend should be spent on each budget item if a person takes home $2,100 a month?

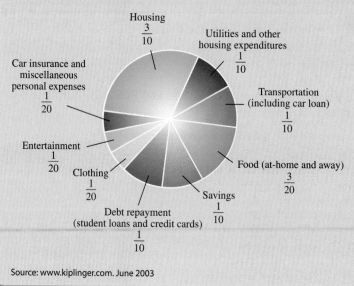

Source: www.kiplinger.com. June 2003

8 Compute with mixed numbers.

A **mixed number** represents the sum of a whole number and a fraction. For example, $5\tfrac{3}{4}$ means $5 + \tfrac{3}{4}$.

Self Check 9

Multiply: $1\tfrac{1}{8} \cdot 9$

Now Try Problem 75

EXAMPLE 9

Divide: $5\tfrac{3}{4} \div 2$

Strategy We begin by writing the mixed number $5\tfrac{3}{4}$ and the whole number 2 as fractions. Then we use the rule for dividing two fractions.

WHY To multiply (or divide) with mixed numbers, we first write them as fractions, and then multiply (or divide) as usual.

Solution

$$5\tfrac{3}{4} \div 2 = \tfrac{23}{4} \div \tfrac{2}{1}$$

Write $5\tfrac{3}{4}$ as an improper fraction by multiplying its whole-number part by the denominator: $5 \cdot 4 = 20$. Then add the numerator to that product: $3 + 20 = 23$. Finally, write the result, 23, over the denominator 4. Write 2 as a fraction: $2 = \tfrac{2}{1}$.

$$= \frac{23}{4} \cdot \frac{1}{2}$$ Multiply by the reciprocal of $\frac{2}{1}$, which is $\frac{1}{2}$.

$$= \frac{23}{8}$$ Multiply the numerators.
Multiply the denominators.

$$= 2\frac{7}{8}$$ Write $\frac{23}{8}$ as a mixed number by dividing the numerator, 23, by the denominator, 8. The quotient, 2, is the whole-number part; the remainder, 7, over the divisor, 8, is the fractional part.

The Language of Algebra Fractions such as $\frac{23}{4}$, with a numerator greater than or equal to the denominator, are called **improper fractions**. In algebra, such fractions are often preferable to their equivalent mixed number form.

EXAMPLE 10 *Freeway Signs* How far apart are the Downtown San Diego and Sea World Drive exits?

Strategy We can find the distance between exits by finding the difference in the mileages on the freeway sign: $6\frac{1}{2} - 1\frac{3}{4}$.

WHY The word *difference* indicates subtraction.

Self Check 10
Subtract: $5\frac{1}{3} - 3\frac{3}{4}$
Now Try Problem 79

Solution

$$\begin{array}{rrrrr} 6\frac{1}{2} = & 6\frac{2}{4} = & 5\frac{2}{4} + \frac{4}{4} = & 5\frac{6}{4} \\ -1\frac{3}{4} = & -1\frac{3}{4} = & -1\frac{3}{4} & = -1\frac{3}{4} \\ \hline & & & 4\frac{3}{4} \end{array}$$

Using vertical form, express $\frac{1}{2}$ as an equivalent fraction with denominator 4. Then, borrow 1 in the form of $\frac{4}{4}$ from 6 to subtract the fractional parts of the mixed numbers.

The Downtown San Diego and Sea World Drive exits are $4\frac{3}{4}$ miles apart.

Success Tip Example 10 could also be solved by writing the mixed numbers $6\frac{1}{2}$ and $1\frac{3}{4}$ as improper fractions and subtracting them.

ANSWERS TO SELF CHECKS

1. $189 = 3 \cdot 3 \cdot 3 \cdot 7$ **2.** $\frac{10}{27}$ **3.** $\frac{12}{25}$ **4.** $\frac{15}{24}$ **5. a.** $\frac{3}{7}$ **b.** in simplest form **6.** $\frac{9}{5}$ **7.** $\frac{13}{240}$ **8. a.** 28 **b.** $\frac{4}{5}$ **9.** $\frac{81}{8} = 10\frac{1}{8}$ **10.** $1\frac{7}{12}$

STUDY SKILLS CHECKLIST

Learning from the Worked Examples

The following checklist will help you become familiar with the example structure in this book. Place a checkmark in each box after you answer the question.

☐ Each section of the book contains worked **Examples** that are numbered. How many worked examples are there in Section 1.3, which begins on page 29?

☐ Each worked example contains a **Strategy**. Fill in the blanks to complete the following strategy for Example 6 on page 36: We will _____ _____ _____ on each side and then compare the values.

☐ Each Strategy statement is followed by an explanation of **Why** that approach is used. Fill in the blanks to complete the following Why for Example 6 on page 36: Simplifying the sides helps us determine _____ _____ _____ _____.

☐ Each worked example has a **Solution**. How many lettered parts are there to the Solution in Example 6 on page 36?

☐ Each example uses magenta **Author notes** to explain the steps of the solution. Fill in the blanks to complete the second author note in the solution of Example 8 on page 23: Multiply the _____.

☐ After reading a worked example, you should work the **Self Check** problem. How many Self Check problems are there for Example 8 on page 23?

☐ At the end of each section, you will find the **Answers to Self Checks**. What is the answer to Self Check problem 4 given on page 25?

☐ After completing a Self Check problem, you can **Now Try** similar problems in the Study Sets. For Example 10 on page 25, which Study Set problem is suggested?

Answers: 6, simplify the expression, which symbol to use, 2, numerators, 2, $\frac{15}{24}$, 79

SECTION 1.2 STUDY SET

VOCABULARY

Fill in the blanks.

1. A factor is a number being _____.
2. Numbers that have only 1 and themselves as factors, such as 23, 37, and 41, are called _____ numbers.
3. When we write 60 as $2 \cdot 2 \cdot 3 \cdot 5$, we say that we have written 60 in _____ form.
4. The _____ of the fraction $\frac{3}{4}$ is 3, and the _____ is 4.
5. Two fractions that represent the same number, such as $\frac{1}{2}$ and $\frac{2}{4}$, are called _____ fractions.
6. $\frac{2}{3}$ is the _____ of $\frac{3}{2}$, because their product is 1.
7. The _____ common denominator for a set of fractions is the smallest number each denominator will divide exactly.
8. The _____ number $7\frac{1}{3}$ represents the sum of a whole number and a fraction: $7 + \frac{1}{3}$.

CONCEPTS

Complete each fact about fractions.

9. a. $\dfrac{a}{a} = \square$ b. $\dfrac{a}{1} = \square$
   c. $\dfrac{a}{b} \cdot \dfrac{c}{d} = \square$ d. $\dfrac{a}{b} \div \dfrac{c}{d} = \square$
   e. $\dfrac{a}{d} + \dfrac{b}{d} = \square$ f. $\dfrac{a}{d} - \dfrac{b}{d} = \square$

10. What two equivalent fractions are shown?

11. Complete each statement.
    a. To simplify a fraction, we remove factors equal to $\square$ in the form of $\frac{2}{2}$, $\frac{3}{3}$, or $\frac{4}{4}$, and so on.
    b. To build a fraction, we multiply it by $\square$ in the form of $\frac{2}{2}$, $\frac{3}{3}$, or $\frac{4}{4}$, and so on.

12. What is the LCD for fractions having denominators of 24 and 36?

NOTATION
Fill in the blanks.

13. **a.** Multiply $\frac{5}{6}$ by a form of 1 to build an equivalent fraction with denominator 30.

    $\frac{5}{6} \cdot \boxed{} = \boxed{}$

    b. Remove common factors to simplify $\frac{12}{42}$.

    $\frac{12}{42} = \frac{2 \cdot \boxed{} \cdot 3}{2 \cdot 3 \cdot \boxed{}} = \boxed{}$

14. **a.** Write $2\frac{15}{16}$ as an improper fraction.
    b. Write $\frac{49}{12}$ as a mixed number.

GUIDED PRACTICE
Find the prime factorization of each number. See Example 1.

15. 75
16. 20
17. 28
18. 54
19. 81
20. 125
21. 117
22. 147
23. 220
24. 270
25. 1,254
26. 1,122

Perform each operation. See Examples 2 and 3.

27. $\frac{5}{6} \cdot \frac{1}{8}$
28. $\frac{2}{3} \cdot \frac{1}{5}$
29. $\frac{7}{11} \cdot \frac{3}{5}$
30. $\frac{8}{19} \cdot \frac{3}{5}$
31. $\frac{3}{4} \div \frac{2}{5}$
32. $\frac{7}{8} \div \frac{6}{13}$
33. $\frac{6}{5} \div \frac{5}{7}$
34. $\frac{4}{3} \div \frac{3}{2}$

Build each fraction or whole number to an equivalent fraction with the indicated denominator. See Example 4.

35. $\frac{1}{3}$, denominator 9
36. $\frac{3}{8}$, denominator 24
37. $\frac{4}{9}$, denominator 54
38. $\frac{9}{16}$, denominator 64
39. 7, denominator 5
40. 15, denominator 3
41. 5, denominator 7
42. 6, denominator 8

Simplify each fraction, if possible. See Examples 5 and 6.

43. $\frac{6}{18}$
44. $\frac{6}{9}$
45. $\frac{24}{28}$
46. $\frac{35}{14}$
47. $\frac{15}{40}$
48. $\frac{15}{21}$
49. $\frac{33}{56}$
50. $\frac{26}{21}$
51. $\frac{26}{39}$
52. $\frac{72}{64}$
53. $\frac{36}{225}$
54. $\frac{175}{490}$

Perform the operations and, if possible, simplify. See Objective 6 and Example 7.

55. $\frac{3}{5} + \frac{3}{5}$
56. $\frac{4}{9} - \frac{1}{9}$
57. $\frac{6}{7} - \frac{2}{7}$
58. $\frac{5}{13} + \frac{6}{13}$
59. $\frac{1}{6} + \frac{1}{24}$
60. $\frac{17}{25} - \frac{2}{5}$
61. $\frac{7}{10} - \frac{1}{14}$
62. $\frac{9}{8} - \frac{5}{6}$
63. $\frac{2}{15} + \frac{7}{9}$
64. $\frac{7}{25} + \frac{3}{10}$
65. $\frac{21}{56} - \frac{9}{40}$
66. $\frac{13}{24} - \frac{3}{40}$

Perform the operations and, if possible, simplify. See Example 8.

67. $16\left(\frac{3}{2}\right)$
68. $30\left(\frac{5}{6}\right)$
69. $18\left(\frac{2}{9}\right)$
70. $14\left(\frac{3}{7}\right)$
71. $\frac{2}{3} - \frac{1}{6} + \frac{5}{18}$
72. $\frac{3}{5} - \frac{7}{10} + \frac{7}{20}$
73. $\frac{5}{12} + \frac{1}{3} - \frac{2}{5}$
74. $\frac{7}{15} + \frac{1}{5} - \frac{4}{9}$

Perform the operations and, if possible, simplify. See Examples 9 and 10.

75. $4\frac{2}{3} \cdot 7$
76. $7 \cdot 1\frac{3}{28}$
77. $8 \div 3\frac{1}{5}$
78. $15 \div 3\frac{1}{3}$
79. $8\frac{2}{9} - 7\frac{2}{3}$
80. $3\frac{4}{5} - 3\frac{1}{10}$
81. $3\frac{3}{16} + 2\frac{5}{24}$
82. $15\frac{5}{6} + 11\frac{5}{8}$

TRY IT YOURSELF
Perform the operations and, if possible, simplify.

83. $\frac{3}{5} + \frac{2}{3}$
84. $\frac{4}{3} + \frac{7}{2}$
85. $21\left(\frac{10}{3}\right)$
86. $27\left(\frac{8}{9}\right)$
87. $6 \cdot 2\frac{7}{24}$
88. $3\frac{1}{2} \cdot \frac{1}{5}$
89. $\frac{2}{3} - \frac{1}{4} + \frac{1}{12}$
90. $\frac{3}{7} - \frac{2}{5} + \frac{2}{35}$

91. $\dfrac{21}{35} \div \dfrac{3}{14}$ 92. $\dfrac{23}{25} \div \dfrac{46}{5}$

93. $\dfrac{4}{3}\left(\dfrac{6}{5}\right)$ 94. $\dfrac{21}{8}\left(\dfrac{2}{15}\right)$

95. $\dfrac{4}{63} + \dfrac{1}{45}$ 96. $\dfrac{5}{18} + \dfrac{1}{99}$

97. $3 - \dfrac{3}{4}$ 98. $4 - \dfrac{7}{3}$

99. $\dfrac{1}{2} \cdot \dfrac{3}{5}$ 100. $\dfrac{3}{4} \cdot \dfrac{5}{7}$

101. $3\dfrac{1}{3} \div 1\dfrac{5}{6}$ 102. $2\dfrac{1}{2} \div 1\dfrac{5}{8}$

103. $\dfrac{11}{21} - \dfrac{8}{21}$ 104. $\dfrac{19}{35} - \dfrac{12}{35}$

APPLICATIONS

105. **BOTANY** To determine the effects of smog on tree development, a scientist cut down a pine tree and measured the width of the growth rings for two five-year time periods.
   a. What was the total growth over the ten-year period?
   b. What is the difference in growth for the two five-year time periods?

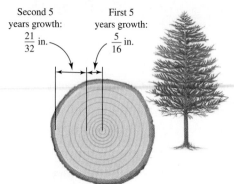

Second 5 years growth: $\dfrac{21}{32}$ in.   First 5 years growth: $\dfrac{5}{16}$ in.

106. **HARDWARE** To secure the bracket to the stock, a bolt and a nut are used. How long should the threaded part of the bolt be?

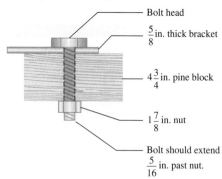

Bolt head
$\dfrac{5}{8}$ in. thick bracket
$4\dfrac{3}{4}$ in. pine block
$1\dfrac{7}{8}$ in. nut
Bolt should extend $\dfrac{5}{16}$ in. past nut.

107. **COOKING** How much butter is left in a $10\dfrac{1}{2}$-pound tub of butter if $4\dfrac{3}{4}$ pounds are used to make a wedding cake?

108. **CALORIES** A company advertises that its mints contain only $3\dfrac{1}{2}$ calories a piece. What is the calorie intake if you eat an entire package of 20 mints?

109. **FRAMES** How many inches of molding are needed to make the square picture frame?

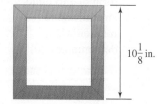

$10\dfrac{1}{8}$ in.

110. **DECORATING** The materials used to make a pillow are shown. Examine the inventory list to decide how many pillows can be manufactured in one production run with the materials in stock.

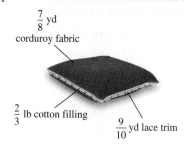

$\dfrac{7}{8}$ yd corduroy fabric

$\dfrac{2}{3}$ lb cotton filling

$\dfrac{9}{10}$ yd lace trim

Factory Inventory List	
Materials	Amount in stock
Lace trim	135 yd
Corduroy fabric	154 yd
Cotton filling	98 lb

WRITING

111. Explain how to add two fractions that have unlike denominators.
112. To multiply two fractions, must they have like denominators? Explain.
113. What are equivalent fractions?
114. Explain the error in the following addition.

$$\dfrac{4}{3} + \dfrac{3}{2} = \dfrac{4+3}{3+2} = \dfrac{7}{5}$$

REVIEW

Use the formula to complete each table.

115. $T = 15g$

Number of gears (g)	Number of teeth (T)
10	
12	

116. $p = r - 200$

Revenue (r)	Profit (p)
1,000	
5,000	

SECTION 1.3
The Real Numbers

Objectives
1. Define the set of integers.
2. Graph sets of numbers on the number line.
3. Define the set of rational numbers.
4. Define the set of irrational numbers.
5. Define the set of real numbers.
6. Find the absolute value of a real number.

We have discussed the set of whole numbers, natural numbers, and the set of prime numbers. In this section, we define other sets of numbers and show that they are part of a larger collection of numbers called **real numbers**.

1 Define the set of integers.

In the table, we see the low temperatures for Rockford during the first week of January. In the left column, we have used the numbers 1, 2, 3, 4, 5, 6, and 7 to denote the calendar days of the month. This collection of numbers is called a **set**, and the members (or **elements**) of the set can be listed within **braces** { }.

Day of the month	Low temperature (°F)
1	4
2	−5
3	−6
4	0
5	3
6	6
7	6

{1, 2, 3, 4, 5, 6, 7}

Each of the numbers 1, 2, 3, 4, 5, 6, and 7 is a member of a basic set of numbers called the **natural numbers**, the numbers that we use for counting.

Natural Numbers

The set of **natural numbers** is {1, 2, 3, 4, 5, 6, 7, 8, 9, 10, … }.

The Language of Algebra The symbol … is called an *ellipsis* and indicates the pattern continues forever.

The natural numbers together with 0 make up another important set of numbers called the **whole numbers**.

Whole Numbers

The set of **whole numbers** is {0, 1, 2, 3, 4, 5, 6, 7, 8, 9, 10, … }.

Caution! Since every natural number is a whole number, we say that the set of natural numbers is a **subset** of the set of whole numbers. However, not all whole numbers are natural numbers. Note that 0 is a whole number but not a natural number.

The table above contains positive and negative temperatures. For example, on the second day of the month, the low temperature was −5° (read as "negative 5 degrees"), and it means 5° below zero. On the fifth day, the low temperature was 3° (3° above zero). The numbers used to represent the temperatures listed in the table are members of a set of numbers called the **integers**.

Integers

The set of **integers** is { … , −4, −3, −2, −1, 0, 1, 2, 3, 4, … }.

Caution! Since every whole number is an integer, we say that the set of whole numbers is a *subset* of the set of integers. The natural numbers are also a subset of the integers. However, not all integers are whole numbers, nor are they all natural numbers. For example, the integer -2 is neither a whole number nor a natural number.

2 Graph sets of numbers on the number line.

We can illustrate sets of numbers with the **number line**. Like a ruler, the number line is straight and has uniform markings, as in the number line below. The arrowheads indicate that the number line continues forever to the left and to the right. Numbers to the right of 0 have values that are greater than 0; they are called **positive numbers**. Numbers to the left of 0 have values that are less than 0; they are called **negative numbers**. The number 0 is neither positive nor negative.

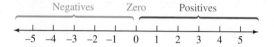

Positive numbers can be written with or without a $+$ sign. For example, $2 = +2$. Negative numbers are always written with a $-$ sign. They can be used to describe amounts that are less than 0, such as a checking account that is $75 overdrawn ($-\$75$), an elevation of 200 feet below sea level (-200 ft), and a loss of 8 points (-8).

Using a process known as **graphing**, a single number or a set of numbers can be represented on a number line. The **graph of a number** is the point on the number line that corresponds to that number. To *graph a number* means to locate its position on the number line and then to highlight it by using a heavy dot.

Self Check 1

Graph the integers between -2 and 2.

Now Try Problem 51

EXAMPLE 1
Graph the integers between -4 and 5.

Strategy We will identify the integers between -4 and 5 and locate their positions on the number line. We will then draw a bold dot for each number and label them.

WHY To *graph a number* means to make a drawing that represents the number.

Solution
The integers between -4 and 5 are $-3, -2, -1, 0, 1, 2, 3$, and 4. To graph each integer, we locate its position on the number line and draw a dot.

Success Tip "Between" means you do not include the endpoints. "From . . . to" means you do include the endpoints as long as they are in the set to be graphed.

As we move to the right on the number line, the values of the numbers increase. As we move to the left, the values of the numbers decrease. In the figure, we know that 5 is greater than -3 because the graph of 5 lies to the right of the graph of -3. We also know that -3 is less than 5 because its graph lies to the left of the graph of 5.

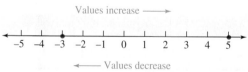

The **inequality symbol** > ("is greater than") can be used to show that 5 is greater than −3, and the inequality symbol < ("is less than") can be used to show that −3 is less than 5.

$5 > -3$ Read as "5 is greater than −3."
$-3 < 5$ Read as "−3 is less than 5."

To distinguish between these two inequality symbols, remember that each one points to the smaller of the two numbers involved.

$5 > -3$ $-3 < 5$
   ↑── Points to the smaller number. ──↑

> **The Language of Algebra** To state that a number x is positive, we can write $x > 0$. To state that a number x is negative, we can write $x < 0$.

EXAMPLE 2
Use one of the symbols > or < to make each of the following statements true: **a.** −4 ☐ 4 **b.** −2 ☐ −3

Strategy To pick the correct inequality sign, we will determine the position of the graph of each number on the number line.

WHY When we compare the positions of two numbers on the number line, the number on the left is the smaller number and the number on the right is the larger.

Solution

a. Since −4 is to the left of 4 on the number line, we have $-4 < 4$.

b. Since −2 is to the right of −3 on the number line, we have $-2 > -3$.

By extending the number line to include negative numbers, we can represent more situations graphically. In the figure to the right, the line graph illustrates the low temperatures listed in the table on page 29. The vertical axis is scaled in units of two degrees Fahrenheit, and temperatures below zero (negative temperatures) are graphed. For example, on the third day of the month, the low was −6°F.

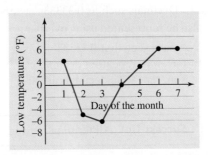

Self Check 2
Use one of the symbols > or < to make each of the following statements true:
a. 1 ☐ −1
b. −5 ☐ −4
Now Try Problems 55 and 57

3 Define the set of rational numbers.

In this course, we will work with positive and negative fractions. For example, the time it takes a motorist to complete the commute home might be $\frac{3}{4}$ of an hour, or a surveyor might indicate that a building's foundation has fallen below finished grade by expressing its elevation as $-\frac{7}{8}$ of an inch.

We will also work with positive and negative mixed numbers. For example, a piece of fabric might be $12\frac{2}{3}$ yards long, or the time before a rocket launch might be expressed as "$-5\frac{1}{2}$ minutes and counting."

Fractions and mixed numbers belong to the set of **rational numbers**, so named because *rational* numbers can be written as the *ratio* (or quotient) of two integers. This means that all rational numbers can be written in fractional form:

$\dfrac{\text{integer}}{\text{nonzero integer}}$ The numerator can be any integer.
                                     The denominator can be any integer except 0.

Fractions such as $\frac{3}{4}$ and $\frac{25}{12}$ are rational numbers, because they have an integer numerator and a nonzero integer denominator. We can use the fact that

$$-\frac{a}{b} = \frac{-a}{b} = \frac{a}{-b}, \quad \text{where } b \neq 0$$

to show that negative fractions are rational numbers. For example, $-\frac{7}{8}$ is a rational number because it can be written as $\frac{-7}{8}$ or as $\frac{7}{-8}$. Positive and negative mixed numbers such as $12\frac{2}{3}$ and $-5\frac{1}{2}$ are also rational numbers because they can be written as the ratios of two integers.

$$12\frac{2}{3} = \frac{38}{3} \quad \text{and} \quad -5\frac{1}{2} = \frac{-11}{2}$$

Many numerical quantities are expressed in decimal notation. For example, a candy bar might cost $0.89, or a dragster might travel at 203.156 mph, or the first-quarter loss of a business might be $-\$2.7$ million. Since these terminating decimals can be written as ratios of two integers, they are rational numbers.

$$0.89 = \frac{89}{100} \qquad 203.156 = 203\frac{156}{1,000} = \frac{203,156}{1,000} \qquad -2.7 = -2\frac{7}{10} = \frac{-27}{10}$$

To find the *decimal equivalent* for a fraction, we divide its numerator by its denominator. For example, to write $\frac{3}{4}$ as an equivalent decimal, we proceed as follows:

```
    0.75
  4)3.0
   -2 8
     20
     20
      0
```

Decimals such as 0.33333 … and 2.161616 … are repeating decimals and can also be represented using **overbar** notation. 0.33333 … = $0.\overline{3}$ and 2.1616 … = $2.\overline{16}$. You have seen that 0.33333 … = $\frac{1}{3}$. It can be shown that 2.161616 … = $2\frac{16}{99} = \frac{214}{99}$. In fact, *any* repeating decimal can be expressed as a ratio of two integers. For this reason, repeating decimals are also rational numbers.

> **Success Tip** All terminating and repeating decimals are rational numbers.

The set of rational numbers is too extensive to be listed in the same way that we have listed other sets in this section. Instead, we use **set-builder notation** to describe it.

The Set of Rational Numbers

$$\left\{ \frac{a}{b} \,\middle|\, a \text{ and } b \text{ are integers and } b \neq 0 \right\}$$

Read as "the set of all numbers of the form $\frac{a}{b}$, where a and b represent integers and $b \neq 0$."

4 Define the set of irrational numbers.

The square root of 2 (denoted $\sqrt{2}$) is the number that, when multiplied by itself, gives 2. That is, $\sqrt{2} \cdot \sqrt{2} = 2$. In illustration (a) on the next page, the anchor wire is the diagonal of a square with sides that are 1 yard in length. It can be shown that the length of the wire is $\sqrt{2}$ yards.

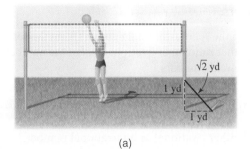

(a)   (b)

The number represented by the Greek letter π (read as "pi") is often used in geometry. In illustration (b), the distance around the hula hoop (circumference) is found by multiplying the diameter of the hoop by π.

Expressed in decimal form,

$$\sqrt{2} = 1.414213562\ldots \quad \text{and} \quad \pi = 3.141592654\ldots$$

The Language of Algebra Since π is irrational, its decimal representation has an infinite number of decimal places that do not repeat a block of digits. In 2011, a Japanese systems engineer and a U.S. computer scientist used a super computer to calculate π to more than 10 trillion decimal places.

These **nonterminating, nonrepeating decimals** cannot be written as the ratio of two integers. Therefore, $\sqrt{2}$ and π are *not* rational numbers—they are called **irrational numbers**.

Irrational Numbers

An **irrational number** is a nonterminating, nonrepeating decimal.

Other examples of irrational numbers are $\sqrt{89}$, $-\sqrt{5}$, $-\pi$, and 3π (this means $3 \cdot \pi$). When doing calculations with irrational numbers, we often approximate them.

Using Your CALCULATOR Approximating Irrational Numbers

We can use the square root key $\boxed{\sqrt{\ }}$ on a calculator to approximate $\boxed{\sqrt{2}}$. To approximate π, we use the $\boxed{\pi}$ key. With a scientific calculator that is reverse entry, we enter these numbers and press these keys.

$2\ \boxed{\sqrt{\ }}$ `1.414213562`

$\boxed{\pi}$ (You may have to use a $\boxed{\text{2nd}}$ or $\boxed{\text{Shift}}$ key first.)   `3.141592654`

We see that $\sqrt{2} \approx 1.414213562$. The symbol $\approx$ means "is approximately equal to." Rounded to the nearest hundredth, $\sqrt{2} \approx 1.41$. Rounded to the nearest thousandth, $\pi \approx 3.142$.

To approximate $\sqrt{2}$ and π using a graphing calculator or direct-entry scientific calculator, we enter these numbers and press these keys.

$\boxed{\text{2nd}}\ \boxed{\sqrt{\ }}\ \boxed{2}\ \boxed{)}\ \boxed{\text{ENTER}}$ `√(2)`
                                                     `1.414213562`

$\boxed{\text{2nd}}\ \boxed{\pi}\ \boxed{\text{ENTER}}$ π
                                   `3.141592654`

5 Define the set of real numbers.

The set of rational numbers together with the set of irrational numbers form the set of **real numbers**. That is, a real number is either rational or irrational. All of the numbers that we have discussed in this section are real numbers.

> **The Real Numbers**
>
> A **real number** is any number that is either a rational or an irrational number.

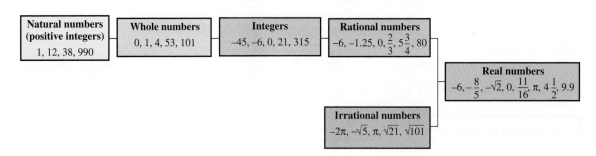

The diagram above shows how the sets of numbers introduced in this section are related. Note that a number can belong to more than one set. For example, -6 is an integer, a rational number, and a real number.

Self Check 3

Use the instructions for Example 3 with the set

$$\left\{0.4, \sqrt{2}, -\tfrac{2}{7}, 45, -2, \tfrac{13}{4}\right\}$$

Now Try Problem 63

EXAMPLE 3 Determine which numbers in the following set are natural numbers, whole numbers, integers, rational numbers, irrational numbers, and real numbers: $\left\{-3.4, \tfrac{2}{5}, 0, -6, 1\tfrac{3}{4}, \pi, 16\right\}$.

Strategy We begin by scanning the given set, looking for any natural numbers. Then we scan it five more times, looking for whole numbers, for integers, for rational numbers, for irrational numbers, and finally, for real numbers.

WHY We need to scan the given set of numbers six times, because numbers in the set can belong to more than one classification.

Solution

Natural numbers: 16   Whole numbers: 0, 16   Integers: 0, -6, 16

Rational numbers: $-3.4, \tfrac{2}{5}, 0, -6, 1\tfrac{3}{4}, 16$ Rational numbers can be expressed as a ratio of two integers: $-3.4 = \tfrac{-34}{10}$, $0 = \tfrac{0}{1}$, $-6 = \tfrac{-6}{1}$, $1\tfrac{3}{4} = \tfrac{7}{4}$, and $16 = \tfrac{16}{1}$.

Irrational numbers: π

Real numbers: $-3.4, \tfrac{2}{5}, 0, -6, 1\tfrac{3}{4}, \pi, 16$

The set of real numbers corresponds to all points on a number line. One and only one point on the number line corresponds to each real number.

Self Check 4

Graph each number in the set:

$$\left\{1.7, \pi, -1\tfrac{3}{4}, -0.333\ldots, \tfrac{5}{2}\right\}$$

EXAMPLE 4 Graph each number in the following set on the number line: $\left\{-3.25, -\tfrac{1}{5}, \sqrt{5}, 3\tfrac{7}{8}, 0.666\ldots, -\tfrac{3}{2}\right\}$.

Strategy We will find decimal approximations for the fractions and irrational numbers.

WHY Decimal approximations help locate the position of the graph of such numbers.

Solution

We locate the position of each number on the number line and then draw a heavy dot. Recall that $0.25 = \frac{1}{4}$, so $-3.25 = -3\frac{1}{4}$. In mixed-number form, $-\frac{3}{2} = -1\frac{1}{2}$. Using a calculator, we see that $\sqrt{5} \approx 2.2$. The repeating decimal $0.666\ldots$ is $\frac{2}{3}$.

Now Try Problem 67

6 Find the absolute value of a real number.

On the number line, -4 and 4 are both a distance of 4 away from 0. Because of this, we say that -4 and 4 are **opposites**.

> **Opposites**
>
> Two numbers represented by points on a number line that are the same distance away from 0, but on opposite sides of it, are called **opposites**.

To write the opposite of a number, a $-$ symbol can be used. For example, the opposite of 6 can be written as -6. The opposite of 0 is 0, so $-0 = 0$. Since the opposite of -6 is 6, we have $-(-6) = 6$. In general, if a represents any real number,

$$-(-a) = a$$

The **absolute value** of a number gives the distance between the number and 0 on the number line. To indicate absolute value, a number is inserted between two vertical bars. For the example, we would write $|-4| = 4$. This notation is read as "The absolute value of negative 4 is 4," and it tells us that the distance between -4 and 0 is 4 units. We also see that $|4| = 4$.

> **Absolute Value**
>
> The **absolute value** of a number is the distance on the number line between the number and 0.

> ***Success Tip*** Since absolute value expresses distance, the absolute value of a number is always positive or zero—never negative.

EXAMPLE 5 Find each absolute value: **a.** $|18|$ **b.** $\left|-\frac{7}{8}\right|$ **c.** $|0|$

Strategy We will find the distance that the number within the absolute value bars is from 0.

WHY The absolute value of a number is the distance between 0 and the number on a number line.

Self Check 5
Find each absolute value:
a. $|100|$
b. $|-4.7|$
c. $|-\sqrt{2}|$

Now Try Problem 72

Solution
a. Since 18 is a distance of 18 from 0 on the number line, $|18| = 18$.
b. Since $-\frac{7}{8}$ is a distance of $\frac{7}{8}$ from 0 on the number line, $\left|-\frac{7}{8}\right| = \frac{7}{8}$.
c. Since 0 is a distance of 0 from 0 on the number line, $|0| = 0$.

Self Check 6
Insert one of the symbols $>$, $<$, or $=$ in each blank:
a. $-(-7)$ ___ -6
b. $3\frac{3}{4}$ ___ $\left|-4\frac{3}{4}\right|$

Now Try Problems 75 and 77

EXAMPLE 6 Insert one of the symbols $>$, $<$, or $=$ in each blank:
a. $-(-3.9)$ ___ 3
b. $-\left|-\frac{4}{5}\right|$ ___ $\left|\sqrt{5}\right|$

Strategy We will simplify the expression on each side and then compare the values.

WHY Simplifying the sides helps us determine which symbol to use.

Solution
a. $-(-3.9) > 3$, because $-(-3.9) = 3.9$.
b. $-\left|-\frac{4}{5}\right| < \left|\sqrt{5}\right|$, because $-\left|-\frac{4}{5}\right| = -\frac{4}{5}$, and $\left|\sqrt{5}\right| = \sqrt{5} \approx 2.2$.

ANSWERS TO SELF CHECKS

1. [number line from −3 to 3 with points marked] 2. a. $>$ b. $<$ 3. natural numbers: 45; whole numbers: 45; integers: 45, −2; rational numbers: 0.4, $-\frac{2}{7}$, 45, −2, $\frac{13}{4}$; irrational numbers: $\sqrt{2}$; real numbers: 0.4, $\sqrt{2}$, $-\frac{2}{7}$, 45, −2, $\frac{13}{4}$ 4. [number line showing $-1\frac{3}{4}$, −0.333..., 1.7, $\frac{5}{2}$, π] 5. a. 100 b. 4.7 c. $\sqrt{2}$ 6. a. $>$ b. $<$

STUDY SKILLS CHECKLIST

Getting the Most from the Study Sets

The following checklist will help you become familiar with the Study Sets in this book. Place a checkmark in each box after you answer the question.

☐ Answers to the odd-numbered **Study Set** problems are located in the appendix beginning on page A-7. On what page do the answers to Study Set 1.2 appear?

☐ Each Study Set begins with **Vocabulary** problems. How many Vocabulary problems appear in Study Set 1.2, which begins on page 26?

☐ Following the Vocabulary problems, you will see **Concepts**. How many Concepts problems appear in Study Set 1.2?

☐ Following the Concepts problems, you will see **Notation** problems. How many Notation problems appear in Study Set 1.2?

☐ After the Notation problems are **Guided Practice** problems, which are linked to similar examples within the section. How many Guided Practice problems appear in Study Set 1.2?

☐ After the Guided Practice problems are **Try It Yourself** problems, which can be used to help you prepare for quizzes. How many Try It Yourself problems appear in Study Set 1.2?

☐ Following the Try It Yourself problems, you will see **Applications**. How many Applications problems appear in Study Set 1.2?

☐ After completing the Application problems, you will see a few **Writing** problems. How many Writing problems appear in Study Set 1.2?

☐ Each Study Set ends with a few **Review** problems. How many Review problems appear in Study Set 1.2?

Answer: A-7, 8, 4, 2, 68, 22, 6, 4, 2

SECTION 1.3 STUDY SET

VOCABULARY
Fill in the blanks.

1. The set of _____ numbers is {1, 2, 3, 4, 5, ...}.
2. The set of _____ numbers is {0, 1, 2, 3, 4, 5, ...}.
3. The set of _____ is {..., −2, −1, 0, 1, 2, ...}.
4. Numbers less than zero are _____, and numbers greater than zero are _____.
5. The symbols < and > are _____ symbols.
6. A _____ number can be written as a quotient (ratio) of two integers.
7. In _____ notation, the set of rational numbers is written as $\left\{\dfrac{a}{b} \,\middle|\, a \text{ and } b \text{ are integers and } b \neq 0\right\}$.
8. An irrational number is a nonterminating, nonrepeating _____.
9. An _____ number cannot be expressed as a quotient (ratio) of two integers.
10. All numbers that can be represented by points on the number line are called _____ numbers.
11. Two numbers represented by points on the number line that are the same distance away from 0, but on opposite sides of it, are called _____.
12. The _____ _____ of a real number is the distance on the number line between the number and 0.

CONCEPTS

13. Show that each of the following numbers is a rational number by expressing it as a fraction with an integer in its numerator and a nonzero integer in its denominator: 6, −9, $-\dfrac{7}{8}$, $3\dfrac{1}{2}$, −0.3, 2.83.

14. Represent each situation using a signed number.
    a. A loss of $15 million
    b. A rainfall total 0.75 inch below average
    c. A score $12\dfrac{1}{2}$ points under the standard
    d. A building foundation $\dfrac{5}{16}$ inch above grade

15. What numbers are a distance of 8 away from 5 on the number line?

16. Suppose the variable m stands for a negative number. Use an inequality to state this fact.

17. The variables a and b represent real numbers. Use an inequality symbol, < or >, to make each statement true.

    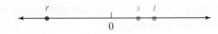

    a. a ___ b b. b ___ a c. b ___ 0 and a ___ 0

18. a. Write the statement −6 < −5 using an inequality symbol that points in the other direction.
    b. Write the statement 16 > −25 using an inequality symbol that points in the other direction.

19. What is the length of the diagonal of the square shown at right?

    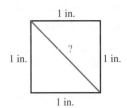

20. The following diagram can be used to show how the natural numbers, whole numbers, integers, rational numbers, and irrational numbers make up the set of real numbers. If the natural numbers are represented as shown, label each of the other sets.

    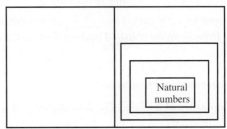

Determine whether each statement is true or false.

21. a. Every whole number is an integer.
    b. Every integer is a natural number.
    c. Every integer is a whole number.
    d. Irrational numbers are nonterminating, nonrepeating decimals.

22. a. Irrational numbers are real numbers.
    b. Every whole number is a rational number.
    c. Every rational number can be written as a fraction.
    d. Every rational number is a whole number.

23. If we begin with the number −4 and find its opposite, and then find the opposite of that result, what number do we obtain?

24. Which number graphed below has the largest absolute value?

Approximate each irrational number to the nearest thousandth.

25. $\sqrt{5}$
26. $\sqrt{19}$
27. $\sqrt{99}$
28. $\sqrt{42}$

29. π
30. 2π
31. $\dfrac{\pi}{2}$
32. $\dfrac{\pi}{6}$

59. $|3.4|$ ▢ $\sqrt{10}$
60. 0.08 ▢ 0.079
61. $-|-1.1|$ ▢ -1
62. $-(-5.5)$ ▢ $-\left(-5\dfrac{1}{2}\right)$

NOTATION
Fill in the blanks.

33. $\sqrt{5}$ is read "the _____ _____ of 5."
34. $|-15|$ is read "the _____ _____ of -15."
35. The symbol $\approx$ means _____.
36. The symbols { }, called _____, are used when writing a set.

Determine which numbers in the set are natural numbers, whole numbers, integers, rational numbers, irrational numbers, and real numbers. See Example 3.

63. $\left\{-\dfrac{5}{6},\ 35.99,\ 0,\ 4\dfrac{3}{8},\ \sqrt{2},\ -50,\ \dfrac{17}{5}\right\}$

64. $\left\{-0.001,\ 10\dfrac{1}{2},\ 6,\ \pi,\ \sqrt{7},\ -23,\ -5.6\right\}$

65. $\left\{-\pi,\ 0,\ 1\dfrac{3}{5},\ 8,\ \pi,\ \sqrt{11},\ -3,\ 2.6,\ 2\right\}$

66. $\left\{3\dfrac{5}{6},\ 9,\ \sqrt{3},\ 0.00023,\ -2.7,\ 0,\ -3\right\}$

For exercises 37 and 38, write each fraction as a decimal. Determine whether the result is a terminating or nonterminating decimal.

37. a. $\dfrac{2}{3}$ b. $\dfrac{4}{5}$
38. a. $\dfrac{5}{8}$ b. $\dfrac{5}{11}$

39. The symbols $<, >, \leq,$ and $\geq$ are called _____.
40. The symbol … is called an _____; it indicates that a set follows an established pattern that continues forever.

Write each expression in simpler form.

41. The opposite of 5
42. The opposite of -9
43. The opposite of $-\dfrac{7}{8}$
44. The opposite of 6.56
45. $-(-10)$
46. $-(-1)$
47. $-(-2.3)$
48. $-\left(-\dfrac{3}{4}\right)$
49. The opposite of the absolute value of 3
50. $-|-5|$

Graph each set of numbers on the number line. See Example 4.

67. $\left\{-\pi,\ 4.25,\ -1\dfrac{1}{2},\ -0.333\ldots,\ \sqrt{2},\ -\dfrac{35}{8},\ 3\right\}$

68. $\left\{\pi,\ -2\dfrac{1}{8},\ 2.75,\ -\sqrt{17},\ \dfrac{17}{4},\ -0.666\ldots,\ -3\right\}$

69. $\left\{-3\dfrac{5}{8},\ 2,\ \sqrt{3},\ \dfrac{17}{5},\ 0.333\ldots,\ 5\right\}$

70. $\left\{-1\dfrac{7}{9},\ 3.5,\ -\sqrt{11},\ \dfrac{9}{4},\ 1.666\ldots,\ 4\right\}$

GUIDED PRACTICE
Graph the integers between the given numbers. See Example 1.

51. -3 and 6
52. -1 and 7
53. -6 and 8

54. -1 and 2

Insert one of the symbols $>, <,$ or $=$ in the blank to make the statement true. See Example 2.

55. 5 ▢ -4
56. -11 ▢ -9
57. -2 ▢ -4
58. 0 ▢ 32

Evaluate each expression. See Example 5.

71. $|-17|$
72. $\left|-\dfrac{3}{5}\right|$
73. $-|-2.5|$
74. $-|\pi|$

Insert one of the symbols >, <, or = in the blank to make the statement true. See Example 6.

75. $-|-2|$ ___ -2

76. $-\left|-\dfrac{2}{3}\right|$ ___ $\dfrac{3}{4}$

77. $-\left|\dfrac{5}{6}\right|$ ___ -2

78. $-\left|-\dfrac{3}{4}\right|$ ___ $\dfrac{3}{4}$

79. $-\left(-\dfrac{5}{8}\right)$ ___ $-\left(-\dfrac{3}{8}\right)$

80. $-19\dfrac{2}{3}$ ___ $-19\dfrac{1}{3}$

81. $\left|-\dfrac{15}{2}\right|$ ___ 7.5

82. $\sqrt{2}$ ___ π

83. $\dfrac{99}{100}$ ___ 0.99

84. $|2|$ ___ $-|-2|$

85. $0.333\ldots$ ___ 0.3

86. $\left|-2\dfrac{2}{3}\right|$ ___ $-\left(-\dfrac{3}{2}\right)$

87. $-(-1)$ ___ $\left|-\dfrac{15}{16}\right|$

88. $-0.666\ldots$ ___ 0

89. $-(-1)$ ___ $\left|-\dfrac{19}{13}\right|$

90. $-0.666\ldots$ ___ -1.34

APPLICATIONS

91. **BANKING** In the table below, which numbers are natural numbers, whole numbers, integers, rational numbers, irrational numbers, and real numbers?

Type of account	Principal	Rate	Time (years)	Interest
Checking	$135.75	0.0275	$\dfrac{31}{365}$	$0.32
Savings	$5,000	0.06	$2\dfrac{1}{2}$	$750

92. **DRAFTING** The drawing below shows the dimensions of an aluminum bracket.
    a. Which numbers shown are natural numbers, whole numbers, integers, rational numbers, irrational numbers, and real numbers?
    b. Approximate all the irrational numbers in the drawing to the nearest thousandth.

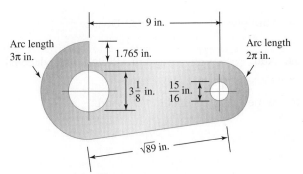

93. **TRADE** Each year from 1994 through 2012, the United States imported more goods and services from Japan than it exported to Japan. This caused trade *deficits*, which are represented by negative numbers on the graph below.
    a. In which year was the deficit the worst? Express that deficit using a signed number.
    b. In which year was the deficit the smallest? Estimate that deficit using a signed number.

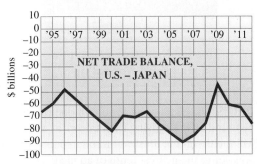

Source: U.S. Bureau of the Census

94. **U.S. BUDGET** A budget *deficit* is a negative number that indicates the government spent more money than it took in that year. A budget *surplus* is a positive number that indicates the government took in more money than it spent that year. The graph below shows the U.S. federal budget deficit/surplus for the years 1980 through 2012.
    a. For how many years was there a budget surplus?
    b. Consider the years there was a budget deficit. For how many of those years was it less than $300 billion?

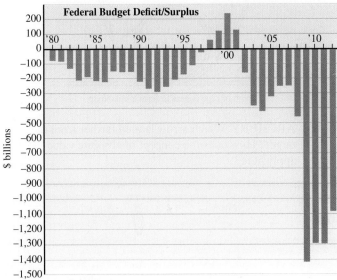

Source: Congressional Budget Office

95. **GEOMETRY** The lengths (in inches) of the three sides of a right triangle are $\sqrt{3}$, 1, and 2. Write the lengths in order from least to greatest.

96. **PHOTOGRAPHY** An exposure screen display on the back of a digital camera is shown below. Where would a photographer set the cursor ☐ on the scale if she wants to underexpose (reduce the light) by $1\frac{1}{3}$ units?

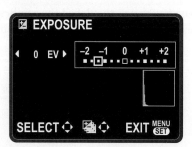

97. **TIRES** The distance a tire rolls in one revolution can be found by multiplying its diameter by π. How far will the tire in the illustration roll in one revolution? Answer to the nearest tenth of an inch.

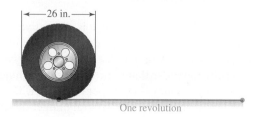

98. **HULA HOOPS** The length of plastic pipe needed to form a hula hoop can be found by multiplying its diameter by π. Find the length of pipe needed to form the hula hoop in the illustration. Answer to the nearest tenth of an inch.

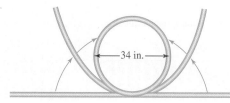

A length of pipe is bent to form a hula hoop.

99. **ARCHERY** Which arrow landed farther from the target? How does the concept of absolute value apply here?

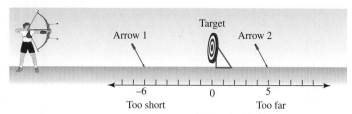

100. Refer to the historical time line.

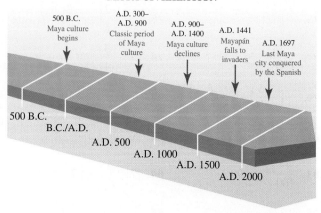

a. What basic unit was used to scale the time line?

b. On the time line, what symbolism is used to represent zero?

c. On the time line, which numbers could be thought of as positive and which could be thought of as negative?

d. Express the dates for the Maya civilization using positive and negative numbers.

WRITING

101. Explain the difference between a rational and an irrational number.

102. Can two different numbers have the same absolute value? Explain.

103. Give two examples each of fractions, mixed numbers, decimals, and negative numbers that you use in your everyday life.

104. In writing courses, students are warned not to use double negatives in their compositions. Identify the double negative in the following sentence. Then rewrite the sentence so that it conveys the same idea without using a double negative. "No one didn't turn in the homework."

REVIEW

105. Simplify: $\dfrac{24}{54}$

106. Prime factor 60.

107. Multiply: $\dfrac{3}{4}\left(\dfrac{8}{5}\right)$

108. Divide: $5\dfrac{2}{3} \div 2\dfrac{5}{9}$

109. Add: $\dfrac{3}{10} + \dfrac{2}{15}$

110. Write $\dfrac{4}{25}$ as a decimal.

111. Write $\dfrac{3}{8}$ as a decimal.

112. Add: $\dfrac{5}{7} + \dfrac{3}{7}$

SECTION 1.4
Adding Real Numbers; Properties of Addition

Objectives
1. Find the sum of two real numbers with the same sign.
2. Find the sum of two real numbers with different signs.
3. Use properties of addition.

Recall that all points on the number line represent the set of real numbers. Real numbers that are greater than zero are *positive real numbers*. See the number line below. Positive numbers can be written with or without a + sign. For example, $2 = +2$ and $4.75 = +4.75$. Real numbers that are less than zero are *negative real numbers*. They must be written with a − sign. For example, *negative* $2 = -2$ and *negative* $4.75 = -4.75$. Positive and negative numbers are called **signed numbers**.

Negatives Zero Positives
$-5\ -4\ -3\ -2\ -1\ \ 0\ \ 1\ \ 2\ \ 3\ \ 4\ \ 5$

Caution! Zero is neither positive nor negative.

We can use signed numbers to describe many situations. Words such as *gain, above, up, to the right,* and *in the future* indicate positive numbers. Words such as *loss, below, down, to the left,* and *in the past* indicate negative numbers.

In words	In symbols	Meaning
16 degrees above 0	$+16°$	positive sixteen degrees
30 seconds after liftoff	30 sec	positive thirty seconds
$10.50 overdrawn	$-\$10.50$	negative ten dollars and fifty cents
$5\frac{1}{2}$ feet below sea level	$-5\frac{1}{2}$ ft	negative five and one-half feet

In the figure to the right, signed numbers are used to denote the 2012 quarterly profits and losses of Toys "R" Us. The fourth-quarter profit of $239 million is represented by the positive number 239. The first-quarter loss of $60 million, the second-quarter loss of $36 million, and the third-quarter loss of $105 million are represented by the negative numbers $-60, -36,$ and -105. To find the 2012 net income of Toys "R" Us, we need to add the following positive and negative numbers.

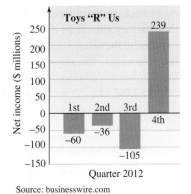

Source: businesswire.com

Net income $= -60 + (-36) + (-105) + 239$

In this section, we will discuss how to perform such additions.

1 Find the sum of two real numbers with the same sign.

We can use a number line to explain the addition of signed numbers. For example, the number line at the top of the next page shows the steps that are used to compute $5 + 2$. We begin at the **origin** (the zero point) and draw an arrow 5 units long, pointing to the right; this represents 5. From that point, we draw an arrow 2 units long, also pointing to the right; this represents 2. We end up at 7; therefore, $5 + 2 = 7$.

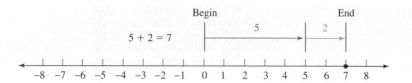

The Language of Algebra   The names of the parts of an addition fact are

Addend   Addend   Sum
↓   ↓   ↓
5   +   2   =   7

To compute $-5 + (-2)$ on a number line, we begin at the origin and draw an arrow 5 units long, pointing to the left; this represents -5. From there, we draw an arrow 2 units long, also pointing to the left; this represents -2. We end up at -7, as shown below. Therefore, $-5 + (-2) = -7$.

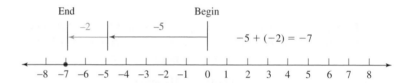

Caution!   To avoid confusion, we write negative numbers within parentheses to separate the negative sign $-$ from the addition symbol $+$: $-5 + (-2)$

As a check, think of this problem in terms of money. If you had a debt of \$5, (-5), and then had another debt of \$2, (-2), you would have a total debt of \$7, (-7).

From these two examples, we observe that when we add two numbers with the same sign, the arrows point in the same direction and build upon each other. Furthermore, the answer that they point to has the same sign as the numbers that are being added. If both numbers are positive, their sum is positive. If both numbers are negative, their sum is negative.

$$5 \;+\; 2 \;=\; 7 \quad \text{and} \quad -5 \;+\; (-2) \;=\; -7$$
positive + positive = positive   negative + negative = negative

These observations suggest the following rule.

Adding Two Real Numbers with the Same (Like) Sign

To add two real numbers with the same (like) sign, add their absolute values and attach their common sign to the sum.

Success Tip   The sum of two positive numbers is always positive. The sum of two negative numbers is always negative.

1.4 Adding Real Numbers; Properties of Addition

EXAMPLE 1 Find the sum: $-25 + (-18)$

Strategy We will use the rule for adding two real numbers that have the same sign.

WHY We use that rule because the addition involves two negative numbers.

Solution
Since both numbers are negative, the answer is negative.

$-25 + (-18) = -43$ Add their absolute values, 25 and 18, to get 43. Attach their common sign (which is a $-$ sign) to 43.

Self Check 1
Find the sum: $-45 + (-12)$
Now Try Problem 27

> **The Language of Algebra** Two negative numbers, as well as two positive numbers, are said to have *like* signs.

2 Find the sum of two real numbers with different signs.

To compute $5 + (-2)$ on the number line, we begin at the origin and draw an arrow 5 units long, pointing to the right; this represents 5. From there, we draw an arrow 2 units long, pointing to the left; this represents -2. We end up at 3, as shown below. Therefore, $5 + (-2) = 3$. In terms of money, if you had $5 (+5), and lost $2 (-2), you would have $3 (+3) left.

To compute $-5 + 2$ on the number line, we start at the origin and draw an arrow 5 units long, pointing to the left; this represents -5. From there, we draw an arrow 2 units long, pointing to the right; this represents 2. We end up at -3, as shown below. Therefore, $-5 + 2 = -3$. In terms of money, if you owed a friend $5 (-5), and paid back $2 (+2), you would still owe your friend $3 (-3).

> **The Language of Algebra** A positive number and a negative number are said to have *unlike* or *different* signs.

From these two examples, we observe that when we add two numbers with different signs, the arrows point in opposite directions. Furthermore, the longer arrow determines the sign of the answer. If the longer arrow represents a positive number, the sum is positive. If the longer arrow represents a negative number, the sum is negative.

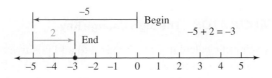

$5 + (-2) = 3$ and $-5 + 2 = -3$
positive + negative = positive   negative + positive = negative

These observations suggest the following rule.

Adding Two Real Numbers with Different (Unlike) Signs

To add two real numbers with different (unlike) signs, subtract their absolute values (the smaller from the larger). To this result, attach the sign of the number with the larger absolute value.

Self Check 2

Add:
a. $63 + (-87)$
b. $-6.27 + 8$
c. $-\dfrac{1}{10} + \dfrac{1}{2}$

Now Try Problem 34

EXAMPLE 2 Add: a. $-17 + 32$ b. $5.4 + (-7.7)$ c. $\dfrac{3}{25} + \left(-\dfrac{19}{25}\right)$

Strategy We will use the rule for adding two real numbers with different signs.

WHY We use that rule because we are asked to add a positive and a negative number.

Solution

a. Since 32 has the larger absolute value, the answer is positive.

$$-17 + 32 = 15 \quad \text{Subtract their absolute values, 17 from 32, to get 15.}$$

b. Since -7.7 has the larger absolute value, the answer is negative.

$$5.4 + (-7.7) = -2.3 \quad \text{Subtract their absolute values, 5.4 from 7.7, to get 2.3. Attach a } - \text{ sign.}$$

c. Since the fractions have the same denominator, we add the numerators and keep the common denominator.

$$\dfrac{3}{25} + \left(-\dfrac{19}{25}\right) = \dfrac{3}{25} + \left(\dfrac{-19}{25}\right) \quad \text{Write } -\tfrac{19}{25} \text{ as } \tfrac{-19}{25}. \left(\text{Recall: } -\tfrac{a}{b} = \tfrac{-a}{b}.\right)$$

$$= -\dfrac{16}{25} \quad \text{Add the numerators: } 3 + (-19) = -16. \text{ Write the sum over the common denominator 25, putting the } - \text{ sign in front of the fraction.}$$

Success Tip The sum of two numbers with different signs may be positive or negative. The sign of the sum is the sign of the number with the greater absolute value.

Using Your CALCULATOR The Sign Change Key

A reverse entry scientific calculator can be used to add positive and negative numbers. For example, to do the addition $-31 + 15 + (-4) + 29$, we do not have to do anything special to enter the positive numbers. To enter negative 31 and negative 4, we must press the *opposite* or *sign change key* $\boxed{+/-}$ after entering 31 and after entering 4.

$$31 \; \boxed{+/-} \; \boxed{+} \; 15 \; \boxed{+} \; 4 \; \boxed{+/-} \; \boxed{+} \; 29 \; \boxed{=} \qquad \boxed{9}$$

Using a graphing calculator or direct-entry scientific calculator, we enter a negative number by first pressing the *negation* key $\boxed{(-)}$. To perform the addition, we enter these numbers and press these keys.

$$\boxed{(-)} \; 31 \; \boxed{+} \; 15 \; \boxed{+} \; \boxed{(-)} \; 4 \; \boxed{+} \; 29 \; \boxed{\text{ENTER}}$$

$$\boxed{\begin{array}{r} -31 + 15 + -4 + 29 \\ 9 \end{array}}$$

As before, the sum is 9.

EXAMPLE 3 Find the 2012 net income of Toys "R" Us from the data given in the graph on page 41.

Strategy We will add the signed numbers that represent the quarterly profits and losses.

WHY The 2012 net income is the sum of the quarterly profits and losses that year.

Solution
One way to find the sum is to perform the additions as they occur from left to right.

$$-60 + (-36) + (-105) + 239 = -96 + (-105) + 239 \quad \text{-60 + (-36) = -96}$$
$$= -201 + 239 \quad \text{-96 + (-105) = -201}$$
$$= 38$$

In 2012, the net income of Toys "R" Us was $38 million.

Self Check 3
Add: $-7 + 13 + (-5) + 10$
Now Try Problem 35

3 Use properties of addition.

The **commutative property of addition** states that two real numbers can be added in either order to get the same result. For example, when adding the numbers 10 and -25, we see that

$$10 + (-25) = -15 \quad \text{and} \quad -25 + 10 = -15$$

To state the *commutative property of addition* concisely, we use variables.

The Commutative Property of Addition

Changing the order when adding does not affect the answer.

For any real numbers a and b,

$$a + b = b + a$$

The Language of Algebra *Commutative* is a form of the word *commute*, meaning to go back and forth. *Commuter* trains take people to and from work.

To find the sum of three numbers, we first add two of them and then add the third to that result. In the following example, we add $-3 + 7 + 5$ in two ways. To show this, we will use grouping symbols ( ), called **parentheses**. Standard practice requires that we perform the operations within parentheses first.

Method 1: Group -3 and 7

$(-3 + 7) + 5 = 4 + 5$ Because of the parentheses, add -3 and 7 first to get 4.
$ = 9$ Then add 4 and 5.

Method 2: Group 7 and 5

$-3 + (7 + 5) = -3 + 12$ Because of the parentheses, add 7 and 5 first to get 12.
$ = 9$ Then add -3 and 12.

Either way, the sum is 9. This illustrates that it doesn't matter how we *group* or *associate* numbers in addition—we get the same result. This property is called the **associative property of addition**.

The Associative Property of Addition

Changing the grouping when adding does not affect the answer.

For any real numbers a, b, and c,

$$(a + b) + c = a + (b + c)$$

The Language of Algebra *Associative* is a form of the word *associate*, meaning to join a group. The NBA (National Basketball *Association*) is a group of professional basketball teams.

Self Check 4
Find the contestant's net gain or loss after the first two questions.
Now Try Problem 43

EXAMPLE 4 *Game Shows*

A contestant on *Jeopardy!* answered the first question correctly to win $100, missed the second question to lose $200, answered the third question correctly to win $300, and answered the fourth question incorrectly to lose $400. Find her net gain or loss after four questions.

Strategy We will add the amounts won and lost after the four questions she had answered.

WHY The net gain or loss after four questions is the sum of the signed amounts from each question.

Solution
"To win $100" can be represented by 100. "To lose $200" can be represented by −200. "To win $300" can be represented by 300, and "to lose $400" can be represented by −400. Her net gain or loss is the sum of these four numbers. We can find the sum by performing the additions from left to right. An alternate method, which uses the commutative and associative properties of addition, is to add the positives, then add the negatives, and add those results.

$$100 + (-200) + 300 + (-400)$$
$$= \mathbf{100 + 300 + (-200) + (-400)} \quad \text{Reorder the numbers.}$$
$$= \mathbf{(100 + 300) + [(-200) + (-400)]} \quad \text{Group the positives together.}$$
$$ \quad \text{Group the negatives together using brackets [ ].}$$
$$= \mathbf{400 + (-600)} \quad \text{Add the positives. Add the negatives.}$$
$$= -200$$

After four questions, she had a net loss of $200.

Whenever we add zero to a number, the number remains the same. For example,

$$0 + 8 = 8, \quad 2.3 + 0 = 2.3, \quad \text{and} \quad -16 + 0 = -16$$

These examples illustrate the **addition property of zero**. Since any number added to 0 remains the same, 0 is called the **identity element** for addition.

Addition Property of Zero (Identity Property of Addition)

When 0 is added to any real number, the result is the same real number.

For any real number a,

$$a + 0 = a \quad \text{and} \quad 0 + a = a$$

> **The Language of Algebra** *Identity* is a form of the word *identical*, meaning the same. You have probably seen *identical* twins.

Recall that two numbers that are the same distance away from the origin, but on opposite sides of it, are called **opposites** or **additive inverses**. For example, 10 is the additive inverse of −10, and −10 is the additive inverse of 10. Whenever we add opposites or additive inverses, the result is 0.

$$10 + (-10) = 0, \quad -\frac{4}{5} + \frac{4}{5} = 0, \quad 56.8 + (-56.8) = 0$$

Addition Property of Opposites (Inverse Property of Addition)

The sum of a real number and its opposite (additive inverse) is 0.

For any real number a,

$$a + (-a) = 0$$

THINK IT THROUGH Calculating Sleep Debt

"63% of the college students in the United States suffer from lack of sleep."

National Sleep Foundation, 2008

Because of our demanding schedules, many of us don't get enough sleep. According to the National Sleep Foundation, sleep deprivation is a common problem among college students. As a result, we build up a *sleep debt*. For example, if you require 8 hours of sleep a night, but only get 7, your sleep debt is −1 hour. It is possible to make up sleep if your sleep debt is not too great. It takes about two hours of weekend sleep to make up for every lost hour of sleep during the week.

How many hours of sleep do you need each night to feel refreshed in the morning? To see whether you get the necessary sleep during the week, complete the following log. On Friday, determine whether you have a sleep debt. Then calculate how many extra hours you need to sleep on the weekend to make up for the sleep debt.

	Bedtime	Awaken	Hours slept	Sleep debt
Sunday night				
Monday night				
Tuesday night				
Wednesday night				
Thursday night				
			Total sleep debt:	

ANSWERS TO SELF CHECKS

1. −57 **2. a.** −24 **b.** 1.73 **c.** $\frac{2}{5}$ **3.** 11 **4.** −$100

STUDY SKILLS CHECKLIST

Get the Most from Your Textbook

The following checklist will help you become familiar with some useful features in this book. Place a checkmark in each box after you answer the question.

☐ Locate the **Definition** for *Absolute Value* on page 35 and the steps for *Simplifying Fractions* on page 19. What color are these boxes?

☐ Find the **Caution** box, the **Success Tip** box, and the **Language of Algebra** box on page 42. What color is used to identify these boxes?

☐ Each chapter begins with **From Campus to Careers** (see page 1). Chapter 1 gives information on how to become a physician assistant. On what page does a related problem appear in Study Set 1.4?

☐ Locate the **Study Skills Workshop** at the beginning of your text beginning on page S-1. How many Objectives appear in the Study Skills Workshop?

Answers: green, red, 49, 7

SECTION 1.4 STUDY SET

VOCABULARY

Fill in the blanks.

1. Real numbers that are greater than zero are called _____ real numbers. Real numbers that are less than zero are called _____ real numbers.
2. The only real number that is neither positive nor negative is _____.
3. The _____ property of addition states that two numbers can be added in either order to get the same result.
4. The grouping symbols ( ) are called _____.
5. The property that allows us to group numbers in addition in any way we want is called the _____ property of addition.
6. Whenever we add _____, or additive _____, the result is 0.

CONCEPTS

Use the following number line to find each sum.

7. $2 + 3$
8. $-3 + (-2)$
9. $4 + (-3)$
10. $-5 + 3$

Fill in the blanks.

11. To add two real numbers with the same sign, _____ their absolute values and attach their common sign to the sum.
12. To add two real numbers with different signs, _____ their absolute values, the _____ from the _____, and attach the sign of the number with the _____ absolute value.

Use the commutative property of addition to complete each statement.

13. $-5 + 1 = $
14. $15 + (-80.5) = $
15. $-20 + (4 + 20) = -20 + $
16. $(5 + 7) + 9 = $ $+ 9$

Use the associative property of addition to complete each statement.

17. $(-6 + 2) + 8 = $ _____
18. $-7 + (7 + 3) = $ _____
19. $-96 + (4 + 200) = $ _____
20. $(-9 + 4) + 15 = $ _____

21. What is the opposite of 7?
22. What is the opposite of -15?
23. What is the opposite of $-\frac{1}{2}$?
24. What is the opposite of $\frac{2}{3}$?

NOTATION

Complete each step.

25. $(-13 + 6) + 4 = $ $+ (6 + 4)$
    $= -13 + $
    $= -3$

26. $-9 + (9 + 43) = ($ $+ 9) + 43$
    $= $ $+ 43$
    $= 43$

1.4 Adding Real Numbers; Properties of Addition

GUIDED PRACTICE

Find each sum. See Example 1.

27. $-65 + (-12)$
28. $-21 + (-12)$
29. $-4.1 + (-5.7)$
30. $-2.5 + (-1.7)$

Find each sum. See Example 2.

31. $6 + (-8)$
32. $4 + (-3)$
33. $15 + (-11)$
34. $27 + (-30)$

Find each sum. See Example 3.

35. $8 + (-5) + 13$
36. $17 + (-12) + (-23)$
37. $21 + (-27) + (-9)$
38. $-32 + 12 + 17$
39. $-27 + (-3) + (-13) + 22$
40. $53 + (-27) + (-32) + (-7)$
41. $57 + (-47) + (-64) + 113$
42. $32 + (-44) + (-37) + (-52)$

Use the associative property of addition to find each sum. See Example 4.

43. $-99 + (99 + 215)$
44. $67 + (-67 + 127)$
45. $(-4 + 15) + (-15)$
46. $(-18 + 37) + (-37)$

TRY IT YOURSELF

Find each sum.

47. $5 + (-5)$
48. $-2.2 + 2.2$
49. $0 + (-6)$
50. $-\dfrac{15}{16} + 0$
51. $-\dfrac{3}{4} + \dfrac{3}{4}$
52. $19 + (-19)$
53. $-6 + 8$
54. $75 + (-13)$
55. $300 + (-335)$
56. $240 + (-340)$
57. $-10.5 + 2.3$
58. $-2.1 + 0.4$
59. $-9.1 + (-11)$
60. $-6.7 + (-7.1)$
61. $-\dfrac{9}{16} + \dfrac{7}{16}$
62. $-\dfrac{3}{4} + \dfrac{1}{4}$
63. $-20 + (-16 + 10)$
64. $-13 + (-16 + 4)$
65. $-\dfrac{1}{4} + \dfrac{2}{3}$
66. $\dfrac{3}{16} + \left(-\dfrac{1}{2}\right)$
67. $4.125 + (-7.341)$
68. $3,718 + (-5,237)$
69. $735 + (-462)$
70. $837 + (-429)$
71. $-5,235 + (-17,235)$
72. $32.137 + (-34.36) + (-32.137)$
73. $736 + 67 + (-736)$
74. $-237.37 + (-315.07) + (-27.4)$
75. $-587.77 + (-1,732.13) + 687.39$
76. $-37.57 + 85.02 + (-77.1)$

APPLICATIONS

77. **MILITARY SCIENCE** During a battle, an army retreated 1,500 meters, regrouped, and advanced 2,400 meters. The next day, it advanced another 1,250 meters. Find the army's net gain.

78. **MEDICAL QUESTIONNAIRES** Find the point total for the six risk factors (in blue) on the medical questionnaire. Then use the table to determine the patient's risk of contracting heart disease in the next 10 years.

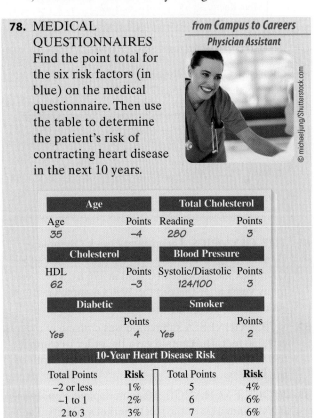

from *Campus to Careers*
Physician Assistant

Age		Total Cholesterol	
Age	Points	Reading	Points
35	−4	280	3

Cholesterol		Blood Pressure	
HDL	Points	Systolic/Diastolic	Points
62	−3	124/100	3

Diabetic		Smoker	
	Points		Points
Yes	4	Yes	2

10-Year Heart Disease Risk			
Total Points	Risk	Total Points	Risk
−2 or less	1%	5	4%
−1 to 1	2%	6	6%
2 to 3	3%	7	6%
4	4%	8	7%

Source: National Heart, Lung, and Blood Institute

79. **GOLF** The illustration on the next page shows the top four finishers in the 1997 Masters Golf Tournament. The scores for each round are related to *par*, the number of strokes an experienced golfer would take to complete the course. A score of −2, for example, indicates that the golfer used two strokes less than par to complete the course. A score of +5 indicates the golfer used five strokes more than par.

  a. Determine the tournament total for each golfer.

b. Tiger Woods won by the largest margin in the history of the Masters. Find the margin.

Leaderboard

	Round 1	2	3	4	Total
Tiger Woods	−2	−6	−7	−3	
Tom Kite	+5	−3	−6	−2	
Tommy Tolles	0	0	0	−5	
Tom Watson	+3	−4	−3	0	

80. **CREDIT CARDS** Refer to the monthly statement. What is the new balance?

Previous Balance	New Purchases, Fees, Advances & Debts	Payments & Credits	New Balance
3,660.66	1,408.78	3,826.58	
04/21/14 Billing Date	05/16/14 Date Payment Due		9,100 Credit Line

81. **THE OLYMPICS** The ancient Greek Olympian Games, which eventually evolved into the modern Olympic Games, were first held in 776 B.C. How many years after this did the 1996 Olympic Games in Atlanta, Georgia, take place?

82. **SUBMARINES** A submarine was cruising at a depth of 1,250 feet. The captain gave the order to climb 550 feet. Relative to sea level, find the new depth of the sub.

83. **STOCK EXCHANGE** Many newspapers publish daily summaries of the stock market's activity. The last entry on the line for June 12 indicates that one share of Walt Disney Co. stock lost $0.81 in value that day. How much did the value of a share of Disney stock rise or fall over the 5-day period shown?

```
June 12  43.88  23.38  Disney  .21  0.5  87  −43  40.75  −.81
June 13  43.88  23.38  Disney  .21  0.5  86  −15  40.19  −.56
June 14  43.88  23.38  Disney  .21  0.5  87  −50  41.00  +.81
June 15  43.88  23.38  Disney  .21  0.5  89  −28  41.81  +.81
June 16  43.88  23.38  Disney              −15  41.19  −.63
```

Based on data from the *Los Angeles Times*

84. **POLITICS** What will be the effect on state government if the following ballot initiative passes?

212 Campaign Spending Limits YES ☐ NO ☐
Limits contributions to $200 in state campaigns. Fiscal impact: Costs of $4.5 million for implementation and enforcement. Increases state revenue by $6.7 million by eliminating tax deductions for lobbying.

85. **MOVIE LOSSES** According to the Numbers Box Office Data website, the movie *Speed Racer*, released in 2008 by Warner Brothers, cost about $120,000,000 to produce, promote, and distribute. It reportedly earned just $93,394,462 worldwide. Use a signed number to express the loss suffered by Warner Brothers.

86. **SAHARA DESERT** From 1980 to 1990, a satellite was used to trace the expansion and contraction of the southern boundary of the Sahara Desert in Africa. If movement southward is represented with a negative number and movement northward with a positive number, use the data in the table to determine the net movement of the Sahara Desert boundary over the 10-year period.

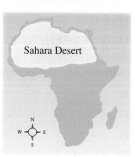

Data from A. Dolgoff, *Physical Geology* (Heath, 1996, p. 496)

Years	Distance/direction
1980–1984	240 km/south
1984–1985	110 km/north
1985–1986	30 km/north
1986–1987	55 km/south
1987–1988	100 km/north
1988–1990	77 km/south

Source: Based on data from A. Dolgoff *Physical Geology* (Heath, 1996, p. 496)

87. **BANKING** On February 1, Marta had $1,704.29 in a checking account. During the month, she made deposits of $713.87 and $1,245.57, made online payments of $813.45, $937.49, and $1,532.79, and had a total of $500 in ATM withdrawals. Find her checking account balance at the end of the month.

88. **CARD GAMES** In the second hand of a card game, Gonzalo was the winner and earned 50 points. Matt and Hydecki had to deduct the value of each of the cards left in their hands from their running point totals. Use the following information to update the score sheet on the next page. (Face cards are counted as 10 points and aces as 1 point.)

Matt   Hydecki

Running point total	Hand 1	Hand 2
Matt	+50	
Gonzalo	−15	
Hydecki	−2	

89. PROFITS AND LOSSES The 2012 quarterly profits and losses of Rite Aid Corporation are shown in the table. Losses are denoted with parentheses. Calculate the company's total net income for 2012.

Quarter	1st	2nd	3rd	4th
Net income $ million	(161)	(28)	(39)	61

Source: www.riteaid.com

90. POLITICS Six months before an election, the incumbent trailed the challenger by 18 points. To overtake her opponent, the incumbent decided to use a four-part strategy. Each part of the plan is shown in the next column, with the expected point gain. With these gains, will the incumbent overtake the challenger on election day?

- TV ads   +10
- Voter mailing   +3
- Union endorsement   +2
- Telephone calls   +1

WRITING

91. Explain why the sum of two positive numbers is always positive and why the sum of two negative numbers is always negative.

92. Explain why we need to subtract the absolute value when we add two real numbers with different signs.

REVIEW

93. True or false: Every real number can be expressed as a decimal.

94. True or false: Irrational numbers are nonterminating, nonrepeating decimals.

95. What two numbers are a distance of 6 away from −3 on the number line?

96. Graph: $\{-2.5, \sqrt{5}, \frac{11}{3}, -0.333\ldots, 0.75\}$

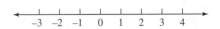

SECTION 1.5
Subtracting Real Numbers

Objectives
1. Use the definition of subtraction.
2. Solve application problems using subtraction.

1 Use the definition of subtraction.

The minus symbol − is used to indicate subtraction. However, this symbol is also used in many other ways, depending on where it appears in an expression.

$4 - 9$ This is read "4 minus 9."

-3 This is usually read "negative three." It can also be read as "the additive inverse of three" or "the opposite of three."

$-(-2)$ This is usually read as "the opposite of negative two" or "the additive inverse of negative 2."

In the expression $-(-2)$, parentheses help us write the opposite of a negative number. To simplify this expression, we find the opposite of the number in the parentheses.

$-(-2) = 2$ Read this equation as "the opposite of negative two is two."

The above equation suggests the following rule.

> **Opposite of an Opposite**
>
> The opposite of the opposite of a number is that number. In symbols:
>
> $-(-a) = a$ Read as "the opposite of the opposite of a is a."

Self Check 1

Simplify each expression:
a. $-(-1)$
b. $-(-y)$
c. $-|-500|$

Now Try Problem 12

EXAMPLE 1
Simplify each expression:
a. $-(-45)$ b. $-(-h)$ c. $-|-10|$

Strategy To simplify each expression, we will use the concept of opposite.

WHY In each case, the outermost $-$ symbol is read as "the opposite."

Solution

a. The number within the parentheses is -45. Its opposite is 45. Therefore, $-(-45) = 45$.

b. The opposite of the opposite of h is h. Therefore, $-(-h) = h$.

c. The notation $-|-10|$ means "the opposite of the absolute value of negative ten." Since $|-10| = 10$, we have:

$$-|-10| = -10$$

The absolute value bars do not affect the $-$ symbol outside them. Therefore, the result is negative.

The subtraction $5 - 2$ can be thought of as taking 2 away from 5. We can use the number line below to illustrate this. Beginning at the origin, we draw an arrow of length 5 units pointing to the right. From that point, we move back 2 units to the left. The result, 3, is called the **difference**.

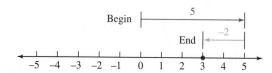

The figure above looks like the illustration for the addition problem $5 + (-2)$. In the problem $5 - 2$, we subtracted 2 from 5. In the problem $5 + (-2)$, we added -2 (which is the opposite of 2) to 5. In each case, the result is 3.

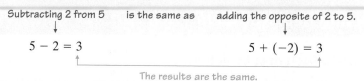

This observation suggests the following rule.

Subtracting Real Numbers

To subtract two real numbers, add the first number to the opposite (additive inverse) of the number to be subtracted.

For any real numbers a and b,

$$a - b = a + (-b)$$

This rule indicates that *subtraction is the same as adding the opposite of the number to be subtracted*. We won't need this rule for every subtraction. For example, $5 - 2$ is obviously 3. However, for more complicated subtractions such as $-8 - (-3)$, the subtraction rule will be helpful.

$-8 - (-3) = -8 + 3$ To subtract -3, add the opposite of -3, which is 3.
$\qquad\qquad\quad = -5$ Do the addition.

The Language of Algebra When we change a number to its opposite, we say we have *changed* (or *reversed*) its sign.

EXAMPLE 2
Subtract: **a.** $-13 - 18$ **b.** $-45 - (-27)$ **c.** $\dfrac{1}{4} - \left(-\dfrac{1}{8}\right)$

Strategy To find each difference, we will use the rule for subtraction: add the first number to the opposite of the second number.

WHY Changing the problem to the addition of the opposite of the second number lessens the chance of making an error.

Solution

a. In $-13 - 18$, the number to be subtracted is 18.

$$-13 - 18 = -13 + (-18)$$ To subtract 18, add the opposite of 18, which is -18.

$$= -31$$ Add their absolute values, 13 and 18, to get 31. Keep their common sign.

Caution! When applying the subtraction rule, *do not* change the first number:

$$-13 - 18 = -13 + (-18)$$

b. In $-45 - (-27)$, the number to be subtracted is -27.

$$-45 - (-27) = -45 + 27$$ To subtract -27, add the opposite of -27, which is 27.

$$= -18$$ Subtract their absolute values, 27 from 45, to get 18. Use the sign of the number with the greater absolute value, which is -45.

c. The lowest common denominator (LCD) for the fractions is 8.

$$\dfrac{1}{4} - \left(-\dfrac{1}{8}\right) = \dfrac{2}{8} - \left(-\dfrac{1}{8}\right)$$ Express $\dfrac{1}{4}$ in terms of eighths: $\dfrac{1}{4} = \dfrac{2}{8}$.

$$= \dfrac{2}{8} + \dfrac{1}{8}$$ The number to be subtracted is $-\dfrac{1}{8}$. Add the opposite of $-\dfrac{1}{8}$, which is $\dfrac{1}{8}$.

$$= \dfrac{3}{8}$$ Add the numerators: $2 + 1 = 3$. Write the sum over the common denominator, 8.

Self Check 2
Subtract:
a. $-32 - 25$
b. $1.7 - (-1.2)$
c. $-\dfrac{1}{2} - \dfrac{1}{8}$

Now Try Problem 20

The Language of Algebra The rule for subtracting real numbers is often stated as:

Subtracting a number is the same as adding its opposite.

EXAMPLE 3
a. Subtract 0.5 from 4.6   **b.** Subtract 4.6 from 0.5

Strategy We will translate each phrase to mathematical symbols and then perform the subtraction. We must be careful when translating the instruction to subtract one number *from* another number.

WHY The order of the numbers in each word phrase must be reversed when we translate it to mathematical symbols.

Self Check 3
a. Subtract 2.2 from 4.9
b. Subtract 4.9 from 2.2

Now Try Problem 27

Solution

a. The number to be subtracted is 0.5.

Subtract 0.5 from 4.6

$4.6 - 0.5 = 4.1$ To translate, reverse the order in which 0.5 and 4.6 appear in the sentence.

b. The number to be subtracted is 4.6.

Subtract 4.6 from 0.5

$0.5 - 4.6 = 0.5 + (-4.6)$ To translate, reverse the order in which 4.6 and 0.5 appear in the sentence. Add the opposite of 4.6.

$= -4.1$

Caution! Notice from parts a and b that $4.6 - 0.5 \neq 0.5 - 4.6$. This result illustrates an important fact: Subtraction is *not* commutative. When subtracting two numbers, it is important that we write them in the correct order, because, in general, $a - b \neq b - a$.

Self Check 4
Perform the operations:
$-40 - (-10) + 7 - (-15)$

Now Try Problem 35

EXAMPLE 4   Perform the operations: $-9 - 15 + 20 - (-6)$

Strategy This expression contains addition and subtraction. We will write each subtraction as addition of the opposite and then evaluate the expression.

WHY It is easy to make an error when subtracting signed numbers. We will probably be more accurate if we write each subtraction as addition of the opposite.

Solution
$-9 - 15 + 20 - (-6) = -9 + (-15) + 20 + 6$

$= -24 + 26$ Add the negatives. Add the positives. Add the results.

$= 2$

2 Solve application problems using subtraction.

Subtraction finds the *difference* between two numbers. When we find the difference between the maximum value and the minimum value of a collection of measurements, we are finding the **range** of the values.

Self Check 5
Find the temperature range for a day that had a low of $-5°F$ and a high of $36°F$.

Now Try Problem 65

EXAMPLE 5   *U.S. Temperatures*   The record high temperature in the United States was 134°F in Death Valley, California, on July 10, 1913. The record low was −80°F at Prospect Creek, Alaska, on January 23, 1971. Find the temperature range for these extremes.

Strategy We will subtract the lowest temperature from the highest temperature.

WHY The *range* of a collection of data indicates the spread of the data. It is the difference between the largest and smallest values.

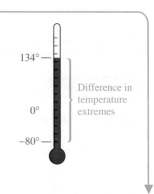

Solution

$134 - (-80) = 134 + 80$ *134° is the higher temperature and −80° is the lower.*

$ = 214$

The temperature range for these extremes is 214°F.

> ### Using Your CALCULATOR   U.S. Temperature Extremes
>
> A calculator can be used to subtract positive and negative real numbers. For instance, to verify the result of 214°F in Example 5, we can find $134 - (-80)$ using a reverse-entry scientific calculator by entering:
>
> 134 [−] 80 [+/−] [=]   ⟶   214
>
> If we use a graphing calculator or a direct-entry scientific calculator, we enter:
>
> 134 [−] [(−)] 80 [ENTER]   ⟶ 134 − ⁻80
>   214
>
> The difference in the record high and low temperatures is 214°F, as we found in Example 5.

Things are constantly changing in our daily lives. The amount of money we have in the bank, the price of gasoline, and our ages are examples. In mathematics, the operation of subtraction is used to measure change. To find the **change** in a quantity, we subtract the earlier value from the later value.

Change = later value − earlier value

EXAMPLE 6   *Water Management*

Lake Mead, on the Nevada-Arizona border, is the largest reservoir in the United States. It is formed by Hoover Dam across the Colorado River. In 2000, the water level in Lake Mead was 89 feet above drought level. By 2010, the water level was 27 feet below drought level. Find the change in the water level over that time span. (Source: Bureau of Reclamation)

Strategy We can represent a water level above drought level using a positive number and a water level below drought level using a negative number. To find the change in the water level, we will subtract.

WHY In general, *to find the change in a quantity, we subtract the earlier value from the later value.*

Solution

$-27 - 89 = -27 + (-89)$ *The earlier water level in 2000 (89 ft) is subtracted from the later water level in 2010 (−27 ft).*

$ = -116$ *Do the addition.*

The negative result indicates that the water level *fell* 116 feet in that time span.

Self Check 6

Find the change in water level for a week that started at 4 feet above normal and went to 7 feet below normal level.

Now Try Problem 71

ANSWERS TO SELF CHECKS

1. a. 1   **b.** y **c.** -500 **2. a.** -57 **b.** 2.9   **c.** $-\dfrac{5}{8}$ **3. a.** 2.7   **b.** -2.7 **4.** -8
5. 41°F   **6.** -11 ft

SECTION 1.5 STUDY SET

VOCABULARY
Fill in the blanks.

1. _____ finds the difference between two numbers.
2. In a subtraction, the result is called the _____.
3. To subtract b from a, add the _____ of b to a.
4. The difference between the maximum and the minimum values of a collection of measurements is called the _____ of the values.

CONCEPTS
Fill in the blanks.

5. $-(-a) = $ ___
6. To subtract two real numbers, change the _____ to addition and take the opposite of the second number: $a - b = a + (-b)$.

NOTATION
Fill in the blanks.

7. The expression -7 is read as _____ 7.
8. The expression $-(-5)$ is read as the _____ of -5.

GUIDED PRACTICE
Simplify each expression. See Example 1.

9. $-(-15)$
10. $-(-p)$
11. $-|7|$
12. $-|-25|$

Find each difference. See Example 2.

13. $8 - (-3)$
14. $17 - (-21)$
15. $-12 - 9$
16. $-25 - 17$
17. $-19 - (-17)$
18. $-30 - (-11)$
19. $-1.5 - 0.8$
20. $-1.5 - (-0.8)$
21. $-\dfrac{1}{8} - \dfrac{3}{8}$
22. $-\dfrac{3}{4} - \dfrac{1}{4}$
23. $-\dfrac{9}{16} - \left(-\dfrac{1}{4}\right)$
24. $-\dfrac{1}{2} - \left(-\dfrac{1}{4}\right)$

Find each difference. See Example 3.

25. Subtract -5 from 17.
26. Subtract 45 from -50.
27. Subtract 1.2 from -1.3.
28. Subtract -1.1 from -2.

Perform the operations. See Example 4.

29. $8 - 9 - 10$
30. $1 - 2 - 3$
31. $-25 - (-50) - 75$
32. $-33 - (-22) - 44$
33. $-6 + 8 - (-1) - 10$
34. $-4 + 5 - (-3) - 13$
35. $61 - (-62) + (-64) - 60$
36. $93 - (-92) + (-94) - 95$

TRY IT YOURSELF
Perform the operations.

37. $2.8 - (-1.8)$
38. $4.7 - (-1.9)$
39. $-44 - 44$
40. $-33 - 33$
41. $0 - (-12)$
42. $0 - 12$
43. $-25 - (-25)$
44. $13 - (-13)$
45. $0 - 4$
46. $0 - (-3)$
47. $8,713 - (-3,753)$
48. $-2,727 - 1,208$
49. $-27,357.875 - 17,213.376$
50. $-45,307.039 - (-27,592.47)$
51. $-62 - 71 - (-37) + 99$
52. $-17 - 32 - (-85) - 51$
53. Subtract 47.5 from 0.
54. Subtract 30.3 from 0.
55. Subtract 5 from -7.
56. Subtract -7 from 5.
57. Subtract -137 from 12.
58. Subtract 512 from -47.
59. Subtract $-\dfrac{1}{3}$ from $\dfrac{5}{3}$.
60. Subtract $\dfrac{2}{5}$ from $\dfrac{4}{5}$.
61. $-\dfrac{5}{6} - \dfrac{3}{4}$
62. $-\dfrac{3}{7} - \dfrac{2}{5}$
63. $-\dfrac{3}{5} - \dfrac{2}{15}$
64. $-\dfrac{4}{11} - \dfrac{1}{2}$

APPLICATIONS

65. **TEMPERATURE RECORDS** Find the difference between the record high temperature of 108°F set in 1926 and the record low of -52°F set in 1979 for New York State.
66. **LIE DETECTORS** A burglar scored -18 on a lie detector test, a score that indicates deception. However, on a second test, he scored $+3$, a score that is inconclusive. Find the difference in the scores.
67. **LAND ELEVATIONS** The elevation of Death Valley, California, is 282 feet below sea level. The elevation of the Dead Sea in Israel is 1,312 feet below sea level. Find the difference in their elevations.

68. THE SUNSHINE STATE Florida's record high temperature of 109°F was set in 1931, and the record low of −2°F was set in 1899. What is the range of these temperature extremes?

69. EYESIGHT Nearsightedness, the condition in which near objects are clear and far objects are blurry, is measured using negative numbers. Farsightedness, the condition in which far objects are clear and near objects are blurry, is measured using positive numbers. Find the range in the measurements shown.

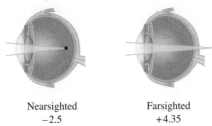

Nearsighted −2.5   Farsighted +4.35

70. WORLD'S COLDEST ICE CREAM Dippin' Dots is an ice cream snack that was invented by Curt Jones in 1987. The tiny multi-colored beads are created by flash freezing ice cream mix in liquid nitrogen at a temperature of −355°F. When they come out of the processor, they are stored at a temperature of −40°F. Find the change in temperature of Dippin' Dots from production to storage. (Source: fundinguniverse.com)

71. RACING To improve handling, drivers often adjust the angle of the wheels of their car. When the wheel leans out, the degree measure is considered positive. When the wheel leans in, the degree measure is considered negative. Find the change in the position of the wheel shown.

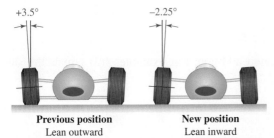

Previous position — Lean outward   New position — Lean inward

72. U.S. JOBS The table lists the three occupations that are predicted to have the largest job declines from 2010 to 2020. Complete the column labeled "Change."

Number of jobs			
Occupation	2010	2020	Change
Farmers/ranchers	1,202,500	1,106,400	
Postal Service mail sorters	142,000	73,000	
Sewing machine operators	163,200	121,100	

Source: Bureau of Labor Statistics

73. GEOGRAPHY The elevation of Denver, Colorado, is 5,183 feet above sea level. The elevation of New Orleans, Louisiana, is 6 feet below sea level. Find the difference in their elevations.

74. PHYSICS The illustration shows an example of a *standing wave*. What is the difference between the height of the crest of the wave and the depth of the trough of the wave? (m stands for meter.)

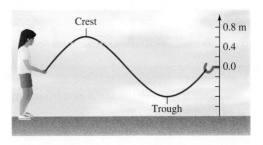

75. FOREIGN POLICY In 2004, Congress forgave $4.1 billion of Iraqi debt owed to the United States. Before that, Iraq's total debt was estimated to be $120.2 billion.

a. Which expression below can be used to find Iraq's total debt after getting debt relief from the United States?

i. $120.2 + 4.1$ ii. $120.2 - (-4.1)$
iii. $-120.2 - (-4.1)$ iv. $-120.2 - 4.1$

b. Find Iraq's total debt after getting the debt relief.

76. HISTORY Plato, a famous Greek philosopher, died in 347 B.C. at the age of 81. When was he born?

77. NASCAR Complete the table below to determine how many points the third, fourth, and fifth place finishers were behind the leader.

2012 NASCAR Sprint Cup series standings			
Rank	Driver	Points	Points behind leader
1	Brad Keselowski	2,400	…
2	Clint Bowyer	2,361	−39
3	Jimmie Johnson	2,360	
4	Kasey Kahne	2,345	
5	Greg Biffle	2,332	

Source: nascar.about.com

78. GAUGES With the engine off, the ammeter on a car reads 0. If the headlights, which draw a current of 7 amps, and the radio, which draws a current of 6 amps, are both turned on, what will be the new reading?

WRITING

79. Is subtracting 2 from 10 the same as subtracting 10 from 2? Explain.

80. Explain why we can subtract by adding the opposite.

81. Explain what it means when we say that *subtraction is not commutative.*

82. Is having a debt of $100 forgiven the same as having a gain of $100? Explain.

REVIEW

83. Find the prime factorization of 30.

84. True or false: $-4 > -5$

85. Use the associative property of addition to simplify $-18 + (18 + 89)$.

86. Multiply: $(4.5)(2.3)$

Objectives

1. Multiply signed numbers.
2. Use properties of multiplication.
3. Divide signed numbers.
4. Use properties of division.

SECTION 1.6
Multiplying and Dividing Real Numbers; Multiplication and Division Properties

In this course, you will often need to multiply or divide positive and negative numbers. For example,

- If the temperature drops 3° per hour for 4 hours, we can find the total drop in temperature by performing the multiplication $4(-3)$.
- If the temperature uniformly drops 15° over a 5-hour period, we can find the number of degrees it drops each hour by performing the division $\frac{-15}{5}$.

In this section, we will show how to perform such multiplications and divisions.

1 Multiply signed numbers.

Multiplication represents repeated addition. For example, 4(3) equals the sum of four 3's.

$$4(3) = 3 + 3 + 3 + 3$$
$$= 12$$

This example illustrates that *the product of two positive numbers is positive.*

To develop a rule for multiplying a positive number and a negative number, we will find $4(-3)$, which is equal to the sum of four -3's.

$$4(-3) = -3 + (-3) + (-3) + (-3)$$
$$= -6 + (-3) + (-3) \quad \text{Work from left to right.}$$
$$= -9 + (-3)$$
$$= -12 \quad \text{The result is negative.}$$

In terms of money, if you lose $3 four times, you have lost a total of $12, which is denoted as $-\$12$.

This example illustrates that *the product of a positive number and a negative number is negative.*

1.6 Multiplying and Dividing Real Numbers; Multiplication and Division Properties

Multiplying Two Numbers That Have Different (Unlike) Signs

To multiply a positive number and a negative number, multiply their absolute values. Then make the final answer negative.

EXAMPLE 1 Multiply: **a.** $-5(7)$ **b.** $8(-12)$ **c.** $-15 \cdot 25$

Strategy To find each product, we will use the rule for multiplying two real numbers with different signs.

WHY In each part, we are asked to multiply a positive number and a negative number.

Solution

a. $-5(7) = -35$ Multiply the absolute values, 5 and 7, to get 35. Since one factor is negative and the other is positive, the answer is negative.

b. $8(-12) = -96$ Multiply the absolute values, 8 and 12, to get 96. Make the answer negative.

c. $-15 \cdot 25 = -375$ Multiply the absolute values, 15 and 25, to get 375. Make the answer negative.

Self Check 1
Multiply:
a. $20(-30)$
b. $-0.4 \cdot 2$

Now Try Problem 25

Success Tip The product of two numbers with unlike signs is *always* negative.

To develop a rule for multiplying two negative numbers, we will find $-4(-3)$. Examine the following pattern, in which we multiply -4 and a series of factors that decrease by 1. After finding the first four products, we graph them on a number line, as shown.

This factor decreases by 1 as you read down the column.   Look for a pattern here.

$$-4(3) = -12$$
$$-4(2) = -8$$
$$-4(1) = -4$$
$$-4(0) = 0$$
$$-4(-1) = ?$$
$$-4(-2) = ?$$
$$-4(-3) = ?$$

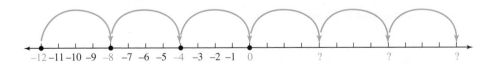

From the pattern, we see that the product increases by 4 each time. Thus,

$$-4(-1) = 4, \quad -4(-2) = 8, \quad \text{and} \quad -4(-3) = 12$$

In terms of money, losing four debts of $3 is the same as gaining $12: $-4(-\$3) = \$12$.

These results illustrate that *the product of two negative numbers is positive.*

Multiplying Two Numbers That Have the Same (Like) Signs

To multiply two numbers that have the same sign, multiply their absolute values. The final answer is positive.

Self Check 2
Multiply:
a. $-15(-8)$
b. $-\dfrac{1}{4}\left(-\dfrac{1}{3}\right)$

Now Try Problem 35

EXAMPLE 2 Multiply: **a.** $-5(-6)$ **b.** $\left(-\dfrac{1}{2}\right)\left(-\dfrac{5}{8}\right)$

Strategy To find each product, we will use the rule for multiplying two real numbers with the same sign.

WHY In each part, we are asked to multiply two negative numbers.

Solution

a. $-5(-6) = 30$ Multiply the absolute values, 5 and 6, to get 30. Since both factors are negative, the product is positive.

b. $\left(-\dfrac{1}{2}\right)\left(-\dfrac{5}{8}\right) = \dfrac{5}{16}$ Multiply the absolute values, $\dfrac{1}{2}$ and $\dfrac{5}{8}$, to get $\dfrac{5}{16}$. The product is positive.

Success Tip The product of two numbers with like signs is *always* positive.

Using Your CALCULATOR Bank Promotions

To attract business, a bank gave a clock radio to each customer who opened a checking account. The radios cost the bank $12.75 each, and 230 new accounts were opened. Each of the 230 radios was given away at a cost of $12.75, which can be expressed as -12.75. To find how much money the promotion cost the bank, we need to find the product of 230 and -12.75.

We can multiply positive and negative numbers with a reverse-entry scientific calculator. To find the product $(230)(-12.75)$, we enter these numbers and press these keys.

230 $\boxed{\times}$ 12.75 $\boxed{+/-}$ $\boxed{=}$ $\boxed{-2932.5}$

Using a graphing calculator or a direct-entry scientific calculator, we enter the following sequence:

230 $\boxed{\times}$ $\boxed{(-)}$ 12.75 $\boxed{\text{ENTER}}$ $\boxed{\begin{array}{r}230* -12.75\\ -2932.5\end{array}}$

The promotion cost the bank $2,932.50.

2 Use properties of multiplication.

A special property of multiplication is that two real numbers can be multiplied in either order to get the same result. For example, when multiplying -6 and 5, we see that

$-6(5) = -30$ and $5(-6) = -30$

This property is called the **commutative property of multiplication**.

The Commutative Property of Multiplication

Changing the order when multiplying does not affect the answer.

For any real numbers a and b,

$$ab = ba$$

To find the product of three numbers, we first multiply two of them, and then we multiply the third by that result. In the following example, we multiply $-3 \cdot 7 \cdot 5$ in two ways. We will use parentheses ( ) to show this. Recall that we perform the operations within parentheses first.

Method 1: Group -3 and 7

$(-3 \cdot 7)5 = (-21)5$ Because of the parentheses, multiply -3 and 7 first.

$ = -105$ Then multiply -21 and 5.

Method 2: Group 7 and 5

$-3(7 \cdot 5) = -3(35)$ Because of the parentheses, multiply 7 and 5 first.

$ = -105$ Then multiply -3 and 35.

Either way, the product is -105, which suggests that it doesn't matter how we *group* or *associate* numbers in multiplication. This property is called the **associative property of multiplication**.

The Associative Property of Multiplication

Changing the grouping when multiplying does not affect the answer.

For any real numbers a, b, and c,

$$(ab)c = a(bc)$$

Success Tip The commutative and associative properties of multiplication are very similar to those of addition. The only change is the operation is now multiplication.

EXAMPLE 3 Multiply: **a.** $-5(-37)(2)$ **b.** $2(-3)(-2)(-3)$

Strategy We will use the commutative and associative properties of multiplication to rewrite the product. Then we will apply the rules for multiplying signed numbers.

WHY Reordering and regrouping the factors make the computations easier, so we are less likely to make an error.

Solution
Using the commutative and associative properties of multiplication, we can reorder and regroup the factors to simplify the computations.

a. $-5(-37)(2) = -10(-37)$ Think of the problem as $-5(2)(-37)$, and then multiply -5 and 2 in your head.

$ = 370$ The product of two negative numbers is positive.

b. $2(-3)(-2)(-3) = -6(6)$ Multiply the first two factors, and then multiply the last two factors.

$ = -36$ The product of two numbers with unlike signs is negative.

Whenever we multiply a number and 0, the product is 0. For example,

$0 \cdot 8 = 0$, $6.5(0) = 0$,   and $0(-12) = 0$

Self Check 3
Multiply:
a. $-25(-3)(-4)$
b. $-1(-2)(-3)(-3)$

Now Try Problem 38

We also see that whenever we multiply a number by 1, the number remains the same. For example,

$$6 \cdot 1 = 6, \quad 4.53(1) = 4.53, \quad \text{and} \quad 1(-9) = -9$$

These examples illustrate the **multiplication properties of 0 and 1**. Since any number multiplied by 1 remains the same (is identical), the number 1 is called the **identity element** for multiplication.

> **Multiplication Properties of 0 and 1**
>
> The product of 0 and any real number is 0. The product of 1 and any real number is that number.
>
> For any real number a,
>
> $$a \cdot 0 = 0 \quad \text{and} \quad 0 \cdot a = 0$$
> $$a \cdot 1 = a \quad \text{and} \quad 1 \cdot a = a$$

Recall that if the product of two numbers is 1, the numbers are **reciprocals**. The numbers are also called **multiplicative inverses** of each other. For example, because $8 \cdot \frac{1}{8} = 1$, the numbers 8 and $\frac{1}{8}$ are reciprocals (or multiplicative inverses). Likewise, $-\frac{3}{4}$ and $-\frac{4}{3}$ are multiplicative inverses because $-\frac{3}{4}\left(-\frac{4}{3}\right) = 1$. All real numbers, except 0, have reciprocals (multiplicative inverses).

> **Multiplicative Inverses or Reciprocals (Inverse Property of Multiplication)**
>
> The product of any nonzero real number and its multiplicative inverse (reciprocal) is 1.
>
> For any nonzero real number a,
>
> $$a\left(\frac{1}{a}\right) = 1$$

> **Caution!** Do not change the sign of a number when finding its reciprocal.

3 Divide signed numbers.

Every division fact containing three numbers can be written as an equivalent multiplication fact containing the same three numbers. For example,

$$\frac{15}{5} = 3 \quad \text{because} \quad 5(3) = 15$$

We will use this relationship between multiplication and division to develop the rules for dividing signed numbers. From the example $\frac{15}{5} = 3$, we see that *the quotient of two positive numbers is positive.*

To determine the quotient of two negative numbers, we consider the division $\frac{-15}{-5} = ?$. We can do the division by examining its related multiplication fact: $-5(?) = -15$. To find the number that should replace the question mark, we use the rules for multiplying signed numbers discussed earlier in this section.

1.6 Multiplying and Dividing Real Numbers; Multiplication and Division Properties

Multiplication fact $-5(?) = -15$
This must be positive 3 if the product is to be negative 15.

Division fact $\dfrac{-15}{-5} = 3$
So the quotient is positive 3.

From this example, we see that *the quotient of two negative numbers is positive.*

To determine the quotient of a positive number and a negative number, we consider $\dfrac{15}{-5} = ?$ and its equivalent multiplication fact $-5(?) = 15$.

Multiplication fact $-5(?) = 15$
This must be negative 3 if the product is to be positive 15.

Division fact $\dfrac{15}{-5} = -3$
So the quotient is negative 3.

From this example, we see that *the quotient of a positive number and a negative number is negative.*

To determine the quotient of a negative number and a positive number, we consider $\dfrac{-15}{5} = ?$ and its equivalent multiplication fact $5(?) = -15$.

Multiplication fact $5(?) = -15$
This must be negative 3 if the product is to be negative 15.

Division fact $\dfrac{-15}{5} = -3$
So the quotient is negative 3.

From this example, we see that *the quotient of a negative number and a positive number is negative.*

We can now summarize the results from the previous discussion. Note that the rules for division are similar to those for multiplication.

Dividing Two Real Numbers

To divide two real numbers, divide their absolute values.

1. The quotient of two numbers with the same (like) signs is positive.
2. The quotient of two numbers with different (unlike) signs is negative.

EXAMPLE 4 Divide: **a.** $\dfrac{66}{11}$ **b.** $\dfrac{-81}{-9}$ **c.** $\dfrac{-45}{9}$ **d.** $\dfrac{28}{-7}$

Strategy To find the quotients, we will use the rules for dividing signed numbers.

WHY The signs of the numbers that we are dividing determine the sign of the result.

Solution
To divide numbers with like signs, we find the quotient of their absolute values and make the quotient positive.

a. $\dfrac{66}{11} = 6$ Divide the absolute values, 66 by 11, to get 6. The answer is positive.

b. $\dfrac{-81}{-9} = 9$ Divide the absolute values, 81 by 9, to get 9. The answer is positive.

Self Check 4
Divide:

a. $\dfrac{48}{12}$

b. $\dfrac{-63}{-9}$

c. $\dfrac{40}{-8}$

d. $\dfrac{-49}{7}$

Now Try Problem 45

To divide numbers with unlike signs, we find the quotient of their absolute values and make the quotient negative.

c. $\dfrac{-45}{9} = -5$ Divide the absolute values, 45 by 9, to get 5. The answer is negative.

d. $\dfrac{28}{-7} = -4$ Divide the absolute values, 28 by 7, to get 4. The answer is negative.

4 Use properties of division.

The examples

$$\dfrac{12}{1} = 12, \qquad \dfrac{-80}{1} = -80, \qquad \text{and} \qquad \dfrac{7.75}{1} = 7.75$$

illustrate that *any number divided by 1 is the number itself.* The examples

$$\dfrac{35}{35} = 1, \qquad \dfrac{-4}{-4} = 1, \qquad \text{and} \qquad \dfrac{0.9}{0.9} = 1$$

illustrate that *any number (except 0) divided by itself is 1.*

> **Division Properties**
>
> Any real number divided by 1 is the number itself. Any number (except 0) divided by itself is 1.
>
> For any real number a,
>
> $$\dfrac{a}{1} = a \qquad \text{and} \qquad \dfrac{a}{a} = 1, \quad \text{where } a \neq 0$$

We will now consider three types of division that involve zero. In the first case, we will examine a division of zero; in the second, a division by zero; in the third case, a division of zero by zero.

Division statement	Related multiplication statement		Result
$\dfrac{0}{2} = ?$	$2(?) = 0$	This must be 0 if the product is to be 0.	$\dfrac{0}{2} = 0$
$\dfrac{2}{0} = ?$	$0(?) = 2$	There is no number that gives 2 when multiplied by 0.	There is no quotient.
$\dfrac{0}{0} = ?$	$0(?) = 0$	Any number times 0 is 0.	Any number can be the quotient.

We see that $\dfrac{0}{2} = 0$. Since $\dfrac{2}{0}$ does not have a quotient, we say that division of 2 by 0 is *undefined*. Since $\dfrac{0}{0}$ can be any number, we say that $\dfrac{0}{0}$ is *indeterminate*. These results suggest the following division facts.

Division Involving 0

1. If a represents a nonzero number, $\dfrac{0}{a} = 0$.

2. If a represents a nonzero number, $\dfrac{a}{0}$ is undefined.

3. $\dfrac{0}{0}$ is indeterminate.

The Language of Algebra When we say a division by 0, like $\dfrac{5}{0}$, is *undefined*, we mean that $\dfrac{5}{0}$ does not represent a number.

Using Your CALCULATOR Depreciation of a House

Over a 17.5-year period, the value of a $124,930 house fell at a uniform rate to $97,105. To find how much the house depreciated per year, we must first find the change in its value by subtracting $124,930 from $97,105. To compute this difference, we enter these numbers and press these keys on a scientific calculator.

97105 [−] 124930 [=]   -27825

-27825 represents a drop in value of $27,825. Since this depreciation occurred in 17.5 years, we divide $-27,825$ by 17.5 to find the amount of depreciation per year. With $-27,825$ on the display, we then enter these numbers and press these keys.

[÷] 17.5 [=]   -1590

If we use a graphing or direct-entry calculator, we enter these numbers and press these keys.

97105 [−] 124930 [ENTER] [÷] 17.5 [ENTER]

```
97105 - 124930
         -27825
Ans/17.5
          -1590
```

The amount of depreciation per year was $1,590.

EXAMPLE 5

Divide, if possible: **a.** $\dfrac{0}{13}$ **b.** $\dfrac{-13}{0}$ **c.** $\dfrac{7}{7}$ **d.** $\dfrac{9}{1}$

Strategy We will determine the appropriate division property to use for each quotient.

WHY Each of these expressions is a special case of division.

Solution

a. $\dfrac{0}{13}$ is division of 0 by a nonzero number. $\dfrac{0}{13} = 0$ Because $13(0) = 0$

b. Since $\dfrac{-13}{0}$ involves division by zero, the division is undefined.

c. $\dfrac{7}{7} = 1$ Because $7(1) = 7$

d. $\dfrac{9}{1} = 9$ Because $1(9) = 9$

Self Check 5

Find each quotient, if possible:

a. $\dfrac{4}{0}$ **b.** $\dfrac{0}{17}$

c. $\dfrac{12}{12}$

Now Try Problems 49 and 51

ANSWERS TO SELF CHECKS

1. a. -600 b. -0.8 2. a. 120 b. $\frac{1}{12}$ 3. a. -300 b. 18 4. a. 4 b. 7 c. -5 d. -7 5. a. undefined b. 0 c. 1

SECTION 1.6 STUDY SET

VOCABULARY
Fill in the blanks.

1. The answer to a multiplication is called a _____. The answer to a division is called a _____.
2. The numbers -4 and -6 are said to have _____ signs. The numbers -10 and 12 are said to have _____ signs.
3. The _____ property of multiplication states that two numbers can be multiplied in either order to get the same result.
4. The statement $(ab)c = a(bc)$ expresses the _____ property of multiplication.
5. Division of a nonzero number by zero is _____.
6. If the product of two numbers is 1, the numbers are called _____ or _____ inverses.

CONCEPTS
Fill in the blanks.

7. The expression $-5 + (-5) + (-5) + (-5)$ can be represented by the multiplication statement _____.
8. The quotient of two numbers with _____ signs is negative.
9. The product of two negative numbers is _____.
10. The product of zero and any number is _____.
11. The product of _____ and any number is that number.
12. The division fact $\frac{25}{-5} = -5$ is related to the multiplication fact _____.
13. a. If we multiply two different numbers and the answer is 0, what is true about one of the numbers?
    b. If we multiply two different numbers and the answer is 1, what is true about the numbers?
14. a. If we divide two nonzero numbers and the answer is 1, what is true about the numbers?
    b. If we divide two numbers and the answer is 0, what is true about the numbers?
15. Which property justifies each statement?
    a. $-5(2 \cdot 17) = (-5 \cdot 2)17$
    b. $5\left(\frac{1}{5}\right) = 1$
    c. $-5 \cdot 2 = 2(-5)$
    d. $-5(1) = -5$
16. a. Find $-1(8)$. In general, what is the result when a number is multiplied by -1?
    b. Find $\frac{8}{-1}$. In general, what is the result when a number is divided by -1?

POS stands for a positive number and NEG stands for a negative number. Determine the sign of each result, if possible.

17. a. POS · NEG   b. POS + NEG
    c. POS − NEG   d. $\frac{\text{POS}}{\text{NEG}}$
18. a. NEG · NEG   b. NEG + NEG
    c. NEG − NEG   d. $\frac{\text{NEG}}{\text{NEG}}$

19. What is wrong with the following statement?

    A negative and a positive is a negative.

20. Give the opposite (additive inverse) and the reciprocal (multiplicative inverse) of each number.
    a. 2 b. $-\frac{4}{5}$
    c. 1.75 d. -5

21. When a calculator was used to compute $16 \div 0$, the message shown appeared on the display screen. Explain what the message means.

    `Error`

22. a. Is 80 divided by -5 the same as -5 divided by 80?
    b. Is 80 times -5 the same as -5 times 80?

NOTATION

Complete each step.

23. $(-37 \cdot 5)2 = -37(\boxed{} \cdot 2)$
    $= -37(\boxed{})$
    $= -370$

24. $-20(5 \cdot 79) = (\boxed{} \cdot 5) \cdot 79$
    $= \boxed{} \cdot 79$
    $= -7,900$

GUIDED PRACTICE

Perform each multiplication. See Example 1.

25. $12(-5)$
26. $(-9)(11)$
27. $-6 \cdot 4$
28. $-8 \cdot 9$
29. $-20(40)$
30. $-10(10)$
31. $(6)(-9)$
32. $(8)(-7)$

Perform each multiplication. See Example 2.

33. $(-6)(-6)$
34. $(-1)(-1)$
35. $-\dfrac{1}{2}\left(-\dfrac{3}{4}\right)$
36. $-\dfrac{1}{3}\left(-\dfrac{5}{16}\right)$

Multiply. See Example 3.

37. $3(-4)(-5)$
38. $(-2)(-4)(-5)$
39. $(-0.4)(0.3)(-0.7)$
40. $0.5(-0.3)(-0.4)$

Perform each division, if possible. See Example 4.

41. $\dfrac{-6}{-2}$
42. $\dfrac{-36}{9}$
43. $\dfrac{4}{-2}$
44. $\dfrac{-9}{3}$
45. $\dfrac{80}{-20}$
46. $\dfrac{-66}{33}$
47. $\dfrac{17}{-17}$
48. $\dfrac{-24}{24}$

Perform each division, if possible. See Example 5.

49. $\dfrac{0}{150}$
50. $\dfrac{225}{0}$
51. $\dfrac{-17}{0}$
52. $\dfrac{0}{-12}$

TRY IT YOURSELF

Perform each operation.

53. $-5.2 \cdot 100$
54. $-1.17 \cdot 1,000$
55. $0(-22)$
56. $-8 \cdot 0$
57. $-3(-4)(0)$
58. $15(0)(-22)$
59. $(-2)(-3)(-4)(-5)$
60. $(-3)(-4)(5)(-6)$
61. $(-23.5)(47.2)$
62. $(-435.7)(-37.8)$
63. $(-6.37)(-7.2)(-9.1)$
64. $(5.2)(-8.2)(7.75)$
65. $-0.6(-4)$
66. $-0.7(-8)$
67. $1.2(-0.4)$
68. $0(-0.2)$
69. $-1\dfrac{1}{4}\left(-\dfrac{3}{4}\right)$
70. $-1\dfrac{1}{8}\left(-\dfrac{3}{8}\right)$
71. $\dfrac{204.6}{-37.2}$
72. $\dfrac{-30.56625}{-4.875}$
73. $\dfrac{-110}{-110}$
74. $\dfrac{-200}{-200}$
75. $\dfrac{-160}{40}$
76. $\dfrac{-250}{-50}$
77. $\dfrac{320}{-16}$
78. $\dfrac{-180}{36}$
79. $\dfrac{0.5}{-100}$
80. $\dfrac{-1.7}{10}$
81. $-\dfrac{1}{3} \div \dfrac{4}{5}$
82. $-\dfrac{1}{8} \div \dfrac{2}{3}$
83. $-\dfrac{3}{16} \div \left(-\dfrac{2}{3}\right)$
84. $-\dfrac{3}{25} \div \left(-\dfrac{2}{3}\right)$
85. $-30 \div (-3)$
86. $-12 \div (-2)$
87. $-42 \div 7$
88. $72 \div (-8)$

Use the associative property of multiplication to help find the product.

89. $-5(2 \cdot 67)$
90. $\left(-\dfrac{5}{16} \cdot \dfrac{1}{7}\right)7$
91. $(-7 \cdot 8) \cdot 5$
92. $(-8 \cdot 7)3$

APPLICATIONS

93. **TEMPERATURE CHANGE** In a lab, the temperature of a fluid was decreased 6° per hour for 12 hours. What signed number indicates the change in temperature?

94. **BACTERIAL GROWTH** To warm a bacterial culture, biologists programmed a heating pad under the culture to increase the temperature 4° every hour for 6 hours. What signed number indicates the change in the temperature of the culture?

95. **FLUID TEMPERATURE** In a lab, the temperature of a fluid was decreased 5° per hour for 14 hours. What signed number indicates the drop in temperature?

96. **REAL ESTATE** A house has depreciated $1,250 each year for 8 years. What signed number indicates its change in value over that time period?

97. **ASTRONOMY** The temperature on Pluto gets as low as −386°F. This is twice the lowest temperature reached on Jupiter. What is the lowest temperature on Jupiter?

98. **CAR RADIATORS** The instructions on the back of a container of antifreeze state, "A 50/50 mixture of antifreeze and water protects against freeze-ups down to −34°F, while a 60/40 mix protects against freeze-ups down to one and one-half times that temperature." To what temperature does the 60/40 mixture protect?

99. **POST OFFICE** The U.S. Postal Service reported an annual loss of $5.1 billion in 2011. For the year 2012, the losses were about triple that. What signed number indicates the Postal Service's financial loss in 2012? (Source: foxnews.com)

100. **ACCOUNTING** In the 2012 income statement for JCPenney, numbers within parentheses represent a loss. Complete the statement given the following facts. The second-quarter loss was worse than the third-quarter loss by a factor of about 1.2. The fourth-quarter loss was approximately $4\frac{2}{5}$ times the third-quarter loss. The first-quarter loss was $\frac{3}{10}$ of the fourth-quarter loss.

JCPenney INCOME STATEMENT				2012
All amounts in millions of dollars	1st Qtr (?)	2nd Qtr (?)	3rd Qtr (125)	4th Qtr (?)

Source: Google Finance

101. **THE *QUEEN MARY*** The ocean liner *Queen Mary* was commissioned in 1936 and cost $22,500,000 to build. In 1967, the ship was purchased for $3,450,000 by the city of Long Beach, California, where it now serves as a hotel and convention center. What signed number indicates the annual average depreciation of the *Queen Mary* over the 31-year period from 1936 to 1967? Round to the nearest dollar.

102. **COMPUTER SPREADSHEETS** The formula = A1*B1*C1 in cell D1 of the spreadsheet instructs the computer to multiply the values in cells A1, B1, and C1 and to print the result *in place of the formula* in cell D1. What values will the computer print in cells D1, D2, and D3?

	A	B	C	D
1	4	−5	−17	= A1*B1*C1
2	22	−30	14	= A2*B2*C2
3	−60	−20	−34	= A3*B3*C3
4				
5				

103. **PHYSICS** An oscilloscope is an instrument that displays electrical signals, which appear as wavy lines on a fluorescent screen. If the magnification setting is switched to × 2, for example, the "height" of the peak and the "depth" of the valley of a graph will be doubled. Use signed numbers to indicate the peak height and the valley depth for each setting of the magnification dial.

a. normal
b. × 0.5
c. × 1.5
d. × 2

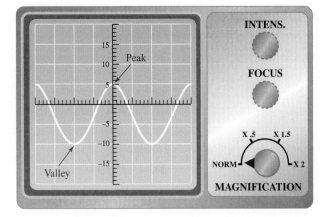

104. **LIGHT** Water acts as a selective filter of light. As shown in the illustration below, red light waves penetrate water only to a depth of about 5 meters. How many times deeper does

a. yellow light penetrate than red light?

b. green light penetrate than orange light?

c. blue light penetrate than yellow light?

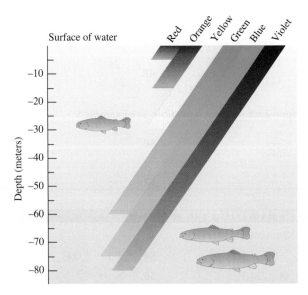

WRITING

105. Explain how you would decide whether the product of several numbers is positive or negative.

106. If the product of five numbers is negative, how many of them could be negative? Explain.

REVIEW

107. Is every integer a rational number?

108. Write the subtraction statement $-3 - (-5)$ as addition of the opposite.

109. Find $\frac{1}{2} + \frac{1}{4} + \frac{1}{3}$ and express the result as a decimal.

110. Describe the balance in a checking account that is overdrawn $65 using a signed number.

111. Find: $0.475(1,000)$

112. Give two examples of irrational numbers.

SECTION 1.7
Exponents and Order of Operations

Objectives
1. Evaluate exponential expressions.
2. Use the order of operations rules.
3. Evaluate expressions with no grouping symbols.
4. Evaluate expressions containing grouping symbols.
5. Find the mean (average).

In this course, we will perform six operations with real numbers: addition, subtraction, multiplication, division, raising to a power, and finding a root. Quite often, we will have to **evaluate** (find the value of) expressions containing more than one operation. In that case, we need to know the order in which the operations are to be performed. That is a topic of this section.

1 Evaluate exponential expressions.

In the expression $3 \cdot 3 \cdot 3 \cdot 3 \cdot 3$, the number 3 is used as a factor 5 times. We call 3 a *repeated factor*. To express a repeated factor, we can use an **exponent**.

Exponent and Base

An **exponent** is used to indicate repeated multiplication. It tells how many times the **base** is used as a factor.

$$\underbrace{3 \cdot 3 \cdot 3 \cdot 3 \cdot 3}_{\text{Five repeated factors of 3.}} = 3^5$$

The exponent is 5. The base is 3.

In the **exponential expression** a^n, a is the base, and n is the exponent. The expression a^n is called a **power of a**. Some examples of powers are

5^2 Read as "5 to the second power" or "5 squared."

9^3 Read as "9 to the third power" or "9 cubed."

$(-2)^5$ Read as "−2 to the fifth power."

The Language of Algebra 5^2 represents the area of a square with sides 5 units long. 4^3 represents the volume of a cube with sides 4 units long.

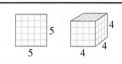

Self Check 1

Write each expression using exponents:
a. (12)(12)(12)(12)(12)(12)
b. $2 \cdot 9 \cdot 9 \cdot 9$
c. fifty squared
d. $(-30)(-30)(-30)$

Now Try Problem 23

EXAMPLE 1
Write each expression using exponents: a. $4 \cdot 4 \cdot 4$
b. $(-5)(-5)(-5)(-5)(-5)$ c. sixteen cubed d. $8 \cdot 8 \cdot 15 \cdot 15 \cdot 15 \cdot 15$

Strategy We will count the number of repeated factors in each expression.

WHY An exponent can be used to represent repeated multiplication.

Solution
a. The factor 4 is repeated 3 times. We can represent this repeated multiplication with an exponential expression having a base of 4 and an exponent of 3: $4 \cdot 4 \cdot 4 = 4^3$.
b. The factor -5 is repeated 5 times: $(-5)(-5)(-5)(-5)(-5) = (-5)^5$.
c. Sixteen cubed can be written as 16^3.
d. $8 \cdot 8 \cdot 15 \cdot 15 \cdot 15 \cdot 15 = 8^2 \cdot 15^4$

In the next example, we use exponents to rewrite expressions involving repeated variable factors.

Self Check 2

Write each product using exponents:
a. $y \cdot y \cdot y \cdot y$
b. $12 \cdot b \cdot b \cdot b \cdot c$

Now Try Problem 27

EXAMPLE 2
Write each product using exponents:
a. $a \cdot a \cdot a \cdot a \cdot a \cdot a$ b. $4 \cdot \pi \cdot r \cdot r$

Strategy We will count the number of repeated factors in each expression.

WHY An exponent can be used to represent repeated multiplication.

Solution
a. $a \cdot a \cdot a \cdot a \cdot a \cdot a = a^6$ a is repeated as a factor 6 times.
b. $4 \cdot \pi \cdot r \cdot r = 4\pi r^2$ r is repeated as a factor 2 times.

Self Check 3

Find each power:
a. 2^5
b. $(-6)^2$
c. $(-5)^3$

Now Try Problem 35

EXAMPLE 3
Find each power: a. 5^3 b. 10^1 c. $(-3)^4$ d. $(-3)^5$

Strategy We will identify the base to determine the repeated factor and identify the exponent to determine the number of times the factor is repeated. Then we will multiply to evaluate the expression.

WHY Exponents represent repeated multiplication.

Solution
We write the base as a factor the number of times indicated by the exponent. Then we perform the multiplication.

a. $5^3 = 5 \cdot 5 \cdot 5 = 125$ The base is 5, the exponent is 3.

b. $10^1 = 10$ The base is 10, the exponent is 1.

c. $(-3)^4 = (-3)(-3)(-3)(-3)$ Write -3 as a factor 4 times.
$= 9(-3)(-3)$ Work from left to right: $(-3)(-3) = 9$.
$= -27(-3)$ Work from left to right: $9(-3) = -27$.
$= 81$

d. $(-3)^5 = (-3)(-3)(-3)(-3)(-3)$ Write -3 as a factor 5 times.
$= 9(-3)(-3)(-3)$ Work from left to right: $(-3)(-3) = 9$.
$= -27(-3)(-3)$ Work from left to right: $9(-3) = -27$.
$= 81(-3)$ Work from left to right: $-27(-3) = 81$.
$= -243$

> **Caution!** Don't make the mistake of multiplying the base and the exponent.
>
> | Incorrect | Correct |
> |---|---|
> | ~~$5^3 = 5 \cdot 3$~~ | $5^3 = 5 \cdot 5 \cdot 5$ |
> | | $= 125$ |

We can now make some observations about raising a negative number to an *even power* (2, 4, 6, 8, and so on) and raising a negative number to an *odd power* (1, 3, 5, 7, and so on). In part c of Example 3, we raised -3 to an even power, and the result was positive. In part d, we raised -3 to an odd power, and the result was negative. These results illustrate the following general rule.

> **Even and Odd Powers of a Negative Number**
>
> When a negative number is raised to an even power, the result is positive.
>
> When a negative number is raised to an odd power, the result is negative.

> **Caution!** Although the expressions -4^2 and $(-4)^2$ look alike, they are not. In -4^2, the base is 4 and the exponent is 2. The $-$ sign in front of 4^2 means the opposite of 4^2. In $(-4)^2$, the base is -4 and the exponent is 2. When we find the value of each expression, it becomes clear that they are not equivalent.
>
> | $-4^2 = -(4 \cdot 4)$ | Write 4 as a factor 2 times. | $(-4)^2 = (-4)(-4)$ | Write -4 as a factor 2 times. |
> |---|---|---|---|
> | $= -16$ | Multiply within the parentheses. | $= 16$ | The product of two negative numbers is positive. |
>
> Different results

EXAMPLE 4

Find each power: **a.** $\left(-\dfrac{2}{3}\right)^3$ **b.** $(0.6)^2$ **c.** -2^6

Strategy We will write each exponential expression as a product and multiply the repeated factors.

WHY Exponents represent repeated multiplication.

Solution

a. $\left(-\dfrac{2}{3}\right)^3 = \left(-\dfrac{2}{3}\right)\left(-\dfrac{2}{3}\right)\left(-\dfrac{2}{3}\right)$ Since $-\dfrac{2}{3}$ is the base and 3 is the exponent, we write $-\dfrac{2}{3}$ as a factor 3 times.

$= \dfrac{4}{9}\left(-\dfrac{2}{3}\right)$ Multiply: $\left(-\dfrac{2}{3}\right)\left(-\dfrac{2}{3}\right) = \dfrac{4}{9}$.

$= -\dfrac{8}{27}$ Do the multiplication.

b. $(0.6)^2 = (0.6)(0.6)$ Since 0.6 is the base and 2 is the exponent, we write 0.6 as a factor 2 times.

$= 0.36$ Do the multiplication.

c. $-2^6 = -(2 \cdot 2 \cdot 2 \cdot 2 \cdot 2 \cdot 2)$ Since 2 is the base and 6 is the exponent, we write 2 as a factor 6 times. We use the opposite of the final value.

$= -64$ Do the multiplication.

Self Check 4

Find each power:

a. $\left(-\dfrac{3}{4}\right)^3$

b. $(-0.3)^2$

c. -5^2

Now Try Problem 37

Using Your CALCULATOR   Finding a Power

On a scientific calculator, we can use the squaring key $\boxed{x^2}$ to find the square of a number, and we can use the exponential key $\boxed{y^x}$ (on some calculators labeled x^y) to raise a number to a power. For example, to evaluate 125^2 and 2^{10} using a scientific calculator, we enter these numbers and press these keys.

125 $\boxed{x^2}$ 15625

2 $\boxed{y^x}$ 10 $\boxed{=}$ 1024

Using a graphing or direct-entry calculator, we can evaluate 125^2 and 2^{10} by pressing these keys.

125 $\boxed{x^2}$ $\boxed{\text{ENTER}}$ 125^2
                                   15625

2 $\boxed{\wedge}$ 10 $\boxed{\text{ENTER}}$ $2 \wedge 10$
                                   1024

We have found that $125^2 = 15{,}625$ and $2^{10} = 1{,}024$.

2 Use the order of operations rules.

Suppose you have been asked to contact a friend if you see a Rolex watch for sale when you are traveling in Europe. While in Switzerland, you find the watch and send the text message shown on the left. The next day, you get the response shown on the right.

You sent this message.   You get this response.

    Something is wrong. The first part of the response (No price too high!) says to buy the watch at any price. The second part (No! Price too high.) says not to buy it, because it's too expensive. The placement of the exclamation point makes us read the two parts of the response differently, resulting in different meanings. When reading a mathematical statement, the same kind of confusion is possible. For example, consider the expression

$$2 + 3 \cdot 6$$

which contains two operations: addition and multiplication. We can consider doing the calculations in two ways. We can add first and then multiply. Or we can multiply first and then add. However, we get different results.

Method 1: Add first		Method 2: Multiply first	
$2 + 3 \cdot 6 = 5 \cdot 6$	Add 2 and 3 first.	$2 + 3 \cdot 6 = 2 + 18$	Multiply 3 and 6 first.
$= 30$	Multiply 5 and 6.	$= 20$	Add 2 and 18.

1.7 Exponents and Order of Operations

If we don't establish a uniform order of operations, the expression $2 + 3 \cdot 6$ has two different values. To avoid this possibility, we always use the following set of priority rules.

> **Order of Operations**
>
> 1. Perform all calculations within parentheses and other grouping symbols following the order listed in steps 2–4 below, working from the innermost pair to the outermost pair.
> 2. Evaluate all exponential expressions.
> 3. Perform all multiplications and divisions as they occur from left to right.
> 4. Perform all additions and subtractions as they occur from left to right.
>
> When grouping symbols have been removed, repeat steps 2–4 to complete the calculation.
>
> If a fraction is present, evaluate the expression above and the expression below the bar separately. Then do the division indicated by the fraction bar, if possible.

It isn't necessary to apply all of these steps in every problem. For example, the expression $2 + 3 \cdot 6$ does not contain any parentheses, and there are no exponential expressions. So we look for multiplications and divisions to perform. To evaluate $2 + 3 \cdot 6$ correctly, we proceed as follows:

$$2 + 3 \cdot 6 = 2 + 18 \quad \text{Multiply first: } 3 \cdot 6 = 18.$$
$$= 20 \quad \text{Add.}$$

Therefore, the correct result when evaluating $2 + 3 \cdot 6$ is 20.

3 Evaluate expressions with no grouping symbols.

EXAMPLE 5 Evaluate: $3 \cdot 2^3 - 4$

Strategy We will scan the expression to determine what operations need to be performed. Then we will perform those operations, one at a time, following the order of operations rules.

WHY The order of operations gives us the steps needed to find the correct result.

Solution
To find the value of this expression, we must perform the operations of multiplication, raising to a power, and subtraction. The rules for the order of operations tell us to begin by evaluating the exponential expression.

$$3 \cdot 2^3 - 4 = 3 \cdot 8 - 4 \quad \text{Evaluate the exponential expression: } 2^3 = 8.$$
$$= 24 - 4 \quad \text{Multiply: } 3 \cdot 8 = 24.$$
$$= 20 \quad \text{Subtract.}$$

Self Check 5
Evaluate: $2 \cdot 3^2 + 17$
Now Try Problem 43

> **The Language of Algebra** Sometimes, for problems like these, the instruction *Simplify* is used instead of *Evaluate*.

Self Check 6

Evaluate: $-40 - 9 \cdot 4 + 10$

Now Try Problem 49

EXAMPLE 6 Evaluate: $-30 - 4 \cdot 5 + 9$

Strategy We will scan the expression to determine what operations need to be performed. Then we will perform those operations, one at a time, following the order of operations rules.

WHY The order of operations gives us the steps needed to find the correct result.

Solution
To evaluate this expression, we must perform the operations of subtraction, multiplication, and addition. The rules for the order of operations tell us to begin with the multiplication.

$$-30 - \mathbf{4 \cdot 5} + 9 = -30 - \mathbf{20} + 9 \quad \text{Multiply: } 4 \cdot 5 = 20.$$
$$= -50 + 9 \quad \text{Working from left to right, subtract } -30 - 20 = -30 + (-20) = -50.$$
$$= -41 \quad \text{Add.}$$

Caution! Some students think that additions are always done before subtractions. As you saw in Example 6, this is not true. Working from left to right, we do the additions or subtractions in the order in which they occur. The same is true for multiplications and divisions.

Self Check 7

Evaluate:
$240 \div (-8)(3) - 3(-2)4$

Now Try Problem 54

EXAMPLE 7 Evaluate: $160 \div (-4)(3) - 6(-2)3$

Strategy We will scan the expression to determine what operations need to be performed. Then we will perform those operations, one at a time, following the order of operations rules.

WHY The order of operations gives us the steps needed to find the correct result.

Solution
Although this expression contains parentheses, there are no operations to perform within them. Since there are no exponents, we perform multiplications and divisions as they occur from left to right.

$$\mathbf{160 \div (-4)}(3) - 6(-2)3 = \mathbf{-40}(3) - 6(-2)3 \quad \text{Divide: } 160 \div (-4) = -40$$
$$= -120 - 6(-2)3 \quad \text{Multiply: } -40(3) = -120$$
$$= -120 - (-12)3 \quad \text{Multiply: } 6(-2) = -12.$$
$$= -120 - (-36) \quad \text{Multiply: } (-12)3 = -36.$$
$$= -120 + 36 \quad \text{Write the subtraction as addition of the opposite.}$$
$$= -84 \quad \text{Add.}$$

Caution! A common mistake is to forget to work from left to right and incorrectly perform the multiplication before the division.

4 Evaluate expressions containing grouping symbols.

Grouping symbols are mathematical punctuation marks. They help determine the order in which an expression is to be evaluated. Examples of grouping symbols are parentheses ( ), brackets [ ], absolute value symbols | |, and the fraction bar —.

EXAMPLE 8 Evaluate: $(6 - 3)^2$

Strategy We will perform the operation(s) within the parentheses first. When there is more than one operation to perform within the parentheses, we follow the order of operations rules.

WHY This is the first step of the order of operations.

Solution
This expression contains parentheses. By the rules for the order of operations, we must perform the operation within the parentheses first.

$(6 - 3)^2 = 3^2$ Subtract within the parentheses: $6 - 3 = 3$.

$ = 9$ Evaluate the exponential expression.

Self Check 8
Evaluate: $(12 - 6)^3$
Now Try Problem 59

EXAMPLE 9 Evaluate: $5^3 + 2(-8 - 3 \cdot 2)$

Strategy We will perform the operation(s) within the parentheses first. When there is more than one operation to perform within the parentheses, we follow the order of operations rules.

WHY This is the first step of the order of operations.

Solution
First, we perform the operations within the parentheses in the proper order.

$5^3 + 2(-8 - 3 \cdot 2) = 5^3 + 2(-8 - 6)$ Multiply within the parentheses: $3 \cdot 2 = 6$.

$ = 5^3 + 2(-14)$ Subtract within the parentheses: $-8 - 6 = -8 + (-6) = -14$.

$ = 125 + 2(-14)$ Evaluate the exponential expression: $5^3 = 125$.

$ = 125 + (-28)$ Multiply: $2(-14) = -28$.

$ = 97$ Add.

Self Check 9
Evaluate: $1^3 + 6(-6 - 3 \cdot 0)$
Now Try Problem 63

> **Success Tip** Multiplication is indicated when a number is next to a parentheses or bracket.

Expressions can contain two or more pairs of grouping symbols. To evaluate the following expression, we begin by working within the innermost pair of grouping symbols. Then we work within the outermost pair.

$$-4[-2 - 3(4 - 8^2)] - 2$$

with the inner $(4 - 8^2)$ labeled "Innermost pair" and the outer $[\ldots]$ labeled "Outermost pair".

> **The Language of Algebra** When one pair of grouping symbols is inside another pair, we say that those grouping symbols are *nested*, or *embedded*.

Chapter 1 An Introduction to Algebra

Self Check 10

Evaluate:
$-5[2(5^2 - 15) + 4] - 10$

Now Try Problem 74

EXAMPLE 10 Evaluate: $-4[-2 - 3(4 - 8^2)] - 2$

Strategy We will work within the parentheses first and then within the brackets. At each stage, we follow the order of operations rules.

WHY By the order of operations, we must work from the *innermost* pair of grouping symbols to the *outermost*.

Solution

We work within the innermost grouping symbols (the parentheses) first.

$-4[-2 - 3(4 - \mathbf{8^2})] - 2$

$= -4[-2 - 3(4 - \mathbf{64})] - 2$ Evaluate the exponential expression within the parentheses: $8^2 = 64$.

$= -4[-2 - 3(-60)] - 2$ Subtract within the parentheses: $4 - 64 = 4 + (-64) = -60$.

$= -4[-2 - (-180)] - 2$ Multiply within the brackets: $3(-60) = -180$.

$= -4(178) - 2$ Subtract within the brackets: $-2 - (-180) = -2 + 180 = 178$.

$= -712 - 2$ Multiply.

$= -714$ Subtract: $-712 - 2 = -712 + (-2) = -714$.

Self Check 11

Evaluate: $\dfrac{-4(-2 + 8) + 6}{8 - 5(-2)}$

Now Try Problem 75

EXAMPLE 11 Evaluate: $\dfrac{-3(3 + 2) + 5}{17 - 3(-4)}$

Strategy We will evaluate the expression above and the expression below the fraction bar separately. Then we will simplify the fraction, if possible.

WHY Fraction bars are grouping symbols. They group the numerator and denominator. The expression could be written as $[-3(3 + 2) + 5] \div [17 - 3(-4)]$.

Solution

We simplify the numerator and the denominator separately.

$\dfrac{-3(\mathbf{3 + 2}) + 5}{17 - \mathbf{3(-4)}} = \dfrac{-3(\mathbf{5}) + 5}{17 - (\mathbf{-12})}$ In the numerator, add within the parentheses. In the denominator, multiply.

$= \dfrac{-15 + 5}{17 + 12}$ In the numerator, multiply. In the denominator, write the subtraction as addition of the opposite of -12, which is 12.

$= \dfrac{-10}{29}$ Do the additions.

$= -\dfrac{10}{29}$ Write the $-$ sign in front of the fraction: $\dfrac{-10}{29} = -\dfrac{10}{29}$.

Success Tip The order of operations is built into most calculators. A left parenthesis key (and a right parenthesis key) should be used when grouping symbols, including a fraction bar, are in the problem.

Self Check 12

Evaluate: $10^3 + 3|24 - 25|$

Now Try Problem 86

EXAMPLE 12 Evaluate: $10|9 - 15| - 2^5$

Strategy The absolute value bars are grouping symbols. We will perform the calculation within them first.

WHY By the order of operations, we must perform all calculations within parentheses and other grouping symbols (such as absolute value bars) first.

Solution
Since the absolute value bars are grouping symbols, we perform the calculation within them first.

$$10|9 - 15| - 2^5 = 10|-6| - 2^5 \quad \text{Subtract: } 9 - 15 = 9 + (-15) = -6.$$
$$= 10(6) - 2^5 \quad 10|-6| \text{ means 10 times } |-6|. \text{ Find the absolute value: } |-6| = 6.$$
$$= 10(6) - 32 \quad \text{Evaluate the exponential expression: } 2^5 = 32.$$
$$= 60 - 32 \quad \text{Multiply.}$$
$$= 28 \quad \text{Subtract.}$$

> **Caution!** When a number is next to an absolute value symbol, multiplication is indicated.

5 Find the mean (average).

The **arithmetic mean** (or **average**) of a set of numbers is a value around which the values of the numbers are grouped.

> **Finding an Arithmetic Mean**
>
> To find the **mean** of a set of values, divide the sum of the values by the number of values.

EXAMPLE 13 *Hotel Reservations* In an effort to improve customer service, a hotel electronically recorded the number of times the reservation desk telephone rang before it was answered by a receptionist. The results of the week-long survey are shown in the table. Find the average number of times the phone rang before a receptionist answered.

Number of rings	Number of calls
1	11
2	46
3	45
4	28
5	20

Strategy First, we will determine the total number of times the reservation desk telephone rang during the week. Then we will divide that result by the total number of calls received.

WHY To find the *average* value of a set of values, we divide the sum of the values by the number of values.

Solution
To find the total number of rings, we multiply each *number of rings* (1, 2, 3, 4, and 5 rings) by the respective number of occurrences and add those subtotals.

Total number of rings = 11(1) + 46(2) + 45(3) + 28(4) + 20(5)

The total number of calls received was 11 + 46 + 45 + 28 + 20. To find the average, we divide the total number of rings by the total number of calls.

Self Check 13
On an evaluation, students are to mark 1 for *strongly agree*, 2 for *agree*, 3 for *disagree*, and 4 for *strongly disagree*. If on a question 17 students marked 1, 5 students marked 2, and 2 students marked 4, find the average response for this question on the survey.

Now Try Problem 133

$$\text{Average} = \frac{11(1) + 46(2) + 45(3) + 28(4) + 20(5)}{11 + 46 + 45 + 28 + 20}$$

$$= \frac{11 + 92 + 135 + 112 + 100}{150}$$

In the numerator, do the multiplications. In the denominator, do the additions.

$$= \frac{450}{150}$$

Do the addition.

$$= 3$$

Simplify the fraction.

The average number of times the phone rang before it was answered was 3.

> **ANSWERS TO SELF CHECKS**
> 1. a. 12^6 b. $2 \cdot 9^3$ c. 50^2 d. $(-30)^3$ 2. a. y^4 b. $12b^3c$ 3. a. 32 b. 36 c. -125
> 4. a. $-\dfrac{27}{64}$ b. 0.09 c. -25 5. 35 6. -66 7. -66 8. 216 9. -35 10. -130
> 11. -1 12. 1,003 13. 1.46

SECTION 1.7 STUDY SET

VOCABULARY

Fill in the blanks.

1. In the exponential expression 3^2, 3 is the _____, and 2 is the _____.

2. 10^2 can be read as ten _____, and 10^3 can be read as ten _____.

3. 7^5 is the fifth _____ of seven.

4. An _____ is used to represent repeated multiplication.

5. The rules for the _____ of operations guarantee that an evaluation of a numerical expression will result in a single answer.

6. The arithmetic _____ or average of a set of numbers is a value around which the values of the numbers are grouped.

CONCEPTS

7. Given: $4 + 5 \cdot 6$

   a. What operations does this expression contain?

   b. Evaluate the expression in two different ways, and state the two possible results.

   c. Which result from part b is correct, and why?

8. a. What repeated multiplication does 5^3 represent?

   b. Write a multiplication statement in which the factor x is repeated 4 times. Then write the expression in simpler form using an exponent.

   c. How can we represent the repeated addition $3 + 3 + 3 + 3 + 3$ in a simpler form?

9. a. How is the mean (or average) of a set of scores found?

   b. Find the average of 75, 81, 47, and 53.

10. In the expression $-8 + 2[15 - (-6 + 1)]$, which grouping symbols are innermost and which are outermost?

11. a. What operations does the expression $12 + 5^2(-3)$ contain?

    b. In what order should they be performed?

12. a. What operations does the expression $20 - (-2)^2 + 3(-1)$ contain?

    b. In what order should they be performed?

13. Consider the expression $\dfrac{36 - 4(7)}{2(10 - 8)}$. In the numerator, what operation should be done first? In the denominator, what operation should be done first?

14. Explain the differences in evaluating $4 \cdot 2^2$ and $(4 \cdot 2)^2$.

1.7 Exponents and Order of Operations

15. To evaluate each expression, what operation should be performed first?
  a. $-80 - 3 + 5 - 2^2$
  b. $-80 - (3 + 5) - 2^2$
  c. $-80 + 3 + (5 - 2)^2$

16. To evaluate each expression, what operation should be performed first?
  a. $(65 - 3)^3$
  b. $65 - 3^3$
  c. $6(5) - (3)^3$

NOTATION

17. Write an exponential expression with a base of 12 and an exponent of 6.

18. Give the name of each grouping symbol: ( ), [ ], | |, and —.

Complete each step.

19. $50 + 6 \cdot 3^2 = 50 + 6 \cdot \square$
$= 50 + \square$
$= 104$

20. $-100 - (25 - 8 \cdot 2) = -100 - (25 - \square)$
$= -100 - \square$
$= -109$

21. $-19 - 2[(1 + 2) \cdot 3] = -19 - 2[\square \cdot 3]$
$= -19 - 2(\square)$
$= -19 - \square$
$= -37$

22. $\dfrac{46 - 2^3}{-3(5) - 4} = \dfrac{46 - \square}{\square - 4}$
$= \dfrac{\square}{\square}$
$= -2$

GUIDED PRACTICE

Write each product using exponents. See Example 1.

23. $3 \cdot 3 \cdot 3 \cdot 3$
24. $(-7)(-7)(-7)(-7)(-7)(-7)$
25. $10 \cdot 10 \cdot 12 \cdot 12 \cdot 12$
26. $5(5)(5)(11)(11)$

Write each product using exponents. See Example 2.

27. $8 \cdot \pi \cdot r \cdot r \cdot r$
28. $4 \cdot \pi \cdot r \cdot r$
29. $6(x)(x)(y)(y)(y)$
30. $76 \cdot s \cdot s \cdot s \cdot s \cdot t$

Find each power. See Examples 3–4.

31. 7^2
32. 11^3
33. $(-6)^2$
34. $(-4)^4$

35. $(-2)^3$
36. -5^3
37. $\left(-\dfrac{2}{5}\right)^3$
38. $\left(-\dfrac{1}{4}\right)^3$
39. $(-0.4)^2$
40. $(-0.5)^2$
41. -6^2
42. -4^4

Evaluate each expression. See Example 5.

43. $3 \cdot 8^2 - 5$
44. $3 \cdot 4^2 - 8$
45. $3 - 5 \cdot 4^2$
46. $-4 \cdot 6^2 + 5$

Evaluate each expression. See Examples 6–7.

47. $8 \cdot 5 - 4 \div 2$
48. $9 \cdot 5 - 6 \div 3$
49. $100 - 8(10) + 60$
50. $50 - 2(5) - 7$
51. $-22 - 15(-3)$
52. $-33 - 8(-10)$
53. $-2(9) - 2(5)$
54. $18 \div 9(-2) - 4(-3)$
55. $5^2 + 13^2$
56. $3^3 - 2^3$
57. $2 \cdot 3^2 + 5 \cdot 2^3$
58. $4 \cdot 2^5 - 3 \cdot 5^2$

Evaluate each expression. See Example 8.

59. $(-5 - 2)^2$
60. $(-3 - 5)^2$
61. $(12 - 2)^3$
62. $(10 - 3)^2$

Evaluate each expression. See Example 9.

63. $175 - 2 \cdot 3^4$
64. $75 - 3 \cdot 1^2$
65. $200 - (-6 + 5)^3$
66. $19 - (-45 + 41)^3$
67. $-6(130 - 4^3)$
68. $-5(150 - 3^3)$
69. $5 \cdot 2^2 \cdot 4 - 30$
70. $2 + (3 \cdot 2^2 \cdot 4)$

Evaluate each expression. See Example 10.

71. $-3[5^2 - (7 - 3)^2]$
72. $3 - [3^3 + (3 - 1)^3]$
73. $5 + (4^2 - 2^3)^2$
74. $(-5)^3[4(2^3 - 3^2)]^2$

Evaluate each expression. See Example 11.

75. $\dfrac{5 \cdot 50 - 160}{-9}$
76. $\dfrac{5(68 - 32)}{-9}$
77. $\dfrac{(4^3 - 10) + (-4)}{5^2 - (-4)(-5)}$
78. $\dfrac{(6 - 5)^4 - (-21)}{(-9)(-3) - 4^2}$
79. $\dfrac{72 - (2 - 2 \cdot 1)}{10^2 - (90 + 2^2)}$
80. $\dfrac{13^2 - 5^2}{-3(5 - 9)}$
81. $\dfrac{40 \div 2 - 5 \cdot 2}{3^2 - (-1)}$
82. $\dfrac{(5 - 2)^2 - (2 - (-1))}{5 \cdot 2 + (-7)}$

Evaluate each expression. See Example 12.

83. $-2|4 - 8|$
84. $-5|1 - 8|$
85. $|7 - 8(4 - 7)|$
86. $|9 - 5(1 - 8)|$
87. $\dfrac{|6 - 4| + |2 - 4|}{26 - 2^4}$
88. $\dfrac{4|9 - 7| + |-7|}{3^2 - 2^2}$
89. $\dfrac{(3 + 5)^2 + |-2|}{-2(5 - 8)}$
90. $\dfrac{|-25| - 8(-5)}{2^4 - 29}$

TRY IT YOURSELF

Evaluate each expression.

91. $-(-6)^4$
92. $-(-7)^2$
93. $-4(6+5)$
94. $-3(5-4)$
95. $4^2-(-2)^2$
96. $3+(-5)^2$
97. $12+2\left(-\dfrac{9}{3}\right)-(-2)$
98. $2+3\left(-\dfrac{25}{5}\right)-(-4)$
99. $1(2)(3)(-4)$
100. $3(4)(5)(-6)$
101. $[6(5)-5(5)]4$
102. $5[9(2)-2(8)]$
103. $(17-5\cdot 2)^3$
104. $(4+2\cdot 3)^4$
105. $-5(-2)^3(3)^2$
106. $-3(-2)^5(2)^2$
107. $-2\left(\dfrac{15}{-5}\right)-\dfrac{6}{2}+9$
108. $-6\left(\dfrac{25}{-5}\right)-\dfrac{36}{9}+1$
109. $5(10+2)-1$
110. $14+3(7-5)$
111. $64-6[15+(-3)3]$
112. $4+2[26+5(-3)]$
113. $(-2)^3\left(\dfrac{-6}{2}\right)(-1)$
114. $(-3)^3\left(\dfrac{-4}{2}\right)(-1)$
115. $\dfrac{-7-3^2}{2\cdot 4}$
116. $\dfrac{-5-3^3}{2^3}$
117. $\dfrac{1}{2}\left(\dfrac{1}{8}\right)+\left(-\dfrac{1}{4}\right)^2$
118. $-\dfrac{1}{9}\left(\dfrac{1}{4}\right)+\left(-\dfrac{1}{6}\right)^2$
119. $3+2[-1-4(5)]$
120. $4+2[-7-3(9)]$
121. $-(2\cdot 3-4)^3$
122. $-(3\cdot 5-2\cdot 6)^2$
123. $\dfrac{2[-4-2(3-1)]}{3(-3)(-2)}$
124. $\dfrac{3[-9+2(7-3)]}{(5-8)(7-9)}$
125. $-\left(\dfrac{40-1^3-2^4}{3(2+5)+2}\right)$
126. $-\left(\dfrac{8^2-10}{2(3)(4)-5(3)}\right)$
127. $\dfrac{3(3,246-1,111)}{561-546}$
128. $54^3-16^4+19(3)$
129. $(23.1)^2-(14.7)(-61)^3$
130. $12-7\left(-\dfrac{85.684}{34.55}\right)^3$

APPLICATIONS

131. **LIGHT** The illustration shows that the light energy that passes through the first unit of area, 1 yard away from the bulb, spreads out as it travels away from the source. How much area does that light energy cover 2 yards, 3 yards, and 4 yards from the bulb? Express each answer using exponents.

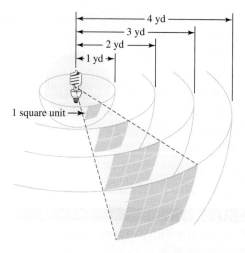

132. **CHAIN LETTERS** A store owner sent two friends a letter advertising her store's low prices. The ad closed with the following request: "Please send a copy of this letter to two of your friends."

    a. Assume that all those receiving letters respond and that everyone in the chain receives just one letter. Complete the table.

    b. How many letters will be circulated in the tenth level of the mailing?

Level	Numbers of letters circulated
1st	$2=2^1$
2nd	$=2$
3rd	$=2$
4th	$=2$

133. **AUTO INSURANCE** See the premium comparison in the table. What is the average 6-month insurance premium?

Allstate	$2,672	Mercury	$1,370
Auto Club	$1,680	State Farm	$2,737
Farmers	$2,485	20th Century	$1,692

    Criteria: Six-month premium. Husband, 45, drives a 2007 Explorer, 12,000 annual miles. Wife, 43, drives a 2008 Dodge Caravan, 12,000 annual miles. Son, 17, is an occasional operator. All have clean driving records.

134. **SWEEPS WEEK** During sweeps week, television networks make a special effort to gain viewers by showing unusually flashy programming. Use the information in the illustration on the next page to determine the average daily gain (or loss) of ratings points by a network for the 7-day sweeps period.

1.7 Exponents and Order of Operations

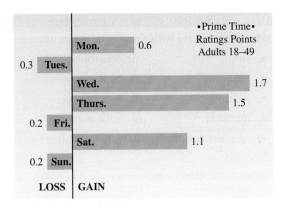

135. YOUTUBE VIDEO CONTEST A video contest is to be part of a promotional kickoff for a new sports drink. The prizes to be awarded are shown.

YouTube Video Contest

Grand prize:
Disney World vacation plus $2,500

Four 1st place prizes of $500
Thirty-five 2nd place prizes of $150
Eighty-five 3rd place prizes of $25

a. How much money will be awarded in the promotion?
b. What is the average cash prize?

136. ENERGY USAGE Refer to the illustration below. Find the average number of therms of natural gas used per month. Then draw a dashed line across the graph showing the average.

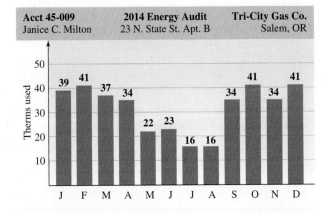

137. SCRABBLE Illustration (a) in the next column shows a portion of the game board before and illustration (b) shows it after the word *QUARTZY* is played. Determine the score. (The number on each tile gives the point value of the letter.)

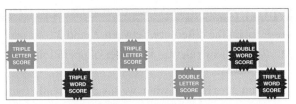
(a) Before the tiles are played

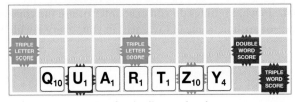
(b) After the tiles are played

138. WRAPPING GIFTS How much ribbon is needed to wrap the package shown if 15 inches of ribbon are needed to make the bow?

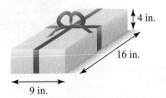

139. SPREADSHEETS This spreadsheet contains data collected by a chemist. For each row, the sum of the values in columns A and B is to be subtracted from the product of 6 and the value in column C. That result is then to be divided by 12 and entered in column D. Use this information to complete the spreadsheet.

	A	B	C	D
1	20	4	8	
2	9	3	16	
3	1	5	11	

140. DOG SHOWS The final score for each dog competing in a toy breeds competition is computed by dividing the sum of the judges' marks, after the highest and lowest have been dropped, by 6. See the table.

a. What was their order of finish?
b. Did any judge rate all the dogs the same?

Judge	1	2	3	4	5	6	7	8
Terrier	14	11	11	10	12	12	13	13
Pekingese	10	9	8	11	11	12	9	10
Pomeranian	15	14	13	11	14	12	10	14

WRITING

141. Explain the difference between 2^3 and 3^2.

142. Explain why rules for the order of operations are necessary.

143. What does it mean when we say perform all additions and subtractions *as they occur from left to right*?

144. In what settings do you encounter or use the concept of arithmetic mean (average) in your everyday life?

REVIEW

145. Match each term with the proper operation.

a. sum i. division
b. difference ii. addition
c. product iii. subtraction
d. quotient iv. multiplication

146. What is the result when we add a number and its opposite?

147. What is the result when we divide a nonzero number by itself?

148. What is wrong with the following statement? Subtraction is the same as adding.

SECTION 1.8
Algebraic Expressions

Objectives

1. Identify terms and coefficients of terms.
2. Write word phrases as algebraic expressions.
3. Analyze problems to determine hidden operations.
4. Evaluate algebraic expressions.

Since problems in algebra are often presented in words, the ability to interpret what you read is important. In this section, we will introduce several strategies that will help you translate English words into mathematical symbols.

1 Identify terms and coefficients of terms.

Recall that variables and/or numbers can be combined with the operations of arithmetic to create **algebraic expressions**. Addition symbols separate expressions into parts called *terms*. For example, the expression $x + 8$ has two terms.

$$\underset{\text{First term}}{x} + \underset{\text{Second term}}{8}$$

Since subtraction can be written as addition of the opposite, the expression $a^2 - 3a - 9$ has three terms.

$$a^2 - 3a - 9 = \underset{\text{First term}}{a^2} + \underset{\text{Second term}}{(-3a)} + \underset{\text{Third term}}{(-9)}$$

In general, a **term** is a product or quotient of numbers and/or variables. A single number or variable is also a term. Examples of terms are:

$$4, \quad y, \quad 6r, \quad -w^3, \quad 3.7x^5, \quad \frac{3}{n}, \quad -15ab^2$$

> **The Language of Algebra** By the commutative property of multiplication, $r6 = 6r$ and $-15b^2a = -15ab^2$. However, we usually write the numerical factor first and the variable factors in alphabetical order.

The numerical factor of a term is called the **coefficient** of the term. For instance, the term $6r$ has a coefficient of 6 because $6r = 6 \cdot r$. The coefficient of $-15ab^2$ is -15 because $-15ab^2 = -15 \cdot ab^2$. More examples are shown on the next page.

A term such as 4, that consists of a single number, is called a **constant term.**

Term	Coefficient	
$8y^2$	8	
$-0.9pq$	-0.9	
$\frac{3}{4}b$	$\frac{3}{4}$	This term could be written $\frac{3b}{4}$.
$-\frac{x}{6}$	$-\frac{1}{6}$	Because $-\frac{x}{6} = -\frac{1x}{6} = -\frac{1}{6} \cdot x$
x	1	Because $x = 1x$
$-t$	-1	Because $-t = -1t$
27	27	

> **The Language of Algebra** Terms such as x and y have *implied* coefficients of 1. *Implied* means suggested without being specifically expressed.

EXAMPLE 1 Identify the coefficient of each term in the following expression: $7x^2 - x + 6$

Strategy We will begin by writing the subtraction as addition of the opposite. Then we will determine the numerical factor of each term.

WHY Addition symbols separate expressions into terms.

Solution
If we write $7x^2 - x + 6$ as $7x^2 + (-x) + 6$, we see that it has three terms: $7x^2$, $-x$, and 6. The numerical factor of each term is its coefficient.

The coefficient of $7x^2$ is **7** because $7x^2$ means $7 \cdot x^2$.

The coefficient of $-x$ is **-1** because $-x$ means $-1 \cdot x$.

The coefficient of the constant 6 is 6.

It is important to be able to distinguish between the *terms* of an expression and the *factors* of a term.

Self Check 1
Identify the coefficient of each term in the expression:
$p^3 - 12p^2 + 3p - 4$

Now Try Problem 15

EXAMPLE 2 Is m used as a *factor* or a *term* in each expression?

a. $m + 6$ **b.** $8m$

Strategy We will begin by determining whether m is involved in an addition or a multiplication.

WHY Addition symbols separate expressions into *terms*. A *factor* is a number being multiplied.

Solution
a. Since m is added to 6, m is a term of $m + 6$.
b. Since m is multiplied by 8, m is a factor of $8m$.

Self Check 2
Is b used as a *factor* or a *term* in each expression?
a. $-27b$
b. $5a + b$

Now Try Problem 21

2 Write word phrases as algebraic expressions.

In the following tables, we list some words and phrases that are used to indicate addition, subtraction, multiplication, and division, and we show how they can be translated to form algebraic expressions.

Addition

The phrase	Translates to
the sum of a and 8	$a + 8$
4 plus c	$4 + c$
16 added to m	$m + 16$
4 more than t	$t + 4$
20 greater than F	$F + 20$
T increased by r	$T + r$
exceeds y by 35	$y + 35$

Subtraction

The phrase	Translates to
the difference of 23 and P	$23 - P$
550 minus h	$550 - h$
18 less than w	$w - 18$
7 decreased by j	$7 - j$
M reduced by x	$M - x$
12 subtracted from L	$L - 12$
5 less f	$5 - f$

Caution! Be careful when translating subtraction. Order is important. For example, when translating the phrase "18 less than w," the terms are reversed.

Multiplication

The phrase	Translates to
the product of 4 and x	$4x$
20 times B	$20B$
twice r	$2r$
triple the profit P	$3P$
$\frac{3}{4}$ of m	$\frac{3}{4}m$

Division

The phrase	Translates to
the quotient of R and 19	$\frac{R}{19}$
s divided by d	$\frac{s}{d}$
the ratio of c to d	$\frac{c}{d}$
k split into 4 equal parts	$\frac{k}{4}$

Caution! The phrase *greater than* is used to indicate addition. The phrase *is greater than* refers to the symbol $>$. Similarly, the phrase *less than* indicates subtraction, and the phrase *is less than* refers to the symbol $<$.

Self Check 3
Write each phrase as an algebraic expression:
a. 80 cents less than t cents
b. $\frac{2}{3}$ of the time T
c. the difference of twice a and 15

Now Try Problems 23, 25, 35

EXAMPLE 3
Write each phrase as an algebraic expression.
a. The sum of the length l and the width 20
b. 5 less than the capacity c
c. The product of the weight w and 2,000, increased by 300

Strategy We will read each phrase and pay close attention to key words that can be translated to mathematical operations. We will refer to the tables as a guide if needed.

WHY Key phrases can be translated to mathematical symbols.

Solution
a. **Key word:** *sum*   **Translation:** add

The phrase translates to $l + 20$.

b. **Key phrase:** *less than*   **Translation:** subtract

The capacity c is to be made less, so we subtract 5 from it: $c - 5$.

c. **Key word:** *product*   **Translation:** multiply

Key phrase: *increased by*   **Translation:** add

The weight w is to be multiplied by 2,000, and then 300 is to be added to the product: $2{,}000w + 300$.

When solving problems, we often begin by letting a variable stand for an unknown quantity.

EXAMPLE 4   *Food Preparation*   A butcher trims 4 ounces of fat from a roast that originally weighed x ounces. Write an algebraic expression that represents the weight of the roast after it is trimmed.

Strategy We will start by letting x represent the original weight of the roast. Then we will look for a key word or phrase to write an expression that represents the trimmed weight of the roast.

WHY The weight after trimming is related to the original weight of the roast.

Solution

We let x = the original weight of the roast (in ounces).

Key word: *trimmed*   **Translation:** subtract

After 4 ounces of fat have been trimmed, the weight of the roast is $(x - 4)$ ounces.

Self Check 4

When a secretary rides the bus to work, it takes her m minutes. If she drives her own car, her travel time exceeds this by 15 minutes. How can we represent the time it takes her to get to work by car?

Now Try Problem 42

EXAMPLE 5   *Competitive Swimming*   The swimming pool to the right is x feet wide. If it is to be sectioned into 8 equally wide swimming lanes, write an algebraic expression that represents the width of each lane.

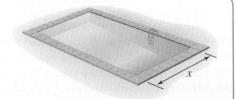

Strategy We start by letting x represent the width of the swimming pool. Then we will look for key words or phrases to write an expression that represents the width of each lane.

WHY The width of each lane is related to the width of the pool.

Solution

We let x = the width of the swimming pool (in feet).

Key phrase: *sectioned into 8 equally wide lanes*   **Translation:** divide

The width of each lane is $\frac{x}{8}$ feet.

Self Check 5

A handyman estimates that it will take the same amount of time to sand as it will to paint some kitchen cabinets. If the entire job takes x hours, how can we express the time it will take him to do the painting?

Now Try Problem 45

When we are solving problems, the variable to be used is rarely specified. We must decide what the unknown quantities are and how they will be represented using variables. The following examples illustrate how to approach these situations.

Self Check 6

A candy bar has twice the number of calories as a serving of pears. Write an expression that represents the number of calories in a candy bar.

Now Try Problem 47

EXAMPLE 6 *Collectibles*
The value of a collectible doll is three times that of an antique toy truck. Write an expression that represents the value of the doll.

Strategy We start by letting x represent the value of the toy truck. Then we will look for key words or phrases to write an expression that represents the value of the antique doll.

WHY The value of the doll is related to the value of the toy truck.

Solution
There are two unknown quantities. Since the doll's value is related to the truck's value, we will let x = the value of the toy truck in dollars.

    Key phrase: 3 *times*   **Translation:** multiply by 3

The value of the doll is $\$3x$.

> **Caution!** A variable is used to represent an unknown number. Therefore, in Example 6, it would be incorrect to write, "Let x = toy truck," because the truck is not a number. We need to write, "Let x = the *value* of the toy truck."

Self Check 7

Part of a $900 donation to a college went to the scholarship fund, the rest to the building fund. Choose a variable to represent the amount donated to one of the funds. Then write an expression that represents the amount donated to the other fund.

Now Try Problem 52

EXAMPLE 7 *Painting*
A 10-inch-long paintbrush has two parts: a handle and bristles. Choose a variable to represent the length of one of the parts. Then write an expression to represent the length of the other part.

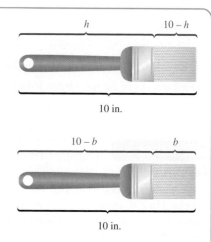

Strategy There are two approaches. We can let h = the length of the handle or we can let b = the length of the bristles.

WHY Both the length of the handle and the length of the bristles are unknown, however we do know the entire length of the paintbrush.

Solution
Refer to the drawing on the top. If we let h = the length of the handle (in inches), then the length of the bristles is $10 - h$.

Now refer to the drawing on the bottom. If we let b = the length of the bristles (in inches), then the length of the handle is $10 - b$.

Self Check 8

The number of votes received by the incumbent in an election was 55 fewer than three times the number the challenger received. Write an expression that represents the number of votes received by the incumbent.

Now Try Problem 53

EXAMPLE 8 *Student Enrollments*
In the second semester, student enrollment in a retraining program at a college was 32 more than twice that of the first semester. Write an expression that represents the student enrollment in the second semester.

Strategy We start by letting x represent the enrollment in the first semester. Then we will look for a key word or phrase to write an expression to represent the second-semester enrollment.

WHY The second-semester enrollment is related to the first-semester enrollment.

Solution
Since the second-semester enrollment is expressed in terms of the first-semester enrollment, we let x = the enrollment in the first semester.

Key phrase: *more than*   **Translation:** add

Key word: *twice*   **Translation:** multiply by 2

The enrollment for the second semester is $2x + 32$.

3 Analyze problems to determine hidden operations.

When analyzing problems, we aren't always given key words or key phrases to help establish what mathematical operation to use. Sometimes a careful reading of the problem is needed to determine the hidden operations.

EXAMPLE 9   **Disney Theme Parks**   Disneyland, located in Anaheim, California, was in operation 16 years before the opening of Walt Disney World in Orlando, Florida. Euro Disney, in Paris, France, was constructed 21 years after Disney World. Use algebraic expressions to express the ages (in years) of each of these Disney attractions.

Strategy   We start by letting x represent the age of Disney World.

WHY   The ages of Disneyland and Euro Disney are related to the age of Disney World.

Solution
The ages of Disneyland and Euro Disney are both related to the age of Disney World. Therefore, we will let x = the age of Disney World.

In carefully reading the problem, we find that Disneyland was built 16 years *before* Disney World, so its age is more than that of Disney World.

Key phrase: *more than*   **Translation:** add

In years, the age of Disneyland is $x + 16$. Euro Disney was built 21 years *after* Disney World, so its age is less than that of Disney World.

Key phrase: *less than*   **Translation:** subtract

In years, the age of Euro Disney is $x - 21$. The results are summarized in the table.

Attraction	Age
Disneyland	$x + 16$
Disney World	x
Euro Disney	$x - 21$

Self Check 9

Kayla worked 5 more hours preparing her tax return than she did on her daughter's return. Kayla's son's return took her 2 more hours to prepare than her daughter's. Write an expression to represent the hours she spent on each return.

Now Try Problem 57

EXAMPLE 10   How many months are in x years?

Strategy   There are no key words, so we must carefully analyze the problem to write an expression that represents the number of months in x years. We will begin by considering some specific cases.

WHY   When no key words are present, it is helpful to work with specifics to get a better understanding of the relationship between the two quantities.

Solution
Let's calculate the number of months in 1 year, 2 years, and 3 years. When we write the results in a table, a pattern is apparent.

Self Check 10

Complete the table. How many days is h hours?

Number of hours	Number of days
24	
48	
72	
h	

Now Try Problem 59

Number of years	Number of months
1	12
2	24
3	36
x	$12x$

We multiply the number of years by 12 to find the number of months.

Therefore, if $x =$ the number of years, the number of months is $12 \cdot x$ or $12x$.

Some problems deal with quantities that have value. In these problems, we must distinguish between *the number of* and *the value of* the unknown quantity. For example, to find the value of 3 quarters, we multiply the number of quarters by the value (in cents) of one quarter. Therefore, the value of 3 quarters is $3 \cdot 25$ cents $= 75$ cents.

The same distinction must be made if the number is unknown. For example, the value of n nickels is not n cents. The value of n nickels is $n \cdot 5$ cents $= (5n)$ cents. For problems of this type, we will use the relationship

Number $\cdot$ value $=$ total value

Self Check 11

Find the value of
a. six $50 savings bonds
b. t $100 savings bonds
c. $x - 4$ $1,000 savings bonds

Now Try Problem 62

EXAMPLE 11 Find the total value of
a. five dimes   b. q quarters   c. $x + 1$ half-dollars

Strategy We will find the total value (in cents) of each collection of coins by multiplying the number of coins by the value of one coin.

WHY Number $\cdot$ value $=$ total value

Solution
To find the total value (in cents) of each collection of coins, we multiply the number of coins by the value (in cents) of one coin, as shown in the table.

Type of Coin	Number	$\cdot$ Value	$=$ Total Value
Dime	5	10	50
Quarter	q	25	$25q$
Half-dollar	$x + 1$	50	$50(x + 1)$

$\leftarrow$ $q \cdot 25$ is written $25q$.

4 Evaluate algebraic expressions.

To **evaluate an algebraic expression**, we replace each variable with a given number value. (When we replace a variable with a number, we say we are **substituting** for the variable.) Then we do the necessary calculations following the rules for the order of operations. For example, to evaluate $x^2 - 2x + 1$ for $x = 3$, we begin by substituting 3 for x.

$$x^2 - 2x + 1 = 3^2 - 2(3) + 1 \quad \text{Substitute 3 for } x.$$
$$= 9 - 2(3) + 1 \quad \text{Evaluate the exponential expression: } 3^2 = 9.$$
$$= 9 - 6 + 1 \quad \text{Do the multiplication: } 2(3) = 6.$$
$$= 4 \quad \text{Working left to right, do the subtraction and then the addition.}$$

We say that 4 is the **value** of this expression when $x = 3$.

Caution! When replacing a variable with its numerical value, use parentheses around the replacement number to avoid possible misinterpretation. For example, when substituting 5 for x in $2x + 1$, we show the multiplication using parentheses: $2(5) + 1$. If we don't show the multiplication, we could misread the expression as $25 + 1$.

EXAMPLE 12
Evaluate each expression for $x = 3$ and $y = -4$:
a. $-y$ **b.** $-3(y + x^2)$

Strategy We will replace x with 3 and y with -4 and then evaluate the expression using the order of operations.

WHY To evaluate an expression means to find its numerical value, once we know the value(s) of the variable(s).

Solution

a. $-y = -(-4)$ Substitute -4 for y.
    $ = 4$ The opposite of -4 is 4.

b. $-3(y + x^2) = -3(-4 + 3^2)$ Substitute 3 for x and -4 for y.
    $ = -3(-4 + 9)$ Work within the parentheses first. Evaluate the exponential expression.
    $ = -3(5)$ Do the addition within the parentheses.
    $ = -15$ Do the multiplication.

Self Check 12
Evaluate each expression for $x = -2$ and $y = 3$:   **a.** $-x$ **b.** $5(x - y)$

Now Try Problem 63

EXAMPLE 13   *Temperature Conversion*

The expression $\frac{9C + 160}{5}$ converts a temperature in degrees Celsius (represented by C) to a temperature in degrees Fahrenheit. Convert $-170°C$, the coldest temperature on the moon, to degrees Fahrenheit.

Strategy We will replace C in the expression with -170 and evaluate it using the order of operations.

WHY The expression evaluated for $C = -170$ converts $-170°C$ to degrees Fahrenheit.

Self Check 13
On January 22, 1943, the temperature in Spearfish, South Dakota, changed from $-20°C$ to $7.2°C$ in two minutes. Convert $-20°C$ to degrees Fahrenheit.

Now Try Problem 75

Solution
To convert $-170°C$ to degrees Fahrenheit, we evaluate the algebraic expression for $C = -170$.

$\frac{9C + 160}{5} = \frac{9(-170) + 160}{5}$ Substitute -170 for C.

$\phantom{\frac{9C + 160}{5}} = \frac{-1{,}530 + 160}{5}$ Do the multiplication.

$\phantom{\frac{9C + 160}{5}} = \frac{-1{,}370}{5}$ Do the addition.

$\phantom{\frac{9C + 160}{5}} = -274$ Do the division.

In degrees Fahrenheit, the coldest temperature on the moon is $-274°$.

Using Your CALCULATOR Evaluating Algebraic Expressions

The rotating drum of a clothes dryer is a cylinder. To find the capacity of the dryer, we can find its volume by evaluating the algebraic expression $\pi r^2 h$, where r represents the radius and h represents the height of the drum. (Here, the cylinder is lying on its side.) If we substitute 13.5 for r and 20 for h, we obtain $\pi(13.5)^2(20)$. Using a scientific calculator, we can evaluate the expression by entering these numbers and pressing these keys.

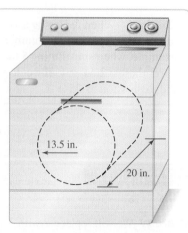

$\boxed{\pi}$ $\boxed{\times}$ 13.5 $\boxed{x^2}$ $\boxed{\times}$ 20 $\boxed{=}$

```
11451.10522
```

Using a graphing or direct-entry calculator, we can evaluate the expression by entering these numbers and pressing these keys.

$\boxed{\text{2nd}}$ $\boxed{\pi}$ $\boxed{\times}$ 13.5 $\boxed{x^2}$ $\boxed{\times}$ 20 $\boxed{\text{ENTER}}$

```
π*13.5²*20
           11451.10522
```

To the nearest cubic inch, the capacity of the dryer is 11,451 in.³

Self Check 14

In Example 14, suppose the initial velocity is 112 feet per second, so the height of the rocket is given by $112t - 16t^2$. Complete the table to find out how many seconds after launch it will hit the ground.

t	$112t - 16t^2$
1	
3	
5	
7	

Now Try Problem 82

EXAMPLE 14 Rocketry

If a toy rocket is shot into the air with an initial velocity of 80 feet per second, its height (in feet) after t seconds in flight is given by the algebraic expression

$$80t - 16t^2$$

How many seconds after the launch will it hit the ground?

Strategy We will substitute positive values for t, the time in flight, until we find the one that gives a height of 0.

WHY When the toy rocket is on the ground, its height above the ground is 0.

Solution

We can substitute positive values for t, the time in flight, until we find the one that gives a height of 0. At that time, the rocket will be on the ground. We will begin by finding the height after the rocket has been in flight for 1 second ($t = 1$) and record the result in a table.

$$80t - 16t^2 = 80(\mathbf{1}) - 16(\mathbf{1})^2 \quad \text{Substitute 1 for } t.$$
$$= 64$$

After 1 second in flight, the height of the rocket is 64 feet. We continue to pick more values of t until we find out when the height is 0.

As we evaluate $80t - 16t^2$ for various values of t, we can show the results in a **table of values**. In the column headed "t," we list each value of the variable to be used in the evaluations. In the column headed "$80t - 16t^2$," we write the result of each evaluation.

t	$80t - 16t^2$
1	64
2	96
3	96
4	64
5	0

Evaluate for $t = 2$:
$80t - 16t^2 = 80(2) - 16(2)^2 = 96$

Evaluate for $t = 3$:
$80t - 16t^2 = 80(3) - 16(3)^2 = 96$

Evaluate for $t = 4$:
$80t - 16t^2 = 80(4) - 16(4)^2 = 64$

Evaluate for $t = 5$:
$80t - 16t^2 = 80(5) - 16(5)^2 = 0$

Since the height of the rocket is 0 when $t = 5$, the rocket will hit the ground in 5 seconds.

The two columns of a table of values are sometimes headed with the terms **input** and **output**, as shown. The t values are the inputs into the expression $80t - 16t^2$, and the resulting values are thought of as the outputs.

Input	Output
1	64
2	96
3	96
4	64
5	0

ANSWERS TO SELF CHECKS

1. 1, −12, 3, −4   **2. a.** factor   **b.** term   **3. a.** $t - 80$ **b.** $\frac{2}{3}T$ **c.** $2a - 15$
4. $(m + 15)$ minutes   **5.** $\frac{x}{2}$ hours   **6.** $x =$ the number of calories in a serving of pears, $2x =$ the number of calories in a candy bar   **7.** $s =$ amount donated to scholarship fund in dollars, $900 - s =$ amount donated to building fund   **8.** $x =$ the number of votes received by the challenger, $3x - 55 =$ the number of votes received by the incumbent
9. Daughter's: x, Kayla's: $x + 5$, son's: $x + 2$ **10.** 1, 2, 3; $\frac{h}{24}$ **11. a.** $300   **b.** $100t
c. $1,000(x - 4)$ **12. a.** 2   **b.** −25   **13.** −4°F   **14.** 7 sec (the heights are 96, 192, 160, and 0)

SECTION 1.8 STUDY SET

VOCABULARY

Fill in the blanks.

1. To _____ an algebraic expression, we substitute the values for the variables and then apply the rules for the order of operations.

2. Variables and/or numbers can be combined with the operation symbols of addition, subtraction, multiplication, and division to create algebraic _____.

3. $2x + 5$ is an example of an algebraic _____, whereas $2x + 5 = 7$ is an example of an _____.

4. When we evaluate an algebraic expression, such as $5x - 8$, for several values of x, we can keep track of the results in an input/output _____.

CONCEPTS

5. Write two algebraic expressions that contain the variable x and the numbers 6 and 20.

6. **a.** Complete the table to determine how many days are in w weeks.

Number of weeks	Number of days
1	
2	
3	
w	

b. Complete the table to answer this question: s seconds is how many minutes?

Number of seconds	Number of minutes
60	
120	
180	
s	

7. When evaluating $3x - 6$ for $x = 4$, what misunderstanding can occur if we don't write parentheses around 4 when it is substituted for the variable?

8. If the knife shown is 12 inches long, write an expression for the length of the blade.

9. a. In the illustration, the weight of the van is 500 pounds less than twice the weight of the car. If the car weighs x pounds, write an expression that represents the weight of the van.

b. If the actual weight of the car is 2,000 pounds, what is the weight of the van?

10. See the illustration.

a. If we let $b =$ the length of the beam, write an expression for the length of the pipe.

b. If we let $p =$ the length of the pipe, write an expression for the length of the beam.

11. Complete the table.

Type of coin	Number ·	Value in cents =	Total value in cents
Nickel	6		
Dime	d		
Half dollar	$x + 5$		

12. If $x = -9$, find the value of

a. $-x$ **b.** $-(-x)$

c. $-x^2$ **d.** $(-x)^2$

NOTATION

Complete each step.

13. Evaluate the expression $9a - a^2$ for $a = 5$.

$$9a - a^2 = 9(\square) - (\square)^2$$
$$= 9(5) - \square$$
$$= \square - 25$$
$$= 20$$

14. Evaluate $\dfrac{4x^2 - 3y}{9(x - y)}$ for $x = 4$ and $y = -3$.

$$\dfrac{4x^2 - 3\square}{9(x - \square)} = \dfrac{4(4)^2 - 3(\square)}{9[4 - (\square)]}$$
$$= \dfrac{4(\square) - 3(\square)}{9(\square)}$$
$$= \dfrac{(\square) - (\square)}{\square}$$
$$= \dfrac{73}{63}$$

GUIDED PRACTICE

Identify the coefficient of each term in the expression. See Example 1.

15. $4x^2 - 5x + 7$ **16.** $-8x^2 + 3x - 2$

17. $9x^2 - 4x$ **18.** $-5x^2 + 6$

Is n used as a factor or a term in each expression? See Example 2.

19. $n - 4$ **20.** $3n - 4$

21. $-5n^2 - 4n + 3$ **22.** $5m^2 + n$

Write each phrase as an algebraic expression. If no variable is given, use x as the variable. See Example 3.

23. The sum of the length l and 15

24. The difference of a number and 10

25. The product of a number and 50

26. Three-fourths of the population p

27. The ratio of the amount won w and lost l

28. The tax t added to c

29. P increased by p

30. 21 less than the total height h

31. The square of k minus 2,005

32. s subtracted from S

33. J reduced by 500

34. Twice the attendance a

35. 1,000 split n equal ways

36. Exceeds the cost c by 25,000

37. 90 more than the current price p
38. 64 divided by the cube of y

Write an algebraic expression that represents each quantity.
See Example 4.

39. A model's skirt is x inches long. The designer then lets the hem down 2 inches. How can we express the length (in inches) of the altered skirt?
40. A caterer always prepares food for 10 more people than the order specifies. If p people are to attend a reception, write an expression for the number of people she should prepare for.
41. Last year a club sold x candy bars for a fund-raiser. This year they want to sell 150 more than last year. Write an expression for the number of candy bars they want to sell this year.
42. The tag on a new pair of 36-inch-long jeans warns that after washing, they will shrink x inches in length. Express the length (in inches) of the jeans after they are washed.

Write an algebraic expression that represents each quantity.
See Example 5.

43. A soft-drink manufacturer produced c cans of cola during the morning shift. Write an expression for how many six-packs of cola can be assembled from the morning shift's production.
44. A student has a paper to type that contains x words. If the student can type 60 words per minute, write an expression for the number of minutes it will take for her to type the paper.
45. A walking path is x feet wide and is striped down the middle. Write an expression for the width of each lane of the path.
46. Tickets to a musical cost a total of $\$t$ for 5 tickets. Write an expression for the cost of one ticket to the musical.

Write an algebraic expression that represents each quantity.
See Example 6.

47. A caravan of b cars, each carrying 5 people, traveled to the state capital for a political rally. Express how many people were in the car caravan.
48. Tickets to a circus cost $5 each. Express how much tickets will cost for a family of x people if they also pay for two of their neighbors.
49. A rectangle is twice as long as it is wide. If the rectangle's width is w, write an expression for the length.
50. If each egg is worth e¢, express the value (in cents) of a dozen eggs.

Write an algebraic expression that represents each quantity.
See Examples 7–8.

51. A 12-foot board is to be cut into two pieces. Choose a variable to represent the length of one piece. Then write an expression that represents the other piece.
52. Part of a $10,000 investment is to be invested in an account paying 2% interest, and the rest in an account paying 3%. Choose a variable to represent the amount invested at 2%. Then write an expression that represents the amount invested at 3%.
53. The number of runners in a marathon this year is 25 more than twice the number that participated last year. Write an expression that represents the number of marathon runners this year.
54. In the second year of operation, a bakery sold 31 more than three times the number of cakes it sold the first year. Write an expression that represents the number of cakes the bakery sold the second year of operation.

Write an algebraic expression that represents each quantity.
See Example 9.

55. IBM was founded 80 years before Apple Computer. Dell Computer Corporation was founded 9 years after Apple.
    a. Let x represent the age (in years) of one of the companies. Write algebraic expressions to represent the ages (in years) of the other two companies.
    b. On April 1, 2008, Apple Computer Company was 32 years old. How old were the other two companies then?
56. Abraham Lincoln was inaugurated 60 years after Thomas Jefferson. Barack Obama was inaugurated 208 years after Jefferson. Write algebraic expressions to represent the year of inauguration of each of these presidents.
57. Florida became a state 27 years after Illinois. California became a state 32 years after Illinois. Write algebraic expressions to represent the year of statehood of each of these states.
58. Minnesota became a state 13 years after Texas. Arizona became a state 67 years after Texas. Write algebraic expressions to represent the year of statehood of each of these states.

Write an algebraic expression that represents each quantity. See Examples 10–11.

59. How many minutes are there in
  a. 5 hours?
  b. h hours?

60. A woman watches television x hours a day. Express the number of hours she watches TV
  a. in a week
  b. in a year

61. a. How many feet are in y yards?
  b. How many yards are in f feet?

62. A sales clerk earns $\$x$ an hour. How much does he earn in
  a. an 8-hour day?
  b. a 40-hour week?

Evaluate each expression, for $x = 3$, $y = -2$, and $z = -4$. See Example 12.

63. $(3 + x)y$
64. $(4 + z)y$
65. $3y^2 - 6y - 4$
66. $-z^2 - z - 12$
67. $(4x)^2 + 3y^2$
68. $4x^2 + (3y)^2$
69. $(x + y)^2 - |z + y|$
70. $[(z - 1)(z + 1)]^2$
71. $-\dfrac{2x + y^3}{y + 2z}$
72. $-\dfrac{2z^2 - y}{2x - y^2}$
73. $\dfrac{yz + 4x}{2z + y}$
74. $\dfrac{5y + z}{z - x}$

Evaluate each formula for the given values. See Example 13.

75. $b^2 - 4ac$ for $a = -1$, $b = 5$, and $c = -2$
76. $a^2 + 2ab + b^2$ for $a = -5$ and $b = -1$
77. $\dfrac{n}{2}[2a + (n - 1)d]$ for $n = 10$, $a = -4$, and $d = 6$
78. $\dfrac{a(1 - r^n)}{1 - r}$ for $a = -5$, $r = 2$, and $n = 3$

Complete each table. See Example 14.

79.

x	$x^3 - 1$
0	
-1	
-3	

80.

g	$g^2 - 7g + 1$
0	
7	
-10	

81.

s	$\dfrac{5s + 36}{s}$
1	
6	
-12	

82.

a	$2{,}500a + a^3$
2	
4	
-5	

83.

Input	Output
x	$2x - \dfrac{x}{2}$
100	
-300	

84.

Input	Output
x	$\dfrac{x}{3} + \dfrac{x}{4}$
12	
-36	

85.

x	$(x + 1)(x + 5)$
-1	
-5	
-6	

86.

x	$\dfrac{1}{x + 8}$
-7	
-9	
-8	

TRY IT YOURSELF

Translate each phrase into an algebraic expression.

87. The total of 35, h, and 300
88. x decreased by 17
89. 680 fewer than the entire population p
90. Triple the number of expected participants x
91. The product of d and 4, decreased by 15
92. Forty-five more than the quotient of y and 6
93. Twice the sum of 200 and t
94. The square of the quantity 14 less than x
95. The absolute value of the difference of a and 2
96. The absolute value of a, decreased by 2
97. Four more than twice x
98. Five less than twice w

APPLICATIONS

99. ROCKETRY The algebraic expression $64t - 16t^2$ gives the height of a toy rocket (in feet) t seconds after being launched. Find the height of the rocket for each of the times shown in the table.

t	h
0	
1	
2	
3	
4	

100. GROWING SOD To determine the number of square feet of sod *remaining* in a field after filling an order, the manager of a sod farm uses the expression $20{,}000 - 3s$ (where s is the number of 1-foot-by-3-foot strips the customer has ordered). To sod a soccer field, a city orders 7,000 strips of sod. Evaluate the expression for this value of s and explain the result. (See the illustration on the next page.)

Strips of sod, cut and ready to be loaded on a truck for delivery

101. ANTIFREEZE The expression

$$\frac{5(F - 32)}{9}$$

converts a temperature in degrees Fahrenheit (given as F) to degrees Celsius. Convert the temperatures listed on the container of antifreeze below to degrees Celsius. Round to the nearest degree.

102. TEMPERATURE ON MARS On Mars, maximum summer temperatures can reach 20°C. However, daily temperatures average −33°C. Convert each of these temperatures to degrees Fahrenheit. See Example 13 (page 89). Round to the nearest degree.

103. TOOLS The utility knife blade shown is in the shape of a trapezoid. Find the area of the front face of the blade. The expression $\frac{1}{2}h(b + d)$ gives the area of a trapezoid.

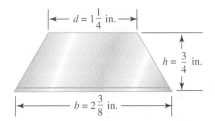

104. TRUMPET MUTES The expression

$$\pi[b^2 + d^2 + (b + d)s]$$

can be used to find the total surface area of the trumpet mute shown. Evaluate the expression for the given dimensions to find the number of square inches of cardboard (to the nearest tenth) used to make the mute.

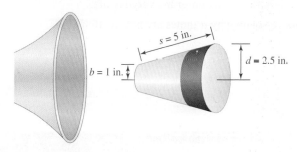

105. LANDSCAPING A grass strip is to be planted around a tree, as shown. Find the number of square feet of sod to order by evaluating the expression $\pi(R^2 - r^2)$. Round to the nearest square foot.

106. ENERGY CONSERVATION A fiberglass blanket wrapped around a water heater helps prevent heat loss. Find the number of square feet of heater surface the blanket covers by evaluating the algebraic expression $2\pi rh$, where r is the radius and h is the height. Round to the nearest square foot.

WRITING

107. What is an algebraic expression? Give some examples.

108. What is a variable? How are variables used in this section?

109. In this section, we substituted a number for a variable. List some other uses of the word *substitute* that you encounter in everyday life.

110. Explain why d dimes are not worth d cents.

REVIEW

111. Simplify: -0

112. Is the statement $-5 > -4$ true or false?

113. Evaluate: $\left|-\frac{2}{3}\right|$

114. Evaluate: $2^3 \cdot 3^2$

115. Write $c \cdot c \cdot c \cdot c$ in exponential form.

116. Evaluate: $15 + 2[15 - (12 - 10)]$

117. Find the mean (average) of the three test scores 84, 93, and 72.

118. Fill in the blanks: In the multiplication statement $5 \cdot x = 5x$, 5 and x are called _____, and $5x$ is called the _____.

Objectives

1 Simplify products.

2 Use the distributive property.

3 Identify like terms.

4 Combine like terms.

SECTION 1.9
Simplifying Algebraic Expressions Using Properties of Real Numbers

In algebra, we often simplify algebraic expressions. To **simplify an algebraic expression**, we use properties of algebra to write the expression in an equivalent, less complicated form.

1 Simplify products.

Two properties that are often used to simplify algebraic expressions are the associative and commutative properties of multiplication. Recall that the associative property of multiplication enables us to change the *grouping of factors* involved in a multiplication. The commutative property of multiplication enables us to change the *order of the factors*.

As an example, let's consider the expression $8(4x)$ and simplify it as follows:

$8(4x) = 8 \cdot (4 \cdot x)$ $4x = 4 \cdot x$

$ = (8 \cdot 4) \cdot x$ Use the associative property of multiplication to group 4 with 8 instead of with x.

$ = 32x$ Do the multiplication within the parentheses: $8 \cdot 4 = 32$.

Since $8(4x) = 32x$, we say that $8(4x)$ simplifies to $32x$. To verify that $8(4x)$ and $32x$ are **equivalent expressions** (represent the same number), we can evaluate each expression for several choices of x. For each value of x, the results should be the same.

If $x = 10$

$8(4x) = 8[4(10)] \qquad 32x = 32(10)$
$ = 8(40) \qquad\qquad = 320$
$ = 320$

If $x = -3$

$8(4x) = 8[4(-3)] \qquad 32x = 32(-3)$
$ = 8(-12) \qquad\quad = -96$
$ = -96$

Self Check 1

Simplify each expression:

a. $9 \cdot 6s$ **b.** $8\left(\frac{7}{8}h\right)$

c. $21p(-3q)$ **d.** $-4(6m)(-2m)$

EXAMPLE 1 Simplify each expression:

a. $15a(-7)$ **b.** $5\left(\frac{4}{5}x\right)$ **c.** $-5r(-6s)$ **d.** $3(7p)(-5p)$

Strategy We will use the commutative and associative properties of multiplication to reorder and regroup the factors.

WHY We want to group the numerical factors of the expression together so that we can find their product.

Now Try Problem 33

Solution

a. $15a(-7) = 15(-7)a$ Use the commutative property of multiplication to change the order of the factors.

$ = -105a$ Working left to right, perform the multiplications.

b. $5\left(\dfrac{4}{5}x\right) = \left(5 \cdot \dfrac{4}{5}\right)x$ Use the associative property of multiplication to group the numbers.

$\phantom{5\left(\dfrac{4}{5}x\right)} = 4x$ Multiply: $5 \cdot \dfrac{4}{5} = \dfrac{5}{1} \cdot \dfrac{4}{5} = \dfrac{\cancel{5} \cdot 4}{1 \cdot \cancel{5}} = 4$.

c. We note that the expression contains two variables.

$-5r(-6s) = [-5(-6)][r \cdot s]$ Use the commutative and associative properties of multiplication to group the numbers and group the variables.

$ = 30rs$ Perform the multiplications within the brackets: $-5(-6) = 30$ and $r \cdot s = rs$.

d. $3(7p)(-5p) = [3(7)(-5)](p \cdot p)$ Use the commutative and associative properties of multiplication to change the order and to regroup the factors.

$ = -105p^2$ Perform the multiplication within the grouping symbols: $3(7)(-5) = -105$ and $p \cdot p = p^2$.

> **Success Tip** By the commutative property of multiplication, we can change the order of the factors.

2 Use the distributive property.

To introduce the **distributive property**, we will consider the expression $4(5 + 3)$, which can be evaluated in two ways.

Method 1. Rules for the order of operations: We compute the sum within the parentheses first.

$4(5 + 3) = 4(8)$ Do the addition within the parentheses first.

$ = 32$ Do the multiplication.

Method 2. The distributive property: We multiply both 5 and 3 by 4, and then we add the results.

$4(5 + 3) = 4(5) + 4(3)$ Distribute the multiplication by 4.

$ = 20 + 12$ Do the multiplications.

$ = 32$ Do the addition.

Notice that each method gives a result of 32.

We can interpret the distributive property geometrically. The figure on the next page shows three rectangles that are divided into squares. Since the area of the rectangle on the left-hand side of the equals sign can be found by multiplying its width by its length, its area is $4(5 + 3)$ square units. We can evaluate this expression, or we can count squares; either way, we see that the area is 32 square units.

The area shown on the right-hand side is the sum of the areas of two rectangles: 4(5) + 4(3). Either by evaluating this expression or by counting squares, we see that this area is also 32 square units. Therefore,

4(5 + 3) = 4(5) + 4(3)

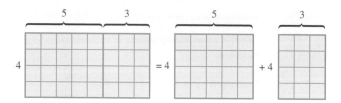

The following figure shows the general case where the width is a and the length is $b + c$.

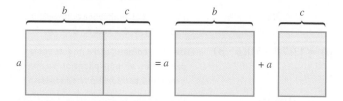

Using the figure as a basis, we can now state the distributive property in symbols.

The Distributive Property

For any real numbers, a, b, and c,

$a(b + c) = ab + ac$

Since subtraction is the same as adding the opposite, the distributive property also holds for subtraction.

The Distributive Property

For any real numbers, a, b, and c,

$a(b - c) = ab - ac$

To illustrate one use of the distributive property, let's consider the expression $5(x + 2)$. Since we are not given the value of x, we cannot add x and 2 within the parentheses. However, we can distribute the multiplication by the factor of 5 that is outside the parentheses to x and to 2 and add those products.

$5(x + 2) = 5(x) + 5(2)$ Distribute the multiplication by 5.
$ = 5x + 10$ Do the multiplications.

Caution! Since the expression $5(x + 2)$ contains parentheses, some students are tempted to perform the addition within the parentheses first. However, we cannot add x and 2, because we do not know the value of x. Instead, we should multiply $x + 2$ by 5, which requires the use of the distributive property.

1.9 Simplifying Algebraic Expressions Using Properties of Real Numbers

EXAMPLE 2 Multiply: **a.** $3(x - 8)$ **b.** $-12(a + 1)$ **c.** $-6(-3y - 8)$

Strategy We will use the distributive property to multiply each term within the parentheses by the factor outside the parentheses.

WHY In each case, we cannot simplify the expression within the parentheses.

Solution

a. $3(x - 8) = 3(x) - 3(8)$ Distribute the multiplication by 3.
          $= 3x - 24$ Do the multiplications.

b. $-12(a + 1) = -12(a) + (-12)(1)$ Distribute the multiplication by -12.
              $= -12a + (-12)$ Do the multiplications.
              $= -12a - 12$ Write the addition of -12 as subtraction of 12.

c. $-6(-3y - 8) = -6(-3y) - (-6)(8)$ Distribute the multiplication by -6.
               $= 18y - (-48)$ Do the multiplications.
               $= 18y + 48$ Add the opposite of -48, which is 48.

Self Check 2
Multiply:
a. $5(p + 2)$
b. $4(t - 1)$
c. $-8(2x - 4)$
Now Try Problem 39

Caution! A common mistake is to forget to distribute the multiplication over each of the terms within the parentheses.

$3(3b - 4) = 9b - 4$

Caution! The fact that an expression contains parentheses does not necessarily mean that the distributive property can be applied. For example, the distributive property does not apply to the expressions:

$6(5x)$ or $6(-7 \cdot y)$ Here a product is multiplied by 6. Simplifying, we have $6(5x) = 30x$ and $6(-7 \cdot y) = -42y$.

However, the distributive property does apply to the expressions:

$6(5 + x)$ or $6(-7 - y)$ Here a sum and a difference are multiplied by 6. Distributing the 6, we have $6(5 + x) = 30 + 6x$ and $6(-7 - y) = -42 - 6y$.

To use the distributive property to simplify $-(x + 10)$, we note that the negative sign in front of the parentheses represents -1.

The $-$ sign represents -1.

$-(x + 10) = -1(x + 10)$
         $= -1(x) + (-1)(10)$ Distribute the multiplication by -1.
         $= -x + (-10)$ Multiply: $-1(x) = -x$ and $(-1)(10) = -10$.
         $= -x - 10$ Write the addition of -10 as a subtraction.

EXAMPLE 3 Simplify: $-(-12 - 3p)$

Strategy We will use the distributive property to multiply each term within the parentheses by -1.

WHY The "$-$" symbol outside the parentheses represents a factor of -1.

Self Check 3
Simplify: $-(-5x + 18)$
Now Try Problem 49

Solution

$$-(-12 - 3p)$$
$$= -1(-12 - 3p) \quad \text{Change the } - \text{ sign in front of the parentheses to } -1.$$
$$= -1(-12) - (-1)(3p) \quad \text{Distribute the multiplication by } -1.$$
$$= 12 - (-3p) \quad \text{Multiply: } -1(-12) = 12 \text{ and } (-1)(3p) = -3p.$$
$$= 12 + 3p \quad \text{To subtract } -3p, \text{ add the opposite of } -3p, \text{ which is } 3p.$$

> **Success Tip** Notice that distributing the multiplication by -1 changes the sign of each term within the parentheses.

Since multiplication is commutative, we can write the distributive property in the following forms.

$$(b + c)a = ba + ca \qquad (b - c)a = ba - ca$$

Self Check 4

Multiply: $(-6x - 24y)\dfrac{1}{3}$

Now Try Problem 54

EXAMPLE 4 Multiply: $(6x + 4y)\dfrac{1}{2}$

Strategy We will use the distributive property to multiply each term within the parentheses by the factor outside the parentheses.

WHY In each case, we cannot simplify the expression within the parentheses.

Solution

$$(6x + 4y)\dfrac{1}{2} = (6x)\dfrac{1}{2} + (4y)\dfrac{1}{2} \quad \text{Distribute the multiplication by } \tfrac{1}{2}.$$
$$= 3x + 2y \quad \text{Multiply: } (6x)\tfrac{1}{2} = \left(6 \cdot \tfrac{1}{2}\right)x = 3x \text{ and } (4y)\tfrac{1}{2} = \left(4 \cdot \tfrac{1}{2}\right)y = 2y.$$

The distributive property can be extended to situations in which there are more than two terms within parentheses.

> **The Extended Distributive Property**
>
> For any real numbers, a, b, c, and d,
>
> $$a(b + c + d) = ab + ac + ad \quad \text{and} \quad a(b - c - d) = ab - ac - ad$$

Self Check 5

Multiply: $-0.7(2r + 5s - 8)$

Now Try Problem 57

EXAMPLE 5 Multiply: $-0.3(3a - 4b + 7)$

Strategy We will use the distributive property to multiply each term within the parentheses by the factor outside the parentheses.

WHY We cannot simplify the expression within the parentheses.

Solution

$$-0.3(3a - 4b + 7)$$
$$= -0.3(3a) - (-0.3)(4b) + (-0.3)(7) \quad \text{Distribute the multiplication by } -0.3.$$
$$= -0.9a - (-1.2b) + (-2.1) \quad \text{Do the three multiplications.}$$

$$= -0.9a + 1.2b + (-2.1) \quad \text{To subtract } -1.2b, \text{ add its opposite, which is } 1.2b.$$
$$= -0.9a + 1.2b - 2.1 \quad \text{Write the addition of } -2.1 \text{ as a subtraction.}$$

3 Identify like terms.

The expression $5p + 7q - 3p + 12$, which can be written $5p + 7q + (-3p) + 12$, contains four terms, $5p$, $7q$, $-3p$, and 12. Since the variable of $5p$ and $-3p$ are the same, we say that these terms are **like** or **similar terms**.

> **Like Terms (Similar Terms)**
>
> **Like terms** (or **similar terms**) are terms with exactly the same variables raised to exactly the same powers. Any numbers (called **constants**) in an expression are considered to be like terms.

Like terms
$4x$ and $7x$
↑   ↑
Same variable

$-10p^2, 25p^2,$ and $150p^2$
Same variable to the same power

Unlike terms
$4x$ and $3y$
↑   ↑
Different variables

$15p$ and $23p^2$
Different exponents on the variable p

> *Caution!* It is important to be able to distinguish between a *term* of an expression and a *factor* of a term. Terms are separated by + symbols. Factors are numbers and/or variables that are multiplied together. For example, x is a term of the expression $18 + x$, because x and 18 are separated by a + symbol. In the expression $18x + 9$, x is a factor of the term $18x$, because x and 18 are multiplied together.

EXAMPLE 6 List like terms:
a. $7r + 5 + 3r$ **b.** $x^4 - 6x^2 - 5$ **c.** $-7m + 7 - 2 + m$

Strategy First we will identify each term of the expression. Then we will look for terms that contain the same variable factors raised to exactly the same powers.

WHY If terms contain the same variables raised to the same powers, they are like terms.

Solution
a. $7r + 5 + 3r$ contains the like terms $7r$ and $3r$.
b. $x^4 - 6x^2 - 5$ contains no like terms.
c. $-7m + 7 - 2 + m$ contains two pairs of like terms: $-7m$ and m are like terms, and the constants, 7 and -2, are like terms.

Self Check 6
List like terms:
a. $5x - 2y + 7y$
b. $-5pq + 17p - 12q - 2pq$

Now Try Problem 59

4 Combine like terms.

To add (or subtract) objects, they must have the same units. For example, we can add dollars to dollars and inches to inches, but we cannot add dollars to inches. The same is true when we work with terms of an algebraic expression. They can be added or subtracted only when they are like terms.

This expression can be simplified, because it contains like terms.	This expression cannot be simplified, because its terms are not like terms.
$3x + 4x$	$3x + 4y$
Like terms	Unlike terms
The variable parts are identical.	The variable parts are not identical.

To simplify an expression containing like terms, we use the distributive property. For example, we can simplify $3x + 4x$ as follows:

$3x + 4x = (3 + 4)x$ Use the distributive property.

$ = 7x$ Perform the addition within the parentheses: $3 + 4 = 7$.

We have simplified the expression $3x + 4x$ by **combining like terms**. The result is the equivalent expression $7x$. This example suggests the following general rule.

Combining Like Terms

To add or subtract like terms, combine their coefficients and keep the same variables with the same exponents.

Self Check 7

Simplify by combining like terms:
a. $5n + (-8n)$
b. $-1.2a^3 + 1.4a^3$

Now Try Problem 65

EXAMPLE 7   Simplify by combining like terms:
a. $-8p + (-12p)$ b. $0.5s^2 - 0.3s^2$

Strategy We will use the distributive property in reverse to add (or subtract) the coefficients of the like terms. We will keep the variable factors raised to the same powers.

WHY To *combine like terms* means to add or subtract the like terms in an expression.

Solution

a. $-8p + (-12p) = -20p$ Add the coefficients of the like terms: $-8 + (-12) = -20$. Keep the variable p.

b. $0.5s^2 - 0.3s^2 = 0.2s^2$ Subtract: $0.5 - 0.3 = 0.2$. Keep the variable part s^2.

Self Check 8

Simplify: $8R + 7r - 14R - 21r$

Now Try Problem 71

EXAMPLE 8   Simplify: $7P - 8p - 12P + 25p$

Strategy We will use the commutative property of addition to write the like terms next to each other. Keep in mind that an uppercase P and a lowercase p are different variables.

WHY To *simplify* an expression, we use properties of real numbers to write an equivalent expression in simpler form.

Solution
The uppercase P and the lowercase p are different variables. We can use the commutative property of addition to write like terms next to each other.

$7P - 8p - 12P + 25p$

$= 7P + (-8p) + (-12P) + 25p$ Rewrite each subtraction as the addition of the opposite.

$= 7P + (-12P) + (-8p) + 25p$ Use the commutative property of addition to write the like terms together.

$= -5P + 17p$ Combine like terms: $7P + (-12P) = -5P$ and $-8p + 25p = 17p$.

The expression in Example 8 contained two sets of like terms, and we rearranged the terms so that like terms were next to each other. With practice, you will be able to combine like terms without having to write them next to each other.

EXAMPLE 9 Simplify: $4(x + 5) - 3(2x - 4)$

Strategy First we will use the distributive property to remove the parentheses. Then we will identify any like terms and combine them.

WHY To *simplify* an expression, we use properties of real numbers, such as the distributive property, to write an equivalent expression in simpler form.

Solution
$4(x + 5) - 3(2x - 4)$
$= 4x + 20 - 6x + 12$ Use the distributive property twice.
$= -2x + 32$ Combine like terms: $4x - 6x = -2x$ and $20 + 12 = 32$.

Self Check 9
Simplify: $-5(y - 4) + 2(4y + 6)$
Now Try Problem 83

ANSWERS TO SELF CHECKS
1. a. $54s$ b. $7h$ c. $-63pq$ d. $48m^2$ 2. a. $5p + 10$ b. $4t - 4$ c. $-16x + 32$
3. $5x - 18$ 4. $-2x - 8y$ 5. $-1.4r - 3.5s + 5.6$ 6. a. $-2y$ and $7y$
b. $-5pq$ and $-2pq$ 7. a. $-3n$ b. $0.2a^3$ 8. $-6R - 14r$ 9. $3y + 32$

SECTION 1.9   STUDY SET

VOCABULARY
Fill in the blanks.

1. To _____ the expression $5(6x)$ means to write it in the simpler form $5(6x) = 30x$.
2. $5(6x)$ and $30x$ are _____ expressions because for each value of x, they represent the same number.
3. To perform the multiplication $2(x + 8)$, we use the _____ property to remove parentheses.
4. Terms such as $7x^2$ and $5x^2$, which have the same variables raised to exactly the same powers, are called _____ terms.

CONCEPTS
5. What property does the equation $a(b + c) = ab + ac$ illustrate?
6. The illustration shows an application of the distributive property. Fill in the blanks.

$2(\square + \square) = 2(\square) + 2(\square)$

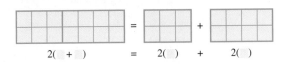

$2(\square + \square)$ $=$ $2(\square)$ $+$ $2(\square)$

Fill in the blanks.

7. $a(b + c + d) =$ _____
8. a. $2(x + 4) = 2x \;\square\; 8$
   b. $2(x - 4) = 2x \;\square\; 8$
9. a. $2(-x + 4) = -2x \;\square\; 8$
   b. $2(-x - 4) = -2x \;\square\; 8$
10. a. $-2(x + 4) = -2x \;\square\; 8$
    b. $-2(x - 4) = -2x \;\square\; 8$
11. a. $-2(-x + 4) = 2x \;\square\; 8$
    b. $-2(-x - 4) = 2x \;\square\; 8$
12. To add or subtract like terms, combine their _____ and keep the same variables and _____.
13. A board was cut into two pieces, as shown. Add the lengths of the two pieces. How long was the original board?

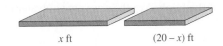

x ft $(20 - x)$ ft

Unless otherwise noted, all content on this page is © Cengage Learning.

14. Let x = the number of miles driven on the first day of a 2-day driving trip. Translate the verbal model to mathematical symbols, and simplify by combining like terms.

   | The miles driven one day | plus | 100 miles more than the miles driven on day 1. |

15. Two angles are called **complementary angles** when the sum of their measures is 90°. Add the measures of the angles in illustration (a). Are they complementary angles?

16. Two angles are called **supplementary angles** if the sum of their measures is 180°. Add the measures of the angles in illustration (b). Are they supplementary angles?

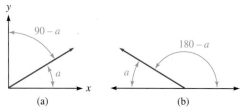

(a)　　(b)
All angle measures are in degrees.

Simplify each expression, if possible.

17. $5(2x)$ and $5 + 2x$
18. $6(-7x)$ and $6 - 7x$
19. $2(3x)(3)$ and $2 + 3x + 3$
20. $-3(2x)(4)$ and $5 - 3x + 2$

NOTATION

Complete each step.

21. $7(a + 2) = \boxed{} \cdot a + \boxed{} \cdot 2$
    $= 7a + \boxed{}$

22. $6(b - 5) + 12b + 7 = 6(\boxed{}) - 6(\boxed{}) + 12b + 7$
    $= 6b - \boxed{} + 12b + 7$
    $= 6b + \boxed{}b - \boxed{} + 7$
    $= \boxed{}b - 23$

23. a. Are $2K$ and $3k$ like terms?
    b. Are $-d$ and d like terms?

24. Fill in the blank: $-(x + 10) = -\boxed{}(x + 10)$

25. Write each expression using fewer symbols.
    a. $5x - (-1)$　　b. $16t + (-6)$

26. In the table in the next column, a student's answers to five problems are compared to the answers in the back of the book. Are the answers equivalent?

Student's answer	Book's answer	Equivalent?
$10x$	$10 + x$	
$3 + y$	$y + 3$	
$5 - 8a$	$8a - 5$	
$3(x) + 4$	$3(x + 4)$	
$2x$	x^2	

GUIDED PRACTICE

Simplify each expression. See Example 1.

27. $9(7m)$　　28. $12n(8)$
29. $5(-7q)$　　30. $-7(5t)$
31. $12\left(\dfrac{5}{12}x\right)$　　32. $15\left(-\dfrac{4}{15}w\right)$
33. $(-5p)(-4b)$　　34. $(-7d)(-7c)$
35. $-5(4r)(-2r)$　　36. $7t(-4t)(-2)$
37. $8q(-2q)(-3)$　　38. $-3m(-5m)(-2m)$

Multiply. See Example 2.

39. $5(x + 3)$　　40. $4(x + 2)$
41. $-2(b - 1)$　　42. $-7(p - 5)$
43. $8(3t - 2)$　　44. $9(2q + 1)$
45. $3(-5t - 4)$　　46. $2(5x - 4)$

Multiply. See Example 3.

47. $-(r - 10)$　　48. $-(h + 4)$
49. $-(x - 7)$　　50. $-(y + 1)$

Multiply. See Example 4.

51. $(3w - 6)\left(-\dfrac{2}{3}\right)$　　52. $(2y - 8)\dfrac{1}{2}$
53. $(9x - 3y)\dfrac{2}{3}$　　54. $(4p + 3q)\dfrac{3}{4}$

Multiply. See Example 5.

55. $17(2x - y + 2)$　　56. $-12(3a + 2b - 1)$
57. $-0.1(-14 + 3p - t)$　　58. $-1.5(-x - y + 5)$

List all like terms, if any. See Example 6.

59. $8p + 7 - 5p$　　60. $-7m - 3m + 5m$
61. $a^4 + 5a^2 - 7$　　62. $6q^2 + 3q - 5q^2 - 2q$

Simplify each expression by combining like terms. See Example 7.

63. $3x + 17x$
64. $12y - 15y$
65. $8x^2 - 5x^2$
66. $17x^2 + 3x^2$
67. $-4x + 4x$
68. $-16y + 16y$
69. $-7b + 7b$
70. $-2c + 2c$

Simplify each expression by combining like terms. See Example 8.

71. $1.8h - 0.7h + p - 3p$
72. $-5.7m + 4.3m + 3n - 1.2n$
73. $a + a + b$
74. $-t - t - T - T$
75. $3x + 5x - 7x + 3y$
76. $-x + 3y + 2y$
77. $-13x^2 + 2x^2 - 5y^2 + 2y^2$
78. $-8x^3 - x^3 + 3y + 5y$

Simplify each expression by combining like terms. See Example 9.

79. $(a + 2) - (a - b)$
80. $3z + 2(Z - z) + Z$
81. $x(x + 3) - 3x^2$
82. $2x + x(x - 3)$
83. $-(c + 7) - 2(c - 3)$
84. $-(z + 2) + 5(3 - z)$
85. $-(c - 6) + 3(c + 1)$
86. $-2(m - 1) - 4(-2 + m)$

TRY IT YOURSELF

Simplify.

87. $0.4(x - 4)$
88. $-2.2(2q + 1)$
89. $2x + 4(X - x) + 3X$
90. $3p - 6(p + z) + p$
91. $0 - 3x$
92. $0 - 4a$
93. $0 - (-t)$
94. $0 - (-2y)$
95. $\frac{3}{5}t + \frac{1}{5}t$
96. $\frac{3}{16}x - \frac{5}{16}x$
97. $(2y - 1)6$
98. $(3w - 5)5$
99. $3(y - 3) + 4(y + 1)$
100. $-5(a - 2) - 4(a + 1)$
101. $8\left(\frac{3}{4}y\right)$
102. $27\left(\frac{2}{3}x\right)$
103. $-0.2r - (-0.6r)$
104. $-1.1m - (-2.4m)$
105. $2z + 5(z - 3)$
106. $12(m + 11) - 11$

APPLICATIONS

107. THE AMERICAN RED CROSS In 1891, Clara Barton founded the Red Cross. Its symbol is a white flag bearing a red cross. If each side of the cross in the illustration has length x, write an algebraic expression for the perimeter (the total distance around the outside) of the cross.

108. BILLIARDS Billiard tables vary in size, but all tables are twice as long as they are wide.

    a. If the following billiard table is x feet wide, write an expression involving x that represents its length.

    b. Write an expression for the perimeter of the table.

109. PING-PONG Write an expression for the perimeter of the table shown in the illustration.

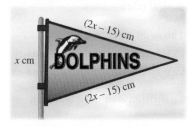

110. SEWING Write an expression for the length of the blue trim needed to outline a pennant with the given side lengths.

111. COUNTER SPACE Write an expression for the perimeter of the customer service desk shown in the illustration below.

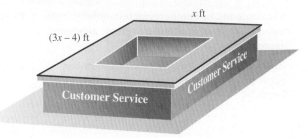

112. **POWER TOOLS** Find the sum of the measures of the two angles that the drill makes with the piece of wood.

WRITING

113. Explain why the distributive property applies to $2(3 + x)$ but not to $2(3x)$.

114. Explain why $3x^2y$ and $5x^2y$ are like terms, and explain why $3x^2y$ and $5xy^2$ are not like terms.

115. Distinguish between a *factor* and a *term* of an algebraic expression. Give examples.

116. Describe how to combine like terms.

REVIEW

Evaluate each expression for $x = -3$, $y = -5$, and $z = 0$.

117. $x^2z(y^3 - z)$

118. $|y^3 - z|$

119. $\dfrac{x - y^2}{2y - 1 + x}$

120. $\dfrac{2y + 1}{x} - x$

CHAPTER 1 SUMMARY AND REVIEW

SECTION 1.1 The Language of Algebra

DEFINITIONS AND CONCEPTS	EXAMPLES
Tables, **bar graphs**, and **line graphs** are used to describe numerical relationships.	See pages 2–3 for examples of tables and graphs.
The result of an addition is called a **sum**. The result of a subtraction is called a **difference**. The result of a multiplication is called a **product**. The result of a division is called a **quotient**.	$6 + 12 = 18$ Sum $21 - 17 = 4$ Difference $8 \times 13 = 104$ Product $\dfrac{35}{7} = 5$ Quotient
An **equation** is a mathematical sentence that contains an = symbol. **Variables** are letters used to stand for numbers. Variables and/or numbers can be combined with the operations of arithmetic to create **algebraic expressions**. Equations that express a known relationship between two or more variables are called **formulas**.	Equations: $3x + 4 = 12$ and $\dfrac{t}{9} = 12$ Variables: x, a, and y Expressions: $5y + 2$, $\dfrac{12x}{5}$, and $a(b - 3)$ Formula: $A = lw$ (the formula for the area of a rectangle)

REVIEW EXERCISES

Consider the following line graph that shows the number of cars parked in a mall parking structure from 6 P.M. to 12 midnight on a Saturday.

1. How many cars were in the parking structure at 11 P.M.?

2. At what time did the parking structure have 500 cars in it?

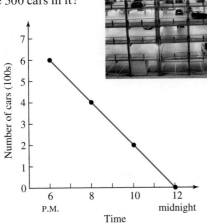

Express each statement in words.

3. $15 - 3 = 12$

4. $15 + 3 = 18$

5. $15 \div 3 = 5$

6. $15 \cdot 3 = 45$

Classify each item as either an algebraic expression or an equation.

7. $5 = 2x + 3$

8. $2x + 3$

9. $\dfrac{t + 6}{12}$

10. $P = 2l + 2w$

11. Use the formula $n = b + 5$ to complete the table.

Number of brackets (b)	Number of nails (n)
5	
10	
20	

12. Use the data in the table to write a formula that mathematically describes the relationship between the two quantities, then state the relationship in words.

Number of children	Total fees (dollars)
1	50
2	100
4	200

SECTION 1.2 Fractions

DEFINITIONS AND CONCEPTS	EXAMPLES
A **prime number** is a natural number greater than 1 that is only divisible by itself and 1.	Prime numbers: {2, 3, 5, 7, 11, 13, 17, … }
A **composite number** is a natural number greater than 1 that is not prime.	Composite numbers: {4, 6, 8, 9, 10, 12, 14, 15, … }
Every composite number can be factored as a product of two or more prime numbers.	$15 = 3 \cdot 5$, $35 = 5 \cdot 7$, $30 = 2 \cdot 3 \cdot 5$
To **multiply two fractions**, multiply their numerators and multiply their denominators.	Multiply: $\dfrac{5}{8} \cdot \dfrac{3}{4} = \dfrac{15}{32}$
One number is the **reciprocal** of another if their product is 1.	The reciprocal of $\dfrac{7}{11}$ is $\dfrac{11}{7}$ because $\dfrac{7}{11} \cdot \dfrac{11}{7} = 1$.
To **divide two fractions**, multiply the first fraction by the reciprocal of the second fraction.	Divide: $\dfrac{4}{7} \div \dfrac{5}{8} = \dfrac{4}{7} \cdot \dfrac{8}{5} = \dfrac{32}{35}$
Multiplication property of 1: the product of 1 and any number is that number.	$1 \cdot 9 = 9$ and $\dfrac{4}{7} \cdot 1 = \dfrac{4}{7}$

To **build a fraction**, multiply it by a form of 1 such as $\frac{2}{2}, \frac{3}{3}, \frac{4}{4}, \ldots$	Write $\frac{3}{7}$ as an equivalent fraction with a denominator of 35. $\frac{3}{7} = \frac{3}{7} \cdot \frac{5}{5} = \frac{15}{35}$
To **simplify a fraction**, remove pairs of factors common to the numerator and the denominator. A fraction is in **simplest form**, or **lowest terms**, when the numerator and denominator have no common factors other than 1.	Simplify: $\frac{39}{52} = \frac{3 \cdot \cancel{13}}{4 \cdot \cancel{13}} = \frac{3}{4}$ $\frac{13}{13} = 1$
To find the **least common denominator (LCD)** of two fractions, prime factor each denominator and find the product of the prime factors, using each factor the greatest number of times it appears in any one factorization.	Find the LCD of $\frac{2}{9}$ and $\frac{5}{6}$. $\left.\begin{array}{l} 9 = 3 \cdot 3 \\ 6 = 2 \cdot 3 \end{array}\right\}$ LCD $= 2 \cdot 3 \cdot 3$
To **add (or subtract) fractions with the same denominator**, add (or subtract) the numerators and keep the common denominator. To **add (or subtract) fractions with unlike denominators**, write each fraction as an equivalent fraction with a denominator that is the LCD. Then add (or subtract) the numerators and keep the common denominator.	Add: $\frac{5}{12} + \frac{7}{8} = \frac{5}{12} \cdot \frac{2}{2} + \frac{7}{8} \cdot \frac{3}{3}$ The LCD is 24. Build each fraction. $= \frac{10}{24} + \frac{21}{24}$ The denominators are now the same. $= \frac{31}{24}$ The result does not simplify.

REVIEW EXERCISES

13. Write 24 as the product of two factors.

14. Write 24 as the product of three factors.

Give the prime factorization of each number, if possible.

15. 54   **16.** 147

17. 385   **18.** 41

Simplify each fraction to lowest terms.

19. $\frac{20}{35}$ **20.** $\frac{24}{18}$

Build each fraction or whole number to an equivalent fraction with the indicated denominator.

21. $\frac{5}{8}$, denominator 64   **22.** 12, denominator 3

Perform each operation.

23. $\frac{16}{35} \cdot \frac{25}{48}$ **24.** $5\frac{3}{5}\left(1\frac{11}{14}\right)$

25. $\frac{1}{3} \div \frac{15}{16}$ **26.** $16\frac{1}{4} \div 5$

27. $\frac{17}{25} - \frac{7}{25}$ **28.** $\frac{17}{12} + \frac{7}{12}$

Perform each operation.

29. $\frac{8}{11} - \frac{1}{2}$ **30.** $\frac{1}{4} + \frac{2}{3}$

31. $61\frac{7}{8} + 19\frac{2}{3}$ **32.** $34\frac{1}{9} - 13\frac{5}{6}$

33. MACHINE SHOPS How much must be milled off the $\frac{17}{24}$-inch-thick steel rod so that the collar will slip over it?

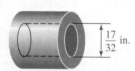

Steel rod

34. DELIVERY TRUCKS A truck can safely carry a one-ton load. Should it be used to deliver one-half ton of sand, one-third ton of gravel, and one-fifth ton of cement in one trip to a job site?

SECTION 1.3 The Real Numbers

DEFINITIONS AND CONCEPTS

Set of numbers:

Natural numbers

Whole numbers

Integers

Inequality symbols

> "is greater than"   < "is less than"

A **rational number** is any number that can be written as a fraction with an integer numerator and a nonzero integer denominator. Rational numbers are either *terminating* or *repeating* decimals.

An **irrational number** is a nonterminating, nonrepeating decimal. Irrational numbers cannot be written as the ratio of two integers.

A **real number** is any number that is either a rational or an irrational number.

Two numbers represented by points on a number line that are the same distance away from 0, but on opposite sides of it, are called **opposites**.

The **absolute value** of a number is the distance on the number line between the number and 0.

EXAMPLES

Natural numbers: $\{1, 2, 3, 4, 5, 6, \ldots\}$
Whole numbers: $\{0, 1, 2, 3, 4, 5, 6, \ldots\}$
Integers: $\{\ldots, -3, -2, -1, 0, 1, 2, 3, \ldots\}$

$11 > 8 \quad 4 < 7$

Rational numbers: $3, \ \frac{2}{3}, \ 0.75, \ 0, \ 0.\overline{25}$

Irrational numbers: $\sqrt{3} = 1.732050808\ldots, \quad \pi = 3.141592654\ldots$

Real numbers: $3, \ \frac{2}{3}, \ 0.75, \ 0, \ 0.\overline{25}, \ \sqrt{3}, \ \pi$

Opposites: 6 and -6

$|8| = 8, \quad |-7.5| = 7.5, \quad \left|-\frac{5}{8}\right| = \frac{5}{8}$

REVIEW EXERCISES

Represent each of these situations with a signed number.

35. A budget deficit of $65 billion

36. 206 feet below sea level

Use one of the symbols > or < to make each statement true.

37. 0 5   **38.** -12 -13

Show that each of the following numbers is a rational number by expressing it as a fraction.

39. 5   **40.** -12

41. 0.7   **42.** $4\frac{2}{3}$

43. Graph each member of the set
$$\left\{\pi, 0.333\ldots, 3.75, -\frac{17}{4}, \frac{7}{8}, -2\right\}$$
on the number line.

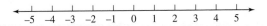

44. Use a calculator to approximate $\sqrt{2}$ to the nearest hundredth.

Determine whether each statement is true or false.

45. All integers are whole numbers.

46. π is a rational number.

47. The set of real numbers corresponds to all points on the number line.

48. 0 is a whole number but not a natural number.

49. A real number is either rational or irrational.

50. Determine which numbers in the given set are natural numbers, whole numbers, integers, rational numbers, irrational numbers, and real numbers.

$$\left\{-\frac{4}{5},\ 99.99,\ 0,\ \sqrt{2},\ -12,\ 4\frac{1}{2},\ 0.666\ldots,\ 8\right\}$$

Write the expression in simplest form.

51. $-\left(-\frac{9}{16}\right)$ **52.** -0

Insert one of the symbols >, <, or = in the blank to make each statement true.

53. $|-6|$ ___ $|5|$ **54.** -9 ___ $-|-10|$

SECTION 1.4 Adding Real Numbers; Properties of Addition

DEFINITIONS AND CONCEPTS	EXAMPLES
To **add two real numbers with like signs**, add their absolute values and attach their common sign to the sum.	Add: $3 + 5 = 8$ Add: $-5 + (-11) = -16$
To **add two real numbers with unlike signs**, subtract their absolute values, the smaller from the larger. To that result, attach the sign of the number with the larger absolute value.	Add: $-8 + 6 = -2$ Add: $18 + (-3) = 15$
The commutative property of addition: $a + b = b + a$	$3 + 8 = 8 + 3$ Reorder
The associative property of addition: $(a + b) + c = a + (b + c)$	$(3 + 5) + 9 = 3 + (5 + 9)$ Regroup

REVIEW EXERCISES

Add.

55. $12 + 33$
56. $-45 + (-37)$
57. $-15 + 37$
58. $25 + (-13)$
59. $12 + (-8) + (-15)$
60. $-25 + (-14) + 35$
61. $-9.9 + (-2.4)$
62. $\frac{5}{16} + \left(-\frac{1}{2}\right)$
63. $35 + (-13) + (-17) + 6$
64. $-21 + (-11) + 32 + (-45)$

Determine what property of addition guarantees that the quantities are equal.

65. $-2 + 5 = 5 + (-2)$
66. $(-2 + 5) + 1 = -2 + (5 + 1)$

SECTION 1.5 Subtracting Real Numbers

DEFINITIONS AND CONCEPTS	EXAMPLES
To *subtract* real numbers, add the opposite: $a - b = a + (-b)$	Subtract: $8 - 5 = 8 + (-5) = 3$

REVIEW EXERCISES

Subtract.

67. $45 - 64$ **68.** $-17 - 32$
69. $-27 - (-12)$ **70.** $3.6 - (-2.1)$

71. ASTRONOMY *Magnitude* is a term used in astronomy to designate the brightness of celestial objects as viewed from Earth. Smaller magnitudes are associated with brighter objects, and larger magnitudes refer to fainter objects. For each of the following pairs of objects, by how many magnitudes do their brightnesses differ?

  a. A full moon and the sun
  b. The star Beta Crucis and a full moon

Object	Magnitude
Sun	-26.5
Full moon	-12.5
Beta Crucis	1.28

Source: Based on data from Abell, Morrison, and Wolf, *Exploration of the Universe* (Saunders, 1987)

72. GEOGRAPHY The tallest peak on Earth is Mt. Everest, at 29,028 feet. The greatest ocean depth is the Mariana Trench, at $-36,205$ feet. Find the difference in the two elevations.

SECTION 1.6 Multiplying and Dividing Real Numbers; Multiplication and Division Properties

DEFINITIONS AND CONCEPTS	EXAMPLES
To **multiply two real numbers**, multiply their absolute values. 1. The product of two real numbers with **like signs** is positive. 2. The product of two real numbers with **unlike signs** is negative.	Multiply: $3(5) = 15$ and $(-3)(-5) = 15$ Multiply: $3(-5) = -15$ and $(-3)(5) = -15$
The commutative property of multiplication: $ab = ba$	$3(-4) = (-4)(3) = -12$
The associative property of multiplication: $(ab)c = a(bc)$	$[3(-4)](-5) = 3[(-4)(-5)] = 60$
To **divide two real numbers**, divide their absolute values. 1. The quotient of two real numbers with **like signs** is positive. 2. The quotient of two real numbers with **unlike signs** is negative.	Divide: $\frac{15}{3} = 5$ and $\frac{-15}{-3} = 5$ Divide: $\frac{-15}{3} = -5$ and $\frac{15}{-3} = -5$
Division of zero by a nonzero number is 0.	Divide: $\frac{0}{5} = 0$
Division by zero is undefined.	Divide: $\frac{5}{0}$ is undefined

REVIEW EXERCISES

Multiply.

73. $-8 \cdot 7$ **74.** $(-9)(-6)$
75. $2(-3)(-2)$ **76.** $(-3)(4)(2)$
77. $(-3)(-4)(-2)$ **78.** $(-4)(-1)(-3)(-3)$
79. $-1.2(-5.3)$ **80.** $0.002(-1,000)$
81. $-\frac{2}{3}\left(\frac{1}{5}\right)$ **82.** $2\frac{1}{4}\left(-\frac{1}{3}\right)$

What property of multiplication is illustrated?

83. $(2 \cdot 3)5 = 2(3 \cdot 5)$

84. $(-5)(-6) = (-6)(-5)$

85. What is the additive inverse of -3?

86. What is the multiplicative inverse of -3?

Perform each division, if possible.

87. $\dfrac{88}{44}$ **88.** $\dfrac{-100}{25}$

89. $\dfrac{-81}{-27}$ **90.** $\dfrac{0}{37}$

91. $-\dfrac{3}{5} \div \dfrac{1}{2}$ **92.** $\dfrac{-60}{0}$

93. $\dfrac{-4.5}{1}$ **94.** $\dfrac{-5}{-5}$

95. The product of two real numbers is 0. What conclusion can be drawn about the numbers?

96. MAGNIFICATION

a. Find the high and low readings displayed on the screen of the emissions-testing device shown below.

b. The picture on the screen can be magnified by switching a setting on the monitor. Find the new high and low readings if every value is doubled.

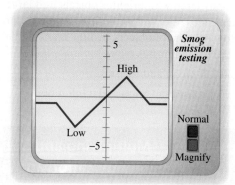

SECTION 1.7 Exponents and Order of Operations

DEFINITIONS AND CONCEPTS	EXAMPLES
An **exponent** is used to represent repeated multiplication. In the **exponential expression** a^n, a is the **base** and n is the **exponent**.	In 5^4, 5 is the base and 4 is the exponent. Thus, $5^4 = 5 \cdot 5 \cdot 5 \cdot 5 = 625$
Order of operations: 1. Perform all calculations within parentheses and other grouping symbols following the order listed in steps 2–4 below, working from the innermost pair to the outermost pair. 2. Evaluate all exponential expressions. 3. Perform all multiplications and divisions, working from left to right. 4. Perform all additions and subtractions, working from left to right. When grouping symbols have been removed, repeat steps 2–4 to complete the calculation. If a fraction is present, evaluate the expression above and the expression below the bar separately. Then do the division indicated by the fraction bar, if possible.	To evaluate $3 + 2[4 - (5^2 + 1)]$, proceed as follows: $3 + 2[4 - (5^2 + 1)]$ $= 3 + 2[4 - (25 + 1)]$ Work within the parentheses and evaluate the exponential expression. $= 3 + 2[4 - 26]$ Add within the parentheses. $= 3 + 2[-22]$ Subtract within the brackets. $= 3 - 44$ Multiply. $= -41$ Subtract.
The **mean (or average)** is a value around which number values are grouped. $$\text{Mean} = \dfrac{\text{sum of values}}{\text{number of values}}$$	To find the mean of 11, 46, 46, 45, and 22, evaluate: $$\text{Mean} = \dfrac{11 + 46 + 46 + 45 + 22}{5} = \dfrac{170}{5} = 34$$

REVIEW EXERCISES

Write each expression using exponents.

97. $8 \cdot 8 \cdot 8 \cdot 8 \cdot 8$
98. $5 \cdot 5 \cdot 5 \cdot 9 \cdot 9$
99. $a(a)(a)(a)$
100. $9 \cdot \pi \cdot r \cdot r$
101. $x \cdot x \cdot x \cdot y \cdot y \cdot y \cdot y$
102. the sixth power of one

Evaluate each expression.

103. 9^2
104. 2 cubed
105. 2^5
106. 50^1

Evaluate each expression.

107. $24 - 3(6)(4)$
108. $-(6-3)^2$
109. $4^3 + 2(-6 - 2 \cdot 2)$
110. $10 - 5[-3 - 2(5 - 7^2)] - 5$
111. $\dfrac{-4(4+2) - 4}{|-18 - 4(5)|}$
112. $(-3)^3 \left(\dfrac{-8}{2}\right) + 5$
113. $\dfrac{|-35| - 2(-7)}{2^4 - 23}$
114. $-9^2 + (-9)^2$

115. **WALK-A-THONS** Use the data in the table to find the average (mean) donation to a charity walk-a-thon.

Donation	Number received
$5	20
$10	65
$20	25
$50	5
$100	10

116. Give the name of each grouping symbol:

( )   [ ] | | —

SECTION 1.8 Algebraic Expressions

DEFINITIONS AND CONCEPTS	EXAMPLES
Addition symbols separate algebraic expressions into *terms*. In a term, the numerical factor is called the *coefficient*.	Since $a^2 + 3a - 5$ can be written as $a^2 + 3a + (-5)$, it has three terms. The coefficient of a^2 is 1, the coefficient of $3a$ is 3, and the coefficient of -5 is -5.
In order to describe numerical relationships, we need to translate the words of a problem into mathematical symbols. Sometimes we must rely on common sense and insight to find **hidden operations**.	5 *more than* x can be expressed as $x + 5$. 25 *less than twice* y can be expressed as $2y - 25$. *One-half of* c can be expressed as $\tfrac{1}{2}c$.
Number · value = total value	The value of 5 dimes is $5 \cdot 10$ cents = 50 cents.
When we replace the variable, or variables, in an algebraic expression with specific numbers and then apply the rules for the order of operations, we are **evaluating the algebraic expression**.	To evaluate $\dfrac{x+y}{x-y}$ for $x = 7$ and $y = 2$, substitute 7 for x and 2 for y and simplify. $\dfrac{x+y}{x-y} = \dfrac{7+2}{7-2} = \dfrac{9}{5}$

REVIEW EXERCISES

Write each phrase as an algebraic expression.

117. 25 more than the height h
118. 15 less than the cutoff score s
119. $\dfrac{1}{2}$ of the time t
120. the product of 6 and x

121. See the illustration.

a. If we let n = the length of the nail in inches, write an algebraic expression for the length of the bolt (in inches).

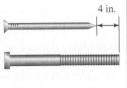

b. If we let b = the length of the bolt in inches, write an algebraic expression for the length of the nail (in inches).

122. Complete the table.

Type of coin	Number	Value (¢)	Total value (¢)
Nickel	6		
Dime	d		

123. Complete the table.

x	$20x - x^3$
0	
1	
−4	

124. Find the volume, to the nearest tenth of a cubic inch, of the ice cream waffle cone by evaluating the algebraic expression
$$\frac{\pi r^2 h}{3}$$

Evaluate each algebraic expression for the given value(s) of the variable(s).

125. $7x^2 - \dfrac{x}{2}$ for $x = 4$

126. $b^2 - 4ac$ for $b = -10$, $a = 3$, and $c = 5$

127. $2(24 - 2c)^3$ for $c = 9$

128. $\dfrac{x + y}{-x - z}$ for $x = 19$, $y = 17$, and $z = -18$

SECTION 1.9 Simplifying Algebraic Expressions Using Properties of Real Numbers

DEFINITIONS AND CONCEPTS	EXAMPLES
We often use the *commutative property of multiplication* to reorder factors and the *associative property of multiplication* to regroup factors when **simplifying expressions**.	Simplify: $-5(3y) = (-5 \cdot 3)y = -15y$ Simplify: $-45b\left(\dfrac{5}{9}\right) = -45\left(\dfrac{5}{9}b\right) = \left(-45 \cdot \dfrac{5}{9}\right)b = -25b$
The **distributive property** can be used to remove parentheses: $a(b + c) = ab + ac$ $a(b - c) = ab - ac$ $a(b + c + d) = ab + ac + ad$	Multiply: $7(x + 3) = 7 \cdot x + 7 \cdot 3 = 7x + 21$ Multiply: $-0.2(4m - 5n - 7)$ $\qquad = -0.2(4m) - (-0.2)(5n) - (-0.2)(7)$ $\qquad = -0.8m + n + 1.4$
Like terms are terms with exactly the same variables raised to exactly the same powers.	$3x$ and $-5x$ are like terms. $-4t^3$ and $6t^4$ are unlike terms because they have different exponents. $0.5xyz$ and $3.7xy$ are unlike terms because they have different variables.
Simplifying the sum or difference of like terms is called **combining like terms**. Like terms are combined by adding or subtracting the coefficients of the terms and keeping the same variables with the same exponents.	Simplify: $4a + 2a = 6a$ Think $(4 + 2)a = 6a$. Simplify: $5p^2 + p - p^2 - 9p = 4p^2 - 8p$ Think $(5 - 1)p^2 = 4p^2$ and $(1 - 9)p = -8p$. Simplify: $2(k - 1) - 3(k + 2) = 2k - 2 - 3k - 6 = -k - 8$

REVIEW EXERCISES

Simplify each expression.

129. $-4(7w)$ **130.** $3(-2x)(-4)$

131. $0.4(5.2f)$ **132.** $\frac{7}{2} \cdot \frac{2}{7}r$

Multiply.

133. $5(x + 3)$ **134.** $-(2x + 3 - y)$

135. $\frac{3}{4}(4c - 8)$ **136.** $-2(-3c - 7)(2.1)$

Simplify each expression by combining like terms.

137. $8p + 5p - 4p$ **138.** $-5m + 2 - 2m - 2$

139. $n + n + n + n$ **140.** $5(p - 2) - 2(3p + 4)$

141. $55.7k^2 - 55.6k^2$

142. $8a^2 + 4a^2 + 2a - 4a^2 - 2a - 1$

143. $\frac{3}{5}w - \left(-\frac{2}{5}w\right)$

144. $-19m + 19m$

Write an equivalent expression for the given expression using fewer symbols.

145. $1x$ **146.** $-1x$

147. $4x - (-1)$ **148.** $4x + (-1)$

149. Write an expression for the perimeter of the rectangle shown in the illustration.

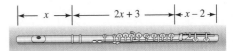

$(7x + 3)$ in.

150. FLUTES Write an expression that represents the total length of the flute shown below.

CHAPTER 1 TEST

1. Fill in the blanks.
   a. Two fractions, such as $\frac{3}{4}$ and $\frac{6}{8}$, that represent the same number are called _____ fractions.
   b. The result of an addition is called a _____.
   c. -5 is the _____ of 5 because $-5 + 5 = 0$.

The following graph shows the cost to hire a security guard. Use the graph to answer Problems 2 and 3.

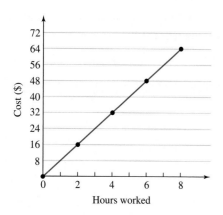

2. What will it cost to hire a security guard for 3 hours?
3. If a school was billed $40 for hiring a security guard for a dance, for how long did the guard work?
4. Use the formula $f = \frac{a}{5}$ to complete the table.

Area in square miles (a)	Number of fire stations (f)
15	
100	
350	

5. Give the prime factorization of 180.
6. Simplify: $\frac{42}{105}$
7. Write $\frac{5}{7}$ as an equivalent fraction with a denominator of 42.

8. SHOPPING Refer to the illustration. What is the cost of the amount of fruit on the scale?

Oranges
84 cents a pound

9. Divide: $\frac{15}{16} \div \frac{5}{8}$
10. Subtract: $\frac{11}{12} - \frac{2}{9}$
11. Add: $11\frac{2}{3} + 8\frac{2}{5}$
12. Multiply: $0.49 \cdot 100$

13. QUALITY CONTROL An electronics company has strict specifications for silicon chips used in a computer. The company will install only chips that are within 0.05 centimeters of the specified thickness. The following table gives that specification for two types of chip. Fill in the blanks to complete the chart.

Chip type	Thickness specification	Acceptable range Low	High
A	0.78 cm		
B	0.643 cm		

14. Write $\frac{5}{6}$ as a decimal.
15. Graph each member of the set on the number line.
$$\left\{-1\frac{1}{4},\ \sqrt{2},\ -3.75,\ \frac{7}{2},\ 0.5,\ -3\right\}$$

16. Determine whether each statement is true or false.
   a. Every integer is a rational number.
   b. Every rational number is an integer.
   c. π is an irrational number.
   d. 0 is a whole number.

Chapter 1 Test

17. iPHONES You can get a more accurate reading of an iPhone's signal strength by dialing *3001#12345#*. Field test mode is then activated and the standard signal strength bars (in the upper left corner of the display) are replaced by a negative number. The closer the negative number is to zero, the stronger the signal. Which iPhone shown below is receiving the strongest signal?.

(i)　　(ii)　　(iii)　　(iv)

18. Insert > or < in the blank to make each statement true.

　a. -2 ___ -3　　b. $-|-7|$ ___ 8

　c. $|-4|$ ___ $-(-5)$　　d. $\left|-\dfrac{7}{8}\right|$ ___ 0.5

19. TELEVISION During sweeps week, networks try to gain viewers by showing their most exciting episodes. Use the data to determine the average daily gain (or loss) of ratings points by a network for the 7-day sweeps period.

Day	Mon.	Tues.	Wed.	Thurs.	Fri.	Sat.	Sun.
Point loss/gain	0.6	−0.3	1.7	1.5	−0.2	1.1	−0.2

20. Add: $(-6) + 8 + (-4)$

21. Add: $-\dfrac{1}{2} + \dfrac{7}{8}$

22. Subtract: $-10 - (-4)$

23. Multiply: $(-2)(-3)(-5)$

24. Divide: $\dfrac{-22}{-11}$

25. Perform each operation.

　a. $0(-3)$　　b. $\dfrac{0}{-3}$

　c. $0 + (-3)$　　d. $3 + (-3)$

26. What property of real numbers is illustrated?

　a. $(-12 + 97) + 3 = -12 + (97 + 3)$

　b. $5 \cdot 2 = 2 \cdot 5$

27. Rewrite each product using exponents.

　a. $9(9)(9)(9)(9)$

　b. $3 \cdot x \cdot x \cdot z \cdot z \cdot z$

28. Evaluate: $8 + 2 \cdot 3^4$

29. Evaluate: $9^2 - 3[45 - 3(6 + 4)]$

30. Evaluate: $\dfrac{3(40 - 2^3)}{-2(6 - 4)^2}$

31. Evaluate $3(x - y) - 5(x + y)$ for $x = 2$ and $y = -5$.

32. Complete the table.

x	$2x - \dfrac{30}{x}$
5	
10	
−30	

33. A band recorded x songs for an album. Technicians had to delete two songs from the album because of poor sound quality. Express the number of songs on the album using an algebraic expression.

34. What is the value of q quarters in cents?

35. a. Explain the difference between an expression and an equation. Give an example of each.

　b. Explain this statement: $a - b = a + (-b)$

　c. Describe the set of real numbers.

36. ACCOUNTING The 2012 quarterly profits for Avis Budget Group Inc. are shown in black, and the losses are shown in red, in the table below. Calculate the company's total net income for 2012.

Quarter	1st	2nd	3rd	4th
Net income ($ million)	(23)	79	280	(46)

Source: Google Financial

Simplify each expression.

37. $5(-4x)$ **38.** $-8(-7t)(4)$

39. $\dfrac{4}{5}(15a + 5) - 16a$

40. $-1.1d^2 - 3.8d^2 - d^2$

41. $9x^2 + 2(7x - 3) - 9(x^2 - 1)$

42. Write an expression that represents the perimeter of the rectangle.

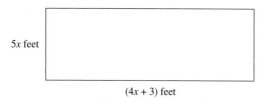

43. TRIATHLONS Write an expression that represents the total length of the triathlon course shown below.

44. FLOOR MATS Find an expression that represents the length of plastic trim used around the floor mat shown below.

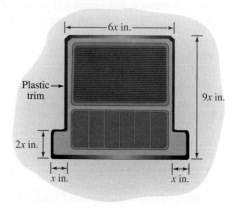

2

Equations, Inequalities, and Problem Solving

2.1 Solving Equations Using Properties of Equality
2.2 More about Solving Equations
2.3 Applications of Percent
2.4 Formulas
2.5 Problem Solving
2.6 More about Problem Solving
2.7 Solving Inequalities
   Chapter Summary and Review
   Chapter Test
   Cumulative Review

from Campus to Careers
Automotive Service Technician

Anyone whose car has ever broken down appreciates the talents of automotive service technicians. To work on today's high-tech cars and trucks, a person needs strong diagnostic and problem-solving skills. Courses in automotive repair, electronics, physics, chemistry, English, computers, and mathematics provide a good educational background for a career as a service technician.

Service technicians must be knowledgeable about the repair and maintenance of automobiles and the fuels that power them. In **Problem 75** of **Study Set 2.4**, you will see how the octane ratings of three familiar grades of gasoline, unleaded, unleaded plus, and premium, are calculated using a formula.

JOB TITLE: Automotive Service Technician
EDUCATION: Formal training at a vocational school or community college is strongly recommended.
JOB OUTLOOK: Demand for technicians will grow as the number of vehicles in operation increases.
ANNUAL EARNINGS: The mean annual salary in 2012 was $39,060.
FOR MORE INFORMATION:
www.bls.gov/oes/current/oes493023.htm

119

Objectives

1. Determine whether a number is a solution.
2. Use the addition property of equality.
3. Use the subtraction property of equality.
4. Use the multiplication property of equality.
5. Use the division property of equality.

SECTION 2.1
Solving Equations Using Properties of Equality

In this section, we introduce four fundamental properties of equality that are used to solve equations.

1 Determine whether a number is a solution.

An **equation** is a statement indicating that two expressions are equal. An example is $x + 5 = 15$. The equal symbol $=$ separates the equation into two parts: The expression $x + 5$ is the **left side** and 15 is the **right side**. The letter x is the **variable** (or the **unknown**). The sides of an equation can be reversed, so we can write $x + 5 = 15$ or $15 = x + 5$.

- An equation can be true: $6 + 3 = 9$
- An equation can be false: $2 + 4 = 7$
- An equation can be neither true nor false. For example, $x + 5 = 15$ is neither true nor false because we don't know what number x represents.

An equation that contains a variable is made true or false by substituting a number for the variable. If we substitute 10 for x in $x + 5 = 15$, the resulting equation is true: $10 + 5 = 15$. If we substitute 1 for x, the resulting equation is false: $1 + 5 = 15$. A number that makes an equation true when substituted for the variable is called a **solution**; it is said to **satisfy** the equation. Therefore, 10 is a solution of $x + 5 = 15$, and 1 is not. The **solution set** of an equation is the set of all numbers that make the equation true.

> **The Language of Algebra** It is important to know the difference between an equation and an expression. An equation contains an $=$ symbol; an expression does not.

Self Check 1

Is 25 a solution of $10 - x = 35 - 2x$?

Now Try Problem 19

EXAMPLE 1 Is 9 a solution of $3y - 1 = 2y + 7$?

Strategy We will substitute 9 for each y in the equation and evaluate the expression on the left side and the expression on the right side separately.

WHY If a true statement results, 9 is a solution of the equation. If we obtain a false statement, 9 is not a solution.

Solution

Evaluate the expression on the left side.

$$3y - 1 = 2y + 7$$
$$3(9) - 1 \stackrel{?}{=} 2(9) + 7$$
$$27 - 1 \stackrel{?}{=} 18 + 7$$
$$26 = 25$$

Evaluate the expression on the right side.

Since $26 = 25$ is false, 9 is not a solution of $3y - 1 = 2y + 7$.

2 Use the addition property of equality.

To **solve an equation** means to find all values of the variable that make the equation true. We can develop an understanding of how to solve equations by referring to the scales shown on the next page.

The first scale represents the equation $x - 2 = 3$. The scale is in balance because the weights on the left side and right side are equal. To find x, we must add 2 to the left side. To keep the scale in balance, we must also add 2 to the right side. After doing this, we see that x grams is balanced by 5 grams. Therefore, x must be 5. We say that we have solved the equation $x - 2 = 3$ and that the solution is 5.

In this example, we solved $x - 2 = 3$ by transforming it to a simpler *equivalent equation*, $x = 5$.

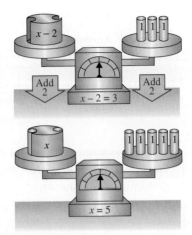

Equivalent Equations

Equations with the same solutions are called **equivalent equations**.

The procedure that we used suggests the following property of equality.

Addition Property of Equality

Adding the same number to both sides of an equation does not change its solution.

For any real numbers a, b, and c,

if $a = b$, then $a + c = b + c$

When we use this property, the resulting equation is *equivalent to the original one*. We will now show how it is used to solve $x - 2 = 3$ algebraically.

EXAMPLE 2 Solve: $x - 2 = 3$

Strategy We will use a property of equality to isolate the variable on one side of the equation.

WHY To solve the original equation, we want to find a simpler equivalent equation of the form $x =$ **a number**, whose solution is obvious.

Solution
We will use the addition property of equality to isolate x on the left side of the equation. We can undo the subtraction of 2 by adding 2 to both sides.

$x - 2 = 3$ This is the equation to solve.
$x - 2 + 2 = 3 + 2$ Add 2 to both sides.
$x + 0 = 5$ The sum of a number and its opposite is zero: $-2 + 2 = 0$.
$x = 5$ When 0 is added to a number, the result is the same number.

Since 5 is obviously the solution of the equivalent equation $x = 5$, the solution of the original equation, $x - 2 = 3$, is also 5. To check this result, we substitute 5 for x in the original equation and simplify.

Self Check 2
Solve: $n - 16 = 33$
Now Try Problem 37

$$x - 2 = 3$$
$$5 - 2 \stackrel{?}{=} 3 \quad \text{Substitute 5 for x.}$$
$$3 = 3 \quad \text{True}$$

Since the statement is true, 5 is the solution. A more formal way to present this result is to write the solution within braces as a solution set: {5}.

> **The Language of Algebra** We solve equations by writing a series of steps that result in an equivalent equation of the form
>
> $x = $ a number
>
> or
>
> a number $= x$
>
> We say the variable is *isolated* on one side of the equation. *Isolated* means alone or by itself.

Self Check 3

Solve: **a.** $-5 = b - 38$
b. $-20 + n = 29$

Now Try Problems 39 and 43

EXAMPLE 3 Solve: **a.** $-19 = y - 7$ **b.** $-27 + y = -3$

Strategy We will use a property of equality to isolate the variable on one side of the equation.

WHY To solve the original equation, we want to find a simpler equivalent equation of the form $y = $ **a number** or **a number** $= y$, whose solution is obvious.

Solution

a. To isolate y on the right side, we use the addition property of equality. We can undo the subtraction of 7 by adding 7 to both sides.

$$-19 = y - 7 \quad \text{This is the equation to solve.}$$
$$-19 + 7 = y - 7 + 7 \quad \text{Add 7 to both sides.}$$
$$-12 = y \quad \text{The sum of a number and its opposite is zero:}$$
$$\quad -7 + 7 = 0.$$

Check: $-19 = y - 7$ This is the original equation.
$-19 \stackrel{?}{=} -12 - 7$ Substitute -12 for y.
$-19 = -19$ True

Since the statement is true, the solution is -12. The solution set is $\{-12\}$.

b. To isolate y, we use the addition property of equality. We can eliminate -27 on the left side by adding its opposite (additive inverse) to both sides.

$$-27 + y = -3 \quad \text{The equation to solve.}$$
$$-27 + y + 27 = -3 + 27 \quad \text{Add 27 to both sides.}$$
$$y = 24 \quad \text{The sum of a number and its opposite is zero:}$$
$$\quad -27 + 27 = 0.$$

Check: $-27 + y = -3$ This is the original equation.
$-27 + 24 \stackrel{?}{=} -3$ Substitute 24 for y.
$-3 = -3$ True

The solution is 24. The solution set is {24}.

Caution! After checking a result, be careful when stating your conclusion. Here, it would be incorrect to say:

The solution is −3.

The number we were checking was 24, not −3.

3 Use the subtraction property of equality.

Since any subtraction can be written as an addition by adding the opposite of the number to be subtracted, the following property is an extension of the addition property of equality.

Subtraction Property of Equality

Subtracting the same number from both sides of an equation does not change its solution.

For any real numbers a, b, and c,

$$\text{if } a = b, \text{ then } a - c = b - c$$

When we use this property, the resulting equation is equivalent to the original one.

EXAMPLE 4 Solve: **a.** $x + \dfrac{1}{8} = \dfrac{7}{4}$ **b.** $54.9 + x = 45.2$

Strategy We will use a property of equality to isolate the variable on one side of the equation.

WHY To solve the original equation, we want to find a simpler equivalent equation of the form $x = $ **a number**, whose solution is obvious.

Solution

a. To isolate x, we use the subtraction property of equality. We can undo the addition of $\frac{1}{8}$ by subtracting $\frac{1}{8}$ from both sides.

$x + \dfrac{1}{8} = \dfrac{7}{4}$ This is the equation to solve.

$x + \dfrac{1}{8} - \dfrac{1}{8} = \dfrac{7}{4} - \dfrac{1}{8}$ Subtract $\frac{1}{8}$ from both sides.

$x = \dfrac{7}{4} - \dfrac{1}{8}$ On the left side, $\frac{1}{8} - \frac{1}{8} = 0$.

$x = \dfrac{7}{4} \cdot \dfrac{2}{2} - \dfrac{1}{8}$ Build $\frac{7}{4}$ so that it has a denominator of 8.

$x = \dfrac{14}{8} - \dfrac{1}{8}$ Multiply the numerators and multiply the denominators.

$x = \dfrac{13}{8}$ Subtract the numerators. Write the result over the common denominator 8.

Verify that $\frac{13}{8}$ is the solution by substituting it for x in the original equation and simplifying.

Self Check 4

Solve: **a.** $x + \dfrac{4}{15} = \dfrac{11}{5}$

b. $0.7 + a = 0.2$

Now Try Problems 49 and 51

b. To isolate x, we use the subtraction property of equality. We can undo the addition of 54.9 by subtracting 54.9 from both sides.

$$54.9 + x = 45.2$$ This is the equation to solve.
$$54.9 + x - \mathbf{54.9} = 45.2 - \mathbf{54.9}$$ Subtract 54.9 from both sides.
$$x = -9.7$$ On the left side, 54.9 − 54.9 = 0.

Check:
$$54.9 + x = 45.2$$ This is the original equation.
$$54.9 + (-9.7) \stackrel{?}{=} 45.2$$ Substitute −9.7 for x.
$$45.2 = 45.2$$ True

The solution is -9.7. The solution set is $\{-9.7\}$.

4 Use the multiplication property of equality.

The first scale shown below represents the equation $\frac{x}{3} = 25$. The scale is in balance because the weights on the left side and right side are equal. To find x, we must triple (multiply by 3) the weight on the left side. To keep the scale in balance, we must also triple the weight on the right side. After doing this, we see that x is balanced by 75. Therefore, x must be 75.

The procedure that we used suggests the following property of equality.

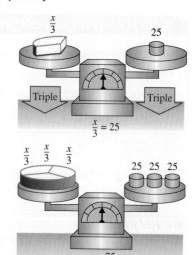

Multiplication Property of Equality

Multiplying both sides of an equation by the same nonzero number does not change its solution.

For any real numbers a, b, and c, where c is not 0,

    if $a = b$,   then $ca = cb$

When we use this property, the resulting equation is equivalent to the original one. We will now show how it is used to solve $\frac{x}{3} = 25$ algebraically.

Self Check 5

Solve: $\dfrac{b}{24} = 3$

Now Try Problem 53

EXAMPLE 5   Solve: $\dfrac{x}{3} = 25$

Strategy We will use a property of equality to isolate the variable on one side of the equation.

WHY To solve the original equation, we want to find a simpler equivalent equation of the form $x = $ **a number**, whose solution is obvious.

Solution

To isolate x, we use the multiplication property of equality. We can undo the division by 3 by multiplying both sides by 3.

$$\frac{x}{3} = 25 \quad \text{This is the equation to solve.}$$

$$3 \cdot \frac{x}{3} = 3 \cdot 25 \quad \text{Multiply both sides by 3.}$$

$$\frac{3x}{3} = 75 \quad \text{Do the multiplications.}$$

$$1x = 75 \quad \text{Simplify } \tfrac{3x}{3} \text{ by removing the common factor of 3 in the numerator and denominator: } \tfrac{3}{3} = 1.$$

$$x = 75 \quad \text{The coefficient 1 need not be written since } 1x = x.$$

If we substitute 75 for x in $\frac{x}{3} = 25$, we obtain the true statement $25 = 25$. This verifies that 75 is the solution. The solution set is {75}.

Since the product of a number and its reciprocal (or multiplicative inverse) is 1, we can solve equations such as $\frac{2}{3}x = 6$, where the coefficient of the variable term is a fraction, as follows.

EXAMPLE 6

Solve: **a.** $\frac{2}{3}x = 6$ **b.** $-\frac{5}{4}x = 3$

Strategy We will use a property of equality to isolate the variable on one side of the equation.

WHY To solve the original equation, we want to find a simpler equivalent equation of the form $x = $ **a number**, whose solution is obvious.

Solution

a. Since the coefficient of x is $\frac{2}{3}$, we can isolate x by multiplying both sides of the equation by the reciprocal of $\frac{2}{3}$, which is $\frac{3}{2}$.

$$\frac{2}{3}x = 6 \quad \text{This is the equation to solve.}$$

$$\frac{3}{2} \cdot \frac{2}{3}x = \frac{3}{2} \cdot 6 \quad \text{To undo the multiplication by } \tfrac{2}{3}\text{, multiply both sides by the reciprocal of } \tfrac{2}{3}.$$

$$\left(\frac{3}{2} \cdot \frac{2}{3}\right)x = \frac{3}{2} \cdot 6 \quad \text{Use the associative property of multiplication to group } \tfrac{3}{2} \text{ and } \tfrac{2}{3}.$$

$$1x = 9 \quad \text{On the left, } \tfrac{3}{2} \cdot \tfrac{2}{3} = 1. \text{ On the right, } \tfrac{3}{2} \cdot 6 = \tfrac{18}{2} = 9.$$

$$x = 9 \quad \text{The coefficient 1 need not be written since } 1x = x.$$

Check:
$$\frac{2}{3}x = 6 \quad \text{This is the original equation.}$$

$$\frac{2}{3}(9) \stackrel{?}{=} 6 \quad \text{Substitute 9 for x in the original equation.}$$

$$6 = 6 \quad \text{On the left side, } \tfrac{2}{3}(9) = \tfrac{18}{3} = 6.$$

Since the statement is true, 9 is the solution. The solution set is {9}.

The Language of Algebra Variable terms with fractional coefficients can be written in two ways. For example:

$$\frac{2x}{3} = \frac{2}{3}x \quad \text{and} \quad -\frac{5a}{4} = -\frac{5}{4}a$$

b. To isolate x, we multiply both sides by the reciprocal of $-\frac{5}{4}$, which is $-\frac{4}{5}$.

Self Check 6

Solve: **a.** $\frac{7}{2}x = 21$

b. $-\frac{3}{8}b = 2$

Now Try Problems 61 and 67

$$-\frac{5}{4}x = 3 \quad \text{This is the equation to solve.}$$

$$-\frac{4}{5}\left(-\frac{5}{4}x\right) = -\frac{4}{5}(3) \quad \text{To undo the multiplication by } -\frac{5}{4}, \text{ multiply both sides by the reciprocal of } -\frac{5}{4}.$$

$$1x = -\frac{12}{5} \quad \begin{array}{l}\text{On the left side, } -\frac{4}{5}\left(-\frac{5}{4}\right) = 1.\\ \text{On the right side, } -\frac{4}{5}(3) = -\frac{12}{5}.\end{array}$$

$$x = -\frac{12}{5} \quad \text{The coefficient 1 need not be written since } 1x = x.$$

The solution is $-\frac{12}{5}$, and the solution set is $\left\{-\frac{12}{5}\right\}$. Verify that this is correct by checking.

5 Use the division property of equality.

Since any division can be rewritten as a multiplication by multiplying by the reciprocal, the following property is a natural extension of the multiplication property.

> **Division Property of Equality**
>
> Dividing both sides of an equation by the same nonzero number does not change its solution.
>
> For any real numbers a, b, and c, where c is not 0,
>
> $$\text{if } a = b, \text{ then } \frac{a}{c} = \frac{b}{c}$$

When we use this property, the resulting equation is equivalent to the original one.

Self Check 7

Solve: **a.** $16x = 176$
b. $10.04 = -0.4r$

Now Try Problems 69 and 79

EXAMPLE 7 Solve: **a.** $2t = 80$ **b.** $-6.02 = -8.6t$

Strategy We will use a property of equality to isolate the variable on one side of the equation.

WHY To solve the original equation, we want to find a simpler equivalent equation of the form $t = $ **a number** or **a number** $= t$, whose solution is obvious.

Solution
a. To isolate t on the left side, we use the division property of equality. We can undo the multiplication by 2 by dividing both sides of the equation by 2.

$$2t = 80 \quad \text{This is the equation to solve.}$$

$$\frac{2t}{2} = \frac{80}{2} \quad \text{Use the division property of equality: Divide both sides by 2.}$$

$$1t = 40 \quad \text{Simplify } \frac{2t}{2} \text{ by removing the common factor of 2 in the numerator and denominator: } \frac{2}{2} = 1.$$

$$t = 40 \quad \text{The product of 1 and any number is that number: } 1t = t.$$

If we substitute 40 for t in $2t = 80$, we obtain the true statement $80 = 80$. This verifies that 40 is the solution. The solution set is $\{40\}$.

The Language of Algebra Since division by 2 is the same as multiplication by $\frac{1}{2}$, we can also solve $2t = 80$ using the multiplication property of equality. That is, we can isolate t by multiplying both sides by the *multiplicative inverse* of 2, which is $\frac{1}{2}$:

$$\frac{1}{2} \cdot 2t = \frac{1}{2} \cdot 80$$

b. To isolate t on the right side, we use the division property of equality. We can undo the multiplication by -8.6 by dividing both sides by -8.6.

$-6.02 = -8.6t$ This is the equation to solve.

$\dfrac{-6.02}{-8.6} = \dfrac{-8.6t}{-8.6}$ Use the division property of equality: Divide both sides by -8.6.

$0.7 = t$ Do the division: $8.6\overline{)6.02}$. The quotient of two negative numbers is positive.

The solution is 0.7. Verify that this is correct by checking.

Success Tip It is usually easier to multiply on each side if the coefficient of the variable term is a *fraction* and to divide on each side if the coefficient is an *integer* or a *decimal*.

EXAMPLE 8 Solve: $-x = 3$

Strategy The variable x is not isolated, because there is a $-$ sign in front of it. Since the term $-x$ has an understood coefficient of -1, the equation can be written as $-1x = 3$. We need to select a property of equality and use it to isolate the variable on one side of the equation.

WHY To find the solution of the original equation, we want to find a simpler equivalent equation of the form $x =$ **a number**, whose solution is obvious.

Solution
To isolate x, we can either multiply or divide both sides by -1.

Multiply both sides by -1:

$-x = 3$ The equation to solve
$-1x = 3$ Write: $-x = -1x$
$(-1)(-1x) = (-1)3$
$1x = -3$
$x = -3$ $1x = x$

Divide both sides by -1:

$-x = 3$ The equation to solve
$-1x = 3$ Write: $-x = -1x$
$\dfrac{-1x}{-1} = \dfrac{3}{-1}$
$1x = -3$ On the left side, $\dfrac{-1}{-1} = 1$.
$x = -3$ $1x = x$

Check: $-x = 3$ This is the original equation.
$-(-3) \stackrel{?}{=} 3$ Substitute -3 for x.
$3 = 3$ On the left side, the opposite of -3 is 3.

Since the statement is true, -3 is the solution. The solution set is $\{-3\}$.

Self Check 8
Solve: $-h = -12$
Now Try Problem 81

ANSWERS TO SELF CHECKS
1. yes **2.** 49 **3. a.** 33 **b.** 49 **4. a.** $\dfrac{29}{15}$ **b.** -0.5 **5.** 72 **6. a.** 6 **b.** $-\dfrac{16}{3}$
7. a. 11 **b.** -25.1 **8.** 12

SECTION 2.1 STUDY SET

VOCABULARY

Fill in the blanks.

1. An _____, such as $x + 1 = 7$, is a statement indicating that two expressions are equal.
2. Any number that makes an equation true when substituted for the variable is said to _____ the equation. Such numbers are called _____.
3. To _____ an equation means to find all values of the variable that make the equation true.
4. To solve an equation, we _____ the variable on one side of the equal symbol.
5. Equations with the same solutions are called _____ equations.
6. To _____ the solution of an equation, we substitute the value for the variable in the original equation and determine whether the result is a true statement.

CONCEPTS

7. Given $x + 6 = 12$:
   a. What is the left side of the equation?
   b. Is this equation true, false, or neither?
   c. Is 5 the solution?
   d. Does 6 satisfy the equation?
8. For each equation, determine what operation is performed on the variable. Then explain how to undo that operation to isolate the variable.
   a. $x - 8 = 24$
   b. $x + 8 = 24$
   c. $\frac{x}{8} = 24$
   d. $8x = 24$
9. Complete the following properties of equality.
   a. If $a = b$, then
   $$a + c = b + \boxed{} \quad \text{and} \quad a - c = b - \boxed{}$$
   b. If $a = b$, then $ca = \boxed{} b$ and $\frac{a}{c} = \frac{b}{\boxed{}}$ $(c \neq 0)$
10. a. To solve $\frac{h}{10} = 20$, do we multiply both sides of the equation by 10 or 20?
    b. To solve $4k = 16$, do we subtract 4 from both sides of the equation or divide both sides by 4?

11. Simplify each expression.
    a. $x + 7 - 7$
    b. $y - 2 + 2$
    c. $\frac{5t}{5}$
    d. $6 \cdot \frac{h}{6}$
12. a. To solve $-\frac{4}{5}x = 8$, we can multiply both sides by the reciprocal of $-\frac{4}{5}$. What is the reciprocal of $-\frac{4}{5}$?
    b. What is $-\frac{5}{4}\left(-\frac{4}{5}\right)$?

NOTATION

Complete each step to solve the equation and check the result.

13. $x - 5 = 45$ Check: $x - 5 = 45$
    $x - 5 + \boxed{} = 45 + \boxed{}$ $\boxed{} - 5 \quad 45$
    $x = \boxed{}$ $\boxed{} = 45$ True
       $\boxed{}$ is the solution.

14. $8x = 40$ Check: $8x = 40$
    $\frac{8x}{\boxed{}} = \frac{40}{\boxed{}}$ $8() \stackrel{?}{=} 40$
    $x = \boxed{}$ $\boxed{} = 40$ True
       $\boxed{}$ is the solution.

15. a. What does the symbol $\stackrel{?}{=}$ mean?
    b. If you solve an equation and obtain $50 = x$, can you write $x = 50$?
16. Fill in the blank: $-x = \boxed{} x$

GUIDED PRACTICE

Check to determine whether the number in red is a solution of the equation. See Example 1.

17. $6, x + 12 = 28$
18. $110, x - 50 = 60$
19. $-8, 2b + 3 = -15$
20. $-2, 5t - 4 = -16$
21. $5, 0.5x = 2.9$
22. $3.5, 1.2 + x = 4.7$
23. $-6, 33 - \frac{x}{2} = 30$
24. $-8, \frac{x}{4} + 98 = 100$
25. $-2, |c - 8| = 10$
26. $-45, |30 - r| = 15$
27. $12, 3x - 2 = 4x - 5$
28. $5, 5y + 8 = 3y - 2$
29. $-3, x^2 - x - 6 = 0$
30. $-2, y^2 + 5y - 3 = 0$
31. $1, \frac{2}{a + 1} + 5 = \frac{12}{a + 1}$
32. $4, \frac{2t}{t - 2} - \frac{4}{t - 2} = 1$

33. $\frac{3}{4}$, $x - \frac{1}{8} = \frac{5}{8}$
34. $\frac{7}{3}$, $-4 = a + \frac{5}{3}$
35. -3, $(x - 4)(x + 3) = 0$
36. 5, $(2x + 1)(x - 5) = 0$

Use a property of equality to solve each equation. Then check the result. See Examples 2–4.

37. $a - 5 = 66$
38. $x - 34 = 19$
39. $9 = p - 9$
40. $3 = j - 88$
41. $x - 1.6 = -2.5$
42. $y - 1.2 = -1.3$
43. $-3 + a = 0$
44. $-1 + m = 0$
45. $d - \frac{1}{9} = \frac{7}{9}$
46. $\frac{7}{15} = b - \frac{1}{15}$
47. $x + 7 = 10$
48. $y + 15 = 24$
49. $s + \frac{1}{5} = \frac{4}{25}$
50. $\frac{1}{6} = h + \frac{4}{3}$
51. $3.5 + f = 1.2$
52. $9.4 + h = 8.1$

Use a property of equality to solve each equation. Then check the result. See Example 5.

53. $\frac{x}{15} = 3$
54. $\frac{y}{7} = 12$
55. $0 = \frac{v}{11}$
56. $\frac{d}{49} = 0$
57. $\frac{d}{-7} = -3$
58. $\frac{c}{-2} = -11$
59. $\frac{y}{0.6} = -4.4$
60. $\frac{y}{0.8} = -2.9$

Use a property of equality to solve each equation. Then check the result. See Example 6.

61. $\frac{4}{5}t = 16$
62. $\frac{11}{15}y = 22$
63. $\frac{2}{3}c = 10$
64. $\frac{9}{7}d = 81$
65. $-\frac{7}{2}r = 21$
66. $-\frac{4}{5}s = 36$
67. $-\frac{5}{4}h = -5$
68. $-\frac{3}{8}t = -3$

Use a property of equality to solve each equation. Then check the result. See Example 7.

69. $4x = 16$
70. $5y = 45$
71. $63 = 9c$
72. $40 = 5t$
73. $23b = 23$
74. $16 = 16h$
75. $-8h = 48$
76. $-9a = 72$
77. $-100 = -5g$
78. $-80 = -5w$
79. $-3.4y = -1.7$
80. $-2.1x = -1.26$

Use a property of equality to solve each equation. Then check the result. See Example 8.

81. $-x = 18$
82. $-y = 50$
83. $-n = \frac{4}{21}$
84. $-w = \frac{11}{16}$

TRY IT YOURSELF

Solve each equation. Then check the result.

85. $8.9 = -4.1 + t$
86. $7.7 = -3.2 + s$
87. $-2.5 = -m$
88. $-1.8 = -b$
89. $-\frac{9}{8}x = 3$
90. $-\frac{14}{3}c = 7$
91. $\frac{3}{4} = d + \frac{1}{10}$
92. $\frac{5}{9} = r + \frac{1}{6}$
93. $-15x = -60$
94. $-14x = -84$
95. $-10 = n - 5$
96. $-8 = t - 2$
97. $\frac{h}{-40} = 5$
98. $\frac{x}{-7} = 12$
99. $a - 93 = 2$
100. $18 = x - 3$

APPLICATIONS

101. **SYNTHESIZERS** To find the unknown angle measure, which is represented by x, solve the equation $x + 115 = 180$.

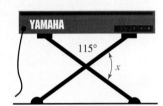

102. **STOP SIGNS** To find the measure of one angle of the stop sign, which is represented by x, solve the equation $8x = 1,080$.

103. **SHARING THE WINNING TICKET** When a Pennsylvania Powerball jackpot was won by a group of 15 coworkers at Quad Graphics, each received $66,666.66. To find the amount of the jackpot, which is represented by x, solve the equation $\frac{x}{15} = 66,666.66$. Round the solution to the nearest dollar.

104. **SOCIAL NETWORKS** In 2012, the annual revenue generated per employee at Twitter was $388,888. This is $712,865 less than the annual revenue generated per employee at Facebook. To find the annual revenue generated per employee at Facebook, which is represented by x, solve the equation $x - 388,888 = 712,865$. (source: web-strategist.com)

WRITING

105. What does it mean to solve an equation?

106. When solving an equation, we *isolate* the variable on one side of the equation. Write a sentence in which the word *isolate* is used in a different context.

107. Explain the error in the following work.

$$\text{Solve:} \quad \cancel{x} + 2 = 40$$
$$x + 2 - 2 = 40$$
$$x = 40$$

108. After solving an equation, how do we check the result?

REVIEW

109. Evaluate $-9 - 3x$ for $x = -3$.

110. Evaluate: $-5^2 + (-5)^2$

111. Translate to symbols: Subtract x from 45

112. Evaluate: $\dfrac{2^3 + 3(5-3)}{15 - 4 \cdot 2}$

Objectives

1. Use more than one property of equality to solve equations.
2. Simplify expressions to solve equations.
3. Clear equations of fractions and decimals.
4. Identify identities and contradictions.

SECTION 2.2
More about Solving Equations

We have solved simple equations by using properties of equality. We will now expand our equation-solving skills by considering more complicated equations. We want to develop a general strategy that can be used to solve any kind of *linear equation in one variable*.

> **Linear Equation in One Variable**
>
> A **linear equation in one variable** can be written in the form
> $$ax + b = c$$
> where a, b and c are real numbers and $a \neq 0$.

1 Use more than one property of equality to solve equations.

Sometimes we must use several properties of equality to solve an equation. For example, on the left side of $2x + 6 = 10$, the variable x is multiplied by 2, and then 6 is added to that product. To isolate x, we use the order of operations rules in reverse. First, we undo the addition of 6, and then we undo the multiplication by 2.

$2x + 6 = 10$	This is the equation to solve.
$2x + 6 - 6 = 10 - 6$	To undo the addition of 6, subtract 6 from both sides.
$2x = 4$	Do the subtractions.
$\dfrac{2x}{2} = \dfrac{4}{2}$	To undo the multiplication by 2, divide both sides by 2.
$x = 2$	Do the divisions.

The solution is 2.

> **The Language of Algebra** We subtract 6 from both sides to isolate the *variable term*, $2x$. Then we divide both sides by 2 to isolate the *variable*, x.

Self Check 1
Solve: $8x - 13 = 43$
Now Try Problem 15

EXAMPLE 1 Solve: $-12x + 5 = 17$

Strategy First we will use a property of equality to isolate the *variable term* on one side of the equation. Then we will use a second property of equality to isolate the *variable* itself.

WHY To solve the original equation, we want to find a simpler equivalent equation of the form $x =$ **a number**, whose solution is obvious.

Solution

- To isolate the variable term, $-12x$, we subtract 5 from both sides to undo the addition of 5.
- To isolate the variable, x, we divide both sides by -12 to undo the multiplication by -12.

$-12x + 5 = 17$	This is the equation to solve.
$-12x + 5 - 5 = 17 - 5$	Use the subtraction property of equality: Subtract 5 from both sides to isolate the variable term $-12x$.
$-12x = 12$	Do the subtractions: $5 - 5 = 0$ and $17 - 5 = 12$.
$\dfrac{-12x}{-12} = \dfrac{12}{-12}$	Use the division property of equality: Divide both sides by -12 to isolate x.
$x = -1$	Do the divisions.

Check:

$-12x + 5 = 17$	This is the original equation.
$-12(-1) + 5 \stackrel{?}{=} 17$	Substitute -1 for x.
$12 + 5 \stackrel{?}{=} 17$	Do the multiplication on the left side.
$17 = 17$	True

The solution is -1. The solution set is $\{-1\}$.

> **Caution!** When checking solutions, always use the original equation.

EXAMPLE 2   Solve: $\dfrac{5}{8}m - 2 = -12$

Strategy We will use properties of equality to isolate the variable on one side of the equation.

WHY To solve the original equation, we want to find a simpler equivalent equation of the form $m = \textbf{a number}$, whose solution is obvious.

Solution
We note that the coefficient of m is $\dfrac{5}{8}$ and proceed as follows.

- To isolate the variable term $\dfrac{5}{8}m$, we add 2 to both sides to undo the subtraction of 2.
- To isolate the variable, m, we multiply both sides by $\dfrac{8}{5}$ to undo the multiplication by $\dfrac{5}{8}$.

$\dfrac{5}{8}m - 2 = -12$	This is the equation to solve.
$\dfrac{5}{8}m - 2 + 2 = -12 + 2$	Use the addition property of equality: Add 2 to both sides to isolate the variable term $\dfrac{5}{8}m$.
$\dfrac{5}{8}m = -10$	Do the additions: $-2 + 2 = 0$ and $-12 + 2 = -10$.
$\dfrac{8}{5}\left(\dfrac{5}{8}m\right) = \dfrac{8}{5}(-10)$	Use the multiplication property of equality: Multiply both sides by $\dfrac{8}{5}$ (which is the reciprocal of $\dfrac{5}{8}$) to isolate m.
$m = -16$	On the left side: $\dfrac{8}{5}\left(\dfrac{5}{8}\right) = 1$ and $1m = m$. On the right side: $\dfrac{8}{5}(-10) = -\dfrac{8 \cdot 2 \cdot \overset{1}{5}}{\underset{1}{5}} = -16$.

The solution is -16. Verify this by substituting -16 into the original equation. The solution set is $\{-16\}$.

Self Check 2

Solve: $\dfrac{7}{12}a - 6 = -27$

Now Try Problem 21

Self Check 3

Solve: $-6.6 - m = -2.7$

Now Try Problem 35

EXAMPLE 3 Solve: $-0.2 = -0.8 - y$

Strategy First, we will use a property of equality to isolate the variable term on one side of the equation. Then we will use a second property of equality to isolate the variable itself.

WHY To solve the original equation, we want to find a simpler equivalent equation of the form **a number** $= y$, whose solution is obvious.

Solution

To isolate the variable term $-y$ on the right side, we eliminate -0.8 by adding 0.8 to both sides.

$$-0.2 = -0.8 - y \qquad \text{This is the equation to solve.}$$
$$-0.2 + \mathbf{0.8} = -0.8 - y + \mathbf{0.8} \qquad \text{Add 0.8 to both sides to isolate } -y.$$
$$0.6 = -y \qquad \text{Do the additions.}$$

Since the term $-y$ has an understood coefficient of -1, the equation can be written as $0.6 = -1y$. To isolate y, we can either multiply both sides or divide both sides by -1. If we choose to divide both sides by -1, we proceed as follows.

$$0.6 = -1y$$
$$\frac{0.6}{-1} = \frac{-1y}{-1} \qquad \text{To undo the multiplication by } -1, \text{ divide both sides by } -1.$$
$$-0.6 = y$$

The solution is -0.6, and the solution set is $\{-0.6\}$. Verify this by substituting -0.6 into the original equation.

2 Simplify expressions to solve equations.

When solving equations, we should simplify the expressions that make up the left and right sides before applying any properties of equality. Often, that involves removing parentheses and/or combining like terms.

Self Check 4

Solve: **a.** $4(a + 2) - a = 11$
b. $9x - 5(x - 9) = 1$

Now Try Problems 45 and 47

EXAMPLE 4 Solve: **a.** $3(k + 1) - 5k = 0$ **b.** $8a - 2(a - 7) = 68$

Strategy We will use the distributive property along with the process of combining like terms to simplify the left side of each equation.

WHY It's best to simplify each side of an equation before using a property of equality.

Solution

a.
$$3(k + 1) - 5k = 0 \qquad \text{This is the equation to solve.}$$
$$3k + 3 - 5k = 0 \qquad \text{Distribute the multiplication by 3.}$$
$$-2k + 3 = 0 \qquad \text{Combine like terms: } 3k - 5k = -2k.$$
$$-2k + 3 - 3 = 0 - 3 \qquad \text{To undo the addition of 3, subtract 3 from both sides. This isolates the variable term } -2k.$$
$$-2k = -3 \qquad \text{Do the subtractions: } 3 - 3 = 0 \text{ and } 0 - 3 = -3.$$
$$\frac{-2k}{-2} = \frac{-3}{-2} \qquad \text{To undo the multiplication by } -2, \text{ divide both sides by } -2. \text{ This isolates the variable } k.$$
$$k = \frac{3}{2} \qquad \text{Simplify: } \frac{-3}{-2} = \frac{3}{2}.$$

Check: $3(k + 1) - 5k = 0$ This is the original equation.

$3\left(\dfrac{3}{2} + 1\right) - 5\left(\dfrac{3}{2}\right) \stackrel{?}{=} 0$ Substitute $\dfrac{3}{2}$ for k.

$3\left(\dfrac{5}{2}\right) - 5\left(\dfrac{3}{2}\right) \stackrel{?}{=} 0$ Do the addition within the parentheses. Think of 1 as $\dfrac{2}{2}$ and then add: $\dfrac{3}{2} + \dfrac{2}{2} = \dfrac{5}{2}$.

$\dfrac{15}{2} - \dfrac{15}{2} \stackrel{?}{=} 0$ Do the multiplications.

$0 = 0$ True

The solution is $\dfrac{3}{2}$ and the solution set is $\left\{\dfrac{3}{2}\right\}$.

Caution! To check a result, we evaluate each side of the equation following the order of operations rules.

b.
$8a - 2(a - 7) = 68$ This is the equation to solve.
$8a - 2a + 14 = 68$ Distribute the multiplication by -2.
$6a + 14 = 68$ Combine like terms: $8a - 2a = 6a$.
$6a + 14 - 14 = 68 - 14$ To undo the addition of 14, subtract 14 from both sides. This isolates the variable term $6a$.
$6a = 54$ Do the subtractions.
$\dfrac{6a}{6} = \dfrac{54}{6}$ To undo the multiplication by 6, divide both sides by 6. This isolates the variable a.
$a = 9$ Do the divisions.

The solution is 9, and the solution set is {9}. Verify this by substituting 9 into the original equation.

When solving an equation, if variables appear on both sides, we can use the addition (or subtraction) property of equality to get all variable terms on one side and all constant terms on the other.

EXAMPLE 5   Solve: $3x - 15 = 4x + 36$

Strategy There are variable terms ($3x$ and $4x$) on both sides of the equation. We will eliminate $3x$ from the left side of the equation by subtracting $3x$ from both sides.

WHY To solve for x, all the terms containing x must be on the same side of the equation.

Solution
$3x - 15 = 4x + 36$ This is the equation to solve.
$3x - 15 - 3x = 4x + 36 - 3x$ Subtract $3x$ from both sides to isolate the variable term on the right side.
$-15 = x + 36$ Combine like terms: $3x - 3x = 0$ and $4x - 3x = x$.
$-15 - 36 = x + 36 - 36$ To undo the addition of 36, subtract 36 from both sides.
$-51 = x$ Do the subtractions.

Self Check 5
Solve: $30 + 6n = 4n - 2$
Now Try Problem 57

Check:

$$3x - 15 = 4x + 36 \quad \text{The original equation.}$$
$$3(-51) - 15 \stackrel{?}{=} 4(-51) + 36 \quad \text{Substitute } -51 \text{ for } x.$$
$$-153 - 15 \stackrel{?}{=} -204 + 36 \quad \text{Do the multiplications.}$$
$$-168 = -168 \quad \text{True}$$

The solution is -51 and the solution set is $\{-51\}$.

3 Clear equations of fractions and decimals.

Equations are usually easier to solve if they don't involve fractions. We can use the multiplication property of equality to clear an equation of fractions by multiplying both sides of the equation by the least common denominator.

Self Check 6

Solve: $\frac{1}{4}x + \frac{1}{2} = -\frac{1}{8}$

Now Try Problem 63

EXAMPLE 6 Solve: $\frac{1}{6}x + \frac{5}{2} = \frac{1}{3}$

Strategy To clear the equations of fractions, we will multiply both sides by their LCD.

WHY It's easier to solve an equation that involves only integers.

Solution

$$\frac{1}{6}x + \frac{5}{2} = \frac{1}{3} \quad \text{This is the equation to solve.}$$

$$6\left(\frac{1}{6}x + \frac{5}{2}\right) = 6\left(\frac{1}{3}\right) \quad \text{Multiply both sides by the LCD of } \frac{1}{6}, \frac{5}{2}, \text{ and } \frac{1}{3}, \text{ which is 6. Don't forget the parentheses.}$$

$$6\left(\frac{1}{6}x\right) + 6\left(\frac{5}{2}\right) = 6\left(\frac{1}{3}\right) \quad \text{On the left side, distribute the multiplication by 6.}$$

$$x + 15 = 2 \quad \text{Do each multiplication: } 6\left(\frac{1}{6}\right) = 1, \, 6\left(\frac{5}{2}\right) = \frac{30}{2} = 15, \text{ and } 6\left(\frac{1}{3}\right) = \frac{6}{3} = 2.$$

$$x + 15 - 15 = 2 - 15 \quad \text{To undo the addition of 15, subtract 15 from both sides.}$$

$$x = -13$$

The solution is -13. Check this by substituting -13 for x in $\frac{1}{6}x + \frac{5}{2} = \frac{1}{3}$.

Caution! Before multiplying both sides of an equation by the LCD, enclose the left and right sides with parentheses or brackets.

$$\left(\frac{1}{6}x + \frac{5}{2}\right) = \left(\frac{1}{3}\right)$$

If an equation contains decimals, it is often convenient to multiply both sides by a power of 10 to change the decimals in the equation to integers.

Self Check 7

Solve:
$(15{,}000 - x)0.08 + 0.07x = 1{,}110$

Now Try Problem 71

EXAMPLE 7 Solve: $0.04(12) + 0.01x = 0.02(12 + x)$

Strategy To clear the equations of decimals, we will multiply both sides by a carefully chosen power of 10.

WHY It's easier to solve an equation that involves only integers.

Solution

The equation contains the decimals 0.04, 0.01, and 0.02. Since the greatest number of decimal places in any one of these numbers is two, we multiply both sides of the equation by 10^2 or 100. This changes 0.04 to 4, and 0.01 to 1, and 0.02 to 2.

$$0.04(12) + 0.01x = 0.02(12 + x)$$

$$100[0.04(12) + 0.01x] = 100[0.02(12 + x)] \quad \text{Multiply both sides by 100.}$$
$$\text{Don't forget the brackets.}$$

$$100 \cdot 0.04(12) + 100 \cdot 0.01x = 100 \cdot 0.02(12 + x) \quad \text{Distribute the multiplication by 100.}$$

$$4(12) + 1x = 2(12 + x) \quad \text{Multiply each decimal by 100 by moving its decimal point 2 places to the right.}$$

$$48 + x = 24 + 2x \quad \text{Distribute the multiplication by 2.}$$

$$48 + x - 24 - x = 24 + 2x - 24 - x \quad \text{Subtract 24 and x from both sides.}$$

$$24 = x \quad \text{Simplify each side.}$$

$$x = 24$$

The solution is 24. Check by substituting 24 for x in the original equation.

The previous examples suggest the following strategy for solving equations. It is important to note that not every step is needed to solve every equation.

Strategy for Solving Linear Equations in One Variable

1. **Clear the equation of fractions or decimals:** Multiply both sides by the LCD to clear fractions or multiply both sides by a power of 10 to clear decimals.
2. **Simplify each side of the equation:** Use the distributive property to remove parentheses, and then combine like terms on each side.
3. **Isolate the variable term on one side:** Add (or subtract) to get the variable term on one side of the equation and a number on the other using the addition (or subtraction) property of equality.
4. **Isolate the variable:** Multiply (or divide) to isolate the variable using the multiplication (or division) property of equality.
5. **Check the result:** Substitute the possible solution for the variable in the *original* equation to see if a true statement results.

EXAMPLE 8

Solve: $\dfrac{7m + 5}{5} = -4m + 1$

Strategy We will follow the steps of the equation-solving strategy to solve the equation.

WHY This is the most efficient way to solve a linear equation in one variable.

Solution

$$\dfrac{7m + 5}{5} = -4m + 1 \quad \text{This is the equation to solve.}$$

Step 1 $\quad 5\left(\dfrac{7m + 5}{5}\right) = 5(-4m + 1) \quad \text{Clear the equation of the fraction by multiplying both sides by 5.}$

Self Check 8

Solve: $6c + 2 = \dfrac{18 - c}{9}$

Now Try Problem 79

Step 2 $\quad 7m + 5 = -20m + 5 \quad$ On the left side, remove the common factor 5 in the numerator and denominator. On the right side, distribute the multiplication by 5.

Step 3 $\quad 7m + 5 + 20m = -20m + 5 + 20m \quad$ To eliminate the term $-20m$ on the right side, add $20m$ to both sides.

$\quad\quad\quad 27m + 5 = 5 \quad$ Combine like terms: $7m + 20m = 27m$ and $-20m + 20m = 0$.

$\quad\quad\quad 27m + 5 - 5 = 5 - 5 \quad$ To isolate the term $27m$, undo the addition of 5 by subtracting 5 from both sides.

$\quad\quad\quad 27m = 0 \quad$ Do the subtractions.

Step 4 $\quad \dfrac{27m}{27} = \dfrac{0}{27} \quad$ To isolate m, undo the multiplication by 27 by dividing both sides by 27.

$\quad\quad\quad m = 0 \quad$ 0 divided by any nonzero number is 0.

The solution is 0.

Step 5 Substitute 0 for m in $\dfrac{7m+5}{5} = -4m + 1$ to check that the solution is 0.

> **Caution!** Remember that when you multiply one side of an equation by a nonzero number, you must multiply the other side of the equation by the same number.

4 Identify identities and contradictions.

Each of the equations in Examples 1 through 8 had exactly one solution. However, some equations have no solutions, and others have infinitely many solutions.

A linear equation in one variable that is true for all values of the variable is an **identity**. One example is the equation

$x + x = 2x \quad$ If we substitute -10 for x, we get the true statement $-20 = -20$. If we substitute 7 for x, we get $14 = 14$, and so on.

Since we can replace x with any number and the equation will be true, all real numbers are solutions of $x + x = 2x$. This equation has infinitely many solutions. Its solution set is written as {all real numbers}.

An equation that is not true for any values of its variable is called a **contradiction**. One example is

$x = x + 1 \quad$ No number is 1 greater than itself.

Since $x = x + 1$ has no solutions, its solution set is the **empty set**, or **null set**, and is written as $\emptyset$.

Self Check 9

Solve:
$3(x + 5) - 4(x + 4) = -x - 1$

Now Try Problem 87

EXAMPLE 9 Solve: $3(x + 8) + 5x = 2(12 + 4x)$

Strategy We will follow the steps of the equation-solving strategy to solve the equation.

WHY This is the most efficient way to solve a linear equation in one variable.

Solution

$3(x + 8) + 5x = 2(12 + 4x) \quad$ This is the equation to solve.

$3x + 24 + 5x = 24 + 8x \quad$ Distribute the multiplication by 3 and by 2.

$8x + 24 = 24 + 8x$	Combine like terms: $3x + 5x = 8x$. Note that the sides of the equation are identical.
$8x - 8x + 24 = 24 + 8x - 8x$	To eliminate the term $8x$ on the right side, subtract $8x$ from both sides.
$24 = 24$	Combine like terms on both sides: $8x - 8x = 0$.

In this case, the terms involving x drop out and the result is true. This means that any number substituted for x in the original equation will give a true statement. Therefore, *all real numbers* are solutions and this equation is an identity.

Success Tip Note that at the step $8x + 24 = 24 + 8x$ we know that the equation is an identity.

EXAMPLE 10 Solve: $3(d + 7) - d = 2(d + 10)$

Strategy We will follow the steps of the equation-solving strategy to solve the equation.

WHY This is the most efficient way to solve a linear equation in one variable.

Solution

$3(d + 7) - d = 2(d + 10)$	This is the equation to solve.
$3d + 21 - d = 2d + 20$	Distribute the multiplication by 3 and by 2.
$2d + 21 = 2d + 20$	Combine like terms: $3d - d = 2d$.
$2d + 21 - 2d = 2d + 20 - 2d$	To eliminate the term $2d$ on the right side, subtract $2d$ from both sides.
$21 = 20$	Combine like terms on both sides: $2d - 2d = 0$.

In this case, the terms involving d drop out and the result is false. This means that any number that is substituted for d in the original equation will give a false statement. Since this equation has *no solution*, it is a contradiction.

The Language of Algebra *Contradiction* is a form of the word *contradict*, meaning conflicting ideas. During a trial, evidence might be introduced that *contradicts* the testimony of a witness.

Self Check 10
Solve:
$-4(c - 3) + 2c = 2(10 - c)$

Now Try Problem 89

ANSWERS TO SELF CHECKS

1. 7 **2.** -36 **3.** -3.9 **4. a.** 1 **b.** -11 **5.** -16 **6.** $-\dfrac{5}{2}$ **7.** 9,000 **8.** 0 **9.** All real numbers; the equation is an identity. **10.** No solution; the equation is a contradiction.

SECTION 2.2   STUDY SET

VOCABULARY

Fill in the blanks.

1. $3x + 8 = 10$ is an example of a linear _____ in one variable.

2. To solve $\dfrac{s}{3} + \dfrac{1}{4} = -\dfrac{1}{2}$, we can _____ the equation of the fractions by multiplying both sides by 12.

3. A linear equation in one variable that is true for all values of the variable is an _____.

4. An equation that is not true for any values of its variable is called a _____.

CONCEPTS

Fill in the blanks.

5. To solve $3x - 5 = 1$, we first undo the _____ of 5 by adding 5 to both sides. Then we undo the _____ by 3 by dividing both sides by 3.

6. To solve $\frac{x}{2} + 3 = 5$, we can undo the _____ of 3 by subtracting 3 from both sides. Then we can undo the _____ by 2 by multiplying both sides by 2.

7. a. Combine like terms on the left side of $6x - 8 - 8x = -24$.
   b. Distribute and then combine like terms on the right side of $-20 = 4(3x - 4) - 9x$.

8. Use a check to determine whether -2 is a solution of the equation.
   a. $6x + 5 = 7$
   b. $8(x + 3) = 8$

9. Multiply.
   a. $20\left(\frac{3}{5}x\right)$
   b. $100 \cdot 0.02x$

10. By what must you multiply both sides of $\frac{2}{3} - \frac{1}{2}b = -\frac{4}{3}$ to clear it of fractions?

11. By what must you multiply both sides of $0.7x + 0.3(x - 1) = 0.5x$ to clear it of decimals?

12. a. Simplify: $3x + 5 - x$
    b. Solve: $3x + 5 = 9$
    c. Evaluate $3x + 5 - x$ for $x = 9$.
    d. Check: Is -1 a solution of $3x + 5 - x = 9$?

NOTATION

Complete the steps to solve the equation and check the result.

13. Solve: $2x - 7 = 21$
    $2x - 7 + \boxed{} = 21 + \boxed{}$
    $2x = 28$
    $\dfrac{2x}{\boxed{}} = \dfrac{28}{\boxed{}}$
    $x = 14$

    Check: $2x - 7 = 21$
    $2(\boxed{}) - 7 \boxed{} 21$
    $\boxed{} - 7 \stackrel{?}{=} 21$
    $\boxed{} = 21$
    $\boxed{}$ is the solution.

14. A student multiplied both sides of $\frac{3}{4}t + \frac{5}{8} = \frac{1}{2}t$ by 8 to clear it of fractions, as shown below. Explain his error in showing this step.

    $8 \cdot \frac{3}{4}t + \frac{5}{8} = 8 \cdot \frac{1}{2}t$

GUIDED PRACTICE

Solve each equation and check the result. See Examples 1–2.

15. $2x + 5 = 17$
16. $3x - 5 = 13$
17. $5q - 2 = 23$
18. $4p + 3 = 43$
19. $-33 = 5t + 2$
20. $-55 = 3w + 5$
21. $\frac{5}{6}k - 5 = 10$
22. $\frac{2}{5}c - 12 = 2$
23. $-\frac{7}{16}h + 28 = 21$
24. $-\frac{5}{8}h + 25 = 15$
25. $\frac{t}{3} + 2 = 6$
26. $\frac{x}{5} - 5 = -12$
27. $-3p + 7 = -3$
28. $-2r + 8 = -1$
29. $-5 - 2d = 0$
30. $-8 - 3c = 0$
31. $2(-3) + 4y = 14$
32. $4(-1) + 3y = 8$
33. $0.7 - 4y = 1.7$
34. $0.3 - 2x = -0.9$

Solve each equation and check the result. See Example 3.

35. $1.2 - x = -1.7$
36. $0.6 = 4.1 - x$
37. $-6 - y = -2$
38. $-1 - h = -9$

Solve each equation and check the result. See Example 4.

39. $3(2y - 2) - y = 5$
40. $2(-3a + 2) + a = 2$
41. $4(5b) + 2(6b - 1) = -34$
42. $9(x + 11) + 5(13 - x) = 0$
43. $-(4 - m) = -10$
44. $-(6 - t) = -12$
45. $10.08 = 4(0.5x + 2.5)$
46. $-3.28 = 8(1.5y - 0.5)$
47. $6a - 3(3a - 4) = 30$
48. $16y - 8(3y - 2) = -24$
49. $-(19 - 3s) - (8s + 1) = 35$
50. $2(3x) - 5(3x + 1) = 58$

Solve each equation and check the result. See Example 5.

51. $5x = 4x + 7$
52. $3x = 2x + 2$
53. $8y + 44 = 4y$
54. $9y + 36 = 6y$
55. $60r - 50 = 15r - 5$
56. $100f - 75 = 50f + 75$
57. $8y - 2 = 4y + 16$
58. $7 + 3w = 4 + 9w$
59. $2 - 3(x - 5) = 4(x - 1)$
60. $2 - (4x + 7) = 3 + 2(x + 2)$
61. $3(A + 2) = 2(A - 7)$
62. $9(T - 1) = 6(T + 2) - T$

Solve each equation and check the result. See Example 6.

63. $\frac{1}{8}y - \frac{1}{2} = \frac{1}{4}$
64. $\frac{1}{15}x - \frac{4}{5} = \frac{2}{3}$
65. $\frac{1}{3} = \frac{5}{6}x + \frac{2}{9}$
66. $\frac{2}{3} = -\frac{2}{3}x + \frac{3}{4}$

67. $\frac{1}{6}y + \frac{1}{4}y = -1$ **68.** $\frac{1}{3}x + \frac{1}{4}x = -2$

69. $\frac{2}{3}y + 2 = \frac{1}{5} + y$ **70.** $\frac{2}{5}x + 1 = \frac{1}{3} + x$

Solve each equation and check the result. See Example 7.

71. $0.06(s + 9) - 1.24 = -0.08s$
72. $0.08(x + 50) - 0.16x = 0.04(50)$
73. $0.09(t + 50) + 0.15t = 52.5$
74. $0.08(x - 100) = 44.5 - 0.07x$
75. $0.06(a + 200) + 0.1a = 172$
76. $0.03x + 0.05(6{,}000 - x) = 280$
77. $0.4b - 0.1(b - 100) = 70$
78. $0.105x + 0.06(20{,}000 - x) = 1{,}740$

Solve each equation and check the result. See Example 8.

79. $\frac{10 - 5s}{3} = s$ **80.** $\frac{40 - 8s}{5} = -2s$

81. $\frac{7t - 9}{16} = t$ **82.** $\frac{11r + 68}{3} = -3$

83. $\frac{5(1 - x)}{6} = -x + 1$ **84.** $\frac{3(14 - u)}{8} = -3u + 6$

85. $\frac{3(d - 8)}{4} = \frac{2(d + 1)}{3}$ **86.** $\frac{3(c - 2)}{2} = \frac{2(2c + 3)}{5}$

Solve each equation, if possible. See Examples 9–10.

87. $8x + 3(2 - x) = 5x + 6$
88. $5(x + 2) = 5x - 2$
89. $-3(s + 2) = -2(s + 4) - s$
90. $21(b - 1) + 3 = 3(7b - 6)$
91. $2(3z + 4) = 2(3z - 2) + 13$
92. $x + 7 = \frac{2x + 6}{2} + 4$
93. $4(y - 3) - y = 3(y - 4)$
94. $5(x + 3) - 3x = 2(x + 8)$

TRY IT YOURSELF

Solve each equation, if possible. Check the result.

95. $3x - 8 - 4x - 7x = -2 - 8$
96. $-6t - 7t - 5t - 1 = 12 - 3$
97. $0.05a + 0.01(90) = 0.02(a + 90)$
98. $0.03x + 0.05(2{,}000 - x) = 99.5$
99. $\frac{3(b + 2)}{2} = \frac{4b - 10}{4}$

100. $\frac{2(5a - 7)}{4} = \frac{9(a - 1)}{3}$

101. $4(a - 3) = -2(a - 6) + 6a$
102. $9(t + 2) = -6(t - 3) + 15t$
103. $10 - 2y = 8$ **104.** $7 - 7x = -21$

105. $2n - \frac{3}{4}n = \frac{1}{2}n + \frac{13}{3}$ **106.** $\frac{5}{6}n + 3n = -\frac{1}{3}n - \frac{11}{9}$

107. $-\frac{2}{3}z + 4 = 8$ **108.** $-\frac{7}{5}x + 9 = -5$

109. $-2(9 - 3s) - (5s + 2) = -25$
110. $4(x - 5) - 3(12 - x) = 7$

WRITING

111. To solve $3x - 4 = 5x + 1$, one student began by subtracting $3x$ from both sides. Another student solved the same equation by first subtracting $5x$ from both sides. Will the students get the same solution? Explain why or why not.

112. What does it mean to clear an equation such as $\frac{1}{4} + \frac{1}{2}x = \frac{3}{8}$ of the fractions?

113. Explain the error in the following work.

Solve: $2x + 4 = 30$
$\frac{2x}{2} + 4 = \frac{30}{2}$
$x + 4 = 15$
$x + 4 - 4 = 15 - 4$
$x = 11$

114. Write an equation that is an identity. Explain why every real number is a solution.

REVIEW

Name the property that is used.

115. $x \cdot 9 = 9x$

116. $4 \cdot \frac{1}{4} = 1$

117. $(x + 1) + 2 = x + (1 + 2)$

118. $2(30y) = (2 \cdot 30)y$

Objectives

1. Change percents to decimals and decimals to percents.
2. Solve percent problems by direct translation.
3. Solve applied percent problems.
4. Find percent of increase and decrease.
5. Solve discount and commission problems.

SECTION 2.3
Applications of Percent

In this section, we will use translation skills from Chapter 1 and equation-solving skills from the first two sections of Chapter 2 to solve problems involving percents.

1 Change percents to decimals and decimals to percents.

The word **percent** means parts per one hundred. We can think of the percent symbol % as representing a denominator of 100. Thus, $93\% = \frac{93}{100}$. Since the fraction $\frac{93}{100}$ is equal to the decimal 0.93, it is also true that $93\% = 0.93$.

When solving percent problems, we must often convert percents to decimals and decimals to percents. To change a percent to a decimal, we *divide by 100 by moving the decimal point 2 places to the left and dropping the % symbol*. For example,

$$31\% = 31.0\% = 0.31$$

To change a decimal to a percent, we *multiply the decimal by 100 by moving the decimal point 2 places to the right, and inserting a % symbol*. For example,

$$0.678 = 67.8\%$$

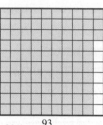

93% or $\frac{93}{100}$ or 0.93 of the figure is shaded.

2 Solve percent problems by direct translation.

There are three basic types of percent problems. Examples of these are:

Type 1 What number is 8% of 215?
Type 2 102 is 21.3% of what number?
Type 3 31 is what percent of 500?

Every percent problem has three parts: the *amount*, the *percent*, and the *base*. For example, in the question *What number is 8% of 215?*, the words "what number" represent the **amount**, 8% represents the **percent**, and 215 represents the **base**. In these problems, the word "is" means "is equal to," and the word "of" means "multiplication."

What number	is	8%	of	215?
↓	↓	↓	↓	↓
Amount	=	Percent	·	base

Self Check 1
What number is 5.6% of 40?
Now Try Problem 13

EXAMPLE 1 What number is 8% of 215?

Strategy We will translate the words of this problem into an equation and then solve the equation.

WHY The variable in the translation equation represents the unknown number that we are asked to find.

Solution
In this problem, the phrase "what number" represents the amount, 8% is the percent, and 215 is the base.

What number	is	8%	of	215?
↓	↓	↓	↓	↓
x	$=$	0.08	$\cdot$	215

Change the percent to a decimal: $8\% = 0.08$.

$x = 17.2$ Do the multiplication.

Thus, 8% of 215 is 17.2.

To check, we note that 17.2 out of 215 is $\frac{17.2}{215} = 0.08 = 8\%$.

The Language of Algebra Translate the word
- *is* to an equal symbol =
- *of* to multiplication
- *what* to a variable

We will illustrate the other two types of percent problems with application problems.

3 Solve applied percent problems.

One method for solving applied percent problems is to use the given facts to write a **percent sentence** of the form

_____ is _____ % of _____ ?

We enter the appropriate numbers in two of the blanks and the words "what" or "what number" in the remaining blank. As before, we translate the words into an equation and then solve it.

EXAMPLE 2   Aging Populations

By the year 2060, the U.S. Bureau of the Census predicts that about 92 million residents will be age 65 or older. The **circle graph** (or **pie chart**) indicates that age group will make up 21.9% of the population. If this prediction is correct, find the population of the United States in 2060. (Round to the nearest million.)

Strategy To find the predicted U.S. population in 2060, we will translate the words of the problem into an equation and then solve the equation.

WHY The variable in the translation equation represents the unknown population in 2060 that we are asked to find.

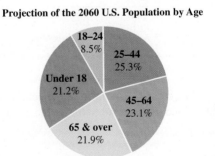

Projection of the 2060 U.S. Population by Age

- 18–24 8.5%
- 25–44 25.3%
- Under 18 21.2%
- 45–64 23.1%
- 65 & over 21.9%

Source: U.S. Bureau of the Census (2012).

Self Check 2

By the year 2050, it is predicted that 86 million, or 21.5%, of U.S. residents will be under 18 years old. If the prediction is correct, find the population in 2050. (Round to the nearest million.)

Now Try Problem 17

Solution

In this problem, 92 is the amount, 21.9% is the percent, and the words "what number" represent the base. The units are in millions.

$$92 \text{ is } 21.9\% \text{ of what number?}$$

$$92 = 0.219 \cdot x$$

$$\frac{92}{0.219} = \frac{0.219x}{0.219} \quad \text{To undo the multiplication by 0.219, divide both sides by 0.219.}$$

$$420.09 \approx x \quad \text{Do the divisions.}$$

$$420 \approx x \quad \text{Round 420.09. The units are millions.}$$

The U.S. population is predicted to be about 420 million in the year 2060. We can check using estimation: 92 million out of a population of 420 million is approximately $\frac{90 \text{ million}}{450 \text{ million}}$, or $\frac{1}{5}$, which is 20%. Since this is close to 21.9%, the answer 420 million seems reasonable.

We pay many types of taxes in our daily lives, such as sales tax, gasoline tax, income tax, and Social Security tax. **Tax rates** are usually expressed as percents.

Self Check 3

The maid mentioned in Example 3 also has $7.25 of Medicare tax deducted from her weekly paycheck. Find her Medicare tax rate.

Now Try Problem 21

EXAMPLE 3
Taxes A maid makes $500 a week. One of the deductions from her weekly paycheck is a Social Security tax of $31. Find her Social Security tax rate.

Strategy To find the tax rate, we will translate the words of the problem into an equation and then solve the equation.

WHY The variable in the translation equation represents the unknown tax rate that we are asked to find.

Solution

$$31 \text{ is what percent of } 500?$$

$$31 = x \cdot 500 \quad \text{31 is the amount, } x \text{ is the percent, and 500 is the base.}$$

$$\frac{31}{500} = \frac{500x}{500} \quad \text{To undo the multiplication by 500, divide both sides by 500.}$$

$$0.062 = x \quad \text{Do the divisions.}$$

$$6.2\% = x \quad \text{Change the decimal 0.062 to a percent.}$$

The Social Security tax rate is 6.2%.

We can use estimation to check: $31 out of $500 is about $\frac{30}{500}$ or $\frac{6}{100}$, which is 6%. Since this is close to 6.2%, the answer seems reasonable.

4 Find percent of increase and decrease.

Percents are often used to describe how a quantity has changed. For example, a health care provider might increase the cost of medical insurance by 3%, or a police department might decrease the number of officers assigned to street patrols by 10%. To describe such changes, we use **percent of increase** or **percent of decrease**.

EXAMPLE 4 Identity Theft
The Federal Trade Commission receives complaints involving the theft of someone's identity information, such as a credit card, Social Security number, or cell phone account. Refer to the data in the table. What was the percent of increase in the number of complaints from 2002 to 2012? (Round to the nearest percent.)

Year	2002	2012
Number of complaints	161,977	369,132

Source: Consumer Sentinel Network Data Book, 2012

Self Check 4
In 2004, there were 246,909 complaints of identity theft. Find the percent increase from 2004 to 2012. (Round to the nearest percent.)

Now Try Problem 43

Strategy First, we will subtract to find the *amount of increase* in the number of complaints. Then we will translate the words of the problem into an equation and solve it.

WHY A percent of increase problem involves finding the *percent of change*, and the change in a quantity is found using subtraction.

Solution
To find the *amount of increase*, we subtract the earlier value (161,977) from the later value (369,132).

$$369,132 - 161,977 = 207,155$$

We know that an increase of 207,155 is some unknown percent of the number of complaints in 2002, which was 161,977.

207,155	is	what percent	of	161,977
↓	↓	↓	↓	↓
207,155	=	x	·	161,977

207,155 is the amount, x is the percent, and 161,977 is the base.

$$\frac{207,155}{161,977} = \frac{161,977x}{161,977}$$ To undo the multiplication by 161,977, divide both sides by 161,977.

$$1.2789 \approx x$$ Do the divisions.

$$127.89\% \approx x$$ Change 1.2789 to a percent.

Rounding 127.89% to the nearest percent, we find that the number of identity theft complaints increased by about 128% from 2002 to 2012.

A 100% increase would be twice the number of complaints: 2(161,977) = 323,954. It seems reasonable that 369,132 complaints is a 128% increase.

Caution! The percent of increase (or decrease) is a percent of the *original* number—that is, the number before the change occurred.

5 Solve discount and commission problems.

When the price of an item is reduced, we call the amount of the reduction a **discount**. If a discount is expressed as a percent, it is called the **rate of discount**.

Self Check 5

A shopper saved $6 on a pen that was discounted 5%. Find the original cost.

Now Try Problem 51

EXAMPLE 5 Health Club Discounts

A 30% discount on a 1-year membership for a fitness center amounted to a $90 savings. Find the cost of a 1-year membership before the discount.

Strategy We will translate the words of the problem into an equation and then solve the equation.

WHY The variable in the translation equation represents the unknown cost of a 1-year membership before the discount that we are asked to find.

Solution

We are told that $90 is 30% of some unknown membership cost.

90	is	30%	of	what number?
↓	↓	↓	↓	↓
90	=	0.30	·	x

90 is the amount, 30% is the percent, and x is the base.

$$\frac{90}{0.30} = \frac{0.30x}{0.30}$$ To undo the multiplication by 0.30, divide both sides by 0.30.

$$300 = x$$ Do the divisions.

A one-year membership cost $300 before the discount.

Instead of working for a salary or at an hourly rate, many salespeople are paid on **commission**. The person is paid a percent of the value of the goods or services that he or she sells. We call that percent the **rate of commission**.

Self Check 6

A jewelry store clerk receives a 4.5% commission on all sales. What was the price of a gold necklace sold by the clerk if his commission was $15.75?

Now Try Problem 53

EXAMPLE 6 Commissions

A real estate agent earned $14,025 for selling a house. If she received a $5\frac{1}{2}$% commission, what was the selling price?

Strategy We will translate the words of the problem into an equation and then solve the equation.

WHY The variable in the translation equation represents the unknown selling price of the house that we are asked to find.

Solution

We are told that $14,025 is $5\frac{1}{2}$% of some unknown selling price of a house.

$14,025	is	5.5%	of	what number?	Write $5\frac{1}{2}$% as 5.5%.
↓	↓	↓	↓	↓	
14,025	=	0.055	·	x	14,025 is the amount, 5.5% is the percent, and x is the base.

$$\frac{14,025}{0.055} = \frac{0.055x}{0.055}$$ To undo the multiplication by 0.055, divide both sides by 0.055.

$$255,000 = x$$ Do the divisions.

The selling price of the house was $255,000.

ANSWERS TO SELF CHECKS

1. 2.24 **2.** 400 million **3.** 1.45% **4.** 50% **5.** $120 **6.** $350

THINK IT THROUGH   Percent of Increases in College Costs

"The evidence is overwhelming that college is a better investment for most graduates than in the past. A new study even shows that a bachelor's degree pays off for jobs that don't require one: secretaries, plumbers and cashiers. And, beyond money, education seems to make people happier and healthier."

"Even for Cashiers, College Pays Off" by David Leonhardt, *The New York Times Sunday Review*, June 25, 2011.

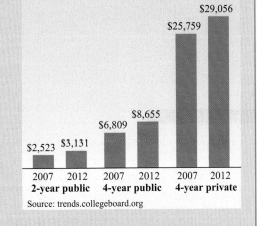

Source: trends.collegeboard.org

Like most everything else, college costs continue to rise. The illustration shows how tuition and fees at 2-year public, 4-year public, and 4-year private institutions have increased from 2007 to 2012. Find the percent of increase in tuition and fees for each type of institution. Round to the nearest one percent.

SECTION 2.3   STUDY SET

VOCABULARY

Fill in the blanks.

1. _____ means parts per one hundred.
2. In the statement "10 is 50% of 20," 10 is the _____, 50% is the percent, and 20 is the _____.
3. In percent questions, the word *of* means _____, and ___ means equals.
4. An employee who is paid a _____ is paid a percent of the value of the goods or services that he or she sells.

CONCEPTS

5. Represent the amount of the figure that is shaded using a fraction, a decimal, and a percent.

6. Fill in the blank: To solve a percent problem, we translate the words of the problem into an _____ and solve it.

7.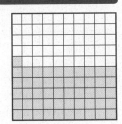

   Source: National Federation of State High School Associations

   a. Find the *amount* of increase in participation.
   b. Fill in the blanks to find the percent of increase in participation: _____ is _____ % of _____?

8. Fill in the blanks using the words *percent*, *amount*, and *base*.

    = ·

9. Translate each sentence into an equation. **Do not solve.**
   a. 12 is 40% of what number?
   b. 99 is what percent of 200?
   c. What is 66% of 3?

10. Use estimation to determine if each statement is reasonable.
    a. 18 is 48% of 93.
    b. 47 is 6% of 206.

NOTATION

11. Change each percent to a decimal.
    a. 35%   b. 8.5%
    c. 150%   d. $2\frac{3}{4}$%

12. Change each decimal to a percent.
    a. 0.9   b. 9
    c. 0.999

GUIDED PRACTICE

See Examples 1–3.

13. What number is 48% of 650?
14. What number is 60% of 200?
15. 78 is what percent of 300?
16. 143 is what percent of 325?
17. 75 is 25% of what number?
18. 78 is 6% of what number?
19. What number is 92.4% of 50?
20. What number is 2.8% of 220?
21. 0.42 is what percent of 16.8?
22. 199.92 is what percent of 2,352?
23. 128.1 is 8.75% of what number?
24. 1.12 is 140% of what number?

APPLICATIONS

25. ANTISEPTICS Use the facts on the label to determine the amount of pure hydrogen peroxide in the bottle.

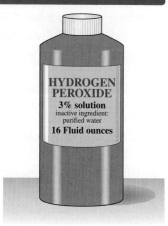

26. DINING OUT Refer to the sales receipt in the next column. Compute the 15% tip (*rounded up* to the nearest dollar). Then find the total cost of the meal.

27. U.S. FEDERAL BUDGET The circle graph shows how the government spent $3,603 billion in 2011. How much was spent on
    a. Social Security/Medicare?
    b. Defense/Veterans?

Based on 2012 Federal Income Tax Form 1040

28. TAX TABLES Use the table to compute the amount of federal income tax to be paid on an income of $39,910.

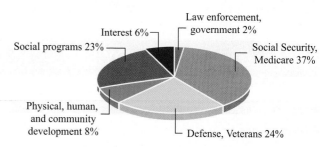

29. PAYPAL Many e-commerce businesses use PayPal to process payments for them. For certain transactions, merchants are charged a fee of 2.9% of the selling price of the item plus $0.30. What would PayPal charge an online art store to collect payment on a painting selling for $350?

30. eBAY When a student sold an Xbox 360 on eBay for $153, she was charged a final value fee of 10% of the total amount of the sale. What profit will the student make from the sale of the Xbox 360?

31. **PRICE GUARANTEES** Home Club offers a "10% Plus" guarantee: If the customer finds the same item selling for less somewhere else, he or she receives the difference in price plus 10% of the difference. A woman bought miniblinds at the Home Club for $120 but later saw the same blinds on sale for $98 at another store. How much can she expect to be reimbursed?

32. **ROOM TAXES** A guest at the San Antonio Hilton Airport Hotel paid $180 for a room plus a 9% city room tax, a $1\frac{3}{4}$% county room tax, and a 6% state room tax. Find the total amount of tax that the guest paid on the room.

33. **COMPUTER MEMORY** The My Computer screen on a student's computer is shown. What percent of the memory on the hard drive Local Disk (C:) of his computer is used? What percent is free? (GB stands for gigabytes.)

34. **GENEALOGY** Through an extensive computer search, a genealogist determined that worldwide, 180 out of every 10 million people had his last name. What percent is this?

35. **DENTISTRY** Refer to the dental record. What percent of the patient's teeth have fillings? Round to the nearest percent.

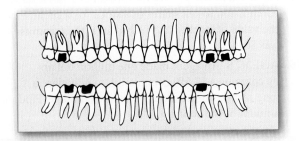

36. **TEST SCORES** The score 175/200 was written by an algebra instructor at the top of a student's test paper. Write the test score as a percent.

37. **DMV WRITTEN TEST** To obtain a learner's permit to drive in Nevada, a score of 80% (or better) on a 50-question multiple-choice test is required. If a teenager answered 33 questions correctly, did he pass the test?

38. **iPHONE** The About screen of an Apple iPhone is shown in the next column. What percent of the capacity is still available? Round to the nearest percent. (GB stands for gigabytes.)

39. **CHILD CARE** After the first day of registration, 84 children had been enrolled in a day-care center. That represented 70% of the available slots. Find the maximum number of children the center could enroll.

40. **RACING PROGRAMS** One month before a stock car race, the sale of ads for the official race program was slow. Only 12 pages, or just 30% of the available pages, had been sold. Find the total number of pages devoted to advertising in the program.

41. **NUTRITION** The Nutrition Facts label from a can of clam chowder is shown.

    a. Find the number of grams of saturated fat in one serving. What percent of a person's recommended daily intake is this?

    b. Determine the recommended number of grams of saturated fat that a person should consume daily.

Nutrition Facts
Serving Size 1 cup (240mL)
Servings Per Container about 2

Amount per serving	
Calories 240 Calories from Fat 140	
	% Daily Value*
Total Fat 15 g	23%
Saturated Fat 5 g	25%
Cholesterol 10 mg	3%
Sodium 980 mg	41%
Total Carbohydrate 21 g	7%
Dietary Fiber 2 g	8%
Sugars 1 g	
Protein 7 g	

42. **COMMERCIALS** Jared Fogle credits his tremendous weight loss to exercise and a diet of low-fat Subway sandwiches. His current weight (about 187 pounds) is 44% of his maximum weight (reached in March of 1998). What did he weigh then?

43. **EXPORTS** According to the graph, between what two years was there a decrease in U.S. exports to Mexico? Find the percent of decrease.

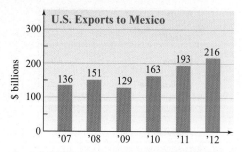

Based on data from www.census.gov/foreign-trade/balance

44. **AUCTIONS** A pearl necklace of former First Lady Jacqueline Kennedy Onassis, originally valued at $700, was sold at auction for $211,500. Find the percent of increase in the value of the necklace. (Round to the nearest percent.)

45. **INSURANCE COSTS** A college student's good grades earned her a student discount on her car insurance premium. Find the percent of decrease to the nearest percent if her annual premium was lowered from $1,050 to $925.

46. **U.S. LIFE EXPECTANCY** Use the following life expectancy data for 1900 and 2011 to find the percent of increase for males and for females. Round to the nearest percent.

	Years of life expected at birth	
	Male	Female
1900	46.3	48.3
2011	76.3	81.1

Source: National Center for Health Statistics

47. **TALK RADIO** Refer to the table and find the percent increase in the number of news/talk radio stations from 2004 to 2012. Round to the nearest percent.

Number of U.S. news/talk radio stations	
2004: 1,285	**2012:** 1,503

Source: *The World Almanac and Book of Facts, 2013*

48. **FOOD LABELS** To be labeled "Reduced Fat," foods must contain at least 25% less fat per serving than the regular product. One serving of the original Jif peanut butter has 16 grams of fat per serving. The new Jif Reduced Fat product contains 12 grams of fat per serving. Does it meet the labeling requirement?

49. **TV SHOPPING** Jan bought a toy from the QVC home shopping network that was discounted 20%. If she saved $15, what was the original price of the toy?

50. **DISCOUNTS** A 12% discount on a watch saved a shopper $48. Find the price of the watch before the discount.

51. **SALES** The price of a certain model patio set was reduced 35% because it was being discontinued. A shopper purchased two of them and saved a total of $210. Find the price of a patio set before the discount.

52. **TV SALES** The price of a plasma screen television was reduced $800 because it was used as a floor model. If this was a 40% savings, find the original price of the TV.

53. **REAL ESTATE** The $3\frac{1}{2}$% commission paid to a real estate agent on the sale of a condominium earned her $3,325. Find the selling price of the condo.

54. **CONSIGNMENT** An art gallery agreed to sell an artist's sculpture for a commission of 45%. What must be the selling price of the sculpture if the gallery would like to make $13,500?

55. **STOCKBROKERS** A stockbroker charges a 2.5% commission to sell shares of a stock for a client. Find the value of stock sold by a broker if the commission was $640.

56. **AGENTS** An agent made one million dollars by charging a 12.5% commission to negotiate a long-term contract for a professional athlete. Find the amount of the contract.

WRITING

57. Explain the error:

What is 5% of 8?
$x = 5 \cdot 8$
$x = 40$
40 is 5% of 8.

58. Write a real-life situation that could be described by "9 is what percent of 20?"
59. Explain why 150% of a number is more than the number.
60. Why is the problem "What is 9% of 100?" easy to solve?

REVIEW

61. Divide: $-\dfrac{16}{25} \div \left(-\dfrac{4}{15}\right)$
62. What two numbers are a distance of 8 away from 4 on the number line?
63. Is -34 a solution of $x + 15 = -49$?
64. Evaluate: $2 + 3[24 - 2(2 - 5)]$

SECTION 2.4
Formulas

Objectives
1. Use formulas from business.
2. Use formulas from science.
3. Use formulas from geometry.
4. Solve for a specified variable.

A **formula** is an equation that states a relationship between two or more variables. Formulas are used in fields such as business, science, and geometry.

1 Use formulas from business.

A formula for retail price: To make a profit, a merchant must sell an item for more than he or she paid for it. The price at which the merchant sells the product, called the **retail price**, is the *sum* of what the item cost the merchant plus the **markup**. Using r to represent the retail price, c the cost, and m the markup, we can write this formula as

$$r = c + m \quad \text{Retail price = cost + markup}$$

A formula for profit: The **profit** a business makes is the *difference* between the **revenue** (the money it takes in) and the cost. Using p to represent the profit, r the revenue, and c the cost, we can write this formula as

$$p = r - c \quad \text{Profit = revenue − cost}$$

If we are given the values of all but one of the variables in a formula, we can use our equation-solving skills to find the value of the remaining variable.

EXAMPLE 1 *Films* Estimates are that Warner Brothers made a $219 million profit on the film *Harry Potter and the Half-Blood Prince*. If the studio received $469 million in worldwide box office revenue, find the cost to make and distribute the film. (Source: www.thenumbers.com, June 2010)

Strategy To find the cost to make and distribute the film, we will substitute the given values in the formula $p = r - c$ and solve for c.

WHY The variable c in the formula represents the unknown cost.

Solution
The film made $219 million (the profit p) and the studio took in $469 million (the revenue r). To find the cost c, we proceed as follows.

$$p = r - c \quad \text{This is the formula for profit.}$$
$$219 = 469 - c \quad \text{Substitute 219 for } p \text{ and 469 for } r.$$
$$219 - 469 = 469 - c - 469 \quad \text{To eliminate 469 on the right side, subtract 469 from both sides.}$$

Self Check 1

A PTA spaghetti dinner made a profit of $275.50. If the cost to host the dinner was $1,235, how much revenue did it generate?

Now Try Problem 11

$$-250 = -c \quad \text{Do the subtractions.}$$
$$\frac{-250}{-1} = \frac{-c}{-1} \quad \text{To solve for } c, \text{ divide (or multiply) both sides by } -1.$$
$$250 = c \quad \text{The units are millions of dollars.}$$

It cost $250 million to make and distribute the film.

A formula for simple interest: When money is borrowed, the lender expects to be paid back the amount of the loan plus an additional charge for the use of the money, called **interest**. When money is deposited in a bank, the depositor is paid for the use of the money. The money the depositor earns is also called interest.

Interest is computed in two ways: either as **simple interest** or as **compound interest**. Simple interest is the *product* of the principal (the amount of money that is invested, deposited, or borrowed), the annual interest rate, and the length of time in years. Using I to represent the simple interest, P the principal, r the annual interest rate, and t the time in years, we can write this formula as

$$I = Prt \quad \text{Interest} = \text{principal} \cdot \text{rate} \cdot \text{time}$$

> **The Language of Algebra** The word *annual* means occurring once a year. An *annual* interest rate is the interest rate paid per year.

Self Check 2

A father loaned his daughter $12,200 at a 2% annual simple interest rate for a down payment on a house. If the interest on the loan amounted to $610, for how long was the loan?

Now Try Problem 15

EXAMPLE 2 *Retirement Income* One year after investing $15,000, a retired couple received a check for $1,125 in interest. Find the interest rate their money earned that year.

Strategy To find the interest rate, we will substitute the given values in the formula $I = Prt$ and solve for r.

WHY The variable r represents the unknown interest rate.

Solution

The couple invested $15,000 (the principal P) for 1 year (the time t) and made $1,125 (the interest I). To find the annual interest rate r, we proceed as follows.

$$I = Prt \quad \text{This is the formula for simple interest.}$$
$$1{,}125 = 15{,}000r(1) \quad \text{Substitute 1,125 for } I, \text{ 15,000 for } P, \text{ and 1 for } t.$$
$$1{,}125 = 15{,}000r \quad \text{Simplify the right side.}$$
$$\frac{1{,}125}{15{,}000} = \frac{15{,}000r}{15{,}000} \quad \text{To solve for } r, \text{ undo the multiplication by 15,000 by dividing both sides by 15,000.}$$
$$0.075 = r \quad \text{Do the divisions.}$$
$$7.5\% = r \quad \text{To write 0.075 as a percent, multiply 0.075 by 100 by moving the decimal point two places to the right and inserting a \% symbol.}$$

The couple received an annual rate of 7.5% that year on their investment. We can display the facts of the problem in a table.

	P	$\cdot$ r	$\cdot t =$	I
Investment	15,000	0.075	1	1,125

Caution! When using the formula $I = Prt$, always write the interest rate r (which is given as a percent) as a decimal (or fraction) before performing any calculations.

2 Use formulas from science.

A formula for distance traveled: If we know the average rate (of speed) at which we will be traveling and the time we will be traveling at that rate, we can find the distance traveled. Using d to represent the distance, r the average rate, and t the time, we can write this formula as

$$d = rt \quad \text{Distance} = \text{rate} \cdot \text{time}$$

EXAMPLE 3 **Whales** As they migrate from the Bering Sea to Baja California, gray whales swim for about 20 hours each day, covering a distance of approximately 70 miles. Estimate their average swimming rate in miles per hour (mph).

Strategy To find the swimming rate, we will substitute the given values in the formula $d = rt$ and solve for r.

WHY The variable r represents the unknown average swimming rate.

Solution
The whales swam 70 miles (the distance d) in 20 hours (the time t). To find their average swimming rate r, we proceed as follows.

$d = rt$ This is the formula for distance traveled.

$70 = r(20)$ Substitute 70 for d and 20 for t.

$\dfrac{70}{20} = \dfrac{20r}{20}$ To solve for r, undo the multiplication by 20 by dividing both sides by 20.

$3.5 = r$ Do the divisions.

The whales' average swimming rate is 3.5 mph. The facts of the problem can be shown in a table.

	r	t	= d
Gray whale	3.5	20	70

Self Check 3
An elevator travels at an average rate of 288 feet per minute. How long will it take the elevator to climb 30 stories, a distance of 360 feet?

Now Try Problem 19

A formula for converting temperatures: In the American system, temperature is measured on the Fahrenheit scale. The Celsius scale is used to measure temperature in the metric system. The formula that relates a Fahrenheit temperature F to a Celsius temperature C is:

$$C = \frac{5}{9}(F - 32)$$

EXAMPLE 4 Convert the temperature shown on the City Savings sign to degrees Fahrenheit.

Strategy To find the temperature in degrees Fahrenheit, we will substitute the given Celsius temperature in the formula $C = \frac{5}{9}(F - 32)$ and solve for F.

Self Check 4
Change $-175°C$, the temperature on Saturn, to degrees Fahrenheit.

Now Try Problem 25

WHY The variable F represents the unknown temperature in degrees Fahrenheit.

Solution

The temperature in degrees Celsius is 30°. To find the temperature in degrees Fahrenheit F, we proceed as follows.

$$C = \frac{5}{9}(F - 32)$$ This is the formula for temperature conversion.

$$30 = \frac{5}{9}(F - 32)$$ Substitute 30 for C, the Celsius temperature.

$$\frac{9}{5} \cdot 30 = \frac{9}{5} \cdot \frac{5}{9}(F - 32)$$ To undo the multiplication by $\frac{5}{9}$, multiply both sides by the reciprocal of $\frac{5}{9}$.

$$54 = F - 32$$ Do the multiplications.

$$54 + 32 = F - 32 + 32$$ To isolate F, undo the subtraction of 32 by adding 32 to both sides.

$$86 = F$$

30°C is equivalent to 86°F.

3 Use formulas from geometry.

To find the **perimeter** of a plane (two-dimensional, flat) geometric figure, such as a rectangle or triangle, we find the distance around the figure by computing the sum of the lengths of its sides. Perimeter is measured in American units, such as inches, feet, yards, and in metric units such as millimeters, meters, and kilometers.

Perimeter Formulas

Rectangle: $P = 2l + 2w$

Square: $P = 4s$

Triangle: $P = a + b + c$

The Language of Algebra When you hear the word *perimeter*, think of the distance around the "rim" of a flat figure.

Self Check 5

The largest flag that consistently flies is the flag of Brazil in Brasilia, the country's capital. It has a perimeter of 1,116 feet and a length of 328 feet. Find its width.

Now Try Problem 27

EXAMPLE 5 *Flags* The largest flag ever flown was an American flag that had a perimeter of 1,520 feet and a length of 505 feet. It was hoisted on cables across Hoover Dam to celebrate the 1996 Olympic Torch Relay. Find the width of the flag.

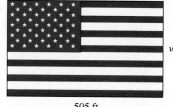

505 ft

Strategy To find the width of the flag, we will substitute the given values in the formula $P = 2l + 2w$ and solve for w.

WHY The variable w represents the unknown width of the flag.

Solution

The perimeter P of the rectangular flag is 1,520 ft, and the length l is 505 ft. To find the width w, we proceed as follows.

2.4 Formulas

$P = 2l + 2w$ This is the formula for the perimeter of a rectangle.
$1{,}520 = 2(505) + 2w$ Substitute 1,520 for P and 505 for l.
$1{,}520 = 1{,}010 + 2w$ Do the multiplication.
$510 = 2w$ To undo the addition of 1,010, subtract 1,010 from both sides.
$255 = w$ To isolate w, undo the multiplication by 2 by dividing both sides by 2.

The width of the flag is 255 feet. If its length is 505 feet and its width is 255 feet, its perimeter is $2(505) + 2(255) = 1{,}010 + 510 = 1{,}520$ feet, as given.

The **area** of a plane (two-dimensional, flat) geometric figure is the amount of surface that it encloses. Area is measured in square units, such as square inches, square feet, square yards, and square meters (written as in.², ft², yd², and m², respectively).

Area Formulas

Rectangle: $A = lw$

Square: $A = s^2$

Triangle: $A = \frac{1}{2}bh$

Trapezoid: $A = \frac{1}{2}h(B + b)$

Circle Formulas

Diameter: $D = 2r$

Radius: $r = \frac{1}{2}D$

Circumference: $C = \pi D$ or $C = 2\pi r$

Area: $A = \pi r^2$

EXAMPLE 6
a. What is the circumference of a circle with diameter of 14 feet? Round to the nearest tenth of a foot. **b.** What is the area of the circle? Round to the nearest tenth of a square foot.

Strategy To find the circumference and area of the circle, we will substitute the proper values into the formulas $C = \pi D$ and $A = \pi r^2$ and find C and A.

WHY The variable C represents the unknown circumference of the circle and A represents the unknown area.

Solution
a. Recall that the circumference of a circle is the distance around it. To find the circumference C of a circle with diameter D equal to 14 ft, we proceed as follows.

$C = \pi D$ This is the formula for the circumference of a circle. πD means $\pi \cdot D$.

$C = \pi(14)$ Substitute 14 for D, the diameter of the circle.

$C = 14\pi$ The exact circumference of the circle is 14π.

Self Check 6
Find the circumference of a circle with a radius of 10 inches. Round to the nearest hundredth of an inch.

Now Try Problem 28

$C \approx 43.98229715$ To use a scientific calculator to approximate the circumference, enter π × 14 = . If you do not have a calculator, use 3.14 as an approximation of π. (Answers may vary slightly depending on which approximation of π is used.)

The circumference is exactly 14π ft. Rounded to the nearest tenth, this is 44.0 ft.

b. The radius r of the circle is one-half the diameter, or 7 feet. To find the area A of the circle, we proceed as follows.

$A = \pi r^2$ This is the formula for the area of a circle. πr^2 means $\pi \cdot r^2$.

$A = \pi(7)^2$ Substitute 7 for r, the radius of the circle.

$A = 49\pi$ Evaluate the exponential expression: $7^2 = 49$. The exact area is 49π ft².

$A \approx 153.93804$ To use a calculator to approximate the area, enter 49 × π = .

The area is exactly 49π ft². To the nearest tenth, the area is 153.9 ft².

Caution! When an approximation of π is used in a calculation, it produces an approximate answer. Remember to use an *is approximately equal to* symbol $\approx$ in your solution to show that.

The **volume** of a three-dimensional geometric solid is the amount of space it encloses. Volume is measured in cubic units, such as cubic inches, cubic feet, and cubic meters (written as in.³, ft³, and m³, respectively). Several volume formulas are given at the top of page 155.

Self Check 7

Find the volume of a cone whose base has a radius of 12 meters and whose height is 9 meters. Round to the nearest tenth of a cubic meter. Use the formula $V = \frac{1}{3}\pi r^2 h$.

Now Try Problem 29

EXAMPLE 7   Find the volume of the cylinder. Round to the nearest tenth of a cubic centimeter.

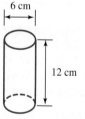

Strategy To find the volume of the cylinder, we will substitute the proper values into the formula $V = \pi r^2 h$ and find V.

WHY The variable V represents the unknown volume.

Solution

Since the radius of a circle is one-half its diameter, the radius r of the circular base of the cylinder is $\frac{1}{2}(6 \text{ cm}) = 3$ cm. The height h of the cylinder is 12 cm. To find volume V of the cylinder, we proceed as follows.

$V = \pi r^2 h$ This is the formula for the volume of a cylinder. $\pi r^2 h$ means $\pi \cdot r^2 \cdot h$.

$V = \pi(3)^2(12)$ Substitute 3 for r and 12 for h.

$V = \pi(9)(12)$ Evaluate the exponential expression.

$V = 108\pi$ Multiply. The exact volume is 108π cm³.

$V \approx 339.2920066$ Use a calculator to approximate the volume.

To the nearest tenth, the volume is 339.3 cubic centimeters. This can be written as 339.3 cm³.

Volume Formulas

Rectangular solid: $V = lwh$

Cube: $V = s^3$

Sphere: $V = \frac{4}{3}\pi r^3$

Cylinder: $V = \pi r^2 h$

Cone: $V = \frac{1}{3}\pi r^2 h$

Pyramid: $V = \frac{1}{3}Bh$

(where B is the area of the base)

4 Solve for a specified variable.

Suppose a shopper wishes to calculate the markup m on several items, knowing their retail price r and their cost c to the merchant. It would take a lot of time to substitute values for r and c into the formula for retail price $r = c + m$ and then repeatedly solve for m. A better way is to solve the formula for m first, substitute values for r and c, and then compute m directly.

To **solve a formula for a specified variable** means to isolate that variable on one side of the equation, with all other variables and constants on the opposite side.

EXAMPLE 8 Solve the formula for retail price, $r = c + m$ for m.

Strategy To solve for m, we will focus on it as if it is the only variable in the equation. We will use a strategy similar to that used to solve linear equations in one variable to isolate m on one side. (See page 135 if you need to review the strategy.)

WHY We can solve the formula as if it were an equation in one variable because all the other variables are treated as if they were numbers (constants).

Solution

$r = c + m$ To solve for m, we will isolate m on this side of the equation.

$r - c = c + m - c$ To isolate m, undo the addition of c by subtracting c from both sides.

$r - c = m$ Simplify the right side: $c - c = 0$.

$m = r - c$ Reverse the sides of the equation so that m is on the left.

Self Check 8

Solve the formula $p = r - c$ for r.

Now Try Problem 31

The Language of Algebra We say that the formula is *solved for m* because m is alone on one side of the equation and the other side does not contain m.

EXAMPLE 9 Solve $A = \frac{1}{2}bh$ for b.

Strategy To solve for b, we will treat b as the only variable in the equation and use properties of equality to isolate it on one side. We will treat the other variables as if they were numbers (constants).

WHY To solve for a specified variable means to isolate it on one side of the equation.

Solution

We use the same steps to solve an equation for a specified variable that we use to solve equations with only one variable.

Self Check 9

Solve $A = \frac{1}{2}r^2 a$ for a.

Now Try Problem 37

Chapter 2 Equations, Inequalities, and Problem Solving

$$A = \frac{1}{2}bh$$

To solve for b, we will isolate b on this side of the equation.

$$2 \cdot A = 2 \cdot \frac{1}{2}bh \quad \text{To clear the equation of the fraction, multiply both sides by 2.}$$

$$2A = bh \quad \text{Simplify.}$$

$$\frac{2A}{h} = \frac{bh}{h} \quad \text{To isolate } b, \text{ undo the multiplication by } h \text{ by dividing both sides by } h.$$

$$\frac{2A}{h} = b \quad \text{On the right side, remove the common factor of } h: \frac{b\cancel{h}}{\cancel{h}} = b.$$

$$b = \frac{2A}{h} \quad \text{Reverse the sides of the equation so that } b \text{ is on the left.}$$

Self Check 10
Solve $P = 2l + 2w$ for w.
Now Try Problem 45

EXAMPLE 10
Solve $P = 2l + 2w$ for l.

Strategy To solve for l, we will treat l as the only variable in the equation and use properties of equality to isolate it on one side. We will treat the other variables as if they were numbers (constants).

WHY To solve for a specified variable means to isolate it on one side of the equation.

Solution

To solve for l, we will isolate l on this side of the equation.

$$P = 2l + 2w$$

$$P - 2w = 2l + 2w - 2w \quad \text{To undo the addition of } 2w, \text{ subtract } 2w \text{ from both sides.}$$

$$P - 2w = 2l \quad \text{Combine like terms: } 2w - 2w = 0.$$

$$\frac{P - 2w}{2} = \frac{2l}{2} \quad \text{To isolate } l, \text{ undo the multiplication by 2 by dividing both sides by 2.}$$

$$\frac{P - 2w}{2} = l \quad \text{Simplify the right side.}$$

We can write the result as $l = \frac{P - 2w}{2}$.

Caution! Do not try to simplify the result this way:

$$l = \frac{P - \cancel{2}w}{\cancel{2}}$$

This step is incorrect because 2 is not a factor of the entire numerator.

Self Check 11
Solve $x + 3y = 12$ for y.
Now Try Problem 47

EXAMPLE 11
In Chapter 3, we will work with equations that involve the variables x and y, such as $3x + 2y = 4$. Solve this equation for y.

Strategy To solve for y, we will treat y as the only variable in the equation and use properties of equality to isolate it on one side.

WHY To solve for a specified variable means to isolate it on one side of the equation.

Solution

$$3x + 2y = 4$$

To solve for y, we will isolate y on this side of the equation.

$$3x + 2y - 3x = 4 - 3x$$

To eliminate 3x on the left side, subtract 3x from both sides.

$$2y = 4 - 3x$$

Combine like terms: $3x - 3x = 0$.

$$\frac{2y}{2} = \frac{4 - 3x}{2}$$

To isolate y, undo the multiplication by 2 by dividing both sides by 2.

$$y = \frac{4}{2} - \frac{3x}{2}$$

Write $\frac{4 - 3x}{2}$ as the difference of two fractions with like denominators, $\frac{4}{2}$ and $\frac{3x}{2}$.

$$y = 2 - \frac{3}{2}x$$

Simplify: $\frac{4}{2} = 2$. Write $\frac{3x}{2}$ as $\frac{3}{2}x$.

$$y = -\frac{3}{2}x + 2$$

On the right side, write the x term first.

Success Tip When solving for a specified variable, there is often more than one way to express the result.

EXAMPLE 12 Solve $V = \pi r^2 h$ for r^2.

Self Check 12
Solve $V = lwh$ for w.
Now Try Problem 55

Strategy To solve for r^2, we will treat it as the only variable expression in the equation and isolate it on one side.

WHY To solve for a specified variable means to isolate it on one side of the equation.

Solution

$$V = \pi r^2 h$$

To solve for r^2, we will isolate r^2 on this side of the equation.

$$\frac{V}{\pi h} = \frac{\pi r^2 h}{\pi h}$$

$\pi r^2 h$ means $\pi \cdot r^2 \cdot h$. To isolate r^2, undo the multiplication by π and h on the right side by dividing both sides by πh.

$$\frac{V}{\pi h} = r^2$$

On the right side, remove the common factors of π and h: $\frac{\cancel{\pi} r^2 \cancel{h}}{\cancel{\pi} \cancel{h}} = r^2$.

$$r^2 = \frac{V}{\pi h}$$

Reverse the sides of the equation so that r^2 is on the left.

ANSWERS TO SELF CHECKS

1. $1,510.50 2. 2.5 yr 3. 1.25 min 4. $-283°$F 5. 230 ft 6. 62.83 in.
7. 1,357.2 m^3 8. $r = p + c$ 9. $a = \frac{2A}{r^2}$ 10. $w = \frac{P - 2l}{2}$
11. $y = 4 - \frac{1}{3}x$ or $y = -\frac{1}{3}x + 4$ 12. $w = \frac{V}{lh}$

SECTION 2.4 STUDY SET

VOCABULARY

Fill in the blanks.

1. A _____ is an equation that is used to state a known relationship between two or more variables.

2. The distance around a plane geometric figure is called its _____, and the amount of surface that it encloses is called its _____.

3. The _____ of a three-dimensional geometric solid is the amount of space it encloses.

Chapter 2 Equations, Inequalities, and Problem Solving

4. The formula $a = P - b - c$ is _____ for a because a is isolated on one side of the equation and the other side does not contain a.

CONCEPTS

5. Use variables to write the formula relating:
   a. Time, distance, rate
   b. Markup, retail price, cost
   c. Costs, revenue, profit
   d. Interest rate, time, interest, principal

6. Complete the table.

	Principal · rate · time = interest		
Account 1	$2,500	5%	2 yr
Account 2	$15,000	4.8%	1 yr

7. Complete the table to find how far light and sound travel in 60 seconds. (*Hint:* mi/sec means miles per second.)

	Rate	· time =	distance
Light	186,282 mi/sec	60 sec	
Sound	1,088 ft/sec	60 sec	

8. Determine which concept (perimeter, area, or volume) should be used to find each of the following. Then determine which unit of measurement, ft, ft^2, or ft^3, would be appropriate.
   a. The amount of storage in a freezer
   b. The amount of ground covered by a sleeping bag lying on the floor
   c. The distance around a dance floor

NOTATION

Complete each step.

9. Solve $Ax + By = C$ for y.

   $$Ax + By = C$$
   $$Ax + By - \boxed{} = C - \boxed{}$$
   $$By = C - Ax$$
   $$\frac{By}{\boxed{}} = \frac{C - Ax}{\boxed{}}$$
   $$y = \frac{C - Ax}{\boxed{}}$$

10. a. Approximate 98π to the nearest hundredth.
    b. In the formula $V = \pi r^2 h$, what does r represent? What does h represent?
    c. What does 45°C mean?
    d. What does 15°F mean?

GUIDED PRACTICE

Use a formula to solve each problem. **See Example 1.**

11. **HOLLYWOOD** As of January 2013, the movie *Zero Dark Thirty* had brought in $231 million worldwide and made a gross profit of $121 million. What did it cost to make the movie? (Source: deadline.com)

12. **VALENTINE'S DAY** Find the markup on a dozen roses if a florist buys them wholesale for $12.95 and sells them for $47.50.

13. **SERVICE CLUBS** After expenses of $55.15 were paid, a Rotary Club donated $875.85 in proceeds from a pancake breakfast to a local health clinic. How much did the pancake breakfast gross?

14. **NEW CARS** The factory invoice for a minivan shows that the dealer paid $16,264.55 for the vehicle. If the sticker price of the van is $18,202, how much over factory invoice is the sticker price?

See Example 2.

15. **ENTREPRENEURS** To start a mobile dog-grooming service, a woman borrowed $2,500. If the loan was for 2 years and the amount of interest was $175, what simple interest rate was she charged?

16. **SAVINGS** A man deposited $5,000 in a credit union paying 6% simple interest. How long will the money have to be left on deposit to earn $6,000 in interest?

17. **LOANS** A student borrowed some money from his father at 2% simple interest to buy a car. If he paid his father $360 in interest after 3 years, how much did he borrow?

18. **BANKING** Three years after opening an account that paid simple interest of 6.45% annually, a depositor withdrew the $3,483 in interest earned. How much money was left in the account?

See Example 3.

19. **SWIMMING** In 1930, a man swam down the Mississippi River from Minneapolis to New Orleans, a total of 1,826 miles. He was in the water for 742 hours. To the nearest tenth, what was his average swimming rate?

20. **PARADES** Rose Parade floats travel down the 5.5-mile-long parade route at a rate of 2.5 mph. How long will it take a float to complete the route if there are no delays?

21. **HOT-AIR BALLOONS** If a hot-air balloon travels at an average of 37 mph, how long will it take to fly 166.5 miles?

22. **AIR TRAVEL** An airplane flew from Chicago to San Francisco in 3.75 hours. If the cities are 1,950 miles apart, what was the average speed of the plane?

See Example 4.

23. **FRYING FOODS** One of the most popular cookbooks in U.S. history, *The Joy of Cooking*, recommends frying foods at 365°F for best results. Convert this to degrees Celsius.

24. **FREEZING POINTS** Saltwater has a much lower freezing point than freshwater does. For saltwater that is as saturated as it can possibly get (23.3% salt by weight), the freezing point is −5.8°F. Convert this to degrees Celsius.

25. **BIOLOGY** Cryobiologists freeze living matter to preserve it for future use. They can work with temperatures as low as −270°C. Change this to degrees Fahrenheit.

26. **METALLURGY** Change 2,212°C, the temperature at which silver boils, to degrees Fahrenheit. Round to the nearest degree.

See Examples 5–7. *If you do not have a calculator, use 3.14 as an approximation of π. Answers may vary slightly depending on which approximation of π is used.*

27. **ENERGY SAVINGS** One hundred inches of foam weather stripping tape was placed around the perimeter of a rectangular window. If the length of the window is 30 inches, what is its width?

28. **RUGS** Find the amount of floor area covered by a circular throw rug that has a radius of 15 inches. Round to the nearest square inch.

29. **STRAWS** Find the volume of a 150 millimeter-long drinking straw that has an inside diameter of 4 millimeters. Round to the nearest cubic millimeter.

30. **RUBBER BANDS** The world's largest rubber band ball is $5\frac{1}{2}$ ft tall and was made in 2006 by Steve Milton of Eugene, Oregon. Find the volume of the ball. Round to the nearest cubic foot. (*Hint:* The formula for the volume of a sphere is $V = \frac{4}{3}\pi r^3$.)

Solve each formula for the specified variable. **See Example 8.**

31. $r = c + m$ for c
32. $p = r - c$ for c
33. $P = a + b + c$ for b
34. $a + b + c = 180$ for a

Solve each formula for the specified variable. **See Example 9.**

35. $E = IR$ for R
36. $d = rt$ for t
37. $V = lwh$ for l
38. $I = Prt$ for r
39. $C = 2\pi r$ for r
40. $V = \pi r^2 h$ for h

41. $V = \frac{1}{3}Bh$ for h
42. $C = \frac{1}{7}Rt$ for R
43. $w = \frac{s}{f}$ for f
44. $P = \frac{ab}{c}$ for c

Solve each formula for the specified variable. **See Examples 10 and 11.**

45. $T = 2r + 2t$ for r
46. $y = mx + b$ for x
47. $Ax + By = C$ for x
48. $A = P + Prt$ for t
49. $K = \frac{1}{2}mv^2$ for m
50. $V = \frac{1}{3}\pi r^2 h$ for h
51. $A = \frac{a+b+c}{3}$ for c
52. $x = \frac{a+b}{2}$ for b
53. $2E = \frac{T-t}{9}$ for t
54. $D = \frac{C-s}{n}$ for s

Solve each equation for the specified variable (or expression). **See Example 12.**

55. $s = 4\pi r^2$ for r^2
56. $E = mc^2$ for c^2
57. $Kg = \frac{wv^2}{2}$ for v^2
58. $c^2 = a^2 + b^2$ for a^2

TRY IT YOURSELF

Solve each equation for the specified variable (or expression).

59. $V = \frac{4}{3}\pi r^3$ for r^3
60. $A = \frac{\pi r^2 S}{360}$ for r^2
61. $\frac{M}{2} - 9.9 = 2.1B$ for M
62. $\frac{G}{0.5} + 16r = -8t$ for G
63. $S = 2\pi rh + 2\pi r^2$ for h
64. $c = bn + 16t^2$ for t^2
65. $3x + y = 9$ for y
66. $-5x + y = 4$ for y
67. $-x + 3y = 9$ for y
68. $5y - x = 25$ for y
69. $4y + 16 = -3x$ for y
70. $6y + 12 = -5x$ for y
71. $A = \frac{1}{2}h(b + d)$ for b
72. $C = \frac{1}{4}s(t - d)$ for t
73. $\frac{7}{8}c + w = 9$ for c
74. $\frac{3}{4}m - t = 5b$ for m

APPLICATIONS

75. If your automobile engine is making a knocking sound, a service technician will probably tell you that the octane rating of the gasoline that you are using is too low. Octane rating numbers are printed on the yellow decals on gas pumps. The formula used to calculate them is

from Campus to Careers
Automotive Service Technician

$$\text{Pump octane number} = \frac{(R + M)}{2}$$

where R is the *research octane number*, which is determined with a test engine running at a low speed and M is the *motor octane number*, which is determined with a test engine running at a higher speed. Calculate the octane rating for the following three grades of gasoline.

Gasoline grade	R	M	Octane rating
Unleaded	92	82	
Unleaded plus	95	83	
Premium	97	85	

76. PROPERTIES OF WATER The boiling point and the freezing point of water are to be given in both degrees Celsius and degrees Fahrenheit on the thermometer. Find the missing degree measures.

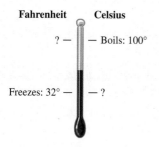

77. AVON PRODUCTS Complete the financial statement.

Income statement (dollar amounts in millions)	Quarter ending Dec '11	Quarter ending Dec '12
Revenue	11,291.6	10,717.1
Cost of goods sold	4,137.4	4,164.8
Operating profit		

Source: Avon Products, Inc.

78. CREDIT The finance charge that a student pays on his credit card is 19.8% APR (annual percentage rate). Determine the finance charges (interest) the student would have to pay if the account's average balance for the year was $2,500.

79. CAMPERS The perimeter of the window of the camper shell is 140 in. Find the length of one of the shorter sides of the window.

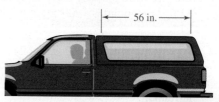

80. FLAGS The flag of Eritrea, a country in east Africa, is shown. The perimeter of the flag is 160 inches.

a. What is the width of the flag?

b. What is the area of the red triangular region of the flag?

81. KITES 650 in.^{2} of nylon cloth were used to make the kite shown. If its height is 26 inches, what is the wingspan?

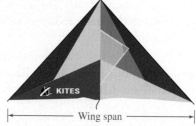

82. MEMORIALS The Vietnam Veterans Memorial is a black granite wall recognizing the more than 58,000 Americans who lost their lives or remain missing. Find the total area of the two triangular surfaces on which the names are inscribed.

83. HDTVs The general rules for high-definition television viewing are given by the formulas:

Minimum viewing distance = $1.5d$
Maximum viewing distance = $3d$

where d is the diagonal measurement of the screen. Fill in the blanks below to give the minimum and maximum viewing distance for a 42-inch HDTV screen. (*Recall:* 12 in. = 1 foot)

Minimum viewing distance: ___ feet ___ inches

Maximum viewing distance: ___ feet ___ inches

84. WHEELCHAIRS Find the diameter of the rear wheel and the radius of the front wheel.

85. ARCHERY The diameter of a standard archery target used in the Olympics is 48.8 inches. Find the area of the target. Round to the nearest square inch.

86. BULLS-EYE See Exercise 85. The diameter of the center yellow ring of a standard archery target is 4.8 inches. What is the area of the bulls-eye? Round to the nearest tenth of a square inch.

87. GEOGRAPHY The circumference of the Earth is about 25,000 miles. Find its diameter to the nearest mile.

88. HORSES A horse trots in a circle around its trainer at the end of a 28-foot-long rope. Find the area of the circle that is swept out. Round to the nearest square foot.

89. YO-YOS How far does a yo-yo travel during one revolution of the "around the world" trick if the length of the string is 21 inches?

90. WORLD HISTORY The Inca Empire (1438–1533) was centered in what is now Peru. A special feature of Inca architecture was the trapezoid-shaped windows and doorways. A standard Inca window was 70 cm (centimeters) high, 50 cm at the base and 40 cm at the top. Find the area of a window opening. The formula for the area of a trapezoid is $A = \frac{1}{2}(\text{height})(\text{upperbase} + \text{lowerbase})$.

91. HAMSTER HABITATS Find the amount of space in the tube.

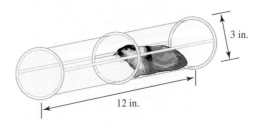

92. TIRES The road surface footprint of a sport truck tire is approximately rectangular. If the area of the footprint is 45 in.², about how wide is the tire?

93. SOFTBALL The strike zone in fast-pitch softball is between the batter's armpit and the top of her knees, as shown. If the area of the strike zone for this batter is 442 in.², what is the width of home plate?

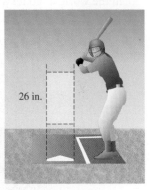

94. FIREWOOD The cord of wood shown occupies a volume of 128 ft³. How long is the stack?

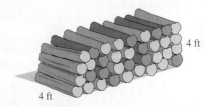

95. TEEPEES The teepees constructed by the Blackfoot Indians were cone-shaped tents about 10 feet high and about 15 feet across at the ground. Estimate the volume of a teepee with these dimensions, to the nearest cubic foot.

96. **IGLOOS** During long journeys, some Canadian Eskimos built winter houses of snow blocks stacked in the dome shape shown. Estimate the volume of an igloo having an interior height of 5.5 feet to the nearest cubic foot.

97. **PYRAMIDS** The Great Pyramid at Giza in northern Egypt is one of the most famous works of architecture in the world. Find its volume to the nearest cubic foot.

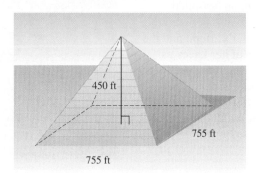

98. **COOKING** If the fish shown in the illustration is 18 inches long, what is the area of the grill? Round to the nearest square inch.

99. **SKATEBOARDING** Refer to the illustration below. A half-pipe ramp is in the shape of a semicircle with a radius of 8 feet. To the nearest tenth of a foot, what is the length of the arc that the rider travels on the ramp?

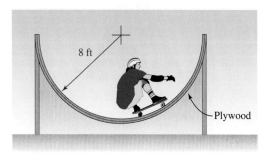

100. **PULLEYS** The approximate length L of a belt joining two pulleys of radii r and R feet with centers D feet apart is given by the formula $L = 2D + 3.25(r + R)$. Solve the formula for D.

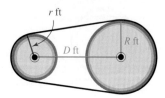

101. **THERMODYNAMICS** The Gibbs free-energy function is given by $G = U - TS + pV$. Solve this formula for the pressure p.

102. **ELECTRICITY** The formula $E = IR + Ir$ can be used to find the internal resistance of a battery.
    a. Solve the formula for r.
    b. Solve the formula for R.

WRITING

103. After solving $A = B + C + D$ for B, a student compared her answer with that at the back of the textbook. Could this problem have two different-looking answers? Explain why or why not.

    Student's answer: $B = A - C - D$
    Book's answer: $B = A - D - C$

104. A student solved $x + 5c = 3c + a$ for c. His answer was $c = \dfrac{3c + a - x}{5}$ for c. Explain why the equation is not solved.

105. Explain the difference between what perimeter measures and what area measures.

106. Explain the error made below.

    $$y = \dfrac{\cancel{3x} + \overset{1}{\cancel{2}}}{\underset{1}{\cancel{2}}}$$

REVIEW

107. Find 82% of 168.
108. 29.05 is what percent of 415?
109. What percent of 200 is 30?
110. A woman bought a coat for $98.95 and some gloves for $7.95. If the sales tax was 6%, how much did the purchase cost her?

SECTION 2.5
Problem Solving

Objectives
1. Apply the steps of a problem-solving strategy.
2. Solve consecutive integer problems.
3. Solve geometry problems.

In this section, you will see that algebra is a powerful tool that can be used to solve a wide variety of real-world problems.

1 Apply the steps of a problem-solving strategy.

To become a good problem solver, you need a plan to follow, such as the following six-step strategy.

> **The Language of Algebra** A *strategy* is a careful plan or method. For example, a businessman might develop a new advertising *strategy* to increase sales, or a long distance runner might have a *strategy* to win a marathon.

Strategy for Problem Solving

1. **Analyze the problem** by reading it carefully to understand the given facts. What information is given? What are you asked to find? What vocabulary is given? Often, a diagram or table will help you understand the facts of the problem.
2. **Assign a variable** to represent an unknown value in the problem. This means, in most cases, to let $x = $ what you are asked to find. If there are other unknown values, represent each of them using an algebraic expression that involves the variable.
3. **Form an equation** by translating the words of the problem into mathematical symbols.
4. **Solve the equation** formed in Step 3.
5. **State the conclusion** clearly. Be sure to include the units (such as feet, seconds, or pounds) in your answer.
6. **Check the result** using the original wording of the problem, not the equation that was formed in Step 3.

EXAMPLE 1 *California Coastline*

The first part of California's magnificent 17-Mile Drive begins at the Pacific Grove entrance and continues to Seal Rock. It is 1 mile longer than the second part of the drive, which extends from Seal Rock to the Lone Cypress, as shown in the map on the next page. The third and final part of the tour winds through the Monterey Peninsula, eventually returning to the entrance. This part of the drive is 1 mile longer than four times the length of the second part. How long is each part of 17-Mile Drive?

Self Check 1
The Mountain-Bay State Park Bike Trail in Northeast Wisconsin is 76 miles long. A couple rode the trail in four days. Each day they rode 2 miles more than the previous day. How many miles did they ride each day?

Now Try Problem 16

Analyze The drive is composed of three parts. We need to find the length of each part. We can straighten out the winding 17-Mile Drive and model it with a line segment.

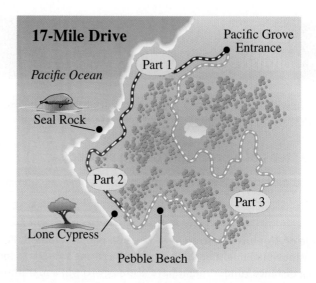

Assign Since the lengths of the first part and of the third part of the drive are related to the length of the second part, we will let x represent the length of the second part. We then express the other lengths in terms of x.

x = the length of the second part of the drive (in miles)

$x + 1$ = the length of the first part of the drive (in miles)

$4x + 1$ = the length of the third part of the drive (in miles)

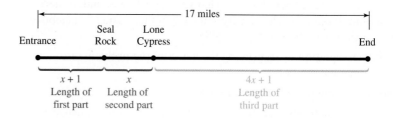

Success Tip When there is more than one unknown value in a problem, let the variable represent the unknown value on which any other unknown values are based. In this case, the three parts of the drive are compared in pairs: The first is compared to the *second*, and the third compared to the *second*. We let x represent the part that the comparisons have in common, which is the second part.

Form Now we translate the words of the problem into an equation.

The length of part 1	plus	the length of part 2	plus	the length of part 3	equals	the total length of the drive.
$x + 1$	$+$	x	$+$	$4x + 1$	$=$	17

Solve

$x + 1 + x + 4x + 1 = 17$

$6x + 2 = 17$ Combine like terms: $x + x + 4x = 6x$ and $1 + 1 = 2$.

$6x = 15$ To undo the addition of 2, subtract 2 from both sides.

$\dfrac{6x}{6} = \dfrac{15}{6}$ To isolate x, undo the multiplication by 6 by dividing both sides by 6.

$x = 2.5$ Do the divisions.

Recall that x represents the length of the second part of the drive. To find the lengths of the first and third parts, we evaluate $x + 1$ and $4x + 1$ for $x = 2.5$.

First part of drive
$x + 1 = \mathbf{2.5} + 1$
$ = 3.5$

Third part of drive
$4x + 1 = 4(\mathbf{2.5}) + 1$ Substitute 2.5 for x.
$ = 11$

State The first part of the drive is 3.5 miles long, the second part is 2.5 miles long, and the third part is 11 miles long.

Check Since 3.5 mi + 2.5 mi + 11 mi = 17 mi, the answers check.

EXAMPLE 2 Advertising
A trucking company had their logo embroidered on the front of baseball caps. They were charged $8.90 per hat plus a one-time setup fee of $25. If the project cost $559, how many hats were embroidered?

Analyze
- It cost $8.90 to have a logo embroidered on a hat. *Given*
- The setup charge was $25. *Given*
- The project cost $559. *Given*
- We are to find the number of hats that were embroidered. *Find*

Assign Let x = the number of hats that were embroidered. If x hats are embroidered, at a cost of $8.90 per hat, the cost to embroider all of the hats is $x \cdot \$8.90$ or $\$8.90x$.

Form

The cost to embroider a hat	times	the number of hats	plus	the setup charge	equals	the total cost.
8.90	·	x	+	25	=	559

Solve
$$8.90x + 25 = 559$$
$$8.90x = 534 \quad \text{To undo the addition of 25, subtract 25 from both sides.}$$
$$\frac{8.90x}{8.90} = \frac{534}{8.90} \quad \text{To isolate } x, \text{ undo the multiplication by 8.90 by dividing both sides by 8.90.}$$
$$x = 60 \quad \text{Do the divisions.}$$

State The company had 60 hats embroidered.

Check The cost to embroider 60 hats is 60($8.90) = $534. When the $25 setup charge is added, we get $534 + $25 = $559. The answer checks.

> **Success Tip** The *Form* step is often the hardest. To help, write a **verbal model** of the situation (shown above in blue) and then translate it into an equation.

Self Check 2
A school club had their motto screenprinted on the front of T-shirts. They were charged $5 per shirt plus a one-time setup fee of $20. If the project cost $255, how many T-shirts were printed?

Now Try Problem 21

EXAMPLE 3 Auctions
A classic car owner is going to sell his 1960 Chevy Impala at an auction. He wants to make $46,000 after paying an 8% commission to the auctioneer. At what selling price (called the "hammer price") will the car owner make this amount of money?

Analyze When the commission is subtracted from the selling price of the car, the owner wants to have $46,000 left.

Assign Let x = the selling price of the car. The amount of the commission is 8% of x, or $0.08x$.

Form

The selling price of the car	minus	the auctioneer's commission	should be	$46,000.
x	−	$0.08x$	=	46,000

Self Check 3
A farmer is going to sell one of his Black Angus cattle at an auction and would like to make $2,597 after paying a 6% commission to the auctioneer. At what selling price will the farmer make this amount of money?

Now Try Problem 27

Solve

$$x - 0.08x = 46{,}000$$
$$0.92x = 46{,}000 \quad \text{Combine like terms: } 1.00x - 0.08x = 0.92x.$$
$$\frac{0.92x}{0.92} = \frac{46{,}000}{0.92} \quad \text{To isolate } x, \text{ undo the multiplication by 0.92 by dividing both sides by 0.92.}$$
$$x = 50{,}000 \quad \text{Do the divisions.}$$

State The owner will make $46,000 if the 1960 Impala sells for $50,000.

Check An 8% commission on $50,000 is 0.08($50,000) = $4,000. The owner will keep $50,000 − $4,000 = $46,000. The answer checks.

2 Solve consecutive integer problems.

Integers that follow one another, such as 15 and 16, are called **consecutive integers**. They are 1 unit apart. **Consecutive even integers** are even integers that differ by 2 units, such as 12 and 14. Similarly, **consecutive odd integers** differ by 2 units, such as 9 and 11. When solving consecutive integer problems, if we let $x =$ the first integer, then

- two consecutive integers are x and $x + 1$
- two consecutive *even* integers are x and $x + 2$
- two consecutive *odd* integers are x and $x + 2$

Self Check 4

The definitions of the words *little* and *lobby* are on back-to-back pages in a dictionary. If the sum of the page numbers is 1,159, on what page can the definition of *little* be found?

Now Try Problem 33

EXAMPLE 4 ***U.S. History*** The year George Washington was chosen president and the year the Bill of Rights went into effect are consecutive odd integers whose sum is 3,580. Find the years.

Analyze We need to find two consecutive odd integers whose sum is 3,580. From history, we know that Washington was elected president first and the Bill of Rights went into effect later.

Assign Let $x =$ the first odd integer (the date when Washington was chosen president). The next odd integer is 2 *greater than* x, therefore $x + 2 =$ the next larger odd integer (the date when the Bill of Rights went into effect).

Form

The first odd integer	plus	the second odd integer	is	3,580.
x	$+$	$x + 2$	$=$	3,580

Solve

$$x + x + 2 = 3{,}580$$
$$2x + 2 = 3{,}580 \quad \text{Combine like terms: } x + x = 2x.$$
$$2x = 3{,}578 \quad \text{To undo the addition of 2, subtract 2 from both sides.}$$
$$x = 1{,}789 \quad \text{To isolate } x, \text{ undo the multiplication by 2 by dividing both sides by 2.}$$

State George Washington was chosen president in the year 1789. The Bill of Rights went into effect in 1789 + 2 = 1791.

Check 1,789 and 1,791 are consecutive odd integers whose sum is 1,789 + 1,791 = 3,580. The answers check.

> ***The Language of Algebra*** *Consecutive* means following one after the other in order. Elton John holds the record for the most *consecutive* years with a song on the Top 50 music chart: 31 years (1970 to 2000).

3 Solve geometry problems.

EXAMPLE 5 *Crime Scenes* Police used 400 feet of yellow tape to fence off a rectangular lot for an investigation. Each width used 50 feet less tape than each length. Find the dimensions of the lot.

Self Check 5

A rectangular counter for the customer service department of a store is 6 feet longer than it is wide. If the perimeter is 32 feet, find the outside dimensions of the counter.

Now Try Problem 39

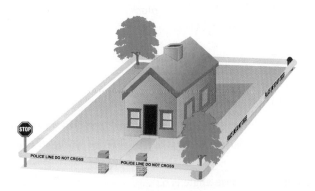

Analyze Since the yellow tape surrounded the lot, the concept of perimeter applies. Recall that the formula for the perimeter of a rectangle is $P = 2l + 2w$. We also know that the width of the lot is 50 feet less than the length.

Assign Since the width of the lot is given in terms of the length, we let l = the length of the lot. Then $l - 50$ = the width.

Form Using the perimeter formula, we have:

2	times	the length	plus	2	times	the width	is	the perimeter.
2	·	l	+	2	·	$(l - 50)$	=	400

Solve

$2l + 2(l - 50) = 400$ Write the parentheses so that the entire expression $l - 50$ is multiplied by 2.

$2l + 2l - 100 = 400$ Distribute the multiplication by 2.

$4l - 100 = 400$ Combine like terms: $2l + 2l = 4l$.

$4l = 500$ To undo the subtraction of 100, add 100 to both sides.

$l = 125$ To isolate l, undo the multiplication by 4 by dividing both sides by 4.

State The length of the lot is 125 feet and width is $125 - 50 = 75$ feet.

Check The width (75 feet) is 50 less than the length (125 feet). The perimeter of the lot is $2(125) + 2(75) = 250 + 150 = 400$ feet. The answers check.

EXAMPLE 6 *Isosceles Triangles* If the vertex angle of an isosceles triangle is 56°, find the measure of each base angle.

Analyze An **isosceles triangle** has two sides of equal length, which meet to form the **vertex angle.** In this case, the measurement of the vertex angle is 56°. We can sketch the triangle as shown. The **base angles** opposite the equal sides are also equal. We need to find their measure.

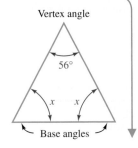

Self Check 6

The perimeter of an isosceles triangle is 32 cm. If the base is 8 cm, find the length of each remaining side.

Now Try Problem 43

Assign If we let x = the measure of one base angle, the measure of the other base angle is also x.

Form Since the sum of the angles of any triangle is 180°, the sum of the base angles and the vertex angle is 180°. We can use this fact to form the equation.

One base angle	plus	the other base angle	plus	the vertex angle	is	180°.
x	+	x	+	56	=	180

Solve

$$x + x + 56 = 180$$
$$2x + 56 = 180 \quad \text{Combine like terms: } x + x = 2x.$$
$$2x = 124 \quad \text{To undo the addition of 56, subtract 56 from both sides.}$$
$$x = 62 \quad \text{To isolate x, undo the multiplication by 2 by dividing both sides by 2.}$$

State The measure of each base angle is 62°.

Check Since $62° + 62° + 56° = 180°$, the answer checks.

ANSWERS TO SELF CHECKS

1. 16 mi, 18 mi, 20 mi, 22 mi   2. 47   3. $2,762.77   4. 579   5. 5 ft by 11 ft   6. 12 cm, 12 cm

---

SECTION 2.5 STUDY SET

VOCABULARY

Fill in the blanks.

1. Integers that follow one another, such as 7 and 8, are called _____ integers.
2. An _____ triangle is a triangle with two sides of the same length.
3. The equal sides of an isosceles triangle meet to form the _____ angle. The angles opposite the equal sides are called _____ angles, and they have equal measures.
4. When asked to find the dimensions of a rectangle, we are to find its _____ and _____.

CONCEPTS

5. A 17-foot pipe is cut into three sections. The longest section is three times as long as the shortest, and the middle-sized section is 2 feet longer than the shortest. Complete the diagram.

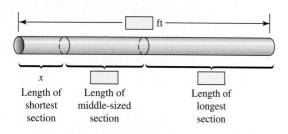

6. It costs $28 per hour to rent a trailer. Write an expression that represents the cost to rent the trailer for x hours.
7. A realtor is paid a 3% commission on the sale of a house. Write an expression that represents the amount of the commission if a house sells for x.
8. The perimeter of the rectangle is 15 feet. Fill in the blanks: 2( ) + 2x =

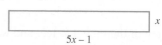

9. What is the sum of the measures of the angles of any triangle?
10. What is x?

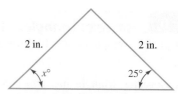

NOTATION

11. **a.** If x represents an integer, write an expression for the next largest integer.

b. If x represents an odd integer, write an expression for the next largest odd integer.

12. What does 45° mean?

GUIDED PRACTICE

Problems with more than one unknown See Example 1.

13. CARPENTRY A 12-foot board has been cut into two sections, one twice as long as the other. How long is each section?

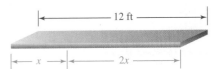

14. ROBOTS The robotic arm shown below will extend a total distance of 18 feet. Find the length of each of the three parts of the arm.

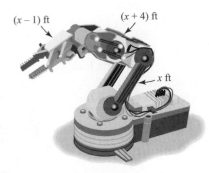

15. NATIONAL PARKS The Natchez Trace Parkway is a historical 444-mile route from Natchez, Mississippi, to Nashville, Tennessee. A couple drove the Trace in four days. Each day they drove 6 miles more than the previous day. How many miles did they drive each day?

16. TOURING A rock group plans to travel for a total of 38 weeks, making three concert stops. They will be in Japan for 4 more weeks than they will be in Australia. Their stay in Sweden will be 2 weeks shorter than that in Australia. How many weeks will they be in each country?

17. SOLAR HEATING Refer to the illustration in the next column. One solar panel is 3.4 feet wider than the other. Find the width of each panel.

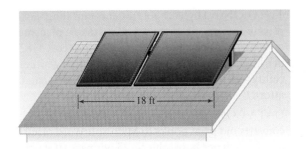

18. ACCOUNTING Determine the 2012 income of Urban Outfitters for each quarter from the data in the graph.

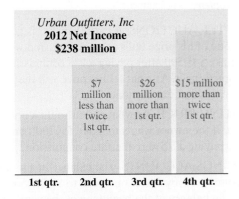

19. COUNTING CALORIES A slice of pie with a scoop of ice cream has 850 calories. The calories in the pie alone are 100 more than twice the calories in the ice cream alone. How many calories are in each food?

20. WASTE DISPOSAL Two tanks hold a total of 45 gallons of a toxic solvent. One tank holds 6 gallons more than twice the amount in the other. How many gallons does each tank hold?

Basic charge/rate problems See Example 2.

21. CONCERTS The fee to rent a concert hall is $2,250 plus $150 per hour to pay for the support staff. For how many hours can an orchestra rent the hall and stay within a budget of $3,300?

22. TRUCK MECHANICS An engine repair cost a truck owner $1,185 in parts and labor. If the parts were $690 and the mechanic charged $45 per hour, how many hours did the repair take?

23. FIELD TRIPS It costs a school $65 a day plus $0.25 per mile to rent a 15-passenger van. If the van is rented for two days, how many miles can be driven on a $275 budget?

24. DECORATIONS A party supply store charges a set-up fee of $80 plus 35¢ per balloon to make a balloon arch. A business has $150 to spend on decorations for their grand opening. How many balloons can they have in the arch? (*Hint:* 35¢ = $0.35.)

25. **TUTORING** High school students enrolling in a private tutoring program must first take a placement test (cost $25) before receiving tutoring (cost $18.75 per hour). If a family has set aside $400 to get their child extra help, how many hours of tutoring can they afford?

26. **DATA CONVERSION** The Books2Bytes service converts old print books to Microsoft Word electronic files for $20 per book plus $2.25 per page. If it cost $1,201.25 to convert a novel, how many pages did the novel have?

Percent problems See Example 3.

27. **CATTLE AUCTIONS** A cattle rancher is going to sell one of his prize bulls at an auction and would like to make $45,500 after paying a 9% commission to the auctioneer. At what selling price will the rancher make this amount of money?

28. **LISTING PRICE** At what price should a home be listed if the owner wants to make $567,000 on its sale after paying a 5.5% real estate commission?

29. **SAVINGS ACCOUNTS** The balance in a savings account grew by 5% in one year, to $5,512.50. What was the balance at the beginning of the year?

30. **ALUMINUM CANS** Today's aluminum cans are much thinner and lighter than those of the past. From 1973 to 2011, the number of empty cans produced from one pound of aluminum increased by about 59%. If 35 cans could be produced from one pound of aluminum in 2011, how many cans could be produced from one pound in 1973? Round to the nearest can. (Source: cancentral.com)

Consecutive integer problems See Example 4.

31. **SOCCER** Ronaldo of Brazil and Gerd Mueller of Germany rank 1 and 2, respectively, with the most goals scored in World Cup play. The number of goals Ronaldo and Mueller have scored are consecutive integers that total 29. Find the number of goals scored by each man.

32. **DICTIONARIES** The definitions of the words *job* and *join* are on back-to-back pages in a dictionary. If the sum of those page numbers is 1,411, on what page can the definition of *job* be found?

33. **TV HISTORY** *Friends* and *Leave It to Beaver* are two of the most popular television shows of all time. The numbers of episodes of each show are consecutive even integers whose sum is 470. If there are more episodes of *Friends,* how many episodes of each were there?

34. **VACATIONS** Use the information in the table below to find the number of paid days off each year a worker in Greece, Japan, and Brazil receives. (The numbers of days are listed in descending order.)

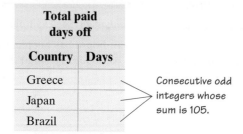

Total paid days off	
Country	Days
Greece	
Japan	
Brazil	

Consecutive odd integers whose sum is 105.

Source: *The World Almanac and Book of Facts, 2013*

35. **CELEBRITY BIRTHDAYS** Selena Gomez, Jennifer Lopez, and Sandra Bullock have birthdays (in that order) on consecutive even-numbered days in July. The sum of the calendar dates of their birthdays is 72. Find each birthday.

36. **LOCKS** The three numbers of the combination for a lock are consecutive integers, and their sum is 81. Find the combination.

Geometry problems See Example 5.

37. **TENNIS** The perimeter of a regulation singles tennis court is 210 feet. The length is 3 feet less than three times the width. What are the dimensions of the court?

38. **SWIMMING POOLS** The seawater Orthlieb Pool in Casablanca, Morocco, is the largest swimming pool in the world. With a perimeter of 1,110 meters, this rectangular pool is 30 meters longer than 6 times its width. Find its dimensions.

39. **ART** The *Mona Lisa* was completed by Leonardo da Vinci in 1506. The length of the picture is 11.75 inches less than twice the width. If the perimeter of the picture is 102.5 inches, find its dimensions.

40. **NEW YORK CITY** Central Park, which lies in the middle of Manhattan, is rectangular in shape and has a 6-mile perimeter. The length is 5 times the width. What are the dimensions of the park?

41. ENGINEERING A truss is in the form of an isosceles triangle. Each of the two equal sides is 4 feet shorter than the third side. If the perimeter is 25 feet, find the lengths of the sides.

42. FIRST AID A sling is in the shape of an isosceles triangle with a perimeter of 144 inches. The longest side of the sling is 18 inches longer than either of the other two sides. Find the lengths of each side.

Triangle problems See Example 6.

43. TV TOWERS The two guy wires supporting a tower form an isosceles triangle with the ground. Each of the base angles of the triangle is 4 times the third angle (the vertex angle). Find the measure of the vertex angle.

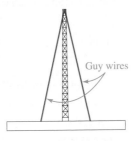

44. CLOTHESLINES A pair of damp jeans are hung in the middle of a clothesline to dry. Find $x°$, the angle that the clothesline makes with the horizontal.

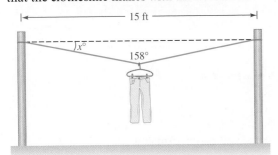

45. MOUNTAIN BICYCLES For the bicycle frame shown in the next column, the angle that the horizontal crossbar makes with the seat support is 15° less than twice the angle at the steering column. The angle at the pedal gear is 25° more than the angle at the steering column. Find these three angle measures.

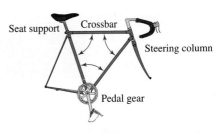

46. TRIANGLES The measure of $\angle 1$ (read as angle 1) of a triangle is one-half that of $\angle 2$. The measure of $\angle 3$ is equal to the sum of the measures of $\angle 1$ and $\angle 2$. Find each angle measure.

47. COMPLEMENTARY ANGLES Two angles are called *complementary angles* when the sum of their measures is 90°. Find the measures of the complementary angles shown in the illustration.

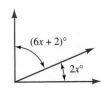

48. SUPPLEMENTARY ANGLES Two angles are called *supplementary angles* when the sum of their measures is 180°. Find the measures of the supplementary angles shown in the illustration.

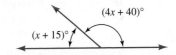

WRITING

49. Create a geometry problem that could be answered by solving the equation $2w + 2(w + 5) = 26$.

50. What information do you need to know to answer the following question?

A business rented a copy machine for $85 per month plus 4¢ for every copy made. How many copies can be made each month?

51. Make a list of words and phrases that translate to an equal symbol $=$.

52. Define the word *strategy*.

REVIEW

Solve.

53. $\dfrac{5}{8}x = -15$

54. $\dfrac{12x + 24}{13} = 36$

55. $\dfrac{3}{4}y = \dfrac{2}{5}y - \dfrac{3}{2}y - 2$

56. $6 + 4(1 - x) = 3(x + 1)$

57. $4.2(y - 4) - 0.6y = -13.2$

58. $16 - 8(b + 4) = 24b + 64$

Objectives

1. Solve investment problems.
2. Solve uniform motion problems.
3. Solve liquid mixture problems.
4. Solve dry mixture problems.
5. Solve number-value problems.

SECTION 2.6
More about Problem Solving

In this section, we will solve problems that involve money, motion, and mixtures. Tables are a helpful way to organize the information given in these problems.

1 Solve investment problems.

To find the amount of *simple interest I* an investment earns, we use the formula $I = Prt$, where P is the principal (the amount invested), r is the annual interest rate, and t is the time in years.

Self Check 1

A student invested $4,200 in two certificates of deposit, one at 2% and the other at 3%. Find the amount invested at each rate if the first-year combined interest income from the two investments was $102.

Now Try Problem 17

EXAMPLE 1 *Paying Tuition* A college student invested the $12,000 inheritance he received and decided to use the annual interest earned to pay his tuition cost of $945. The highest rate offered by a bank at that time was 6% annual simple interest. At this rate, he could not earn the needed $945, so he invested some of the money in a riskier, but more profitable, investment offering a 9% return. How much did he invest at each rate?

Analyze We know that $12,000 was invested for 1 year at two rates: 6% and 9%. We are asked to find the amount invested at each rate so that the total return would be $945.

Assign Let $x =$ the amount invested at 6%. Then $12{,}000 - x =$ the amount invested at 9%.

Form To organize the facts of the problem, we enter the principal, rate, time, and interest earned in a table.

Step 1: List each investment in a row of the table.

Bank			
Riskier investment			

Step 2: Label the columns using $I = Prt$ reversed, and also write Total.

	P	$\cdot\ r$	$\cdot\ t =$	I
Bank				
Riskier investment				
				Total:

Step 3: Enter the rates, times, and total interest.

	P	$\cdot\ r$	$\cdot\ t =$	I
Bank		0.06	1	
Riskier investment		0.09	1	
				Total: **945**

Step 4: Enter each unknown principal.

	P · r · t =			I
Bank	x	0.06	1	
Riskier investment	12,000 − x	0.09	1	
				Total: 945

Step 5: In the last column, multiply P, r, and t to obtain expressions for the interest earned.

	P · r · t =			I
Bank	x	0.06	1	**0.06x** ← This is x · 0.06 · 1.
Riskier investment	12,000 − x	0.09	1	**0.09(12,000 − x)** ← This is (12,000 − x) · 0.09 · 1.
				Total: 945

Use the information in this column to form an equation.

The interest earned at 6%	plus	the interest earned at 9%	equals	the total interest.
0.06x	+	[0.09(12,000 − x)]	=	945

Solve

$$0.06x + 0.09(12,000 - x) = 945$$

$$100[0.06x + 0.09(12,000 - x)] = 100(945)$$ Multiply both sides by 100 to clear the equation of decimals.

$$100(0.06x) + 100(0.09)(12,000 - x) = 100(945)$$ Distribute the multiplication by 100.

$$6x + 9(12,000 - x) = 94{,}500$$ Do the multiplications by 100.

$$6x + 108{,}000 - 9x = 94{,}500$$ Use the distributive property.

$$-3x + 108{,}000 = 94{,}500$$ Combine like terms.

$$-3x = -13{,}500$$ Subtract 108,000 from both sides.

$$x = 4{,}500$$ To isolate x, divide both sides by −3.

State The student invested $4,500 at 6% and $12,000 − $4,500 = $7,500 at 9%.

Check The first investment earned 0.06($4,500), or $270. The second earned 0.09($7,500), or $675. Since the total return was $270 + $675 = $945, the answers check.

2 Solve uniform motion problems.

If we know the rate r at which we will be traveling and the time t we will be traveling at that rate, we can find the distance d traveled by using the formula $d = rt$.

Self Check 2

Two search-and-rescue teams leave base at the same time looking for a lost boy. The first team, on foot, heads north at 2 mph, and the other, on horseback, heads south at 4 mph. How long will it take them to search a distance of 21 miles between them?

Now Try Problem 27

EXAMPLE 2 *Rescues at Sea*

A cargo ship, heading into port, radios the Coast Guard that it is experiencing engine trouble and that its speed has dropped to 3 knots (this is 3 sea miles per hour). Immediately, a Coast Guard cutter leaves port and speeds at a rate of 25 knots directly toward the disabled ship, which is 56 sea miles away. How long will it take the Coast Guard to reach the ship? (Sea miles are also called nautical miles.)

Analyze We know the *rate* of each ship (25 knots and 3 knots), and we know that they must close a *distance* of 56 sea miles between them. We don't know the *time* it will take to do this.

Assign Let $t =$ the time it takes the Coast Guard to reach the cargo ship. During the rescue, the ships don't travel at the same rate, but they do travel for the same amount of time. Therefore, t also represents the travel time for the cargo ship.

Form We enter the rates, the variable t for each time, and the total distance traveled by the ships (56 sea miles) in the table. To fill in the last column, we use the formula $r \cdot t = d$ twice to find an expression for each distance traveled: $25 \cdot t = 25t$ and $3 \cdot t = 3t$.

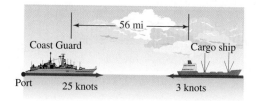

	r	$\cdot$ t	$=$ d
Coast Guard cutter	25	t	$25t$
Cargo ship	3	t	$3t$
		Total:	56

Multiply $r \cdot t$ to obtain an expression for each distance traveled.

Use the information in this column to form an equation.

The distance the cutter travels	plus	the distance the ship travels	equals	the original distance between the ships.
$25t$	$+$	$3t$	$=$	56

Solve

$25t + 3t = 56$

$28t = 56$ Combine like terms: $25t + 3t = 28t$.

$t = \dfrac{56}{28}$ To isolate t, divide both sides by 28.

$t = 2$ Do the division.

State The ships will meet in 2 hours.

Check In 2 hours, the Coast Guard cutter travels $25 \cdot 2 = 50$ sea miles, and the cargo ship travels $3 \cdot 2 = 6$ sea miles. Together, they travel $50 + 6 = 56$ sea miles. Since this is the original distance between the ships, the answer checks.

Success Tip A sketch is helpful when solving uniform motion problems.

2.6 More about Problem Solving

EXAMPLE 3   Concert Tours
While on tour, a country music star travels by bus. Her musical equipment is carried in a truck. How long will it take her bus, traveling 60 mph, to overtake the truck, traveling at 45 mph, if the truck had a $1\frac{1}{2}$-hour head start to her next concert location?

Analyze We know the rate of each vehicle (60 mph and 45 mph) and that the truck began the trip $1\frac{1}{2}$ or 1.5 hours earlier than the bus. We need to determine how long it will take the bus to catch up to the truck.

Assign Let t = the time it takes the bus to overtake the truck. With a 1.5-hour head start, the truck is on the road longer than the bus. Therefore, $t + 1.5$ = the truck's travel time.

Form We enter each rate and time in the table, and use the formula $r \cdot t = d$ twice to fill in the distance column.

Self Check 3
A couple leaves a vacation spot by car traveling at 50 mph. Half an hour later, their friends leave the same spot in a second car traveling at 60 mph. How long will it take the second car to catch up with the first?

Now Try Problem 31

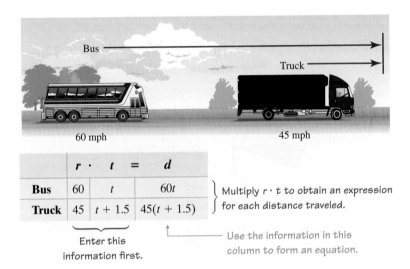

	r ·	t =	d
Bus	60	t	$60t$
Truck	45	$t + 1.5$	$45(t + 1.5)$

Multiply $r \cdot t$ to obtain an expression for each distance traveled.

Enter this information first.

Use the information in this column to form an equation.

When the bus overtakes the truck, they will have traveled the same distance.

The distance traveled by the bus	is the same as	the distance traveled by the truck.
$60t$	=	$45(t + 1.5)$

Solve

$60t = 45(t + 1.5)$

$60t = 45t + 67.5$ Distribute the multiplication by 45: $45(1.5) = 67.5$.

$15t = 67.5$ Subtract $45t$ from both sides: $60t - 45t = 15t$.

$t = 4.5$ To isolate t, divide both sides by 15: $\frac{67.5}{15} = 4.5$.

State The bus will overtake the truck in 4.5 or $4\frac{1}{2}$ hours.

Check In 4.5 hours, the bus travels $60(4.5) = 270$ miles. The truck travels for $1.5 + 4.5 = 6$ hours at 45 mph, which is $45(6) = 270$ miles. Since the distances traveled are the same, the answer checks.

> **Success Tip** We used 1.5 hrs for the head start because it is easier to solve $60t = 45(t + 1.5)$ than $60t = 45\left(t + 1\frac{1}{2}\right)$.

Chapter 2 Equations, Inequalities, and Problem Solving

3 Solve liquid mixture problems.

We now discuss how to solve mixture problems. In the first type, a liquid mixture of a desired strength is made from two solutions with different concentrations.

Self Check 4

How many gallons of a 3% salt solution must be mixed with a 7% salt solution to obtain 25 gallons of a 5.4% salt solution?

Now Try Problem 39

EXAMPLE 4 *Mixing Solutions* A chemistry experiment calls for a 30% sulfuric acid solution. If the lab supply room has only 50% and 20% sulfuric acid solutions, how much of each should be mixed to obtain 12 liters of a 30% acid solution?

Analyze The 50% solution is too strong, and the 20% solution is too weak. We must find how much of each should be combined to obtain 12 liters of a 30% solution.

Assign If x = the number of liters of the 50% solution used in the mixture, the remaining $(12 - x)$ liters must be the 20% solution.

Form The amount of pure sulfuric acid in each solution is given by

Amount of solution · strength of the solution = amount of pure sulfuric acid

A table and sketch are helpful in organizing the facts of the problem.

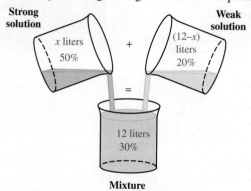

	Amount	· Strength	=	Amount of pure sulfuric acid
Strong	x	0.50		$0.50x$
Weak	$12 - x$	0.20		$0.20(12 - x)$
Mixture	12	0.30		$12(0.30)$

Multiply amount · strength three times to fill in this column.

Enter this information first.

Use the information in this column to form an equation.

The sulfuric acid in the 50% solution	plus	the sulfuric acid in the 20% solution	equals	the sulfuric acid in the mixture.
$0.50x$	+	$0.20(12 - x)$	=	$12(0.30)$

Solve

$0.50x + 0.20(12 - x) = 12(0.30)$

$0.5x + 2.4 - 0.2x = 3.6$ Distribute the multiplication by 0.20.

$0.3x + 2.4 = 3.6$ Combine like terms: $0.5x - 0.2x = 0.3x$.

$0.3x = 1.2$ Subtract 2.4 from both sides.

$x = 4$ To isolate x, undo the multiplication by 0.3 by dividing both sides by 0.3: $\frac{1.2}{0.3} = 4$.

State 4 liters of 50% solution and 12 − 4 = 8 liters of 20% solution should be used.

Check The amount of acid in 4 liters of the 50% solution is 0.50(4) = 2.0 liters and the amount of acid in 8 liters of the 20% solution is 0.20(8) = 1.6 liters. Thus, the amount of acid in these two solutions is 2.0 + 1.6 = 3.6 liters. The amount of acid in 12 liters of the 30% mixture is also 0.30(12) = 3.6 liters. Since the amounts of acid are equal, the answers check.

> **Success Tip** The strength *(concentration)* of a mixture is always between the strengths of the two solutions used to make it.

4 Solve dry mixture problems.

In another type of mixture problem, a dry mixture of a specified value is created from two differently priced ingredients.

EXAMPLE 5 *Snack Foods* Because cashews priced at $9 per pound were not selling, a produce clerk decided to combine them with less expensive peanuts and sell the mixture for $7 per pound. How many pounds of peanuts, selling at $6 per pound, should be mixed with 50 pounds of cashews to obtain such a mixture?

Analyze We need to determine how many pounds of peanuts to mix with 50 pounds of cashews to obtain a mixture worth $7 per pound.

Assign Let x = the number of pounds of peanuts to use in the mixture. Since 50 pounds of cashews will be combined with the peanuts, the mixture will weigh $50 + x$ pounds.

Form The value of the mixture and of each of its ingredients is given by

   Amount · the price = the total value

We can organize the facts of the problem in a table.

	Amount	· Price	= Total value
Peanuts	x	6	$6x$
Cashews	50	9	450
Mixture	$50 + x$	7	$7(50 + x)$

Multiply amount · price three times to fill in this column.

Enter this information first.

Use the information in this column to form an equation.

The value of the peanuts	plus	the value of the cashews	equals	the value of the mixture.
$6x$	+	450	=	$7(50 + x)$

Solve

$6x + 450 = 7(50 + x)$

$6x + 450 = 350 + 7x$ Distribute the multiplication by 7.

$450 = 350 + x$ To eliminate the term $6x$ on the left side, subtract $6x$ from both sides: $7x - 6x = x$.

$100 = x$ To isolate x, subtract 350 from both sides.

State 100 pounds of peanuts should be used in the mixture.

Self Check 5

Candy worth $1.90 per pound is to be mixed with 60 lb of a second candy selling for $1.20 per pound. How many pounds of the $1.90 per pound mixture should be used to make a mixture worth $1.48 per pound?

Now Try Problem 45

Check The value of 100 pounds of peanuts at $6 per pound is $100(6) = \$600$, and the value of 50 pounds of cashews at $9 per pound is $50(9) = \$450$. The total of these two amounts is $1,050. Since the value of 150 pounds of the mixture at $7 per pound is $150(7) = \$1,050$, the answer checks.

5 Solve number-value problems.

When problems deal with collections of different items having different values, we must distinguish between the *number of* and the *value of* the items. For these problems, we will use the fact that

$$\text{Number} \cdot \text{value} = \text{total value}$$

Self Check 6

A small electronics store buys iPods for $189, iPod skins for $32, and iTunes cards for $15. If they place an order for three times as many iPods as skins and 6 times as many iTunes cards as skins, how many of each did they order if the total before shipping and tax totaled $2,756?

Now Try Problem 53

EXAMPLE 6 *Dining Area Improvements*

A restaurant owner needs to purchase some tables, chairs, and dinner plates for the dining area of her establishment. She plans to buy four chairs and four plates for each new table. She also plans to buy 20 additional plates in case of breakage. If a table costs $100, a chair $50, and a plate $5, how many of each can she buy if she takes out a loan for $6,500 to pay for the new items?

Analyze We know the *value* of each item: Tables cost $100, chairs cost $50, and plates cost $5 each. We need to find the *number* of tables, chairs, and plates she can purchase for $6,500.

Assign The number of chairs and plates she needs depends on the number of tables she buys. So we let $t =$ the number of tables to be purchased. Since every table requires four chairs and four plates, she needs to order $4t$ chairs. Because 20 additional plates are needed, she should order $(4t + 20)$ plates.

Form We can organize the facts of the problem in a table.

	Number	· Value	= Total value
Tables	t	100	$100t$
Chairs	$4t$	50	$50(4t)$
Plates	$4t + 20$	5	$5(4t + 20)$
			Total: 6,500

Multiply number · value three times to fill in this column.

Enter this information first.

Use the information in this column to form an equation.

The value of the tables	plus	the value of the chairs	plus	the value of the plates	equals	the value of the purchase.
$100t$	$+$	$50(4t)$	$+$	$5(4t + 20)$	$=$	$6,500$

Solve

$$100t + 50(4t) + 5(4t + 20) = 6,500$$
$$100t + 200t + 20t + 100 = 6,500 \quad \text{Do the multiplications and distribute.}$$
$$320t + 100 = 6,500 \quad \text{Combine like terms: } 100t + 200t + 20t = 320t.$$
$$320t = 6,400 \quad \text{Subtract 100 from both sides.}$$
$$t = 20 \quad \text{To isolate } t, \text{ divide both sides by 320.}$$

To find the number of chairs and plates to buy, we evaluate $4t$ and $4t + 20$ for $t = 20$.

Chairs: $4t = 4(20)$ **Plates:** $4t + 20 = 4(20) + 20$ Substitute 20 for t.
$= 80$ $= 100$

State The owner can buy 20 tables, 80 chairs, and 100 plates.

Check The total value of 20 tables is $20(\$100) = \$2,000$, the total value of 80 chairs is $80(\$50) = \$4,000$, and the total value of 100 plates is $100(\$5) = \$500$. Because the total purchase is $\$2,000 + \$4,000 + \$500 = \$6,500$, the answers check.

ANSWERS TO SELF CHECKS

1. $2,400 at 2%, $1,800 at 3%   2. 3.5 hr   3. 2.5 hr   4. 10 gal   5. 40 lb
6. 12 iPods, 4 skins, 24 cards

SECTION 2.6   STUDY SET

VOCABULARY

Fill in the blanks.

1. Problems that involve depositing money are called _____ problems, and problems that involve moving vehicles are called uniform _____ problems.

2. Problems that involve combining ingredients are called _____ problems, and problems that involve collections of different items having different values are called _____ problems.

CONCEPTS

3. Complete the *principal column* given that part of $30,000 is invested in stocks and the rest in art.

	P	· r · t = I
Stocks	x	
Art		

4. A man made two investments that earned a combined annual interest of $280. Complete the table and then form an equation for this investment problem.

	P	· r	· t =	I
Bank	x	0.04	1	
Stocks	6,000 − x	0.06	1	
			Total:	

5. Complete the *rate column* given that the eastbound plane flew 150 mph slower than the westbound plane.

	r	· t = d
West	r	
East		

6. a. Complete the *time column* given that a runner wants to overtake a walker and the walker had a $\frac{1}{2}$-hour head start.

	r ·	t = d
Runner		t
Walker		

b. Complete the *time column* given that part of a 6-hour drive was in fog and the other part was in clear conditions.

	r ·	t = d
Foggy		t
Clear		

7. A husband and wife drive in opposite directions to work. Their drives last the same amount of time and their workplaces are 80 miles apart. Complete the table and then form an equation for this distance problem.

	r ·	t = d
Husband	35	t
Wife	45	
		Total:

8. a. How many gallons of acetic acid are there in barrel 2? (See the figure on the next page.)

b. Suppose the contents of the two barrels are poured into an empty third barrel. How many gallons of liquid will the third barrel contain?

c. Estimate the strength of the solution in the third barrel: 15%, 35%, or 60% acid?

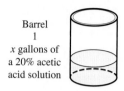 

Barrel 1: x gallons of a 20% acetic acid solution
Barrel 2: 42 gallons of a 40% acetic acid solution

9. a. Two antifreeze solutions are combined to form a mixture. Complete the table and then form an equation for this mixture problem.

	Amount · Strength = Pure antifreeze		
Strong	6	0.50	
Weak	x	0.25	
Mixture		0.30	

b. Two oil-and-vinegar salad dressings are combined to make a new mixture. Complete the table and then form an equation for this mixture problem.

	Amount · Strength = Pure vinegar		
Strong	x	0.06	
Weak		0.03	
Mixture	10	0.05	

10. The value of all the nylon brushes that a paint store carries is $670. Complete the table and then form an equation for this number-value problem.

	Number · Value = Total value		
1-inch	2x	4	
2-inch	x	5	
3-inch	x + 10	7	
		Total:	

NOTATION

11. Write 6% and 15.2% in decimal form.
12. By what power of 10 should each decimal be multiplied to make it a whole number?
   a. 0.08   b.

GUIDED PRACTICE

Solve each equation. See Example 1.

13. $0.18x + 0.45(12 - x) = 0.36(12)$
14. $0.12x + 0.20(4 - x) = 0.6$
15. $0.08x + 0.07(15,000 - x) = 1,110$
16. $0.108x + 0.07(16,000 - x) = 1,500$

APPLICATIONS

Investment problems See Example 1.

17. **CORPORATE INVESTMENTS** The financial board of a corporation invested $25,000 overseas, part at 4% and part at 7% annual interest. Find the amount invested at each rate if the first-year combined income from the two investments was $1,300.

18. **LOANS** A credit union loaned out $50,000, part at an annual rate of 5% and the rest at an annual rate of 8%. They collected combined simple interest of $3,400 from the loans that year. How much was loaned out at each rate?

19. **OLD COINS** A salesperson used her $3,500 year-end bonus to purchase some old coins, earning a 15% annual return on the gold coins and a 12% annual return on the silver coins. If the total return on her investment the first year was $480, how much did she invest in each type of coin?

20. **HIGH-RISK COMPANIES** An investment club used funds totaling $200,000 to invest in a biotech company and an ethanol plant, earning annual returns of 11% and 14%, respectively. The club made a total of $24,250 the first year. How much was invested in each company?

21. **RETIREMENT** A professor wants to supplement her pension with investment interest. If she has $28,000 invested at 6% interest, how much would she also have to invest at 7% to earn a total of $3,500 per year in supplemental income?

22. **EXTRA INCOME** An investor wants to receive $1,000 annually from two investments. He has put $4,500 in a money market account paying 4% annual interest. How much should he invest in a stock fund with a 10% annual return to achieve his goal?

23. **1099 FORMS** The form on the next page shows the interest income Terrell Washington earned in 2012 from two savings accounts. He deposited a total of $15,000 at the first of that year, and made no further deposits or withdrawals. How much money did he deposit in account 822 and in account 721?

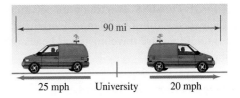

USA HOME SAVINGS — 2012
This is important tax information and is being furnished to the Internal Revenue Service.

RECIPIENT'S name: **TERRELL WASHINGTON**

Account Number	Annual Percent Yield	Interest earned
822	5%	?
721	4.5%	?

FORM 1099 Total Interest Income $720.00

24. **INVESTMENT PLANS** A financial planner recommends a plan for a client who has $65,000 to invest. (See the chart.) At the end of the presentation, the client asks, "How much will be invested at each rate?" Answer this question using the given information.

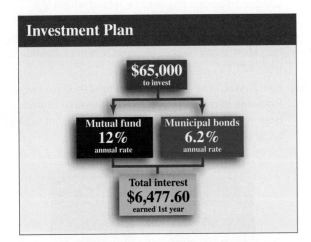

Investment Plan
$65,000 to invest → Mutual fund 12% annual rate / Municipal bonds 6.2% annual rate
Total interest $6,477.60 earned 1st year

25. **INVESTMENTS** Equal amounts are invested in each of three accounts paying 7%, 8%, and 10.5% annually. If one year's combined interest income is $1,249.50, how much is invested in each account?

26. **PERSONAL LOANS** Maggy lent her brother some money at 2% annual interest. She lent her sister twice as much money at half of the interest rate. In one year, Maggy collected combined interest of $200 from her brother and sister. How much did she lend each of them?

Uniform motion problems See Examples 2–3.

27. **TORNADOES** During a storm, two teams of scientists leave a university at the same time in vans to search for tornadoes. The first team travels east at 20 mph and the second travels west at 25 mph. If their radios have a range of up to 90 miles, how long will it be before they lose radio contact? (See the next column.)

28. **UNMANNED AIRCRAFT** Two remotely controlled aircraft are launched in opposite directions. One flies east at 78 mph and the other west at 82 mph. How long will it take the aircraft to fly a combined distance of 560 miles?

29. **HELLO/GOODBYE** A husband and wife work different shifts at the same plant. When the husband leaves from work to make the 20-mile trip home, the wife leaves their home and drives to work. They travel on the same road. The husband's driving rate is 45 mph and the wife's is 35 mph. How long into their drives can they wave at each other when passing on the road?

30. **AIR TRAFFIC CONTROL** An airliner leaves Berlin, Germany, headed for Montreal, Canada, flying at an average speed of 450 mph. At the same time, an airliner leaves Montreal headed for Berlin, averaging 500 mph. If the airports are 3,800 miles apart, when will the air traffic controllers have to make the pilots aware that the planes are passing each other?

31. **CYCLING** A cyclist leaves his training base for a morning workout, riding at the rate of 18 mph. One and one-half hours later, his support staff leaves the base in a car going 45 mph in the same direction. How long will it take the support staff to catch up with the cyclist?

32. **PARENTING** How long will it take a mother, running at 4 feet per second, to catch up with her toddler, running down the sidewalk at 2 feet per second, if the child had a 5-second head start?

33. **ROAD TRIPS** A car averaged 40 mph for part of a trip and 50 mph for the remainder. If the 5-hour trip covered 210 miles, for how long did the car average 40 mph?

34. **CROSS-TRAINING** An athlete runs up a set of stadium stairs at a rate of 2 stairs per second, immediately turns around, and then descends the same stairs at a rate of 3 stairs per second. If the workout takes 90 seconds, how long does it take him to run up the stairs?

35. **WINTER DRIVING** A trucker drove for 4 hours before he encountered icy road conditions. He reduced his speed by 20 mph and continued driving for 3 more hours. Find his average speed during the first part of the trip if the entire trip was 325 miles.

36. **SPEED OF TRAINS** Two trains are 330 miles apart, and their speeds differ by 20 mph. Find the speed of each train if they are traveling toward each other and will meet in 3 hours.

Liquid mixture problems See Example 4.

37. **SALT SOLUTIONS** How many gallons of a 3% salt solution must be mixed with 50 gallons of a 7% solution to obtain a 5% solution?

38. **PHOTOGRAPHY** A photographer wishes to mix 2 liters of a 5% acetic acid solution with a 10% solution to get a 7% solution. How many liters of 10% solution must be added?

39. **MAKING CHEESE** To make low-fat cottage cheese, milk containing 4% butterfat is mixed with milk containing 1% butterfat to obtain 15 gallons of a mixture containing 2% butterfat. How many gallons of each milk must be used?

40. **ANTIFREEZE** How many quarts of a 10% antifreeze solution must be mixed with 16 quarts of a 40% antifreeze solution to make a 30% solution?

41. **PRINTING** A printer has ink that is 8% cobalt blue color and ink that is 22% cobalt blue color. How many ounces of each ink are needed to make 1 gallon (64 ounces) of ink that is 15% cobalt blue color?

42. **FLOOD DAMAGE** One website recommends a 6% chlorine bleach–water solution to remove mildew. A chemical lab has 3% and 15% chlorine bleach–water solutions in stock. How many gallons of each should be mixed to obtain 100 gallons of the mildew spray?

43. **INTERIOR DECORATING** The colors on the paint chip card below are created by adding different amounts of orange tint to a white latex base. How many gallons of Desert Sunrise should be mixed with 1 gallon of Bright Pumpkin to obtain Cool Cantaloupe?

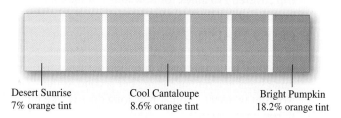

Desert Sunrise Cool Cantaloupe Bright Pumpkin
7% orange tint   8.6% orange tint   18.2% orange tint

44. **ANTISEPTICS** A nurse wants to add water to 30 ounces of a 10% solution of benzalkonium chloride to dilute it to an 8% solution. How much water must she add? (*Hint:* Water is 0% benzalkonium chloride.)

Dry mixture problems See Example 5.

45. **LAWN SEED** A store sells bluegrass seed for $6 per pound and ryegrass seed for $3 per pound. How much ryegrass must be mixed with 100 pounds of bluegrass to obtain a blend that will sell for $5 per pound?

46. **COFFEE BLENDS** A store sells regular coffee for $8 a pound and gourmet coffee for $14 a pound. To get rid of 40 pounds of the gourmet coffee, a shopkeeper makes a blend to put on sale for $10 a pound. How many pounds of regular coffee should he use?

47. **RAISINS** How many scoops of natural seedless raisins costing $3.45 per scoop must be mixed with 20 scoops of golden seedless raisins costing $2.55 per scoop to obtain a mixture costing $3 per scoop?

48. **FERTILIZER** Fertilizer with weed control costing $38 per 50-pound bag is to be mixed with a less expensive fertilizer costing $6 per 50-pound bag to make 16 bags of fertilizer that can be sold for $28 per bag. How many bags of cheaper fertilizer should be used?

49. **PACKAGED SALAD** How many 10-ounce bags of Romaine lettuce must be mixed with fifty 10-ounce bags of Iceberg lettuce to obtain a blend that sells for $2.50 per ten-ounce bag?

50. **MIXING CANDY** Lemon drops worth $3.80 per pound are to be mixed with jelly beans that cost $2.40 per pound to make 100 pounds of a mixture worth $2.96 per pound. How many pounds of each candy should be used?

51. **BRONZE** A pound of tin is worth $1 more than a pound of copper. Four pounds of tin are mixed with 6 pounds of copper to make bronze that sells for $3.65 per pound. How much is a pound of tin worth?

52. **SNACK FOODS** A bag of peanuts is worth $.30 less than a bag of cashews. Equal amounts of peanuts and cashews are used to make 40 bags of a mixture that sells for $1.05 per bag. How much is a bag of cashews worth?

Number-value problems See Example 6.

53. **RENTALS** The owners of an apartment building rent equal numbers of 1-, 2-, and 3-bedroom units. The monthly rent for a 1-bedroom is $550, a 2-bedroom is $700, and a 3-bedroom is $900. If the total monthly income is $36,550, how many of each type of unit are there?

54. **WAREHOUSING** A store warehouses 40 more portables than big-screen TV sets, and 15 more consoles than big-screen sets. The monthly storage cost for a portable is $1.50, a console is $4.00, and a big-screen is $7.50. If storage for all the televisions costs $276 per month, how many big-screen sets are in stock?

55. **SOFTWARE** Three software applications are priced as shown. Spreadsheet and database programs sold in equal numbers, but 15 more word processing applications were sold than the other two combined. If the three applications generated sales of $72,000, how many spreadsheets were sold?

Software	Price
Spreadsheet	$150
Database	$195
Word processing	$210

56. **INVENTORIES** With summer approaching, the number of air conditioners sold is expected to be double that of stoves and refrigerators combined. Stoves sell for $350, refrigerators for $450, and air conditioners for $500, and sales of $56,000 are expected. If stoves and refrigerators sell in equal numbers, how many of each appliance should be stocked?

57. **PIGGY BANKS** When a child emptied his coin bank, he had a collection of pennies, nickels, and dimes. There were 20 more pennies than dimes and the number of nickels was triple the number of dimes. If the coins had a value of $5.40, how many of each type coin were in the bank?

58. **WISHING WELLS** A scuba diver, hired by an amusement park, collected $121 in nickels, dimes, and quarters at the bottom of a wishing well. There were 500 nickels, and 90 more quarters than dimes. How many quarters and dimes were thrown into the wishing well?

59. **BASKETBALL** Epiphanny Prince, of New York, scored 113 points in a high school game on February 1, 2006, breaking a national prep record that was held by Cheryl Miller. Prince made 46 more 2-point baskets than 3-point baskets, and only 1 free throw. How many 2-point and 3-point baskets did she make?

60. **MUSEUM TOURS** The admission prices for the Coca-Cola Museum in Atlanta are shown. A family purchased 3 more children's tickets than adult tickets, and 1 less senior ticket than adult tickets. The total cost of the tickets was $148. How many of each type of ticket did they purchase?

WRITING

61. Create a mixture problem of your own, and solve it.

62. Is it possible to mix a 10% sugar solution with a 20% sugar solution to get a 30% sugar solution? Explain.

REVIEW

Multiply.

63. $-25(2x - 5)$

64. $-12(3a + 4b - 32)$

65. $-(3x - 3)$

66. $\frac{1}{2}(4b - 8)$

67. $(4y - 4)4$

68. $3(5t + 1)2$

SECTION 2.7
Solving Inequalities

Objectives
1. Determine whether a number is a solution of an inequality.
2. Graph solution sets and use interval notation.
3. Solve linear inequalities.
4. Solve compound inequalities.
5. Solve inequality applications.

In our daily lives, we often speak of one value being *greater than* or *less than* another. For example, a sick child might have a temperature *greater than* 98.6°F or a granola bar might contain *less than* 2 grams of fat.

In mathematics, we use *inequalities* to show that one expression is greater than or is less than another expression.

1 Determine whether a number is a solution of an inequality.

An **inequality** is a statement that contains one or more of the following symbols.

Inequality Symbols

< is less than	> is greater than	≠ is not equal to
≤ is less than or equal to	≥ is greater than or equal to	

An inequality can be true, false, or neither true nor false. For example,

- $9 \geq 9$ is true because $9 = 9$.
- $37 < 24$ is false.
- $x + 1 > 5$ is neither true nor false because we don't know what number x represents.

An inequality that contains a variable can be made true or false depending on the number that is substituted for the variable. If we substitute 10 for x in $x + 1 > 5$, the resulting inequality is true: $10 + 1 > 5$. If we substitute 1 for x, the resulting inequality is false: $1 + 1 > 5$. A number that makes an inequality true is called a **solution** of the inequality, and we say that the number *satisfies* the inequality. Thus, 10 is a solution of $x + 1 > 5$ and 1 is not.

The Language of Algebra Because $<$ requires one number to be strictly less than another number and $>$ requires one number to be strictly greater than another number, $<$ and $>$ are called *strict inequalities*.

In this section, we will find the solutions of *linear inequalities in one variable*.

Linear Inequality in One Variable

A **linear inequality in one variable** can be written in one of the following forms where a, b, and c are real numbers and $a \neq 0$.

$$ax + b > c \qquad ax + b \geq c \qquad ax + b < c \qquad ax + b \leq c$$

Self Check 1

Is 2 a solution of $3x - 1 \geq 0$?

Now Try Problem 13

EXAMPLE 1 Is 9 a solution of $2x + 4 \leq 21$?

Strategy We will substitute 9 for x and evaluate the expression on the left side.

WHY If a true statement results, 9 is a solution of the inequality. If we obtain a false statement, 9 is not a solution.

Solution

$$2x + 4 \leq 21$$
$$2(9) + 4 \stackrel{?}{\leq} 21 \quad \text{Substitute 9 for } x. \text{ Read } \stackrel{?}{\leq} \text{ as "is possibly less than or equal to."}$$
$$18 + 4 \stackrel{?}{\leq} 21$$
$$22 \leq 21 \quad \text{This inequality is false.}$$

The statement $22 \leq 21$ is false because neither $22 < 21$ nor $22 = 21$ is true. Therefore, 9 is not a solution.

2 Graph solution sets and use interval notation.

The **solution set** of an inequality is the set of all numbers that make the inequality true. Some solution sets are easy to determine. For example, if we replace the variable in $x > -3$ with a number greater than -3, the resulting inequality will be true. Because there are infinitely many real numbers greater than -3, it follows that $x > -3$ has infinitely many solutions. Since there are too many solutions to list, we use **set-builder notation** to describe the solution set.

$\{x \mid x > -3\}$ Read as "the set of all x such that x is greater than −3."

We can illustrate the solution set by **graphing the inequality** on a number line. To graph $x > -3$, a **parenthesis** or **open circle** is drawn on the endpoint −3 to indicate that −3 is not part of the graph. Then we shade all of the points on the number line to the right of −3. The right arrowhead is also shaded to show that the solutions continue forever to the right.

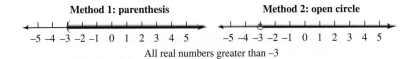

All real numbers greater than −3

The graph of $x > -3$ is an example of an **interval** on the number line. We can write intervals in a compact form called **interval notation**.

The interval notation that represents the graph of $x > -3$ is $(-3, \infty)$. As on the number line, a left parenthesis is written next to −3 to indicate that −3 is not included in the interval. The **positive infinity symbol** ∞ that follows indicates that the interval continues without end to the right. With this notation, *a parenthesis is always used next to an infinity symbol.*

The illustration below shows the relationship between the symbols used to graph an interval and the corresponding interval notation. If we begin at −3 and move to the right, the shaded arrowhead on the graph indicates that the interval approaches positive infinity ∞.

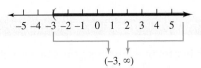

> **Success Tip** Think of interval notation as a way to tell someone how to draw the graph, from left to right, giving them only a "start" and a "stop" instruction.

We now have three ways to describe the solution set of an inequality.

Set-builder notation   **Number line graph**   **Interval notation**
$\{x \mid x > -3\}$ $(-3, \infty)$

> **Success Tip** The *infinity* symbol ∞ does not represent a number. It indicates that an interval extends to the right without end.

EXAMPLE 2   Graph: $x \leq 2$

Strategy We need to determine which real numbers, when substituted for x, would make $x \leq 2$ a true statement.

WHY To graph $x \leq 2$ means to draw a "picture" of all of the values of x that make the inequality true.

Self Check 2

Graph: $x \geq 0$

Now Try Problem 17

Solution

If we replace x with a number less than or equal to 2, the resulting inequality will be true. To graph the solution set, a **bracket** or a **closed circle** is drawn at the endpoint 2 to indicate that 2 is part of the graph. Then we shade all of the points on the number line to the left of 2 and the left arrowhead.

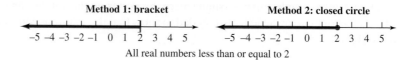

All real numbers less than or equal to 2

The interval is written as $(-\infty, 2]$. The right bracket indicates that 2 is included in the interval. The **negative infinity symbol** $-\infty$ shows that the interval continues forever to the left. The illustration below shows the relationship between the symbols used to graph the interval and the corresponding interval notation.

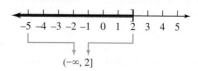

3 Solve linear inequalities.

To **solve an inequality** means to find all values of the variable that make the inequality true. As with equations, there are properties that we can use to solve inequalities.

> **Addition and Subtraction Properties of Inequality**
>
> Adding the same number to, or subtracting the same number from, both sides of an inequality does not change its solutions.
>
> For any real numbers a, b, and c,
>
> If $a < b$, then $a + c < b + c$.
> If $a < b$, then $a - c < b - c$.
>
> Similar statements can be made for the symbols $\leq$, $>$, and $\geq$.

After applying one of these properties, the resulting inequality is equivalent to the original one. **Equivalent inequalities** have the same solution set.

Like equations, inequalities are solved by isolating the variable on one side.

Self Check 3

Solve $x - 3 < -2$. Write the solution set in interval notation and graph it.

Now Try Problem 25

EXAMPLE 3 Solve $x + 3 > 2$. Write the solution set in interval notation and graph it.

Strategy We will use a property of inequality to isolate the variable on one side.

WHY To solve the original inequality, we want to find a simpler equivalent inequality of the form $x >$ **a number** or $x <$ **a number**, whose solution is obvious.

Solution
We will use the subtraction property of inequality to isolate x on the left side of the inequality. We can undo the addition of 3 by subtracting 3 from both sides.

$$x + 3 > 2 \quad \text{This is the inequality to solve.}$$
$$x + 3 - 3 > 2 - 3 \quad \text{Subtract 3 from both sides.}$$
$$x > -1$$

All real numbers greater than -1 are solutions of $x + 3 > 2$. The solution set can be written in set-builder notation as $\{x \mid x > -1\}$ and in interval notation as $(-1, \infty)$. The graph of the solution set is shown below.

Since there are infinitely many solutions, we cannot check all of them.

As an informal check, we can pick some numbers in the graph, say 0 and 30, substitute each number for x in the original inequality, and see whether true statements result.

Check:
$$x + 3 > 2 \qquad\qquad x + 3 > 2$$
$$0 + 3 \stackrel{?}{>} 2 \quad \text{Substitute 0 for x.} \qquad 30 + 3 \stackrel{?}{>} 2 \quad \text{Substitute 30 for x.}$$
$$3 > 2 \quad \text{True} \qquad\qquad 33 > 2 \quad \text{True}$$

The solution set appears to be correct.

Caution! Since we use parentheses and brackets in interval notation, we will use them to graph inequalities. Note that parentheses, not brackets, are written next to ∞ and $-\infty$ because there is no endpoint.

$$(-3, \infty) \qquad (-\infty, 2]$$

As with equations, there are properties for multiplying and dividing both sides of an inequality by the same number. To develop what is called *the multiplication property of inequality,* we consider the true statement $2 < 5$. If both sides are multiplied by a positive number, such as 3, another true inequality results.

$$2 < 5 \quad \text{This inequality is true.}$$
$$3 \cdot 2 < 3 \cdot 5 \quad \text{Multiply both sides by 3.}$$
$$6 < 15 \quad \text{This inequality is true.}$$

However, if we multiply both sides of $2 < 5$ by a negative number, such as -3, the direction of the inequality symbol is reversed to produce another true inequality.

$$2 < 5 \quad \text{This inequality is true.}$$
$$-3 \cdot 2 > -3 \cdot 5 \quad \text{Multiply both sides by } -3 \text{ and reverse the direction of the inequality.}$$
$$-6 > -15 \quad \text{This inequality is true.}$$

The inequality $-6 > -15$ is true because -6 is to the right of -15 on the number line.

Dividing both sides of an inequality by the same negative number also requires that the direction of the inequality symbol be reversed.

$$-4 < 6 \quad \text{This inequality is true.}$$
$$\frac{-4}{-2} > \frac{6}{-2} \quad \text{Divide both sides by } -2 \text{ and change } < \text{ to } >.$$
$$2 > -3 \quad \text{This inequality is true.}$$

These examples illustrate the multiplication and division properties of inequality.

Multiplication and Division Properties of Inequality

Multiplying or dividing both sides of an inequality by the same positive number does not change its solutions.

For any real numbers a, b, and c, where c is positive,

If $a < b$, then $ac < bc$.   If $a < b$, then $\dfrac{a}{c} < \dfrac{b}{c}$.

If we multiply or divide both sides of an inequality by a negative number, the direction of the inequality symbol must be reversed for the inequalities to have the same solutions.

For any real numbers a, b, and c, where c is negative,

If $a < b$, then $ac > bc$.   If $a < b$, then $\dfrac{a}{c} > \dfrac{b}{c}$.

Similar statements can be made for the symbols $\leq$, $>$, and $\geq$.

Self Check 4

Solve each inequality. Write the solution set in interval notation and graph it.

a. $-\dfrac{h}{20} \leq 10$

$\quad \overset{-200 \quad 0 \quad 200}{\longleftrightarrow}$

b. $-12a > -144$

$\quad \overset{10 \quad 11 \quad 12 \quad 13}{\longleftrightarrow}$

Now Try Problems 31 and 33

EXAMPLE 4 Solve each inequality. Write the solution set in interval notation and graph it.   **a.** $-\dfrac{3}{2}t \geq -12$ **b.** $-5t < 55$

Strategy We will use a property of inequality to isolate the variable on one side.

WHY To solve the original inequality, we want to find a simpler equivalent inequality, whose solution is obvious.

Solution

a. To undo the multiplication by $-\dfrac{3}{2}$, we multiply both sides by the reciprocal, which is $-\dfrac{2}{3}$.

$$-\dfrac{3}{2}t \geq -12 \quad \text{This is the inequality to solve.}$$

$$-\dfrac{2}{3}\left(-\dfrac{3}{2}t\right) \leq -\dfrac{2}{3}(-12) \quad \begin{array}{l}\text{Multiply both sides by } -\dfrac{2}{3}. \text{ Since we are multiplying} \\ \text{both sides by a negative number, reverse the direction} \\ \text{of the inequality.}\end{array}$$

$$t \leq 8 \quad \text{Do the multiplications.}$$

The solution set is $(-\infty, 8]$ and it is graphed as shown.

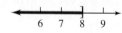

b. To undo the multiplication by -5, we divide both sides by -5.

$$-5t < 55 \quad \text{This is the inequality to solve.}$$

$$\dfrac{-5t}{-5} > \dfrac{55}{-5} \quad \begin{array}{l}\text{To isolate } t, \text{ undo the multiplication by } -5 \text{ by dividing both sides by} \\ -5. \text{ Since we are dividing both sides by a negative number, reverse} \\ \text{the direction of the inequality.}\end{array}$$

$$t > -11$$

The solution set is $(-11, \infty)$ and it is graphed as shown.

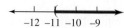

EXAMPLE 5
Solve $-5 > 3x + 7$. Write the solution set in interval notation and graph it.

Strategy First we will use a property of inequality to isolate the *variable term* on one side. Then we will use a second property of inequality to isolate the *variable* itself.

WHY To solve the original inequality, we want to find a simpler equivalent inequality of the form $x >$ **a number** or $x <$ **a number**, whose solution is obvious.

Solution

$-5 > 3x + 7$	This is the inequality to solve.
$-5 - 7 > 3x + 7 - 7$	To isolate the variable term, 3x, undo the addition of 7 by subtracting 7 from both sides.
$-12 > 3x$	Do the subtractions.
$\dfrac{-12}{3} > \dfrac{3x}{3}$	To isolate x, undo the multiplication by 3 by dividing both sides by 3.
$-4 > x$	Do the divisions.

To determine the solution set, it is useful to rewrite the inequality $-4 > x$ in an equivalent form with the variable on the left side. If -4 is greater than x, it follows that x must be less than -4.

$x < -4$

The solution set is $(-\infty, -4)$, whose graph is shown below.

Self Check 5
Solve $-13 < 2r - 7$. Write the solution set in interval notation and graph it.

Now Try Problem 39

EXAMPLE 6
Solve $5.1 - 3k < 19.5$. Write the solution set in interval notation and graph it.

Strategy We will use properties of inequality to isolate the variable on one side.

WHY To solve the original inequality, we want to find a simpler equivalent inequality of the form $k >$ **a number** or $k <$ **a number**, whose solution is obvious.

Solution

$5.1 - 3k < 19.5$	This is the inequality to solve.
$5.1 - 3k - \mathbf{5.1} < 19.5 - \mathbf{5.1}$	To isolate $-3k$ on the left side, subtract 5.1 from both sides.
$-3k < 14.4$	Do the subtractions.
$\dfrac{-3k}{-3} > \dfrac{14.4}{-3}$	To isolate k, undo the multiplication by -3 by dividing both sides by -3 and reverse the direction of the $<$ symbol.
$k > -4.8$	Do the divisions.

The solution set is $(-4.8, \infty)$, whose graph is shown below.

Self Check 6
Solve $-9n + 1.8 > -17.1$. Write the solution set in interval notation and graph it.

Now Try Problem 47

The equation-solving strategy on page 135 can be applied to inequalities. However, when solving inequalities, we must remember to *change the direction of the inequality symbol when multiplying or dividing both sides by a negative number.*

Self Check 7

Solve $5(b - 2) \geq -(b - 3) + 2b$. Write the solution set in interval notation and graph it.

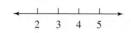

Now Try Problem 53

EXAMPLE 7
Solve $8(y + 1) \geq 2(y - 4) + y$. Write the solution set in interval notation and graph it.

Strategy We will follow the steps of the equation-solving strategy (adapted to inequalities) to solve the inequality.

WHY This is the most efficient way to solve a linear inequality in one variable.

Solution

$8(y + 1) \geq 2(y - 4) + y$	This is the inequality to solve.
$8y + 8 \geq 2y - 8 + y$	Distribute the multiplication by 8 and by 2.
$8y + 8 \geq 3y - 8$	Combine like terms: $2y + y = 3y$.
$8y + 8 - 3y \geq 3y - 8 - 3y$	To eliminate $3y$ from the right side, subtract $3y$ from both sides.
$5y + 8 \geq -8$	Combine like terms on both sides.
$5y + 8 - 8 \geq -8 - 8$	To isolate $5y$, undo the addition of 8 by subtracting 8 from both sides.
$5y \geq -16$	Do the subtractions.
$\dfrac{5y}{5} \geq \dfrac{-16}{5}$	To isolate y, undo the multiplication by 5 by dividing both sides by 5. Do not reverse the direction of the $\geq$ symbol.
$y \geq -\dfrac{16}{5}$	

The solution set is $\left[-\dfrac{16}{5}, \infty\right)$. To graph it, we note that $-\dfrac{16}{5} = -3\dfrac{1}{5}$.

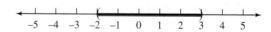

4 Solve compound inequalities.

Two inequalities can be combined into a **compound inequality** to show that an expression lies between two fixed values. For example, $-2 < x < 3$ is a combination of

$$-2 < x \quad \text{and} \quad x < 3$$

It indicates that x is greater than -2 and that x is also less than 3. The solution set of $-2 < x < 3$ consists of all numbers that lie between -2 and 3, and we write it as $(-2, 3)$. The graph of the compound inequality is shown below.

Self Check 8

Graph $-2 \leq x < 1$ and write the solution set in interval notation.

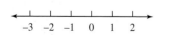

Now Try Problem 19

EXAMPLE 8
Graph: $-4 \leq x < 0$

Strategy We need to determine which real numbers, when substituted for x, would make $-4 \leq x < 0$ a true statement.

WHY To graph $-4 \leq x < 0$ means to draw a "picture" of all of the values of x that make the compound inequality true.

Solution

If we replace the variable in $-4 \leq x < 0$ with a number between -4 and 0, including -4, the resulting compound inequality will be true. Therefore, the solution set is the interval $[-4, 0)$. To graph the interval, we draw a bracket at -4, a parenthesis at 0, and shade in between.

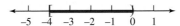

To check, we pick a number in the graph, such as -2, and see whether it satisfies the inequality. Since $-4 \leq -2 < 0$ is true, the answer appears to be correct.

> **Success Tip** Note that the two inequality symbols in $-4 \leq x < 0$ point in the same direction and point to the smaller number.

To solve compound inequalities, we isolate the variable in the middle part of the inequality. To do this, we apply the properties of inequality to all *three* parts of the inequality.

EXAMPLE 9
Solve $-4 < 2(x - 1) \leq 4$. Write the solution set in interval notation and graph it.

Strategy We will use properties of inequality to isolate the variable by itself as the middle part of the inequality.

WHY To solve the original inequality, we want to find a simpler equivalent inequality of the form **a number** $< x \leq$ **a number**, whose solution is obvious.

Solution

$-4 < 2(x - 1) \leq 4$	This is the compound inequality to solve.
$-4 < 2x - 2 \leq 4$	Distribute the multiplication by 2.
$-4 + 2 < 2x - 2 + 2 \leq 4 + 2$	To isolate $2x$, undo the subtraction of 2 by adding 2 to all three parts.
$-2 < 2x \leq 6$	Do the additions.
$\dfrac{-2}{2} < \dfrac{2x}{2} \leq \dfrac{6}{2}$	To isolate x, we undo the multiplication by 2 by dividing all three parts by 2.
$-1 < x \leq 3$	Do the divisions.

The solution set is $(-1, 3]$ and its graph is shown.

Self Check 9
Solve $-6 \leq 3(t + 2) \leq 6$. Write the solution set in interval notation and graph it.

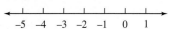

Now Try Problem 69

5 Solve inequality applications.

In some application problems, an inequality, rather than an equation, is needed to find the solution. This is the case when we are asked to determine when one quantity *is more than* (or *is less than*) another. Phrases that call for an inequality are listed in the table on the next page.

The statement	Translates to	The statement	Translates to
a does not exceed b.	$a \leq b$	a will exceed b.	$a > b$
a is at most b.	$a \leq b$	a surpasses b.	$a > b$
a is no more than b.	$a \leq b$	a is at least b.	$a \geq b$
a is between b and c.	$b < a < c$	a is not less than b.	$a \geq b$

Self Check 10

A student has scores of 78%, 82%, and 76% on three exams. What percent score does he need on the last test to earn a grade of no less than a B (80%)?

Now Try Problem 99

EXAMPLE 10 *Grades*
A student has scores of 72%, 74%, and 78% on three exams. What percent score does he need on the last exam to earn a grade of no less than B (80%)?

Analyze We know three scores. We are to find what the student must score on the last exam to earn a grade of B or higher.

Assign We can let x = the score on the fourth (and last) exam.

Form an Inequality To find the average grade, we add the four scores and divide by 4. To earn a grade of *no less than* B, the student's average must be *greater than or equal to* 80%.

The average of the four grades	must be no less than	80.
$\dfrac{72 + 74 + 78 + x}{4}$	$\geq$	80

Solve

$\dfrac{224 + x}{4} \geq 80$ Combine like terms in the numerator: $72 + 74 + 78 = 224$.

$4\left(\dfrac{224 + x}{4}\right) \geq 4(80)$ To clear the inequality of the fraction, multiply both sides by 4.

$224 + x \geq 320$ Simplify each side.

$x \geq 96$ To isolate x, undo the addition of 224 by subtracting 224 from both sides.

State To earn a B, the student must score 96% or better on the last exam. Assuming the student cannot score higher than 100% on the exam, the solution set is written as [96, 100]. The graph is shown below.

Check Pick some numbers in the interval, and verify that the average of the four scores will be 80% or greater.

ANSWERS TO SELF CHECKS

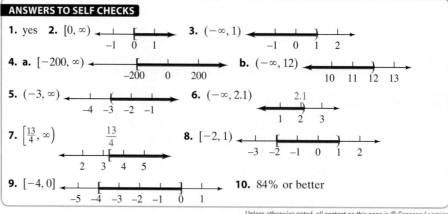

1. yes 2. $[0, \infty)$ 3. $(-\infty, 1)$
4. a. $[-200, \infty)$ b. $(-\infty, 12)$
5. $(-3, \infty)$ 6. $(-\infty, 2.1)$
7. $\left[\dfrac{13}{4}, \infty\right)$ 8. $[-2, 1)$
9. $[-4, 0]$ 10. 84% or better

SECTION 2.7 STUDY SET

VOCABULARY

Fill in the blanks.

1. An _____ is a statement that contains one of the symbols: $>$, $\geq$, $<$, or $\leq$.
2. To _____ an inequality means to find all the values of the variable that make the inequality true.
3. The solution set of $x > 2$ can be expressed in _____ notation as $(2, \infty)$.
4. The inequality $-4 < x \leq 10$ is an example of a _____ inequality.

CONCEPTS

Fill in the blanks.

5. a. Adding the _____ number to both sides of an inequality does not change the solutions.
   b. Multiplying or dividing both sides of an inequality by the same _____ number does not change the solutions.
   c. If we multiply or divide both sides of an inequality by a _____ number, the direction of the inequality symbol must be reversed for the inequalities to have the same solutions.
6. To solve $-4 \leq 2x + 1 < 3$, properties of inequality are applied to all _____ parts of the inequality.
7. Rewrite the inequality $32 < x$ in an equivalent form with the variable on the left side.
8. The solution set of an inequality is graphed below. Which of the four numbers, 3, -3, 2, and 4.5, when substituted for the variable in that inequality, would make it true?

NOTATION

9. Write each symbol.
   a. is less than or equal to
   b. infinity
   c. bracket
   d. is greater than
10. Consider the graph of the interval $[4, 8)$.
    a. Is the endpoint 4 included or not included in the graph?
    b. Is the endpoint 8 included or not included in the graph?

Complete each step to solve the inequality.

11. $\quad 4x - 5 \geq 7$
    $4x - 5 + \square \geq 7 + \square$
    $\quad 4x \geq \square$
    $\quad \dfrac{4x}{\square} \geq \dfrac{12}{\square}$
    $\quad x \geq 3$ \quad Solution set: $[\square, \infty)$

12. $\quad -6x > 12$
    $\quad \dfrac{-6x}{\square} \;\square\; \dfrac{12}{-6}$
    $\quad x < \square$ \quad Solution set: $(\square, -2)$

GUIDED PRACTICE

See Example 1.

13. Determine whether each number is a solution of $3x - 2 > 5$.
    a. 5 \quad b. -4
14. Determine whether each number is a solution of $3x + 7 < 4x - 2$.
    a. 12 \quad b. 9
15. Determine whether each number is a solution of $-5(x - 1) \geq 2x + 12$.
    a. 1 \quad b. -1
16. Determine whether each number is a solution of $\frac{4}{5}a \geq -2$.
    a. $-\dfrac{5}{4}$ \quad b. -15

Graph each inequality and describe the graph using interval notation. See Example 2.

17. $x < 5$
18. $x \geq -2$
19. $-3 < x \leq 1$
20. $-4 \leq x \leq 2$

Write the inequality that is represented by each graph. Then describe the graph using interval notation.

21. ⟵———)———⟶
    $\quad\quad\;\; -1$

22. ⟵———[———⟶
    $\quad\quad\;\; 2$

23. ⟵(—————]⟶
   -7   2

24. ⟵(———]⟶
    4   6

Solve each inequality. Write the solution set in interval notation and graph it. See Examples 3–4.

25. $x + 2 > 5$

26. $x + 5 \geq 2$

27. $g - 30 \geq -20$

28. $h - 18 \leq -3$

29. $8h < 48$

30. $2t > 22$

31. $-\dfrac{3}{16}x \geq -9$

32. $-\dfrac{7}{8}x \leq 21$

33. $-3y \leq -6$

34. $-6y \geq -6$

35. $\dfrac{2}{3}x \geq 2$

36. $\dfrac{3}{4}x < 3$

Solve each inequality. Write the solution set in interval notation and graph it. See Examples 5–6.

37. $9x + 1 > 64$

38. $4x + 8 < 32$

39. $0.5 \geq 2x - 0.3$

40. $0.8 > 7x - 0.04$

41. $\dfrac{x}{8} - (-9) \geq 11$

42. $\dfrac{x}{6} - (-12) > 14$

43. $\dfrac{m}{-42} - 1 > -1$

44. $\dfrac{a}{-25} + 3 < 3$

45. $-x - 3 \leq 7$

46. $-x - 9 > 3$

47. $-3x - 7 > -1$

48. $-5x + 7 \leq 12$

Solve each inequality. Write the solution set in interval notation and graph it. See Example 7.

49. $9a + 4 > 5a - 16$

50. $8t + 1 < 4t - 19$

51. $0.4x \leq 0.1x + 0.45$

52. $0.9s \leq 0.3s + 0.54$

53. $8(5 - x) \leq 10(8 - x)$

54. $17(3 - x) \geq 3 - 13x$

55. $8x + 4 > -(3x - 4)$

56. $7x + 6 \geq -(x - 6)$

57. $\dfrac{1}{2} + \dfrac{n}{5} > \dfrac{3}{4}$

58. $\dfrac{1}{3} + \dfrac{c}{5} > -\dfrac{3}{2}$

59. $\dfrac{6x + 1}{4} \leq x + 1$

60. $\dfrac{3x - 10}{5} \leq x + 4$

Solve each compound inequality. Write the solution set in interval notation and graph it. See Examples 8–9.

61. $2 < x - 5 < 5$

62. $-8 < t - 8 < 8$

63. $0 \leq x + 10 \leq 10$

64. $-9 \leq x + 8 < 1$

65. $-3 \leq \dfrac{c}{2} \leq 5$

66. $-12 < \dfrac{b}{3} < 0$

67. $3 \leq 2x - 1 < 5$
68. $4 < 3x - 5 \leq 7$
69. $-9 < 6x + 9 \leq 45$
70. $-30 \leq 10d + 20 < 90$
71. $6 < -2(x - 1) < 12$
72. $4 \leq -4(x - 2) < 20$

TRY IT YOURSELF

Solve each inequality or compound inequality. Write the solution set in interval notation and graph it.

73. $6 - x \leq 3(x - 1)$
74. $3(3 - x) \geq 6 + x$
75. $\dfrac{y}{4} + 1 \leq -9$
76. $\dfrac{r}{8} - 7 \geq -8$
77. $0 < 5(x + 2) \leq 15$
78. $-18 \leq 9(x - 5) < 27$
79. $-1 \leq -\dfrac{1}{2}n$
80. $-3 \geq -\dfrac{1}{3}t$
81. $-m - 12 > 15$
82. $-t + 5 < 10$
83. $-\dfrac{2}{3} \geq \dfrac{2y}{3} - \dfrac{3}{4}$
84. $-\dfrac{2}{9} \geq \dfrac{5x}{6} - \dfrac{1}{3}$
85. $9x + 13 \geq 2x + 6x$
86. $7x - 16 < 2x + 4x$
87. $7 < \dfrac{5}{3}a + (-3)$
88. $5 < \dfrac{7}{2}a + (-9)$
89. $-8 \leq \dfrac{y}{8} - 4 \leq 2$
90. $6 < \dfrac{m}{16} + 7 < 8$
91. $-2(2x - 3) > 17$
92. $-3(x + 0.2) < 0.3$
93. $\dfrac{5}{3}(x + 1) \geq -x + \dfrac{2}{3}$
94. $\dfrac{5}{2}(7x - 15) \geq \dfrac{11}{2}x - \dfrac{3}{2}$
95. $2x + 9 \leq x + 8$
96. $3x + 7 \leq 4x - 2$
97. $-7x + 1 < -5$
98. $-3x - 10 \geq -5$

APPLICATIONS

99. **GRADES** A student has test scores of 68%, 75%, and 79% in a government class. What must she score on the last exam to earn a B (80% or better) in the course?

100. **OCCUPATIONAL TESTING** An employment agency requires that applicants average at least 70% on a battery of four job skills tests. If an applicant scored 70%, 74%, and 84% on the first three exams, what must he score on the fourth test to maintain a 70% or better average?

101. **GAS MILEAGE** A car manufacturer produces three models in equal quantities. One model has an economy rating of 17 miles per gallon, and the second model is rated at 19 mpg. If government regulations require the manufacturer to have a fleet average that exceeds 21 mpg, what economy rating is required for the third model?

102. **SERVICE CHARGES** When the average daily balance of a customer's checking account falls below $500 in any week, the bank assesses a $5 service charge. The table shows the daily balances of one customer. What must Friday's balance be to avoid the service charge?

Day	Balance
Monday	$540.00
Tuesday	$435.50
Wednesday	$345.30
Thursday	$310.00

103. GEOMETRY The perimeter of an equilateral triangle is at most 57 feet. What could the length of a side be? (*Hint:* All three sides of an equilateral triangle are equal.)

104. GEOMETRY The perimeter of a square is no less than 68 centimeters. How long can a side be?

105. COUNTER SPACE A rectangular counter is being built for the customer service department of a store. Designers have determined that the outside perimeter of the counter (shown in red) needs to exceed 30 feet. Determine the acceptable values for x.

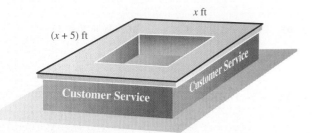

106. NUMBER PUZZLES What numbers satisfy the condition: Four more than three times the number is at most 10?

107. GRADUATIONS It costs a student $18 to rent a cap and gown and 80 cents for each graduation announcement that she orders. If she doesn't want her spending on these graduation costs to exceed $50, how many announcements can she order?

108. TELEPHONES A cellular telephone company has currently enrolled 36,000 customers in a new calling plan. If an average of 1,200 people are signing up for the plan each day, in how many days will the company surpass its goal of having 150,000 customers enrolled?

109. WINDOWS An architect needs to design a triangular bathroom window that has an area no greater than 100 in.^{2}. If the base of the window must be 16 inches long, what window heights will meet this condition?

110. ROOM TEMPERATURES To hold the temperature of a room between 19° and 22° Celsius, what Fahrenheit temperatures must be maintained? (*Hint:* Use the formula $C = \frac{5}{9}(F - 32)$.)

111. EXERCISE The graph below shows the target heartbeat ranges for different ages and exercise intensity levels. If we let b represent the number of beats per minute, then the compound inequality that estimates the heartbeat rate range for a 30-year-old involved in a high-intensity workout is about $168 \leq b \leq 198$. Use a compound inequality to estimate the heartbeat rate ranges for the following ages and zones.

a. 45-year-old, fat-burning zone

b. 70-year-old, high-intensity zone

c. 25-year-old, intermediate zone

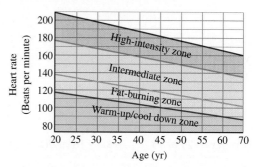

Source: elitefitness.co.nz

112. NUMBER PUZZLES What *whole* numbers satisfy the condition: Twice the number decreased by 1 is between 50 and 60?

WRITING

113. Explain why multiplying both sides of an inequality by a negative number reverses the direction of the inequality.

114. Explain the use of parentheses and brackets for graphing intervals.

REVIEW

Complete each table.

115.

x	$x^2 - 3$
-2	
0	
3	

116.

x	$\dfrac{x}{3} + 2$
-6	
0	
12	

STUDY SKILLS CHECKLIST

Preparing for the Chapter 2 Test

There are common mistakes that students make when working with the topics of Chapter 2. To make sure you are prepared for your test and to help you avoid these common mistakes, study this checklist below.

☐ When solving an equation, if you multiply by the LCD to remove the fractions, remember to multiply the other side by the same number.

$$\frac{2}{3}x + \frac{1}{12} = \frac{1}{4}x$$

$$12\left(\frac{2}{3}x + \frac{1}{12}\right) = 12\left(\frac{1}{4}x\right)$$

☐ When *solving an equation*, find a simpler equivalent equation of the form $x =$ a number. When *simplifying expressions*, remove the parentheses and combine like terms.

☐ Do not try to divide by the coefficient of the variable until that term is isolated.

$$-3x + 9 = 16$$
$$-3x + 9 - 9 = 16 - 9$$
$$-3x = 7$$
$$\frac{-3x}{-3} = \frac{7}{-3}$$
$$x = -\frac{7}{3}$$

☐ When solving inequalities, remember to *change the direction of the inequality symbol* when multiplying or dividing both sides by a negative number.

CHAPTER 2 SUMMARY AND REVIEW

SECTION 2.1 Solving Equations Using Properties of Equality

DEFINITIONS AND CONCEPTS	EXAMPLES
An **equation** is a statement indicating that two expressions are equal. The equal symbol = separates an equation into two parts: the left side and the right side.	$2x + 4 = 10$ $-5(a + 4) = -11a$ $\frac{3}{2}t + 6 = t - \frac{1}{3}$
A number that makes an equation a true statement when substituted for the variable is called a **solution** of the equation.	Determine whether 2 is a solution of $x + 4 = 3x$. **Check:** $x + 4 = 3x$ $2 + 4 \stackrel{?}{=} 3(2)$ Substitute 2 for each x. $6 = 6$ True Since the resulting statement is true, 2 is a solution.
Equivalent equations have the same solutions.	$x - 2 = 6$ and $x = 8$ are equivalent equations because they have the same solution, 8.

To **solve an equation**, isolate the variable on one side of the equation by undoing the operations performed on it using properties of equality. **Addition (subtraction) property of equality:** If the same number is added to (or subtracted from) both sides of an equation, the result is an equivalent equation.	Solve: $x - 5 = 7$ $x - 5 + 5 = 7 + 5$ $x = 12$ The solution is 12.	Solve: $c + 9 = 16$ $c + 9 - 9 = 16 - 9$ $c = 7$ The solution is 7.
Multiplication (division) property of equality: If both sides of an equation are multiplied (or divided) by the same nonzero number, the result is an equivalent equation.	Solve: $\frac{1}{3}m = 2$ $3\left(\frac{1}{3}m\right) = 3(2)$ $m = 6$ The solution is 6.	Solve: $10y = 50$ $\frac{10y}{10} = \frac{50}{10}$ $y = 5$ The solution is 5.

REVIEW EXERCISES

Use a check to determine whether the number in red is a solution of the equation.

1. 84, $x - 34 = 50$
2. 3, $5y + 2 = 12$
3. -30, $\frac{x}{5} = 6$
4. 2, $a^2 - a - 1 = 0$
5. -3, $5b - 2 = 3b - 8$
6. 1, $\frac{2}{y+1} = \frac{12}{y+1} - 5$

Fill in the blanks.

7. An _____ is a statement indicating that two expressions are equal.
8. To solve $x - 8 = 10$ means to find all the values of the variable that make the equation a _____ statement.

Solve each equation and check the result.

9. $x - 9 = 12$
10. $-y = -32$
11. $a + 3.7 = -16.9$
12. $100 = -7 + r$
13. $120 = 5c$
14. $t - \frac{1}{2} = \frac{3}{2}$
15. $\frac{4}{3}t = -12$
16. $3 = \frac{q}{-2.6}$
17. $6b = 0$
18. $\frac{15}{16}s = -3$

SECTION 2.2 More about Solving Equations

DEFINITIONS AND CONCEPTS	EXAMPLES	
A five-step **strategy for solving linear equations**: 1. *Clear* the equation of fractions or decimals. 2. *Simplify* each side. Use the distributive property and combine like terms when necessary. 3. *Isolate the variable term.* Use the addition and subtraction properties of equality. 4. *Isolate the variable.* Use the multiplication and division properties of equality. 5. *Check* the result in the original equation.	Solve: $2(y + 2) + 4y = 11 - y$ $2y + 4 + 4y = 11 - y$ $6y + 4 = 11 - y$ $6y + 4 + y = 11 - y + y$ $7y + 4 = 11$ $7y + 4 - 4 = 11 - 4$ $7y = 7$ $\frac{7y}{7} = \frac{7}{7}$ $y = 1$ The solution is 1, and the solution set is {1}. Verify this using a check.	Distribute the multiplication by 2. Combine like terms: $2y + 4y = 6y$. To eliminate $-y$ on the right, add y to both sides. Combine like terms. To isolate the variable term $7y$, subtract 4 from both sides. Simplify each side of the equation. To isolate y, divide both sides by 7. Do the division.

To clear an equation of fractions, multiply both sides of the equation by the LCD.	To solve $\frac{1}{2} + \frac{x}{3} = \frac{3}{4}$, first multiply both sides by 12: $$12\left(\frac{1}{2} + \frac{x}{3}\right) = 12\left(\frac{3}{4}\right)$$
To clear an equation of decimals, multiply both sides by a power of 10 to change the decimals in the equation to integers.	To solve $0.5(x - 4) = 0.1x + 0.2$, first multiply both sides by 10: $$10[0.5(x - 4)] = 10(0.1x + 0.2)$$
A linear equation in one variable that is true for all values of the variable is called an **identity**.	When we solve $x + 5 + x = 2x + 5$, the variables drop out and we obtain a true statement $5 = 5$. All real numbers are solutions.
An equation that is not true for any value of its variable is called a **contradiction**.	When we solve $y + 2 = y$, the variables drop out and we obtain a false statement $2 = 0$. The equation has no solutions.

REVIEW EXERCISES

Solve each equation. Check the result.

19. $5x + 4 = 14$
20. $98.6 - t = 129.2$
21. $\frac{n}{5} + (-2) = 4$
22. $\frac{b - 5}{4} = -6$
23. $5(2x - 4) - 5x = 0$
24. $-2(x - 5) = 5(-3x + 4) + 3$
25. $\frac{3}{4} = \frac{1}{2} + \frac{d}{5}$
26. $\frac{5(7 - x)}{4} = 2x - 3$
27. $\frac{3(2 - c)}{2} = \frac{-2(2c + 3)}{5}$
28. $\frac{b}{3} + \frac{11}{9} + 3b = -\frac{5}{6}b$
29. $0.15(x + 2) + 0.3 = 0.35x - 0.4$
30. $0.5 - 0.02(y - 2) = 0.16 + 0.36y$
31. $3(a + 8) = 6(a + 4) - 3a$
32. $2(y + 10) + y = 3(y + 8)$

SECTION 2.3   Applications of Percent

DEFINITIONS AND CONCEPTS	EXAMPLES
To solve **percent problems**, use the facts of the problem to write a sentence of the form: ⎕ is ⎕% of ⎕ ? Translate the sentence to mathematical symbols: *is* translates to an = symbol, and *of* means multiply. Then solve the equation.	648 is 30% of what number? $$648 = 30\% \cdot x \quad \text{Translate.}$$ $$648 = 0.30x \quad \text{Change 30\% to a decimal: } 30\% = 0.30.$$ $$\frac{648}{0.30} = \frac{0.30x}{0.30} \quad \text{To isolate } x, \text{ divide both sides by 0.30.}$$ $$2{,}160 = x \quad \text{Do the division.}$$ Thus, 648 is 30% of 2,160.
To find the **percent of increase** or **the percent of decrease**, find what percent the increase or decrease is of the original amount.	SALE PRICES To find the percent of decrease when ground beef prices are reduced from $4.89 to $4.59 per pound, we first find the amount of decrease: $4.89 - 4.59 = 0.30$. Then we determine what percent 0.30 is of 4.89 (the original price). 0.30 is what% of 4.89? $$0.30 = x \cdot 4.89 \quad \text{Translate.}$$

$$0.30 = 4.89x$$
$$\frac{0.30}{4.89} = \frac{4.89x}{4.89} \quad \text{To isolate } x, \text{ divide both sides by 4.89.}$$
$$0.061349693 \approx x \quad \text{Do the division.}$$
$$006.1349693\% \approx x \quad \text{Write the decimal as a percent.}$$

To the nearest tenth of a percent, the percent of decrease is 6.1%.

REVIEW EXERCISES

33. Fill in the blanks.
    a. _____ means parts per one hundred.
    b. When the price of an item is reduced, we call the amount of the reduction a _____.
    c. An employee who is paid a _____ is paid a percent of the goods or services that he or she sells.

34. 4.81 is 2.5% of what number?

35. What number is 15% of 950?

36. What percent of 410 is 49.2?

37. INTERNET USAGE Refer to the graph below.
    a. When, for the first time, does the graph indicate that more than half of Americans age 65+ were using the Internet?
    b. In 2012, the number of Americans age 65+ was estimated to be 42,450,000. About how many were using the Internet in April 2012?

38. COST OF LIVING A retired trucker receives a monthly Social Security check of $764. If she is to receive a 3.5% cost-of-living increase soon, how much larger will her check be?

39. FAMILY BUDGETS It is recommended that a family pay no more than 30% of its monthly income (after taxes) on housing. If a family has an after-tax income of $1,890 per month and pays $625 in housing costs each month, are they within the recommended range?

40. DISCOUNTS A shopper saved $148.50 on a food processor that was discounted 33%. What did it originally cost?

41. TUPPERWARE The hostess of a Tupperware party is paid a 25% commission on her in-home party's sales. What would the hostess earn if sales totaled $600?

42. COLLECTIBLES A collector of football trading cards paid $6 for a 1984 Dan Marino rookie card several years ago. If the card is now worth $100, what is the percent of increase in the card's value? (Round to the nearest percent.)

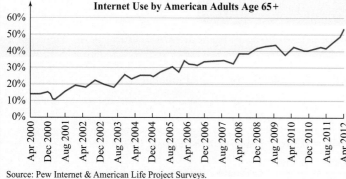

Source: Pew Internet & American Life Project Surveys.

SECTION 2.4 Formulas

DEFINITIONS AND CONCEPTS	EXAMPLES
A **formula** is an equation that states a relationship between two or more variables.	Retail price: $r = c + m$ Profit: $p = r - c$ Simple Interest: $I = Prt$ Distance: $d = rt$ Temperature: $C = \frac{5}{9}(F - 32)$

Chapter 2 Summary and Review 201

The **perimeter** of a plane geometric figure is the distance around it. The **area** of a plane geometric figure is the amount of surface that it encloses. The **volume** of a three-dimensional geometric solid is the amount of space it encloses.	Rectangle: $P = 2l + 2w$ Circle: $C = \pi D$ or $C = 2\pi r$ $\qquad A = lw \qquad\qquad\qquad A = \pi r^2$ Rectangular solid: $V = lwh$ Cylinder: $V = \pi r^2 h$ *See the inside back cover of the text for more geometric formulas.
If we are given the values of all but one of the variables in a formula, we can use our equation-solving skills to find the value of the remaining variable.	BEDDING The area of a standard queen-size bed sheet is 9,180 in.². If the width is 102 inches, what is the length? $A = lw$ This is the formula for the area of a rectangle. $9{,}180 = 102w$ Substitute 9,180 for the area A and 102 for the width w. $\dfrac{9{,}180}{102} = \dfrac{102w}{102}$ To isolate w, divide both sides by 102. $90 = w$ Do the division. The length of a standard queen-size bed sheet is 90 inches.
To solve a formula for a specific variable means to isolate that variable on one side of the equation, with all other variables and constants on the opposite side. Treat the specified variable as if it were the only variable in the equation. Treat the other variables as if they were numbers (constants).	Solve the formula for the volume of a cone for h. $V = \dfrac{1}{3}\pi r^2 h$ This is the formula for the volume of a cone. $3(V) = 3\left(\dfrac{1}{3}\pi r^2 h\right)$ To clear the equation of the fraction, multiply both sides by 3. $3V = \pi r^2 h$ Simplify. $\dfrac{3V}{\pi r^2} = \dfrac{\pi r^2 h}{\pi r^2}$ To isolate h, divide both sides by πr^2. $\dfrac{3V}{\pi r^2} = h$ or $h = \dfrac{3V}{\pi r^2}$

REVIEW EXERCISES

43. SHOPPING Find the markup on a CD player whose wholesale cost is $219 and whose retail price is $395.

44. RESTAURANTS One month, a restaurant had sales of $13,500 and made a profit of $1,700. Find the expenses for the month.

45. SNAILS A typical garden snail travels at an average rate of 2.5 feet per minute. How long would it take a snail to cross a 20-foot-long flower bed?

46. CERTIFICATES OF DEPOSIT A $26,000 investment in a CD earned $1,170 in interest the first year. What was the annual interest rate?

47. JEWELRY Gold melts at about 1,065°C. Change this to degrees Fahrenheit.

48. CAMPING
    a. Find the perimeter of the air mattress in the next column.
    b. Find the amount of sleeping area on the top surface of the air mattress.
    c. Find the approximate volume of the air mattress if it is 3 inches thick.

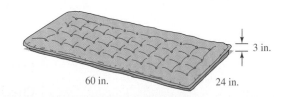

49. Find the area of a triangle with a base 17 meters long and a height of 9 meters.

50. Find the area of a trapezoid with bases 11 inches and 13 inches long and a height of 12 inches.

51. a. Find the circumference of a circle with a radius of 8 centimeters. Round to the nearest hundredth of one centimeter.
    b. Find the area of the circle. Round to the nearest square centimeter.

52. Find the volume of a 12-foot cylinder whose circular base has a radius of 0.5 feet. Give the result to the nearest tenth.

53. Find the volume of a pyramid that has a square base, measuring 6 feet on a side, and a height of 10 feet.

54. HALLOWEEN After being cleaned out, a spherical pumpkin has an inside diameter of 9 inches. To the nearest hundredth, what is its volume?

Solve each formula for the specified variable.

55. $A = 2\pi rh$ for h

56. $A - BC = \dfrac{G - K}{3}$ for G

57. $C = \dfrac{1}{4}s(t - d)$ for t

58. $4y - 3x = 16$ for y

SECTION 2.5 Problem Solving

DEFINITIONS AND CONCEPTS

To solve application problems, use the six-step problem-solving strategy.

1. Analyze the problem.
2. Assign a variable.
3. Form an equation.
4. Solve the equation.
5. State the conclusion.
6. Check the result.

EXAMPLES

INCOME TAXES After taxes, an author kept $85,340 of her total annual earnings. If her earnings were taxed at a 15% rate, how much did she earn that year?

Analyze the Problem The author earned some unknown amount of money. On that amount, she paid 15% in taxes. The difference between her total earnings and the taxes paid was $85,340.

Assign a Variable If we let $x =$ the author's total earnings, the amount of taxes that she paid was 15% of x or $0.15x$.

Form an Equation We can use the words of the problem to form an equation.

Her total earnings	minus	the taxes that she paid	equals	the money that she kept.
x	$-$	$0.15x$	$=$	$85{,}340$

Solve the Equation

$x - 0.15x = 85{,}340$

$0.85x = 85{,}340$ Combine like terms.

$x = 100{,}400$ To isolate x, divide both sides by 0.85.

State the Conclusion The author earned $100,400 that year.

Check the Result The taxes were 15% of $100,400 or $15,060. If we subtract the taxes from her total earnings, we get $100,400 − $15,060 = $85,340. The answer checks.

REVIEW EXERCISES

59. SOUND SYSTEMS A 45-foot-long speaker wire is to be cut into three pieces. One piece is to be 15 feet long. Of the remaining pieces, one must be 2 feet less than 3 times the length of the other. Find the length of the shorter piece.

60. SIGNING PETITIONS A professional signature collector is paid $50 a day plus $2.25 for each verified signature he gets from a registered voter. How many signatures are needed to earn $500 a day?

61. LOTTERY WINNINGS After taxes, a lottery winner was left with a lump sum of $1,800,000. If 28% of the original prize was withheld to pay federal income taxes, what was the original cash prize?

62. NASCAR The car numbers of drivers Paul Menard and Kevin Harvick are consecutive odd integers whose sum is 56. If Menard's number is the smaller, find the number of each car.

Chapter 2 Summary and Review

63. ART HISTORY *American Gothic* was painted in 1930 by Grant Wood. The length of the painting is 5 inches more than the width. Find the dimensions of the painting if it has a perimeter of $109\frac{1}{2}$ inches.

64. GEOMETRY Find the missing angle measures of the triangle.

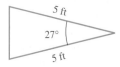

SECTION 2.6 More about Problem Solving

DEFINITIONS AND CONCEPTS	EXAMPLES
To solve application problems, use the six-step problem-solving strategy. 1. Analyze the problem. 2. Assign a variable. 3. Form an equation. 4. Solve the equation. 5. State the conclusion. 6. Check the result. Tables are a helpful way to organize the facts of a problem.	**TRUCKING** Two trucks leave from the same place at the same time traveling in opposite directions. One travels at a rate of 60 mph and the other at 50 mph. How long will it take them to be 165 miles apart? **Analyze the Problem** We know that one truck travels at 60 mph and the other at 50 mph. Together, the trucks will travel a distance of 165 miles. **Assign a Variable** We enter each rate in the table under the heading *r*. Since the trucks travel for the same length of time, say *t* hours, we enter *t* for each truck under the heading *t*. Since $d = r \cdot t$, the first truck will travel $60t$ miles and the second will travel $50t$ miles.

	r	· t =	d
Truck 1	60	t	60t
Truck 2	50	t	50t
		Total:	165

Use the information in this column to form an equation.

Form an Equation

The distance the first truck travels	plus	the distance the second truck travels	is	165 miles.
$60t$	$+$	$50t$	$=$	165

Solve the Equation

$$60t + 50t = 165$$
$$110t = 165 \quad \text{Combine like terms.}$$
$$\frac{110t}{110} = \frac{165}{110} \quad \text{To isolate } t, \text{ divide both sides by 110.}$$
$$t = 1.5 \quad \text{Do the division.}$$

State the Conclusion The trucks will be 165 miles apart in 1.5 hours.

Check the Result If the first truck travels 60 mph for 1.5 hours, it will go $60(1.5) = 90$ miles. If the second truck travels 50 mph for 1.5 hours, it will go $50(1.5) = 75$ miles. Since 90 miles + 75 miles = 165 miles, the result checks.

REVIEW EXERCISES

65. INVESTMENT INCOME A woman has $27,000. Part is invested for 1 year in a certificate of deposit paying 7% interest, and the remaining amount in a cash management fund paying 9%. After 1 year, the total interest on the two investments is $2,110. How much is invested at each rate?

66. WALKING AND BICYCLING A bicycle path is 5 miles long. A man walks from one end at the rate of 3 mph. At the same time, a friend bicycles from the other end, traveling at 12 mph. In how many minutes will they meet?

67. AIRPLANES How long will it take a jet plane, flying at 450 mph, to overtake a propeller plane, flying at 180 mph, if the propeller plane had a $2\frac{1}{2}$-hour head start?

68. AUTOGRAPHS Kesha has collected the autographs of 8 more television celebrities than she has of movie stars. Each TV celebrity autograph is worth $75, and each movie star autograph is worth $250. If her collection is valued at $1,900, how many of each type of autograph does she have?

69. MIXTURES A store manager mixes candy worth 90¢ per pound with gumdrops worth $1.50 per pound to make 20 pounds of a mixture worth $1.20 per pound. How many pounds of each kind of candy does he use?

70. MILK Cream is about 22% butterfat, and low-fat milk is about 2% butterfat. How many gallons of cream must be mixed with 18 gallons of low-fat milk to make whole milk that contains 4% butterfat?

SECTION 2.7 Solving Inequalities

DEFINITIONS AND CONCEPTS	EXAMPLES
An **inequality** is a mathematical statement that contains an $>$, $<$, $\geq$, or $\leq$ symbol.	$3x < 8$ $\quad \frac{1}{2}y - 4 \geq 12 \quad$ $2z + 4 \leq z - 5$
A **solution of an inequality** is any number that makes the inequality true.	Determine whether 3 is a solution of $2x - 7 < 5$. **Check:** $\quad 2x - 7 < 5$ $\quad\quad\quad\quad 2(3) - 7 \stackrel{?}{<} 5 \quad$ Substitute 3 for x. $\quad\quad\quad\quad\quad -1 < 5 \quad$ True Since the resulting statement is true, 3 is a solution.
We **solve inequalities** as we solve equations. However, if we multiply or divide both sides by a negative number, we must reverse the inequality symbol.	Solve: $\quad -3(z - 1) \geq -6$ $\quad\quad\quad\quad -3z + 3 \geq -6 \quad$ Distribute the multiplication by -3. $\quad\quad\quad\quad\quad -3z \geq -9 \quad$ To isolate the variable term $-3z$, subtract 3 from both sides. $\quad\quad\quad\quad \frac{-3z}{-3} \leq \frac{-9}{-3} \quad$ To isolate z, divide both sides by -3. Reverse the inequality symbol. $\quad\quad\quad\quad\quad z \leq 3 \quad$ Do the division.
Interval notation can be used to describe the solution set of an inequality. A **parenthesis** indicates that a number is not in the solution set of an inequality. A **bracket** indicates that a number is included in the solution set.	In interval notation, the solution set is $(-\infty, 3]$, whose graph is shown. $\quad\quad\quad \longleftarrow \mid \quad \mid \quad \mid \quad]\!\!-\!\!-\!\!\mid \longrightarrow$ $\quad\quad\quad\quad 0 \quad 1 \quad 2 \quad 3 \quad 4$

REVIEW EXERCISES

Solve each inequality. Write the solution set in interval notation and graph it.

71. $3x + 2 < 5$

72. $-\dfrac{3}{4}x \geq -9$

73. $\dfrac{3}{4} < \dfrac{d}{5} + \dfrac{1}{2}$

74. $5(3 - x) \leq 3(x - 3)$

75. $\dfrac{t}{-5} - (-1.8) \geq -6.2$

76. $63 < 7a$

77. $8 < x + 2 < 13$

78. $0 \leq 3 - 2x < 10$

79. SPORTS EQUIPMENT The acceptable weight w of Ping-Pong balls used in competition can range from 2.40 to 2.53 grams. Express this range using a compound inequality.

80. SIGNS A large office complex has a strict policy about signs. Any sign to be posted in the building must be rectangular in shape, its width must be 18 inches, and its perimeter is not to exceed 132 inches. What possible sign lengths meet these specifications?

CHAPTER 2 TEST

1. Fill in the blanks.
   a. To _____ an equation means to find all of the values of the variable that make the equation true.
   b. _____ means parts per one hundred.
   c. The distance around a circle is called its _____.
   d. An _____ is a statement that contains one of the symbols $>$, $\geq$, $<$, or $\leq$.
   e. The _____ property of _____ says that multiplying both sides of an equation by the same nonzero number does not change its solution.

2. Use a check to determine whether 3 is a solution of $5y + 2 = 12$.

Solve each equation.

3. $3h + 2 = 8$
4. $\frac{4}{5}t = -4$
5. $-22 = -x$
6. $\frac{11b - 11}{5} = 3b - 2$
7. $0.8(5x - 1,000) + 1.3 = 2.9 - x$
8. $2(y - 7) - 3y = -(y - 3) - 17$
9. $\frac{m}{2} - \frac{1}{3} = \frac{1}{4} + \frac{m}{6}$
10. $9 - 5(2x + 10) = -1$
11. $24t = -6(8 - 4t)$
12. $6a + (-7) = 3a - 7 + 2a$
13. What is 15.2% of 80?

14. **DOWN PAYMENTS** To buy a house, a woman was required to make a down payment of $26,400. What did the house sell for if this was 15% of the purchase price?

15. **IMPROVING HORSEPOWER** The following graph shows how the installation of a special computer chip increases the horsepower of a truck. Find the percent of increase in horsepower for the engine running at 4,000 revolutions per minute (rpm) and round to the nearest tenth of one percent.

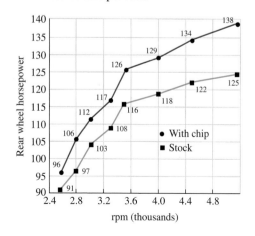

16. **COMMISSIONS** An appliance store salesperson receives a commission of 5% of the price of every item that she sells. How much will she make if she sells a $599.99 refrigerator?

17. **GRAND OPENINGS** On its first night of business, a pizza parlor brought in $445. The owner estimated his profits that night to be $150. What were his costs?

18. Find the Celsius temperature reading if the Fahrenheit reading is 14°.

19. **PETS** The spherical fishbowl is three-quarters full of water. To the nearest cubic inch, find the volume of water in the bowl. (*Hint:* The volume of a sphere is given by $V = \frac{4}{3}\pi r^3$.)

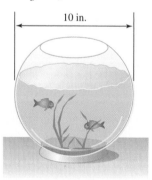

20. Solve $A = P + Prt$ for r.

21. **IRONS** Estimate the area of the soleplate of the iron.

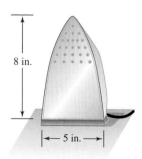

22. **TELEVISION** In a typical 30-minute block of time on TV, the number of programming minutes is 2 less than three times the number of minutes of commercials. How many minutes of programming and commercials are there?

23. **HOME SALES** A condominium owner cleared $114,600 on the sale of his condo after paying a 4.5% real estate commission. What was the selling price?

24. **COLORADO** The border of the state of Colorado is approximately a rectangle with a perimeter of 1,320 miles. Find the length (east to west) and width (north to south), if the length is 100 miles longer than the width.

25. **TEA** How many pounds of green tea, worth $40 a pound, should be mixed with herbal tea, worth $50 a pound, to produce 20 pounds of a blend worth $42 a pound?

26. **READING** A bookmark is inserted between two page numbers whose sum is 825. What are the page numbers?

27. The following table shows the four types of problems an instructor put on a history test and the point value for each type.
    a. Complete the table.
    b. Write an equation that could be used to find x.

Type of question	Number	·	Value	=	Total value
Multiple choice	x		5		
True/false	$3x$		2		
Essay	$x - 2$		10		
Fill-in	x		5		
					Total: 110 points

28. **"LIGHTNING BOLT"** In 2010, Usain Bolt of Jamaica held the world records for the 100-meter and 200-meter sprints. His *maximum stride angle*, shown below, is 5° less than 1.5 times its supplement. Find his maximum stride angle. (Source: somaxsports.com)

Maximum stride angle

29. **TRAVEL TIMES** A car leaves Rockford, Illinois, at the rate of 65 mph, bound for Madison, Wisconsin. At the same time, a truck leaves Madison at the rate of 55 mph, bound for Rockford. If the cities are 72 miles apart, how long will it take for the car and the truck to meet?

30. **PICKLES** To make pickles, fresh cucumbers are soaked in a salt water solution called *brine*. How many liters of a 2% brine solution must be added to 30 liters of a 10% brine solution to dilute it to an 8% solution?

31. **GEOMETRY** If the vertex angle of an isosceles triangle is 44°, find the measure of each base angle.

32. **INVESTMENTS** Part of $13,750 was invested at 9% annual interest, and the rest was invested at 8%. After one year, the accounts paid $1,185 in interest. How much was invested at the lower rate?

Solve each inequality. Write the solution set in interval notation and graph it.

33. $-8x - 20 \leq 4$

34. $-8.1 > \dfrac{t}{2} + (-11.3)$

35. $-12 \leq 2(x + 1) < 10$

36. **AWARDS** A city honors its citizen of the year with a framed certificate. An artist charges $15 for the frame and 75 cents per word for writing out the proclamation. If a city regulation does not allow gifts in excess of $150, what is the maximum number of words that can be written on the certificate?

CHAPTERS 1–2 CUMULATIVE REVIEW

1. Classify each of the following as an equation or an expression. [Section 1.1]
   a. $4m - 3 + 2m$
   b. $4m = 3 + 2m$

2. If $t =$ time and $w =$ weight, use the formula $t = \dfrac{w}{5}$ to complete the table. [Section 1.1]

Weight (lb)	Cooking time (hr)
15	
20	
25	

3. Write each phrase as an algebraic expression. [Section 1.1]
   a. The sum of the width w and 12.
   b. Four less than a number n.

4. Give the prime factorization of 100. [Section 1.2]

5. Simplify: $\dfrac{24}{36}$ [Section 1.2]

6. Multiply: $\dfrac{11}{21}\left(-\dfrac{14}{33}\right)$ [Section 1.2]

7. COOKING A recipe calls for $\frac{3}{4}$ cup of flour, and the only measuring container you have holds $\frac{1}{8}$ of a cup. How many $\frac{1}{8}$ cups of flour would you need to use to follow the recipe? [Section 1.2]

8. Add: $\dfrac{4}{5} + \dfrac{2}{3}$ [Section 1.2]

9. Subtract: $42\dfrac{1}{8} - 29\dfrac{2}{3}$ [Section 1.2]

10. Write $\dfrac{15}{16}$ as a decimal. [Section 1.3]

11. Evaluate each expression. [Section 1.3]
    a. $|-65|$
    b. $-|-12|$

12. Determine whether the statement is true or false. [Section 1.3]
    a. Every natural number is a whole number.
    b. -3 is a rational number.
    c. π is a rational number.

13. Perform each operation.
    a. $-6 + (-12) + 8$ [Section 1.4]
    b. $-15 - (-1)$ [Section 1.5]
    c. $2(-32)$ [Section 1.6]
    d. $\dfrac{0}{35}$ [Section 1.6]

14. Write each product using exponents. [Section 1.7]
    a. $4 \cdot 4 \cdot 4$
    b. $\pi \cdot r \cdot r \cdot h$

15. SICK DAYS Use the data in the table to find the average (mean) number of sick days used by this group of employees this year. [Section 1.7]

Name	Sick days	Name	Sick days
Chung	4	Ryba	0
Cruz	8	Nguyen	5
Damron	3	Tomaka	4
Hammond	2	Young	6

16. Complete the table. [Section 1.8]

x	$x^2 - 3$
-2	
0	
3	

Let $x = -5$, $y = 3$, and $z = 0$. Evaluate each expression. [Section 1.8]

17. $(3x - 2y)z$

18. $\dfrac{x - 3y + |z|}{2 - x}$

19. $x^2 - y^2 + z^2$

20. $\dfrac{x}{y} + \dfrac{y + 2}{3 - z}$

Simplify each expression. [Section 1.9]

21. $-8(4d)$

22. $5(2x - 3y + 1)$

23. $2x + 3x$

24. $3a + 6a - 17a$

25. $q(q - 5) + 7q^2$

26. $5(6t - 4) - 3t$

Solve each equation. [Sections 2.1 and 2.2]

27. $3x - 4 = 23$
28. $\dfrac{x}{5} + 3 = 7$
29. $-5p + 0.7 = 3.7$
30. $\dfrac{y - 4}{5} = 3$
31. $-\dfrac{4}{5}x = 16$
32. $-9(n + 2) - 2(n - 3) = 10$
33. $9y - 3 = 6y$
34. $\dfrac{1}{2} + \dfrac{x}{5} = \dfrac{3}{4}$

35. 45 is 15% of what number? [Section 2.3]

36. What is 35% of 250? [Section 2.3]

37. 65 is what percent of 260? [Section 2.3]

38. **TIPS** A diner at a local restaurant received a bill for $10.75. Compute a 15% tip (rounded up to the nearest dollar). [Section 2.3]

39. Find the area of a rectangle with sides of 5 meters and 13 meters. [Section 2.4]

40. Find the volume of a cone that is 10 centimeters tall and has a circular base whose diameter is 12 centimeters. Round to the nearest hundredth. [Section 2.4]

41. Solve $A = P + Prt$ for t. [Section 2.4]

42. **WORK** Physicists say that *work* is done when an object is moved a distance d by a force F. To find the work done (in foot-pounds), we can use the formula $W = Fd$. Find the work done in lifting the bundle of newspapers onto the workbench shown in the illustration. (*Hint:* The force that must be applied to lift the newspapers is equal to the weight of the newspapers.) [Section 2.4]

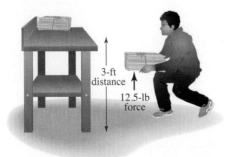

43. **WORK** See Exercise 42. Find the weight of a 1-gallon can of paint if the amount of work done to lift it onto the workbench is 28.35 foot-pounds. [Section 2.4]

44. Find the unknown angle measure represented by x. [Section 2.5]

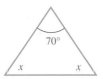

45. **INVESTING** An investment club invested part of $10,000 at 9% annual interest and the rest at 8%. If the annual income from these investments was $860, how much was invested at 8%? [Section 2.6]

46. **GOLDSMITHING** How many ounces of a 40% gold alloy must be mixed with 10 ounces of a 10% gold alloy to obtain an alloy that is 25% gold? [Section 2.6]

Solve each inequality, graph the solution, and use interval notation to describe the solution. [Section 2.7]

47. $x - 4 > -6$

48. $-6x \geq -12$

49. $8x + 4 \geq 5x + 1$

50. $\dfrac{x}{4} - \dfrac{1}{3} > \dfrac{5}{6} + \dfrac{x}{3}$

51. $-1 \leq 2x + 1 < 5$

52. **ICE SKATING** For the free-skating portion of a competition, an ice skater received scores of 5.3, 4.8, 4.7, 4.9, and 5.1 from the first five judges. What score must she receive from the sixth and final judge to average better than 5.0 for her performance? [Section 2.7]

3 Graphs, Linear Equations, and Inequalities in Two Variables; Functions

3.1 Graphing Using the Rectangular Coordinate System
3.2 Equations Containing Two Variables
3.3 Graphing Linear Equations
3.4 Rate of Change and the Slope of a Line
3.5 Slope–Intercept Form
3.6 Point–Slope Form; Writing Linear Equations
3.7 Graphing Linear Inequalities
3.8 Functions
Chapter Summary and Review
Chapter Test
Cumulative Review

from Campus to Careers
Physical Therapist

Physical therapists create treatment plans that improve a patient's ability to move, relieve pain, and prevent or lessen physical disabilities. Physical therapists use specialized equipment, wheelchairs, and crutches and use their own strength to help patients work through sometimes painful exercises. Strong communication skills and a desire to help people in need are also important.

In **Problem 52** of **Study Set 3.1**, you will see how physical therapists can detect scoliosis of the human spine.

JOB TITLE: Physical Therapist
EDUCATION: Physical therapists are required to hold at least an associate's degree and must pass a state licensure test.
JOB OUTLOOK: Employment is expected to grow by 39% through the year 2020.
ANNUAL EARNINGS: In 2012, the mean annual salary for physical therapists was $81,110.
FOR MORE INFORMATION:
www.bls.gov/oes/current/oes291123.htm

Objectives

1. Construct a rectangular coordinate system.
2. Plot ordered pairs and determine the coordinates of a point.
3. Graph paired data.
4. Read line graphs.
5. Read step graphs.

SECTION 3.1
Graphing Using the Rectangular Coordinate System

It is often said, "A picture is worth a thousand words." In this section, we will show how numerical relationships can be described using mathematical pictures called **graphs**. We will also show how graphs are constructed and how we can obtain information from them.

1 Construct a rectangular coordinate system.

When designing the Gateway Arch in St. Louis, architects created a mathematical model called a **rectangular coordinate graph**. This graph, shown below, is drawn on a grid called a **rectangular coordinate system**. This coordinate system is also called a **Cartesian coordinate system**, after the 17th-century French mathematician René Descartes.

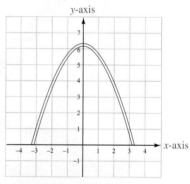

Scale: 1 unit = 100 ft

A rectangular coordinate system is formed by two perpendicular number lines. The horizontal number line is called the ***x*-axis**, and the vertical number line is called the ***y*-axis**. On the *x*-axis, the positive direction is to the right. On the *y*-axis, the positive direction is upward. The scale on each axis should fit the data. For example, the axes of the graph of the arch are scaled in units of 100 feet.

> **Success Tip** If no scale is indicated on the axes, we assume that the axes are scaled in units of 1.

René Descartes
(1596–1650)

The point where the axes intersect is called the **origin**. This is the zero point on each axis. The axes form a **coordinate plane**, and they divide it into four regions called **quadrants**, which are numbered counterclockwise using Roman numerals as shown to the right. The axes are not considered to be in any quadrant.

Each point in a coordinate plane can be identified by an **ordered pair** of real numbers x and y written in the form (x, y).

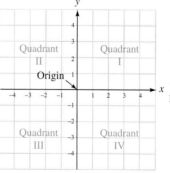

The vertical number line is called the *y*-axis.

The horizontal number line is called the *x*-axis.

The first number, x, in the pair is called the **x-coordinate**, and the second number, y, is called the **y-coordinate**. The numbers in the pair are called the **coordinates** of the point. Some examples of such pairs are $(3, -4)$, $\left(-1, \frac{3}{2}\right)$, and $(0, 2.5)$.

$$(3, -4)$$

The x-coordinate is listed first. The y-coordinate is listed second.

> **Caution!** Do not be confused by this new use of parentheses. The notation $(3, -4)$ represents a point on the coordinate plane, whereas $3(-4)$ indicates multiplication. Also, don't confuse the ordered pair with interval notation.

2 Plot ordered pairs and determine the coordinates of a point.

The process of locating a point in the coordinate plane is called **graphing** or **plotting** the point. In the figure to the right, we use two blue arrows to show how to graph the point with coordinates of $(3, -4)$. Since the x-coordinate, 3, is positive, we start at the origin and move 3 units to the *right* along the x-axis. Since the y-coordinate, -4, is negative, we then move *down* 4 units and draw a dot. The graph of $(3, -4)$ lies in quadrant IV.

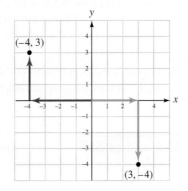

Two red arrows are used to show how to plot the point $(-4, 3)$. We start at the origin, move 4 units to the *left* along the x-axis, and then move *up* 3 units and draw a dot. The graph of $(-4, 3)$ lies in quadrant II.

> **The Language of Algebra** Note that the point with coordinates $(3, -4)$ is not the same as the point with coordinates $(-4, 3)$. Since the order of the coordinates of a point is important, we call the pairs **ordered pairs**.

In the figure to the right, we see that the points $(-4, 0)$, $(0, 0)$, and $(2, 0)$ lie on the x-axis. In fact, all points with a y-coordinate of zero will lie on the x-axis. We also see that the points $(0, -3)$, $(0, 0)$, and $(0, 3)$ lie on the y-axis. All points with an x-coordinate of zero lie on the y-axis. We can also see that the coordinates of the origin are $(0, 0)$.

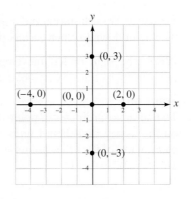

Self Check 1

Plot the points:
a. $(2, -2)$ b. $(-4, 0)$
c. $\left(1.5, \frac{5}{2}\right)$ d. $(0, 5)$

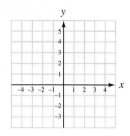

Now Try Problem 21

EXAMPLE 1
Plot each point. Then state the quadrant in which it lies or the axis on which it lies.   a. $(-2, 3)$ b. $\left(-1, -\frac{3}{2}\right)$ c. $(0, 2.5)$ d. $(4, 2)$

Strategy We will start at the origin and move the corresponding number of units right or left for the *x*-coordinate, then move the corresponding number of units up or down for the *y*-coordinate, to locate the point. We will draw a dot at the point.

WHY The coordinates of a point determine its location on the coordinate plane.

Solution

a. Since the *x*-coordinate, -2, is negative, we start at the origin and move 2 units to the *left* along the *x*-axis. Since the *y*-coordinate, 3, is positive, we then move *up* 3 units and draw a dot. The point lies in quadrant II.

b. To plot $\left(-1, -\frac{3}{2}\right)$, we begin at the origin and move 1 unit to the *left* and $\frac{3}{2}$ (or $1\frac{1}{2}$) units *down*. The point lies in quadrant III.

c. To graph $(0, 2.5)$, we begin at the origin and do not move right or left, because the *x*-coordinate is 0. Since the *y*-coordinate is positive, we move 2.5 units *up*. The point lies on the *y*-axis.

d. To graph $(4, 2)$, we begin at the origin and move 4 units to the *right* and 2 units *up*. The point lies in quadrant I.

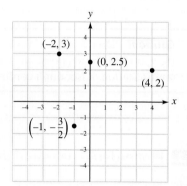

Self Check 2

Find the coordinates of each point in figure (b).

Now Try Problem 25

EXAMPLE 2
Find the coordinates of points A, B, C, D, and E plotted in figure (a) below.

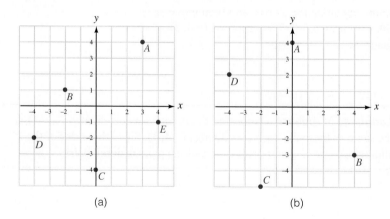

(a) (b)

Strategy We will start at the origin and count to the right or left on the *x*-axis, and then up or down to reach each point.

WHY The right or left movement gives the *x*-coordinate and the up or down movement gives the *y*-coordinate of the point.

Solution

To find the coordinates of point A, we start at the origin, move 3 units to the right on the *x*-axis, and then 4 units up. The coordinates of point A are $(3, 4)$. The coordinates of the other points are found in the same manner: $B(-2, 1)$, $C(0, -4)$, $D(-4, -2)$, $E(4, -1)$.

3 Graph paired data.

Every day, we deal with quantities that are related:

- The time it takes to cook a turkey depends on the weight of the turkey.
- Our weight depends on how much we eat.
- The sales tax that we pay depends on the price of the item purchased.

We can use graphs to visualize such relationships. For example, suppose we know the number of gallons of water that are in a tub at several time intervals after the water has been turned on. We can list that information in a **table**.

The information in the table can be used to construct a graph that shows the relationship between the amount of water in the tub and the time the water has been running. Since the amount of water in the tub depends on the time, we will associate *time* with the *x*-axis and *amount of water* with the *y*-axis.

At various times, the amount of water in the tub was measured and recorded in the table.

Time (min)	Water in tub (gal)
0	0
1	8
3	24
4	32

↑ ↑
x-coordinate y-coordinate

→ (0, 0)
→ (1, 8)
→ (3, 24)
→ (4, 32)

↑
The data in the table can be expressed as ordered pairs (x, y).

To construct the graph below, we plot the four ordered pairs and draw a straight line through the resulting data points. The *y*-axis is scaled in larger units (4 gallons) because the data range from 0 to 32 gallons.

From the graph, we can see that the amount of water in the tub steadily increases as the water is allowed to run. We can also use the graph to make observations about the amount of water in the tub at other times. For example, the dashed line on the graph shows that in 5 minutes, the tub will contain 40 gallons of water.

x	y	(x, y)
0	0	(0, 0)
1	8	(1, 8)
3	24	(3, 24)
4	32	(4, 32)

The data can be listed in a table with headings x, y, and (x, y).

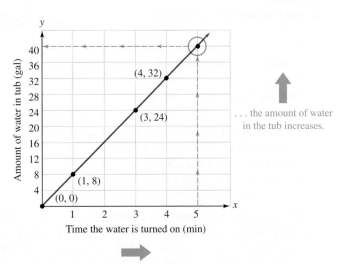

... the amount of water in the tub increases.

As the time increases ...

Self Check 3

Use the graph in Example 3 to answer the following questions.
a. At what times were there exactly 50 people in the audience?
b. What was the size of the audience that watched the taping?
c. How long did it take for the audience to leave the studio after the taping ended?

Now Try Problems 33 and 34

4 Read line graphs.

Since graphs are a popular way to present information, the ability to read and interpret them is very important.

EXAMPLE 3 TV Shows

The graph below shows the number of people in an audience before, during, and after the taping of a television show. On the x-axis, zero represents the time when taping began. Use the graph to answer the following questions and complete the table.

a. How many people were in the audience when taping began?
b. What was the size of the audience 10 minutes before taping began?
c. At what times were there exactly 100 people in the audience?

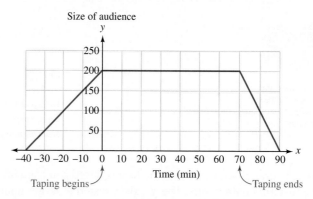

Strategy We will use an ordered pair of the form (*time, size of audience*) to describe each situation mentioned in parts a, b, and c.

WHY The coordinates of specific points on the graph can be used to answer each of these questions.

Solution

a. The time when taping began is represented by 0 on the x-axis. The point on the graph directly above 0 is (0, **200**). The y-coordinate indicates that 200 people were in the audience when the taping began. We enter this result in the table at the right.

b. Ten minutes before taping began is represented by −10 on the x-axis. The point on the graph directly above −10 is (−10, **150**). The y-coordinate indicates that 150 people were in the audience 10 minutes before the taping began. We enter this result in the table.

c. We can draw a horizontal line passing through 100 on the y-axis. Since this line intersects the graph twice, at (**−20**, 100) and at (**80**, 100), there are two times when 100 people were in the audience. The x-coordinates of the points tell us those times: 20 minutes before taping began and 80 minutes after. Enter these results in the table.

Time (min) x	Size of audience y
0	200
−10	150
−20	100
80	100

5 Read step graphs.

The graph below shows the cost of renting a trailer for different periods of time. For example, the cost of renting the trailer for 4 days is $60, which is the y-coordinate of the point (4, 60). The cost of renting the trailer for a period lasting over 4 and up to 5 days jumps to $70. Since the jumps in cost form steps in the graph, we call this graph a **step graph**.

EXAMPLE 4 Use the step graph shown below to answer the following questions. Write the results in a table.

a. Find the cost of renting the trailer for 2 days.

b. Find the cost of renting the trailer for $5\frac{1}{2}$ days.

c. What is the longest amount of time that you can rent the trailer if you have $50?

Self Check 4
Use the graph in Example 4 to answer the following questions.
a. Find the cost of renting the trailer for 1 day.
b. Find the cost of renting the trailer for $4\frac{1}{2}$ days.
c. What is the longest amount of time that you can rent the trailer if you have $40?

Now Try Problems 41 and 43

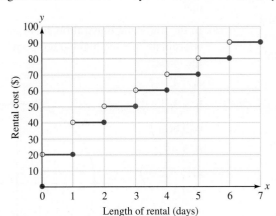

Strategy We will use an ordered pair of the form (*days, rental cost*) to describe each situation mentioned in parts a, b, and c.

WHY The coordinates of specific points on the graph can be used to answer each of these questions.

Solution

a. The solid dot at the end of each step indicates the rental cost for 1, 2, 3, 4, 5, 6, or 7 days. Each open circle indicates that that point is not on the graph. We locate 2 days on the *x*-axis and move up to locate the point on the graph directly above the 2. Since the point has coordinates (2, **40**), a 2-day rental would cost $40. We enter this result in the table below.

b. We locate $5\frac{1}{2}$ days on the *x*-axis and move straight up to locate the point with coordinates $\left(5\frac{1}{2}, \mathbf{80}\right)$, which indicates that a $5\frac{1}{2}$-day rental would cost $80. We then enter this result in the table.

c. We draw a horizontal line through the point labeled 50 on the *y*-axis. Since this line intersects one step in the graph, we can look down to the *x*-axis to find the *x*-values that correspond to a *y*-value of 50. From the graph, we see that the trailer can be rented for more than 2 and up to 3 days for $50. The point has coordinates (**3**, 50). Enter the results in the table.

Length of rental (days) *x*	Cost (dollars) *y*
2	40
$5\frac{1}{2}$	80
3	50

ANSWERS TO SELF CHECKS

1. (graph showing points (0, 5), (1.5, 5/2), (−4, 0), (2, −2))
2. $A(0, 4)$, $B(4, -3)$, $C(-2, -5)$, $D(-4, 2)$
3. a. 30 min before and 85 min after taping began b. 200 c. 20 min 4. a. $20 b. $70 c. 2 days

SECTION 3.1   STUDY SET

VOCABULARY

Fill in the blanks.

1. The point with coordinates (4, 2) can be graphed on a _____ coordinate system.
2. On the rectangular coordinate system, the horizontal number line is called the _____ and the vertical number line is called the _____.
3. On the rectangular coordinate system, the point (0, 0) where the axes cross is called the _____.
4. On the rectangular coordinate system, the axes form the _____ plane.
5. The x- and y-axes divide the coordinate plane into four regions called _____.
6. The pair of numbers (−1, −5) is called an _____ pair.
7. In the ordered pair $\left(-\frac{3}{2}, -5\right)$, $-\frac{3}{2}$ is called the _____ and −5 is called the _____.
8. The process of locating the position of a point on a coordinate plane is called _____ the point.

CONCEPTS

Fill in the blanks.

9. To plot the point with coordinates (−5, 4.5), we start at the _____ and move 5 units to the _____ and then move 4.5 units ___.
10. To plot the point with coordinates $\left(6, -\frac{3}{2}\right)$, we start at the _____ and move 6 units to the _____ and then move $\frac{3}{2}$ units ___.
11. Do (3, 2) and (2, 3) represent the same point?
12. In the ordered pair (4, 5), is the number 4 associated with the horizontal or the vertical axis?
13. In which quadrant do points with a negative x-coordinate and a positive y-coordinate lie?
14. In which quadrant do points with a positive x-coordinate and a negative y-coordinate lie?
15. In the following illustration, fill in the missing coordinate of each highlighted point on the graph of the circle.

   a. $(4, \ )$
   b. $(3, \ )$
   c. $(5, \ )$
   d. $(-3, \ )$
   e. $(-5, \ )$
   f. $(0, \ )$

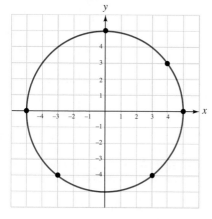

16. In the following illustration, fill in the missing coordinate of each point on the graph of the line.

   a. $(-4, \ )$
   b. $(\ , 0)$
   c. $(\ , 2)$
   d. $(\ , -1)$
   e. $(\ , 1)$

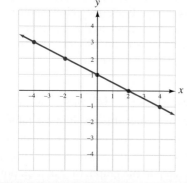

NOTATION

17. Explain the difference between (3, 5), 3(5), and 5(3 + 5).

18. In the table, which column contains values associated with the vertical axis of a graph?

x	y
2	0
5	−2
−1	$-\frac{1}{2}$

19. Do these ordered pairs name the same point?
$(2.5, -\frac{7}{2}), (2\frac{1}{2}, -3.5), (2.5, -3\frac{1}{2})$

20. Do these ordered pairs name the same point?
$(1.25, 4), (1\frac{1}{4}, 4.0), (-\frac{5}{4}, 4)$

GUIDED PRACTICE

Plot each point on the grid provided. See Example 1.

21. $(-3, 4), (4, 3.5),$
$(-2, -\frac{5}{2}), (2, 0)$

22. $(0, -4), (\frac{3}{2}, 0),$
$(3, -4) \; (-3, 4)$

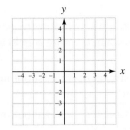 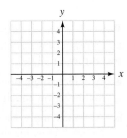

23. $(4, 4), (0.5, -3),$
$(-4, -4), (0, -2)$

24. $(0, 0), (0, 3),$
$(-2, 0), (0, -1)$

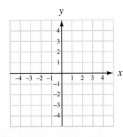 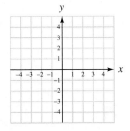

Determine the coordinates of each point. See Example 2.

25. A   **26.** B
27. C   **28.** D
29. E   **30.** F
31. G   **32.** H

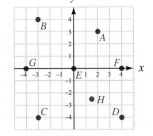

The graph gives the heart rate of a woman before, during, and after an aerobic workout. Use the graph to answer Problems 33–40. See Example 3.

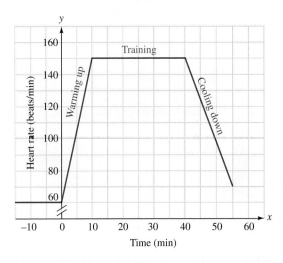

33. What information does the point $(-10, 60)$ give us?

34. After beginning her workout, how long did it take the woman to reach her training-zone heart rate?

35. What was the woman's heart rate half an hour after beginning the workout?

36. For how long did the woman work out at her training zone?

37. At what time was her heart rate 100 beats per minute?

38. How long was her cool-down period?

39. What was the difference in the woman's heart rate before the workout and after the cool-down period?

40. What was her approximate heart rate 8 minutes after beginning?

The graph gives the charges for renting skiing equipment for certain lengths of time. Use the graph to answer Problems 41–44. See Example 4.

41. Find the charge for a 1-day rental.

42. Find the charge for a 2-day rental.

43. What is the charge if the equipment is kept for 5 days?

44. What is the charge if the equipment is kept for a week?

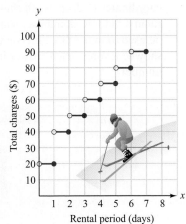

APPLICATIONS

45. BRIDGE CONSTRUCTION Find the coordinates of each rivet, weld, and anchor.

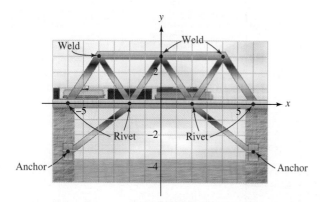

Scale: 1 unit = 8 ft

46. GOLF A golfer is videotaped and then has her swing displayed on a computer monitor so that it can be analyzed. Give the coordinates of the points that are highlighted in red.

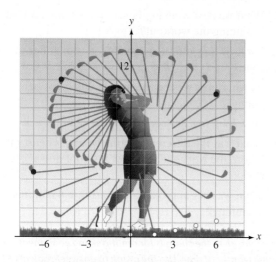

47. GAMES In the game Battleship, the player uses coordinates to drop depth charges from a battleship to hit a hidden submarine. What coordinates should be used to make three hits on the exposed submarine shown? Express each answer in the form (letter, number).

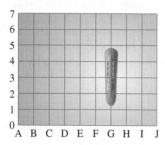

48. MAPS Use coordinates of the form (number, letter) to locate each position on the following map: Rockford, Forreston, Harvard, and the intersection of state Highway 251 and U.S. Highway 30.

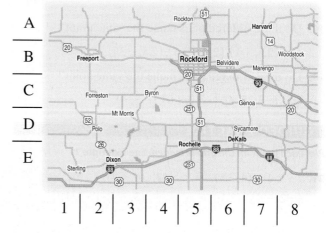

49. WATER PRESSURE The graph shows how the path of a stream of water changes when the hose is held at two different angles.

a. At which angle does the stream of water shoot up higher? How much higher?

b. At which angle does the stream of water shoot out farther? How much farther?

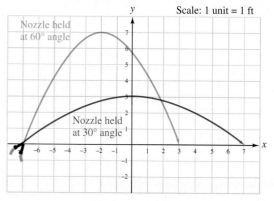

50. LANDMARKS A scale model of the block letter H in the Hollywood sign can be drawn by plotting the following points and connecting them: $(0, 0)$, $(13, 0)$, $(13, 16)$, $(26, 16)$, $(26, 0)$, $(39, 0)$, $(39, 45)$, $(26, 45)$, $(26, 29)$, $(13, 29)$, $(13, 45)$, and $(0, 45)$. The scale is 1 unit on the graph is equal to 1 foot on the actual sign. If a gallon of paint covers 350 square feet, how many gallons are needed to paint the front side of the letter H? Round to the nearest gallon.

3.1 Graphing Using the Rectangular Coordinate System

51. GEOGRAPHY A coordinate system that describes the location of any place on the surface of the Earth uses a series of *latitude* and *longitude* lines, as shown below. Estimate the location of the Deep Water Horizon (the oil drilling rig in the Gulf of Mexico that exploded in 2010) using an ordered pair of the form (latitude, longitude). (Source: sailwx.info)

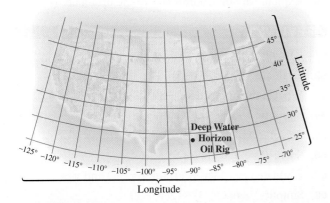

52. Scoliosis is a lateral curvature of the spine that can be detected when a grid is superimposed over an X-ray. In the illustration, find the coordinates of the center points of the indicated vertebrae. Note that T3 means the third thoracic vertebra, L4 means the fourth lumbar vertebra, and so on.

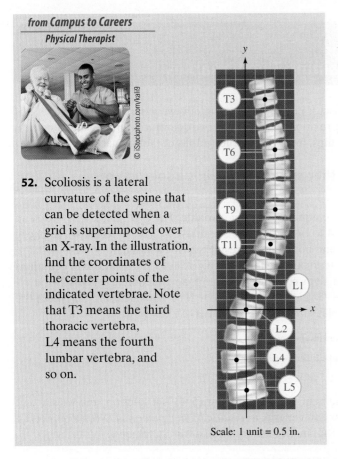

Scale: 1 unit = 0.5 in.

53. DENTISTRY Dentists describe teeth as being located in one of four quadrants as shown below:

a. How many teeth are located in the upper left quadrant?

b. Why would the upper left quadrant appear on the right in the illustration?

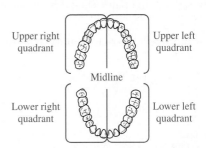

54. GAS MILEAGE The following table gives the number of miles (y) that a truck can be driven on x gallons of gasoline. Plot the ordered pairs and draw a line connecting the points.

x	y
2	10
3	15
5	25

a. Estimate how far the truck can go on 7 gallons of gasoline.

b. How many gallons of gas are needed to travel a distance of 20 miles?

c. How far can the truck go on 6.5 gallons of gasoline?

55. VALUE OF A CAR The following table shows the value y (in thousands of dollars) of a car that is x years old. Plot the ordered pairs and draw a line connecting the points.

x	y
3	7
4	5.5
5	4

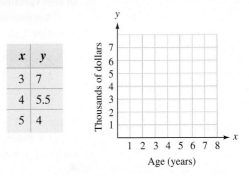

a. What does the point (3, 7) on the graph tell you?

b. Estimate the value of the car when it is 7 years old.

c. After how many years will the car be worth $2,500?

56. BOATING The table below shows the cost to rent a sailboat for a given number of hours. Plot the data in the table as ordered pairs. Then draw a line through the points.

a. How much will it cost to rent the boat for 3 hours?

b. For how long can the boat be rented for $60?

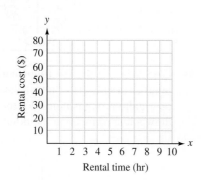

Rental time (hr)	Cost ($)
2	20
4	30
6	40

WRITING

57. Explain why the point $(-3, 3)$ is not the same as the point $(3, -3)$.

58. Explain what is meant when we say that the rectangular coordinate graph of the St. Louis Gateway Arch is made up of *infinitely many* points.

59. Explain how to plot the point $(-2, 5)$.

60. Explain why the coordinates of the origin are $(0, 0)$.

REVIEW

61. Evaluate: $-3 - 3(-5)$

62. Evaluate: $(-5)^2 + (-5)$

63. What is the opposite of -8?

64. Simplify: $|-1 - 9|$

65. Solve: $-4x + 7 = -21$

66. Solve $P = 2l + 2w$ for w.

67. Evaluate $(x + 1)(x + y)^2$ for $x = -2$ and $y = -5$.

68. Simplify: $-6(x - 3) - 2(1 - x)$

SECTION 3.2
Equations Containing Two Variables

Objectives
1. Determine whether an ordered pair is a solution of an equation.
2. Complete ordered-pair solutions of equations.
3. Construct a table of solutions.
4. Graph equations by plotting points.
5. Graph equations that use different variables.

In this section, we will discuss equations that contain two variables. These equations are often used to describe relationships between two quantities. To see a mathematical picture of these relationships, we will construct graphs of their equations.

1 Determine whether an ordered pair is a solution of an equation.

We have previously solved **equations in one variable**. For example, $x - 4 = 3$ is an equation in x. If we add 4 to both sides, we see that $x = 7$ is the solution. To check this, we replace x with 7 and note that the result is a true statement: $3 = 3$.

In this chapter, we extend our equation-solving skills to find solutions of **equations in two variables**. To begin, let's consider $y = x - 1$, an equation in x and y.

A solution of $y = x - 1$ is a pair of values, one for x and one for y, that make the equation true. For example, suppose x is 5 and y is 4. Then we have:

$y = x - 1$ This is the original equation.
$4 \stackrel{?}{=} 5 - 1$ Substitute 5 for x and 4 for y.
$4 = 4$ True

Since $4 = 4$ is a true statement, the ordered pair $(5, 4)$ is a solution, and we say that $(5, 4)$ **satisfies** the equation. In general, a *solution of an equation in two variables* is an ordered pair of numbers that makes the equation a true statement.

Self Check 1
Is $(9, 8)$ a solution of $y = x - 1$?

Now Try Problem 17

EXAMPLE 1 Is the ordered pair $(-1, -3)$ a solution of $y = x - 1$?

Strategy We will substitute -1 for x and -3 for y and see whether the resulting equation is true.

WHY An ordered pair is a *solution* of $y = x - 1$ if replacing the variables with the values of the ordered pair results in a true statement.

Solution

$y = x - 1$ This is the original equation.
$-3 \stackrel{?}{=} -1 - 1$ Substitute −1 for x and −3 for y.
$-3 = -2$ Do the subtraction: −1 − 1 = −2. False

Since $-3 = -2$ is a false statement, $(-1, -3)$ is not a solution of $y = x - 1$.

EXAMPLE 2   Is the ordered pair $(-6, 36)$ a solution of $y = x^2$?

Strategy   We will substitute −6 for x and 36 for y and see whether the resulting equation is true.

WHY   An ordered pair is a *solution* of $y = x^2$ if replacing the variables with the values of the ordered pair results in a true statement.

Solution
We substitute −6 for x and 36 for y and see whether the resulting equation is a true statement.

$y = x^2$ This is the original equation.
$36 \stackrel{?}{=} (-6)^2$ Substitute −6 for x and 36 for y.
$36 = 36$ Find the power: $(-6)^2 = 36$. True

Since the equation $36 = 36$ is true, $(-6, 36)$ is a solution of $y = x^2$.

Self Check 2
Is $(-2, 5)$ a solution of $y = x^2$?
Now Try Problem 19

> **Language of Algebra**   Equations in two variables often involve the variables x and y. However, other letters can be used. For example, $a - 3b = 6$ and $n = 2m + 1$ are equations in two variables.

2 Complete ordered-pair solutions of equations.

If only one of the values of an ordered-pair solution is known, we can substitute it into the equation to determine the other value.

EXAMPLE 3   Complete the solution $(-4, \_\_)$ of the equation $y = -x + 2$.

Strategy   We will substitute the known x-coordinate of the solution into the given equation.

WHY   We can use the resulting equation in one variable to find the unknown y-coordinate of the solution.

Solution
In the ordered pair $(-4, \_\_)$, the x-value is −4; the y-value is not known. To find y, we substitute −4 for x in the equation and evaluate the right side.

$y = -x + 2$ This is the original equation.
$y = -(-4) + 2$ Substitute −4 for x.
$y = 4 + 2$ The opposite of −4 is 4.
$y = 6$ This is the missing y-coordinate of the solution.

The completed ordered pair is $(-4, 6)$.

Self Check 3
Complete the solution $(-3, \_\_)$ of the equation $y = 2x - 4$.
Now Try Problem 26

3 Construct a table of solutions.

To find a solution of an equation in x and y, we can select a number, substitute it for x, and find the corresponding value of y. For example, to find a solution of the equation $y = x - 1$, we can let $x = -4$ (called the **input value**), substitute -4 for x, and solve for y (called the **output value**).

$y = x - 1$ This is the original equation.
$y = -4 - 1$ Substitute the input -4 for x.
$y = -5$ The output is -5.

$y = x - 1$		
x	y	(x, y)
-4	-5	$(-4, -5)$

The ordered pair $(-4, -5)$ is a solution. We list this ordered pair in red in the **table of solutions** (or **table of values**).

To find another solution of $y = x - 1$, we select another value of x, say -2, and find the corresponding y-value.

$y = x - 1$ This is the original equation.
$y = -2 - 1$ Substitute the input -2 for x.
$y = -3$ The output is -3.

$y = x - 1$		
x	y	(x, y)
-4	-5	$(-4, -5)$
-2	-3	$(-2, -3)$

A second solution is $(-2, -3)$, and we list it in the table of solutions.
If we let $x = 0$, we can find a third ordered pair that satisfies $y = x - 1$.

$y = x - 1$ This is the original equation.
$y = 0 - 1$ Substitute the input 0 for x.
$y = -1$ The output is -1.

$y = x - 1$		
x	y	(x, y)
-4	-5	$(-4, -5)$
-2	-3	$(-2, -3)$
0	-1	$(0, -1)$

A third solution is $(0, -1)$, which we also add to the table of solutions.
If we let $x = 2$, we can find a fourth solution.

$y = x - 1$ This is the original equation.
$y = 2 - 1$ Substitute the input 2 for x.
$y = 1$ The output is 1.

$y = x - 1$		
x	y	(x, y)
-4	-5	$(-4, -5)$
-2	-3	$(-2, -3)$
0	-1	$(0, -1)$
2	1	$(2, 1)$

A fourth solution is $(2, 1)$, and we add it to the table of solutions.
If we let $x = 4$, we have

$y = x - 1$ This is the original equation.
$y = 4 - 1$ Substitute the input 4 for x.
$y = 3$ The output is 3.

A fifth solution is $(4, 3)$.

Since we can choose any real number for x, and since any choice of x will give a corresponding value of y, it is apparent that the equation $y = x - 1$ has *infinitely many solutions*. We have found five of them: $(-4, -5)$, $(-2, -3)$, $(0, -1)$, $(2, 1)$, and $(4, 3)$.

$y = x - 1$		
x	y	(x, y)
-4	-5	$(-4, -5)$
-2	-3	$(-2, -3)$
0	-1	$(0, -1)$
2	1	$(2, 1)$
4	3	$(4, 3)$

3.2 Equations Containing Two Variables   225

4 Graph equations by plotting points.

To graph the equation $y = x - 1$, we plot the ordered pairs listed in the table of solutions on a rectangular coordinate system, as shown in figure (a). We can see that the five points lie on a line.

We then use a straightedge to draw a line through the points, because the graph of any solution of $y = x - 1$ will lie on this line. See figure (b). The line is a picture of all the solutions of the equation $y = x - 1$. It is called the **graph of the equation**. The arrowheads show that the line continues forever in both directions. See figure (c).

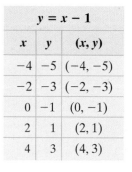

$y = x - 1$		
x	y	(x, y)
-4	-5	$(-4, -5)$
-2	-3	$(-2, -3)$
0	-1	$(0, -1)$
2	1	$(2, 1)$
4	3	$(4, 3)$

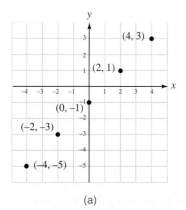 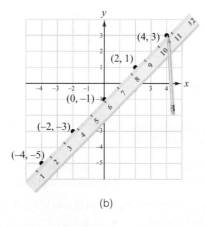 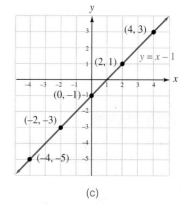

(a) (b) (c)

To graph an equation in x and y, we follow these steps.

Graphing an Equation in x and y

1. Make a table of solutions containing several ordered pairs of numbers (x, y) that satisfy the equation. Do this by selecting values for x and finding the corresponding values for y.
2. Plot each ordered pair on a rectangular coordinate system.
3. Carefully draw a line or smooth curve through the points.

Since we will usually choose a number for x and then find the corresponding value of y, the value of y depends on x. For this reason, we call y the **dependent variable** and x the **independent variable**. The value of the independent variable is the input value, and the value of the dependent variable is the output value.

EXAMPLE 4 Graph: $y = -2x - 2$

Strategy We will find several solutions of the equation, plot them on a rectangular coordinate system, and then draw a graph passing through the points.

WHY To *graph* an equation in two variables means to make a drawing that represents all of its solutions.

Solution
To make a table of solutions, we choose numbers for x and find the corresponding values of y. If $x = -3$, we have

$y = -2x - 2$ This is the original equation.
$y = -2(-3) - 2$ Substitute -3 for x.
$y = 6 - 2$ Do the multiplication: $-2(-3) = 6$.
$y = 4$ Do the subtraction.

Self Check 4
Graph: $y = -3x + 1$

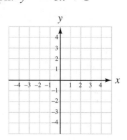

Now Try Problem 33

Thus, $x = -3$ and $y = 4$ is a solution. In a similar manner, we find the corresponding y-values for x-values of -2, -1, 0, and 1 and record the results in the table of solutions. After plotting the ordered pairs, we draw a line through the points to get the graph shown.

$y = -2x - 2$		
x	y	(x, y)
-3	4	$(-3, 4)$
-2	2	$(-2, 2)$
-1	0	$(-1, 0)$
0	-2	$(0, -2)$
1	-4	$(1, -4)$
↑	↑	↑
Select	Find	Plot

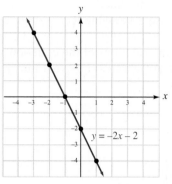

Self Check 5

Graph $y = x^2 - 2$ and compare the result to the graph of $y = x^2$. What do you notice?

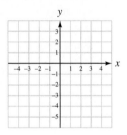

Now Try Problem 37

EXAMPLE 5   Graph: $y = x^2$

Strategy We will find several solutions of the equation, plot them on a rectangular coordinate system, and then draw a graph passing through the points.

WHY To *graph* an equation in two variables means to make a drawing that represents all of its solutions.

Solution

To make a table of solutions, we will choose numbers for x and find the corresponding values of y. If $x = -3$, we have

$y = x^2$ This is the original equation.
$y = (-3)^2$ Substitute the input -3 for x.
$y = 9$ The output is 9.

Thus, $x = -3$ and $y = 9$ is a solution. In a similar manner, we find the corresponding y-values for x-values of -2, -1, 0, 1, 2, and 3. If we plot the ordered pairs listed in the table and join the points with a smooth curve, we get the graph shown in the figure, which is called a **parabola**.

$y = x^2$		
x	y	(x, y)
-3	9	$(-3, 9)$
-2	4	$(-2, 4)$
-1	1	$(-1, 1)$
0	0	$(0, 0)$
1	1	$(1, 1)$
2	4	$(2, 4)$
3	9	$(3, 9)$

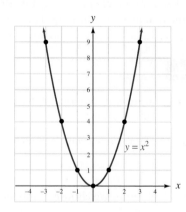

> **Success Tip** When selecting x-values for a table of solutions, a rule of thumb is to choose some negative numbers, some positive numbers, and 0. When $x = 0$, the computations to find y are usually quite simple.

Unless otherwise noted, all content on this page is © Cengage Learning.

3.2 Equations Containing Two Variables

EXAMPLE 6 Graph: $y = |x|$

Strategy We will find several solutions of the equation, plot them on a rectangular coordinate system, and then draw a graph passing through the points.

WHY To *graph* an equation in two variables means to make a drawing that represents all of its solutions.

Solution
To make a table of solutions, we will choose numbers for x and find the corresponding values of y. If $x = -5$, we have

$y = |x|$ This is the original equation.
$y = |-5|$ Substitute the input -5 for x.
$y = 5$ The output is 5.

The ordered pair $(-5, 5)$ satisfies the equation. This pair and several others that satisfy the equation are listed in the table of solutions in the figure. If we plot the ordered pairs in the table, we see that they lie in a "V" shape. We join the points to complete the graph shown in the figure.

Self Check 6
Graph $y = |x| + 2$ and compare the result to the graph of $y = |x|$. What do you notice?

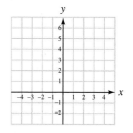

Now Try Problem 43

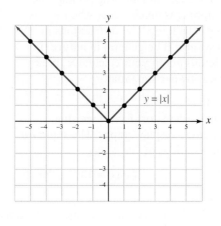

EXAMPLE 7 Graph: $y = x^3$

Strategy We will find several solutions of the equation, plot them on a rectangular coordinate system, and then draw a graph passing through the points.

WHY To *graph* an equation in two variables means to make a drawing that represents all of its solutions.

Solution
If we let $x = -2$, we have

$y = x^3$ This is the original equation.
$y = (-2)^3$ Substitute the input -2 for x.
$y = -8$ The output is -8.

The ordered pair $(-2, -8)$ satisfies the equation. This ordered pair and several others that satisfy the equation are listed in the table of solutions on the next page. Plotting the ordered pairs and joining them with a smooth curve gives us the graph shown.

Self Check 7
Graph $y = (x - 2)^3$ and compare the result to the graph of $y = x^3$. What do you notice?

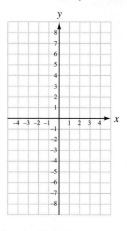

Now Try Problem 45

$y = x^3$		
x	y	(x, y)
-2	-8	$(-2, -8)$
-1	-1	$(-1, -1)$
0	0	$(0, 0)$
1	1	$(1, 1)$
2	8	$(2, 8)$

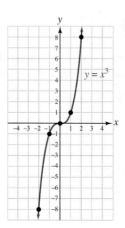

Using Your CALCULATOR Using a Graphing Calculator to Graph an Equation

We have graphed equations by making tables of solutions and plotting points. The task of graphing is much easier when we use a graphing calculator. The instructions in this discussion will be general in nature. For specific details about your calculator, please consult your owner's manual.

The viewing window
All graphing calculators have a viewing **window**, used to display graphs. The **standard window** has settings of

$$\text{Xmin} = -10, \quad \text{Xmax} = 10, \quad \text{Ymin} = -10, \quad \text{and} \quad \text{Ymax} = 10$$

which indicate that the minimum x- and y-coordinates used in the graph will be -10 and that the maximum x- and y-coordinates will be 10.

Graphing an equation
To graph the equation $y = x - 1$ using a graphing calculator, we press the $\boxed{Y=}$ key and enter the right-hand side of the equation after the symbol Y_1. The display will show the equation

$$Y_1 = x - 1$$

Then we press the $\boxed{\text{GRAPH}}$ key to produce the graph shown in figure (a) on the next page.

Next, we will graph the equation $y = |x - 4|$. Since absolute values are always nonnegative, the minimum y-value is zero. To obtain a reasonable viewing window, we press the $\boxed{\text{WINDOW}}$ key and set the Ymin value slightly lower, at Ymin $= -3$. We set Ymax to be 10 units greater than Ymin, at Ymax $= 7$. The minimum value of y occurs when $x = 4$. To center the graph in the viewing window, we set the Xmin and Xmax values 5 units to the left and right of 4. Therefore, Xmin $= -1$ and Xmax $= 9$.

After entering the right-hand side of the equation, we obtain the graph shown in figure (b) on the next page. Consult your owner's manual to learn how to enter an absolute value.

(continued)

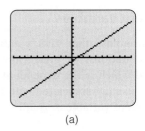 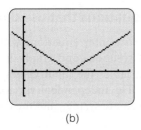

(a) (b)

Changing the viewing window

The choice of viewing windows is extremely important when graphing equations. To show this, let's graph $y = x^2 - 25$ with x-values from -1 to 6 and y-values from -5 to 5.

To graph this equation, we set the x and y window values and enter the right-hand side of the equation. The display will show

$$Y_1 = x^2 - 25$$

Then we press the $\boxed{\text{GRAPH}}$ key to produce the graph shown in figure (c). Although the graph appears to be a straight line, it is not. Actually, we are seeing only part of a parabola. If we pick a viewing window with x-values of -6 to 6 and y-values of -30 to 2, as in figure (d), we can see that the graph is a parabola.

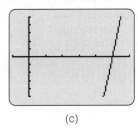 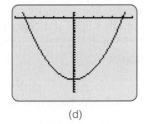

(c) (d)

Use a graphing calculator to graph each equation. Use a viewing window of $x = -5$ to 5 and $y = -5$ to 5.

1. $y = 2.1x - 1.1$ **2.** $y = 1.12x^2 - 1$

3. $y = |x + 0.7|$ **4.** $y = 0.1x^3 + 1$

Graph each equation in a viewing window of $x = -4$ to 4 and $y = -4$ to 4. Each graph is not what it first appears to be. Pick a better viewing window and find a better representation of the true graph.

5. $y = -x^3 - 8.2$ **6.** $y = -|x - 4.01|$

7. $y = x^2 + 5.9$ **8.** $y = -x + 7.95$

5 Graph equations that use different variables.

We will often encounter equations with variables other than x and y. When we make tables of solutions and graph these equations, we must know which is the independent variable (the input values) and which is the dependent variable (the output values). The independent variable is usually associated with the horizontal axis of the coordinate system, and the dependent variable is usually associated with the vertical axis.

Self Check 8

If the maximum speed limit on a rural highway is 55 mph, the formula for the distance traveled in t time is $d = 55t$. Graph the equation.

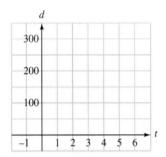

Now Try Problem 49

EXAMPLE 8 *Speed Limits* In some states, the maximum speed limit on a U.S. interstate highway is 75 mph. The distance covered by a vehicle traveling at 75 mph depends on the time the vehicle travels at that speed. This relationship is described by the equation $d = 75t$, where d represents the distance (in miles) and t represents the time (in hours). Graph the equation.

Strategy We will find several solutions of the equation, plot them on a rectangular coordinate system, and then draw a graph passing through the points.

WHY We can use the graph to estimate the distance traveled (in miles) after traveling an amount of time at 75 mph.

Solution

Since d depends on t in the equation $d = 75t$, t is the independent variable (the input) and d is the dependent variable (the output). Therefore, we choose values for t and find the corresponding values of d. Since t represents the time spent traveling at 75 mph, we choose no negative values for t.

If $t = 0$, we have

$d = 75t$ This is the original equation.
$d = 75(0)$ Substitute the input 0 for t.
$d = 0$ Do the multiplication.

The pair $t = 0$ and $d = 0$, or $(0, 0)$, is a solution. This ordered pair and others that satisfy the equation are listed in the table of solutions shown below. If we plot the ordered pairs and draw a line through them, we obtain the graph shown. From the graph, we see (as expected) that the distance covered steadily increases as the traveling time increases.

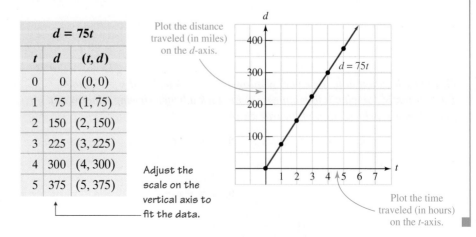

ANSWERS TO SELF CHECKS

1. yes **2.** no **3.** $(-3, -10)$

4.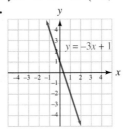

5.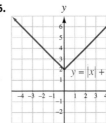

The graph has the same shape but is 2 units lower.

6.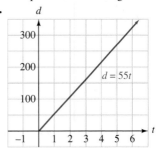

The graph has the same shape but is 2 units higher.

7.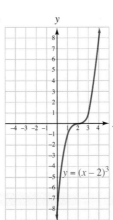

The graph has the same shape but is 2 units to the right.

8.

SECTION 3.2 STUDY SET

VOCABULARY

Fill in the blanks.

1. The equation $7 = -2x + 5$ is an equation in ____ variable. The equation $y = x + 1$ is an equation in ____ variables, x and y.

2. An ordered pair is a _____ of an equation if the numbers in the ordered pair satisfy the equation.

3. When constructing a _____ of solutions, the values of x are the input values and the values of y are the _____ values.

4. In equations containing the variables x and y, x is called the independent _____ and y is called the _____ variable.

CONCEPTS

5. Consider the equation: $y = -2x + 6$
   a. How many variables does the equation contain?
   b. Does the ordered pair $(4, -2)$ satisfy the equation?
   c. Is $(-3, 12)$ a solution of the equation?
   d. How many solutions does this equation have?

6. To graph an equation, five solutions were found, they were plotted (in black), and a straight line was drawn through them, as shown. From the graph, determine three other solutions of the equation.

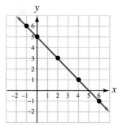

7. Fill in the blanks: The graph of $y = -x + 5$ is shown in Problem 6. Every point on the graph represents an ordered-pair _____ of $y = -x + 5$, and every ordered-pair solution is a _____ on the graph.

8. Consider the graph of an equation shown below.
   a. If the coordinates of point M are substituted into the equation, is the result a true or false statement?
   b. If the coordinates of point N are substituted into the equation, is the result a true or false statement?

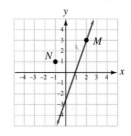

9. Complete the table.

$y = x^3$	
x (inputs)	y (outputs)
0	
−1	
−2	
1	
2	

10. What is wrong with the graph of $y = x - 3$ shown below?

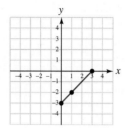

11. To graph $y = -x + 1$, a student constructed a table of solutions and plotted the ordered pairs as shown. Instead of drawing a crooked line through the points, what should he have done?

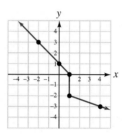

12. To graph $y = x^2 - 4$, a table of solutions is constructed and a graph is drawn, as shown. Explain the error made here.

$y = x^2 - 4$		
x	y	(x, y)
0	−4	(0, −4)
2	0	(2, 0)

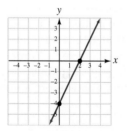

13. Explain the error with the graph of $y = x^2$ shown in the illustration.

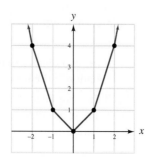

14. Several solutions of an equation are listed in the table of solutions. When graphing them, with what variable should the horizontal and vertical axes of the graph be labeled?

t	s	(t, s)
0	4	(0, 4)
1	5	(1, 5)
2	10	(2, 10)

NOTATION

Complete each step.

15. Verify that $(-2, 6)$ satisfies $y = -x + 4$.

$y = -x + 4$
$\boxed{} \stackrel{?}{=} -\left(\boxed{}\right) + 4$
$6 \stackrel{?}{=} \boxed{} + 4$
$6 = 6$

16. For the equation $y = |x - 2|$, if $x = -3$, find y.

$y = |x - 2|$
$y = \left|\boxed{} - 2\right|$
$y = \left|\boxed{}\right|$
$y = 5$

GUIDED PRACTICE

Determine whether the ordered pair satisfies the equation. See Examples 1–2.

17. $y = 2x - 4$, (4, 4)   **18.** $y = -3x + 1$, (2, −4)

19. $y = x^2$, (8, 48)   **20.** $y = -x^2 + 2$, (1, 1)

21. $y = |x - 2|$, (4, −3)   **22.** $y = |x + 3|$, (0, 3)

23. $y = x^3 + 1$, (−2, −7)   **24.** $y = -x^3 - 1$, (1, −2)

Complete the solution of each equation. See Example 3.

25. $y = 3x - 4$, (1, ?)

26. $y = \frac{1}{2}x - 3$, (2, ?)

27. $y = -5x + 3$, (−3, ?)

28. $y = -\frac{2}{5}x - 1$, (−5, ?)

Complete each table. See Objective 3.

29.

$y = x - 3$	
x	y
0	
1	
−2	

30.

$y = |x - 3|$	
x	y
0	
−1	
3	

31.

Input	Output
0	
2	
−2	

  $y = x^2 - 3$

32.

Input	Output
0	
2	
−1	

$y = x + 1$

Construct a table of solutions and graph each equation. See Example 4.

33. $y = 2x - 3$ **34.** $y = 3x + 1$

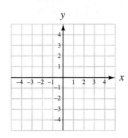 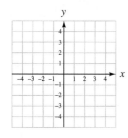

35. $y = -2x + 1$ **36.** $y = -3x + 2$

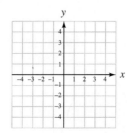 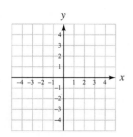

Construct a table of solutions and graph each equation. Compare the result to the graph of $y = x^2$. See Example 5.

37. $y = x^2 + 1$ **38.** $y = -x^2$

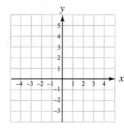 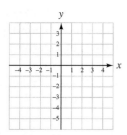

39. $y = (x - 2)^2$ **40.** $y = (x + 2)^2$

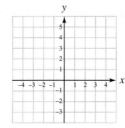 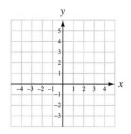

Construct a table of solutions and graph each equation. Compare the result to the graph of $y = |x|$. See Example 6.

41. $y = -|x|$ **42.** $y = |x| - 2$

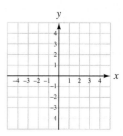 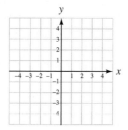

43. $y = |x + 2|$ **44.** $y = |x - 2|$

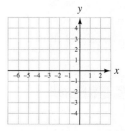 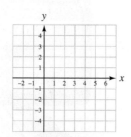

Construct a table of solutions and graph each equation. Compare the result to the graph of $y = x^3$. See Example 7.

45. $y = -x^3$ **46.** $y = x^3 + 2$

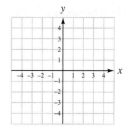 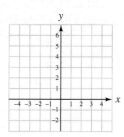

47. $y = x^3 - 2$ **48.** $y = (x + 2)^3$

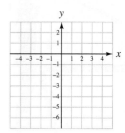 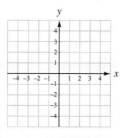

APPLICATIONS

See Example 8.

49. BILLIARDS The path traveled by the black 8 ball is described by the equations $y = 2x - 4$ and $y = -2x + 12$. Construct a table of solutions for $y = 2x - 4$ using the x-values 1, 2, and 4. Do the same for $y = -2x + 12$ using the x-values 4, 6, and 8. Then graph the path of the 8 ball.

x	1	2	4
y			

x	4	6	8
y			

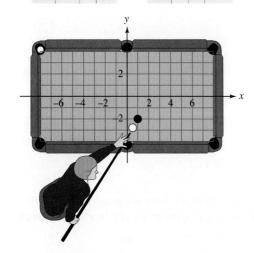

50. TABLE TENNIS The illustration shows the path traveled by a Ping-Pong ball as it bounces off the table. Use the information in the illustration to complete the table.

x	-7	-3	1	3	5
y					

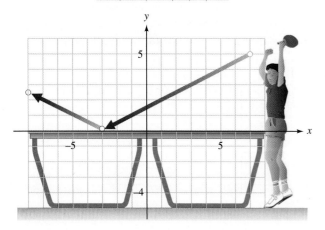

51. SUSPENSION BRIDGES The suspension cables of a bridge hang in the shape of a parabola, as shown in the next column. Use the information in the illustration to complete the table.

x	0	2	4	-2	-4
y					

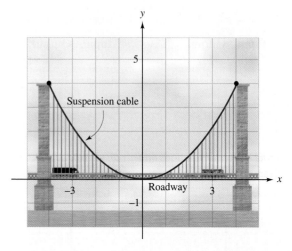

52. FIREBOATS A stream of water from a high-pressure hose on a fireboat travels in the shape of a parabola. Use the information in the graph to complete the table.

x	1	2	3	4
y				

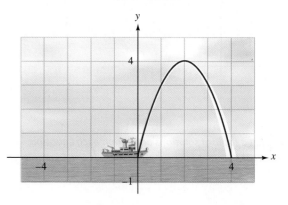

53. MANUFACTURING The graph on the next page shows the relationship between the length l (in inches) of a machine bolt and the cost C (in cents) to manufacture it.

  a. What information does the point (2, 8) on the graph give us?

  b. How much does it cost to make a 7-inch bolt?

  c. What length bolt is the least expensive to make?

  d. Describe how the cost changes as the length of the bolt increases.

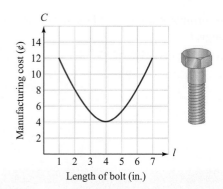

54. SOFTBALL The following graph shows the relationship between the distance d (in feet) traveled by a batted softball and the height h (in feet) it attains.
  a. What information does the point (40, 40) on the graph give us?
  b. At what distance from home plate does the ball reach its maximum height?
  c. Where will the ball land?

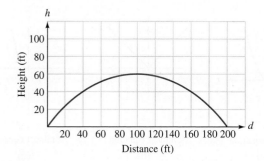

55. MARKET VALUE OF A HOUSE The following graph shows the relationship between the market value v of a house and the time t since it was purchased.

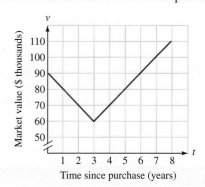

  a. What was the purchase price of the house?
  b. When did the value of the house reach its lowest point?
  c. When did the value of the house begin to surpass the purchase price?
  d. Describe how the market value of the house changed over the 8-year period.

56. POLITICAL SURVEYS The following graph shows the relationship between the percent P of those surveyed who rated their senator's job performance as satisfactory or better and the time t she had been in office.
  a. When did her job performance rating reach a maximum?
  b. When was her job performance rating at or above the 60% mark?
  c. Describe how her job performance rating changed over the 12-month period.

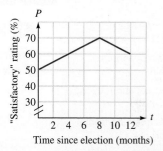

WRITING

57. What is a table of solutions?

58. To graph an equation in two variables, how many solutions of the equation must be found?

59. Give an example of an equation in one variable and an equation in two variables. How do their solutions differ?

60. When we say that $(-2, -6)$ is a solution of $y = x - 4$, what do we mean?

61. On a quiz, students were asked to graph $y = 3x - 1$. One student made the table of solutions on the left. Another student made the one on the right. Which table is incorrect? Or could they both be correct? Explain.

x	y	(x, y)
0	−1	(0, −1)
2	5	(2, 5)
3	8	(3, 8)
4	11	(4, 11)
5	14	(5, 14)

x	y	(x, y)
−2	−7	(−2, −7)
−1	−4	(−1, −4)
1	2	(1, 2)
−3	−10	(−3, −10)
2	5	(2, 5)

62. What does it mean when we say that an equation in two variables has infinitely many solutions?

REVIEW

63. Solve: $\dfrac{x}{8} = -12$

64. Combine like terms: $3t - 4T + 5T - 6t$

65. Is $\dfrac{x+5}{6}$ an expression or an equation?

66. What formula is used to find the perimeter of a rectangle?

67. What number is 0.5% of 250?

68. Solve: $-3x + 5 > -7$

69. Find: $-2.5 - (-2.6)$

70. Evaluate: $(-5)^3$

Objectives

1. Identify linear equations.
2. Complete ordered-pair solutions of linear equations.
3. Graph linear equations by plotting points.
4. Graph linear equations by the intercept method.
5. Identify and graph horizontal and vertical lines.
6. Use graphs of linear equations to solve application problems.

SECTION 3.3
Graphing Linear Equations

1 Identify linear equations.

We have previously graphed the equations shown below. The graph of the equation $y = x - 1$ is a line, and we call it a **linear equation**. Since the graphs of $y = x^2$, $y = |x|$, and $y = x^3$ are *not* lines, they are **nonlinear equations**.

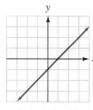
$y = x - 1$
Linear equation
(a)

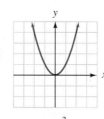
$y = x^2$
Nonlinear equation
(b)

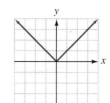
$y = |x|$
Nonlinear equation
(c)

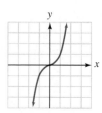
$y = x^3$
Nonlinear equation
(d)

In this section, we will discuss how to graph linear equations and show how to use their graphs to solve problems.

Any equation, such as $y = x - 1$, whose graph is a straight line is called a **linear equation in x and y**. Some other examples of linear equations are

$$y = \dfrac{1}{2}x + 2, \quad 3x - 2y = 8, \quad 5y - x + 2 = 0, \quad y = 4, \quad \text{and} \quad x = -3$$

A linear equation in x and y is any equation that can be written in a special form, called **standard** (or **general**) form.

Standard (General) Form of a Linear Equation

If A, B, and C represent real numbers, the equation

$Ax + By = C$ (both A and B are not zero)

is called the **standard form** (or **general form**) of the equation of a line.

Whenever possible, we will write the standard form $Ax + By = C$ so that A, B, and C are integers and $A \geq 0$. Note that in a linear equation in x and y, the exponents on x and y are 1.

EXAMPLE 1
Which of the following equations are linear equations?

a. $3x = 1 - 2y$ **b.** $y = x^3 + 1$ **c.** $y = -\frac{1}{2}x$

Strategy We will try to write each equation in standard form $Ax + By = C$. We will also note the exponents on x and y.

WHY If we can write an equation in standard form $Ax + By = C$, it is a linear equation. If the exponents on x or y are not 1, the equation is nonlinear.

Solution

a. Since the equation $3x = 1 - 2y$ can be written in $Ax + By = C$ form, it is a linear equation.

$3x = 1 - 2y$	This is the original equation.
$3x + 2y = 1 - 2y + 2y$	Add 2y to both sides.
$3x + 2y = 1$	Simplify the right-hand side: $-2y + 2y = 0$.

Here $A = 3$, $B = 2$, and $C = 1$.

b. Since the exponent on x in $y = x^3 + 1$ is 3, the equation is a nonlinear equation.

c. Since the equation $y = -\frac{1}{2}x$ can be written in $Ax + By = C$ form, it is a linear equation.

$y = -\frac{1}{2}x$	This is the original equation.
$-2(y) = -2\left(-\frac{1}{2}x\right)$	Multiply both sides by -2 so that the coefficient of x will be 1.
$-2y = x$	Simplify the right-hand side: $-2\left(-\frac{1}{2}\right) = 1$.
$0 = x + 2y$	Add 2y to both sides.
$x + 2y = 0$	Write the equation in standard form.

Here $A = 1$, $B = 2$, and $C = 0$.

Self Check 1
Which of the following are linear equations and which are nonlinear?
a. $y = |x|$
b. $-x = 6 - y$
c. $y = x$

Now Try Problems 38 and 40

2 Complete ordered-pair solutions of linear equations.

To find solutions of linear equations, we substitute selected values for one variable and solve for the other.

EXAMPLE 2
Complete the table of solutions for $3x + 2y = 5$.

Strategy In each case, we will substitute the known coordinate of the solution into the given equation.

WHY We can solve the resulting equation in one variable to find the unknown coordinate of the solution.

Solution
In the first row, we are given an x-value of 7. To find the corresponding y-value, we substitute 7 for x and solve for y.

$3x + 2y = 5$	This is the original equation.
$3(7) + 2y = 5$	Substitute 7 for x.
$21 + 2y = 5$	Do the multiplication: $3(7) = 21$.
$2y = -16$	Subtract 21 from both sides: $5 - 21 = -16$.
$y = -8$	Divide both sides by 2.

x	y	(x, y)
7		(7,)
	4	(, 4)

Self Check 2
Complete the table of solutions for $3x + 2y = 5$.

x	y	(x, y)
	−2	(, −2)
5		(5,)

Now Try Problem 42

A solution of $3x + 2y = 5$ is $(7, -8)$.

In the second row, we are given a y-value of 4. To find the corresponding x-value, we substitute 4 for y and solve for x.

$3x + 2y = 5$ This is the original equation.
$3x + 2(4) = 5$ Substitute 4 for y.
$3x + 8 = 5$ Do the multiplication: 2(4) = 8.
$3x = -3$ Subtract 8 from both sides: 5 − 8 = −3.
$x = -1$ Divide both sides by 3.

Another solution is $(-1, 4)$. The completed table is shown on the right.

x	y	(x, y)
7	−8	(7, −8)
−1	4	(−1, 4)

3 Graph linear equations by plotting points.

It is impossible to list the infinitely many solutions of a linear equation. However, to show all of its solutions, we can draw a mathematical "picture" of them. We call this picture the *graph of the equation*.

Graphing Linear Equations

1. Find three pairs (x, y) that satisfy the equation by selecting three numbers for x and finding the corresponding values of y.
2. Plot each resulting pair (x, y) on a rectangular coordinate system. If the three points do not lie on a straight line, check your computations.
3. Draw the straight line passing through the points.

Success Tip When selecting x values for a table of solutions, a rule of thumb is to choose a negative number, a positive number, and 0.

Self Check 3

Graph $y = -3x + 2$ and compare the result to the graph of $y = -3x$. What do you notice?

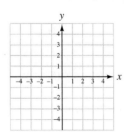

Now Try Problem 45

EXAMPLE 3   Graph: $y = -3x$

Strategy We will find three solutions of the equation, plot them on a rectangular coordinate system, and then draw a straight line passing through the points.

WHY To *graph* a linear equation in two variables means to make a drawing that represents all of its solutions.

Solution
To find three ordered pairs that satisfy the equation, we begin by choosing three x-values: -2, 0, and 2.

If $x = -2$ If $x = 0$ If $x = 2$
$y = -3x$ $y = -3x$ $y = -3x$
$y = -3(-2)$ $y = -3(0)$ $y = -3(2)$
$y = 6$ $y = 0$ $y = -6$

We enter the results in a table of solutions, plot the points, and draw a straight line through the points. The graph appears on the next page.

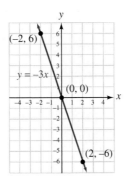

This point will serve as a check.

Success Tip Since two points determine a line, only two points are needed to graph a linear equation. However, we will often plot a third point as a check. If the three points do not lie on a straight line, then at least one of them is in error.

When graphing linear equations, it is often easier to find solutions of the equation if it is first solved for y.

EXAMPLE 4 Graph: $2y = 4 - x$

Strategy We will use properties of equality to solve the given equation for y. Then we will use the point-plotting method of this section to graph the resulting equivalent equation.

WHY The calculations to find several solutions of a linear equation in two variables are usually easier when the equation is solved for y.

Solution
To solve for y, we undo the multiplication of 2 by dividing both sides by 2.

$2y = 4 - x$

$\dfrac{2y}{2} = \dfrac{4}{2} - \dfrac{x}{2}$ On the right-hand side, dividing each term by 2 is equivalent to dividing the entire side by 2: $\dfrac{4-x}{2} = \dfrac{4}{2} - \dfrac{x}{2}$.

$y = 2 - \dfrac{x}{2}$ Simplify: $\dfrac{4}{2} = 2$

Since each value of x will be divided by 2, we will choose values of x that are divisible by 2. Three such choices are -4, 0, and 4. If $x = -4$, we have

$y = 2 - \dfrac{x}{2}$

$y = 2 - \dfrac{-4}{2}$ Substitute -4 for x.

$y = 2 - (-2)$ Divide: $\dfrac{-4}{2} = -2$

$y = 4$ Do the subtraction.

A solution is $(-4, 4)$. This pair and two others satisfying the equation are shown in the table on the next page. If we plot the points and draw a straight line through them, we will obtain the graph also shown on the next page.

Self Check 4
Solve $3y = 3 + x$ for y. Then graph the equation.

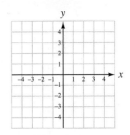

Now Try Problem 53

240   Chapter 3   Graphs, Linear Equations, and Inequalities in Two Variables; Functions

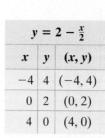
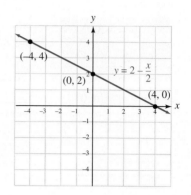

4 Graph linear equations by the intercept method.

In the figure to the right, the graph of $3x + 4y = 12$ intersects the y-axis at the point $(0, 3)$; we call this point the **y-intercept** of the graph. Since the graph intersects the x-axis at $(4, 0)$, the point $(4, 0)$ is the **x-intercept**.

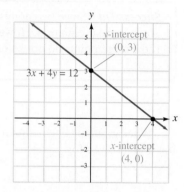

In general, we have the following definitions.

> **y- and x-intercepts**
>
> The **y-intercept** of a line is the point $(0, b)$ where the line intersects the y-axis. To find b, substitute 0 for x in the equation of the line and solve for y.
>
> The **x-intercept** of a line is the point $(a, 0)$ where the line intersects the x-axis. To find a, substitute 0 for y in the equation of the line and solve for x.

Plotting the x- and y-intercepts of a graph and drawing a straight line through them is called the **intercept method of graphing a line**. This method is useful when graphing equations written in standard form $Ax + By = C$.

Self Check 5

Graph $4x + 3y = 6$ using the intercept method.

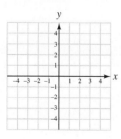

Now Try Problem 59

EXAMPLE 5   Graph: $3x - 2y = 8$

Strategy We will let $y = 0$ to find the x-intercept of the graph. We will then let $x = 0$ to find the y-intercept.

WHY Since two points determine a line, the y-intercept and the x-intercept are enough information to graph this linear equation.

Solution

x-intercept: $y = 0$

$$3x - 2y = 8$$
$$3x - 2(0) = 8 \quad \text{Substitute 0 for } y.$$
$$3x = 8 \quad \text{Simplify the left-hand side: } 2(0) = 0.$$
$$x = \frac{8}{3} \quad \text{Divide both sides by 3.}$$

The x-intercept is $\left(\frac{8}{3}, 0\right)$, which can be written $\left(2\frac{2}{3}, 0\right)$. This ordered pair is entered in the table below.

y-intercept: $x = 0$

$$3x - 2y = 8$$
$$3(0) - 2y = 8 \quad \text{Substitute 0 for x.}$$
$$-2y = 8 \quad \text{Simplify the left-hand side: } 3(0) = 0.$$
$$y = -4 \quad \text{Divide both sides by } -2.$$

The y-intercept is $(0, -4)$. It is entered in the table below. As a check, we find one more point on the line. If $x = 4$, then $y = 2$. We plot these three points and draw a straight line through them. The graph of $3x - 2y = 8$ is shown in the figure.

$3x - 2y = 8$		
x	y	(x, y)
$\frac{8}{3} = 2\frac{2}{3}$	0	$\left(2\frac{2}{3}, 0\right)$
0	-4	$(0, -4)$
4	2	$(4, 2)$

↑ This point serves as a check.

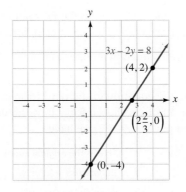

5 Identify and graph horizontal and vertical lines.

Equations such as $y = 4$ and $x = -3$ are linear equations, because they can be written in the standard form $Ax + By = C$. For example, $y = 4$ is equivalent to $0x + 1y = 4$ and $x = -3$ is equivalent to $1x + 0y = -3$. We now discuss how to graph these types of linear equations.

EXAMPLE 6 Graph: $y = 4$

Strategy To find three ordered-pair solutions of this equation to plot, we will select three values for x and use 4 for y each time.

WHY The given equation requires that $y = 4$.

Solution
We can write the equation in standard form as $0x + y = 4$. Since the coefficient of x is 0, the numbers chosen for x have no effect on y. The value of y is always 4. For example, if $x = 2$, we have

$$0x + y = 4 \quad \text{This is the original equation, } y = 4, \text{ written in standard form.}$$
$$0(2) + y = 4 \quad \text{Substitute 2 for x.}$$
$$y = 4 \quad \text{Simplify the left side.}$$

The table of solutions shown on the next page contains three ordered pairs that satisfy the equation $y = 4$. If we plot the points and draw a straight line through them, the result is a horizontal line. The y-intercept is $(0, 4)$, and there is no x-intercept.

Self Check 6
Graph: $y = -2$

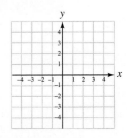

Now Try Problem 62

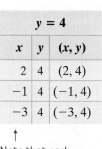

Note that each y-coordinate is 4.

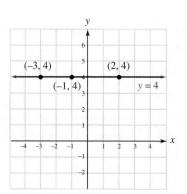

Self Check 7

Graph: $x = 4$

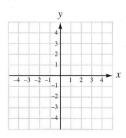

Now Try Problem 64

EXAMPLE 7 Graph: $x = -3$

Strategy To find three ordered-pair solutions of this equation to plot, we will select -3 for x each time and use three different values for y.

WHY The given equation requires that $x = -3$.

Solution
We can write the equation in standard form as $x + 0y = -3$. Since the coefficient of y is 0, the numbers chosen for y have no effect on x. The value of x is always -3. For example, if $y = -2$, we have

$x + 0y = -3$ This is the original equation written in standard form.
$x + 0(-2) = -3$ Substitute -2 for y.
$x = -3$ Simplify the left side.

The table of solutions shown below contains three ordered pairs that satisfy the equation $x = -3$. If we plot the points and draw a line through them, the result is a vertical line. The x-intercept is $(-3, 0)$, and there is no y-intercept.

$x = -3$		
x	y	(x, y)
-3	-2	$(-3, -2)$
-3	0	$(-3, 0)$
-3	3	$(-3, 3)$

Note that each x-coordinate is -3.

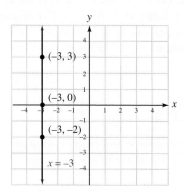

From the results of Examples 6 and 7, we have the following facts.

Equations of Horizontal and Vertical Lines

The equation $y = b$ represents the horizontal line that intersects the y-axis at $(0, b)$. If $b = 0$, the line is the x-axis.

The equation $x = a$ represents the vertical line that intersects the x-axis at $(a, 0)$. If $a = 0$, the line is the y-axis.

6 Use graphs of linear equations to solve application problems.

EXAMPLE 8 *Birthday Parties* A restaurant offers a party package that includes food, drinks, cake, and party favors for a cost of $25 plus $3 per child. Write a linear equation that will give the cost for a party of any size, and then graph the equation.

Strategy We will form an equation and use the plotting points method to graph the equation.

WHY The graph is a picture of all the solutions of the equation.

Solution

We can let c represent the cost of the party. The cost c is the sum of the basic charge of $25 and the cost per child times the number of children attending. If the number of children attending is n, at $3 per child, the total cost for the children is $3n$.

The cost	is	the basic $25 charge	plus	$3	times	the number of children.
c	$=$	25	$+$	3	$\cdot$	n

For the equation $c = 25 + 3n$, the independent variable (input) is n, the number of children. The dependent variable (output) is c, the cost of the party. We will find three points on the graph of the equation by choosing n-values of 0, 5, and 10 and finding the corresponding c-values. The results are shown in the table.

If $n = 0$	If $n = 5$	If $n = 10$
$c = 25 + 3(0)$	$c = 25 + 3(5)$	$c = 25 + 3(10)$
$c = 25$	$c = 25 + 15$	$c = 25 + 30$
	$c = 40$	$c = 55$

$c = 25 + 3n$

n	c
0	25
5	40
10	55

Next, we graph the points and draw a line through them. We don't draw an arrowhead on the left, because it doesn't make sense to have a negative number of children attend a party. Note that the c-axis is scaled in units of $5 to accommodate costs ranging from $0 to $65. We can use the graph to determine the cost of a party of any size. For example, to find the cost of a party with 8 children, we locate 8 on the horizontal axis and then move up to find a point on the graph directly above the 8. Since the coordinates of that point are (8, 49), the cost for 8 children would be $49.

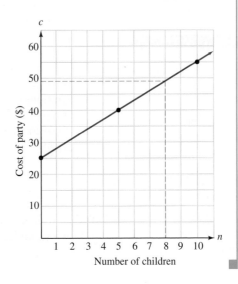

Self Check 8

A laser tag business offers a party package that includes invitations, a party room, and two rounds of laser tag. The cost is $15 plus $10 per child. Write a linear equation that will give the cost for a party of any size, and then graph the equation.

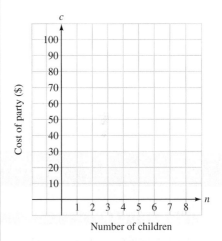

Now Try Problem 83

ANSWERS TO SELF CHECKS

1. a. nonlinear **b.** linear **c.** linear **2.** (3, −2), (5, −5)

3.

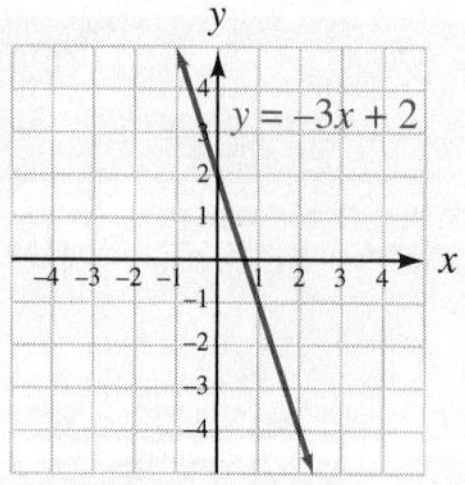

It is a line 2 units above the graph of $y = -3x$.

4.

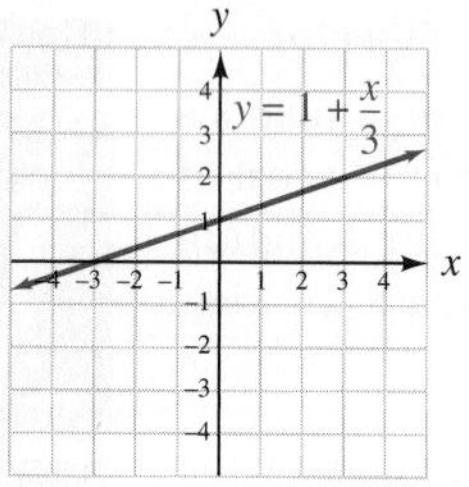

5.

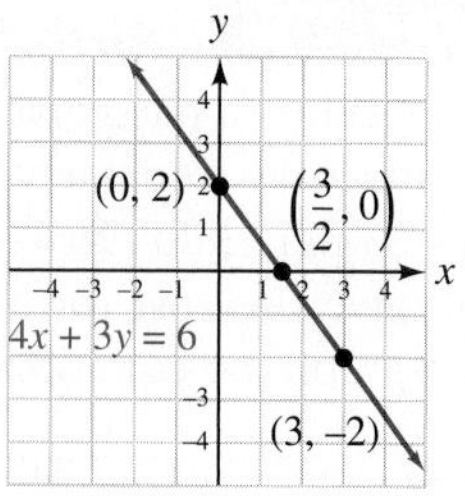

6.

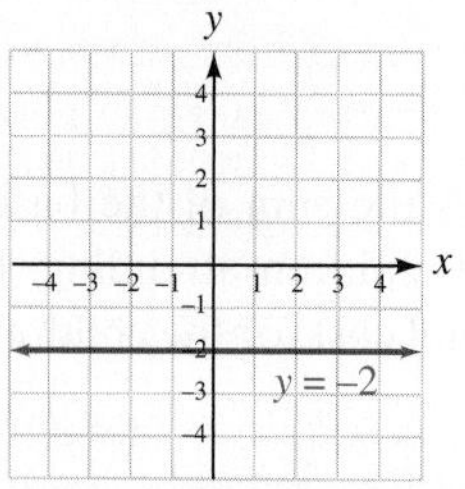

7.

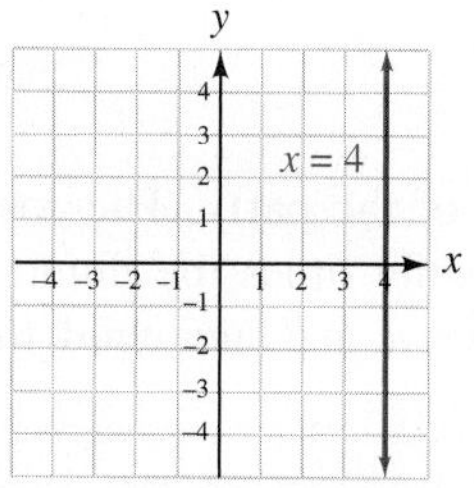

8.

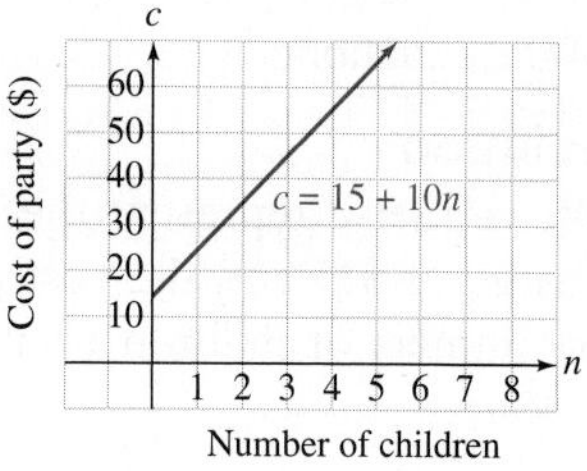

SECTION 3.3 STUDY SET

VOCABULARY

Fill in the blanks.

1. An equation whose graph is a line and whose variables are to the first power is called a ______ equation.

2. The equation $Ax + By = C$ is the ______________ form of the equation of a line.

3. The ________ of a line is the point $(0, b)$ where the line intersects the y-axis.

4. The ________ of a line is the point $(a, 0)$ where the line intersects the x-axis.

CONCEPTS

Fill in the blanks.

5. To find the y-intercept of the graph of a linear equation, let ___ = 0 and solve for ___.

6. To find the x-intercept of the graph of a linear equation, let ___ = 0 and solve for ___.

7. Lines parallel to the y-axis are _______ lines.

8. Lines parallel to the x-axis are ________ lines.

9. What is another name for the line $x = 0$?

10. What is another name for the line $y = 0$?

Find the power of each variable in Problems 11–13.

11. $y = 2x - 6$

12. $y = x^2 - 6$

13. $y = x^3 + 2$

14. In a linear equation in x and y, what are the exponents on x and y?

Consider the graph of a linear equation shown below.

15. Why will the coordinates of point A yield a true statement when substituted into the equation?

16. Why will the coordinates of point B yield a false statement when substituted into the equation?

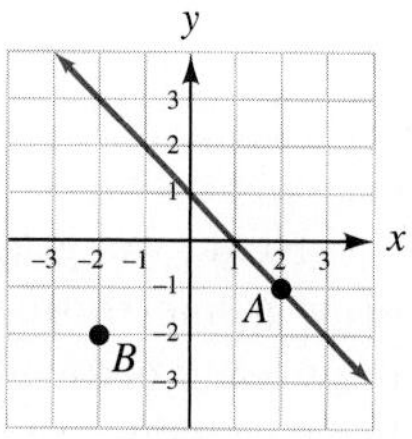

17. A student found three solutions of a linear equation and plotted them as shown. What conclusion can he draw?

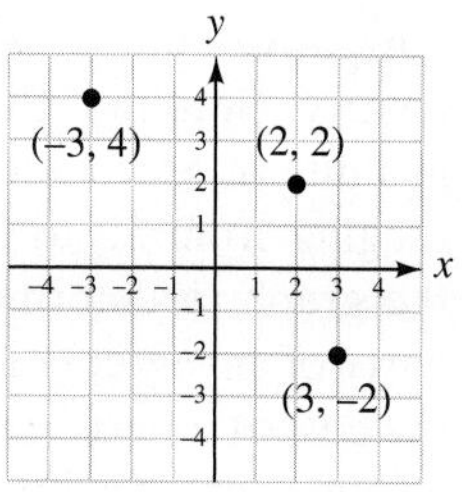

18. How many solutions are there for a linear equation in two variables?

19. a. Give the x-intercept of the graph on the right.
    b. Give the y-intercept of the graph on the right.

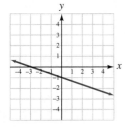

20. a. Give the x-intercept of the graph on the right.
    b. Give the y-intercept of the graph on the right.

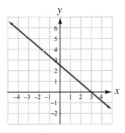

On the coordinate system below, draw the graph of a line...
21. with no x-intercept.
22. with no y-intercept.
23. with an x-intercept of $(2, 0)$.
24. with a y-intercept of $\left(0, -\dfrac{5}{2}\right)$.

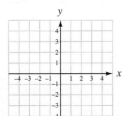

NOTATION
Write each equation in standard form, $Ax + By = C$.
25. $4x = y + 6$
26. $2y = x$
27. $x - 9 = -3y$
28. $x = 12$

Solve each equation for y.
29. $x + y = 8$
30. $2x - y = 8$
31. $3x + \dfrac{y}{2} = 4$
32. $y - 2 = 0$

GUIDED PRACTICE
Classify each of the following as the graph of a linear equation or of a nonlinear equation. See Objective 1.

33.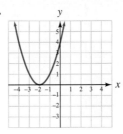

34.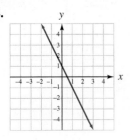

35.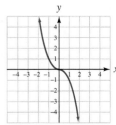

36.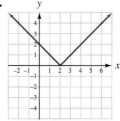

Classify each equation as linear or nonlinear. See Example 1.
37. $y = x^3$
38. $2x + 3y = 6$
39. $y = -2$
40. $y = |x + 2|$

Complete each table of solutions. See Example 2.
41. $5y = 2x + 10$

x	y
10	
	0
5	

42. $2x + 4y = 24$

x	y
4	
	7
-4	

43. $x - 2y = 4$

x	y
0	
	0
1	

44. $5x - y = 3$

x	y
0	
	0
1	

Find three solutions of the equation and draw its graph. See Example 3.

45. $y = -x + 2$

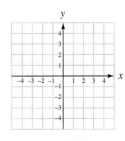

46. $y = -x - 1$

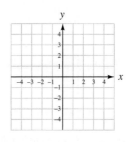

47. $y = 2x + 1$

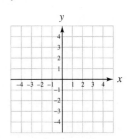

48. $y = 3x - 2$

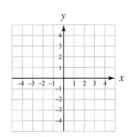

49. $y = x$
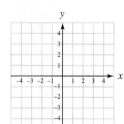

50. $y = 3x$
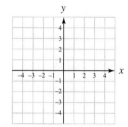

51. $y = -3x$
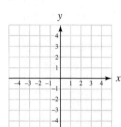

52. $y = -2x$
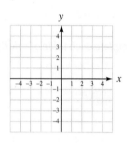

Solve each equation for y, find three solutions of the equation, and then draw its graph. See Example 4.

53. $2y = 4x - 6$
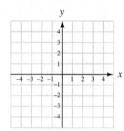

54. $3y = 6x - 3$
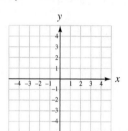

55. $2y = x - 4$
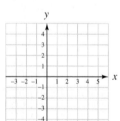

56. $4y = x + 16$
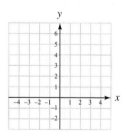

Graph each equation using the intercept method. See Example 5.

57. $2y - 2x = 6$
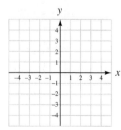

58. $3x - 3y = 9$
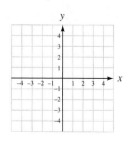

59. $4x + 5y = 20$
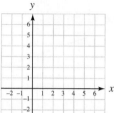

60. $3x + y = -3$
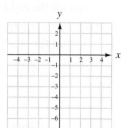

Graph each equation. See Examples 6–7.

61. $y = 4$
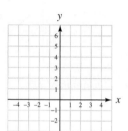

62. $y = -3$
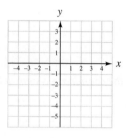

63. $x = -2$
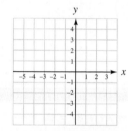

64. $x = 5$
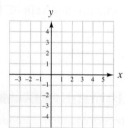

TRY IT YOURSELF

Graph each equation.

65. $15y + 5x = -15$
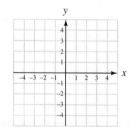

66. $8x + 4y = -24$
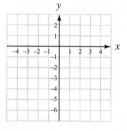

67. $3x + 4y = 8$
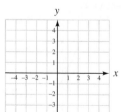

68. $2x + 3y = 9$
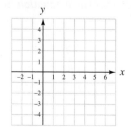

69. $y = \dfrac{x}{3}$

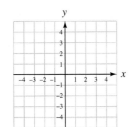

70. $y = -\dfrac{x}{3} - 1$

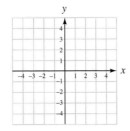

71. $-4y + 9x = -9$

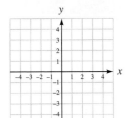

72. $-4y + 5x = -15$

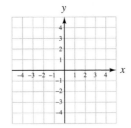

73. $3x + 4y = 12$

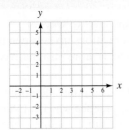

74. $4x - 3y = 12$

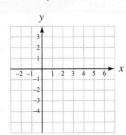

75. $y = -\dfrac{1}{2}$

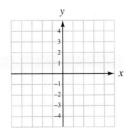

76. $y = \dfrac{5}{2}$

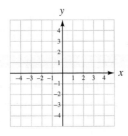

77. $x = \dfrac{4}{3}$

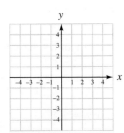

78. $x = -\dfrac{5}{3}$

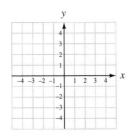

79. $y = -\dfrac{3}{2}x + 2$

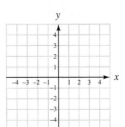

80. $y = \dfrac{2}{3}x - 2$

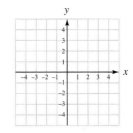

81. $2y + x = -2$

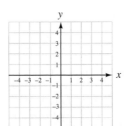

82. $4y + 2x = 8$

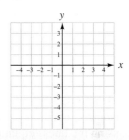

APPLICATIONS

83. EDUCATION COSTS Each semester, a college charges a services fee of $50 plus $25 for each unit taken by a student.

   a. Write a linear equation that gives the total enrollment cost c for a student taking u units.

   b. Complete the table of solutions below and graph the equation.

   c. Use the graph to find the total cost for a student taking 18 units the first semester and 12 units the second semester.

   d. What does the y-intercept of the line tell you?

u	c
4	
8	
14	

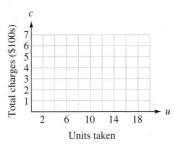

84. GROUP RATES To promote the sale of tickets for a cruise to Alaska, a travel agency reduces the regular ticket price of $3,000 by $5 for each individual traveling in the group.

   a. Write a linear equation that would find the ticket price t for the cruise if a group of p people travel together.

   b. Complete the table of solutions on the next page and graph the equation.

c. As the size of the group increases, what happens to the ticket price?

d. Use the graph to determine the cost of an individual ticket if a group of 25 will be traveling together.

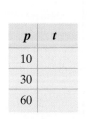

p	t
10	
30	
60	

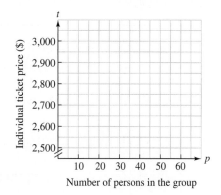

85. **PHYSIOLOGY** Physiologists have found that a woman's height h (in inches) can be approximated using the linear equation $h = 3.9r + 28.9$, where r represents the length of her radius bone in inches.

   a. Complete the table of solutions. Round to the nearest tenth and then graph the equation.

   b. Complete this sentence: From the graph, we see that the longer the radius bone, the …

   c. From the graph, estimate the height of a woman whose radius bone is 7.5 inches long.

r	h
7	
8.5	
9	

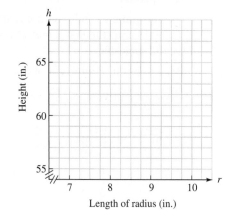

86. **RESEARCH EXPERIMENTS** A psychology major found that the time t (in seconds) that it took a rat to complete a maze was related to the number of trials n the rat had been given. The resulting equation was $t = 25 - 0.25n$.

   a. Complete the table of solutions and graph the equation.

   b. Complete this sentence: From the graph, we see that the more trials the rat had, the …

c. From the graph, estimate the time it will take the rat to complete the maze on its 32nd trial.

n	t
4	
12	
16	

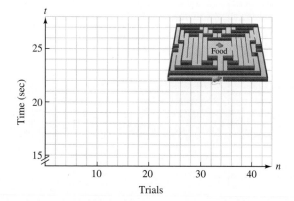

WRITING

87. A linear equation and a graph are two ways of mathematically describing a relationship between two quantities. Which do you think is more informative, and why?

88. From geometry, we know that two points determine a line. Explain why it is a good practice when graphing linear equations to find and plot three points instead of just two.

89. How can we tell by looking at an equation whether its graph will be a straight line?

90. Can the x-intercept and the y-intercept of a line be the same point? Explain.

REVIEW

91. Simplify: $-(-5 - 4c)$

92. Write the set of integers.

93. Solve: $\dfrac{x + 6}{2} = 1$

94. Evaluate: $-2^2 + 2^2$

95. Write a formula that relates profit, revenue, and costs.

96. Find the volume, to the nearest tenth, of a sphere with radius 6 feet.

97. Evaluate: $1 + 2[-3 - 4(2 - 8^2)]$

98. Evaluate $\dfrac{x + y}{x - y}$ for $x = -2$ and $y = -4$.

SECTION 3.4
Rate of Change and the Slope of a Line

Since our world is one of constant change, we must be able to describe change so that we can plan for the future. In this section, we will show how to describe the amount of change of one quantity in relation to the amount of change of another quantity by finding a **rate of change**.

1 Find rates of change.

The line graph in figure (a) below shows the number of business permits issued each month by a city over a 12-month period. From the shape of the graph, we can see that the number of permits issued *increased* each month.

For situations such as the one graphed in (a), it is often useful to calculate a rate of increase (called a **rate of change**). We do so by finding the **ratio** of the change in the number of business permits issued each month to the number of months over which that change took place.

Objectives

1. Find rates of change.
2. Find the slope of a line from its graph.
3. Find the slope of a line given two points.
4. Recognize positive and negative slope.
5. Find slopes of horizontal and vertical lines.
6. Use slope to graph a line.
7. Determine whether lines are parallel or perpendicular using slope.

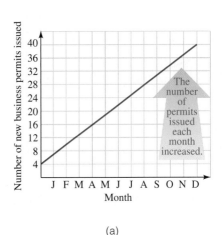

(a)

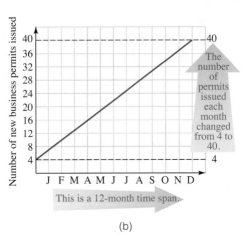

(b)

Ratios and Rates

A **ratio** is the quotient of two numbers or the quotient of two quantities with the same units. In symbols, if a and b represent two numbers, the ratio of a to b is $\dfrac{a}{b}$. Ratios that are used to compare quantities with different units are called **rates**.

In figure (b), we see that the number of permits issued prior to the month of January was 4. By the end of the year, the number of permits issued during the month of December was 40. This is a change of $40 - 4$, or 36, over a 12-month period. So we have

$$\text{Rate of change} = \frac{\text{change in number of permits issued each month}}{\text{change in time}}$$

The rate of change is a ratio.

$$\text{Rate of change} = \frac{36 \text{ permits}}{12 \text{ months}}$$

$$\text{Rate of change} = \frac{\overset{1}{\cancel{12}} \cdot 3 \text{ permits}}{\underset{1}{\cancel{12}} \text{ months}} \qquad \text{Factor 36 as } 12 \cdot 3 \text{ and remove the common factor of 12.}$$

$$\text{Rate of change} = \frac{3 \text{ permits}}{1 \text{ month}}$$

The number of business permits being issued increased at a rate of 3 per month, denoted as 3 permits/month.

> **The Language of Algebra** The preposition *per* means for each, or for every. When we say the rate of change is 3 permits *per* month, we mean 3 permits for each month.

Self Check 1

Find the rate of change in the number of subscribers over the second 5-year period. Write the rate in simplest form.

Now Try Problem 79

EXAMPLE 1

The graph shows the number of subscribers to a newspaper. Find the rate of change in the number of subscribers over the first 5-year period. Write the rate in simplest form.

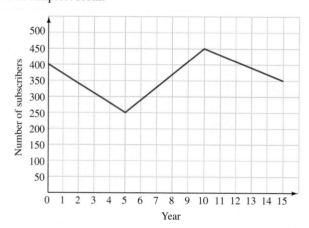

Strategy We will form a ratio of the change in the number of subscribers over the change in the time.

WHY The rate of change is given by this ratio.

Solution

$$\text{Rate of change} = \frac{\text{change in number of subscribers}}{\text{change in time}} \qquad \text{Set up the ratio.}$$

$$\text{Rate of change} = \frac{(250 - 400) \text{ subscribers}}{(5 - 0) \text{ years}} \qquad \text{Subtract the earlier number of subscribers from the later number of subscribers.}$$

$$\text{Rate of change} = \frac{-150 \text{ subscribers}}{5 \text{ years}} \qquad 250 - 400 = -150$$

$$\text{Rate of change} = \frac{-30 \cdot \overset{1}{\cancel{5}} \text{ subscribers}}{\underset{1}{\cancel{5}} \text{ years}} \qquad \text{Factor } -150 \text{ as } -30 \cdot 5 \text{ and remove the common factor of 5.}$$

$$\text{Rate of change} = \frac{-30 \text{ subscribers}}{1 \text{ year}}$$

The number of subscribers for the first 5 years *decreased* by 30 per year, as indicated by the negative sign in the result. We can write this as -30 subscribers/year.

2 Find the slope of a line from its graph.

The **slope of a nonvertical line** is a number that measures the line's steepness. We can calculate the slope by picking two points on the line and writing the ratio of the vertical change (called the **rise**) to the corresponding horizontal change (called the **run**) as we move from one point to the other. As an example, we will find the slope of the line that was used to describe the number of building permits issued and show that it gives the rate of change.

In the following figure, the line passes through points $P(0, 4)$ and $Q(12, 40)$. Moving along the line from point P to point Q causes the value of y to change from $y = 4$ to $y = 40$, an increase of $40 - 4 = 36$ units. We say that the *rise* is 36. Notice that a right triangle, called a **slope triangle**, is created by this process.

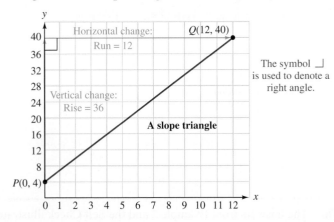

Moving from point P to point Q, the value of x increases from $x = 0$ to $x = 12$, an increase of $12 - 0 = 12$ units. We say that the *run* is 12. The slope of a line, usually denoted with the letter m, is defined as the ratio of the change in y to the change in x.

$m = \dfrac{\text{change in } y\text{-values}}{\text{change in } x\text{-values}}$ Slope is a ratio.

$m = \dfrac{40 - 4}{12 - 0}$ To find the change in y (the rise), subtract the y-values.
To find the change in x (the run), subtract the x-values.

$m = \dfrac{36}{12}$ Do the subtractions.

$m = 3$ Do the division.

This is the same value we obtained when we found the rate of change of the number of business permits issued over the 12-month period. Therefore, by finding the slope of the line, we found the rate of change.

EXAMPLE 2 Find the slope of the line shown in figure (a).

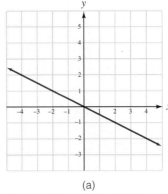
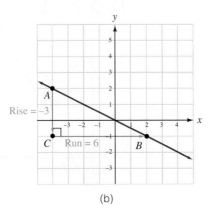

(a)   (b)

Self Check 2
Find the slope of the line using two points different from those used in Example 2.

Now Try Problem 27

Strategy We will pick two points on the line, construct a slope triangle, and find the rise and the run. Then we will write the ratio of the rise to the run.

WHY The slope of a line is the ratio of the rise to the run.

Solution

We begin by choosing two points on the line—call them A and B—as shown in figure (b). One way to move from point A to point B is to start at point A and count *downward* 3 grid squares. Because this movement is downward, the rise is -3. Then, moving right, we count 6 grid squares to reach B. This indicates a run of 6. To find the slope of the line, we write a ratio of rise to run in simplified form. Usually the letter m is used to denote slope, so we have

$$m = \frac{\text{rise}}{\text{run}} \quad \text{The slope of a line is the ratio of the rise to the run.}$$

$$m = \frac{-3}{6} \quad \text{From the slope triangle, the rise is } -3 \text{ and the run is 6.}$$

$$m = -\frac{1}{2} \quad \text{Simplify the fraction.}$$

The slope of the line is $-\dfrac{1}{2}$.

Success Tip The answers from Example 2 and the Self Check illustrate an important fact about slope: *The same value for the slope of a line will result no matter which two points on the line are used to determine the rise and the run.*

3 Find the slope of a line given two points.

We can generalize the graphic method for finding slope to develop a slope formula. To begin, we select points P and Q on the line shown in the figure below. To distinguish between the coordinates of these points, we use **subscript notation**. Point P has coordinates (x_1, y_1), which are read as "x sub 1 and y sub 1." Point Q has coordinates (x_2, y_2), which are read as "x sub 2 and y sub 2."

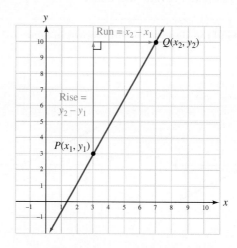

As we move from P to Q, the rise is the difference of the y-coordinates: $y_2 - y_1$. We call this difference the **change in y**. The run is the difference of the x-coordinates: $x_2 - x_1$. This difference is called the **change in x**. Since the slope is the ratio $\frac{\text{rise}}{\text{run}}$, we have the following formula for calculating slope.

Slope of a Nonvertical Line

The **slope** of a nonvertical line passing through points (x_1, y_1) and (x_2, y_2) is

$$m = \frac{\text{vertical change}}{\text{horizontal change}} = \frac{\text{rise}}{\text{run}} = \frac{\text{change in } y}{\text{change in } x} = \frac{y_2 - y_1}{x_2 - x_1} \quad \text{if } x_2 \neq x_1$$

EXAMPLE 3 Find the slope of the line passing through $(1, 2)$ and $(3, 8)$.

Strategy We will use the slope formula to find the slope of the line.

WHY We know the coordinates of two points on the line.

Solution
When using the slope formula, it makes no difference which point you call (x_1, y_1) and which point you call (x_2, y_2). If we let (x_1, y_1) be $(1, 2)$ and (x_2, y_2) be $(3, 8)$, then

$m = \dfrac{y_2 - y_1}{x_2 - x_1}$ This is the slope formula.

$m = \dfrac{8 - 2}{3 - 1}$ Substitute 8 for y_2, 2 for y_1, 3 for x_2, and 1 for x_1.

$m = \dfrac{6}{2}$ Do the subtractions.

$m = 3$ Simplify. Think of this as a $\frac{3}{1}$ rise-to-run ratio.

The slope of the line is 3. The graph of the line, including the slope triangle, is shown here. Note that we obtain the same value for the slope if we let $(x_1, y_1) = (3, 8)$ and $(x_2, y_2) = (1, 2)$.

$m = \dfrac{y_2 - y_1}{x_2 - x_1} = \dfrac{2 - 8}{1 - 3} = \dfrac{-6}{-2} = 3$

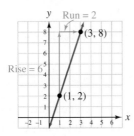

Self Check 3
Find the slope of the line passing through $(2, 1)$ and $(4, 11)$.

Now Try Problem 33

Caution! When finding the slope of a line, always subtract the y-values and the x-values in the same order. Otherwise your answer will have the wrong sign:

$m \neq \dfrac{y_2 - y_1}{x_1 - x_2}$ and $m \neq \dfrac{y_1 - y_2}{x_2 - x_1}$

THINK IT THROUGH Average Rate of Tuition Increase

"Whatever happens in the future to the economy, whether up or down or more of the same, all current predictions point to a continuing rise over the coming decade in the cost of college education."

Daniel Silver in Show Me the Money, News-Tribune

The line graphed below approximates the average cost of tuition and fees at U.S. public two-year academic institutions for the years 2000–2012. Find the average rate of increase in cost over this time period by finding the slope of the line.

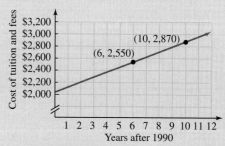

Source: The College Board

Self Check 4

Find the slope of the line that passes through $(-1, -2)$ and $(1, -7)$.

Now Try Problem 39

EXAMPLE 4
Find the slope of the line that passes through $(-2, 4)$ and $(5, -6)$ and draw its graph.

Strategy We will use the slope formula to find the slope of the line.

WHY We know the coordinates of two points on the line.

Solution
Since we know the coordinates of two points on the line, we can find its slope. If (x_1, y_1) is $(-2, 4)$ and (x_2, y_2) is $(5, -6)$, then

$$x_1 = -2 \quad \text{and} \quad x_2 = 5$$
$$y_1 = 4 \qquad\qquad y_2 = -6$$

$$m = \frac{y_2 - y_1}{x_2 - x_1} \quad \text{This is the slope formula.}$$

$$m = \frac{-6 - 4}{5 - (-2)} \quad \text{Substitute } -6 \text{ for } y_2, \ 4 \text{ for } y_1, \ 5 \text{ for } x_2, \text{ and } -2 \text{ for } x_1.$$

$$m = -\frac{10}{7} \quad \begin{array}{l}\text{Simplify the numerator: } -6 - 4 = -10. \\ \text{Simplify the denominator: } 5 - (-2) = 7.\end{array}$$

The slope of the line is $-\frac{10}{7}$. The figure below shows the graph of the line. Note that the line falls from left to right—a fact that is indicated by its negative slope.

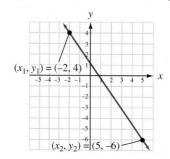

4 Recognize positive and negative slope.

In Example 3, the slope of the line was positive (3). In Example 4, the slope of the line was negative $\left(-\frac{10}{7}\right)$. In general, lines that rise from left to right have a positive slope, and lines that fall from left to right have a negative slope, as shown below.

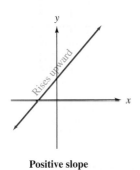
Positive slope

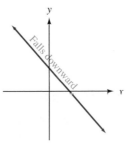
Negative slope

5 Find slopes of horizontal and vertical lines.

In the next two examples, we will calculate the slope of a horizontal line and show that a vertical line has no defined slope.

EXAMPLE 5 Find the slope of the line $y = 3$.

Strategy We will find the coordinates of two points on the line.

WHY We can then use the slope formula to find the slope of the line.

Solution
To find the slope of the line $y = 3$, we need to know two points on the line. Graph the horizontal line $y = 3$ and label two points on the line: $(-2, 3)$ and $(3, 3)$.

If (x_1, y_1) is $(-2, 3)$ and (x_2, y_2) is $(3, 3)$, we have

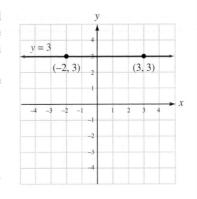

$m = \dfrac{y_2 - y_1}{x_2 - x_1}$ This is the slope formula.

$m = \dfrac{3 - 3}{3 - (-2)}$ Substitute 3 for y_2, 3 for y_1, 3 for x_2, and -2 for x_1.

$m = \dfrac{0}{5}$ Simplify the numerator and the denominator.

$m = 0$

The slope of the line $y = 3$ is 0.

The y-coordinates of any two points on any horizontal line will be the same, and the x-values will be different. Thus, the numerator of $\dfrac{y_2 - y_1}{x_2 - x_1}$ will always be zero, and the denominator will always be nonzero. Therefore, the slope of a horizontal line is zero.

Self Check 5
Find the slope of the line $y = -2$.
Now Try Problem 45

EXAMPLE 6 If possible, find the slope of the line $x = -2$.

Strategy We will find the coordinates of two points on the line.

WHY We can then use the slope formula to try to find the slope of the line.

Self Check 6
Find the slope of $x = 5$.
Now Try Problem 47

Solution

To find the slope of the line $x = -2$, we need to know two points on the line. We graph the vertical line $x = -2$ and label two points on the line: $(-2, -1)$ and $(-2, 3)$.

If (x_1, y_1) is $(-2, -1)$ and (x_2, y_2) is $(-2, 3)$, we have

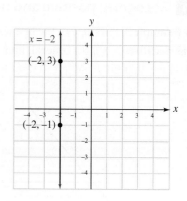

$m = \dfrac{y_2 - y_1}{x_2 - x_1}$ This is the slope formula.

$m = \dfrac{3 - (-1)}{-2 - (-2)}$ Substitute 3 for y_2, -1 for y_1, -2 for x_2, and -2 for x_1.

$m = \dfrac{4}{0}$ Simplify the numerator and the denominator.

Since division by zero is undefined, $\frac{4}{0}$ has no meaning. The slope of the line $x = -2$ is undefined.

The y-values of any two points on a vertical line will be different, and the x-values will be the same. Thus, the numerator of $\dfrac{y_2 - y_1}{x_2 - x_1}$ will always be nonzero, and the denominator will always be zero. Therefore, the slope of a vertical line is undefined.

We now summarize the results from Examples 5 and 6.

Slopes of Horizontal and Vertical Lines

Horizontal lines (lines with equations of the form $y = b$) have a slope of 0.

Vertical lines (lines with equations of the form $x = a$) have undefined slope.

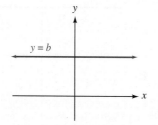

Horizontal line: 0 slope

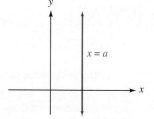

Vertical line: undefined slope

6 Use slope to graph a line.

We can graph a line whenever we know the coordinates of one point on the line and the slope of the line. For example, to graph the line that passes through $P(2, 4)$ and has a slope of 3, we first plot $P(2, 4)$, as in the figure. We can express the slope of 3 as a fraction: $3 = \dfrac{3}{1}$. Therefore, the line *rises* 3 units for every 1 unit it *runs* to the right. We can find a second point on the line by starting at $P(2, 4)$ and moving 3 units up (rise) and 1 unit to the right (run). This brings us to a point that we will call Q with coordinates $(2 + \mathbf{1}, 4 + \mathbf{3})$ or $(3, 7)$. The required line passes through points P and Q.

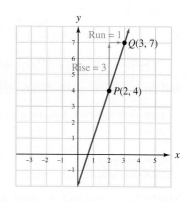

EXAMPLE 7
Graph the line that passes through the point $(-3, 4)$ with slope $-\frac{2}{5}$.

Strategy We will plot the given point and identify the rise and the run of the slope. Then we will start at the plotted point and find a second point on the line by forming a slope triangle.

WHY Once we locate two points on the line, we can draw the graph of the line.

Solution
We plot the point $(-3, 4)$ as shown in the figure to the right. Then, after writing the slope $-\frac{2}{5}$ as $\frac{-2}{5}$, we see that the *rise* is -2 and the *run* is 5. From the point $(-3, 4)$, we can find a second point on the line by moving 2 units down (rise) and then 5 units right (run). (A rise of -2 means to move down 2 units.) This brings us to the point with coordinates of $(2, 2)$. We then draw a line that passes through the two points.

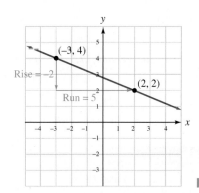

Self Check 7
Graph the line that passes through the point $(-4, 2)$ with slope -4.

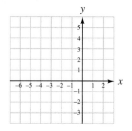

Now Try Problem 49

7 Determine whether lines are parallel or perpendicular using slope.

Two lines that lie in the same plane but do not intersect are called **parallel lines**. Parallel lines have the same slope and different y-intercepts. For example, the lines graphed in figure (a) are parallel because they both have slope $-\frac{2}{3}$.

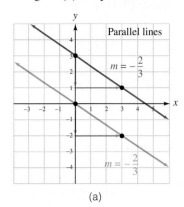
(a)

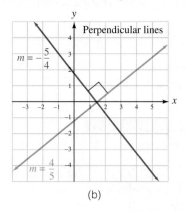
(b)

Lines that intersect to form four right angles (angles with measure $90°$) are called **perpendicular lines**. If the product of the slopes of two lines is -1, the lines are perpendicular. This means that the slopes are **negative** (or **opposite**) **reciprocals**. In figure (b), we know that the lines with slopes $\frac{4}{5}$ and $-\frac{5}{4}$ are perpendicular because

$$\frac{4}{5}\left(-\frac{5}{4}\right) = -\frac{20}{20} = -1 \qquad \frac{4}{5} \text{ and } -\frac{5}{4} \text{ are negative reciprocals.}$$

Slopes of Parallel and Perpendicular Lines

1. Two lines with the same slope are parallel.
2. Two lines are perpendicular if the product of the slopes is -1; that is, if their slopes are negative reciprocals.
3. Any horizontal line and any vertical line are perpendicular.

Self Check 8
Determine whether the line that passes through (2, 1) and (6, 8) and the line that passes through (−1, 0) and (4, 7) are parallel, perpendicular, or neither.

Now Try Problems 57, 59, and 61.

EXAMPLE 8
Determine whether the line that passes through (7, −9) and (10, 2) and the line that passes through (0, 1) and (3, 12) are parallel, perpendicular, or neither.

Strategy We will use the slope formula to find the slope of each line.

WHY If the slopes are equal, the lines are parallel. If the slopes are negative reciprocals, the lines are perpendicular. Otherwise, the lines are neither parallel nor perpendicular.

Solution
To calculate the slope of each line, we use the slope formula.

The line through (7, −9) and (10, 2):   *The line through (0, 1) and (3, 12):*

$$m = \frac{y_2 - y_1}{x_2 - x_1} = \frac{2 - (-9)}{10 - 7} = \frac{11}{3} \qquad m = \frac{y_2 - y_1}{x_2 - x_1} = \frac{12 - 1}{3 - 0} = \frac{11}{3}$$

Since the slopes are the same, the lines are parallel.

Self Check 9
Find the slope of a line perpendicular to the line passing through (−4, 1) and (9, 5).

Now Try Problem 67

EXAMPLE 9
Find the slope of a line perpendicular to the line passing through (1, −4) and (8, 4).

Strategy We will use the slope formula to find the slope of the line passing through (1, −4) and (8, 4).

WHY We can then form the negative reciprocal of the result to produce the slope of a line perpendicular to the given line.

Solution
The slope of the line that passes through (1, −4) and (8, 4) is

$$m = \frac{y_2 - y_1}{x_2 - x_1} = \frac{4 - (-4)}{8 - 1} = \frac{8}{7}$$

A line perpendicular to the given line has a slope that is the negative (or opposite) reciprocal of $\frac{8}{7}$, which is $-\frac{7}{8}$.

ANSWERS TO SELF CHECKS
1. 40 subscribers/year 2. $-\frac{1}{2}$ 3. 5 4. $-\frac{5}{2}$ 5. 0 6. undefined
7. 8. neither 9. $-\frac{13}{4}$

SECTION 3.4 STUDY SET

VOCABULARY
Fill in the blanks.

1. A _____ is the quotient of two numbers.

2. Ratios used to compare quantities with different units are called _____.

3. The _____ of a line is defined as the ratio of the change in y to the change in x.

4. The vertical change between two points on a coordinate system is often called the _____.
5. The horizontal change between two points on a coordinate system is often called the _____.
6. $m = \dfrac{\text{\_\_\_\_\_ change}}{\text{horizontal change}} = \dfrac{\text{rise}}{\text{\_\_\_\_}} = \dfrac{\text{change in \_\_\_\_}}{\text{change in \_\_\_\_}}$
7. Two lines that lie in the same plane but do not intersect are called _____ lines.
8. The rate of _____ of a linear relationship can be found by finding the slope of the graph of the line.

CONCEPTS

Fill in the blanks.

9. _____ lines have a slope of 0.
10. Vertical lines have _____ slope.
11. A line with positive slope _____ from left to right.
12. A line with negative slope _____ from left to right.

In the following illustration, which line has

13. a positive slope?
14. a negative slope?
15. zero slope?
16. undefined slope?

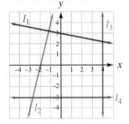

Consider the graph of the line in the following illustration:

17. Find its slope using points A and B.
18. Find its slope using points B and C.
19. Find its slope using points A and C.
20. What observation is suggested by your answers to parts a, b, and c?

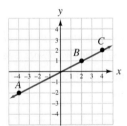

21. The following table shows the coordinates of two points on a line. Use the information to determine the slope of the graph of the line.

x	y
-4	2
5	-7

22. GROWTH RATES Use the graph to find the rate of change of a boy's height during the time shown.

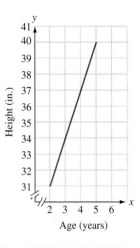

NOTATION

23. Write the formula used to find the slope of the line passing through (x_1, y_1) and (x_2, y_2).
24. Explain the difference between y^2 and y_2.

GUIDED PRACTICE

Find the slope of each line. See Examples 1–2.

25.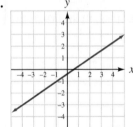

26.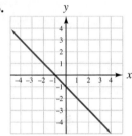

27.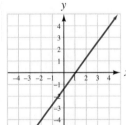

28.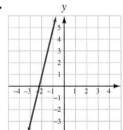

29.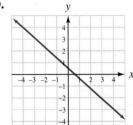

30.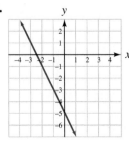

31. 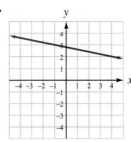 **32.**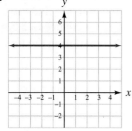

Find the slope of the line passing through the given points, when possible. See Examples 3–4.

33. $(2, 4)$ and $(1, 3)$ **34.** $(1, 3)$ and $(2, 5)$

35. $(3, 4)$ and $(2, 7)$ **36.** $(3, 6)$ and $(5, 2)$

37. $(0, 0)$ and $(4, 5)$ **38.** $(4, 3)$ and $(7, 8)$

39. $(-3, 5)$ and $(-5, 6)$ **40.** $(6, -2)$ and $(-3, 2)$

41. $(5, 7)$ and $(-4, 7)$ **42.** $(-1, -12)$ and $(6, -12)$

43. $(8, -4)$ and $(8, -3)$ **44.** $(-2, 8)$ and $(-2, 15)$

Find the slope of each line, if possible. See Examples 5–6.

45. $y = 5$ **46.** $x = -5$
47. $x = 4$ **48.** $y = -7$

Graph the line that passes through the given point and has the given slope. See Example 7.

49. $(0, 0)$, $m = -4$ **50.** $(0, 0)$, $m = 5$

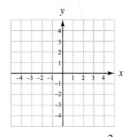

51. $(-3, -3)$, $m = -\dfrac{3}{2}$ **52.** $(-2, -1)$, $m = \dfrac{4}{3}$

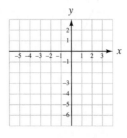 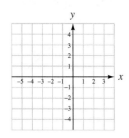

53. $(-5, 1)$, $m = 0$ **54.** $(0, 3)$,

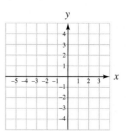 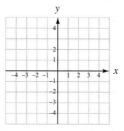

55. $(-1, -4)$, undefined slope **56.** $(-3, -2)$, $m = 0$

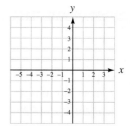 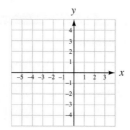

Determine whether the lines through each pair of points are parallel, perpendicular, or neither. See Example 8.

57. $(5, 3)$ and $(1, 4)$ **58.** $(2, 4)$ and $(-1, -1)$
$(-3, -4)$ and $(1, -5)$ $(8, 0)$ and $(11, 5)$

59. $(-4, -2)$ and $(2, -3)$ **60.** $(-2, 4)$ and $(6, -7)$
$(7, 1)$ and $(8, 7)$ $(-6, 4)$ and $(5, 12)$

61. $(2, 2)$ and $(4, -3)$ **62.** $(-1, -3)$ and $(2, 4)$
$(-3, 4)$ and $(-1, 9)$ $(5, 2)$ and $(8, -5)$

63. $(4, 2)$ and $(5, -3)$ **64.** $(8, -3)$ and $(8, -8)$
$(-5, 3)$ and $(-2, 9)$ $(11, 3)$ and $(22, 3)$

Find the slope of a line perpendicular to the line passing through the given two points. See Example 9.

65. $(0, 0)$ and $(5, -9)$ **66.** $(0, 0)$ and $(5, 12)$
67. $(-1, 7)$ and $(1, 10)$ **68.** $(-7, 6)$ and $(0, 4)$

APPLICATIONS

69. POOL DESIGN Find the slope of the bottom of the swimming pool as it drops off from the shallow end to the deep end, as shown.

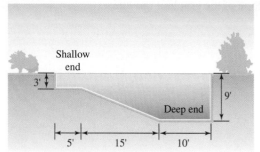

70. **DRAINAGE** To measure the amount of fall (slope) of a concrete patio slab, a 10-foot-long 2-by-4, a 1-foot ruler, and a level were used. Find the amount of fall in the slab. Explain what it means.

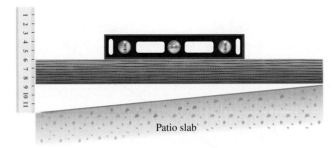

71. **GRADE OF A ROAD** The vertical fall of the road shown is 264 feet for a horizontal run of 1 mile. Find the slope of the decline and use that fact to complete the roadside warning sign for truckers. (*Hint:* 1 mile = 5,280 feet.)

72. **TREADMILLS** For each height setting listed in the table, find the resulting slope of the jogging surface of the treadmill below. Express each incline as a percent.

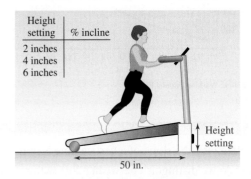

73. **ACCESSIBILITY** The illustration in the next column shows two designs to make the upper level of a stadium wheelchair-accessible.
    a. Find the slope of the ramp in design 1.
    b. Find the slopes of the ramps in design 2.
    c. Give one advantage and one drawback of each design.

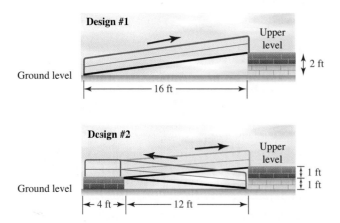

74. **IRRIGATION** The following graph shows the number of gallons of water remaining in a reservoir as water is discharged from it to irrigate a field. Find the rate of change in the number of gallons of water for the time the field was being irrigated.

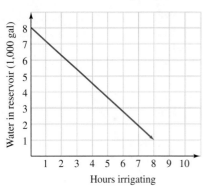

75. **DEPRECIATION** The following graph shows how the value of some sound equipment decreased over the years. Find the rate of change of its value during this time.

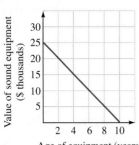

76. **ARCHITECTURE** Locate the coordinates of the peak of the roof if it is to have a pitch of $\frac{2}{5}$ and the roof line is to pass through the two points in black.

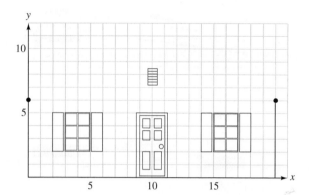

77. **ENGINE OUTPUT** Use the graph below to find the rate of change in the horsepower (hp) produced by an automobile engine for engine speeds in the range of 2,400–4,800 revolutions per minute (rpm).

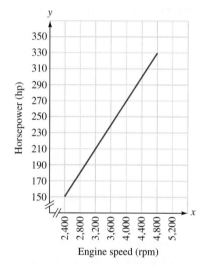

78. **COMMERCIAL JETS** Examine the graph and consider trips of more than 7,000 miles by a Boeing 777. Use a rate of change to estimate how the maximum payload decreases as the distance traveled increases.

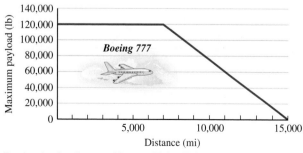

Based on data from Lawrence Livermore National Laboratory and *Los Angeles Times* (October 22, 1998).

79. **MILK PRODUCTION** The following graph approximates the amount of milk produced per cow in the United States for the years 1996–2012. Find the rate of change.

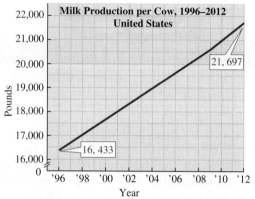

Source: Agricultural and Applied Economics, UW Madison, Brian Gould

80. **WAL-MART** The graph below approximates the net sales of Wal-Mart for the years 1991–2012. Find the rate of change in sales for the years
   a. 1991–1998
   b. 1998–2009
   c. 2009–2012

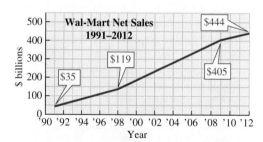

Source: WalMart Annual Reports

WRITING

81. Explain why the slope of a vertical line is undefined.
82. How do we distinguish between a line with positive slope and a line with negative slope?
83. Give an example of a rate of change that government officials might be interested in knowing so they can plan for the future needs of our country.
84. Explain the difference between a rate of change that is positive and one that is negative. Give an example of each.

REVIEW

85. In what quadrant does the point $(-3, 6)$ lie?
86. What is the name given the point $(0, 0)$?
87. Is $(-1, -2)$ a solution of $y = x^2 + 1$?
88. What basic shape does the graph of the equation $y = |x - 2|$ have?
89. Is the equation $y = 2x + 2$ linear or nonlinear?
90. Solve: $-3x \leq 15$

SECTION 3.5
Slope–Intercept Form

Objectives
1. Use slope–intercept form to identify the slope and y-intercept of a line.
2. Write a linear equation in slope–intercept form.
3. Use the slope and y-intercept to graph a linear equation.
4. Recognize parallel and perpendicular lines.
5. Use slope–intercept form to write an equation to model data.

Of all the ways in which a linear equation can be written, one form, called *slope–intercept form*, is probably the most useful. When an equation is written in this form, two important features of its graph are evident.

1 Use slope–intercept form to identify the slope and y-intercept of a line.

The graph of $y = -\frac{2}{3}x + 4$ shown in the figure enables us to see that the slope of the line is $-\frac{2}{3}$ and that the y-intercept is $(0, 4)$.

$y = -\frac{2}{3}x + 4$

x	y	(x, y)
0	4	(0, 4)
3	2	(3, 2)

To find the slope of the line, we pick two points on the line, (0, 4) and (3, 2); draw a slope triangle; and count grid squares

slope = $\frac{\text{rise}}{\text{run}} = \frac{-2}{3} = -\frac{2}{3}$

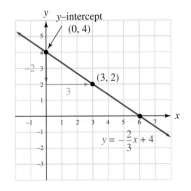

From the equation and the graph, we can make two observations:

- The graph crosses the y-axis at 4. This is the same as the constant term in $y = -\frac{2}{3}x + 4$
- The slope of the line is $-\frac{2}{3}$. This is the same as the coefficient of x in $y = -\frac{2}{3}x + 4$

$$y = -\frac{2}{3}x + 4$$

↑ The slope of the line is $-\frac{2}{3}$.   ↑ The y-intercept is $(0, 4)$.

These observations suggest the following form of an equation of a line.

Slope–Intercept Form of the Equation of a Line

If a linear equation is written in the form

$y = mx + b$

the graph of the equation is a line with slope m and y-intercept $(0, b)$.

Chapter 3 Graphs, Linear Equations, and Inequalities in Two Variables; Functions

Self Check 1

Find the slope and the y-intercept:
a. $y = -5x - 1$
b. $y = \dfrac{7}{8}x$
c. $y = 5 - \dfrac{x}{3}$

Now Try Problems 33 and 36

EXAMPLE 1
Find the slope and the y-intercept of the graph of each equation:

a. $y = 6x - 2$ b. $y = -\dfrac{5}{4}x$ c. $y = \dfrac{x}{2} + 6$

Strategy We will write each equation in slope–intercept form, $y = mx + b$.

WHY When the linear equations are written in slope–intercept form, the slope and the y-intercept of their graphs become apparent.

Solution

a. If we write the subtraction as the addition of the opposite, the equation will be in $y = mx + b$ form:

$$y = 6x + (-2)$$

Since $m = 6$ and $b = -2$, the slope of the line is 6 and the y-intercept is $(0, -2)$.

b. Writing $y = -\dfrac{5}{4}x$ in slope–intercept form, we have

$$y = -\dfrac{5}{4}x + 0$$

Since $m = -\dfrac{5}{4}$ and $b = 0$, the slope of the line is $-\dfrac{5}{4}$ and the y-intercept is $(0, 0)$.

c. Since $\dfrac{x}{2}$ means $\dfrac{1}{2}x$, we can rewrite $y = \dfrac{x}{2} + 6$ as

$$y = \dfrac{1}{2}x + 6$$

We see that $m = \dfrac{1}{2}$ and $b = 6$, so the slope of the line is $\dfrac{1}{2}$ and the y-intercept is $(0, 6)$.

Caution! If a linear equation is written in the form $y = mx + b$, the slope of the graph is the *coefficient* of x, not the term involving x. For example, it would be incorrect to say that the graph of $y = 5x + 1$ has a slope of $m = 5x$. Its graph has slope $m = 5$.

THINK IT THROUGH Prospects for a Teaching Career

"We should keep our best teachers in the classroom—and they should be earning a lot more money—as much as $150,000 a year."

Arne Duncan, Secretary of Education, 2011

Have you ever thought about becoming a teacher? There will be plenty of openings in the future, especially for mathematics and science teachers. The equation

$$y = 1{,}150x + 32{,}850$$

approximates the average beginning teacher salary y, where x is the number of years after 1990. Graph the equation. What information about beginning teacher salaries is given by the slope of the line? By the y-intercept? What is the predicted average beginning teacher salary in 2020? (Source: American Federation of Teachers)

2 Write a linear equation in slope–intercept form.

The equation of any nonvertical line can be written in slope–intercept form. To do so, we apply the properties of equality to solve the equation for y.

EXAMPLE 2 Find the slope and the y-intercept of the line determined by $6x - 3y = 9$.

Strategy We will use the properties of equality to write each equation in slope–intercept form, $y = mx + b$.

WHY When the linear equations are written in slope–intercept form, the slope and the y-intercept of their graphs become apparent.

Solution
To find the slope and the y-intercept of the line, we write the equation in slope–intercept form by solving for y.

$6x - 3y = 9$

$-3y = -6x + 9$ Subtract 6x from both sides.

$\dfrac{-3y}{-3} = \dfrac{-6x}{-3} + \dfrac{9}{-3}$ To undo the multiplication by −3, divide both sides by −3. On the right-hand side, dividing each term by −3 is equivalent to dividing the entire side by −3: $\dfrac{-6x + 9}{-3} = \dfrac{-6x}{-3} + \dfrac{9}{-3}$

$y = 2x - 3$ Do the divisions. Here, $m = 2$ and $b = -3$.

From the equation, we see that the slope is 2 and the y-intercept is $(0, -3)$.

Self Check 2
Find the slope and the y-intercept of the line determined by $8x - 2y = -2$.

Now Try Problem 41

3 Use the slope and y-intercept to graph a linear equation.

To graph $y = 2x - 3$, we plot the y-intercept $(0, -3)$, as shown. Since the slope is $\dfrac{\text{rise}}{\text{run}} = 2 = \dfrac{2}{1}$, the line rises 2 units for every unit it moves to the right. If we begin at $(0, -3)$ and move 2 units up (rise) and then 1 unit to the right (run), we locate the point $(1, -1)$, which is a second point on the line. We then draw a line through $(0, -3)$ and $(1, -1)$.

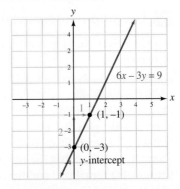

4 Recognize parallel and perpendicular lines.

The slope–intercept form enables us to quickly identify parallel and perpendicular lines.

EXAMPLE 3 Determine whether the graphs of $y = -\dfrac{2}{3}x$ and $y = -\dfrac{2}{3}x + 3$ are parallel, perpendicular, or neither.

Strategy We will find the slope of each line and compare the slopes.

WHY If the slopes are equal, the lines are parallel. If the slopes are negative reciprocals, the lines are perpendicular. Otherwise, the lines are neither parallel nor perpendicular.

Solution
The graph of $y = -\dfrac{2}{3}x$ is a line with slope $-\dfrac{2}{3}$. The graph of $y = -\dfrac{2}{3}x + 3$ is a line with slope of $-\dfrac{2}{3}$. Since the slopes $-\dfrac{2}{3}$ and $-\dfrac{2}{3}$ are the same, the lines are parallel.

Self Check 3
Determine whether the graphs of $y = -\dfrac{3}{2}x + 1$ and $y = \dfrac{3}{2}x + 4$ are parallel, perpendicular, or neither.

Now Try Problems 45 and 46

Unless otherwise noted, all content on this page is © Cengage Learning.

Self Check 4

Determine whether the graphs of $y = 4x + 6$ and $x - 4y = -8$ are parallel, perpendicular, or neither.

Now Try Problem 49

EXAMPLE 4 Are the graphs of $y = -5x + 6$ and $x - 5y = -10$ parallel, perpendicular, or neither?

Strategy We will find the slope of each line and then compare the slopes.

WHY If the slopes are equal, the lines are parallel. If the slopes are negative reciprocals, the lines are perpendicular. Otherwise, the lines are neither parallel nor perpendicular.

Solution

The graph of $y = -5x + 6$ is a line with slope -5. To find the slope of the graph of $x - 5y = -10$, we will write the equation in slope–intercept form.

$$x - 5y = -10$$

$$-5y = -x - 10 \quad \text{To eliminate } x \text{ from the left side, subtract } x \text{ from both sides.}$$

$$\frac{-5y}{-5} = \frac{-x}{-5} - \frac{10}{-5} \quad \text{To isolate } y, \text{ undo the multiplication by } -5 \text{ by dividing both sides by } -5.$$

$$y = \frac{x}{5} + 2 \quad m = \tfrac{1}{5} \text{ because } \tfrac{x}{5} = \tfrac{1}{5}x.$$

The graph of $y = \frac{x}{5} + 2$ is a line with slope $\frac{1}{5}$. Since the slopes -5 and $\frac{1}{5}$ are negative reciprocals, the lines are perpendicular. This is verified by the fact that the product of their slopes is -1.

$$-5\left(\frac{1}{5}\right) = -\frac{5}{5} = -1$$

> **Success Tip** Graphs are not necessary to determine if two lines are parallel, perpendicular, or neither. We simply examine the slopes of the lines.

5 Use slope–intercept form to write an equation to model data.

If we are given the slope and y-intercept of a line, we can write its equation, as in the next example.

Self Check 5

To encourage larger orders, a screen printing service offers a $0.02 per shirt discount from the normal cost of $15 per shirt. Write a linear equation that determines the per shirt cost c of a shirt if n shirts are ordered.

Now Try Problem 77

EXAMPLE 5 *Limo Service*

On weekends, a limousine service charges a fee of $100, plus 50¢ per mile, for the rental of a stretch limo. Write a linear equation that describes the relationship between the rental cost and the number of miles driven. Graph the result.

Strategy We will determine the slope and the y-intercept of the graph of the equation from the given facts about the limo service.

WHY If we know the slope and y-intercept, we can use the slope–intercept form, $y = mx + b$, to write the equation to model the situation.

Solution

To write an equation describing this relationship, we will let x represent the number of miles driven and y represent the cost (in dollars). We can make two observations:

- The cost increases by 50¢ or $0.50 for each mile driven. This is the *rate of change* of the rental cost to miles driven, and it will be the *slope* of the graph of the equation. Thus, $m = 0.50$.
- The basic fee is $100. Before driving any miles (that is, when $x = 0$), the cost y is 100. The ordered pair $(0, 100)$ will be the y-intercept of the graph of the equation. So we know that $b = 100$.

We substitute 0.50 for m and 100 for b in the slope–intercept form to get

$y = 0.50x + 100$ Here the cost y depends on x, the number of miles driven.
   ↑   ↑
$m = 0.50$ $b = 100$

To graph $y = 0.50x + 100$, we plot its y-intercept, $(0, 100)$. Since the slope is $0.50 = \frac{50}{100} = \frac{5}{10}$, we can start at $(0, 100)$ and locate a second point on the line by moving 5 units up (rise) and then 10 units to the right (run). This point will have coordinates $(0 + 10, 100 + 5)$ or $(10, 105)$. We draw a straight line through these two points to get a graph that illustrates the relationship between the rental cost and the number of miles driven. We draw the graph only in quadrant I, because the number of miles driven is always positive.

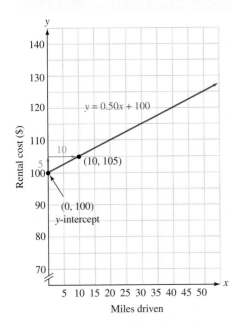

We can also graph $y = 0.50x + 100$ by plotting points. If we pick x-values of 0, 20, and 40, and find the corresponding values of y, we obtain the following table.

x	y
0	100
20	110
40	120

ANSWERS TO SELF CHECKS

1. a. $m = -5, (0, -1)$ b. $m = \frac{7}{8}, (0, 0)$ c. $m = -\frac{1}{3}, (0, 5)$ 2. $m = 4, (0, 1)$
3. neither   4. neither   5. $c = -0.02n + 15$

SECTION 3.5 STUDY SET

VOCABULARY

Fill in the blanks.

1. The equation $y = mx + b$ is called the _____ form for the equation of a line.
2. _____ lines do not intersect. _____ lines meet at right angles.

CONCEPTS

3. The graph of the linear equation $y = mx + b$ has a _____ of $(0, b)$ and a _____ of m.

4. The numbers $\frac{5}{6}$ and $-\frac{6}{5}$ are negative _____ because their product is -1.

Determine whether each equation is in slope–intercept form.

5. $7x + 4y = 2$
6. $5y = 2x - 3$
7. $y = 6x + 1$
8. $x = 4y - 8$

Determine the slope of the graph of each equation.

9. $y = \frac{-2x}{3} - 2$
10. $y = \frac{x}{4} + 1$

11. $y = 2 - 8x$
12. $y = 3x$
13. $y = x$
14. $y = -x$

See the illustration.

15. What is the slope of the line?
16. What is the y-intercept of the line?
17. Write the equation of the line.

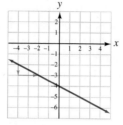

See the illustration.

18. What is the slope of the line?
19. What is the y-intercept of the line?
20. Write the equation of the line.

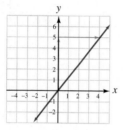

In the illustration, the slope of line l_1 is 2.

21. Determine the slope of line l_2.
22. Determine the slope of line l_3.
23. Determine the slope of line l_4.
24. Which lines have the same y-intercept?

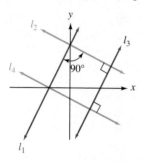

25. See the illustration.
   a. Determine the y-intercept of line l_1.

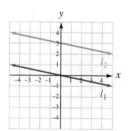

   b. What do lines l_1 and l_2 have in common? How are they different?

26. True or false? The slope of the graph of the equation $y = 4x + 2$ is $4x$.

Without graphing, determine whether the graphs of each pair of lines are parallel, perpendicular, or neither.

27. $y = 0.5x - 3$, $y = \frac{1}{2}x + 3$
28. $y = 0.75x$, $y = -\frac{4}{3}x + 2$
29. $y = -x$, $y = -2x$
30. $y = \frac{2}{3}x - 4$, $y = -\frac{3}{2}x + 4$

NOTATION

Complete each step to solve the equation for y. Then find the slope and the y-intercept of its graph.

31.
$$6x - 2y = 10$$
$$6x - \boxed{} - 2y = -6x + 10$$
$$-2y = \boxed{} + 10$$
$$\frac{-2y}{\boxed{}} = \frac{-6x}{\boxed{}} + \frac{10}{\boxed{}}$$
$$y = \boxed{} - 5$$

The slope is $\boxed{}$ and the y-intercept is $\boxed{}$.

32.
$$2x + 5y = 15$$
$$2x + 5y - \boxed{} = \boxed{} + 15$$
$$\boxed{} = -2x + 15$$
$$\frac{5y}{\boxed{}} = \frac{-2x}{\boxed{}} + \frac{15}{\boxed{}}$$
$$y = -\frac{2}{5}x + 3$$

The slope is $\boxed{}$ and the y-intercept is $\boxed{}$.

GUIDED PRACTICE

Find the slope and the y-intercept of the graph of each equation. See Examples 1–2.

33. $y = 4x + 2$
34. $y = -4x - 2$
35. $y = \frac{x}{4} - \frac{1}{2}$
36. $y = \frac{1}{2}x + 6$
37. $4x - 2 = y$
38. $6 - x = y$
39. $6y = x - 6$
40. $6x - 1 = y$
41. $4x - 3y = 12$
42. $2x + 3y = 6$
43. $10x - 5y = 12$
44. $-4x + 4y = -9$

3.5 Slope–Intercept Form

For each pair of equations, determine whether their graphs are parallel, perpendicular, or neither. See Examples 3–4.

45. $y = 6x + 8$
$y = 6x$

46. $y = 3x - 15$
$y = -\dfrac{1}{3}x + 4$

47. $y = x$
$y = -x$

48. $y = \dfrac{1}{2}x - \dfrac{4}{5}$
$y = 0.5x + 3$

49. $y = -2x - 9$
$2x - y = 9$

50. $y = \dfrac{3}{4}x + 1$
$4x - 3y = 15$

51. $x - y = 12$
$-2x + 2y = -23$

52. $x = 9$
$y = 8$

TRY IT YOURSELF

Find the slope and the y-intercept of the graph of each equation.

53. $x + y = 8$

54. $x - y = -30$

55. $2x + 3y = 6$

56. $3x - 5y = 15$

57. $3y - 13 = 0$

58. $-5y - 2 = 0$

59. $y = -5x$

60. $y = 14x$

Write an equation of the line with the given slope and y-intercept. Then graph it.

61. $m = 5, (0, -3)$

62. $m = -2, (0, 1)$

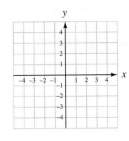
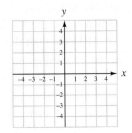

63. $m = \dfrac{1}{4}, (0, -2)$

64. $m = \dfrac{1}{3}, (0, -5)$

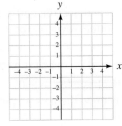
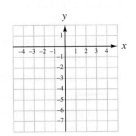

65. $m = -3, (0, 6)$

66. $m = 2, (0, 1)$

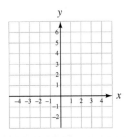
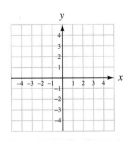

67. $m = -\dfrac{8}{3}, (0, 5)$

68. $m = -\dfrac{7}{6}, (0, 2)$

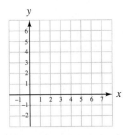
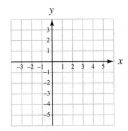

Find the slope and the y-intercept of the graph of each equation. Then graph it.

69. $y = 3x + 3$

70. $y = -3x + 5$

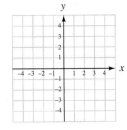
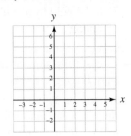

71. $y = -\dfrac{x}{2} + 2$

72. $y = \dfrac{x}{3}$

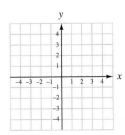
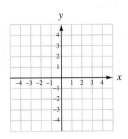

73. $3x + 4y = 16$

74. $2x + 3y = 9$

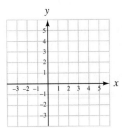
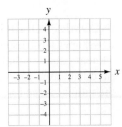

75. $10x - 5y = 5$ **76.** $4x - 2y = 6$

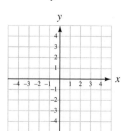 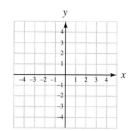

APPLICATIONS

77. PRODUCTION COSTS A television production company charges a basic fee of $5,000 and then $2,000 an hour when filming a commercial.

   a. Write a linear equation that describes the relationship between the total production costs y and the hours of filming x.

   b. Use your answer to part a to find the production costs if a commercial required 8 hours of filming.

78. COLLEGE FEES Each semester, students enrolling at a community college must pay tuition costs of $20 per unit as well as a $40 student services fee.

   a. Write a linear equation that gives the total fees y to be paid by a student enrolling at the college and taking x units.

   b. Use your answer to part a to find the enrollment cost for a student taking 12 units.

79. CHEMISTRY EXPERIMENT The following illustration shows a portion of a student's chemistry lab manual. Use the information to write a linear equation relating the temperature y (in degrees Fahrenheit) of the compound to the time x (in minutes) elapsed during the lab procedure.

> Chem. Lab #1   Aug. 13
> **Step 1:** Removed compound from freezer @ –10° F.
>
> **Step 2:** Used heating unit to raise temperature of compound 5° F every minute.

80. INCOME PROPERTY Use the information in the newspaper advertisement in the next column to write a linear equation that gives the amount of income y (in dollars) the apartment owner will receive when the unit is rented for x months.

> **APARTMENT FOR RENT**
> 1 bedroom/1 bath, with garage
> $500 per month +
> $250 nonrefundable security fee

81. SALAD BAR For lunch, a delicatessen offers a "Salad and Soda" special where customers serve themselves at a well-stocked salad bar. The cost is $2.00 for the drink and 25¢ = $0.25 an ounce for the salad.

   a. Write a linear equation that will find the cost y of a lunch when a salad weighing x ounces is purchased.

   b. Graph the equation using the grid below.

   c. How would the graph from part b change if the delicatessen began charging $3.00 for the drink?

   d. How would the graph from part b change if the cost of the salad changed to 40¢ an ounce?

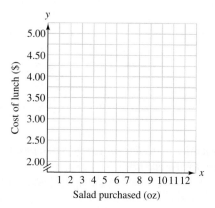

82. iPADS When a student purchased an Apple iPad with Wi-Fi + 3G for $629.99, he also enrolled in a 250 MB data plan that cost $14.95 per month.

   a. Write a linear equation that gives the cost for him to purchase and use the iPad for m months.

   b. Use your answer to part a to find the cost to purchase and use the iPad for 2 years.

83. SEWING COSTS A tailor charges a basic fee of $20 plus $2.50 per letter to sew an athlete's name on the back of a jacket.

   a. Write a linear equation that will find the cost y to have a name containing x letters sewn on the back of a jacket.

   b. Graph the equation on the grid on the next page.

   c. Suppose the tailor raises the basic fee to $30. On your graph from part b, draw the new graph showing the increased cost.

3.6 Point–Slope Form; Writing Linear Equations   271

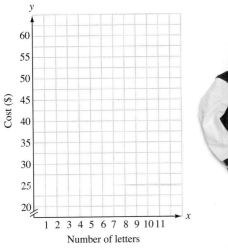

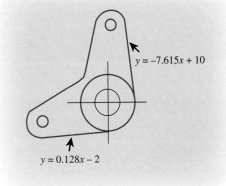

84. **PRINTING PRESSES** Every three minutes, 100 feet of paper is used off of an 8,000-foot roll to print the pages of a magazine. Write a linear equation that relates the number of feet of paper that remain on the roll and the number of minutes the printing press has been operating.

85. **EMPLOYMENT SERVICE** A policy statement of LIZCO, Inc., is shown below. Suppose a secretary had to pay an employment service $500 to get placed in a new job at LIZCO. Write a linear equation that tells the secretary the actual cost y of the employment service to her x months after being hired.

> Policy no. 23452– A new hire will be reimbursed by LIZCO for any employment service fees paid by the employee at the rate of $20 per month.

86. **COMPUTER DRAFTING** The illustration in the next column shows a computer-generated drawing of an engine mount. When the designer clicks on a line of the drawing, the computer determines its equation. Are the two lines selected in the drawing perpendicular?

WRITING

87. Explain the advantages of writing the equation of a line in slope–intercept form ($y = mx + b$) as opposed to standard form ($Ax + By = C$).

88. Why is $y = mx + b$ called the slope–intercept form of the equation of a line?

89. What is the minimum number of points needed to draw the graph of a line? Explain why.

90. List some examples of parallel and perpendicular lines that you see in your daily life.

REVIEW

91. Find the slope of the line passing through the points $(6, -2)$ and $(-6, 1)$.

92. Is $(3, -7)$ a solution of $y = 3x - 2$?

93. Evaluate: $-4 - (-4)$

94. Solve: $2(x - 3) = 3x$

95. To evaluate $[-2(4 - 8) + 4^2]$, which operation should be performed first?

96. Translate to mathematical symbols: four less than twice the price p.

97. What percent of 6 is 1.5?

98. Is -6.75 a solution of $x + 1 > -9$?

SECTION 3.6
Point–Slope Form; Writing Linear Equations

Objectives

1. Use point–slope form to write an equation of a line.
2. Write an equation of a line given two points on the line.
3. Write equations of horizontal and vertical lines.
4. Write linear equations that model data.

If we know the slope of a line and its y-intercept, we can use the slope–intercept form to write the equation of the line. The question that now arises is, can any point on a line be used in combination with its slope to write its equation? In this section we will answer this question.

1 Use point–slope form to write an equation of a line.

Refer to the line graphed on the next page with slope 3 and passing through the point $(2, 1)$. If we pick another point on the line, (x, y), we can find the slope of the line by using the coordinates of points $(2, 1)$ and (x, y). Using the slope formula, we have

$$\frac{y_2 - y_1}{x_2 - x_1} = m \quad \text{This is the slope formula.}$$

$$\frac{y - 1}{x - 2} = m \quad \text{Substitute } y \text{ for } y_2, 1 \text{ for } y_1, x \text{ for } x_2, \text{ and } 2 \text{ for } x_1.$$

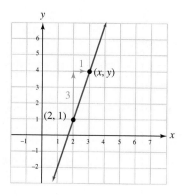

Since the slope of the line is given to be 3, we can substitute 3 for m in the previous equation.

$$\frac{y - 1}{x - 2} = m$$

$$\frac{y - 1}{x - 2} = 3$$

We then multiply both sides by $x - 2$ to get

$$\frac{y - 1}{x - 2}(x - 2) = 3(x - 2) \quad \text{Clear the equation of the fraction.}$$

$$y - 1 = 3(x - 2) \quad \text{Simplify the left-hand side.}$$

The resulting equation displays the slope of the line and the coordinates of one point on the line:

$$y - 1 = 3(x - 2)$$

where $y - 1$ has the y-coordinate of the point, 3 is the slope of the line, and $x - 2$ has the x-coordinate of the point.

In general, suppose we know that the slope of a line is m and that the line passes through the point (x_1, y_1). Then if (x, y) is any other point on the line, we can use the definition of slope to write

$$\frac{y - y_1}{x - x_1} = m$$

If we multiply both sides by $x - x_1$, we have

$$y - y_1 = m(x - x_1)$$

This form of a linear equation is called the **point–slope form**. It can be used to write the equation of a line when the slope and one point on the line are known.

Point–Slope Form of the Equation of a Line

If a line with slope m passes through the point (x_1, y_1), the equation of the line is

$$y - y_1 = m(x - x_1)$$

3.6 Point–Slope Form; Writing Linear Equations

EXAMPLE 1 Write an equation of a line that has a slope of -3 and passes through $(-1, 5)$. Write the answer in slope–intercept form.

Strategy We will use the point–slope form, $y - y_1 = m(x - x_1)$, to write an equation of the line.

WHY We are given the slope of the line and the coordinates of a point that it passes through.

Solution

$y - y_1 = m(x - x_1)$ This is the point–slope form.

$y - 5 = -3[x - (-1)]$ Substitute -3 for m, -1 for x_1, and 5 for y_1.

$y - 5 = -3(x + 1)$ Simplify within the brackets.

We can write this result in slope–intercept form, as follows:

$y - 5 = -3x - 3$ Distribute the multiplication by -3.

$y = -3x + 2$ To undo the subtraction of 5, add 5 to both sides: $-3 + 5 = 2$.

In slope–intercept form, the equation is $y = -3x + 2$.

Self Check 1
Write an equation of a line that has a slope of -2 and passes through $(4, -3)$. Write the answer in slope–intercept form.

Now Try Problems 25 and 31

2 Write an equation of a line given two points on the line.

In the next example, we will show that it is possible to write the equation of a line when we know the coordinates of two points on the line.

EXAMPLE 2 Write an equation of the line passing through $(4, 0)$ and $(6, -8)$.

Strategy We will use the point–slope form, $y - y_1 = m(x - x_1)$, to write an equation of the line.

WHY We can calculate the slope of the line using the coordinates of the two points given and the slope formula. We also know the coordinates of a point that the line passes through (we can choose either point).

Solution
First we find the slope of the line that passes through $(4, 0)$ and $(6, -8)$.

$m = \dfrac{y_2 - y_1}{x_2 - x_1}$ This is the slope formula.

$= \dfrac{-8 - 0}{6 - 4}$ Substitute -8 for y_2, 0 for y_1, 6 for x_2, and 4 for x_1.

$= \dfrac{-8}{2}$ Simplify.

$= -4$

Since the line passes through both $(4, 0)$ and $(6, -8)$, we can choose either point and substitute its coordinates into the point–slope form. If we choose $(4, 0)$, we substitute 4 for x_1, 0 for y_1, and -4 for m and proceed as follows.

$y - y_1 = m(x - x_1)$ This is the point–slope form.

$y - 0 = -4(x - 4)$ Substitute -4 for m, 4 for x_1, and 0 for y_1.

$y = -4x + 16$ Distribute the multiplication by -4.

The equation of the line is $y = -4x + 16$.

Self Check 2
Write an equation of the line passing through $(0, -3)$ and $(2, 1)$.

Now Try Problems 33 and 38

Success Tip In Example 2, either of the given points can be used as (x_1, y_1) when writing the point–slope equation. The results will be the same. We usually choose the point whose coordinates make the computations the easiest.

3 Write equations of horizontal and vertical lines.

We have graphed horizontal and vertical lines. We will now discuss how to write their equations.

Self Check 3

Write an equation of each line.
a. a horizontal line passing through $(3, 2)$
b. a vertical line passing through $(-1, -3)$

Now Try Problems 41 and 44

EXAMPLE 3 Write an equation of each line and then graph it. **a.** a horizontal line passing through $(-2, -4)$ **b.** a vertical line passing through $(1, 3)$.

Strategy We will use the appropriate form, either $y = b$ or $x = a$, to write an equation of each line.

WHY These are the standard forms for the equations of a horizontal and a vertical line.

Solution

a. The equation of a horizontal line can be written in the form $y = b$. Since the y-coordinate of $(-2, -4)$ is -4, the equation of the line is $y = -4$. The graph is shown in the figure.

b. The equation of a vertical line can be written in the form $x = a$. Since the x-coordinate of $(1, 3)$ is 1, the equation of the line is $x = 1$. The graph is shown in the figure.

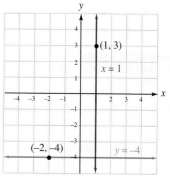

4 Write linear equations that model data.

In the next two examples, we will see how the point–slope form can be used to write linear equations that model certain real-world situations.

Self Check 4

Celsius and Fahrenheit measures of temperature are not the same. There is, however, a linear relationship between the two. Degrees Fahrenheit increases by 9° for each 5° increase in Celsius. If a 212° Fahrenheit temperature measure is the same as 100° Celsius, write a linear equation that relates Fahrenheit measure to Celsius measure.

Now Try Problem 73

EXAMPLE 4 *Men's Shoe Sizes* The length (in inches) of a man's foot is not his shoe size. For example, the smallest adult men's shoe size is 5, and it fits a 9-inch-long foot. There is, however, a linear relationship between the two. It can be stated this way: Shoe size increases by 3 sizes for each 1-inch increase in foot length.

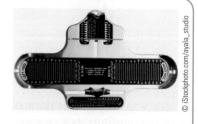

a. Write a linear equation that relates shoe size s to foot length L.

b. Shaquille O'Neal, a famous basketball player, has a foot that is about 14.6 inches long. Find his shoe size.

Strategy We will first find the slope of the line that describes the linear relationship between shoe size and the length of a foot. Then we will determine the coordinates of a point on that line.

WHY Once we know the slope and the coordinates of one point on the line, we can use the point–slope form to write the equation of the line.

Solution

a. Since shoe size s depends on the length L of the foot, ordered pairs have the form (L, s). Because the relationship is linear, the graph of the desired equation is a line.

- The line's slope is the rate of change: $\frac{3 \text{ sizes}}{1 \text{ inch}}$. Therefore, $m = 3$.
- A 9-inch-long foot wears size 5, so the line passes through $(9, 5)$.

We substitute 3 for m and the coordinates of the point into the point–slope form and solve for s.

$s - s_1 = m(L - L_1)$	This is the point–slope form using the variables L and s.
$s - 5 = 3(L - 9)$	Substitute 3 for m, 9 for L_1, and 5 for s_1.
$s - 5 = 3L - 27$	Distribute the multiplication by 3.
$s = 3L - 22$	To isolate s, add 5 to both sides: $-27 + 5 = -22$.

The equation relating men's shoe size and foot length is $s = 3L - 22$.

b. To find Shaquille's shoe size, we substitute 14.6 inches for L in the equation.

$s = 3L - 22$	
$s = 3(\mathbf{14.6}) - 22$	
$s = 43.8 - 22$	Do the multiplication.
$s = 21.8$	Do the subtraction.

Since men's shoes come in only full- and half-sizes, we round 21.8 up to 22. Shaquille O'Neal wears size 22 shoes.

EXAMPLE 5 Market Research

A company that makes a breakfast cereal has found that the number of discount coupons redeemed for its product is linearly related to the coupon's value. In one advertising campaign, 10,000 of the "10¢ off" coupons were redeemed. In another campaign, 45,000 of the "50¢ off" coupons were redeemed. How many coupons can the company expect to be redeemed if it issues a "35¢ off" coupon?

Self Check 5

Orders for awards to be given to math team members were placed on two separate occasions. The first order of 32 awards cost $172 and the second order of 5 awards cost $37. Assuming no price change, write an equation of the line that would give the cost for an order of any number of awards.

Now Try Problem 65

Strategy We will determine two points of the graph of the equation from the given facts about the coupon.

WHY If we know two points, we can use the point–slope form, $y - y_1 = m(x - x_1)$, to write the equation to model the situation.

Solution

If we let x represent the value of a coupon and y represent the number of coupons that will be redeemed, ordered pairs will have the form

(coupon value, number redeemed)

Two points on the graph of the equation are (10, 10,000) and (50, 45,000). These points are plotted on the graph shown to the right. To write the equation of the line passing through the points, we first find the slope of the line.

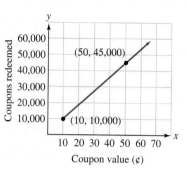

$$m = \frac{y_2 - y_1}{x_2 - x_1} \quad \text{This is the slope formula.}$$

$$m = \frac{45{,}000 - 10{,}000}{50 - 10} \quad \text{Substitute 45,000 for } y_2\text{, 10,000 for } y_1\text{, 50 for } x_2\text{, and 10 for } x_1.$$

$$m = \frac{35{,}000}{40} \quad \text{Simplify.}$$

$$m = 875 \quad \text{Do the division.}$$

We then substitute 875 for m and the coordinates of one known point—say, (10, 10,000)—into the point–slope form of the equation of a line and proceed as follows.

$$y - y_1 = m(x - x_1) \quad \text{This is the point-slope form.}$$
$$y - 10{,}000 = 875(x - 10) \quad \text{Substitute for } m, x_1, \text{ and } y_1.$$
$$y - 10{,}000 = 875x - 8{,}750 \quad \text{Distribute the multiplication by 875.}$$
$$y = 875x + 1{,}250 \quad \text{Add 10,000 to both sides.}$$

To find the expected number of coupons that will be redeemed, we substitute the value of the coupon, 35¢, into the equation $y = 875x + 1{,}250$ and find y.

$$y = 875x + 1{,}250$$
$$y = 875(35) + 1{,}250 \quad \text{Substitute 35 for x.}$$
$$y = 30{,}625 + 1{,}250 \quad \text{Do the multiplication.}$$
$$y = 31{,}875 \quad \text{Do the addition.}$$

The company can expect 31,875 of the 35¢ coupons to be redeemed.

ANSWERS TO SELF CHECKS

1. $y = -2x + 5$ 2. $y = 2x - 3$ 3. a. $y = 2$ b. $x = -1$ 4. $F = \frac{9}{5}C + 32$
5. $c = 5n + 12$

SECTION 3.6   STUDY SET

VOCABULARY

Fill in the blanks.

1. $y - y_1 = m(x - x_1)$ is called the _____ form of the equation of a line.

2. In the illustration, point P has an x-_____ of 2 and a y-_____ of -1.

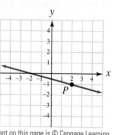

CONCEPTS

3. The linear equation $y = 2x - 3$ is written in *slope-intercept* form. What are the slope and the y-intercept of the graph of this line?

4. The linear equation $y - 4 = 6(x - 5)$ is written in *point-slope* form. What is the slope of the graph of this equation and what point does it pass through?

5. In what form is the equation $y - 4 = 2(x - 5)$ written?

6. In what form is the equation $y = 2x + 15$ written?

Refer to the illustration below.

7. Find two points on the line whose coordinates are integers.

8. Determine the slope of the line.

9. Use the results of Problems 7 and 8 to write the equation of the line in point–slope form.

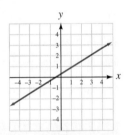

10. Is it true that the equations
$$y - 1 = 2(x - 2), \ y = 2x - 3, \text{ and } 2x - y = 3$$
all describe the same line?

In each case, is enough information given to write an equation of the line?

11. The line passes through $(2, -7)$.

12. The slope of the line is $-\frac{3}{4}$.

13. The line has the following table of solutions:

x	y
2	3
-3	-6

14. The line is horizontal.

15. The line is vertical and passes through $(-1, 1)$.

16. The line has the following table of solutions:

x	y
4	5

NOTATION

Complete each step to write a point-slope equation in slope-intercept form.

17. $y - 2 = -3(x - 4)$
$y - 2 = \boxed{} + \boxed{}$
$y - 2 + \boxed{} = -3x + 12 + \boxed{}$
$y = -3x + 14$

18. $y + 2 = \frac{1}{2}(x + 2)$
$y + 2 = \boxed{} + \boxed{}$
$y + 2 - \boxed{} = \frac{1}{2}x + 1 - \boxed{}$
$y = \frac{1}{2}x - 1$

19. Complete the steps to write the slope–intercept equation of the line with slope -2 that passes through the point $(-1, 5)$.
$y - y_1 = m(x - x_1)$
$y - \boxed{} = -2[x - (\boxed{})]$
$y - 5 = \boxed{} - 2$
$y = -2x + 3$

20. Complete the steps to write the slope–intercept equation of the line with slope 4 that passes through the point $(0, 3)$.
$y - y_1 = m(x - x_1)$
$y - \boxed{} = 4(x - \boxed{})$
$y - 3 = \boxed{}$
$y = 4x + 3$

GUIDED PRACTICE

Use the point–slope form to write an equation of the line with the given slope and point. See Objective 1.

21. $m = 3$, passes through $(2, 1)$

22. $m = 2$, passes through $(4, 3)$

23. $m = -\frac{4}{5}$, passes through $(-5, -1)$

24. $m = -\frac{7}{8}$, passes through $(-2, -9)$

Use the point–slope form to write an equation of the line with the given slope and point. Then write the result in slope–intercept form. See Example 1.

25. $m = \dfrac{1}{5}$, passes through $(10, 1)$
26. $m = \dfrac{1}{4}$, passes through $(8, 1)$
27. $m = -5$, passes through $(-9, 8)$
28. $m = -4$, passes through $(-2, 10)$
29. $m = -\dfrac{4}{3}$,

x	y
6	-4

30. $m = -\dfrac{3}{2}$,

x	y
-2	1

31. $m = -\dfrac{2}{3}$, passes through $(3, 0)$
32. $m = -\dfrac{5}{4}$, passes through $(2, 0)$

Write an equation of the line that passes through the two given points. Write the result in slope–intercept form. See Example 2.

33. $(1, 7)$ and $(-2, 1)$
34. $(-2, 2)$ and $(2, -8)$
35. $(5, 1)$ and $(-5, 0)$
36. $(-3, 0)$ and $(3, 1)$
37. $(5, 5)$ and $(7, 5)$
38. $(-2, 1)$ and $(-2, 15)$
39.

x	y
-4	3
2	0

40.

x	y
-1	-4
1	-2

Write an equation of the line with the given characteristics. See Example 3.

41. Vertical, passes through $(4, 5)$
42. Vertical, passes through $(-2, -5)$
43. Horizontal, passes through $(4, 5)$
44. Horizontal, passes through $(-2, -5)$

TRY IT YOURSELF

Find the equation of the line with the following characteristics. Write the equation in slope–intercept form, if possible.

45. $m = 8$, passes through $(0, 4)$
46. $m = 6$, passes through $(0, -4)$
47. $m = -3$, passes through the origin
48. $m = -1$, passes through the origin
49. Passes through $(-8, 2)$ and $(-8, 17)$
50. Vertical, passes through $(-3, 7)$
51. Vertical, passes through $(12, -23)$
52. Slope 7 and y-intercept $(0, 0)$
53. Slope 3 and y-intercept $(0, 4)$
54. Passes through $(-2, -1)$ and $(-1, -5)$
55. Passes through $(-3, 6)$ and $(-1, -4)$
56. x-intercept $(7, 0)$ and y-intercept $(0, -2)$
57. x-intercept $(-3, 0)$ and y-intercept $(0, 7)$
58. Slope $\dfrac{1}{10}$, passes through the origin
59. Slope $\dfrac{9}{8}$, passes through the origin
60. Undefined slope, passes through $\left(-\dfrac{1}{8}, 12\right)$
61. Undefined slope, passes through $\left(\dfrac{2}{5}, -\dfrac{5}{6}\right)$
62. Horizontal, passes through $(-8, 12)$
63. Horizontal, passes through $(9, -32)$
64. Slope $\dfrac{2}{3}$, x-intercept $(4, 0)$

APPLICATIONS

65. **TOXIC CLEANUP** Three months after cleanup began at a dump site, 800 cubic yards of toxic waste had yet to be removed. Two months later, that number had been lowered to 720 cubic yards.

    a. Write a linear equation that mathematically describes the linear relationship between the length of time x (in months) the cleanup crew has been working and the number of cubic yards y of toxic waste remaining.

    b. Use your answer to part a to predict the number of cubic yards of waste that will still be on the site 1 year after the cleanup project began.

66. **DEPRECIATION** To lower its corporate income tax, accountants of a large company depreciated a word processing system over several years using a linear model, as shown in the worksheet on the next page.

    a. Use the information in the worksheet to write a linear equation relating the years since the system was purchased x and its value y, in dollars.

b. Find the purchase price of the system by substituting $x = 0$ into your answer from part a.

Tax Worksheet

Method of depreciation: Linear

Property	Years after purchase	Value
Word processing system	2	$60,000
"	4	$30,000

67. POLE VAULTING Find the equations of the lines that describe the positions of the pole for parts 1, 3, and 4 of the jump. Write the equations in slope–intercept form, if possible.

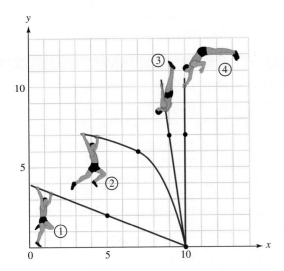

68. FREEWAY DESIGN The graph below shows the route of a proposed freeway.

a. Give the coordinates of the points where the proposed freeway will join Interstate 25 and Highway 40.

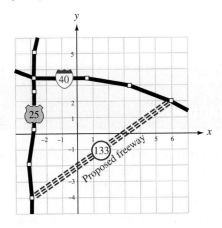

b. Write the equation of the line that mathematically describes the route of the proposed freeway. Answer in slope–intercept form.

69. COUNSELING In the first year of her practice, a family counselor saw 75 clients. In the second year, the number of clients grew to 105. If a linear trend continues, write an equation that gives the number of clients c the counselor will have t years after beginning her practice.

70. GOT MILK The diagram below shows the amount of milk that an average American drank in one year for the years 1980–2011. A straight line can be used to model these data.

a. Use the highlighted two points on the line to find its equation. Write the equation in slope–intercept form.

b. Use your answer to part a to predict the amount of milk that an average American will drink in 2030.

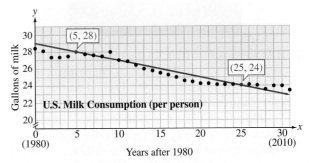

71. CONVERTING TEMPERATURES The relationship between Fahrenheit temperature, F, and Celsius temperature, C, is linear.

a. Use the data in the illustration below to write two ordered pairs of the form (C, F).

b. Use your answer to part a to write a linear equation relating the Fahrenheit and Celsius scales.

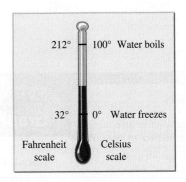

72. TRAMPOLINES The relationship between the circumference of a circle and its radius is linear. For example, the length l of the protective pad that wraps around a trampoline is related to the radius r of the trampoline. Use the data in the illustration to write a linear equation that approximates the length of pad needed for any trampoline radius.

Radius (ft)	Approximate length of padding (ft)
3	19
7	44

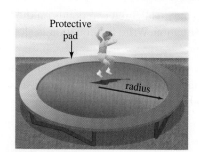

73. RAISING A FAMILY In the report *Expenditures on Children by Families*, the U.S. Department of Agriculture projected the yearly child-rearing expenditures on children from birth through age 17. For a child born in 2010 to a two-parent middle-income family, the report estimated annual expenditures of $10,808 when the child is 6 years old, and $14,570 when the child is 15 years old.

a. Write two ordered pairs of the form (child's age, annual expenditure).

b. Assume the relationship between the child's age a and the annual expenditures E is linear. Use your answers to part a to write an equation in slope–intercept form that models this relationship.

c. What are the projected child-rearing expenses when the child is 17 years old?

74. AUTOMATION An automated production line uses distilled water at a rate of 300 gallons every 2 hours to make shampoo. After the line had run for 7 hours, planners noted that 2,500 gallons of distilled water remained in the storage tank. Write a linear equation relating the time in hours x since the production line began and the number of gallons y of distilled water in the storage tank.

WRITING

75. Why is $y - y_1 = m(x - x_1)$ called the point–slope form of the equation of a line?

76. If we know two points that a line passes through, we can write its equation. Explain how to do this.

77. If we know the slope of a line and a point it passes through, we can write its equation. Explain how to do this.

78. Think of several points on the graph of the horizontal line $y = 4$. What do the points have in common? How do they differ?

REVIEW

79. Find the slope of the line passing through the points $(2, 4)$ and $(-6, 8)$.

80. Is the graph of $y = x^2$ a straight line?

81. Find the area of a circle with a diameter of 12 feet. Round to the nearest tenth.

82. If a 15-foot board is cut into two pieces and we let x represent the length of one piece (in feet), write an expression for the length of the other piece.

83. Evaluate: $(-1)^5$

84. Solve: $\dfrac{x-3}{4} = -4$

85. What is the coefficient of the second term of $-4x^2 + 6x - 13$?

86. Simplify: $(-2p)(-5)(4x)$

Objectives

1. Determine whether an ordered pair is a solution of an inequality.
2. Graph a linear inequality in two variables.
3. Solve application problems involving linear inequalities in two variables.

SECTION 3.7
Graphing Linear Inequalities

Recall that an **inequality** is a statement that contains one of the symbols $<$, $\leq$, $>$, or $\geq$. Inequalities in one variable, such as $x - 5 < 2$ and $3x + 4 > 2x$, were solved in Section 2.7. Because they have an infinite number of solutions, we represented their solution sets graphically, by shading intervals on a number line.

We now extend that concept to linear inequalities *in two variables*, as we introduce a procedure that is used to graph their solution sets.

1. Determine whether an ordered pair is a solution of an inequality.

If the = symbol in a linear equation in two variables is replaced with an inequality symbol, we have a **linear inequality in two variables**. Some examples are

$$2x - y > -3, \quad y < 3, \quad x + 4y \geq 6, \quad \text{and} \quad x \leq -2$$

As with linear equations, an ordered pair (x, y) is **a solution of an inequality** in x and y if a true statement results when the variables in the inequality are replaced by the coordinates of the ordered pair.

EXAMPLE 1 Determine whether each ordered pair is a solution of $x - y \leq 5$. Then graph each solution: **a.** $(4, 2)$ **b.** $(0, -6)$ **c.** $(1, -4)$

Strategy We will substitute each ordered pair of coordinates into the inequality.

WHY If the resulting statement is true, the ordered pair is a solution.

Solution
a. For $(4, 2)$:

$x - y \leq 5$ This is the original inequality.
$4 - 2 \stackrel{?}{\leq} 5$ Substitute 4 for x and 2 for y.
$2 \leq 5$ This is true.

Because $2 \leq 5$ is true, $(4, 2)$ is a solution of $x - y \leq 5$. We say that $(4, 2)$ *satisfies* the inequality. This solution is graphed as shown on the right.

b. For $(0, -6)$:

$x - y \leq 5$ This is the original inequality.
$0 - (-6) \stackrel{?}{\leq} 5$ Substitute 0 for x and -6 for y.
$6 \leq 5$ This is false.

Because $6 \leq 5$ is false, $(0, -6)$ is not a solution.

c. For $(1, -4)$:

$x - y \leq 5$ This is the original inequality.
$1 - (-4) \stackrel{?}{\leq} 5$ Substitute 1 for x and -4 for y.
$5 \leq 5$ This is true.

Because $5 \leq 5$ is true, $(1, -4)$ is a solution, and we graph it as shown.

Self Check 1

Using the inequality in Example 1, determine whether each ordered pair is a solution.
a. $(8, 2)$
b. $(4, -1)$
c. $(-2, 4)$
d. $(-3, -5)$

Now Try Problem 33

In Example 1, we graphed two solutions of the inequality $x - y \leq 5$. Since there are infinitely many more ordered pairs (x, y) that make the inequality true, it would not be reasonable to plot them. Fortunately, there is an easier way to show all of the solutions.

2. Graph a linear inequality in two variables.

The graph of $x - y = 5$ is a line consisting of the points whose coordinates satisfy the equation. The graph of the inequality $x - y \leq 5$ is not a line, but an area bounded by a line, called a **half-plane**. The half-plane consists of the points whose coordinates satisfy the inequality, and we use a two-step procedure to find them.

Self Check 2

Graph: $2x + y \leq 4$

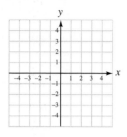

Now Try Problem 37

EXAMPLE 2   Graph: $x - y \leq 5$

Strategy We will graph the related *equation* $x - y = 5$ to establish a boundary line between two regions of the coordinate plane. Then we will determine which region contains points whose coordinates satisfy the given inequality.

WHY The graph of a linear inequality in two variables is a region of the coordinate plane on one side of the boundary line.

Solution

Since the inequality symbol $\leq$ includes an equals sign, the graph of $x - y \leq 5$ includes the graph of $x - y = 5$.

Step 1 Graph the equation $x - y = 5$ using the intercept method, as shown in figure (a).

$x - y = 5$

x	y	(x, y)
0	−5	(0, −5)
5	0	(5, 0)
6	1	(6, 1)

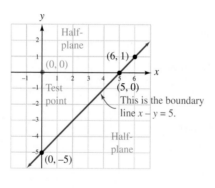

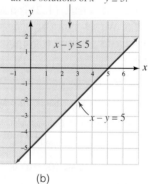

(a) (b)

Step 2 Since the inequality $x - y \leq 5$ allows $x - y$ to be less than 5, the coordinates of points other than those shown on the boundary satisfy the inequality. For example, the coordinates of the origin (0, 0) satisfy the inequality. We can verify this by letting x and y be zero in the given inequality:

$x - y \leq 5$

$0 - 0 \stackrel{?}{\leq} 5$ Substitute 0 for x and 0 for y.

$0 \leq 5$

Because $0 \leq 5$, the coordinates of the origin satisfy the original inequality. In fact, the coordinates of every point on the *same side* of the line as the origin satisfy the inequality. The graph of $x - y \leq 5$ is the half-plane that is shaded in figure (b). Since the **boundary line** $x - y = 5$ is included, we draw it with a solid line.

EXAMPLE 3   Graph: $x + y \geq 3$

Strategy We will graph the related *equation* $x + y = 3$ to establish a boundary line between two regions of the coordinate plane. Then we will determine which region contains points whose coordinates satisfy the given inequality.

WHY The graph of a linear inequality in two variables is a region of the coordinate plane on one side of the boundary line.

Solution
To graph the inequality $x + y \geq 3$, we graph the boundary line whose equation is $x + y = 3$. Since the graph of $x + y \geq 3$ includes the line $x + y = 3$, we draw the boundary with a solid line. See figure (a) below. Note that it divides the coordinate plane into two half-planes.

To decide which half-plane to shade, we substitute the coordinates of some point that lies on one side of the boundary line into the inequality. If we use the origin $(0, 0)$ for the **test point**, we have

$x + y \geq 3$
$0 + 0 \geq 3$ Substitute 0 for x and 0 for y.
$0 \geq 3$ This is false.

Since $0 \geq 3$ is a false statement, the origin is not in the graph. In fact, the coordinates of *every* point on the origin's side of the boundary line will not satisfy the inequality. However, every point on the other side of the boundary line will satisfy the inequality. We shade that half-plane. The graph of $x + y \geq 3$ is the half-plane that is shaded in figure (b).

Self Check 3
Graph: $x - y \leq -2$

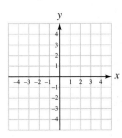

Now Try Problem 39

$x + y = 3$		
x	y	(x, y)
0	3	$(0, 3)$
3	0	$(3, 0)$
1	2	$(1, 2)$

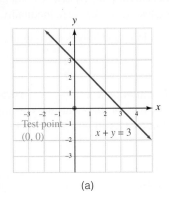
(a)

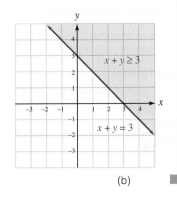
(b)

EXAMPLE 4   Graph: $y > 2x$

Strategy We will graph the related *equation* $y = 2x$ to establish a boundary line between two regions of the coordinate plane. Then we will determine which region contains points whose coordinates satisfy the given inequality.

WHY The graph of a linear inequality in two variables is a region of the coordinate plane on one side of the boundary line.

Solution
To find the boundary line, we graph $y = 2x$. Since the symbol $>$ does not include an equals sign, the points on the graph of $y = 2x$ are not part of the graph of $y > 2x$. We draw the boundary line as a dashed line to show this. See figure (a) on the next page.

To determine which half-plane to shade, we substitute the coordinates of some point that lies on one side of the boundary line into $y > 2x$. Since the origin is on the boundary, we cannot use it as a test point. The point $(2, 0)$, for example, is not on the boundary line. To see whether $(2, 0)$ satisfies $y > 2x$, we substitute 2 for x and 0 for y in the inequality.

$y > 2x$
$0 > 2(2)$ Substitute 2 for x and 0 for y.
$0 > 4$ This is false.

Self Check 4
Graph: $y < 3x$

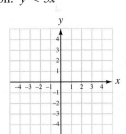

Now Try Problem 42

Since $0 > 4$ is a false statement, the point $(2, 0)$ does not satisfy the inequality, and is not on the side of the dashed line we wish to shade. Instead, we shade the other side of the boundary line. The graph of the solution set of $y > 2x$ is shown in figure (b).

y = 2x		
x	y	(x, y)
0	0	(0, 0)
−1	−2	(−1, −2)
1	2	(1, 2)

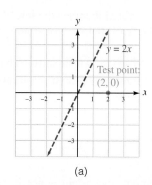
(a)

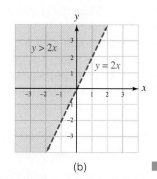
(b)

Self Check 5

Graph: $2x - y < 4$

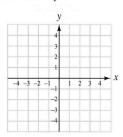

Now Try Problem 45

EXAMPLE 5 Graph: $x + 2y < 6$

Strategy We will graph the related *equation* $x + 2y = 6$ to establish a boundary line between two regions of the coordinate plane. Then we will determine which region contains points whose coordinates satisfy the given inequality.

WHY The graph of a linear inequality in two variables is a region of the coordinate plane on one side of the boundary line.

Solution

We find the boundary by graphing the equation $x + 2y = 6$. We draw the boundary as a dashed line to show that it is not part of the solution. We then choose a test point not on the boundary and see whether its coordinates satisfy $x + 2y < 6$. The origin is a convenient choice.

$$x + 2y < 6$$
$$0 + 2(0) < 6 \quad \text{Substitute 0 for x and 0 for y.}$$
$$0 < 6 \quad \text{This is true.}$$

Since $0 < 6$ is a true statement, we shade the side of the line that includes the origin. The graph is shown in the illustration.

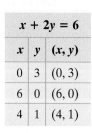

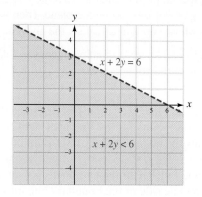

EXAMPLE 6 Graph: $y \geq 0$

Strategy We will graph the related *equation* $y = 0$ to establish a boundary line between two regions of the coordinate plane. Then we will determine which region contains points whose coordinates satisfy the given inequality.

WHY The graph of a linear inequality in two variables is a region of the coordinate plane on one side of the boundary line.

Solution
We find the boundary by graphing the equation $y = 0$. We draw the boundary as a solid line to show that it is part of the solution. We then choose a test point not on the boundary and see whether its coordinates satisfy $y \geq 0$. The point $(0, 1)$ is a convenient choice.

$y \geq 0$
$1 \geq 0$ Substitute 1 for y.

Since $1 > 0$ is a true statement, we shade the side of the line that includes $(0, 1)$. The graph is shown below.

Self Check 6
Graph: $x \geq 2$

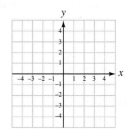

Now Try Problem 51

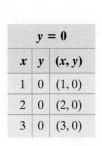
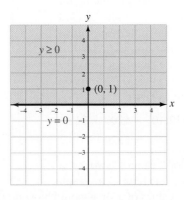

The following is a summary of the procedure for graphing linear inequalities.

Graphing Linear Inequalities in Two Variables

1. Replace the inequality symbol with an equal symbol = and graph the boundary line of the region. If the inequality allows the possibility of equality (the symbol is either $\leq$ or $\geq$), draw the boundary line as a solid line. If equality is not allowed ($<$ or $>$), draw the boundary line as a dashed line.

2. Pick a test point that is on one side of the boundary line. (Use the origin if possible.) Replace x and y in the inequality with the coordinates of that point. If the inequality is satisfied, shade the side that contains that point. If the inequality is not satisfied, shade the other side of the boundary.

3 Solve application problems involving linear inequalities in two variables.

When solving application problems, phrases such as *at least*, *at most*, and *should not exceed* indicate that an inequality should be used.

EXAMPLE 7 *Earning Money* Carlos has two jobs, one paying $10 per hour and one paying $12 per hour. He must earn at least $240 per week to pay his expenses while attending college. Write an inequality that shows the ways he can schedule his time to achieve his goal.

Strategy We will form an inequality using the facts about the two jobs Carlos has. Then we will graph the solution of the inequality.

Self Check 7

Brianna and Ashley pool their resources to purchase some songs and movies on iTunes. If songs cost $1 and movies cost $4, write an inequality to represent the number of songs and movies they can buy if they have a combined spending limit of $25.

Now Try Problem 73

WHY The graph of a linear inequality in two variables gives the ways Carlos can schedule his time to achieve his goal.

Solution

If we let x represent the number of hours Carlos works on the first job and y the number of hours he works on the second job, we have

The hourly rate on the first job	times	the hours worked on the first job	plus	the hourly rate on the second job	times	the hours worked on the second job	is at least	$240.
10	·	x	+	12	·	y	≥	240

The graph of the inequality $10x + 12y \geq 240$ is shown in the figure to the right. Any point in the shaded region indicates a possible way Carlos can schedule his time and earn $240 or more per week. For example, if he works 20 hours on the first job and 10 hours on the second job, he will earn

$$\$10(20) + \$12(10) = \$200 + \$120$$
$$= \$320$$

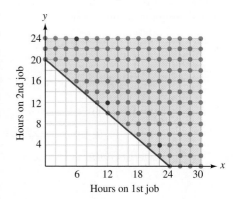

Since Carlos cannot work a negative number of hours, a graph showing negative values of x or y would have no meaning.

ANSWERS TO SELF CHECKS

1. **a.** not a solution **b.** solution **c.** solution **d.** solution

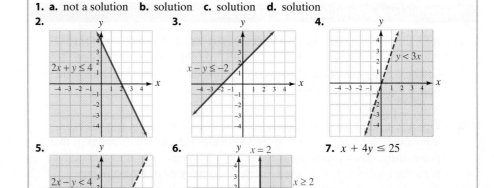

7. $x + 4y \leq 25$

SECTION 3.7 STUDY SET

VOCABULARY

Fill in the blanks.

1. $2x - y \leq 4$ is a linear _____ in x and y.

2. In the illustration on the next page, the line $2x - y = 4$ divides the rectangular coordinate system into two _____.

3.7 Graphing Linear Inequalities   287

3. In the illustration, the graph of the line $2x - y = 4$ is the _____ line.

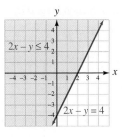

4. When graphing a linear inequality, we determine which half-plane to shade by substituting the coordinates of a test _____ into the inequality.

CONCEPTS

Determine whether each ordered pair is a solution of $5x - 3y \leq 0$.

5. $(1, 1)$ 6. $(-2, -3)$

7. $(0, 0)$ 8. $\left(\dfrac{1}{5}, -\dfrac{4}{3}\right)$

Determine whether each ordered pair is a solution of $x + 4y < -1$.

9. $(3, 1)$ 10. $(-2, 0)$

11. $(-0.5, 0.2)$ 12. $\left(-2, \dfrac{1}{4}\right)$

Determine whether the graph of each linear inequality includes the boundary line. When graphed, is the boundary line solid or dashed?

13. $y > -x$ 14. $5x - 3y \leq -2$

The graph of a linear inequality is shown below. Determine whether each point satisfies the inequality.

15. $(1, -3)$
16. $(-2, -1)$
17. $(2, 3)$
18. $(3, -4)$

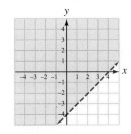

The graph of a linear inequality is shown below. Determine whether each point satisfies the inequality.

19. $(2, 1)$
20. $(-2, -4)$
21. $(4, -2)$
22. $(-3, 4)$

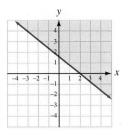

23. The boundary line for the graph of a linear inequality is shown. Why can't the origin be used as a test point to determine which side to shade?

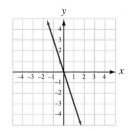

24. To determine how many pallets (x) and barrels (y) a delivery truck can hold, a dispatcher refers to the loading sheet on the right. Can a truck make a delivery of 4 pallets and 10 barrels in one trip?

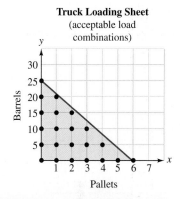

NOTATION

Write the meaning of each symbol.

25. $\leq$ 26. $>$

27. $<$ 28. $\geq$

GUIDED PRACTICE

Determine whether each ordered pair is a solution of the given inequality. See Example 1.

29. $2x + y > 6, (3, 2)$ 30. $4x - 2y \geq -6, (-2, 1)$

31. $-5x - 8y < 8, (-8, 4)$ 32. $x + 3y > 14; (-3, 8)$

33. $4x - y \leq 0, \left(\dfrac{1}{2}, 1\right)$ 34. $9x - y \leq 2, \left(\dfrac{1}{3}, 1\right)$

35. $-5x + 2y > -4, (0.8, 0.6)$

36. $6x - 2y < -7, (-0.2, 1.5)$

Complete each graph by shading the correct side of the boundary. See Examples 2–3.

37. $x + y \leq 1$

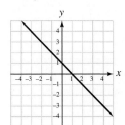

38. $4x + 3y \leq 12$

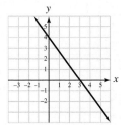

39. $x - 2y \geq 4$

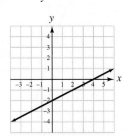

40. $y + 9x \geq 3$

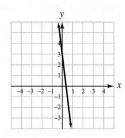

Graph each inequality. See Example 4.

41. $y > x - 3$

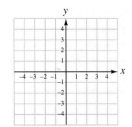

42. $y > 3x$

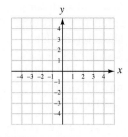

43. $y \leq x + 2$

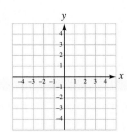

44. $y \geq 2x$

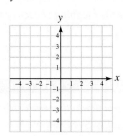

Graph each inequality. See Example 5.

45. $2y - x < 8$

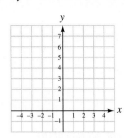

46. $3x + 2y > 12$

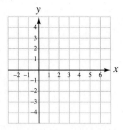

47. $2x + y > 2$

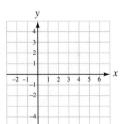

48. $7x - 2y < 21$

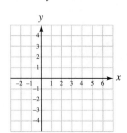

Graph each inequality. See Example 6.

49. $x < 2$

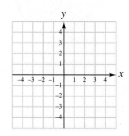

50. $y > -3$

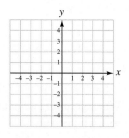

51. $y \leq 1$

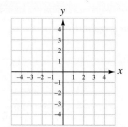

52. $x \geq -4$

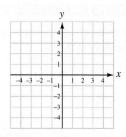

TRY IT YOURSELF

Graph each inequality.

53. $y - x \geq 0$

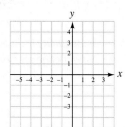

54. $y + x < 0$

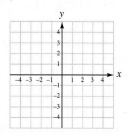

55. $y \geq 3 - x$

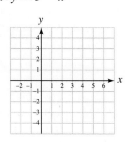

56. $y < 2 - x$

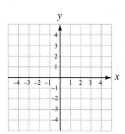

57. $y < 2 - 3x$

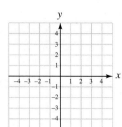

58. $y \geq 5 - 2x$

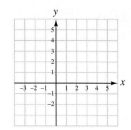

59. $y > 2x - 4$

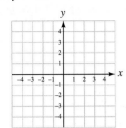

60. $y \leq 4x$

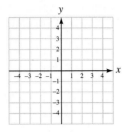

61. $y + 2x < 0$

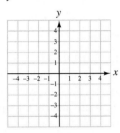

62. $3x - 2y > 6$

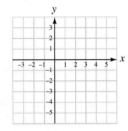

63. $3x - 4y > 12$

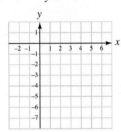

64. $5x + 4y \geq 20$

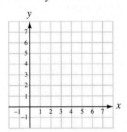

65. $3(x + y) + x < 6$

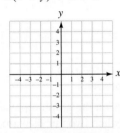

66. $2(x - y) - y \geq 4$

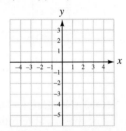

67. $4x - 3(x + 2y) \geq -6y$

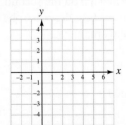

68. $3y + 2(x + y) < 5y$

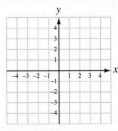

APPLICATIONS

69. NATO See the illustration in the next column. In March 1999, NATO aircraft targeted Serbian military forces that were south of the 44th parallel in Yugoslavia, Montenegro, and Kosovo. Shade the geographic area that NATO was trying to rid of Serbian forces.

Based on data from *Los Angeles Times* (March 24, 1999)

70. U.S. HISTORY When he ran for president in 1844, the campaign slogan of James K. Polk was "54-40 or fight!" It meant that Polk was willing to fight Great Britain for possession of the Oregon Territory north to the 54°40′ parallel, as shown below. In 1846, Polk accepted a compromise to establish the 49th parallel as the permanent boundary of the United States. Shade the area of land that Polk conceded to the British.

In Problems 71–76, write an inequality and graph it for nonnegative values of x and y. Then give three ordered pairs that satisfy the inequality.

71. PRODUCTION PLANNING It costs a bakery $3 to make a cake and $4 to make a pie. Production costs cannot exceed $120 per day. Write and then graph an inequality that shows the possible combinations of cakes (x) and pies (y) that can be made.

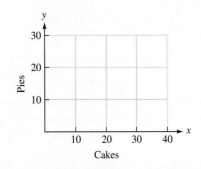

72. **HIRING BABYSITTERS** Mary has a choice of two babysitters. Sitter 1 charges $6 per hour, and sitter 2 charges $7 per hour. Mary can afford no more than $42 per week for sitters. Write and then graph an inequality that shows the possible ways that she can hire sitter 2 (x) and sitter 2 (y).

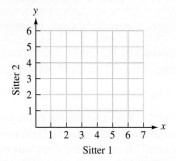

73. **INVENTORIES** A clothing store advertises that it maintains an inventory of at least $4,400 worth of men's jackets. A leather jacket costs $100 and a nylon jacket costs $88. Write and then graph an inequality that shows the possible ways that leather jackets (x) and nylon jackets (y) can be stocked.

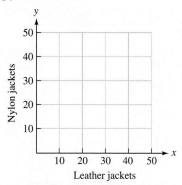

74. **MAKING SPORTING GOODS** A sporting goods manufacturer allocates at least 2,400 units of production time per day to make baseballs and footballs. It takes 20 units of time to make a baseball and 30 units of time to make a football. Write and then graph an inequality that shows the possible ways to schedule the production time to make baseballs (x) and footballs (y).

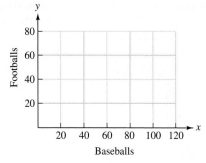

75. **INVESTING** Robert has up to $8,000 to invest in two companies. Stock in Robotronics sells for $40 per share, and stock in Macrocorp sells for $50 per share. Write and then graph an inequality that shows the possible ways that he can buy shares of Robotronics (x) and Macrocorp (y).

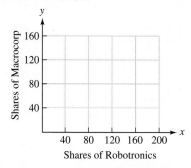

76. **TICKET SALES** Tickets to the Rockford Rox baseball games cost $6 for reserved seats and $4 for general admission. Nightly receipts must average at least $10,200 to meet expenses. Write and then graph an inequality that shows the possible ways that the Rox can sell reserved seats (x) and general admission tickets (y).

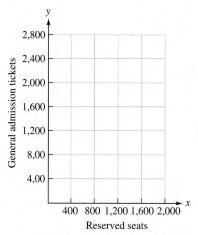

77. **ROLLING DICE** The points on the graph represent all of the possible outcomes when two fair dice are rolled a single time. For example, (5, 2), shown in red, represents a 5 on the first die and a 2 on the second. Which of the following sentences best describes the outcomes that lie in the shaded area?

   (i) Their sum is at most 6.
   (ii) Their sum exceeds 6.
   (iii) Their sum does not exceed 6.
   (iv) Their sum is at least 6.

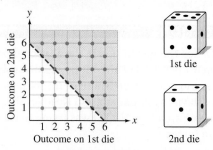

78. **DENTAL ASSISTANT** A dentist's office schedules 1-hour-long appointments for adults and $\frac{3}{4}$-hour-long appointments for children. The appointment times do not overlap. Let c represent the number of appointments scheduled for children and a represent the number of appointments scheduled for adults. Graph $\frac{3}{4}c + a \leq 9$, which shows the possible ways the time for seeing patients can be scheduled so that it does not exceed 9 hours per day. Find three possible combinations of children/adult appointments.

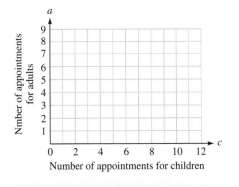

WRITING

79. Explain how to find the boundary for the graph of a linear inequality in two variables.
80. Explain how to determine which side of the boundary line to shade when graphing a linear inequality in two variables.

REVIEW

81. 39.75 is what percent of 265?
82. Solve: $2(x - 4) \leq -12$.
83. Solve $c = d\pi$ for d.
84. Solve: $-2x + 5 = -9$
85. Write a formula relating distance, rate, and time.
86. What is the slope of the line $2x - 3y = 2$?
87. Solve $A = P + Prt$ for t.
88. What is the sum of the measures of the three angles of any triangle?

SECTION 3.8
Functions

Objectives
1. Find the domain and range of a relation.
2. Identify functions and their domains and ranges.
3. Use function notation.
4. Graph functions.
5. Use the vertical line test.

In this section, we will discuss *relations* and *functions*. These two concepts are included in our study of graphing because they involve ordered pairs.

1 Find the domain and range of a relation.

The following table shows the number of medals won by American athletes at several recent Winter Olympics.

USA Winter Olympic Medal Count

Year	1988	1992	1994*	1998	2002	2006	2010
Medals	6	11	13	13	34	25	37
	Calgary CAN	Albertville FRA	Lillehammer NOR	Nagano JPN	Salt Lake City USA	Turin ITA	Vancouver CAN

*The Winter Olympics were moved ahead two years so that the winter and summer games would alternate every two years.

We can display the data in the table as a set of ordered pairs, where the **first component** represents the year and the **second component** represents the number of medals won by American athletes:

{(1988, 6), (1992, 11), (1994, 13), (1998, 13), (2002, 34), (2006, 25)(2010, 37)}

A set of ordered pairs, such as this, is called a **relation**. The set of all first components is called the **domain** of the relation and the set of all second components is called the **range** of a relation.

Self Check 1
Find the domain and range of the relation {(8, 2), (−1, 10), (6, 2), (−5, −5)}.
Now Try Problem 21

EXAMPLE 1 Find the domain and range of the relation {(1, 7), (4, −6), (−3, 1), (2, 7)}.

Strategy We will examine the first and second components of the ordered pairs.

WHY The set of first components is the domain and the set of second components is the range.

Solution
The relation {(**1**, 7), (**4**, −6), (**−3**, 1), (**2**, 7)} has the domain {**−3, 1, 2, 4**}, and the range is {**−6, 1, 7**}. The elements of the domain and range are usually listed in increasing order, and if a value is repeated, it is listed only once.

In everyday life, we see a wide variety of situations in which one quantity depends on another:

- The distance traveled by a car depends on its speed.
- The cost of renting a video depends on the number of days it is rented.
- A state's number of representatives in Congress depends on the state's population.

We will discuss many situations in which one quantity depends on another according to a specific rule, called a *function*. For example, the equation $y = 2x - 3$ sets up a rule where each value of y depends on the choice of some number x. The rule is: *To find y, double the value of x and subtract* 3. In this case, y (the *dependent variable*) depends on x (the *independent variable*).

2 Identify functions and their domains and ranges.

We have previously described relationships between two quantities in different ways.

Using words

| The number of tires to order | is | two | times | the number of bicycles to be manufactured. |

Here words are used to state that the number of bicycle tires to order depends on the number of bicycles to be manufactured.

Using equations
$t = 1{,}500 - d$

This equation describes how the amount of take-home pay t depends on the amount of deductions d.

Using graphs

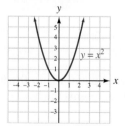

This rectangular coordinate graph shows many ordered pairs (x, y) that satisfy the equation $y = x^2$, where the value of the y-coordinate depends on the value of the x-coordinate.

Using tables

Acres	Schools
400	4
800	8
1,000	10
2,000	20

This table shows that the number of schools needed depends on the size of the housing development.

An **arrow** or **mapping diagram** can also be used to illustrate a relation. The data from the Winter Olympics example is shown below in that form.

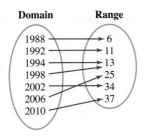

Notice that for each year, there corresponds exactly one medal count. That is, this relation assigns to each member of the domain exactly one member of the range.

Two observations can be made about these examples:

- Each one establishes a relationship between two sets of values. For example, the number of bicycle tires that must be ordered *depends* on the number of bicycles to be manufactured.
- In these relationships, each value in one set is assigned a *single* value of a second set. For example, for each number of bicycles to be manufactured, there is exactly one number of tires to order.

Relationships between two quantities that exhibit both of these characteristics are called **functions**.

Functions

A **function** is a set of ordered pairs (a relation) in which to each first component there corresponds exactly one second component.

Using the variables x and y, we can restate the previous definition as follows.

y Is a Function of x

For y to be a function of x, each value of x must determine exactly one value of y.

EXAMPLE 2 Determine whether the equation, table, and arrow diagram define y to be a function of x.

a. $y = 4x + 1$

b.

x	y
0	6
5	3
9	1
5	7
10	8

c.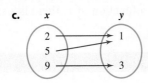

Self Check 2

a. Does $y = 2 - x^2$ define a function?

b. Does the table to the right define a function?

x	y
2	4
1	1
0	0
−1	1
−2	4

Now Try Problems 29 and 31

Strategy We will check to see whether each value of x is assigned exactly one value of y.

WHY If this is true, the equation, table, or arrow diagram defines y to be a function of x.

Solution
a. For each value of the independent variable x, we apply the rule: *Multiply x by 4 and add 1*. Since this arithmetic gives a single value of the dependent variable y, the equation defines a function.
b. Since the table assigns two different values of y, 3 and 7, to the x-value of 5, it does not define y as a function of x.
c. Each value of x is assigned to one value of y: 2→1, 5→1, 9→3. This is a function.

> **Success Tip** The table in Self Check 2 illustrates an important fact about functions. In a function, different values of x can determine the *same* value of y. In the table, x-values of 2 and -2 determine a y-value of 4, and x-values of 1 and -1 determine a y-value of 1. Nevertheless, each value of x determines exactly one value of y, so the table does define a function.

We have seen that functions can be represented by equations in two variables. Some examples of functions are

$$y = 2x - 10, \quad y = x^2 + 2x - 3, \quad \text{and} \quad s = 5 - 16t$$

For a function, the set of all possible values of the independent variable (the inputs) is called the **domain of the function**. The set of all possible values of the dependent variable (the outputs) is called the **range of the function**.

Self Check 3
Find the domain and range of the function $y = -x$.
Now Try Problem 39

EXAMPLE 3 Find the domain and range of $y = |x|$.

Strategy We will determine which real numbers are allowable inputs for x for the domain. Then we will determine which real numbers are possible outputs of y for the range.

WHY The domain is the set of all possible values of the input variable, and the range is the set of all possible values of the output variable.

Solution
To find the domain of $y = |x|$, we determine which real numbers are allowable inputs for x. Since we can find the absolute value of any real number, the domain is the set of all real numbers. Since the absolute value of any real number x is greater than or equal to zero, the range of $y = |x|$ is the set of all real numbers greater than or equal to zero.

3 Use function notation.

When the variable y is a function of x, there is a special notation that we can use to denote the function.

> **Function Notation**
>
> The notation $y = f(x)$ denotes that y is a function of x.

The notation $y = f(x)$ is read as "y equals f of x." Note that y and $f(x)$ are two different notations for the same quantity. Thus, the equations $y = 4x + 1$ and $f(x) = 4x + 1$ represent the same relationship.

This is the variable used
to represent the input value.
↓
$$f(x) = 4x + 1$$
↓ ↓
This is the | This expression shows
name of the | how to obtain an output
function. | for a given input.

> **Caution!** The symbol $f(x)$ denotes a function. It does not mean "f times x."
> Read $f(x)$ as "f of x."

The notation $y = f(x)$ provides a compact way of representing the value of y that corresponds to some number x. For example, if $f(x) = 4x + 1$, the value of y that is determined when $x = 2$ is denoted by $f(2)$.

$f(x) = 4x + 1$	This is the function.
$f(2) = 4(2) + 1$	Replace x with 2.
$f(2) = 8 + 1$	Do the multiplication.
$f(2) = 9$	Do the addition.

Thus, $f(2) = 9$.

The letter f used in the notation $y = f(x)$ represents the word *function*. However, other letters can be used to represent functions. For example, $y = g(x)$ and $y = h(x)$ also denote functions involving the variable x.

EXAMPLE 4 For $g(x) = 3 - 2x$ and $h(x) = x^3 - 1$, find:
a. $g(3)$ **b.** $h(-2)$

Self Check 4
Find $g(0)$ and $h(4)$ using the functions in Example 4.

Now Try Problem 43

Strategy We will substitute 3 for x in $3 - 2x$ and substitute -2 for x in $x^3 - 1$, and then evaluate each expression.

WHY The numbers 3 and -2, which are within the parentheses, are inputs that should be substituted for the variable x. The expression that the value of x is substituted in is determined by the name of the function.

Solution
a. To find $g(3)$, we use the function rule $g(x) = 3 - 2x$ and replace x with 3.

$g(x) = 3 - 2x$	
$g(3) = 3 - 2(3)$	Substitute 3 for x.
$g(3) = 3 - 6$	Do the multiplication.
$g(3) = -3$	Do the subtraction.

Thus, $g(3) = -3$.

b. To find $h(-2)$, we use the function rule $h(x) = x^3 - 1$ and replace x with -2.

$h(x) = x^3 - 1$	
$h(-2) = (-2)^3 - 1$	Substitute -2 for x.
$h(-2) = -8 - 1$	Evaluate the power.
$h(-2) = -9$	Do the subtraction.

Thus, $h(-2) = -9$.

We can think of a function as a machine that takes some input x and turns it into some output $f(x)$, as shown in part (a) of the figure. The machine in part (b) turns the input value of -2 into the output value of -9, and we can write $f(-2) = -9$.

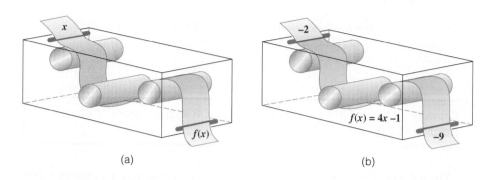

(a) (b)

Using Your CALCULATOR Business Profits

Accountants have found that the function $f(x) = -0.000065x^2 + 12x - 278{,}000$ estimates the profit a bowling alley will make when x games are bowled per year. Suppose that management predicts that 90,000 games will be bowled in the upcoming year. The expected profit for that year can be found by evaluating $f(90{,}000)$ on a reverse-entry scientific calculator.

$$f(90{,}000) = -0.000065(90{,}000)^2 + 12(90{,}000) - 278{,}000$$

.000065 $\boxed{+\backslash -}$ $\boxed{\times}$ 90000 $\boxed{x^2}$ $\boxed{+}$ 12 $\boxed{\times}$ 90000 $\boxed{-}$ 278000 $\boxed{=}$

$\boxed{275500}$

To evaluate $f(90{,}000)$ with a graphing calculator or a direct-entry scientific calculator, we enter these numbers and press these keys.

$\boxed{(-)}$ .000065 $\boxed{\times}$ 90000 $\boxed{x^2}$ $\boxed{+}$ 12 $\boxed{\times}$ 90000 $\boxed{-}$ 278000 $\boxed{\text{ENTER}}$

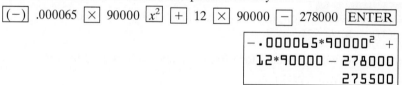

4 Graph functions.

We have seen that a function such as $f(x) = 4x + 1$ assigns to each value of x a single value $f(x)$. The input–output pairs generated by a function can be written in the form $(x, f(x))$. These ordered pairs can be plotted on a rectangular coordinate system to give the graph of the function.

Self Check 5

Graph: $f(x) = -3x - 2$

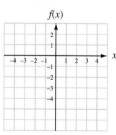

EXAMPLE 5 Graph: $f(x) = 4x + 1$

Strategy We can graph the function by creating a table of function values and plotting the corresponding ordered pairs.

WHY After drawing a line through the plotted points, we will have the graph of the function.

Solution

To make a table, we choose several values for x and find the corresponding values of $f(x)$. If x is -1, we have

$f(x) = 4x + 1$ This is the function to graph.
$f(-1) = 4(-1) + 1$ Substitute -1 for each x.
$f(-1) = -4 + 1$ Do the multiplication.
$f(-1) = -3$ Do the addition.

Thus, $f(-1) = -3$. This means that, when x is -1, $f(x)$ is -3, and it indicates that the ordered pair $(-1, -3)$ lies on the graph of $f(x)$.

Similarly, we find the corresponding values of $f(x)$ for x-values of 0 and 1. Then we plot the resulting ordered pairs and draw a straight line through them to get the graph of $f(x) = 4x + 1$. Since $y = f(x)$, the graph of $f(x) = 4x + 1$ is the same as the graph of the equation $y = 4x + 1$.

Now Try Problem 45

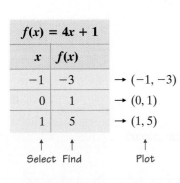
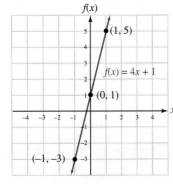

The vertical axis can be labeled y or $f(x)$.

The Language of Algebra A table of function values is similar to a table of solutions, except that the second column is usually labeled $f(x)$ instead of y.

x	$f(x)$

x	y

We call $f(x) = 4x + 1$ from Example 5 a **linear function** because its graph is a nonvertical line. Any linear equation, except those of the form $x = a$, can be written using function notation by writing it in slope–intercept form ($y = mx + b$) and then replacing y with $f(x)$.

EXAMPLE 6   Graph: $f(x) = |x|$

Strategy We can graph the function by creating a table of function values and plotting the corresponding ordered pairs.

WHY After drawing a "V" shape through the plotted points, we will have the graph of the function.

Solution
To create a table of function values, we choose values for x and find the corresponding values of $f(x)$. For $x = -4$ and $x = 3$, we have

$f(x) = |x|$ $f(x) = |x|$
$f(-4) = |-4|$ $f(3) = |3|$
$f(-4) = 4$ $f(3) = 3$

Thus, $f(-4) = 4$ and $f(3) = 3$.

Self Check 6
Graph: $f(x) = |x| + 2$

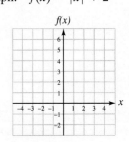

Now Try Problem 46

Similarly, we find the corresponding values of $f(x)$ for several other x-values. When we plot the resulting ordered pairs, we see that they lie in a "V" shape. We join the points to complete the graph as shown. We call $f(x) = |x|$ an **absolute value function**.

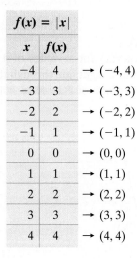
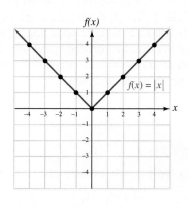

The figure below shows the graphs of four basic functions.

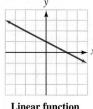
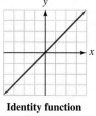
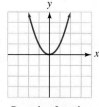
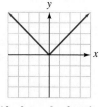

Linear function
$f(x) = mx + b$
(a)

Identity function
$f(x) = x$
(b)

Squaring function
$f(x) = x^2$
(c)

Absolute value function
$f(x) = |x|$
(d)

5 Use the vertical line test.

We can use the **vertical line test** to determine whether a given graph is the graph of a function. If any vertical line intersects a graph more than once, the graph cannot represent a function, because to one value of x, there corresponds more than one value of y. The graph in figure (a), shown in red, is not the graph of a function, because the x-value -1 determines three different y-values: 3, -1, and -4.

The graph shown in figure (b) does represent a function, because every vertical line that can be drawn intersects the graph exactly once.

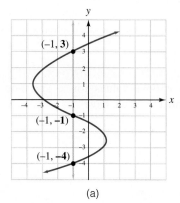
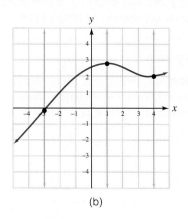

(a)   (b)

EXAMPLE 7   Determine whether each of the following is a graph of a function.

a.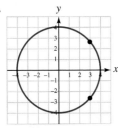

b.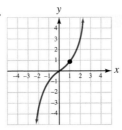

Self Check 7
Determine whether each of the following is a graph of a function.

a.

b.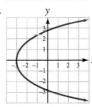

Now Try Problems 49 and 50

Strategy We will check to see whether any vertical line intersects the graph more than once.

WHY If any vertical line does intersect the graph more than once, the graph is not a function.

Solution
a. This graph is not the graph of a function, because the vertical line intersects the graph at more than one point.

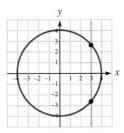

b. This graph is the graph of a function, because no vertical line will intersect the graph at more than one point.

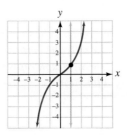

ANSWERS TO SELF CHECKS

1. domain: $\{-5, -1, 6, 8\}$, range: $\{-5, 2, 10\}$ 2. a. yes   b. yes
3. domain: all real numbers, range: all real numbers   4. 3, 63
5. 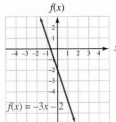   6. 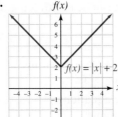   7. a. function   b. not a function

SECTION 3.8 STUDY SET

VOCABULARY

Fill in the blanks.

1. A set of ordered pairs is called a _____.
2. A _____ is a set of ordered pairs in which to each first component there corresponds exactly one second component.
3. In the equation $y = 2x + 8$, x is the _____ variable.
4. In the equation $y = 2x + 8$, y is the _____ variable.
5. The set of all input values for a function is called the _____; the set of all output values is called the _____.
6. $f(x) = 6 - 5x$ is an example of _____ notation.

CONCEPTS

Consider the function $f(x) = x^2$.

7. If positive real numbers are substituted for x, what type of numbers result?
8. If negative real numbers are substituted for x, what type of numbers result?
9. If 0 is substituted for x, what number results?
10. What are the domain and range of the function?

Consider the function $g(x) = x^4$.

11. What type of numbers can be inputs in this function? What is the special name for this set?
12. What type of numbers will be outputs in this function? What is the special name for this set?
13. Complete part b so that the statements say the same thing.
    a. In the equation $y = -5x + 1$, find the value of y when $x = -1$.
    b. In the equation $f(x) = -5x + 1$, find _____.
14. A function can be thought of as a machine that converts inputs into outputs. Use the terms *domain*, *range*, *input*, and *output* to label the diagram of a function machine shown in the next column. Then find $f(2)$.

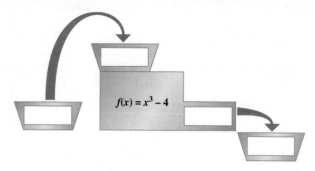

15. See the illustration.
    a. Give the coordinates of the points where the blue vertical line intersects the red graph.
    b. Is the red graph the graph of a function? Explain.

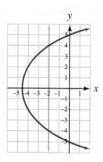

16. A student wrote the following statement about the illustration. What is wrong with his reasoning?

    When I drew a vertical line through the graph, it intersected the graph only once. By the vertical line test, the graph represents a function.

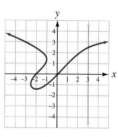

NOTATION

Fill in the blanks to make the statements true.

17. The function notation $f(4) = -5$ states that when 4 is substituted for ____ in function f, the result is ____. This fact can be illustrated graphically by plotting the point (____, ____).
18. $f(x) = 6 - 5x$ is read as "f ____ x is $6 - 5x$."
19. If $f(x) = 6 - 5x$, then $f(0) = 6$ is read as "f ____ zero ____ 6."
20. The equations $y = 3x + 5$ and ____ $= 3x + 5$ are the same.

GUIDED PRACTICE

Find the domain and range of each relation. See Example 1.

21. $\{(6, -1), (-1, -10), (-6, 2), (8, -5)\}$

22. $\{(11, -3), (0, 0), (4, 5), (-3, -7)\}$

23. $\{(0, 9), (-8, 50), (6, 9)\}$

24. $\{(1, -12), (-6, 8), (5, 8)\}$

Determine whether each equation, table, or arrow diagram defines a function. If not, indicate an input value for which there is more than one output value. See Example 2.

25. $y = 2x + 10$ **26.** $y = x - 15$

27. $y = x^2$ **28.** $y = |x|$

29. $y^2 = x$ **30.** $|y| = x$

31.

x	y
1	7
2	15
3	23
4	16
5	8

32.

x	y
-1	1
-3	1
-5	1
-7	1
-9	1

33.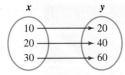

34.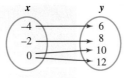

35.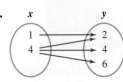

36.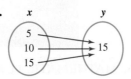

Find the domain and range of each function. See Example 3.

37. $f(x) = x + 1$ **38.** $f(x) = 3x - 2$

39. $y = x^2$ **40.** $y = -|x|$

Find each value. See Example 4.

41. $f(x) = 4x - 1$
  a. $f(1)$ b. $f(-2)$
  c. $f\left(\dfrac{1}{4}\right)$ d. $f(50)$

42. $g(x) = 1 - 5x$
  a. $g(0)$ b. $g(-75)$
  c. $g(0.2)$ d. $g\left(-\dfrac{4}{5}\right)$

43. $h(t) = 2t^2$
  a. $h(0.4)$ b. $h(-3)$
  c. $h(1,000)$ d. $h\left(\dfrac{1}{8}\right)$

44. $v(t) = 6 - t^2$
  a. $v(30)$ b. $v(6)$
  c. $v(-1)$ d. $v(0.5)$

Complete each table and graph the function. See Examples 5-6.

45. $f(x) = -2 - 3x$

x	f(x)
0	
1	
-1	
-2	

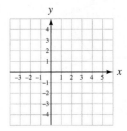

46. $h(x) = |1 - x|$

x	h(x)
0	
1	
2	
3	
-1	
-2	

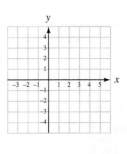

47. $f(x) = \tfrac{1}{2}x - 2$

x	f(x)
-2	
0	
2	

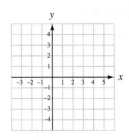

48. $f(x) = -\frac{2}{3}x + 3$

x	f(x)
0	
3	
6	

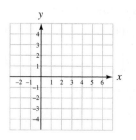

Determine whether each of the following graphs is the graph of a function. See Example 7.

49.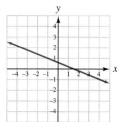

50.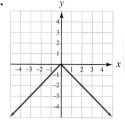

51.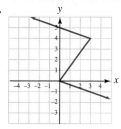

52.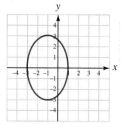

59.
t	d
3	4
3	−4
4	3
4	−3

60.
x	y
1	1
2	2
3	3
4	4

61.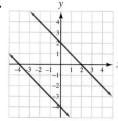

62.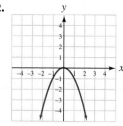

63. {(3, 4), (5, 1), (6, 2)}
64. {(3, −2), (2, 4), (3, −1), (5, 6)}

APPLICATIONS

65. **REFLECTIONS** When a beam of light hits a mirror, it is reflected off the mirror at the same angle that the incoming beam struck the mirror, as shown. What type of function could serve as a mathematical model for the path of the light beam shown here?

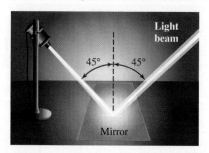

TRY IT YOURSELF

Determine whether each equation, table, or graph defines a function. If not, indicate an input value for which there is more than one output value.

53. $y = x^3$
54. $y = -x$
55. $x = 3$
56. $y = 3$

57.
x	y
−4	6
−1	0
0	−3
2	4
−1	2

58.
x	y
30	2
30	4
30	6
30	8
30	10

66. **MATHEMATICAL MODELS** The illustration below shows the path of a basketball shot taken by a player. What type of function could be used to mathematically model the path of the basketball?

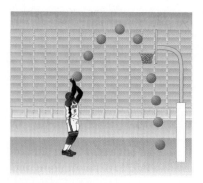

67. TIDES The illustration below shows the graph of a function f, which gives the height of the tide for a 24-hour period in Seattle, Washington. (Note that military time is used on the x-axis: 3 A.M. = 3, noon = 12, 3 P.M. = 15, 9 P.M. = 21, and so on.)

a. Find the domain of the function.
b. Find: $f(3)$
c. Find: $f(6)$
d. Estimate: $f(15)$
e. What information does $f(12)$ give?

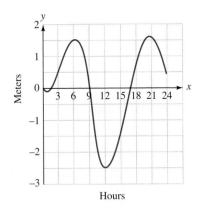

68. RENTALS The function $C(d) = 500 + 100(d - 3)$ gives the cost in dollars to rent an RV motor home for d days. Find the cost of renting the motor home for a vacation that will last 7 days.

(3-day minimum)

69. LAWN SPRINKLERS The function $A(r) = \pi r^2$ can be used to determine the area that will be watered by a rotating sprinkler that sprays out a stream of water r feet. Find $A(5)$, $A(10)$, and $A(20)$. Round to the nearest tenth.

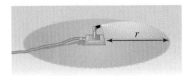

70. BALLISTICS The height of a toy rocket shot from the ground straight upward is given by the function $f(t) = -16t^2 + 256t$.

a. Find the height of the rocket 3 seconds after it is shot.
b. Find $f(16)$. Interpret the result.

71. PLATFORM DIVING The number of feet a diver is above the surface of the water is given by the function $h(t) = -16t^2 + 16t + 32$, where t is the elapsed time in seconds after the diver jumped. Find the height of the diver for the times shown in the table.

t	$h(t)$
0	
0.5	
1.5	
2.0	

72. PARTS The function $f(r) = 2.30 + 3.25(r + 0.40)$ approximates the length (in feet) of the belt that joins the two pulleys shown. r is the radius (in feet) of the smaller pulley. Find the belt length needed for each pulley in the parts list.

Parts List

Pulley	r	Belt length
P-45M	0.32	
P-08D	0.24	
P-00A	0.18	
P-57X	0.38	

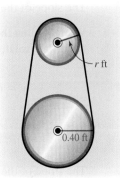

WRITING

73. In the function $y = -5x + 2$, why do you think x is called the *independent* variable and y the *dependent* variable?

74. Explain what a politician meant when she said, "The speed at which the downtown area will be redeveloped is a function of the number of low-interest loans made available to the property owners."

REVIEW

75. Write the equation of the horizontal line passing through $(-3, 6)$.
76. Is -3 a solution of $t^2 - t + 1 = 13$?
77. Write the formula that relates profit, revenue, and costs.
78. What is the word used to represent the perimeter of a circle?
79. Use the distributive property: $-3(2x - 4)$
80. Evaluate $r^2 - r$ for $r = -0.5$.

STUDY SKILLS CHECKLIST

Preparing for the Chapter 3 Test

There are common difficulties that students have when working with the topics of Chapter 3. To make sure you are prepared for the test and to help you overcome these difficulties, study the checklist below.

☐ When finding the slope of a line, if you are given two ordered pairs, use the formula:

$m = \dfrac{y_2 - y_1}{x_2 - x_1}$ Find the slope of the line that passes through $(-2, 5)$, $(6, -1)$.

$m = \dfrac{-1 - 5}{6 - (-2)}$

$m = \dfrac{-6}{8}$

$m = -\dfrac{3}{4}$

The slope of the line is $-\dfrac{3}{4}$.

☐ When finding the slope of a line, if you are given an equation, isolate y and the slope is the *coefficient* of x.

$2x + 3y = 6$

$3y = -2x + 6$ Subtract 2x from both sides.

$y = -\dfrac{2}{3}x + 2$ Divide both sides by 3.

The slope of the line is $-\dfrac{2}{3}$.

☐ To graph a linear equation in two variables:
1. Find three ordered pairs that are solutions of the equation by selecting three values for x and calculating the corresponding values of y.
2. Plot these points on a rectangular coordinate system.
3. Draw a straight line passing through them.

☐ To graph a nonlinear equation in two variables:
1. Find at least *seven* ordered pairs that are solutions of the equation by selecting *seven* values for x and calculating the corresponding values of y.
2. Plot these points on a rectangular coordinate system.
3. Draw a curve passing through them.

☐ To write an equation of a line, you:
1. need the slope of the line, m.
2. need a point that line passes through (x_1, y_1).
3. input these values in the equation, $y - y_1 = m(x - x_1)$.

☐ To graph a linear inequality in two variables:
1. Graph the boundary line. Draw a solid line if the inequality contains an $\leq$ or an $\geq$ symbol. Draw a dashed line if the inequality contains an $<$ or an $>$ symbol.
2. Pick a test point on one side of the boundary. Use the origin if possible. If the test point is a solution of the inequality, shade the side of the boundary that contains the point. If the test point is not a solution of the inequality, shade the other side.

CHAPTER 3 SUMMARY AND REVIEW

SECTION 3.1 Graphing Using the Rectangular Coordinate System

DEFINITIONS AND CONCEPTS

A **rectangular coordinate system** is composed of a horizontal number line called the **x-axis** and a vertical number line called the **y-axis**.

The coordinates of the **origin** are $(0, 0)$.

To plot or **graph** ordered pairs means to locate their position on a coordinate system.

The x- and y-axes divide the coordinate plane into four distinct regions, called **quadrants**.

EXAMPLES

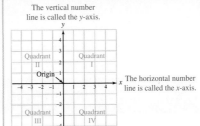
The vertical number line is called the y-axis.
The horizontal number line is called the x-axis.

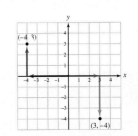

REVIEW EXERCISES

1. **a.** Graph the points with coordinates $(-1, 3)$, $(0, 1.5)$ $(-4, -4)$, $\left(2, \frac{7}{2}\right)$, and $(4, 0)$.

   b. In what quadrant does the point $(-3, -4)$ lie?

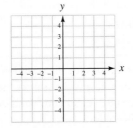

2. Use the graph in the illustration to complete the table.

x	y
3	
	0
−3	

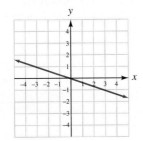

3. HAWAII Extimate the coordinates of Oahu using an ordered pair of the form (longitude, latitiude).

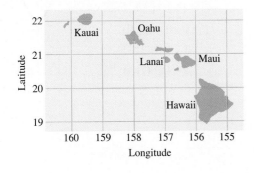

4. SNOWFALL The graph below gives the amount of snow on the ground at a mountain resort as measured once each day over a 7-day period.

   a. On the first day, how much snow was on the ground?

   b. What was the difference in the amount of snow on the ground when the measurements were taken on the second and third day?

   c. How much snow was on the ground on the sixth day?

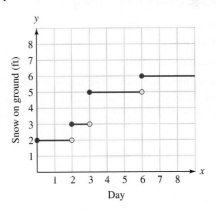

5. **COLLEGE ENROLLMENTS** The graph at the right gives the number of students enrolled at a college for the period from 4 weeks before to 5 weeks after the semester began.
   a. What was the maximum enrollment and when did it occur?
   b. How many students had enrolled 2 weeks before the semester began?
   c. When was enrollment 2,250?

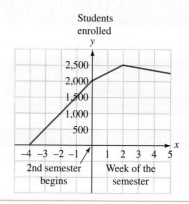

SECTION 3.2 Equations Containing Two Variables

DEFINITIONS AND CONCEPTS	EXAMPLES		
An **ordered pair** is a **solution** of an equation if, after substituting the values of the ordered pair for the variables in the equation, the result is a true statement.	The ordered pair $(1, 2)$ is a solution of $x + 2y = 5$ because the values of the ordered pair satisfy the equation: $$x + 2y = 5$$ $$1 + 2(2) \stackrel{?}{=} 5$$ $$5 = 5$$		
Solutions of an equation can be shown in a **table of solutions**. In an equation in x and y, x is called the **independent variable**, or **input**, and y is called the **dependent variable**, or **output**.	A table of solutions for the equation $x + 2y = 5$ includes ordered pairs that satisfy the equation. The ordered pair found above and others are shown in the table.	x | y | (x, y) -1 | 3 | $(-1, 3)$ 1 | 2 | $(1, 2)$ 5 | 0 | $(5, 0)$	
To graph an equation in two variables: 1. Make a table of solutions that contains several solutions written as ordered pairs. 2. Plot each ordered pair. 3. Draw a line or smooth curve through the points.	To graph the equation $x + 2y = 5$, make a table of solutions, plot the points, and draw the graph as in the illustration.	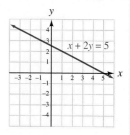	
	A table and a graph for the equation $y = x^2$ is shown in the illustration.	x | y | (x, y) -2 | 4 | $(-2, 4)$ -1 | 1 | $(-1, 1)$ 0 | 0 | $(0, 0)$ 1 | 1 | $(1, 1)$ 2 | 4 | $(2, 4)$	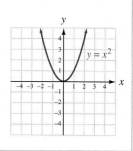

REVIEW EXERCISES

6. Determine whether $(-3, 5)$ is a solution of $y = |2 + x|$.

7. a. Complete the following table of solutions and graph the equation $y = -x^3$.

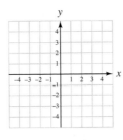

x	y	(x, y)
-2		
-1		
0		
1		
2		

b. How would the graph of $y = -x^3 + 2$ compare to the graph of the equation given in part a?

8. The graph below shows the relationship between the number of oranges O an acre of land will yield if t orange trees are planted on it.

a. If $t = 70$, what is O?

b. What importance does the point $(40, 18)$ on the graph have?

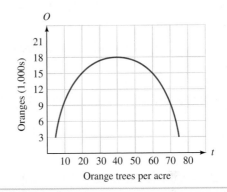
Orange trees per acre

SECTION 3.3 Graphing Linear Equations

DEFINITIONS AND CONCEPTS	EXAMPLES
An equation whose graph is a straight line and whose variables are raised to the first power is called a **linear equation**.	The equations $y = -3x - 2$ and $3x + 2y = 6$ are linear equations, because their graphs are straight lines. The equations $y = x^2$ and $y = |x|$ are not linear equations, because the graphs of these equations are not straight lines.
The **standard** or **general form** of a linear equation is $Ax + By = C$, where A, B, and C are real numbers and A and B are not both zero.	To write the equation $y = -3x - 2$ in standard or general form, add $3x$ to both sides to get $3x + y = -2$.
To graph a linear equation: **1.** Find three (x, y) pairs that satisfy the equation by selecting three x-values and finding their corresponding y-values. **2.** Plot each ordered pair. **3.** Draw a straight line through the points.	A table of solutions and the graph of $3x + 2y = 6$ are shown in the illustration. | x	y	(x, y)
---	---	---
0	3	(0, 3)
2	0	(2, 0)
4	-3	(4, -3)	 ↑ ↑ ↑ Select Find Plot 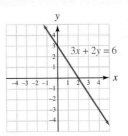
To find the y-intercept of a linear equation, substitute 0 for x in the equation of the line and solve for y.	The y-intercept of the line in the previous graph is $(0, 3)$.
To find the x-intercept of a linear equation, substitute 0 for y in the equation of the line and solve for x.	The x-intercept of the line in the previous graph is $(2, 0)$.

The equation $y = b$ represents the **horizontal line** that intersects the y-axis at $(0, b)$.

The graph of $y = 3$ is a horizontal line.

The equation $x = a$ represents the **vertical line** that intersects the x-axis at $(a, 0)$.

The graph of $x = 3$ is a vertical line.

REVIEW EXERCISES

Classify each equation as either linear or nonlinear.

9. $y = |x + 2|$
10. $3x + 4y = 12$
11. $y = 2x - 3$
12. $y = x^2 - x$
13. The equation $5x + 2y = 10$ is in standard form; what are A, B, and C?
14. Complete the table of solutions for the equation $3x + 2y = -18$.

x	y	(x, y)
-2		$(-2, \ \ )$
	3	$(\ \ , 3)$

15. Solve the equation $x + 2y = 6$ for y, find three solutions, and graph it.

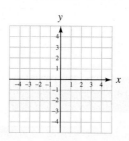

16. Graph $-4x + 2y = 8$ by finding its x- and y-intercepts.

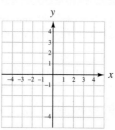

Graph each equation.

17. $y = 4$

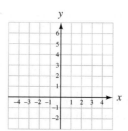

18. $x = -1$

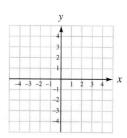

19. Since two points determine a line, only two points are needed to graph a linear equation. Why is it a good idea to plot a third point?

20. The graph of a linear equation is shown.

    a. If the coordinates of point M are substituted into the equation, will the result be true or false?

    b. If the coordinates of point N are substituted into the equation, will the result be true or false?

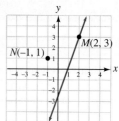

SECTION 3.4 Rate of Change and the Slope of a Line

DEFINITIONS AND CONCEPTS

Slope of a nonvertical line:

$$m = \frac{\text{change in the } y\text{-values}}{\text{change in the } x\text{-values}} = \frac{\text{rise}}{\text{run}}$$

The **slope** of a nonvertical line passing through points (x_1, y_1) and (x_2, y_2) is

$$m = \frac{y_2 - y_1}{x_2 - x_1}$$

EXAMPLES

If the change in the y-values between two points is 8 and the change in the x-values between the same two points is 5, the slope of the line is $m = \dfrac{8}{5}$.

To find the slope of the line passing through the points $(-3, 2)$ and $(1, -5)$, substitute into the slope formula:

$$m = \frac{y_2 - y_1}{x_2 - x_1} = \frac{-5 - 2}{1 - (-3)} = \frac{-7}{4} = -\frac{7}{4}$$

Chapter 3 Summary and Review

Lines that rise from left to right have a **positive slope**, and lines that fall from left to right have a **negative slope**.	In the illustration, line l_1 has a positive slope. Line l_2 has a negative slope.
Horizontal lines have **zero slope**. Vertical lines have **undefined slope**.	Line 3 in the illustration has a slope of 0. Line 4 has no defined slope. 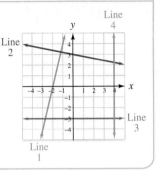

REVIEW EXERCISES

In Problems 21–25, find the slope of the line.

21.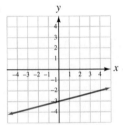

22.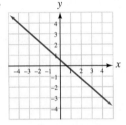

23. The line with the table of solutions shown here.

x	y	(x, y)
2	−3	(2, −3)
4	−17	(4, −17)

24. The line passing through the points $(2, -5)$ and $(5, -5)$.

25. The line passing through the points $(1, -4)$ and $(3, -7)$.

26. CARPENTRY If a truss like the one shown below is used to build the roof of a shed, find the slope (pitch) of the roof.

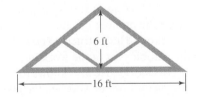

27. Graph the line that passes through $(-2, 4)$ and has slope $m = -\frac{4}{5}$.

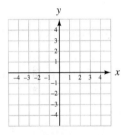

28. TOURISM The graph shows the number of international travelers to the United States in 2-year increments.

a. Between 2000 and 2002, the largest decline in the number of visitors occurred. What was the rate of change?

b. Between 2006 and 2008, the largest increase in the number of visitors occurred. What was the rate of change?

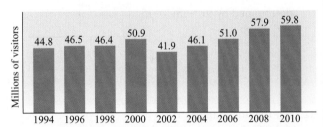

Based on data from *World Almanac and Book of Facts*, 2013.

SECTION 3.5 Slope–Intercept Form

DEFINITIONS AND CONCEPTS	EXAMPLES
If a linear equation is written in **slope–intercept form**, $$y = mx + b$$ the graph of the equation is a line with slope m and y-intercept $(0, b)$.	The slope of the graph of the equation $y = 4x - 8$ is 4. The y-intercept is the point $(0, -8)$. To find the slope and y-intercept of the graph of the equation $3x + 2y = 12$, solve the equation for y. $$3x + 2y = 12$$ $$2y = -3x + 12 \quad \text{Subtract 3x from both sides.}$$ $$y = -\frac{3}{2}x + 6 \quad \text{Divide both sides by 2.}$$ The slope is $-\frac{3}{2}$ and the y-intercept is $(0, 6)$.
Two lines with the same slope are **parallel**.	Since the graphs of $y = -2x + 5$ and $y = -2x + 7$ both have a slope of -2, the graphs will be parallel.
The product of the slopes of **perpendicular** lines is -1.	The graph of $y = -2x + 5$ has a slope of -2. The graph of $y = \frac{1}{2}x + 7$ has a slope of $\frac{1}{2}$. Since the product of the slopes is -1, the graphs are perpendicular.

REVIEW EXERCISES

Find the slope and the y-intercept of each line.

29. $y = \dfrac{3}{4}x - 2$

30. $y = -4x$

31. Find the slope and the y-intercept of the line determined by $9x - 3y = 15$ and graph it.

32. For the line graphed at right, find the slope and the y-intercept. Then write the equation of the line. Express your answer in slope–intercept form.

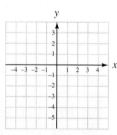

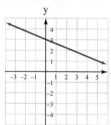

Without graphing, determine whether the graphs of the given pairs of lines would be parallel, perpendicular, or neither.

33. $\begin{cases} y = -\dfrac{2}{3}x + 6 \\ y = -\dfrac{2}{3}x - 6 \end{cases}$

34. $\begin{cases} x + 5y = -10 \\ y = 5x \end{cases}$

35. **COPIERS** A business buys a used copy machine that, when purchased, has already produced 75,000 copies.
   a. If the business plans to run 300 copies a week, write a linear equation that would find the number of copies c the machine has made in its lifetime after the business has used it for w weeks.
   b. Use your result to part a to predict the total number of copies that will have been made on the machine 1 year, or 52 weeks, after being purchased by the business.

36. **PHYSICAL FITNESS** A fitness instructor wants to determine the number of calories a client burns during a workout. The instructor knows that during the aerobic part of the workout, the client will burn 220 calories. He also knows that during the swimming part of the workout, the client will burn 7.8 calories per minute. Write a linear model in slope–intercept form that gives the total number of calories c the client will burn if she concludes a workout with m minutes of swimming.

Chapter 3 Summary and Review

SECTION 3.6 Point–Slope Form; Writing Linear Equations

DEFINITIONS AND CONCEPTS

If a line with slope m passes through the point (x_1, y_1), the equation of the line in **point–slope form** is

$$y - y_1 = m(x - x_1)$$

EXAMPLES

If a line has a slope of **5** and passes through the point (**3, 4**), its equation in point–slope form is

$$y - 4 = 5(x - 3)$$

REVIEW EXERCISES

Write an equation of a line with the given slope that passes through the given point. Express the result in point–slope form. Then change it to slope–intercept form and graph the equation.

37. $m = 3, (1, 5)$

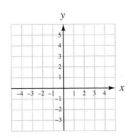

38. $m = -\dfrac{1}{2}, (-4, -1)$

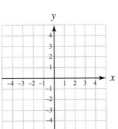

Write an equation of the line with the following characteristics. Express the result in slope–intercept form.

39. passing through $(3, 7)$ and $(-6, 1)$
40. horizontal, passing through $(6, -8)$
41. **CAR REGISTRATION** When it was 2 years old, the annual registration fee for a Dodge van was $380. When it was 4 years old, the registration fee dropped to $310. If the relationship is linear, write an equation that gives the registration fee f in dollars for the van when it is x years old.
42. **SALVAGE VALUE** A truck was purchased for $19,984. Its salvage value at the end of 8 years is expected to be $1,600. Find the depreciation equation.

SECTION 3.7 Graphing Linear Inequalities

DEFINITIONS AND CONCEPTS

An ordered pair (x, y) is a **solution of an inequality** in x and y if a true statement results when the variables are replaced by the coordinates of the ordered pair.

EXAMPLES

To determine whether $(-2, 5)$ is a solution of $x + 3y > 6$, substitute the coordinates into the inequality and see whether a true statement results.

$$x + 3y > 6$$
$$-2 + 3(5) \overset{?}{>} 6$$
$$13 > 6 \quad \text{True}$$

Since a true statement results, $(-2, 5)$ is a solution.

To graph a linear inequality:

1. Graph the boundary line. Draw a solid line if the inequality contains an $\leq$ or an $\geq$ symbol. Draw a dashed line if the inequality contains an $<$ or an $>$ symbol.

To graph $2x - y \leq 4$, proceed as follows:

1. Graph the boundary line $2x - y = 4$ and draw it as a solid line because the inequality symbol is $\leq$.

Unless otherwise noted, all content on this page is © Cengage Learning.

2. Pick a test point on one side of the boundary. Use the origin if possible. Replace x and y with the coordinates of that point. If the inequality is satisfied, shade the side of the boundary that contains the point. If the inequality is not satisfied, shade the other side.

2. Test the point $(0, 0)$:
$$2x - y \leq 4$$
$$2(0) - 0 \stackrel{?}{\leq} 4$$
$$0 \leq 4 \quad \text{True}$$

Since the coordinates of the test point satisfy the inequality, shade the side of a boundary that contains $(0, 0)$.

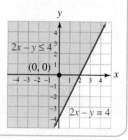

REVIEW EXERCISES

Use a check to determine whether each ordered pair is a solution of $2x - y \leq -4$.

43. $(0, 5)$ **44.** $(2, 8)$

45. $(-3, -2)$ **46.** $\left(\dfrac{1}{2}, -5\right)$

Graph each inequality.

47. $x - y < 5$ **48.** $2x - 3y \geq 6$

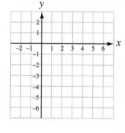 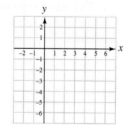

49. $y \leq -2x$ **50.** $y < -4$

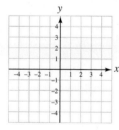 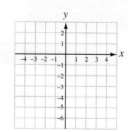

51. The graph of a linear inequality is shown in the next column. Would a true or a false statement occur if the coordinates of

  a. point A were substituted into the inequality?

  b. point B were substituted into the inequality?

  c. point C were substituted into the inequality?

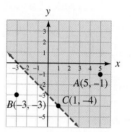

52. WORK SCHEDULES A student told her employer that during the school year, she would be available for up to 30 hours a week, working either 3- or 5-hour shifts. If x represents the number of 3-hour shifts and y represents the number of 5-hour shifts she works, the inequality $3x + 5y \leq 30$ shows the possible combinations of shifts she can work. Graph the inequality and find three possible combinations.

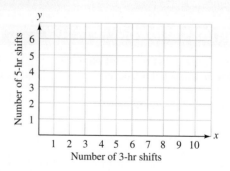

SECTION 3.8 Functions

DEFINITIONS AND CONCEPTS	EXAMPLES
A **relation** is a set of ordered pairs. The set of all **first components** is called the **domain** of the relation, and the set of all **second components** is called the **range** of a relation.	The relation $\{(4, 7), (0, -3), (-3, 8), (1, 7)\}$ has the domain $\{-3, 0, 1, 4\}$, and the range is $\{-3, 7, 8\}$.
A **function** is a set of ordered pairs in which to each first component there corresponds exactly one second component. For a function, the set of all possible values of the independent variable x (the inputs) is called the **domain**. The set of all possible values of the dependent variable y (the outputs) is called the **range**.	The equation $y = \|2x\|$ defines a function because each value of x determines exactly one value of y. For example, if $x = 1$, then $y = 2$, and if $x = 3$, then $y = 6$. Since x can be any real number in the function $y = \|2x\|$, the domain is the set of real numbers. Because of the absolute value in the function $y = \|2x\|$, y must be either positive or 0. Thus the range is the set of all real numbers that are greater than or equal to 0.
The notation $y = f(x)$ denotes that y is a function of x. Four basic functions are Linear: $f(x) = mx + b$ Identity: $f(x) = x$ Squaring: $f(x) = x^2$ Absolute value: $f(x) = \|x\|$	If $f(x) = 3x + 5$, we can find $f(2)$ and $f(-3)$ as follows: $f(x) = 3x + 5 \qquad f(x) = 3x + 5$ $f(2) = 3(2) + 5 \qquad f(-3) = 3(-3) + 5$ $f(2) = 11 \qquad f(-3) = -4$ Linear function $f(x) = mx + b$ — Identity function $f(x) = x$ — Squaring function $f(x) = x^2$ — Absolute value function $f(x) = \|x\|$
The vertical line test: If a vertical line intersects a graph in more than one point, the graph is not the graph of a function.	A function — Not a function

REVIEW EXERCISES

Find the domain and range of each relation.

53. $\{(7, -3), (-5, 9), (4, 4), (0, -11)\}$

54. $\{(2, -2), (15, -8), (-6, 9) (1, -8)\}$

In each case, determine whether a function is defined.

55. $y = 3x - 2$ **56.** $y^2 = x$

57.

x	2	2	3	4	5	6
y	-2	2	3	-4	5	-6

58.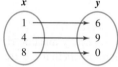

Find the domain and range of each function.

59. $f(x) = x + 10$

60. $y = x^2$

For the function $g(x) = 1 - 6x$, find each value.

61. $g(1)$ **62.** $g(-6)$

63. $g(0.5)$ **64.** $g\left(\dfrac{3}{2}\right)$

Complete the table and graph the function.

65. $h(x) = 1 - |x|$

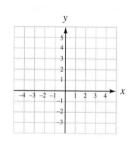

66. $f(x) = x^2 + 2$

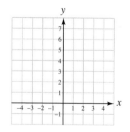

Determine whether each graph is the graph of a function.

67. 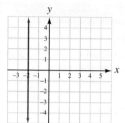   **68.**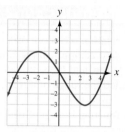

69. The function $f(r) = 15.7r^2$ estimates the volume in cubic inches of a can 5 inches tall with a radius of r inches. Find the volume of the can in the illustration. Round to the nearest tenth.

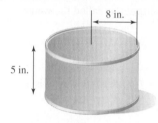

70. EARTH'S ATMOSPHERE The illustration below shows a graph of the temperatures of the atmosphere at various altitudes above Earth's surface. The temperature is expressed in degrees Kelvin, a scale widely used in scientific work.

a. Estimate the coordinates of three points on the graph that have an x-coordinate of 200.

b. Explain why this is not the graph of a function.

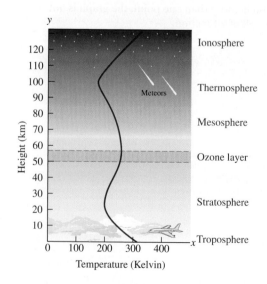

CHAPTER 3 TEST

1. The graph shows the number of dogs being boarded in a kennel over a 3-day holiday weekend.

   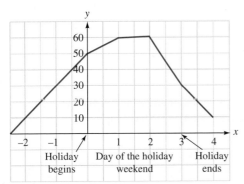

   a. How many dogs were in the kennel 2 days before the holiday?

   b. What is the maximum number of dogs that were boarded on the holiday weekend?

   c. When were there 30 dogs in the kennel?

   d. What information does the y-intercept of the graph give?

2. Plot each point on a rectangular coordinate system: $(1, 3), (-2, 4), (-3, -2), (3, -2), (-1, 0), (0, -1)$, and $\left(-\frac{1}{2}, \frac{7}{2}\right)$.

   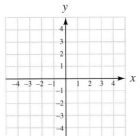

3. Find the coordinates of each point shown in the graph.

   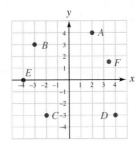

4. In which quadrant is each point located?

   a. $(-1, -5)$
   b. $\left(6, -2\frac{3}{4}\right)$

5. Graph: $y = x^2 - 4$

   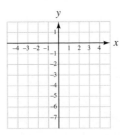

6. Graph: $8x + 4y = -24$

   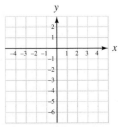

7. Is $(-3, -4)$ a solution of $3x - 4y = 7$?

8. Complete the table of solutions for the linear equation $x + 4y = 6$.

   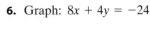

x	y	(x, y)
2		
	3	

9. Is $y = x^3$ a linear equation?

10. What are the x- and y-intercepts of the graph of $2x - 3y = 6$?

11. Find the slope and the y-intercept of $x + 2y = 8$.

12. Graph: $x = -4$

    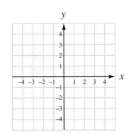

13. Graph the line passing through $(-2, -4)$ having a slope of $\frac{2}{3}$.

    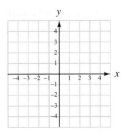

14. Find the slope of the line passing through $(-1, 3)$ and $(3, -1)$.

15. What is the slope of a vertical line?

16. Find the slope of the line.

    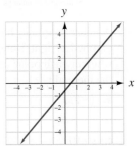

Unless otherwise noted, all content on this page is © Cengage Learning.

17. Find the slope of a line that is perpendicular to a line with slope $-\frac{2}{3}$.

18. When graphed, are the lines $y = 2x + 6$ and $6x - 3y = 0$ parallel, perpendicular, or neither?

Refer to the graph in the illustration below, which shows the elevation changes in a 26-mile marathon course.

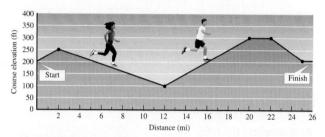

19. Find the rate of change of the decline on which the woman is running.

20. Find the rate of change of the incline on which the man is running.

21. DEPRECIATION After it is purchased, a $15,000 computer loses $1,500 in resale value every year. Write a linear equation that gives the resale value v of the computer x years after being purchased.

22. Write an equation of the line passing through $(-2, 5)$ and $(-3, -2)$. Answer in slope–intercept form.

23. WATER HEATERS The scatter diagram shows how excessively high temperatures affect the life of a water heater. Write an equation of the line that models the data for water temperatures between 140° and 180°. Let T represent the temperature of the water in degrees Fahrenheit and y represent the expected life of the heater in years. Give the answer in slope–intercept form.

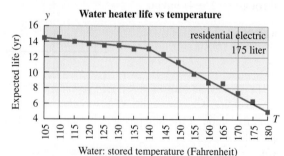

24. A linear inequality has been graphed at right. Determine whether each point satisfies the inequality.

    a. $(-2, 3)$
    b. $(3, -4)$
    c. $(0, 0)$

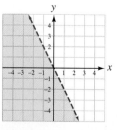

25. Does the point with coordinates $(1, -2)$ satisfy the inequality $x - y > -2$?

26. Graph: $x - y > -2$

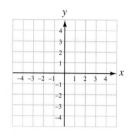

27. Find the domain and range of the relation: $\{(6, 1), (-2, 4), (7, 5), (0, 6)\}$

28. Is the circle the graph of a function?

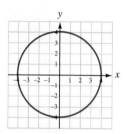

Determine whether the table, arrow diagram, or equation defines y to be a function of x. If a function is defined, give its domain and range. If it does not define a function, find ordered pairs that show a value of x that corresponds to more than one value of y.

29.
x	y
1	4
2	3
3	2
4	1

30.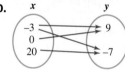

31. $y = 2x - 8$

32. Find the domain and range of the function $f(x) = -|x|$.

33. If $f(x) = 2x - 7$, find $f(-3)$.

34. If $g(x) = 3.5x^3 + 2x^2 - x$, find $g(6)$.

35. TELEPHONE CALLS The function $C(n) = 0.30n + 15$ gives the cost C per month in dollars for making n phone calls. Find $C(45)$ and explain what it means.

36. Graph: $f(x) = |x| - 1$

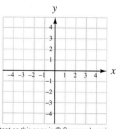

CHAPTERS 1–3 CUMULATIVE REVIEW

1. Give the prime factorization of 108. [Section 1.2]

2. Write $\dfrac{1}{250}$ as a decimal. [Section 1.3]

3. Determine whether each statement is true or false. [Section 1.3]
   a. Every whole number is an integer.
   b. Every integer is a real number.
   c. 0 is a whole number, an integer, and a rational number.

4. Add $-3 + (-15)$ [Section 1.4]

Evaluate each expression. [Section 1.7]

5. $12 - 2 \cdot 3$

6. $\dfrac{(6-5)^4 - (-21)}{-27 + 4^2}$

7. $19 - 2[(-3.1 + 6.1) \cdot 3]$

8. $64 - 6[15 - (3)3]$

9. Evaluate $b^2 - 4ac$ for $a = 2$, $b = -8$, and $c = 4$. [Section 1.8]

10. Suppose x sheets from a 500-sheet ream of paper have been used. Write an algebraic expression to represent the number of sheets that are left. [Section 1.8]

Consider the algebraic expression $3x^2 - 2x + 1$. [Section 1.8]

11. How many terms does the expression have?

12. What is the coefficient of the second term?

Use the distributive property to remove parentheses. [Section 1.9]

13. $2(x + 4)$

14. $2(x - 4)$

15. $-2(x + 4)$

16. $-2(x - 4)$

Simplify each expression. [Section 1.9]

17. $5a + 10 - a$

18. $-2b^2 + 6b^2$

19. $(a + 2) - (a - 2)$

20. $-y - y - y$

Solve each equation. [Sections 2.1 and 2.2]

21. $3x - 5 = 13$

22. $1.2 - x = -1.7$

23. $\dfrac{2x}{3} - 2 = 4$

24. $\dfrac{y - 2}{7} = -3$

25. $-3(2y - 2) - y = 5$

26. $9y - 3 = 6y$

27. $\dfrac{1}{3} + \dfrac{c}{5} = -\dfrac{3}{2}$

28. $5(x + 2) = 5x - 2$

29. PENNIES A recent telephone survey of adults asked whether the penny should be discontinued from the national currency. The results are shown in the circle graph. If 869 people favored keeping the penny, how many took part in the survey? [Section 2.3]

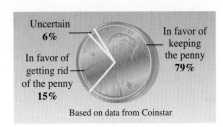

30. Solve the equation $y = mx + b$ for x. [Section 2.4]

31. Find the perimeter and the area of the gauze pad of the bandage shown. [Section 2.4]

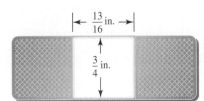

32. If the vertex of an isosceles triangle is 22°, find the measure of each base angle. [Section 2.5]

33. Complete the table. [Section 2.6]

Solution	Liters	% acid	Amount of acid
50% solution	x	0.50	
25% solution	$13 - x$	0.25	
30% mixture	13	0.30	

34. INVESTMENT A student saving for a class trip to Europe invests $2,500 in two accounts. One account pays 3% annual interest and the other, riskier account pays 8%. Find the amount invested at each rate if the first year's combined interest from the two accounts is $105. [Section 2.6]

35. ROAD TRIPS A bus, carrying the members of a marching band, and a truck, carrying their instruments, leave a high school at the same time. The bus travels at 60 mph and the truck at 50 mph. In how many hours will they be 75 miles apart? [Section 2.6]

36. MIXING CANDY Candy corn worth $1.90 per pound is to be mixed with black gumdrops that cost $1.20 per pound to make 200 pounds of a mixture worth $1.48 per pound. How many pounds of each candy should be used? [Section 2.6]

Solve each inequality, graph the solution set, and write the solution in interval notation. [Section 2.7]

37. $-\dfrac{3}{16}x \geq -9$

38. $8x + 4 > 3x + 4$

39. MEDICATION Dosages for a certain medication are shown below. Find the dosage for: [Section 3.1]
  a. a 5-year-old child
  b. a 9-year-old child

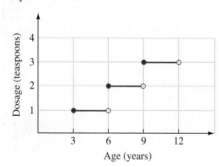

40. Is $(-2, 4)$ a solution of $y = 2x - 8$? [Section 3.2]

Graph each equation. [Section 3.2]

41. $y = |x - 2|$ **42.** $4y + 2x = -8$

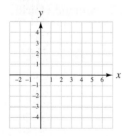 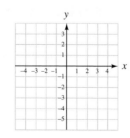

43. ROOFING Find the slope of the roof. [Section 3.4]

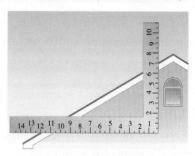

44. What is the slope of the graph of the line $y = 5$? [Section 3.4]

45. What is the slope of the line passing through $(-2, 4)$ and $(5, -6)$? [Section 3.4]

46. Find the slope and the y-intercept of the graph of the line described by $4x - 6y = -12$. [Section 3.5]

47. Write an equation of the line that has slope -2 and y-intercept of $(0, 1)$. [Section 3.5]

48. Write an equation of the line that has slope $-\dfrac{7}{8}$ and passes through $(2, -9)$. Express the answer in point–slope form. [Section 3.6]

Graph each inequality. [Section 3.7]

49. $y \geq x + 1$ **50.** $x < 4$

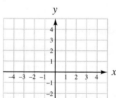 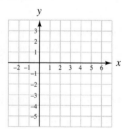

51. If $f(x) = x^2 - 3x$, find $f(-2)$. [Section 3.8]

52. Is this the graph of a function? Explain why or why not. [Section 3.8]

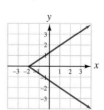

4 Solving Systems of Equations and Inequalities

4.1 Solving Systems of Equations by Graphing

4.2 Solving Systems of Equations by Substitution

4.3 Solving Systems of Equations by Addition (Elimination)

4.4 Problem Solving Using Systems of Equations

4.5 Solving Systems of Linear Inequalities

Chapter Summary and Review

Chapter Test

Cumulative Review

from Campus to Careers
Air Traffic Controller

The air traffic control system is a vast network of people and equipment that ensures the safe operation of commercial and private aircraft. Air traffic controllers coordinate the movement of air traffic to make certain that planes stay a safe distance apart. Their immediate concern is safety, but controllers also must direct planes efficiently to minimize delays. Some regulate airport traffic, and others regulate airport arrivals and departures.

In **Problem 73** of **Study Set 4.1**, you will see one way that air traffic controllers can guard against an air collision.

JOB TITLE: Air Traffic Controller

EDUCATION: Completion of an FAA-approved program, a passing score on an FAA-authorized pre-employment test, and a thorough medical exam are required.

JOB OUTLOOK: There is expected to be a modest decline of 3% in employment from 2010 to 2020.

ANNUAL EARNINGS: In 2012, the median salary was $122,530.

FOR MORE INFORMATION:
www.bls.gov/oes/current/oes532021.htm

Chapter 4 Solving Systems of Equations and Inequalities

Objectives

1. Determine whether a given ordered pair is a solution of a system.
2. Solve systems of linear equations by graphing.
3. Use graphing to identify inconsistent systems and dependent equations.
4. Identify the number of solutions of a linear system without graphing.

SECTION 4.1
Solving Systems of Equations by Graphing

The following illustration shows the average amounts of chicken and beef eaten per person each year in the United States from 1985 to 2013. Plotting both graphs on the same coordinate system makes it easy to compare recent trends. The point of intersection of the graphs indicates that Americans ate equal amounts of chicken and beef in 1992—about 66 pounds of each, per person.

In this section, we will use a similar graphical approach to solve systems of equations.

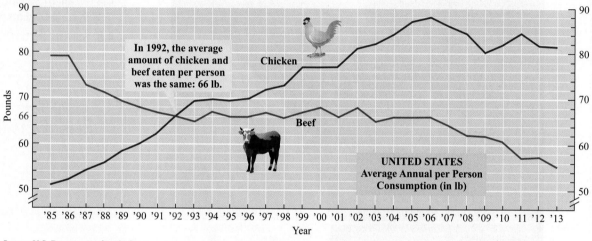

Source: U.S. Department of Agriculture

1 Determine whether a given ordered pair is a solution of a system.

We have previously discussed equations in two variables, such as $x + y = 3$. Because there are infinitely many pairs of numbers whose sum is 3, there are infinitely many pairs (x, y) that satisfy this equation. Some of these pairs are listed in table (a).

Now consider the equation $x - y = 1$. Because there are infinitely many pairs of numbers whose difference is 1, there are infinitely many pairs (x, y) that satisfy $x - y = 1$. Some of these pairs are listed in table (b).

$x + y = 3$		
x	y	(x, y)
0	3	(0, 3)
1	2	(1, 2)
2	1	(2, 1)
3	0	(3, 0)

(a)

$x - y = 1$		
x	y	(x, y)
0	−1	(0, −1)
1	0	(1, 0)
2	1	(2, 1)
3	2	(3, 2)

(b)

From the two tables, we see that (2, 1) satisfies both equations.

Unless otherwise noted, all content on this page is © Cengage Learning.

When two equations with the same variables are considered simultaneously (at the same time), we say that they form a **system of equations**. Using a left brace {, we can write the equations from the previous example as a system:

$$\begin{cases} x + y = 3 \\ x - y = 1 \end{cases}$$ Read as "the system of equations $x + y = 3$ and $x - y = 1$."

Because the ordered pair (2, 1) satisfies both of these equations, it is called a **solution of the system**. In general, a system of linear equations can have exactly one solution, no solution, or infinitely many solutions.

> **The Language of Algebra** We say that (2, 1) *satisfies* $x + y = 3$, because the *x*-coordinate, 2, and the *y*-coordinate, 1, make the equation true when substituted for *x* and *y*: $2 + 1 = 3$. To *satisfy* means to make content, as in *satisfy* your thirst or a *satisfied* customer.

EXAMPLE 1 Determine whether $(-2, 5)$ is a solution of each system of equations.

a. $\begin{cases} 3x + 2y = 4 \\ x - y = -7 \end{cases}$ b. $\begin{cases} 4y = 18 - x \\ y = 2x \end{cases}$

Strategy We will substitute the *x*- and *y*-coordinates of $(-2, 5)$ for the corresponding variables in both equations of the system.

WHY If both equations are satisfied (made true) by the *x*- and *y*-coordinates, then the ordered pair is a solution of the system.

Solution

a. Recall that in an ordered pair, the first number is the *x*-coordinate and the second number is the *y*-coordinate. To determine whether $(-2, 5)$ is a solution, we substitute -2 for *x* and 5 for *y* in each equation.

Check: $3x + 2y = 4$ The first equation. $x - y = -7$ The second equation.
$3(-2) + 2(5) \stackrel{?}{=} 4$ $-2 - 5 \stackrel{?}{=} -7$
$-6 + 10 \stackrel{?}{=} 4$ $-7 = -7$ True
$4 = 4$ True

Since $(-2, 5)$ satisfies both equations, it is a solution of the system.

b. Again, we substitute -2 for *x* and 5 for *y* in each equation.

Check: $4y = 18 - x$ The first equation. $y = 2x$ The second equation.
$4(5) \stackrel{?}{=} 18 - (-2)$ $5 \stackrel{?}{=} 2(-2)$
$20 \stackrel{?}{=} 18 + 2$ $5 = -4$ False
$20 = 20$ True

Although $(-2, 5)$ satisfies the first equation, it does not satisfy the second. Because it does not satisfy both equations, $(-2, 5)$ is not a solution of the system.

Self Check 1
Determine whether $(4, -1)$ is a solution of: $\begin{cases} x - 2y = 6 \\ y = 3x - 11 \end{cases}$

Now Try Problems 21 and 26

> **The Language of Algebra** A system of equations is two (or more) equations that we consider *simultaneously*—at the same time. Some professional sports teams *simulcast* their games. That is, the announcer's play-by-play description is broadcast on radio and television at the same time.

2 Solve systems of linear equations by graphing.

To use the graphing method to solve

$$\begin{cases} x + y = 3 \\ 3x - y = 1 \end{cases}$$

we graph both equations on one set of coordinate axes using the intercept method, as shown below.

$x + y = 3$		
x	y	(x, y)
0	3	(0, 3)
3	0	(3, 0)
2	1	(2, 1)

$3x - y = 1$		
x	y	(x, y)
0	−1	(0, −1)
$\frac{1}{3}$	0	$(\frac{1}{3}, 0)$
2	5	(2, 5)

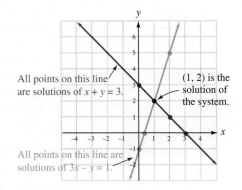

All points on this line are solutions of $x + y = 3$.

(1, 2) is the solution of the system.

All points on this line are solutions of $3x - y = 1$.

Although there are infinitely many pairs (x, y) that satisfy $x + y = 3$, and infinitely many pairs (x, y) that satisfy $3x - y = 1$, only the coordinates of the point where their graphs intersect satisfy both equations simultaneously. Thus, the solution of the system is $(1, 2)$.

To check this result, we substitute 1 for x and 2 for y in each equation and verify that the pair $(1, 2)$ satisfies each equation.

Check: **First equation**
$x + y = 3$
$1 + 2 \stackrel{?}{=} 3$
$3 = 3$ True

Second equation
$3x - y = 1$
$3(1) - 2 \stackrel{?}{=} 1$
$3 - 2 \stackrel{?}{=} 1$
$1 = 1$ True

When the graphs of two equations in a system are different lines, the equations are called **independent equations**. When a system of equations has a solution, the system is called a **consistent system**.

To solve a system of equations in two variables by graphing, we follow these steps.

The Graphing Method

1. Carefully graph each equation on the same rectangular coordinate system.
2. If the lines intersect, determine the coordinates of the point of intersection of the graphs. That ordered pair is the solution of the system.
3. Check the proposed solution in each equation of the original system.

EXAMPLE 2

Solve the system of equations by graphing: $\begin{cases} 2x + 3y = 2 \\ 3x = 2y + 16 \end{cases}$

Strategy We will graph both equations on the same coordinate system.

WHY Recall that the graph of a linear equation is a "picture" of its solutions. If both equations are graphed on the same coordinate system, we can see whether they have any common solutions.

Solution
Using the intercept method, we graph both equations on one set of coordinate axes.

Although there are infinitely many pairs (x, y) that satisfy $2x + 3y = 2$, and infinitely many pairs (x, y) that satisfy $3x = 2y + 16$, only the coordinates of the point where the graphs intersect satisfy both equations at the same time. From the graph, the solution appears to be $(4, -2)$.

$2x + 3y = 2$		
x	y	(x, y)
0	$\frac{2}{3}$	$(0, \frac{2}{3})$
1	0	$(1, 0)$
-2	2	$(-2, 2)$

$3x = 2y + 16$		
x	y	(x, y)
0	-8	$(0, -8)$
$\frac{16}{3}$	0	$(\frac{16}{3}, 0)$
2	-5	$(2, -5)$

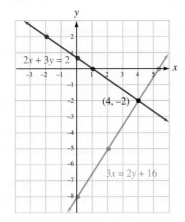

Self Check 2

Solve the system of equations by graphing: $\begin{cases} 2x = y - 5 \\ x + y = -1 \end{cases}$

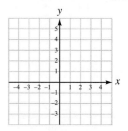

Now Try Problem 33

To check, we substitute 4 for x and -2 for y in each equation and verify that the pair $(4, -2)$ satisfies each equation.

Check:
$2x + 3y = 2$ *This is the first equation.*
$2(4) + 3(-2) \stackrel{?}{=} 2$
$8 - 6 \stackrel{?}{=} 2$
$2 = 2$ *True*

$3x = 2y + 16$ *This is the second equation.*
$3(4) \stackrel{?}{=} 2(-2) + 16$
$12 \stackrel{?}{=} -4 + 16$
$12 = 12$ *True*

The equations in this system are independent equations, and the system is a consistent system of equations.

EXAMPLE 3

Solve the system of equations by graphing: $\begin{cases} -\dfrac{x}{2} - 1 = \dfrac{y}{2} \\ \dfrac{1}{3}x - \dfrac{1}{2}y = -4 \end{cases}$

Strategy We will multiply both sides of each equation by a value that will remove the fractions. Then we will graph both equations of the equivalent system on the same coordinate plane.

Self Check 3

Solve the system of equations by graphing: $\begin{cases} -\dfrac{x}{2} = \dfrac{y}{4} \\ \dfrac{1}{4}x - \dfrac{3}{8}y = -2 \end{cases}$

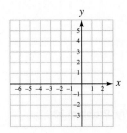

Now Try Problem 39

WHY Recall that the graph of a linear equation is a "picture" of its solutions. It is easier to graph equations that have integer coefficients. If both equations are graphed on the same coordinate system, we can see whether they have any common solutions.

Solution

We can multiply both sides of the first equation by 2 to clear it of fractions.

$$-\frac{x}{2} - 1 = \frac{y}{2}$$

$$2\left(-\frac{x}{2} - 1\right) = 2\left(\frac{y}{2}\right)$$

(1) $\qquad -x - 2 = y \qquad$ We will label this Equation 1.

We then multiply both sides of the second equation by 6 to clear it of fractions.

$$\frac{1}{3}x - \frac{1}{2}y = -4$$

$$6\left(\frac{1}{3}x - \frac{1}{2}y\right) = 6(-4)$$

(2) $\qquad 2x - 3y = -24 \qquad$ We will label this Equation 2.

Equations 1 and 2 form the following **equivalent system**, which has the same solutions as the original system:

$$\begin{cases} -x - 2 = y \\ 2x - 3y = -24 \end{cases}$$

We graph $-x - 2 = y$ by plotting the y-intercept $(0, -2)$ and then drawing a slope of -1. We graph $2x - 3y = -24$ using the intercept method. It appears that the point of intersection is $(-6, 4)$.

A check will show that when the coordinates of $(-6, 4)$ are substituted into the two original equations, true statements result. Therefore, the equations are independent and the system is consistent.

$y = -x - 2$

So $m = -1 = \dfrac{-1}{1}$ and $b = -2$.

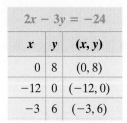

$2x - 3y = -24$		
x	y	(x, y)
0	8	$(0, 8)$
-12	0	$(-12, 0)$
-3	6	$(-3, 6)$

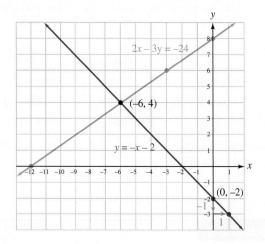

Caution! When solving a system of equations, always check your answer by substituting into the *original* equations. Do not check by substituting into the equations of an equivalent system. If an algebraic error was made while finding the equivalent system, an answer that would not satisfy the original system might appear to be correct.

THINK IT THROUGH Bridging the Gender Gap in Education

"The woman of the 20th century will be the peer of man. In education, in art and science, in literature, in the home, the church, the state, everywhere she will be his acknowledged equal."

Susan B. Anthony, 1900

The following graph shows the percent of associate degrees awarded in the U.S. by gender for the years 1970–2012. Determine the point of intersection of the lines and explain its importance.

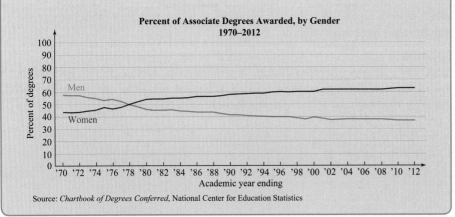

Percent of Associate Degrees Awarded, by Gender 1970–2012

Source: *Chartbook of Degrees Conferred*, National Center for Education Statistics

Using Your CALCULATOR   Solving Systems with a Graphing Calculator

We can use a graphing calculator to solve a system of equations, such as

$$\begin{cases} 2x + y = 12 \\ 2x - y = -2 \end{cases}$$

However, before we can enter the equations into the calculator, we must solve them for y.

$$2x + y = 12 \qquad\qquad 2x - y = -2$$
$$y = -2x + 12 \qquad\quad -y = -2x - 2$$
$$\qquad\qquad\qquad\qquad\quad y = 2x + 2$$

We enter the resulting equations and graph them on the same coordinate axes. If we use the standard window settings, their graphs will look like figure (a) below.

To find the solution of the system, we use the INTERSECT feature that is found on most graphing calculators. With this option, the cursor automatically moves to the point of intersection of the graphs and displays the coordinates of that point. In figure (b), we see that the solution is (2.5, 7). Consult your owner's manual for specific keystrokes to use INTERSECT.

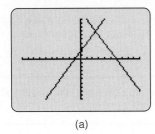
(a)

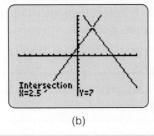
(b)

3 Use graphing to identify inconsistent systems and dependent equations.

Sometimes a system of equations has no solution. Such systems are called **inconsistent systems**.

Self Check 4

Solve the system of equations by graphing: $\begin{cases} y = \dfrac{3}{2}x \\ 3x - 2y = 6 \end{cases}$

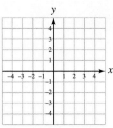

Now Try Problem 41

EXAMPLE 4 Solve the system of equations by graphing: $\begin{cases} y = -2x - 6 \\ 4x + 2y = 8 \end{cases}$

Strategy We will graph both equations on the same coordinate system.

WHY If both equations are graphed on the same coordinate system, we can see whether they have any common solutions.

Solution
Since $y = -2x - 6$ is written in slope–intercept form, we can graph it by plotting the y-intercept $(0, -6)$ and then drawing a slope of -2. (The rise is -2, and the run is 1.) We graph $4x + 2y = 8$ using the intercept method.

$y = -2x - 6$

So, $m = -2 = \dfrac{-2}{1}$

and

$b = -6$.

$4x + 2y = 8$		
x	y	(x, y)
0	4	$(0, 4)$
2	0	$(2, 0)$
1	2	$(1, 2)$

The system is graphed below. Since the lines in the figure are parallel, they have the same slope. We can verify this by writing the second equation in slope–intercept form and observing that the coefficients of x in each equation are equal and the y-intercepts are different, $(0, -6)$ and $(0, 4)$.

$y = -2x - 6 \qquad 4x + 2y = 8$
$\qquad\qquad\qquad\qquad 2y = -4x + 8$
$\qquad\qquad\qquad\qquad\; y = -2x + 4$

Different y-intercepts
Same slope

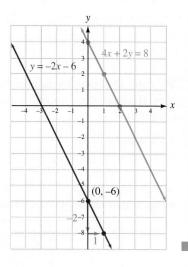

Because parallel lines do not intersect, this system has no solution and is inconsistent. Since the graphs are different lines, the equations of the system are independent.

Caution! A common error is to graph the parallel lines but forget to answer with the words *no solution*.

Some systems of equations have infinitely many solutions, as we will see in the next example.

EXAMPLE 5

Solve the system of equations by graphing: $\begin{cases} y - 4 = 2x \\ 4x + 8 = 2y \end{cases}$

Strategy We will graph both equations on the same coordinate system.

WHY If both equations are graphed on the same coordinate system, we can see whether they have any common solutions.

Solution
We graph both equations using the intercept method.

$y - 4 = 2x$		
x	y	(x, y)
0	4	(0, 4)
−2	0	(−2, 0)
−1	2	(−1, 2)

$4x + 8 = 2y$		
x	y	(x, y)
0	4	(0, 4)
−2	0	(−2, 0)
−3	−2	(−3, −2)

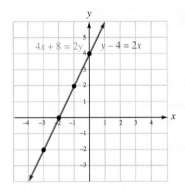

The lines in the figure coincide (they are the same line). Because the lines intersect at infinitely many points, the system has infinitely many solutions. From the graph, we can see that some of the solutions are (0, 4), (−1, 2), and (−3, −2). Equations that have the same graph are called **dependent equations**. Therefore, this system is consistent and its equations are dependent.

Self Check 5

Solve the system of equations by graphing: $\begin{cases} 6x - 2y = 4 \\ y + 2 = 3x \end{cases}$

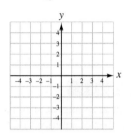

Now Try Problem 43

There are three possible outcomes when we solve a system of two linear equations using the graphing method.

Consistent system Independent equations	Inconsistent system Independent equations	Consistent system Dependent equations
The two lines intersect at one point.	The two lines are parallel.	The two lines are identical.

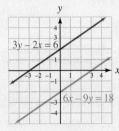
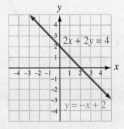

Exactly one solution (the point of intersection) — No solution — Infinitely many solutions (any point on the line is a solution)

4 Identify the number of solutions of a linear system without graphing.

We can determine the number of solutions that a system of two linear equations has by writing each equation in slope–intercept form.

- If the lines have different slopes, they intersect, and the system has one solution. (See Example 2.)
- If the lines have the same slope and different y-intercepts, they are parallel, and the system has no solution. (See Example 4.)
- If the lines have the same slope and the same y-intercept, they are the same line, and the system has infinitely many solutions. (See Example 5.)

Self Check 6

Without graphing, determine the number of solutions of:
$$\begin{cases} 3x + 6y = 1 \\ 2x + 4y = 0 \end{cases}$$

Now Try Problems 45 and 48

EXAMPLE 6

Without graphing, determine the number of solutions of:
$$\begin{cases} 5x + y = 5 \\ 3x + 2y = 8 \end{cases}$$

Strategy We will write both equations in slope–intercept form.

WHY We can determine the number of solutions of a linear system by comparing the slopes and y-intercepts of the graphs of the equations.

Solution

To write each equation in slope–intercept form, we solve for y.

$5x + y = 5$ The first equation. $\qquad$ $3x + 2y = 8$ The second equation.
$y = -5x + 5$ $\qquad\qquad\qquad\qquad\qquad\quad$ $2y = -3x + 8$
$\qquad\qquad\qquad\qquad\qquad\qquad\qquad\qquad\quad$ $y = -\dfrac{3}{2}x + 4$

————— Different slopes —————

Since the slopes are different, the lines are neither parallel nor identical. Therefore, they will intersect at one point and the system has one solution.

ANSWERS TO SELF CHECKS

1. no **2.** $(-2, 1)$ 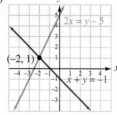 **3.** $(-2, 4)$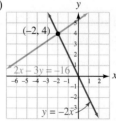

4. no solution 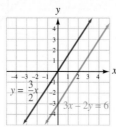 **5.** infinitely many solutions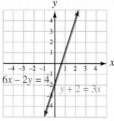

6. no solution

SECTION 4.1 STUDY SET

VOCABULARY

Fill in the blanks.

1. The pair of equations $\begin{cases} x - y = -1 \\ 2x - y = 1 \end{cases}$ is called a _____ of equations.

2. Because the ordered pair (2, 3) satisfies both equations in Problem 1, it is called a _____ of the system of equations.

3. When the graphs of two equations in a system are different lines, the equations are called _____ equations.

4. When a system of equations has a solution, the system is called a _____ system.

5. Systems of equations that have no solution are called _____ systems.

6. Equations that have the same graph are called _____ equations.

CONCEPTS

Refer to the following illustration. Determine whether a true or false statement would result if the coordinates of each point were substituted into the equation for the indicated line.

7. point A, line l_1
8. point B, line l_2
9. point A, line l_2
10. point B, line l_1
11. point C, line l_1
12. point C, line l_2

13. The following tables were created to graph the two linear equations in a system. What is the solution of the system?

Equation 1

x	y
0	−5
−5	0
−4	−1
1	−6
2	−7

Equation 2

x	y
0	3
−3	0
−2	1
−4	−1
1	4

14. **a.** To graph $5x - 2y = 10$, we can use the intercept method. Complete the table.

x	y
0	
	0

b. To graph $y = 3x - 2$, we can use the slope and y-intercept. Fill in the blanks.

slope: ▢ = ▢/1   y-intercept: ▢

15. How many solutions does the system of equations graphed in the illustration have? Is the system consistent or inconsistent?

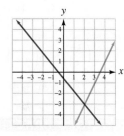

16. How many solutions does the system of equations graphed in the illustration have? Are the equations dependent or independent?

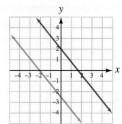

17. The solution of the system of equations graphed on the right is $\left(\frac{2}{5}, -\frac{1}{3}\right)$. Knowing this, can you see any disadvantages to the graphing method?

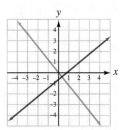

18. Draw the graphs of two linear equations so that the system has

a. one solution (−3, −2).   **b.** infinitely many solutions, three of which are (−2, 0), (1, 2), and (4, 4).

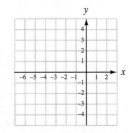
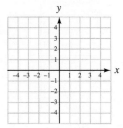

NOTATION

Fill in the blanks to clear each equation of fractions.

19. $\dfrac{1}{6}x - \dfrac{1}{3}y = \dfrac{11}{2}$

   $\boxed{}\left(\dfrac{1}{6}x - \dfrac{1}{3}y\right) = \boxed{}\left(\dfrac{11}{2}\right)$

   $\boxed{}\left(\dfrac{1}{6}x\right) - 6\left(\boxed{}\right) = 6\left(\dfrac{11}{2}\right)$

   $x - 2y = 33$

20. $\dfrac{3x}{5} - \dfrac{4y}{5} = -1$

   $\boxed{}\left(\dfrac{3x}{5} - \dfrac{4y}{5}\right) = \boxed{}(-1)$

   $\boxed{}\left(\dfrac{3x}{5}\right) - 5\left(\boxed{}\right) = 5(-1)$

   $3x - 4y = -5$

GUIDED PRACTICE

Use a check to determine whether the ordered pair is a solution of the given system of equations. See Example 1.

21. $(1, 1)$, $\begin{cases} x + y = 2 \\ 2x - y = 1 \end{cases}$

22. $(1, 3)$, $\begin{cases} 2x + y = 5 \\ 3x - y = 0 \end{cases}$

23. $(3, -2)$, $\begin{cases} 2x + y = 4 \\ y = 1 - x \end{cases}$

24. $(-2, 4)$, $\begin{cases} 2x + 2y = 4 \\ 3y = 10 - x \end{cases}$

25. $(-2, -4)$, $\begin{cases} 4x + 5y = -23 \\ -3x + 2y = 0 \end{cases}$

26. $(-5, 2)$, $\begin{cases} -2x + 7y = 17 \\ 3x - 4y = -19 \end{cases}$

27. $\left(\dfrac{1}{2}, 3\right)$, $\begin{cases} 2x + y = 4 \\ 4x - 11 = 3y \end{cases}$

28. $\left(2, \dfrac{1}{3}\right)$, $\begin{cases} x - 3y = 1 \\ -2x + 6 = -6y \end{cases}$

Solve each system of equations by graphing. See Example 2.

29. $\begin{cases} 2x + 3y = 12 \\ 2x - y = 4 \end{cases}$

30. $\begin{cases} 5x + y = 5 \\ 5x + 3y = 15 \end{cases}$

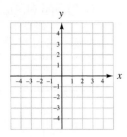
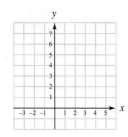

31. $\begin{cases} 3x + 2y = -8 \\ 2x - 3y = -1 \end{cases}$

32. $\begin{cases} x + 4y = -2 \\ y = -x - 5 \end{cases}$

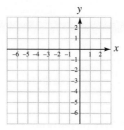
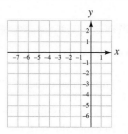

33. $\begin{cases} x + y = 2 \\ y = x - 4 \end{cases}$

34. $\begin{cases} x + y = 1 \\ y = x + 5 \end{cases}$

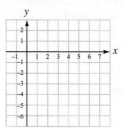
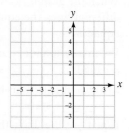

35. $\begin{cases} x = 4 \\ 2y = 12 - 4x \end{cases}$

36. $\begin{cases} x = 3 \\ 3y = 6 - 2x \end{cases}$

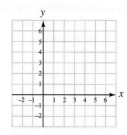
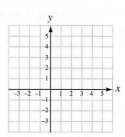

Solve each system of the equations by graphing. See Example 3.

37. $\begin{cases} x + 2y = -4 \\ x - \dfrac{1}{2}y = 6 \end{cases}$

38. $\begin{cases} \dfrac{2}{3}x - y = -3 \\ 3x + y = 3 \end{cases}$

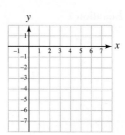
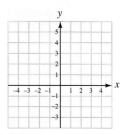

39. $\begin{cases} -\dfrac{3}{4}x + y = 3 \\ \dfrac{1}{4}x + y = -1 \end{cases}$

40. $\begin{cases} \dfrac{1}{3}x + y = 7 \\ \dfrac{2x}{3} - y = -4 \end{cases}$

49. $\begin{cases} x + y = 6 \\ x + y = 8 \end{cases}$

50. $\begin{cases} 5x + y = 0 \\ 5x + y = 6 \end{cases}$

51. $\begin{cases} 6x + y = 0 \\ 2x + 2y = 0 \end{cases}$

52. $\begin{cases} x + y = 1 \\ 2x - 2y = 5 \end{cases}$

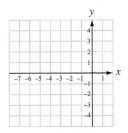
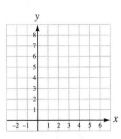

TRY IT YOURSELF

Use a check to determine whether the ordered pair is a solution of the given system of equations.

53. $\left(-\dfrac{2}{5}, \dfrac{1}{4}\right)$, $\begin{cases} x - 4y = -6 \\ 8y = 10x + 12 \end{cases}$

54. $\left(-\dfrac{1}{3}, \dfrac{3}{4}\right)$, $\begin{cases} 3x + 4y = 2 \\ 12y = 3(2 - 3x) \end{cases}$

55. $(0.2, 0.3)$, $\begin{cases} 20x + 10y = 7 \\ 20y = 15x + 3 \end{cases}$

56. $(2.5, 3.5)$, $\begin{cases} 4x - 3 = 2y \\ 4y + 1 = 6x \end{cases}$

Solve each system of equations by graphing. If the equations of the system are dependent or if the system is inconsistent, so indicate. See Examples 4–5.

41. $\begin{cases} y = -\dfrac{1}{3}x - 4 \\ x + 3y = 6 \end{cases}$

42. $\begin{cases} y = -x + 1 \\ 4x + 4y = 4 \end{cases}$

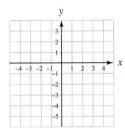
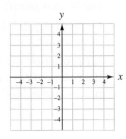

Solve each system of equations by graphing. If the equations of the system are dependent or if the system is inconsistent, so indicate.

57. $\begin{cases} 2x - 3y = -18 \\ 3x + 2y = -1 \end{cases}$

58. $\begin{cases} -x + 3y = -11 \\ 3x - y = 17 \end{cases}$

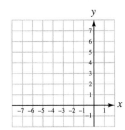
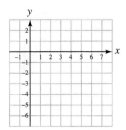

43. $\begin{cases} y = x - 1 \\ 3x - 3y = 3 \end{cases}$

44. $\begin{cases} y = -\dfrac{1}{2}x - 3 \\ x + 2y = 2 \end{cases}$

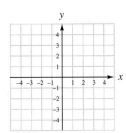
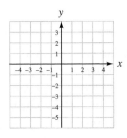

59. $\begin{cases} \dfrac{1}{3}x - \dfrac{1}{2}y = \dfrac{1}{6} \\ \dfrac{2x}{5} + \dfrac{y}{2} = \dfrac{13}{10} \end{cases}$

60. $\begin{cases} \dfrac{3x}{4} + \dfrac{2y}{3} = -\dfrac{19}{6} \\ 3y = -x \end{cases}$

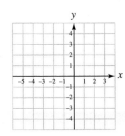
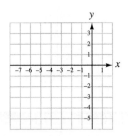

Find the slope and y-intercept of the graph of lines in each system of equations. Then use that information to determine the number of solutions of each system. See Example 6.

45. $\begin{cases} y = 6x - 7 \\ y = -2x + 1 \end{cases}$

46. $\begin{cases} y = \dfrac{1}{2}x + 8 \\ y = 4x - 10 \end{cases}$

47. $\begin{cases} 3x - y = -3 \\ y - 3x = 3 \end{cases}$

48. $\begin{cases} x + 4y = 4 \\ 12y = 12 - 3x \end{cases}$

61. $\begin{cases} 3x - 6y = 18 \\ x = 2y + 3 \end{cases}$
62. $\begin{cases} 2y = 3x + 2 \\ \dfrac{3}{2}x - y = 3 \end{cases}$

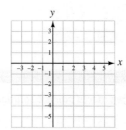

63. $\begin{cases} 4x - 2y = 8 \\ y = 2x - 4 \end{cases}$
64. $\begin{cases} -\dfrac{3}{5}x - \dfrac{1}{5}y = \dfrac{6}{5} \\ x + \dfrac{y}{3} = -2 \end{cases}$

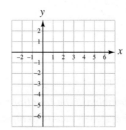

65. $\begin{cases} y = 4 - x \\ y = 2 + x \end{cases}$
66. $\begin{cases} 3x - y = 4 \\ 2x + y = 1 \end{cases}$

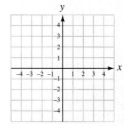

67. $\begin{cases} 6x - 2y = 5 \\ 3x = y + 10 \end{cases}$
68. $\begin{cases} x - 3y = -2 \\ 5x + y = 6 \end{cases}$

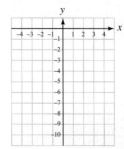

APPLICATIONS

69. **TORNADOES** The illustration in the next column shows the paths of two devastating tornadoes that swept through Moore, Oklahoma—one in 1999 and the other in 2013. Identify the location of the intersection of the tornadoes' paths as an ordered pair of the form (street, avenue).

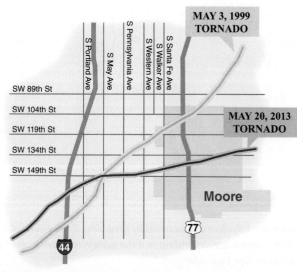

70. **DAILY TRACKING POLLS** Use the graph below to answer the following.
    a. Which political candidate was ahead on October 28, and by how much?
    b. On what day did the challenger pull even with the incumbent?
    c. If the election was held November 4, who did the poll predict would win, and by how many percentage points?

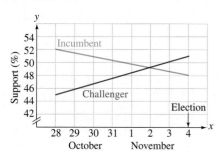

71. **LATITUDE AND LONGITUDE** Refer to the map below.
    a. Name three American cities that lie on a latitude line of 30° north.
    b. Name three American cities that lie on a longitude line of 90° west.
    c. What city lies on both lines?

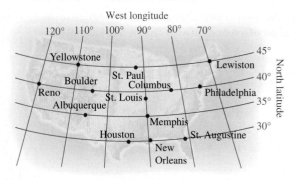

Unless otherwise noted, all content on this page is © Cengage Learning.

72. ECONOMICS The following graph illustrates the law of supply and demand.

  a. Complete this sentence: As the price of an item increases, the *supply* of the item _____.

  b. Complete this sentence: As the price of an item increases, the *demand* for the item _____.

  c. At what price will the supply equal the demand? How many items will be supplied at this price?

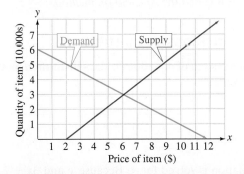

73. The equations describing the paths of two airplanes are $y = -\frac{1}{2}x + 3$ and $3y = 2x + 2$. Graph each equation on the radar screen shown. Is there a possibility of a midair collision? If so, where?

from Campus to Careers
Air Traffic Controller

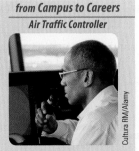

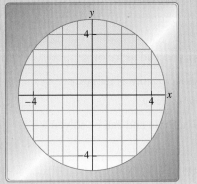

74. TV COVERAGE A television camera is located at $(-2, 0)$ and will follow the launch of a rocket, as shown. (Each unit in the illustration is 1 mile.) As the rocket rises vertically on a path described by $x = 2$, the farthest the camera can tilt back is a line of sight given by $y = \frac{5}{2}x + 5$. For how many miles of the rocket's flight will it be in view of the camera?

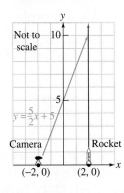

75. DEMOGRAPHICS The illustration below compares the projected median population age for the United States and China for the years 2015 through 2050. Identify the point of intersection of the graphs as an ordered pair of the form (year, age).

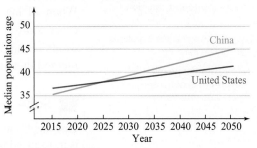

Source: United Nations Population Division

76. BEVERAGES Refer to the graph below. In what year was the average number of gallons of milk and carbonated soft drinks consumed per person the same? Estimate the number of gallons.

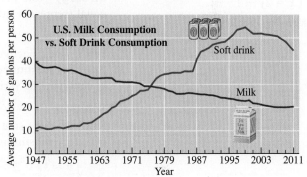

Source: USDA, Economic Research Service

WRITING

77. Look up the word *simultaneous* in a dictionary and give its definition. In mathematics, what is meant by a simultaneous solution of a system of equations?

78. Suppose the solution of a system is $\left(\frac{1}{3}, -\frac{3}{5}\right)$. Do you think you would be able to find the solution using the graphing method? Explain.

REVIEW

79. Are the graphs of $y = 2x - 3$ and $4x - 2y = 8$ parallel, perpendicular, or neither?

80. Are the graphs of the lines $y = 5x$ and $y = -\frac{1}{5}x$ parallel, perpendicular, or neither?

81. If $f(x) = -4x - x^2$, find $f(3)$.

82. In what quadrant does $(-12, 15)$ lie?

83. Write the equation for the *y*-axis.

84. Does $(1, 2)$ lie on the line $2x + y = 4$?

85. What point does the line with equation $y - 2 = 7(x - 5)$ pass through?

86. Is the word *domain* associated with the inputs or the outputs of a function?

Objectives

1. Solve systems of linear equations by substitution.
2. Find a substitution equation.
3. Solve systems of linear equations that contain fractions.
4. Use substitution to identify inconsistent systems and dependent equations.

SECTION 4.2
Solving Systems of Equations by Substitution

When solving a system of equations by graphing, it is often difficult to determine the exact coordinates of the point of intersection. For example, a solution of $\left(\frac{1}{16}, -\frac{3}{5}\right)$ would be almost impossible to identify. In this section, we will introduce an algebraic method that finds *exact* solutions. It is called the *substitution method*.

1 Solve systems of linear equations by substitution.

One algebraic method for solving a system of equations is the **substitution method**. It is introduced in the following example.

Self Check 1

Solve the system:
$$\begin{cases} 3x - 5y = 5 \\ x = 2y - 1 \end{cases}$$

Now Try Problem 17

EXAMPLE 1 Solve the system: $\begin{cases} y = 3x - 2 \\ 2x + y = 8 \end{cases}$

Strategy Note that the first equation is solved for y. Because y and $3x - 2$ are equal (represent the same value), we will substitute $3x - 2$ for y in the second equation.

WHY The objective is to obtain one equation containing only one unknown. When $3x - 2$ is substituted for y in the second equation, the result will be just that—an equation in one variable, x.

Solution
Since the right side of $y = 3x - 2$ is used to make a substitution, $y = 3x - 2$ is called the **substitution equation**.

$$\begin{cases} y = \boxed{3x - 2} \\ 2x + y = 8 \end{cases}$$

To find the solution of the system, we proceed as follows:

$2x + y = 8$ This is the second equation of the system.
$2x + 3x - 2 = 8$ Substitute $3x - 2$ for y.

The resulting equation has only one variable and can be solved for x.

$2x + 3x - 2 = 8$
$5x - 2 = 8$ Combine like terms: $2x + 3x = 5x$.
$5x = 10$ Add 2 to both sides.
$x = 2$ Divide both sides by 5. This is the x-value of the solution.

We can find the y-value by substituting 2 for x in either equation of the original system. Because $y = 3x - 2$ is already solved for y, it is easier to substitute into this equation.

$y = 3x - 2$ This is the first equation of the system.
$ = 3(2) - 2$ Substitute 2 for x.
$ = 6 - 2$
$ = 4$ This is the y-value of the solution. We would have obtained the same result if we had substituted 2 for x in $2x + y = 8$ and solved for y.

The solution to the given system is $(2, 4)$. To check, we substitute 2 for x and 4 for y in each equation.

Check:

First equation	Second equation
$y = 3x - 2$	$2x + y = 8$
$4 \stackrel{?}{=} 3(2) - 2$	$2(2) + 4 \stackrel{?}{=} 8$
$4 \stackrel{?}{=} 6 - 2$	$4 + 4 \stackrel{?}{=} 8$
$4 = 4$ True	$8 = 8$ True

If we graphed the lines represented by the equations of the given system, they would intersect at the point (2, 4). The equations of this system are independent, and the system is consistent.

> **Caution!** When using the substitution method, a common error is to find the value of one of the variables, say x, and forget to find the value of the other. Remember that a solution of a linear system of two equations is an ordered pair (x, y).

To solve a system of equations in x and y by the substitution method, we follow these steps.

The Substitution Method

1. Solve one of the equations for either x or y. If this is already done, go to step 2. (We call this equation the **substitution equation**.)
2. Substitute the expression for x or for y obtained in step 1 into the other equation, and solve that equation.
3. Substitute the value of the variable found in step 2 into the substitution equation to find the value of the remaining variable.
4. Check the proposed solution in each equation of the original system. Write the solution as an ordered pair.

EXAMPLE 2   Solve the system: $\begin{cases} 2x + y = -10 \\ x = -3y \end{cases}$

Strategy We will use the substitution method to solve this system.

WHY The substitution method works well when one of the equations of the system (in this case, $x = -3y$) is solved for a variable.

Solution

Step 1 The second equation, $x = -3y$, tells us that x and $-3y$ have the same value. Therefore, we can substitute $-3y$ for x in the first equation.

$$\begin{cases} 2x + y = -10 \\ x = -3y \end{cases}$$ This is the substitution equation.

Step 2 When we substitute $-3y$ for x in the first equation, the resulting equation contains only one variable, and it can be solved for y.

$2x + y = -10$	This is the first equation of the system.
$2(-3y) + y = -10$	Replace x with $-3y$.
$-6y + y = -10$	Do the multiplication.
$-5y = -10$	Combine like terms.
$y = 2$	Divide both sides by -5.

Self Check 2

Solve the system:
$\begin{cases} x = -2y \\ 3x - 2y = 8 \end{cases}$

Now Try Problem 21

Step 3 To find x, substitute 2 for y in the equation $x = -3y$.

$x = -3y$ This is the second equation of the system.
$x = -3(2)$ Substitute 2 for y.
$x = -6$

The solution is $(-6, 2)$.

Step 4 Verify the solution by checking it in both equations.

Check: **First equation**   **Second equation**

$2x + y = -10$ $x = -3y$
$2(-6) + 2 \stackrel{?}{=} -10$ $-6 \stackrel{?}{=} -3(2)$
$-12 + 2 \stackrel{?}{=} -10$ $-6 = -6$ True
$-10 = -10$ True

2 Find a substitution equation.

To find a substitution equation, solve one of the equations of the system for one of its variables. If possible, solve for a variable whose coefficient is 1 or -1 to avoid working with fractions.

Self Check 3

Solve the system:
$\begin{cases} 2x - 3y = 13 \\ 3x + y = 3 \end{cases}$

Now Try Problem 27

EXAMPLE 3   Solve the system: $\begin{cases} 2x + y = -5 \\ 3x + 5y = -4 \end{cases}$

Strategy Since the system does not contain an equation solved for x or y, we must choose an equation and solve it for x or y. We will solve for y in the first equation, because y has a coefficient of 1 in that equation. Then we will use the substitution method to solve the system.

WHY Solving $2x + y = -5$ for x or $3x + 5y = -4$ for x or y would involve working with cumbersome fractions.

Solution

Step 1 To find a substitution equation, solve the first equation for y.

$2x + y = -5$ This is the first equation of the system.
$y = -5 - 2x$ Subtract 2x from both sides to isolate y. This is the substitution equation.

Step 2 We then substitute $-5 - 2x$ for y in the second equation and solve for x.

$3x + 5y = -4$ This is the second equation of the system.
$3x + 5(-5 - 2x) = -4$ Substitute $-5 - 2x$ for y. Don't forget to write $-5 - 2x$ within parentheses.
$3x - 25 - 10x = -4$ Distribute the multiplication by 5.
$-7x - 25 = -4$ Combine like terms: $3x - 10x = -7x$.
$-7x = 21$ Add 25 to both sides.
$x = -3$ Divide both sides by -7.

Step 3 To find y, substitute -3 for x in the equation $y = -5 - 2x$.

$y = -5 - 2x$
$y = -5 - 2(-3)$ Substitute -3 for x.
$y = -5 + 6$
$y = 1$

Step 4 The solution is $(-3, 1)$. Check it in the original equations.

Systems of equations are sometimes written in variables other than x and y. For example, the system

$$\begin{cases} 3a - 3b = 5 \\ 3 - a = -2b \end{cases}$$

is written in a and b. Regardless of the variables used, the procedures used to solve the system remain the same. Unless told otherwise, list the values of the variables of a solution in alphabetical order. Here, the solution should be expressed in the form (a, b).

EXAMPLE 4 Solve the system: $\begin{cases} 3a - 3b = 5 \\ 3 - a = -2b \end{cases}$

Strategy Since the coefficient of a in the second equation is -1, we will solve that equation for a. Then we will use the substitution method to solve the system.

WHY If we solve for the variable with a numerical coefficient of -1 or 1, we can avoid having to work with fractions.

Solution

Step 1 Solve the second equation for a.

$3 - a = -2b$ This is the second equation of the system.
$-a = -2b - 3$ Subtract 3 from both sides.

To obtain a on the left-hand side, we can multiply (or divide) both sides of the equation by -1.

$-1(-a) = -1(-2b - 3)$ Multiply both sides by -1.
$a = 2b + 3$ Do the multiplications. This is the substitution equation.

Step 2 We then substitute $2b + 3$ for a in the first equation and proceed as follows:

$3a - 3b = 5$ This is the first equation of the system.
$3(2b + 3) - 3b = 5$ Substitute: $2b + 3$ for a. Don't forget the parentheses.
$6b + 9 - 3b = 5$ Distribute the multiplication by 3.
$3b + 9 = 5$ Combine like terms: $6b - 3b = 3b$.
$3b = -4$ Subtract 9 from both sides: $5 - 9 = -4$.
$b = -\dfrac{4}{3}$ Divide both sides by 3.

Step 3 To find a, we substitute $-\dfrac{4}{3}$ for b in $a = 2b + 3$ and simplify.

$a = 2b + 3$

$a = 2\left(-\dfrac{4}{3}\right) + 3$ Substitute: $-\dfrac{4}{3}$ for b.

$a = -\dfrac{8}{3} + \dfrac{9}{3}$ Do the multiplication: $2\left(-\dfrac{4}{3}\right) = -\dfrac{8}{3}$. Write 3 as $\dfrac{9}{3}$.

$a = \dfrac{1}{3}$ Add. This is the a-value of the solution.

Step 4 The solution is $\left(\dfrac{1}{3}, -\dfrac{4}{3}\right)$. Check it in the original equations.

Self Check 4

Solve the system: $\begin{cases} 2s - t = 4 \\ 3s - 5t = 2 \end{cases}$

Now Try Problem 29

3 Solve systems of linear equations that contain fractions.

It is usually helpful to clear any equations of fractions and combine any like terms before performing a substitution.

Self Check 5

Solve the system:
$$\begin{cases} \frac{1}{3}x - \frac{1}{6}y = -\frac{1}{3} \\ x + y = -3 - 2x - y \end{cases}$$

Now Try Problem 31

EXAMPLE 5

Solve the system: $\begin{cases} \frac{x}{2} + \frac{y}{4} = -\frac{1}{4} \\ 2x - y = 2 + y - x \end{cases}$

Strategy We will use properties of algebra to write each equation of the system in simpler form. Then we will use the substitution method to solve the resulting equivalent system.

WHY The first equation will be easier to work with if we clear it of fractions. The second equation will be easier to work with if we eliminate the variable terms on the right side.

Solution

We begin by clearing the first equation of fractions by multiplying both sides by the LCD.

$$\frac{x}{2} + \frac{y}{4} = -\frac{1}{4}$$

$$4\left(\frac{x}{2} + \frac{y}{4}\right) = 4\left(-\frac{1}{4}\right) \quad \text{Multiply both sides by the LCD, which is 4.}$$

$$2x + y = -1$$

We can write the second equation in standard form ($Ax + By = C$) by adding x and subtracting y from both sides.

$$2x - y = 2 + y - x$$
$$2x - y + x - y = 2 + y - x + x - y$$
$$3x - 2y = 2 \quad \text{Combine like terms.}$$

The two results form the following equivalent system, which has the same solution as the original one.

(1) $\begin{cases} 2x + y = -1 \\ 3x - 2y = 2 \end{cases}$
(2)

Step 1 To solve this system, we solve Equation 1 for y.

$$2x + y = -1$$
$$2x + y - 2x = -1 - 2x \quad \text{Subtract 2x from both sides.}$$
(3) $\qquad y = -1 - 2x \quad \text{Combine like terms. This is the substitution equation.}$

Step 2 To find x, we substitute $-1 - 2x$ for y in Equation 2 and proceed as follows:

$$3x - 2y = 2$$
$$3x - 2(-1 - 2x) = 2 \quad \text{Substitute.}$$
$$3x + 2 + 4x = 2 \quad \text{Distribute the multiplication by } -2.$$
$$7x + 2 = 2 \quad \text{Combine like terms.}$$
$$7x = 0 \quad \text{Subtract 2 from both sides.}$$
$$x = 0 \quad \text{Divide both sides by 7.}$$

Step 3 To find y, we substitute 0 for x in Equation 3.

$$y = -1 - 2x$$
$$y = -1 - 2(0)$$
$$y = -1$$

Step 4 The solution is $(0, -1)$. Check it in the original equations.

4 Use substitution to identify inconsistent systems and dependent equations.

In the previous section, we solved inconsistent systems and systems of dependent equations graphically. We can also solve these systems using the substitution method.

EXAMPLE 6 Solve the system: $\begin{cases} 0.01x = 0.12 - 0.04y \\ 2x = 4(3 - 2y) \end{cases}$

Self Check 6

Solve the system:
$\begin{cases} 0.1x - 0.4 = 0.1y \\ -2y = 2(2 - x) \end{cases}$

Now Try Problem 33

Strategy We will use properties of algebra to write the first equation of the system in simpler form. Then we will use the substitution method to solve the resulting equivalent system.

WHY The first equation will be easier to work with if we clear it of decimals.

Solution
We can clear the first equation of decimals by multiplying both sides by 100.

$$\begin{cases} x = 12 - 4y \\ 2x = 4(3 - 2y) \end{cases}$$

Since $x = 12 - 4y$, we can substitute $12 - 4y$ for x in the second equation and solve for y.

$2x = 4(3 - 2y)$	This is the second equation.
$2(12 - 4y) = 4(3 - 2y)$	Substitute.
$24 - 8y = 12 - 8y$	Distribute.
$24 \neq 12$	Add $8y$ to both sides.

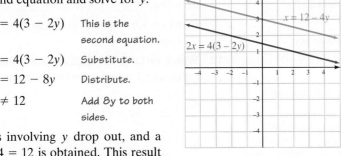

Here, the terms involving y drop out, and a false result of $24 = 12$ is obtained. This result indicates that the equations are independent and also that the system is inconsistent. When the equations are graphed, the graphs are parallel lines. This system has no solution.

EXAMPLE 7 Solve the system: $\begin{cases} x = -3y + 6 \\ 2x + 6y = 12 \end{cases}$

Self Check 7

Solve the system: $\begin{cases} y = 2 - x \\ 3x + 3y = 6 \end{cases}$

Now Try Problem 34

Strategy We will use the substitution method to solve this system.

WHY The substitution method works well when one of the equations of the system (in this case, $x = -3y + 6$) is solved for a variable.

Solution
We can substitute $-3y + 6$ for x in the second equation and proceed as follows:

$2x + 6y = 12$	This is the second equation of the system.
$2(-3y + 6) + 6y = 12$	Substitute.

$-6y + 12 + 6y = 12$ Distribute the multiplication by 2.
$12 = 12$ True

Here, the terms involving y drop out, and we get $12 = 12$. This true statement indicates that the equations are dependent. When these equations are graphed, their graphs are identical.

Because any ordered pair that satisfies one equation of the system also satisfies the other, the system has infinitely many solutions. To find some, we substitute 0, 3, and 6 for x in either equation and solve for y. The pairs $(0, 2)$, $(3, 1)$, and $(6, 0)$ are some of the solutions.

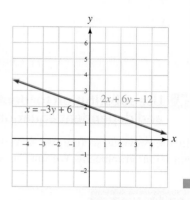

ANSWERS TO SELF CHECKS

1. $(15, 8)$ 2. $(2, -1)$ 3. $(2, -3)$ 4. $\left(\dfrac{18}{7}, \dfrac{8}{7}\right)$ 5. $(-1, 0)$ 6. no solution
7. infinitely many solutions

SECTION 4.2 STUDY SET

VOCABULARY

Fill in the blanks.

1. We say that the equation $y = 2x + 4$ is solved for _____ or that y is expressed in _____ of x.
2. "To _____ a solution of a system" means to see whether the coordinates of the ordered pair satisfy both equations.
3. When we write $2(x - 6)$ as $2x - 12$, we are applying the _____ property.
4. In mathematics, "to _____" means to replace an expression with one that is equivalent to it.
5. A dependent system has _____ many solutions.
6. In the term y, the coefficient is understood to be _____.

CONCEPTS

7. Consider the system: $\begin{cases} 2x + 3y = 12 \\ y = 2x + 4 \end{cases}$

    a. How many variables does each equation of the system contain?
    b. Substitute $2x + 4$ for y in the first equation. How many variables does the resulting equation contain?

8. For each equation, solve for y.
    a. $y + 2 = x$
    b. $2 + x + y = 0$
    c. $2 - y = x$

9. Given the equation $x - 2y = -10$,
    a. solve it for x.
    b. solve it for y.
    c. which variable was easier to solve for, x or y? Explain.

10. Which variable in which equation should be solved for in step 1 of the substitution method?
    a. $\begin{cases} x - 2y = 2 \\ 2x + 3y = 11 \end{cases}$
    b. $\begin{cases} 2x - 3y = 2 \\ 2x - y = 11 \end{cases}$

11. a. Find the error when $x - 4$ is substituted for y.

    $x + 2y = 5$ This is the first equation of the system.
    $x + 2x - 4 = 5$ Substitute for y: $y = x - 4$.
    $3x - 4 = 5$ Combine like terms.
    $3x = 9$ Add 4 to both sides.
    $x = 3$ Do the divisions.

    b. Rework the problem to find the correct value of x.

12. A student uses the substitution method to solve the system $\begin{cases} 4a + 5b = 2 \\ b = 3a - 11 \end{cases}$ and she finds that $a = 3$. What is the easiest way for her to determine the value of b?

4.2 Solving Systems of Equations by Substitution

13. Consider the system: $\begin{cases} y = \frac{1}{2}x \\ 2x + y = 2 \end{cases}$

   a. Graph the equations on the same coordinate system. Why is it difficult to determine the solution of the system?

   b. Solve the system by the substitution method.

14. Suppose the equation $-2 = 1$ is obtained when a system is solved by the substitution method.

   a. Does the system have a solution?

   b. Which of the following is a possible graph of the system?

   i. 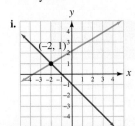   ii.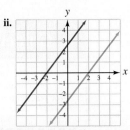

NOTATION

Complete each step to find the solution of the system.

15. Solve: $\begin{cases} y = 3x \\ x - y = 4 \end{cases}$

   $x - y = 4$ This is the second equation.
   $x - (\ \ ) = 4$
   $-2x = \boxed{}$
   $x = -2$
   $y = 3x$ This is the first equation.
   $y = 3(\ \ )$
   $y = -6$

   The solution is $\boxed{}$.

16. Solve: $\begin{cases} 2x + y = -5 \\ 2 - 2y = x \end{cases}$

   $2x + y = -5$ This is the first equation.
   $2(\ \ ) + y = -5$
   $4 - \boxed{} + y = -5$
   $\boxed{} - 3y = -5$
   $-3y = \boxed{}$
   $y = 3$

   $2 - 2y = x$ This is the second equation.
   $2 - 2(\ \ ) = x$
   $2 - 6 = x$
   $-4 = x$

   The solution is $\boxed{}$.

GUIDED PRACTICE

Use the substitution method to solve each system.
See Example 1.

17. $\begin{cases} y = 2x \\ x + y = 6 \end{cases}$ 18. $\begin{cases} y = 3x \\ x + y = 4 \end{cases}$

19. $\begin{cases} y = 2x - 6 \\ 2x + y = 6 \end{cases}$ 20. $\begin{cases} y = 2x - 9 \\ x + 3y = 8 \end{cases}$

Use the substitution method to solve each system.
See Examples 2–3.

21. $\begin{cases} 6x - 3y = 5 \\ x = -2y \end{cases}$ 22. $\begin{cases} 5x + 10y = 3 \\ x = -\frac{1}{2}y \end{cases}$

23. $\begin{cases} r + 3s = 9 \\ 3r + 2s = 13 \end{cases}$ 24. $\begin{cases} x - 2y = 2 \\ 2x + 3y = 11 \end{cases}$

25. $\begin{cases} 3x + 4y = -7 \\ 2y - x = -1 \end{cases}$ 26. $\begin{cases} 4x + 5y = -2 \\ x + 2y = -2 \end{cases}$

27. $\begin{cases} -2x + y = 5 \\ x + 2y = -5 \end{cases}$ 28. $\begin{cases} 2x + y = 0 \\ 3x + 2y = -1 \end{cases}$

Use the substitution method to solve each system.
See Examples 4–5.

29. $\begin{cases} 2a = 3b - 13 \\ -b = -2a - 7 \end{cases}$ 30. $\begin{cases} a = 3b - 1 \\ -b = -2a - 2 \end{cases}$

31. $\begin{cases} 5m = \frac{1}{2}n - 1 \\ \frac{1}{4}n = 10m - 1 \end{cases}$ 32. $\begin{cases} \frac{2}{3}m = 1 - 2n \\ 2(5n - m) + 11 = 0 \end{cases}$

Use the substitution method to solve each system, if possible.
See Examples 6–7.

33. $\begin{cases} 2a + 4b = -24 \\ a = 20 - 2b \end{cases}$ 34. $\begin{cases} 3a + 6b = -15 \\ a = -2b - 5 \end{cases}$

35. $\begin{cases} 9x = 3y + 12 \\ 4 = 3x - y \end{cases}$ 36. $\begin{cases} 8y = 15 - 4x \\ x + 2y = 4 \end{cases}$

Work area:

$2 - 2y = x$ This is the second equation.
$2 - 2(\ \ ) = x$
$2 - 6 = x$
$-4 = x$

The solution is $\boxed{}$.

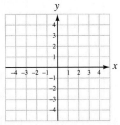

TRY IT YOURSELF

Use the substitution method to solve each system. If the equations of a system are dependent or if a system is inconsistent, so indicate.

37. $\begin{cases} 0.4x + 0.5y = 0.2 \\ 3x - y = 11 \end{cases}$

38. $\begin{cases} 0.5u + 0.3v = 0.5 \\ 4u - v = 4 \end{cases}$

39. $\begin{cases} 0.02x + 0.05y = -0.02 \\ -\dfrac{x}{2} = y \end{cases}$

40. $\begin{cases} y = -\dfrac{x}{2} \\ 0.02x - 0.03y = -0.07 \end{cases}$

41. $\begin{cases} y - x = 3x \\ 2x + 2y = 14 - y \end{cases}$

42. $\begin{cases} y + x = 2x + 2 \\ 6x - 4y = 21 - y \end{cases}$

43. $\begin{cases} 2x - y = x + y \\ -2x + 4y = 6 \end{cases}$

44. $\begin{cases} x = -3y + 6 \\ 2x + 4y = 6 + x + y \end{cases}$

45. $\begin{cases} 3(x - 1) + 3 = 8 + 2y \\ 2(x + 1) = 8 + y \end{cases}$

46. $\begin{cases} 4(x - 2) = 19 - 5y \\ 3(x - 2) - 2y = -y \end{cases}$

47. $\begin{cases} \dfrac{1}{2}x + \dfrac{1}{2}y = -1 \\ \dfrac{1}{3}x - \dfrac{1}{2}y = -4 \end{cases}$

48. $\begin{cases} \dfrac{2}{3}y + \dfrac{1}{5}z = 1 \\ \dfrac{1}{3}y - \dfrac{2}{5}z = 3 \end{cases}$

49. $\begin{cases} b = \dfrac{2}{3}a \\ 8a - 3b = 3 \end{cases}$

50. $\begin{cases} a = \dfrac{2}{3}b \\ 9a + 4b = 5 \end{cases}$

51. $\begin{cases} \dfrac{6x - 1}{3} - \dfrac{5}{3} = \dfrac{3y + 1}{2} \\ \dfrac{1 + 5y}{4} + \dfrac{x + 3}{4} = \dfrac{17}{2} \end{cases}$

52. $\begin{cases} \dfrac{5x - 2}{4} + \dfrac{1}{2} = \dfrac{3y + 2}{2} \\ \dfrac{7y + 3}{3} = \dfrac{x}{2} + \dfrac{7}{3} \end{cases}$

APPLICATIONS

53. **GEOMETRY** In the illustration, x and y represent the degree measures of angles. If $x + y = 90$ and $y = 3x$, find x and y.

54. **GEOMETRY** In the illustration, x and y represent the degree measures of angles. If $x + y = 180$ and $x = 4y$, find x and y.

55. **DINING** See the following breakfast menu. What substitution from the a la carte menu will the restaurant owner allow customers to make if they don't want hash browns with their country breakfast? Why?

56. **DISCOUNT COUPONS** In mathematics, the substitution property states: *If $a = b$, then a may replace b or b may replace a in any statement.* Where on the following coupon is there an application of the substitution property? Explain.

Golden Spur — $16.95 Value

Buy one roast beef dinner and get a second roast beef dinner (or other entree of equal value) free!

Valid anytime

57. **HIGH SCHOOL SPORTS** The equations shown model the number of boys and girls taking part in high school soccer programs. In both models, x is the number of years after 2012, and y is the number of participants. If the trends continue, the graphs will intersect. Use the substitution method to predict the year when the number of boys and girls participating in high school soccer will be the same. How many of each gender will be participating at that time?

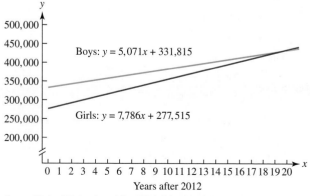

Source: National Federation of State High School Associations

58. OFF ROADING The *angle of approach* indicates how steep an incline a vehicle can drive up without damaging the front bumper. The *angle of departure* indicates a vehicle's ability to exit an incline without damaging the rear bumper. The angle of approach *a* and the angle of departure *d* for an H3 Hummer are described by the system:

$$\begin{cases} a + d = 77 \\ a = d + 3 \end{cases}$$

Use substitution to solve the system. (Each angle is measured in degrees.)

WRITING

59. Explain how to use substitution to solve a system of equations.

60. If the equations of a system are written in standard form, why is it to your advantage to solve for a variable whose coefficient is 1 when using the substitution method?

61. When solving a system, what advantages and disadvantages are there with the graphing method? With the substitution method?

62. In this section, the substitution method for solving a system of two equations was discussed. List some other uses of the word *substitution*, or *substitute*, that you encounter in everyday life.

REVIEW

63. What is the slope of the line $y = -\frac{5}{8}x - 12$?

64. If $g(x) = -3x + 9$, find: $g(-3)$

65. Find the y-intercept of $2x - 3y = 18$.

66. Write the equation of the line passing through $(-1, 5)$ with a slope of -3.

67. Can a circle represent the graph of a function?

68. What is the range of the function $f(x) = |x|$?

69. On what axis does $(3, 0)$ lie?

70. On what axis does $(0, -2)$ lie?

SECTION 4.3
Solving Systems of Equations by Addition (Elimination)

Objectives

1 Solve systems of linear equations by the addition (elimination) method.

2 Use multiplication to eliminate a variable.

3 Use the addition (elimination) method to identify inconsistent systems and dependent equations.

4 Determine the most efficient method to use to solve a linear system.

In step 1 of the substitution method for solving a system of equations, we solve one equation for one of the variables. At times, this can be difficult—especially if neither variable has a coefficient of 1 or −1. In cases such as these, we can use another algebraic method called the **addition** or **elimination method** to find the exact solution of the system. This method is based on the addition property of equality: *When equal quantities are added to both sides of an equation, the results are equal.*

1 Solve systems of linear equations by the addition (elimination) method.

EXAMPLE 1
Solve the system: $\begin{cases} x + y = 8 \\ x - y = -2 \end{cases}$

Strategy Since the coefficients of the y-terms are opposites, we will add the left sides and the right sides of the given equations.

WHY When we add the equations in this way, the y-terms will drop out and the result will be an equation that contains only one variable, x.

Solution
Since $x - y$ and -2 are equal quantities, we can add $x - y$ to the left side and -2 to the right side of the first equation, $x + y = 8$.

Self Check 1
Solve the system:
$\begin{cases} 3x + 2y = 8 \\ -3x + 4y = -2 \end{cases}$

Now Try Problem 15

$$x + y = 8 \quad \text{To add the equations, add the like terms, column by column.}$$
$$\underline{x - y = -2} \quad \text{Combine like terms: } x + x = 2x, y + (-y) = 0, \text{ and } 8 + (-2) = 6.$$
$$2x = 6 \quad \text{Write each result here.}$$

We can then solve the resulting equation for x.

$$2x = 6$$
$$x = 3 \quad \text{Divide both sides by 2.}$$

To find y, we substitute 3 for x in either equation and solve it for y.

$$x + y = 8 \quad \text{This is the first equation of the system.}$$
$$3 + y = 8 \quad \text{Substitute 3 for x.}$$
$$y = 5 \quad \text{Subtract 3 from both sides.}$$

We check the solution by verifying that (3, 5) satisfies each equation of the system.

To solve a system of equations in x and y by the addition method, we follow these steps.

The Addition (Elimination) Method

1. Write both equations of the system in standard form: $Ax + By = C$.
2. If necessary, multiply one or both of the equations by a nonzero number chosen to make the coefficients of x (or the coefficients of y) opposites.
3. Add the equations to eliminate the terms involving x (or y).
4. Solve the equation resulting from step 3.
5. Find the value of the remaining variable by substituting the solution found in step 4 into any equation containing both variables. Or repeat steps 2–4 to eliminate the other variable.
6. Check the proposed solution in each equation of the original system. Write the solution as an ordered pair.

Self Check 2

Solve the system:
$$\begin{cases} x + 3y = 7 \\ 2x - 3y = -22 \end{cases}$$

Now Try Problem 17

EXAMPLE 2 Solve the system: $\begin{cases} 5x + y = -4 \\ -5x + 2y = 7 \end{cases}$

Strategy Since the coefficients of the x-terms are opposites, we will add the left sides and the right sides of the given equations.

WHY When we add the equations in this way, the x-terms will drop out and the result will be an equation that contains only one variable, y.

Solution

Step 1 Both equations are in standard $Ax + By = C$ form.

Step 2 The coefficients of x are opposites.

Step 3 Add the equations to eliminate x.

$$5x + y = -4 \quad \text{Combine like terms: } 5x + (-5x) = 0, y + 2y = 3y, \text{ and}$$
$$\underline{-5x + 2y = 7} \quad -4 + 7 = 3.$$
$$3y = 3$$

Step 4 Since the result of the addition is an equation in one variable, we can solve for y.

$$3y = 3$$
$$y = 1 \quad \text{Divide both sides by 3.}$$

Step 5 To find x, we substitute 1 for y in either equation. If we use $5x + y = -4$, we have

$5x + y = -4$ This is the first equation of the system.
$5x + 1 = -4$ Substitute 1 for y.
$5x = -5$ Subtract 1 from both sides.
$x = -1$ Divide both sides by 5.

Step 6 Verify that $(-1, 1)$ satisfies each original equation. ∎

2 Use multiplication to eliminate a variable.

In Example 1, the coefficients of the terms y in the first equation and $-y$ in the second equation were opposites. When we added the equations, the variable y was eliminated. For many systems, however, we are not able to immediately eliminate a variable by adding. In such cases, we use the multiplication property of equality to create coefficients of x or y that are opposites.

EXAMPLE 3 Solve the system: $\begin{cases} 3x + y = 7 \\ x + 2y = 4 \end{cases}$

Self Check 3
Solve the system:
$\begin{cases} 3x - 4y = -7 \\ 2x + y = 10 \end{cases}$
Now Try Problem 23

Strategy We will multiply the second equation by -3 to make the coefficients of x opposites.

WHY Neither the coefficients of x nor the coefficients of y are opposites in the original system.

Solution
Step 1 Both equations are in standard $Ax + By = C$ form.

Step 2 If we add the equations as they are, neither variable will be eliminated. We must write the equations so that the coefficients of one of the variables are opposites. To eliminate x, we can multiply both sides of the second equation by -3 to get

$\begin{cases} 3x + y = 7 \\ x + 2y = 4 \end{cases} \xrightarrow{\text{Unchanged}} \begin{cases} 3x + y = 7 \\ -3(x + 2y) = -3(4) \end{cases} \xrightarrow{\text{Unchanged}} \begin{cases} 3x + y = 7 \\ -3x - 6y = -12 \end{cases}$
                  Multiply by -3 Simplify

Step 3 The coefficients of the terms $3x$ and $-3x$ are now opposites. When the equations are added, x is eliminated.

$3x + y = 7$
$\underline{-3x - 6y = -12}$
$ -5y = -5$

Step 4 Solve the resulting equation to find y.

$-5y = -5$
$y = 1$ Divide both sides by -5.

Step 5 To find x, we substitute 1 for y in the equation $x + 2y = 4$.

$x + 2y = 4$ This is the second equation of the original system.
$x + 2(1) = 4$ Substitute 1 for y.
$x + 2 = 4$ Do the multiplication.
$x = 2$ Subtract 2 from both sides.

Step 6 Check the solution $(2, 1)$ in the original system of equations.

Self Check 4

Solve the system:
$$\begin{cases} 2a + 3b = 7 \\ 5a + 2b = 1 \end{cases}$$

Now Try Problem 27

EXAMPLE 4

Solve the system: $\begin{cases} 2a - 5b = 10 \\ 3a - 2b = -7 \end{cases}$

Strategy We will use the addition method to solve this system.

WHY Since none of the variables has a coefficient of 1 or -1, it would be difficult to solve this system using substitution.

Solution

Step 1 Both equations are in standard $Ax + By = C$ form. We see that neither the coefficients of a nor b are opposites. Adding these equations as written does not eliminate a variable.

Step 2 To eliminate a, we can multiply the first equation by 3 and the second equation by -2 to get

$$\begin{cases} 2a - 5b = 10 \\ 3a - 2b = -7 \end{cases} \xrightarrow{\text{Multiply by 3}} \begin{cases} 3(2a - 5b) = 3(10) \\ -2(3a - 2b) = -2(-7) \end{cases} \xrightarrow{\text{Simplify}} \begin{cases} 6a - 15b = 30 \\ -6a + 4b = 14 \end{cases}$$

Step 3 When these equations are added, the terms $6a$ and $-6a$ are eliminated.

$$\begin{array}{r} 6a - 15b = 30 \\ \underline{-6a + 4b = 14} \\ -11b = 44 \end{array}$$

Step 4 Solve the resulting equation to find b.

$-11b = 44$

$b = -4$ Divide both sides by -11.

Step 5 To find a, we substitute -4 for b in the equation $2a - 5b = 10$.

$2a - 5b = 10$ This is the first equation of the original system.

$2a - 5(-4) = 10$ Substitute -4 for b.

$2a + 20 = 10$ Simplify.

$2a = -10$ Subtract 20 from both sides.

$a = -5$ Divide both sides by 2.

Step 6 Check the solution $(-5, -4)$ in the original equations.

Self Check 5

Solve the system:
$$\begin{cases} \dfrac{1}{3}x + \dfrac{1}{6}y = 1 \\ \dfrac{1}{2}x - \dfrac{1}{4}y = 0 \end{cases}$$

Now Try Problem 29

EXAMPLE 5

Solve the system: $\begin{cases} \dfrac{5}{6}x + \dfrac{2}{3}y = \dfrac{7}{6} \\ \dfrac{10}{7}x - \dfrac{4}{9}y = \dfrac{17}{21} \end{cases}$

Strategy We will begin by clearing each equation of fractions. Then we will use the addition method to solve the resulting equivalent system.

WHY It is easier to create a pair of terms that are opposites if their coefficients are integers rather than fractions.

Solution

Step 1 To clear the equations of fractions, we multiply both sides of the first equation by 6 and both sides of the second equation by 63. This gives the equivalent system

(1) $\begin{cases} 5x + 4y = 7 \\ 90x - 28y = 51 \end{cases}$
(2)

Step 2 We can solve for x by eliminating the terms involving y. To do so, we multiply Equation 1 by 7.

$$\begin{cases} 5x + 4y = 7 \\ 90x - 28y = 51 \end{cases} \xrightarrow{\text{Multiply by 7}} \begin{cases} 7(5x + 4y) = 7(7) \\ 90x - 28y = 51 \end{cases} \xrightarrow{\text{Simplify}} \begin{cases} 35x + 28y = 49 \\ 90x - 28y = 51 \end{cases}$$

Step 3 Add the equations.

$$\begin{array}{r} 35x + 28y = 49 \\ 90x - 28y = 51 \\ \hline 125x = 100 \end{array}$$

Step 4 Solve the resulting equation for x.

$125x = 100$

$x = \dfrac{100}{125}$ Divide both sides by 125.

$x = \dfrac{4}{5}$ Simplify $\dfrac{100}{125}$: Remove the common factor of 25.

Step 5 To solve for y, we substitute $\dfrac{4}{5}$ for x in Equation 1 and simplify.

$5x + 4y = 7$

$5\left(\dfrac{4}{5}\right) + 4y = 7$

$4 + 4y = 7$ Simplify.

$4y = 3$ Subtract 4 from both sides.

$y = \dfrac{3}{4}$ Divide both sides by 4.

Step 6 Check the solution of $\left(\dfrac{4}{5}, \dfrac{3}{4}\right)$ in the original equations.

EXAMPLE 6

Solve the system: $\begin{cases} 2(2x + y) = 13 \\ 8x = 2y - 16 \end{cases}$

Strategy We will use properties of algebra to write the first equation of the system in simpler form and the second equation in standard form. Then we will use the addition method to solve the resulting equivalent system.

WHY Writing the equations in standard form is the first step of the addition method.

Solution

Step 1 We begin by writing each equation in $Ax + By = C$ form. For the first equation, we need only apply the distributive property. To write the second equation in standard form, we subtract $2y$ from both sides.

$2(2x + y) = 13 \qquad\qquad 8x = 2y - 16$

$4x + 2y = 13 \qquad\qquad 8x - 2y = 2y - 16 - 2y$

$8x - 2y = -16$

The two resulting equations form the following system.

(1) $\begin{cases} 4x + 2y = 13 \\ 8x - 2y = -16 \end{cases}$
(2)

Self Check 6

Solve the system:
$\begin{cases} -3y = -5 - x \\ 3(x - y) = -11 \end{cases}$

Now Try Problem 33

Step 2 The coefficients of y are opposites.

Step 3 When the equations are added, the terms involving y are eliminated.

$$4x + 2y = 13$$
$$8x - 2y = -16$$
$$\overline{12x = -3}$$

Step 4 Solve the resulting equation for x.

$$12x = -3$$
$$x = -\frac{1}{4} \quad \text{Divide both sides by 12 and simplify the fraction: } -\frac{3}{12} = -\frac{1}{4}.$$

Step 5 We can use Equation 1 to find y.

$$4x + 2y = 13$$
$$4\left(-\frac{1}{4}\right) + 2y = 13 \quad \text{Substitute } -\frac{1}{4} \text{ for } x.$$
$$-1 + 2y = 13 \quad \text{Do the multiplication.}$$
$$2y = 14 \quad \text{Add 1 to both sides.}$$
$$y = 7 \quad \text{Divide both sides by 2.}$$

Step 6 Verify that $\left(-\frac{1}{4}, 7\right)$ satisfies each original equation.

3 Use the addition (elimination) method to identify inconsistent systems and dependent equations.

We have solved inconsistent systems and systems of dependent equations by substitution and by graphing. We can also solve these systems using the addition method.

Self Check 7

Solve the system:
$$\begin{cases} 2t - 7v = 5 \\ -2t + 7v = 3 \end{cases}$$

Now Try Problem 37

EXAMPLE 7 Solve the system, if possible: $\begin{cases} 3x - 2y = 8 \\ -3x + 2y = -12 \end{cases}$

Strategy We will use the addition method to solve this system.

WHY The terms $3x$ and $-3x$ are immediately eliminated.

Solution
We can add the equations to eliminate the term involving x.

$$3x - 2y = 8$$
$$-3x + 2y = -12$$
$$\overline{ \, 0 = -4}$$

Here the terms involving both x and y drop out, and a false result of $0 = -4$ is obtained. This indicates that the equations of the system are independent and that the system is inconsistent. This system has no solution.

Self Check 8

Solve the system:
$$\begin{cases} \dfrac{3x + y}{6} = \dfrac{1}{3} \\ -0.3x - 0.1y = -0.2 \end{cases}$$

Now Try Problem 39

EXAMPLE 8 Solve the system: $\begin{cases} \dfrac{2x - 5y}{2} = \dfrac{19}{2} \\ -0.2x + 0.5y = -1.9 \end{cases}$

Strategy We will begin by clearing the equations of fractions and decimals. Then we will use the addition method to solve the resulting equivalent system.

WHY Writing the equations in standard form is the first step of the addition method.

Solution
We can multiply both sides of the first equation by **2** to clear it of fractions and both sides of the second equation by **10** to clear it of decimals.

$$\begin{cases} 2\left(\dfrac{2x-5y}{2}\right) = 2\left(\dfrac{19}{2}\right) \\ 10(-0.2x + 0.5y) = 10(-1.9) \end{cases} \xrightarrow{\text{Simplify}} \begin{cases} 2x - 5y = 19 \\ -2x + 5y = -19 \end{cases}$$

We add the resulting equations to get

$$\begin{array}{r} 2x - 5y = 19 \\ -2x + 5y = -19 \\ \hline 0 = 0 \end{array}$$

As in Example 7, both x and y drop out. However, this time a true result is obtained. This indicates that the equations are dependent and that the system has infinitely many solutions.

Any ordered pair that satisfies one equation also satisfies the other equation. Some solutions are $(2, -3)$, $(12, 1)$, and $\left(0, -\dfrac{19}{5}\right)$.

4 Determine the most efficient method to use to solve a linear system.

If no method is specified for solving a particular linear system, the following guidelines can be helpful in determining whether to use graphing, substitution, or addition.

1. If you want to show trends and see the point that the two graphs have in common, then use the **graphing method**. However, this method is not exact and can be lengthy.
2. If one of the equations is solved for one of the variables, or easily solved for one of the variables, use the **substitution method**.
3. If both equations are in standard $Ax + By = C$ form, and no variable has a coefficient of 1 or -1, use the **addition method**.
4. If the coefficient of one of the variables is 1 or -1, you have a choice. You can write each equation in standard $Ax + By = C$ form and use addition, or you can solve for the variable with coefficient 1 or -1 and use substitution.

Here are some examples of suggested approaches:

$$\underbrace{\begin{cases} 2x + 3y = 1 \\ y = 4x - 3 \end{cases}}_{\text{Substitution}} \quad \underbrace{\begin{cases} 5x + 3y = 9 \\ 8x + 4y = 3 \end{cases}}_{\text{Addition}} \quad \underbrace{\begin{cases} 4x - y = -6 \\ 3x + 2y = 1 \end{cases}}_{\text{Addition}} \quad \underbrace{\begin{cases} x - 23 = 6y \\ 7x - 9y = -3 \end{cases}}_{\text{Substitution}}$$

Each method that we use to solve systems of equations has advantages and disadvantages.

ANSWERS TO SELF CHECKS

1. $(2, 1)$ **2.** $(-5, 4)$ **3.** $(3, 4)$ **4.** $(-1, 3)$ **5.** $\left(\dfrac{3}{2}, 3\right)$ **6.** $\left(-3, \dfrac{2}{3}\right)$ **7.** no solution **8.** infinitely many solutions

---

SECTION 4.3 STUDY SET

VOCABULARY

Fill in the blanks.

1. The _____ of the term $-3x$ is -3.
2. The _____ of 4 is -4.
3. $Ax + By = C$ is the _____ form of the equation of a line.
4. If we add the equations $\begin{array}{r} 5x - 6y = 10 \\ -3x + 6y = 24 \end{array}$ the variable y will be _____.

CONCEPTS

5. If the addition (elimination) method is to be used to solve this system, what is wrong with the form in which it is written?
$$\begin{cases} 2x - 5y = -3 \\ -2y + 3x = 10 \end{cases}$$

6. Can the system
$$\begin{cases} 2x + 5y = -13 \\ -2x - 3y = -5 \end{cases}$$
be solved more easily using the addition method or the substitution method? Explain.

7. What algebraic step should be performed to clear this equation of fractions?
$$\frac{2}{3}x + 4y = -\frac{4}{5}$$

8. If the addition method is used to solve
$$\begin{cases} 3x + 12y = 4 \\ 6x - 4y = 8 \end{cases}$$
   a. By what would we multiply the first equation to eliminate x?
   b. By what would we multiply the second equation to eliminate y?

9. Solve: $\begin{cases} 4x + 2y = 2 \\ 3x - 2y = 12 \end{cases}$
   a. by the graphing method.
   b. by the addition method.

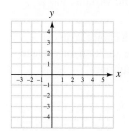

10. a. Suppose $0 = 0$ is obtained when a system is solved by the addition method. Does the system have a solution? Which of the following is a possible graph of the system?
    b. Suppose $0 = 2$ is obtained when a system is solved by the addition method. Does the system have a solution? Which of the following is a possible graph of the system?

i   ii   iii

NOTATION

Fill in the blanks to solve the system.

11. Solve: $\begin{cases} x + y = 5 \\ x - y = -3 \end{cases}$

    $x + y = 5$
    $\underline{x - y = -3}$
    $ \square = 2$
    $x = \square$

    $x + y = 5$ This is the first equation.
    $(\square) + y = 5$
    $y = 4$
    The solution is $\square$.

12. Solve: $\begin{cases} x - 2y = 8 \\ -x + 5y = -17 \end{cases}$

    $x - 2y = 8$
    $\underline{-x + 5y = -17}$
    $ \square = -9$
    $y = \square$

    $x - 2y = 8$ This is the first equation.
    $x - 2(\square) = 8$
    $x + 6 = 8$
    $x = 2$
    The solution is $\square$.

GUIDED PRACTICE

Use the addition (elimination) method to solve each system. See Examples 1–2.

13. $\begin{cases} x - y = -5 \\ x + y = 1 \end{cases}$
14. $\begin{cases} x + y = 1 \\ x - y = 5 \end{cases}$

15. $\begin{cases} 2r + s = -1 \\ -2r + s = 3 \end{cases}$
16. $\begin{cases} 3m + n = -6 \\ m - n = -2 \end{cases}$

17. $\begin{cases} 2x + y = -2 \\ -2x - 3y = -6 \end{cases}$
18. $\begin{cases} 3x + 4y = 8 \\ 5x - 4y = 24 \end{cases}$

19. $\begin{cases} 4x + 3y = 24 \\ 4x - 3y = -24 \end{cases}$
20. $\begin{cases} 5x - 4y = 8 \\ -5x - 4y = 8 \end{cases}$

Use the addition (elimination) method to solve each system. See Examples 3–4.

21. $\begin{cases} x + y = 5 \\ x + 2y = 8 \end{cases}$
22. $\begin{cases} x + 2y = 0 \\ x - y = -3 \end{cases}$

23. $\begin{cases} 2x + y = 4 \\ 2x + 3y = 0 \end{cases}$
24. $\begin{cases} 2x + 5y = -13 \\ 2x - 3y = -5 \end{cases}$

25. $\begin{cases} 3x - 5y = -29 \\ 3x + 4y = 34 \end{cases}$
26. $\begin{cases} 3x - 5y = 16 \\ 4x + 5y = 33 \end{cases}$
27. $\begin{cases} 8x - 4y = 18 \\ 3x - 2y = 8 \end{cases}$
28. $\begin{cases} 4x + 6y = 5 \\ 8x - 9y = 3 \end{cases}$

Use the addition (elimination) method to solve each system. See Example 5.

29. $\begin{cases} \dfrac{3}{5}s + \dfrac{4}{5}t = 1 \\ -\dfrac{1}{4}s + \dfrac{3}{8}t = 1 \end{cases}$
30. $\begin{cases} \dfrac{1}{2}s - \dfrac{1}{4}t = 1 \\ \dfrac{1}{3}s + t = 3 \end{cases}$
31. $\begin{cases} \dfrac{3}{5}x + y = 1 \\ \dfrac{4}{5}x - y = -1 \end{cases}$
32. $\begin{cases} \dfrac{1}{2}x + \dfrac{4}{7}y = -1 \\ 5x - \dfrac{4}{5}y = -10 \end{cases}$

Use the addition (elimination) method to solve each system. See Example 6.

33. $\begin{cases} -3(x - 2y) = -9 \\ 5x = 4y + 15 \end{cases}$
34. $\begin{cases} -4x + 13 = -3y \\ 2(4y - 3x) = -16 \end{cases}$
35. $\begin{cases} 4(x + 1) = 17 - 3(y - 1) \\ 2(x + 2) + 3(y - 1) = 9 \end{cases}$
36. $\begin{cases} 5(x - 1) = 8 - 3(y + 2) \\ 4(x + 2) - 7 = 3(2 - y) \end{cases}$

Use the addition (elimination) method to solve each system. If the equations of a system are dependent or if a system is inconsistent, so indicate. See Examples 7–8.

37. $\begin{cases} 2a - 3b = -6 \\ 2a - 3b = 8 \end{cases}$
38. $\begin{cases} 3a - 4b = 6 \\ 2(2b + 3) = 3a \end{cases}$
39. $\begin{cases} -2(x + 1) = 3y - 6 \\ 3(y + 2) = 10 - 2x \end{cases}$
40. $\begin{cases} 3x + 2y + 1 = 5 \\ 3(x - 1) = -2y - 4 \end{cases}$

TRY IT YOURSELF

Solve each system, if possible.

41. $\begin{cases} 3x + 4y = 12 \\ 4x + 5y = 17 \end{cases}$
42. $\begin{cases} 2x + 11y = -10 \\ 5x + 4y = 22 \end{cases}$
43. $\begin{cases} 2x + y = 10 \\ 0.1x + 0.2y = 1.0 \end{cases}$
44. $\begin{cases} 0.3x + 0.2y = 0 \\ 2x - 3y = -13 \end{cases}$
45. $\begin{cases} 2x - y = 16 \\ 0.03x + 0.02y = 0.03 \end{cases}$
46. $\begin{cases} -5y + 2x = 4 \\ -0.02y + 0.03x = 0.04 \end{cases}$
47. $\begin{cases} 6x + 3y = 0 \\ 5y = 2x + 12 \end{cases}$
48. $\begin{cases} 0 = 4x - 3y \\ 5x = 4y - 2 \end{cases}$

49. $\begin{cases} 3x + 12y = -12 \\ x = 3y + 10 \end{cases}$
50. $\begin{cases} 3x + 2y = 3 \\ y = 2(x - 8) \end{cases}$
51. $\begin{cases} 4x - 8y = 36 \\ 3x - 6y = 27 \end{cases}$
52. $\begin{cases} 2x + 4y = 15 \\ 3x = 8 - 6y \end{cases}$
53. $\begin{cases} x = y \\ 0.1x + 0.2y = 1.0 \end{cases}$
54. $\begin{cases} x = y \\ 0.4x - 0.8y = -0.5 \end{cases}$
55. $\begin{cases} \dfrac{m}{4} + \dfrac{m}{3} = -\dfrac{1}{12} \\ \dfrac{m}{2} - \dfrac{5n}{4} = \dfrac{7}{4} \end{cases}$
56. $\begin{cases} \dfrac{x}{2} - \dfrac{y}{3} = -2 \\ \dfrac{x}{3} + \dfrac{2y}{3} = \dfrac{4}{3} \end{cases}$

APPLICATIONS

57. **EDUCATION** The graph shows educational trends during the years 1980–2011 for persons 25 years or older. The equation $9x + 11y = 352$ approximates the percent y that had less than high school completion x years after 1980. The equation $5x - 11y = -198$ approximates the percent y that had a bachelor's or higher degree x years after 1980. Use the addition method to determine in what year the percents were equal.

Source: U.S. Department of Commerce, Census Bureau

58. **CFL BULBS** The graph below shows how a more expensive, but more energy-efficient, compact fluorescent light bulb eventually costs less to use than an incandescent light bulb. The equation $60c - d = 96$ approximates the cost c (in dollars) to purchase and use a CFL bulb 8 hours a day for d days. The equation $15c - d = 6$ does the same for an incandescent bulb. Use the elimination method to determine after how many days the upgrade to a CFL bulb begins to save money.

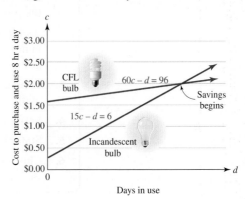

Source: Whitesites.com

WRITING

59. Why is it usually to your advantage to write the equations of a system in standard form before using the addition method to solve the system?

60. How would you decide whether to use substitution or addition to solve a system of equations?

61. In this section, we discussed the addition method for solving a system of two equations. Some instructors call it the *elimination method*. Why do you think it is known by this name?

62. Explain the error in the following work.

Solve: $\begin{cases} x + y = 1 \\ x - y = 5 \end{cases}$

$$x + y = 1$$
$$\underline{x - y = 5}$$
$$2x = 6$$
$$\frac{2x}{2} = \frac{6}{2}$$
$$\boxed{x = 3} \text{ The solution is 3.}$$

REVIEW

63. Solve: $8(3x - 5) - 12 = 4(2x + 3)$

64. Solve: $3y + \dfrac{y + 2}{2} = \dfrac{2(y + 3)}{3} + 16$

65. Simplify: $x - x$

66. Simplify: $3.2m - 4.4 + 2.1m + 16$

67. Find the area of a triangular sign with a base of 4 feet and a height of 3.75 feet.

68. Translate to mathematical symbols: *the product of the sum of x and y and the difference of x and y*.

69. What is 10 less than x?

70. Solve $y = mx + b$ for m.

Objectives

1. Solve problems using two unknowns.
2. Use systems to solve geometry problems.
3. Use systems to solve break-point problems.
4. Use systems to solve interest problems.
5. Use systems to solve uniform motion problems.
6. Use systems to solve mixture problems.

SECTION 4.4
Problem Solving Using Systems of Equations

We have previously formed equations involving one variable to solve problems. In this section, we consider ways to solve problems using two variables and the six-step problem-solving strategy discussed in Chapter 2.

1 Solve problems using two unknowns.

We can use the following steps to solve problems involving two unknown quantities.

Problem-Solving Strategy

1. **Analyze the problem** by reading it carefully to understand the given facts. What information is given? What are you asked to find? What vocabulary is given? Often a diagram or table will help you understand the facts of the problem.

2. **Assign variables** to represent unknown values in the problem. This means, in most cases, to let $x =$ one of the unknowns that you are asked to find, and $y =$ the other unknown.

3. **Form a system of equations** by translating the words of the problem into mathematical symbols.

4. **Solve the system** of equations using graphing, substitution, or elimination.

5. **State the conclusion clearly.** Be sure to include the units.

6. **Check the results** using the words of the problem, not the equations that were formed in step 3.

EXAMPLE 1 Photography
At a school, two picture packages are available, as shown in the illustration. Find the cost of a class picture and the cost of an individual wallet-size picture.

Analyze
- Package 1 contains 1 class picture and 10 wallet-size pictures.
- Package 2 contains 2 class pictures and 15 wallet-size pictures.
- Find the cost of a class picture and the cost of a wallet-size picture.

Assign Let c = the cost of 1 class picture and w = the cost of 1 wallet-size picture.

Form To write an equation that models the first package, we note that (in dollars) the cost of 1 class picture is c and the cost of 10 wallet-size pictures is $10 \cdot w = 10w$.

The cost of 1 class picture	plus	the cost of 10 wallet-size pictures	is	$19.
c	$+$	$10w$	$=$	19

To write an equation that models the second package, we note that (in dollars) the cost of 2 class pictures is $2 \cdot c = 2c$, and the cost of 15 wallet-size pictures is $15 \cdot w = 15w$.

The cost of 2 class pictures	plus	the cost of 15 wallet-size pictures	is	$31.
$2c$	$+$	$15w$	$=$	31

The resulting system is $\begin{cases} c + 10w = 19 \\ 2c + 15w = 31 \end{cases}$

Solve Since both equations are written in standard $Ax + By = C$ form, we will use the addition method to solve this system. To eliminate c, we proceed as follows:

$-2c - 20w = -38$ Multiply both sides of $c + 10w = 19$ by -2.
$\underline{2c + 15w = 31}$
$ -5w = -7$ Add the equations to eliminate c.
$w = 1.4$ Divide both sides by -5 to isolate w. This is the cost of a wallet-size picture.

To find c, substitute 1.4 for w in the first equation of the original system.

$c + 10w = 19$
$c + 10(1.4) = 19$ Substitute 1.4 for w.
$c + 14 = 19$ Multiply.
$c = 5$ Subtract 14 from both sides. This is the cost of a class picture.

State A class picture costs $5, and a wallet-size picture costs $1.40.

Self Check 1
Team Sports Photography offers two packages for their school sports pictures. The first package includes a team photo and 8 individual wallets for $29. The second package offers two team photos and one individual wallet for $35.50. Find the cost of a team photo and an individual wallet photo.

Now Try Problem 27

Check Package 1 has 1 class picture and 10 wallets: $5 + 10($1.40) = $5 + $14 = $19. Package 2 has 2 class pictures and 15 wallets: 2($5) + 15($1.40) = $10 + $21 = $31. The results check.

> **Caution!** In Example 1 we are to find two unknowns, the cost of a class picture and the cost of a wallet-size picture. Remember to give both answers in the state-the-conclusion step of the solution.

Self Check 2

A carpenter wants to cut a 12-foot board into two pieces. The longer piece is to be twice as long as the shorter piece. Find the length of each piece.

Now Try Problem 21

EXAMPLE 2 **Lawn Care** An installer of underground irrigation systems wants to cut a 20-foot length of plastic tubing into two pieces. The longer piece is to be 2 feet longer than twice the shorter piece. Find the length of each piece.

Analyze Refer to the figure, which shows the 20-foot-long pipe cut into two pieces.

Assign We can let s = the length of the shorter piece and ℓ = the length of the longer piece.

Form

The length of the shorter piece	plus	the length of the longer piece	is	20 feet.
s	$+$	ℓ	$=$	20

Since the longer piece is 2 feet longer than twice the shorter piece, we have

The length of the longer piece	is	2	times	the length of the shorter piece	plus	2 feet.
ℓ	$=$	2	$\cdot$	s	$+$	2

Solve Since the second equation is solved for ℓ, we will use the substitution method to solve the system.

(1) $\begin{cases} s + \ell = 20 \\ (2) \quad \ell = 2s + 2 \end{cases}$

$s + 2s + 2 = 20$ Substitute $2s + 2$ for ℓ in Equation 1.
$3s + 2 = 20$ Combine like terms.
$3s = 18$ Subtract 2 from both sides.
$s = 6$ Divide both sides by 3.

The shorter piece should be 6 feet long. To find the length of the longer piece, we substitute 6 for s in Equation 2 and find ℓ.

$\ell = 2s + 2$
$\ell = 2(6) + 2$ Substitute.
$\ell = 12 + 2$ Simplify.
$\ell = 14$

State The longer piece should be 14 feet long, and the shorter piece 6 feet long.

Check The sum of 6 and 14 is 20, and 14 is 2 more than twice 6. The results check.

THINK IT THROUGH College Students and Television Viewing

"Many college students no longer plan their lives around TV network line-ups, but they still watch hours and hours of television—often on their laptops."

Jenna Johnson, Post Local, 2011

According to Nielsen Media Research, the typical high school graduate will have spent 6,000 more hours in front of a TV set than in the classroom. Combined, the television viewing and classroom time totals about 30,000 hours. Use two equations in two variables to find the number of hours spent in the classroom and the number of hours spent watching television by the typical high school graduate.

2 Use systems to solve geometry problems.

EXAMPLE 3 Gardening
Tom has 150 feet of fencing to enclose a rectangular garden. If the garden's length is to be 5 feet less than 3 times its width, find the area of the garden.

Analyze To find the area of a rectangle, we need to know its length and width.

Assign We can let ℓ = the length of the garden and w = its width, as shown in the figure.

Form Since the perimeter of a rectangle is two lengths plus two widths, we have

2 times	the length of the garden	plus	2 times	the width of the garden	is	150 feet.
2 ·	ℓ	+	2 ·	w	=	150

Since the length is 5 feet less than 3 times the width,

The length of the garden	is	3 times	the width of the garden	minus	5 feet.
ℓ	=	3 ·	w	−	5

Solve Since the second equation is solved for ℓ, we will use the substitution method to solve this system.

(1) $\begin{cases} 2\ell + 2w = 150 \\ \ell = 3w - 5 \end{cases}$
(2)

$2(3w - 5) + 2w = 150$ Substitute $3w - 5$ for ℓ in Equation 1.
$6w - 10 + 2w = 150$ Distribute the multiplication by 2.
$8w - 10 = 150$ Combine like terms.
$8w = 160$ Add 10 to both sides.
$w = 20$ Divide both sides by 8.

The width of the garden is 20 feet. To find the length, we substitute 20 for w in Equation 2 and simplify.

$\ell = 3w - 5$
$\ell = 3(20) - 5$ Substitute.
$\ell = 55$ The length of the garden is 55 feet.

Self Check 3
In 1917, James Montgomery Flagg created the classic *I Want You* poster to help recruiting for World War I. The perimeter of the poster is 114 inches, and its length is 9 inches less than twice its width. Find the area of the poster.

Now Try Problem 34

Now we find the area of the rectangle with dimensions 55 feet by 20 feet.

$A = \ell w$ This is the formula for the area of a rectangle.
$A = 55 \cdot 20$ Substitute 55 for ℓ and 20 for w.
$A = 1{,}100$

State The garden covers an area of 1,100 square feet.

Check Because the dimensions of the garden are 55 feet by 20 feet, the perimeter is

$P = 2\ell + 2w$
$P = 2(55) + 2(20)$ Substitute 55 for ℓ and 20 for w.
$P = 110 + 40$
$P = 150$

It is also true that 55 feet is 5 feet less than 3 times 20 feet. The results check.

3 Use systems to solve break-point problems.

Self Check 4

A personal copy machine costs $100 with a per copy cost of $0.10. A larger copy machine costs $250 with a per copy cost of $0.07. How many copies will need to be made for the purchase of the larger copier to be worthwhile?

Now Try Problem 35

EXAMPLE 4   *Manufacturing* The setup cost of a machine that mills brass plates is $750. After setup, it costs $0.25 to mill each plate. Management is considering the purchase of a larger machine that can produce the same plate at a cost of $0.20 per plate. If the setup cost of the larger machine is $1,200, how many plates would the company have to produce to make the purchase worthwhile?

Analyze We need to find the number of plates (called the **break point**) that will cost equal amounts to produce on either machine.

Assign We can let c = the cost of milling p plates.

Form If we call the machine currently being used machine 1, and the new, larger one machine 2, we can form two equations.

The cost of making p plates on machine 1	is	the setup cost of machine 1	plus	the cost per plate on machine 1	times	the number of plates p to be made.
c	=	750	+	0.25	·	p

The cost of making p plates on machine 2	is	the setup cost of machine 2	plus	the cost per plate on machine 2	times	the number of plates p to be made.
c	=	1,200	+	0.20	·	p

Solve Since the costs are equal at the break point, we can use the substitution method to solve the system

(1) $\begin{cases} c = 750 + 0.25p \\ (2) \quad c = 1{,}200 + 0.20p \end{cases}$

$750 + 0.25p = 1{,}200 + 0.20p$ Substitute for c in the second equation.
$0.25p = 450 + 0.20p$ Subtract 750 from both sides.
$0.05p = 450$ Subtract $0.20p$ from both sides.
$p = 9{,}000$ Divide both sides by 0.05.

State If 9,000 plates are milled, the cost will be the same on either machine. If more than 9,000 plates are milled, the cost will be lower on the larger machine, because it mills the plates less expensively than the smaller machine.

Check We check the solution by substituting 9,000 for p in Equations 1 and 2 and verifying that 3,000 is the value of c in both cases.

If we graph the two equations, we can illustrate the break point.

Machine 1

$c = 750 + 0.25p$

p	c
0	750
1,000	1,000
5,000	2,000

Machine 2

$c = 1,200 + 0.20p$

p	c
0	1,200
4,000	2,000
12,000	3,600

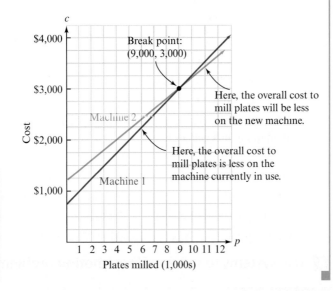

4 Use systems to solve interest problems.

EXAMPLE 5 **White-Collar Crime** A company secretly moved $150,000 out of the country to avoid paying corporate income tax on it. Some of the money was invested in a Swiss bank account that paid 8% interest annually. The rest was deposited in a Cayman Islands account paying 7% annual interest. The combined interest earned the first year was $11,500. How much money was invested in each account?

Analyze We are told that an unknown part of the $150,000 was invested at an annual rate of 8% and the rest at 7%. Together, the accounts earned $11,500 in interest.

Assign We can let x = the amount invested in the Swiss bank account and y = the amount invested in the Cayman Islands account.

Form Because the total investment was $150,000, we have

The amount invested in the Swiss account	+	the amount invested in the Cayman Is. Account	is	$150,000.
x	+	y	=	150,000

Since the annual income on x dollars invested at 8% is $0.08x$, the income on y dollars invested at 7% is $0.07y$, and the combined income is $11,500, we have

The income on the 8% investment	+	the income on the 7% investment	is	$11,500.
$0.08x$	+	$0.07y$	=	11,500

The resulting system is

(1) $\begin{cases} x + y = 150,000 \\ 0.08x + 0.07y = 11,500 \end{cases}$
(2)

Self Check 5

A woman invested $10,000, some at 9% and the rest at 10%. The annual income from these two investments was $975. How much was invested at each rate?

Now Try Problem 37

Solve To solve the system, we use the addition method to eliminate x.

$$-8x - 8y = -1{,}200{,}000 \quad \text{Multiply both sides of Equation 1 by } -8.$$
$$\underline{8x + 7y = 1{,}150{,}000} \quad \text{Multiply both sides of Equation 2 by 100.}$$
$$-y = -50{,}000$$
$$y = 50{,}000 \quad \text{Multiply (or divide) both sides by } -1.$$

To find x, we substitute 50,000 for y in Equation 1 and simplify.

$$x + y = 150{,}000$$
$$x + \mathbf{50{,}000} = 150{,}000 \quad \text{Substitute.}$$
$$x = 100{,}000 \quad \text{Subtract 50,000 from both sides.}$$

State $100,000 was invested in the Swiss bank account, and $50,000 was invested in the Cayman Islands account.

Check

$$\$100{,}000 + \$50{,}000 = \$150{,}000 \quad \text{The two investments total \$150,000.}$$
$$0.08(\$100{,}000) = \$8{,}000 \quad \text{The Swiss bank account earned \$8,000.}$$
$$0.07(\$50{,}000) = \$3{,}500 \quad \text{The Cayman Islands account earned \$3,500.}$$

The combined interest is $8,000 + $3,500 = $11,500. The results check.

5 Use systems to solve uniform motion problems.

Self Check 6
It takes a salmon 40 minutes to swim 10,000 feet upstream and 8 minutes to swim that same portion of a river downstream. Find the speed of the salmon in still water and the speed of the current.

Now Try Problem 42

EXAMPLE 6 *Boating* A boat traveled 30 miles downstream in 3 hours and made the return trip in 5 hours. Find the speed of the boat in still water and the speed of the current.

Analyze Traveling downstream, the speed of the boat will be faster than it would be in still water. Traveling upstream, the speed of the boat will be slower than it would be in still water.

Assign We can let s = the speed of the boat in still water and c = the speed of the current. Then the rate of the boat going downstream is $s + c$, and its rate going upstream is $s - c$. We can organize the information as shown.

Form Since each trip is 30 miles long, the Distance column of the table gives two equations in two variables. To write each equation in standard form, distribute.

	Rate	· Time	= Distance
Downstream	$s + c$	3	$3(s + c)$
Upstream	$s - c$	5	$5(s - c)$

$$\begin{cases} 3(s + c) = 30 \\ 5(s - c) = 30 \end{cases} \xrightarrow{\text{Distribute}} \begin{cases} 3s + 3c = 30 \quad (1) \\ 5s - 5c = 30 \quad (2) \end{cases}$$

Solve To solve this system by addition, we multiply Equation 1 by 5, multiply Equation 2 by 3, add the equations, and solve for s.

$$15s + 15c = 150$$
$$\underline{15s - 15c = 90}$$
$$30s = 240$$
$$s = 8 \quad \text{Divide both sides by 30.}$$

To find c, it appears that the calculations will be easiest if we use $3s + 3c = 30$.

$$3s + 3c = 30$$
$$3(8) + 3c = 30 \quad \text{Substitute 8 for } s.$$
$$24 + 3c = 30 \quad \text{Multiply.}$$
$$3c = 6 \quad \text{Subtract 24 from both sides.}$$
$$c = 2 \quad \text{Divide both sides by 3. This is the speed of the current.}$$

State The speed of the boat in still water is 8 miles per hour, and the speed of the current is 2 miles per hour.

Check With a 2-mph current, the boat's downstream speed will be 8 + 2 = 10 mph. In 3 hours, it will travel 10 · 3 = 30 miles. With a 2-mph current, the boat's upstream speed will be 8 − 2 = 6 mph. In 5 hours, it will cover 6 · 5 = 30 miles. The results check.

6 Use systems to solve mixture problems.

EXAMPLE 7 *Medical Technology*

A laboratory technician has one batch of antiseptic that is 40% alcohol and a second batch that is 60% alcohol. She would like to make 8 liters of solution that is 55% alcohol. How many liters of each batch should she use?

Analyze Some 60% solution must be added to some 40% solution to make a 55% solution.

Assign We can let x = the number of liters to be used from batch 1 and y = the number of liters to be used from batch 2. We then organize the information as shown.

	Number of liters of solution	·	% of concentration	=	Number of liters of alcohol
Batch 1	x		0.40		$0.40x$
Batch 2	y		0.60		$0.60y$
Mixture	8		0.55		$0.55(8)$

↑ One equation comes from information in this column.
↑ 40%, 60%, and 55% have been expressed as decimals.
↑ Another equation comes from information in this column.

Self Check 7

How much 1% milk must be added to 4% milk to obtain 60 liters of 2% milk?

Now Try Problem 43

Form The information in the table provides two equations.

(1) $\begin{cases} x + y = 8 \end{cases}$ The number of liters of batch 1 plus the number of liters of batch 2 equals the total number of liters in the mixture.

(2) $\phantom{\begin{cases}}0.40x + 0.60y = 0.55(8)$ The amount of alcohol in batch 1 plus the amount of alcohol in batch 2 equals the amount of alcohol in the mixture.

Solve We can use addition to solve this system.

$-40x - 40y = -320$ Multiply both sides of Equation 1 by -40.
$\underline{40x + 60y = 440}$ Multiply both sides of Equation 2 by 100.
$20y = 120$
$y = 6$ Divide both sides by 20.

To find x, we substitute 6 for y in Equation 1 and simplify.

$x + y = 8$
$x + 6 = 8$ Substitute.
$x = 2$ Subtract 6 from both sides.

State The technician should use 2 liters of the 40% solution and 6 liters of the 60% solution.

Check The check is left to the reader.

ANSWERS TO SELF CHECKS

1. team photo: $17.00, wallet: $1.50 2. 4 ft, 8 ft 3. 770 in.² 4. 5,000 copies
5. $2,500 at 9%, $7,500 at 10% 6. salmon: 750 ft/min, current: 500 ft/min 7. 40 liters

SECTION 4.4 STUDY SET

VOCABULARY

Fill in the blanks.

1. A _____ is a letter that stands for a number.
2. An _____ is a statement indicating that two quantities are equal.
3. $\begin{cases} a + b = 20 \\ a = 2b + 4 \end{cases}$ is a _____ of linear equations.
4. A _____ of a system of linear equations satisfies both equations simultaneously.

CONCEPTS

5. For each case in the illustration, write an algebraic expression that represents the speed of the canoe in mph if its speed in still water is x mph.

6. See the illustration.
   a. If the contents of the two test tubes are poured into a third test tube, how much solution will the third test tube contain? (mL means milliliters.)
   b. Which is the best estimate of the concentration of the solution in the third test tube: 25%, 35%, or 45% acid solution?

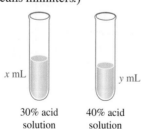

7. Use the information in the table to answer the questions about two investments.

	Principal ·	Rate ·	Time =	Interest
City Bank	x	5%	1 yr	
USA Savings	y	11%	1 yr	

   a. How much money was deposited in the USA Savings account?
   b. What interest rate did the City Bank account earn?
   c. Complete the table.

8. Use the information in the table to answer the questions about a plane flying in windy conditions.

	Rate ·	Time =	Distance
With	$x + y$	3 hr	450 mi
Against	$x - y$	5 hr	450 mi

   a. For how long did the plane fly against the wind?
   b. At what rate did the plane travel when flying with the wind?
   c. Write two equations that could be used to solve for x and y.

9. a. If a problem contains two unknowns, and if two variables are used to represent them, how many equations must be written to find the unknowns?
   b. Name three methods that can be used to solve a system of linear equations.

10. Put the steps of the six-step problem-solving strategy listed below in the correct order.

    State the conclusion.   Form two equations.
    Analyze the problem.   Check the result.
    Solve the system.   Assign variables.

NOTATION

Write a formula that relates the given quantities.

11. length, width, area of a rectangle
12. length, width, perimeter of a rectangle
13. rate, time, distance traveled
14. principal, rate, time, interest earned

Translate each verbal model into mathematical symbols. Use variables to represent any unknowns.

15. 2 · length of pool + 2 · width of pool is 90 yards.

16. $6 · [number of adults] + $2 · [number of children] is $26.

APPLICATIONS

Use a system of two equations in two variables to solve each problem.

17. One integer is twice another. Their sum is 96.
18. The sum of two integers is 38. Their difference is 12.
19. Three times one integer plus another integer is 29. The first integer plus twice the second is 18.
20. Twice one integer plus another integer is 21. The first integer plus 3 times the second is 33.
21. TREE TRIMMING When fully extended, the arm on the tree service truck shown is 51 feet long. If the upper part of the arm is 7 feet shorter than the lower part, how long is each part of the arm?

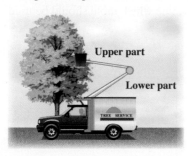

22. TV PROGRAMMING The producer of a 30-minute TV documentary about World War I divided it into two parts. Four times as much program time was devoted to the causes of the war as to the outcome. How long is each part of the documentary?
23. ALASKA Most of the 1,422-mile-long Alaska Highway is actually in Canada. Find the length of the highway that is in Canada and the length that is in Alaska if it is known that the *difference* in the lengths is 1,020 miles.

24. EXECUTIVE BRANCH The annual salaries of the president and vice president of the United States total $630,700 a year. If the president makes $169,300 more than the vice president, find each of their annual salaries.
25. CAUSES OF DEATH According to the *National Vital Statistics Reports*, in 2011, the number of Americans who died from heart disease was about 5 times the number who died from accidents. If the total number of deaths from these two causes was 720,000, how many Americans died from each cause in 2011?
26. BUYING PAINTING SUPPLIES Two partial receipts for paint supplies are shown. How much does each gallon of paint and each brush cost?

27. WEDDING PICTURES A photographer sells the two wedding picture packages shown in the illustration. How much does a 10 × 14 photo cost? An 8 × 10 photo?

28. BUYING TICKETS If receipts for the movie advertised below were $1,440 for an audience of 190 people, how many senior citizens attended?

29. **SELLING ICE CREAM** At a store, ice cream cones cost $1.80 and sundaes cost $3.30. One day a total of 148 cones and sundaes were sold, and the receipts totaled $360.90. How many cones were sold?

30. **THE MARINE CORPS** The Marine Corps War Memorial in Arlington, Virginia, portrays the raising of the U.S. flag on Iwo Jima during World War II. Find the measures of the two angles shown below if the measure of one of the angles is 15° less than twice the other.

31. **PHYSICAL THERAPY** To rehabilitate her knee, an athlete does leg extensions. Her goal is to regain a full 90° range of motion in this exercise. Use the information in the illustration to determine her current range of motion in degrees.

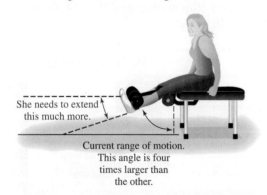

32. **THEATER SCREENS** At an IMAX theater, the giant rectangular movie screen has a width 26 feet less than its length. If its perimeter is 332 feet, find the area of the screen.

33. **ART** In 1770, Thomas Gainsborough painted *The Blue Boy*. The sum of the length and width of the painting is 118 in. The difference between the length and width is 22 in. Find the length and the width of the painting.

34. **GEOMETRY** A 50-meter-long path surrounds the rectangular garden shown below. If the width of the garden is two-thirds its length, find its area.

35. **MAKING TIRES** A company has two molds to form tires. One mold has a setup cost of $1,000, and the other has a setup cost of $3,000. The cost to make each tire with the first mold is $15, and the cost to make each tire with the second mold is $10.

   a. Find the break point.

   b. Check your result by graphing both equations on the coordinate system in the illustration.

   c. If a production run of 500 tires is planned, determine which mold should be used.

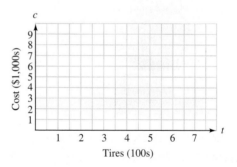

36. **CHOOSING A FURNACE** A high-efficiency 90+ furnace can be purchased for $2,250 and costs an average of $412 per year to operate in Big Bear Lake, California. An 80+ furnace can be purchased for only $1,715, but it costs $466 per year to operate.

   a. Find the break point.

   b. If you intended to live in a Big Bear Lake house for 7 years, which furnace would you choose?

37. **STUDENT LOANS** A college used a $5,000 gift from an alumnus to make two student loans. The first was at 5% annual interest to a nursing student. The second was at 7% to a business major. If the college collected $310 in interest the first year, how much was loaned to each student?

38. **FINANCIAL PLANNING** In investing $6,000 of a couple's money, a financial planner put some of it into a savings account paying 6% annual interest. The rest was invested in a riskier mini-mall development plan paying 12% annually. The combined interest earned for the first year was $540. How much money was invested at each rate?

39. THE GULF STREAM The Gulf Stream is a warm ocean current of the North Atlantic Ocean that flows northward, as shown. Heading north with the Gulf Stream, a cruise ship traveled 300 miles in 10 hours. Against the current, it took 15 hours to make the return trip. Find the speed of the current.

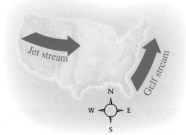

40. THE JET STREAM The jet stream is a strong wind current that flows across the United States, as shown in the illustration above. Flying with the jet stream, a plane flew 3,000 miles in 5 hours. Against the same wind, the trip took 6 hours. Find the airspeed of the plane (the speed in still air).

41. AVIATION An airplane can fly downwind a distance of 600 miles in 2 hours. However, the return trip against the same wind takes 3 hours. Find the speed of the wind.

42. BOATING A boat can travel 24 miles downstream in 2 hours and can make the return trip in 3 hours. Find the speed of the boat in still water.

43. MARINE BIOLOGY A marine biologist wants to set up an aquarium containing 3% salt water. He has two tanks on hand that contain 6% and 2% salt water. How much water from each tank must he use to fill a 16-liter aquarium with a 3% saltwater mixture?

44. COMMEMORATIVE COINS A foundry has been commissioned to make souvenir coins. The coins are to be made from an alloy that is 40% silver. The foundry has on hand two alloys, one with 50% silver content and one with a 25% silver content. How many kilograms of each alloy should be used to make 20 kilograms of the 40% silver alloy?

45. MIXING NUTS A merchant wants to mix peanuts with cashews, as shown in the illustration in the next column, to get 48 pounds of mixed nuts that will be sold at $4 per pound. How many pounds of each should the merchant use?

46. COFFEE SALES A coffee supply store waits until the orders for its special coffee blend reach 100 pounds before making up a batch. Coffee selling for $8.75 a pound is blended with coffee selling for $3.75 a pound to make a product that sells for $6.35 a pound. How much of each type of coffee should be used to make the blend that will fill the orders?

47. MARKDOWN A set of golf clubs has been marked down 40% to a sale price of $384. Let r represent the retail price and d the discount. Then use the following equations to find the original retail price.

$$\text{Retail price} - \text{discount} = \text{sale price}$$

$$\text{Discount} = \text{discount rate} \cdot \text{retail price}$$

48. MARKUP A stereo system retailing at $565.50 has been marked up 45% from wholesale. Let w represent the wholesale cost and m the markup. Then use the following equations to find the wholesale cost.

$$\text{Wholesale cost} + \text{markup} = \text{retail price}$$

$$\text{Markup} = \text{markup rate} \cdot \text{wholesale cost}$$

WRITING

49. When solving a problem using two variables, why isn't one equation sufficient to find the two unknown quantities?

50. Describe an everyday situation in which you might need to make a mixture.

REVIEW

Graph the solution set of each inequality.

51. $x < 4$

52. $x \geq -3$

53. $-1 < x \leq 2$

54. $-2 \leq x \leq 0$

Solve each equation.

55. $\dfrac{x}{5} - 4 = 2$

56. $\dfrac{x-5}{4} + 6 = 0$

57. $3(x + 8) - 6(x + 4) = -3x$

58. $2(x + 10) + x = 3(x + 8)$

SECTION 4.5
Solving Systems of Linear Inequalities

Objectives
1. Solve a system of linear inequalities by graphing.
2. Use graphing to identify systems of inequalities with no solutions.
3. Solve application problems involving systems of linear inequalities.

In Section 4.1, we solved systems of linear *equations* by the graphing method. The solution of such a system is the point of intersection of the straight lines. We now consider how to solve systems of linear *inequalities* graphically. When the solution of a linear inequality is graphed, the result is a half-plane. Therefore, we would expect to find the graphical solution of a system of inequalities by looking for the intersection, or overlap, of shaded half-planes.

1 Solve a system of linear inequalities by graphing.

Self Check 1

Graph the solutions of the system:
$$\begin{cases} x - y \leq 2 \\ x + y \geq -1 \end{cases}$$

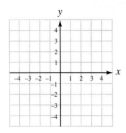

Now Try Problem 26

EXAMPLE 1 Graph the solutions of the system: $\begin{cases} x + y \geq 1 \\ x - y \geq 1 \end{cases}$

Strategy We will graph the solutions of $x + y \geq 1$ in one color and the solutions of $x - y \geq 1$ in another color on the same coordinate system.

WHY We need to see where the graphs of the two inequalities intersect (overlap).

Solution

We first graph each inequality. For this first example, we will graph each inequality on a separate set of axes, although in practice we will draw them on the same axes.

To graph $x + y \geq 1$, we begin by graphing the boundary line $x + y = 1$. Since the inequality contains an $\geq$ symbol, the boundary is a solid line. Because the coordinates of the test point $(0, 0)$ do not satisfy $x + y \geq 1$, we shade (in red) the side of the boundary that does **not** contain $(0, 0)$. See figure (a) on the next page.

To graph $x - y \geq 1$, we begin by graphing the boundary line $x - y = 1$. Since the inequality contains an $\geq$ symbol, the boundary is a solid line. Because the coordinates of the test point $(0, 0)$ do not satisfy $x - y \geq 1$, we shade (in blue) the side of the boundary that does **not** contain $(0, 0)$. See figure (b).

$x + y = 1$		
x	y	(x, y)
0	1	$(0, 1)$
1	0	$(1, 0)$
2	-1	$(2, -1)$

$x - y = 1$		
x	y	(x, y)
0	-1	$(0, -1)$
1	0	$(1, 0)$
2	1	$(2, 1)$

Unless otherwise noted, all content on this page is © Cengage Learning.

Shading: Check the test point (0, 0).

$x + y \geq 1$

$0 + 0 \overset{?}{\geq} 1$ Substitute

$0 \geq 1$ False

(0, 0) is not a solution of $x + y \geq 1$.

Shading: Check the test point (0, 0).

$x - y \geq 1$

$0 - 0 \overset{?}{\geq} 1$ Substitute

$0 \geq 1$ False

(0, 0) is not a solution of $x - y \geq 1$.

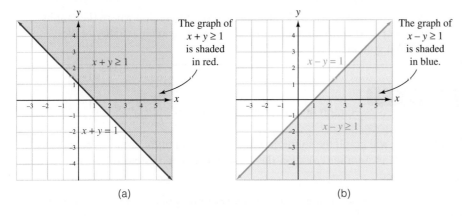

(a) The graph of $x + y \geq 1$ is shaded in red.

(b) The graph of $x - y \geq 1$ is shaded in blue.

In the following figure, we show the result when the inequalities $x + y \geq 1$ and $x - y \geq 1$ are graphed one at a time on the same coordinate axes. The area that is shaded twice represents the set of simultaneous solutions of the given system of inequalities. Any point in the doubly shaded region (shown in purple) has coordinates that satisfy both inequalities of the system.

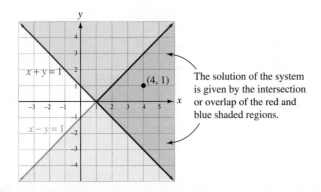

The solution of the system is given by the intersection or overlap of the red and blue shaded regions.

To see whether this is true, we can pick a point, such as (4, 1), that lies in the doubly shaded region and show that its coordinates satisfy both inequalities.

Check: $x + y \geq 1$ and $x - y \geq 1$

$4 + 1 \geq 1$ $4 - 1 \geq 1$

$5 \geq 1$ True $3 \geq 1$ True

The resulting true statements verify that (4, 1) is a solution of the system. If we pick a point that is not in the doubly shaded region, its coordinates will fail to satisfy at least one of the inequalities.

> **The Language of Algebra**   To solve a system of linear inequalities we *superimpose* the graphs of the inequalities. That is, we place one graph on top of the other. Most cameras *superimpose* the date and time over the picture.

In general, to solve systems of linear inequalities, we will follow these steps.

> **Solving Systems of Inequalities**
>
> 1. Graph each inequality on the same rectangular coordinate system.
> 2. Use shading to highlight the intersection of the graphs (the region where the graphs overlap). The points in this region are the solutions of the system.
> 3. As an informal check, pick a test point from the region where the graphs intersect, and verify that its coordinates satisfy each inequality of the original system.

Self Check 2

Graph the solutions of the system:
$$\begin{cases} x + 3y < 3 \\ -x + 3y > 3 \end{cases}$$

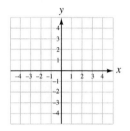

Now Try Problem 31

EXAMPLE 2 Graph the solutions of the system: $\begin{cases} 2x + y < 4 \\ y > 2x + 2 \end{cases}$

Strategy We will graph the solutions of $2x + y < 4$ in one color and the solutions of $y > 2x + 2$ in another color on the same coordinate system to see where the graphs intersect.

WHY The solution set of the system is the set of all points in the intersection of the two graphs.

Solution

First, we graph each inequality on one set of axes, as shown in the figure.

Boundary for $2x + y < 4$

$2x + y = 4$

x	y	(x, y)
0	4	(0, 4)
2	0	(2, 0)
1	2	(1, 2)

Boundary for $y > 2x + 2$

$y = 2x + 2$

So $m = 2 = \dfrac{2}{1}$

and $b = 2$.

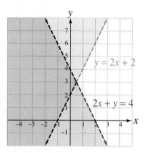

Shading for $2x + y < 4$

Check the test point $(0, 0)$.

$2x + y < 4$

$2(0) + 0 \overset{?}{<} 4$ Substitute

$0 < 4$ True

Shading for $y > 2x + 2$

Check the test point $(0, 0)$.

$y > 2x + 2$

$0 \overset{?}{>} 2(0) + 2$

$0 > 2$ False

We note that

- The graph of $2x + y < 4$ includes all points below the boundary line $2x + y = 4$. Since the boundary is not included, we draw it as a dashed line.

- The graph of $y > 2x + 2$ includes all points above the boundary line $y = 2x + 2$. Since the boundary is not included, we draw it as a dashed line.

The area that is shaded twice (the region in purple) is the solution of the given system of inequalities. Any point in the doubly shaded region has coordinates that will satisfy both inequalities of the system.

Pick a point in the doubly shaded region and show that it satisfies both inequalities.

EXAMPLE 3

Graph the solutions of the system: $\begin{cases} x \leq 2 \\ y > 3 \end{cases}$

Strategy We will graph the solutions of $x \leq 2$ in one color and the solutions of $y > 3$ in another color on the same coordinate system to see where the graphs intersect.

WHY The solution set of the system is the set of all points in the intersection of the two graphs.

Solution
We graph each inequality on one set of axes, as shown in the figure.

Boundary for $x \leq 2$

	$x = 2$	
x	y	(x, y)
2	0	(2, 0)
2	2	(2, 2)
2	4	(2, 4)

Boundary for $y > 3$

	$y = 3$	
x	y	(x, y)
0	3	(0, 3)
1	3	(1, 3)
4	3	(4, 3)

Shading for $x \leq 2$

Check the test point $(0, 0)$.

$x \leq 2$
$0 \stackrel{?}{\leq} 2$ True

Shading for $y > 3$

Check the test point $(0, 0)$.

$y > 3$
$0 \stackrel{?}{>} 3$ False

We note that

- The graph of $x \leq 2$ includes all points to the left of the boundary line $x = 2$. Since the boundary is included, we draw it as a solid line.
- The graph $y > 3$ includes all points above the boundary line $y = 3$. Since the boundary is not included, we draw it as a dashed line.

The area that is shaded twice is the solution of the given system of inequalities. Any point in the doubly shaded region (purple) has coordinates that will satisfy both inequalities of the system. Pick a point in the doubly shaded region and show that this is true.

Self Check 3

Graph the solutions of the system:
$\begin{cases} y \leq 1 \\ x > 2 \end{cases}$

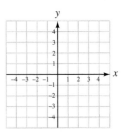

Now Try Problem 34

2 Use graphing to identify systems of inequalities with no solutions.

EXAMPLE 4

Graph the solutions of the system: $\begin{cases} y < 3x - 1 \\ y \geq 3x + 1 \end{cases}$

Strategy We will graph the solutions of $y < 3x - 1$ in one color and the solutions of $y \geq 3x + 1$ in another color on the same coordinate system to see where or if the graphs intersect.

WHY The solution set of the system is the set of all points in the intersection of the two graphs.

Self Check 4

Graph the solutions of the system:

$\begin{cases} y \geq -\dfrac{1}{2}x + 1 \\ y < -\dfrac{1}{2}x - 1 \end{cases}$

(continued)

368 Chapter 4 Solving Systems of Equations and Inequalities

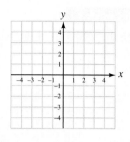

Now Try Problem 37

Solution
We graph each inequality as shown in the figure and make the following observations:

- The graph of $y < 3x - 1$ includes all points below the dashed line $y = 3x - 1$.
- The graph of $y \geq 3x + 1$ includes all points on and above the solid line $y = 3x + 1$.

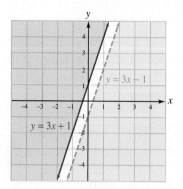

Because the graphs of these inequalities do not intersect, the solution set is empty. There are no solutions.

Self Check 5
Graph the solutions of the system:
$$\begin{cases} x \leq 1 \\ y \leq 2 \\ 2x - y \leq 4 \end{cases}$$

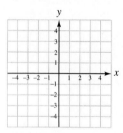

Now Try Problem 39

EXAMPLE 5
Graph the solutions of the system: $\begin{cases} x \geq 0 \\ y \geq 0 \\ x + 2y \leq 6 \end{cases}$

Strategy We will graph the solutions of $x \geq 0$, $y \geq 0$, and $x + 2y \leq 6$ on the same coordinate system to see where all three graphs intersect (overlap).

WHY The solution set of the system is the set of all points in the intersection of the three graphs.

Solution
We graph each inequality as shown in the figure and make the following observations:

- The graph of $x \geq 0$ includes all points on the y-axis and to the right.
- The graph of $y \geq 0$ includes all points on the x-axis and above.
- The graph of $x + 2y \leq 6$ includes all points on the line $x + 2y = 6$ and below.

The solution is the region that is shaded three times. This includes triangle OPQ and the triangular region it encloses.

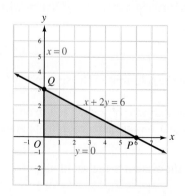

3 Solve application problems involving systems of linear inequalities.

EXAMPLE 6 Landscaping

A homeowner budgets from $300 to $600 for trees and bushes to landscape his yard. After shopping around, he finds that good trees cost $150 and mature bushes cost $75. What combinations of trees and bushes can he afford to buy?

Analyze

- The homeowner wants to spend *at least* $300 but *not more than* $600 for trees and bushes.
- Trees cost $150 each and bushes cost $75 each.
- What combination of trees and bushes can he buy?

Assign We can let x represent the number of trees purchased and y the number of bushes purchased.

Form

The cost of a tree	times	the number of trees purchased	plus	the cost of a bush	times	the number of bushes purchased	should at least be	$300.
$150	·	x	+	$75	·	y	$\geq$	$300

The cost of a tree	times	the number of trees purchased	plus	the cost of a bush	times	the number of bushes purchased	should not be more than	$600.
$150	·	x	+	$75	·	y	$\leq$	$600

Solve The graph of the following system is shown below.

$$\begin{cases} 150x + 75y \geq 300 \\ 150x + 75y \leq 600 \end{cases}$$

State The coordinates of each point in the shaded region represent a possible combination of the number of trees (x) and the number of bushes (y) that can be purchased. These possibilities are

(0, 4), (0, 5), (0, 6), (0, 7), (0, 8)
(1, 2), (1, 3), (1, 4), (1, 5), (1, 6)
(2, 0), (2, 1), (2, 2), (2, 3), (2, 4)
(3, 0), (3, 1), (3, 2), (4, 0)

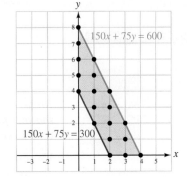

Only these points can be used, because the homeowner cannot buy a portion of a tree or a bush.

Check Suppose the homeowner picks a combination of 2 trees and 4 bushes, as represented by (2, 4). Show that this point satisfies both inequalities of the system.

Self Check 6

BUYING CDs AND DVDs An electronics store sells CDs for $10 and DVDs for $20. Carly wants to spend at least $100 but no more than $200 on ($x$) CDs and ($y$) DVDs. What combinations of CDs and DVDs can she afford to buy?

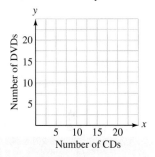

Now Try Problem 51

ANSWERS TO SELF CHECKS

1.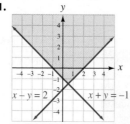

2.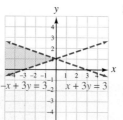

3.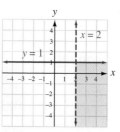

4. No solution.

5.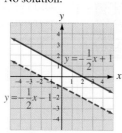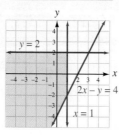

6. Any ordered pair in the shaded region with whole number coordinates.

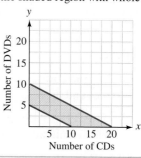

SECTION 4.5 STUDY SET

VOCABULARY

Fill in the blanks.

1. $\begin{cases} x + y > 2 \\ x + y < 4 \end{cases}$ is a system of linear _____.

2. The _____ of a system of linear inequalities is all the ordered pairs that make all inequalities of the system true at the same time.

3. Any point in the doubly _____ region of the graph of the solution of a system of two linear inequalities has coordinates that satisfy both inequalities of the system.

4. To graph a linear inequality such as $x + y > 2$, first graph the boundary. Then pick a test _____ to determine which half-plane to shade.

CONCEPTS

In the illustration, the solution of linear inequality 1 was shaded in red, and the solution of linear inequality 2 was shaded in blue. The overlap of the red and the blue regions is shown in purple. Determine whether a true or a false statement results when the coordinates of the given point are substituted into the given inequality.

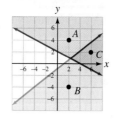

5. A, inequality 1
6. A, inequality 2
7. B, inequality 1
8. B, inequality 2
9. C, inequality 1
10. C, inequality 2

Match each equation, inequality, or system with the graph of its solution.

11. $x + y = 2$
12. $x + y \geq 2$
13. $\begin{cases} x + y = 2 \\ x - y = 2 \end{cases}$
14. $\begin{cases} x + y \geq 2 \\ x - y \leq 2 \end{cases}$

i 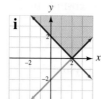 ii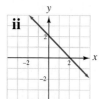

iii 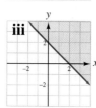 iv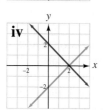

The graph of the solution of a system of two linear inequalities is shown in purple. Determine whether each point is a part of the solution set.

15. $(4, -2)$
16. $(1, 3)$
17. the origin
18. $(3, 2)$

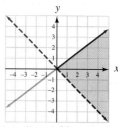

19. Use a system of inequalities to describe the purple shaded region in the illustration.

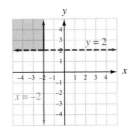

NOTATION

20. Fill in the blank: The graph of the solution of a system of linear inequalities shown on the right can be described as the triangle _____ and the triangular region it encloses.

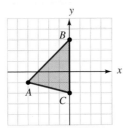

Represent each phrase using either $>$, $<$, $\geq$, or $\leq$.

21. is not more than

22. must be at least

23. should not surpass

24. cannot go below

GUIDED PRACTICE

Graph the solutions of each system of inequalities. See Example 1.

25. $\begin{cases} 3x + 4y \geq -7 \\ 2x - 3y \geq 1 \end{cases}$

26. $\begin{cases} 2x - 4y \geq -6 \\ 3x + y \geq 5 \end{cases}$

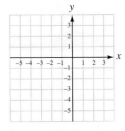 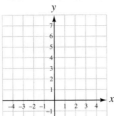

27. $\begin{cases} x + 2y \leq 3 \\ 2x - y \geq 1 \end{cases}$

28. $\begin{cases} 2x + y \geq 3 \\ x - 2y \leq -1 \end{cases}$

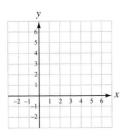 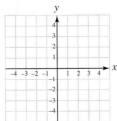

Graph the solutions of each system of inequalities. See Example 2.

29. $\begin{cases} x + y < -1 \\ x - y > -1 \end{cases}$

30. $\begin{cases} x + y > 2 \\ x - y < -2 \end{cases}$

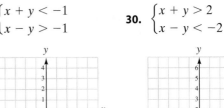 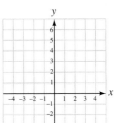

31. $\begin{cases} 2x - 3y \leq 0 \\ y \geq x - 1 \end{cases}$

32. $\begin{cases} y > 2x - 4 \\ y \geq -x - 1 \end{cases}$

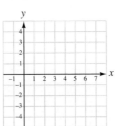 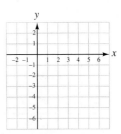

Graph the solutions of each system of inequalities. See Example 3.

33. $\begin{cases} x > 2 \\ y \leq 3 \end{cases}$

34. $\begin{cases} x \geq -1 \\ y > -2 \end{cases}$

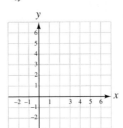 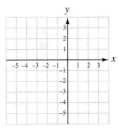

35. $\begin{cases} x > 0 \\ y > 0 \end{cases}$

36. $\begin{cases} x \leq 0 \\ y < 0 \end{cases}$

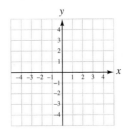 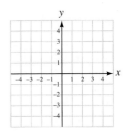

Graph the solutions of each system of inequalities, when possible. See Examples 4–5.

37. $\begin{cases} y < -x + 1 \\ y > -x + 3 \end{cases}$

38. 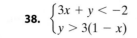 $\begin{cases} 3x + y < -2 \\ y > 3(1 - x) \end{cases}$

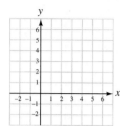
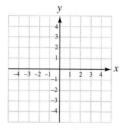

39. $\begin{cases} x \geq 0 \\ y \geq 0 \\ x + y \leq 3 \end{cases}$

40. $\begin{cases} x - y \leq 6 \\ x + 2y \leq 6 \\ x \geq 0 \end{cases}$

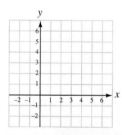
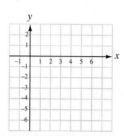

TRY IT YOURSELF

Graph the solutions of each system of inequalities.

41. $\begin{cases} 2x + y < 7 \\ y > 2(1 - x) \end{cases}$

42. 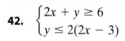 $\begin{cases} 2x + y \geq 6 \\ y \leq 2(2x - 3) \end{cases}$

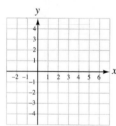
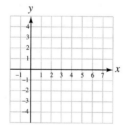

43. $\begin{cases} 3x + y \leq 1 \\ 4x - y > -8 \end{cases}$

44. $\begin{cases} 2x - 3y < 0 \\ 2x + 3y \geq 12 \end{cases}$

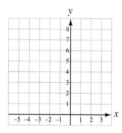
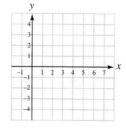

45. 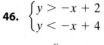 $\begin{cases} 3x - y \leq -4 \\ 3y > -2(x + 5) \end{cases}$

46. $\begin{cases} y > -x + 2 \\ y < -x + 4 \end{cases}$

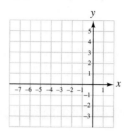
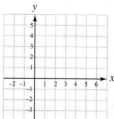

47. $\begin{cases} \dfrac{x}{2} + \dfrac{y}{3} \geq 2 \\ \dfrac{x}{2} - \dfrac{y}{2} < -1 \end{cases}$

48. $\begin{cases} \dfrac{x}{3} - \dfrac{y}{2} < -3 \\ \dfrac{x}{3} + \dfrac{y}{2} > -1 \end{cases}$

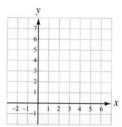
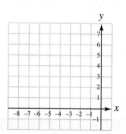

APPLICATIONS

49. **BIRDS OF PREY** Parts a and b of the illustration show the individual fields of vision for each eye of an owl. In part c, shade the area where the fields of vision overlap—that is, the area that is seen by both eyes.

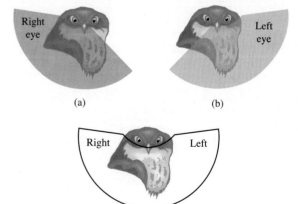

50. **EARTH SCIENCE** Shade the area of Earth's surface that is north of the Tropic of Capricorn and south of the Tropic of Cancer.

In Exercises 51–54, graph each system of inequalities and give two possible solutions.

51. BUYING COMPACT DISCS Melodic Music has compact discs on sale for either $10 or $15. If a customer wants to spend at least $30 but no more than $60 on CDs, use the illustration to graph a system of inequalities showing the possible combinations of $10 CDs ($x$) and $15 CDs ($y$) that the customer can buy.

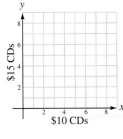

52. BUYING BOATS Dry Boatworks wholesales aluminum boats for $800 and fiberglass boats for $600. Northland Marina wants to make a purchase totaling at least $2,400, but no more than $4,800. Use the illustration to graph a system of inequalities showing the possible combinations of aluminum boats (x) and fiberglass boats (y) that can be ordered.

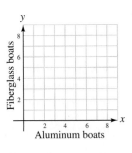

53. BUYING FURNITURE A distributor wholesales desk chairs for $150 and side chairs for $100. Best Furniture wants its order to total no more than $900; Best also wants to order more side chairs than desk chairs. Use the illustration to graph a system of inequalities showing the possible combinations of desk chairs (x) and side chairs (y) that can be ordered.

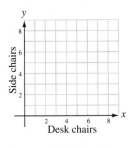

54. ORDERING FURNACE EQUIPMENT J. Bolden Heating Company wants to order no more than $2,000 worth of electronic air cleaners and humidifiers from a wholesaler that charges $500 for air cleaners and $200 for humidifiers. If Bolden wants more humidifiers than air cleaners, use the illustration to graph a system of inequalities showing the possible combinations of air cleaners (x) and humidifiers (y) that can be ordered.

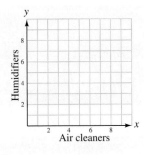

55. PESTICIDES To eradicate a fruit fly infestation, helicopters sprayed an area of a city that can be described by $y \geq -2x + 1$ (within the city limits). Two weeks later, more spraying was ordered over the area described by $y \geq \frac{1}{4}x - 4$ (within the city limits). In the illustration, show the part of the city that was sprayed twice.

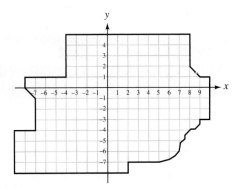

56. REDEVELOPMENT Refer to the following diagram. A government agency has declared an area of a city east of First Street, north of Second Avenue, south of Sixth Avenue, and west of Fifth Street as eligible for federal redevelopment funds. Describe this area of the city mathematically using a system of four inequalities, if the corner of Central Avenue and Main Street is considered the origin.

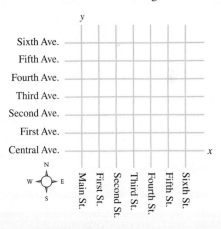

WRITING

57. Explain how to use graphing to solve a system of inequalities.

58. Explain when a system of inequalities will have no solutions.

59. Describe how the graphs of the solutions of these systems are similar and how they differ.

$$\begin{cases} x + y = 4 \\ x - y = 4 \end{cases} \quad \text{and} \quad \begin{cases} x + y \geq 4 \\ x - y \geq 4 \end{cases}$$

60. When a solution of a system of linear inequalities is graphed, what does the shading represent?

REVIEW

Complete each table.

61. $y = 2x^2$

x	y
8	
-2	

62. $t = -|s + 2|$

s	t
-3	
-10	

63. $f(x) = 4 + x^3$

Input	Output
0	
-3	

64. $g(x) = 2x - x^2$

x	g(x)
5	
-5	

STUDY SKILLS CHECKLIST

Preparing for the Chapter 4 Test

In Chapter 4 you learned three methods to solve a system of linear equations. You also learned how to graph the solutions of a system of linear inequalities. As you prepare for the exam on this material, make sure you also review the following checklist.

☐ To check a proposed solution of a system of equations, be sure the coordinates of the ordered pair satisfies *both* equations.

Is $(3, -2)$ a solution of the system $\begin{cases} 3x + 4y = 1 \\ x + 2y = -1 \end{cases}$?

$3(3) + 4(-2) \stackrel{?}{=} 1$ Substitute 3 for x and −2 for y.

$9 - 8 \stackrel{?}{=} 1$

$1 = 1$ True

$3 + 2(-2) \stackrel{?}{=} -1$ Substitute 3 for x and −2 for y.

$3 - 4 \stackrel{?}{=} -1$

$-1 = -1$ True

Yes, $(3, -2)$ is a solution of the system.

☐ When solving a system of equations by graphing, you must determine the coordinates of the point at the intersection of the graphs. That ordered pair is the solution of the system.

☐ When using the substitution or the addition (elimination) method, remember to find the value of *both* the variables.

For the system of linear equations $\begin{cases} x = 2y - 3 \\ x + 4y = 3 \end{cases}$ the y-coordinate of the solution is $y = 1$.

To find the *x*-value, substitute 1 for *y* in either equation:

$x = 2y - 3$

$x = 2(\mathbf{1}) - 3$

$x = 2 - 3$

$x = -1$

The solution is $(-1, 1)$.

CHAPTER 4   SUMMARY AND REVIEW

SECTION 4.1   Solving Systems of Equations by Graphing

DEFINITIONS AND CONCEPTS	EXAMPLES
When two equations are considered at the same time, we say that they form a **system of equations**.	Is $(4, 3)$ a solution of the system $\begin{cases} x + y = 7 \\ x - y = 5 \end{cases}$?

Chapter 4 Summary and Review

A **solution of a system** of equations in two variables is an ordered pair that satisfies both equations of the system.	To answer this question, we substitute 4 for x and 3 for y in each equation. $x + y = 7 \qquad\qquad x - y = 5$ $4 + 3 \stackrel{?}{=} 7 \qquad\quad 4 - 3 \stackrel{?}{=} 5$ $7 = 7 \quad \text{True} \qquad\quad 1 = 5 \quad \text{False}$ Although $(4, 3)$ satisfies the first equation, it does not satisfy the second. Because it does not satisfy both equations, it is not a solution of the system.
To solve a system graphically: 1. Graph each equation on the same rectangular coordinate system. 2. If the lines intersect, determine the coordinates of the point of intersection of the graphs. That ordered pair is the solution. 3. Check the proposed solution in each equation of the original system.	Use graphing to solve the system: $\begin{cases} y = -2x + 3 \\ x - 2y = 4 \end{cases}$ **Step 1** Graph each equation as shown below. $y = -2x + 3$ $m = \dfrac{-2}{1}$ $b = 3$ $x - 2y = 4$ | x	y	(x, y)
---	---	---
0	-2	$(0, -2)$
4	0	$(4, 0)$	 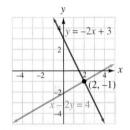 **Step 2** It appears that the graphs intersect at the point $(2, -1)$. To verify that this is the solution of the system, substitute 2 for x and -1 for y in each equation. **Step 3** Check $x - 2y = 4 \qquad\qquad y = -2x + 3$ $2 - 2(-1) \stackrel{?}{=} 4 \qquad\quad -1 \stackrel{?}{=} -2(2) + 3$ $4 = 4 \quad \text{True} \qquad\qquad -1 = -1 \quad \text{True}$ Since $(2, -1)$ makes both equations true, it is the solution of the system.
A system of equations that has at least one solution is called a **consistent system**. If the graphs of the equations of the system are parallel lines, the system has no solution and is called an **inconsistent system**. Equations with different graphs are called **independent equations**. If the graphs of the equations in a system are the same line, the system has infinitely many solutions. The equations are called **dependent equations**. We can determine the **number of solutions** that a system of two linear equations has by writing each equation in slope-intercept form, $y = mx + b$, and comparing the slopes and y-intercepts.	There are three possible outcomes when solving a system of two linear equations by graphing. *Consistent system* *Independent equations* • Exactly one solution • The lines have different slopes. *Inconsistent system* *Independent equations* • No solution • The lines have the same slope but different y-intercepts. *Consistent system* *Dependent equations* • Infinitely many solutions • The lines have the same slope and same y-intercept.

REVIEW EXERCISES

1. Use a check to determine whether $(2, -3)$ is a solution of the system.

$$\begin{cases} 3x - 2y = 12 \\ 2x + 3y = -5 \end{cases}$$

2. BACHELOR'S DEGREES Estimate the point of intersection of the graph below. Explain its significance.

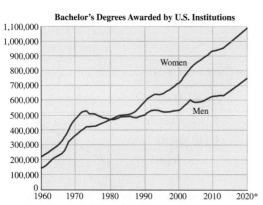

Use the graphing method to solve each system. If the equations of a system are dependent or if a system is inconsistent, so indicate.

3. $\begin{cases} x + y = 7 \\ 2x - y = 5 \end{cases}$

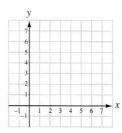

4. $\begin{cases} y = -\frac{x}{3} \\ 2x + y = 5 \end{cases}$

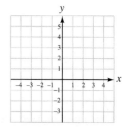

5. $\begin{cases} 3x + 6y = 6 \\ x + 2y = 2 \end{cases}$

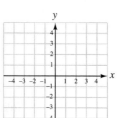

6. $\begin{cases} 6x + 3y = 12 \\ 2x + y = 2 \end{cases}$

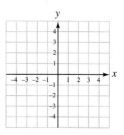

SECTION 4.2 Solving Systems of Equations by Substitution

DEFINITIONS AND CONCEPTS	EXAMPLES
To solve a system of equations in x and y by the **substitution method**: 1. Solve one of the equations for either x or y. If this is already done, go to step 2. (We call this equation the substitution equation.) 2. Substitute the expression for x or for y obtained in step 1 into the other equation, and solve that equation. 3. Substitute the value of the variable found in step 2 in any equation to find the value of the remaining variable. 4. Check the proposed solution in each equation of the original system. Write the solution as an ordered pair.	To solve the system $\begin{cases} x + 2y = -4 \\ 3x - 2y = 12 \end{cases}$ by substitution, begin by solving the first equation for x. $x + 2y = -4$ This is the first equation. $x = -2y - 4$ Subtract 2y from both sides. This is the substitution equation. Then substitute the result for x in the second equation, and solve for y. $3x - 2y = 12$ This is the second equation. $3(-2y - 4) - 2y = 12$ Substitute $-2y - 4$ for x. $-6y - 12 - 2y = 12$ Distribute. $-8y = 24$ Combine terms and add 12 to both sides. $y = -3$ Divide both sides by -8. To find x, substitute -3 for y in Equation 1. $x = -2y - 4$ This is Equation 1. $x = -2(-3) - 4$ Substitute -3 for y. $x = 2$ Simplify. The solution $(2, -3)$ was checked in the previous section.

REVIEW EXERCISES

Use the substitution method to solve each system.

7. $\begin{cases} x = y \\ 5x - 4y = 3 \end{cases}$

8. $\begin{cases} y = 15 - 3x \\ 7y + 3x = 15 \end{cases}$

9. $\begin{cases} 0.2x + 0.2y = 0.6 \\ 3x = 2 - y \end{cases}$

10. $\begin{cases} 6(r + 2) = s - 1 \\ r - 5s = -7 \end{cases}$

11. $\begin{cases} 9x + 3y = 5 \\ 3x + y = \dfrac{5}{3} \end{cases}$

12. $\begin{cases} \dfrac{x}{6} + \dfrac{y}{10} = 3 \\ \dfrac{5x}{16} - \dfrac{3y}{16} = \dfrac{15}{8} \end{cases}$

13. In solving a system using the substitution method, suppose you obtain the result of $8 = 9$.
    a. How many solutions does the system have?
    b. Describe the graph of the system.
    c. What term is used to describe the system?

14. Fill in the blank. With the substitution method, the objective is to use an appropriate substitution to obtain one equation in _____ variable.

SECTION 4.3 Solving Systems of Equations by Addition (Elimination)

DEFINITIONS AND CONCEPTS

To solve a system of equations using the **addition (elimination) method**:

1. Write both equations of the system in standard form: $Ax + By = C$.
2. If necessary, multiply one or both of the equations by a nonzero number chosen to make the coefficients of x (or the coefficients of y) opposites.
3. Add the equations to eliminate the terms involving x (or y).
4. Solve the equation resulting from step 3.
5. Find the value of the remaining variable by substituting the solution found in step 4 into any equation containing both variables. Or repeat steps 2–4 to eliminate the other variable.
6. Check the proposed solution in each equation of the original system. Write the solution as an ordered pair.

EXAMPLES

To solve the system $\begin{cases} x + 2y = -4 \\ 3x - 2y = 12 \end{cases}$ by addition (elimination), note that both equations are in standard $Ax + By = C$ form. Since the coefficients of y are opposites, add the equations to eliminate the variable y.

$$\begin{aligned} x + 2y &= -4 \\ 3x - 2y &= 12 \\ \hline 4x &= 8 \\ x &= 2 \quad \text{Divide both sides by 4.} \end{aligned}$$

To find y using addition, we multiply both sides of the first equation by -3 and add the equations to eliminate the variable x.

$$\begin{aligned} -3x - 6y &= 12 \\ 3x - 2y &= 12 \\ \hline -8y &= 24 \\ y &= -3 \quad \text{Divide both sides by } -8. \end{aligned}$$

We could also find y by substituting 2 for x in equation 1.

$$\begin{aligned} x + 2y &= -4 \\ 2 + 2y &= -4 \\ 2y &= -6 \\ \frac{2y}{2} &= \frac{-6}{2} \\ y &= -3 \end{aligned}$$

The solution $(2, -3)$ was checked in Section 4.1.

REVIEW EXERCISES

Solve each system using the addition (elimination) method.

15. $\begin{cases} 2x + y = 1 \\ 5x - y = 20 \end{cases}$

16. $\begin{cases} x + 8y = 7 \\ x - 4y = 1 \end{cases}$

17. $\begin{cases} 5a + b = 2 \\ 3a + 2b = 11 \end{cases}$

18. $\begin{cases} 11x + 3y = 27 \\ 8x + 4y = 36 \end{cases}$

19. $\begin{cases} 9x + 3y = 15 \\ 3x = 5 - y \end{cases}$

20. $\begin{cases} \dfrac{x}{3} + \dfrac{y + 2}{2} = 1 \\ \dfrac{x + 8}{8} + \dfrac{y - 3}{3} = 0 \end{cases}$

21. $\begin{cases} 0.02x + 0.05y = 0 \\ 0.3x - 0.2y = -1.9 \end{cases}$

22. $\begin{cases} -\dfrac{1}{4}x = 1 - \dfrac{2}{3}y \\ 6(x - 3y) + 2y = 5 \end{cases}$

For each system, determine which method, substitution or addition (elimination), would be easier to use to solve the system and explain why.

23. $\begin{cases} 6x + 2y = 5 \\ 3x - 3y = -4 \end{cases}$

24. $\begin{cases} x = 5 - 7y \\ 3x - 3y = -4 \end{cases}$

SECTION 4.4 Problem Solving Using Systems of Equations

DEFINITIONS AND CONCEPTS

To solve problems involving two unknown quantities:

1. **Analyze the problem** by reading it carefully to understand the given facts. What information is given? What are you asked to find? What vocabulary is given? Often a diagram or table will help you understand the facts of the problem.

2. **Assign variables** to represent unknown values in the problem. This means, in most cases, to let $x =$ one of the unknowns that you are asked to find, and $y =$ the other unknown.

3. **Form a system of equations** by translating the words of the problem into mathematical symbols.

4. **Solve the system** of equations using graphing, substitution, or elimination.

5. **State the conclusion clearly.** Be sure to include the units.

6. **Check the results** using the words of the problem, not the equations that were formed in step 3.

EXAMPLE

PLUMBING A plumber plans to cut a 20-foot pipe into two pieces so that one piece is 4 feet longer than the other.

Analyze Refer to the figure, which shows the 20-foot-long pipe cut into two pieces.

Assign Let $\ell =$ the length of the longer piece and $s =$ the length of the shorter piece.

Form Since the sum of the lengths is to be 20 feet, you have

The length of the longer piece	plus	the length of the shorter piece	is	20 feet.
ℓ	$+$	s	$=$	20

Since the longer piece is to be 4 feet longer than the shorter piece, we have

The length of the longer piece	is	the length of the shorter piece	plus	4 feet.
ℓ	$=$	s	$+$	4

Solve To find ℓ and s, solve the system $\begin{cases} \ell + s = 20 \\ \ell = s + 4 \end{cases}$ as follows

$s + 4 + s = 20$ Substitute $s + 4$ in the second equation for ℓ in the first equation.

$2s + 4 = 20$ Combine like terms.

$2s = 16$ Subtract 4 from both sides.

$s = 8$ Divide both sides by 2.

State The shorter piece should be 8 feet long. Because the longer piece is to be 4 feet longer, the longer piece should be 12 feet long.

Check This result checks because the sum of 8 and 12 is 20, and 12 is 4 greater than 8.

REVIEW EXERCISES

Use a system of two equations in two variables to solve each problem.

25. ELEVATIONS The elevation of Las Vegas, Nevada, is 20 times greater than that of Baltimore, Maryland. The sum of their elevations is 2,100 feet. Find the elevation of each city.

26. PAINTING EQUIPMENT When fully extended, the ladder shown is 35 feet in length. If the extension is 7 feet shorter than the base, how long is each part of the ladder?

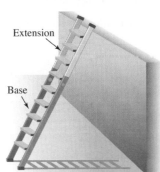

27. CRASH INVESTIGATIONS In an effort to protect evidence, investigators used 420 yards of yellow "Police Line—Do Not Cross" tape to seal off a large rectangular area around an airplane crash site. How much area will the investigators have to search if the width of the rectangle is three-fourths of the length? (*Hint:* Find the length and width of the rectangle first.)

28. ENDORSEMENTS A company selling a home juicing machine is contemplating hiring either an athlete or an actor to serve as the spokesperson for the product. The terms of each contract would be as follows:

Celebrity	Base pay	Commission per item sold
Athlete	$30,000	$5
Actor	$20,000	$10

  a. For each celebrity, write an equation giving the money ($y) the celebrity would earn if x juicers were sold.

  b. For what number of juicers would the athlete and the actor earn the same amount?

  c. Using the illustration, graph the equations from part a. The company expects to sell more than 3,000 juicers. Which celebrity would cost the company less money to serve as a spokesperson?

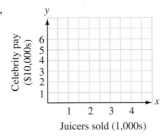

29. Complete each table.

  a.

	Amount · Strength = Amount of pesticide		
Weak	x	0.02	
Strong	y	0.09	
Mixture	100	0.08	

  b.

		Rate · Time = Distance		
With the wind		$s + w$	5	
Against the wind		$s - w$	7	

  c.

		P · r · t = I			
Mack Financial		x	0.11	1	
Union Savings		y	0.06	1	

  d.

	Amount · Price = Total value		
Caramel corn	x	4	
Peanuts	y	8	
Mixture	10	5	

30. CANDY STORES A merchant wants to mix gummy worms worth $3 per pound and gummy bears worth $1.50 per pound to make 30 pounds of a mixture worth $2.10 per pound. How many pounds of each should he use?

31. BOATING It takes a motorboat 4 hours to travel 56 miles down a river, and 3 hours longer to make the return trip. Find the speed of the current.

32. SHOPPING Packages containing 2 bottles of contact lens cleaner and 3 bottles of soaking solution cost $63.40, and packages containing 3 bottles of cleaner and 2 bottles of soaking solution cost $69.60. Find the cost of a bottle of cleaner and the cost of a bottle of soaking solution.

33. INVESTING Carlos invested part of $3,000 in a 10% certificate account and the rest in a 6% passbook account. The total annual interest from both accounts is $270. How much did he invest at 6%?

34. ANTIFREEZE How much of a 40% antifreeze solution must a mechanic mix with a 70% antifreeze solution if he needs 20 gallons of a 50% antifreeze solution? (*Hint:* The answers are mixed numbers.)

SECTION 4.5 Solving Systems of Linear Inequalities

DEFINITIONS AND CONCEPTS

Solving a system of linear inequalities by graphing:

1. Graph each inequality on the same rectangular coordinate system.

2. Use shading to highlight the intersection of the graphs. The points in this region are the solutions of the system.

3. As an informal check, pick a test point from the region where the graphs intersect and verify that its coordinates satisfy each inequality of the original system.

EXAMPLE

Graph the solution of the system: $\begin{cases} y \leq x + 1 \\ y > -1 \end{cases}$

Step 1 Graph each inequality on the same coordinate system as shown.

Step 2 Use shading to highlight where the graphs intersect.

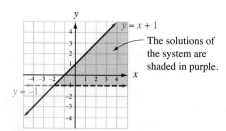

The solutions of the system are shaded in purple.

Step 3 Pick a point from the solution region, such as $(1, 0)$, and verify that it satisfies both inequalities.

REVIEW EXERCISES

Solve each system of inequalities by graphing.

35. $\begin{cases} 5x + 3y < 15 \\ 3x - y > 3 \end{cases}$

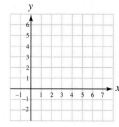

36. $\begin{cases} x \geq 3y \\ y > 3x \end{cases}$

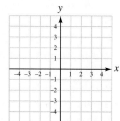

37. Use a check to determine whether each ordered pair is a solution of the system: $\begin{cases} x + 2y \leq 3 \\ 2x - y > 1 \end{cases}$

  a. $(5, -4)$ **b.** $(-1, -3)$

38. GIFT SHOPPING A grandmother wants to spend at least $40 but no more than $60 on school clothes for her grandson. If T-shirts sell for $10 and pants sell for $20, write a system of inequalities that describes the possible combinations of T-shirts (x) and pants (y) she can buy. Graph the system in the illustration. Give two possible solutions.

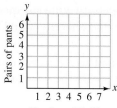

CHAPTER 4 TEST

1. Fill in the blanks.
   a. A _____ of a system of linear equations is an ordered pair that satisfies each equation.
   b. A system of equations that has at least one solution is called a _____ system.
   c. A system of equations that has no solution is called an _____ system.
   d. Equations with different graphs are called _____ equations.
   e. A system of _____ equations has an infinite number of solutions.

2. The following graph shows two different ways in which a salesperson can be paid according to the number of items he or she sells.

   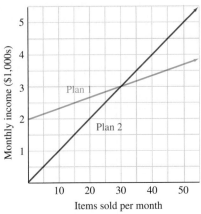

   a. What is the point of intersection of the graph? Explain its significance.
   b. If a salesperson expects to sell more than 30 items per month, which plan is more profitable?

Use a check to determine whether the given ordered pair is a solution of the given system.

3. $(5, 3)$, $\begin{cases} 3x + 2y = 21 \\ x + y = 8 \end{cases}$

4. $(-2, -1)$, $\begin{cases} 4x + y = -9 \\ 2x - 3y = -7 \end{cases}$

5. Solve the system by graphing: $\begin{cases} 3x + y = 7 \\ x - 2y = 0 \end{cases}$

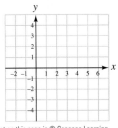

6. To solve a system of two linear equations in x and y, a student used a graphing calculator. From the calculator display shown, determine whether the system has a solution. Explain your answer.

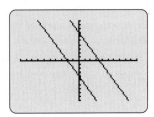

Solve each system by substitution.

7. $\begin{cases} y = x - 1 \\ 2x + y = -7 \end{cases}$

8. $\begin{cases} 3a + 4b = -7 \\ 2b - a = -1 \end{cases}$

Solve each system by addition (elimination).

9. $\begin{cases} 3x - y = 2 \\ 2x + y = 8 \end{cases}$

10. $\begin{cases} 4x + 3y = -3 \\ -3x = -4y + 21 \end{cases}$

Classify each system as consistent or inconsistent, and classify the equations as dependent or independent.

11. $\begin{cases} x + y = 4 \\ x + y = 6 \end{cases}$

12. $\begin{cases} \dfrac{x}{3} + y = 4 \\ x + 3y = 12 \end{cases}$

Solve each system using substitution or elimination.

13. $\begin{cases} 3x - 5y - 16 = 0 \\ \dfrac{x}{2} - \dfrac{5}{6}y = \dfrac{1}{3} \end{cases}$

14. $\begin{cases} 3a + 4b = -7 \\ 2b - a = -1 \end{cases}$

15. $\begin{cases} y = 3x - 1 \\ y = 2x + 4 \end{cases}$

16. $\begin{cases} 0.6c + 0.5d = 0 \\ 0.02c + 0.09d = 0 \end{cases}$

17. Find the slope and the y-intercept of the graph of each line in the system $\begin{cases} y = 4x - 10 \\ x - 2y = -16 \end{cases}$. Then use that information to determine the *number of solutions* of the system. **Do not solve the system.**

18. How many solutions does the system of two linear equations graphed on the right have? Give three of the solutions.

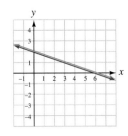

19. Which method would be most efficient to solve the following system?

$$\begin{cases} 5x - 3y = 5 \\ 3x + 3y = 3 \end{cases}$$

Explain your answer. (You do not need to solve the system.)

20. Match each equation, inequality, or system with the graph of its solution.

a. $2x + y = 2$ b. $2x + y \geq 2$

c. $\begin{cases} 2x + y = 2 \\ 2x - y = 2 \end{cases}$ d. $\begin{cases} 2x + y \geq 2 \\ 2x - y \leq 2 \end{cases}$

i.   ii.

iii.   iv.

21. CHILD CARE On a mother's 22-mile commute to work, she drops her daughter off at a child care center. The first part of the trip is 6 miles less than the second part. How long is each part of her morning commute?

22. VACATIONING It cost a family of 7 a total of $268 for general admission tickets to the San Diego Zoo. How many adult tickets and how many child tickets were purchased?

Use a system of two equations in two variables to solve each problem.

23. FINANCIAL PLANNING A woman invested some money at 8% and some at 9%. The interest for 1 year on the combined investment of $10,000 was $840. How much was invested at 9%?

24. TAILWINDS/HEADWINDS Flying with a tailwind, a pilot flew an airplane 450 miles in 2.5 hours. Flying into a headwind, the return trip took 3 hours. Find the speed of the plane in calm air and the speed of the wind.

25. ANTIFREEZE How many pints of a 5% antifreeze solution and how many pints of a 20% antifreeze solution must be mixed to obtain 12 pints of a 15% solution?

26. SUNSCREEN A sunscreen selling for $1.50 per ounce is to be combined with another sunscreen selling for $0.80 per ounce. How many ounces of each are needed to make 10 ounces of a sunscreen mix that sells for $1.01 per ounce?

27. Solve the system by graphing: $\begin{cases} 2x + 3y \leq 6 \\ x \geq 2 \end{cases}$

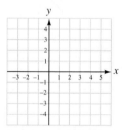

28. CLOTHES SHOPPING This system of inequalities describes the number of $20 shirts, x, and $40 pairs of pants, y, a person can buy if he or she plans to spend not less than $80 but not more than $120. Graph the system. Then give three solutions.

$$\begin{cases} 20x + 40y \geq 80 \\ 20x + 40y \leq 120 \end{cases}$$

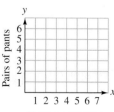

CHAPTERS 1–4 CUMULATIVE REVIEW

1. Fill in the blanks. The answer to an addition problem is called a _____. The answer to a subtraction problem is called a _____. The answer to a multiplication problem is called a _____. The answer to a division problem is called a _____. [Section 1.1]

2. Give the prime factorization of 100. [Section 1.2]

3. Divide: $\frac{3}{4} \div \frac{6}{5}$ [Section 1.2]

4. Subtract: $\frac{7}{10} - \frac{1}{14}$ [Section 1.2]

5. Is π a rational or irrational number? [Section 1.3]

6. Graph each member of the set on the number line. [Section 1.3]

$$\left\{-2\frac{1}{4}, \sqrt{2}, -1.75, \frac{7}{2}, 0.5\right\}$$

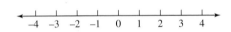

7. Write $\frac{2}{3}$ as a decimal. [Section 1.3]

8. What property of real numbers is illustrated? [Section 1.6]

$$3(2x) = (3 \cdot 2)x$$

Evaluate each expression.

9. $-3^2 + |4^2 - 5^2|$ [Section 1.7]

10. $(4-5)^{20}$ [Section 1.7]

11. $\dfrac{-3-(-7)}{2^2-3}$ [Section 1.7]

12. $12 - 2[1-(-8+2)]$ [Section 1.7]

13. RACING Suppose a driver has completed x laps of a 250-lap race. Write an expression for how many more laps he must make to finish the race. [Section 1.8]

14. What is the value of d dimes? [Section 1.8]

Simplify each expression. [Section 1.9]

15. $13r - 12r$

16. $27\left(\dfrac{2}{3}x\right)$

17. $4(d-3) - (d-1)$

18. $(13c - 3)(-6)$

Solve each equation. Check each result. [Section 2.2]

19. $3(x-5) + 2 = 2x$

20. $\dfrac{x-5}{3} - 5 = 7$

21. $\dfrac{2}{5}x + 1 = \dfrac{1}{3} + x$

22. $-\dfrac{5}{8}h = 15$

23. GYMNASTICS After the first day of registration, 119 children had been enrolled in a Gymboree class. That represented 85% of the available slots. Find the maximum number of children the center could enroll. [Section 2.3]

24. Solve $A = \dfrac{1}{2}h(b+B)$ for h. [Section 2.4]

25. MIXING CANDY The owner of a candy store wants to make a 30-pound mixture of two candies to sell for $4 per pound. If red licorice bits sell for $3.80 per pound and lemon gumdrops sell for $4.40 per pound, how many pounds of each should be used? [Section 2.6]

26. Solve $8(4+x) > 10(6+x)$. Write the solution set in interval notation and graph it. [Section 2.7]

27. In what quadrant does $(-3.5, 6)$ lie? [Section 3.1]

28. Use a check to determine whether $(-2, 8)$ is a solution of $y = -2x + 3$. [Section 3.2]

Graph each equation.

29. $x = 4$ [Section 3.3]

30. $4x - 3y = 12$ [Section 3.3]

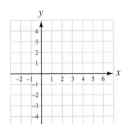 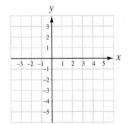

Find the slope of the line with the given properties.

31. Passing through $(-2, 4)$ and $(6, 8)$ [Section 3.4]

32. A line that is horizontal [Section 3.4]

33. An equation of $2x - 3y = 12$ [Section 3.5]

34. Are the graphs of the lines parallel or perpendicular? [Section 3.5]
$$y = -\frac{3}{4}x + \frac{15}{4} \qquad 4x - 3y = 25$$

Find an equation of the line with the following properties. Write the equation in slope–intercept form.

35. Slope $= \frac{2}{3}$, y-intercept $= (0, 5)$ [Section 3.5]

36. Passing through $(-2, 4)$ and $(6, 10)$ [Section 3.6]

37. A horizontal line passing through $(2, 4)$ [Section 3.6]

38. Graph: $y < \frac{x}{3} - 1$ [Section 3.7]

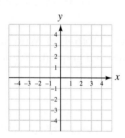

39. If $f(x) = -2x^2 - 3x^3$, find $f(-1)$. [Section 3.8]

40. Refer to the graph on the right. Is this the graph of a function? If it is, give the domain and range. If it is not, find ordered pairs that show a value of x that is assigned more than one value of y. [Section 3.8]

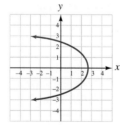

Solve each system by graphing.

41. $\begin{cases} x + 4y = -2 \\ y = -x - 5 \end{cases}$ [Section 4.1]

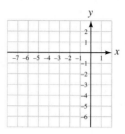

42. $\begin{cases} 2x - 3y < 0 \\ y > x - 1 \end{cases}$ [Section 4.5]

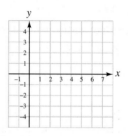

43. Solve $\begin{cases} x - 2y = 2 \\ 2x + 3y = 11 \end{cases}$ by substitution. [Section 4.2]

44. Solve $\begin{cases} \dfrac{3}{2}x - \dfrac{2}{3}y = 0 \\ \dfrac{3}{4}x + \dfrac{4}{3}y = \dfrac{5}{2} \end{cases}$ by elimination. [Section 4.3]

Use a system of two equations in two variables to solve each problem.

45. NUTRITION The table shows per serving nutritional information for egg noodles and rice pilaf. How many servings of each food should be eaten to consume exactly 22 grams of protein and 21 grams of fat? [Section 4.4]

	Protein (g)	Fat (g)
Egg noodles	5	3
Rice pilaf	4	5

46. INVESTMENT CLUBS Part of $8,000 was invested by an investment club at 10% interest and the rest at 12%. If the annual income from these investments is $900, how much was invested at each rate? [Section 4.4]

5

Exponents and Polynomials

5.1 Natural-Number Exponents; Rules for Exponents
5.2 Zero and Negative Integer Exponents
5.3 Scientific Notation
5.4 Polynomials
5.5 Adding and Subtracting Polynomials
5.6 Multiplying Polynomials
5.7 Dividing Polynomials by Monomials
5.8 Dividing Polynomials by Polynomials

Chapter Summary and Review
Chapter Test
Cumulative Review

from Campus to Careers
Police Officer

People depend on the police to protect their lives and property. The job can be dangerous because police officers must arrest suspects and respond to emergencies. The daily activities of police officers can vary greatly depending on their specialty, such as patrol officer, game warden, or detective. Regardless of their duties, they must write reports and maintain records that will be needed if they testify in court.

In **Problem 80** of **Study Set 5.4**, you will see how police officers can compute the stopping distance of a car.

JOB TITLE: Police Officer
EDUCATION: For many departments, two years of college or a college degree may be required. Physical education courses are helpful. Foreign language skills are desirable.
JOB OUTLOOK: Employment opportunities are expected to grow 7 percent through 2020.
ANNUAL EARNINGS: Median annual salary in 2012 was $55,270. Earnings often exceed the base salary because of payments for overtime.
FOR MORE INFORMATION:
www.bls.gov/oco/ocos160.htm

Objectives

1. Identify bases and exponents.
2. Multiply exponential expressions that have like bases.
3. Divide exponential expressions that have like bases.
4. Raise exponential expressions to a power.
5. Find powers of products and quotients.

SECTION 5.1
Natural-Number Exponents; Rules for Exponents

In this section, we will use the definition of exponent to develop some rules for simplifying expressions that contain exponents.

1 Identify bases and exponents.

We have used natural-number exponents to indicate repeated multiplication. For example,

$9^2 = 9 \cdot 9 = 81$ Write 9 as a factor 2 times.

$7^3 = 7 \cdot 7 \cdot 7 = 343$ Write 7 as a factor 3 times.

$(-2)^4 = (-2)(-2)(-2)(-2) = 16$ Write -2 as a factor 4 times.

$-2^4 = -(2 \cdot 2 \cdot 2 \cdot 2) = -16$ The $-$ sign in front of 2^4 means the opposite of 2^4.

These examples illustrate a definition for x^n, where n is a natural number.

Natural-Number Exponents

A natural-number exponent tells how many times its base is to be used as a factor. For any number x and any natural number n,

$$x^n = \underbrace{x \cdot x \cdot x \cdot \cdots \cdot x}_{n \text{ factors of } x}$$

In the **exponential expression** x^n, x is called the **base** and n is called the **exponent**. The entire expression is called a **power of** x.

$$\text{base} \rightarrow x^n \leftarrow \text{exponent}$$

If an exponent is a natural number, it tells how many times its base is to be used as a factor. An exponent of 1 indicates that its base is to be used one time as a factor, an exponent of 2 indicates that its base is to be used two times as a factor, and so on. The base of an exponential expression can be a number, a variable, or a combination of numbers and variables.

$$x^1 = x \qquad (y+1)^2 = (y+1)(y+1) \qquad (-5s)^3 = (-5s)(-5s)(-5s)$$

Self Check 1

Identify the base and the exponent:
a. $5x^4$
b. $(5x)^4$

Now Try Problems 25, 26, and 31

EXAMPLE 1
Identify the base and the exponent in each expression:

a. 7^6 b. $4x^3$ c. $-x^5$ d. $(2x)^4$

Strategy To identify the base and exponent, we will look for the exponent first. Then we will look for the base. We will report the base first, then the exponent.

WHY The exponent is the small raised number. The base is the number or variable directly in front of the exponent, unless there are parentheses.

Solution

a. In 7^6, the base is 7 and the exponent is 6.

b. $4x^3$ means $4 \cdot x^3$. The base is x and the exponent is 3.

c. $-x^5$ means $-1 \cdot x^5$. The base is x and the exponent is 5.

d. Because of the parentheses in $(2x)^4$, the base is $2x$ and the exponent is 4.

5.1 Natural-Number Exponents; Rules for Exponents

EXAMPLE 2 Write each expression in an equivalent form using an exponent:

a. $\dfrac{x}{5} \cdot \dfrac{x}{5} \cdot \dfrac{x}{5} \cdot \dfrac{x}{5}$ **b.** $4 \cdot y \cdot y \cdot y \cdot y \cdot y$

Strategy We will look for repeated factors and count the number of times each appears.

WHY We can use an exponent to represent repeated multiplication.

Solution

a. Since there are four repeated factors of $\dfrac{x}{5}$ in $\dfrac{x}{5} \cdot \dfrac{x}{5} \cdot \dfrac{x}{5} \cdot \dfrac{x}{5}$, the expression can be written as $\left(\dfrac{x}{5}\right)^4$.

b. Since there are five repeated factors of y, the expression can be written $4y^5$.

Self Check 2
Write as an exponential expression:
$(2c + d)(2c + d)(2c + d)$
Now Try Problems 33, 38, and 44

2 Multiply exponential expressions that have like bases.

To develop a rule for multiplying exponential expressions that have the same base, we consider the product $x^2 \cdot x^3$. Since the expression x^2 means that x is to be used as a factor two times, and the expression x^3 means that x is to be used as a factor three times, we have

$$x^2 \cdot x^3 = \overbrace{x \cdot x}^{\text{2 factors of } x} \cdot \overbrace{x \cdot x \cdot x}^{\text{3 factors of } x}$$

$$= \overbrace{x \cdot x \cdot x \cdot x \cdot x}^{\text{5 factors of } x}$$

$$= x^5$$

This example suggests the following rule:

Product Rule for Exponents

To multiply two exponential expressions that have the same base, keep the common base and add the exponents. If m and n represent natural numbers, then

$$x^m x^n = x^{m+n}$$

EXAMPLE 3 Simplify each expression:

a. $9^5(9^6)$ **b.** $x^3 \cdot x^4$ **c.** $y^2 y^4 y$ **d.** $(c^2 d^3)(c^4 d^5)$ **e.** $(a+b)^4 (a+b)^3$

Strategy We will identify exponential expressions that have the same base in each product. Then we will use the product rule for exponents to simplify the expression.

WHY The product rule for exponents is used to multiply exponential expressions that have the same base.

Solution

a. To simplify $9^5(9^6)$ means to write it in an equivalent form using one base and one exponent.

$9^5(9^6) = 9^{5+6}$ Use the product rule for exponents: Keep the common base, which is 9, and add the exponents.

$ = 9^{11}$ Do the addition. Since 9^{11} is a very large number, we will leave it in this form. We won't evaluate it.

Self Check 3
Simplify:
a. $7^8(7^7)$
b. $z \cdot z^3$
c. $x^2 x^3 x^6$
d. $(s^4 t^3)(s^4 t^4)$
e. $(r + s)^2 (r + s)^5$

Now Try Problems 45, 49, 57, and 59

b. $x^3 \cdot x^4 = x^{3+4}$ Keep the common base x and add the exponents.
$ = x^7$ Do the addition.

c. $y^2 y^4 y = y^{2+4} y$ Working from left to right, keep the common base y and add the exponents.
$ = y^6 y^1$ Do the addition. Write y as y^1.
$ = y^{6+1}$ Keep the common base and add the exponents.
$ = y^7$ Do the addition.

> **Success Tip** When simplifying exponential expressions, it is often helpful to write any simple variable factors with an exponent of 1. In this case, we wrote y as y^1.

d. $(c^2 d^3)(c^4 d^5) = c^2 d^3 c^4 d^5$ Use the associative property of multiplication.
$ = c^2 c^4 d^3 d^5$ Change the order of the factors.
$ = c^{2+4} d^{3+5}$ Keep the common base c and add the exponents. Keep the common base d and add the exponents.
$ = c^6 d^8$ Do the additions.

e. $(a+b)^4 (a+b)^3 = (a+b)^{4+3}$ Keep the common base $(a+b)$ and add the exponents.
$ = (a+b)^7$ Do the addition.

> **Caution!** We cannot use the product rule to simplify expressions like $3^2 \cdot 2^3$, where the bases are not the same. However, we can simplify this expression by doing the arithmetic: $3^2 \cdot 2^3 = 9 \cdot 8 = 72$.

> **THINK IT THROUGH** **PIN Code Choices**
>
> "While college students are using credit cards less and less, debit card use is on the rise—more than twice as many students carry debit cards than credit cards."
> *New York State Higher Education Services Corporation*
>
> Every year, there are more than 100 billion ATM transactions in the United States. Before each transaction, the card owner is required to enter his or her PIN (personal identification number). When an ATM card is issued, many financial institutions have the applicant select a four-digit PIN. There are 10 possible choices for the first digit of the PIN, 10 possible choices for the second digit, and so on. Write the total number of possible choices of a PIN as an exponential expression. Then evaluate the expression.

3 Divide exponential expressions that have like bases.

To develop a rule for dividing exponential expressions that have the same base, we now consider the fraction

$$\frac{4^5}{4^2}$$

where the exponent in the numerator is greater than the exponent in the denominator.

We can simplify this fraction as follows:

$$\frac{4^5}{4^2} = \frac{4 \cdot 4 \cdot 4 \cdot 4 \cdot 4}{4 \cdot 4} \quad \text{Write each expression without using exponents.}$$

$$= \frac{\overset{1}{\cancel{4}} \cdot \overset{1}{\cancel{4}} \cdot 4 \cdot 4 \cdot 4}{\underset{1}{\cancel{4}} \cdot \underset{1}{\cancel{4}}} \quad \text{Remove the common factors of 4.}$$

$$= 4^3$$

We can quickly find this result if we keep the common base 4 and subtract the exponents on 4^5 and 4^2.

$$\frac{4^5}{4^2} = 4^{5-2} = 4^3$$

This example suggests another rule for exponents.

Quotient Rule for Exponents

To divide exponential expressions that have the same base, keep the common base and subtract the exponents. If m and n represent natural numbers, $m > n$, and $x \neq 0$, then

$$\frac{x^m}{x^n} = x^{m-n}$$

EXAMPLE 4 Simplify. Assume that there are no divisions by 0.

a. $\dfrac{20^{16}}{20^9}$ **b.** $\dfrac{x^4}{x^3}$ **c.** $\dfrac{a^3 b^8}{ab^5}$

Strategy We will identify exponential expressions that have the same base in each quotient. Then we will use the quotient rule for exponents to simplify the expression.

WHY The quotient rule for exponents is used to divide exponential expressions that have the same base.

Solution

a. To simplify $\dfrac{20^{16}}{20^9}$ means to write it in an equivalent form using one base and one exponent.

$$\frac{20^{16}}{20^9} = 20^{16-9} \quad \text{Use the quotient rule for exponents: Keep the common base, which is 20, and subtract the exponents.}$$

$$= 20^7 \quad \text{Do the subtraction.}$$

b. $\dfrac{x^4}{x^3} = x^{4-3}$ Keep the common base x and subtract the exponents.

$\phantom{\dfrac{x^4}{x^3}} = x^1$ Do the subtraction.

$\phantom{\dfrac{x^4}{x^3}} = x$

c. $\dfrac{a^3 b^8}{ab^5} = \dfrac{a^3}{a^1} \cdot \dfrac{b^8}{b^5}$ Group the like bases together. Write a as a^1.

$\phantom{\dfrac{a^3 b^8}{ab^5}} = a^{3-1} b^{8-5}$ Keep the common base a and subtract the exponents. Keep the common base b and subtract the exponents.

$\phantom{\dfrac{a^3 b^8}{ab^5}} = a^2 b^3$ Do the subtractions.

Self Check 4

Simplify:

a. $\dfrac{55^{30}}{55^5}$

b. $\dfrac{a^5}{a^3}$

c. $\dfrac{b^{15} c^4}{b^4 c}$

Now Try Problems 61, 65, and 69

Caution! Don't make the mistake of dividing the bases when using the quotient rule. Keep the *same* base and subtract the *exponents*.

$$\cancel{\frac{20^{16}}{20^9} = 1^7}$$

Self Check 5

Simplify: $\dfrac{b^2 b^6 b}{b^4 b^4}$

Now Try Problem 73

EXAMPLE 5 Simplify: $\dfrac{a^3 a^5 a^7}{a^4 a}$

Strategy We will use the product rule and quotient rule to write an equivalent expression using one base and one exponent.

WHY The expression involves multiplication and division of exponential expressions that have the same base.

Solution
We use the product rule for exponents to simplify the numerator and denominator separately and proceed as follows.

$\dfrac{a^3 a^5 a^7}{a^4 a} = \dfrac{a^{15}}{a^4 a^1}$ In the numerator, keep the common base a and add the exponents. In the denominator, write a as a^1.

$= \dfrac{a^{15}}{a^5}$ In the denominator, keep the common base a and add the exponents.

$= a^{15-5}$ Use the quotient rule for exponents: Keep the common base a and subtract the exponents.

$= a^{10}$ Do the subtraction.

Success Tip Sometimes more than one rule for exponents is needed to simplify an expression.

It is important to pay close attention to the operation between the exponential expressions and then use the appropriate rules. To add or subtract exponential expressions, they must be like terms. To multiply or divide exponential expressions, only the bases need to be the same.

$x^2 + x^3$ The operation is addition and these are not like terms because the exponents are different. We cannot simplify the expression any further.

$x^4 + x^4 = 2x^4$ The operation is addition and these are like terms. We can simplify the expression by adding the numerical coefficients and keeping the variable expression.

$x^7 \cdot x^3 = x^{10}$ The operation is multiplication and the bases are the same. We keep the base and add the exponents.

$\dfrac{x^4}{x^3} = x$ The operation is division and the bases are the same. We keep the base and subtract the exponents. An exponent of 1 need not be written.

4 Raise exponential expressions to a power.

To find another rule for exponents, we consider the expression $(x^3)^4$, which can be written as $x^3 \cdot x^3 \cdot x^3 \cdot x^3$. Because each of the four factors of x^3 contains three factors of x, there are $4 \cdot 3$ (or 12) factors of x. This product can be written as x^{12}.

5.1 Natural-Number Exponents; Rules for Exponents   391

$$(x^3)^4 = x^3 \cdot x^3 \cdot x^3 \cdot x^3$$

$$= \underbrace{x \cdot x \cdot x \cdot x \cdot x \cdot x \cdot x \cdot x \cdot x \cdot x \cdot x \cdot x}_{\text{12 factors of } x}$$

$$= x^{12}$$

We can quickly find this result if we keep the base of x and multiply the exponents.

$$(x^3)^4 = x^{3 \cdot 4} = x^{12}$$

This illustrates the following rule for exponents.

Power Rule for Exponents

To raise an exponential expression to a power, keep the base and multiply the exponents. If m and n represent natural numbers, then

$$(x^m)^n = x^{m \cdot n} \quad \text{or, more simply,} \quad (x^m)^n = x^{mn}$$

EXAMPLE 6 Simplify each expression: **a.** $(2^3)^7$ **b.** $(z^8)^8$

Strategy In each case, we want to write an equivalent expression using one base and one exponent. We will use the power rule for exponents to do this.

WHY Each expression is a power of a power.

Solution

a. To simplify $(2^3)^7$ means to write it in an equivalent form using one base and one exponent.

$(2^3)^7 = 2^{3 \cdot 7}$ Keep the base of 2 and multiply the exponents.

$= 2^{21}$ Do the multiplication.

b. $(z^8)^8 = z^{8 \cdot 8}$ Keep the base and multiply the exponents.

$= z^{64}$ Do the multiplication.

Self Check 6
Simplify each expression:
a. $(5^4)^6$
b. $(y^5)^2$
Now Try Problems 77 and 83

EXAMPLE 7 Simplify each expression: **a.** $(x^2 x^5)^2$ **b.** $(z^2)^4 (z^3)^3$

Strategy In each case, we want to write an equivalent expression using one base and one exponent. We will use the product and power rules for exponents to do this.

WHY The expressions involve products and powers of exponential expressions that have the same base.

Solution

a. We begin by using the product rule for exponents. Then we use the power rule.

$(x^2 x^5)^2 = (x^7)^2$ Within the parentheses, keep the base x and add the exponents.

$= x^{14}$ Keep the base x and multiply the exponents.

Self Check 7
Simplify each expression:
a. $(a^4 a^3)^3$
b. $(a^3)^3 (a^4)^2$
Now Try Problems 85 and 89

b. We begin by using the power rule for exponents twice. Then we use the product rule.

$$(z^2)^4(z^3)^3 = z^8 z^9 \quad \text{For each power of } z \text{ raised to a power, keep the base } z \text{ and multiply the exponents.}$$

$$= z^{17} \quad \text{Keep the base } z \text{ and add the exponents.}$$

5 Find powers of products and quotients.

To develop two more rules for exponents, we consider the expression $(2x)^3$, which is a *power of the product* of 2 and x, and the expression $\left(\frac{2}{x}\right)^3$, which is a *power of the quotient* of 2 and x.

$$(2x)^3 = (2x)(2x)(2x)$$
$$= (2 \cdot 2 \cdot 2)(x \cdot x \cdot x)$$
$$= 2^3 x^3$$
$$= 8x^3$$

$$\left(\frac{2}{x}\right)^3 = \left(\frac{2}{x}\right)\left(\frac{2}{x}\right)\left(\frac{2}{x}\right) \quad \text{Assume } x \neq 0.$$

$$= \frac{2 \cdot 2 \cdot 2}{x \cdot x \cdot x} \quad \text{Multiply the numerators. Multiply the denominators.}$$

$$= \frac{2^3}{x^3}$$

$$= \frac{8}{x^3} \quad \text{Evaluate: } 2^3 = 8.$$

These examples illustrate the following rules for exponents.

Powers of a Product and a Quotient

To raise a product to a power, we raise each factor of the product to that power. To raise a fraction to a power, we raise both the numerator and the denominator to that power. If n represents a natural number, then

$$(xy)^n = x^n y^n \quad \text{and if} \quad y \neq 0, \quad \text{then} \quad \left(\frac{x}{y}\right)^n = \frac{x^n}{y^n}$$

Self Check 8

Simplify:
a. $(2t)^4$
b. $(c^3 d^4)^6$
c. $(-3ab^5)^3$

Now Try Problems 93, 95, and 99

EXAMPLE 8 Simplify: **a.** $(3c)^3$ **b.** $(x^2 y^3)^5$ **c.** $(-2a^3 b)^2$

Strategy We will use the power of a product rule for exponents to write the simplified expression.

WHY Within each set of parentheses is a product, and each of those products is raised to a power.

Solution

a. Since $3c$ is the product of 3 and c, the expression $(3c)^3$ is a power of a product.

$$(3c)^3 = 3^3 c^3 \quad \text{Use the power rule for products: Raise each factor of the product } 3c \text{ to the 3rd power.}$$
$$= 27c^3 \quad \text{Evaluate: } 3^3 = 27.$$

b. $(x^2 y^3)^5 = (x^2)^5 (y^3)^5 \quad$ Raise each factor of the product $x^2 y^3$ to the 5th power.
$\qquad\qquad\; = x^{10} y^{15} \quad$ For each power of a power, keep the base and multiply the exponents.

c. $(-2a^3 b)^2 = (-2)^2 (a^3)^2 b^2 \quad$ Raise each of the three factors of the product $-2a^3 b$ to the 2nd power.

$\qquad\qquad\quad\; = 4a^6 b^2 \quad$ Evaluate: $(-2)^2 = 4$. Keep the base a and multiply the exponents.

EXAMPLE 9

Simplify: $\dfrac{(a^3b^4)^2}{ab^5}$

Strategy We will use the power of a product rule and the quotient rule for exponents to write the simplified expression.

WHY The expression involves a power of a product and a quotient of exponential expressions that have the same base.

Solution

$\dfrac{(a^3b^4)^2}{ab^5} = \dfrac{(a^3)^2(b^4)^2}{a^1b^5}$ In the numerator, raise each factor within the parentheses to the 2nd power. In the denominator, write a as a^1.

$= \dfrac{a^6b^8}{a^1b^5}$ In the numerator, for each power of a power, keep the base and multiply the exponents.

$= a^{6-1}b^{8-5}$ Keep each of the bases, a and b, and subtract the exponents.

$= a^5b^3$ Do the subtractions.

Self Check 9

Simplify: $\dfrac{(c^4d^5)^3}{c^2d^3}$

Now Try Problem 101

EXAMPLE 10

Simplify: **a.** $\left(\dfrac{4}{k}\right)^3$ **b.** $\left(\dfrac{3x^2}{2y^3}\right)^5$

Strategy In each case, we will use the power of a quotient rule for exponents to simplify the expression.

WHY Each expression is a quotient raised to a power.

Solution

a. Since $\dfrac{4}{k}$ is the quotient of 4 and k, the expression $\left(\dfrac{4}{k}\right)^3$ is a power of a quotient.

$\left(\dfrac{4}{k}\right)^3 = \dfrac{4^3}{k^3}$ Use the power rule for quotients: Raise the numerator and denominator to the 3rd power.

$= \dfrac{64}{k^3}$ Evaluate: $4^3 = 64$.

b. $\left(\dfrac{3x^2}{2y^3}\right)^5 = \dfrac{(3x^2)^5}{(2y^3)^5}$ Raise the numerator and denominator to the 5th power.

$= \dfrac{3^5(x^2)^5}{2^5(y^3)^5}$ In the numerator and denominator, raise each factor within the parentheses to the 5th power.

$= \dfrac{243x^{10}}{32y^{15}}$ Evaluate 3^5 and 2^5. For each power of a power, keep the base and multiply the exponents.

Self Check 10

Simplify:

a. $\left(\dfrac{x}{7}\right)^3$

b. $\left(\dfrac{2x^3}{3y^2}\right)^4$

Now Try Problems 105 and 107

EXAMPLE 11

Simplify: $\dfrac{(5b)^9}{(5b)^6}$

Strategy We will use the quotient rule for exponents and then the power of a product rule.

WHY The expression involves division of exponential expressions that have the same base, $5b$.

Solution

$\dfrac{(5b)^9}{(5b)^6} = (5b)^{9-6}$ Keep the common base $5b$, and subtract the exponents.

$= (5b)^3$ Do the subtraction.

$= 5^3b^3$ Raise each factor within the parentheses to the 3rd power.

$= 125b^3$ Evaluate: $5^3 = 125$.

Self Check 11

Simplify: $\dfrac{(-2h)^{20}}{(-2h)^{14}}$

Now Try Problem 109

The rules for natural-number exponents are summarized below.

Rules for Exponents

If n represents a natural number, then

$$x^n = \overbrace{x \cdot x \cdot x \cdot \cdots \cdot x}^{n \text{ factors of } x}$$

If m and n represent natural numbers and there are no divisions by zero, then

Exponent of 1

$$x^1 = x$$

Product Rule

$$x^m x^n = x^{m+n}$$

Quotient Rule

$$\frac{x^m}{x^n} = x^{m-n}$$

Power Rule

$$(x^m)^n = x^{mn}$$

Power of a Product

$$(xy)^n = x^n y^n$$

Power of a Quotient

$$\left(\frac{x}{y}\right)^n = \frac{x^n}{y^n}$$

ANSWERS TO SELF CHECKS

1. a. base: x, exponent: 4 b. base: $5x$, exponent: 4   2. $(2c+d)^3$ 3. a. 7^{15} b. z^4 c. x^{11} d. $s^8 t^7$ e. $(r+s)^7$ 4. a. 55^{25} b. a^2 c. $b^{11} c^3$ 5. b 6. a. 5^{24} b. y^{10}
7. a. a^{21} b. a^{17} 8. a. $16t^4$ b. $c^{18} d^{24}$ c. $-27 a^3 b^{15}$ 9. $c^{10} d^{12}$ 10. a. $\dfrac{x^3}{343}$ b. $\dfrac{16 x^{12}}{81 y^8}$
11. $64 h^6$

SECTION 5.1   STUDY SET

VOCABULARY

Fill in the blanks.

1. The _____ of the exponential expression $(-5)^3$ is -5. The _____ is 3.
2. The exponential expression x^4 represents a repeated multiplication where x is to be written as a _____ four times.
3. x^n is called a _____ of x.
4. The expression $(2x^2 b)^5$ is a power of a _____, and $\left(\dfrac{2x^2}{b}\right)^5$ is a power of a _____.

CONCEPTS

Fill in the blanks.

5. $(3x)^4$ means ☐ · ☐ · ☐ · ☐
6. Using an exponent, $(-5y)(-5y)(-5y)$ can be written as ☐.
7. $x^m x^n =$ ☐
8. $(xy)^n =$ ☐
9. $\left(\dfrac{a}{b}\right)^n =$ ☐
10. $(a^b)^c =$ ☐
11. $\dfrac{x^m}{x^n} =$ ☐
12. $x = x^{\square}$
13. $(x^m)^n =$ ☐
14. $(t^3)^2 =$ ☐ · ☐
15. Write a power of a product that has two factors.
16. Write a power of a quotient.
17. To simplify $(2y^3 z^2)^4$, how many factors within the parentheses must be raised to the fourth power?
18. To simplify $\left(\dfrac{y^3}{z^2}\right)^4$ what two expressions must be raised to the fourth power?

NOTATION

Complete each step.

19. $(x^4 x^2)^3 = (\square)^3$
    $= x^{18}$

20. $\dfrac{a^3 a^4}{a^2} = \dfrac{\square}{a^2}$
    $= a^{\square - 2}$
    $= a^5$

Evaluate each expression.

21. $(-4)^2$
22. $(-5)^2$
23. -4^2
24. -5^2

GUIDED PRACTICE

Identify the base and the exponent in each expression. See Example 1.

25. 4^3
26. $(-8)^2$
27. r^5
28. $\left(\dfrac{5}{x}\right)^3$
29. $(-3x)^2$
30. $-x^4$
31. $-\dfrac{1}{3}y^6$
32. $3.14r^4$

Write each expression in an equivalent form using an exponent. See Example 2.

33. $\dfrac{a}{3} \cdot \dfrac{a}{3} \cdot \dfrac{a}{3}$
34. $\dfrac{y}{4} \cdot \dfrac{y}{4} \cdot \dfrac{y}{4} \cdot \dfrac{y}{4}$
35. $-\dfrac{b}{6} \cdot \dfrac{b}{6} \cdot \dfrac{b}{6} \cdot \dfrac{b}{6} \cdot \dfrac{b}{6}$
36. $-\dfrac{c}{5} \cdot \dfrac{c}{5} \cdot \dfrac{c}{5}$
37. $6 \cdot x \cdot x \cdot x \cdot x \cdot x$
38. $-3 \cdot y \cdot y \cdot y$
39. $(4t)(4t)(4t)(4t)$
40. $(-5u)(-5u)$
41. $-4 \cdot t \cdot t \cdot t$
42. $-5 \cdot u \cdot u \cdot u$
43. $(x+y)(x+y)$
44. $(a+b)(a+b)(a+b)(a+b)$

Simplify each expression. See Example 3.

45. $12^3 \cdot 12^4$
46. $3^4 \cdot 3^6$
47. $(2^3)(2^2)$
48. $(5^5)(5^3)$
49. $a^3 \cdot a^3$
50. $m^7 \cdot m^7$
51. $x^4 x^3$
52. $y^5 y^2$
53. $a^3 a a^5$
54. $b b^2 b^3$
55. $y^3(y^2 y^4)$
56. $(y^4 y) y^6$
57. $(a^2 b^3)(a^3 b^3)$
58. $(u^3 v^5)(u^4 v^5)$
59. $(m+n)^2 (m+n)^3$
60. $(c-d)^3 (c-d)^4$

Simplify each expression. Assume there are no divisions by 0. See Example 4.

61. $\dfrac{8^{12}}{8^4}$
62. $\dfrac{10^4}{10^2}$
63. $\dfrac{12^{15}}{12^{12}}$
64. $\dfrac{15^{17}}{15^{10}}$
65. $\dfrac{x^{15}}{x^3}$
66. $\dfrac{y^6}{y^3}$
67. $\dfrac{c^{10}}{c^9}$
68. $\dfrac{h^{20}}{h^{10}}$
69. $\dfrac{x^2 y^8}{xy^4}$
70. $\dfrac{p^3 q^8}{p^2 q^6}$
71. $\dfrac{c^3 d^7}{cd}$
72. $\dfrac{r^8 s^9}{rs}$

Simplify each expression. Assume there are no divisions by 0. See Example 5.

73. $\dfrac{m^2 m^5 m^7}{m^2 m}$
74. $\dfrac{b^4 b^5}{b^2 b^3}$
75. $\dfrac{a^2 a^3 a^4}{a^4 a^4}$
76. $\dfrac{a a^2 a^6}{a^2 a^3}$

Simplify each expression. See Example 6.

77. $(3^2)^4$
78. $(4^3)^3$
79. $(2^3)^5$
80. $(5^4)^2$
81. $(y^5)^3$
82. $(b^3)^6$
83. $(m^{50})^{10}$
84. $(n^{25})^4$

Simplify each expression. See Example 7.

85. $(x^2 x^3)^5$
86. $(y^3 y^4)^4$
87. $(b^3 b^4)^2$
88. $(m^4 m^2)^3$
89. $(y^3)^2 (y^2)^4$
90. $(b^4)^3 (b^2)^2$
91. $(p^2)^3 (p^3)^3$
92. $(q^4)^2 (q^2)^3$

Simplify each expression. See Example 8.

93. $(4p^3)^3$
94. $(3n^2)^4$
95. $(a^3 b^2)^3$
96. $(r^3 s^2)^2$
97. $(p^2 q^4)^2$
98. $(r^3 s^5)^3$
99. $(-2r^2 s^3)^3$
100. $(-3x^2 y^4)^2$

Simplify each expression. See Example 9.

101. $\dfrac{(x^2 y^3)^2}{xy^2}$
102. $\dfrac{(a^3 b^4)^3}{a^3 b^4}$
103. $\dfrac{(p^5 q^2)^3}{p^{10} q^4}$
104. $\dfrac{(m^5 n^2)^3}{m^{12} n^3}$

Simplify each expression. See Example 10.

105. $\left(\dfrac{a}{b}\right)^3$
106. $\left(\dfrac{r}{s}\right)^4$
107. $\left(\dfrac{2x^2}{y^3}\right)^5$
108. $\left(\dfrac{3u^4}{2v^2}\right)^6$

Simplify each expression. See Example 11.

109. $\dfrac{(3z)^5}{(3z)^2}$

110. $\dfrac{(2t)^{10}}{(2t)^5}$

111. $\dfrac{(6k)^7}{(6k)^4}$

112. $\dfrac{(-3a)^{12}}{(-3a)^{10}}$

TRY IT YOURSELF

Simplify each expression. Assume there are no divisions by 0.

113. $xy^2 \cdot x^2 y$
114. $s^8 t^2 s^2 t^7$
115. $(3zz^2 z^3)^5$
116. $(4t^3 t^6 t^2)^2$
117. $(x^5)^2 (x^7)^3$
118. $(y^3 y)^2 (y^2)^2$
119. $(uv)^4$
120. $(xy)^3$
121. $\left(\dfrac{-2a}{b}\right)^5$
122. $\left(\dfrac{-2t}{3}\right)^4$
123. $\dfrac{(a^2 b)^{15}}{(a^2 b)^9}$
124. $\dfrac{(s^2 t^3)^4}{(s^2 t^3)^2}$
125. $\dfrac{(ab^2)^3}{(ab)^2}$
126. $\dfrac{(m^3 n^4)^3}{(mn^2)^3}$
127. $\dfrac{(r^4 s^3)^4}{(rs^3)^3}$
128. $\dfrac{(x^2 y^5)^5}{(x^3 y)^2}$
129. $\left(\dfrac{y^3 y}{2yy^2}\right)^3$
130. $\left(\dfrac{2y^3 y}{yy^2}\right)^3$
131. $\left(\dfrac{3t^3 t^4 t^5}{4t^2 t^6}\right)^3$
132. $\left(\dfrac{4t^3 t^4 t^5}{3t^2 t^6}\right)^3$
133. $\left(\dfrac{t^2}{2}\right)\left(\dfrac{t^2}{2}\right)\left(\dfrac{t^2}{2}\right)$
134. $c \cdot c \cdot c \cdot d \cdot d$
135. $(cd^4)(cd)$
136. $ab^3 c^4 \cdot ab^4 c^2$
137. $\dfrac{y^3 y^4}{yy^2}$
138. $\dfrac{b^4 b^5}{b^2 b^3}$

APPLICATIONS

Find each area or volume. You may leave π in your answer. Refer to the area and volume formulas in Section 2.4.

139.
a^5 mi by a^5 mi

140.
$4y^3$ yd

141.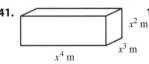
x^2 m, x^3 m, x^4 m

142.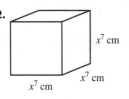
x^7 cm

143. **ART HISTORY** Leonardo da Vinci's drawing relating a human figure to a square and a circle is shown. Find an expression for the following:

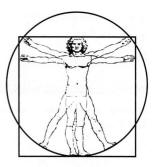

   a. The area of the square if the man's height is $5x$ feet.

   b. The area of the circle if the waist-to-feet distance is $3a$ feet. Leave π in your answer.

144. **PACKAGING** A bowling ball fits tightly against all sides of a cardboard box that it is packaged in. Find expressions for the volume of the ball and box. Leave π in your answer.

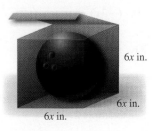
$6x$ in., $6x$ in., $6x$ in.

145. **BOUNCING BALLS** A ball is dropped from a height of 32 feet. Each rebound is one-half of its previous height.

   a. Draw a diagram of the path of the ball, showing four bounces.

   b. Explain why the expressions $32\left(\tfrac{1}{2}\right)$, $32\left(\tfrac{1}{2}\right)^2$, $32\left(\tfrac{1}{2}\right)^3$, and $32\left(\tfrac{1}{2}\right)^4$ represent the height of the ball on the first, second, third, and fourth bounces, respectively. Find the heights of the first four bounces.

146. **CHILDBIRTH** Mr. and Mrs. Emory Harrison, of Johnson City, Tennessee, had 13 sons in a row during the 1940s and 1950s. The **probability** of a family of 13 children all being male is $\left(\tfrac{1}{2}\right)^{13}$. Evaluate this expression.

147. COMPUTERS Text is stored by computers using a sequence of eight 0's and 1's. Such a sequence is called a **byte**. An example of a byte is 10101110.

a. Write four other bytes, all ending in 1.

b. Each of the eight digits of a byte can be one of *two* possibilities (either 0 or 1). The total number of different bytes can be represented by an exponential expression with base 2. What is it?

148. INVESTING Guess the answer to the following problem. Then use a calculator to find the correct answer. Were you close?

If the value of 1¢ is to double every day, what will the penny be worth after 31 days?

WRITING

149. Explain the mistake.

$$2^3 \cdot 2^2 = 4^5$$
$$= 1{,}024$$

150. Are the expressions $2x^3$ and $(2x)^3$ equivalent? Explain.

151. Is the operation of raising to a power commutative? That is, is $a^b = b^a$? Explain why or why not.

152. When a number is raised to a power, is the result always larger than the original number? Support your answer with some examples.

REVIEW

Match each equation with its graph below.

153. $y = 2x - 1$ **154.** $y = 3x - 1$

155. $y = 3$ **156.** $x = 3$

a.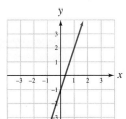

b.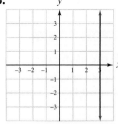

c.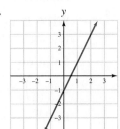

d.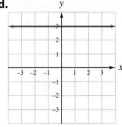

SECTION 5.2
Zero and Negative Integer Exponents

Objectives
1. Use the zero exponent rule.
2. Use the negative integer exponent rule.
3. Use exponent rules to write equivalent expressions with positive exponents.
4. Use all exponent rules to simplify expressions.
5. Use exponent rules with variable exponents.

In the previous section, we discussed natural-number exponents. We now extend the discussion to include exponents that are zero and exponents that are negative integers.

1 Use the zero exponent rule.

When we discussed the quotient rule for exponents in the previous section, the exponent in the numerator was always greater than the exponent in the denominator. We now consider what happens when the exponents are equal. To develop the definition of a zero exponent, we will simplify the expression $\dfrac{5^3}{5^3}$ in two ways and compare the results. If we use the quotient rule for exponents, where the exponents in the numerator and denominator are equal, we obtain 5^0. However, by removing the common factors of 5, we obtain 1.

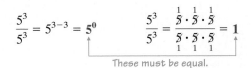

These must be equal.

For this reason, we conclude that $5^0 = 1$. This example suggests the following rule.

> **Zero Exponents**
>
> Any nonzero real number raised to the 0 power is 1. For any nonzero real number x,
>
> $$x^0 = 1$$

Self Check 1

Write each expression without using exponents:
a. $(-0.115)^0$
b. $-5a^0 b$

Now Try Problems 21, 23, and 24

EXAMPLE 1 Write each expression without using exponents:

a. $\left(\dfrac{1}{13}\right)^0$ b. $3x^0$ c. $(3x)^0$

Strategy We will note the base and exponent in each case. Since each expression has an exponent that is zero, we will use the zero exponent rule.

WHY If an expression contains a nonzero base raised to the 0 power, we can replace it with 1.

Solution

a. $\left(\dfrac{1}{13}\right)^0 = 1$ The base is $\tfrac{1}{13}$; the exponent is 0.

b. $3x^0 = 3(1)$ The base is x; the exponent is 0.
$= 3$

c. $(3x)^0 = 1$ The base is $3x$; the exponent is 0.

Parts b and c point out that $3x^0 \neq (3x)^0$.

> **Caution!** Remember, the base is only that which is directly in front of the exponent unless there are parentheses. For $3x^0$, the base is x and the exponent is 0. For $(3x)^0$, the base is $3x$ and the exponent 0.

2 Use the negative integer exponent rule.

To develop the definition of a negative exponent, we will simplify the expression $\dfrac{6^2}{6^5}$ in two ways. If we use the quotient rule for exponents, where the exponent in the numerator is less than the exponent in the denominator, we obtain 6^{-3}. However, by removing the common factors of 6, we obtain $\dfrac{1}{6^3}$.

$$\dfrac{6^2}{6^5} = 6^{2-5} = 6^{-3} \qquad \dfrac{6^2}{6^5} = \dfrac{\overset{1}{\cancel{6}} \cdot \overset{1}{\cancel{6}}}{\underset{1}{\cancel{6}} \cdot \underset{1}{\cancel{6}} \cdot 6 \cdot 6 \cdot 6} = \dfrac{1}{6^3}$$

These must be equal.

For this reason, we conclude that $6^{-3} = \dfrac{1}{6^3}$. In general, we have the following rule.

Negative Exponents

If x represents any nonzero number and n represents a natural number, then

$$x^{-n} = \dfrac{1}{x^n}$$

In words, x^{-n} is the reciprocal of x^n.

The definition of a negative exponent states that another way to write x^{-n} is to write its reciprocal, changing the sign of the exponent. We can use this definition to write expressions that contain negative exponents as expressions without negative exponents.

The Language of Algebra The *negative integers* are: $-1, -2, -3, -4, -5, \ldots$

EXAMPLE 2 Simplify by using the definition of negative exponents:
a. 3^{-5} **b.** $(-2)^{-3}$

Strategy Since each expression has an exponent that is negative, we will use the negative exponent rule.

WHY This rule enables us to write an exponential expression that has a negative exponent in an equivalent form using a positive exponent.

Solution

a. $3^{-5} = \dfrac{1}{3^5}$ Write the reciprocal of 3^{-5} and change the exponent from -5 to 5.

$= \dfrac{1}{243}$ Evaluate 3^5.

b. $(-2)^{-3} = \dfrac{1}{(-2)^3}$ Write the reciprocal of $(-2)^{-3}$ and change the exponent from -3 to 3.

$= -\dfrac{1}{8}$ Evaluate $(-2)^3$.

Self Check 2
Simplify by using the definition of negative exponents:
a. 4^{-4}
b. $(-5)^{-3}$
Now Try Problems 27 and 29

Caution! A negative exponent does not indicate a negative number. It indicates a reciprocal. For example:

$$4^{-2} = \dfrac{1}{4^2} = \dfrac{1}{16} \qquad 4^{-2} \neq -16 \qquad 4^{-2} \neq -\dfrac{1}{4^2}$$

3 Use exponent rules to write equivalent expressions with positive exponents.

Negative exponents can appear in the numerator and/or the denominator of a fraction. To develop rules for such situations, we consider the following example.

$$\dfrac{a^{-4}}{b^{-3}} = \dfrac{\dfrac{1}{a^4}}{\dfrac{1}{b^3}} = \dfrac{1}{a^4} \div \dfrac{1}{b^3} = \dfrac{1}{a^4} \cdot \dfrac{b^3}{1} = \dfrac{b^3}{a^4}$$

We can obtain this result in a simpler way. In $\dfrac{a^{-4}}{b^{-3}}$, we can move a^{-4} from the numerator to the denominator and change the sign of the exponent, and we can move b^{-3} from the denominator to the numerator and change the sign of the exponent.

$$\dfrac{a^{-4}}{b^{-3}} = \dfrac{b^{3}}{a^{4}}$$

> **The Language of Algebra** *Factors* of a numerator or denominator may be moved *across the fraction bar* if we change the sign of their exponent.

This example suggests the following rules.

Changing from Negative to Positive Exponents

A factor can be moved from the denominator to the numerator or from the numerator to the denominator of a fraction if the sign of its exponent is changed.

For any nonzero real numbers x and y, and any integers m and n,

$$\dfrac{1}{x^{-n}} = x^{n} \quad \text{and} \quad \dfrac{x^{-m}}{y^{-n}} = \dfrac{y^{n}}{x^{m}}$$

These rules streamline the process when simplifying fractions involving negative exponents.

Self Check 3

Simplify by using the definition of negative exponents:

a. $\dfrac{1}{9^{-1}}$

b. $\dfrac{8^{-2}}{7^{-1}}$

Now Try Problems 31 and 33

EXAMPLE 3 Simplify by using the definition of negative exponents:

a. $\dfrac{1}{5^{-2}}$ b. $\dfrac{2^{-3}}{3^{-4}}$

Strategy Since the exponents are negative numbers, we will use the negative exponent rule.

WHY It is usually easier to simplify exponential expressions if the exponents are positive.

Solution

a. $\dfrac{1}{5^{-2}} = 5^{2}$ Move 5^{-2} to the numerator and change the sign of the exponent.

$= 25$ Evaluate 5^{2}.

b. $\dfrac{2^{-3}}{3^{-4}} = \dfrac{3^{4}}{2^{3}}$ Move 2^{-3} to the denominator and change the sign of the exponent. Move 3^{-4} to the numerator and change the sign of the exponent.

$= \dfrac{81}{8}$ Evaluate 3^{4} and 2^{3}.

EXAMPLE 4 Simplify by using the definition of negative exponents. Assume that no denominators are zero.

a. x^{-4} b. $\dfrac{x^{-3}}{y^{-7}}$ c. $(-2x)^{-2}$ d. $-2x^{-2}$

5.2 Zero and Negative Integer Exponents

Strategy We will note the base and exponent in each case. Since each expression has exponents that are negative numbers, we will use the negative exponent rule.

WHY The negative exponent rule enables us to write an exponential expression that has negative exponents in an equivalent form using positive exponents.

Solution

a. $x^{-4} = \dfrac{1}{x^4}$

b. $\dfrac{x^{-3}}{y^{-7}} = \dfrac{y^7}{x^3}$

c. $(-2x)^{-2} = \dfrac{1}{(-2x)^2} = \dfrac{1}{4x^2}$

d. $-2x^{-2} = -2\left(\dfrac{1}{x^2}\right) = -\dfrac{2}{x^2}$

> **Self Check 4**
> Simplify by using the definition of negative exponents:
> **a.** a^{-5} **b.** $\dfrac{r^{-4}}{s^{-5}}$ **c.** $3y^{-3}$
>
> **Now Try** Problems 35, 37, 39, and 41

When a fraction base is raised to a negative power, we can use rules for exponents to change the sign of the exponent. For example,

$$\left(\dfrac{x}{y}\right)^{-2} = \dfrac{x^{-2}}{y^{-2}} = \dfrac{y^2}{x^2} = \left(\dfrac{y}{x}\right)^2$$

The exponent is the opposite of -2. The base is the reciprocal of $\dfrac{x}{y}$.

This example suggests the following rule.

Negative Exponents and Reciprocals

A fraction raised to a power is equal to the reciprocal of the fraction raised to the opposite power.

For any nonzero real numbers x and y, and any integer n,

$$\left(\dfrac{x}{y}\right)^{-n} = \left(\dfrac{y}{x}\right)^n$$

EXAMPLE 5
Write $\left(\dfrac{5}{16}\right)^{-1}$ without using exponents.

Strategy We will use the negative exponent and reciprocal rule.

WHY The expression involves a fraction base to a negative exponent. It is often easier to simplify exponential expressions if the exponents are positive.

Solution

$\left(\dfrac{5}{16}\right)^{-1} = \left(\dfrac{16}{5}\right)^1$ The base is $\dfrac{5}{16}$ and the exponent is -1. Write the reciprocal of the base and change the sign of the exponent.

$= \dfrac{16^1}{5^1}$ To raise a fraction to a power, we raise both the numerator and the denominator to that power.

$= \dfrac{16}{5}$

> **Self Check 5**
> Write $\left(\dfrac{3}{7}\right)^{-2}$ without using exponents.
>
> **Now Try** Problem 45

4 Use all exponent rules to simplify expressions.

The rules for exponents involving products, powers, and quotients are also true for zero and negative exponents.

> **Summary of Exponent Rules**
>
> If m and n represent integers and there are no divisions by zero, then
>
> | Product rule | Power rule | Power of a product |
> |---|---|---|
> | $x^m \cdot x^n = x^{m+n}$ | $(x^m)^n = x^{mn}$ | $(xy)^n = x^n y^n$ |
>
> | Quotient rule | Power of a quotient | Exponents of 0 and 1 |
> |---|---|---|
> | $\dfrac{x^m}{x^n} = x^{m-n}$ | $\left(\dfrac{x}{y}\right)^n = \dfrac{x^n}{y^n}$ | $x^0 = 1$ and $x^1 = x$ |
>
> | Negative exponent | Negative exponents appearing in fractions |
> |---|---|
> | $x^{-n} = \dfrac{1}{x^n}$ | $\dfrac{1}{x^{-n}} = x^n \qquad \dfrac{x^{-m}}{y^{-n}} = \dfrac{y^n}{x^m} \qquad \left(\dfrac{x}{y}\right)^{-n} = \left(\dfrac{y}{x}\right)^n$ |

The rules for exponents are used to simplify expressions involving products, quotients, and powers. In general, **an expression involving exponents is simplified when**

- Each base occurs only once
- There are no parentheses
- There are no negative or zero exponents

Self Check 6

Simplify and write the result without using negative exponents:

a. $(x^4)^{-3}$

b. $\dfrac{a^4}{a^8}$

c. $\dfrac{a^{-4} a^{-5}}{a^{-3}}$

Now Try Problems 47, 49, 51, and 53

EXAMPLE 6 Simplify and write the result without using negative exponents. Assume that no denominators are zero.

a. $(x^{-3})^2$ **b.** $\dfrac{x^3}{x^7}$ **c.** $(x^3 x^2)^{-3}$ **d.** $\dfrac{y^{-4} y^{-3}}{y^{-20}}$

Strategy In each case, we want to use the exponent rules to write an equivalent expression that uses each base with a positive exponent only once.

WHY The expressions are not in simplest form. In each case, either the bases occur as a factor more than once, there are parentheses, or there are negative exponents.

Solution

a. $(x^{-3})^2 = x^{-6}$ Use the power rule. Keep the base and multiply the exponents.

$\qquad = \dfrac{1}{x^6}$ Write the reciprocal of x^{-6} and change the sign of the exponent.

b. $\dfrac{x^3}{x^7} = x^{3-7}$ Use the quotient rule. Keep the base and subtract the exponents.

$\qquad = x^{-4}$ Do the subtraction: $3 - 7 = -4$.

$\qquad = \dfrac{1}{x^4}$ Write the reciprocal of x^{-4} and change the sign of the exponent.

c. $(x^3 x^2)^{-3} = (x^5)^{-3}$ Use the product rule. Keep the base and add the exponents.

$\qquad = x^{-15}$ Use the power rule. Keep the base and multiply the exponents.

$\qquad = \dfrac{1}{x^{15}}$ Write the reciprocal of x^{-15} and change the sign of the exponent.

d. $\dfrac{y^{-4} y^{-3}}{y^{-20}} = \dfrac{y^{-7}}{y^{-20}}$ Use the product rule in the numerator.

$\qquad = y^{-7-(-20)}$ Use the quotient rule.

$\qquad = y^{13}$ Do the subtraction: $-7 - (-20) = -7 + 20 = 13$.

EXAMPLE 7
Simplify and write the answer without negative exponents. Assume no denominators are zero.

a. $\dfrac{12a^3b^4}{4a^5b^2}$ **b.** $\left(-\dfrac{x^3y^2}{xy^{-3}}\right)^{-2}$

Strategy In each case, we want to use the exponent rules to write an equivalent expression that uses each base with a positive exponent only once.

WHY The expressions are not in simplest form. In each case, either the bases occur as factors more than once, there are parentheses, or there are negative exponents.

Solution

a. $\dfrac{12a^3b^4}{4a^5b^2} = 3a^{3-5}b^{4-2}$ Simplify the numerical coefficients. Use the quotient rule twice.

$\phantom{\dfrac{12a^3b^4}{4a^5b^2}} = 3a^{-2}b^2$ Do the subtractions.

$\phantom{\dfrac{12a^3b^4}{4a^5b^2}} = \dfrac{3b^2}{a^2}$ Move a^{-2} to the denominator and change the sign of the exponent.

b. $\left(-\dfrac{x^3y^2}{xy^{-3}}\right)^{-2} = \left(-\dfrac{xy^{-3}}{x^3y^2}\right)^2$ Write the reciprocal of the base and change the sign of the exponent.

$= (-x^{1-3}y^{-3-2})^2$ Use the quotient rule for exponents twice.

$= (-x^{-2}y^{-5})^2$ Do the subtractions.

$= x^{-4}y^{-10}$ Raise each factor to the second power.

$= \dfrac{1}{x^4y^{10}}$ Move x^{-4} and y^{-10} to the denominator and change the sign of the exponents.

Self Check 7
Simplify and write the result without using negative exponents:

a. $\dfrac{20x^5y^3}{15x^2y^8}$

b. $\left(\dfrac{x^5y^3}{xy^{-3}}\right)^{-3}$

Now Try Problems 55 and 57

5 Use exponent rules with variable exponents.

We can apply the rules for exponents to simplify expressions involving variable exponents.

EXAMPLE 8
Simplify. Assume that there are no divisions by 0.

a. $\dfrac{6^n}{6^n}$ **b.** $x^{2m}x^{3m}$ **c.** $\dfrac{y^{2m}}{y^{4m}}$

Strategy We will use the rules for exponents and the rules for adding, subtracting, and multiplying variable expressions.

WHY The exponents are variables.

Solution

a. $\dfrac{6^n}{6^n} = 6^{n-n}$ Keep the common base and subtract the exponents.

$\phantom{\dfrac{6^n}{6^n}} = 6^0$ Combine like terms: $n - n = 0$.

$\phantom{\dfrac{6^n}{6^n}} = 1$ Evaluate: $6^0 = 1$.

b. $x^{2m}x^{3m} = x^{2m+3m}$ Keep the common base and add the exponents.

$\phantom{x^{2m}x^{3m}} = x^{5m}$ Combine like terms: $2m + 3m = 5m$.

c. $\dfrac{y^{2m}}{y^{4m}} = y^{2m-4m}$ Keep the base and subtract the exponents.

$\phantom{\dfrac{y^{2m}}{y^{4m}}} = y^{-2m}$ Combine like terms: $2m - 4m = -2m$.

$\phantom{\dfrac{y^{2m}}{y^{4m}}} = \dfrac{1}{y^{2m}}$ Write the reciprocal of y^{-2m} and change the exponent to $2m$.

Self Check 8
Simplify each expression:

a. $\dfrac{7^m}{7^m}$

b. $z^{3n}z^{2n}$

c. $\dfrac{z^{3n}}{z^{5n}}$

Now Try Problems 59, 61, and 65

Using Your CALCULATOR Finding Present Value

As a gift for their newborn grandson, the grandparents want to deposit enough money in the bank now so that when he turns 18, the young man will have a college fund of $20,000 waiting for him. How much should they deposit now if the money will earn 6% annually?

To find how much money P must be invested at an annual rate i (expressed as a decimal) to have $\$A$ in n years, we use the formula $P = A(1 + i)^{-n}$. If we substitute 20,000 for A, 0.06 (6%) for i, and 18 for n, we have

$$P = A(1 + i)^{-n} \qquad \text{P is called the present value.}$$
$$P = 20{,}000(1 + 0.06)^{-18}$$

To find P with a reverse-entry calculator, we enter these numbers and press these keys.

$$\boxed{(} \; 1 \; \boxed{+} \; .06 \; \boxed{)} \; \boxed{y^x} \; 18 \; \boxed{+/-} \; \boxed{\times} \; 20000 \; \boxed{=} \qquad \boxed{7006.875823}$$

To evaluate the expression with a graphing or a direct-entry calculator, we use the following keystrokes.

$$20000 \; \boxed{\times} \; \boxed{(} \; 1 \; \boxed{+} \; .06 \; \boxed{)} \; \boxed{\wedge} \; \boxed{(-)} \; 18 \; \boxed{\text{ENTER}}$$

```
20000 × (1 + .06)^ - 1
8
           7006.875823
```

They must invest approximately $7,006.88 to have $20,000 in 18 years.

ANSWERS TO SELF CHECKS

1. a. 1 b. $-5b$ 2. a. $\dfrac{1}{256}$ b. $-\dfrac{1}{125}$ 3. a. 9 b. $\dfrac{7}{64}$ 4. a. $\dfrac{1}{a^5}$ b. $\dfrac{s^5}{r^4}$ c. $\dfrac{3}{y^3}$

5. $\dfrac{49}{9}$ 6. a. $\dfrac{1}{x^{12}}$ b. $\dfrac{1}{a^4}$ c. $\dfrac{1}{a^6}$ 7. a. $\dfrac{4x^3}{3y^5}$ b. $\dfrac{1}{x^{12}y^{18}}$ 8. a. 1 b. z^{5n} c. $\dfrac{1}{z^{2n}}$

SECTION 5.2 STUDY SET

VOCABULARY

Fill in the blanks.

1. In the exponential expression 8^{-3}, 8 is the _____ and -3 is the _____.
2. In the exponential expression 5^{-1}, the exponent is a _____ integer.
3. Another way to write 2^{-3} is to write its _____ and to change the sign of the exponent: $2^{-3} = \dfrac{1}{2^3}$.
4. In the expression 6^m, the _____ is a variable.

CONCEPTS

5. In parts a and b, fill in the blanks as you simplify the fraction in two different ways. Then complete the sentence in part c.

   a. $\dfrac{6^4}{6^4} = 6^{}$

   $= 6^{}$

   b. $\dfrac{6^4}{6^4} = \dfrac{\cdot\cdot\cdot}{6\cdot 6\cdot 6\cdot 6}$

   $= $

   c. So we define 6^0 to be _____, and, in general, if x is any nonzero real number, then $x^0 = $ _____.

6. In parts a and b, fill in the blanks as you simplify the fraction in two different ways. Then complete the sentence in part c.

   a. $\dfrac{8^3}{8^5} = 8^{}$

   $= 8^{}$

   b. $\dfrac{8^3}{8^5} = \dfrac{\cdot\cdot}{8\cdot 8\cdot 8\cdot 8\cdot 8}$

   $= \dfrac{1}{8^2}$

   c. We define 8^{-2} to be _____, and, in general, if x is any nonzero real number, then $x^{-n} = $ _____.

Complete each table.

7.

x	3^x
2	
1	
0	
−1	
−2	

8.

x	4^x
2	
1	
0	
−1	
−2	

9.

x	$(-9)^x$
2	
1	
0	
−1	
−2	

10.

x	$(-5)^x$
2	
1	
0	
−1	
−2	

Use the graph to determine the missing y-coordinates in the table and express each y-coordinate as a power of 2.

11.

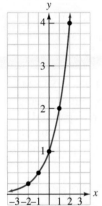

x	y	y as a power of 2
2	2	
1	2	
0	2	
−1	2	
−2	2	

12.

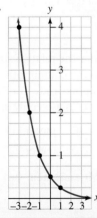

x	y	y as a power of 2
1	2	
0	2	
−1	2	
−2	2	
−3	2	

NOTATION

Complete each step.

13. $(y^5 y^3)^{-5} = ()^{-5}$
$= y^{}$
$= \dfrac{1}{y^{40}}$

14. $\left(\dfrac{a^2 b^3}{a^{-3} b}\right)^{-3} = \left(a^{2-(-3)} b^{-1}\right)^{-3}$
$= \left(a^{} b^{}\right)^{-3}$
$= \dfrac{1}{(a^5 b^2)^{}}$
$= \dfrac{1}{a^{15} b^6}$

15. In the expression $3x^{-2}$, what is the base and what is the exponent?

16. In the expression $-3x^{-2}$, what is the base and what is the exponent?

17. Determine the base and the exponent and evaluate each expression.
  a. -4^2
  b. 4^{-2}
  c. -4^{-2}

18. Determine the base and the exponent and evaluate each expression.
  a. $(-7)^2$
  b. $(-7)^{-2}$
  c. -7^{-2}

GUIDED PRACTICE

Write each expression without using exponents. See Example 1.

19. 7^0 **20.** 9^0

21. $\left(\dfrac{1}{4}\right)^0$ **22.** $\left(\dfrac{3}{8}\right)^0$

23. $2x^0$ **24.** $(2x)^0$

25. $(-x)^0$ **26.** $-x^0$

Simplify each expression by using the definition of negative exponents. See Example 2.

27. 12^{-2} **28.** 7^{-2}

29. $(-4)^{-1}$ **30.** $(-8)^{-2}$

Simplify each expression by using the definition of negative exponents. See Example 3.

31. $\dfrac{1}{5^{-3}}$ **32.** $\dfrac{1}{3^{-3}}$

33. $\dfrac{2^{-4}}{3^{-1}}$ **34.** $\dfrac{7^{-2}}{2^{-3}}$

Write each expression without using negative exponents. See Example 4.

35. x^{-2} **36.** y^{-3}

37. $\dfrac{a^{-2}}{b^{-3}}$ **38.** $\dfrac{m^{-7}}{n^{-5}}$

39. $(-4y)^{-2}$ **40.** $(-5d)^{-3}$

41. $-2b^{-5}$ **42.** $-3c^{-4}$

Write each expression without using exponents. See Example 5.

43. $\left(\dfrac{7}{8}\right)^{-1}$
44. $\left(\dfrac{16}{5}\right)^{-1}$
45. $\left(\dfrac{3}{4}\right)^{-2}$
46. $\left(\dfrac{2}{3}\right)^{-2}$

Write each expression without using negative exponents. See Example 6.

47. $(a^{-4})^3$
48. $(b^{-3})^5$
49. $\dfrac{a^3}{a^8}$
50. $\dfrac{t^5}{t^{12}}$
51. $(ab^2)^{-3}$
52. $(m^2n^3)^{-2}$
53. $\dfrac{y^{-4}y^{-3}}{y^{-10}}$
54. $\dfrac{x^{-4}x^{-5}}{x^{-8}x^{-4}}$

Write each expression without using negative exponents. See Example 7.

55. $\dfrac{15p^2q^3}{5p^3q^2}$
56. $\dfrac{27m^3n^5}{6m^5n^3}$
57. $\left(\dfrac{a^2b^5}{a^2b^{-2}}\right)^{-2}$
58. $\left(\dfrac{a^3b^{-2}}{a^2b^3}\right)^{-3}$

Simplify each expression. Assume that there are no divisions by 0. See Example 8.

59. $\dfrac{7^n}{7^n}$
60. $\dfrac{8^p}{8^p}$
61. $x^{2m}x^m$
62. $y^{3m}y^{2m}$
63. $u^{2m}u^{-3m}$
64. $r^{5m}r^{-6m}$
65. $\dfrac{y^{3m}}{y^{2m}}$
66. $\dfrac{z^{4m}}{z^{2m}}$

TRY IT YOURSELF

Write each answer without using parentheses or negative exponents.

67. $\left(\dfrac{a^2b^3}{ab^4}\right)^0$
68. $\dfrac{2}{3}\left(\dfrac{xyz}{x^2y}\right)^0$
69. $\dfrac{5}{2x^0}$
70. $\dfrac{4}{3a^0}$
71. -4^{-3}
72. -6^{-3}
73. $-(-4)^{-3}$
74. $-(-4)^{-2}$
75. $\dfrac{y^4y^3}{y^4y^{-2}}$
76. $\dfrac{x^{12}x^{-7}}{x^3x^4}$
77. $\dfrac{1}{c^{-5}}$
78. $\dfrac{3}{a^{-7}}$
79. $\dfrac{3^{-2}}{2^{-3}}$
80. $\dfrac{5^{-3}}{3^{-4}}$
81. $(2y)^{-4}$
82. $(-3x)^{-1}$
83. $2^5 \cdot 2^{-2}$
84. $10^2 \cdot 10^{-4}$
85. $4^{-3} \cdot 4^{-2} \cdot 4^5$
86. $3^{-4} \cdot 3^5 \cdot 3^{-3}$
87. $\dfrac{3^5 \cdot 3^{-2}}{3^3}$
88. $\dfrac{6^2 \cdot 6^{-3}}{6^{-2}}$
89. $\dfrac{y^4}{y^5}$
90. $\dfrac{t^7}{t^{10}}$
91. $\dfrac{(r^2)^3}{(r^3)^4}$
92. $\dfrac{(b^3)^4}{(b^5)^4}$
93. $\dfrac{10a^4a^{-2}}{5a^2a^0}$
94. $\dfrac{9b^0b^3}{3b^{-3}b^4}$
95. $(ab^2)^{-2}$
96. $(c^2d^3)^{-2}$
97. $(x^2y)^{-3}$
98. $(-xy^2)^{-4}$
99. $(x^{-4}x^3)^3$
100. $(y^{-2}y)^3$
101. $(a^{-2}b^3)^{-4}$
102. $(y^{-3}z^5)^{-6}$
103. $(-2x^3y^{-2})^{-5}$
104. $(-3u^{-2}v^3)^{-3}$
105. $\left(\dfrac{a^3}{a^{-4}}\right)^2$
106. $\left(\dfrac{a^4}{a^{-3}}\right)^3$
107. $\left(\dfrac{4x^2}{3x^{-5}}\right)^4$
108. $\left(\dfrac{-3r^4r^{-3}}{r^{-3}r^7}\right)^3$
109. $\left(\dfrac{12y^3z^{-2}}{3y^{-4}z^3}\right)^2$
110. $\left(\dfrac{6xy^3}{3x^{-1}y}\right)^3$

APPLICATIONS

111. **THE DECIMAL NUMERATION SYSTEM** Decimal numbers are written by putting digits into place-value columns that are separated by a decimal point. Express the place value of each of the columns shown using a power of 10.

112. **UNIT COMPARISONS** Consider the relative sizes of the items listed in the table. In the column titled "measurement," write the most appropriate number from the following list. Each number is used only once.

10^0 meter 10^{-1} meter 10^{-2} meter
10^{-3} meter 10^{-4} meter 10^{-5} meter

Item	Measurement (m)
Thickness of a dime	
Height of a bathroom sink	
Length of a pencil eraser	
Thickness of soap bubble film	
Diameter of a hamburger patty	
Thickness of a piece of paper	

113. RETIREMENT YEARS How much money should a young married couple invest now at an 8% annual rate if they want to have $100,000 in the bank when they reach retirement age in 40 years? (*Hint:* See the Using Your Calculator box in this section.)

114. BACTERIA During bacterial reproduction, the time required for a population to double is called the **generation time**. If b bacteria are introduced into a medium, then after the generation time has elapsed, there will be $2b$ bacteria. After n generations, there will be $b \cdot 2^n$ bacteria. Explain what this expression represents when $n = 0$.

115. ELECTRONICS The total resistance R of a certain circuit is given by

$$R = \left(\frac{1}{R_1} + \frac{1}{R_2}\right)^{-1} + R_3$$

Find R if $R_1 = 4$, $R_2 = 2$, and $R_3 = 1$.

116. BIOLOGY Refer to the illustration below in which the size of a bacterium is expressed as a fraction of a meter. Fill in the exponent to express the fraction as a power of 10.

Bacterium $\dfrac{1}{1,000,000}$ meter = $10^{\boxed{}}$ meter

WRITING

117. Explain how you would help a friend understand that 2^{-3} is not equal to -8.

118. Describe how you would verify on a calculator that

$$2^{-3} = \frac{1}{2^3}$$

REVIEW

119. IQ TESTS An IQ (intelligence quotient) is a score derived from the formula

$$\text{IQ} = \frac{\text{mental age}}{\text{chronological age}} \cdot 100$$

Find the mental age of a 10-year-old girl if she has an IQ of 135.

120. DIVING When you are under water, the pressure in your ears is given by the formula

Pressure = depth · density of water

Find the density of water (in lb/ft^3) if, at a depth of 9 feet, the pressure on your eardrum is 561.6 lb/ft^2.

121. Write the equation of the line having slope $\frac{3}{4}$ and y-intercept -5.

122. Find $f(-6)$ if $f(x) = x^2 - 3x + 1$.

SECTION 5.3
Scientific Notation

Objectives

1. Define scientific notation.
2. Write numbers in scientific notation.
3. Convert from scientific notation to standard notation.
4. Perform computations with scientific notation.

Scientists often deal with extremely large and extremely small numbers. Two examples are shown below.

The distance from Earth to the sun is approximately 150,000,000 kilometers.

The influenza virus, which causes "flu" symptoms of cough, sore throat, headache, and congestion, has a diameter of 0.00000256 inch.

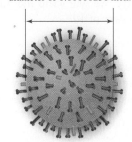

The large number of zeros in 150,000,000 and 0.00000256 makes them difficult to read and hard to remember. In this section, we will discuss a notation that will make such numbers easier to use.

1 Define scientific notation.

Scientific notation provides a compact way of writing large numbers, such as 5,213,000,000,000, and small numbers, such as 0.000000000000914.

> **Scientific Notation**
>
> A number is written in **scientific notation** when it is written as the product of a number between 1 (including 1) and 10, denoted N, and an integer power of 10, denoted n. In symbols, scientific notation has the form $N \times 10^n$.

These numbers are written in scientific notation:

$$3.9 \times 10^6, \quad 2.24 \times 10^{-4}, \quad \text{and} \quad 9.875 \times 10^{22}$$

Every number written in scientific notation has the following form:

$$\underbrace{\blacksquare.\blacksquare}_{\text{A decimal between 1 and 10}} \times 10^{\overbrace{}^{\text{An integer exponent}}}$$

2 Write numbers in scientific notation.

To write a number in scientific notation ($N \times 10^n$), we first determine N then n.

Self Check 1

The distance from Earth to the sun is approximately 93,000,000 miles. Write this number in scientific notation.

Now Try Problem 19

EXAMPLE 1 Change to scientific notation: 150,000,000

Strategy We will write the number as a product of a number between 1 and 10 and a power of 10.

WHY Numbers written in scientific notation have the form $N \times 10^n$.

Solution
We note that 1.5 lies between 1 and 10. To obtain 150,000,000, the decimal point in 1.5 must be moved eight places to the right.

$$1.\underbrace{50000000}_{\text{8 places to the right}}$$

Because multiplying a number by 10 moves the decimal point one place to the right, we can accomplish this by multiplying 1.5 by 10 eight times. We can show the multiplication of 1.5 by 10 eight times using the notation 10^8. Thus, 150,000,000 written in scientific notation is 1.5×10^8.

Self Check 2

The *Salmonella* bacterium, which causes food poisoning, is 0.00009055 inch long. Write this number in scientific notation.

Now Try Problem 25

EXAMPLE 2 Change to scientific notation: 0.00000256

Strategy We will write the number as a product of a number between 1 and 10 and a power of 10.

WHY Numbers written in scientific notation have the form $N \times 10^n$.

Solution
We note that 2.56 is between 1 and 10. To obtain 0.00000256, the decimal point in 2.56 must be moved six places to the left.

$$\underbrace{000002}_{\text{6 places to the left}}.56$$

We can accomplish this by dividing 2.56 by 10^6, which is equivalent to multiplying 2.56 by $\frac{1}{10^6}$ (or by 10^{-6}). Thus, 0.00000256 written in scientific notation is 2.56×10^{-6}.

5.3 Scientific Notation

EXAMPLE 3 Write in scientific notation: **a.** 235,000 **b.** 0.0000073

Strategy We will write each number as a product of a number between 1 and 10 and a power of 10.

WHY Numbers written in scientific notation have the form $N \times 10^n$.

Solution

a. $235{,}000 = 2.35 \times 10^5$ Because $2.35 \times 10^5 = 235{,}000$ and 2.35 is between 1 and 10

b. $0.0000073 = 7.3 \times 10^{-6}$ Because $7.3 \times 10^{-6} = 0.0000073$ and 7.3 is between 1 and 10

Self Check 3
Write in scientific notation:
a. 17,500
b. 0.657
Now Try Problems 27 and 29

> **Sucess Tip** From Examples 1, 2, and 3, we see that when a real number greater than 10 is written in scientific notation, the exponent on 10 is a positive number. When a real number between 0 and 1 is written in scientific notation, the exponent on 10 is a negative number.

Using Your CALCULATOR Calculators and Scientific Notation

When displaying a very large or a very small number as an answer, most scientific calculators express it in scientific notation. To show this, we will find the values of $(453.46)^5$ and $(0.0005)^{12}$. We enter these numbers and press these keys.

453.46 $\boxed{y^x}$ 5 $\boxed{=}$ $\boxed{1.917321395 \quad 13}$

.0005 $\boxed{y^x}$ 12 $\boxed{=}$ $\boxed{2.44140625 \quad -40}$

Since the answers in standard notation require more space than the calculator display has, the calculator gives each result in scientific notation. The first display represents $1.917321395 \times 10^{13}$, and the second represents $2.44140625 \times 10^{-40}$.

If using a graphing or direct-entry calculator, we see that the letter E is used when displaying a number in scientific notation.

453.46 $\boxed{\wedge}$ 5 $\boxed{\text{ENTER}}$ $\boxed{\begin{array}{l}453.46\wedge5 \\ 1.917321395\text{E}13\end{array}}$

.0005 $\boxed{\wedge}$ 12 $\boxed{\text{ENTER}}$ $\boxed{\begin{array}{l}.0005\wedge12 \\ 2.44140625\text{E}-40\end{array}}$

> **Caution!** When reading an answer such as $\boxed{1.917321395 \quad 13}$ off the calculator, be careful to write $1.917321395 \times 10^{13}$, not $1.917321395 \; ^{13}$.

EXAMPLE 4 Write in scientific notation: 432.0×10^5

Strategy We will write the number as a product of a number between 1 and 10 and a power of 10.

WHY Numbers written in scientific notation have the form $N \times 10^n$.

Solution
The number 432.0×10^5 is not written in scientific notation, because 432.0 is not a number between 1 and 10. To write this number in scientific notation, we proceed as follows:

$432.0 \times 10^5 = 4.32 \times 10^2 \times 10^5$ Write 432.0 in scientific notation.

$\qquad\qquad\quad = 4.32 \times 10^7$ $10^2 \times 10^5 = 10^{2+5} = 10^7$.

Self Check 4
Write in scientific notation:
85×10^{-3}
Now Try Problem 39

3 Convert from scientific notation to standard notation.

We can change a number written in scientific notation to **standard notation**. For example, to write 9.3×10^7 in standard notation, we multiply 9.3 by 10^7.

$9.3 \times \mathbf{10^7} = 9.3 \times \mathbf{10{,}000{,}000}$ 10^7 is equal to 1 followed by 7 zeros.
$ = 93{,}000{,}000$

The following numbers are written in both scientific and standard notation. In each case, the exponent gives the number of places that the decimal point moves, and the sign of the exponent indicates the direction that it moves.

$5.32 \times 10^5 = 5\,3\,2\,0\,0\,0.$ The decimal point moves 5 places to the right.
$8.95 \times 10^{-4} = 0.0\,0\,0\,8\,9\,5$ The decimal point moves 4 places to the left.
$9.77 \times 10^0 = 9.77$ There is no movement of the decimal point.

The following summarizes our observations.

> **Converting from Scientific to Standard Notation**
>
> 1. If the exponent on 10 is positive, move the decimal point the same number of places to the right as the exponent.
> 2. If the exponent on 10 is negative, move the decimal point the same number of places to the left as the absolute value of the exponent.

Self Check 5

Convert to standard notation:
a. 4.76×10^5
b. 9.8×10^{-3}

Now Try Problems 49 and 51

EXAMPLE 5   Convert to standard notation: a. 3.4×10^5 b. 2.1×10^{-4}

Strategy We will identify the exponent on the 10 and consider its sign.

WHY The exponent gives the number of decimal places that we should move the decimal point. The sign of the exponent indicates whether it should be moved to the right or the left.

Solution

a. $3.4 \times \mathbf{10^5} = 3.4 \times \mathbf{100{,}000}$
$ = 340{,}000$

b. $2.1 \times 10^{-4} = 2.1 \times \dfrac{1}{10^4}$
$\phantom{2.1 \times 10^{-4}} = 2.1 \times \dfrac{1}{10{,}000}$
$\phantom{2.1 \times 10^{-4}} = 2.1 \times \mathbf{0.0001}$
$\phantom{2.1 \times 10^{-4}} = 0.00021$

4 Perform computations with scientific notation.

Another advantage of scientific notation becomes apparent when we evaluate products or quotients that contain very large or very small numbers.

Self Check 6

Use scientific notation to evaluate:
$(2{,}540{,}000{,}000{,}000)\,(0.00041)$

Now Try Problem 55

EXAMPLE 6   *Astronomy*   Except for the sun, the nearest star visible to the naked eye from most parts of the United States is Sirius. Light from Sirius reaches Earth in about 70,000 hours. If light travels at approximately 670,000,000 mph, how far from Earth is Sirius?

Strategy We will use the formula $d = rt$ to find the distance from Sirius to Earth.

WHY We know the *rate* at which light travels and the *time* it takes to travel from Sirius to Earth. We want to know the distance.

Solution

We are given the rate at which light travels (670,000,000 mph) and the time it takes the light to travel from Sirius to Earth (70,000 hr). We can find the distance the light travels using the formula $d = rt$.

$$d = rt$$
$$d = 670{,}000{,}000(70{,}000) \quad \text{Substitute 670,000,000 for } r \text{ and 70,000 for } t.$$
$$= (6.7 \times 10^8)(7.0 \times 10^4) \quad \text{Write each number in scientific notation.}$$
$$= (6.7 \cdot 7.0) \times (10^8 \cdot 10^4) \quad \text{Group the numbers together and the powers of 10 together.}$$
$$= (6.7 \cdot 7.0) \times 10^{8+4} \quad \text{Keep the base and add the exponents.}$$
$$= 46.9 \times 10^{12} \quad \text{Perform the multiplication. Perform the addition.}$$

We note that 46.9 is not between 0 and 1, so 46.9×10^{12} is not written in scientific notation. To answer in scientific notation, we proceed as follows.

$$= 4.69 \times 10^1 \times 10^{12} \quad \text{Write 46.9 in scientific notation as } 4.69 \times 10^1.$$
$$= 4.69 \times 10^{13} \quad \text{Keep the base of 10 and add the exponents.}$$

Sirius is approximately 4.69×10^{13} or 46,900,000,000,000 miles from Earth.

THINK IT THROUGH STEM Majors and Space Travel

"Today, jobs in STEM fields—science, technology, engineering, and math—go unfilled for lack of qualified workers, even in the current economy."

The Hill, 2013

Many educators believe that manned flights to Mars would ignite a passion for science, technology, engineering, and math studies among young people. The minimum distance Mars is from Earth is 135 times further than the moon is from Earth. Traveling such a long way poses many problems. If the average distance from Earth to the moon is about 2.4×10^5 miles, what is the distance between Earth and Mars? Express the result in scientific notation.

EXAMPLE 7 Atoms

Scientific notation is used in chemistry. As an example, we can approximate the weight (in grams) of one atom of the heaviest naturally occurring element, uranium, by evaluating the following expression.

$$\frac{2.4 \times 10^2}{6.0 \times 10^{23}}$$

Self Check 7

Find the approximate weight (in grams) of one atom of gold by evaluating: $\dfrac{1.98 \times 10^2}{6.0 \times 10^{23}}$

Now Try Problem 57

Strategy We will divide the numbers and the powers of 10 separately.

WHY We can use the quotient rule for exponents to simplify the calculations.

Solution

$$\frac{2.4 \times 10^2}{6.0 \times 10^{23}} = \frac{2.4}{6.0} \times \frac{10^2}{10^{23}} \quad \text{Divide the numbers and the powers of 10 separately.}$$
$$= \frac{2.4}{6.0} \times 10^{2-23} \quad \text{For the powers of 10, keep the base and subtract the exponents.}$$
$$= 0.4 \times 10^{-21} \quad \text{Perform the division. Then subtract the exponents.}$$
$$= 4.0 \times 10^{-1} \times 10^{-21} \quad \text{Write 0.4 in scientific notation as } 4.0 \times 10^{-1}.$$
$$= 4.0 \times 10^{-22} \quad \text{Keep the base and add the exponents.}$$

One atom of uranium weighs 4.0×10^{-22} gram. Written in standard notation, this is 0.00000000000000000000004 g.

Using Your CALCULATOR Entering Numbers in Scientific Notation

We can evaluate the expression from Example 7 by entering the numbers written in scientific notation, using the $\boxed{EE}$ key on a scientific calculator.

$$2.4 \; \boxed{EE} \; 2 \; \boxed{\div} \; 6 \; \boxed{EE} \; 23 \; \boxed{=} \qquad \boxed{4.^{-22}}$$

The result shown in the display means 4.0×10^{-22}.

If we use a graphing calculator, the keystrokes are similar.

$$2.4 \; \boxed{2\text{nd}} \; \boxed{EE} \; 2 \; \boxed{\div} \; 6 \; \boxed{2\text{nd}} \; \boxed{EE} \; 23 \; \boxed{ENTER} \qquad \boxed{\begin{array}{r} 2.4E2/6E23 \\ 4 \; E - 22 \end{array}}$$

ANSWERS TO SELF CHECKS
1. 9.3×10^7 2. 9.055×10^{-5} 3. a. 1.75×10^4 b. 6.57×10^{-1} 4. 8.5×10^{-2}
5. a. $476,000$ b. 0.0098 6. 1.0414×10^9 7. 3.3×10^{-22} g

SECTION 5.3 STUDY SET

VOCABULARY

Fill in the blanks.

1. A number is written in _____ notation when it is written as the product of a number between 1 (including 1) and 10 and an integer power of 10.
2. The number 125,000 is written in _____ notation.

CONCEPTS

Fill in the blanks by writing the number in standard notation.

3. $2.5 \times 10^2 =$ ____
4. $2.5 \times 10^{-2} =$ ____
5. $2.5 \times 10^{-5} =$ ____
6. $2.5 \times 10^5 =$ ____

Fill in the blanks with a power of 10.

7. $387,000 = 3.87 \times$ ____
8. $38.7 = 3.87 \times$ ____
9. $0.00387 = 3.87 \times$ ____
10. $0.000387 = 3.87 \times$ ____
11. When we multiply a decimal by 10^5, the decimal point moves ___ places to the ____.
12. When we multiply a decimal by 10^{-7}, the decimal point moves ___ places to the ____.
13. Dividing a decimal by 10^4 is equivalent to multiplying it by ____.
14. Multiplying a decimal by 10^0 does not move the decimal point, because $10^0 =$ ____.

15. When a real number greater than 10 is written in scientific notation, the exponent on 10 is a _____ number.
16. When a real number between 0 and 1 is written in scientific notation, the exponent on 10 is a _____ number.

NOTATION

Complete each step.

17. Write in scientific notation: 63.7×10^5

$$63.7 \times 10^5 = \underline{} \times 10^5$$
$$= 6.37 \times 10^{\underline{}+5}$$
$$= 6.37 \times 10^6$$

18. Simplify: $\dfrac{64{,}000}{0.00004}$

$$\frac{64{,}000}{0.00004} = \frac{6.4 \times \underline{}}{4 \times \underline{}}$$
$$= \frac{\underline{}}{} \times \frac{10^4}{10^{-5}}$$
$$= 1.6 \times 10^{\underline{}-(-5)}$$
$$= 1.6 \times 10^9$$

GUIDED PRACTICE

Write each number in scientific notation. See Example 1.

19. 23,000
20. 4,750
21. 625,000
22. 320,000

Write each answer in scientific notation. See Example 2.

23. 0.062
24. 0.75
25. 0.00073
26. 0.000057

Write each number in scientific notation. See Example 3.

27. 543,000
28. 17,000,000
29. 0.00000875
30. 0.000002
31. 1,700,000
32. 290,000
33. 909,000,000
34. 7,007,000,000
35. 0.00502
36. 0.00073
37. 0.0000051
38. 0.04

Write each number in scientific notation. See Example 4.

39. 42.5×10^2
40. 25.2×10^{-3}
41. 0.25×10^{-2}
42. 0.3×10^3
43. 201.8×10^{15}
44. 154.3×10^{17}
45. 0.073×10^{-3}
46. 0.0017×10^{-4}

Write each number in standard notation. See Example 5.

47. 2.3×10^2
48. 3.75×10^4
49. 8.12×10^5
50. 1.2×10^3
51. 1.15×10^{-3}
52. 4.9×10^{-2}
53. 9.76×10^{-4}
54. 7.63×10^{-5}

Use scientific notation and the rules for exponents to simplify each expression. Give all answers in standard notation. See Examples 6–7.

55. $(3.4 \times 10^2)(2.1 \times 10^3)$
56. $(4.1 \times 10^{-3})(3.4 \times 10^4)$
57. $\dfrac{9.3 \times 10^2}{3.1 \times 10^{-2}}$
58. $\dfrac{7.2 \times 10^6}{1.2 \times 10^8}$

TRY IT YOURSELF

Simplify if necessary, then write the answer in standard notation.

59. 25×10^6
60. 0.07×10^3
61. 0.51×10^{-3}
62. 2.37×10^{-4}
63. 617×10^{-2}
64. $5,280 \times 10^{-3}$
65. 0.699×10^3
66. 0.012×10^4
67. $\dfrac{0.00000129}{0.0003}$
68. $\dfrac{169,000,000,000}{26,000,000}$
69. $\dfrac{96,000}{(12,000)(0.00004)}$
70. $\dfrac{(0.48)(14,400,000)}{96,000,000}$

71. $(456.4)^6$
72. $(0.053)^4$
73. $(0.009)^{-6}$
74. 225^{-3}
75. $\left(\dfrac{1}{3}\right)^{-25}$
76. $\left(\dfrac{8}{5}\right)^{50}$

APPLICATIONS

77. **ASTRONOMY** The distance from Earth to Alpha Centauri (the nearest star outside our solar system) is about 25,700,000,000,000 miles. Express this number in scientific notation.

78. **SPEED OF SOUND** The speed of sound at sea level under standard conditions is 33,100 centimeters per second. Express this number in scientific notation.

79. **GEOGRAPHY** The largest ocean in the world is the Pacific Ocean, which covers 6.38×10^7 square miles. Express this number in standard notation.

80. **ATOMS** The number of atoms in 1 gram of iron is approximately 1.08×10^{22}. Express this number in standard notation.

81. **LENGTH OF A METER** One meter is approximately 0.00622 mile. Use scientific notation to express this number.

82. **ANGSTROM** One angstrom is 1.0×10^{-7} millimeter. Express this number in standard notation.

83. **WAVELENGTHS** Transmitters, vacuum tubes, and lights emit energy that can be modeled as a wave, as shown. Examples of the most common types of electromagnetic waves are given in the table. List the wavelengths in order from shortest to longest.

This distance between the two crests of the wave is called the wavelength.

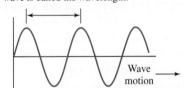

Wave motion

Type	Use	Wavelength (m)
visible light	lighting	9.3×10^{-6}
infrared	photography	3.7×10^{-5}
x-ray	medical	2.3×10^{-11}
radio wave	communication	3.0×10^2
gamma ray	treating cancer	8.9×10^{-14}
microwave	cooking	1.1×10^{-2}
ultraviolet	sun lamp	6.1×10^{-8}

84. **MARS EXPLORATION** On August 6, 2012, *Curiosity,* a rover vehicle, landed on Mars to perform a scientific investigation of the planet. The distance from Mars to Earth is approximately 3.5×10^7 miles. Use scientific notation to express this distance in feet. (*Hint:* 5,280 feet = 1 mile.)

85. **PROTONS** The mass of one proton is approximately 1.7×10^{-24} gram. Use scientific notation to express the mass of 1 million protons.

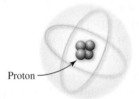

86. **SPEED OF SOUND** The speed of sound in air is approximately 3.3×10^4 centimeters per second. Use scientific notation to express this speed in kilometers per second. (*Hint:* 100 centimeters = 1 meter and 1,000 meters = 1 kilometer.)

87. **LIGHT YEARS** One light year is about 5.87×10^{12} miles. Use scientific notation to express this distance in feet. (*Hint:* 5,280 feet = 1 mile.)

88. **OIL RESERVES** In 2013, Saudi Arabia had crude oil reserves of about 2.67×10^{11} barrels. A barrel contains 42 gallons of oil. Use scientific notation to express Saudi Arabia oil reserves in gallons. (Source: U.S. Energy Information Administration)

89. **INSURED DEPOSITS** In 2013, the total insured deposits in U.S. banks and savings and loans was approximately 5.37×10^{12} dollars. If all of this money were invested at 4% simple annual interest, how much would it earn in 1 year? Use scientific notation to express the answer. (Source: bloomberg.com)

90. **CURRENCY** In 2013, the number of $20 bills in circulation was approximately 7.4×10^9. Find the total value of the currency. Use scientific notation to express the answer. (Source: federalreserve.gov)

91. **THE DOD** The graph shows the number of Department of Defense personnel for 1960–2010. Estimate each of the following and express your answers in scientific and standard notation.

    a. The number of personnel in 1968
    b. The number of personnel in 2010

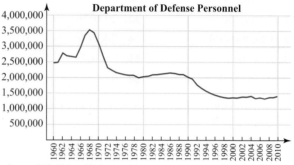

Source: U.S. Census Bureau

92. **SUPERCOMPUTERS** In 2013, the world's fastest computer was IBM's BlueGene/L System. If it could make 3.386×10^{16} calculations in one second, how many could it make in one minute? Give the answer in scientific notation.

WRITING

93. In what situations would scientific notation be more convenient than standard notation?

94. To multiply a number by a power of 10, we move the decimal point. Which way, and how far? Explain.

95. 2.3×10^{-3} contains a negative sign but represents a positive number. Explain.

96. Is this a true statement? $2.0 \times 10^3 = 2 \times 10^3$ Explain.

REVIEW

97. If $y = -1$, find the value of $-5y^{55}$.

98. What is the *y*-intercept of the graph of $y = -3x - 5$?

Which property of real numbers is illustrated by each statement?

99. $5 + z = z + 5$

100. $7(u + 3) = 7u + 7 \cdot 3$

Solve each equation.

101. $3(x - 4) - 6 = 0$

102. $8(3x - 5) - 4(2x + 3) = 12$

SECTION 5.4
Polynomials

Objectives

1. Know the vocabulary for polynomials.
2. Evaluate polynomials.

In arithmetic, we learned how to add, subtract, multiply, divide, and find powers of numbers. In algebra, we will learn how to perform these operations on *polynomials*. In this section, we will introduce polynomials, classify them into groups, define their degrees, and show how to evaluate them at specific values of their variables.

1 Know the vocabulary for polynomials.

Recall that a **term** is a number or a product of a number and one or more variables, which may be raised to powers. Examples of terms are

$$3x, \quad -4y^2, \quad \tfrac{1}{2}a^2b^3, \quad t, \quad \text{and} \quad 25$$

The **numerical coefficients**, or simply **coefficients**, of the first four of these terms are $3, -4, \tfrac{1}{2}$, and 1, respectively. Because $25 = 25x^0$, 25 is considered to be the numerical coefficient of the term 25.

> **Polynomials**
>
> A **polynomial** is a term or a sum of terms in which all variables have whole-number exponents. No variable appears in a denominator.

Here are some examples of polynomials:

$$3x + 2, \quad 4y^2 - 2y - 3, \quad -8xy^2, \quad \text{and} \quad a^3 + 3a^2b + 3ab^2 + b^3$$

The polynomial $3x + 2$ has two terms, $3x$ and 2, and we say it is a **polynomial in one variable**, x. A single number is called a **constant**, so its last term, 2, is called the **constant term**.

Since $4y^2 - 2y - 3$ can be written as $4y^2 + (-2y) + (-3)$, it is the sum of three terms, $4y^2, -2y$, and -3. It is written in **decreasing** or **descending powers** of y, because the powers on y decrease from left to right.

$-8xy^2$ is a polynomial with just one term. We say that it is a **polynomial in two variables**, x and y.

The four-term polynomial $a^3 + 3a^2b + 3ab^2 + b^3$ is written in descending powers of a and **ascending powers** of b.

> **Caution!** The expression $2x^3 - 3x^{-2} + 5$ is not a polynomial, because the second term contains a variable with an exponent that is not a whole number. Similarly, $y^2 - \dfrac{7}{y}$ is not a polynomial, because $\dfrac{7}{y}$ has a variable in the denominator.

Chapter 5 Exponents and Polynomials

Self Check 1

Determine whether each expression is a polynomial:
a. $3x^{-4} + 2x^2 - 3$
b. $7.5p^3 - 4p^2 - 3p + 4$

Now Try Problems 17 and 18

EXAMPLE 1
Determine whether each expression is a polynomial.

a. $x^2 + 2x + 1$ b. $3a^{-1} - 2a - 3$ c. $\frac{1}{2}x^3 - 2.3x$ d. $\frac{p+3}{p-1}$

Strategy We will note the exponents on the variable bases. We will also identify each denominator.

WHY The expression is a polynomial when all the variables have whole-number exponents and no variable appears in a denominator of a fraction.

Solution

a. $x^2 + 2x + 1$ is a polynomial.

b. $3a^{-1} - 2a - 3$ is not a polynomial. In the first term, the exponent on the variable is not a whole number.

c. $\frac{1}{2}x^3 - 2.3x$ is a polynomial, because it can be written as the sum $\frac{1}{2}x^3 + (-2.3x)$.

d. $\frac{p+3}{p-1}$ is not a polynomial. For a polynomial, variables cannot be in the denominator of a fraction.

A polynomial with one term is called a **monomial**. A polynomial with two terms is called a **binomial**. A polynomial with three terms is called a **trinomial**. Here are some examples.

Monomials	Binomials	Trinomials
$-6x$	$3u^3 - 4u^2$	$-5t^2 + 4t + 3$
$5x^2y$	$18a^2b + 4ab$	$27x^3 - 6x - 2$
29	$-29z^{17} - 1$	$a^2 + 2ab + b^2$

Self Check 2

Classify each polynomial as a monomial, a binomial, or a trinomial:
a. $5x$
b. $-5x^2 + 2x - 0.5$
c. $16x^2 - 9y^2$

Now Try Problems 26, 29, and 33

EXAMPLE 2
Classify each polynomial as a monomial, a binomial, or a trinomial:

a. $5.2x^4 + 3.1x$ b. $7g^4 - 5g^3 - 2$ c. $-5x^2y^3$

Strategy We will count the number of terms in each polynomial.

WHY The number of terms determines the type of polynomial.

Solution

a. The polynomial $5.2x^4 + 3.1x$ has two terms, $5.2x^4$ and $3.1x$, so it is a binomial.

b. The polynomial $7g^4 - 5g^3 - 2$ has three terms, $7g^4$, $-5g^3$, and -2, so it is a trinomial.

c. The polynomial $-5x^2y^3$ has one term, so it is a monomial.

Success Tip Recall that terms are separated by + or − symbols and the numerical coefficient is the numerical factor of the term.

The monomial $7x^6$ is called a **monomial of sixth degree** or a **monomial of degree 6**, because the variable x occurs as a factor six times. The monomial $3x^3y^4$ is a monomial of seventh degree, because the variables x and y occur as factors a total of seven times. Here are some more examples:

$2.7a$ is a monomial of degree 1.

$-2x^3$ is a monomial of degree 3.

$47x^2y^3$ is a monomial of degree 5.

8 is a monomial of degree 0, because $8 = 8x^0$.

These examples illustrate the following definition.

> **Degree of a Monomial**
>
> If a represents a nonzero constant, the **degree of the monomial** ax^n is n.
>
> The **degree of a monomial** in several variables is the sum of the exponents on those variables.

> **Caution!** Note that the degree of ax^n is not defined when $a = 0$. Since $ax^n = 0$ when $a = 0$, the constant 0 has no defined degree.

Because each term of a polynomial is a monomial, we define the degree of a polynomial by considering the degrees of each of its terms.

> **Degree of a Polynomial**
>
> The **degree of a polynomial** is determined by the term with the largest degree.

Here are some examples:

- $x^2 + 2x$ is a binomial of degree 2, because the degree of its first term is 2 and the degree of its second term is less than 2.
- $1 + d^3 - 3d^2$ is a trinomial of degree 3, because the degree of its second term is 3 and the degree of each of its other terms is less than 3.
- $25y^{13} - 15y^8z^{10} - 32y^{10}z^8 + 4$ is a polynomial of degree 18, because its second and third terms are of degree 18. Its other terms have degree less than 18.

EXAMPLE 3 Find the degree of each polynomial:

a. $-4x^3 - 5x^2 + 3x$ **b.** $1.6w - 1.6$ **c.** $-17a^2b^3 + 12ab^6$

Strategy We will find the degree of each term and compare them.

WHY The degree of the polynomial is the same as the highest-degree term.

Solution

a. The trinomial $-4x^3 - 5x^2 + 3x$ has terms of degree 3, 2, and 1. Therefore, its degree is 3.

b. The first term of $1.6w - 1.6$ has degree 1 and the second term has degree 0, so the binomial has degree 1.

c. The degree of the first term of $-17a^2b^3 + 12ab^6$ is 5 and the degree of the second term is 7, so the binomial has degree 7.

Self Check 3

Find the degree of each polynomial:

a. $15p^3 - 15p^2 - 3p + 4$
b. $-14st^4 + 12s^3t$

Now Try Problems 39 and 44

If written in descending powers of the variable, the **leading term** of a polynomial is the term of highest degree. For example, the leading term of $-4x^3 - 5x^2 + 3x$ is $-4x^3$. The coefficient of the leading term (in this case, -4) is called the **leading coefficient**.

2 Evaluate polynomials.

Each of the equations below defines a function, because each input x-value determines exactly one output value. Since the right side of each equation is a polynomial, these functions are called **polynomial functions**.

$$f(x) = 6x + 4 \qquad g(x) = 3x^2 + 4x - 5 \qquad h(x) = -x^3 + x^2 - 2x + 3$$

This polynomial has two terms. Its degree is 1.   This polynomial has three terms. Its degree is 2.   This polynomial has four terms. Its degree is 3.

To *evaluate a polynomial function* for a specific value, we replace the variable in the defining equation with the input value. Then we simplify the resulting expression to find the output. For example, suppose we wish to evaluate the polynomial function $f(x) = 6x + 4$ for $x = 1$. Then $f(1)$ represents the value of $f(x) = 6x + 4$ when $x = 1$. We find $f(1)$ as follows.

$f(x) = 6x + 4$ This is the given function.
$f(1) = 6(1) + 4$ Substitute 1 for x. The number 1 is the input.
$f(1) = 6 + 4$ Do the multiplication.
$f(1) = 10$ Do the addition. 10 is the output.

Thus, $f(1) = 10$.

Self Check 4
Consider the function
$h(x) = -x^3 + x - 2x + 3$
Find:
a. $h(0)$
b. $h(-3)$

Now Try Problems 51 and 59

EXAMPLE 4 Consider the function $g(x) = 3x^2 + 4x - 5$. Find:
a. $g(0)$ **b.** $g(-2)$

Strategy We will substitute the value in the parentheses on the left side of the equation for the letter on the right side. Then we will follow the rules for the order of operations to simplify the right side.

WHY To *evaluate a polynomial* means to find its numerical value, once we know the value of the variable.

Solution
a. $g(x) = 3x^2 + 4x - 5$ This is the given function.
$g(0) = 3(0)^2 + 4(0) - 5$ To find g(0), substitute 0 for x.
$g(0) = 3(0) + 4(0) - 5$ Evaluate the power.
$g(0) = 0 + 0 - 5$ Do the multiplications.
$g(0) = -5$

b. $g(x) = 3x^2 + 4x - 5$ This is the given function.
$g(-2) = 3(-2)^2 + 4(-2) - 5$ To find g(−2), substitute −2 for x.
$g(-2) = 3(4) + 4(-2) - 5$ Evaluate the power.
$g(-2) = 12 + (-8) - 5$ Do the multiplications.
$g(-2) = -1$ Do the addition and subtraction.

Self Check 5
Find the number of cans used in a display having a square base formed by 5 cans per side.

Now Try Problem 80

EXAMPLE 5 **Supermarket Displays** The polynomial function $f(c) = \frac{1}{3}c^3 + \frac{1}{2}c^2 + \frac{1}{6}c$ gives the number of cans used in a display shaped like a square pyramid, having a square base formed by c cans per side. Find the number of cans of soup used in the display shown in the figure on the next page.

Strategy We will count the number of cans along one side of the square base of the display. Then we will evaluate the function at that number.

WHY The function gives the number of cans in the display based on the number of cans along one side of the square base.

Solution
Since each side of the square base of the display is formed by 4 cans, $c = 4$. We can find the number of cans used in the display by finding $f(4)$.

$f(c) = \frac{1}{3}c^3 + \frac{1}{2}c^2 + \frac{1}{6}c$ This is the given function.

$f(4) = \frac{1}{3}(4)^3 + \frac{1}{2}(4)^2 + \frac{1}{6}(4)$ Substitute 4 for c.

$f(4) = \frac{1}{3}(64) + \frac{1}{2}(16) + \frac{1}{6}(4)$ Find the powers.

$f(4) = \frac{64}{3} + 8 + \frac{2}{3}$ Multiply, and then simplify: $\frac{4}{6} = \frac{2}{3}$.

$f(4) = \frac{66}{3} + 8$ Add the fractions.

$f(4) = 22 + 8$ Simplify the fraction.

$f(4) = 30$ Do the addition.

30 cans of soup were used in the display.

ANSWERS TO SELF CHECKS
1. a. no **b.** yes **2. a.** monomial **b.** trinomial **c.** binomial **3. a.** 3 **b.** 5
4. a. 3 **b.** 33 **5.** 55

SECTION 5.4 STUDY SET

VOCABULARY

Fill in the blanks.

1. A _____ is a term or a sum of terms in which all variables have whole-number exponents.
2. The numerical _____ of the term $-25x^2y^3$ is -25.
3. The _____ of the monomial $3x^7$ is 7.
4. The degree of a polynomial is the same as the _____ of its term with the largest degree.
5. A _____ is a polynomial with one term.
6. A _____ is a polynomial with two terms.
7. A _____ is a polynomial with three terms.
8. For the polynomial $6x^2 + 3x - 1$, the _____ term is $6x^2$, and the leading _____ is 6. The _____ term is -1.
9. The notation $f(x)$ is read as f ___ x.
10. $f(2)$ represents the _____ of a function when $x = 2$.

CONCEPTS

Fill in the blanks.

11. $4x^3 + 7x^2 - 3x - 15$ is a polynomial in x. It is written in _____ powers of x.
12. $-7 + 2y + 3y^2 - 8y^3$ is a polynomial in y. It is written in _____ powers of y.

13. Write $x - 9 + 3x^2$ in descending powers of x.

14. Write $-2xy + y^2 + x^2$ in descending powers of x.

NOTATION

Complete each step.

15. If $f(x) = -2x^2 + 3x - 1$, find $f(2)$.

$f(2) = -2()^2 + 3() - 1$

$f(2) = -2() + - 1$

$f(2) = -8 + 6 - $

$f(2) = - 1$

$f(2) = -3$

16. If $f(x) = -2x^2 + 3x - 1$, find $f(-2)$.

$f(-2) = -2()^2 + 3() - 1$

$f(-2) = -2() + 3() - 1$

$f(-2) = + (-6) - 1$

$f(-2) = - 1$

$f(-2) = -15$

GUIDED PRACTICE

Determine whether each expression is a polynomial. See Example 1.

17. $x^3 - 5x^2 - 2$ **18.** $x^{-4} - 5x$

19. $\dfrac{1}{2x} + 3$ **20.** $x^3 - 1$

21. $x^2 - y^2$ **22.** $a^4 + a^3 + a^2 + a$

23. $a^3 + 2a^2 - a + 2$ **24.** $\dfrac{1}{x^2 + x - 7}$

Classify each polynomial as a monomial, a binomial, a trinomial, or none of these. See Example 2.

25. $3x + 7$ **26.** $3y - 5$

27. $y^2 + 4y + 3$ **28.** $3xy$

29. $3z^2$ **30.** $3x - 2x^3 + 3x - 1$

31. $t - 32$ **32.** $9x^2y^3z^4$

33. $s^2 - 23s + 31$ **34.** $2x^3 - 5x^2 + 6x - 3$

35. $3x^5 - x^4 - 3x^3 + 7$ **36.** x^3

Find the degree of each polynomial. See Example 3.

37. $3x^4$ **38.** $3x^5$

39. $-2x^2 + 3x + 1$ **40.** $-3x + 3x^2 - 5x^4$

41. $3x - 5$ **42.** $y^3 + 4y^2$

43. $-5r^2s^2 - r^3s + 3$ **44.** $4r^2s^3 - 5r^2s^8$

45. $x^{12} + 3x^2y^3$ **46.** $17ab^5 - 12a^3b$

47. 38 **48.** -24

Let $f(x) = 5x - 3$ and find each value. See Example 4.

49. $f(2)$ **50.** $f(0)$

51. $f(-1)$ **52.** $f(-2)$

53. $f\left(\dfrac{1}{5}\right)$ **54.** $f\left(\dfrac{4}{5}\right)$

55. $f(-0.9)$ **56.** $f(-1.2)$

Let $g(x) = -x^2 - 4$ and find each value.

57. $g(0)$ **58.** $g(1)$

59. $g(-1)$ **60.** $g(-2)$

61. $g(1.3)$ **62.** $g(2.4)$

63. $g(-13.6)$ **64.** $g(-25.3)$

TRY IT YOURSELF

Let $h(x) = x^3 - 2x + 3$ and find each value.

65. $h(0)$ **66.** $h(3)$

67. $h(-2)$ **68.** $h(-1)$

69. $h(0.9)$ **70.** $h(0.4)$

71. $h(-8.1)$ **72.** $h(-7.7)$

Let $f(x) = -x^4 - x^3 + x^2 + x - 1$ and find each value.

73. $f(1)$ **74.** $f(-1)$

75. $f(-2)$ **76.** $f(2)$

APPLICATIONS

77. PACKAGING To make boxes, a manufacturer cuts equal-sized squares from each corner of a 10-inch × 12-inch piece of cardboard, and then folds up the sides. The polynomial function

$$f(x) = 4x^3 - 44x^2 + 120x$$

gives the volume (in cubic inches) of the resulting box when a square with sides x inches long is cut from each corner. Find the volume of a box if 3-inch squares are cut out.

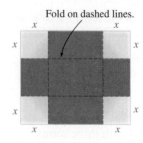

78. MAXIMIZING REVENUE The revenue (in dollars) that a manufacturer of office desks receives is given by the polynomial function

$$f(d) = -0.08d^2 + 100d$$

where d is the number of desks manufactured.

a. Find the total revenue if 625 desks are manufactured.

b. Does increasing the number of desks being manufactured to 650 increase the revenue?

79. WATER BALLOONS Some college students launched water balloons from the balcony of their dormitory on unsuspecting sunbathers. The height in feet of the balloons at a time t seconds after being launched is given by the polynomial function

$$f(t) = -16t^2 + 12t + 20$$

What was the height of the balloons 0.5 second and 1.5 seconds after being launched?

80. STOPPING DISTANCE *from Campus to Careers* — Police Officer

The number of feet that a car travels before stopping depends on the driver's reaction time and the braking distance, as shown below. For one driver, the stopping distance is given by the polynomial function

$$f(v) = 0.04v^2 + 0.9v$$

where v is the velocity of the car. Find the stopping distance when the driver is traveling at 30 mph.

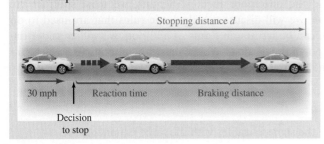

81. SUSPENSION BRIDGES The function

$$f(s) = 400 + 0.0066667s^2 - 0.0000001s^4$$

approximates the length of the cable between the two vertical towers of a suspension bridge, where s is the sag in the cable. Estimate the length of the cable of the bridge in the next column if the sag is 24.6 feet.

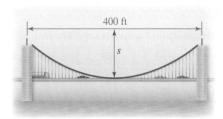

82. PRODUCE DEPARTMENTS Suppose a grocer is going to set up a pyramid-shaped display of cantaloupes like that shown in the figure in Example 5. If each side of the square base of the display is made of six cantaloupes, how many will be used in the display?

83. DOLPHINS At a marine park, three trained dolphins jump in unison over an arching stream of water whose path can be described by the polynomial function

$$f(x) = -0.05x^2 + 2x$$

Given the takeoff points for each dolphin, how high must each dolphin jump to clear the stream of water?

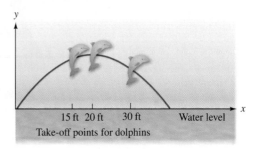

84. TUNNELS The height of an arch at the entrance to a tunnel is described by the polynomial function

$$f(x) = -0.25x^2 + 23$$

where x is the distance from the centerline of the pavement. What is the height of the arch at the edge of the pavement?

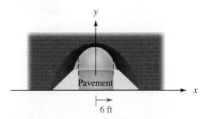

WRITING

85. Describe how to determine the degree of a polynomial.

86. List some words that contain the prefixes *mono*, *bi*, or *tri*.

REVIEW

Solve each inequality and graph its solution set.

87. $-4(3y + 2) \leq 28$ 88. $-5 < 3t + 4 \leq 13$

Write each expression without using parentheses or negative exponents.

89. $(x^2 x^4)^3$
90. $(a^2)^3 (a^3)^2$
91. $\left(\dfrac{y^2 y^5}{y^4}\right)^3$
92. $\left(\dfrac{2t^2}{t}\right)^{-4}$

Objectives

1. Add monomials.
2. Subtract monomials.
3. Add polynomials.
4. Subtract polynomials.
5. Add and subtract multiples of polynomials.

SECTION 5.5
Adding and Subtracting Polynomials

In figure (a), the heights of the Seattle Space Needle and the Eiffel Tower in Paris are given. Using rules from arithmetic, we can find the difference in the heights of the towers by subtracting two numbers.

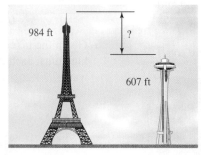

Arithmetic

$984 - 607 = 377$

The difference in height is 377 feet.

(a)

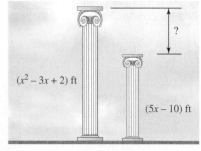

Algebra

$(x^2 - 3x + 2) - (5x - 10) = ?$

(b)

In figure (b), the heights of two types of classical Greek columns are expressed using *polynomials*. To find the difference in their heights, we must subtract the polynomials. In this section, we will discuss the algebraic rules that are used to do this. Since any subtraction can be written in terms of addition, we will consider the procedures used to add polynomials first. We begin with monomials, which are polynomials having just one term.

1 Add monomials.

Recall that like terms have the same variables with the same exponents:

Like terms	**Unlike terms**
$-7x$ and $15x$	$-7x$ and $15a$
$4y^3$ and $16y^3$	$4y^3$ and $16y^2$
$\dfrac{1}{2}xy^2$ and $-\dfrac{1}{3}xy^2$	$\dfrac{1}{2}xy^2$ and $-\dfrac{1}{3}x^2y$

Also recall that to combine like terms, we combine their coefficients and keep the same variables with the same exponents. For example,

$$4y + 5y = (4 + 5)y \quad \text{and} \quad 8x^2 + x^2 = (8 + 1)x^2$$
$$= 9y \qquad\qquad\qquad\qquad\qquad = 9x^2$$

Unless otherwise noted, all content on this page is © Cengage Learning.

Likewise,

$$3a + 4b + 6a + 3b = 9a + 7b \quad \text{and} \quad 4cd^3 + 9cd^3 = 13cd^3$$

These examples suggest that to add like monomials, we simply combine like terms.

> **The Language of Algebra** Simplifying the sum or difference of like terms is called *combining like terms*.

EXAMPLE 1 Perform the following additions.

a. $4x^4 + 81x^4$ **b.** $8x^2y^2 + 6x^2y^2 + x^2y^2$ **c.** $32c^2 + 10c + 4c^2$

Strategy We will note the terms that have the same variables with the same exponents. Then we will combine their coefficients and keep the same variables with the same exponents.

WHY Only like terms can be simplified with addition.

Solution
a. $4x^4 + 81x^4 = 85x^4$ Think: $(4 + 81)x^4 = 85x^4$.
b. $8x^2y^2 + 6x^2y^2 + x^2y^2 = 15x^2y^2$ Think: $(8 + 6 + 1)x^2y^2 = 15x^2y^2$.
c. $32c^2 + 10c + 4c^2 = 32c^2 + 4c^2 + 10c$ Write the like terms together.
                    $= 36c^2 + 10c$ Think: $(32 + 4)c^2 = 36c^2$.

Self Check 1
Perform the following additions:
a. $27x^6 + 8x^6$
b. $12pq^2 + 5pq^2 + 8pq^2$
c. $6a^3 + 15a + a^3$

Now Try Problem 19

> **Caution!** When combining like terms, the exponents on the variables *stay the same*. Don't incorrectly add the exponents.

> **Success Tip** When performing operations on polynomials, we usually write the terms of the solution in decreasing (or descending) powers of one variable. For instance, in Example 1, part c, the solution was written as $36c^2 + 10c$ instead of $10c + 36c^2$.

2 Subtract monomials.

To subtract one monomial from another, we add the opposite of the monomial that is to be subtracted.

EXAMPLE 2 Find each difference.

a. $8x^2 - 3x^2$ **b.** $6xy - 9xy$ **c.** $-3r - 5 - 4r$ **d.** $0.9x^2 - 1.2x - 0.5x^2 - 0.4x$

Strategy We will note the terms that have the same variables with the same exponents. Then we will combine their coefficients and keep the same variables with the same exponents.

WHY Only like terms can be simplified with subtraction.

Solution
a. $8x^2 - 3x^2 = 8x^2 + (-3x^2)$ Add the opposite of $3x^2$, which is $-3x^2$.
             $= 5x^2$ Combine like terms.
b. $6xy - 9xy = 6xy + (-9xy)$ Add the opposite of $9xy$, which is $-9xy$.
            $= -3xy$ Combine like terms.

Self Check 2
Find each difference:
a. $12m^3 - 7m^3$
b. $-4pq - 27p - 8pq$
c. $-2.5x^3 - 0.3x^3$

Now Try Problems 25 and 30

c. $-3r - 5 - 4r = -3r + (-5) + (-4r)$ Add the opposite of 5 and 4r.
   $= -3r + (-4r) + (-5)$ Write like terms together.
   $= -7r - 5$ Combine like terms. Write the addition of -5 as a subtraction of 5.

d. $0.9x^2 - 1.2x - 0.5x^2 - 0.4x = 0.9x^2 + (-1.2x) + (-0.5x^2) + (-0.4x)$
   $= 0.9x^2 + (-0.5x^2) + (-1.2x) + (-0.4x)$
   $= 0.4x^2 - 1.6x$

3 Add polynomials.

Because of the distributive property, we can remove parentheses enclosing several terms when the sign preceding the parentheses is a $+$ sign. We simply drop the parentheses.

$+ (3x^2 + 3x - 2) = +1(3x^2 + 3x - 2)$
$= 1(3x^2) + 1(3x) + 1(-2)$ Distribute the multiplication by 1.
$= 3x^2 + 3x + (-2)$ Multiplicative identity property.
$= 3x^2 + 3x - 2$ Write the addition of -2 as a subtraction of 2.

We can add polynomials by removing parentheses, if necessary, and then combining any like terms that are contained within the polynomials.

Self Check 3

Add: $(2a^2 - a + 4) + (5a^2 + 6a - 5)$

Now Try Problem 35

EXAMPLE 3 Add: $(3x^2 - 3x + 2) + (2x^2 + 7x - 4)$

Strategy We will reorder and write the like terms together. Then we will combine like terms.

WHY To add polynomials means to combine their like terms.

Solution
$(3x^2 - 3x + 2) + (2x^2 + 7x - 4)$
$= 3x^2 - 3x + 2 + 2x^2 + 7x - 4$ Drop the parentheses.
$= 3x^2 + 2x^2 - 3x + 7x + 2 - 4$ Write like terms together.
$= 5x^2 + 4x - 2$ Combine like terms.

Problems such as Example 3 are often written with like terms aligned vertically. We can then add the polynomials column by column.

$3x^2 - 3x + 2$
$+\ 2x^2 + 7x - 4$
$\overline{5x^2 + 4x - 2}$

Self Check 4

Add $4q^2 - 7$ and $2q^2 - 8q + 9$ using the vertical form.

Now Try Problem 39

EXAMPLE 4 Add $4x^2 - 3$ and $3x^2 - 8x + 8$ using the vertical form.

Strategy We will write one polynomial underneath the other, aligning the like terms and drawing a horizontal line beneath them. Then we will add the like terms, column by column, and write each result under the line.

WHY *Vertical form* means to arrange the like terms in columns.

Solution
Since the first polynomial does not have an x-term, we leave a space so that the constant terms can be aligned.

$4x^2 - 3$
$+\ 3x^2 - 8x + 8$
$\overline{7x^2 - 8x + 5}$

5.5 Adding and Subtracting Polynomials

4 Subtract polynomials.

Because of the distributive property, we can remove parentheses enclosing several terms when the sign preceding the parentheses is a $-$ sign. We simply drop the minus sign and the parentheses, and *change the sign of every term within the parentheses.*

$$-(3x^2 + 3x - 2) = -1(3x^2 + 3x - 2)$$
$$= -1(3x^2) + (-1)(3x) + (-1)(-2)$$
$$= -3x^2 + (-3x) + 2$$
$$= -3x^2 - 3x + 2$$

This suggests that the way to subtract polynomials is to remove parentheses, change the sign of each term of the second polynomial, and combine like terms.

EXAMPLE 5 Find each difference.

a. $(3x - 4) - (5x + 7)$ b. $(3x^2 - 4x - 6) - (2x^2 - 6x)$
c. $(-t^3 - 2t^2 - 1) - (-t^3 - 2t^2 + 1)$

Strategy We will change the signs of the terms of the polynomial being subtracted, drop the parentheses, and combine like terms.

WHY This is the method for subtracting two polynomials.

Solution

a. $(3x - 4) - (5x + 7) = 3x - 4 - 5x - 7$ Change the sign of each term inside $(5x + 7)$ and drop the parentheses.
$$= -2x - 11$$ Combine like terms.

b. $(3x^2 - 4x - 6) - (2x^2 - 6x)$
$$= 3x^2 - 4x - 6 - 2x^2 + 6x$$ Change the sign of each term of $2x^2 - 6x$ and drop the parentheses.
$$= x^2 + 2x - 6$$ Combine like terms.

c. $(-t^3 - 2t^2 - 1) - (-t^3 - 2t^2 + 1)$
$$= -t^3 - 2t^2 - 1 + t^3 + 2t^2 - 1$$ Change the sign of each term of $-t^3 - 2t^2 + 1$ and drop the parentheses.
$$= -2$$ Combine like terms.

Self Check 5
Find the difference:
$(-2a^2 + 5) - (-5a^2 - 7)$
Now Try Problems 41 and 43

To subtract polynomials in vertical form, we add the opposite of the **subtrahend** (the bottom polynomial) to the **minuend** (the top polynomial).

EXAMPLE 6 Subtract $3x^2 - 2x$ from $2x^2 + 4x$.

Strategy Since $3x^2 - 2x$ is to be subtracted from $2x^2 + 4x$, we will write $3x^2 - 2x$ below $2x^2 + 4x$ in vertical form. Then we will change the signs of the terms of $3x^2 - 2x$ and add column by column.

WHY *Vertical form* means to align the like terms in columns.

Solution

```
              Change signs
   2x² + 4x                 2x² + 4x
 − 3x² − 2x    ────→     + −3x² + 2x
              and add      −x² + 6x
```

Self Check 6
Subtract $2p^2 + 2p - 8$ from $5p^2 - 6p + 7$.
Now Try Problem 45

Self Check 7

Subtract $-2q^2 - 2q$ from the sum of $q^2 - 6q$ and $3q^2 + q$.

Now Try Problem 49

EXAMPLE 7 Subtract $12a - 7$ from the sum of $6a + 5$ and $4a - 10$.

Strategy We will use brackets to show that $(12a - 7)$ is to be subtracted from the sum of $(6a + 5)$ and $(4a - 10)$.

WHY The key words of the problem *subtract from* and *sum* indicate mathematical operations.

Solution

$$[(6a + 5) + (4a - 10)] - (12a - 7)$$

Next, we remove the grouping symbols to obtain

$$= 6a + 5 + 4a - 10 - 12a + 7 \quad \text{Change the sign of each term in } (12a - 7).$$
$$= -2a + 2 \quad \text{Combine like terms.}$$

5 Add and subtract multiples of polynomials.

Because of the distributive property, we can remove parentheses enclosing several terms when a monomial precedes the parentheses. We simply multiply every term within the parentheses by that monomial. For example, to add $3(2x + 5)$ and $2(4x - 3)$, we proceed as follows:

$$3(2x + 5) + 2(4x - 3) = 6x + 15 + 8x - 6 \quad \text{Distribute the multiplication by 3 and by 2.}$$
$$= 6x + 8x + 15 - 6 \quad \text{Use the commutative property of addition to reorder terms.}$$
$$= 14x + 9 \quad \text{Combine like terms.}$$

Self Check 8

Remove parentheses and simplify: $2(a^2 - 3a) + 5(a^2 + 2a)$

Now Try Problems 53 and 55

EXAMPLE 8 Remove the parentheses and simplify.

a. $3(x^2 + 4x) + 2(x^2 - 4)$ **b.** $-8(y^2 - 2y + 3) - 4(2y^2 + y - 6)$

Strategy We will use the distributive property to remove parentheses. Then we will combine like terms.

WHY This is what it means to simplify a polynomial.

Solution

a. $3(x^2 + 4x) + 2(x^2 - 4) = 3x^2 + 12x + 2x^2 - 8 \quad \text{Use the distributive property to remove parentheses.}$

$$= 5x^2 + 12x - 8 \quad \text{Combine like terms.}$$

b. $-8(y^2 - 2y + 3) - 4(2y^2 + y - 6) = -8y^2 + 16y - 24 - 8y^2 - 4y + 24$
$$= -16y^2 + 12y$$

Self Check 9

Two warning flares are fired upward at the same time from different parts of a ship. After t seconds, the height of the first flare is $(-16t^2 + 115t + 25)$ feet and the height of the higher-traveling second flare is $(-16t^2 + 130t + 30)$ feet. Find a polynomial that represents the difference in their heights.

Now Try Problem 96

EXAMPLE 9 **Property Values** A condo purchased for \$95,000 is expected to appreciate according to the polynomial function $f(x) = 2,500x + 95,000$, where $f(x)$ is the value of the condo after x years. A second condo purchased for \$125,000 is expected to appreciate according to the equation $f(x) = 4,500x + 125,000$. Find one polynomial function that will give the total value of both properties after x years.

Strategy To find the polynomial function that will give the total value of both properties, we will add the two polynomials.

WHY To *find a total* means to add.

Solution

The value of the first condo after x years is given by the polynomial $2,500x + 95,000$. The value of the second condo after x years is given by the polynomial

$4,500x + 125,000$. The value of both properties will be the sum of these two polynomials.

$$(2,500x + 95,000) + (4,500x + 125,000) = 7,000x + 220,000$$

The total value of the properties is given by the polynomial function $f(x) = 7,000x + 220,000$.

ANSWERS TO SELF CHECKS

1. a. $35x^6$ b. $25pq^2$ c. $7a^3 + 15a$ 2. a. $5m^3$ b. $-12pq - 27p$ c. $-2.8x^3$
3. $7a^2 + 5a - 1$ 4. $6q^2 - 8q + 2$ 5. $3a^2 + 12$ 6. $3p^2 - 8p + 15$ 7. $6q^2 - 3q$
8. $7a^2 + 4a$ 9. $(15t + 5)$ ft

SECTION 5.5   STUDY SET

VOCABULARY

Fill in the blanks.

1. The expression $(x^2 - 3x + 2) + (x^2 - 4x)$ is the sum of two _____.
2. The expression $(x^2 - 3x + 2) - (x^2 - 4x)$ is the _____ of two polynomials.
3. _____ terms have the same variables and the same exponents.
4. "To add or subtract like terms" means to combine their _____ and keep the same variables with the same exponents.

CONCEPTS

Fill in the blanks.

5. To add like monomials, combine like _____.
6. $a - b = a +$ _____
7. To add two polynomials, combine any _____ terms contained in the polynomials.
8. To subtract two polynomials, change the _____ of each term in the second polynomial, and combine like terms.
9. When the sign preceding parentheses is a $-$ sign, we can remove the parentheses by dropping the sign and the parentheses, and _____ the sign of every term within the parentheses.
10. When a monomial precedes parentheses, we can remove the parentheses by _____ every term within the parentheses by that monomial.
11. $-(-2x^2 - 3x + 4) =$ _____
12. $-3(-2x^2 - 3x + 4) =$ _____

13. What is the result of addition in the *x*-column?

$$\begin{array}{r} 4x^2 + x - 12 \\ + 5x^2 - 8x + 23 \end{array}$$

14. What is the result of the subtraction in the *x*-column?

$$\begin{array}{r} 8x^2 - 7x - 1 \\ - (4x^2 + 6x - 9) \end{array} \longrightarrow \begin{array}{r} 8x^2 - 7x - 1 \\ + -4x^2 - 6x + 9 \end{array}$$

NOTATION

Complete each step.

15. $(5x^2 + 3x) - (7x^2 - 2x)$
$= 5x^2 +$ ____ $- 7x^2 +$ ____
$= 5x^2 -$ ____ $+ 3x + 2x$
$= -2x^2 + 5x$

16. $4(3x^2 - 2x) - (2x + 4)$
$= 12x^2 -$ ____ $-$ ____ $- 4$
$= 12x^2 - 10x - 4$

GUIDED PRACTICE

Perform the following additions. See Example 1.

17. $4y + 5y$
18. $2x + 3x$
19. $8t^2 + 4t^2$
20. $15x^2 + 10x^2$
21. $4r + 3r + 7r$
22. $2b + 7b + 3b$
23. $4ab + 4ab + ab$
24. $xy + 4xy + 2xy$

Find each difference. See Example 2.

25. $7a^3 - 4a^3$
26. $12ab - 5ab$
27. $-32u^3 - 16u^3$
28. $-25x^3 - 7x^3$

29. $-3m - 6 - 4m$
30. $6c - 10 - 3c - 2$
31. $0.8p^2 - 3.1p - 2.7p^2 - 1.4p$
32. $1.7x - 3.2y - 2.5x - 7.5y$

Find each sum. See Examples 3–4.

33. $(3x + 7) + (4x - 3)$
34. $(2y - 3) + (4y + 7)$
35. $(3x^2 - 3x - 2) + (3x^2 + 4x - 3)$
36. $(4c^2 + 3c - 2) + (3c^2 + 4c + 2)$
37. $3x^2 + 4x + 5$
    $+\ \underline{2x^2 - 3x + 6}$
38. $2x^3 + 2x^2 - 3x + 5$
    $+\ \underline{3x^3 - 4x^2 - x - 7}$
39. $2x^3 - 3x^2 + 4x - 7$
    $+\ \underline{-9x^3 - 4x^2 - 5x + 6}$
40. $-3x^3 + 4x^2 - 4x + 9$
    $+\ \underline{2x^3 + 9x - 3}$

Find each difference. See Examples 5–6.

41. $(4a + 3) - (2a - 4)$
42. $(5b - 7) - (3b - 5)$
43. $(3a^2 - 2a + 4) - (a^2 - 3a + 7)$
44. $(2b^2 + 3b - 5) - (2b^2 - 4b - 9)$
45. $3x^2 + 4x - 5$
    $-\ \underline{-2x^2 - 2x + 3}$
46. $3y^2 - 4y + 7$
    $-\ \underline{6y^2 - 6y - 13}$
47. $4x^3 + 4x^2 - 3x + 10$
    $-\ \underline{5x^3 - 2x^2 - 4x - 4}$
48. $3x^3 + 4x^2 + 7x + 12$
    $-\ \underline{-4x^3 + 6x^2 + 9x - 3}$

Perform the operations. See Example 7.

49. Find the difference when $t^3 - 2t^2 + 2$ is subtracted from the sum of $3t^3 + t^2$ and $-t^3 + 6t - 3$.
50. Find the difference when $-3z^3 - 4z + 7$ is subtracted from the sum of $2z^2 + 3z - 7$ and $-4z^3 - 2z - 3$.
51. Find the sum when $3x^2 + 4x - 7$ is added to the sum of $-2x^2 - 7x + 1$ and $-4x^2 + 8x - 1$.
52. Find the difference when $32x^2 - 17x + 45$ is subtracted from the sum of $23x^2 - 12x - 7$ and $-11x^2 + 12x + 7$.

Simplify each expression. See Example 8.

53. $2(x + 3) + 4(x - 2)$
54. $3(y - 4) - 5(y + 3)$
55. $-2(x^2 + 7x - 1) - 3(x^2 - 2x + 2)$
56. $-5(y^2 - 2y - 6) + 6(2y^2 + 2y - 5)$
57. $2(2y^2 - 2y + 2) - 4(3y^2 - 4y - 1) + 4(y^2 - y - 1)$
58. $-4(z^2 - 5z) - 5(4z^2 - 1) + 6(2z - 3)$
59. $2(ab^2 - b) - 3(a + 2ab) + (b - a + a^2b)$
60. $3(xy + y) - 2(x - 4 + y) + 2(y^3 + y^2)$

TRY IT YOURSELF

Perform the operations and simplify.

61. $1.8x - 1.9x$
62. $1.7y - 2.2y$
63. $\dfrac{1}{2}st + \dfrac{3}{2}st$
64. $\dfrac{2}{5}at + \dfrac{1}{5}at$
65. $(3x)^2 - 4x^2 + 10x^2$
66. $(2x)^4 - (3x^2)^2$
67. $(2x + 3y) + (5x - 10y)$
68. $(5x - 8y) - (-2x + 5y)$
69. $(-8x - 3y) - (-11x + y)$
70. $(-4a + b) + (5a - b)$
71. $(2x^2 - 3x + 1) - (4x^2 - 3x + 2) + (2x^2 + 3x + 2)$
72. $(-3z^2 - 4z + 7) + (2z^2 + 2z - 1) - (2z^2 - 3z + 7)$
73. $(4.52x^2 + 1.13x - 0.89) + (9.02x^2 - 7.68x + 7.04)$
74. $(0.891a^4 - 0.442a^2 + 0.121a) - (-0.160a^4 + 0.249a^2 + 0.789a)$
75. $2(5a + 3b - c) - 5(-2a + 4b + 4c)$
76. $-3(4p - 3q + r) + 4(-2p - 3q - 2r)$
77. $-3x^2 + 4x + 25$
    $+\ \underline{5x^2 - 12}$

78. $-6x^3 - 4x^2 + 7$
    $+ \underline{-7x^3 + 9x^2}$

79. $-2x^2y^2 + 12y^2$
    $- \underline{10x^2y^2 + 9xy - 24y^2}$

80. $25x^3 + 31xz^2$
    $- \underline{12x^3 + 27x^2z - 17xz^2}$

Find the polynomial that represents the perimeter of each figure.

81.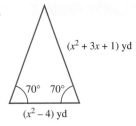

82.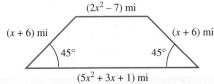

APPLICATIONS

83. **GREEK ARCHITECTURE** Find the difference in the heights of the columns shown in figure (b) on page 422 at the beginning of this section.

84. **CLASSICAL GREEK COLUMNS** If the columns shown in figure (b) on page 422 at the beginning of this section were stacked one atop the other, to what height would they reach?

85. **AUTO MECHANICS** Find the polynomial representing the length of the fan belt shown below. The dimensions are in inches. Your answer will involve π.

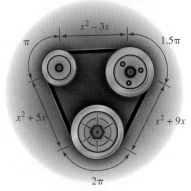

86. **READING BLUEPRINTS**
    a. What is the difference in the length and width of the one-bedroom apartment shown below?
    b. Find the perimeter of the apartment.

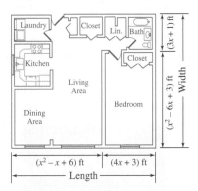

If a house is purchased for $305,000 and is expected to appreciate $900 per year, its value after x years is given by the polynomial function $f(x) = 900x + 305{,}000$.

87. **VALUE OF A HOUSE** Find the expected value of the house in 10 years.

88. **VALUE OF A HOUSE** A second house is purchased for $220,000 and is expected to appreciate $1,000 per year.
    a. Find a polynomial function that will give the value of the house in x years.
    b. Find the value of this second house after 12 years.

89. **VALUE OF TWO HOUSES** Find one polynomial function that will give the combined value of both houses, one from the directions and the other from Problem 88, after x years.

90. **VALUE OF TWO HOUSES** Find the value of the two houses after 20 years by
    a. substituting 20 into the polynomial functions $f(x) = 900x + 305{,}000$ and $f(x) = 1{,}000x + 220{,}000$ and adding.
    b. substituting 20 into the result of Problem 89.

A business purchases two computers, one for $6,600 and the other for $9,200. The first computer is expected to depreciate $1,100 per year and the second $1,700 per year.

91. **VALUE OF A COMPUTER** Write a polynomial function that gives the value of the first computer after x years.

92. **VALUE OF A COMPUTER** Write a polynomial function that gives the value of the second computer after x years.

93. VALUE OF TWO COMPUTERS Find one polynomial function that gives the combined value of both computers whose functions were found in Problems 91 and 92 after x years.

94. VALUE OF TWO COMPUTERS In two ways, find the combined value of the two computers after 3 years.

95. NAVAL OPERATIONS Two warning flares are simultaneously fired upward from different parts of a ship. After t seconds, the height of the first flare is $(-16t^2 + 128t + 20)$ feet and the height of the higher-traveling second flare is $(-16t^2 + 150t + 40)$ feet.

a. Find a polynomial that represents the difference in the heights of the flares.

b. In 4 seconds, the first flare reaches its peak, explodes, and lights up the sky. How much higher is the second flare at that time?

96. PIÑATAS Find the polynomial that represents the length of the rope used to hold up the piñata.

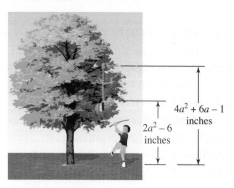

97. JETS Find the polynomial representing the length of the passenger jet.

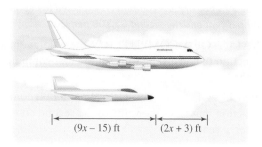

98. WATER SKIING Find the polynomial representing the distance of the water skier from the boat.

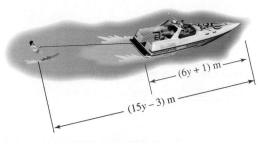

WRITING

99. How do you recognize like terms?

100. How do you add like terms?

101. Explain the concept that is illustrated by the statement

$$-(x^2 + 3x - 1) = -1(x^2 + 3x - 1)$$

102. Explain the mistake made in the following simplification:

$$(12x - 4) - (3x - 1) = 12x - 4 - 3x - 1$$
$$= 9x - 5$$

REVIEW

103. What is the sum of the measures of the angles of a triangle?

104. What is the sum of the measures of two complementary angles?

105. Solve the inequality $-4(3x - 3) \geq -12$ and graph the solution.

106. CURLING IRONS A curling iron is plugged into a 110-volt electrical outlet and used for $\frac{1}{4}$ hour. If its resistance is 10 ohms, find the electrical power (in kilowatt hours, kwh) used by the curling iron by applying the formula:

$$\text{kwh} = \frac{(\text{volts})^2}{1{,}000 \cdot \text{ohms}} \cdot \text{hours}$$

SECTION 5.6
Multiplying Polynomials

Objectives
1. Multiply monomials.
2. Multiply a polynomial by a monomial.
3. Multiply binomials.
4. Use special product formulas.
5. Multiply polynomials.
6. Multiply polynomials to solve equations.

We now discuss multiplying polynomials. We will begin with the simplest case—finding the product of two monomials.

1 Multiply monomials.

To multiply two monomials, such as $8x^2$ and $-3x^4$, we use the commutative and associative properties of multiplication to group the numerical factors and the variable factors. Then we multiply the numerical factors and multiply the variable factors.

$$8x^2(-3x^4) = 8(-3)x^2x^4$$
$$= -24x^6$$

This example suggests the following rule.

> **Multiplying Monomials**
>
> To multiply two monomials, multiply the numerical factors (the coefficients) and then multiply the variable factors.

EXAMPLE 1 Multiply:

a. $3x^4(2x^5)$ **b.** $-2a^2b^3(5ab^2)$ **c.** $-4y^5z^2(2y^3z^3)(3yz)$ **d.** $\left(\frac{3}{4}x^2y^3\right)\left(\frac{8}{3}xy^2\right)$

Strategy We will multiply the numerical factors and then multiply the variable factors.

WHY The commutative and associative properties of multiplication enable us to reorder and regroup the factors.

Solution

a. $3x^4(2x^5) = 3(2)\,x^4x^5$ Reorder the factors.

$= 6x^9$ Multiply the numerical factors, 3 and 2. Multiply the variable factors: $x^4x^5 = x^{4+5} = x^9$.

b. $-2a^2b^3(5ab^2) = -10a^3b^5$ Think: $-2(5) = -10$, $a^2 \cdot a = a^3$, and $b^3 \cdot b^2 = b^5$.

c. $-4y^5z^2(2y^3z^3)(3yz) = -24y^9z^6$ Think: $-4(2)(3) = -24$, $y^5 \cdot y^3 \cdot y = y^9$, and $z^2 \cdot z^3 \cdot z = z^6$.

d. $\left(\frac{3}{4}x^2y^3\right)\left(\frac{8}{3}xy^2\right) = 2x^3y^5$ Think: $\frac{3}{4}\left(\frac{8}{3}\right) = 2$, $x^2 \cdot x = x^3$, and $y^3 \cdot y^2 = y^5$.

Self Check 1
Multiply:
a. $(5a^2b^3)(6a^3b^4)$
b. $(-15p^3q^2)(5p^3q^2)$
c. $\left(\frac{2}{3}x^2y\right)(9y^2)$

Now Try Problems 13 and 18

> **Success Tip** Notice that we *multiply* the numerical coefficients. To multiply the variable factors with like bases, keep the base and *add* the exponents.

2 Multiply a polynomial by a monomial.

To find the product of a polynomial (with more than one term) and a monomial, we use the distributive property. To multiply $2x + 4$ by $5x$, for example, we proceed as follows:

$$5x(2x + 4) = 5x(2x) + 5x(4)$$ Distribute the multiplication by $5x$.
$$= 10x^2 + 20x$$ Multiply the monomials: $5x(2x) = 10x^2$ and $5x(4) = 20x$.

This example suggests the following rule.

> **Multiplying Polynomials by Monomials**
>
> To multiply a polynomial with more than one term by a monomial, multiply each term of the polynomial by the monomial and simplify.

Self Check 2

Multiply:
a. $2p^3(3p^2 - 5p)$
b. $-5a^2b(3a + 2b - 4ab)$
c. $(3ab - 4b + 2)(1.5a^4b)$

Now Try Problems 21 and 31

EXAMPLE 2 Multiply:

a. $3a^2(3a^2 - 5a)$ b. $-2xz^2(2x - 3z + 2z^2)$ c. $(2.1b^2 - 3b)(0.1b^3)$

Strategy We will multiply each term of the polynomial by the monomial.

WHY We use the distributive property to multiply a monomial and a polynomial.

Solution

a. $3a^2(3a^2 - 5a) = 3a^2(3a^2) - 3a^2(5a)$ Distribute the multiplication by $3a^2$.
$= 9a^4 - 15a^3$ Multiply the monomials.

b. $-2xz^2(2x - 3z + 2z^2)$
$= -2xz^2(2x) - (-2xz^2)(3z) + (-2xz^2)(2z^2)$ Use the distributive property.
$= -4x^2z^2 - (-6xz^3) + (-4xz^4)$ Multiply the monomials.
$= -4x^2z^2 + 6xz^3 - 4xz^4$

c. $(2.1b^2 - 3b)(0.1b^3) = 2.1b^2(0.1b^3) - 3b(0.1b^3)$ Distribute the multiplication by $0.1b^3$.
$= 0.21b^5 - 0.3b^4$ Multiply the monomials.

> **Success Tip** $(2.1b^2 - 3b)(0.1b^3)$ could also be rewritten as $(0.1b^3)(2.1b^2 - 3b)$ using the commutative property of multiplication. Then use the distributive property to find the product $0.21b^5 - 0.3b^4$.

3 Multiply binomials.

To multiply two binomials, we must use the distributive property more than once. For example, to multiply $2a - 4$ by $3a + 5$, we proceed as follows.

$(2a - 4)(3a + 5) = (2a - 4)(3a) + (2a - 4)(5)$ Distribute the multiplication by $(2a - 4)$.
$= 3a(2a - 4) + 5(2a - 4)$ Use the commutative property of multiplication.
$= 3a(2a) - 3a(4) + 5(2a) - 5(4)$ Distribute the multiplication by $3a$ and by 5.
$= 6a^2 - 12a + 10a - 20$ Do the multiplications.
$= 6a^2 - 2a - 20$ Combine like terms.

This example suggests the following rule.

> **Multiplying Two Binomials**
>
> To multiply two binomials, multiply each term of one binomial by each term of the other binomial and combine like terms.

EXAMPLE 3
Multiply: $(2x - y)(3x + 4y)$

Strategy We will multiply each term of $3x + 4y$ by $2x$ and $-y$.

WHY To multiply two binomials, each term of one binomial must be multiplied by each term of the other binomial.

Solution

$(2x - y)(3x + 4y) = 2x(3x + 4y) - y(3x + 4y)$ Multiply $3x + 4y$ by $2x$ and by $-y$.

$\qquad = 6x^2 + 8xy - 3xy - 4y^2$ Distribute the multiplication by $2x$. Distribute the multiplication by $-y$.

$\qquad = 6x^2 + 5xy - 4y^2$ Combine like terms.

Self Check 3
Multiply: $(7x + 2y)(5x - 2y)$
Now Try Problem 37

We can use a shortcut method, called the **FOIL** method, to multiply binomials. FOIL is an acronym for **F**irst terms, **O**uter terms, **I**nner terms, and **L**ast terms. To use the FOIL method to multiply $2a - 4$ by $3a + 5$:

1. Multiply the **F**irst terms $2a$ and $3a$ to obtain $6a^2$.
2. Multiply the **O**uter terms $2a$ and 5 to obtain $10a$.
3. Multiply the **I**nner terms -4 and $3a$ to obtain $-12a$.
4. Multiply the **L**ast terms -4 and 5 to obtain -20.

Then we simplify the resulting polynomial, if possible.

$(2a - 4)(3a + 5) = 2a(3a) + 2a(5) + (-4)(3a) + (-4)(5)$

$\qquad = 6a^2 + 10a - 12a - 20$ Do the multiplications.

$\qquad = 6a^2 - 2a - 20$ Combine like terms.

EXAMPLE 4
Find each product: **a.** $(x + 5)(x + 7)$ **b.** $(3x + 4)(2x - 3)$
c. $(a - 7b)(a - 4b)$ **d.** $\left(\dfrac{3}{4}r - 3s\right)\left(\dfrac{1}{2}r + 4t\right)$

Strategy We will use the FOIL method.

WHY In each case we are to find the product of two binomials, and the FOIL method is a shortcut for multiplying two binomials.

Solution

a. $(x + 5)(x + 7) = x(x) + x(7) + 5(x) + 5(7)$
$\qquad = x^2 + 7x + 5x + 35$ Multiply the monomials.
$\qquad = x^2 + 12x + 35$ Combine like terms.

b. $(3x + 4)(2x - 3) = 3x(2x) + 3x(-3) + 4(2x) + 4(-3)$
$\qquad = 6x^2 - 9x + 8x - 12$ Multiply the monomials.
$\qquad = 6x^2 - x - 12$ Combine like terms.

Self Check 4
Find each product:
a. $(y + 3)(y + 1)$
b. $(2a - 1)(3a + 2)$
c. $(5y - 2z)(2y + 3z)$
d. $\left(4r + \dfrac{1}{2}s\right)\left(2r - \dfrac{1}{2}s\right)$

Now Try Problems 40 and 41

c. $(a - 7b)(a - 4b) = a(a) + a(-4b) + (-7b)(a) + (-7b)(-4b)$
$= a^2 - 4ab - 7ab + 28b^2$ Multiply the monomials.
$= a^2 - 11ab + 28b^2$ Combine like terms.

d. $\left(\frac{3}{4}r - 3s\right)\left(\frac{1}{2}r + 4t\right) = \frac{3}{4}r\left(\frac{1}{2}r\right) + \frac{3}{4}r(4t) - 3s\left(\frac{1}{2}r\right) - 3s(4t)$
$= \frac{3}{8}r^2 + 3rt - \frac{3}{2}rs - 12st$ There are no like terms.

Self Check 5

Simplify: $(x + 3)(2x - 1) + 2x(x - 1)$.

Now Try Problems 45 and 51

EXAMPLE 5   Simplify each expression.

a. $3(2x - 3)(x + 1)$ **b.** $(x + 1)(x - 2) - 3x(x + 3)$

Strategy We will use the FOIL method and the distributive property to remove the parentheses. Then we will combine like terms.

WHY To simplify an expression means to remove parentheses and combine like terms.

Solution

a. $3(2x - 3)(x + 1) = 3(2x^2 + 2x - 3x - 3)$ Multiply the binomials.
$= 3(2x^2 - x - 3)$ Combine like terms.
$= 6x^2 - 3x - 9$ Distribute the multiplication by 3.

b. $(x + 1)(x - 2) - 3x(x + 3) = x^2 - 2x + x - 2 - 3x^2 - 9x$
$= -2x^2 - 10x - 2$ Combine like terms.

4 Use special product formulas.

Certain products of binomials occur so frequently in algebra that it is worthwhile to learn formulas for computing them. To develop a rule to find the *square of a sum*, we consider $(x + y)^2$.

$(x + y)^2 = (x + y)(x + y)$ In $(x + y)^2$, the base is $(x + y)$ and the exponent is 2.
$= x^2 + xy + xy + y^2$ Multiply the binomials.
$= x^2 + 2xy + y^2$ Combine like terms: $xy + xy = 2xy$.

We note that the terms of this result are related to the terms of the original expression. That is, $(x + y)^2$ is equal to the square of its first term (x^2), plus twice the product of both its terms $(2xy)$, plus the square of its last term (y^2).

To develop a rule to find the *square of a difference*, we consider $(x - y)^2$.

$(x - y)^2 = (x - y)(x - y)$
$= x^2 - xy - xy + y^2$ Multiply the binomials.
$= x^2 - 2xy + y^2$ Combine like terms: $-xy - xy = -2xy$.

Again, the terms of the result are related to the terms of the original expression. When we find $(x - y)^2$, the product is composed of the square of its first term (x^2), twice the product of both its terms $(-2xy)$, and the square of its last term (y^2).

The final special product is the product of two binomials that differ only in the signs of the last terms. To develop a rule to find the product of a *sum and a difference*, we consider $(x + y)(x - y)$.

$(x + y)(x - y) = x^2 - xy + xy - y^2$ Multiply the binomials.
$\qquad\qquad\quad = x^2 - y^2$ Combine like terms: $-xy + xy = 0$.

The product is the square of the first term (x^2) minus the square of the second term (y^2). The expression $x^2 - y^2$ is called a **difference of two squares**.

Because these special products occur so often, it is wise to memorize their forms.

Special Products

$(x + y)^2 = x^2 + 2xy + y^2$ The square of a sum
$(x - y)^2 = x^2 - 2xy + y^2$ The square of a difference
$(x + y)(x - y) = x^2 - y^2$ The product of a sum and difference

EXAMPLE 6 Find: **a.** $(t + 9)^2$ **b.** $(8a - 5)^2$ **c.** $(3y + 4z)(3y - 4z)$

Strategy We will identify the special product and apply the appropriate rule.

WHY The rules for special products enables us to compute them quickly.

Solution

a. This is the square of a sum. The terms of the binomial being squared are t and 9.

$(t + 9)^2 = \underbrace{t^2}_{\substack{\text{The square} \\ \text{of the first} \\ \text{term, } t.}} + \underbrace{2(t)(9)}_{\substack{\text{Twice the} \\ \text{product of} \\ \text{both terms.}}} + \underbrace{9^2}_{\substack{\text{The square} \\ \text{of the last} \\ \text{term, 9.}}}$

$= t^2 + 18t + 81$

b. This is the square of a difference. The terms of the binomial being squared are $8a$ and -5.

$(8a - 5)^2 = \underbrace{(8a)^2}_{\substack{\text{The square} \\ \text{of the first} \\ \text{term, } 8a.}} - \underbrace{2(8a)(5)}_{\substack{\text{Twice the} \\ \text{product of} \\ \text{both terms.}}} + \underbrace{5^2}_{\substack{\text{The square} \\ \text{of the last} \\ \text{term, 5.}}}$

$= 64a^2 - 80a + 25$ Use the power rule for products: $(8a)^2 = 8^2 a^2 = 64a^2$.

c. The binomials differ only in the signs of the last terms. This is the product of a sum and a difference.

$(3y + 4z)(3y - 4z) = (3y)^2 - (4z)^2$ This is the square of the first term minus the square of the second term.

$\qquad\qquad\qquad\quad = 9y^2 - 16z^2$ Use the power rule for products twice.

Self Check 6
Find:
a. $(r + 6)^2$
b. $(7g - 2)^2$
c. $(5m - 9n)(5m + 9n)$

Now Try Problems 53, 56, and 57

> **Caution!** A common error when squaring a binomial is to forget the middle term of the product. For example, $(x + 2)^2 \neq x^2 + 4$ and $(x - 2)^2 \neq x^2 - 4$. Applying the special product formulas, we have $(x + 2)^2 = x^2 + \mathbf{4x} + 4$ and $(x - 2)^2 = x^2 - \mathbf{4x} + 4$.

5 Multiply polynomials.

We must use the distributive property more than once to multiply a polynomial by a binomial. For example, to multiply $3x^2 + 3x - 5$ by $2x + 3$, we proceed as follows:

$$(2x + 3)(3x^2 + 3x - 5) = (2x + 3)3x^2 + (2x + 3)3x - (2x + 3)5$$
$$= 3x^2(2x + 3) + 3x(2x + 3) - 5(2x + 3)$$
$$= 6x^3 + 9x^2 + 6x^2 + 9x - 10x - 15$$
$$= 6x^3 + 15x^2 - x - 15$$

This example suggests the following rule.

Multiplying Polynomials

To multiply one polynomial by another, multiply each term of one polynomial by each term of the other polynomial and combine like terms.

It is often convenient to organize the work vertically.

Self Check 7
Multiply using vertical form:
a. $(3x + 2)(2x^2 - 4x + 5)$
b. $(-2x^2 + 3)(2x^2 - 4x - 1)$
Now Try Problem 63

EXAMPLE 7   Multiply using vertical form:

a. $(3a^2 - 4a + 7)(2a + 5)$ b. $(3y^2 - 5y + 4)(-4y^2 - 3)$

Strategy First, we will write one polynomial underneath the other and draw a horizontal line beneath them. Then, we will multiply each term of the upper polynomial by each term of the lower polynomial, making sure to line up like terms.

WHY *Vertical form* means to use an approach similar to that used in arithmetic to multiply two numbers.

Solution
a. Multiply:

$$\begin{array}{r} 3a^2 - 4a + 7 \\ 2a + 5 \\ \hline 15a^2 - 20a + 35 \\ 6a^3 - 8a^2 + 14a \\ \hline 6a^3 + 7a^2 - 6a + 35 \end{array}$$

Multiply $3a^2 - 4a + 7$ by 5.
Multiply $3a^2 - 4a + 7$ by $2a$.
In each column, combine like terms.

b. Multiply:

$$\begin{array}{r} 3y^2 - 5y + 4 \\ -4y^2 - 3 \\ \hline -9y^2 + 15y - 12 \\ -12y^4 + 20y^3 - 16y^2 \\ \hline -12y^4 + 20y^3 - 25y^2 + 15y - 12 \end{array}$$

Multiply $3y^2 - 5y + 4$ by -3.
Multiply $3y^2 - 5y + 4$ by $-4y^2$.
Combine like terms.

When finding the product of three polynomials, we begin by multiplying any two of them, and then we multiply that result by the third polynomial.

Self Check 8
Find the product:
$-2y(y + 3)(3y - 2)$
Now Try Problem 67

EXAMPLE 8   Find the product: $-3a(4a + 1)(a - 7)$

Strategy We will find the product of $4a + 1$ and $a - 7$ and then multiply that result by $-3a$.

WHY It is better to find the most difficult product first. Save the simpler multiplication by $-3a$ for last.

Solution
$$-3a(4a + 1)(a - 7) = -3a(4a^2 - 28a + a - 7) \quad \text{Multiply the two binomials.}$$
$$= -3a(4a^2 - 27a - 7) \quad \text{Combine like terms.}$$
$$= -12a^3 + 81a^2 + 21a \quad \text{Distribute the multiplication by } -3a.$$

6 Multiply polynomials to solve equations.

To solve an equation involving products of polynomials, we can do the multiplication on each side and proceed as follows.

EXAMPLE 9   Solve: $(x + 2)(x + 3) = x(x + 7)$

Strategy We will multiply the binomials on the left side of the equation and use the distributive property on the right side.

WHY The first step in solving equations is to remove parentheses.

Solution
$(x + 2)(x + 3) = x(x + 7)$

$$x^2 + 3x + 2x + 6 = x^2 + 7x$$
$$x^2 + 5x + 6 = x^2 + 7x \quad \text{Combine like terms: } 3x + 2x = 5x.$$
$$x^2 + 5x + 6 - x^2 = x^2 + 7x - x^2 \quad \text{Subtract } x^2 \text{ from both sides.}$$
$$5x + 6 = 7x \quad \text{Combine like terms: } x^2 - x^2 = 0.$$
$$6 = 2x \quad \text{Subtract } 5x \text{ from both sides.}$$
$$3 = x \quad \text{Divide both sides by 2. The solution is 3.}$$

Check:
$$(x + 2)(x + 3) = x(x + 7)$$
$$(3 + 2)(3 + 3) \stackrel{?}{=} 3(3 + 7) \quad \text{Replace } x \text{ with 3.}$$
$$5(6) \stackrel{?}{=} 3(10) \quad \text{Do the additions within parentheses.}$$
$$30 = 30 \quad \text{Do the multiplications.}$$

Self Check 9
Solve:
$(x + 5)(x - 4) = x(x + 3)$

Now Try Problem 71

EXAMPLE 10   **Dimensions of a Painting**   A square painting is surrounded by a border 2 inches wide. If the area of the border is 96 square inches, find the dimensions of the painting.

Analyze Refer to the figure, which shows a square painting surrounded by a border 2 inches wide. We know that the area of this border is 96 square inches, and we are to find the dimensions of the painting.

Assign Let $x =$ the length of a side of the square painting. Since the border is 2 inches wide, the length and the width of the outer rectangle are both $(x + 2 + 2)$ inches. Then the outer rectangle is also a square, and its dimensions are $(x + 4)$ by $(x + 4)$ inches.

Form Since the area of a square is the product of its length and width, the area of the larger square is $(x + 4)(x + 4)$, and the area of the painting is $x \cdot x$. If we subtract the area of the painting from the area of the larger square, the difference is 96.

Self Check 10
A square painting is surrounded by a border 1 inch wide. If the area of the border is 84 square inches, find the dimensions of the painting.

Now Try Problem 116

The area of the large square	minus	the area of the square painting	is	the area of the border.
$(x+4)(x+4)$	$-$	$x \cdot x$	$=$	96

Solve

$$(x+4)(x+4) - x^2 = 96 \quad x \cdot x = x^2.$$
$$x^2 + 8x + 16 - x^2 = 96 \quad (x+4)(x+4) = (x+4)^2 = x^2 + 8x + 16.$$
$$8x + 16 = 96 \quad \text{Combine like terms: } x^2 - x^2 = 0.$$
$$8x = 80 \quad \text{Subtract 16 from both sides.}$$
$$x = 10 \quad \text{Divide both sides by 8.}$$

State The dimensions of the painting are 10 inches by 10 inches.

Check Verify that the 2-inch-wide border of a 10-inch-square painting would have an area of 96 square inches.

ANSWERS TO SELF CHECKS

1. a. $30a^5b^7$ **b.** $-75p^6q^4$ **c.** $6x^2y^3$ **2. a.** $6p^5 - 10p^4$ **b.** $-15a^3b - 10a^2b^2 + 20a^3b^2$ **c.** $4.5a^5b^2 - 6a^4b^2 + 3a^4b$ **3.** $35x^2 - 4xy - 4y^2$ **4. a.** $y^2 + 4y + 3$ **b.** $6a^2 + a - 2$ **c.** $10y^2 + 11yz - 6z^2$ **d.** $8r^2 - rs - \frac{1}{4}s^2$ **5.** $4x^2 + 3x - 3$ **6. a.** $r^2 + 12r + 36$ **b.** $49g^2 - 28g + 4$ **c.** $25m^2 - 81n^2$ **7. a.** $6x^3 - 8x^2 + 7x + 10$ **b.** $-4x^4 + 8x^3 + 8x^2 - 12x - 3$ **8.** $-6y^3 - 14y^2 + 12y$ **9.** $x = -10$ **10.** 20 in. by 20 in.

SECTION 5.6 STUDY SET

VOCABULARY

Fill in the blanks.

1. The expression $(2a - 4)(3a + 5)$ is the product of two _____.

2. The expression $(2a - 4)(3a^2 + 5a - 1)$ is the product of a _____ and a _____.

3. In the acronym FOIL, F stands for _____ terms, O for _____ terms, I for _____ terms, and L for _____ terms.

4. $(x + 5)^2$ is the square of a _____, and $(x - 5)^2$ is the square of a _____.

CONCEPTS

Consider the product $(2x + 5)(3x - 4)$.

5. The product of the first terms is _____.
6. The product of the outer terms is _____.
7. The product of the inner terms is _____.
8. The product of the last terms is _____.

9. Label each arrow using one of the letters, F, O, I, or L. Then fill in the blanks.

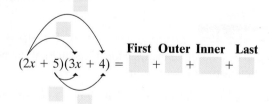

First Outer Inner Last
$(2x + 5)(3x + 4) =$ ____ + ____ + ____ + ____

10. $(3a)(2a^2)$ can be classified as a monomial · monomial. Classify the following products by identifying the types of polynomial factors.

a. $6x(x - 7)$
b. $(9a + 1)(5a - 3)$
c. $(c - d)(c^2 - c + d)$
d. $6m(m^2 + 1)(m^2 - 1)$

NOTATION

Complete each step.

11. $7x(3x^2 - 2x + 5) = \boxed{}(3x^2) - \boxed{}(2x) + \boxed{}(5)$
    $= 21x^3 - 14x^2 + 35x$

12. $(2x + 5)(3x - 2) = 2x(3x) - \boxed{}(2) + \boxed{}(3x) - \boxed{}(2)$
    $= 6x^2 - \boxed{} + \boxed{} - 10$
    $= 6x^2 + 11x - 10$

GUIDED PRACTICE

Multiply. See Example 1.

13. $(3x^2)(4x^3)$
14. $(-2a^3)(3a^2)$
15. $(-5x^3y^6)(x^2y^2)$
16. $(3x^2y)(2xy^2)$
17. $(x^2y^5)(x^2z^5)(-3z^3)$
18. $(3ab^2)(-2ab)(4ab^3)$
19. $\left(\dfrac{1}{2}x^2y^3\right)\left(\dfrac{2}{3}x^3y^2\right)$
20. $\left(-\dfrac{3}{4}r^4st^2\right)(2r^2st)\left(-\dfrac{2}{3}rst\right)$

Multiply. See Example 2.

21. $-4t(t + 7)$
22. $-8c(2c - 3)$
23. $6s^2(s^2 - 3s)$
24. $3a^3(2a^2 + 5a)$
25. $3y(x + 4y)$
26. $-3a^2b(a - b)$
27. $2x^2(3x^2 + 4x - 7)$
28. $3y^3(2y^2 - 7y - 8)$
29. $2ab^2(2a + 3b - 2a^2)$
30. $-2p^2q(3p - 2q - 3pq)$
31. $(3.1p^2 - 4q)(0.2p^2)$
32. $(1.5m^2 + 5.1n)(1.2m^3)$

Find each product. See Examples 3–4.

33. $(a + 4)(a + 5)$
34. $(y - 3)(y + 5)$
35. $(3x - 2)(x + 4)$
36. $(t + 4)(2t - 3)$
37. $(2a + 4)(3a - 5)$
38. $(2b - 1)(3b + 4)$
39. $(3x - 5)(2x + 1)$
40. $(2y - 5)(3y + 7)$
41. $(2t + 3s)(3t - s)$
42. $(3a - 2b)(4a + b)$
43. $\left(\dfrac{1}{4}t - u\right)\left(-\dfrac{1}{2}t + u\right)$
44. $\left(-\dfrac{1}{3}t + 2s\right)\left(\dfrac{2}{3}t - 3s\right)$

Simplify each expression. See Example 5.

45. $4(2x + 1)(x - 2)$
46. $-5(3a - 2)(2a + 3)$
47. $3a(a + b)(a - b)$
48. $-2r(r + s)(r + s)$
49. $2t(t + 2) + 3t(t - 5)$
50. $3y(y + 2) + (y + 1)(y - 1)$
51. $(x + y)(x - y) + x(x + y)$
52. $(3x + 4)(2x - 2) - (2x + 1)(x + 3)$

Find each product. See Example 6.

53. $(x + 4)^2$
54. $(y + 7)^2$
55. $(t - 3)^2$
56. $(z - 5)^2$
57. $(4x + 5)(4x - 5)$
58. $(5z + 1)(5z - 1)$
59. $(x - 2y)^2$
60. $(3a + 2b)^2$

Find each product. See Example 7.

61. $x^2 - 2x + 1$
    $\underline{x + 2}$

62. $5r^2 + r + 6$
    $\underline{2r - 1}$

63. $4x^2 + 3x - 4$
    $\underline{3x + 2}$

64. $x^2 - x + 1$
    $\underline{x + 1}$

Find each product. See Example 8.

65. $3x(-2x^2)(x + 4)$
66. $-2a^2(-3a^3)(3a - 2)$
67. $-2x(x + 3)(2x - 3)$
68. $5x^2(2x + 3)(2x - 5)$

Solve each equation. See Example 9.

69. $(s - 4)(s + 1) = s^2 + 5$
70. $(y - 5)(y - 2) = y^2 - 4$
71. $z(z + 2) = (z + 4)(z - 4)$
72. $(z + 3)(z - 3) = z(z - 3)$
73. $(x + 4)(x - 4) = (x - 2)(x + 6)$
74. $(y - 1)(y + 6) = (y - 3)(y - 2) + 8$
75. $(a - 3)^2 = (a + 3)^2$
76. $(b + 2)^2 = (b - 1)^2$

TRY IT YOURSELF

Find each product.

77. $3x(x - 2)$
78. $4y(y + 5)$
79. $-2x^2(3x^2 - x)$
80. $4b^3(2b^2 - 2b)$

81. $3xy(x + y)$
82. $-4x^2z(3x^2 - z)$
83. $(r + 4)(r - 4)$
84. $(b + 2)(b - 2)$
85. $(2s + 1)(2s + 1)$
86. $(3t - 2)(3t - 2)$
87. $(x + y)(x + z)$
88. $(a - b)(x + y)$
89. $(x + 2)(x^2 - 2x + 3)$
90. $(x - 5)(x^2 + 2x - 3)$
91. $(4t + 3)(t^2 + 2t + 3)$
92. $(3x + 1)(2x^2 - 3x + 1)$
93. $(-3x + y)(x^2 - 8xy + 16y^2)$
94. $(3x - y)(x^2 + 3xy - y^2)$
95. $(x + 5)^2$
96. $(m - 6)^2$
97. $(2a - 3b)^2$
98. $(2x + 5y)^2$
99. $(4x + 5y)^2$
100. $(6p - 5q)^2$
101. $2a(a + 2) - 3a(5 - a)$
102. $(3b + 4)(2b - 2) + (2b + 1)(b + 3)$

APPLICATIONS

Find the area of each figure. You may leave π in your answer.

103.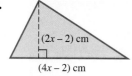
$(2x - 2)$ cm
$(4x - 2)$ cm

104.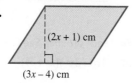
$(2x + 1)$ cm
$(3x - 4)$ cm

105.
$(x + 3)$ in.

106.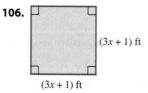
$(3x + 1)$ ft
$(3x + 1)$ ft

107. **TOYS** Find polynomials that represent the perimeter and the area of the screen of the Etch A Sketch.

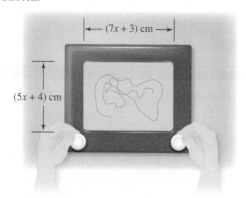
$(7x + 3)$ cm
$(5x + 4)$ cm

108. **SUNGLASSES** An ellipse is an oval-shaped curve. The area of an ellipse is approximately $0.785lw$, where l is its length and w is its width. Find a polynomial that represents the approximate area of one of the elliptical-shaped lenses of the sunglasses.

$(x - 1)$ in.
$(x + 1)$ in.

109. **STAMPS** Find a polynomial that represents the area of the stamp.

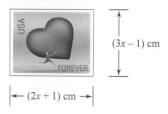
$(3x - 1)$ cm
$(2x + 1)$ cm

110. **LUGGAGE** Find a polynomial that represents the volume of the garment bag.

$(2x + 2)$ in.
$(x - 3)$ in.
x in.

111. **GARDENING** See the illustration below.
   a. What is the area of the region planted with corn? tomatoes? beans? carrots? Use your answers to find the total area of the garden.

   b. What is the length of the garden? What is its width? Use your answers to find its area.

   c. How do the answers from parts a and b for the area of the garden compare?

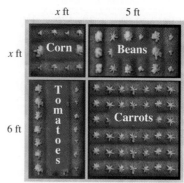

112. **PAINTING** See the illustration below. To purchase the correct amount of enamel to paint these two garage doors, a painter must find their areas. Find a polynomial that gives the number of square feet to be painted. All dimensions are in feet, and the windows are squares with sides of x feet.

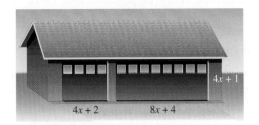

113. **INTEGER PROBLEM** The difference between the squares of two consecutive positive integers is 11. Find the integers. (*Hint:* Let x and $x + 1$ represent the consecutive integers.)

114. **INTEGER PROBLEM** If 3 less than a certain integer is multiplied by 4 more than the integer, the product is 6 less than the square of the integer. Find the integer.

115. **STONE-GROUND FLOUR** The radius of a millstone is 3 meters greater than the radius of another, and their areas differ by 15π square meters. Find the radius of the larger millstone.

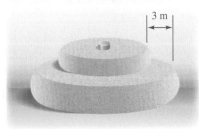

116. **BOOKBINDING** Two square sheets of cardboard used for making book covers differ in area by 44 square inches. An edge of the larger square is 2 inches greater than an edge of the smaller square. Find the length of an edge of the smaller square.

117. **BASEBALL** In major league baseball, the distance between bases is 30 feet greater than it is in softball. The bases in major league baseball mark the corners of a square that has an area 4,500 square feet greater than for softball. Find the distance between the bases in baseball.

118. **PULLEY DESIGNS** The radius of one pulley in the illustration in the next column is 1 inch greater than the radius of the second pulley, and their areas differ by 4π square inches. Find the radius of the smaller pulley.

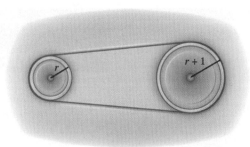

119. **SIGNS** Find a polynomial that represents the area of the sign.

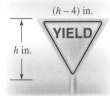

120. **BASEBALL** Find a polynomial that represents the amount of space within the batting cage.

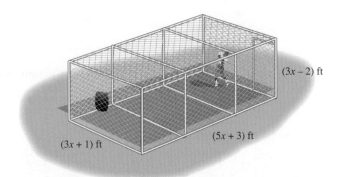

WRITING

121. Describe the steps involved in finding the product of $x + 2$ and $x - 2$.

122. Writing $(x + y)^2$ as $x^2 + y^2$ illustrates a common error. Explain.

REVIEW

Refer to the illustration.

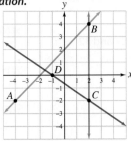

123. Find the slope of line AB.
124. Find the slope of line BC.
125. Find the slope of line CD.
126. Find the slope of the x-axis.
127. Find the y-intercept of line AB.
128. Find the x-intercept of line AB.

Objectives

1. Divide a monomial by a monomial.
2. Divide a polynomial by a monomial.
3. Write equivalent forms of formulas.

SECTION 5.7
Dividing Polynomials by Monomials

In this section, we will discuss how to divide polynomials by monomials. We will first divide monomials by monomials and then divide polynomials with more than one term by monomials.

1 Divide a monomial by a monomial.

Recall that to simplify a fraction, we write both its numerator and denominator as the product of several factors and then remove all common factors.

$$\frac{4}{6} = \frac{2 \cdot 2}{2 \cdot 3} \quad \text{Factor 4 and 6.} \qquad \frac{20}{25} = \frac{4 \cdot 5}{5 \cdot 5} \quad \text{Factor 20 and 25.}$$

$$= \frac{\overset{1}{\cancel{2}} \cdot 2}{\underset{1}{\cancel{2}} \cdot 3} \quad \text{Remove the common factor 2.} \qquad = \frac{4 \cdot \overset{1}{\cancel{5}}}{\underset{1}{\cancel{5}} \cdot 5} \quad \text{Remove the common factor 5.}$$

$$= \frac{2}{3} \qquad\qquad\qquad\qquad\qquad = \frac{4}{5}$$

We can use the same method to simplify algebraic fractions that contain variables.

$$\frac{3p^2}{6p} = \frac{3 \cdot p \cdot p}{2 \cdot 3 \cdot p} \quad \text{Factor } p^2 \text{ and 6.}$$

$$= \frac{\overset{1}{\cancel{3}} \cdot \cancel{p} \cdot p}{2 \cdot \underset{1}{\cancel{3}} \cdot \underset{1}{\cancel{p}}} \quad \text{Remove the common factors 3 and } p.$$

$$= \frac{p}{2}$$

To divide monomials, we can use either the preceding method for simplifying arithmetic fractions or the rules for exponents.

Self Check 1

Divide: $\dfrac{-5p^2q^3}{10pq^4}$

Now Try Problems 27 and 32

EXAMPLE 1

Divide: **a.** $\dfrac{x^2y}{xy^2}$ **b.** $\dfrac{-8a^3b^2}{4ab^3}$

Strategy We will use the rules for simplifying fractions and/or the quotient rule for exponents.

WHY We need to make sure that the numerator and denominator have no common factors other than 1. If that is the case, then the fraction is in *simplest form*.

Solution

By simplifying fractions

a. $\dfrac{x^2y}{xy^2} = \dfrac{x \cdot x \cdot y}{x \cdot y \cdot y}$

$= \dfrac{\overset{1}{\cancel{x}} \cdot x \cdot \overset{1}{\cancel{y}}}{\underset{1}{\cancel{x}} \cdot y \cdot \underset{1}{\cancel{y}}}$

$= \dfrac{x}{y}$

Using the rules for exponents

$\dfrac{x^2y}{xy^2} = x^{2-1}y^{1-2}$

$= x^1 y^{-1}$

$= \dfrac{x}{y}$

b. $\dfrac{-8a^3b^2}{4ab^3} = \dfrac{-2 \cdot 4 \cdot a \cdot a \cdot a \cdot b \cdot b}{4 \cdot a \cdot b \cdot b \cdot b}$

$= -\dfrac{2a^2}{b}$

$\dfrac{-8a^3b^2}{4ab^3} = \dfrac{-2^3 a^3 b^2}{2^2 ab^3}$

$= -2^{3-2} a^{3-1} b^{2-3}$

$= -2^1 a^2 b^{-1}$

$= -\dfrac{2a^2}{b}$

EXAMPLE 2

Simplify $\dfrac{25(s^2t^3)^2}{15(st^3)^3}$ and write the result using positive exponents only.

Strategy We will use the rules for exponents to remove the parentheses. Then we will simplify the fraction.

WHY We need to make sure that the numerator and denominator have no common factors other than 1. If that is the case, then the fraction is in *simplest form*.

Solution

$\dfrac{25(s^2t^3)^2}{15(st^3)^3} = \dfrac{25 s^4 t^6}{15 s^3 t^9}$ Use the power rules for exponents: $(xy)^n = x^n y^n$ and $(x^m)^n = x^{mn}$.

$= \dfrac{5 \cdot 5 \cdot s^{4-3} t^{6-9}}{5 \cdot 3}$ Factor 25 and 15. Use the quotient rule for exponents: $\dfrac{x^m}{x^n} = x^{m-n}$.

$= \dfrac{5 \cdot 5 \cdot s^1 t^{-3}}{5 \cdot 3}$ Remove the common factors of 5. Do the subtractions.

$= \dfrac{5s}{3t^3}$ Use the negative integer exponent rule: $t^{-3} = \dfrac{1}{t^3}$.

Self Check 2

Simplify: $\dfrac{-24(h^3 p)^5}{20(h^2 p^2)^3}$

Now Try Problems 35 and 39

2 Divide a polynomial by a monomial.

In Section 1.2, we used the following rules to add and subtract fractions with like denominators.

Adding and Subtracting Fractions with Like Denominators

To add (or subtract) fractions with like denominators, we add (or subtract) their numerators and keep the common denominator. In symbols, if a, b, and d represent numbers, and d is not 0,

$\dfrac{a}{d} + \dfrac{b}{d} = \dfrac{a+b}{d}$ and $\dfrac{a}{d} - \dfrac{b}{d} = \dfrac{a-b}{d}$

We can use this rule in reverse to divide polynomials by monomials.

Dividing a Polynomial by a Monomial

To divide a polynomial by a monomial, divide each term of the polynomial by the monomial.

If A, B and D represent monomials, where $D \neq 0$, then

$\dfrac{A+B}{D} = \dfrac{A}{D} + \dfrac{B}{D}$

Chapter 5 Exponents and Polynomials

Self Check 3

Divide: $\dfrac{4 - 8b}{4}$

Now Try Problem 41

EXAMPLE 3 Divide: $9x + 6$ by 3

Strategy We will set up the fraction to represent the division. Then we will divide each term of the numerator by the denominator.

WHY This is the rule for dividing a polynomial by a monomial.

Solution

$$\dfrac{9x + 6}{3} = \dfrac{9x}{3} + \dfrac{6}{3} \quad \text{Divide each term of the numerator by the denominator.}$$

$$= 3x + 2 \quad \text{Simplify each fraction.}$$

Self Check 4

Divide: $\dfrac{9a^2b - 6ab^2 + 3ab}{3ab}$

Now Try Problem 49

EXAMPLE 4 Divide: $\dfrac{6x^2y^2 + 4x^2y - 2xy}{2xy}$

Strategy We will divide each term of the polynomial in the numerator by the monomial in the denominator.

WHY This is the rule for dividing a polynomial by a monomial.

Solution

$$\dfrac{6x^2y^2 + 4x^2y - 2xy}{2xy}$$

$$= \dfrac{6x^2y^2}{2xy} + \dfrac{4x^2y}{2xy} - \dfrac{2xy}{2xy} \quad \text{Divide each term of the numerator by the denominator.}$$

$$= 3xy + 2x - 1 \quad \text{Simplify each fraction.}$$

Self Check 5

Divide: $\dfrac{14p^3q + pq^2 - p}{7p^2q}$

Now Try Problem 51

EXAMPLE 5 Divide: $\dfrac{12a^3b^2 - 4a^2b + a}{6a^2b^2}$

Strategy We will divide each term of the polynomial in the numerator by the monomial in the denominator.

WHY This is the rule for dividing a polynomial by a monomial.

Solution

$$\dfrac{12a^3b^2 - 4a^2b + a}{6a^2b^2}$$

$$= \dfrac{12a^3b^2}{6a^2b^2} - \dfrac{4a^2b}{6a^2b^2} + \dfrac{a}{6a^2b^2} \quad \text{Divide each term of the numerator by the denominator.}$$

$$= 2a - \dfrac{2}{3b} + \dfrac{1}{6ab^2} \quad \text{Simplify each fraction.}$$

Self Check 6

Divide: $\dfrac{(x + y)^2 - (x - y)^2}{xy}$

Now Try Problem 55

EXAMPLE 6 Divide: $\dfrac{(x - y)^2 - (x + y)^2}{xy}$

Strategy We will remove the parentheses in the numerator and combine like terms. Then we will divide each term of the polynomial in the numerator by the monomial in the denominator.

WHY This is the rule for dividing a polynomial by a monomial.

Solution

$$\frac{(x-y)^2 - (x+y)^2}{xy}$$

$$= \frac{x^2 - 2xy + y^2 - (x^2 + 2xy + y^2)}{xy} \quad \text{Use the special product rules to square the binomials in the numerator.}$$

$$= \frac{x^2 - 2xy + y^2 - x^2 - 2xy - y^2}{xy} \quad \text{Change the sign of each term within } (x^2 + 2xy + y^2).$$

$$= \frac{-4xy}{xy} \quad \text{Combine like terms in the numerator.}$$

$$= -4 \quad \text{Remove the common factors } x \text{ and } y.$$

3 Write equivalent forms of formulas.

The area of a trapezoid is given by the formula $A = \frac{1}{2}h(B+b)$, where B and b are its bases and h is its height. To solve the formula for b, we proceed as follows.

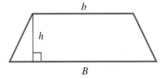

$$A = \frac{1}{2}h(B+b)$$

$$2 \cdot A = 2 \cdot \frac{1}{2}h(B+b) \quad \text{Multiply both sides by 2 to clear the equation of the fraction.}$$

$$2A = h(B+b) \quad \text{Simplify: } 2 \cdot \frac{1}{2} = \frac{2}{2} = 1.$$

$$2A = hB + hb \quad \text{Distribute the multiplication by } h.$$

$$2A - hB = hB + hb - hB \quad \text{Subtract } hB \text{ from both sides.}$$

$$2A - hB = hb \quad \text{Combine like terms: } hB - hB = 0.$$

$$\frac{2A - hB}{h} = \frac{hb}{h} \quad \text{To undo the multiplication by } h, \text{ divide both sides by } h.$$

$$\frac{2A - hB}{h} = b \quad \text{On the right side, remove the common factor of } h.$$

EXAMPLE 7
Another student worked the previous problem in a different way and got a result of $b = \frac{2A}{h} - B$. Is this result correct?

Strategy We will use the rule for dividing a polynomial by a monomial on the right side of $b = \frac{2A - hB}{h}$.

WHY If the right side of $b = \frac{2A - hB}{h}$ is the same as the right side of $b = \frac{2A}{h} - B$ after applying the rule for dividing a polynomial by a monomial, then the answer is also correct.

Solution
To determine whether this result is correct, we must show that

$$\frac{2A - hB}{h} = \frac{2A}{h} - B$$

We can do this by dividing $2A - hB$ by h.

$$\frac{2A - hB}{h} = \frac{2A}{h} - \frac{hB}{h} \quad \text{Divide each term of } 2A - hB \text{ by the denominator, which is } h.$$

$$= \frac{2A}{h} - B \quad \text{Simplify the second fraction: } \frac{\overset{1}{\cancel{h}}B}{\underset{1}{\cancel{h}}} = B.$$

The results are the same.

Self Check 7
Suppose another student got $b = 2A - \frac{hB}{h}$. Is this result correct?

Now Try Problem 77

446 Chapter 5 Exponents and Polynomials

ANSWERS TO SELF CHECKS

1. $-\dfrac{p}{2q}$ 2. $-\dfrac{6h^9}{5p}$ 3. $1-2b$ 4. $3a-2b+1$ 5. $2p+\dfrac{q}{7p}-\dfrac{1}{7pq}$ 6. 4 7. no

SECTION 5.7 STUDY SET

VOCABULARY

Fill in the blanks.

1. A _____ is an algebraic expression that is the sum of one or more terms containing whole-number exponents on the variables.
2. A _____ is a polynomial with one term.
3. A binomial is a polynomial with _____ terms.
4. A trinomial is a polynomial with _____ terms.
5. $\dfrac{x^m}{x^n} = x^{m-n}$ is the _____ rule for exponents.
6. To _____ a fraction, we remove common factors of the numerator and denominator.

CONCEPTS

Fill in the blanks.

7. $\dfrac{18x+9}{9} = \dfrac{18x}{\square} + \dfrac{9}{\square}$

8. $\dfrac{30x^2+12x-24}{6} = \dfrac{30x^2}{6}\,\square\,\dfrac{12x}{6}\,\square\,\dfrac{24}{6}$

9. $\dfrac{x^m}{x^n} = x^{\square}$

10. $x^{-n} = \dfrac{1}{\square}$

11. a. Solve the formula $d = rt$ for t.
    b. Use your answer from part a to complete the table.

	$r \cdot t = d$	
Motorcycle	$2x$	$6x^3$

12. a. Solve the formula $I = Prt$ for r.
    b. Use your answer from part a to complete the table.

	$P \cdot r \cdot t = I$		
Savings account	$8x^3$	$2x$	$24x^6$

13. How many nickels would have a value of $(10x+35)$ cents?
14. How many twenty-dollar bills would have a value of $\$(60x-100)$?

NOTATION

Complete each step.

15. $\dfrac{a^2b^3}{a^3b^2} = \dfrac{a \cdot a \cdot \square \cdot \square \cdot \square}{\square \cdot \square \cdot \square \cdot b \cdot b}$

    $= \dfrac{\overset{1}{\cancel{a}} \cdot \overset{1}{\cancel{a}} \cdot \overset{1}{\cancel{b}} \cdot \overset{1}{\cancel{b}} \cdot \square}{\underset{1}{\cancel{a}} \cdot \underset{1}{\cancel{a}} \cdot \square \cdot \underset{1}{\cancel{b}} \cdot \underset{1}{\cancel{b}}}$

    $= \dfrac{b}{a}$

16. $\dfrac{6pq^2 - 9p^2q^2 + pq}{3p^2q}$

    $= \dfrac{6pq^2}{\square} - \dfrac{9p^2q^2}{\square} + \dfrac{pq}{\square}$

    $= \dfrac{6 \cdot p \cdot q \cdot q}{3 \cdot p \cdot p \cdot q} - \dfrac{\square}{3 \cdot p \cdot p \cdot q} + \dfrac{p \cdot q}{3 \cdot p \cdot p \cdot q}$

    $= \dfrac{2q}{p} - 3q + \dfrac{1}{\square}$

GUIDED PRACTICE

Simplify each fraction. See Objective 1.

17. $\dfrac{5}{15}$ 18. $\dfrac{64}{128}$

19. $\dfrac{-125}{75}$ 20. $\dfrac{-98}{21}$

21. $\dfrac{120}{160}$ 22. $\dfrac{70}{420}$

23. $\dfrac{-3{,}612}{-3{,}612}$ 24. $\dfrac{-288}{-112}$

Perform each division. See Example 1.

25. $\dfrac{x^5}{x^2}$ 26. $\dfrac{a^{12}}{a^8}$

27. $\dfrac{r^3s^2}{rs^3}$ 28. $\dfrac{y^4z^3}{y^2z^2}$

29. $\dfrac{8x^3y^2}{4xy^3}$ 30. $\dfrac{-3y^3z}{6yz^2}$

31. $\dfrac{12u^5v}{-4u^2v^3}$ 32. $\dfrac{16rst^2}{-8rst^3}$

5.7 Dividing Polynomials by Monomials

Perform each division. See Example 2.

33. $\dfrac{(a^3b^4)^3}{ab^4}$

34. $\dfrac{(a^2b^3)^3}{a^6b^6}$

35. $\dfrac{15(r^2s^3)^2}{-5(rs^5)^3}$

36. $\dfrac{-5(a^2b)^3}{10(ab^2)^3}$

37. $\dfrac{-32(x^3y)^3}{128(x^2y^2)^3}$

38. $\dfrac{68(a^6b^7)^2}{-96(abc^2)^3}$

39. $\dfrac{-(4x^3y^3)^2}{(x^2y^4)^3}$

40. $\dfrac{(2r^3s^2)^2}{-(4r^2s^2)^2}$

Perform each division. See Examples 3–5.

41. $\dfrac{6x+9}{3}$

42. $\dfrac{8x+12y}{4}$

43. $\dfrac{5x-10y}{25xy}$

44. $\dfrac{2x-32}{16x}$

45. $\dfrac{3x^2+6y^3}{3x^2y^2}$

46. $\dfrac{4a^2-9b^2}{12ab}$

47. $\dfrac{15a^3b^2-10a^2b^3}{5a^2b^2}$

48. $\dfrac{9a^4b^3-16a^3b^4}{12a^2b}$

49. $\dfrac{4x-2y+8z}{4xy}$

50. $\dfrac{5a^2+10b^2-15ab}{5ab}$

51. $\dfrac{12x^3y^2-8x^2y-4x}{4xy}$

52. $\dfrac{12a^2b^2-8a^2b-4ab}{4ab}$

Perform each division. See Example 6.

53. $\dfrac{5x(4x-2y)}{2y}$

54. $\dfrac{9y(x^2-3xy)}{3x^2}$

55. $\dfrac{(x-y)^2-(x+y)^2}{2xy}$

56. $\dfrac{(2m-n)(3m-2n)}{-3m^2n^2}$

TRY IT YOURSELF

Simplify each expression.

57. $\dfrac{-16r^3y^2}{-4r^2y^4}$

58. $\dfrac{35xyz^2}{-7x^2yz}$

59. $\dfrac{-65rs^2t}{15r^2s^3t}$

60. $\dfrac{112u^3z^6}{-42u^3z^6}$

61. $\dfrac{x^2x^3}{xy^6}$

62. $\dfrac{x^2y^2}{x^2y^3}$

63. $\dfrac{(a^2a^3)^4}{(a^4)^3}$

64. $\dfrac{(b^3b^4)^5}{(bb^2)^2}$

65. $\dfrac{-25x^2y+30xy^2-5xy}{-5xy}$

66. $\dfrac{-30a^2b^2-15a^2b-10ab^2}{-10ab}$

67. $\dfrac{(-2x)^3+(3x^2)^2}{6x^2}$

68. $\dfrac{(-3x^2y)^3+(3xy^2)^3}{27x^3y^4}$

69. $\dfrac{4x^2y^2-2(x^2y^2+xy)}{2xy}$

70. $\dfrac{-5a^3b-5a(ab^2-a^2b)}{10a^2b^2}$

71. $\dfrac{(a+b)^2-(a-b)^2}{2ab}$

72. $\dfrac{(x-y)^2+(x+y)^2}{2x^2y^2}$

APPLICATIONS

73. **POOL** The rack shown is used to set up the balls when beginning a game of pool. If the perimeter of the rack, in inches, is given by the polynomial $6x^2-3x+9$, what is the length of one side?

74. **CHECKERBOARDS** If the perimeter (in inches) of the checkerboard is $12x^2-8x+32$, find an expression that represents the length of one side.

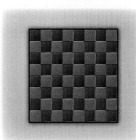

75. **AIR CONDITIONING** If the volume occupied by the air conditioning unit shown is $(36x^3-24x^2)$ cubic feet, find an expression that represents its height.

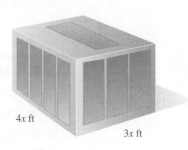

4x ft

3x ft

Unless otherwise noted, all content on this page is © Cengage Learning.

76. MINIBLINDS The area covered by the miniblinds is $(3x^3 - 6x)$ square feet. Find an expression that represents the length of the blinds.

3x ft

77. CONFIRMING FORMULAS Are these formulas the same?

$$l = \frac{P - 2w}{2} \quad \text{and} \quad l = \frac{P}{2} - w$$

78. CONFIRMING FORMULAS Are these formulas the same?

$$r = \frac{G + 2b}{2b} \quad \text{and} \quad r = \frac{G}{2b} + b$$

79. ELECTRIC BILLS On an electric bill, the following two formulas are used to compute the average cost of x kwh of electricity. Are the formulas equivalent?

$$\frac{0.08x + 5}{x} \quad \text{and} \quad 0.08x + \frac{5}{x}$$

80. PHONE BILLS On a phone bill, the following two formulas are used to compute the average cost per minute of x minutes of phone usage. Are the formulas equivalent?

$$\frac{0.15x + 12}{x} \quad \text{and} \quad 0.15 + \frac{12}{x}$$

WRITING

81. Explain the error.

$$\frac{3x + 5}{5} = \frac{3x + \overset{1}{\cancel{5}}}{\underset{1}{\cancel{5}}} = 3x$$

82. Explain how to perform this division.

$$\frac{4x^2y + 8xy^2}{4xy}$$

REVIEW

Identify each polynomial as a monomial, a binomial, a trinomial, or none of these.

83. $5a^2b + 2ab^2$

84. $-3x^3y$

85. $-2x^3 + 3x^2 - 4x + 12$

86. $17t^2 - 15t + 27$

87. What is the degree of the trinomial $3x^2 - 2x + 4$?

88. What is the coefficient of the second term of the trinomial $-7t^2 + 5t + 17$?

SECTION 5.8
Dividing Polynomials by Polynomials

Objectives
1. Divide polynomials by polynomials.
2. Write powers in descending order.
3. Divide polynomials that have missing terms.

In this section, we will discuss how to divide one polynomial by another.

1 Divide polynomials by polynomials.

To divide one polynomial by another, we use a method similar to long division in arithmetic. We illustrate the method with several examples.

Self Check 1
Divide $x^2 + 7x + 12$ by $x + 3$.
Now Try Problem 17

EXAMPLE 1 Divide $x^2 + 5x + 6$ by $x + 2$.

Strategy We will use the long division method. The dividend is $x^2 + 5x + 6$ and the divisor is $x + 2$.

WHY Since the divisor has more than one term, we must use the long division method to divide the polynomials.

Solution

We write the division using the symbol $\overline{)}$ and proceed as follows:

Step 1
$$x+2\overline{)x^2+5x+6}$$
with x above.

Divide the first term of $x^2 + 5x + 6$ by the first term of $x + 2$. $\frac{x^2}{x} = x$. Write x above the division symbol.

Step 2
$$\begin{array}{r} x \\ x+2\overline{)x^2+5x+6} \\ x^2+2x \end{array}$$

Multiply each term in the divisor by x. Write the result, $x^2 + 2x$, under $x^2 + 5x$, and draw a line. Be sure to align like terms.

Step 3
$$\begin{array}{r} x \\ x+2\overline{)x^2+5x+6} \\ \underline{-(x^2+2x)} \downarrow \\ 3x+6 \end{array}$$

Subtract $x^2 + 2x$ from $x^2 + 5x$. Work vertically, column by column: $x^2 - x^2 = 0$ and $5x - 2x = 3x$.
Bring down the 6.

Step 4
$$\begin{array}{r} x+3 \\ x+2\overline{)x^2+5x+6} \\ \underline{-(x^2+2x)} \\ 3x+6 \end{array}$$

Divide the first term of $3x + 6$ by the first term of the divisor. $\frac{3x}{x} = +3$. Write $+3$ above the division symbol to form the second term of the quotient.

Step 5
$$\begin{array}{r} x+3 \\ x+2\overline{)x^2+5x+6} \\ \underline{-(x^2+2x)}+6 \\ 3x+6 \\ 3x+6 \end{array}$$

Multiply each term in the divisor by 3. Write the product under $3x + 6$ and draw a line.

Step 6
$$\begin{array}{r} x+3 \\ x+2\overline{)x^2+5x+6} \\ \underline{-(x^2+2x)} \\ 3x+6 \\ \underline{-(3x+6)} \\ 0 \end{array}$$

Subtract $3x + 6$ from $3x + 6$. Work vertically: $3x - 3x = 0$ and $6 - 6 = 0$. This is the remainder.

The quotient is $x + 3$ and the remainder is 0.

Step 7 Check the work by verifying that $(x + 2)(x + 3)$ is $x^2 + 5x + 6$.

$$(x+2)(x+3) = x^2 + 3x + 2x + 6$$
$$= x^2 + 5x + 6$$

The answer checks.

EXAMPLE 2

Divide: $\dfrac{6x^2 - 7x - 2}{2x - 1}$

Strategy We will use the long division method. The dividend is $6x^2 - 7x - 2$ and the divisor is $2x - 1$.

WHY Since the divisor has more than one term, we must use the long division method to divide the polynomials.

Solution
We write the division using a long division symbol $\overline{)}$ and proceed as follows:

Step 1
$$2x-1\overline{)6x^2-7x-2}$$
with $3x$ above.

Divide the first term of the dividend by the first term of the divisor. $\frac{6x^2}{2x} = 3x$. Write the $3x$ above the division symbol.

Self Check 2

Divide: $\dfrac{8x^2 + 6x - 3}{2x + 3}$

Now Try Problems 21 and 27

Step 2
$$\begin{array}{r} 3x \\ 2x-1\overline{)6x^2-7x-2} \\ 6x^2-3x \end{array}$$

Multiply each term in the divisor by $3x$. Write the product under $6x^2 - 7x$ and draw a line.

Step 3
$$\begin{array}{r} 3x \\ 2x-1\overline{)6x^2-7x-2} \\ \underline{-(6x^2-3x)} \\ -4x-2 \end{array}$$

Subtract $6x^2 - 3x$ from $6x^2 - 7x$. Work vertically: $6x^2 - 6x^2 = 0$ and $-7x - (-3x) = -7x + 3x = -4x$.
Bring down the -2.

Step 4
$$\begin{array}{r} 3x-2 \\ 2x-1\overline{)6x^2-7x-2} \\ \underline{-(6x^2-3x)} \\ -4x-2 \end{array}$$

Divide the first term of $-4x - 2$ by the first term of the divisor. $\frac{-4x}{2x} = -2$. Write -2 above the division symbol.

Step 5
$$\begin{array}{r} 3x-2 \\ 2x-1\overline{)6x^2-7x-2} \\ \underline{-(6x^2-3x)} \\ -4x-2 \\ -4x+2 \end{array}$$

Multiply each term in the divisor by -2. Write the product under $-4x - 2$ and draw a line.

Step 6
$$\begin{array}{r} 3x-2 \\ 2x-1\overline{)6x^2-7x-2} \\ \underline{-(6x^2-3x)} \\ -4x-2 \\ \underline{-(-4x+2)} \\ -4 \end{array}$$

Subtract $-4x + 2$ from $-4x - 2$. Work vertically: $-4x - (-4x) = -4x + 4x = 0$ and $-2 - 2 = -4$.

Here the quotient is $3x - 2$ and the remainder is -4. It is common to write the answer as either

$$3x - 2 + \frac{-4}{2x-1} \quad \text{or} \quad 3x - 2 - \frac{4}{2x-1} \quad \text{Quotient} + \frac{\text{remainder}}{\text{divisor}}.$$

Step 7 To check, we multiply

$$3x - 2 + \frac{-4}{2x-1} \quad \text{by} \quad 2x - 1$$

The product should be the dividend.

$$(2x-1)\left(3x - 2 + \frac{-4}{2x-1}\right) = (2x-1)(3x-2) + (2x-1)\left(\frac{-4}{2x-1}\right)$$
$$= (2x-1)(3x-2) - 4$$
$$= 6x^2 - 4x - 3x + 2 - 4$$
$$= 6x^2 - 7x - 2$$

Because the result is the dividend, the answer checks.

2 Write powers in descending order.

The division method works best when the terms of the divisor and the dividend are written in descending powers of the variable. This means that the term involving the highest power of x appears first, the term involving the second-highest power of x appears second, and so on. For example, the terms in

$$3x^3 + 2x^2 - 7x + 5$$

have their exponents written in descending order.

If the powers in the dividend or divisor are not in descending order, we use the commutative property of addition to write them that way.

5.8 Dividing Polynomials by Polynomials

EXAMPLE 3
Divide: $(4x^2 + 2x^3 + 12 - 2x) \div (x + 3)$

Strategy We will write the dividend in descending powers of x and use the long division method to divide the polynomials.

WHY It is easier to carry out the division when the powers of the variables are written in descending order.

Solution
We write the dividend so that the exponents are in descending order.

$$\begin{array}{r} 2x^2 - 2x + 4 \\ x+3\overline{\smash{\big)}2x^3 + 4x^2 - 2x + 12} \\ \underline{-(2x^3 + 6x^2)} \\ -2x^2 - 2x \\ \underline{-(-2x^2 - 6x)} \\ 4x + 12 \\ \underline{-(4x + 12)} \\ 0 \end{array}$$

The first division: $\frac{2x^3}{x} = 2x^2$.

The second division: $\frac{-2x^2}{x} = -2x$.

The third division: $\frac{4x}{x} = 4$.

Check: $(x + 3)(2x^2 - 2x + 4) = 2x^3 - 2x^2 + 4x + 6x^2 - 6x + 12$
$= 2x^3 + 4x^2 - 2x + 12$

Self Check 3
Divide:
$(x^2 - 10x + 6x^3 + 4) \div (2x - 1)$

Now Try Problem 29

3 Divide polynomials that have missing terms.

When we write the terms of a dividend in descending powers of x, we must determine whether some powers of the variable are missing. When this happens, we should write such terms with a coefficient of 0 or leave a blank space for them.

EXAMPLE 4
Divide: $\dfrac{x^2 - 4}{x + 2}$

Strategy The dividend $x^2 - 4$ is missing an x-term. We will insert a $0x$ term as a placeholder and use the long division method.

WHY We insert placeholder terms so that like terms will be aligned in the same column when we subtract.

Solution

$$\begin{array}{r} x - 2 \\ x+2\overline{\smash{\big)}x^2 + 0x - 4} \\ \underline{-(x^2 + 2x)} \\ -2x - 4 \\ \underline{-(-2x - 4)} \\ 0 \end{array}$$

The first division: $\frac{x^2}{x} = x$.

The second division: $\frac{-2x}{x} = -2$.

Check: $(x + 2)(x - 2) = x^2 - 2x + 2x - 4$
$= x^2 - 4$

Self Check 4
Divide: $\dfrac{x^2 - 9}{x - 3}$

Now Try Problem 35

ANSWERS TO SELF CHECKS

1. $x + 4$ 2. $4x - 3 + \dfrac{6}{2x + 3}$ 3. $3x^2 + 2x - 4$ 4. $x + 3$

SECTION 5.8 STUDY SET

VOCABULARY

Fill in the blanks.

1. In the division $x + 1 \overline{)x^2 + 2x + 1}$, the expression $x + 1$ is called the _____ and $x^2 + 2x + 1$ is called the _____.
2. The answer to a division problem is called the _____.
3. If a division does not come out even, the leftover part is called a _____.
4. The powers of x in $2x^4 + 3x^3 + 4x^2 - 7x - 2$ are said to be written in _____ order.

CONCEPTS

Write each polynomial with the powers in descending order.

5. $4x^3 + 7x - 2x^2 + 6$
6. $5x^2 + 7x^3 - 3x - 9$
7. $9x + 2x^2 - x^3 + 6x^4$
8. $7x^5 + x^3 - x^2 + 2x^4$

Identify the missing terms in each polynomial.

9. $5x^4 + 2x^2 - 1$
10. $-3x^5 - 2x^3 + 4x - 6$
11. a. Solve $d = rt$ for r.
    b. Use your answer to part a and the long division method to complete the table.

	r · t = d	
Subway	$x + 4$	$x^2 + x - 12$

12. a. Solve $I = Prt$ for P.
    b. Use your answer to part a and the long division method to complete the table.

	P · r · t = I		
Bonds	$x + 4$	1	$x^2 + 7x + 12$

13. Using long division, a student found that
$$\frac{3x^2 + 8x + 4}{3x + 2} = x + 2$$
Use multiplication to see whether the result is correct.

14. Using long division, a student found that
$$\frac{x^2 + 4x - 21}{x - 3} = x - 7$$
Use multiplication to see whether the result is correct.

NOTATION

Complete each long division.

15.
$$\begin{array}{r} \square + 2 \\ x + 2 \overline{)x^2 + 4x + 4} \\ \underline{x^2 + \square} \\ \square + 4 \\ \underline{2x + 4} \\ 0 \end{array}$$

16.
$$\begin{array}{r} \square + x - 2 + \square \\ 2x + 1 \overline{)2x^3 + 3x^2 - 3x + 5} \\ \underline{\square + x^2} \\ 2x^2 - 3x \\ \underline{2x^2 + \square} \\ \square + 5 \\ \underline{-4x - \square} \\ 7 \end{array}$$

GUIDED PRACTICE

Perform each division. See Example 1.

17. Divide $x^2 + 4x - 12$ by $x - 2$.
18. Divide $x^2 - 5x + 6$ by $x - 2$.
19. Divide $y^2 + 13y + 12$ by $y + 1$.
20. Divide $z^2 - 7z + 12$ by $z - 3$.

Perform each division. See Example 2.

21. $\dfrac{6a^2 + 5a - 6}{2a + 3}$
22. $\dfrac{8a^2 + 2a - 3}{2a - 1}$
23. $\dfrac{3b^2 + 11b + 6}{3b + 2}$
24. $\dfrac{3b^2 - 5b + 2}{3b - 2}$
25. $\dfrac{2x^2 + 5x + 2}{2x - 3}$
26. $\dfrac{3x^2 - 8x + 8}{3x - 2}$
27. $\dfrac{4x^2 + 6x - 1}{2x + 1}$
28. $\dfrac{6x^2 - 11x + 2}{3x - 1}$

Write the terms so that the powers of x are in descending order. Then perform each division. See Example 3.

29. $5x + 3 \overline{)11x + 10x^2 + 3}$

30. $2x - 7 \overline{)-x - 21 + 2x^2}$

31. $4 + 2x \overline{)-10x - 28 + 2x^2}$

32. $1 + 3x \overline{)9x^2 + 1 + 6x}$

Perform each division. See Example 4.

33. $\dfrac{x^2 - 1}{x - 1}$

34. $\dfrac{x^2 - 9}{x + 3}$

35. $\dfrac{4x^2 - 9}{2x + 3}$

36. $\dfrac{25x^2 - 16}{5x - 4}$

TRY IT YOURSELF

Perform each division. If there is a remainder, write the answer in quotient $+ \dfrac{\text{remainder}}{\text{divisor}}$ *form.*

37. $2x - 1 \overline{)x - 2 + 6x^2}$

38. $2 + x \overline{)3x + 2x^2 - 2}$

39. $3 + x \overline{)2x^2 - 3 + 5x}$

40. $x - 3 \overline{)2x^2 - 3 - 5x}$

41. $2x + 3 \overline{)2x^3 + 7x^2 + 4x - 3}$

42. $2x - 1 \overline{)2x^3 - 3x^2 + 5x - 2}$

43. $3x + 2 \overline{)6x^3 + 10x^2 + 7x + 2}$

44. $4x + 3 \overline{)4x^3 - 5x^2 - 2x + 3}$

45. $2x + 1 \overline{)2x^3 + 3x^2 + 3x + 1}$

46. $3x - 2 \overline{)6x^3 - x^2 + 4x - 4}$

47. $\dfrac{x^3 + 3x^2 + 3x + 1}{x + 1}$

48. $\dfrac{x^3 + 6x^2 + 12x + 8}{x + 2}$

49. $\dfrac{2x^3 + 7x^2 + 4x + 3}{2x + 3}$

50. $\dfrac{6x^3 + x^2 + 2x + 1}{3x - 1}$

51. $\dfrac{2x^3 + 4x^2 - 2x + 3}{x - 2}$

52. $\dfrac{3y^3 - 4y^2 + 2y + 3}{y + 3}$

53. $\dfrac{x^3 + 1}{x + 1}$

54. $\dfrac{x^3 - 8}{x - 2}$

55. $\dfrac{a^3 + a}{a + 3}$

56. $\dfrac{y^3 - 50}{y - 5}$

57. $3x - 4 \overline{)15x^3 - 23x^2 + 16x}$

58. $2y + 3 \overline{)21y^2 + 6y^3 - 20}$

APPLICATIONS

59. **FURNACE FILTERS** The area of the furnace filter shown is $(x^2 - 2x - 24)$ square inches.
    a. Find an expression for its length.
    b. Find an expression for its perimeter.

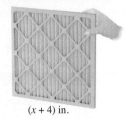

$(x + 4)$ in.

60. **ADVERTISING** Find the polynomial that represents the length of one of the longer sides of the rectangular billboard if its area is represented by $x^3 - 4x^2 + x + 6$.

61. **COMMUNICATIONS** See the illustration. Telephone poles were installed every $(2x - 3)$ feet along a stretch of railroad track $(8x^3 - 6x^2 + 5x - 21)$ feet long. How many poles were used?

$(2x - 3)$ ft

62. **CONSTRUCTION COSTS** Find the price per square foot to remodel each of the three rooms listed in the table.

Room	Remodeling cost	Area (ft²)	Cost (per ft²)
Bathroom	$(2x^2 + x - 6)$	$2x - 3$	
Bedroom	$(x^2 + 9x + 20)$	$x + 4$	
Kitchen	$(3x^3 - 9x - 6)$	$3x + 3$	

WRITING

63. Explain how the following are related: *dividend*, *divisor*, *quotient*, and *remainder*.

64. How would you check the results of a division?

REVIEW

Simplify each expression.

65. $(x^5 x^6)^2$

66. $(a^2)^3 (a^3)^4$

67. $3(2x^2 - 4x + 5) + 2(x^2 + 3x - 7)$

68. $-2(y^3 + 2y^2 - y) - 3(3y^3 + y)$

69. What can be said about the slopes of two parallel lines?

70. What is the slope of a line perpendicular to a line with a slope of $\frac{3}{4}$?

STUDY SKILLS CHECKLIST

Preparing for the Chapter 5 Test

While preparing for the Chapter 5 exam, it is important to keep the following in mind:

☐ When simplifying an exponential expression with negative exponents, write the reciprocal of the base and change the sign of the exponent.

$$7^{-2} = \frac{1}{7^2} = \frac{1}{49}$$

☐ Any nonzero number raised to the zero power is 1.

$$37^0 = 1, \quad 8x^0 = 8(1) = 8, \quad (-3x)^0 = 1$$

☐ To multiply a polynomial by a monomial, use the distributive property.

$$6xy^3(3xy - y^2) = 18x^2y^4 - 6xy^5$$

☐ The square of a binomial is a *trinomial*. A common error when squaring a binomial is to forget the middle term of the product.

$$(5x - 7)^2 = (5x - 7)(5x - 7)$$
$$= 25x^2 - 35x - 35x + 49$$
$$= 25x^2 - 70x + 49$$

$(5x - 7)^2 \ne 25x^2 + 49$

☐ To divide a polynomial by a *monomial,* do not use long division. Divide each term of the polynomial by the monomial in the denominator. Then simplify each fraction.

$$\frac{15x^2 + 3xy - 5y^2}{10xy} = \frac{15x^2}{10xy} + \frac{3xy}{10xy} - \frac{5y^2}{10xy}$$
$$= \frac{3x}{2y} + \frac{3}{10} - \frac{y}{2x}$$

☐ To divide a polynomial by a *polynomial*, use long division.

Divide: $\dfrac{x^2 - 6x + 5}{x - 5}$

$$\begin{array}{r} x - 1 \\ x-5 \overline{\smash{)}\, x^2 - 6x + 5} \\ \underline{-(x^2 - 5x)} \\ -x + 5 \\ \underline{-x + 5} \\ 0 \end{array}$$

CHAPTER 5 SUMMARY AND REVIEW

SECTION 5.1 Natural-Number Exponents; Rules for Exponents

DEFINITIONS AND CONCEPTS

If n represents a natural number, then

$$x^n = \underbrace{x \cdot x \cdot x \cdot \cdots \cdot x}_{n \text{ factors of } x}$$

where x is called the **base** and n is called the **exponent**.

EXAMPLES

$5^4 = 5 \cdot 5 \cdot 5 \cdot 5 = 625$ 5 is the base, 4 is the exponent

$a^3 = a \cdot a \cdot a$ a is the base, 3 is the exponent

$(-y)^4 = (-y)(-y)(-y)(-y)$ $-y$ is the base, 4 is the exponent

$-y^4 = -y \cdot y \cdot y \cdot y$ y is the base, 4 is the exponent

Rules for exponents:

If m and n represent integers, then

Product rule: $x^m x^n = x^{m+n}$

Power rule: $(x^m)^n = x^{mn}$

Power of a product rule: $(xy)^n = x^n y^n$

Power of a quotient rule: $\left(\dfrac{x}{y}\right)^n = \dfrac{x^n}{y^n}$ $(y \neq 0)$

Quotient rule: $\dfrac{x^m}{x^n} = x^{m-n}$ $(x \neq 0)$

$a^3 a^4 = a^{3+4} = a^7$

$(a^3)^4 = a^{3 \cdot 4} = a^{12}$

$(ab)^3 = (ab)(ab)(ab) = a \cdot a \cdot a \cdot b \cdot b \cdot b = a^3 b^3$

$\left(\dfrac{a}{b}\right)^3 = \dfrac{a}{b} \cdot \dfrac{a}{b} \cdot \dfrac{a}{b} = \dfrac{a^3}{b^3}$

$\dfrac{a^7}{a^3} = a^{7-3} = a^4$

REVIEW EXERCISES

Write each expression without using exponents.

1. $-3x^4$
2. $\left(\dfrac{1}{2}pq\right)^3$

Evaluate each expression.

3. 5^3
4. $(-8)^2$
5. -8^2
6. $(5-3)^2$

Simplify each expression.

7. $x^3 x^2$
8. $-3y(y^5)$
9. $(y^7)^3$
10. $(3x)^4$
11. $b^3 b^4 b^5$
12. $-z^2(z^3 y^2)$
13. $(-16s)^2 s$
14. $(2x^2 y)^2$

15. $(x^2 x^3)^3$
16. $\left(\dfrac{x^2 y}{xy^2}\right)^2$
17. $\dfrac{x^7}{x^3}$
18. $\dfrac{(5y^2 z^3)^3}{(yz)^5}$

Find the area or the volume of each figure.

19.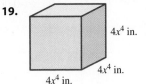
$4x^4$ in.
$4x^4$ in.
$4x^4$ in.

20.
y^2 m
y^2 m

SECTION 5.2 Zero and Negative Integer Exponents

DEFINITIONS AND CONCEPTS

Rules for exponents: For any nonzero real numbers x and y and any integers m and n,

Zero exponent: $x^0 = 1$

Negative exponents: $x^{-n} = \dfrac{1}{x^n}$

Negative to positive rules:

$$\dfrac{1}{x^{-n}} = x^n \qquad \dfrac{x^{-m}}{y^{-n}} = \dfrac{y^n}{x^m}$$

Negative exponents and reciprocals:

$$\left(\dfrac{x}{y}\right)^{-n} = \left(\dfrac{y}{x}\right)^n$$

EXAMPLES

$5^0 = 1 \qquad \left(\dfrac{a}{4}\right)^0 = 1 \qquad (x+4)^0 = 1$

$5^{-2} = \dfrac{1}{5^2} = \dfrac{1}{25}$

$\dfrac{1}{5^{-2}} = 5^2 = 25 \qquad \dfrac{4^{-3}}{5^{-2}} = \dfrac{5^2}{4^3} = \dfrac{25}{64}$

$\left(\dfrac{2}{3}\right)^{-3} = \left(\dfrac{3}{2}\right)^3 = \dfrac{3^3}{2^3} = \dfrac{27}{8}$

REVIEW EXERCISES

Write each expression without using negative or zero exponents or parentheses.

21. x^0

22. $(3x^2y^2)^0$

23. $(3x^0)^2$

24. 10^{-3}

25. $\left(\dfrac{3}{4}\right)^{-1}$

26. -5^{-2}

27. x^{-5}

28. $-6y^4y^{-5}$

29. $\dfrac{x^{-3}}{x^7}$

30. $(x^{-3}x^{-4})^{-2}$

31. $\left(\dfrac{x^2}{x}\right)^{-5}$

32. $\left(\dfrac{3z^4}{z^3}\right)^{-3}$

Write each expression with a single exponent.

33. $y^{3n}y^{4n}$

34. $\dfrac{z^{8c}}{z^{10c}}$

SECTION 5.3 Scientific Notation

DEFINITIONS AND CONCEPTS

A number is written in **scientific notation** if it is written as the product of a number between 1 (including 1) and 10 and an integer power of 10.

EXAMPLES

Standard Notation	Scientific Notation
93,000,000	9.3×10^7
0.000375	3.75×10^{-4}

REVIEW EXERCISES

Write each number in scientific notation.

35. 728

36. 9,370,000

37. 0.0136

38. 0.00942

39. 0.018×10^{-2}

40. 753×10^3

Write each number in standard notation.

41. 7.26×10^5

42. 3.91×10^{-4}

43. 2.68×10^0

44. 5.76×10^1

Simplify each fraction by first writing each number in scientific notation. Then perform the arithmetic. Express the result in standard notation.

45. $\dfrac{(0.00012)(0.00004)}{0.00000016}$

46. $\dfrac{(4,800)(20,000)}{600,000}$

47. WORLD POPULATION In the year 2013, the world's population was estimated to be 7.098 billion. Write this number in standard notation and in scientific notation. (Source: United States Census Bureau)

48. ATOMS The illustration shows a cross section of an atom. How many nuclei, placed end to end, would it take to stretch across the atom?

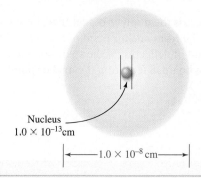

Nucleus 1.0×10^{-13} cm

1.0×10^{-8} cm

SECTION 5.4   Polynomials

DEFINITIONS AND CONCEPTS	EXAMPLES
A **polynomial** is a term or a sum of terms in which all variables have whole-number exponents. No variable appears in a denominator.	Polynomials: $3x$, $2x^2 - x + 7$, $5x^3 + 7x^4y - 3y^4$ Not polynomials: $y^2 - y^{-5}$, $4a - \dfrac{3}{a^2} + 5$
The **degree of a monomial** ax^n is n. The **degree of a monomial** in several variables is the sum of the exponents on those variables. The **degree of a polynomial** is the same as the degree of its term with the largest degree.	The polynomials above have degrees of 1, 2, and 5, respectively.
If $f(x)$ is a polynomial function in x, then $f(3)$ is the value of the function when $x = 3$.	If $f(x) = 3x^2 + x - 2$, then $f(2) = 3(2)^2 + 2 - 2$ $f(-1) = 3(-1)^2 + (-1) - 2$ $f(2) = 3(4) + 2 - 2$ $f(-1) = 3(1) - 1 - 2$ $f(2) = 12 + 0$ $f(-1) = 3 - 1 - 2$ $f(2) = 12$ $f(-1) = 0$

REVIEW EXERCISES

Determine whether each expression is a polynomial.

49. $x^3 - x^2 - x - 1$

50. $x^{-2} - x^{-1} - 1$

51. $\dfrac{11}{y} + 4y$

52. $-16x^2y + 5xy^2$

Consider the polynomial $3x^3 - x^2 + x + 10$.

53. How many terms does the polynomial have?

54. What is the leading term?

55. What is the coefficient of the second term?

56. What is the constant term?

Find the degree of each polynomial and classify it as a monomial, binomial, trinomial, or none of these.

57. $13x^7$

58. $-16a^2b$

59. $5^3x + x^2$

60. $-3x^5 + x - 1$

61. $9xy^2 + 21x^3y^3$

62. $4s^4 - 3s^2 + 5s + 4$

Let $f(x) = 3x^2 + 2x + 1$. Find each value.

63. $f(3)$

64. $f(0)$

65. $f(-2)$

66. $f(-0.2)$

67. DIVING The number of inches that the woman deflects the diving board is given by the function

$$f(x) = 0.1875x^2 - 0.0078125x^3$$

where x is the number of feet that she stands from the front anchor point of the board. Find the amount of deflection if she stands on the end of the diving board, 8 feet from the anchor point.

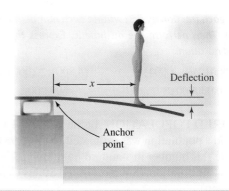

SECTION 5.5 Adding and Subtracting Polynomials

DEFINITIONS AND CONCEPTS	EXAMPLES	
To add (or subtract) polynomials, remove parentheses, and add (or subtract) like terms by combining the numerical coefficients and using the same variables and the same exponents.	Add: $(3x^2 + 2x - 4) + (2x^2 - 5x + 8)$ $= 3x^2 + 2x - 4 + 2x^2 - 5x + 8$ $= 3x^2 + 2x^2 + 2x - 5x - 4 + 8$ $= 5x^2 - 3x + 4$	Subtract: $(5a^2 - a + 2) - (a^2 + a - 3)$ $= 5a^2 - a + 2 - a^2 - a + 3$ $= 5a^2 - a^2 - a - a + 2 + 3$ $= 4a^2 - 2a + 5$
To add (or subtract) polynomials when a monomial precedes parentheses, we can use the distributive property to remove the parentheses and add (or subtract) the polynomials.	Simplify: $-2(4a^2 + 3a + 5) - 4(3a^2 - 7)$ $= -8a^2 - 6a - 10 - 12a^2 + 28$ $= -8a^2 - 12a^2 - 6a - 10 + 28$ $= -20a^2 - 6a + 18$	Distribute the multiplication of -2 and -4. Write like terms together. Combine like terms.

REVIEW EXERCISES

Simplify each expression.

68. $3x^6 + 5x^5 - x^6$

69. $x^2y^2 - 3x^2y^2$

70. $(3x^2 + 2x) + (5x^2 - 8x)$

71. $3(9x^2 + 3x + 7) - 2(11x^2 - 5x + 9)$

Perform the operations.

72. $\quad 2x^2 + 5x + 2$
$\quad\underline{+\ \ x^2 - 3x + 6}$

73. $\quad 20x^3 \qquad\quad + 12x$
$\quad\underline{-\ 12x^3 + 7x^2 - \ \ 7x}$

SECTION 5.6 Multiplying Polynomials

DEFINITIONS AND CONCEPTS	EXAMPLES
To multiply two monomials, multiply the numerical factors and then multiply the variable factors.	Multiply: $(3a^3b^2)(-2a^2b^3) = (3)(-2)a^3a^2b^2b^3$ $= -6a^5b^5$
To multiply a polynomial with more than one term by a monomial, multiply each term of the polynomial by the monomial and simplify.	Multiply: $-5a^4b(2a^2 - 3a + 2) = (-5a^4b)(2a^2) - (-5a^4b)(3a) + (-5a^4b)(2)$ $= -10a^6b + 15a^5b - 10a^4b$

Chapter 5 Summary and Review

To multiply two binomials, use the FOIL method.	Multiply: $(2a + 3b)(3a - 4b) = (2a)(3a) + (2a)(-4b) + (3b)(3a) + (3b)(-4b)$ $= 6a^2 - 8ab + 9ab - 12b^2$ $= 6a^2 + ab - 12b^2$
Special products: $(x + y)^2 = x^2 + 2xy + y^2$ $(x - y)^2 = x^2 - 2xy + y^2$ $(x + y)(x - y) = x^2 - y^2$	Multiply: $(a + 2b)^2 = a^2 + 2(a)(2b) + (2b)^2 = a^2 + 4ab + 4b^2$ $(a - 2b)^2 = a^2 + 2(a)(-2b) + (-2b)^2 = a^2 - 4ab + 4b^2$ $(a + 2b)(a - 2b) = a^2 - (2b)^2 = a^2 - 4b^2$
To multiply one polynomial by another, multiply each term of one polynomial by each term of the other polynomial, and simplify.	Multiply: $(a + 2)(a^2 + 3a - 4)$ $= a(a^2) + a(3a) + a(-4) + 2(a^2) + 2(3a) + 2(-4)$ $= a^3 + 3a^2 - 4a + 2a^2 + 6a - 8$ $= a^3 + 5a^2 + 2a - 8$

REVIEW EXERCISES

Find each product.

74. $(2x^2)(5x)$
75. $(-6x^4z^3)(x^6z^2)$
76. $(2rst)(-3r^2s^3t^4)$
77. $5b^3 \cdot 6b^2 \cdot 4b^6$
78. $5(x + 3)$
79. $x^2(3x^2 - 5)$
80. $x^2y(y^2 - xy)$
81. $-2y^2(y^2 - 5y)$
82. $2x(3x^4)(x + 2)$
83. $-3x(x^2 - x + 2)$
84. $(x + 3)(x + 2)$
85. $(2x + 1)(x - 1)$
86. $(3a - 3)(2a + 2)$
87. $6(a - 1)(a + 1)$
88. $(a - b)(2a + b)$
89. $(-3x - y)(2x + y)$
90. $(x + 3)(x + 3)$
91. $(x + 5)(x - 5)$
92. $(a - 3)^2$
93. $(x + 4)^2$
94. $(-2y + 1)^2$
95. $(y^2 + 1)(y^2 - 1)$
96. $(3x + 1)(x^2 + 2x + 1)$
97. $(2a - 3)(4a^2 + 6a + 9)$

Solve each equation.

98. $x^2 + 3 = x(x + 3)$
99. $x^2 + x = (x + 1)(x + 2)$
100. $(x + 2)(x - 5) = (x - 4)(x - 1)$
101. $(x + 5)(3x + 1) = x^2 + (2x - 1)(x - 5)$
102. APPLIANCE Find polynomials that represent the perimeter of the base, the area of the base, and the volume occupied by the dishwasher shown below.

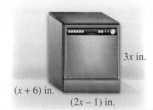

$3x$ in.
$(x + 6)$ in.
$(2x - 1)$ in.

SECTION 5.7 Dividing Polynomials by Monomials

DEFINITIONS AND CONCEPTS	EXAMPLES
To divide monomials, use the method for simplifying fractions or use the rules for exponents.	$\dfrac{10}{15} = \dfrac{5 \cdot 2}{5 \cdot 3} = \dfrac{2}{3}$ $\dfrac{12a^2b^3}{8a^4b^2} = \dfrac{3 \cdot 4 \cdot a \cdot a \cdot b \cdot b \cdot b}{2 \cdot 4 \cdot a \cdot a \cdot a \cdot a \cdot b \cdot b} = \dfrac{3b}{2a^2}$ Remove common factors.

To **divide a polynomial by a monomial**, divide each term of the numerator by the denominator.

$$\frac{6a^2b + 8ab^2}{2ab} = \frac{6a^2b}{2ab} + \frac{8ab^2}{2ab}$$ Divide each term of the polynomial in the numerator by the denominator, $2ab$.

$$= \frac{3 \cdot 2 \cdot a \cdot a \cdot b}{2 \cdot a \cdot b} + \frac{4 \cdot 2 \cdot a \cdot b \cdot b}{2 \cdot a \cdot b} = 3a + 4b$$

REVIEW EXERCISES

Simplify each expression.

103. $\dfrac{-14x^2y}{21xy^3}$

104. $\dfrac{(x^2)^2}{xx^4}$

107. $\dfrac{15a^2b + 20ab^2 - 25ab}{5ab}$

108. $\dfrac{(x+y)^2 + (x-y)^2}{-2xy}$

Perform each division. All variables represent positive numbers.

105. $\dfrac{8x+6}{2}$

106. $\dfrac{14xy - 21x}{7xy}$

109. **SAVINGS BONDS** How many $50 savings bonds would have a total value of $(50x + 250)$?

SECTION 5.8 Dividing Polynomials by Polynomials

DEFINITIONS AND CONCEPTS	EXAMPLES
Long division is used to **divide one polynomial by another**. When a division has a remainder, write the answer in the form $$\text{Quotient} + \frac{\text{remainder}}{\text{divisor}}$$	Divide: $\quad\quad\quad\quad x + 2 + \frac{1}{x+1}$ $x+1\overline{)x^2 + 3x + 3}$ $\quad\;\underline{-(x^2 + x)}$ $\quad\quad\quad\; 2x + 3$ $\quad\quad\;\;\underline{-(2x + 2)}$ $\quad\quad\quad\quad\quad 1$ The first division: $\frac{x^2}{x} = x$. The second division: $\frac{2x}{x} = 2$. The remainder is 1.
The division method works best when the exponents of the terms of the divisor and the dividend are **written in descending order**.	$\quad\quad\quad\quad\quad\quad\quad\quad a - 2b$ $a+b\overline{)a^2 - 2b^2 - ab} \longrightarrow a+b\overline{)a^2 - ab - 2b^2}$ $\quad\quad\quad\quad\quad\quad\quad\;\underline{-(a^2 + ab)}$ $\quad\quad\quad\quad\quad\quad\quad\quad\; -2ab - 2b^2$ $\quad\quad\quad\quad\quad\quad\quad\;\underline{-(-2ab - 2b^2)}$ $\quad\quad\quad\quad\quad\quad\quad\quad\quad\quad\quad 0$ The first division: $\frac{a^2}{a} = a$. The second division: $\frac{-2ab}{a} = -2b$.
When the dividend has a **missing term**, write it with a coefficient of zero or leave a blank space.	$\quad\quad\quad\quad\quad\quad\quad\quad\quad\quad\quad\quad 4x - 1$ $4x+1\overline{)16x^2 - 1} \longrightarrow 4x+1\overline{)16x^2 + 0x - 1}$ $\quad\quad\quad\quad\quad\quad\quad\quad\quad\;\underline{-(16x^2 + 4x)}$ $\quad\quad\quad\quad\quad\quad\quad\quad\quad\quad\quad -4x - 1$ $\quad\quad\quad\quad\quad\quad\quad\quad\quad\;\underline{-(-4x - 1)}$ $\quad\quad\quad\quad\quad\quad\quad\quad\quad\quad\quad\quad\quad 0$ The first division: $\frac{16x^2}{4x} = 4x$. The second division: $\frac{-4x}{4x} = -1$.

REVIEW EXERCISES

Perform each division.

110. $x + 2\overline{)x^2 + 3x + 5}$

111. $x - 1\overline{)x^2 - 6x + 5}$

112. $\dfrac{2x^2 + 3 + 7x}{x + 3}$

113. $\dfrac{3x^2 + 14x - 2}{3x - 1}$

114. $2x - 1\overline{)6x^3 + x^2 + 1}$

115. $3x + 1\overline{)-13x - 4 + 9x^3}$

116. Use multiplication to show that the answer when dividing $3y^2 + 11y + 6$ by $y + 3$ is $3y + 2$.

117. **ZOOLOGY** The distance in inches traveled by a certain type of snail in $(2x - 1)$ minutes is given by the polynomial $8x^2 + 2x - 3$. At what rate did the snail travel?

CHAPTER 5 TEST

1. Fill in the blanks.
   a. In the expression y^{10}, the _____ is y and the _____ is 10.
   b. We call a polynomial with exactly one term a _____, with exactly two terms a _____, and with exactly three terms a _____.
   c. The _____ of a term of a polynomial in one variable is the value of the exponent on the variable.
   d. $(x+y)^2$, $(x-y)^2$, and $(x+y)(x-y)$ are called _____ products.

2. Use exponents to rewrite $2xxxyyyy$.

Write each expression as an expression containing only one exponent.

3. $y^2(yy^3)$

4. $(2x^3)^5(x^2)^3$

Simplify each expression. Write answers without using parentheses or negative exponents.

5. $3x^0$

6. $2y^{-5}y^2$

7. $\dfrac{y^2}{yy^{-2}}$

8. $\left(\dfrac{a^2b^{-1}}{4a^3b^{-2}}\right)^{-3}$

9. $\left(\dfrac{8}{m^6}\right)^{-2}$

10. $\dfrac{-6a}{b^{-9}}$

11. What is the volume of a cube that has sides of length $10y^4$ inches?

12. Rewrite 4^{-2} using a positive exponent and then evaluate the result.

13. a. ELECTRICITY One ampere (amp) corresponds to the flow of 6,250,000,000,000,000,000 electrons per second past any point in a direct current (DC) circuit. Write this number in scientific notation.

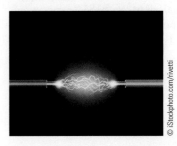

   b. BIOLOGY The length of a paramecium is 2.5×10^{-4} meters. Write that number in standard notation.

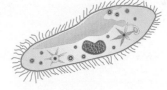

14. SPEED OF LIGHT Light travels 1.86×10^5 miles per second. How far does it travel in a minute? Express the answer in scientific notation.

15. Identify $3x^2 + 2$ as a monomial, binomial, or trinomial.

16. Identify $x^4 + 8x^2 - 12$ as a monomial, binomial, or trinomial. Then complete the table.

Term	Coefficient	Degree

Degree of the polynomial: ▊

17. Find the degree of the polynomial: $3x^2y^3 + 2x^3y - 5x^2y$

18. FREE FALL A visitor standing on the rim of the Grand Canyon drops a rock over the side. The distance (in feet) that the rock is from the canyon floor t seconds after being dropped is given by the polynomial $-16t^2 + 5{,}184$. Find the position of the rock 18 seconds after being dropped. Explain your answer.

19. Let $f(x) = x^2 + x - 2$. Find: $f(-2)$.

20. Simplify: $(xy)^2 + 5x^2y^2 - (3x)^2y^2$

21. Simplify: $-6(x - y) + 2(x + y) - 3(x + 2y)$

22. Subtract:
$$\begin{array}{r} 2x^2 - 7x + 3 \\ -\underline{3x^2 - 2x - 1} \end{array}$$

23. Find a polynomial that represents the perimeter of the rectangle.

$(5a^2 + 3a - 1)$ in.

$(a - 9)$ in.

24. ICE CHESTS. Write a polynomial that represents
  a. The area of the base of the ice chest.
  b. The volume of the ice chest.

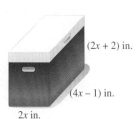

Find each product.

25. $(-2x^3)(2x^2y)$

26. $3y^2(y^2 - 2y + 3)$

27. $(x - 9)(x + 9)$

28. $(3y - 4)^2$

29. $(2x - 5)(3x + 4)$

30. $(2x - 3)(x^2 - 2x + 4)$

31. Solve: $(a + 2)^2 = (a - 3)^2$

32. Simplify: $\dfrac{8x^2y^3z^4}{16x^3y^2z^4}$

33. Simplify: $\dfrac{6a^2 - 12b^2}{24ab}$

34. Divide: $2x + 3 \overline{)2x^2 - x - 6}$

35. Divide: $2x - 1 \overline{)1 + x^2 + 6x^3}$

36. Divide: $\dfrac{x^2 - 4}{x + 2}$

37. In your own words, explain this rule for exponents:
$$x^{-n} = \dfrac{1}{x^n}$$

38. A rectangle has an area of $(x^2 - 6x + 5)$ ft² and a length of $(x - 1)$ feet. Show how long division can be used to find the width of the rectangle. Explain your steps.

CHAPTERS 1–5 CUMULATIVE REVIEW

1. Use exponents to write the prime factorization of 270. [Section 1.2]

2. **a.** Use the variables a and b to state the commutative property of addition. [Section 1.4]
   b. Use the variables x, y, and z to state the associative property of multiplication. [Section 1.6]

Evaluate each expression.

3. $3 - 4[-10 - 4(-5)]$ [Section 1.7]

4. $\dfrac{|-45| - 2(-5) + 1^5}{2 \cdot 9 - 2^4}$ [Section 1.7]

Simplify each expression.

5. $27\left(\dfrac{2}{3}x\right)$ [Section 1.9]

6. $3x^2 + 2x^2 - 5x^2$ [Section 1.9]

Solve each equation.

7. $2 - (4x + 7) = 3 + 2(x + 2)$ [Section 2.2]

8. $\dfrac{2}{5}y + 3 = 9$ [Section 2.2]

9. CANDY SALES The circle graph shows how $7.1 billion in seasonal candy sales for 2012 was spent. Find the candy sales for Halloween. [Section 2.3]

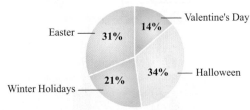

Source: National Confectioners Association

10. AIR CONDITIONING Find the volume of air contained in the duct. Round to the nearest tenth of a cubic foot. [Section 2.4]

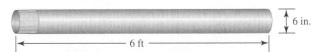

11. ANGLE OF ELEVATION Find x. [Section 2.5]

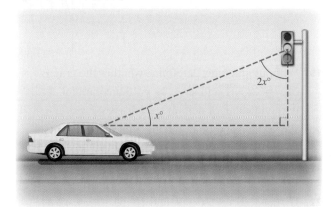

12. LIVESTOCK AUCTION A farmer is going to sell one of her prize hogs at an auction and would like to make $6,000 after paying a 4% commission to the auctioneer. At what selling price will the farmer make this amount of money? [Section 2.5]

13. STOCK MARKET An investment club invested part of $45,000 in a high-yield mutual fund that earned 12% annual simple interest. The remainder of the money was invested in Treasury bonds that earned 6.5% simple annual interest. The two investments earned $4,300 in one year. How much was invested in each account? [Section 2.6]

14. Solve $-4x + 6 > 17$ and graph the solution set. Then describe the graph using interval notation. [Section 2.7]

Graph each equation.

15. $y = 3x$ [Section 3.2]   16. $x = -2$ [Section 3.3]

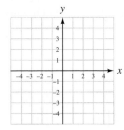 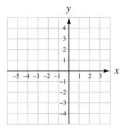

17. Find the slope of the line passing through $(6, -2)$ and $(-3, 2)$. [Section 3.4]

18. Find the slope and y-intercept of the line. Then write the equation of the line. [Section 3.5]

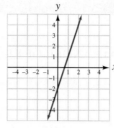

19. Without graphing, determine whether the graphs of $y = \frac{3}{2}x - 1$ and $2x + 3y = 10$ are parallel, perpendicular, or neither. [Section 3.5]

20. Write the equation of the line that passes through $(-2, 10)$ with slope -4. Write the result in slope–intercept form. [Section 3.6]

21. Is $(-2, 1)$ a solution of $2x - 3y \geq -6$? [Section 3.7]

22. If $f(x) = 2x^2 + 3x - 9$, find $f(-5)$. [Section 3.8]

23. Is $\left(\frac{2}{3}, -1\right)$ a solution of the system $\begin{cases} y = -3x + 1 \\ 3x + 3y = -2 \end{cases}$? [Section 4.1]

24. Solve the system by graphing:
$\begin{cases} 3x + 2y = 14 \\ y = \frac{1}{4}x \end{cases}$
[Section 4.1]

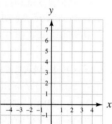

25. Solve the system $\begin{cases} 2b - 3a = 18 \\ a + 3b = 5 \end{cases}$ by substitution. [Section 4.2]

26. Solve $\begin{cases} 8s + 10t = 24 \\ 11s - 3t = -34 \end{cases}$ by addition (elimination). [Section 4.3]

27. VACATIONS One-day passes to Universal Studios Hollywood cost a family of 5 (2 adults and 3 children) $396. A family of 6 (3 adults and 3 children) paid $480 for their one-day passes. Find the cost of an adult one-day pass and a child's one-day pass to Universal Studios. [Section 4.4]

28. Graph: $\begin{cases} y \leq 2x - 1 \\ x + 3y > 6 \end{cases}$
[Section 4.5]

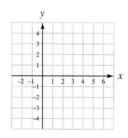

Simplify. Do not use negative exponents in the answer.

29. $(-3x^2y^4)^2$ [Section 5.1]

30. $(v^5)^2(v^3)^4$ [Section 5.1]

31. $ab^3c^4 \cdot ab^4c^2$ [Section 5.1]

32. $\left(\dfrac{4t^3t^4t^5}{3t^2t^6}\right)^3$ [Section 5.1]

33. $(2y)^{-4}$ [Section 5.2]

34. $\dfrac{a^4b^0}{a^{-3}}$ [Section 5.2]

35. -5^{-2} [Section 5.2]

36. $\left(\dfrac{a}{x}\right)^{-10}$ [Section 5.2]

Write each number in scientific notation.

37. 615,000 [Section 5.3]

38. 0.0000013 [Section 5.3]

39. Graph: $y = x^2$ [Section 5.4]

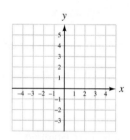

40. MUSICAL INSTRUMENTS The amount of deflection of the horizontal beam (in inches) is given by the polynomial $0.01875x^4 - 0.15x^3 + 1.2x$, where x is the distance (in feet) that the gong is hung from one end of the beam. Find the deflection if the gong is hung in the middle of the support. [Section 5.4]

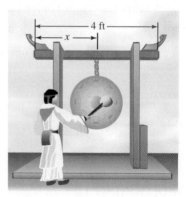

Perform the operations.

41. $(4c^2 + 3c - 2) + (3c^2 + 4c + 2)$ [Section 5.5]

42. Subtract: $\quad 17x^4 - 3x^2 - 65x - 12$
$\underline{\quad\; -23x^4 + 14x^2 + \;\;3x - 23}$
[Section 5.5]

43. $(2t + 3s)(3t - s)$ [Section 5.6]

44. $3x(2x + 3)^2$ [Section 5.7]

45. $\dfrac{2x - 32}{16x}$ [Section 5.8]

46. $5x + 3\overline{)11x + 10x^2 + 3}$ [Section 5.8]

Factoring and Quadratic Equations

6

6.1 The Greatest Common Factor; Factoring by Grouping
6.2 Factoring Trinomials of the Form $x^2 + bx + c$
6.3 Factoring Trinomials of the Form $ax^2 + bx + c$
6.4 Factoring Perfect-Square Trinomials and the Difference of Two Squares
6.5 Factoring the Sum and Difference of Two Cubes
6.6 A Factoring Strategy
6.7 Solving Quadratic Equations by Factoring
6.8 Applications of Quadratic Equations
Chapter Summary and Review
Chapter Test
Cumulative Review

© iofoto/Shutterstock.com

from Campus to Careers
Elementary School Teacher

It has been said that a teacher takes a hand, opens a mind, and touches a heart. That is certainly true for the thousands of dedicated elementary school teachers across the country. Elementary school teachers use their training in mathematics in many ways. Besides teaching math on a daily basis, they calculate student grades, analyze test results, and order instructional materials and supplies. They use measurement and geometry for designing bulletin board displays and they construct detailed schedules so that the classroom time is used wisely.

In **Problem 28** of **Study Set 6.8**, you will determine the maximum dimensions of a bulletin board so that it meets the fire code requirements of an elementary school classroom.

JOB TITLE: Elementary School Teacher
EDUCATION: A bachelor's degree and completion of an approved teacher training program
JOB OUTLOOK: Job opportunities should increase by 17% from 2010 to 2020.
ANNUAL EARNINGS: The U.S. median salary in 2012 was $53,400*.
*Can vary greatly by region and experience.
FOR MORE INFORMATION:
www.bls.gov/oes/current/oes252021.htm

Objectives

1. Find the greatest common factor of a list of terms.
2. Factor out the greatest common factor.
3. Factor by grouping.

SECTION 6.1
The Greatest Common Factor; Factoring by Grouping

Recall that in Chapter 5 we used the distributive property to multiply a monomial and a binomial. For example,

$$4y(3y + 5) = 4y \cdot 3y + 4y \cdot 5$$
$$= 12y^2 + 20y$$

In this section, we will reverse the operation of multiplication. Given a polynomial such as $12y^2 + 20y$, we will ask, "What factors were multiplied to obtain $12y^2 + 20y$?" The process of finding the factors of a known product is called **factoring**.

Multiplication: Given the factors, we find the polynomial. →
$$4y(3y + 5) = 12y^2 + 20y$$
← Factoring: Given a polynomial, we find the factors.

To **factor a polynomial** means to express it as a product of two (or more) polynomials. The first step when factoring a polynomial is to determine whether its terms have any common factors.

1 Find the greatest common factor of a list of terms.

To determine whether two or more integers have common factors, it is helpful to write them as a product of prime numbers. For example, the prime factorizations of 90 and 42 are given below.

$$90 = \mathbf{2} \cdot \mathbf{3} \cdot 3 \cdot 5$$
$$42 = \mathbf{2} \cdot \mathbf{3} \cdot 7$$

The color highlighting indicates that 90 and 42 have one prime factor of 2 and one prime factor of 3 in common. We can conclude that $2 \cdot 3 = 6$ is the largest natural number that divides 90 and 42 exactly, and we say that 6 is their *greatest common factor (GCF)*.

$$\frac{90}{6} = 15 \quad \text{and} \quad \frac{42}{6} = 7$$

The Greatest Common Factor (GCF)

The **greatest common factor (GCF)** of a list of integers is the largest common factor of those integers.

Self Check 1

Find the GCF of:
a. 45, 60, and 75
b. 16, 28, and 35

Now Try Problems 23 and 25

EXAMPLE 1 Find the GCF of each list of numbers:

a. 24, 60, and 96   b. 6, 35, and 50

Strategy We will prime factor each number in the list. Then we will identify the common prime factors and find their product.

WHY The product of the common prime factors is the GCF of the numbers in the list.

Solution

a. The prime factors of the three numbers are shown:

$24 = \mathbf{2 \cdot 2} \cdot 2 \cdot \mathbf{3}$ This can be written as $2^3 \cdot 3$.
$60 = \mathbf{2 \cdot 2 \cdot 3} \cdot 5$ This can be written as $2^2 \cdot 3 \cdot 5$.
$96 = \mathbf{2 \cdot 2} \cdot 2 \cdot 2 \cdot 2 \cdot \mathbf{3}$ This can be written as $2^5 \cdot 3$.

The highlighting shows that 24, 60, and 96 each have two factors of 2 and one factor of 3; their greatest common factor is $\mathbf{2 \cdot 2 \cdot 3} = 12$.

b. The prime factorization of each number is shown:

$6 = 2 \cdot 3$
$35 = 5 \cdot 7$
$50 = 2 \cdot 5 \cdot 5$

Since there are no prime factors common to 6, 35, and 50, their GCF is 1.

> **Success Tip** The exponent on any factor in a GCF is the *smallest* exponent that appears on that factor in all of the numbers under consideration.

Strategy for Finding the GCF

1. Write each coefficient as a product of prime factors.
2. Identify the numerical and variable factors common to each term.
3. Multiply the common numerical and variable factors identified in step 2 to obtain the GCF. If there are no common factors, the GCF is 1.

EXAMPLE 2   Find the GCF of each list of terms:
a. $12x^2$ and $20x$ **b.** $9a^5b^2$, $15a^4b^2$, and $90a^3b^3$

Strategy We will prime factor each coefficient of each term in the list. Then we will identify the numerical and variable factors common to each term and find their product.

WHY The product of the common factors is the GCF of the terms in the list.

Solution

a. Step 1 We write each coefficient, 12 and 20, as a product of prime factors. Recall that an exponent, as in x^2, indicates repeated multiplication.

$12x^2 = \mathbf{2 \cdot 2} \cdot 3 \cdot \mathbf{x} \cdot x$ This can be written as $2^2 \cdot 3 \cdot x^2$.
$20x = \mathbf{2 \cdot 2} \cdot 5 \cdot \mathbf{x}$ This can be written as $2^2 \cdot 5 \cdot x$.

Step 2 There are two common factors of 2 and one common factor of x.

Step 3 We multiply the common factors, 2, 2, and x, to obtain the GCF.

$\text{GCF} = \mathbf{2 \cdot 2 \cdot x} = 2^2 \cdot x = 4x$

b. Step 1 We write the coefficients, 9, 15, and 90, as products of primes. The exponents on the variables represent repeated multiplication.

$9a^5b^2 = \mathbf{3 \cdot 3} \cdot \mathbf{a \cdot a \cdot a} \cdot a \cdot a \cdot \mathbf{b \cdot b}$ This can be written as $3^2 \cdot a^5 \cdot b^2$.
$15a^4b^2 = \mathbf{3} \cdot 5 \cdot \mathbf{a \cdot a \cdot a} \cdot a \cdot \mathbf{b \cdot b}$ This can be written as $3 \cdot 5 \cdot a^4 \cdot b^2$.
$90a^3b^3 = 2 \cdot \mathbf{3 \cdot 3} \cdot 5 \cdot \mathbf{a \cdot a \cdot a \cdot b \cdot b} \cdot b$ This can be written as $2 \cdot 3^2 \cdot 5 \cdot a^3 \cdot b^3$.

Self Check 2

Find the GCF of each list of terms:
a. $33c$ and $22c^4$
b. $42s^3t^2$, $63s^2t^4$, and $21s^3t^3$

Now Try Problems 29 and 31

Step 2 The highlighting shows one common factor of 3, three common factors of a, and two common factors of b.

Step 3 GCF = $3 \cdot a \cdot a \cdot a \cdot b \cdot b = 3a^3b^2$

2 Factor out the greatest common factor.

To factor $12y^2 + 20y$, we find the GCF of $12y^2$ and $20y$ (which is $4y$) and use the distributive property in reverse: $ab + ac = a(b + c)$.

$12y^2 + 20y = \mathbf{4y} \cdot 3y + \mathbf{4y} \cdot 5$ Write each term of the polynomial as the product of the GCF, 4y, and one other factor.

$= \mathbf{4y}(3y + 5)$ 4y is a common factor of both terms.

This process is called **factoring out the greatest common factor**.

Self Check 3

Factor: $18x - 24$

Now Try Problem 37

EXAMPLE 3   Factor: $25 - 5m$

Strategy We will prime factor each coefficient of each term in the polynomial. Then we will write each term of the polynomial as a product of the GCF and one other factor.

WHY We can then use the distributive property to factor out the GCF.

Solution
The prime factorizations are shown:

$\left. \begin{array}{l} 25 = \mathbf{5} \cdot 5 \\ 5m = \mathbf{5} \cdot m \end{array} \right\}$ GCF = 5

We can use the distributive property in reverse to factor out the GCF.

$25 - 5m = \mathbf{5} \cdot 5 - \mathbf{5} \cdot m$ Factor each monomial using 5 and one other factor.

$= 5(5 - m)$ Factor out the common factor of 5.

To check, we multiply: $5(5 - m) = 5 \cdot 5 - 5 \cdot m = 25 - 5m$.

Self Check 4

Factor: $32x^2y^4 + 12x^3y^3$

Now Try Problem 45

EXAMPLE 4   Factor: $35a^3b^2 + 14a^2b^3$

Strategy We will prime factor each coefficient of each term in the polynomial. Then we will write each term of the polynomial as a product of the GCF and one other factor.

WHY We can then use the distributive property to factor out the GCF.

Solution
The prime factorizations of $35a^3b^2$ and $14a^2b^3$ are shown:

$\left. \begin{array}{l} 35a^3b^2 = 5 \cdot \mathbf{7} \cdot \mathbf{a} \cdot \mathbf{a} \cdot a \cdot \mathbf{b} \cdot \mathbf{b} \\ 14a^2b^3 = 2 \cdot \mathbf{7} \cdot \mathbf{a} \cdot \mathbf{a} \cdot \mathbf{b} \cdot \mathbf{b} \cdot b \end{array} \right\}$ GCF = $7 \cdot a \cdot a \cdot b \cdot b = 7a^2b^2$

We factor out the GCF, $7a^2b^2$.

$35a^3b^2 + 14a^2b^3 = \mathbf{7a^2b^2} \cdot 5a + \mathbf{7a^2b^2} \cdot 2b$

$= 7a^2b^2(5a + 2b)$ Factor out the GCF $7a^2b^2$.

To check, we multiply: $7a^2b^2(5a + 2b) = 35a^3b^2 + 14a^2b^3$.

Caution! If the GCF of the terms of a polynomial is the same as one of the terms, leave a 1 in place of that term when factoring out the GCF.

EXAMPLE 5   Factor: $4x^3y^2z - 2x^2yz + xz$

Strategy We will prime factor each coefficient of each term in the polynomial. Then we will write each term of the polynomial as a product of the GCF and one other factor.

WHY We can then use the distributive property to factor out the GCF.

Solution
The expression has three terms. We factor out the GCF, which is xz.

$$4x^3y^2z - 2x^2yz + xz = xz \cdot 4x^2y^2 - xz \cdot 2xy + xz \cdot 1$$
$$= xz(4x^2y^2 - 2xy + 1) \quad \text{Factor out the GCF } xz.$$

The last term of $4x^3y^2z - 2x^2yz + xz$ has an implied coefficient of 1. When xz is factored out, we must write this coefficient of 1, as shown in blue. To check, we multiply: $xz(4x^2y^2 - 2xy + 1) = 4x^3y^2z - 2x^2yz + xz$.

Self Check 5
Factor: $2ab^2c + 4a^2bc - ab$
Now Try Problem 51

EXAMPLE 6   Factor: $x(x + 4) + 3(x + 4)$

Strategy We will identify the terms of the expression and find their GCF.

WHY We can then use the distributive property in reverse to factor out the GCF.

Solution
The given expression has two terms:

$\underbrace{x(x + 4)}_{\text{The first term}} + \underbrace{3(x + 4)}_{\text{The second term}}$

The GCF of the terms is $x + 4$, which can be factored out.

$$x(x + 4) + 3(x + 4) = x(x + 4) + 3(x + 4)$$
$$= (x + 4)(x + 3) \quad \text{Factor out the GCF } (x + 4).$$

Self Check 6
Factor: $2y(y - 1) - 7(y - 1)$
Now Try Problem 55

It is often useful to factor out a common factor having a negative coefficient.

EXAMPLE 7   Factor -1 out of $-a^3 + 2a^2 - 4$.

Strategy We will write each term of the polynomial as the product of -1 and one other factor.

WHY We can then use the distributive property in reverse to factor out the -1.

Solution
$$-a^3 + 2a^2 - 4 = (-1)a^3 + (-1)(-2a^2) + (-1)4$$
$$= -1(a^3 - 2a^2 + 4) \quad \text{Factor out } -1.$$
$$= -(a^3 - 2a^2 + 4) \quad \text{The coefficient of 1 need not be written.}$$

We check by multiplying: $-(a^3 - 2a^2 + 4) = -a^3 + 2a^2 - 4$.

Self Check 7
Factor -1 out of $-b^4 - 3b^2 + 2$.
Now Try Problem 61

Self Check 8

Factor out the opposite of the GCF in $-27xy^2 - 18x^2y + 36x^2y^2$.

Now Try Problem 66

EXAMPLE 8 Factor out the opposite of the GCF in $-18a^2b + 6ab^2 - 12a^2b^2$.

Strategy First we will determine the GCF of the terms of the polynomial. Then we will write each term of the polynomial as the product of the opposite of the GCF and one other factor.

WHY We can then use the distributive property to factor out the opposite of the GCF.

Solution

The GCF is $6ab$, the opposite of $6ab$ is $-6ab$. We write each term of the polynomial as the product of $-6ab$ and another factor. Then we factor out $-6ab$.

$$-18a^2b + 6ab^2 - 12a^2b^2 = (-6ab)3a - (-6ab)b + (-6ab)2ab$$

$$= -6ab(3a - b + 2ab) \quad \text{Factor out } -6ab.$$

We check by multiplying: $-6ab(3a - b + 2ab) = -18a^2b + 6ab^2 - 12a^2b^2$.

> **Success Tip** When the first coefficient of a polynomial is negative, factor out the opposite of the GCF.

3 Factoring by grouping.

When a polynomial has four or more terms, we see if we can factor it by arranging the terms in groups that have common factors. This method is called **factoring by grouping**.

> **Factoring by Grouping**
>
> 1. Group the terms of the polynomial so that the first two terms have a common factor and the last two terms have a common factor.
> 2. Factor out the common factor from each group.
> 3. Factor out the resulting common binomial factor. If there is no common binomial factor, regroup the terms of the polynomial and repeat steps 2 and 3.

Self Check 9

Factor: $7x - 7y + xy - y^2$

Now Try Problem 70

EXAMPLE 9 Factor: $2c - 2d + cd - d^2$

Strategy We note that there is no common factor of all four terms. Then we will factor out a common factor from the first two terms and a common factor from the last two terms.

WHY Often this will produce a common binomial factor that can be factored out.

Solution

Since 2 is a common factor of the first two terms and d is a common factor of the last two terms, we have

$$2c - 2d + cd - d^2 = 2(c - d) + d(c - d) \quad \text{Factor out 2 from } 2c - 2d \text{ and } d \text{ from } cd - d^2.$$

$$= (c - d)(2 + d) \quad \text{Factor out } c - d.$$

We check by multiplying:

$$(c - d)(2 + d) = 2c + cd - 2d - d^2$$

$$= 2c - 2d + cd - d^2 \quad \text{Rearrange the terms.}$$

EXAMPLE 10
Factor: **a.** $x^3 + 6x^2 + x + 6$ **b.** $x^2 - ax - x + a$

Strategy We will follow the steps for factoring a four-termed polynomial.

WHY Since the terms of the polynomials do not have a common factor (other than 1), the only option is to attempt to factor these polynomials by grouping.

Solution

a. The first two terms, x^3 and $6x^2$, have a common factor of x^2. The only common factor of the last two terms, x and 6, is 1.

$$x^3 + 6x^2 + x + 6 = x^2(x+6) + 1(x+6)$$
Factor out x^2 from $x^3 + 6x^2$.
Factor out 1 from $x + 6$.
Don't forget the blue $+$ sign.

$$= (x+6)(x^2+1)$$
Factor out the common binomial factor, $x + 6$. Caution: Don't write $(x+6)^2$.

Check the factorization by multiplying.

b. Since x is a common factor of the first two terms, we can factor it out.

$$x^2 - ax - x + a = x(x-a) - x + a$$ Factor out x from $x^2 - ax$.

When factoring four terms by grouping, if the coefficient of the 3rd term is negative, we often factor out a negative coefficient from the last two terms. If we factor -1 from $-x + a$, a common binomial factor $x - a$ appears within the second set of parentheses, which we can factor out.

$$x^2 - ax - x + a = x(x-a) - 1(x-a)$$
Factor out -1, change the sign of $-x$ and a, and write -1 in front of the parentheses.

$$= (x-a)(x-1)$$
Factor out the common factor, $x - a$. Caution: Don't write $(x-a)^2$.

Check by multiplying.

The next example illustrates that when factoring a polynomial, we should always look for a common factor first.

Self Check 10
Factor:
a. $a^5 + 11a^4 + a + 11$
b. $b^2 - bc - b + c$

Now Try Problems 71 and 73

EXAMPLE 11
Factor: $10k + 10m - 2km - 2m^2$

Strategy Since all four terms have a common factor of 2, we factor it out first. Then we will factor the resulting polynomial by grouping.

WHY Factoring out the GCF first makes the factoring process easier.

Solution
$$10k + 10m - 2km - 2m^2 = 2(5k + 5m - km - m^2)$$ Factor out the GCF 2.
$$= 2[5(k+m) - m(k+m)]$$ Factor out 5 from $5k + 5m$. Factor out $-m$ from $-km - m^2$.
$$= 2[(k+m)(5-m)]$$ Factor out $(k+m)$.
$$= 2(k+m)(5-m)$$

Use multiplication to check the result.

Self Check 11
Factor: $-4t - 4s - 4tz - 4sz$

Now Try Problem 77

ANSWERS TO SELF CHECKS

1. a. 15 **b.** 1 **2. a.** 11c **b.** $21s^2t^2$ **3.** $6(3x-4)$ **4.** $4x^2y^3(8y+3x)$
5. $ab(2bc + 4ac - 1)$ **6.** $(y-1)(2y-7)$ **7.** $-(b^4 + 3b^2 - 2)$
8. $-9xy(3y + 2x - 4xy)$ **9.** $(x-y)(7+y)$ **10. a.** $(a+11)(a^4+1)$
b. $(b-c)(b-1)$ **11.** $-4(t+s)(1+z)$

SECTION 6.1 STUDY SET

VOCABULARY

Fill in the blanks.

1. The process of finding the individual factors of a known product is called _____.
2. To _____ a polynomial means to express the polynomial as a product of two (or more) polynomials.
3. The _____ factorization of 12 is $2 \cdot 2 \cdot 3$.
4. The GCF of several integers is the _____ common factor of those integers.
5. When we write $15x^2 - 25x$ as $5x(3x - 5)$, we say that we have _____ _____ the greatest common factor.
6. When a polynomial has four or more terms, we can attempt to factor it by rearranging its terms in groups that have common factors. This process is called factoring by _____.

CONCEPTS

Explain what is wrong with each solution.

7. Factor: $6a + 9b + 3$

$$6a + 9b + 3 = 3(2a + 3b + 0)$$
$$= 3(2a + 3b)$$

8. Factor out the GCF: $30a^3 - 12a^2$

$$30a^3 - 12a^2 = 6a(5a^2 - 2a)$$

9. Factor: $ab + b + a + 1$.

$$ab + b + a + 1 = b(a + 1) + (a + 1)$$
$$= (a + 1)b$$

10. What algebraic concept is illustrated in the work shown below?

$$4 \cdot 5x + 4 \cdot 3 = 4(5x + 3)$$

11. The prime factorizations of three monomials are shown here. Find their GCF.

$$3 \cdot 3 \cdot 5 \cdot x \cdot x$$
$$2 \cdot 3 \cdot 5 \cdot x \cdot y$$
$$2 \cdot 2 \cdot 3 \cdot x \cdot y \cdot y$$

Consider the polynomial $2k - 8 + hk - 4h$.

12. Is there a common factor of all the terms?
13. What is the common factor of the first two terms?
14. What is the common factor of the last two terms?

Complete each factorization.

15. $4a + 12 = \boxed{}(a + 3)$
16. $r^4 + r^2 = r^2(\boxed{} + 1)$
17. $4y^2 + 8y - 2xy = 2y(2y + \boxed{} - \boxed{})$
18. $3x^2 - 6xy + 9xy^2 = \boxed{}(\boxed{} - 2y + 3y^2)$

NOTATION

Complete each factorization.

19. Factor: $b^3 - 6b^2 + 2b - 12$

$$b^3 - 6b^2 + 2b - 12 = \boxed{}(b - 6) + 2\boxed{}$$
$$= (b - 6)\boxed{}$$

20. Factor: $12b^3 - 6b^2 + 2b - 2$

$$12b^3 - 6b^2 + 2b - 2 = \boxed{}(6b^3 - 3b^2 + b - 1)$$

21. In the expression $4x^2y + xy$, what is the coefficient of the last term?
22. Is the following statement true?

$$-(x^2 - 3x + 1) = -1(x^2 - 3x + 1)$$

GUIDED PRACTICE

Find the GCF of each list of numbers. See Example 1.

23. 6, 8, 10
24. 28, 35, 21
25. 30, 45, 60
26. 78, 104, 156

Find the GCF of each list of terms. See Example 2.

27. $25y^3, 35y$
28. $36a^2, 48a$
29. $20p^2q, 40pq^2$
30. $36m^2n^2, 54mn$
31. $6t^3, 12t^2, 18t$
32. $28r^3, 14r^2, 35r^4$
33. $30a^3b^3, 45a^2b^2, 60ab$
34. $28u^4v^3, 35u^3v^2, 49u^2v$

Factor each expression. See Example 3.

35. $3x + 6$
36. $2y - 10$
37. $36 - 6x$
38. $48 + 12y$

Factor each expression. See Example 4.

39. $t^3 + 2t^2$
40. $b^3 - 3b^2$
41. $a^3 - a^2$
42. $r^3 + r^2$
43. $24x^2y^3 + 8xy^2$
44. $3x^2y^3 - 9x^4y^3$
45. $12uv - 18uv^2$
46. $14xy - 16x^2y^2$

Factor each expression. See Example 5.

47. $12x^2 - 6 - 24a$
48. $27a^2 - 9 + 45b$
49. $3 + 3y - 6z$
50. $2 - 4y + 8z$
51. $ab + ac - a$
52. $rs - rt + r$
53. $12r^2 - 3r + 9r^2s^2$
54. $6a - 12a^3b + 36ab$

Factor each expression. See Example 6.

55. $3(x + 2) - x(x + 2)$
56. $t(5 - s) + 4(5 - s)$
57. $h^2(14 + r) + 2(14 + r)$
58. $k^2(14 + v) - 7(14 + v)$

Factor −1 out of each expression. See Example 7.

59. $-a - b$
60. $-x - 2y$
61. $-3m - 4n + 1$
62. $-3r + 2s - 3$

In each expression, factor out the negative of the GCF. See Example 8.

63. $-3x^2 - 6x$
64. $-4a^2 + 6a$
65. $-4a^2b^3 + 12a^3b^2 + 4a^2b^2$
66. $-30x^4y^3 + 24x^3y^2 - 60x^2y$

Factor each expression. See Example 9.

67. $2x + 2y + ax + ay$
68. $bx + bz + 5x + 5z$
69. $7r + 7s - kr - ks$
70. $3xy + 3xz - 5y - 5z$

Factor each expression. See Example 10.

71. $ab + ac + b + c$
72. $xy + 3y^2 + x + 3y$
73. $mp - np - mq + nq$
74. $9p - 9q - mp + mq$

Factor each expression completely. See Example 11.

75. $ax^3 + bx^3 + 2ax^2y + 2bx^2y$
76. $x^3y^2 - 2x^2y^2 + 3xy^2 - 6y^2$
77. $4a^2b + 12a^2 - 8ab - 24a$
78. $-4abc - 4ac^2 + 2bc + 2c^2$

TRY IT YOURSELF

Completely factor each expression (including −1, if necessary).

79. $\pi R^2 - \pi ab$
80. $\frac{1}{3}\pi R^2 h - \frac{1}{3}\pi rh$
81. $-2x + 5y$
82. $-3x + 8z$
83. $-3ab - 5ac + 9bc$
84. $-6yz + 12xz - 5xy$
85. $-4a^2b^2c^2 + 14a^2b^2c - 10ab^2c^2$
86. $-10x^4y^3z^2 + 8x^3y^2z - 20x^2y$
87. $2ab + 2ac + 3b + 3c$
88. $3ac + a + 3bc + b$
89. $6x^2 - 2x - 15x + 5$
90. $6x^2 + 2x + 9x + 3$
91. $9mp + 3mq - 3np - nq$
92. $ax + bx - a - b$
93. $2xy + y^2 - 2x - y$
94. $2xy - 3y^2 + 2x - 3y$
95. $8z^5 + 12z^2 - 10z^3 - 15$
96. $2a^4 + 2a^3 - 4a - 4$
97. $x^3y - x^2y - xy^2 + y^2$
98. $2x^3z - 4x^2z + 32xz - 64z$

APPLICATIONS

99. **PICTURE FRAMING** The dimensions of a family portrait and the frame in which it is mounted are shown. Write an algebraic expression that describes
    a. the area enclosed by the picture frame.
    b. the area of the portrait.
    c. the area of the mat used in the framing. Express the result in factored form.

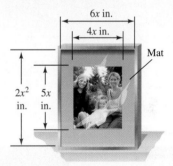

100. **REARVIEW MIRRORS** The dimensions of the three rearview mirrors on an automobile are given in the illustration below. Write an algebraic expression that gives
    a. the area of the rearview mirror mounted on the windshield.
    b. the total area of the two side mirrors.
    c. the total area of all three mirrors. Express the result in factored form.

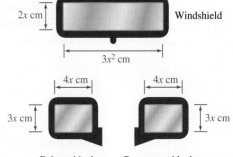

101. COOKING See the illustration below.
  a. What is the length of a side of the square griddle, in terms of r? What is the area of the cooking surface of the griddle, in terms of r?
  b. How many square inches of the cooking surface do the pancakes cover, in terms of r?
  c. Find the amount of cooking surface that is not covered by the pancakes. Express the result in factored form.

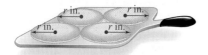

102. AIRCRAFT CARRIERS The rectangular landing area of $(x^3 + 4x^2 + 5x + 20)$ ft² is shaded. The dimensions of the landing area can be found by factoring. What are the length and width of the landing area?

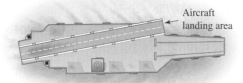

Aircraft landing area

WRITING

103. To add $5x$ and $7x$, we combine like terms: $5x + 7x = 12x$. Explain how this is related to factoring out a common factor.

104. One student commented, "Factoring undoes the distributive property." What do you think she meant? Give an example.

105. If asked to write $ax + ay - bx - by$ in factored form, explain why $a(x + y) - b(x + y)$ is not an acceptable answer.

106. When asked to factor $rx - sy + ry - sx$, a student wrote the expression as $rx + ry - sx - sy$. Then she factored it by grouping. Can the terms be rearranged in this manner? Explain your answer.

REVIEW

107. Simplify: $\left(\dfrac{y^3 y}{2yy^2}\right)^3$

108. Find the slope of the line passing through the points $(3, 5)$ and $(-2, -7)$.

109. Does the point $(3, 5)$ lie on the graph of the line $4x - y = 7$?

110. Simplify: $-5(3a - 2)(2a + 3)$

Objectives

1. Factor trinomials of the form $x^2 + bx + c$.
2. Factor trinomials of the form $x^2 + bx + c$ after factoring out the GCF.
3. Factor trinomials of the form $x^2 + bx + c$ using the grouping method.

SECTION 6.2

Factoring Trinomials of the Form $x^2 + bx + c$

In Chapter 5, we learned how to multiply binomials. For example, to multiply $x + 2$ and $x + 3$, we proceed as follows:

$$(x + 2)(x + 3) = x^2 + 3x + 2x + 6$$
$$= x^2 + 5x + 6$$

To *factor the trinomial* $x^2 + 5x + 6$, we will reverse the multiplication process and determine what factors were multiplied to obtain $x^2 + 5x + 6$. Since the product of two binomials is often a trinomial, many trinomials factor into the product of two binomials.

Multiplication: Given the binomial factors, we find the trinomial. →

$$(x + 2)(x + 3) = x^2 + 5x + 6$$

← Factoring: Given the trinomial, we find the binomial factors.

We will now consider how to factor trinomials of the form $ax^2 + bx + c$, where a (called the **leading coefficient**) is 1.

1 Factor trinomials of the form $x^2 + bx + c$.

To develop a method for factoring trinomials, we multiply $(x + 4)$ and $(x + 5)$.

$$(x + 4)(x + 5) = x \cdot x + x \cdot 5 + 4 \cdot x + 4 \cdot 5 \quad \text{Use the FOIL method.}$$
$$= x^2 + 5x + 4x + 20$$
$$= x^2 + 9x + 20$$

First term, Middle term, Last term

The result has three terms. We can see that

- the first term, x^2, is the product of x and x,
- the last term, 20, is the product of 4 and 5, and
- the coefficient of the middle term, 9, is the sum of 4 and 5.

We can use these facts to factor trinomials with lead coefficients of 1.

EXAMPLE 1 Factor: $x^2 + 5x + 6$

Strategy We will assume that $x^2 + 5x + 6$ is the product of two binomials and we will use a systematic method to find their terms.

WHY Since the terms of $x^2 + 5x + 6$ do not have a common factor (other than 1), the only option available is to try to factor it as the product of two binomials.

Solution
Since the first term of the trinomial is x^2, the first term of each binomial factor must be x. To fill in the blanks, we must find two integers whose product is $+6$ and whose sum is $+5$.

$$x^2 + 5x + 6 = (x \quad)(x \quad) \quad \text{Because } x \cdot x \text{ will give } x^2.$$

The positive factorizations of 6 and the sum of the factors are shown in the table.

Factors of 6	Sum of the factors of 6
1(6)	1 + 6 = 7
2(3)	2 + 3 = 5

The last row contains the integers 2 and 3, whose product is 6 and whose sum is 5. To complete the factorization, we enter 2 and 3 as the second terms of the binomial factors.

$$x^2 + 5x + 6 = (x + 2)(x + 3)$$

To check the result, we verify that $(x + 2)(x + 3)$ is $x^2 + 5x + 6$.

$$(x + 2)(x + 3) = x^2 + 3x + 2x + 6 \quad \text{Use the FOIL method.}$$
$$= x^2 + 5x + 6 \quad \text{This is the original trinomial.}$$

Self Check 1
Factor: $y^2 + 7y + 6$
Now Try Problem 25

Success Tip When factoring trinomials, the binomial factors can be written in either order. In Example 1, an equivalent factorization is $x^2 + 5x + 6 = (x + 3)(x + 2)$.

Chapter 6 Factoring and Quadratic Equations

Self Check 2

Factor: $p^2 - 5p + 6$

Now Try Problem 29

EXAMPLE 2 Factor: $y^2 - 7y + 12$

Strategy We will assume that $y^2 - 7y + 12$ is the product of two binomials and we will use a systematic method to find their terms.

WHY Since the terms of $y^2 - 7y + 12$ do not have a common factor (other than 1), the only option available is to try to factor it as the product of two binomials.

Solution
Since the first term of the trinomial is y^2, the first term of each binomial factor must be y. To fill in the blanks, we must find two integers whose product is 12 and whose sum is -7.

$$y^2 - 7y + 12 = (y \quad)(y \quad) \quad \text{Because } y \cdot y \text{ will give } y^2.$$

The two-integer factorizations of 12 and the sums of the factors are shown in the following table.

Factors of 12	Sum of the factors of 12
1(12)	$1 + 12 = 13$
2(6)	$2 + 6 = 8$
3(4)	$3 + 4 = 7$
$-1(-12)$	$-1 + (-12) = -13$
$-2(-6)$	$-2 + (-6) = -8$
$-3(-4)$	$-3 + (-4) = -7$

The last row contains the integers -3 and -4, whose product is 12 and whose sum is -7. To complete the factorization, we enter -3 and -4 as the second terms of the binomial factors.

$$y^2 - 7y + 12 = (y - 3)(y - 4)$$

To check the result, we verify that $(y - 3)(y - 4)$ is $y^2 - 7y + 12$.

$$(y - 3)(y - 4) = y^2 - 4y - 3y + 12 \quad \text{Use the FOIL method.}$$
$$= y^2 - 7y + 12 \quad \text{This is the original trinomial.}$$

Self Check 3

Factor: $p^2 + 3p - 18$

Now Try Problem 35

EXAMPLE 3 Factor: $a^2 + 2a - 15$

Strategy We will assume that $a^2 + 2a - 15$ is the product of two binomials and we will use a systematic method to find their terms.

WHY Since the terms of $a^2 + 2a - 15$ do not have a common factor (other than 1), the only option available is to try to factor it as the product of two binomials.

Solution
Since the first term of the trinomial is a^2, the first term of each binomial factor must be a. To fill in the blanks, we must find two integers whose product is -15 and whose sum is 2.

$$a^2 + 2a - 15 = (a \quad)(a \quad) \quad \text{Because } a \cdot a \text{ will give } a^2.$$

Unless otherwise noted, all content on this page is © Cengage Learning.

The possible factorizations of −15 and the sum of the factors are shown in the following table.

Factors of −15	Sum of the factors of −15
1(−15)	1 + (−15) = −14
3(−5)	3 + (−5) = −2
5(−3)	5 + (−3) = 2
15(−1)	15 + (−1) = 14

The third row contains the integers 5 and −3, whose product is −15 and whose sum is 2. To complete the factorization, we enter 5 and −3 as the second binomial factors.

$$a^2 + 2a - 15 = (a + 5)(a - 3)$$

We can check by multiplying.

$(a + 5)(a - 3) = a^2 - 3a + 5a - 15$ Use the FOIL method.
$ = a^2 + 2a - 15$ This is the original trinomial.

EXAMPLE 4 Factor: $z^2 - 4z - 21$

Strategy We will assume that $z^2 - 4z - 21$ is the product of two binomials and we will use a systematic method to find their terms.

WHY Since the terms of $z^2 - 4z - 21$ do not have a common factor (other than 1), the only option available is to try to factor it as the product of two binomials.

Solution
Since the first term of the trinomial is z^2, the first term of each binomial factor must be z. To fill in the blanks, we must find two integers whose product is −21 and whose sum is −4.

$$z^2 - 4y - 21 = (z)(z)$$ Because $z \cdot z$ will give z^2.

The factorizations of −21 and the sums of the factors are shown in the following table.

Factors of −21	Sum of the factors of −21
1(−21)	1 + (−21) = −20
3(−7)	3 + (−7) = −4
7(−3)	7 + (−3) = 4
21(−1)	21 + (−1) = 20

The second row contains the integers 3 and −7, whose product is −21 and whose sum is −4. To complete the factorization, we enter 3 and −7 as the second terms of the binomial factors.

$$z^2 - 4z - 21 = (z + 3)(z - 7)$$

We can check by multiplying.

$(z + 3)(z - 7) = z^2 - 7z + 3z - 21$ Use the FOIL method.
$ = z^2 - 4z - 21$ This is the original trinomial.

Self Check 4
Factor: $q^2 - 2q - 24$
Now Try Problem 37

The following sign patterns can be helpful when factoring trinomials.

> **Factoring Trinomials Whose Leading Coefficient Is 1**
>
> To factor $x^2 + bx + c$, find two integers whose product is c and whose sum is b.
>
> 1. If c is positive, the integers have the same sign.
> 2. If c is negative, the integers have opposite signs.

When factoring out trinomials of the form $ax^2 + bx + c$, where $a = -1$, we begin by factoring out -1.

Self Check 5

Factor: $-x^2 + 11x - 28$

Now Try Problem 41

EXAMPLE 5 Factor: $-h^2 + 2h + 15$

Strategy We will factor out -1 and then factor the resulting trinomial.

WHY It is easier to factor trinomials whose leading coefficient is positive.

Solution
We factor out -1 and then factor $h^2 - 2h - 15$.

$$-h^2 + 2h + 15 = -\mathbf{1}(h^2 - 2h - 15) \quad \text{Factor out } -1.$$
$$= -(h^2 - 2h - 15) \quad \text{The 1 need not be written.}$$
$$= -(h - 5)(h + 3) \quad \text{Use the integers } -5 \text{ and } 3, \text{ because their product is } -15 \text{ and their sum is } -2.$$

We can check by multiplying.

$$-(h - 5)(h + 3) = -(h^2 + 3h - 5h - 15) \quad \text{Multiply the binomials first.}$$
$$= -(h^2 - 2h - 15) \quad \text{Combine like terms.}$$
$$= -h^2 + 2h + 15 \quad \text{This is the original trinomial.}$$

The trinomial in the next example is of a form similar to $x^2 + bx + c$, and we can use the methods of this section to factor it.

Self Check 6

Factor: $s^2 + 6st - 7t^2$

Now Try Problem 45

EXAMPLE 6 Factor: $x^2 - 4xy - 5y^2$

Strategy We will assume that $x^2 - 4xy - 5y^2$ is the product of two binomials and we will use a systematic method to find their terms.

WHY Since the terms of $x^2 - 4xy - 5y^2$ do not have a common factor (other than 1), the only option available is to try to factor it as the product of two binomials.

Solution
The trinomial has two variables, x and y. Since the first term is x^2, the first term of each factor must be x.

$$x^2 - 4xy - 5y^2 = (x \quad\quad)(x \quad\quad) \quad \text{Because } x \cdot x \text{ will give } x^2.$$

To fill in the blanks, we must find two *expressions* whose product is the last term, $-5y^2$, and that will give a middle term of $-4xy$. Two such expressions are $-5y$ and y.

$$x^2 - 4xy - 5y^2 = (x - \mathbf{5y})(x + \mathbf{y})$$

Check: $(x - 5y)(x + y) = x^2 + xy - 5xy - 5y^2 \quad \text{Use the FOIL method.}$
$$= x^2 - 4xy - 5y^2 \quad \text{This is the original trinomial.}$$

2 Factor trinomials of the form $x^2 + bx + c$ after factoring out the GCF.

If the terms of a trinomial have a common factor, the GCF should always be factored out before any of the factoring techniques of this section are used. A trinomial is **factored completely** when no factor can be factored further. Always factor completely when you are asked to factor.

EXAMPLE 7 Factor: $2x^4 + 26x^3 + 80x^2$

Strategy We will factor out the GCF, $2x^2$, first. Then we will factor the resulting trinomial.

WHY The first step in factoring any polynomial is to factor out the GCF. Factoring out the GCF first makes factoring by any method easier.

Solution
We begin by factoring out the GCF, $2x^2$.

$$2x^4 + 26x^3 + 80x^2 = \mathbf{2x^2}(x^2 + 13x + 40)$$

Next, we factor $x^2 + 13x + 40$. The integers 8 and 5 have a product of 40 and a sum of 13, so the completely factored form of the given trinomial is

$$2x^4 + 26x^3 + 80x^2 = 2x^2(x + 8)(x + 5) \quad \text{The complete factorization must include } 2x^2.$$

Check by multiplying $2x^2$, $x + 8$, and $x + 5$.

Self Check 7
Factor: $4m^5 + 8m^4 - 32m^3$
Now Try Problem 51

EXAMPLE 8 Factor completely: $-13g^2 + 36g + g^3$

Strategy We will write the terms of the trinomial in descending powers of g.

WHY It is easier to factor a trinomial if its terms are written in descending powers of one variable.

Solution
Before factoring the trinomial, we write its terms in descending powers of g.

$$-13g^2 + 36g + g^3 = g^3 - 13g^2 + 36g \quad \text{Rearrange the terms.}$$
$$= g(g^2 - 13g + 36) \quad \text{Factor out } g, \text{ which is the GCF.}$$
$$= g(g - 9)(g - 4) \quad \text{Factor the trinomial.}$$

Check by multiplying.

Self Check 8
Factor: $-12t + t^3 + 4t^2$
Now Try Problem 55

If a trinomial cannot be factored using only integers, it is called a **prime polynomial**, or more specifically, a **prime trinomial**.

EXAMPLE 9 Factor, if possible: $x^2 + 2x + 3$

Strategy We will assume that $x^2 + 2x + 3$ is the product of two binomials and we will use a systematic method to find their terms.

WHY Since the terms of $x^2 + 2x + 3$ do not have a common factor (other than 1), the only option available is to try to factor it as the product of two binomials.

Self Check 9
Factor, if possible: $x^2 - 4x + 6$
Now Try Problem 58

Solution
To factor the trinomial, we must find two integers whose product is 3 and whose sum is 2. The possible factorizations of 3 and the sums of the factors are shown in the following table.

Factors of 3	Sum of the factors of 3
1(3)	1 + 3 = 4
−1(−3)	−1 + (−3) = −4

Since two integers whose product is 3 and whose sum is 2 do not exist, $x^2 + 2x + 3$ cannot be factored. It is a *prime trinomial*.

3 Factor trinomials of the form $x^2 + bx + c$ using the grouping method.

Another way to factor trinomials of the form $x^2 + bx + c$ is to write them as equivalent four-termed polynomials and factor by grouping. To factor $x^2 + 8x + 15$ using this method, we proceed as follows.

1. First, identify b as the coefficient of the x-term, and c as the last term. For trinomials of the form $x^2 + bx + c$, we call c the **key number**.

$$\left.\begin{array}{c} x^2 + bx + c \\ \downarrow \quad \downarrow \quad \downarrow \\ x^2 + 8x + 15 \end{array}\right\} b = 8 \text{ and } c = 15$$

2. Now find two integers whose product is the key number, 15, and whose sum is $b = 8$. Since the integers must have a positive product and a positive sum, we consider only positive factors of 15.

Key number = 15	$b = 8$
Positive factors of 15	Sum of the positive factors of 15
1 · 15 = 15	1 + 15 = 16
3 · 5 = 15	3 + 5 = 8

The second row of the table contains the correct pair of integers 3 and 5, whose product is the key number 15 and whose sum is $b = 8$.

3. Express the middle term, $8x$, of the trinomial as the *sum of two terms*, using the integers 3 and 5 found in step 2 as coefficients of the two terms.

$$x^2 + 8x + 15 = x^2 + 3x + 5x + 15 \quad \text{Express 8x as 3x + 5x.}$$

4. Factor the equivalent four-term polynomial by grouping:

$$x^2 + 3x + 5x + 15 = x(x + 3) + 5(x + 3) \quad \text{Factor x out of } x^2 + 3x \text{ and 5 out of 5x + 15.}$$
$$= (x + 3)(x + 5) \quad \text{Factor out x + 3.}$$

Check the factorization by multiplying.

The grouping method is an alternative to the method for factoring trinomials discussed earlier in this section. It is especially useful when the constant term, c, has many factors.

Factoring Trinomials of the Form $x^2 + bx + c$ Using Grouping

To factor a trinomial that has a leading coefficient of 1:

1. Identify b and the key number, c.
2. Find two integers whose product is the key number and whose sum is b.
3. Express the middle term, bx, as the sum (or difference) of two terms. Enter the two numbers found in step 2 as coefficients of x in the form shown below. Then factor the equivalent four-term polynomial by grouping.

   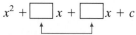

   The product of these numbers must be c, and their sum must be b.

4. Check the factorization using multiplication.

EXAMPLE 10
Factor by grouping: $a^2 + a - 20$

Strategy We will express the middle term, a, of the trinomial as the difference of two carefully chosen terms.

WHY We want to produce an equivalent four-term polynomial that can be factored by grouping.

Solution
Since $a^2 + a - 20 = a^2 + 1a - 20$, we identify b as **1** and the key number c as **−20**. We must find two integers whose product is −20 and whose sum is 1. Since the integers must have a negative product, their signs must be different.

Key number = −20	$b = 1$
Factors of −20	Sum of the factors of −20
$1(-20) = -20$	$1 + (-20) = -19$
$2(-10) = -20$	$2 + (-10) = -8$
$4(-5) = -20$	$4 + (-5) = -1$
$5(-4) = -20$	$5 + (-4) = 1$
$10(-2) = -20$	$10 + (-2) = 8$
$20(-1) = -20$	$20 + (-1) = 19$

The fourth row of the table contains the correct pair of integers 5 and −4, whose product is −20 and whose sum is 1. They serve as the coefficients of $5a$ and $-4a$, the two terms that we use to represent the middle term, a, of the trinomial.

$$a^2 + a - 20 = a^2 + 5a - 4a - 20 \quad \text{Express the middle term, } a, \text{ as } 5a - 4a.$$
$$= a(a + 5) - 4(a + 5) \quad \text{Factor } a \text{ out of } a^2 + 5a \text{ and } -4 \text{ out of } -4a - 20.$$
$$= (a + 5)(a - 4) \quad \text{Factor out } a + 5.$$

Check the factorization by multiplying.

Self Check 10
Factor by grouping:
$m^2 + m - 42$

Now Try Problem 61

> **Success Tip** We could also express the middle term as $-4a + 5a$. We obtain the same binomial factors, but in reverse order.
>
> $a^2 - 4a + 5a - 20$
> $= a(a - 4) + 5(a - 4)$
> $= (a - 4)(a + 5)$

Self Check 11

Factor by grouping:
$q^2 - 2qt - 24t^2$

Now Try Problem 63

EXAMPLE 11 Factor by grouping: $x^2 - 4xy - 5y^2$

Strategy We will express the middle term, $-4xy$, of the trinomial as the sum of two carefully chosen terms.

WHY We want to produce an equivalent four-term polynomial that can be factored by grouping.

Solution
In $x^2 - 4xy - 5y^2$, we identify b as -4 and the key number c as -5. We must find two integers whose product is -5 and whose sum is -4. Such a pair is -5 and 1. They serve as the coefficients of $-5xy$ and $1xy$, the two terms that we use to represent the middle term, $-4xy$, of the trinomial.

Key number = -5	$b = -4$
Factors	**Sum**
$-5(1) = -5$	$-5 + 1 = -4$

$x^2 - 4xy - 5y^2 = x^2 - 5xy + 1xy - 5y^2$ Express the middle term, $-4xy$, as $-5xy + 1xy$. ($1xy - 5xy$ could also be used.)

$= x(x - 5y) + y(x - 5y)$ Factor x out of $x^2 - 5xy$ and y out of $1xy - 5y^2$.

$= (x - 5y)(x + y)$ Factor out $x - 5y$.

Check the factorization by multiplying.

Self Check 12

Factor completely:
$3m^3 - 27m^2 + 24m$

Now Try Problem 65

EXAMPLE 12 Factor completely: $2x^3 - 20x^2 + 18x$

Strategy We will factor out the GCF, $2x$, first. Then we will factor the resulting trinomial using the grouping method.

WHY The first step in factoring any polynomial is to factor out the GCF.

Solution
We begin by factoring out the GCF, $2x$, from $2x^3 - 20x^2 + 18x$.

$2x^3 - 20x^2 + 18x = 2x(x^2 - 10x + 9)$

Key number = 9	$b = -10$
Factors	**Sum**
$-9(-1) = 9$	$-9 + (-1) = -10$

To factor $x^2 - 10x + 9$ by grouping, we must find two integers whose product is the key number 9 and whose sum is $b = -10$. Such a pair is -9 and -1.

$x^2 - 10x + 9 = x^2 - 9x - 1x + 9$ Express $-10x$ as $-9x - 1x$. ($-1x - 9x$ could also be used.)

$= x(x - 9) - 1(x - 9)$ Factor x out of $x^2 - 9x$ and -1 out of $-1x + 9$.

$= (x - 9)(x - 1)$ Factor out $x - 9$.

The complete factorization of the original trinomial is

$2x^3 - 20x^2 + 18x = \mathbf{2x(x - 9)(x - 1)}$ Don't forget to write the GCF, 2x.

Check the factorization by multiplying.

ANSWERS TO SELF CHECKS

1. $(y + 1)(y + 6)$ **2.** $(p - 3)(p - 2)$ **3.** $(p + 6)(p - 3)$ **4.** $(q - 6)(q + 4)$
5. $-(x - 4)(x - 7)$ **6.** $(s + 7t)(s - t)$ **7.** $4m^3(m + 4)(m - 2)$ **8.** $t(t - 2)(t + 6)$
9. not possible; prime trinomial **10.** $(m + 7)(m - 6)$ **11.** $(q + 4t)(q - 6t)$
12. $3m(m - 8)(m - 1)$

SECTION 6.2 STUDY SET

VOCABULARY

Fill in the blanks.

1. A polynomial with three terms is called a _____.
2. In the polynomial $x^2 - x - 6$, x^2 is the _____ term, $-x$ is the middle _____, and ___ is the last term.
3. The statement $x^2 - x - 12 = (x - 4)(x + 3)$ shows that $x^2 - x - 12$ _____ into the product of two binomials.
4. A trinomial is said to be factored _____ when no factor can be factored further.
5. A _____ polynomial cannot be factored by using only integer coefficients.
6. When factoring trinomials of the form $x^2 + bx + c$ by the grouping method, the number c is called the _____ number.

CONCEPTS

Fill in the blanks.

7. Two factorizations of 4 that involve only positive numbers are ___ · ___ and ___ · ___. Two factorizations of 4 that involve only negative numbers are ___(___) and ___(___).
8. Before attempting to factor a trinomial, be sure that the exponents are written in _____ order.
9. Before attempting to factor a trinomial into two binomials, always factor out any _____ factors first.
10. To factor $x^2 + x - 56$, we must find two integers whose _____ is -56 and whose _____ is 1.
11. Two factors of 18 whose sum is -9 are ___ and ___.

12. Complete the table.

Factors of 8	Sum of the factors of 8
1(8)	
2(4)	
−1(−8)	
−2(−4)	

13. Complete the table.

Factors of −9	Sum of the factors of −9
1(−9)	
3(−3)	
−1(9)	

14. Find two integers whose
   a. product is 10 and whose sum is 7.
   b. product is 8 and whose sum is −6.
   c. product is −6 and whose sum is 1.
   d. product is −9 and whose sum is −8.

15. Given: $x^2 + 8x + 15$
   a. What is the coefficient of the x^2 term?
   b. What is the last term? What is the coefficient of the middle term?
   c. What two integers have a product of 15 and a sum of 8?

16. What trinomial has the factorization of $(x + 8)(x - 2)$?

Complete each factorization.

17. $x^2 + 3x + 2 = (x \;\; 2)(x \;\; 1)$
18. $y^2 + 4y + 3 = (y \;\; 3)(y \;\; 1)$
19. $t^2 - 9t + 14 = (t \;\; 7)(t \;\; 2)$
20. $c^2 - 9c + 8 = (c \;\; 8)(c \;\; 1)$
21. $a^2 + 6a - 16 = (a \;\; 8)(a \;\; 2)$
22. $x^2 - 3x - 40 = (x \;\; 8)(x \;\; 5)$

NOTATION

Complete each factorization.

23. $6 + 5x + x^2 = x^2 + \underline{} + 6$
    $= (x + 3)(x + \underline{})$
24. $-a^2 - a + 20 = \underline{}(a^2 + a - 20)$
    $= -(a + 5)(a - \underline{})$

GUIDED PRACTICE

Factor each trinomial. See Example 1.

25. $z^2 + 12z + 11$
26. $x^2 + 7x + 10$
27. $p^2 + 9p + 14$
28. $q^2 + 11q + 24$

Factor each trinomial. See Example 2.

29. $m^2 - 5m + 6$
30. $n^2 - 7n + 10$
31. $y^2 - 13y + 30$
32. $r^2 - 10r + 24$

Factor each trinomial. See Example 3.

33. $b^2 + 6b - 7$
34. $x^2 + 5x - 24$
35. $a^2 + 6a - 16$
36. $b^2 + 2b - 99$

Factor each trinomial. See Example 4.

37. $a^2 - 4a - 5$
38. $t^2 - 5t - 50$
39. $z^2 - 3z - 18$
40. $s^2 - 2s - 120$

Factor each trinomial. See Example 5.

41. $-x^2 - 7x - 10$
42. $-x^2 + 9x - 20$
43. $-t^2 - 15t + 34$
44. $-t^2 - t + 30$

Factor each trinomial. See Example 6.

45. $x^2 + 4xy + 4y^2$
46. $a^2 + 10ab + 9b^2$
47. $m^2 + 3mn - 10n^2$
48. $m^2 - mn - 12n^2$

Factor each trinomial. See Example 7.

49. $2x^2 + 10x + 12$
50. $3y^2 - 21y + 18$
51. $5p^3 + 25p^2 - 70p$
52. $3m^4 - 9m^3 - 54m^2$

Factor each trinomial. See Example 8.

53. $4rx + r^2 + 3x^2$
54. $a^2 + 5b^2 + 6ab$
55. $-3a^2b + a^3 + 2ab^2$
56. $-13yz^2 + y^2z - 14z^3$

Factor each trinomial, if possible. See Example 9.

57. $r^2 - 9r - 12$
58. $u^2 + 10u + 15$
59. $r^2 - 2rs + 4s^2$
60. $m^2 + 3mn - 20n^2$

Factor each trinomial by grouping. See Examples 10–12.

61. $p^2 + p - 30$
62. $q^2 - 10q + 24$
63. $m^2 - 3mn - 4n^2$
64. $r^2 - 2rs - 15s^2$
65. $3x^3 - 27x^2 + 60x$
66. $2x^3 + 4x^2 - 70x$
67. $4y^3 - 28y^2 + 40y$
68. $5y^3 + 45y^2 + 100y$

TRY IT YOURSELF

Completely factor each of the following expressions. If an expression is prime, so indicate.

69. $a^2 - 10a - 39$
70. $v^2 + 9v + 15$
71. $s^2 + 11s - 26$
72. $y^2 + 8y + 12$
73. $r^2 - 2r + 4$
74. $m^2 + 3m - 10$
75. $a^2 - 4ab - 12b^2$
76. $p^2 + pq - 6q^2$
77. $-r^2 + 14r - 40$
78. $-r^2 + 14r - 45$
79. $-a^2 - 4ab - 3b^2$
80. $-a^2 - 6ab - 5b^2$
81. $4 - 5x + x^2$
82. $y^2 + 5 + 6y$
83. $10y + 9 + y^2$
84. $x^2 - 13 - 12x$
85. $-r^2 + 2 + r$
86. $u^2 - 3 + 2u$
87. $-5a^2 + 25a - 30$
88. $-2b^2 + 20b - 18$
89. $-4x^2y - 4x^3 + 24xy^2$
90. $3x^2y^3 + 3x^3y^2 - 6xy^4$

Choose the correct method from Section 6.1 or 6.2 to factor completely each expression. If an expression is prime, so indicate.

91. $m^2 - m - 12$ **92.** $u^2 + u - 42$

93. $3a^2b + 3ab^2$ **94.** $3p^2 - 12p + 6$

95. $-4a^2 - 8a$ **96.** $3p + p^2 + 3q + pq$

97. $-x^2 + 6xy + 7y^2$ **98.** $-x^2 - 10xy + 11y^2$

99. $12xy + 4x^2y - 72y$ **100.** $48xy + 6xy^2 + 96x$

101. $3ap + 2p + 3aq + 2q$

102. $-9abc - 9ac^2 + 3bc + 3c^2$

103. $3z^2 - 15z + 12$ **104.** $5m^2 + 45m - 50$

APPLICATIONS

105. PETS The cage shown is used for transporting dogs. Its volume is $(x^3 + 12x^2 + 27x)$ in.³. The dimensions of the cage can be found by factoring this expression. If the cage is longer than it is tall and taller than it is wide, write expressions that represent its length, height, and width.

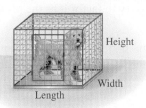

106. PHOTOGRAPHY A picture cube is a clever way to display 6 photographs in a small amount of space. Suppose the surface area of the entire cube is given by the polynomial $(6s^2 + 12s + 6)$ in.². Find the polynomial that represents the length of an edge of the cube.

WRITING

107. Explain what it means when we say that a trinomial is the product of two binomials. Give an example.

108. Are $2x^2 - 12x + 16$ and $x^2 - 6x + 8$ factored in the same way? Explain why or why not.

109. When factoring $x^2 - 2x - 3$, one student got $(x - 3)(x + 1)$, and another got $(x + 1)(x - 3)$. Are both answers acceptable? Explain.

110. Explain how to use multiplication to check the factorization of a trinomial.

111. A student begins to factor a trinomial as shown below. Explain why the student is off to a bad start.

$$x^2 - 2x - 63 = (x -)(x -)$$

112. Explain why the given trinomial is not factored completely.

$$3x^2 - 3x - 60 = 3(x^2 - x - 20)$$

REVIEW

Graph the solution of each inequality on the number line.

113. $x - 3 > 5$ **114.** $x + 4 \leq 3$

115. $-3x - 5 \geq 4$ **116.** $2x - 3 < 7$

SECTION 6.3
Factoring Trinomials of the Form $ax^2 + bx + c$

Objectives
1. Factor trinomials using the trial-and-check method.
2. Factor trinomials after factoring out the GCF.
3. Factor trinomials using the grouping method.

In this section, we will factor trinomials with leading coefficients other than 1, such as $2x^2 + 5x + 3$ and $6a^2 - 17a + 5$. Two methods are used to factor these trinomials. With the first method, educated guesses are made. These guesses are checked by multiplication. The correct factorization is determined by a process of elimination. The second method is an extension of factoring by grouping.

1 Factor trinomials using the trial-and-check method.

In the work below, we find the products $(2x + 1)(x + 3)$ and $(2x + 3)(x + 1)$. There are several observations that can be made when we compare the results.

$$(2x + 1)(x + 3) = 2x^2 + 6x + x + 3 \qquad (2x + 3)(x + 1) = 2x^2 + 2x + 3x + 3$$
$$= 2x^2 + 7x + 3 \qquad \qquad \qquad = 2x^2 + 5x + 3$$

In each case, the result is a trinomial, and

- the first terms are the same $(2x^2)$,
- the last terms are the same (3), and
- the middle terms are different ($7x$ and $5x$).

These observations indicate that when the last terms in $(2x + 1)(x + 3)$ are interchanged to form $(2x + 3)(x + 1)$, only the middle terms of the products are different. This fact is helpful when factoring trinomials using the *trial-and-check method*.

Self Check 1

Factor: $3x^2 + 7x + 2$

Now Try Problem 27

EXAMPLE 1   Factor: $2x^2 + 5x + 3$

Strategy We will assume that $2x^2 + 5x + 3$ is the product of two binomials and we will use a systematic method to find their terms.

WHY Since the terms of $2x^2 + 5x + 3$ do not have a common factor (other than 1), the only option available is to try to factor it as the product of two binomials.

Solution

Since the first term is $2x^2$, the first terms of the binomial factors must be $2x$ and x. To fill in the blanks, we must find two factors of 3 that will give a middle term of $5x$.

$$(2x \quad\quad)(x \quad\quad) \qquad \text{Because } 2x \cdot x \text{ will give } 2x^2.$$

Because each term of the trinomial is positive, we need only consider positive factors of the last term. Since the positive factors of 3 are 1 and 3, there are two possible factorizations.

$$(2x + 1)(x + 3) \qquad \text{or} \qquad (2x + 3)(x + 1)$$

The first possibility is incorrect: When we find the outer and inner products and combine like terms, we obtain an incorrect middle term of $7x$.

Outer: $6x$

$(2x + 1)(x + 3)$ Multiply and add to find the middle term: $6x + x = 7x$.

Inner: x

The second possibility is correct, because it gives a middle term of $5x$.

Outer: $2x$

$(2x + 3)(x + 1)$ Multiply and add to find the middle term: $2x + 3x = 5x$.

Inner: $3x$

Thus,

$$2x^2 + 5x + 3 = (2x + 3)(x + 1)$$

EXAMPLE 2   Factor: $6a^2 - 17a + 5$

Strategy We will assume that $6a^2 - 17a + 5$ is the product of two binomials and we will use a systematic method to find their terms.

WHY Since the terms of $6a^2 - 17a + 5$ do not have a common factor (other than 1), the only option available is to try to factor it as the product of two binomials.

Solution
Since the first term is $6a^2$, the first terms of the binomial factors must be $6a$ and a or $3a$ and $2a$. To fill in the blanks, we must find two factors of 5 that will give a middle term of $-17a$.

$$(6a\ \_\_\_)(a\ \_\_\_) \quad \text{or} \quad (3a\ \_\_\_)(2a\ \_\_\_)$$

Because the sign of the last term is positive and the sign of the middle term is negative, we need only consider negative factors of the last term. Since the negative factors of 5 are -1 and -5, there are four possible factorizations.

$(6a - 1)(a - 5) \quad -30a - a = -31a.$

$(6a - 5)(a - 1) \quad -6a - 5a = -11a.$

$(3a - 1)(2a - 5) \quad -15a - 2a = -17a.$

$(3a - 5)(2a - 1) \quad -3a - 10a = -13a.$

Only the possibility shown in blue gives the correct middle term of $-17a$. Thus,

$$6a^2 - 17a + 5 = (3a - 1)(2a - 5)$$

Self Check 2
Factor: $6x^2 - 7x + 2$
Now Try Problem 31

EXAMPLE 3   Factor: $3y^2 - 7y - 6$

Strategy We will assume that $3y^2 - 7y - 6$ is the product of two binomials and we will use a systematic method to find their terms.

WHY Since the terms of $3y^2 - 7y - 6$ do not have a common factor (other than 1), the only option available is to try to factor it as the product of two binomials.

Solution
Since the first term is $3y^2$, the first terms of the binomial factors must be $3y$ and y.

$(3y\ \_\_\_)(y\ \_\_\_)$ Because $3y \cdot y$ will give $3y^2$.

The second terms of the binomials must be two integers whose product is -6. There are four such pairs: $-1(6), 1(-6), -2(3),$ and $2(-3)$. When these pairs are entered, and then reversed, as second terms of the binomials, there are eight possibilities to consider. Four of them can be discarded because they include a binomial whose terms have a common factor. If $3y^2 - 7y - 6$ does not have a common factor, neither can any of its binomial factors.

For the factors -1 and 6: $(3y - 1)(y + 6)$ or $(3y + 6)(y - 1)$
$18y - y = 17y$ A common factor of 3

Self Check 3
Factor: $5a^2 - 23a - 10$
Now Try Problem 35

For the factors 1 and −6: $(3y + 1)(y − 6)$ or $(3y − 6)(y + 1)$

$-18y + y = -17y$

A common factor of 3

For the factors −2 and 3: $(3y − 2)(y + 3)$ or $(3y + 3)(y − 2)$

$9y − 2y = 7y$

A common factor of 3

For the factors 2 and −3: $(3y + 2)(y − 3)$ or $(3y − 3)(y + 2)$

$-9y + 2y = -7y$

A common factor of 3

Only the possibility shown in blue gives the correct middle term of $-7y$. Thus,

$$3y^2 − 7y − 6 = (3y + 2)(y − 3)$$

Check the factorization by multiplication.

> **Success Tip** If a trinomial does not have a common factor, the terms of each of its binomial factors will not have a common factor.

Self Check 4

Factor: $4x^2 + 4xy − 3y^2$

Now Try Problem 39

EXAMPLE 4 Factor: $4b^2 + 8bc − 45c^2$

Strategy We will assume that $4b^2 + 8bc − 45c^2$ is the product of two binomials and we will use a systematic method to find their terms.

WHY Since the terms of $4b^2 + 8bc − 45c^2$ do not have a common factor (other than 1), the only option available is to try to factor it as the product of two binomials.

Solution

Since the first term is $4b^2$, the first terms of the factors must be $4b$ and b or $2b$ and $2b$.

$(4b\ \ )(b\ \ )$ or $(2b\ \ )(2b\ \ )$ Because $4b \cdot b$ or $2b \cdot 2b$ will give $4b^2$.

To fill in the blanks, we must find two factors of $-45c^2$ that will give a middle term of $8bc$.

Since $-45c^2$ has many factors, there are many possible combinations for the last terms of the binomial factors. The signs of the factors must be different, because the last term of the trinomial is negative.

If we pick factors of $4b$ and b for the first terms, and $-c$ and $45c$ for the last terms, the multiplication gives an incorrect middle term of $179bc$. So the factorization is incorrect.

$(4b − c)(b + 45c)$ $180bc − bc = 179bc$

If we pick factors of $4b$ and b for the first terms and $15c$ and $-3c$ for the last terms, the multiplication gives an incorrect middle term of $3bc$.

$$(4b + 15c)(b - 3c) \qquad -12bc + 15bc = 3bc$$

If we pick factors of $2b$ and $2b$ for the first terms and $-5c$ and $9c$ for the last terms, we have

$$(2b - 5c)(2b + 9c) \qquad 18bc - 10bc = 8bc$$

which gives the correct middle term of $8bc$. Thus,

$$4b^2 + 8bc - 45c^2 = (2b - 5c)(2b + 9c)$$

Check by multiplication.

Because some guesswork is often necessary, it is difficult to give specific rules for factoring trinomials with a leading coefficient other than 1. However, the following hints are helpful.

Factoring $ax^2 + bx + c$ $(a \neq 1)$

1. Write the trinomial in descending powers of the variable and factor out any GCF (including -1 if that is necessary to make the leading coefficient positive).
2. Attempt to write the trinomial as *the product of two binomials*. The coefficients of the first terms of each binomial factor must be factors of a, and the last terms must be factors of c.

$$(\boxed{} x + \boxed{})(\boxed{} x + \boxed{})$$

   with Factors of a on the first terms and Factors of c on the last terms.

3. If the sign of the last term of the trinomial is positive, the signs between the terms of the binomial factors are the same as the sign of the middle term. If the sign of the last term is negative, the signs between the terms of the binomial factors are opposite.
4. Try combinations of coefficients of the first terms and last terms until you find one that gives the middle term of the trinomial. If no combination works, the trinomial is prime.
5. Check the factorization by multiplying.

Chapter 6 Factoring and Quadratic Equations

Self Check 5

Factor: $12y - 2y^3 - 2y^2$

Now Try Problem 43

2 Factor trinomials after factoring out the GCF.

EXAMPLE 5 Factor: $2x^2 - 8x^3 + 3x$

Strategy We will write the trinomial in descending powers of x and factor out the common factor, $-x$.

WHY It is easier to factor trinomials that have a positive leading coefficient.

Solution
We write the trinomial in descending powers of x

$$-8x^3 + 2x^2 + 3x$$

and we factor out the negative of the GCF, which is $-x$.

$$-8x^3 + 2x^2 + 3x = -x(8x^2 - 2x - 3)$$

We must now factor $8x^2 - 2x - 3$. Its factorization has the form

$(x \quad)(8x \quad)$ or $(2x \quad)(4x \quad)$ Because $x \cdot 8x$ or $2x \cdot 4x$ will give $8x^2$.

To fill in the blanks, we find two factors of the last term of the trinomial (-3) that will give a middle term of $-2x$. Because the sign of the last term is negative, the signs within its binomial factors will be different. If we pick factors of $2x$ and $4x$ for the first terms and 1 and -3 for the last terms, we have

$(2x + 1)(4x - 3)$ $-6x + 4x = -2x$.

which gives the correct middle term of $-2x$, so it is correct.

$$8x^2 - 2x - 3 = (2x + 1)(4x - 3)$$

We can now give the complete factorization.

$$-8x^3 + 2x^2 + 3x = -x(8x^2 - 2x - 3)$$
$$= -x(2x + 1)(4x - 3)$$

Check by multiplication.

3 Factor trinomials using the grouping method.

The method of factoring by grouping can be used to help factor trinomials of the form $ax^2 + bx + c$. For example, to factor $2x^2 + 5x + 3$, we proceed as follows.

1. We find the product ac: In $2x^2 + 5x + 3$, $a = 2$, $b = 5$, and $c = 3$, so $ac = 2(3) = 6$. This number is called the **key number**.
2. Next, find two numbers whose product is $ac = 6$ and whose sum is $b = 5$. Since the numbers must have a positive product and a positive sum, we consider only positive factors of 6. The correct factors are 2 and 3.

Positive factors of 6	Sum of the factors of 6
$1 \cdot 6 = 6$	$1 + 6 = 7$
$2 \cdot 3 = 6$	$2 + 3 = 5$

3. Use the factors 2 and 3 as coefficients of two terms to be placed between $2x^2$ and 3:

$$2x^2 + 5x + 3 = 2x^2 + 2x + 3x + 3$$ Express $5x$ as $2x + 3x$.

Unless otherwise noted, all content on this page is © Cengage Learning.

4. Factor by grouping:

$$2x^2 + 2x + 3x + 3 = 2x(x + 1) + 3(x + 1) \quad \text{Factor 2x out of } 2x^2 + 2x \text{ and 3 out of } 3x + 3.$$

$$= (x + 1)(2x + 3) \quad \text{Factor out } x + 1.$$

So $2x^2 + 5x + 3 = (x + 1)(2x + 3)$. Verify this factorization by multiplication.

Factoring $ax^2 + bx + c$ by Grouping

1. Write the trinomial in descending powers of the variable and factor out any GCF (including -1 if that is necessary to make the leading coefficient positive).
2. Calculate the key number ac.
3. Find two numbers whose product is the key number found in step 2 and whose sum is the coefficient of the middle term of the trinomial.
4. Write the numbers in the blanks of the form shown below, and then factor the polynomial by grouping.

$$ax^2 + \boxed{}x + \boxed{}x + c$$

The product of these numbers must be ac and their sum must be b.

5. Check the factorization using multiplication.

EXAMPLE 6 Factor: $10x^2 + 13x - 3$

Strategy We will express the middle term, $13x$, of the trinomial as the sum of two carefully chosen terms.

WHY We want to produce an equivalent four-term polynomial that can be factored by grouping.

Solution
Since $a = 10$ and $c = -3$ in the trinomial, $ac = -30$. We now find two factors of -30 whose sum is 13. Two such factors are 15 and -2. We use these factors as coefficients of two terms to be placed between $10x^2$ and -3.

$$10x^2 + 13x - 3 = 10x^2 + 15x - 2x - 3 \quad \text{Express 13x as 15x} - 2x.$$

Finally, we factor by grouping.

$$10x^2 + 15x - 2x - 3 = 5x(2x + 3) - 1(2x + 3) \quad \text{Factor out 5x from } 10x^2 + 15x.$$
$$\text{Factor out } -1 \text{ from } -2x - 3.$$

$$= (2x + 3)(5x - 1) \quad \text{Factor out } (2x + 3).$$

So $10x^2 + 13x - 3 = (2x + 3)(5x - 1)$. Check the result.

Self Check 6
Factor: $15a^2 + 17a - 4$
Now Try Problem 51

EXAMPLE 7 Factor: $12x^5 - 17x^4 + 6x^3$

Strategy We will factor out the GCF, x^3, first. Then we will factor the resulting trinomial using the grouping method.

WHY The first step in factoring any polynomial is to factor out the GCF.

Solution
First, we factor out the GCF, which is x^3.

$$12x^5 - 17x^4 + 6x^3 = x^3(12x^2 - 17x + 6)$$

Self Check 7
Factor: $21a^4 - 13a^3 + 2a^2$
Now Try Problem 55

To factor $12x^2 - 17x + 6$, we need to find two integers whose product is $12(6) = 72$ and whose sum is -17. Two such numbers are -8 and -9.

$$12x^2 - 17x + 6 = 12x^2 - 8x - 9x + 6 \quad \text{Express } -17x \text{ as } -8x - 9x.$$
$$= 4x(3x - 2) - 3(3x - 2) \quad \text{Factor out } 4x \text{ and factor out } -3.$$
$$= (3x - 2)(4x - 3) \quad \text{Factor out } 3x - 2.$$

The complete factorization is

$$12x^5 - 17x^4 + 6x^3 = x^3(3x - 2)(4x - 3) \quad \text{Do not forget to write the GCF, } x^3.$$

Check the result.

ANSWERS TO SELF CHECKS

1. $(3x + 1)(x + 2)$ 2. $(3x - 2)(2x - 1)$ 3. $(5a + 2)(a - 5)$ 4. $(2x + 3y)(2x - y)$
5. $-2y(y + 3)(y - 2)$ 6. $(3a + 4)(5a - 1)$ 7. $a^2(7a - 2)(3a - 1)$

SECTION 6.3 STUDY SET

VOCABULARY

Fill in the blanks.

1. The trinomial $3x^2 - x - 12$ has a _____ coefficient of 3. The ___ term is -12.
2. The numbers 3 and 2 are _____ of the first term of the trinomial $6x^2 + x - 12$.
3. Consider $(x - 2)(5x - 1)$. The product of the _____ terms is $-x$ and the product of the _____ terms is $-10x$.
4. When we write $2x^2 + 7x + 3$ as $(2x + 1)(x + 3)$, we say that we have _____ the trinomial—it has been expressed as the product of two _____.
5. The _____ term of $4x^2 - 7x + 13$ is $-7x$.
6. The ___ of the middle terms of the polynomial $4a^2 - 12a - a + 3$ is $-13a$.
7. The _____ of the terms of the trinomial $6b^3 - 3b^2 - 12b$ is $3b$.
8. When factoring the trinomial $ax^2 + bx + c$ by grouping, the product ac is called the ___ number.

CONCEPTS

Complete each sentence.

9.
These coefficients must be factors of ▯.

$$5x^2 + 6x - 8 = (\ \ x + \ \ )(\ \ x + \ \ )$$

These numbers must be factors of ▯.

10.
The product of these coefficients must be ▯.

$$3x^2 + 16x + 5 = 3x^2 + \ \ x + \ \ x + 5$$

The sum of these coefficients must be ▯.

A trinomial has been partially factored. Complete each statement that describes the type of integers we should consider for the blanks.

11. $5y^2 - 13y + 6 = (5y\ \square)(y\ \square)$
    Since the last term of the trinomial is _____ and the middle term is _____, the integers must be _____ factors of 6.

12. $5y^2 + 13y + 6 = (5y\ \square)(y\ \square)$
    Since the last term of the trinomial is _____ and the middle term is _____, the integers must be _____ factors of 6.

13. $5y^2 + 7y - 6 = (5y\ \square)(y\ \square)$
    Since the last term of the trinomial is _____, the signs of the integers will be _____.

14. $5y^2 - 7y - 6 = (5y\ \square)(y\ \square)$
    Since the last term of the trinomial is _____, the signs of the integers will be _____.

Complete each factorization.

15. $3a^2 + 13a + 4 = (3a\ \ 1)(a\ \ 4)$
16. $2b^2 + 7b + 6 = (2b\ \ 3)(b\ \ 2)$

17. $4z^2 - 13z + 3 = (z \;\; 3)(4z \;\; 1)$
18. $4t^2 - 4t + 1 = (2t \;\; 1)(2t \;\; 1)$
19. $2m^2 + 5m - 12 = (2m \;\; 3)(m \;\; 4)$
20. $10u^2 - 13u - 3 = (2u \;\; 3)(5u \;\; 1)$

A trinomial is to be factored by the grouping method. Complete each statement that describes the type of integers we should consider for the blanks.

21. $8c^2 - 11c + 3 = 8c^2 + \boxed{}c + \boxed{}c + 3$
    We need to find two integers whose product is ___ and whose sum is ___.

22. $15c^2 + 4c - 4 = 15c^2 + \boxed{}c + \boxed{}c - 4$
    We need to find two integers whose product is ___ and whose sum is ___.

Complete each step of the factorization by grouping.

23. $12t^2 + 17t + 6 = 12t^2 + \boxed{}t + 8t + 6$
    $= \boxed{}(4t + 3) + \boxed{}(4t + 3)$
    $= (\boxed{})(3t + 2)$

24. $35t^2 - 11t - 6 = 35t^2 + \boxed{}t - 21t - 6$
    $= 5t(7t + 2) - 3(7t \;\; 2)$
    $= (\boxed{})(5t - 3)$

NOTATION

25. Write a trinomial of the form: $ax^2 + bx + c$
    a. where $a = 1$
    b. where $a \neq 1$

26. Write the terms of the trinomial $40 - t - 4t^2$ in descending powers of the variable.

GUIDED PRACTICE

Factor each trinomial. See Example 1.

27. $3a^2 + 13a + 4$
28. $2b^2 + 7b + 6$
29. $4z^2 + 13z + 3$
30. $6y^2 + 7y + 2$

Factor each trinomial. See Example 2.

31. $4t^2 - 4t + 1$
32. $10x^2 - 9x + 2$
33. $2x^2 - 3x + 1$
34. $2y^2 - 7y + 3$

Factor each trinomial. See Example 3.

35. $8u^2 - 2u - 15$
36. $2x^2 - 3x - 2$
37. $12y^2 - y - 1$
38. $10a^2 - 3a - 4$

Factor each trinomial. See Example 4.

39. $6r^2 + rs - 2s^2$
40. $4a^2 - 4ab + b^2$
41. $2b^2 - 5bc + 2c^2$
42. $3m^2 + 5mn + 2n^2$

Factor each trinomial. See Example 5.

43. $4x^2 + 10x - 6$
44. $9x^2 + 21x - 18$
45. $-y^3 - 13y^2 - 12y$
46. $-2xy^2 - 8xy + 24x$
47. $6x^3 - 15x^2 - 9x$
48. $9y^3 + 3y^2 - 6y$
49. $30r^5 + 63r^4 - 30r^3$
50. $6s^5 - 26s^4 - 20s^3$

Factor each trinomial by grouping. See Example 6.

51. $10y^2 - 3y - 1$
52. $6m^2 + 19m + 3$
53. $12y^2 - 5y - 2$
54. $10x^2 + 21x - 10$

Factor each trinomial by grouping. See Example 7.

55. $12y^4 + y^3 - y^2$
56. $36p^3 - 3p^2 - 18p$
57. $-16m^3n - 20m^2n^2 - 6mn^3$
58. $-84x^4 - 100x^3y - 24x^2y^2$

TRY IT YOURSELF

Completely factor each of the following expressions. If an expression is prime, so indicate.

59. $4x^2 + 8x + 3$
60. $15t^2 - 34t + 8$
61. $7x^2 - 9x + 2$
62. $8m^2 + 5m - 10$
63. $10u^2 - 13u - 6$
64. $-5t^2 - 13t - 6$
65. $-16y^2 - 10y - 1$
66. $-16m^2 + 14m - 3$
67. $-16x^2 - 16x - 3$
68. $13x^2 + 24xy + 11y^2$
69. $4b^2 + 15bc - 4c^2$
70. $6r^2 + rs - 2s^2$
71. $12x^2 + 5xy - 3y^2$
72. $-13x + 3x^2 - 10$
73. $-14 + 3a^2 - a$
74. $15 + 8a^2 - 26a$
75. $16 - 40a + 25a^2$
76. $6a^2 + 6b^2 - 13ab$
77. $11uv + 3u^2 + 6v^2$
78. $pq + 6p^2 - q^2$

Choose the correct method from Sections 6.1, 6.2, or 6.3 to factor completely each of the following expressions. If an expression is prime, so indicate.

79. $12y^2 + 12 - 25y$
80. $15c^2d^3 - 25c^3d^2 - 10c^4d^4$
81. $6x^2 - 15x + 2xy - 5y$
82. $12t^2 - 1 - 4t$
83. $6a^2 - 10 - 11a$
84. $3x^2 + 6 + x$
85. $12p^2 + 5pq - 2q^2$
86. $25 + 2u^2 + 3u$
87. $-3a^3 - 6a^2 + 9a$
88. $3m^2 + 4m - 6mn - 8n$
89. $-28u^3v^3 + 26u^2v^4 - 6uv^5$
90. $9t^3 + 33t^2 - 12t$
91. $-16x^4y^3 + 30x^3y^4 + 4x^2y^5$
92. $22pq + 6p^2 + 12q^2$
93. $-11mn + 12m^2 + 2n^2$
94. $-18b + 36b^3 - 3b^2$

APPLICATIONS

95. **OFFICE FURNITURE** The area of the desktop shown below is given by the expression $(4x^2 + 20x - 11)$ in.^{2}. Factor this expression to find the expressions that represent its length and width. Then determine the difference in the length and width of the desktop.

96. **STORAGE** The volume of the 8-foot-wide portable storage container shown below is given by the expression $(72x^2 + 120x - 400)$ ft^3. If its dimensions can be determined by factoring the expression, find the height and the length of the container.

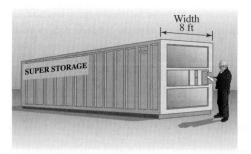

WRITING

97. A student begins to factor a trinomial as shown below. Explain why the student is off to a bad start.

$$3x^2 - 5x - 2 = (3x - \square)(x - \square)$$

98. Two students factor $2x^2 + 20x + 42$ and get two different answers:

$$(2x + 6)(x + 7) \quad \text{and} \quad (x + 3)(2x + 14)$$

Do both answers check? Why don't they agree? Is either answer completely correct? Explain.

99. Why is the process of factoring $6x^2 - 5x - 6$ more complicated than the process of factoring $x^2 - 5x - 6$?

100. How can the factorization shown below be checked?

$$6x^2 - 5x - 6 = (3x + 2)(2x - 3)$$

REVIEW

101. Simplify: $(x^2x^5)^2$
102. Simplify: $\dfrac{(a^3b^4)^2}{ab^5}$
103. Evaluate: $\dfrac{1}{2^{-3}}$
104. Evaluate: 7^0

Objectives

1. Recognize perfect-square trinomials.
2. Factor perfect-square trinomials.
3. Factor the difference of two squares.

SECTION 6.4

Factoring Perfect-Square Trinomials and the Difference of Two Squares

In this section, we will discuss a method that can be used to factor two types of trinomials, called *perfect-square trinomials*. We also develop techniques for factoring a type of binomial called the *difference of two squares*.

1 Recognize perfect-square trinomials.

We have seen that the square of a binomial is a trinomial. We have also seen that the special product rules shown below can be used to quickly find the square of a sum and the square of a difference. The terms of the resulting trinomial are related to the terms of the binomial that was squared.

$$(A + B)^2 = A^2 + 2AB + B^2$$

- A^2: This is the square of the first term of the binomial.
- $2AB$: This is twice the product of the terms of the binomial.
- B^2: This is the square of the last term of the binomial.

$$(A - B)^2 = A^2 - 2AB + B^2$$

Trinomials that are squares of a binomial are called **perfect-square trinomials**. Some examples are

$y^2 + 6y + 9$ Because it is the square of $(y + 3)$: $(y + 3)^2 = y^2 + 6y + 9$
$t^2 - 14t + 49$ Because it is the square of $(t - 7)$: $(t - 7)^2 = t^2 - 14t + 49$
$4m^2 - 20m + 25$ Because it is the square of $(2m - 5)$: $(2m - 5)^2 = 4m^2 - 20m + 25$

EXAMPLE 1   Determine whether the following are perfect-square trinomials: **a.** $x^2 + 10x + 25$ **b.** $c^2 - 12c - 36$ **c.** $25y^2 - 30y + 9$ **d.** $4t^2 + 18t + 81$

Strategy We will compare each trinomial, term by term, to one of the special product forms discussed above.

WHY If a trinomial matches one of these forms, it is a perfect-square trinomial.

Solution

a. To determine whether this is a perfect-square trinomial, we note that

$$x^2 + 10x + 25$$

- The first term is the square of x.
- The middle term is twice the product of x and 5: $2 \cdot x \cdot 5 = 10x$.
- The last term is the square of 5.

Thus, $x^2 + 10x + 25$ is a perfect-square trinomial.

b. To determine whether this is a perfect-square trinomial, we note that

$$c^2 - 12c - 36$$

The last term, -36, is not the square of a real number.

Since the last term is negative, $c^2 - 12c - 36$ is not a perfect-square trinomial.

c. To determine whether this is a perfect-square trinomial, we note that

$$25y^2 - 30y + 9$$

- The first term is the square of $5y$.
- The middle term is twice the product of $5y$ and -3: $2(5y)(-3) = -30y$.
- The last term is the square of -3.

Thus, $25y^2 - 30y + 9$ is a perfect-square trinomial.

Self Check 1

Determine whether the following are perfect-square trinomials:
a. $y^2 + 4y + 4$
b. $b^2 - 6b - 9$
c. $4z^2 + 4z + 4$
d. $49x^2 - 28x + 16$

Now Try Problems 25, 29, and 32

The Language of Algebra The expressions $25y^2$ and 9 are called *perfect squares* because $25y^2 = (5y)^2$ and $9 = 3^2$.

d. To determine whether this is a perfect-square trinomial, we note that

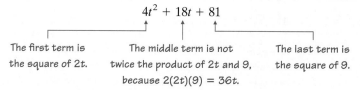

The first term is the square of $2t$.

The middle term is not twice the product of $2t$ and 9, because $2(2t)(9) = 36t$.

The last term is the square of 9.

Thus, $4t^2 + 18t + 81$ is not a perfect-square trinomial.

2 Factor perfect-square trinomials.

We can factor perfect-square trinomials using the methods previously discussed in Sections 6.2 and 6.3. However, in many cases, we can factor them more quickly by inspecting their terms and applying the special product rules in reverse.

> **Factoring Perfect-Square Trinomials**
>
> $A^2 + 2AB + B^2 = (A + B)^2$ Each of these trinomials factors as the square of a binomial.
>
> $A^2 - 2AB + B^2 = (A - B)^2$

When factoring perfect-square trinomials, it is helpful to know the integers that are perfect squares. The number 400, for example, is a perfect integer square, because $400 = 20^2$.

$1 = 1^2$	$25 = 5^2$	$81 = 9^2$	$169 = 13^2$	$289 = 17^2$
$4 = 2^2$	$36 = 6^2$	$100 = 10^2$	$196 = 14^2$	$324 = 18^2$
$9 = 3^2$	$49 = 7^2$	$121 = 11^2$	$225 = 15^2$	$361 = 19^2$
$16 = 4^2$	$64 = 8^2$	$144 = 12^2$	$256 = 16^2$	$400 = 20^2$

Self Check 2

Factor:
a. $x^2 + 18x + 81$
b. $16x^2 - 8xy + y^2$

Now Try Problems 33 and 35

EXAMPLE 2 Factor: **a.** $x^2 + 20x + 100$ **b.** $9x^2 - 30xy + 25y^2$

Strategy The terms of each trinomial do not have a common factor (other than 1). We will determine whether each is a perfect-square trinomial.

WHY If it is, we can factor it using a special product rule in reverse.

Solution

a. $x^2 + 20x + 100$ is a perfect-square trinomial, because:

- The first term x^2 is the square of x.
- The last term 100 is the square of 10.
- The middle term is twice the product of x and 10: $2(x)(10) = 20x$.

To find the factorization, we match $x^2 + 20x + 100$ to the proper rule for factoring a perfect-square trinomial.

$$A^2 + 2 \cdot A \cdot B + B^2 = (A + B)^2$$
$$x^2 + 20x + 10^2 = x^2 + 2 \cdot x \cdot 10 + 10^2 = (x + 10)^2$$

Therefore, $x^2 + 20x + 10^2 = (x + 10)^2$. Check by finding $(x + 10)^2$.

b. $9x^2 - 30xy + 25y^2$ is a perfect-square trinomial, because:

- The first term $9x^2$ is the square of $3x$: $(3x)^2 = 9x^2$.
- The last term $25y^2$ is the square of $-5y$: $(-5y)^2 = 25y^2$.
- The middle term is twice the product of $3x$ and $-5y$: $2(3x)(-5y) = -30xy$.

We can use these observations to write the trinomial in one of the special product forms that then leads to its factorization.

$$9x^2 - 30xy + 25y^2 = (3x)^2 - 2(3x)(5y) + (-5y)^2 \quad -2(3x)(5y) = 2(3x)(-5y)$$
$$= (3x - 5y)^2$$

Therefore, $9x^2 - 30xy + 25y^2 = (3x - 5y)^2$. Check by finding $(3x - 5y)^2$.

> **Success Tip** The sign of the middle term of a perfect-square trinomial is the same as the sign of the second term of the squared binomial.
>
> $$A^2 + 2AB + B^2 = (A + B)^2$$
> $$A^2 - 2AB + B^2 = (A - B)^2$$

EXAMPLE 3
Factor completely: $4a^3 - 4a^2 + a$

Strategy We will factor out the GCF, a, first. Then we will factor the resulting perfect-square trinomial using a special product rule in reverse.

WHY The first step in factoring any polynomial is to factor out the GCF.

Solution
The terms of $4a^3 - 4a^2 + a$ have the common factor a, which should be factored out first. Within the parentheses, we recognize $4a^2 - 4a + 1$ as a perfect-square trinomial of the form $A^2 - 2AB + B^2$, and factor it as such.

$$4a^3 - 4a^2 + a = a(4a^2 - 4a + 1) \quad \text{Factor out } a.$$
$$= a(2a - 1)^2 \quad 4a^2 = (2a)^2, 1 = (-1)^2, \text{ and } -4a = 2(2a)(-1).$$

Self Check 3
Factor completely:
$49x^3 - 14x^2 + x$

Now Try Problem 41

3 Factor the difference of two squares.

Recall the special product rule for multiplying the sum and difference of the same two terms:

$$(A + B)(A - B) = A^2 - B^2$$

The binomial $A^2 - B^2$ is called a **difference of two squares**, because A^2 is the square of A and B^2 is the square of B. If we reverse this rule, we obtain a method for factoring a difference of two squares.

Factoring ⟶
$$A^2 - B^2 = (A + B)(A - B)$$

This pattern is easy to remember if we think of a difference of two squares as the square of a **First** quantity minus the square of a **Last** quantity.

> **The Language of Algebra** The expression $A^2 - B^2$ is a *difference of two squares*, whereas $(A - B)^2$ is the *square of a difference*. They are not equivalent because $(A - B)^2 \neq A^2 - B^2$.

Factoring a Difference of Two Squares

To factor the square of a First quantity minus the square of a Last quantity, multiply the First plus the Last by the First minus the Last.

$$F^2 - L^2 = (F + L)(F - L)$$

Self Check 4

Factor, if possible:
a. $c^2 - 4$
b. $121 - t^2$
c. $x^2 - 24$
d. $s^2 + 36$

Now Try Problems 45 and 53

EXAMPLE 4 Factor, if possible:
a. $x^2 - 9$ b. $16 - b^2$ c. $n^2 - 45$ d. $a^2 + 81$

Strategy The terms of each binomial do not have a common factor (other than 1). The only option available is to attempt to factor each as a difference of two squares.

WHY If a binomial is a difference of two squares, we can factor it using a special product rule in reverse.

Solution

a. $x^2 - 9$ is the difference of two squares because it can be written as $x^2 - 3^2$. We can match it to the rule for factoring a difference of two squares to find the factorization.

$$\mathbf{F}^2 - \mathbf{L}^2 = (\mathbf{F} + \mathbf{L})(\mathbf{F} - \mathbf{L})$$
$$x^2 - 3^2 = (x + 3)(x - 3) \quad \text{9 is a perfect integer square: } 9 = 3^2.$$

Therefore, $x^2 - 9 = (x + 3)(x - 3)$.

Check by multiplying: $(x + 3)(x - 3) = x^2 - 9$.

b. $16 - b^2$ is the difference of two squares because $16 - b^2 = 4^2 - b^2$. Therefore,

$$16 - b^2 = (4 + b)(4 - b) \quad \text{16 is a perfect integer square: } 16 = 4^2.$$

Check by multiplying.

c. Since 45 is not a perfect integer square, $n^2 - 45$ cannot be factored using integers. It is prime.

d. $a^2 + 81$ can be written $a^2 + 9^2$, and is, therefore, the **sum of two squares**. We might attempt to factor $a^2 + 81$ as $(a + 9)(a + 9)$ or $(a - 9)(a - 9)$. However, the following checks show that neither product is $a^2 + 81$.

$$(a + 9)(a + 9) = a^2 + 18a + 81 \qquad (a - 9)(a - 9) = a^2 - 18a + 81$$

In general, the sum of two squares (with no common factor other than 1) cannot be factored using real numbers. Thus, $a^2 + 81$ is prime.

Terms containing variables such as $25x^2$ and $4y^4$ are perfect squares, because they can be written as the square of a quantity. For example:

$$25x^2 = (5x)^2 \quad \text{and} \quad 4y^4 = (2y^2)^2$$

Self Check 5

Factor:
a. $16y^2 - 9$
b. $9m^2 - 64n^4$

Now Try Problems 57 and 59

EXAMPLE 5 Factor: a. $25x^2 - 49$ b. $4y^4 - 121z^2$

Strategy In each case, the terms of the binomial do not have a common factor (other than 1). To factor them, we will write each binomial in a form that clearly shows it is a difference of two squares.

WHY We can then use a special product rule in reverse to factor it.

Solution

a. We can write $25x^2 - 49$ in the form $(5x)^2 - 7^2$ and match it to the rule for factoring the difference of two squares:

$$F^2 - L^2 = (F + L)(F - L)$$
$$(5x)^2 - 7^2 = (5x + 7)(5x - 7)$$

Therefore, $25x^2 - 49 = (5x + 7)(5x - 7)$. Check by multiplying.

b. We can write $4y^4 - 121z^2$ in the form $(2y^2)^2 - (11z)^2$ and match it to the rule for factoring the difference of two squares:

$$F^2 - L^2 = (F + L)(F - L)$$
$$(2y^2)^2 - (11z)^2 = (2y^2 + 11z)(2y^2 - 11z)$$

Therefore, $4y^4 - 121z^2 = (2y^2 + 11z)(2y^2 - 11z)$. Check by multiplying.

> **Success Tip** Remember that a *difference of two squares* is a binomial. Each term is a square and the terms have different signs. The powers of the variables in the terms must be even.

EXAMPLE 6 Factor completely: $8x^2 - 8$

Strategy We will factor out the GCF, 8, first. Then we will factor the resulting difference of two squares.

WHY The first step in factoring any polynomial is to factor out the GCF.

Solution

$8x^2 - 8 = 8(x^2 - 1)$ The GCF is 8.

$\qquad = 8(x + 1)(x - 1)$ Think of $x^2 - 1$ as $x^2 - 1^2$ and factor the difference of two squares.

Check: $8(x + 1)(x - 1) = 8(x^2 - 1)$ Multiply the binomials first.

$\qquad\qquad\qquad\qquad = 8x^2 - 8$ Distribute the multiplication by 8.

Self Check 6

Factor completely: $4x^2 - 400$

Now Try Problem 65

ANSWERS TO SELF CHECKS

1. a. yes **b.** no **c.** no **d.** no **2. a.** $(x + 9)^2$ **b.** $(4x - y)^2$ **3.** $x(7x - 1)^2$
4. a. $(c + 2)(c - 2)$ **b.** $(11 + t)(11 - t)$ **c.** prime **d.** prime
5. a. $(4y + 3)(4y - 3)$ **b.** $(3m + 8n^2)(3m - 8n^2)$ **6.** $4(x + 10)(x - 10)$

SECTION 6.4   STUDY SET

VOCABULARY

Fill in the blanks.

1. $x^2 + 6x + 9$ is a _____-square trinomial because it is the square of the binomial $x + 3$.

2. The binomial $x^2 - 25$ is called a _____ of two squares.

CONCEPTS

Fill in the blanks.

3. Consider: $25x^2 + 30x + 9$
   a. The first term is the square of ___.
   b. The last term is the square of ___.
   c. The middle term is twice the product of ___ and ___.

4. Consider: $49x^2 - 28xy + 4y^2$
   a. The first term is the square of ___.
   b. The last term is the square of ___.
   c. The middle term is twice the product of ___ and ___.

5. If a trinomial is the square of one quantity, plus the square of a second quantity, plus ___ the product of the quantities, it factors into the square of the ___ of the quantities.

6. Explain why each trinomial is not a perfect-square trinomial.
   a. $9h^2 - 6h + 7$
   b. $j^2 - 8j - 16$
   c. $25r^2 + 20r + 16$

7. List the first ten perfect integer squares.

8. To factor the square of a First quantity minus the square of a Last quantity, we multiply the ___ plus the ___ by the ___ minus the ___.

9. a. $36x^2 = ()^2$ b. $100x^4 = ()^2$
   c. $4x^2 - 9 = ()^2 - ()^2$

10. a. Three incorrect factorizations of $x^2 + 36$ are given below. Explain why each one is wrong.
    $(x + 6)(x - 6)$
    $(x + 6)(x + 6)$
    $(x - 6)(x - 6)$
    b. Can $x^2 + 36$ be factored using only integer coefficients?

Complete each factorization.

11. $a^2 - 6a + 9 = (a -)^2$
12. $t^2 + 2t + 1 = (t 1)^2$
13. $4x^2 + 4x + 1 = (2x 1)^2$
14. $9y^2 - 12y + 4 = (3y -)^2$
15. $y^2 - 49 = (y +)(y -)$
16. $p^4 - q^2 = (p^2 + q)(-)$
17. $t^2 - w^2 = (+)(t - w)$
18. $49u^2 - 64v^2 = (+ 8v)(7u 8v)$

NOTATION

Write each expression as a polynomial in simpler form.

19. $(3a)^2 - 2(3a)(5b) + (5b)^2$
20. $(2s)^2 + 2(2s)(9t) + (9t)^2$
21. $(6x)^2 - (5y)^2$
22. $(4x)^2 - (9y)^2$

Use an exponent to write each expression in simpler form.

23. $(x + 8)(x + 8)$
24. $(x - 8)(x - 8)$

GUIDED PRACTICE

Determine whether the following expressions are perfect-square trinomials. See Example 1.

25. $x^2 + 18x + 81$
26. $x^2 + 14x + 49$
27. $y^2 + 2y + 4$
28. $y^2 + 4y + 16$
29. $9n^2 - 30n - 25$
30. $9a^2 - 48a - 64$
31. $4y^2 - 12y + 9$
32. $9x^2 - 30x + 25$

Factor each polynomial. See Example 2.

33. $x^2 + 6x + 9$
34. $x^2 + 10x + 25$
35. $t^2 - 20t + 100$
36. $r^2 + 24r + 144$
37. $a^2 + 2ab + b^2$
38. $a^2 - 2ab + b^2$
39. $16x^2 - 8xy + y^2$
40. $25x^2 + 20xy + 4y^2$

Factor each polynomial. See Example 3.

41. $y^3 - 8y^2 + 16y$
42. $u^4 - 18u^3 + 81u^2$
43. $8x^3 + 24x^2 + 18x$
44. $108x^3 + 36x^2 + 3x$

Factor each polynomial. If a polynomial is prime, so indicate. See Example 4.

45. $x^2 - 16$
46. $x^2 - 25$
47. $t^2 - 49$
48. $m^2 - 121$
49. $49 - c^2$
50. $81 - t^2$
51. $144 - 25a^2$
52. $169 - 9t^2$
53. $p^2 - 54$
54. $q^2 - 20$
55. $a^2 + b^2$
56. $121a^2 + 144b^2$

Factor each polynomial. See Example 5.

57. $4y^2 - 1$
58. $9z^2 - 1$
59. $49a^2 - 169$
60. $16b^2 - 225$
61. $9x^2 - y^2$
62. $4x^2 - z^2$
63. $16a^2 - 25b^2$
64. $36a^2 - 121b^2$

Factor each polynomial. See Example 6.

65. $8a^2 - 32$ **66.** $2p^2 - 200$

67. $7 - 7a^2$ **68.** $5 - 20x^2$

TRY IT YOURSELF

Factor each expression completely.

69. $z^2 - 2z + 1$ **70.** $v^2 - 14v + 49$

71. $4x^2 - 4x + 1$ **72.** $4x^2 - 20x + 25$

73. $a^4 - 144b^2$ **74.** $81y^4 - 100z^2$

75. $t^2z^2 - 64$ **76.** $900 - B^2C^2$

77. $6x^4 - 6x^2y^2$ **78.** $4b^2y - 16c^2y$

79. $a^2b^2 - 144$ **80.** $20m^2 + 100m + 125$

81. $16 - 40z + 25z^2$ **82.** $49p^2 + 28pq + 4q^2$

83. $8a^2x^3y - 2b^2xy$ **84.** $16x^2 - 40x^3 + 25x^4$

Choose the correct method from Section 6.1, 6.2, 6.3, or 6.4 to factor completely each of the following expressions:

85. $8x^2 - 32y^2$ **86.** $2a^2 - 200b^2$

87. $8m^2n^3 - 24mn^4$ **88.** $3rs + 6r^2 - 18s^2$

89. $x^2 + 7x + 1$ **90.** $14t^3 - 40t^2 + 6t^4$

91. $-9x^2y^2 + 6xy - 1$ **92.** $2c^2 - 5cd - 3d^2$

93. $2ac + 4ad + bc + 2bd$ **94.** $10r^2 - 13r - 4$

95. $6x^2 - x - 16$ **96.** $4x^2 + 9y^2$

97. $6a^3 + 35a^2 - 6a$ **98.** $21t^3 - 10t^2 + t$

99. $70p^4q^3 - 35p^4q^2 + 49p^5q^2$
100. $2ab^2 + 8ab - 24a$
101. $a^2c + a^2d^2 + bc + bd^2$
102. $-8p^3q^7 - 4p^2q^3$

APPLICATIONS

103. GENETICS The Hardy–Weinberg equation, one of the fundamental concepts in population genetics, is $p^2 + 2pq + q^2 = 1$ where p represents the frequency of a certain dominant gene and q represents the frequency of a certain recessive gene. Factor the left side of the equation.

104. CHECKERS The area of a square checkerboard is represented by the polynomial $25x^2 - 40x + 16$. Use factoring to find an expression that represents the length of a side.

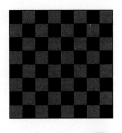

105. PHYSICS The illustration shows a time-sequence picture of a falling apple. Factor the expression, which gives the distance the apple falls during the time interval from t_1 to t_2 seconds.

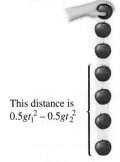

This distance is $0.5gt_1^2 - 0.5gt_2^2$

106. SIGNAL FLAGS The maritime signal flag for the letter X is shown. Find the polynomial that represents the area of the shaded region and express it in factored form.

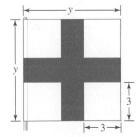

107. MOVIE STUNTS The function that gives the distance a stuntwoman is above the ground t seconds after she falls over the side of a 144-foot-tall building is $h(t) = 144 - 16t^2$. Factor the right side.

108. DARTS A circular dartboard has a series of rings around a solid center, called the bull's-eye. To find the area of the outer black ring, we can use the formula

$$A = \pi R^2 - \pi r^2$$

Factor the expression on the right side of the equation.

WRITING

109. When asked to factor $x^2 - 25$, one student wrote $(x + 5)(x - 5)$, and another student wrote $(x - 5)(x + 5)$. Are both answers correct? Explain.

110. Explain the error in the factorization shown below:
$$4x^2 - 16y^2 = (2x + 4y)(2x - 4y)$$

REVIEW

Perform each division.

111. $\dfrac{5x^2 + 10y^2 - 15xy}{5xy}$

112. $\dfrac{-30c^2d^2 - 15c^2d - 10cd^2}{-10cd}$

113. $2a - 1 \overline{)a - 2 + 6a^2}$

114. $4b + 3 \overline{)4b^3 - 5b^2 - 2b + 3}$

Objective

1 Factor the sum and difference of two cubes.

SECTION 6.5
Factoring the Sum and Difference of Two Cubes

In this section we will discuss how to factor two types of binomials, called the *sum* and the *difference of two cubes*.

1 Factor the sum and difference of two cubes.

We have seen that the sum of two squares, such as $x^2 + 4$ or $25a^2 + 9b^2$, cannot be factored. However, the sum of two cubes and the difference of two cubes can be factored.

The sum of two cubes
$x^3 + 8$

This is x cubed.   This is 2 cubed: $2^3 = 8$.

The difference of two cubes
$a^3 - 64b^3$

This is a cubed.   This is $4b$ cubed: $(4b)^3 = 64b^3$.

To find rules for factoring the sum of two cubes and the difference of two cubes, we need to find the products shown below. Note that each term of the trinomial is multiplied by each term of the binomial.

$$(x + y)(x^2 - xy + y^2) = x^3 - x^2y + xy^2 + x^2y - xy^2 + y^3$$
$$= x^3 + y^3 \qquad \text{Combine like terms: } -x^2y + x^2y = 0 \text{ and } xy^2 - xy^2 = 0.$$

> **The Language of Algebra** The expression $x^3 + y^3$ is a *sum of two cubes*, whereas $(x + y)^3$ is the *cube of a sum*. If you expand $(x + y)^3$, you will see that $(x + y)^3 \neq x^3 + y^3$.

$$(x - y)(x^2 + xy + y^2) = x^3 + x^2y + xy^2 - x^2y - xy^2 - y^3$$
$$= x^3 - y^3 \qquad \text{Combine like terms.}$$

These results justify the rules for factoring the **sum and difference of two cubes**. They are easier to remember if we think of a sum (or a difference) of two cubes as the cube of a **F**irst quantity plus (or minus) the cube of the **L**ast quantity.

Factoring the Sum and Difference of Two Cubes

To factor the cube of a First quantity plus the cube of a Last quantity, multiply the First plus the Last by the First squared, minus the First times the Last, plus the Last squared.

$$F^3 + L^3 = (F + L)(F^2 - FL + L^2)$$

To factor the cube of a First quantity minus the cube of a Last quantity, multiply the First minus the Last by the First squared, plus the First times the Last, plus the Last squared.

$$F^3 - L^3 = (F - L)(F^2 + FL + L^2)$$

To factor the sum or difference of two cubes, it is helpful to know the cubes of integers from 1 to 10. The number 216, for example, is a **perfect integer cube**, because $216 = 6^3$.

$1 = 1^3$	$27 = 3^3$	$125 = 5^3$	$343 = 7^3$	$729 = 9^3$
$8 = 2^3$	$64 = 4^3$	$216 = 6^3$	$512 = 8^3$	$1{,}000 = 10^3$

EXAMPLE 1 Factor: $x^3 + 8$

Strategy We will write $x^3 + 8$ in a form that clearly shows it is the sum of two cubes.

WHY We can then use the rule for factoring the sum of two cubes.

Solution
$x^3 + 8$ is the sum of two cubes because it can be written as $x^3 + 2^3$. We can match it to the rule for factoring the sum of two cubes to find its factorization.

$$\mathbf{F}^3 + \mathbf{L}^3 = (\mathbf{F} + \mathbf{L})(\mathbf{F}^2 - \mathbf{F}\ \mathbf{L} + \mathbf{L}^2)$$ To write the trinomial factor:
- Square the first term of the binomial factor.
- Multiply the terms of the binomial factor.

$$x^3 + 2^3 = (x + 2)(x^2 - x \cdot 2 + 2^2)$$ · Square the last term of the binomial factor.

$$= (x + 2)(x^2 - 2x + 4) \qquad x^2 - 2x + 4 \text{ does not factor.}$$

Therefore, $x^3 + 8 = (x + 2)(x^2 - 2x + 4)$. We can check by multiplying.

$$(x + 2)(x^2 - 2x + 4) = x^3 - 2x^2 + 4x + 2x^2 - 4x + 8$$
$$= x^3 + 8 \qquad \text{This is the original binomial.}$$

Caution! A common error is to try to factor $x^2 - 2x + 4$. It is not a perfect square trinomial, because the middle term needs to be $-4x$. Furthermore, it cannot be factored by the methods of Section 6.2. It is prime.

Terms containing variables such as $64b^3$ and m^6 are also perfect cubes, because they can be written as the cube of a quantity:

$$64b^3 = (4b)^3 \quad \text{and} \quad m^6 = (m^2)^3$$

EXAMPLE 2 Factor: $a^3 - 64b^3$

Strategy We will write $a^3 - 64b^3$ in a form that clearly shows it is the difference of two cubes.

WHY We can then use the rule for factoring the difference of two cubes.

Self Check 1
Factor: $h^3 + 27$
Now Try Problem 17

Self Check 2
Factor: $8c^3 - 1$
Now Try Problem 31

Solution

$a^3 - 64b^3$ is the difference of two cubes because it can be written as $a^3 - (4b)^3$. We can match it to the rule for factoring the difference of two cubes to find its factorization.

$$F^3 - L^3 = (F - L)(F^2 + F \cdot L + L^2)$$
$$a^3 - (4b)^3 = (a - 4b)[a^2 + a \cdot 4b + (4b)^2]$$
$$= (a - 4b)(a^2 + 4ab + 16b^2) \quad \text{$a^2 + 4ab + 16b^2$ does not factor.}$$

Therefore, $a^3 - 64b^3 = (a - 4b)(a^2 + 4ab + 16b^2)$. Check by multiplying.

You should memorize the rules for factoring the sum and the difference of two cubes. Note that the right side of each rule has the form

(a binomial)(a trinomial)

and that there is a relationship between the signs that appear in these forms.

The Sum of Two Cubes

$$F^3 + L^3 = (F + L)(F^2 - FL + L^2)$$

The same sign (first), Opposite signs, Always plus

The Difference of Two Cubes

$$F^3 - L^3 = (F - L)(F^2 + FL + L^2)$$

The same sign (first), Opposite signs, Always plus

If the terms of a binomial have a common factor, the GCF (or the opposite of the GCF) should always be factored out first.

Self Check 3
Factor: $4c^3 + 4d^3$

Now Try Problems 33 and 39

EXAMPLE 3
Factor: $-2t^5 + 250t^2$

Strategy We will factor out the common factor, $-2t^2$. We can then factor the resulting binomial as a difference of two cubes.

WHY The first step in factoring any polynomial is to factor out the GCF, or its opposite.

Solution

$$-2t^5 + 250t^2 = -2t^2(t^3 - 125) \quad \text{Factor out the opposite of the GCF, } -2t^2.$$
$$= -2t^2(t - 5)(t^2 + 5t + 25) \quad \text{Factor } t^3 - 125.$$

Therefore, $-2t^5 + 250t^2 = -2t^2(t - 5)(t^2 + 5t + 25)$. Check by multiplying.

ANSWERS TO SELF CHECKS
1. $(h + 3)(h^2 - 3h + 9)$ 2. $(2c - 1)(4c^2 + 2c + 1)$ 3. $4(c + d)(c^2 - cd + d^2)$

SECTION 6.5   STUDY SET

VOCABULARY

1. The binomial $x^3 + 27$ is called a sum of two _____.
2. The binomial $x^3 - 8$ is called a _____ of two cubes.

CONCEPTS

Fill in the blanks.

3. $F^3 + L^3 = (\boxed{} + \boxed{})(F^2 - FL + L^2)$
4. $F^3 - L^3 = (F - L)(\boxed{} + FL + \boxed{})$

6.5 Factoring the Sum and Difference of Two Cubes

5. $(x - 2)(x^2 + 2x + 4)$ is the factorization of what binomial?
6. $(x + 3)(x^2 - 3x + 9)$ is the factorization of what binomial?
7. List the first five positive perfect integer cubes.
8. Use multiplication to determine whether the factorization is correct.
$$b^3 + 64 = (b + 4)(b^2 + 4b + 16)$$
9. $27m^3 = (\quad)^3$
10. $a^6 = (\quad)^3$
11. $8x^3 - 27 = (\quad)^3 - (\quad)^3$
12. $x^3 + 64y^3 = (\quad)^3 + (\quad)^3$

NOTATION

Complete each factorization.

13. $a^3 + 8 = (a + 2)(a^2 - \quad + 4)$
14. $x^3 - 1 = (x - 1)(x^2 + \quad + 1)$
15. $b^3 + 27 = (\quad)(b^2 - 3b + 9)$
16. $z^3 - 125 = (\quad)(z^2 + 5z + 25)$

GUIDED PRACTICE

Factor each polynomial. See Example 1.

17. $y^3 + 1$
18. $b^3 + 125$
19. $y^3 + 343$
20. $b^3 + 216$
21. $a^3 + 64$
22. $n^3 + 1$
23. $m^3 + 512$
24. $t^3 + 729$

Factor each polynomial. See Example 2.

25. $x^3 - 8$
26. $a^3 - 27$
27. $z^3 - 343$
28. $c^3 - 1{,}000$
29. $s^3 - 8t^3$
30. $27a^3 - b^3$
31. $s^3 - 64t^3$
32. $64x^3 - 27y^3$

Factor each polynomial. See Example 3.

33. $-x^3 + 216$
34. $-x^3 - 125$
35. $2x^3 + 54$
36. $2x^3 - 2$
37. $64m^3x - 8n^3x$
38. $16r^4 + 128rs^3$
39. $x^4y + 216xy^4$
40. $16a^5 - 54a^2b^3$

TRY IT YOURSELF

Completely factor each expression.

41. $8 + x^3$
42. $27 - y^3$
43. $8u^3 + w^3$
44. $a^3 + 8b^3$
45. $64x^3 - 27$
46. $27x^3 + 125$
47. $a^6 - b^3$
48. $a^3 + b^6$
49. $x^9 + y^6$
50. $x^3 - y^9$
51. $81r^4s^2 - 24rs^5$
52. $4m^5n + 500m^2n^4$
53. $3a^3 + 24b^3$
54. $-2x^5 + 128x^2$
55. $8p^6 - 27q^6$
56. $125p^3 - 64y^3$

Choose the correct method from Sections 6.1–6.5 to factor completely each of the following expressions:

57. $x^2 - 81$
58. $y^2 - 625$
59. $a^2 - 16$
60. $b^2 - 256$
61. $81r^2 - 256s^2$
62. $16y^2 - 81z^2$
63. $a^2(x - a) - b^2(x - a)$
64. $at^2 - 16a$
65. $x^2y^2 - 2x^2 - y^2 + 2$
66. $5x^3y^3z^4 + 25x^2y^3z^2 - 35x^3y^2z^5$
67. $81p^4 - 16q^4$
68. $30a^4 + 5a^3 - 200a^2$
69. $54x^3 + 250y^6$
70. $-16x^4y^2z + 24x^5y^3z^4 - 15x^2y^3z^7$

APPLICATIONS

71. **MAILING BREAKABLES** Write a polynomial that describes the amount of space in the larger box that must be filled with styrofoam chips if the smaller box containing a glass tea cup is to be placed within the larger box for mailing. Then factor the polynomial.

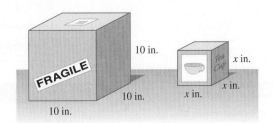

506 Chapter 6 Factoring and Quadratic Equations

72. MELTING ICE In one hour, the block of ice shown below had melted to the size shown on the right. Write a polynomial that describes the volume of ice that melted away. Then factor the polynomial.

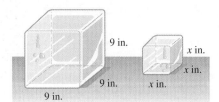

WRITING

73. Explain why $x^3 - 25$ is not the difference of two cubes.

74. Explain why $x^6 - 1$ can be thought of as a difference of two squares or as a difference of two cubes.

REVIEW

75. Solve: $x + 20 = 4x - 1 + 2x$

76. Evaluate $2x^2 + 5x - 3$ for $x = -3$.

77. Is 4 a solution of $3(m - 8) + 2m = 4 - (m + 2)$?

78. When expressed as a decimal, is $\frac{7}{9}$ a terminating or repeating decimal?

Objective

1 Factor randomly chosen polynomials.

SECTION 6.6
A Factoring Strategy

The factoring methods discussed so far will be used in the remaining chapters to simplify expressions and solve equations. In such cases, we must determine the factoring method—it will not be specified. This section will give you practice in selecting the appropriate factoring method to use given a randomly chosen polynomial.

1 Factor randomly chosen polynomials.

The following strategy is helpful when factoring polynomials.

> **Steps for Factoring a Polynomial**
>
> 1. Is there a common factor? If so, factor out the GCF, or the opposite of the GCF, so that the leading coefficient is positive.
> 2. How many terms does the polynomial have?
>
>   If it has *two terms*, look for the following problem types:
>
>   **a.** The difference of two squares
>
>   **b.** The sum of two cubes
>
>   **c.** The difference of two cubes
>
>   If it has *three terms*, look for the following problem types:
>
>   **a.** A perfect-square trinomial
>
>   **b.** If the trinomial is not a perfect square, use the trial-and-check method or the grouping method.
>
>   If it has *four or more terms*, try to factor by grouping.
> 3. Can any factors be factored further? If so, factor them completely.
> 4. Does the factorization check? Check by multiplying.

Self Check 1

Factor: $11a^6 - 11a^2$

Now Try Problem 21

EXAMPLE 1 Factor: $2x^4 - 162$

Strategy We will answer the four questions listed in the *Steps for Factoring a Polynomial*.

WHY The answers help us determine which factoring techniques to use.

Solution

Is there a common factor? Yes. Factor out the GCF, which is 2.

$$2x^4 - 162 = 2(x^4 - 81)$$

How many terms does it have? The polynomial within the parentheses, $x^4 - 81$, has two terms. It is a difference of two squares.

$$2x^4 - 162 = 2(x^4 - 81) \quad \text{Think of } x^4 - 81 \text{ as } (x^2)^2 - 9^2.$$
$$= 2(x^2 + 9)(x^2 - 9)$$

Is it factored completely? No. $x^2 - 9$ is also the difference of two squares and can be factored.

$$2x^4 - 162 = 2(x^4 - 81)$$
$$= 2(x^2 + 9)(x^2 - 9) \quad \text{Think of } x^2 - 9 \text{ as } x^2 - 3^2.$$
$$= 2(x^2 + 9)(x + 3)(x - 3) \quad x^2 + 9 \text{ is a sum of two squares and does not factor.}$$

Therefore, $2x^4 - 162 = 2(x^2 + 9)(x + 3)(x - 3)$.

> **The Language of Algebra** Remember that the instruction to *factor* means to *factor completely*. A polynomial is *factored completely* when no factor can be factored further.

Does it check? Yes.

$$2(x^2 + 9)(x + 3)(x - 3) = 2(x^2 + 9)(x^2 - 9) \quad \text{Multiply } (x + 3)(x - 3) \text{ first.}$$
$$= 2(x^4 - 81) \quad \text{Multiply } (x^2 + 9)(x^2 - 9).$$
$$= 2x^4 - 162 \quad \text{This is the original polynomial.}$$

EXAMPLE 2 Factor: $-4c^5d^2 - 12c^4d^3 - 9c^3d^4$

Strategy We will answer the four questions listed in the *Steps for Factoring a Polynomial*.

WHY The answers help us determine which factoring techniques to use.

Self Check 2

Factor: $-32h^4 - 80h^3 - 50h^2$

Now Try Problem 33

Solution

Is there a common factor? Yes. Factor out the opposite of the GCF, $-c^3d^2$, so that the leading coefficient is positive.

$$-4c^5d^2 - 12c^4d^3 - 9c^3d^4 = -c^3d^2(4c^2 + 12cd + 9d^2)$$

How many terms does it have? The polynomial within the parentheses has three terms. It is a perfect-square trinomial because $4c^2 = (2c)^2$, $9d^2 = (3d)^2$, and $12cd = 2 \cdot 2c \cdot 3d$.

$$-4c^5d^2 - 12c^4d^3 - 9c^3d^4 = -c^3d^2(\mathbf{4c^2 + 12cd + 9d^2})$$
$$= -c^3d^2(\mathbf{2c + 3d})^2$$

Is it factored completely? Yes. The binomial $2c + 3d$ does not factor further.

Therefore, $-4c^5d^2 - 12c^4d^3 - 9c^3d^4 = -c^3d^2(2c + 3d)^2$.

Does it check? Yes.

$$-c^3d^2(2c + 3d)^2 = -c^3d^2(4c^2 + 12cd + 9d^2) \quad \text{Use a special product rule.}$$
$$= -4c^5d^2 - 12c^4d^3 - 9c^3d^4 \quad \text{This is the original polynomial.}$$

EXAMPLE 3 Factor: $y^4 - 3y^3 + y - 3$

Strategy We will answer the four questions listed in the *Steps for Factoring a Polynomial*.

Self Check 3

Factor: $b^4 + b^3 + 8b + 8$

Now Try Problem 37

WHY The answers help us determine which factoring techniques to use.

Solution

Is there a common factor? No. There is no common factor (other than 1).

How many terms does it have? Since the polynomial has four terms, we will try factoring by grouping.

$y^4 - 3y^3 + y - 3 = y^3(y - 3) + 1(y - 3)$ Factor out y^3 from $y^4 - 3y^3$. Factor out 1 from $y - 3$.

$= (y - 3)(y^3 + 1)$

Is it factored completely? No. We can factor $y^3 + 1$ as a sum of two cubes.

$y^4 - 3y^3 + y - 3 = y^3(y - 3) + 1(y - 3)$

$= (y - 3)(y^3 + 1)$ Think of $y^3 + 1$ as $y^3 + 1^3$.

$= (y - 3)(y + 1)(y^2 - y + 1)$ $y^2 - y + 1$ does not factor further.

Therefore, $y^4 - 3y^3 + y - 3 = (y - 3)(y + 1)(y^2 - y + 1)$.

Does it check? Yes.

$(y - 3)(y + 1)(y^2 - y + 1) = (y - 3)(y^3 + 1)$ Multiply the last two factors.

$= y^4 + y - 3y^3 - 3$ Use the FOIL method.

$= y^4 - 3y^3 + y - 3$ This is the original polynomial.

Self Check 4

Factor: $6m^2 - 54m + 6m^3$

Now Try Problem 45

EXAMPLE 4 Factor: $32n - 4n^2 + 4n^3$

Strategy We will answer the four questions listed in the *Steps for Factoring a Polynomial*.

WHY The answers help us determine which factoring techniques to use.

Solution

Is there a common factor? Yes. When we write the terms in descending powers of n, we see that the GCF is $4n$.

$4n^3 - 4n^2 + 32n = 4n(n^2 - n + 8)$

How many terms does it have? The polynomial within the parentheses has three terms. It is not a perfect-square trinomial because the last term, 8, is not a perfect integer square.

To factor the trinomial $n^2 - n + 8$, we must find two integers whose product is 8 and whose sum is -1. As we see in the table, there are no such integers. Thus, $n^2 - n + 8$ is prime.

Negative factors of 8	Sum of the negative factors of 8
$-1(-8) = 8$	$-1 + (-8) = -9$
$-2(-4) = 8$	$-2 + (-8) = -10$

Is it factored completely? Yes.

Therefore, $4n^3 - 4n^2 + 32n = 4n(n^2 - n + 8)$. Remember to write the GCF, $4n$, from the first step.

Does it check? Yes.

$4n(n^2 - n + 8) = 4n^3 - 4n^2 + 32n$ This is the original polynomial.

EXAMPLE 5 Factor: $3y^3 - 4y^2 - 4y$

Strategy We will answer the four questions listed in the *Steps for Factoring a Polynomial*.

WHY The answers help us determine which factoring techniques to use.

Solution
Is there a common factor? Yes. The GCF is y.

$$3y^3 - 4y^2 - 4y = y(3y^2 - 4y - 4)$$

How many terms does it have? The polynomial within the parentheses has three terms. It is not a perfect-square trinomial because the first term, $3y^2$, is not a perfect square.

If we use grouping to factor $3y^2 - 4y - 4$, the key number is $ac = 3(-4) = -12$. We must find two integers whose product is -12 and whose sum is $b = -4$.

Key number $= -12$	$b = -4$
Factors of -12	Sum of the factors of -12
$2(-6) = -12$	$2 + (-6) = -4$

From the table, the correct pair is 2 and -6. They serve as the coefficients of $2y$ and $-6y$, the two terms that we use to represent the middle term, $-4y$, of the trinomial.

$$\begin{aligned} 3y^2 - 4y - 4 &= 3y^2 + 2y - 6y - 4 && \text{Express } -4y \text{ as } 2y - 6y. \\ &= y(3y + 2) - 2(3y + 2) && \text{Factor } y \text{ from the first two terms and factor } -2 \text{ from the last two terms.} \\ &= (3y + 2)(y - 2) && \text{Factor out } 3y + 2. \end{aligned}$$

The trinomial $3y^2 - 4y - 4$ factors as $(3y + 2)(y - 2)$.

Is it factored completely? Yes. Because $3y + 2$ and $y - 2$ do not factor.

Therefore, $3y^3 - 4y^2 - 4y = y(3y + 2)(y - 2)$. Remember to write the GCF, y, from the first step.

Does it check? Yes.

$$\begin{aligned} y(3y + 2)(y - 2) &= y(3y^2 - 4y - 4) && \text{Multiply the binomials.} \\ &= 3y^3 - 4y^2 - 4y && \text{This is the original polynomial.} \end{aligned}$$

Self Check 5

Factor: $6y^3 + 21y^2 - 12y$

Now Try Problem 67

ANSWERS TO SELF CHECKS
1. $11a^2(a^2 + 1)(a + 1)(a - 1)$ 2. $-2h^2(4h + 5)^2$ 3. $(b + 1)(b + 2)(b^2 - 2b + 4)$
4. $6m(m^2 + m - 9)$ 5. $3y(2y - 1)(y + 4)$

SECTION 6.6 STUDY SET

VOCABULARY

Fill in the blanks.

1. To factor a polynomial means to express it as a _____ of two (or more) polynomials.

2. A polynomial is factored _____ when each factor is prime.

CONCEPTS

For each of the following polynomials, which factoring method would you use first?

3. $2x^5y - 4x^3y$
4. $9b^2 + 12y - 5$
5. $x^2 + 18x + 81$
6. $ax + ay - x - y$
7. $x^3 + 27$
8. $y^3 - 64$
9. $m^2 + 3mn + 2n^2$
10. $16 - 25z^2$

11. What is the first question that should be asked when using the strategy of this section to factor a polynomial?
12. Use multiplication to determine whether the factorization is correct.
$$5c^3d^2 - 40c^2d^3 + 35cd^4 = 5cd^2(c - 7d)(c - d)$$

NOTATION

Complete each factorization.

13. $6m^3 - 28m^2 + 16m = 2m(3m^2 - \boxed{} + 8)$
    $= 2m(3m - 2)(\boxed{} - 4)$

14. $2a^3 + 3a^2 - 2a - 3$
    $= \boxed{}(2a + 3) - 1(\boxed{} + 3)$
    $= (\boxed{})(a^2 - 1)$
    $= (2a + 3)(a + 1)(\boxed{})$

TRY IT YOURSELF

The following is a list of random factoring problems. Factor each expression completely. If an expression is not factorable, write "prime." See Examples 1–5.

15. $2b^2 + 8b - 24$
16. $32 - 2t^4$
17. $8p^3q^7 + 4p^2q^3$
18. $8m^2n^3 - 24mn^4$
19. $2 + 24y + 40y^2$
20. $6r^2 + 3rs - 18s^2$
21. $8x^4 - 8$
22. $t - 90 + t^2$
23. $14c - 147 + c^2$
24. $ab^2 - 4a + 3b^2 - 12$
25. $x^2 + 7x + 1$
26. $3a^3 + 24b^3$
27. $-2x^5 + 128x^2$
28. $16 - 40z + 25z^2$
29. $a^2c + a^2d^2 + bc + bd^2$
30. $6t^4 + 14t^3 - 40t^2$
31. $-9x^2y^2 + 6xy - 1$
32. $x^2y^2 - 2x^2 - y^2 + 2$
33. $-20m^3 - 100m^2 - 125m$
34. $5x^3y^3z^4 + 25x^2y^4z^2 - 35x^3y^2z^5$
35. $2c^2 - 5cd - 3d^2$
36. $125p^3 - 64y^3$
37. $p^4 - 2p^3 - 8p + 16$
38. $a^2 + 8a + 3$
39. $a^2(x - a) - b^2(x - a)$
40. $a^2b^2 - 144$
41. $70p^4q^3 - 35p^4q^2 + 49p^5q^2$
42. $-16x^4y^2z + 24x^5y^3z^4 - 15x^2y^3z^7$
43. $2x^3 + 10x^2 + x + 5$
44. $u^2 - 18u + 81$
45. $8v^2 - 14v^3 + v^4$
46. $28 - 3m - m^2$
47. $x^4 - 13x^2 + 36$
48. $81r^4 - 256$
49. $8a^2x^3 - 2b^2x$
50. $12x^2 + 14x - 6$
51. $6x^2 - 14x + 8$
52. $12x^2 - 12$
53. $4x^2y^2 + 4xy^2 + y^2$
54. $81r^4s^2 - 24rs^5$
55. $4m^5 + 500m^2$
56. $ae + bf + af + be$
57. $x^4 - 2x^2 - 8$
58. $6x^2 - x - 16$
59. $4x^2 + 9y^2$
60. $x^4y + 216xy^4$
61. $16a^5 - 54a^2$
62. $25x^2 - 16y^2$
63. $27x - 27y - 27z$
64. $12x^2 + 52x + 35$
65. $xy - ty + xs - ts$
66. $bc + b + cd + d$
67. $35x^8 - 2x^7 - x^6$
68. $x^3 - 25$
69. $5(x - 2) + 10y(x - 2)$
70. $16x^2 - 40x^3 + 25x^4$
71. $49p^2 + 28pq + 4q^2$
72. $x^2y^2 - 6xy - 16$
73. $4t^2 + 36$
74. $r^5 + 3r^3 + 2r^2 + 6$
75. $m^2n^2 - 9m^2 + 3n^2 - 27$
76. $z^2 + 6yz^2 + 9y^2z^2$

WRITING

77. Which factoring method do you find the most difficult? Why?
78. What four questions make up the factoring strategy for polynomials discussed in this section?

79. What does it mean to factor a polynomial?
80. How is a factorization checked?

REVIEW

81. Graph the real numbers $-3, 0, 2,$ and $-\frac{3}{2}$ on a number line.

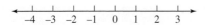

82. Graph the interval $(-2, 3]$ on a number line.

83. Graph: $y = \frac{1}{2}x + 1$

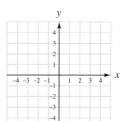

84. Graph: $y < 2 - 3x$

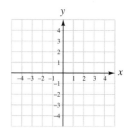

SECTION 6.7
Solving Quadratic Equations by Factoring

Objectives
1. Define quadratic equations.
2. Solve quadratic equations using the zero-factor property.
3. Solve quadratic equations by factoring.
4. Solve third-degree equations by factoring.

Equations that involve first-degree polynomials, such as $9x - 6 = 0$, are called *linear equations*. Equations that involve second-degree polynomials, such as $9x^2 - 6x = 0$, are called **quadratic equations**. In this section, we will define quadratic equations and learn how to solve many of them by factoring.

1 Define quadratic equations.

If a polynomial contains one variable with an exponent to the second (but no higher) power, it is called a **second-degree polynomial**. A quadratic, or second-degree equation, has a term in which the exponent on the variable is 2, and has no other terms of higher degree. Some examples are

$$9x^2 - 6x = 0, \quad x^2 - 2x - 63 = 0, \quad \text{and} \quad 2x^2 + 3x - 2 = 0$$

Quadratic Equation

A **quadratic equation** is an equation that can be written in the **standard form**

$$ax^2 + bx + c = 0$$

where a, b, and c represent real numbers, and a is not 0.

To write a quadratic equation such as $21x = 10 - 10x^2$ in $ax^2 + bx + c = 0$ form (called **standard form**), we use the addition and subtraction properties of equality to get 0 on the right-hand side.

$$21x = 10 - 10x^2$$
$$\mathbf{10x^2} + 21x = 10 - 10x^2 + \mathbf{10x^2} \quad \text{Add } 10x^2 \text{ to both sides.}$$
$$10x^2 + 21x = 10 \quad \text{Combine like terms: } -10x^2 + 10x^2 = 0.$$
$$10x^2 + 21x - 10 = 0 \quad \text{Subtract 10 from both sides.}$$

2 Solve quadratic equations using the zero-factor property.

To **solve a quadratic equation**, we find all the values of the variable that make the equation true.

The method that we have used to solve linear equations cannot be used to solve a quadratic equation, because those techniques cannot isolate the variable on one side of the equation. However, we can often solve quadratic equations using factoring and the following property of real numbers.

> **The Zero-Factor Property**
>
> If a and b represent real numbers, and
>
> if $ab = 0$, then $a = 0$ or $b = 0$.

In words, the zero-factor property states that when the product of two numbers is zero, at least one of them must be zero.

Self Check 1

Solve: $(4x - 3)(5x - 4) = 0$

Now Try Problem 15

EXAMPLE 1   Solve: $(4y - 1)(y + 6) = 0$

Strategy We will set $4y - 1$ equal to 0 and $y + 6$ equal to 0 and solve each equation.

WHY The product of $4y - 1$ and $y + 6$ is equal to 0. By the zero-factor property, $4y - 1$ must equal 0, or $y + 6$ must equal 0.

Solution
The left side of the equation is $(4y - 1)(y + 6)$. By the zero-factor property, one of these factors must be 0.

$$4y - 1 = 0 \quad \text{or} \quad y + 6 = 0$$

We can solve each of the linear equations.

$4y - 1 = 0$ or $y + 6 = 0$ Set each factor equal to zero.
$4y = 1$ $\quad\quad\quad y = -6$ Solve each equation.
$y = \dfrac{1}{4}$

The equation has two solutions, $\frac{1}{4}$ and -6. The solution set is $\left\{-6, \frac{1}{4}\right\}$. To check, we substitute the results for y in the original equation and simplify.

For $y = \frac{1}{4}$:
$(4y - 1)(y + 6) = 0$
$\left[4\left(\dfrac{1}{4}\right) - 1\right]\left[\dfrac{1}{4} + 6\right] \stackrel{?}{=} 0$
$(1 - 1)\left(6\dfrac{1}{4}\right) \stackrel{?}{=} 0$
$0\left(6\dfrac{1}{4}\right) \stackrel{?}{=} 0$
$0 = 0$ True

For $y = -6$:
$(4y - 1)(y + 6) = 0$
$[4(-6) - 1](-6 + 6) \stackrel{?}{=} 0$
$(-24 - 1)(0) \stackrel{?}{=} 0$
$-25(0) \stackrel{?}{=} 0$
$0 = 0$ True

3 Solve quadratic equations by factoring.

In Example 1, the right side of the equation was zero and the left side was in factored form, so we were able to use the zero-factor property immediately. However, to solve many quadratic equations, we must first do the factoring.

EXAMPLE 2   Solve: $9x^2 - 6x = 0$

Strategy We will factor the binomial on the left side of the equation and use the zero-factor property.

WHY To use the zero-factor property, we need one side of the equation to be factored completely and the other side to be 0.

Solution
We begin by factoring the left side of the equation.

$9x^2 - 6x = 0$ This is the equation to solve.
$3x(3x - 2) = 0$ Factor out the GCF of 3x.

By the zero-factor property, we have

$3x = 0$ or $3x - 2 = 0$

We can solve each of the linear equations to get

$x = 0$ or $x = \dfrac{2}{3}$

To check, we substitute the results for x in the original equation and simplify.

For $x = 0$
$9x^2 - 6x = 0$
$9(0)^2 - 6(0) \stackrel{?}{=} 0$
$0 - 0 \stackrel{?}{=} 0$
$0 = 0$ True

For $x = \dfrac{2}{3}$
$9x^2 - 6x = 0$
$9\left(\dfrac{2}{3}\right)^2 - 6\left(\dfrac{2}{3}\right) \stackrel{?}{=} 0$
$9\left(\dfrac{4}{9}\right) - 6\left(\dfrac{2}{3}\right) \stackrel{?}{=} 0$
$4 - 4 \stackrel{?}{=} 0$
$0 = 0$ True

The solutions of $9x^2 - 6x = 0$ are 0 and $\dfrac{2}{3}$. The solution set is $\left\{0, \dfrac{2}{3}\right\}$.

We can use the following steps to solve a quadratic equation by factoring.

> **Solving Quadratic Equations by Factoring**
>
> 1. Write the equation in standard form: $ax^2 + bx + c = 0$ or $0 = ax^2 + bx + c$.
> 2. Factor completely.
> 3. Use the zero-factor property to set each factor equal to zero.
> 4. Solve each resulting linear equation.
> 5. Check the results in the original equation.

Self Check 2
Solve: $5x^2 + 10x = 0$
Now Try Problem 23

EXAMPLE 3   Solve: $x^2 = 9$

Strategy We will subtract 9 from both sides of the equation to get 0 on the right side. Then we will factor the resulting binomial and use the zero-factor property.

WHY To use the zero-factor property, we need one side of the equation to be factored completely and the other side to be 0.

Solution
Before we can use the zero-factor property, we must subtract 9 from both sides to make the right side zero.

Self Check 3
Solve: $x^2 = 25$
Now Try Problem 25

$$x^2 = 9 \quad \text{This is the equation to solve.}$$
$$x^2 - 9 = 0 \quad \text{Subtract 9 from both sides.}$$
$$(x + 3)(x - 3) = 0 \quad \text{Factor the difference of two squares.}$$
$$x + 3 = 0 \quad \text{or} \quad x - 3 = 0 \quad \text{Set each factor equal to zero.}$$
$$x = -3 \quad | \quad x = 3 \quad \text{Solve each linear equation.}$$

Check each possible solution by substituting it into the original equation.

For $x = -3$
$$x^2 = 9$$
$$(-3)^2 \stackrel{?}{=} 9$$
$$9 = 9$$

For $x = 3$
$$x^2 = 9$$
$$(3)^2 \stackrel{?}{=} 9$$
$$9 = 9$$

The solutions of $x^2 = 9$ are -3 and 3. The solution set is $\{-3, 3\}$.

Self Check 4

Solve: $x^2 + 5x + 6 = 0$

Now Try Problem 31

EXAMPLE 4 Solve: $x^2 - 2x - 63 = 0$

Strategy We will factor the trinomial on the left side of the equation and use the zero-factor property.

WHY To use the zero-factor property, we need one side of the equation to be factored completely and the other side to be 0.

Solution
$$x^2 - 2x - 63 = 0 \quad \text{This is the equation to solve.}$$
$$(x + 7)(x - 9) = 0 \quad \text{Factor the trinomial } x^2 - 2x - 63.$$
$$x + 7 = 0 \quad \text{or} \quad x - 9 = 0 \quad \text{Set each factor equal to zero.}$$
$$x = -7 \quad | \quad x = 9 \quad \text{Solve each linear equation.}$$

The solutions are -7 and 9. The solution set is $\{-7, 9\}$. Check each one.

Self Check 5

Solve: $3x^2 - 6 = -7x$

Now Try Problem 35

EXAMPLE 5 Solve: $2x^2 + 3x = 2$

Strategy We will subtract 2 from both sides of the equation to get 0 on the right side. Then we will factor the resulting trinomial and use the zero-factor property.

WHY To use the zero-factor property, we need one side of the equation to be factored completely and the other side to be 0.

Solution
The equation is not in $ax^2 + bx + c = 0$ form. To get 0 on the right side, we proceed as follows:

$$2x^2 + 3x = 2 \quad \text{This is the equation to solve.}$$
$$2x^2 + 3x - 2 = 0 \quad \text{Subtract 2 from both sides so that the right-hand side is zero.}$$
$$(2x - 1)(x + 2) = 0 \quad \text{Factor } 2x^2 + 3x - 2.$$
$$2x - 1 = 0 \quad \text{or} \quad x + 2 = 0 \quad \text{Set each factor equal to zero.}$$
$$2x = 1 \quad | \quad x = -2 \quad \text{Solve each linear equation.}$$
$$x = \frac{1}{2}$$

The solution set is $\{-2, \frac{1}{2}\}$. Use a check to verify that $\frac{1}{2}$ and -2 are solutions.

EXAMPLE 6 Solve: $x(9x - 12) = -4$

Strategy To write the equation in standard form, we will distribute the multiplication by x and add 4 to both sides. Then we will factor the resulting trinomial and use the zero-factor property.

WHY To use the zero-factor property, we need one side of the equation to be factored completely and the other side to be 0.

Solution
First, we need to write the equation in the form $ax^2 + bx + c = 0$.

$x(9x - 12) = -4$	This is the equation to solve.
$9x^2 - 12x = -4$	Distribute the multiplication by x.
$9x^2 - 12x + 4 = 0$	To get 0 on the right side, add 4 to both sides.
$(3x - 2)(3x - 2) = 0$	Factor the trinomial.
$3x - 2 = 0$ or $3x - 2 = 0$	Set each factor equal to zero.
$3x = 2 \qquad\qquad 3x = 2$	Add 2 to both sides.
$x = \dfrac{2}{3} \qquad\qquad x = \dfrac{2}{3}$	Divide both sides by 3.

The equation has two solutions that are the same. We call $\frac{2}{3}$ a *repeated solution*. The solution set is $\{\frac{2}{3}\}$. Check by substituting it into the original equation.

Self Check 6
Solve: $x(4x + 12) = -9$
Now Try Problem 37

4 Solve third-degree equations by factoring.

EXAMPLE 7 Solve: $6x^3 + 12x = 17x^2$

Strategy This equation is not quadratic, because it contains a term involving x^3. However, we can solve it by using factoring. First we get 0 on the right side by subtracting $17x^2$ from both sides. Then we factor the polynomial on the left side and use an extension of the zero-factor property.

WHY To use the zero-factor property, we need one side of the equation to be factored completely and the other side to be 0.

Solution

$6x^3 + 12x = 17x^2$	This is the equation to solve.
$6x^3 - 17x^2 + 12x = 0$	Subtract $17x^2$ from both sides to get 0 on the right-hand side.
$x(6x^2 - 17x + 12) = 0$	Factor out the GCF, x.
$x(2x - 3)(3x - 4) = 0$	Factor $6x^2 - 17x + 12$.
$x = 0$ or $2x - 3 = 0$ or $3x - 4 = 0$	Set each factor equal to zero.
$\qquad\qquad 2x = 3 \qquad\qquad 3x = 4$	Solve the linear equations.
$\qquad\qquad x = \dfrac{3}{2} \qquad\qquad x = \dfrac{4}{3}$	

This equation has three solutions, 0, $\frac{3}{2}$, and $\frac{4}{3}$. The solution set is $\{0, \frac{3}{2}, \frac{4}{3}\}$.

Self Check 7
Solve: $10x^3 + x^2 - 2x = 0$
Now Try Problem 43

ANSWERS TO SELF CHECKS

1. $\dfrac{3}{4}, \dfrac{4}{5}$ 2. $0, -2$ 3. $-5, 5$ 4. $-2, -3$ 5. $\dfrac{2}{3}, -3$ 6. $-\dfrac{3}{2}, -\dfrac{3}{2}$ 7. $0, \dfrac{2}{5}, -\dfrac{1}{2}$

SECTION 6.7 STUDY SET

VOCABULARY

Fill in the blanks.

1. Any equation that can be written in the form $ax^2 + bx + c = 0$ is called a _____ equation.
2. To _____ a binomial or trinomial means to write it as a product.

CONCEPTS

Fill in the blanks.

3. When the product of two numbers is 0, at least one of them is ☐. Symbolically, we can state this: If $ab = 0$, then $a = $ ☐ or $b = $ ☐.
4. We can often use _____ and the zero-factor property to solve quadratic equations.
5. To write a quadratic equation in standard form means that one side of the equation must be _____ and the other side must be in the form $ax^2 + bx + c$.
6. Classify each equation as quadratic or linear.
   a. $3x^2 + 4x + 2 = 0$ b. $3x + 7 = 0$
   c. $2 = -16 - 4x$ d. $-6x + 2 = x^2$
7. Check to see whether the given number is a solution of the given quadratic equation.
   a. $x^2 - 4x = 0$; $x = 4$
   b. $x^2 + 2x - 4 = 0$; $x = -2$
8. a. Evaluate $x^2 + 6x - 16$ for $x = 0$.
   b. Factor: $x^2 + 6x - 16$
9. The equation $3x^2 - 4x + 5 = 0$ is written in $ax^2 + bx + c = 0$ form. What are a, b, and c?
10. a. How many solutions does the linear equation $2a + 3 = 2$ have?
    b. How many solutions does the quadratic equation $2a^2 + 3a = 2$ have?

NOTATION

Complete each step to solve the equation.

11. $7y^2 + 14y = 0$
    ☐$(y + 2) = 0$
    $7y = 0$ or ☐ $= 0$
    $y = 0$ | $y = -2$

12. $12p^2 - p - 6 = 0$
    (☐ $- 3$)($3p + $ ☐) $= 0$
    ☐ $= 0$ or $3p + 2 = $ ☐
    $4p = $ ☐ $3p = $ ☐
    $p = \dfrac{3}{4}$ | $p = -\dfrac{2}{3}$

GUIDED PRACTICE

Solve each equation. See Example 1.

13. $(x - 2)(x + 3) = 0$ 14. $(x - 3)(x - 2) = 0$
15. $(2s - 5)(s + 6) = 0$ 16. $(3h - 4)(h + 1) = 0$
17. $x(x - 3) = 0$ 18. $x(x + 5) = 0$
19. $(x - 1)(x + 2)(x - 3) = 0$
20. $(x + 2)(x + 3)(x - 4) = 0$

Solve each equation. See Example 2.

21. $w^2 - 7w = 0$ 22. $p^2 + 5p = 0$
23. $3x^2 + 8x = 0$ 24. $5x^2 - x = 0$

Solve each equation. See Example 3.

25. $x^2 = 100$ 26. $z^2 = 25$
27. $4x^2 = 81$ 28. $9y^2 = 64$

Solve each equation. See Example 4.

29. $x^2 - 4x - 21 = 0$ 30. $x^2 + 2x - 15 = 0$
31. $x^2 - 13x + 12 = 0$ 32. $x^2 + 7x + 6 = 0$

Solve each equation. See Example 5.

33. $4r^2 + 4r = -1$ 34. $9m^2 + 6m = -1$
35. $-15x^2 + 2 = -7x$ 36. $-8x^2 - 10x = -3$

Solve each equation. See Example 6.

37. $x(2x - 3) = 20$ 38. $x(2x - 3) = 14$
39. $(d + 1)(8d + 1) = 18d$ 40. $4h(3h + 2) = h + 12$

Solve each equation. See Example 7.

41. $x^3 + 3x^2 + 2x = 0$ 42. $x^3 - 7x^2 + 10x = 0$
43. $k^3 - 27k - 6k^2 = 0$ 44. $j^3 - 22j - 9j^2 = 0$

TRY IT YOURSELF

Solve each equation.

45. $x(2x - 5) = 0$
46. $x(5x + 7) = 0$
47. $8s^2 - 16s = 0$
48. $15s^2 - 20s = 0$
49. $x^2 - 25 = 0$
50. $x^2 - 36 = 0$
51. $4x^2 - 1 = 0$
52. $9y^2 - 1 = 0$
53. $9y^2 - 4 = 0$
54. $16z^2 - 25 = 0$
55. $x^2 - 9x + 8 = 0$
56. $x^2 - 14x + 45 = 0$
57. $a^2 + 8a = -15$
58. $a^2 - a = 56$
59. $2y - 8 = -y^2$
60. $-3y + 18 = y^2$
61. $2x^2 - 5x + 2 = 0$
62. $2x^2 + x - 3 = 0$
63. $5x^2 - 6x + 1 = 0$
64. $6x^2 - 5x + 1 = 0$
65. $(x - 1)(x^2 + 5x + 6) = 0$
66. $(x - 2)(x^2 - 8x + 7) = 0$
67. $2x(3x^2 + 10x) = -6x$
68. $2x^3 = 2x(x + 2)$
69. $x^3 + 7x^2 = x^2 - 9x$
70. $x^2(x + 10) = 2x(x - 8)$

WRITING

71. Explain the zero-factor property.
72. Find the error in the following solution:

$$x(x + 1) = 6$$
$$x = 6 \quad \text{or} \quad x + 1 = 6$$
$$x = 5$$

The solutions are 6 and 5.

REVIEW

Perform the operations and simplify.

73. $5b(2b - 3) + 2b(b - 5)$
74. $3x(a + x) - 2(x + 3)$
75. $(2b - 3)(2b - 5)$
76. $(a + x)(x + 3)$
77. $\dfrac{a + 1}{2} + \dfrac{a - 1}{2}$
78. $\dfrac{x + 2}{3} - \dfrac{2x + 1}{2}$
79. $3z - 2 \overline{)6z^2 + 5z - 6}$
80. $2a - 1 \overline{)6a^3 + a^2 + 4a - 2}$

SECTION 6.8
Applications of Quadratic Equations

Objectives

1. Solve problems given the quadratic equation model.
2. Solve problems involving consecutive integers.
3. Solve problems involving geometric figures.
4. Solve problems involving the Pythagorean Theorem.

In Chapter 2, we solved mixture, investment, and uniform motion problems. To model those equations, we used *linear equations* in one variable. We will now consider situations that are modeled by *quadratic equations*.

1 Solve problems given the quadratic equation model.

EXAMPLE 1 **Softball** A pitcher can throw a fastball underhand at 63 feet per second (about 45 mph). If she throws a ball into the air with that velocity, its height h in feet, t seconds after being released, is given by the formula

$$h = -16t^2 + 63t + 4$$

After the ball is thrown, in how many seconds will it hit the ground?

Solution
When the ball hits the ground, its height will be 0 feet. To find the time that it will take for the ball to hit the ground, we set h equal to 0, and solve the quadratic equation for t.

$h = -16t^2 + 63t + 4$
$0 = -16t^2 + 63t + 4$ Substitute 0 for the height h.
$0 = -(16t^2 - 63t - 4)$ Factor out -1.
$0 = -(16t + 1)(t - 4)$ Factor $16t^2 - 63t - 4$.

Self Check 1

A student uses rubber tubing to launch a water balloon from the roof of his dormitory. The height h (in feet) of the balloon, t seconds after being launched, is given by $h = -16t^2 + 48t + 64$. After how many seconds will the balloon hit the ground?

Now Try Problem 11

$16t + 1 = 0$	or $\quad t - 4 = 0$	Set each factor that contains a variable equal to 0.
$16t = -1$	$t = 4$	Solve each equation.
$t = -\dfrac{1}{16}$		

Since time cannot be negative, we discard the solution $-\frac{1}{16}$. The second solution indicates that the ball hits the ground 4 seconds after being released. Check this answer by substituting 4 for t in $h = -16t^2 + 63t + 4$. You should get $h = 0$.

2 Solve problems involving consecutive integers.

Consecutive integers are integers that follow one another, such as 15 and 16. When solving consecutive integer problems, if we let $x =$ the first integer, then:

- two consecutive integers are x and $x + 1$
- two consecutive even integers are x and $x + 2$
- two consecutive odd integers are x and $x + 2$

Self Check 2

The product of two consecutive integers is 552. Find the integers.

Now Try Problem 19

EXAMPLE 2 **Women's Tennis** In the 1998 Australian Open, sisters Venus and Serena Williams played against each other for the first time as professionals. Venus was victorious over her younger sister. At that time, their ages were consecutive integers whose product was 272. How old were Venus and Serena when they met in this match?

Analyze
- Venus is older than Serena.
- Their ages were consecutive integers.
- The product of their ages was 272.
- Find Venus' and Serena's age when they played this match.

Assign Let $x =$ Serena's age when she played in the 1998 Australian Open. Since their ages were consecutive integers, and since Venus is older, we let $x + 1 =$ Venus' age.

Form The word *product* indicates multiplication.

Serena's age	times	Venus' age	was	272.
x	$\cdot$	$(x + 1)$	$=$	272

Solve

$x(x + 1) = 272$	
$x^2 + x = 272$	Distribute the multiplication by x. Note that this is a quadratic equation.
$x^2 + x - 272 = 0$	Subtract 272 from both sides to make the right side 0.
$(x + 17)(x - 16) = 0$	Factor $x^2 + x - 272$. Two numbers whose product is -272 and whose sum is 1 are 17 and -16.
$x + 17 = 0 \quad$ or $\quad x - 16 = 0$	Set each factor equal to 0.
$x = -17 \qquad\qquad\quad x = 16$	Solve each equation.

State The solutions of the equation are -17 and 16. Since x represents Serena's age, and it cannot be negative, we discard -17. Thus, Serena Williams was 16 years old and Venus Williams was $16 + 1 = 17$ years old when they played against each other for the first time as professionals.

> **Success Tip** The prime factorization of 272 is helpful in determining that $272 = 17 \cdot 16$.
>
> $$16 \begin{cases} 2\,\overline{|272} \\ 2\,\overline{|136} \\ 2\,\overline{|68} \\ 2\,\overline{|34} \\ 17 \end{cases}$$

3 Solve problems involving geometric figures.

EXAMPLE 3 Perimeter of a Rectangle

Assume that the rectangle has an area of 52 square centimeters and that its length is 1 centimeter more than 3 times its width. Find the perimeter of the rectangle.

$3w + 1$

$w \quad A = 52 \text{ cm}^2$

Analyze The area of the rectangle is 52 square centimeters. Recall that the formula that gives the area of a rectangle is $A = lw$. To find the perimeter of the rectangle, we need to know its length and width. We are told that its length is related to its width; the length is 1 centimeter more than 3 times the width.

Assign Let w represent the width of the rectangle. Then $3w + 1$ represents its length.

Form Because the area is 52 square centimeters, we substitute 52 for A and $3w + 1$ for l in the formula $A = lw$.

$A = lw$
$52 = (3w + 1)w$

Solve Now we solve the equation for w.

$52 = (3w + 1)w$ This is the equation to solve.
$52 = 3w^2 + w$ Distribute the multiplication by w.
$0 = 3w^2 + w - 52$ Subtract 52 from both sides to make the left side zero.
$0 = (3w + 13)(w - 4)$ Factor the trinomial.
$3w + 13 = 0 \quad \text{or} \quad w - 4 = 0$ Set each factor equal to zero.
$3w = -13 \quad\quad\quad\quad w = 4$ Solve each linear equation.
$w = -\dfrac{13}{3}$

State Since the width cannot be negative, we discard the solution $-\dfrac{13}{3}$. Thus, the width of the rectangle is 4, and the length is given by

$3w + 1 = 3(4) + 1$ Substitute 4 for w.
$ = 12 + 1$
$ = 13$

The dimensions of the rectangle are 4 centimeters by 13 centimeters. We find the perimeter by substituting 13 for l and 4 for w in the formula for the perimeter of a rectangle.

Self Check 3

A rectangle has an area of 55 square meters. Its length is 1 meter more than twice its width. Find the perimeter of the rectangle.

Now Try Problem 23

$P = 2l + 2w$
$P = 2(13) + 2(4)$
$P = 26 + 8$
$P = 34$

The perimeter of the rectangle is 34 centimeters.

Check A rectangle with dimensions of 13 centimeters by 4 centimeters does have an area of 52 square centimeters, and the length is 1 centimeter more than 3 times the width. A rectangle with these dimensions has a perimeter of 34 centimeters.

4 Solve problems involving the Pythagorean theorem.

The next example involves a right triangle. A **right triangle** is a triangle that contains a 90° angle. The longest side of a right triangle is the **hypotenuse**, which is the side opposite the right angle. The remaining two sides are the **legs** of the triangle. The **Pythagorean theorem** provides a formula relating the lengths of the three sides of a right triangle.

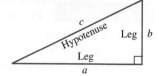

THINK IT THROUGH Pythagorean Triples

"Fraternity and sorority members form the largest network of volunteers in the U.S. Nationally, fraternity and sorority members volunteer approximately 10 million hours of community service annually."

University of Missouri–Kansas City National Statistics

The first college social fraternity, Phi Beta Kappa, was founded in 1776 on the campus of The College of William and Mary. However, secret societies have existed since ancient times, and the essence of today's fraternities and sororities comes from those roots. Pythagoras, the Greek mathematician of the sixth century B.C., was the leader of a secret society called the Pythagoreans. They were a community of men and women that studied mathematics, and in particular, the "magic 3-4-5 triangle." This right triangle is special because the sum of the squares of the lengths of its legs is equal to the square of the length of its hypotenuse: $3^2 + 4^2 = 5^2$ or $9 + 16 = 25$. Today, we call a set of three natural numbers a, b, and c that satisfy $a^2 + b^2 = c^2$ a Pythagorean triple. Show that each list of numbers is a Pythagorean triple.

1. 5, 12, 13
2. 7, 24, 25
3. 8, 15, 17
4. 9, 40, 41
5. 11, 60, 61
6. 12, 35, 37

Pythagorean Theorem

If the length of the hypotenuse of a right triangle is c and the lengths of the two legs are a and b, then

$$c^2 = a^2 + b^2$$

EXAMPLE 4 Right Triangles
The longer leg of a right triangle is 3 units longer than the shorter leg. If the hypotenuse is 6 units longer than the shorter leg, find the lengths of the sides of the triangle.

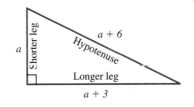

Analyze We begin by drawing a right triangle and labeling the legs and the hypotenuse.

Assign We let a = the length of the shorter leg. Then the length of the hypotenuse is $a + 6$ and the length of the longer leg is $a + 3$.

Form By the Pythagorean theorem, we have

$$\begin{pmatrix}\text{The length of}\\ \text{the shorter leg}\end{pmatrix}^2 \text{ plus } \begin{pmatrix}\text{The length of}\\ \text{the longer leg}\end{pmatrix}^2 \text{ equals } \begin{pmatrix}\text{The length of}\\ \text{the hypotenuse}\end{pmatrix}^2$$

$$a^2 \quad + \quad (a+3)^2 \quad = \quad (a+6)^2$$

Solve

$$a^2 + (a+3)^2 = (a+6)^2$$
$$a^2 + a^2 + 6a + 9 = a^2 + 12a + 36 \quad \text{Find } (a+3)^2 \text{ and } (a+6)^2.$$
$$2a^2 + 6a + 9 = a^2 + 12a + 36 \quad \text{Combine like terms on the left side.}$$
$$a^2 - 6a - 27 = 0 \quad \text{Subtract } a^2, 12a, \text{ and } 36 \text{ from both sides to make the right side 0.}$$

Now solve the quadratic equation for a.

$$a^2 - 6a - 27 = 0$$
$$(a-9)(a+3) = 0 \quad \text{Factor.}$$
$$a - 9 = 0 \quad \text{or} \quad a + 3 = 0 \quad \text{Set each factor to zero.}$$
$$a = 9 \quad \quad \quad a = -3 \quad \text{Solve each equation.}$$

State Since a side cannot have a negative length, we discard the solution -3. Thus, the shorter leg is 9 units long, the hypotenuse is $9 + 6 = 15$ units long, and the longer leg is $9 + 3 = 12$ units long.

Check The longer leg, with length 12, is 3 units longer than the shorter leg, with length 9. The hypotenuse, with length 15, is 6 units longer than the shorter leg. Since these lengths satisfy the Pythagorean theorem, the results check.

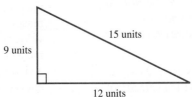

$$9^2 + 12^2 \stackrel{?}{=} 15^2$$
$$81 + 144 \stackrel{?}{=} 225$$
$$225 = 225$$

Self Check 4
The longer leg of a right triangle is 7 inches longer than the shorter leg. If the hypotenuse is 9 units longer than the shorter leg, find the lengths of the sides of the triangle.

Now Try Problem 29

ANSWERS TO SELF CHECKS

1. 4 sec **2.** 23 and 24 **3.** 32 m **4.** 8 in., 15 in., and 17 in.

Unless otherwise noted, all content on this page is © Cengage Learning.

SECTION 6.8 STUDY SET

VOCABULARY

Fill in the blanks.

1. Integers that follow one another, such as 6 and 7, are called _____ integers.
2. A _____ triangle is a triangle that contains a 90° angle.
3. The longest side of a right triangle is called the _____.
4. The _____ theorem is a formula that relates the lengths of the three sides of a right triangle.

CONCEPTS

Fill in the blanks.

5. The formula for the area of a rectangle is $A =$ ___.
6. If a and b are legs of a right triangle and c is the hypotenuse, then ___ = ___.

NOTATION

Complete each step.

7. $0 = -16t^2 + 32t + 48$
   $0 = (t^2 - 2t - 3)$
   $0 = -16(t-3)(t +)$
   $t - 3 = $ or $t + 1 = $
   $t = $ | $t = $

8. $6 = w(w+1)$
   $6 = + w$
   $0 = w^2 + w $
   $0 = (w)(w -)$
   $w + 3 = $ or $w - 2 = $
   $w = $ | $w = $

APPLICATIONS

An object has been thrown straight up into the air. The formula $h = vt - 16t^2$ gives the height h of the object above the ground after t seconds, when it is thrown upward with an initial velocity v. See Example 1.

9. **TIME OF FLIGHT** After how many seconds will the object hit the ground if it is thrown with a velocity of 144 feet per second?

10. **TIME OF FLIGHT** After how many seconds will the object hit the ground if it is thrown with a velocity of 160 feet per second?

11. **OFFICIATING** Before a football game, a coin toss is used to determine which team will kick off. The height h (in feet) of a coin above the ground t seconds after being flipped up into the air is given by

    $$h = -16t^2 + 22t + 3$$

    How long will the coin be in the air?

12. **DOLPHINS** The height h in feet reached by a dolphin t seconds after breaking the surface of the water is given by

    $$h = -16t^2 + 32t$$

    How long will it take the dolphin to jump out of the water and touch the trainer's hand?

    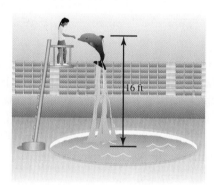

13. **EXHIBITION DIVING** In Acapulco, Mexico, men diving from a cliff to the water 64 feet below are a tourist attraction. A diver's height h above the water t seconds after diving is given by $h = -16t^2 + 64$. How long does a dive last?

14. **FORENSIC MEDICINE** The kinetic energy E of a moving object is given by $E = \frac{1}{2}mv^2$, where m is the mass of the object (in kilograms) and v is the object's velocity (in meters per second). Kinetic energy is measured in joules. Examining the damage done to a victim, a police pathologist determines that the energy of a 3-kilogram mass at impact was 54 joules. Find the velocity at impact. (*Hint:* Multiply both sides of the equation by 2.)

15. CHOREOGRAPHY For the finale of a musical, 36 dancers are to assemble in a triangular series of rows, where each successive row has one more dancer than the previous row. The illustration shows the beginning of such a formation. The relationship between the number of rows r and the number of dancers d is given by

$$d = \frac{1}{2}r(r + 1)$$

Determine the number of rows in the formation. (*Hint:* Multiply both sides of the equation by 2.)

16. CRAFTS The illustration shows how a geometric wall hanging can be created by stretching yarn from peg to peg across a wooden ring. The relationship between the number of pegs p placed evenly around the ring and the number of yarn segments s that crisscross the ring is given by the formula

$$s = \frac{p(p - 3)}{2}$$

How many pegs are needed if the designer wants 27 segments to crisscross the ring? (*Hint:* Multiply both sides of the equation by 2.)

See Example 2.

17. NASCAR The car numbers of drivers Marcos Ambrose and Danica Patrick are consecutive positive integers whose product is 90. If Ambrose's car number is the smaller, find the number of each car.

18. BASEBALL Catcher Thurman Munson and pitcher Whitey Ford are two of the sixteen New York Yankees who have had their uniform numbers retired. These numbers are consecutive integers whose product is 240. If Munson's was the smaller number, determine the uniform number of each player.

19. CUSTOMER SERVICE At a pharmacy, customers take a number to reserve their place in line. If the product of the ticket number now being served and the next ticket number to be served is 156, what number is now being served?

20. HISTORY Delaware was the first state to enter the Union and Hawaii was the 50th. If we order the positions of entry for the rest of the states, we find that Kentucky entered the Union right after Vermont, and the product of their order-of-entry numbers is 210. Use the given information to complete each statement:

Kentucky was the ___ th state to enter the Union.
Vermont was the ___ th state to enter the Union.

21. PLOTTING POINTS The x- and y-coordinates of a point in quadrant I are consecutive odd integers whose product is 143. Find the coordinates of the point.

22. PRESIDENTS George Washington was born on 2-22-1732. He died in 1799 at the age of 67. The month in which he died and the day of the month on which he died are consecutive even integers whose product is 168. When did Washington die?

See Example 3.

23. INSULATION The area of the rectangular slab of foam insulation in the illustration is 36 square meters. Find the dimensions of the slab.

24. FLAGS The length of the flag of Australia is twice its width. If the area of an Australian flag is 18 ft², find its dimensions.

25. SHIPPING PALLETS The length of a rectangular shipping pallet is 2 feet less than 3 times its width. Its area is 21 square feet. Find the dimensions of the pallet.

26. BILLIARDS Pool tables are rectangular and their length is twice their width. Find the dimensions of a pool table if it occupies 50 ft² of floor space.

27. FURNITURE A rectangular kitchen table has an area of 15 square feet. Find the dimensions of the table if its length is 2 ft longer than its width.

28. BULLETIN BOARDS *from Campus to Careers — Elementary School Teacher*
Suppose you are an elementary school teacher. You want to order a rectangular bulletin board to mount on a classroom wall that has an area of 90 square feet. Fire code requirements allow for no more than 30% of a classroom wall to be covered by a bulletin board. If the length of the board is to be three times its width, what are the dimensions of the largest bulletin board that meets the fire code?

See Example 4.

29. BOATING The inclined ramp of the boat launch shown is 8 meters longer than the rise of the ramp. The run is 7 meters longer than the rise. How long are the three sides of the ramp?

30. CONSTRUCTION Trusses are triangular wooden structures used to support roofs. For the truss shown below, the right side is 1 meter longer than the left side, and the span is 2 meters longer than the left side. Find the length of each side of the truss.

31. GARDENING TOOLS The dimensions (in millimeters) of the teeth of a pruning saw blade are given in the illustration. Find each length.

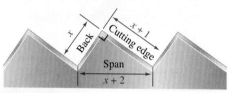

32. HARDWARE An aluminum brace used to support a wooden shelf has a length that is 2 inches less than twice the width of the shelf. The brace is anchored to the wall 8 inches below the shelf. Find the width of the shelf and the length of the brace.

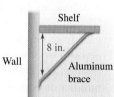

33. DESIGNING A TENT The length of the base of the triangular sheet of canvas above the door of a tent is 2 feet more than twice its height. The area is 30 square feet. Find the height and the length of the base of the triangle.

34. DIMENSIONS OF A TRIANGLE The height of a triangle is 2 inches less than 5 times the length of its base. The area is 36 square inches. Find the length of the base and the height of the triangle.

More problems that are modeled by quadratic equations.

35. TUBING A piece of cardboard in the shape of a parallelogram is twisted to form the tube for a roll of paper towels. The parallelogram has an area of 60 square inches. If its height h is 7 inches more than the length of the base b, what is the length of the base? (*Hint:* The formula for the area of a parallelogram is $A = bh$.)

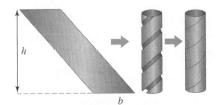

36. SWIMMING POOL BORDERS The owners of the rectangular swimming pool in the illustration want to surround the pool with a crushed-stone border of uniform width. They have enough stone to cover 74 square meters. How wide should they make the border? (*Hint:* The area of the larger rectangle minus the area of the smaller is the area of the border.)

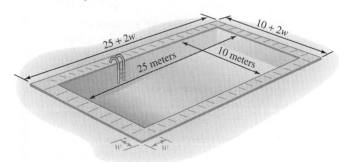

37. HOUSE CONSTRUCTION The formula for the area of a trapezoid is

$$A = \frac{h(B + b)}{2}$$

The area of the trapezoidal truss in the illustration is 24 square meters. Find the height of the trapezoid if one base is 8 meters and the other base is the same as the height. (*Hint:* Multiply both sides of the equation by 2.)

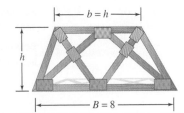

38. VOLUME OF A PYRAMID The volume of a pyramid is given by the formula

$$V = \frac{Bh}{3}$$

where B is the area of its base and h is its height. The volume of the following pyramid is 192 cubic centimeters. Find the dimensions of its rectangular base if one edge of the base is 2 centimeters longer than the other and the height of the pyramid is 12 centimeters.

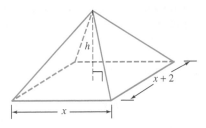

39. THRILL RIDES At the peak of a roller coaster ride, a rider's wristwatch flew off his wrist. The height h (in feet) of the watch, t seconds after he lost it, is given by $h = -16t^2 + 64t + 80$. After how many seconds will the watch hit the ground?

40. PARADES A celebrity on the top of a parade float is tossing pieces of candy to children on the street below. The height h (in feet) of a piece of candy, t seconds after being thrown, is given by $h = -16t^2 + 16t + 32$. After how many seconds will the candy hit the ground?

WRITING

41. Suppose that to find the length of the base of a triangle, you write a quadratic equation and solve it to find $b = 6$ or $b = -8$. Explain why one solution should be discarded.

42. What error is apparent in the following illustration?

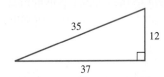

REVIEW

Find each special product.

43. $(5b - 2)^2$

44. $(2a + 3)^2$

45. $(s^2 + 4)^2$

46. $(m^2 - 1)^2$

47. $(9x + 6)(9x - 6)$

48. $(5b + 2c)(5b - 2c)$

STUDY SKILLS CHECKLIST

Preparing for the Chapter 6 Test

In the first five sections of Chapter 6, different factoring methods were discussed. In Section 6.6, an overall factoring strategy was presented on page 506. This strategy is helpful when factoring randomly chosen polynomials. As you study the material for the test on this chapter, review the following checklist for factoring and for solving quadratic equations.

☐ Always factor out the greatest common factor (GCF) first. If this step is not done, it is easy to miss factorizations that need to be done to factor the expression completely.

Factor: $2x^3 - 50x$

$$2x^3 - 50x = 2x(x^2 - 25) \quad \text{Factor out the GCF } 2x.$$
$$= 2x(x^2 - 5^2) \quad x^2 - 25 \text{ is the difference of two squares.}$$
$$= 2x(x + 5)(x - 5)$$

☐ Although $x^2 - 9x$ and $x^2 - 9$ look very similar, they have entirely different factorizations. Be sure to look closely at each term, and remember the *Steps for Factoring a Polynomial* on page 506 of this text.

$$x^2 - 9x = x(x - 9) \quad \text{Factor out the GCF } x.$$
$$x^2 - 9 = (x + 3)(x - 3) \quad \text{There are no common factors. It is the difference of two squares and factors as the product of two binomials.}$$

☐ To factor the sum or difference of two cubes, write the two terms as the base to the third power that is equivalent to the original polynomial. Then follow the rule for factoring the sum or difference of two cubes.

Factor: $x^3 + 8y^3$

$$x^3 + 8y^3 = x^3 + (2y)^3 \quad \text{This polynomial is the sum of two cubes. Write each term as a base to the third power that is equivalent to the original polynomial.}$$
$$= (x + 2y)[x^2 - x(2y) + (2y)^2]$$
$$= (x + 2y)(x^2 - 2xy + 4y^2)$$

☐ Although there is a factorization for the sum of two *cubes*, in general the sum of two squares is a prime polynomial.

$$x^3 + y^3 = (x + y)(x^2 - xy + y^2)$$
$$x^2 + y^2 \quad \text{Is a prime polynomial.}$$

☐ It is important to remember that the instruction to factor means to factor completely. A polynomial is factored completely when no factor can be factored further.

Factor: $x^4 - 16y^4$

$$x^4 - 16y^4 = (x^2 + 4y^2)(x^2 - 4y^2) \quad \text{Think of } x^4 - 16y^4 \text{ as } (x^2)^2 - (4y^2)^2.$$
$$= (x^2 + 4y^2)(x^2 - 4y^2) \quad \text{Think of } x^2 - 4y^2 \text{ as } x^2 - (2y)^2.$$
$$= (x^2 + 4y^2)(x + 2y)(x - 2y)$$
$x^2 + 4y^2$ is a sum of two squares and does not factor.

☐ To solve a quadratic equation by factoring, write the equation in $ax^2 + bx + c = 0$ form, factor the polynomial completely, use the zero-factor property to set each factor equal to zero, and solve the resulting linear equations.

Solve: $x(x + 5) = 14$

$$x(x + 5) = 14 \quad \text{This is the equation to solve.}$$
$$x^2 + 5x = 14 \quad \text{Distribute the multiplication by } x.$$
$$x^2 + 5x - 14 = 0 \quad \text{Subtract 14 from both sides to get 0 on the right side of the equation.}$$
$$(x + 7)(x - 2) = 0 \quad \text{Factor } x^2 + 5x - 14.$$
$$x + 7 = 0 \quad \text{or} \quad x - 2 = 0 \quad \text{Set each factor equal to 0.}$$
$$x = -7 \quad \text{or} \quad x = 2 \quad \text{Solve each linear equation.}$$

CHAPTER 6   SUMMARY AND REVIEW

SECTION 6.1   The Greatest Common Factor; Factoring by Grouping

DEFINITIONS AND CONCEPTS

The **greatest common factor** of a list of integers is the largest common factor of those integers.

To **find the greatest common factor (GCF)** of several monomials:

1. Write each coefficient as a product of prime factors.
2. Identify the numerical and variable factors common to each term.
3. Multiply the common numerical and variable factors identified in step 2 to obtain the GCF. If there are no common factors, the GCF is 1.

EXAMPLES

To find the GCF of 24, 36, and 60, prime factor each number in the list:

$24 = 4 \cdot 6 = \mathbf{2 \cdot 2} \cdot 2 \cdot \mathbf{3}$
$36 = 4 \cdot 9 = \mathbf{2 \cdot 2 \cdot 3} \cdot 3$
$60 = 4 \cdot 15 = \mathbf{2 \cdot 2 \cdot 3} \cdot 5$

Since 24, 36, and 60 each have two factors of 2 and one factor of 3, their greatest common factor is $2 \cdot 2 \cdot 3 = 12$.

To find the GCF of $18a^2b$ and $24a^2b^2$, prime factor each coefficient in the list:

$18a^2b = \mathbf{2} \cdot 3 \cdot \mathbf{3 \cdot a \cdot a \cdot b}$
$24a^2b^2 = \mathbf{2} \cdot 2 \cdot 2 \cdot \mathbf{3 \cdot a \cdot a \cdot b} \cdot b$

Since $18a^2b$ and $24a^2b^2$ each have one factor of 2, one factor of 3, two factors of a, and one factor of b, their greatest common factor is $2 \cdot 3 \cdot a \cdot a \cdot b = 6a^2b$.

To **factor a polynomial** means to express it as a product of two (or more) polynomials.

Factor:

$81 - 9x = 9(9 - x)$	Factor out the GCF of 9.
$18a^2b + 24a^2b^2 = 6a^2b(3 + 4b)$	Factor out the GCF of $6a^2b$.
$-15x^2y + 20xy^2 = -5xy(3x - 4y)$	Factor out the GCF of $-5xy$.

If a polynomial has four terms, try **factoring by grouping**.

1. Group the terms of the polynomial so that the first two terms have a common factor and the last two terms have a common factor.
2. Factor out the common factor from each group.
3. Factor out the resulting common binomial factor. If there is no common binomial factor, regroup the terms of the polynomial and repeat steps 2 and 3.

Factor:

$a^2 + 2a + ab + 2b = a(a + 2) + b(a + 2)$	Factor out a from $a^2 + 2a$ and b from $ab + 2b$.
$= (a + 2)(a + b)$	Factor out $(a + 2)$.

REVIEW EXERCISES

Find the GCF of each list of numbers.

1. 35, 45
2. 45, 54
3. 12, 30, 42
4. 30, 45, 60
5. $36p^2q^2$, $54pq$
6. $28p^4q^3$, $35p^3q^2$, $63p^2q$

Factor each polynomial completely.

7. $3x + 9y$
8. $5ax^2 + 15a$
9. $7s^2 + 14s$
10. $\pi ab - \pi ac$
11. $2x^3 + 4x^2 - 8x$
12. $x^2yz + xy^2z + xyz$
13. $-5ab^2 + 10a^2b - 15ab$
14. $4(x - 2) - x(x - 2)$

Factor out -1 from each polynomial.

15. $-a - 7$
16. $-4t^2 + 3t - 1$

Factor by grouping.

17. $2c + 2d + ac + ad$
18. $3xy + 9x - 2y - 6$
19. $2a^3 - a + 2a^2 - 1$
20. $4m^2n + 12m^2 - 8mn - 24m$

SECTION 6.2 Factoring Trinomials of the Form $x^2 + bx + c$

DEFINITIONS AND CONCEPTS	EXAMPLES
Many trinomials factor as the product of two binomials. To **factor a trinomial** of the form $x^2 + bx + c$, find two integers whose product is c, and whose sum is b. $(x \quad)(x \quad)$ The product of these numbers must be c and their sum must be b.	Factor: $p^2 + 7p + 12 = (p \quad)(p \quad)$ The product of these numbers must be 12 and their sum must be 7. Since $3 \cdot 4 = 12$ and $3 + 4 = 7$, we have: $p^2 + 7p + 12 = (p + 3)(p + 4)$
Before factoring a trinomial, write it in **descending powers** of one variable. Also, factor out -1 if that is necessary to make the **leading coefficient positive**.	Factor: $7q - q^2 - 6$ $= -q^2 + 7q - 6$ Write the terms in descending powers of q. $= -(q^2 - 7q + 6)$ Factor out -1. $= -(q - 1)(q - 6)$ Factor the trinomial.
If a trinomial cannot be factored using only integer coefficients, it is called a **prime polynomial**.	$q^2 + 2q - 5$ is a prime trinomial because there are no two integers whose product is -5 and whose sum is 2.
The GCF should always be factored out first. A trinomial is **factored completely** when it is expressed as a product of prime polynomials.	Factor: $3p^3 - 6p^2 - 24p$ $= 3p(p^2 - 2p - 8)$ Factor out $3p$. $= 3p(p + 2)(p - 4)$ Factor the trinomial.
To factor a trinomial of the form $x^2 + bx + c$ by **grouping**, write it as an equivalent four-term polynomial. $x^2 + \quad x + \quad x + c$ The product of these numbers must be c, and their sum must be b.	Factor by grouping: $p^2 + 7p + 12$ Find two numbers whose product is 12 and whose sum is 7. $= p^2 + 4p + 3p + 12$ Write $7p$ as $4p + 3p$. $= p(p + 4) + 3(p + 4)$ Factor p out of $p^2 + 4p$ and 3 out of $3p + 12$. $= (p + 4)(p + 3)$ Factor out $p + 4$.

REVIEW EXERCISES

Factor each trinomial, if possible.

21. $x^2 + 2x - 24$
22. $x^2 - 4x - 12$
23. $n^2 - 7n + 10$
24. $t^2 + 10t + 15$
25. $-y^2 + 9y - 20$
26. $10y + 9 + y^2$
27. $c^2 + 3cd - 10d^2$
28. $-3mn + m^2 + 2n^2$

Completely factor each trinomial.

29. $5a^2 + 45a - 50$
30. $-4x^2y - 4x^3 + 24xy^2$

Use grouping to factor each trinomial completely.

31. $p^2 + p - 20$
32. $-4q^3 + 4q^2 + 24q$

SECTION 6.3 Factoring Trinomials of the Form $ax^2 + bx + c$

DEFINITIONS AND CONCEPTS	EXAMPLES
To use the **trial-and-check method** to factor $ax^2 + bx + c$, we must determine four integers. The product of these numbers must be a. $ax^2 + bx + c = (\;\;x\;\;)(\;\;x\;\;)$ The product of these numbers must be c. Use the FOIL method to check the factorization.	Factor: $2p^2 - 5p - 12$ The product of these numbers must be 2. $2p^2 - 5p - 12 = (2p + 3)(1p - 4)$ The product of these numbers must be -12. The numbers must also give the correct middle term when we use the FOIL method to check. To check, verify that $(2p + 3)(p - 4) = 2p^2 - 5p - 12$.
To use **grouping** to factor $ax^2 + bx + c$, write it as an equivalent four-term polynomial: $ax^2 + bx + c = ax^2 + \;\;x + \;\;x + c$ The product of these numbers must be ac, and their sum must be b. Then factor the four-term polynomial by grouping. Use the FOIL method to check your work.	Factor by grouping: $2p^2 - 5p - 12$ Find two numbers whose product is $2(-12) = -24$ and whose sum is -5. Two such numbers are -8 and 3. $= 2p^2 - 8p + 3p - 12$ Write $-5p$ as $-8p + 3p$. $= 2p(p - 4) + 3(p - 4)$ Factor the first two terms and the last two terms. $= (p - 4)(2p + 3)$ Factor out $p - 4$. To check, verify that $(p - 4)(2p + 3) = 2p^2 - 5p - 12$.

REVIEW EXERCISES

Factor each trinomial completely, if possible.

33. $2x^2 - 5x - 3$ **34.** $10y^2 + 21y - 10$

35. $-3x^2 + 14x + 5$ **36.** $-9p^2 - 6p + 6p^3$

37. $4b^2 - 17bc + 4c^2$ **38.** $3y^2 + 7y - 11$

Use grouping to factor each trinomial completely.

39. $12p^2 - 2 - 5p$ **40.** $12q^3 - q^2 - 6q$

41. $-16p^2 - 24p - 4pq - 6q$

42. ENTERTAINING The rectangular area occupied by the table setting shown is $(12x^2 - x - 1)$ square inches. Factor the expression to find the binomials that represent the length and width of the table setting.

SECTION 6.4   Factoring Perfect-Square Trinomials and the Difference of Two Squares

DEFINITIONS AND CONCEPTS	EXAMPLES
Special product formulas can be used to factor **perfect-square trinomials**. $A^2 + 2AB + B^2 = (A + B)^2$ $A^2 - 2AB + B^2 = (A - B)^2$	Factor: $p^2 + 8p + 16 = p^2 + 2 \cdot p \cdot 4 + 4^2 = (p + 4)^2$ $m^2 - 18mn + 81n^2 = m^2 - 2 \cdot m \cdot 9n + (9n)^2 = (m - 9n)^2$
To factor the **difference of two squares**, use the formula $F^2 - L^2 = (F + L)(F - L)$	Factor: $4p^2 - 25q^2$ $= (2p)^2 - (5q)^2$ $= (2p + 5q)(2p - 5q)$ This is the difference of two squares.
In general, the sum of two squares (with no common factor other than 1) cannot be factored using real numbers.	$x^2 + 25$ and $81x^2 + 49$ are prime polynomials.

REVIEW EXERCISES

Factor each polynomial completely.

43. $x^2 + 10x + 25$ **44.** $9y^2 - 24y + 16$

45. $-z^2 + 2z - 1$ **46.** $25a^2 + 20ab + 4b^2$

Factor each polynomial completely, if possible.

47. $x^2 - 9$ **48.** $49t^2 - 25y^2$

49. $x^2y^2 - 400$ **50.** $8at^2 - 32a$

51. $4c^3 - 64c$ **52.** $h^2 + 36$

SECTION 6.5 Factoring the Sum and Difference of Two Cubes

DEFINITIONS AND CONCEPTS

To factor the **sum and difference of two cubes**, use the formulas

$$F^3 + L^3 = (F + L)(F^2 - FL + L^2)$$

$$F^3 - L^3 = (F - L)(F^2 + FL + L^2)$$

EXAMPLES

Factor: $p^3 + 64 = p^3 + 4^3$ This is the sum of two cubes.
$$= (p + 4)(p^2 - p \cdot 4 + 4^2)$$
$$= (p + 4)(p^2 - 4p + 16)$$

Factor: $m^3 - 27n^3 = m^3 - (3n)^3$ This is the difference of two cubes.
$$= (m - 3n)[m^2 + m \cdot (3n) + (3n)^2]$$
$$= (m - 3n)(m^2 + 3mn + 9n^2)$$

REVIEW EXERCISES

Factor each polynomial completely, if possible.

53. $h^3 + 1$ **54.** $125p^3 + q^3$ **55.** $x^3 - 27$ **56.** $16x^5 - 54x^2y^3$

SECTION 6.6 A Factoring Strategy

DEFINITIONS AND CONCEPTS

To **factor a random polynomial**, use the **factoring strategy** discussed in the section.

Remember that the instruction to factor means to factor completely. A polynomial is factored completely when no factor can be factored further.

EXAMPLES

Factor: $a^5 + 8a^2 + 4a^3 + 32$

Is there a common factor other than 1?   No.

How many terms does it have?   Since the polynomial has four terms, try factoring by grouping.

$$a^5 + 8a^2 + 4a^3 + 32 = a^2(a^3 + 8) + 4(a^3 + 8)$$
$$= (a^3 + 8)(a^2 + 4)$$

Is it factored completely?   No. We can factor $a^3 + 8$ as the sum of two cubes.

$$a^5 + 8a^2 + 4a^3 + 32 = a^2(a^3 + 8) + 4(a^3 + 8)$$
$$= (a^3 + 8)(a^2 + 4)$$
$$= (a + 2)(a^2 - 2a + 4)(a^2 + 4)$$

Does it check?   Check by multiplication.

REVIEW EXERCISES

Factor each polynomial completely, if possible.

57. $14y^3 + 6y^4 - 40y^2$ **58.** $s^2t + s^2u^2 + tv + u^2v$ **61.** $12w^2 - 36w + 27$ **62.** $121p^2 + 36q^2$

59. $j^4 - 16$ **60.** $-3j^3 - 24k^3$ **63.** $15x^2y - 5xy^2 + 5xy$ **64.** $2x^3 + 12$

Chapter 6 Factoring and Quadratic Equations

SECTION 6.7 Solving Quadratic Equations by Factoring

DEFINITIONS AND CONCEPTS	EXAMPLES
A **quadratic equation** is an equation of the form $ax^2 + bx + c = 0$ $(a \neq 0)$, where a, b, and c represent real numbers.	Quadratic equations: $$x^2 - 3x - 4 = 0$$ $$3a^2 + 5a - 6 = 0$$
The Zero-Factor Property If a and b represent real numbers, and if $ab = 0$, then $a = 0$ or $b = 0$.	Solve: $(4a - 3)(2a + 3) = 0$ $\quad 4a - 3 = 0 \quad$ or $\quad 2a + 3 = 0 \quad$ Set each factor equal to 0. $\quad 4a = 3 \quad\mid\quad 2a = -3 \quad$ Solve each equation. $\quad a = \dfrac{3}{4} \quad\mid\quad a = -\dfrac{3}{2}$ The solutions are $\dfrac{3}{4}$ and $-\dfrac{3}{2}$. The solution set is $\left\{-\dfrac{3}{2}, \dfrac{3}{4}\right\}$.
To use the **factoring method to solve a quadratic equation**: 1. Write the equation in $ax^2 + bx + c = 0$ form. 2. Factor the left side. 3. Use the *zero-factor property* (if $ab = 0$, then $a = 0$ or $b = 0$) and set each factor equal to zero. 4. Solve each resulting linear equation. 5. Check the results in the original equation.	Solve: $\quad 9p^2 - 3p = 0$ $\quad 3p(3p - 1) = 0$ $\quad 3p = 0 \quad$ or $\quad 3p - 1 = 0 \quad$ Set each factor equal to 0. $\quad p = 0 \quad\mid\quad p = \dfrac{1}{3} \quad$ Solve each equation. The solutions are 0 and $\dfrac{1}{3}$. The solution set is $\left\{0, \dfrac{1}{3}\right\}$. Solve: $\quad 2a^2 + 1 = 3a$ $\quad 2a^2 - 3a + 1 = 0$ $\quad (2a - 1)(a - 1) = 0$ $\quad 2a - 1 = 0 \quad$ or $\quad a - 1 = 0 \quad$ Set each factor equal to 0. $\quad a = \dfrac{1}{2} \quad\mid\quad a = 1 \quad$ Solve each equation. The solutions are $\dfrac{1}{2}$ and 1. The solution set is $\left\{\dfrac{1}{2}, 1\right\}$.

REVIEW EXERCISES

Solve each quadratic equation by factoring.

65. $x^2 + 2x = 0$
66. $x(x - 6) = 0$
67. $x^2 - 9 = 0$
68. $a^2 - 7a + 12 = 0$
69. $t^2 + 4t + 4 = 0$
70. $2x - x^2 + 24 = 0$
71. $5a^2 - 6a + 1 = 0$
72. $2p^3 = 2p(p + 2)$

SECTION 6.8 Applications of Quadratic Equations

DEFINITIONS AND CONCEPTS	EXAMPLES
To solve many application problems, use the **six-step problem-solving strategy**: 1. Analyze the problem. 2. Assign a variable. 3. Form an equation. 4. Solve the equation. 5. State the conclusion. 6. Check the result.	Find two consecutive positive integers whose product is 72. **Analyze** *Consecutive integers* are integers that follow each other. The word *product* indicates multiplication. **Assign** Let $x =$ the smaller positive integer. Then $x + 1 =$ the larger integer.

Form

| The smaller integer | times | the larger integer | equals | 72. |

$$x \cdot (x+1) = 72$$

Solve

$$x(x+1) = 72$$
$$x^2 + x = 72 \quad \text{Remove parentheses.}$$
$$x^2 + x - 72 = 0 \quad \text{Subtract 72 from both sides.}$$
$$(x+9)(x-8) = 0 \quad \text{Factor the trinomial.}$$
$$x + 9 = 0 \quad \text{or} \quad x - 8 = 0 \quad \text{Set each factor equal to 0.}$$
$$x = -9 \quad | \quad x = 8 \quad \text{Solve each equation.}$$

State Since we are looking for a positive integer, -9 must be discarded. Thus, the smaller integer is 8. The larger integer is $x + 1 = 9$.

Check The integers 8 and 9 are consecutive positive integers and their product is 72.

The Pythagorean Theorem
If the length of the hypotenuse of a right triangle is c and the lengths of the two legs are a and b, then $c^2 = a^2 + b^2$.

To show that a triangle with sides of 5, 12, and 13 units is a right triangle, we verify that $5^2 + 12^2 = 13^2$.

$$5^2 + 12^2 \stackrel{?}{=} 13^2$$
$$25 + 144 \stackrel{?}{=} 169$$
$$169 = 169 \quad \text{True}$$

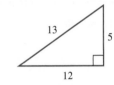

REVIEW EXERCISES

73. BALLOONING A hot-air balloonist dropped his camera overboard while traveling at a height of 1,600 ft. The height h (in feet) of the camera t seconds after being dropped is given by $h = -16t^2 + 1,600$. In how many seconds will the camera hit the ground?

74. Find two consecutive positive integers whose product is 42.

75. CONSTRUCTION The face of the triangular preformed concrete panel shown has an area of 45 square meters, and its base is 3 meters longer than twice its height. How long is its base?

76. GARDENING A rectangular flower bed occupies 27 square feet and is 3 feet longer than twice its width. Find its dimensions.

77. TIGHTROPE WALKERS A circus performer intends to walk up a taut cable to a platform atop a pole, as shown. How high above the ground is the platform?

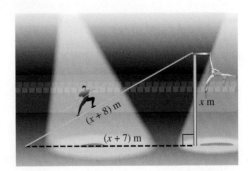

CHAPTER 6 TEST

1. Fill in the blanks.
   a. The letters GCF stand for _____ _____ _____.
   b. To factor a polynomial means to express it as a _____ of two (or more) polynomials.
   c. The _____ theorem provides a formula relating the lengths of the three sides of a right triangle.
   d. $y^2 - 25$ is a _____ of two squares.
   e. The trinomial $x^2 + x - 6$ factors as the product of two _____ : $(x + 3)(x - 2)$.

Find the prime factorization of each number.

2. a. 196 b. 111

Factor each polynomial completely. If a polynomial cannot be factored, write "prime."

3. $4x + 16$
4. $30a^2b^3 - 20a^3b^2 + 5abc$
5. $q^2 - 81$
6. $x^2 + 9$
7. $16x^4 - 81$
8. $x^2 + 4x + 3$
9. $-x^2 + 9x + 22$
10. $9a - 9b + ax - bx$
11. $2a^2 + 5a - 12$
12. $18x^2 - 60xy + 50y^2$
13. $x^3 + 8$
14. $2a^3 - 54$

15. **LANDSCAPING** The combined area of the portions of the square lot that the sprinkler doesn't reach is given by $4r^2 - \pi r^2$, where r is the radius of the circular spray. Factor this expression.

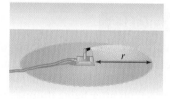

16. **CHECKERS** The area of the square checkerboard is represented by the expression $16x^2 - 24x + 9$. Find an expression that represents the length of a side.

17. What is the greatest common factor of $4a^3b^2$ and $18ab^2$?

18. Factor: $x^2 - 3x - 54$. Show a check of your answer.

Solve each equation.

19. $(x + 3)(x - 2) = 0$

20. $x^2 - 25 = 0$

21. $6x^2 - x = 0$

22. $x^2 + 6x + 9 = 0$

23. $6x^2 + x - 1 = 0$

24. $x^2 + 7x = -6$

25. $x^3 + 7x^2 = -6x$

26. DRIVING SAFETY Virtually all cars have a blind spot where it is difficult for the driver to see a car behind and to the right. The area of the blind spot shown is 54 square feet. Find the width and length of the blind spot.

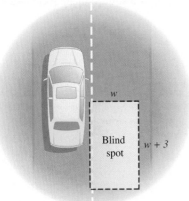

27. ROCKETRY The height h, in feet, of a toy rocket t seconds after being launched is given by $h = -16t^2 + 80t$. After how many seconds will the rocket hit the ground?

28. ATVs The area of a triangular safety flag on an all-terrain vehicle is 33 in.² Its height is 1 inch less than twice the length of the base. Find the length of the base and the height of the flag.

29. Find two consecutive positive integers whose product is 156.

30. Find the length of the hypotenuse of the following right triangle.

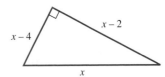

31. What is a quadratic equation? Give an example.

32. If the product of two numbers is 0, what conclusion can be drawn about the numbers?

CHAPTERS 1–6 CUMULATIVE REVIEW

1. **HEART RATES** Refer to the graph. Determine the difference in the maximum heart beat rate for a 70-year-old as compared to someone half that age. [Section 1.1]

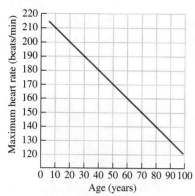

Based on data from *Cardiopulmonary Anatomy and Physiology: Essentials for Respiratory Care*, 2nd ed.

2. Find the prime factorization of 250. [Section 1.2]

3. Find the quotient: $\dfrac{16}{5} \div \dfrac{10}{3}$ [Section 1.2]

4. Write $\dfrac{124}{125}$ as a decimal. [Section 1.3]

5. Determine whether each statement is true or false. [Section 1.3]
   a. Every integer is a whole number.
   b. Every integer is a rational number.
   c. π is a real number.

6. Which division is undefined, $\dfrac{0}{5}$ or $\dfrac{5}{0}$? [Section 1.6]

Evaluate each expression.

7. $3 + 2[-1 - 4(5)]$ [Section 1.7]

8. $\dfrac{|-25| - 2(-5)}{9 - 2^4}$ [Section 1.7]

9. What is -3 cubed? [Section 1.7]

10. What is the value of x twenty-dollar bills? [Section 1.8]

11. Evaluate $\dfrac{-x - a}{y - b}$ for $x = -2$, $y = 1$, $a = 5$, and $b = 2$. [Section 1.8]

12. Identify the coefficient of each term in the expression $8x^2 - x + 9$. [Section 1.8]

Simplify each expression.

13. $-8y^2 - 5y^2 + 6$ [Section 1.9]

14. $3z + 2(y - z) + y$ [Section 1.9]

Solve each equation.

15. $-(3a + 1) + a = 2$ [Section 2.2]

16. $2 - (4x + 7) = 3 + 2(x + 2)$ [Section 2.2]

17. $\dfrac{3t - 21}{2} = t - 6$ [Section 2.2]

18. $-\dfrac{1}{3} - \dfrac{x}{5} = \dfrac{3}{2}$ [Section 2.2]

19. **WATERMELONS** The heaviest watermelon on record weighed 270 pounds. If watermelon is 92% water by weight, what was its water weight? (Round to the nearest pound.) [Section 2.3]

20. Find the distance traveled by a truck traveling for $5\tfrac{1}{2}$ hours at a rate of 60 miles per hour. [Section 2.4]

21. What is the formula for simple interest? [Section 2.4]

22. **GEOMETRY TOOLS** A compass is adjusted so that the distance between the pointed ends is 2 inches. Then a circle is drawn. What will the area of the circle be? Round to the nearest tenth of a square inch. [Section 2.4]

23. Solve $A = P + Prt$ for t. [Section 2.4]

24. HISTORY George Washington was the first president of the United States. John Adams was the second, Thomas Jefferson was the third, and so on. Grover Cleveland was president two *different* times, as shown in the illustration. The sum of the numbers of Cleveland's presidencies is 46. Find these two numbers. [Section 2.5]

Grover Cleveland   Benjamin Harrison   Grover Cleveland

25. PHOTOGRAPHIC CHEMICALS A photographer wishes to mix 6 liters of a 5% acetic acid solution with a 10% solution to get a 7% solution. How many liters of 10% solution must be added? [Section 2.6]

26. Solve $-\dfrac{x}{2} + 4 > 5$. Write the solution set in interval notation and graph it. [Section 2.7]

27. Is $(-2, 5)$ a solution of $3x + 2y = 4$? [Section 3.2]

28. Graph: $y = 2x - 3$ [Section 3.2]

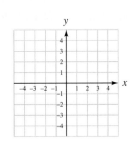

29. Is the graph of $x = 3$ a vertical or horizontal line? [Section 3.3]

30. If two lines are parallel, what can be said about their slopes? [Section 3.4]

31. Find the rate of change of the temperature for the period of time shown in the graph below.

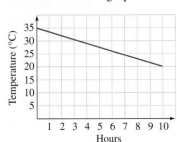

32. Find the slope and the y-intercept of the graph of $3x - 3y = 6$. [Section 3.5]

33. Graph the line passing through $(-4, 1)$ that has slope -3. [Section 3.5]

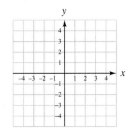

34. Find an equation of the line passing through $(-2, 5)$ and $(-3, -2)$. Write the equation in slope–intercept form. [Section 3.6]

35. Graph: $8x + 4y \geq -24$ [Section 3.7]

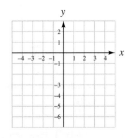

36. If $f(x) = 3x^2 - 2x + 1$, find $f(-2)$. [Section 3.8]

37. Is $\left(\dfrac{1}{2}, 1\right)$ a solution of the system $\begin{cases} 4x - y = 1 \\ 2x + y = 2 \end{cases}$? [Section 4.1]

38. Solve the system $\begin{cases} 3x - 2y = 6 \\ x - y = 1 \end{cases}$ by graphing. [Section 4.1]

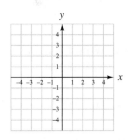

39. Solve the system $\begin{cases} y = -4x + 1 \\ 4x - y = 5 \end{cases}$ by substitution. [Section 4.2]

40. Solve the system $\begin{cases} 5a + 3b = -8 \\ 2a + 9b = 2 \end{cases}$ by addition (elimination). [Section 4.3]

41. FUND-RAISING A Rotary Club held a citywide recycling drive. They collected a total of 14 tons of newspaper and cardboard that earned them $356. They were paid $31 per ton for the newspaper and $18 per ton for the cardboard. How many tons of each did they collect? [Section 4.4]

42. Graph: $\begin{cases} 4x + 3y \geq 12 \\ y < 4 \end{cases}$ [Section 4.5]

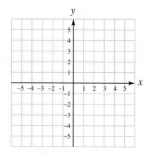

Simplify each expression. Write each answer without negative exponents.

43. $-y^2(4y^3)$ [Section 5.1]

44. $\dfrac{(x^2y^5)^5}{(x^3y)^2}$ [Section 5.1]

45. $\left(\dfrac{b^5}{b^{-2}}\right)$ [Section 5.2]

46. $2x^0$ [Section 5.2]

47. Write 0.00009011 in scientific notation. [Section 5.3]

48. Write 1,700,000 in scientific notation. [Section 5.3]

49. Find the degree of $7y^3 + 4y^2 + y + 3$. [Section 5.4]

50. Graph: $y = x^3 + 2$ [Section 5.4]

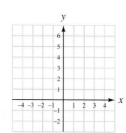

Perform the operations.

51. $(x^2 - 3x + 8) - (3x^2 + x + 3)$ [Section 5.5]

52. $4b^3(2b^2 - 2b)$ [Section 5.6]

53. $(3x - 2)(x + 4)$ [Section 5.6]

54. $(y - 6)^2$ [Section 5.6]

55. $\dfrac{12a^2b^2 - 8a^2b - 4ab}{4ab}$ [Section 5.7]

56. $x - 3\overline{)2x^2 - 5x - 3}$ [Section 5.8]

57. PLAYPENS Find an expression that represents the
  a. perimeter of the playpen. [Section 5.5]
  b. area of the floor of the playpen. [Section 5.6]
  c. volume of the playpen. [Section 5.6]

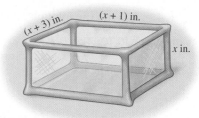

58. Find the GCF of $24x^5y^8$ and $54x^6y$. [Section 6.1]

Factor completely.

59. $9b^3 - 27b^2$ [Section 6.1]

60. $ax + bx + ay + by$ [Section 6.1]

61. $u^2 - 3 + 2u$ [Section 6.2]

62. $10x^2 + x - 2$ [Section 6.3]

63. $4a^2 - 12a + 9$ [Section 6.4]

64. $9z^2 - 1$ [Section 6.4]

65. $t^3 - 8$ [Section 6.5]

66. $3a^2b^2 - 6a^2 - 3b^2 + 6$ [Section 6.6]

Solve each equation.

67. $15s^2 - 20s = 0$ [Section 6.7]

68. $2x^2 - 5x = -2$ [Section 6.7]

Rational Expressions and Equations

7

- **7.1** Simplifying Rational Expressions
- **7.2** Multiplying and Dividing Rational Expressions
- **7.3** Adding and Subtracting with Like Denominators; Least Common Denominators
- **7.4** Adding and Subtracting with Unlike Denominators
- **7.5** Simplifying Complex Fractions
- **7.6** Solving Rational Equations
- **7.7** Problem Solving Using Rational Equations
- **7.8** Proportions and Similar Triangles
- **7.9** Variation

  Chapter Summary and Review
  Chapter Test
  Cumulative Review

from Campus to Careers
Recreation Director

People of all ages enjoy participating in activities, such as arts and crafts, camping, sports, and the performing arts. Recreation directors plan, organize, and oversee these activities in local playgrounds, camps, community centers, religious organizations, theme parks, and tourist attractions. The job of recreation director requires mathematical skills such as budgeting, scheduling, and forecasting trends.

Problem 21 of **Study Set 7.7** involves an area of responsibility for many recreation directors—swimming pools.

JOB TITLE: Recreation Director
EDUCATION: An associate's or bachelor's degree in parks and recreation is preferred.
JOB OUTLOOK: Employment is expected to increase by 19% through the year 2020.
ANNUAL EARNINGS: Average annual salary in 2011 for entry-level positions was $36,000 to $40,000.
FOR MORE INFORMATION:
http://stats.bls.gov/oco/ocos058.htm

Objectives

1. Evaluate rational expressions.
2. Find numbers that cause a rational expression to be undefined.
3. Simplify rational expressions.
4. Simplify rational expressions that have factors that are opposites.

SECTION 7.1
Simplifying Rational Expressions

Fractions such as $\frac{1}{2}$ and $\frac{3}{4}$ that are the quotient of two integers are *rational numbers*. Fractions such as

$$\frac{3}{2y}, \quad \frac{x}{x+2}, \quad \text{and} \quad \frac{5a^2 + 6ab + b^2}{25a^2 - b^2}$$

where the numerators and denominators are polynomials are called *rational expressions*.

> **Rational Expressions**
>
> A **rational expression** is an expression of the form $\frac{A}{B}$, where A and B are polynomials and B does not equal zero.

1 Evaluate rational expressions.

To evaluate a rational expression, we replace each variable with a given number value and simplify.

Self Check 1

Evaluate $\dfrac{3a - 7}{a^3 + 1}$ for $a = 1$.

Now Try Problem 15

EXAMPLE 1   Evaluate $\dfrac{2x - 1}{x^2 + 1}$ for $x = -3$.

Strategy We will replace each x in the rational expression with -3. Then we will evaluate the expression following the order of operations.

WHY Recall from Chapter 1 that to *evaluate an expression* means to find its numerical value, once we know the value of its variable.

Solution

$$\frac{2x-1}{x^2+1} = \frac{2(-3)-1}{(-3)^2+1} \quad \text{Substitute } -3 \text{ for } x.$$

$$= \frac{-6-1}{9+1} \quad \text{In the numerator, do the multiplication. In the denominator, evaluate the exponential expression.}$$

$$= \frac{-7}{10} \quad \text{In the numerator, do the subtraction. In the denominator, do the addition.}$$

$$= -\frac{7}{10} \quad \text{Write the negative symbol in front of the fraction: } \frac{-a}{b} = -\frac{a}{b}.$$

2 Find numbers that cause a rational expression to be undefined.

Since rational expressions indicate division, we must make sure that the denominator of a rational expression is not equal to 0.

EXAMPLE 2   Find all real numbers for which each rational expression is undefined: **a.** $\dfrac{7x}{x-5}$ **b.** $\dfrac{x-1}{x^2-x-6}$

Strategy To find the real numbers for which each rational expression is undefined, we will find the values of the variable that make the *denominator* 0.

WHY A denominator of 0 makes a rational expression undefined, because a denominator of 0 indicates division by 0. We don't need to examine the numerator of the rational expression; it can be any value, including 0.

Solution

a. The expression $\dfrac{7x}{x-5}$ will be undefined if the denominator is zero. To find the value, we solve

$x - 5 = 0$ Set the denominator equal to zero.
$x = 5$ Add 5 to both sides to solve for x.

Check: $\dfrac{7x}{x-5} = \dfrac{7(5)}{5-5} = \dfrac{35}{0}$

Since $\dfrac{35}{0}$ is undefined, the rational expression $\dfrac{7x}{x-5}$ is undefined for $x = 5$.

b. The expression $\dfrac{x-1}{x^2-x-6}$ will be undefined for values of x that make the denominator 0. To find these values, we solve $x^2 - x - 6 = 0$.

$x^2 - x - 6 = 0$ Set the denominator of the rational expression equal to 0.

$(x - 3)(x + 2) = 0$ Factor the trinomial.

$x - 3 = 0$ or $x + 2 = 0$ Set each factor equal to 0.

$x = 3$ | $x = -2$ Solve each linear equation.

Since the values 3 and -2 make the denominator 0, the expression $\dfrac{x-1}{x^2-x-6}$ is undefined for $x = 3$ or $x = -2$.

> **Self Check 2**
> Find the values for x for which each expression is undefined.
> **a.** $\dfrac{x}{x+9}$
> **b.** $\dfrac{x+7}{x^2-25}$
>
> **Now Try** Problems 17 and 21

3 Simplify rational expressions.

To simplify a fraction we remove a factor equal to 1. This can be accomplished in two ways. For example, to simplify $\dfrac{6}{15}$, we proceed as follows:

Method 1

$\dfrac{6}{15} = \dfrac{2 \cdot 3}{5 \cdot 3}$ Factor the numerator and the denominator.

$= \dfrac{2}{5} \cdot \dfrac{3}{3}$ From Section 1.2, we know that $\dfrac{a \cdot c}{b \cdot d} = \dfrac{a}{b} \cdot \dfrac{c}{d}$.

$= \dfrac{2}{5} \cdot 1$ A number divided by itself is equal to 1: $\dfrac{3}{3} = 1$.

$= \dfrac{2}{5}$ Any number multiplied by 1 remains the same.

Method 2

$\dfrac{6}{15} = \dfrac{2 \cdot 3}{5 \cdot 3}$ Factor the numerator and the denominator.

$= \dfrac{2 \cdot \cancel{3}}{5 \cdot \cancel{3}}$ Remove $\dfrac{3}{3} = 1$.

$= \dfrac{2}{5}$ Multiply to find the numerator: $2 \cdot 1 = 2$. Multiply to find the denominator: $5 \cdot 1 = 5$.

When all pairs of factors common to the numerator and denominator of a fraction have been removed, the fraction is **expressed in simplified form**. Since it usually requires fewer steps, we will use method 2 in the following examples. The generalization of method 2 is called the **fundamental property of fractions** and can be applied to rational expressions as well.

Fundamental Property of Fractions

If A, B, and C are polynomials, and B and C are not 0,

$$\dfrac{AC}{BC} = \dfrac{A}{B}$$

Chapter 7 Rational Expressions and Equations

Simplifying rational expressions is similar to simplifying fractions. We write the rational expression so that the numerator and denominator have no common factors other than 1.

> **Simplifying Rational Expressions**
>
> 1. Factor the numerator and denominator completely to determine their common factors.
> 2. Remove factors equal to 1 by replacing each pair of factors common to the numerator and denominator with the equivalent fraction $\frac{1}{1}$.
> 3. Multiply the remaining factors in the numerator and in the denominator.

Self Check 3

Simplify: $\dfrac{32a^3b^2}{24ab^4}$

Now Try Problem 39

EXAMPLE 3 Simplify: $\dfrac{21x^2y}{14xy^2}$

Strategy We will write the numerator and denominator in factored form and then remove pairs of factors that are equal to 1.

WHY The rational expression is simplified when the numerator and denominator have no common factor other than 1.

Solution

$$\frac{21x^2y}{14xy^2} = \frac{3 \cdot 7 \cdot x \cdot x \cdot y}{2 \cdot 7 \cdot x \cdot y \cdot y}$$ Factor the numerator and denominator.

$$= \frac{3 \cdot \overset{1}{\cancel{7}} \cdot \overset{1}{\cancel{x}} \cdot x \cdot \overset{1}{\cancel{y}}}{2 \cdot \underset{1}{\cancel{7}} \cdot \underset{1}{\cancel{x}} \cdot y \cdot \underset{1}{\cancel{y}}}$$ Replace $\frac{7}{7}$, $\frac{x}{x}$, and $\frac{y}{y}$ with the equivalent fraction $\frac{1}{1}$. This removes the factor $\frac{7 \cdot x \cdot y}{7 \cdot x \cdot y}$, which is equal to 1.

$$= \frac{3x}{2y}$$ Multiply the remaining factors in the numerator and in the denominator: $3 \cdot 1 \cdot 1 \cdot x \cdot 1 = 3x$ and $2 \cdot 1 \cdot 1 \cdot y \cdot 1 = 2y$.

To simplify rational expressions, we often make use of the factoring techniques discussed in Chapter 6.

Self Check 4

Simplify: $\dfrac{x^2 - 5x}{5x - 25}$

Now Try Problem 43

EXAMPLE 4 Simplify: $\dfrac{x^2 + 3x}{3x + 9}$

Strategy We will begin by factoring the numerator and denominator. Then we will remove any factors common to the numerator and denominator.

WHY We need to make sure that the numerator and denominator have no common factors other than 1. When this is true, the rational expression is simplified.

Solution
We note that the terms of the numerator have a common factor of x and the terms of the denominator have a common factor of 3.

$$\frac{x^2 + 3x}{3x + 9} = \frac{x(x + 3)}{3(x + 3)}$$ Factor the numerator and the denominator.

$$= \frac{x\cancel{(x + 3)}}{3\cancel{(x + 3)}}$$ Remove a factor equal to 1 by replacing $\frac{x + 3}{x + 3}$ with $\frac{1}{1}$.

$$= \frac{x}{3}$$ Multiply in the numerator: $x \cdot 1 = x$.
Multiply in the denominator: $3 \cdot 1 = 3$.

7.1 Simplifying Rational Expressions

EXAMPLE 5 Simplify: $\dfrac{x^2 + 13x + 12}{x^2 - 144}$

Strategy We will begin by factoring the numerator and denominator. Then we will remove any factors common to the numerator and denominator.

WHY We need to make sure that the numerator and denominator have no common factors other than 1. When this is true, the rational expression is simplified.

Solution
The numerator is a trinomial, and the denominator is a difference of two squares.

$$\dfrac{x^2 + 13x + 12}{x^2 - 144} = \dfrac{(x+1)(x+12)}{(x+12)(x-12)} \quad \text{Factor the numerator and the denominator.}$$

$$= \dfrac{(x+1)\cancel{(x+12)}}{\cancel{(x+12)}(x-12)} \quad \text{Remove a factor equal to 1 by replacing } \dfrac{x+12}{x+12} \text{ with } \dfrac{1}{1}.$$

$$= \dfrac{x+1}{x-12}$$

Self Check 5
Simplify: $\dfrac{3x^2 - 8x - 3}{x^2 - 9}$

Now Try Problem 45

Caution! When simplifying a fraction, remember that only *factors* that are common to the *entire numerator* and the *entire denominator* can be removed. For example, consider the correct simplification

$$\dfrac{5+8}{5} = \dfrac{13}{5}$$

It would be incorrect to remove the common *term* of 5 in this simplification. Doing so gives an incorrect answer of 9.

$$\dfrac{5+8}{5} = \dfrac{\cancel{5}+8}{\cancel{5}} = \dfrac{1+8}{1} = 9$$

When simplifying rational expressions, *it is incorrect to remove terms common to both the numerator and denominator.*

$$\dfrac{\cancel{x}+5}{\cancel{x}+6} \qquad \dfrac{a^2 - \cancel{3a} + \cancel{2}}{\cancel{a} + \cancel{2}} \qquad \dfrac{\cancel{y^2} - 36}{\cancel{y^2} - y - 7}$$

Any number or algebraic expression divided by 1 remains unchanged. For example,

$$\dfrac{37}{1} = 37, \quad \dfrac{5x}{1} = 5x, \quad \text{and} \quad \dfrac{3x+y}{1} = 3x+y$$

In general, we have the following.

Division by 1

For any real number a, $\dfrac{a}{1} = a$.

Self Check 6

Simplify: $\dfrac{a^2 + a - 2}{a - 1}$

Now Try Problem 49

EXAMPLE 6 Simplify: $\dfrac{x^3 + x^2}{x + 1}$

Strategy We will begin by factoring the numerator and denominator. Then we will remove any factors common to the numerator and denominator.

WHY We need to make sure that the numerator and denominator have no common factors other than 1. When this is true, the rational expression is simplified.

Solution

$$\dfrac{x^3 + x^2}{x + 1} = \dfrac{x^2(x + 1)}{x + 1} \qquad \text{Factor the numerator.}$$

$$= \dfrac{x^2\cancel{(x+1)}}{\cancel{x+1}} \qquad \text{Remove a factor equal to 1 by replacing } \tfrac{x+1}{x+1} \text{ with } \tfrac{1}{1}.$$

$$= \dfrac{x^2}{1}$$

$$= x^2 \qquad \text{Denominators of 1 need not be written.}$$

Self Check 7

Simplify: $\dfrac{4(x - 2) + 4}{3(x - 2) + 3}$

Now Try Problem 55

EXAMPLE 7 Simplify: $\dfrac{5(x + 3) - 5}{7(x + 3) - 7}$

Strategy We will begin by simplifying the numerator, $5(x + 3) - 5$, and the denominator, $7(x + 3) - 7$, separately. Then we will factor each result and remove any common factors.

WHY We cannot immediately remove $x + 3$ because it is not a factor of the *entire* numerator and the *entire* denominator.

Solution

$$\dfrac{5(x + 3) - 5}{7(x + 3) - 7} = \dfrac{5x + 15 - 5}{7x + 21 - 7} \qquad \text{Use the distributive property in the numerator and the denominator.}$$

$$= \dfrac{5x + 10}{7x + 14} \qquad \text{Combine like terms: } 15 - 5 = 10 \text{ and } 21 - 7 = 14.$$

$$= \dfrac{5(x + 2)}{7(x + 2)} \qquad \text{Factor the numerator and the denominator.}$$

$$= \dfrac{5\cancel{(x+2)}}{7\cancel{(x+2)}} \qquad \text{Remove the common factor of } (x + 2) \text{ in the numerator and denominator.}$$

$$= \dfrac{5}{7}$$

Self Check 8

Simplify: $\dfrac{a(a + 2) - 2(a - 1)}{a^2 + 2}$

Now Try Problem 57

EXAMPLE 8 Simplify: $\dfrac{x(x + 3) - 3(x - 1)}{x^2 + 3}$

Strategy We will begin by simplifying the numerator, $x(x + 3) - 3(x - 1)$. Then we will look for any common factors that can be removed in the numerator and the denominator.

WHY We need to make sure that the numerator and denominator have no common factors other than 1. When this is true, the rational expression is simplified.

Solution

We simplify the numerator and look for any common factors to remove in the numerator and denominator.

$$\frac{x(x+3) - 3(x-1)}{x^2+3} = \frac{x^2 + 3x - 3x + 3}{x^2+3}$$ Use the distributive property twice in the numerator.

$$= \frac{x^2+3}{x^2+3}$$ Combine like terms in the numerator: $3x - 3x = 0$.

$$= \frac{\overset{1}{\cancel{x^2+3}}}{\underset{1}{\cancel{x^2+3}}}$$ Remove the common factor of x^2+3 in the numerator and denominator.

$$= 1$$

Sometimes a fraction does not simplify. For example, to attempt to simplify

$$\frac{x^2+x-2}{x^2+x}$$

we factor the numerator and the denominator.

$$\frac{x^2+x-2}{x^2+x} = \frac{(x+2)(x-1)}{x(x+1)}$$

Because there are no factors common to the numerator and denominator, this fraction is already in simplest terms.

4 Simplify rational expressions that have factors that are opposites.

If the terms of two polynomials are the same, except that they are opposite in sign, the polynomials are called **opposites (negatives)**. For example, the following pairs of polynomials are opposites of each other:

$x - y$	and	$-x + y$	Compare terms: x and $-x$; $-y$ and y.
$2a - 1$	and	$-2a + 1$	Compare terms: $2a$ and $-2a$; -1 and 1.
$-3x^2 - 2x + 5$	and	$3x^2 + 2x - 5$	Compare terms: $-3x^2$ and $3x^2$; $-2x$ and $2x$; 5 and -5.

Example 9 shows why the quotient of two binomials that are opposites is equal to -1.

EXAMPLE 9

Simplify: $\dfrac{2a-1}{1-2a}$

Strategy We will rearrange the terms of the numerator $2a - 1$, and factor out -1.

WHY This step is useful when the numerator and denominator contain factors that are opposites, such as $2a - 1$ and $1 - 2a$. It produces a common factor that can be removed.

Solution

$$\frac{2a-1}{1-2a} = \frac{-1+2a}{1-2a}$$ In the numerator, think of $2a - 1$ as $2a + (-1)$. Then change the order of the terms: $2a + (-1) = -1 + 2a$.

$$= \frac{-(1-2a)}{1-2a}$$ In the numerator, factor out -1: $-1 + 2a = -(1-2a)$.

Self Check 9

Simplify: $\dfrac{3p-2}{2-3p}$

Now Try Problem 61

$$= \frac{-(1 - 2a)}{1 - 2a} \quad \text{Remove the common factor of } 1 - 2a \text{ in the numerator and denominator.}$$

$$= -1$$

In general, we have this important fact.

The Quotient of Opposites

The quotient of any nonzero expression and its opposite is -1.

Caution! Only apply the preceding rule to expressions that are opposites. For example, it would be incorrect to use this rule to simplify $\frac{x+1}{1+x}$. Since $x + 1$ equals $1 + x$ by the commutative property of addition, this is the quotient of a number and itself. The result is 1, not -1.

$$\frac{x+1}{1+x} = \frac{x+1}{x+1} = 1$$

Self Check 10

Simplify, if possible:

a. $\dfrac{m^2 - 100}{10m - m^2}$

b. $\dfrac{2x - 3}{2x + 3}$

Now Try Problem 63

EXAMPLE 10 Simplify, if possible: a. $\dfrac{y^2 - 1}{3 - 3y}$ b. $\dfrac{t + 8}{t - 8}$

Strategy We will begin by factoring the numerator and denominator. Then we look for common factors, or factors that are opposites, and remove them.

WHY We need to make sure that the numerator and denominator have no common factor (or opposite factors) other than 1. When this is the case, then the rational expression is simplified.

Solution

a. $\dfrac{y^2 - 1}{3 - 3y} = \dfrac{(y+1)(y-1)}{3(1-y)}$ Factor the numerator. Factor the denominator.

$$= \dfrac{(y+1)(y-1)}{3(1-y)}$$ Since $y - 1$ and $1 - y$ are opposites, simplify by replacing $\frac{y-1}{1-y}$ with the equivalent fraction $\frac{-1}{1}$. This removes the factor $\frac{y-1}{1-y} = -1$.

$$= \dfrac{-(y+1)}{3}$$

This result may be written in several other equivalent forms.

$$\dfrac{-(y+1)}{3} = -\dfrac{y+1}{3} \quad \text{The } - \text{ symbol in } -(y+1) \text{ can be written in the front of the fraction, and the parentheses can be dropped.}$$

$$\dfrac{-(y+1)}{3} = \dfrac{-y-1}{3} \quad \text{The } - \text{ symbol in } -(y+1) \text{ represents a factor of } -1. \text{ Distribute the multiplication by } -1 \text{ in the numerator.}$$

$$\dfrac{-(y+1)}{3} = \dfrac{y+1}{-3} \quad \text{The } - \text{ symbol in } -(y+1) \text{ can be applied to the denominator. However, we don't usually use this form.}$$

Caution! A − symbol in front of a fraction may be applied to the numerator or to the denominator, but not to both:

$$-\frac{y+1}{3} \neq \frac{-(y+1)}{-3}$$

b. The binomials $t + 8$ and $t - 8$ are not opposites because their first terms do not have opposite signs. Thus, $\frac{t+8}{t-8}$ does not simplify.

ANSWERS TO SELF CHECKS

1. -2 2. a. -9 b. 5 or -5 3. $\frac{4a^2}{3b^2}$ 4. $\frac{x}{5}$ 5. $\frac{3x+1}{r+3}$ 6. $a+2$ 7. $\frac{4}{3}$ 8. 1
9. -1 10. a. $-\frac{m+10}{m}$ b. does not simplify

SECTION 7.1 STUDY SET

VOCABULARY

Fill in the blanks.

1. In a fraction, the part above the fraction bar is called the _____, and the part below is called the _____.

2. A fraction that has polynomials in its numerator and denominator, such as $\frac{x+2}{x-3}$, is called a _____ expression.

3. Division by 0 is _____.

4. A fraction is in _____ form when all common factors of the numerator and denominator have been removed.

5. To _____ a rational expression means we remove factors common to the numerator and denominator.

6. If the terms of two polynomials are the same, except for sign, the polynomials are called _____ of each other.

CONCEPTS

7. What value of x makes each rational expression undefined?

   a. $\frac{x+2}{x}$ b. $\frac{x+2}{x-6}$ c. $\frac{x+2}{x+6}$

8. In the following work, what common factor has been removed?

$$\frac{x^2+2x+1}{x^2+4x+3} = \frac{\overset{1}{(x+1)}(x+1)}{(x+3)\underset{1}{(x+1)}} = \frac{x+1}{x+3}$$

9. Simplify each rational expression.

   a. $\frac{x-8}{x-8}$ b. $\frac{x-8}{8-x}$ c. $\frac{x-8}{1}$

10. Explain the error in the following work.

$$\frac{x}{x+2} = \frac{\overset{1}{\cancel{x}}}{\cancel{x}+2} = \frac{1}{3}$$

NOTATION

Complete each step.

11. $\dfrac{x^2+5x-6}{x^2-1} = \dfrac{(x+\;\;)(x-1)}{(x+1)(x-\;\;)}$

$$= \frac{x+6}{\;\;}$$

12. $\dfrac{5(x+2)-5}{4(x+2)-4} = \dfrac{5x+\;\;-5}{4x+\;\;-4}$

$$= \frac{5x+\;\;}{4x+\;\;}$$

$$= \frac{5\;\;}{4(x+1)}$$

$$= \frac{5}{\;\;}$$

GUIDED PRACTICE

Evaluate each expression for $x = 6$. See Example 1.

13. $\dfrac{x-2}{x-5}$ 14. $\dfrac{3x-2}{x-2}$

15. $\dfrac{-2x-3}{x^2-1}$ 16. $\dfrac{x^2-11}{-x-4}$

Which value(s) of x, if any, make each rational expression undefined? See Example 2.

17. $\dfrac{15}{x-2}$
18. $\dfrac{5x}{x+5}$
19. $\dfrac{15x+2}{16}$
20. $\dfrac{x^2-4x}{25}$
21. $\dfrac{30}{x^2-36}$
22. $\dfrac{2x-15}{x^2-49}$
23. $\dfrac{15}{x^2+x-2}$
24. $\dfrac{x-20}{x^2+2x-8}$

Simplify each fraction. See Objective 3.

25. $\dfrac{28}{35}$
26. $\dfrac{14}{20}$
27. $\dfrac{9}{27}$
28. $\dfrac{15}{45}$
29. $-\dfrac{36}{48}$
30. $-\dfrac{32}{40}$
31. $-\dfrac{49}{35}$
32. $-\dfrac{36}{52}$

Simplify each expression. If an expression cannot be simplified, so indicate. Assume that no denominators are 0. See Examples 3–4.

33. $\dfrac{45}{9a}$
34. $\dfrac{48}{16y}$
35. $\dfrac{2x}{3x}$
36. $\dfrac{5y}{7y}$
37. $\dfrac{2x^2}{3y}$
38. $\dfrac{5x^2}{2y^2}$
39. $\dfrac{15x^2y}{5xy^2}$
40. $\dfrac{12xz}{4xz^2}$
41. $\dfrac{6x+3}{3y}$
42. $\dfrac{4x+12}{2y}$
43. $\dfrac{a^2+4a}{4a+16}$
44. $\dfrac{9c-27}{bc-3b}$

Simplify each expression. Assume that no denominators are 0. See Examples 5–6.

45. $\dfrac{3x^2-14x+8}{x^2-16}$
46. $\dfrac{4x^2+22x+10}{x^2-25}$
47. $\dfrac{x^2+3x+2}{x^2+x-2}$
48. $\dfrac{x^2+x-6}{x^2-x-2}$
49. $\dfrac{x^4-x^3}{x-1}$
50. $\dfrac{a^3+a^2}{a^2+a}$
51. $\dfrac{x^2+3x+2}{x^3+x^2}$
52. $\dfrac{x^2-13x+30}{x^4-3x^3}$

Simplify each expression. Assume that no denominators are 0. See Examples 7–8.

53. $\dfrac{4(x+3)+4}{3(x+2)+6}$
54. $\dfrac{4+2(x-5)}{3x-5(x-2)}$
55. $\dfrac{6+2(x+2)}{3(x-2)+21}$
56. $\dfrac{-5(a+2)+15}{-12+4(a+2)}$
57. $\dfrac{a(a+2)-2(a+1)}{2a^2-4}$
58. $\dfrac{2b(b-3)+6(b-1)}{4b^2-12}$
59. $\dfrac{x^2-9}{(2x+3)-(x+6)}$
60. $\dfrac{x^2+5x+4}{2(x+3)-(x+2)}$

Simplify each expression. Assume that no denominators are 0. See Examples 9–10.

61. $\dfrac{x-7}{7-x}$
62. $\dfrac{18-d}{d-18}$
63. $\dfrac{6x-30}{5-x}$
64. $\dfrac{3-4t}{8t-6}$
65. $\dfrac{2-a}{a^2-a-2}$
66. $\dfrac{4-b}{b^2-5b+4}$
67. $\dfrac{a+b-c}{c-a-b}$
68. $\dfrac{x-y-z}{z+y-x}$

TRY IT YOURSELF

Simplify each expression. If it is already in simplified form, so indicate. Assume that no denominators are 0.

69. $\dfrac{x^2-5x}{x^3-25x}$
70. $\dfrac{4x}{8x+12}$
71. $\dfrac{6xy}{15x^2y^3}$
72. $\dfrac{7-a}{a^2-49}$
73. $\dfrac{5+5}{5z}$
74. $\dfrac{x+x}{2}$
75. $\dfrac{(3+4)a}{24-3}$
76. $\dfrac{(3-18)k}{25}$
77. $\dfrac{6x^2}{4x^2}$
78. $\dfrac{9x^3}{6x}$
79. $\dfrac{x+3}{3x+9}$
80. $\dfrac{2x+14}{x-7}$
81. $\dfrac{2x^2-8x}{x^2-6x+8}$
82. $\dfrac{3y^2-15y}{y^2-3y-10}$
83. $\dfrac{6x^2-13x+6}{3x^2+x-2}$
84. $\dfrac{7x^2+20x-3}{3x^2+8x-3}$
85. $\dfrac{3x^2-27}{2x^2-5x-3}$
86. $\dfrac{2x^2-8}{3x^2-5x-2}$
87. $\dfrac{m^2-2mn+n^2}{2m^2-2n^2}$
88. $\dfrac{c^2-d^2}{c^2-2cd+d^2}$

89. $\dfrac{5x^2 + 2x - 3}{6x^2 - x - 1}$ 90. $\dfrac{3x^2 - 10x - 77}{x^2 - 4x - 21}$

91. $\dfrac{x(x - 8) + 16}{16 - x^2}$ 92. $\dfrac{x^2 - 3(2x - 3)}{9 - x^2}$

93. $\dfrac{y - xy}{xy - x}$ 94. $\dfrac{x^2 + y^2}{x + y}$

95. $\dfrac{6a - 6b + 6c}{9a - 9b + 9c}$ 96. $\dfrac{3a - 3b - 6}{2a - 2b - 4}$

97. $\dfrac{15x - 3x^2}{25y - 5xy}$ 98. $\dfrac{rz - 2x}{yz - 2y}$

99. $\dfrac{12 - 3x^2}{x^2 - x - 2}$ 100. $\dfrac{-5x + 10}{x^2 - 4x + 4}$

101. $\dfrac{x^2 - 8x + 15}{x^2 - x - 6}$ 102. $\dfrac{x^2 - 6x - 7}{x^2 + 8x + 7}$

APPLICATIONS

103. **ORGAN PIPES** The number of vibrations n per second of an organ pipe is given by the formula

$n = \dfrac{512}{L}$, where L is the length of the pipe in feet.

How many times per second will a 6-foot pipe vibrate?

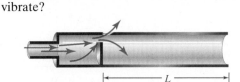

104. **WORD PROCESSORS** For the word processor shown, the number of words w that can be typed on a piece of paper is given by the formula

$w = \dfrac{8{,}000}{x}$

where x is the font size used. Find the number of words that can be typed on a page for each font size choice shown.

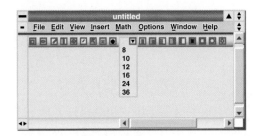

105. **ROOFING** Refer to the illustration in the next column. The *pitch* of a roof is a measure of how steep or how flat the roof is. If pitch = $\dfrac{\text{rise}}{\text{run}}$, find the pitch of the roof of the cabin shown. Express the result in simplified form.

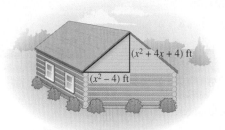

$(x^2 + 4x + 4)$ ft

$(x^2 - 4)$ ft

106. **GRAPHIC DESIGN** A physical activity pyramid, in the shape of an equilateral triangle, is to be enlarged and distributed to schools for display in their health classes. What is the length of a side of the original design divided by the length of a side of the enlargement? Express the result in lowest terms.

 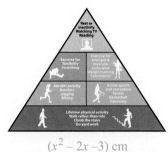

$(2x - 6)$ cm $(x^2 - 2x - 3)$ cm
Original design Enlargement

WRITING

107. Explain why $\dfrac{x - 7}{7 - x} = -1$.

108. Explain the difference between a factor and a term. Give several examples.

109. Explain the error.

$$\dfrac{3(\overset{1}{\cancel{x + 1}}) - x}{\underset{1}{\cancel{x + 1}}} = 3 - x$$

110. Explain why there are no values for x for which $\dfrac{x - 7}{x^2 + 49}$ is undefined.

REVIEW

111. State the associative property of addition using the variables a, b, and c.

112. State the distributive property using the variables x, y, and z.

113. If $ab = 0$, what must be true about a or b?

114. What is the product of a number and 1?

115. Find the opposite of $-\dfrac{5}{3}$.

116. Find the cube of 2, squared.

Objectives

1. Multiply rational expressions.
2. Multiply a rational expression by a polynomial.
3. Divide rational expressions.
4. Convert units of measurement.

SECTION 7.2
Multiplying and Dividing Rational Expressions

In this section, we extend the rules for multiplying and dividing numerical fractions to problems involving multiplication and division of rational expressions.

1 Multiply rational expressions.

Recall that to multiply fractions, we multiply their numerators and multiply their denominators. For example,

$$\frac{4}{7} \cdot \frac{3}{5} = \frac{4 \cdot 3}{7 \cdot 5}$$ Multiply the numerators and multiply the denominators.

$$= \frac{12}{35}$$ Do the multiplication in the numerator: $4 \cdot 3 = 12$.
Do the multiplication in the denominator: $7 \cdot 5 = 35$.

We use the same procedure to multiply rational expressions.

Multiplying Rational Expressions

To multiply rational expressions, multiply their numerators and their denominators. Then, if possible, factor and simplify.

For any two rational expressions, $\frac{A}{B}$ and $\frac{C}{D}$,

$$\frac{A}{B} \cdot \frac{C}{D} = \frac{AC}{BD}$$

Self Check 1

Multiply: $\dfrac{3x}{4} \cdot \dfrac{x-3}{5}$

Now Try Problems 11, 15, and 21

EXAMPLE 1 Multiply: **a.** $\dfrac{x}{3} \cdot \dfrac{2}{5}$ **b.** $\dfrac{7}{9} \cdot \dfrac{-5}{3x}$ **c.** $\dfrac{x^2}{2} \cdot \dfrac{3}{y^2}$ **d.** $\dfrac{8}{t} \cdot \dfrac{t-1}{t-2}$

Strategy We will use the rule for multiplying rational expressions. In the process, we must be ready to factor the numerators and denominators so that all common factors can be removed. If no common factors are present, the fraction is already in simplest form.

WHY We want to give the result in simplified form, which requires that the numerator and denominator have no common factors other than 1.

Solution

a. $\dfrac{x}{3} \cdot \dfrac{2}{5} = \dfrac{x \cdot 2}{3 \cdot 5}$

$= \dfrac{2x}{15}$

b. $\dfrac{7}{9} \cdot \dfrac{-5}{3x} = \dfrac{7(-5)}{9 \cdot 3x}$

$= \dfrac{-35}{27x}$

$= -\dfrac{35}{27x}$

c. $\dfrac{x^2}{2} \cdot \dfrac{3}{y^2} = \dfrac{x^2 \cdot 3}{2 \cdot y^2}$

$= \dfrac{3x^2}{2y^2}$

d. $\dfrac{8}{t} \cdot \dfrac{t-1}{t-2} = \dfrac{8(t-1)}{t(t-2)}$

$= \dfrac{8t-8}{t(t-2)}$

Caution! In Example 1d, it is not necessary to multiply in the denominator. When we add and subtract rational expressions (see Section 7.3), it is usually more convenient to leave the denominator in factored form as shown.

7.2 Multiplying and Dividing Rational Expressions

EXAMPLE 2

Multiply: $\dfrac{35x^2y}{7y^2z} \cdot \dfrac{z}{5xy}$

Strategy We will use the rule for multiplying rational expressions. In the process, we must be ready to factor the numerators and denominators so that all common factors can be removed.

WHY We want to give the result in simplified form, which requires that the numerator and denominator have no common factors other than 1.

Solution

$$\dfrac{35x^2y}{7y^2z} \cdot \dfrac{z}{5xy} = \dfrac{35x^2y \cdot z}{7y^2z \cdot 5xy}$$ Multiply the numerators and multiply the denominators.

$$= \dfrac{5 \cdot 7 \cdot x \cdot x \cdot y \cdot z}{7 \cdot y \cdot y \cdot z \cdot 5 \cdot x \cdot y}$$ Factor $35x^2$ and factor y^2.

$$= \dfrac{\overset{1}{\cancel{5}} \cdot \overset{1}{\cancel{7}} \cdot \cancel{x} \cdot x \cdot \overset{1}{\cancel{y}} \cdot \overset{1}{\cancel{z}}}{\underset{1}{\cancel{7}} \cdot \underset{1}{\cancel{y}} \cdot y \cdot \underset{1}{\cancel{z}} \cdot \underset{1}{\cancel{5}} \cdot \underset{1}{\cancel{x}} \cdot y}$$ Simplify by removing the common factors.

$$= \dfrac{x}{y^2}$$ Do the multiplications in the numerator and the denominator.

Self Check 2

Multiply: $\dfrac{a^2b^2}{2a} \cdot \dfrac{9a^3}{3b^3}$

Now Try Problem 17

EXAMPLE 3

Multiply: $\dfrac{x^2 - x}{2x + 4} \cdot \dfrac{x + 2}{x}$

Strategy We will use the rule for multiplying rational expressions. In the process, we must be ready to factor the monomials, binomials, or trinomials in the numerators and denominators so that all common factors can be removed.

WHY We want to give the result in simplified form, which requires that the numerator and denominator have no common factors other than 1.

Solution

$$\dfrac{x^2 - x}{2x + 4} \cdot \dfrac{x + 2}{x} = \dfrac{(x^2 - x)(x + 2)}{(2x + 4)(x)}$$ Multiply the numerators and multiply the denominators.

We now factor the numerator and denominator to see whether this product can be simplified.

$$= \dfrac{x(x - 1)(x + 2)}{2(x + 2)x}$$ Factor the numerator: $(x^2 - x) = x(x - 1)$. Factor the denominator: $(2x + 4) = 2(x + 2)$.

$$= \dfrac{\overset{1}{\cancel{x}}(x - 1)\overset{1}{\cancel{(x + 2)}}}{2\underset{1}{\cancel{(x + 2)}}\underset{1}{\cancel{x}}}$$ Simplify by removing the common factors.

$$= \dfrac{x - 1}{2}$$

Self Check 3

Multiply: $\dfrac{x^2 + x}{3x + 6} \cdot \dfrac{x + 2}{x + 1}$

Now Try Problem 25

EXAMPLE 4

Multiply: $\dfrac{x^2 - 3x}{x^2 - x - 6} \cdot \dfrac{x^2 + x - 2}{x^2 - x}$

Strategy We will use the rule for multiplying rational expressions. In the process, we must be ready to factor the monomials, binomials, or trinomials in the numerators and denominators so that all common factors can be removed.

WHY We want to give the result in simplified form, which requires that the numerator and denominator have no common factors other than 1.

Self Check 4

Multiply: $\dfrac{a^2 + a}{a^2 - 4} \cdot \dfrac{a^2 - a - 2}{a^2 + 2a + 1}$

Now Try Problem 31

Solution

$$\frac{x^2 - 3x}{x^2 - x - 6} \cdot \frac{x^2 + x - 2}{x^2 - x}$$

$$= \frac{(x^2 - 3x)(x^2 + x - 2)}{(x^2 - x - 6)(x^2 - x)} \quad \text{Multiply the numerators and multiply the denominators.}$$

$$= \frac{x(x - 3)(x + 2)(x - 1)}{(x + 2)(x - 3)x(x - 1)} \quad \text{Factor the numerator and denominator to see whether the result can be simplified.}$$

$$= \frac{\overset{1}{\cancel{x}}(\overset{1}{\cancel{x - 3}})(\overset{1}{\cancel{x + 2}})(\overset{1}{\cancel{x - 1}})}{(\underset{1}{\cancel{x + 2}})(\underset{1}{\cancel{x - 3}})\underset{1}{\cancel{x}}(\underset{1}{\cancel{x - 1}})} \quad \text{Simplify by removing the common factors.}$$

$$= 1$$

2 Multiply a rational expression by a polynomial.

Since any number divided by 1 remains unchanged, we can write any polynomial as a fraction by inserting a denominator of 1.

Self Check 5

Multiply:

a. $\dfrac{9}{7y} \cdot 7y$

b. $36b\left(\dfrac{1}{6b}\right)$

c. $4x\left(\dfrac{x + 3}{x}\right)$

Now Try Problems 35 and 39

EXAMPLE 5 Multiply: **a.** $\dfrac{4}{x} \cdot x$ **b.** $63x\left(\dfrac{1}{7x}\right)$ **c.** $5a\left(\dfrac{3a - 1}{a}\right)$

Strategy We will write each of the monomials, x, $63x$, and $5a$, as rational expressions with a denominator of 1. (Remember, any number divided by 1 remains unchanged.) Then we will use the rule for multiplying rational expressions.

WHY Writing x, $63x$, and $5a$ over 1 is helpful during the multiplication process when we multiply numerators and multiply denominators.

Solution

a. $\dfrac{4}{x} \cdot x = \dfrac{4}{x} \cdot \dfrac{x}{1}$ Write x as a fraction: $x = \dfrac{x}{1}$.

$= \dfrac{4 \cdot \overset{1}{\cancel{x}}}{\underset{1}{\cancel{x}} \cdot 1}$ Multiply the fractions and remove the common factor in the numerator and denominator.

$= 4$ Simplify.

b. $63x\left(\dfrac{1}{7x}\right) = \dfrac{63x}{1}\left(\dfrac{1}{7x}\right)$ Write $63x$ as a fraction: $63x = \dfrac{63x}{1}$.

$= \dfrac{63x \cdot 1}{1 \cdot 7 \cdot x}$ Multiply the fractions.

$= \dfrac{9 \cdot \overset{1}{\cancel{7}} \cdot \overset{1}{\cancel{x}} \cdot 1}{1 \cdot \underset{1}{\cancel{7}} \cdot \underset{1}{\cancel{x}}}$ Write $63x$ in factored form as $9 \cdot 7 \cdot x$ and remove the common factors.

$= 9$ Simplify.

c. $5a\left(\dfrac{3a - 1}{a}\right) = \dfrac{5a}{1}\left(\dfrac{3a - 1}{a}\right)$ Write $5a$ as a fraction: $5a = \dfrac{5a}{1}$.

$= \dfrac{5\overset{1}{\cancel{a}}(3a - 1)}{1 \cdot \underset{1}{\cancel{a}}}$ Multiply the fractions and remove the common factor of a.

$= 5(3a - 1)$ Simplify.

$= 15a - 5$ Distribute the multiplication by 5.

EXAMPLE 6

Multiply: $\dfrac{x^2 + x}{x^2 + 8x + 7} \cdot (x + 7)$

Strategy We will write the binomial factor over 1. Then we will use the rule for multiplying rational expressions.

WHY Writing $(x + 7)$ over 1 is helpful during the multiplication process when we multiply numerators and denominators.

Solution

$\dfrac{x^2 + x}{x^2 + 8x + 7} \cdot (x + 7)$

$= \dfrac{x^2 + x}{x^2 + 8x + 7} \cdot \dfrac{x + 7}{1}$ Write $x + 7$ as a fraction with a denominator of 1.

$= \dfrac{x(x + 1)(x + 7)}{(x + 1)(x + 7)1}$ Multiply the fractions and factor where possible.

$= \dfrac{x\cancel{(x + 1)}\cancel{(x + 7)}}{\cancel{(x + 1)}\cancel{(x + 7)}1}$ Simplify by removing the common factors.

$= x$

Self Check 6

Multiply: $(a - 7) \cdot \dfrac{a^2 - a}{a^2 - 8a + 7}$

Now Try Problem 45

3 Divide rational expressions.

Division by a nonzero number is equivalent to multiplying by its reciprocal. Thus, to divide two fractions, we can invert the divisor (the fraction following the $\div$ sign) and multiply. For example,

$\dfrac{4}{7} \div \dfrac{3}{5} = \dfrac{4}{7} \cdot \dfrac{5}{3}$ Invert $\tfrac{3}{5}$ and change the division to a multiplication.

$= \dfrac{20}{21}$ Multiply the numerators and denominators.

We use the same procedures to divide rational expressions.

Dividing Rational Expressions

To divide two rational expressions, multiply the first by the reciprocal of the second. Then, if possible, factor and simplify.

For any two rational expressions $\dfrac{A}{B}$ and $\dfrac{C}{D}$, where $\dfrac{C}{D} \neq 0$,

$\dfrac{A}{B} \div \dfrac{C}{D} = \dfrac{A}{B} \cdot \dfrac{D}{C}$

EXAMPLE 7

Divide: **a.** $\dfrac{a}{13} \div \dfrac{17}{26}$ **b.** $-\dfrac{9x}{35y} \div \dfrac{15x^2}{14}$

Strategy We will use the rule for dividing rational expressions. After multiplying by the reciprocal, we will factor the monomials that are not prime and remove any common factors of the numerator and denominator.

Self Check 7

Divide: $-\dfrac{8a}{3b} \div \dfrac{16a^2}{9b^2}$

Now Try Problems 47 and 55

WHY We want to give the result in simplified form, which requires that the numerator and denominator have no common factor other than 1.

Solution

a. $\dfrac{a}{13} \div \dfrac{17}{26} = \dfrac{a}{13} \cdot \dfrac{26}{17}$ Invert the divisor, which is $\dfrac{17}{26}$, and change the division to a multiplication.

$= \dfrac{a \cdot 2 \cdot 13}{13 \cdot 17}$ Multiply the fractions and factor where possible.

$= \dfrac{a \cdot 2 \cdot \cancel{13}}{\cancel{13} \cdot 17}$ Simplify by removing the common factors.

$= \dfrac{2a}{17}$

b. $-\dfrac{9x}{35y} \div \dfrac{15x^2}{14} = -\dfrac{9x}{35y} \cdot \dfrac{14}{15x^2}$ Multiply by the reciprocal of $\dfrac{15x^2}{14}$.

$= -\dfrac{3 \cdot 3 \cdot x \cdot 2 \cdot 7}{5 \cdot 7 \cdot y \cdot 3 \cdot 5 \cdot x \cdot x}$ Multiply the fractions and factor where possible.

$= -\dfrac{3 \cdot \cancel{3} \cdot \cancel{x} \cdot 2 \cdot \cancel{7}}{5 \cdot \cancel{7} \cdot y \cdot \cancel{3} \cdot 5 \cdot \cancel{x} \cdot x}$ Simplify by removing the common factors.

$= -\dfrac{6}{25xy}$ Multiply the remaining factors.

Self Check 8

Divide:
$\dfrac{z^2 - 1}{z^2 + 4z + 3} \div \dfrac{z - 1}{z^2 + 2z - 3}$

Now Try Problem 63

EXAMPLE 8   Divide: $\dfrac{x^2 + x}{3x - 15} \div \dfrac{x^2 + 2x + 1}{6x - 30}$

Strategy We will use the rule for dividing rational expressions. After multiplying by the reciprocal, we will factor the binomials and trinomials that are not prime. Then we will remove any common factors of the numerator and denominator.

WHY We want to give the result in simplified form, which requires that the numerator and denominator have no common factor other than 1.

Solution

$\dfrac{x^2 + x}{3x - 15} \div \dfrac{x^2 + 2x + 1}{6x - 30}$

$= \dfrac{x^2 + x}{3x - 15} \cdot \dfrac{6x - 30}{x^2 + 2x + 1}$ Invert the divisor and change the division to multiplication.

$= \dfrac{x(x + 1) \cdot 2 \cdot 3(x - 5)}{3(x - 5)(x + 1)(x + 1)}$ Multiply the fractions and factor.

$= \dfrac{x\cancel{(x + 1)} \cdot 2 \cdot \cancel{3}\cancel{(x - 5)}}{\cancel{3}\cancel{(x - 5)}\cancel{(x + 1)}(x + 1)}$ Simplify by removing the common factors.

$= \dfrac{2x}{x + 1}$

To divide a rational expression by a polynomial, we write the polynomial as a fraction by inserting a denominator of 1, and then we divide the fractions.

EXAMPLE 9

Divide: $\dfrac{2x^2 - 3x - 2}{2x + 1} \div (4 - x^2)$

Strategy We begin by writing $4 - x^2$ as a rational expression by writing it over 1. Then we will use the rule for dividing rational expressions.

WHY Writing $4 - x^2$ over 1 is helpful when we invert its numerator and denominator to find its reciprocal.

Solution

$$\dfrac{2x^2 - 3x - 2}{2x + 1} \div (4 - x^2)$$

$$= \dfrac{2x^2 - 3x - 2}{2x + 1} \div \dfrac{4 - x^2}{1} \quad \text{Write } 4 - x^2 \text{ as a fraction with a denominator of 1.}$$

$$= \dfrac{2x^2 - 3x - 2}{2x + 1} \cdot \dfrac{1}{4 - x^2} \quad \text{Invert the divisor and change the division to multiplication.}$$

$$= \dfrac{(2x + 1)(x - 2) \cdot 1}{(2x + 1)(2 + x)(2 - x)} \quad \text{Multiply the fractions and factor where possible.}$$

$$= \dfrac{\overset{1}{\cancel{(2x+1)}}\overset{-1}{\cancel{(x-2)}} \cdot 1}{\underset{1}{\cancel{(2x+1)}}(2+x)\underset{1}{\cancel{(2-x)}}} \quad \text{Remove the common factors. The binomials } x - 2 \text{ and } 2 - x \text{ are opposites: } \tfrac{x-2}{2-x} = -1.$$

$$= \dfrac{-1}{2 + x}$$

$$= -\dfrac{1}{2 + x}$$

Self Check 9

Divide: $\dfrac{a^2 - b^2}{a^2 + ab} \div (b - a)$

Now Try Problem 67

4 Convert units of measurement.

We can use the concepts discussed in this section to make conversions from one unit of measure to another. *Unit conversion factors* play an important role in this process. A **unit conversion factor** is a fraction that has a value of 1. For example, we can use the fact that 1 square yard = 9 square feet to form two unit conversion factors:

$\dfrac{1 \text{ yd}^2}{9 \text{ ft}^2} = 1$ Read as "1 square yard per 9 square feet."

$\dfrac{9 \text{ ft}^2}{1 \text{ yd}^2} = 1$ Read as "9 square feet per 1 square yard."

> **Success Tip** Remember that unit conversion factors are equal to 1. Some examples are:
>
> $\dfrac{12 \text{ in.}}{1 \text{ ft}} = 1 \qquad \dfrac{60 \text{ min}}{1 \text{ hr}} = 1$

Since a unit conversion factor is equal to 1, multiplying a measurement by a unit conversion factor does not change the measurement, it only changes the units of measure.

EXAMPLE 10

Carpeting A roll of carpeting is 12 feet wide and 150 feet long. Find the number of square yards of carpeting on the roll.

Strategy We will begin by determining the number of square feet of carpeting on the roll. Then we will multiply that result by a unit conversion factor.

Self Check 10

Convert 5,400 ft² to square yards.

Now Try Problem 71

WHY A properly chosen unit conversion factor can convert the number of square feet of carpeting on the roll to the number of square yards on the roll.

Solution

When unrolled, the carpeting forms a rectangular shape with an area of $12 \cdot 150 = 1,800$ square feet. We will multiply 1,800 ft² by a unit conversion factor such that the units of ft² are removed and the units of yd² are introduced. Since $1 \text{ yd}^2 = 9 \text{ ft}^2$, we will use $\frac{1 \text{ yd}^2}{9 \text{ ft}^2}$.

$$\frac{1,800 \text{ ft}^2}{1 \text{ roll}} = \frac{1,800 \text{ ft}^2}{1 \text{ roll}} \cdot \boxed{\frac{1 \text{ yd}^2}{9 \text{ ft}^2}}$$
Multiply by a unit conversion factor that relates yd² to ft².

$$= \frac{1,800 \text{ ft}^2}{1 \text{ roll}} \cdot \frac{1 \text{ yd}^2}{9 \text{ ft}^2}$$
Remove the units of ft² that are common to the numerator and denominator.

$$= \frac{200 \text{ yd}^2}{1 \text{ roll}}$$
Divide 1,800 by 9 to get 200.

There are 200 yd² of carpeting on the roll.

Self Check 11

A mosquito flaps its wings about 600 times per second. How many times per minute does a mosquito flap its wings?

Now Try Problem 75

EXAMPLE 11 *The Speed of Light* The speed with which light moves through space is about 186,000 miles per second. Express this speed in miles per minute.

Strategy The speed of light can be expressed as $\frac{186,000 \text{ mi}}{1 \text{ sec}}$. We will multiply that fraction by a unit conversion factor.

WHY A properly chosen unit conversion factor can convert the number of miles traveled per second to the number of miles traveled per minute.

Solution

We will multiply $\frac{186,000 \text{ mi}}{1 \text{ sec}}$ by a unit conversion factor such that the units of seconds are removed and the units of minutes are introduced. Since 60 seconds = 1 minute, we will use $\frac{60 \text{ sec}}{1 \text{ min}}$.

$$\frac{186,000 \text{ mi}}{1 \text{ sec}} = \frac{186,000 \text{ mi}}{1 \text{ sec}} \cdot \boxed{\frac{60 \text{ sec}}{1 \text{ min}}}$$
Multiply by a unit conversion factor that relates seconds to minutes.

$$= \frac{186,000 \text{ mi}}{1 \text{ sec}} \cdot \frac{60 \text{ sec}}{1 \text{ min}}$$
Remove the units of seconds that are common to the numerator and denominator.

$$= \frac{11,160,000 \text{ mi}}{1 \text{ min}}$$
Multiply 186,000 and 60 to get 11,160,000.

The speed of light is about 11,160,000 miles per minute.

ANSWERS TO SELF CHECKS

1. $\frac{3x^2 - 9x}{20}$ 2. $\frac{3a^4}{2b}$ 3. $\frac{x}{3}$ 4. $\frac{a}{a+2}$ 5. a. 9 b. 6 c. $4x + 12$ 6. a 7. $-\frac{3b}{2a}$ 8. $z - 1$ 9. $-\frac{1}{a}$ 10. 600 yd² 11. 36,000 flaps per minute

SECTION 7.2 STUDY SET

VOCABULARY

Fill in the blanks.

1. The _____ of $\frac{3}{x+1}$ is $\frac{x+1}{3}$.

2. A _____ conversion factor is a fraction containing units that is equal to 1, such as $\frac{3 \text{ ft}}{1 \text{ yd}}$.

7.2 Multiplying and Dividing Rational Expressions

CONCEPTS

Fill in the blanks.

3. To multiply rational expressions, multiply their _____ and multiply their _____.

4. $\dfrac{A}{B} \cdot \dfrac{C}{D} =$ ☐

5. To divide rational expressions, multiply the first by the _____ of the second.

6. $\dfrac{A}{B} \div \dfrac{C}{D} = \dfrac{A}{B} \cdot$ ☐

7. The product of $\dfrac{x}{x+2}$ and its reciprocal $\dfrac{x+2}{x}$ is ___.

8. Use the fact that 1 tablespoon = 3 teaspoons to write two unit conversion factors.

NOTATION

Complete each step.

9. $\dfrac{x^2 + x}{3x - 6} \cdot \dfrac{x - 2}{x + 1} = \dfrac{(x^2 + x)\cdot\;\boxed{}}{\boxed{}\cdot(x+1)}$

$= \dfrac{x\;\boxed{}(x-2)}{3\;\boxed{}(x+1)}$

$= \dfrac{x}{\boxed{}}$

10. $\dfrac{x^2 - x}{4x + 12} \div \dfrac{x - 1}{x + 3} = \dfrac{x^2 - x}{4x + 12} \cdot \dfrac{\boxed{}}{\boxed{}}$

$= \dfrac{\boxed{}(x+3)}{(4x+12)\boxed{}}$

$= \dfrac{x\;\boxed{}(x+3)}{4\;\boxed{}(x-1)}$

$= \dfrac{x}{\boxed{}}$

GUIDED PRACTICE

Perform each multiplication. Simplify answers if possible. See Examples 1–2.

11. $\dfrac{3}{y} \cdot \dfrac{y}{2}$

12. $\dfrac{2}{z} \cdot \dfrac{z}{3}$

13. $\dfrac{7z}{9z} \cdot \dfrac{-4z}{2z}$

14. $\dfrac{-8}{2x} \cdot \dfrac{16x}{3x}$

15. $\dfrac{2x^2 y}{3xy} \cdot \dfrac{3xy^2}{2}$

16. $\dfrac{2x^2 z}{z} \cdot \dfrac{5x}{z}$

17. $\dfrac{8x^2 y^2}{4x^2} \cdot \dfrac{2xy}{2y}$

18. $\dfrac{9x^2 y}{3x} \cdot \dfrac{3xy}{3y}$

19. $-\dfrac{2xy}{x^2} \cdot \dfrac{3xy}{2}$

20. $-\dfrac{3x}{x^2} \cdot \dfrac{2xz}{3}$

21. $\dfrac{4}{7} \cdot \dfrac{z + 2}{z}$

22. $\dfrac{2}{a} \cdot \dfrac{a + 3}{5}$

Perform each multiplication. Simplify answers if possible. See Examples 3–4.

23. $\dfrac{x - 2}{2} \cdot \dfrac{2x}{x - 2}$

24. $\dfrac{y + 3}{y} \cdot \dfrac{3y}{y + 3}$

25. $\dfrac{x + 5}{5} \cdot \dfrac{x}{x + 5}$

26. $\dfrac{y - 9}{y + 9} \cdot \dfrac{y}{9}$

27. $\dfrac{(x+1)^2}{x+1} \cdot \dfrac{x+2}{x+1}$

28. $\dfrac{(y-3)^2}{y-3} \cdot \dfrac{y-3}{y-3}$

29. $\dfrac{2x + 6}{x + 3} \cdot \dfrac{3}{4x}$

30. $\dfrac{3y - 9}{y - 3} \cdot \dfrac{y}{3y^2}$

31. $\dfrac{x^2 + x - 6}{5x} \cdot \dfrac{5x - 10}{x + 3}$

32. $\dfrac{z^2 + 4z - 5}{5z - 5} \cdot \dfrac{5z}{z + 5}$

33. $\dfrac{m^2 - 2m - 3}{2m + 4} \cdot \dfrac{m^2 - 4}{m^2 + 3m + 2}$

34. $\dfrac{p^2 - p - 6}{3p - 9} \cdot \dfrac{p^2 - 9}{p^2 + 6p + 9}$

Perform each multiplication. Simplify answers if possible. See Examples 5–6.

35. $\dfrac{5}{m} \cdot m$

36. $p \cdot \dfrac{10}{p}$

37. $4d \cdot \dfrac{3}{2d}$

38. $9x \cdot \dfrac{25}{3x}$

39. $15x\left(\dfrac{x + 1}{15x}\right)$

40. $30t\left(\dfrac{t - 7}{30t}\right)$

41. $12y\left(\dfrac{y + 8}{6y}\right)$

42. $16x\left(\dfrac{3x + 8}{4x}\right)$

43. $(x + 8)\dfrac{x + 5}{x + 8}$

44. $(y - 2)\dfrac{y + 3}{y - 2}$

45. $(10h + 90)\dfrac{h - 3}{h + 9}$

46. $(r^2 - 25r)\dfrac{r + 4}{r - 25}$

Perform each division. Simplify answers when possible. See Example 7.

47. $\dfrac{2}{y} \div \dfrac{4}{3}$

48. $\dfrac{3}{a} \div \dfrac{a}{9}$

49. $\dfrac{3x}{2} \div \dfrac{x}{2}$

50. $\dfrac{y}{6} \div \dfrac{2}{3y}$

51. $\dfrac{3x}{y} \div \dfrac{2x}{4}$

52. $\dfrac{3y}{8} \div \dfrac{2y}{4y}$

53. $\dfrac{4x}{3x} \div \dfrac{2y}{9y}$

54. $\dfrac{14}{7y} \div \dfrac{10}{5z}$

55. $-\dfrac{x^2}{3} \div \dfrac{2x}{4}$

56. $-\dfrac{z^2}{z} \div \dfrac{z}{3z}$

57. $\dfrac{x^2 y}{3xy} \div \dfrac{xy^2}{6y}$

58. $\dfrac{2xz}{z} \div \dfrac{4x^2}{z^2}$

Perform each division. Simplify answers when possible.
See Example 8.

59. $\dfrac{x^2 - 4}{3x + 6} \div \dfrac{x - 2}{x + 2}$

60. $\dfrac{x^2 - 9}{5x + 15} \div \dfrac{x - 3}{x + 3}$

61. $\dfrac{x^2 - 1}{3x - 3} \div \dfrac{x + 1}{3}$

62. $\dfrac{x^2 - 16}{x - 4} \div \dfrac{3x + 12}{x}$

63. $\dfrac{x^2 - 2x - 35}{3x^2 + 27x} \div \dfrac{x^2 + 7x + 10}{6x^2 + 12x}$

64. $\dfrac{x^2 - x - 6}{2x^2 + 9x + 10} \div \dfrac{x^2 - 25}{2x^2 + 15x + 25}$

65. $\dfrac{2d^2 + 8d - 42}{d - 3} \div \dfrac{2d^2 + 14d}{d^2 + 5d}$

66. $\dfrac{5x^2 + 13x - 6}{x + 3} \div \dfrac{5x^2 - 17x + 6}{x - 2}$

Perform each division. Simplify answers when possible.
See Example 9.

67. $\dfrac{x^2 + 2x - 3}{x - 1} \div (9 - x^2)$

68. $\dfrac{x^2 - 1}{3x - 3} \div (x + 1)$

69. $\dfrac{x^2 - 10x + 9}{x - 9} \div (x - 1)$

70. $\dfrac{3m + n}{18} \div (9m^2 + 6mn + n^2)$

Complete each unit conversion. See Examples 10–11.

71. $\dfrac{150 \text{ yards}}{1} \cdot \dfrac{3 \text{ feet}}{1 \text{ yard}} = ?$

72. $\dfrac{60 \text{ inches}}{1} \cdot \dfrac{1 \text{ foot}}{12 \text{ inches}} = ?$

73. $\dfrac{6 \text{ pints}}{1} \cdot \dfrac{1 \text{ gallon}}{8 \text{ pints}} = ?$

74. $\dfrac{4 \text{ cups}}{1} \cdot \dfrac{1 \text{ gallon}}{16 \text{ cups}} = ?$

75. $\dfrac{30 \text{ miles}}{1 \text{ hour}} \cdot \dfrac{1 \text{ hour}}{60 \text{ minutes}} = ?$

76. $\dfrac{300 \text{ meters}}{3 \text{ months}} \cdot \dfrac{12 \text{ months}}{1 \text{ year}} = ?$

77. $\dfrac{30 \text{ meters}}{1 \text{ second}} \cdot \dfrac{60 \text{ seconds}}{1 \text{ minute}} = ?$

78. $\dfrac{288 \text{ inches}^2}{1 \text{ yr}} \cdot \dfrac{1 \text{ foot}^2}{144 \text{ inches}^2} = ?$

TRY IT YOURSELF

Perform each operation and simplify.

79. $\dfrac{5y}{7} \cdot \dfrac{7}{5}$

80. $\dfrac{4x}{3y} \cdot \dfrac{3y}{7x}$

81. $\dfrac{x^2 - x}{x} \cdot \dfrac{3x - 6}{3x - 3}$

82. $\dfrac{5z - 10}{z + 2} \cdot \dfrac{3}{3z - 6}$

83. $\dfrac{7y - 14}{y - 2} \cdot \dfrac{x^2}{7x}$

84. $\dfrac{y^2 + 3y}{9} \cdot \dfrac{3x}{y + 3}$

85. $\dfrac{b^2 - 5b + 6}{b^2 - 10b + 16} \div \dfrac{b^2 + 2b}{b^2 - 6b - 16}$

86. $\dfrac{m^2 + m - 6}{m^2 - 6m + 9} \div \dfrac{m^2 - 4}{m^2 - 9}$

87. $\dfrac{x + 2}{3x} \div \dfrac{x + 2}{2}$

88. $\dfrac{z - 3}{3z} \div \dfrac{z + 3}{z}$

89. $\dfrac{(z - 2)^2}{3z^2} \div \dfrac{z - 2}{6z}$

90. $\dfrac{(x + 7)^2}{x + 7} \div \dfrac{(x - 3)^2}{x + 7}$

91. $\dfrac{(z - 7)^2}{z + 2} \div \dfrac{z(z - 7)}{5z^2}$

92. $\dfrac{y(y + 2)}{y^2(y - 3)} \div \dfrac{y^2(y + 2)}{(y - 3)^2}$

93. $\dfrac{2r - 3s}{12} \div (4r^2 - 12rs + 9s^2)$

94. $\dfrac{r^2 - 11r + 18}{r - 9} \div (r - 2)$

APPLICATIONS

95. **INTERNATIONAL ALPHABET** The symbols representing the letters A, B, C, D, E, and F in an international code used at sea are printed six to a sheet and then cut into separate cards. If each card is a square, find the area of the large printed sheet shown in the illustration.

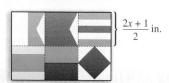

$\dfrac{2x + 1}{2}$ in.

96. **PHYSICS** The following table contains algebraic expressions for the rate an object travels, and the time traveled at that rate, in terms of a constant k. Complete the table.

Rate (mph)	Time (hr)	Distance (mi)
$\dfrac{k^2 + k - 6}{k - 3}$	$\dfrac{k^2 - 9}{k^2 - 4}$	

WRITING

97. Explain how to multiply two fractions and how to simplify the result.

98. Explain why any mathematical expression can be written as a fraction.

99. To divide fractions, you must first know how to multiply fractions. Explain.

100. Explain how to do the division: $\dfrac{a}{b} \div \dfrac{c}{d} \div \dfrac{e}{f}$

REVIEW

Simplify each expression. Write all answers without using negative exponents.

101. $2x^3y^2(-3x^2y^4)$

102. $\dfrac{8x^4y^5}{-2x^3y^2}$

103. $(3y)^{-4}$

104. $x^{3m} \cdot x^{4m}$

Perform the operations.

105. $-4(y^3 - 4y^2 + 3y - 2) - 4(-2y^3 - y)$

106. $y - 5\overline{)5y^3 - 3y^2 + 4y - 1}$

SECTION 7.3
Adding and Subtracting with Like Denominators; Least Common Denominators

Objectives

1. Add and subtract rational expressions that have the same denominator.
2. Find the least common denominator.
3. Build rational expressions into equivalent expressions.

In this section, we extend the rules for adding and subtracting numerical fractions to problems involving addition and subtraction of rational expressions.

1 Add and subtract rational expressions that have the same denominator.

To add (or subtract) fractions with a common denominator, we add (or subtract) their numerators and keep the common denominator. For example,

$$\dfrac{3}{7} + \dfrac{2}{7} = \dfrac{3+2}{7} \qquad \text{and} \qquad \dfrac{3}{7} - \dfrac{2}{7} = \dfrac{3-2}{7}$$

$$= \dfrac{5}{7} \qquad\qquad\qquad\qquad = \dfrac{1}{7}$$

We use the same procedure to add and subtract rational expressions with like denominators.

Adding and Subtracting Fractions with Like Denominators

If $\dfrac{A}{D}$ and $\dfrac{B}{D}$ are rational expressions,

$$\dfrac{A}{D} + \dfrac{B}{D} = \dfrac{A+B}{D} \qquad \text{and} \qquad \dfrac{A}{D} - \dfrac{B}{D} = \dfrac{A-B}{D}$$

EXAMPLE 1 Add: **a.** $\dfrac{x}{8} + \dfrac{3x}{8}$ **b.** $\dfrac{3x+y}{5x} + \dfrac{x+y}{5x}$

Strategy Because the rational expressions have the same denominator, we will add the numerators and write the sum over the common denominator. Then, if possible, we will factor and simplify.

WHY This is the rule for adding rational expressions such as these that have the same denominator.

Self Check 1

Add:

a. $\dfrac{x}{7} + \dfrac{4x}{7}$

b. $\dfrac{3x}{7y} + \dfrac{4x}{7y}$

Now Try Problems 11 and 17

Solution

a. $\dfrac{x}{8} + \dfrac{3x}{8} = \dfrac{x + 3x}{8}$ Add the numerators and keep the common denominator.

$= \dfrac{4x}{8}$ Combine like terms: $x + 3x = 4x$.

$= \dfrac{\overset{1}{\cancel{4}} \cdot x}{\underset{1}{\cancel{4}} \cdot 2}$ Factor the numerator and denominator and remove the common factor, 4.

$= \dfrac{x}{2}$ Simplify.

b. $\dfrac{3x + y}{5x} + \dfrac{x + y}{5x} = \dfrac{3x + y + x + y}{5x}$ Add the numerators and keep the common denominator.

$= \dfrac{4x + 2y}{5x}$ Combine like terms in the numerator.

Self Check 2

Add: $\dfrac{x + 4}{6x - 12} + \dfrac{x - 8}{6x - 12}$

Now Try Problem 19

EXAMPLE 2

Add: $\dfrac{3x + 21}{5x + 10} + \dfrac{8x + 1}{5x + 10}$

Strategy We will add the numerators and write the sum over the common denominator. Then, if possible, we will factor and simplify.

WHY This is the rule for adding rational expressions that have the same denominator.

Solution

$\dfrac{3x + 21}{5x + 10} + \dfrac{8x + 1}{5x + 10} = \dfrac{3x + 21 + 8x + 1}{5x + 10}$ Add the numerators and keep the common denominator.

$= \dfrac{11x + 22}{5x + 10}$ Combine like terms in the numerator.

$= \dfrac{11\overset{1}{\cancel{(x + 2)}}}{5\underset{1}{\cancel{(x + 2)}}}$ Simplify the result by factoring the numerator and denominator. Remove the common factor of $x + 2$.

$= \dfrac{11}{5}$

Self Check 3

Subtract: $\dfrac{2y + 1}{y + 5} - \dfrac{y - 4}{y + 5}$

Now Try Problems 23 and 29

EXAMPLE 3

Subtract: **a.** $\dfrac{5x}{3} - \dfrac{2x}{3}$ **b.** $\dfrac{5x + 1}{x - 3} - \dfrac{4x - 2}{x - 3}$

Strategy We will use the rule for subtracting rational expressions that have the same denominators. In part b, it is important to note that the numerator of the second fraction has *two* terms.

WHY We must make sure that the entire numerator (not just the first term) of the second fraction is subtracted.

Solution

a. $\dfrac{5x}{3} - \dfrac{2x}{3} = \dfrac{5x - 2x}{3}$

$= \dfrac{3x}{3}$ Combine like terms: $5x - 2x = 3x$.

$= \dfrac{x}{1}$ Remove the common factor of 3.

$= x$ Denominators of 1 need not be written.

b. $\dfrac{5x + 1}{x - 3} - \dfrac{4x - 2}{x - 3} = \dfrac{5x + 1 - (4x - 2)}{x - 3}$ The second numerator, $4x - 2$, is written within parentheses to make sure that we subtract both of its terms.

$= \dfrac{5x + 1 - 4x + 2}{x - 3}$ Distribute the multiplication by -1: $-(4x - 2) = -4x + 2$.

$= \dfrac{x + 3}{x - 3}$ Combine like terms in the numerator.

To add and/or subtract three or more rational expressions, we follow the rules for the order of operations.

EXAMPLE 4

Simplify: $\dfrac{3x + 1}{x^2 + x + 1} - \dfrac{5x + 2}{x^2 + x + 1} + \dfrac{2x + 1}{x^2 + x + 1}$

Strategy We note that all of the denominators of the rational expressions to be added and subtracted have the same denominator. Then we will perform the additions and subtractions from left to right.

WHY We must follow the rule for adding and/or subtracting rational expressions as well as the rules for order of operations.

Solution
This example combines addition and subtraction. Unless parentheses indicate otherwise, we perform additions and subtractions from left to right.

$\dfrac{3x + 1}{x^2 + x + 1} - \dfrac{5x + 2}{x^2 + x + 1} + \dfrac{2x + 1}{x^2 + x + 1}$

$= \dfrac{3x + 1 - (5x + 2) + 2x + 1}{x^2 + x + 1}$ Combine the numerators and keep the common denominator.

$= \dfrac{3x + 1 - 5x - 2 + 2x + 1}{x^2 + x + 1}$ Distribute the multiplication by -1: $-(5x + 2) = -5x - 2$.

$= \dfrac{0}{x^2 + x + 1}$ Combine like terms in the numerator.

$= 0$ If the numerator of a fraction is zero and the denominator is not zero, the fraction's value is zero.

Self Check 4

Simplify:

$\dfrac{2a^2 - 3}{a - 5} + \dfrac{3a^2 + 2}{a - 5} - \dfrac{5a^2}{a - 5}$

Now Try Problem 35

2 Find the least common denominator.

Since the denominators of the fractions in the addition $\dfrac{4}{7} + \dfrac{3}{5}$ are different, we cannot add the fractions in their present form.

four-sevenths + three-fifths
              ⎣— Different denominators —⎦

Chapter 7 Rational Expressions and Equations

To add these fractions, we need to express them as equivalent expressions with a common denominator. The **least common denominator (LCD)** is usually the easiest one to use. The least common denominator of several rational expressions can be found as follows.

> **Finding the LCD**
>
> 1. Factor each denominator completely.
> 2. The LCD is a product that uses each different factor obtained in step 1 the greatest number of times it appears in any one factorization.

Self Check 5

Find the LCD of each pair of rational expressions:

a. $\dfrac{y+7}{6y^3}$ and $\dfrac{7}{75y}$

b. $\dfrac{a-3}{a+3}$ and $\dfrac{21}{a}$

Now Try Problems 41 and 43

EXAMPLE 5

Find the LCD of each pair of rational expressions:

a. $\dfrac{11}{8x}$ and $\dfrac{7}{18x^2}$

b. $\dfrac{20}{x}$ and $\dfrac{4x}{x-9}$

Strategy We will begin by factoring completely the denominator of each rational expression. Then we will form a product using each factor the greatest number of times it appears in any one factorization.

WHY Since the LCD must contain the factors of each denominator, we need to write each denominator in factored form.

Solution

a. $8x = 2 \cdot 2 \cdot 2 \cdot x$ Prime factor 8.
$18x^2 = 2 \cdot 3 \cdot 3 \cdot x \cdot x$ Prime factor 18. Factor x^2.

The factorizations of $8x$ and $18x^2$ contain the factors 2, 3, and x. The LCD of $\dfrac{11}{8x}$ and $\dfrac{7}{18x^2}$ should contain each factor of $8x$ and $18x^2$ the greatest number of times it appears in any one factorization.

The greatest number of times the factor 2 appears is three times.
The greatest number of times the factor 3 appears is twice.
The greatest number of times the factor x appears is twice.

$$\text{LCD} = 2 \cdot 2 \cdot 2 \cdot 3 \cdot 3 \cdot x \cdot x$$
$$= 72x^2$$

The LCD for $\dfrac{11}{8x}$ and $\dfrac{7}{18x^2}$ is $72x^2$.

Success Tip The factorizations can be written:

$$8x = 2^3 \cdot x$$
$$18x^2 = 2 \cdot 3^2 \cdot x^2$$

Note that the highest power of each factor is used to form the LCD.

$$\text{LCD} = 2^3 \cdot 3^2 \cdot x^2 = 72x^2$$

b. Since the denominators of $\dfrac{20}{x}$ and $\dfrac{4x}{x-9}$ are completely factored, the factor x appears once and the factor $x-9$ appears once. Thus, the LCD is $x(x-9)$.

EXAMPLE 6
Find the LCD of each pair of rational expressions:

a. $\dfrac{x}{7x+7}$ and $\dfrac{x-2}{5x+5}$ **b.** $\dfrac{6-x}{x^2+8x+16}$ and $\dfrac{15x}{x^2-16}$

Strategy We will begin by factoring completely each binomial and trinomial in the denominators of the rational expressions. Then we will form a product using each factor the greatest number of times it appears in any one factorization.

WHY Since the LCD must contain the factors of each denominator, we need to write each denominator in factored form.

Solution

a. Factor each denominator completely.

$$7x + 7 = 7(x+1) \quad \text{The GCF is 7.}$$
$$5x + 5 = 5(x+1) \quad \text{The GCF is 5.}$$

The factorizations of $7x+7$ and $5x+5$ contain the factors 7, 5, and $x+1$.

The LCD of $\dfrac{x}{7x+7}$ and $\dfrac{x-2}{5x+5}$ should contain each factor of $7x+7$ and $5x+5$ the greatest number of times it appears in any one factorization.

- The greatest number of times the factor 7 appears is once.
- The greatest number of times the factor 5 appears is once.
- The greatest number of times the factor $x+1$ appears is once.

$$\text{LCD} = 7 \cdot 5 \cdot (x+1) = 35(x+1)$$

Success Tip Rather than performing the multiplication, it is often better to leave an LCD in factored form:

$$\text{LCD} = 35(x+1)$$

b. Factor each denominator completely.

$$x^2 + 8x + 16 = (x+4)(x+4) \quad \text{Factor the trinomial.}$$
$$x^2 - 16 = (x+4)(x-4) \quad \text{Factor the difference of two squares.}$$

The factorizations of $x^2 + 8x + 16$ and $x^2 - 16$ contain the factors $x+4$ and $x-4$.

- The greatest number of times the factor $x+4$ appears is twice.
- The greatest number of times the factor $x-4$ appears is once.

$$\text{LCD} = (x+4)(x+4)(x-4) = (x+4)^2(x-4)$$

Self Check 6

Find the LCD of each pair of rational expressions:

a. $\dfrac{x^3}{x^2-6x}$ and $\dfrac{25x}{2x-12}$

b. $\dfrac{m+1}{m^2-9}$ and $\dfrac{6m^2}{m^2-6m+9}$

Now Try Problems 47 and 49

3 Build rational expressions into equivalent expressions.

Recall from Chapter 1 that writing a fraction as an equivalent fraction with a larger denominator is called **building the fraction**. For example, to write $\dfrac{3}{5}$ as an equivalent fraction with a denominator of 35, we multiply it by 1 in the form of $\dfrac{7}{7}$:

$$\dfrac{3}{5} = \dfrac{3}{5} \cdot \dfrac{7}{7} = \dfrac{21}{35} \quad \begin{array}{l}\text{Multiply the numerators.}\\ \text{Multiply the denominators.}\end{array}$$

To add and subtract rational expressions with different denominators, we must write them as equivalent expressions having a common denominator. To do so, we build rational expressions.

Chapter 7 Rational Expressions and Equations

> **Building Rational Expressions**
>
> To build a rational expression, multiply it by 1 in the form of $\dfrac{c}{c}$, where c is any nonzero number or expression.

Self Check 7

Write each rational expression as an equivalent expression with the indicated denominator:

a. $\dfrac{7}{20m^2}$, denominator $60m^3$

b. $\dfrac{2c}{c+1}$, denominator $(c+1)(c+3)$

Now Try Problems 53 and 57

EXAMPLE 7 Write each rational expression as an equivalent expression with the indicated denominator:

a. $\dfrac{7}{15n}$, denominator $30n^3$

b. $\dfrac{6x}{x+4}$, denominator $(x+4)(x-4)$

Strategy We will begin by asking, "By what must we multiply the given denominator to get the required denominator?"

WHY The answer to that question helps us determine the form of 1 to be used to build an equivalent rational expression.

Solution

a. We need to multiply the denominator of $\dfrac{7}{15n}$ by $2n^2$ to obtain a denominator of $30n^3$. It follows that $\dfrac{2n^2}{2n^2}$ is the form of 1 that should be used to build an equivalent expression.

$$\dfrac{7}{15n} = \dfrac{7}{15n} \cdot \dfrac{2n^2}{2n^2}$$ Multiply the given rational expression by 1, in the form of $\dfrac{2n^2}{2n^2}$.

$$= \dfrac{14n^2}{30n^3}$$ Multiply the numerators.
Multiply the denominators.

b. We need to multiply the denominator of $\dfrac{6x}{x+4}$ by $x-4$ to obtain a denominator of $(x+4)(x-4)$. It follows that $\dfrac{x-4}{x-4}$ is the form of 1 that should be used to build an equivalent expression.

$$\dfrac{6x}{x+4} = \dfrac{6x}{x+4} \cdot \dfrac{x-4}{x-4}$$ Multiply the given rational expression by 1, in the form of $\dfrac{x-4}{x-4}$.

$$= \dfrac{6x(x-4)}{(x+4)(x-4)}$$ Write the product of the numerators.
Write the product of the denominators.

$$= \dfrac{6x^2 - 24x}{(x+4)(x-4)}$$ In the numerator, distribute the multiplication by $6x$.
Leave the denominator in factored form.

Success Tip To get this answer, we multiplied the factors in the numerator to obtain a polynomial in unfactored form: $6x^2 - 24x$. However, we left the denominator in factored form. This approach is beneficial in the next section when we add and subtract rational expressions with unlike denominators.

Self Check 8

Write $\dfrac{x-3}{x^2-4x}$ as an equivalent expression with a denominator of $x(x-4)(x+8)$.

Now Try Problem 59

EXAMPLE 8 Write $\dfrac{x+1}{x^2+6x}$ as an equivalent expression with a denominator of $x(x+6)(x+2)$.

Strategy We will begin by factoring the denominator of $\dfrac{x+1}{x^2+6x}$. Then we will compare the factors of x^2+6x to those of $x(x+6)(x+2)$.

WHY This comparison will enable us to answer the question, "By what must we multiply x^2+6x to obtain $x(x+6)(x+2)$?"

Solution

We factor the denominator to determine what factors are missing.

$$\frac{x+1}{x^2+6x} = \frac{x+1}{x(x+6)} \quad \text{Factor out the GCF, } x, \text{ from } x^2+6x.$$

It is now apparent that we need to multiply the denominator by $x+2$ to obtain a denominator of $x(x+6)(x+2)$. It follows that $\dfrac{x+2}{x+2}$ is the form of 1 that should be used to build an equivalent expression.

$$\frac{x+1}{x^2+6x} = \frac{x+1}{x(x+6)} \cdot \frac{x+2}{x+2} \quad \text{Multiply the given rational expression by 1, in the form of } \tfrac{x+2}{x+2}.$$

$$= \frac{(x+1)(x+2)}{x(x+6)(x+2)} \quad \text{Write the product of the numerators. Write the product of the denominators.}$$

$$= \frac{x^2+3x+2}{x(x+6)(x+2)} \quad \text{In the numerator, use the FOIL method to multiply } (x+1)(x+2). \text{ Leave the denominator in factored form.}$$

> **Success Tip** When building rational expressions, write the numerator of the result as a polynomial in unfactored form. Write the denominator in factored form.

ANSWERS TO SELF CHECKS

1. a. $\dfrac{5x}{7}$ b. $\dfrac{x}{y}$ 2. $\dfrac{1}{3}$ 3. 1 4. $-\dfrac{1}{a-5}$ 5. a. $150y^3$ b. $a(a+3)$ 6. a. $2x(x-6)$
b. $(m+3)(m-3)^2$ 7. a. $\dfrac{21m}{60m^3}$ b. $\dfrac{2c^2+6c}{(c+1)(c+3)}$ 8. $\dfrac{x^2+5x-24}{x(x-4)(x+8)}$

SECTION 7.3 STUDY SET

VOCABULARY

Fill in the blanks.

1. The rational expressions $\dfrac{7}{6n}$ and $\dfrac{n+1}{6n}$ have a common _____ of $6n$.

2. The _____ of $\dfrac{x-8}{x+6}$ and $\dfrac{6-5x}{x}$ is $x(x+6)$.

3. To _____ a rational expression, multiply it by a form of 1. For example, $\dfrac{2}{n^2} \cdot \dfrac{8}{8} = \dfrac{16}{8n^2}$.

4. The expressions $\dfrac{2}{n^2}$ and $\dfrac{16}{8n^2}$ are called _____ expressions because they have the same value for all values of n, except for $n=0$.

CONCEPTS

Fill in the blanks.

5. To add two fractions with like denominators, add their _____ and keep the _____ _____.

6. To subtract two fractions with _____ denominators, subtract their numerators and keep the common denominator.

7. To build the rational expression $\dfrac{3y}{4x}$ into $\dfrac{9y^2}{12xy}$, multiply both the numerator and the denominator by ▢.

8. The sum of two rational expressions is $\dfrac{4x+4}{5(x+1)}$. After factoring the numerator and simplifying the result, you will have ▢.

NOTATION

Complete each step.

9. $\dfrac{6a-1}{4a+1} + \dfrac{2a+3}{4a+1} = \dfrac{6a-1+▢}{4a+1}$

$= \dfrac{8a+▢}{4a+1}$

$= \dfrac{2▢}{4a+1}$

$= ▢$

The type of multiplication that is used to build rational expressions is shown below. Fill in the blanks.

10. a. $\dfrac{4x}{5} \cdot \dfrac{2}{2} = \dfrac{}{10}$ b. $\dfrac{3}{t} \cdot \dfrac{t-2}{t-2} = \dfrac{}{t(t-2)}$

c. $\dfrac{m+1}{m-3} \cdot \dfrac{m-5}{m-5} = \dfrac{}{(m-3)(m-5)}$

GUIDED PRACTICE

Perform each addition. Simplify answers, if possible. See Example 1.

11. $\dfrac{x}{9} + \dfrac{2x}{9}$
12. $\dfrac{5x}{7} + \dfrac{9x}{7}$
13. $\dfrac{2x}{y} + \dfrac{2x}{y}$
14. $\dfrac{4y}{3x} + \dfrac{2y}{3x}$
15. $\dfrac{4}{7y} + \dfrac{10}{7y}$
16. $\dfrac{x^2}{4y} + \dfrac{x^2}{4y}$
17. $\dfrac{y+2}{10z} + \dfrac{y+4}{10z}$
18. $\dfrac{x+3}{2x^2} + \dfrac{x+5}{2x^2}$

Perform each addition. Simplify answers, if possible. See Example 2.

19. $\dfrac{3x-5}{x-2} + \dfrac{6x-13}{x-2}$
20. $\dfrac{8x-7}{x+3} + \dfrac{2x+37}{x+3}$
21. $\dfrac{b}{b^2-4} + \dfrac{2}{b^2-4}$
22. $\dfrac{a}{a^2+5a+6} + \dfrac{3}{a^2+5a+6}$

Perform each subtraction. Simplify answers, if possible. See Example 3.

23. $\dfrac{35y}{72} - \dfrac{44y}{72}$
24. $\dfrac{13t}{99} - \dfrac{35t}{99}$
25. $\dfrac{2x}{y} - \dfrac{x}{y}$
26. $\dfrac{7y}{5t} - \dfrac{4y}{5t}$
27. $\dfrac{7x+7}{5y} - \dfrac{2x+7}{5y}$
28. $\dfrac{3y-2}{2y+6} - \dfrac{2y-5}{2y+6}$
29. $\dfrac{5x+8}{3x+15} - \dfrac{3x-2}{3x+15}$
30. $\dfrac{2c}{c^2-d^2} - \dfrac{2d}{c^2-d^2}$

Perform the operations. Simplify answers, if possible. See Example 4.

31. $-\dfrac{x}{y} + \dfrac{2x}{y} - \dfrac{x}{y}$
32. $\dfrac{5y}{8x} + \dfrac{4y}{8x} - \dfrac{9y}{8x}$
33. $\dfrac{3x}{y+2} - \dfrac{3y}{y+2} + \dfrac{x+y}{y+2}$
34. $\dfrac{3y}{x-5} + \dfrac{x}{x-5} - \dfrac{y-x}{x-5}$

35. $\dfrac{2a}{a^2+2a+1} - \dfrac{3a-1}{a^2+2a+1} + \dfrac{a+2}{a^2+2a+1}$
36. $\dfrac{b+2}{b^2-3b-2} - \dfrac{b}{b^2-3b-2} + \dfrac{b-2}{b^2-3b-2}$
37. $\dfrac{3p+1}{p^2-3p+2} + \dfrac{2p-1}{p^2-3p+2} - \dfrac{p}{p^2-3p+2}$
38. $\dfrac{2q+1}{q^2-q+1} + \dfrac{q-1}{q^2-q+1} - \dfrac{3q+2}{q^2-q+1}$

Find the LCD of each pair of rational expressions. See Examples 5–6.

39. $\dfrac{1}{2x}, \dfrac{9}{6x}$
40. $\dfrac{4}{9y}, \dfrac{11}{3y}$
41. $\dfrac{33}{15a^3}, \dfrac{9}{10a}$
42. $\dfrac{m-21}{12m^4}, \dfrac{m+1}{18m}$
43. $\dfrac{8}{c}, \dfrac{8-c}{c+2}$
44. $\dfrac{d^2-5}{d+9}, \dfrac{d-3}{d}$
45. $\dfrac{3x+1}{3x-1}, \dfrac{3x}{6x+2}$
46. $\dfrac{-2x}{x^2-1}, \dfrac{5x}{x+1}$
47. $\dfrac{b-9}{4b+8}, \dfrac{b}{6}$
48. $\dfrac{5m+6}{4m+12}, \dfrac{7}{6m}$
49. $\dfrac{8-b}{b^2+6b+9}, \dfrac{12b}{b^2-9}$
50. $\dfrac{2a+3}{a^2-4a+4}, \dfrac{a-2}{a^2-4}$

Write each rational expression as an equivalent fraction with the indicated denominator. See Examples 7–8.

51. $\dfrac{25}{4}$; $20x$
52. $\dfrac{5}{y}$; y^2
53. $\dfrac{8}{x}$; x^2y
54. $\dfrac{7}{y}$; xy^2
55. $\dfrac{2y}{x}$; $x(x+1)$
56. $\dfrac{3x}{y}$; $y(y-1)$
57. $\dfrac{3x}{x+1}$; $(x+1)^2$
58. $\dfrac{5y}{y-2}$; $(y-2)^2$
59. $\dfrac{z+2}{z^2-6z}$; $z(z-6)(z+1)$
60. $\dfrac{y+1}{y^2+2y}$; $y(y+2)(y-2)$
61. $\dfrac{3}{2x+4}$; $2x^2+6x+4$
62. $\dfrac{4}{3x-3}$; $3x^2+3x-6$

TRY IT YOURSELF

Perform the operations. Then simplify, if possible.

63. $\dfrac{9y}{3x} - \dfrac{6y}{3x}$

64. $\dfrac{5r^2}{2r} - \dfrac{r^2}{2r}$

65. $\dfrac{6x - 5}{3xy} - \dfrac{3x - 5}{3xy}$

66. $\dfrac{3t}{t^2 - 8t + 7} - \dfrac{3}{t^2 - 8t + 7}$

67. $\dfrac{13x}{15} + \dfrac{12x}{15} - \dfrac{5x}{15}$

68. $\dfrac{13y}{32} + \dfrac{13y}{32} - \dfrac{10y}{32}$

69. $\dfrac{x + 1}{x - 2} - \dfrac{2(x - 3)}{x - 2} + \dfrac{3(x + 1)}{x - 2}$

70. $\dfrac{3xy}{x - y} - \dfrac{x(3y - x)}{x - y} - \dfrac{x(x - y)}{x - y}$

71. $\dfrac{3t}{t^2 - 8t + 7} - \dfrac{21}{t^2 - 8t + 7}$

72. $\dfrac{10t}{t^2 - 2t + 1} - \dfrac{10}{t^2 - 2t + 1}$

73. $\dfrac{1}{(t - 1)(t + 1)} - \dfrac{6 - t}{(t + 1)(t - 1)}$

74. $\dfrac{5r - 27}{3r^2 - 9r} + \dfrac{4r}{3r^2 - 9r}$

75. $\dfrac{9a}{5a^2 + 25a} + \dfrac{a + 50}{5a^2 + 25a}$

76. $\dfrac{a^2 + a}{4a^2 - 8a} + \dfrac{2a^2 - 7a}{4a^2 - 8a}$

77. $\dfrac{3b^2}{b + 1} - \dfrac{-b + 2}{b + 1}$

78. $\dfrac{7}{(t - 1)(t + 1)} - \dfrac{6 - t}{(t + 1)(t - 1)}$

APPLICATIONS

Refer to the illustration of the funnel in the next column.

79. Find the total height of the funnel.

80. Refer to the illustration in the next column. What is the difference between the diameter of the opening at the top of the funnel and the diameter of its spout?

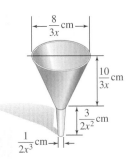

WRITING

81. Explain how to add fractions with the same denominator.

82. Explain how to find a least common denominator.

Explain what is wrong with the work shown below.

83. $\dfrac{2x + 3}{x + 5} - \dfrac{x + 2}{x + 5} = \dfrac{2x + 3 - x + 2}{x + 5}$

$= \dfrac{x + 5}{x + 5}$

$= 1$

84. $\dfrac{5x + 4}{y} + \dfrac{x}{y} = \dfrac{5x - 4 + x}{y + y}$

$= \dfrac{6x - 4}{2y}$

$= \dfrac{2(3x - 2)}{2y}$

$= \dfrac{3x - 2}{y}$

REVIEW

Write each number in prime-factored form.

85. 49

86. 64

87. 136

88. 315

SECTION 7.4
Adding and Subtracting with Unlike Denominators

Objectives

1. Add and subtract rational expressions that have unlike denominators.
2. Add and subtract rational expressions that have denominators that are opposites.

We have discussed a method for finding the least common denominator (LCD) of two rational expressions. We have also built rational expressions into equivalent expressions having a given denominator. We will now use these skills to add and subtract rational expressions with unlike denominators.

1 Add and subtract rational expressions that have unlike denominators.

The following steps summarize how to add (or subtract) rational expressions that have different denominators.

> **Adding or Subtracting Rational Expressions That Have Unlike Denominators**
>
> 1. Find the LCD.
> 2. Rewrite each rational expression as an equivalent expression with the LCD as the denominator. To do this, build each fraction using a form of 1 that involves any factor(s) needed to obtain the LCD.
> 3. Add (or subtract) the numerators and write the sum (or difference) over the LCD.
> 4. Simplify the result, if possible.

Self Check 1

Add: $\dfrac{y}{2} + \dfrac{6y}{7}$

Now Try Problem 13

EXAMPLE 1 Add: $\dfrac{4x}{7} + \dfrac{3x}{5}$

Strategy We will use the procedure for adding rational expressions that have unlike denominators. The first step is to determine the LCD.

WHY If we are to add (or subtract) fractions, their denominators must be the same. Since the denominators of these rational expressions are different, we cannot add them in their present form.

Solution

Step 1 The denominators are 7 and 5. The LCD is $7 \cdot 5 = 35$.

Step 2 We need to multiply the denominator of $\dfrac{4x}{7}$ by 5 and the denominator of $\dfrac{3x}{5}$ by 7 to obtain the LCD, 35. It follows that $\dfrac{5}{5}$ and $\dfrac{7}{7}$ are the forms of 1 that should be used to write the equivalent rational expressions.

$$\dfrac{4x}{7} + \dfrac{3x}{5} = \dfrac{4x}{7} \cdot \dfrac{5}{5} + \dfrac{3x}{5} \cdot \dfrac{7}{7}$$ Build the rational expressions so that each has a denominator of 35.

$$= \dfrac{20x}{35} + \dfrac{21x}{35}$$ Do the multiplications.

Step 3 $$= \dfrac{20x + 21x}{35}$$ Add the numerators and keep the common denominator.

$$= \dfrac{41x}{35}$$ Combine like terms in the numerator.

Step 4 Since 41 and 35 have no common factors, the result cannot be simplified.

Self Check 2

Add: $\dfrac{3}{28z} + \dfrac{5}{21z}$

Now Try Problem 19

EXAMPLE 2 Add: $\dfrac{5}{24b} + \dfrac{11}{18b}$

Strategy We will use the procedure for adding rational expressions that have unlike denominators. The first step is to determine the LCD.

WHY If we are to add (or subtract) fractions, their denominators must be the same. Since the denominators of these rational expressions are different, we cannot add them in their present form.

Solution

Step 1 To find the LCD, we form a product that uses each different factor of $24b$ and $18b$ the greatest number of times it appears in any one factorization.

$$\left.\begin{array}{l}24b = \mathbf{2 \cdot 2 \cdot 2 \cdot 3} \cdot b \\ 18b = \mathbf{2 \cdot 3 \cdot 3} \cdot b\end{array}\right\} \quad \text{LCD} = \mathbf{2 \cdot 2 \cdot 2 \cdot 3 \cdot 3} \cdot b = 72b$$

Step 2 We need to multiply the denominator of $\dfrac{5}{24b}$ by 3 and the denominator of $\dfrac{11}{18b}$ by 4 to obtain the LCD, $72b$. It follows that $\dfrac{3}{3}$ and $\dfrac{4}{4}$ are the forms of 1 that should be used to write the equivalent rational expressions.

$$= \frac{5}{24b} \cdot \frac{3}{3} + \frac{11}{18b} \cdot \frac{4}{4} \quad \text{Build each fraction to get the common denominator.}$$

$$= \frac{15}{72b} + \frac{44}{72b} \quad \text{Do the multiplications.}$$

Step 3 $= \dfrac{15 + 44}{72b}$ Add the numerators and keep the common denominator.

$$= \frac{59}{72b} \quad \text{Combine like terms in the numerators.}$$

Step 4 Since 59 and 72 have no common factors, the result cannot be simplified.

EXAMPLE 3 Add: $\dfrac{x + 4}{x^2} + \dfrac{x - 5}{4x}$

Strategy We will use the procedure for adding rational expressions when the denominators are different. The first step is to find the LCD.

WHY Since the denominators are different, we cannot add these rational expressions in their present form.

Solution
First we find the LCD.

$$\left.\begin{array}{l}x^2 = x \cdot x \\ 4x = 2 \cdot 2 \cdot x\end{array}\right\} \quad \text{LCD} = x \cdot x \cdot 2 \cdot 2 = 4x^2$$

$$\frac{x + 4}{x^2} + \frac{x - 5}{4x} = \frac{x + 4}{x^2} \cdot \frac{4}{4} + \frac{x - 5}{4x} \cdot \frac{x}{x} \quad \text{Build the fractions to get the common denominator, } 4x^2.$$

$$= \frac{4x + 16}{4x^2} + \frac{x^2 - 5x}{4x^2} \quad \text{Do the multiplications.}$$

$$= \frac{4x + 16 + x^2 - 5x}{4x^2} \quad \text{Add the numerators and keep the common denominator.}$$

$$= \frac{x^2 - x + 16}{4x^2} \quad \text{Combine like terms in the numerator.}$$

Self Check 3
Add: $\dfrac{a - 1}{9a} + \dfrac{2 - a}{a^2}$

Now Try Problem 23

EXAMPLE 4 Subtract: $\dfrac{x}{x + 1} - \dfrac{3}{x}$

Strategy We will use the procedure for subtracting rational expressions when the denominators are different. The first step is to find the LCD.

Self Check 4
Subtract: $\dfrac{a}{a - 1} - \dfrac{5}{a}$

Now Try Problem 31

WHY Since the denominators are different, we cannot subtract these rational expressions in their present form.

Solution
By inspection, the least common denominator is $(x + 1)x$.

$$\frac{x}{x+1} - \frac{3}{x} = \frac{x}{x+1} \cdot \frac{x}{x} - \frac{3}{x} \cdot \frac{x+1}{x+1} \quad \text{Build the fractions to get the common denominator.}$$

$$= \frac{x(x) - 3(x+1)}{x(x+1)} \quad \text{Subtract the numerators and keep the common denominator.}$$

$$= \frac{x^2 - 3x - 3}{x(x+1)} \quad \text{Do the multiplications in the numerator. Leave the denominator in factored form.}$$

Self Check 5

Subtract:
$$\frac{b}{b-2} - \frac{8}{b^2 - 4}$$

Now Try Problem 35

EXAMPLE 5 Subtract: $\dfrac{a}{a-1} - \dfrac{2}{a^2 - 1}$

Strategy We use the procedure for subtracting rational expressions when the denominators are binomials. The first step is to find the LCD.

WHY Since the denominators are different, we cannot subtract these rational expressions in their present form.

Solution
We factor $a^2 - 1$ to see that the LCD is $(a + 1)(a - 1)$.

$$\frac{a}{a-1} - \frac{2}{a^2 - 1}$$

$$= \frac{a}{(a-1)} \cdot \frac{a+1}{a+1} - \frac{2}{(a+1)(a-1)} \quad \text{Build the first fraction to get the LCD.}$$

$$= \frac{a(a+1) - 2}{(a-1)(a+1)} \quad \text{Subtract the numerators and keep the common denominator.}$$

$$= \frac{a^2 + a - 2}{(a-1)(a+1)} \quad \text{Distribute the multiplication by } a.$$

$$= \frac{(a+2)\cancel{(a-1)}}{\cancel{(a-1)}(a+1)} \quad \text{Simplify the result by factoring } a^2 + a - 2. \text{ Remove the common factor of } a - 1.$$

$$= \frac{a+2}{a+1}$$

Self Check 6

Subtract:
$$\frac{a}{a^2 - 2a + 1} - \frac{1}{6a - 6}$$

Now Try Problem 41

EXAMPLE 6 Subtract: $\dfrac{2a}{a^2 + 4a + 4} - \dfrac{1}{2a + 4}$

Strategy We use the procedure for subtracting rational expressions when the denominators are binomials and/or trinomials. The first step is to find the LCD.

WHY Since the denominators are different, we cannot subtract these rational expressions in their present form.

Solution
Find the least common denominator by factoring each denominator.

$$\left.\begin{array}{l} a^2 + 4a + 4 = (a+2)(a+2) \\ 2a + 4 = 2(a+2) \end{array}\right\} \quad \text{LCD} = (a+2)(a+2)2$$

We build each fraction into a new fraction with a denominator of $2(a + 2)(a + 2)$.

$$\frac{2a}{a^2 + 4a + 4} - \frac{1}{2a + 4}$$

$$= \frac{2a}{(a + 2)(a + 2)} - \frac{1}{2(a + 2)} \qquad \text{Write the denominators in factored form.}$$

$$= \frac{2a}{(a + 2)(a + 2)} \cdot \frac{2}{2} - \frac{1}{2(a + 2)} \cdot \frac{a + 2}{a + 2} \qquad \text{Build each fraction to get a common denominator.}$$

$$= \frac{4a - 1(a + 2)}{2(a + 2)^2} \qquad \text{Subtract the numerators and keep the common denominator. Write } (a + 2)(a + 2) \text{ as } (a + 2)^2.$$

$$= \frac{4a - a - 2}{2(a + 2)^2} \qquad \text{Distribute the multiplication by } -1. \text{ Leave the denominator in factored form.}$$

$$= \frac{3a - 2}{2(a + 2)^2} \qquad \text{Combine like terms in the numerator.}$$

EXAMPLE 7
Add: $\dfrac{4b}{a - 5} + b$

Self Check 7
Add: $\dfrac{10y}{n + 4} + y$

Now Try Problem 49

Strategy We will begin by writing the second addend, b, as $\dfrac{b}{1}$ and then find the LCD.

WHY To add b to the rational expression, $\dfrac{4b}{a - 5}$, we must rewrite b as a rational expression.

Solution
The LCD of $\dfrac{4b}{a - 5}$ and $\dfrac{b}{1}$ is $1(a - 5)$, or simply $a - 5$. Since we must multiply the denominator of $\dfrac{b}{1}$ by $a - 5$ to obtain the LCD, we will use $\dfrac{a - 5}{a - 5}$ to write an equivalent rational expression.

$$\frac{4b}{a - 5} + b = \frac{4b}{a - 5} + \frac{b}{1} \cdot \frac{a - 5}{a - 5} \qquad \text{Build } \tfrac{b}{1} \text{ so that it has a denominator of } a - 5.$$

$$= \frac{4b}{a - 5} + \frac{ab - 5b}{a - 5} \qquad \text{Multiply numerators: } b(a - 5) = ab - 5b. \text{ Multiply denominators: } 1(a - 5) = a - 5.$$

$$= \frac{4b + ab - 5b}{a - 5} \qquad \text{Add the numerators. Write the sum over the common denominator.}$$

$$= \frac{ab - b}{a - 5} \qquad \text{Combine like terms in the numerator: } 4b - 5b = -b.$$

Although the numerator factors as $b(a - 1)$, the numerator and denominator do not have a common factor. Therefore, the result is in simplest form.

2 Add and subtract rational expressions that have denominators that are opposites.

Recall that two polynomials are **opposites** if their terms are the same but they are opposite in sign. For example, $x - 5$ and $5 - x$ are opposites. If we multiply one of these binomials by -1, the subtraction is reversed, and the result is the other binomial.

$-1(x - 5) = -x + 5$ $\qquad\qquad$ $-1(5 - x) = -5 + x$

$\qquad\quad = 5 - x$ $\quad$ Write the expression $\qquad\qquad = x - 5$ $\quad$ Write the expression
$\qquad\qquad\qquad\qquad$ with 5 first. $\qquad\qquad\qquad\qquad\qquad\qquad$ with x first.

Chapter 7 Rational Expressions and Equations

> **Multiplying by −1**
>
> When a polynomial is multiplied by −1, the result is its opposite.

Self Check 8

Subtract: $\dfrac{5}{a-b} - \dfrac{2}{b-a}$

Now Try Problem 57

EXAMPLE 8 Subtract: $\dfrac{3}{x-y} - \dfrac{x}{y-x}$

Strategy We note that the denominators are opposites. Either can serve as the LCD; we will choose $x - y$. To obtain a common denominator, we will multiply $\dfrac{x}{y-x}$ by $\dfrac{-1}{-1}$.

WHY When $y - x$ is multiplied by −1, the subtraction is reversed, and the result is $x - y$.

Solution
We note that the second denominator is the opposite (negative) of the first. So we can multiply the numerator and denominator of the second fraction by −1 to get

$$\dfrac{3}{x-y} - \dfrac{x}{y-x} = \dfrac{3}{x-y} - \dfrac{x}{y-x} \cdot \dfrac{-1}{-1}$$ Multiply numerator and denominator by −1.

$$= \dfrac{3}{x-y} - \dfrac{-x}{-y+x}$$ Distribute the multiplication by −1: $-1(y-x) = -y + x$.

$$= \dfrac{3}{x-y} - \dfrac{-x}{x-y}$$ $-y + x = x - y$. The fractions now have a common denominator of $x - y$.

$$= \dfrac{3-(-x)}{x-y}$$ Subtract the numerators and keep the common denominator.

$$= \dfrac{3+x}{x-y}$$ $-(-x) = x$.

To add and/or subtract three or more rational expressions, we follow the rules for the order of operations.

Self Check 9

Perform the operations:
$\dfrac{5}{ab^2} - \dfrac{b}{a} + \dfrac{a}{b}$

Now Try Problem 63

EXAMPLE 9 Perform the operations: $\dfrac{3}{x^2y} + \dfrac{2}{xy} - \dfrac{1}{xy^2}$

Strategy We note that the denominators of the rational expressions to be added and subtracted do not have the same denominator. We will find the LCD of all three rational expressions, build to get equivalent rational expressions with the LCD, and then perform the additions and subtractions from left to right.

WHY We must follow the rule for adding and/or subtracting rational expressions as well as the rules for order of operations.

Solution
Find the least common denominator.

$$\left.\begin{array}{l} x^2y = x \cdot x \cdot y \\ xy = x \cdot y \\ xy^2 = x \cdot y \cdot y \end{array}\right\} \text{Factor each denominator.}$$

In any one of these denominators, the factor x occurs at most twice, and the factor y occurs at most twice. Thus,

$$\text{LCD} = x \cdot x \cdot y \cdot y$$
$$\text{LCD} = x^2y^2$$

We build each fraction into one with a denominator of x^2y^2.

$$\frac{3}{x^2y} + \frac{2}{xy} - \frac{1}{xy^2}$$

$$= \frac{3}{x \cdot x \cdot y} \cdot \frac{y}{y} + \frac{2}{x \cdot y} \cdot \frac{x \cdot y}{x \cdot y} - \frac{1}{x \cdot y \cdot y} \cdot \frac{x}{x}$$

Factor each denominator and build each fraction.

$$= \frac{3y + 2xy - x}{x^2y^2}$$

Do the multiplications and combine the numerators. Write the result over the LCD.

ANSWERS TO SELF CHECKS

1. $\dfrac{19y}{14}$ 2. $\dfrac{29}{84z}$ 3. $\dfrac{a^2 - 10a + 18}{9a^2}$ 4. $\dfrac{a^2 - 5a + 5}{a(a - 1)}$ 5. $\dfrac{b + 4}{b + 2}$ 6. $\dfrac{5a + 1}{6(a - 1)^2}$

7. $\dfrac{ny + 14y}{n + 4}$ 8. $\dfrac{7}{a - b}$ 9. $\dfrac{5 - b^3 + a^2b}{ab^2}$

SECTION 7.4 STUDY SET

VOCABULARY

1. $\dfrac{x}{x - 7}$ and $\dfrac{1}{x - 7}$ have _____ denominators. $\dfrac{x - 1}{2x + 3}$ and $\dfrac{x}{2x - 3}$ have _____ denominators.

2. Two polynomials are _____ if their terms are the same, but are opposite in sign.

CONCEPTS

3. Write the denominator of $\dfrac{x + 1}{20x^2}$ in factored form.

4. Write the denominator of $\dfrac{3x^2 - 4}{x^2 + 4x - 12}$ in factored form.

5. Factor each denominator and find the LCD of $\dfrac{1}{12a}$ and $\dfrac{1}{18a^2}$.

6. Factor each denominator and find the LCD of $\dfrac{1}{x^2 - 36}$ and $\dfrac{1}{3x - 18}$.

7. Find the LCD of $\dfrac{x - 1}{x + 6}$ and $\dfrac{1}{x + 3}$.

8. Find the LCD of $\dfrac{x + 3}{x^2 - 4}$ and $\dfrac{x - 1}{x^2 + 2x}$.

NOTATION

Complete each step.

9. $\dfrac{2}{5} + \dfrac{7}{3x} = \dfrac{2}{5} \cdot \dfrac{\boxed{}}{3x} + \dfrac{7}{3x} \cdot \dfrac{\boxed{}}{5}$

$$= \dfrac{6x}{\boxed{}} + \dfrac{35}{\boxed{}}$$

$$= \dfrac{6x + \boxed{}}{15x}$$

10. Are the student's answers and the book's answers equivalent?

Student's answer	Book's answer	Equivalent?
$\dfrac{m^2 + 2m}{(m - 1)(m - 4)}$	$\dfrac{m^2 + 2m}{(m - 4)(m - 1)}$	
$\dfrac{-5x^2 - 7}{4x(x + 3)}$	$-\dfrac{5x^2 + 7}{4x(x + 3)}$	
$\dfrac{-2x}{x - y}$	$-\dfrac{2x}{y - x}$	

GUIDED PRACTICE

Add the rational expressions. Simplify the result, if possible. See Example 1.

11. $\dfrac{2y}{9} + \dfrac{y}{3}$

12. $\dfrac{7y}{6} + \dfrac{10y}{9}$

13. $\dfrac{x}{3} + \dfrac{2x}{7}$

14. $\dfrac{y}{4} + \dfrac{3y}{5}$

15. $\dfrac{5a}{6} + \dfrac{5a}{3}$

16. $\dfrac{4x}{3} + \dfrac{x}{6}$

17. $\dfrac{21x}{14} + \dfrac{5x}{21}$

18. $\dfrac{8a}{15} + \dfrac{5a}{12}$

Add the rational expressions. Simplify the result, if possible. See Example 2.

19. $\dfrac{4}{15a} + \dfrac{9}{20a}$

20. $\dfrac{5}{14b} + \dfrac{2}{21b}$

21. $\dfrac{6}{25p} + \dfrac{7}{15p}$

22. $\dfrac{3}{54q} + \dfrac{2}{63q}$

Add the rational expressions. Simplify the result, if possible. See Example 3.

23. $\dfrac{y+2}{5y^2} + \dfrac{y+4}{15y}$

24. $\dfrac{x+3}{x^2} + \dfrac{x+5}{2x}$

25. $\dfrac{x-1}{x} + \dfrac{y+1}{y}$

26. $\dfrac{a+2}{b} + \dfrac{b-2}{a}$

Subtract the rational expressions. Simplify the result, if possible. See Example 4.

27. $\dfrac{21x}{14} - \dfrac{5x}{21}$

28. $\dfrac{8a}{15} - \dfrac{5a}{12}$

29. $\dfrac{7}{8} - \dfrac{4}{t}$

30. $\dfrac{5}{a} - \dfrac{3}{b}$

31. $\dfrac{a}{a+2} - \dfrac{2}{a}$

32. $\dfrac{2n}{n-3} - \dfrac{1}{n}$

33. $\dfrac{5}{d} - \dfrac{d}{d+1}$

34. $\dfrac{8}{b} - \dfrac{2b}{b-2}$

Perform the operations. Simplify the result, if possible. See Examples 5–6.

35. $\dfrac{y}{y+1} - \dfrac{2}{y^2-1}$

36. $\dfrac{b}{b+2} - \dfrac{8}{b^2-4}$

37. $\dfrac{d}{d^2+6d+5} - \dfrac{3}{d^2+5d+4}$

38. $\dfrac{r}{r^2+5r+6} - \dfrac{2}{r^2+3r+2}$

39. $\dfrac{4}{s^2+5s+4} + \dfrac{s}{s^2+2s+1}$

40. $\dfrac{3}{t^2+t-6} + \dfrac{1}{t^2+3t-10}$

41. $\dfrac{3b}{b^2+2b+1} - \dfrac{1}{2b+2}$

42. $\dfrac{2a}{a^2+6a+9} - \dfrac{5}{3a+9}$

43. $\dfrac{1}{b+3} - \dfrac{1}{b^2+7b+12}$

44. $\dfrac{1}{c+6} - \dfrac{-4}{c^2+8c+12}$

45. $\dfrac{b}{b+1} - \dfrac{b+1}{2b+2}$

46. $\dfrac{2x}{x+2} + \dfrac{x+1}{x-3}$

Perform the operations. Simplify the result, if possible. See Example 7.

47. $\dfrac{8}{x} + 6$

48. $\dfrac{2}{y} + 7$

49. $\dfrac{9}{x-4} + x$

50. $\dfrac{9}{m+4} + 9$

51. $b - \dfrac{3}{a^2}$

52. $c - \dfrac{5}{3b}$

53. $\dfrac{x+2}{x+1} - 5$

54. $\dfrac{y+8}{y-8} - 4$

Perform the operations. Simplify the result, if possible. See Example 8.

55. $\dfrac{2}{a-b} - \dfrac{a}{b-a}$

56. $\dfrac{m}{m-n} - \dfrac{n}{n-m}$

57. $\dfrac{5}{a-4} + \dfrac{7}{4-a}$

58. $\dfrac{p}{p-q} + \dfrac{q}{q-p}$

59. $\dfrac{2x+2}{x-2} - \dfrac{2x}{2-x}$

60. $\dfrac{y+3}{y-1} - \dfrac{y+4}{1-y}$

61. $\dfrac{r+2}{r^2-4} + \dfrac{4}{4-r^2}$

62. $\dfrac{h}{h^2-49} + \dfrac{7}{49-h^2}$

Perform the operations. Simplify the result, if possible. See Example 9.

63. $\dfrac{3}{a^2b} + \dfrac{4}{ab} - \dfrac{2}{ab^2}$

64. $\dfrac{2}{pq} - \dfrac{a}{pq^2} + \dfrac{b}{p^2q^2}$

65. $\dfrac{2}{x-1} + \dfrac{3}{x+1} - \dfrac{1}{x^2-1}$

66. $\dfrac{3}{2y+2} - \dfrac{2}{3y+3} + \dfrac{1}{y+1}$

67. $\dfrac{2x}{x^2-3x+2} + \dfrac{2x}{x-1} - \dfrac{x}{x-2}$

68. $\dfrac{4a}{a-2} - \dfrac{3a}{a-3} + \dfrac{4a}{a^2-5a+6}$

69. $\dfrac{2x}{x-1} + \dfrac{3x}{x+1} - \dfrac{x+3}{x^2-1}$

70. $\dfrac{a}{a-1} - \dfrac{2}{a+2} + \dfrac{3(a-2)}{a^2+a-2}$

TRY IT YOURSELF

Perform the operations and simplify, if possible.

71. $\dfrac{2y}{5x} - \dfrac{y}{2}$

72. $\dfrac{2}{x} \cdot 3x$

73. $\dfrac{7}{m^2} + \dfrac{2}{m}$

74. $\dfrac{3}{8a} + \dfrac{11}{12a}$

75. $\dfrac{4x}{3} + \dfrac{2x}{y}$

76. $14 + \dfrac{10}{y^2}$

77. $\dfrac{x}{x+1} + \dfrac{x-1}{x}$

78. $\dfrac{3x}{xy} + \dfrac{x+1}{y-1}$

79. $\dfrac{x+5}{xy} - \dfrac{x-1}{x^2y}$

80. $\dfrac{y+7}{2y} - \dfrac{y^2+5y+14}{2y^2}$

81. $\dfrac{x}{x-2} + \dfrac{4+2x}{x^2-4}$

82. $\dfrac{y}{y+3} - \dfrac{2y-6}{y^2-9}$

83. $\dfrac{4}{b-6} - \dfrac{b}{6-b}$

84. $\dfrac{t+1}{t-7} - \dfrac{t+1}{7-t}$

85. $\dfrac{4x+1}{8x-12} + \dfrac{x-3}{2x-3}$

86. $\dfrac{x+1}{x-1} + \dfrac{x-1}{x+1}$

87. $\dfrac{2}{a^2+4a+3} + \dfrac{1}{a+3}$

88. $\dfrac{3}{s-8} + t$

89. $\dfrac{x+1}{2x+4} - \dfrac{x^2}{2x^2-8}$

90. $\dfrac{x+1}{x+2} - \dfrac{x^2+1}{x^2-x-6}$

91. $\dfrac{7}{3a} + \dfrac{1}{a-2}$

92. $\dfrac{5}{9x} + \dfrac{4}{x+6}$

93. $\dfrac{3}{x+2} + \dfrac{x}{x-1} - \dfrac{3x}{x^2+x-2}$

94. $\dfrac{2x}{x^2-3x+2} + \dfrac{2x}{x-1} - \dfrac{x}{x-2}$

WRITING

95. Explain the error:

$$\dfrac{3}{x} + \dfrac{8}{y} = \dfrac{3+8}{x+y} = \dfrac{11}{x+y}$$

96. Explain how to add two rational expressions with unlike denominators.

REVIEW

97. Find the slope and y-intercept of the graph of $y = 8x + 2$.

98. Find the slope and y-intercept of the graph of $3x + 4y = -36$.

99. What is the slope of the graph of $y = 2$?

100. Is the graph of the equation $x = 0$ the x-axis or the y-axis?

SECTION 7.5
Simplifying Complex Fractions

Objectives

1. Simplify complex fractions using division.
2. Simplify complex fractions using the LCD.

A rational expression whose numerator and/or denominator contain fractions is called a **complex rational expression** or a **complex fraction**. The expression above the main fraction bar of a complex fraction is the numerator, and the expression below the main fraction bar is the denominator. Two examples of complex fractions are:

$$\dfrac{\dfrac{5x}{3}}{\dfrac{2x}{9}} \quad \begin{array}{l} \leftarrow \text{Numerator of complex fraction} \\ \leftarrow \text{Main fraction bar} \\ \leftarrow \text{Denominator of complex fraction} \end{array} \quad \dfrac{\dfrac{1}{2} - \dfrac{1}{x}}{\dfrac{x}{3} + \dfrac{1}{5}}$$

Chapter 7 Rational Expressions and Equations

In this section, we will discuss two methods for simplifying complex fractions. To **simplify a complex fraction** means to write it in the form $\dfrac{A}{B}$, where A and B are polynomials that have no common factors.

1 Simplify complex fractions using division.

One method for simplifying complex fractions uses the fact that the main fraction bar indicates division.

> **Simplifying Complex Fractions Method 1: Using Division**
>
> 1. Add or subtract in the numerator and/or denominator so that the numerator is a single fraction and the denominator is a single fraction.
> 2. Perform the indicated division by multiplying the numerator of the complex fraction by the reciprocal of the denominator.
> 3. Simplify the result, if possible.

Self Check 1

Simplify: $\dfrac{\dfrac{7y^3}{8}}{\dfrac{21y^2}{20}}$

Now Try Problem 15

EXAMPLE 1

Simplify: $\dfrac{\dfrac{5x^2}{3}}{\dfrac{2x^3}{9}}$

Strategy We will perform the division indicated by the main fraction bar using the procedure for dividing rational expressions from Section 7.2.

WHY We can skip the first step of method 1 and immediately divide because the numerator and the denominator of the complex fraction are already single fractions.

Solution

$$\dfrac{\dfrac{5x^2}{3}}{\dfrac{2x^3}{9}} = \dfrac{5x^2}{3} \div \dfrac{2x^3}{9} \quad \text{Write the division indicated by the main fraction bar using a } \div \text{ symbol.}$$

$$= \dfrac{5x^2}{3} \cdot \dfrac{9}{2x^3} \quad \text{To divide rational expressions, multiply the first by the reciprocal of the second.}$$

$$= \dfrac{5x^2 \cdot 9}{3 \cdot 2x^3} \quad \text{Write the product of the numerators. Write the product of the denominators.}$$

$$= \dfrac{5 \cdot \cancel{x} \cdot \cancel{x} \cdot \cancel{3} \cdot 3}{\cancel{3} \cdot 2 \cdot \cancel{x} \cdot \cancel{x} \cdot x} \quad \text{Factor 9 as 3 · 3. Then simplify by removing factors equal to 1.}$$

$$= \dfrac{15}{2x} \quad \text{Multiply the remaining factors in the numerator. Multiply the remaining factors in the denominator.}$$

Self Check 2

Simplify: $\dfrac{\dfrac{1}{3} + \dfrac{1}{x}}{\dfrac{x}{5} - \dfrac{1}{2}}$

Now Try Problem 21

EXAMPLE 2

Simplify: $\dfrac{\dfrac{1}{2} - \dfrac{1}{x}}{\dfrac{x}{3} + \dfrac{1}{5}}$

Strategy We will simplify the expressions above and below the main fraction bar separately to write $\dfrac{1}{2} - \dfrac{1}{x}$ and $\dfrac{x}{3} + \dfrac{1}{5}$ as single fractions. Then we will perform the indicated division.

WHY The numerator and the denominator of the complex fraction must be written as single fractions before dividing.

Solution
To write the numerator as a single fraction, we build $\frac{1}{2}$ and $\frac{1}{x}$ to have an LCD of $2x$, and then subtract. To write the denominator as a single fraction, we build $\frac{x}{3}$ and $\frac{1}{5}$ to have an LCD of 15, and then add.

$$\frac{\frac{1}{2} - \frac{1}{x}}{\frac{x}{3} + \frac{1}{5}} = \frac{\frac{1}{2} \cdot \frac{x}{x} - \frac{1}{x} \cdot \frac{2}{2}}{\frac{x}{3} \cdot \frac{5}{5} + \frac{1}{5} \cdot \frac{3}{3}}$$

← The LCD for the numerator is $2x$. Build each fraction so that each has a denominator of $2x$.

← The LCD for the denominator is 15. Build each fraction so that each has a denominator of 15.

$$= \frac{\frac{x}{2x} - \frac{2}{2x}}{\frac{5x}{15} + \frac{3}{15}}$$

Multiply the numerators and multiply the denominators.

$$= \frac{\frac{x-2}{2x}}{\frac{5x+3}{15}}$$

Subtract in the numerator and add in the denominator of the complex fraction.

Now that the numerator and the denominator of the complex fraction are single fractions, we perform the indicated division.

$$\frac{\frac{x-2}{2x}}{\frac{5x+3}{15}} = \frac{x-2}{2x} \div \frac{5x+3}{15}$$

Write the division indicated by the main fraction bar using a ÷ symbol.

$$= \frac{x-2}{2x} \cdot \frac{15}{5x+3}$$

Multiply by the reciprocal of $\frac{5x+3}{15}$.

$$= \frac{15(x-2)}{2x(5x+3)}$$

Write the product of the numerators.
Write the product of the denominators.

Since the numerator and denominator have no common factor, the result does not simplify.

> **Success Tip** The result after simplifying a complex fraction can often have several equivalent forms. The result for Example 2 could be written:
>
> $$\frac{15x - 30}{2x(5x+3)}$$

EXAMPLE 3

Simplify: $\dfrac{\frac{6}{x} + y}{\frac{6}{y} + x}$

Strategy We will simplify the expressions above and below the main fraction bar separately to write $\frac{6}{x} + y$ and $\frac{6}{y} + x$ as single fractions. Then we will perform the indicated division.

WHY The numerator and the denominator of the complex fraction must be written as single fractions before dividing.

Self Check 3

Simplify: $\dfrac{\frac{2}{a} - b}{\frac{2}{b} - a}$

Now Try Problems 33 and 35

Solution

To write $\frac{6}{x} + y$ as a single fraction, we build y into a fraction with a denominator of x and add. To write $\frac{6}{y} + x$ as a single fraction, we build x into a fraction with a denominator of y and add.

$$\frac{\frac{6}{x} + y}{\frac{6}{y} + x} = \frac{\frac{6}{x} + \frac{y}{1} \cdot \frac{x}{x}}{\frac{6}{y} + \frac{x}{1} \cdot \frac{y}{y}}$$

← Write y as $\frac{y}{1}$. The LCD for the numerator is x. Build $\frac{y}{1}$ so that it has a denominator of x.

← Write x as $\frac{x}{1}$. The LCD for the denominator is y. Build $\frac{x}{1}$ so that it has a denominator of y.

$$= \frac{\frac{6}{x} + \frac{xy}{x}}{\frac{6}{y} + \frac{xy}{y}}$$

Multiply in the numerator and multiply in the denominator.

$$= \frac{\frac{6 + xy}{x}}{\frac{6 + xy}{y}}$$

Add in the numerator and in the denominator of the complex fraction.

Now that the numerator and the denominator of the complex fraction are single fractions, we can perform the division.

$$\frac{\frac{6 + xy}{x}}{\frac{6 + xy}{y}} = \frac{6 + xy}{x} \div \frac{6 + xy}{y}$$

Write the division indicated by the main fraction bar using a $\div$ symbol.

$$= \frac{6 + xy}{x} \cdot \frac{y}{6 + xy}$$

Multiply by the reciprocal of $\frac{6 + xy}{y}$.

$$= \frac{y(6 + xy)}{x(6 + xy)}$$

Write the product of the numerators.
Write the product of the denominators.

$$= \frac{y\cancel{(6 + xy)}}{x\cancel{(6 + xy)}}$$

Simplify the result by removing a factor equal to 1.

$$= \frac{y}{x}$$

2 Simplify complex fractions using the LCD.

A second method for simplifying complex fractions uses the concepts of LCD and multiplication by a form of 1. The multiplication by 1 produces a simpler, equivalent expression, which will not contain fractions in its numerator or denominator.

Simplifying Complex Fractions Method 2: Multiplying by the LCD

1. Find the LCD of all fractions within the complex fraction.
2. Multiply the complex fraction by 1 in the form $\frac{\text{LCD}}{\text{LCD}}$.
3. Perform the operations in the numerator and denominator. No fractional expressions should remain within the complex fraction.
4. Simplify the result, if possible.

We will use method 2 to rework Example 2.

7.5 Simplifying Complex Fractions

EXAMPLE 4

Simplify: $\dfrac{\dfrac{1}{2} - \dfrac{1}{x}}{\dfrac{x}{3} + \dfrac{1}{5}}$

Strategy Using method 1 to simplify this complex fraction, we worked with $\dfrac{1}{2} - \dfrac{1}{x}$ and $\dfrac{x}{3} + \dfrac{1}{5}$ separately. With method 2, we will use the LCD of *all four* fractions within the complex fraction.

WHY Multiplying a complex fraction by 1 in the form of $\dfrac{\text{LCD}}{\text{LCD}}$ clears its numerator and denominator of fractions.

Solution
The denominators of all the fractions within the complex fraction are 2, x, 3, and 5. Thus, their LCD is $2 \cdot x \cdot 3 \cdot 5 = 30x$.

We now multiply the complex fraction by a factor equal to 1, using the LCD: $\dfrac{30x}{30x} = 1$.

$\dfrac{\dfrac{1}{2} - \dfrac{1}{x}}{\dfrac{x}{3} + \dfrac{1}{5}} = \dfrac{\dfrac{1}{2} - \dfrac{1}{x}}{\dfrac{x}{3} + \dfrac{1}{5}} \cdot \dfrac{30x}{30x}$

$= \dfrac{\left(\dfrac{1}{2} - \dfrac{1}{x}\right)30x}{\left(\dfrac{x}{3} + \dfrac{1}{5}\right)30x}$ ← Write the product of the numerators.
← Write the product of the denominators.

$= \dfrac{\dfrac{1}{2}(30x) - \dfrac{1}{x}(30x)}{\dfrac{x}{3}(30x) + \dfrac{1}{5}(30x)}$ ← In the numerator, distribute the multiplication by 30x.
← In the denominator, distribute the multiplication by 30x.

$= \dfrac{15x - 30}{10x^2 + 6x}$ Do each of the four multiplications by 30x. Notice that no fractional expressions remain within the complex fraction.

To attempt to simplify the result, factor the numerator and denominator. Since they do not have a common factor, the result is in simplest form.

$\dfrac{15x - 30}{10x^2 + 6x} = \dfrac{15(x - 2)}{2x(5x + 3)}$

Success Tip When simplifying a complex fraction, the same result will be obtained regardless of the method used. See Example 2.

Self Check 4
Use method 2 to simplify:
$\dfrac{\dfrac{1}{4} - \dfrac{1}{x}}{\dfrac{x}{5} + \dfrac{1}{3}}$

Now Try Problem 43

EXAMPLE 5

Simplify: $\dfrac{\dfrac{1}{8} - \dfrac{1}{y}}{\dfrac{8 - y}{4y^2}}$

Strategy Using method 1, we would work with $\dfrac{1}{8} - \dfrac{1}{y}$ and $\dfrac{8 - y}{4y^2}$ separately. With method 2, we use the LCD of all three fractions within the complex fraction.

Self Check 5
Simplify: $\dfrac{\dfrac{10 - n}{5n^2}}{\dfrac{1}{10} - \dfrac{1}{n}}$

Now Try Problem 53

Chapter 7 Rational Expressions and Equations

WHY Multiplying a complex fraction by 1 in the form of $\frac{\text{LCD}}{\text{LCD}}$ clears its numerator and denominator of fractions.

Solution
The denominators of all fractions within the complex fraction are 8, y, and $4y^2$. Therefore, the LCD is $8y^2$ and we multiply the complex fraction by a factor equal to 1, using the LCD: $\frac{8y^2}{8y^2} = 1$.

$$\frac{\frac{1}{8} - \frac{1}{y}}{\frac{8-y}{4y^2}} = \frac{\frac{1}{8} - \frac{1}{y}}{\frac{8-y}{4y^2}} \cdot \frac{8y^2}{8y^2}$$

$$= \frac{\left(\frac{1}{8} - \frac{1}{y}\right)8y^2}{\left(\frac{8-y}{4y^2}\right)8y^2} \quad \longleftarrow \text{Write the product of the numerators.}$$
$$\quad \longleftarrow \text{Write the product of the denominators.}$$

$$= \frac{\frac{1}{8}(8y^2) - \frac{1}{y}(8y^2)}{\left(\frac{8-y}{4y^2}\right)(8y^2)} \quad \text{In the numerator, distribute the multiplication by } 8y^2.$$

$$= \frac{y^2 - 8y}{(8-y)2} \quad \text{Do each of the three multiplications by } 8y^2.$$

$$= \frac{y(y-8)^{-1}}{(8-y)2}_{1} \quad \text{In the numerator, factor out the GCF, } y. \text{ Since } y - 8 \text{ and } 8 - y \text{ are opposites, simplify by replacing } \frac{y-8}{8-y} \text{ with } \frac{-1}{1}.$$

$$= -\frac{y}{2}$$

Self Check 6
Simplify: $\dfrac{2}{\dfrac{1}{x+2} + 2}$

Now Try Problem 55

EXAMPLE 6 Simplify: $\dfrac{1}{1 + \dfrac{1}{x+1}}$

Strategy Although either method can be used, we will use method 2 to simplify this complex fraction.

WHY Method 2 is often easier when the complex fraction contains a sum or difference.

Solution
The only fraction within the complex fraction has the denominator $x + 1$. Therefore, the LCD is $x + 1$. We multiply the complex fraction by a factor equal to 1, using the LCD: $\frac{x+1}{x+1} = 1$.

$$\frac{1}{1 + \dfrac{1}{x+1}} = \frac{1}{1 + \dfrac{1}{x+1}} \cdot \frac{x+1}{x+1}$$

$$= \frac{1(x+1)}{\left(1 + \dfrac{1}{x+1}\right)(x+1)} \quad \text{Write the product of the numerators.}$$
$$\quad \text{Write the product of the denominators.}$$

$$= \frac{1(x+1)}{1(x+1) + \dfrac{1}{x+1}(x+1)} \quad \text{In the denominator, distribute the multiplication by } x + 1.$$

$$= \frac{x+1}{x+1+1} \quad \text{Do each of the three multiplications by } x+1.$$

$$= \frac{x+1}{x+2} \quad \text{Combine like terms in the denominator.}$$

The result does not simplify.

> **Success Tip** Simplifying using the LCD (method 2) works well when the complex fraction has sums and/or differences in the numerator and/or denominator.

ANSWERS TO SELF CHECK

1. $\dfrac{5y}{6}$ 2. $\dfrac{10(x+3)}{3x(2x-5)}$ 3. $\dfrac{b}{a}$ 4. $\dfrac{15(x-4)}{4x(3x+5)}$ 5. $-\dfrac{2}{n}$ 6. $\dfrac{2(x+2)}{2x+5}$

SECTION 7.5 STUDY SET

VOCABULARY

Fill in the blanks.

1. If a fraction has a fraction in its numerator or denominator, it is called a _____ _____ .

2. The denominator of the complex fraction $\dfrac{\dfrac{3}{x}+\dfrac{x}{y}}{\dfrac{1}{x}+2}$ is _____ .

CONCEPTS

Fill in the blanks.

3. Method 1: To simplify a complex fraction, we write the numerator and the denominator of a complex fraction as _____ fractions. Then perform the indicated _____ by multiplying the numerator of the complex fraction by the _____ of the denominator.

4. Method 2: To simplify a complex fraction, we multiply the numerator and denominator of the complex fraction by the _____ of the fractions in its numerator and denominator.

NOTATION

Complete each step.

5. $\dfrac{\dfrac{2b-a}{ab}}{\dfrac{b+2a}{ab}} = \dfrac{2b-a}{ab} \boxed{} \dfrac{b+2a}{ab}$

$= \dfrac{2b-a}{ab} \cdot \dfrac{\boxed{}}{b+2a}$

$= \dfrac{(2b-a)\boxed{}}{ab\boxed{}}$

$= \dfrac{\boxed{}}{b+2a}$

6. $\dfrac{\dfrac{2}{a}-\dfrac{1}{b}}{\dfrac{1}{a}+\dfrac{2}{b}} = \dfrac{\boxed{}\left(\dfrac{2}{a}-\dfrac{1}{b}\right)}{\boxed{}\left(\dfrac{1}{a}+\dfrac{2}{b}\right)}$

$= \dfrac{\boxed{} - a}{b+2a}$

GUIDED PRACTICE

Simplify each complex fraction. See Example 1.

7. $\dfrac{\dfrac{2}{3}}{\dfrac{3}{4}}$ 8. $\dfrac{\dfrac{3}{5}}{\dfrac{2}{7}}$ 9. $\dfrac{\dfrac{4}{5}}{\dfrac{32}{15}}$

10. $\dfrac{\dfrac{7}{8}}{\dfrac{49}{4}}$ 11. $\dfrac{\dfrac{x}{2}}{\dfrac{6}{5}}$ 12. $\dfrac{\dfrac{9}{4}}{\dfrac{7}{x}}$

13. $\dfrac{\dfrac{x}{y}}{\dfrac{1}{x}}$ 14. $\dfrac{\dfrac{y}{x}}{\dfrac{x}{xy}}$

15. $\dfrac{\dfrac{5t^2}{9x^2}}{\dfrac{3t}{x^2 t}}$ 16. $\dfrac{\dfrac{5w^2}{4tz}}{\dfrac{15wt}{z^2}}$

17. $\dfrac{\dfrac{4t-8}{t^2}}{\dfrac{8t-16}{t^5}}$ 18. $\dfrac{\dfrac{9m-27}{m^6}}{\dfrac{2m-6}{m^8}}$

Simplify each complex fraction. See Example 2.

19. $\dfrac{\frac{1}{2}+\frac{3}{4}}{\frac{3}{2}+\frac{1}{4}}$

20. $\dfrac{\frac{2}{3}-\frac{5}{2}}{\frac{2}{3}-\frac{3}{2}}$

21. $\dfrac{\frac{3}{5}+\frac{1}{y}}{\frac{y}{3}-\frac{2}{5}}$

22. $\dfrac{\frac{2}{x}-\frac{1}{3}}{\frac{2}{3}+\frac{x}{5}}$

23. $\dfrac{\frac{1}{y}-\frac{5}{2}}{\frac{3}{y}+\frac{1}{2}}$

24. $\dfrac{\frac{1}{6}-\frac{5}{s}}{\frac{2}{s}}$

25. $\dfrac{\frac{4}{c}-\frac{c}{6}}{\frac{2}{c}}$

26. $\dfrac{\frac{10}{n}-\frac{n}{4}}{\frac{8}{n}}$

27. $\dfrac{\frac{2}{s}-\frac{2}{s^2}}{\frac{4}{s^2}+\frac{4}{s^3}}$

28. $\dfrac{\frac{3}{a}-\frac{2}{b}}{\frac{1}{a}+\frac{5}{b}}$

29. $\dfrac{\frac{2}{a}+\frac{1}{b}}{\frac{1}{a}-\frac{3}{b}}$

30. $\dfrac{\frac{1}{4}+\frac{1}{y}}{\frac{y}{3}-\frac{1}{2}}$

Simplify each complex fraction. See Example 3.

31. $\dfrac{\frac{2}{3}+1}{\frac{1}{3}+1}$

32. $\dfrac{\frac{3}{5}-2}{\frac{2}{5}-2}$

33. $\dfrac{\frac{1}{x}-3}{\frac{5}{x}+2}$

34. $\dfrac{\frac{1}{y}+3}{\frac{3}{y}-2}$

35. $\dfrac{\frac{2}{x}+2}{\frac{4}{x}+2}$

36. $\dfrac{\frac{3}{x}-3}{\frac{9}{x}-3}$

37. $\dfrac{\frac{3y}{x}-y}{y-\frac{y}{x}}$

38. $\dfrac{\frac{y}{x}+3y}{y+\frac{2y}{x}}$

39. $\dfrac{4-\frac{1}{8h}}{12+\frac{3}{4h}}$

40. $\dfrac{12+\frac{1}{3b}}{12-\frac{1}{b^2}}$

41. $\dfrac{1-\frac{9}{d^2}}{2+\frac{6}{d}}$

42. $\dfrac{1-\frac{16}{a^2}}{\frac{12}{a}+3}$

Simplify each complex fraction by using the LCD. See Example 4.

43. $\dfrac{\frac{1}{6}-\frac{2}{x}}{\frac{1}{6}+\frac{1}{x}}$

44. $\dfrac{\frac{3}{4}+\frac{1}{y}}{\frac{5}{6}-\frac{1}{y}}$

45. $\dfrac{\frac{a}{7}-\frac{7}{a}}{\frac{1}{a}+\frac{1}{7}}$

46. $\dfrac{\frac{t}{9}-\frac{9}{t}}{\frac{1}{t}+\frac{1}{9}}$

47. $\dfrac{\frac{m}{n}+\frac{n}{m}}{\frac{m}{n}-\frac{n}{m}}$

48. $\dfrac{\frac{2a}{b}-\frac{b}{a}}{\frac{2a}{b}+\frac{b}{a}}$

49. $\dfrac{\frac{1}{4}+\frac{1}{y}}{\frac{y}{3}-\frac{1}{2}}$

50. $\dfrac{\frac{3}{5}+\frac{2}{x}}{\frac{1}{3}-\frac{4}{y}}$

Simplify each complex fraction by using the LCD. See Example 5.

51. $\dfrac{\frac{1}{c}+\frac{5}{4}}{\frac{2}{c^2}}$

52. $\dfrac{\frac{1}{s}+\frac{10}{3}}{\frac{7}{s^2}}$

53. $\dfrac{\frac{2}{x}}{\frac{2}{y}-\frac{4}{x}}$

54. $\dfrac{\frac{2y}{3}}{\frac{2y}{3}-\frac{8}{y}}$

Simplify each complex fraction by using the LCD. See Example 6.

55. $\dfrac{\frac{5}{x+1}}{1+\frac{1}{x+1}}$

56. $\dfrac{\frac{1}{x-4}}{1-\frac{1}{x-4}}$

57. $\dfrac{\frac{4}{x-4}}{\frac{x}{x-4}+3}$

58. $\dfrac{\frac{2}{x-2}}{\frac{2}{x-2}-1}$

59. $\dfrac{3+\frac{3}{x-1}}{3-\frac{3}{x-1}}$

60. $\dfrac{2-\frac{2}{x+1}}{2+\frac{2}{x+1}}$

61. $\dfrac{m-\frac{1}{2m+1}}{1-\frac{m}{2m+1}}$

62. $\dfrac{1-\frac{r}{2r+1}}{r-\frac{1}{2r+1}}$

TRY IT YOURSELF

Simplify each complex fraction.

63. $\dfrac{\dfrac{3y}{x} - y}{y - \dfrac{y}{x}}$

64. $\dfrac{\dfrac{y}{x} + 3y}{y + \dfrac{2y}{x}}$

65. $\dfrac{\dfrac{3}{x+1}}{5 + \dfrac{1}{x+1}}$

66. $\dfrac{\dfrac{1}{x-1}}{1 - \dfrac{1}{x-1}}$

67. $\dfrac{\dfrac{3x}{3x+2}}{\dfrac{x}{3x+2} + x}$

68. $\dfrac{2 - \dfrac{5}{x}}{7 + \dfrac{2}{x}}$

69. $\dfrac{\dfrac{1}{y} + 3}{\dfrac{3}{y} - 2}$

70. $\dfrac{\dfrac{1}{x-5}}{\dfrac{2}{x-5} - 1}$

71. $\dfrac{1}{\dfrac{1}{x} + \dfrac{1}{y}}$

72. $\dfrac{1}{\dfrac{b}{a} - \dfrac{a}{b}}$

73. $\dfrac{\dfrac{2}{x}}{\dfrac{2}{y} - \dfrac{4}{x}}$

74. $\dfrac{\dfrac{3x}{5}}{\dfrac{3x}{5} - \dfrac{6}{x}}$

75. $\dfrac{3 + \dfrac{3}{x-1}}{3 - \dfrac{3}{x}}$

76. $\dfrac{2 - \dfrac{2}{x+1}}{2 + \dfrac{2}{x}}$

77. $\dfrac{\dfrac{3}{x} + \dfrac{4}{x+1}}{\dfrac{2}{x+1} - \dfrac{3}{x}}$

78. $\dfrac{\dfrac{5}{y-3} - \dfrac{2}{y}}{\dfrac{1}{y} + \dfrac{2}{y-3}}$

79. $\dfrac{\dfrac{2}{x} - \dfrac{3}{x+1}}{\dfrac{2}{x+1} - \dfrac{3}{x}}$

80. $\dfrac{\dfrac{5}{y} + \dfrac{4}{y+1}}{\dfrac{4}{y} - \dfrac{5}{y+1}}$

81. $\dfrac{\dfrac{1}{y^2+y} - \dfrac{1}{xy+x}}{\dfrac{1}{xy+x} - \dfrac{1}{y^2+y}}$

82. $\dfrac{\dfrac{2}{b^2-1} - \dfrac{3}{ab-a}}{\dfrac{3}{ab-a} - \dfrac{2}{b^2-1}}$

APPLICATIONS

83. **GARDENING TOOLS** In the illustration in the next column, what is the result when the opening of the cutting blades is divided by the opening of the handles? Express the result in simplest form.

84. **EARNED RUN AVERAGE** The earned run average (ERA) is a statistic that gives the average number of earned runs a pitcher allows. For a softball pitcher, this is based on a six-inning game. The formula for ERA is

$$\text{ERA} = \dfrac{\dfrac{\text{earned runs}}{\text{innings pitched}}}{6}$$

Simplify the complex fraction on the right side of the equation.

85. **ELECTRONICS** In electronic circuits, resistors oppose the flow of an electric current. To find the total resistance of a parallel combination of two resistors, we can use the formula

$$\text{Total resistance} = \dfrac{1}{\dfrac{1}{R_1} + \dfrac{1}{R_2}}$$

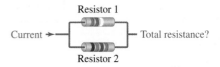

where R_1 is the resistance of the first resistor and R_2 is the resistance of the second. Simplify the complex fraction on the right side of the formula.

86. **DATA ANALYSIS** Use the data in the table to find the average measurement for the three-trial experiment.

	Trial 1	Trial 2	Trial 3
Measurement	$\dfrac{k}{2}$	$\dfrac{k}{3}$	$\dfrac{k}{2}$

WRITING

87. Explain how to use method 1 to simplify

$$\dfrac{1 + \dfrac{1}{x}}{3 - \dfrac{1}{x}}$$

88. Explain how to use method 2 to simplify the expression in Problem 87.

REVIEW

Write each expression as an expression involving only one exponent.

89. $t^3 t^4 t^2$

90. $(a^0 a^2)^3$

91. $-2r(r^3)^2$

92. $(s^3)^2(s^4)^0$

Write each expression without using parentheses or negative exponents.

93. $\left(\dfrac{3r}{4r^3}\right)^4$

94. $\left(\dfrac{12y^{-3}}{3y^2}\right)^{-2}$

95. $\left(\dfrac{6r^{-2}}{2r^3}\right)^{-2}$

96. $\left(\dfrac{4x^3}{5x^{-3}}\right)^{-2}$

Objectives

1. Solve rational equations.
2. Solve for a specified variable in a formula.

SECTION 7.6
Solving Rational Equations

In Chapter 2, we solved equations such as $\dfrac{1}{6}x + \dfrac{5}{2} = \dfrac{1}{3}$ by multiplying both sides by the LCD. With this approach, the equation that results is equivalent to the original equation, but easier to solve because it is cleared of fractions.

In this section, we will extend the fraction-clearing strategy to solve another type of equation, called a *rational equation*.

Rational Equations

A **rational equation** is an equation that contains one or more rational expressions.

Rational equations often have a variable in a denominator. Some examples are:

$$\dfrac{2}{3} = \dfrac{1}{3x} + \dfrac{3}{x} \qquad \dfrac{2}{x} + \dfrac{1}{4} = \dfrac{5}{2x} \qquad \dfrac{11x}{x-5} = 6 + \dfrac{55}{x-5}$$

1 Solve rational equations.

To **solve a rational equation**, we find all the values of the variable that make the equation true. Any value of the variable that makes a denominator in a rational equation equal to 0 cannot be a solution of the equation. Such a number must be rejected, because division by 0 is undefined.

The following steps can be used to solve rational equations.

Strategy for Solving Rational Equations

1. Determine which numbers cannot be solutions of the equation.
2. Multiply both sides of the equation by the LCD of all rational expressions in the equation. This clears the equation of fractions.
3. Solve the resulting equation.
4. Check all possible solutions in the original equation.

EXAMPLE 1

Solve: $\dfrac{x}{6} + \dfrac{5}{2} = \dfrac{1}{3}$

Strategy We will use the multiplication property of equality to clear this rational equation of fractions by multiplying both sides by the LCD.

WHY Equations that contain only integers are usually easier to solve than equations that contain fractions.

Solution
There are no restrictions on x, because no value of x ever makes a denominator 0. Since the denominators are 3, 6, and 2, we multiply both sides of the equation by the LCD, 6.

$\dfrac{x}{6} + \dfrac{5}{2} = \dfrac{1}{3}$ This is the equation to solve.

$6\left(\dfrac{x}{6} + \dfrac{5}{2}\right) = 6\left(\dfrac{1}{3}\right)$ Multiply both sides of the equation by the LCD of $\dfrac{x}{6}$, $\dfrac{5}{2}$, and $\dfrac{1}{3}$, which is 6.

$6 \cdot \dfrac{x}{6} + 6 \cdot \dfrac{5}{2} = 6 \cdot \dfrac{1}{3}$ Distribute the multiplication by 6.

$x + 15 = 2$ Do the multiplications.

$x + 15 - 15 = 2 - 15$ To undo the addition of 15, subtract 15 from both sides.

$x = -13$ Do the subtractions.

The solution is -13 and the solution set is $\{-13\}$. Verify this using a check.

Self Check 1

Solve: $\dfrac{2x}{3} = \dfrac{x}{6} + \dfrac{3}{2}$

Now Try Problem 25

This method can be used to solve rational equations. It is important to note that if we multiply both sides of an equation by an expression that involves a variable, we must check the apparent solutions.

EXAMPLE 2

Solve: $\dfrac{4}{x} + 1 = \dfrac{6}{x}$

Strategy This equation contains two rational expressions that have a variable in their denominator. We begin by asking, "What value(s) of x make either denominator 0?" Then we will use the multiplication property of equality to clear this rational equation of fractions by multiplying both sides by the LCD, x.

WHY If a number makes the denominator of a rational expression 0, that number cannot be a solution of the equation because division by 0 is undefined.

Solution
If x is 0, the denominators of $\dfrac{4}{x}$ and $\dfrac{6}{x}$ are 0 and the rational expressions would be undefined. Therefore, 0 cannot be a solution.

To clear the equation of fractions, we multiply both sides by the LCD of $\dfrac{4}{x}$ and $\dfrac{6}{x}$, which is x.

$\dfrac{4}{x} + 1 = \dfrac{6}{x}$ This is the equation to solve.

$x\left(\dfrac{4}{x} + 1\right) = x\left(\dfrac{6}{x}\right)$ Write each side of the equation within parentheses, and then multiply both sides by x.

$\overset{1}{\cancel{x}} \cdot \dfrac{4}{\underset{1}{\cancel{x}}} + x \cdot 1 = \overset{1}{\cancel{x}} \cdot \dfrac{6}{\underset{1}{\cancel{x}}}$ Distribute the multiplication by x.

$4 + x = 6$ Do each multiplication.

$x = 2$ Subtract 4 from both sides.

Self Check 2

Solve: $\dfrac{6}{x} - 1 = \dfrac{3}{x}$

Now Try Problem 31

The solution is 2 and the solution set is {2}. We verify this using a check.

Check: $\dfrac{4}{x} + 1 = \dfrac{6}{x}$ This is the original equation.

$\dfrac{4}{2} + 1 \stackrel{?}{=} \dfrac{6}{2}$ Substitute 2 for x.

$2 + 1 \stackrel{?}{=} 3$ Simplify.

$3 = 3$ True.

Self Check 3

Solve: $\dfrac{7}{6} - \dfrac{2r - 11}{r} = \dfrac{1}{r}$

Now Try Problem 33

EXAMPLE 3   Solve: $\dfrac{22}{5} - \dfrac{3a - 1}{a} = \dfrac{8}{a}$

Strategy This equation contains two rational expressions that have a variable in their denominator. We begin by asking, "What value(s) of a make either denominator 0?" Then we will use the multiplication property of equality to clear this rational equation of fractions by multiplying both sides by the LCD, $5a$.

WHY If a number makes the denominator of a rational expression 0, that number cannot be a solution of the equation because division by 0 is undefined.

Solution

We multiply both sides by $5a$, the LCD of the rational expressions in the equation.

$\dfrac{22}{5} - \dfrac{3a - 1}{a} = \dfrac{8}{a}$ This is the equation to solve.

$5a\left(\dfrac{22}{5} - \dfrac{3a - 1}{a}\right) = 5a\left(\dfrac{8}{a}\right)$ Write each side of the equation within parentheses, and then multiply both sides by $5a$.

$\overset{1}{\cancel{5}} \cdot a\left(\dfrac{22}{\cancel{5}}\right) - 5 \cdot \overset{1}{\cancel{a}}\left(\dfrac{3a - 1}{\cancel{a}}\right) = 5 \cdot \overset{1}{\cancel{a}}\left(\dfrac{8}{\cancel{a}}\right)$ Distribute the multiplication by $5a$.

$22a - 5(3a - 1) = 40$ Simplify. Note that $3a - 1$ must be written within parentheses.

$22a - 15a + 5 = 40$ Distribute the multiplication by -5.

$7a + 5 = 40$ Combine like terms: $22a - 15a = 7a$.

$7a = 35$ Subtract 5 from both sides.

$a = 5$ Divide both sides by 7.

The solution is 5 and the solution set is {5}. We verify this using a check.

Check: $\dfrac{22}{5} - \dfrac{3a - 1}{a} = \dfrac{8}{a}$ This is the original equation.

$\dfrac{22}{5} - \dfrac{3(5) - 1}{5} \stackrel{?}{=} \dfrac{8}{5}$ Substitute 5 for a.

$\dfrac{22}{5} - \dfrac{14}{5} \stackrel{?}{=} \dfrac{8}{5}$

$\dfrac{8}{5} = \dfrac{8}{5}$ True.

Caution! After multiplying both sides by the LCD and simplifying, the equation should not contain any fractions. If it does, check for an algebraic error, or perhaps your LCD is incorrect.

EXAMPLE 4

Solve: $\dfrac{x+2}{x+3} + \dfrac{1}{x^2+2x-3} = 1$

Strategy We will multiply both sides by the LCD of the two rational expressions in the equation. To find the LCD, we must factor the second denominator.

WHY To find the restrictions of the variable and to find the LCD, we need to write each denominator in factored form.

Solution
To find the LCD, we must factor the second denominator.

$\dfrac{x+2}{x+3} + \dfrac{1}{x^2+2x-3} = 1$ This is the given equation.

$\dfrac{x+2}{x+3} + \dfrac{1}{(x+3)(x-1)} = 1$ Factor $x^2 + 2x - 3$.

We see that -3 and 1 cannot be solutions of the equation.
   To clear the equation of fractions, we multiply both sides by the LCD, which is $(x+3)(x-1)$.

$(x+3)(x-1)\left[\dfrac{x+2}{x+3} + \dfrac{1}{(x+3)(x-1)}\right] = (x+3)(x-1)1$

Next, we distribute the multiplication by $(x+3)(x-1)$.

$(x+3)(x-1)\dfrac{x+2}{x+3} + (x+3)(x-1)\dfrac{1}{(x+3)(x-1)} = (x+3)(x-1)1$

$(x-1)(x+2) + 1 = (x+3)(x-1)$	Simplify.
$x^2 + x - 2 + 1 = x^2 + 2x - 3$	Multiply the pairs of binomials.
$x^2 + x - 1 = x^2 + 2x - 3$	Combine like terms.
$x - 1 = 2x - 3$	Subtract x^2 from both sides.
$-x - 1 = -3$	Subtract $2x$ from both sides.
$-x = -2$	Add 1 to both sides.
$x = 2$	Multiply (or divide) both sides by -1.

The solution is 2 and the solution set is {2}. Verify this using a check.

Self Check 4

Solve: $\dfrac{1}{x+3} + \dfrac{1}{x-3} = \dfrac{10}{x^2-9}$

Now Try Problem 39

EXAMPLE 5

Solve: $\dfrac{4}{5} + y = \dfrac{4y-50}{5y-25}$

Strategy We will multiply both sides by the LCD of the two rational expressions in the equation. To find the LCD, we must factor the second denominator.

WHY To find the restrictions of the variable and to find the LCD, we need to write each denominator in factored form.

Solution
To find the LCD, we must factor the second denominator.

$\dfrac{4}{5} + y = \dfrac{4y-50}{5y-25}$ This is the given equation.

$\dfrac{4}{5} + y = \dfrac{4y-50}{5(y-5)}$ Factor $5y - 25$.

We see that 5 cannot be a solution of the equation. To clear the equation of fractions, we multiply both sides by the LCD, which is $5(y-5)$.

Self Check 5

Solve: $\dfrac{x-6}{3x-9} - \dfrac{1}{3} = \dfrac{x}{2}$

Now Try Problem 47

$$5(y-5)\left[\frac{4}{5}+y\right] = 5(y-5)\left[\frac{4y-50}{5(y-5)}\right] \quad \text{Multiply both sides by the LCD, which is } 5(y-5).$$

$$\overset{1}{\cancel{5}}(y-5)\left(\frac{4}{\cancel{5}}\right) + 5(y-5)y = \overset{1}{\cancel{5}}\overset{1}{\cancel{(y-5)}}\left[\frac{4y-50}{\cancel{5}\cancel{(y-5)}}\right] \quad \text{Distribute } 5(y-5).$$

$$4(y-5) + 5y(y-5) = 4y - 50$$
$$4y - 20 + 5y^2 - 25y = 4y - 50 \quad \text{Distribute 4 and 5y.}$$
$$5y^2 - 25y - 20 = -50 \quad \text{Subtract 4y from both sides and rearrange terms.}$$
$$5y^2 - 25y + 30 = 0 \quad \text{Add 50 to both sides.}$$
$$y^2 - 5y + 6 = 0 \quad \text{Divide both sides by 5.}$$
$$(y-3)(y-2) = 0 \quad \text{Factor } y^2 - 5y + 6.$$
$$y - 3 = 0 \quad \text{or} \quad y - 2 = 0 \quad \text{Set each factor equal to zero.}$$
$$y = 3 \quad | \quad y = 2 \quad \text{Solve each equation.}$$

The solutions are 3 and 2 and the solution set is {2, 3}. Verify that both satisfy the original equation.

Success Tip Don't confuse procedures. To simplify the expression $\frac{4}{5} + y$, we build each fraction to have the LCD 5, add the numerators, and write the sum over the LCD. To solve the equation $\frac{4}{5} + y = \frac{4y-50}{5y-25}$ we multiply both sides by the LCD $5y - 25$ to eliminate the denominators.

Self Check 6

Solve: $\dfrac{x+5}{x-2} = \dfrac{7}{x-2}$

Now Try Problem 51

EXAMPLE 6 Solve: $\dfrac{x+3}{x-1} = \dfrac{4}{x-1}$

Strategy We will multiply both sides of the equation by the LCD, $x - 1$.

WHY This will clear the equation of fractions.

Solution
If $x = 1$, the denominators of $\dfrac{x+3}{x-1}$ and $\dfrac{4}{x-1}$ are 0, and the rational expressions are undefined. Therefore, 1 cannot be a solution of the equation.

To clear the equation of fractions, we multiply both sides by the LCD, which is $x - 1$.

$$\frac{x+3}{x-1} = \frac{4}{x-1} \quad \text{This is the equation to solve.}$$

$$(x-1)\frac{x+3}{x-1} = (x-1)\frac{4}{x-1} \quad \text{Multiply both sides by } x - 1.$$

$$x + 3 = 4 \quad \text{Simplify.}$$

$$x = 1 \quad \text{Subtract 3 from both sides.}$$

Earlier, we determined that 1 makes both denominators in the original equation 0. Therefore 1 cannot be a solution. Since 1 is the only possible solution, and it must be rejected, it follows that $\dfrac{x+3}{x-1}$ and $\dfrac{4}{x-1}$ has *no solution*. The solution set is written as { } or ∅.

Here is how a check of the possible solution, 1, would look.

Check: $\dfrac{x+3}{x-1} = \dfrac{4}{x-1}$ This is the original equation.

$\dfrac{1+3}{1-1} \stackrel{?}{=} \dfrac{4}{1-1}$ Substitute 1 for x.

$\dfrac{4}{0} \stackrel{?}{=} \dfrac{4}{0}$ Simplify.

When solving an equation, a possible solution that does not satisfy the original equation is called an **extraneous solution**. In this example, 1 is an extraneous solution.

> **The Language of Algebra** Extraneous means not a vital part. Mathematicians speak of extraneous solutions. Rock groups don't want extraneous sounds (like feedback) coming from their amplifiers. Artists erase extraneous marks on their sketches.

2 Solve for a specified variable in a formula.

Many formulas are equations that contain rational expressions. To solve these formulas for a specified variable, we use the same steps, in the same order, as we do when solving rational equations having only one variable.

EXAMPLE 7

The formula $\dfrac{1}{r} = \dfrac{1}{r_1} + \dfrac{1}{r_2}$ is used in electronics to calculate parallel resistances. Solve it for r.

Self Check 7
Solve the formula in Example 7 for r_1.

Now Try Problem 53

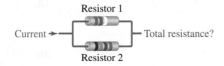

Strategy We will begin by multiplying both sides of the equation by the LCD, rr_1r_2.

WHY This will clear the equation of fractions.

Solution

$\dfrac{1}{r} = \dfrac{1}{r_1} + \dfrac{1}{r_2}$ This is the given formula.

$rr_1r_2\left(\dfrac{1}{r}\right) = rr_1r_2\left(\dfrac{1}{r_1} + \dfrac{1}{r_2}\right)$ Multiply both sides by rr_1r_2.

$\dfrac{rr_1r_2}{r} = \dfrac{rr_1r_2}{r_1} + \dfrac{rr_1r_2}{r_2}$ Distribute the multiplication by rr_1r_2.

$r_1r_2 = rr_2 + rr_1$ Simplify each fraction.

$r_1r_2 = r(r_2 + r_1)$ Factor out r.

$\dfrac{r_1r_2}{r_2 + r_1} = r$ To isolate r, divide both sides by $r_2 + r_1$.

or

$r = \dfrac{r_1r_2}{r_2 + r_1}$ Write the equation with r on the left side.

ANSWERS TO SELF CHECKS

1. 3 2. 3 3. 12 4. 5 5. 1, 2 6. No solution; 2 is extraneous. 7. $r_1 = \dfrac{rr_2}{r_2 - r}$

SECTION 7.6 STUDY SET

VOCABULARY

Fill in the blanks.

1. Equations that contain one or more rational expressions, such as $\dfrac{x+2}{x+3} + \dfrac{1}{x^2 + 2x - 3} = 1$ are called _____ _____.

2. To clear an equation of fractions, we multiply both sides by the _____ of the fractions in the equation.

3. If you multiply both sides of an equation by an expression that involves a variable, you must _____ the solution.

4. False solutions that result from multiplying both sides of an equation by a variable are called _____ solutions.

CONCEPTS

Use a check to determine whether 5 is a solution of the following equations.

5. $\dfrac{1}{x-1} = 1 - \dfrac{3}{x-1}$

6. $\dfrac{x}{x-5} = 3 + \dfrac{5}{x-5}$

By what should we multiply both sides of each equation to clear it of fractions?

7. $\dfrac{1}{x} + 2 = \dfrac{5}{x}$

8. $\dfrac{x}{x-2} - \dfrac{x}{x-1} = 5$

9. A student was asked to solve a rational equation. The first step of his solution is as follows:

$$12x\left(\dfrac{5}{x} + \dfrac{2}{3}\right) = 12x\left(\dfrac{7}{4x}\right)$$

   a. What equation was he asked to solve?
   b. What LCD is used to clear the equation of fractions?

10. Consider the rational equation $\dfrac{x}{x-3} = \dfrac{1}{x} + \dfrac{2}{x-3}$.

   a. What values of x make a denominator 0?
   b. What values of x make a rational expression undefined?
   c. What numbers can't be solutions of the equation?

11. Perform each multiplication.

   a. $4x\left(\dfrac{3}{4x}\right)$ b. $(x+6)(x-2)\left(\dfrac{3}{x-2}\right)$

12. Fill in the blanks.

$$\underbrace{8x\left(\dfrac{3}{4x}\right)}_{} = \underbrace{8x\left(\dfrac{1}{8x}\right)}_{} + \underbrace{8x\left(\dfrac{5}{4}\right)}_{}$$

$$ = + $$

By what should both sides of the equation be multiplied to clear it of fractions?

13. $\dfrac{2}{3} = \dfrac{1}{2} + \dfrac{x}{6}$

14. $\dfrac{7}{4} = \dfrac{x}{8} + \dfrac{5}{2}$

15. $\dfrac{5}{3} - \dfrac{5}{2} = \dfrac{5s}{4}$

16. $\dfrac{n}{9} - \dfrac{n}{6} = \dfrac{4n}{3}$

17. $\dfrac{1}{y} = 20 - \dfrac{5}{y}$

18. $\dfrac{x}{x^2 - 4} = \dfrac{4}{x-2}$

19. $\dfrac{x}{5} = \dfrac{3x}{10} + \dfrac{7}{2x}$

20. $\dfrac{2x}{x-6} = 4 + \dfrac{1}{x-6}$

NOTATION

Complete each step.

21. $\dfrac{2}{a} + \dfrac{1}{2} = \dfrac{7}{2a}$

$\left(\dfrac{2}{a} + \dfrac{1}{2}\right) = \left(\dfrac{7}{2a}\right)$

$ \cdot \dfrac{2}{a} + \cdot \dfrac{1}{2} = \cdot \dfrac{7}{2a}$

$ + a = 7$

$4 + a - = 7 - $

$a = 3$

7.6 Solving Rational Equations

22.
$$\frac{3}{5} + \frac{7}{a+2} = 2$$
$$\boxed{}\left(\frac{3}{5} + \frac{7}{a+2}\right) = \boxed{} \cdot 2$$
$$\boxed{} \cdot \frac{3}{5} + \boxed{} \cdot \frac{7}{a+2} = \boxed{} \cdot 2$$
$$3(a+2) + \boxed{} = 10(a+2)$$
$$3a + \boxed{} + 35 = 10a + \boxed{}$$
$$3a + \boxed{} = 10a + 20$$
$$-7a = \boxed{}$$
$$a = 3$$

23. The following work shows both sides of an equation being multiplied by the LCD to clear it of fractions. What was the original equation?

$$5(x-1)\left(\frac{3}{5}\right) + 5(x-1)\left(\frac{7}{x-1}\right) = 5(x-1) \cdot 2$$

24. After solving a rational equation, a student checked her answer and obtained the following:

$$\frac{-1}{0} + \frac{1}{0} = 0$$

What conclusion can be drawn?

GUIDED PRACTICE

Solve each equation and check the result. See Example 1.

25. $\frac{x}{2} + 4 = \frac{3x}{2}$ **26.** $\frac{2y}{5} - 8 = \frac{4y}{5}$

27. $\frac{x+1}{3} + \frac{x-1}{5} = \frac{2}{15}$

28. $\frac{3x-1}{6} - \frac{x+3}{2} = \frac{3x+4}{3}$

Solve each equation and check the result. See Example 2.

29. $\frac{3}{x} + 2 = 3$ **30.** $\frac{2}{x} + 9 = 11$

31. $\frac{3}{4h} + \frac{2}{h} = 1$ **32.** $\frac{5}{3k} + 2 = -\frac{1}{k}$

Solve each equation and check the result. See Example 3.

33. $\frac{15}{4} - \frac{a+3}{2a} = \frac{10}{2a}$ **34.** $\frac{7}{3} + \frac{a+1}{a} = \frac{11}{a}$

35. $\frac{3r}{2} - \frac{3}{r} = \frac{3r}{2} + 3$ **36.** $\frac{2p}{3} - \frac{1}{p} = \frac{2p-1}{3}$

Solve each equation and check the result. See Example 4.

37. $\frac{2}{y+1} + 5 = \frac{12}{y+1}$ **38.** $\frac{3}{p+6} - 2 = \frac{7}{p+6}$

39. $\frac{x+4}{x+3} - \frac{4x}{x^2+6x+9} = 1$

40. $\frac{1}{x^2+x-2} + \frac{x+1}{x+2} = 1$

Solve each equation and check the result. See Example 5.

41. $\frac{a}{4} - \frac{4}{a} = 0$ **42.** $0 = \frac{t}{3} - \frac{12}{t}$

43. $\frac{2x}{x^2+x-2} + \frac{2}{x+2} = 1$

44. $\frac{4x}{x^2+2x-3} + \frac{3}{x+3} = 1$

45. $\frac{3}{b-1} = \frac{2b}{b+4}$ **46.** $\frac{a}{a+4} = \frac{a-4}{6}$

47. $\frac{b+2}{b+3} + 1 = \frac{-7}{b-5}$

48. $1 - \frac{3}{b} = \frac{-8b}{b^2+3b}$

Solve each equation and check the result. See Example 6.

49. $\frac{2x+3}{x-2} = \frac{7}{x-2}$ **50.** $\frac{5}{a} - \frac{4}{a} = 8 + \frac{1}{a}$

51. $\frac{x}{x-5} - \frac{5}{x-5} = 3$ **52.** $\frac{3}{y-2} + 1 = \frac{3}{y-2}$

Solve each formula for the indicated variable. See Example 7.

53. $\frac{1}{a} + \frac{1}{b} = 1$ for a **54.** $\frac{1}{a} + \frac{1}{b} = 1$ for b

55. $I = \frac{E}{R+r}$ for r **56.** $h = \frac{2A}{b+d}$ for A

TRY IT YOURSELF

Solve each equation and check the result. If an equation has no solution, so indicate.

57. $\frac{3}{4} + \frac{x}{4} = \frac{2x+2}{x+3}$ **58.** $\frac{11}{b} + \frac{13}{b} = 12$

59. $\frac{1}{3} + \frac{2}{x-3} = 1$ **60.** $\frac{3}{5} + \frac{7}{x+2} = 2$

61. $\frac{z-4}{z-3} = \frac{z+2}{z+1}$ **62.** $\frac{a+2}{a+8} = \frac{a-3}{a-2}$

63. $\frac{v}{v+2} + \frac{1}{v-1} = 1$ **64.** $\frac{x}{x-2} = 1 + \frac{1}{x-3}$

65. $\dfrac{a^2}{a+2} - \dfrac{4}{a+2} = a$

66. $\dfrac{z^2}{z+1} + 2 = \dfrac{1}{z+1}$

67. $\dfrac{7}{q^2-q-2} + \dfrac{1}{q+1} = \dfrac{3}{q-2}$

68. $\dfrac{3}{x-1} - \dfrac{1}{x+9} = \dfrac{18}{x^2+8x-9}$

69. $\dfrac{u}{u-1} + \dfrac{1}{u} = \dfrac{u^2+1}{u^2-u}$

70. $\dfrac{3}{x-2} + \dfrac{1}{x} = \dfrac{2(3x+2)}{x^2-2x}$

71. $\dfrac{n}{n^2-9} + \dfrac{n+8}{n+3} = \dfrac{n-8}{n-3}$

72. $\dfrac{7}{x-5} - \dfrac{3}{x+5} = \dfrac{40}{x^2-25}$

73. $\dfrac{5}{x+4} + \dfrac{1}{x+4} = x-1$

74. $\dfrac{7}{x-3} + \dfrac{1}{x-3} = x-5$

75. $\dfrac{3}{x+1} - \dfrac{x-2}{2} = \dfrac{x-2}{x+1}$

76. $\dfrac{2}{x-1} + \dfrac{x-2}{3} = \dfrac{4}{x-1}$

77. $\dfrac{x-4}{x-3} + \dfrac{x-2}{x-3} = x-3$

78. $\dfrac{5}{4y+12} - \dfrac{3}{4} = \dfrac{5}{4y+12} - \dfrac{y}{4}$

79. $\dfrac{x}{x-1} - \dfrac{12}{x^2-x} = \dfrac{-1}{x-1}$

80. $y + \dfrac{2}{3} = \dfrac{2y-12}{3y-9}$

Solve each formula for the indicated variable or expression.

81. $\dfrac{a}{b} = \dfrac{c}{d}$ for d

82. $F = \dfrac{L^2}{6d} + \dfrac{d}{2}$ for L^2

APPLICATIONS

83. **OPTICS** The focal length f of a lens is given by the formula
$$\dfrac{1}{f} = \dfrac{1}{d_1} + \dfrac{1}{d_2}$$
where d_1 is the distance from the object to the lens and d_2 is the distance from the lens to the image. Solve the formula for f.

84. **OPTICS** Solve the formula in Problem 83 for d_1.

85. **MEDICINE** Radioactive tracers are used for diagnostic work in nuclear medicine. The **effective half-life** H of a radioactive material in an organism is given by the formula
$$H = \dfrac{RB}{R+B}$$
where R is the radioactive half-life and B is the biological half-life of the tracer. Solve the formula for R.

86. **CHEMISTRY** Charles's law describes the relationship between the volume and the temperature of a gas that is kept at a constant pressure. It states that as the temperature of the gas increases, the volume of the gas will increase:
$$\dfrac{V_1}{V_2} = \dfrac{T_1}{T_2}$$
Solve the formula for V_2.

WRITING

87. Why is it important to check your solutions of an equation that contains fractions with variables in the denominator?

88. Explain the difference between the procedures used to simplify $\dfrac{1}{x} + \dfrac{1}{3}$ and the procedure used to solve $\dfrac{1}{x} + \dfrac{1}{3} = \dfrac{1}{2}$.

REVIEW

Factor each expression.

89. $x^2 + 4x$

90. $x^2 - 16y^2$

91. $2x^2 + x - 3$

92. $6a^2 - 5a - 6$

93. $x^4 - 16$

94. $4x^2 + 10x - 6$

95. $a^2b^2 - a$

96. $m^3n^2 - m$

SECTION 7.7
Problem Solving Using Rational Equations

Objectives
1. Solve number problems.
2. Solve shared-work problems.
3. Solve uniform motion problems.
4. Solve investment problems.

1 Solve number problems.

We will now use the six-step problem-solving strategy to solve application problems. We begin with an example in which we find an unknown number.

EXAMPLE 1 **A Number Problem** If the same number is added to both the numerator and the denominator of the fraction $\frac{3}{5}$, the result is $\frac{4}{5}$. Find the number.

Analyze We are asked to find a number. If we add it to both the numerator and the denominator of $\frac{3}{5}$, we will get $\frac{4}{5}$.

Assign Let n = the unknown number.

Form We add n to both the numerator and the denominator of $\frac{3}{5}$. Then set the result equal to $\frac{4}{5}$ to get the equation

$$\frac{3+n}{5+n} = \frac{4}{5}$$

Solve To solve the equation, we proceed as follows:

$\frac{3+n}{5+n} = \frac{4}{5}$ This is the equation to solve.

$5(5+n)\frac{3+n}{5+n} = 5(5+n)\frac{4}{5}$ Multiply both sides by $5(5+n)$, which is the LCD of the fractions appearing in the equation.

$5(3+n) = (5+n)4$ Simplify.

$15 + 5n = 20 + 4n$ Distribute the multiplications by 5 and by 4.

$15 + n = 20$ Subtract $4n$ from both sides.

$n = 5$ Subtract 15 from both sides.

State The number is 5.

Check When we add 5 to both the numerator and denominator of $\frac{3}{5}$, we get

$$\frac{3+5}{5+5} = \frac{8}{10} = \frac{4}{5}$$

The result checks.

Self Check 1
If the same number is added to both the numerator and denominator of the fraction $\frac{7}{9}$, the result is $\frac{8}{9}$. Find the number.

Now Try Problem 11

2 Solve shared-work problems.

We can use rational equations to model shared-work problems. In this case, we assume that the work is being performed at a constant rate by all of those involved.

EXAMPLE 2 **Filling an Oil Tank** An inlet pipe can fill an oil tank in 7 days, and a second inlet pipe can fill the same tank in 9 days. If both pipes are used, how long will it take to fill the tank?

Analyze The key is to determine what each pipe can do in 1 day. If we add what the first pipe can do in 1 day to what the second pipe can do in 1 day, the sum is what they can do in 1 day, working together.

Chapter 7 Rational Expressions and Equations

Self Check 2

A school secretary can prepare a mass mailing of an informational flyer in 6 hours. A student worker would take 8 hours to prepare the mailing. How long will it take to prepare the mailing if they work together?

Now Try Problem 21

Since the first pipe can fill the tank in 7 days, it can do $\frac{1}{7}$ of the job in 1 day. Since the second pipe can fill the tank in 9 days, it can do $\frac{1}{9}$ of the job in 1 day. If it takes x days for both pipes to fill the tank, together they can do $\frac{1}{x}$ of the job in 1 day.

Assign Let $x =$ the number of days it will take to fill the tank if both inlet pipes are used.

Form

What the first inlet pipe can do in 1 day	plus	what the second inlet pipe can do in 1 day	equals	what they can do together in 1 day.
$\frac{1}{7}$	+	$\frac{1}{9}$	=	$\frac{1}{x}$

Solve To solve the equation, we proceed as follows:

$$\frac{1}{7} + \frac{1}{9} = \frac{1}{x} \quad \text{This is the equation to solve.}$$

$$63x\left(\frac{1}{7} + \frac{1}{9}\right) = 63x\left(\frac{1}{x}\right) \quad \text{Multiply both sides by 63x, which is the LCD, to clear the equation of fractions.}$$

$$63x\left(\frac{1}{7}\right) + 63x\left(\frac{1}{9}\right) = 63x\left(\frac{1}{x}\right) \quad \text{Distribute the multiplication by 63x.}$$

$$9x + 7x = 63 \quad \text{Do the multiplications.}$$

$$16x = 63 \quad \text{Combine like terms.}$$

$$x = \frac{63}{16} \quad \text{Divide both sides by 16.}$$

State If both inlet pipes are used, it will take $\frac{63}{16}$ or $3\frac{15}{16}$ days to fill the tank.

Check In $\frac{63}{16}$ days, the first pipe fills $\frac{1}{7}\left(\frac{63}{16}\right)$ of the tank and the second pipe fills $\frac{1}{9}\left(\frac{63}{16}\right)$ of the tank. The sum of these efforts, $\frac{9}{16} + \frac{7}{16}$, is equal to one full tank.

3 Solve uniform motion problems.

Recall that we use the distance formula $d = rt$ to solve motion problems. The relationship between distance, rate, and time can be expressed in another way by solving for t.

$$d = rt \quad \text{Distance} = \text{rate} \cdot \text{time.}$$

$$\frac{d}{r} = \frac{rt}{r} \quad \text{To undo the multiplication by } r \text{ and isolate } t, \text{ divide both sides by } r.$$

$$\frac{d}{r} = t \quad \text{Simplify the right side by removing the common factor of } r: \frac{\overset{1}{\cancel{r}}t}{\underset{1}{\cancel{r}}} = t.$$

$$t = \frac{d}{r}$$

Self Check 3

A cyclist can ride 24 miles in the same amount of time that her friend can walk 8 miles. If the cyclist travels 8 mph faster than the walker, find the speed of the walker.

Now Try Problem 29

EXAMPLE 3 *Track and Field* A coach can run 10 miles in the same amount of time as his best student-athlete can run 12 miles. If the student can run 1 mile per hour faster than the coach, how fast can the student run?

Analyze
- The coach runs 10 miles in the same time that the student runs 12 miles.
- The student runs 1 mph faster than the coach.
- Find the speed that each runs.

Assign Since the student's speed is 1 mph faster than the coach's, let $r =$ the speed that the coach can run. Then, $r + 1 =$ the speed that the student can run.

Form The expressions for the rates are entered in the Rate column of the table. The distances run by the coach and by the student are entered in the Distance column of the table.

Using $t = \frac{d}{r}$, we find that the time it takes the coach to run 10 miles, at a rate of r mph, is $\frac{10}{r}$ hours. Similarly, we find that the time it takes the student to run 12 miles, at a rate of $(r + 1)$ mph, is $\frac{12}{r+1}$ hours. These expressions are entered in the Time column of the table.

	Rate	· Time	= Distance
Coach	r	$\frac{10}{r}$	10
Student	$r + 1$	$\frac{12}{r+1}$	12

Enter the information in these two columns first.

To get these entries, divide the distance by the rate to obtain an expression for the time: $t = \frac{d}{r}$.

The time it takes the coach to run 10 miles **is the same time** the time it takes the student to run 12 miles.

$$\frac{10}{r} = \frac{12}{r+1}$$

Solve We can solve the equation as follows:

$$\frac{12}{r+1} = \frac{10}{r} \quad \text{This is the equation to solve.}$$

$$r(r+1)\frac{12}{r+1} = r(r+1)\frac{10}{r} \quad \text{Multiply both sides by } r(r+1).$$

$$12r = 10(r+1) \quad \text{Simplify.}$$
$$12r = 10r + 10 \quad \text{Distribute the multiplication by 10.}$$
$$2r = 10 \quad \text{Subtract } 10r \text{ from both sides.}$$
$$r = 5 \quad \text{Divide both sides by 2.}$$

State The coach can run 5 mph. The student, running 1 mph faster, can run 6 mph.

Check If the coach runs at a rate of **5** mph for $\frac{10}{5} = 2$ hrs, he will travel a distance of $5 \cdot 2 = 10$ miles. If the student runs at a rate of 6 mph for $\frac{12}{5+1} = \frac{12}{6} = 2$ hrs, he will travel a distance of $6 \cdot 2 = 12$ miles. These results check.

4 Solve investment problems.

EXAMPLE 4 *Comparing Investments* An amount of money invested for 1 year in bonds will earn $120. At a bank, that same amount of money will only earn $75 interest, because the interest rate paid by the bank is 3% less than that paid by the bonds. Find the rate of interest paid by each investment.

Analyze
- The investment in bonds earns $120 in 1 year.
- The same amount of money, invested in a bank, earns $75 in 1 year.
- The interest rate paid by the bank is 3% less than that paid by the bonds.
- Find the bond's rate of interest and the bank's rate of interest.

Assign Since the interest rate paid by the bank is 3% less than that paid by the bonds, let $r =$ the bond's rate of interest, and $r - 0.03 =$ the bank's interest rate. (Recall that $3\% = 0.03$.)

Self Check 4
An amount of money invested for 1 year in a certificate of deposit will earn $210. The same amount of money in a savings account will earn $70. If the certificate of deposit's interest rate is 2% more than the savings account's rate, find the interest rate of the savings account.

Now Try Problem 41

Form If an investment earns $120 interest in 1 year at some rate r, we can use $P = \frac{I}{rt}$ to find that the principal invested was $\frac{120}{r}$ dollars. Similarly, if another investment earns $75 interest in 1 year at some rate $r - 0.03$, the principal invested was $\frac{75}{r - 0.03}$ dollars. We can organize the facts of the problem in a table.

	Principal ·	Rate ·	Time =	Interest
Bonds	$\frac{120}{r}$	r	1	120
Bank	$\frac{75}{r - 0.03}$	$r - 0.03$	1	75

Divide to get each of these entries: $P = \frac{I}{rt}$. Enter the information first.

The amount invested in the bonds equals the amount invested in the bank.

$$\frac{120}{r} = \frac{75}{r - 0.03}$$

Solve

$$\frac{120}{r} = \frac{75}{r - 0.03}$$ This is a rational equation.

$$r(r - 0.03)\left(\frac{120}{r}\right) = r(r - 0.03)\left(\frac{75}{r - 0.03}\right)$$ Multiply both sides by the LCD, $r(r - 0.03)$. Then remove common factors of the numerator and denominator.

$(r - 0.03)120 = 75r$ Simplify. The fractions have been cleared.

$120r - 3.6 = 75r$ On the left side, distribute the multiplication by 120.

$45r - 3.6 = 0$ To isolate the variable term on the left side, subtract $75r$ from both sides.

$45r = 3.6$ To undo the subtraction of 3.6, add 3.6 to both sides.

$r = 0.08$ To undo the multiplication by 45 and isolate r, divide both sides by 45.

If $r = 0.08$, then the bank's interest rate is given by $r - 0.03 = 0.05$.

State The bonds pay 0.08, or 8%, interest. The bank's interest rate is 5%.

Check The amount invested at 8% that will earn $120 interest in 1 year is $\frac{120}{(0.08)1} = \$1,500$. The amount invested at 5% that will earn $75 interest in 1 year is $\frac{75}{(0.05)1} = \$1,500$. The amounts invested in the bonds and the bank are the same. The results check.

ANSWERS TO SELF CHECKS

1. 9   2. $3\frac{3}{7}$ hr   3. 4 mph   4. 1%

SECTION 7.7 STUDY SET

VOCABULARY

Fill in the blanks.

1. Problems that involve people or things completing jobs are called shared-_____ problems.

2. Problems that involve moving vehicles are called uniform _____ problems.

3. Problems that involve depositing money are called _____ problems.

4. The amount of money invested is called the _____.

CONCEPTS

5. Write the formula that relates distance, rate, and time.
6. Write the formula that relates interest, principal, rate, and time.

NOTATION

7. Write $\frac{55}{9}$ days as a mixed number.
8. Write $32\frac{2}{5}$ days as an improper fraction.
9. Write 9% as a decimal.
10. Write 0.035 as a percent.

GUIDED PRACTICE

Solve each number problem. See Example 1.

11. If the same number is added to both the numerator and the denominator of $\frac{2}{5}$, the result is $\frac{2}{3}$. Find the number.
12. If the same number is subtracted from both the numerator and the denominator of $\frac{11}{13}$, the result is $\frac{3}{4}$. Find the number.
13. If the denominator of $\frac{3}{4}$ is increased by a number and the numerator of the fraction is doubled, the result is 1. Find the number.
14. If a number is added to the numerator of $\frac{7}{8}$ and the same number is subtracted from the denominator, the result is 2. Find the number.
15. If a number is added to the numerator of $\frac{3}{4}$ and twice the number is added to the denominator, the result is $\frac{4}{7}$. Find the number.
16. If a number is added to the numerator of $\frac{5}{7}$ and twice the number is subtracted from the denominator, the result is 8. Find the number.
17. The sum of a number and its reciprocal is $\frac{13}{6}$. Find the number and its reciprocal.
18. The sum of the reciprocals of two consecutive even integers is $\frac{7}{24}$. Find the integers. (*Hint:* Let $x =$ the first integer and $x + 2 =$ the second integer.)

APPLICATIONS

See Example 1.

19. COOKING If the same number is added to both the numerator and the denominator of the amount of butter used in the following recipe, the result is the amount of brown sugar to be used. Find the number.

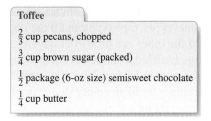

Toffee
$\frac{2}{3}$ cup pecans, chopped
$\frac{3}{4}$ cup brown sugar (packed)
$\frac{1}{2}$ package (6-oz size) semisweet chocolate
$\frac{1}{4}$ cup butter

20. TAPE MEASURES If the same number is added to both the numerator and the denominator of the first measurement, the result is the second measurement. Find the number.

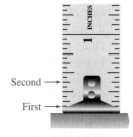

See Example 2.

21. Suppose you are a recreation director at a summer camp. The water in the camp swimming pool was drained out for the winter and it is now time to refill the pool. One pipe can fill the empty pool in 12 hours and another can fill the empty pool in 18 hours. Suppose both pipes are opened at 8:00 A.M. and you have scheduled a swimming activity for 2:00 P.M. that day. Will the pool be filled by then?

from Campus to Careers
Recreation Director

22. ROOFING HOUSES A homeowner estimates that it will take him 7 days to roof his house. A professional roofer estimates that he could roof the house in 4 days. How long will it take if the homeowner helps the roofer?
23. HOLIDAY DECORATING One crew can put up holiday decorations in the mall in 8 hours. If a slower crew can put up the decorations in 10 hours, how long will it take if both crews work together?
24. GROUNDSKEEPING It takes a groundskeeper 45 minutes to prepare a softball field for a game. If it takes his assistant 55 minutes to prepare the same field, how long will it take if they work together?
25. FILLING POOLS One inlet pipe can fill an empty pool in 4 hours, and a drain can empty the pool in 8 hours. How long will it take the pipe to fill the pool if the drain is left open?

26. **SEWAGE TREATMENT** A sludge pool is filled by two inlet pipes. One pipe can fill the pool in 15 days, and the other can fill it in 21 days. However, if no sewage is added, continuous waste removal will empty the pool in 36 days. How long will it take the two inlet pipes to fill an empty sludge pool?

27. **GRADING PAPERS** On average, it takes a teacher 30 minutes to grade a set of quizzes. If it takes her aide twice as long to grade the quizzes, how long will it take if they work together?

28. **DOG KENNELS** It takes the owner of a dog kennel 6 hours to clean all of the cages. If it takes her assistant 2 hours longer, how long will it take if they work together?

See Example 3.

29. **CYCLING** Maurice Garin of France won the first Tour de France road race in 1903. Chris Froome of Great Britain won the Tour de France in 2013. Froome's average speed in 2013 was 10 mph faster than Garin's in 1903. In the time it took Garin to ride 80 miles, Froome could have ridden 130 miles. Find the average speed of each cyclist.

30. **PHYSICAL FITNESS** A woman can bike 28 miles in the same time it takes her to walk 8 miles. If she can bike 10 mph faster than she can walk, how fast can she walk?

31. **PACKAGING FRUIT** The illustration shows how apples are processed for market. Although the second conveyor belt is shorter, an apple spends the same amount of time on each belt because the second conveyor belt moves 1 foot per second slower than the first. Determine the speed of each conveyor belt.

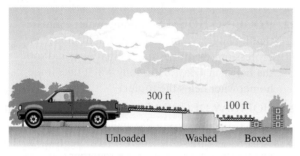

32. **COMPARING TRAVEL** A plane can fly 300 miles in the same time as it takes a car to go 120 miles. If the car travels 90 mph slower than the plane, find the speed of the plane.

33. **BIRDS IN FLIGHT** On average, a Canada goose can fly 10 mph faster than a great blue heron. In the same time that a Canada goose can fly 120 miles, a heron can fly 80 miles. Find the flying speed of each bird.

34. **MUSCLE CARS** The top speed of a Dodge Charger SRT8 is 33 mph less than the top speed of a Chevrolet Corvette Z06. At top speeds, a Corvette can travel 6 miles in the same time that a Charger can travel 5 miles. Find the speed of each car.

35. **TRAVEL TIMES** A company president flew 680 miles one way in the corporate jet but returned in a smaller plane that could fly only half as fast. If the total travel time was 6 hours, find the speed of each plane.

36. **TRAVEL TIMES** A car can travel 280 miles in the same time as a truck can travel 240 miles. If the car travels at a speed that is 10 mph more than the speed of the truck, find the speeds of the car and the truck.

37. **WIND SPEED** When a plane flies downwind, the wind pushes the plane so that its speed is the *sum* of the speed of the plane in still air and the speed of the wind. Traveling upwind, the wind pushes against the plane so that its speed is the *difference* of the speed of the plane in still air and the speed of the wind. Suppose a plane that can travel 255 mph in still air can travel 300 miles downward in the *same time* as it takes to travel 210 miles upwind. Complete the following table and find the speed of the wind, represented by x.

	Rate · Time = Distance		
Downwind	$255 + x$		300
Upwind	$255 - x$		210

38. **BOATING** A boat that travels 18 mph in still water can travel 22 miles downstream in the same time as it takes to travel 14 miles upstream. Find the speed of the current in the river.

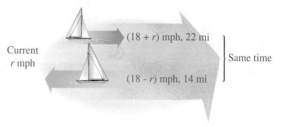

39. **RIVER TOURS** A river boat tour begins by going 60 miles upstream against a 5-mph current. There, the boat turns around and returns with the current. What

still-water speed should the captain use to complete the tour in 5 hours?

40. **BOAT TRAVEL** A boat can cruise at 12 mph in still water. If it cruises 6 miles upstream on the Mississippi River in the same amount of time it cruises 10 miles downstream, find the current in that portion of the river.

See Example 4.

41. **COMPARING INVESTMENTS** An amount of money invested for 1 year in tax-free bonds will earn $300. In a certain credit union account, that same amount of money will only earn $200 interest in a year, because the interest paid is 2% less than that paid by the bonds. Find the rate of interest paid by each investment.

42. **COMPARING INVESTMENTS** An amount of money invested for 1 year in a savings account will earn $1,500. That same amount of money invested in a mini-mall development will earn $6,500 interest in a year, because the interest paid is 10% more than that paid by the savings account. Find the rate of interest paid by each investment.

43. **COMPARING INVESTMENTS** Two certificates of deposit (CDs) pay interest at rates that differ by 1%. Money invested for 1 year in the first CD earns $175 interest. The same principal invested in the second CD earns $200. Find the two rates of interest.

44. **COMPARING INTEREST RATES** Two bond funds pay interest at rates that differ by 2%. Money invested for 1 year in the first fund earns $315 interest. The same amount invested in the second fund earns $385. Find the lower rate of interest.

45. **SALES** A dealer bought some radios for a total of $1,200. She gave away 6 radios as gifts, sold the rest for $10 more than she paid for each radio, and broke even. How many radios did she buy?

46. **FURNACE REPAIR** A repairman purchased several furnace blower motors for a total cost of $210. If his cost per motor had been $5 less, he could have purchased one additional motor. How many motors did he buy at the regular rate?

WRITING

In Example 2, one inlet pipe could fill an oil tank in 7 days, and another could fill the same tank in 9 days. We were asked to find how long it would take if both pipes were used. Explain why each of the approaches in Problems 47–49 is incorrect.

47. The time it would take to fill the tank is the *sum* of the lengths of time it takes each pipe to fill the tank:

    7 days + 9 days = 16 days.

48. The time it would take to fill the tank is the *difference* in the lengths of time it takes each pipe to fill the tank:

    9 days − 7 days = 2 days.

49. The time it would take to fill the tank is the *average* of the lengths of time it takes each pipe to fill the tank:

    $$\frac{7 \text{ days} + 9 \text{ days}}{2} = \frac{16 \text{ days}}{2} = 8 \text{ days}$$

50. Write a shared-work problem that can be modeled by the equation $\frac{1}{3} + \frac{1}{4} = \frac{1}{x}$.

REVIEW

51. Solve using substitution: $\begin{cases} x + y = 4 \\ y = 3x \end{cases}$

52. Solve using addition (elimination): $\begin{cases} 5x - 4y = 19 \\ 3x + 2y = 7 \end{cases}$

53. Use a check to determine whether $\frac{21}{5}$ is a solution of $x + 20 = 4x - 1 + 2x$.

54. Solve: $4x^2 + 8x = 0$

55. Evaluate $2x^2 + 5x - 3$ for $x = -3$.

56. Solve $T - R = ma$ for R.

SECTION 7.8
Proportions and Similar Triangles

Objectives
1. Write ratios and rates in simplest form.
2. Solve proportions.
3. Use proportions to solve problems.
4. Use proportions to solve problems that involve similar triangles.

In this section, we will discuss a problem-solving tool called a *proportion*. A proportion is a type of rational equation that involves two *ratios* or two *rates*.

1 Write ratios and rates in simplest form.

Ratios are used to compare two numbers or two quantities measured in the same units. Here are some examples.

- To prepare fuel for a lawnmower, gasoline must be mixed with oil in a 50-to-1 ratio.
- Gold is combined with other metals in the ratio of 14 to 10 to make 14-karat jewelry.
- The ratio of the width to the height of an HDTV screen is 16 to 9.

Ratios

A **ratio** is the quotient of two numbers or the quotient of two quantities that have the same units.

There are three common ways to write a ratio: with the word *to*, with a colon, or as a fraction. For example, the ratio comparing the width to the height of an HDTV screen can be written as

$$16 \text{ to } 9, \quad 16:9, \quad \text{or} \quad \frac{16}{9}$$

Each of these forms can be read as "the ratio of 16 to 9."

Self Check 1

Translate each phrase into a ratio written in fractional form:

a. The ratio of 15 to 2
b. 12 hours to 2 days

Now Try Problems 19 and 25

EXAMPLE 1 Translate each phrase into a ratio written in fractional form:

a. The ratio of 5 to 9 b. 12 ounces to 2 pounds

Strategy To translate, we need to identify the number (or quantity) before the word *to* and the number (or quantity) after it.

WHY The number before the word *to* is the numerator of the ratio and the number after it is the denominator.

Solution

a. The ratio of 5 *to* 9 is written $\frac{5}{9}$ (numerator/denominator).

b. To write a ratio of two quantities with the same units, we must express 2 pounds in terms of ounces. Since 1 pound = 16 ounces, 2 pounds = 32 ounces. The ratio of 12 ounces to 32 ounces can be simplified so that no units appear in the final form.

$$\frac{12 \text{ ounces}}{32 \text{ ounces}} = \frac{3 \cdot \cancel{4} \; \cancel{\text{ounces}}}{\cancel{4} \cdot 8 \; \cancel{\text{ounces}}} = \frac{3}{8}$$

The Language of Algebra A ratio that is the quotient of two quantities having the same units should be simplified so that no units appear in the final answer.

When quotients are used to compare quantities with different units, they are called *rates*. For example, if the 495-mile drive from New Orleans to Dallas takes 9 hours, the average rate of speed is the quotient of the miles driven to the length of time the trip takes.

$$\text{Average rate of speed} = \frac{495 \text{ miles}}{9 \text{ hours}} = \frac{\cancel{9} \cdot 55 \text{ miles}}{\cancel{9} \cdot 1 \text{ hour}} = \frac{55 \text{ miles}}{1 \text{ hour}}$$

Rates

A **rate** is a quotient of two quantities that have different units.

7.8 Proportions and Similar Triangles

THINK IT THROUGH *Student Loan Calculations*

"Nearly half of high school seniors in the U.S. don't know how much money they will need for college, and even more don't understand basic student loan terms, according to the Credit Union National Association's first annual High School Student Borrowing Survey."

cuna.org

Many student loan programs calculate a *debt-to-income ratio* to assist them in determining whether the borrower has sufficient income to repay the loan. A debt-to-income ratio compares an applicant's monthly debt payments (mortgages, credit cards, auto loans, etc.) to their gross monthly income. Most education lenders require borrower debt-to-income ratios of $\frac{2}{5}$ or less, according to the Nellie Mae Debt Management Edvisor. Calculate the debt-to-income ratio for each loan applicant shown below. Then determine whether it makes them eligible for a student loan.

	Applicant #1	Applicant #2	Applicant #3
Monthly debt payments	$250	$1,000	$1,200
Gross monthly income	$1,000	$2,000	$3,000
Debt-to-income ratio			
Is the ratio $\leq \frac{2}{5}$?			

2 Solve proportions.

Consider the following table, in which we are given the costs of various numbers of gallons of gasoline.

Number of gallons	Cost
2	$7.50
5	$18.75
8	$30.00
12	$45.00

If we divide the costs by the numbers of gallons purchased, we see that the quotients are equal. In this example, each quotient represents the cost of 1 gallon of gasoline, which is $3.75.

$$\frac{\$7.50}{2} = \$3.75, \quad \frac{\$18.75}{5} = \$3.75, \quad \frac{\$30.00}{8} = \$3.75, \quad \text{and} \quad \frac{\$45.00}{12} = \$3.75$$

When two ratios or rates $\left(\text{such as } \frac{\$18.75}{5} \text{ and } \frac{\$7.50}{2}\right)$ are equal, they form a *proportion*.

Proportions

A **proportion** is a mathematical statement that two ratios or two rates are equal.

Some examples of proportions are

$$\frac{1}{2} = \frac{3}{6}, \qquad \frac{3 \text{ waiters}}{7 \text{ tables}} = \frac{9 \text{ waiters}}{21 \text{ tables}}, \qquad \text{and} \qquad \frac{a}{b} = \frac{c}{d}$$

- The proportion $\frac{1}{2} = \frac{3}{6}$ can be read as "1 is to 2 as 3 is to 6."
- The proportion $\frac{3 \text{ waiters}}{7 \text{ tables}} = \frac{9 \text{ waiters}}{21 \text{ tables}}$ can be read as "3 waiters is to 7 tables as 9 waiters is to 21 tables."
- The proportion $\frac{a}{b} = \frac{c}{d}$ can be read as "a is to b as c is to d."

In the proportion $\frac{a}{b} = \frac{c}{d}$, a and d are called the **extremes**, and b and c are called the **means**. We can show that the product of the extremes (ad) is equal to the product of the means (bc) by multiplying both sides of the proportion by bd and observing that $ad = bc$.

$$\frac{a}{b} = \frac{c}{d}$$

$$\overset{1}{\cancel{b}d} \cdot \frac{a}{\cancel{b}} = b\overset{1}{\cancel{d}} \cdot \frac{c}{\cancel{d}} \qquad \text{To clear the equation of fractions, multiply both sides by the LCD, which is } bd.$$

$$ad = bc \qquad \text{Do each multiplication and simplify.}$$

Since $ad = bc$, the product of the extremes equals the product of the means.

The Fundamental Property of Proportions

In a proportion, the product of the extremes is equal to the product of the means. If $\frac{a}{b} = \frac{c}{d}$, then $ad = bc$, and if $ad = bc$, then $\frac{a}{b} = \frac{c}{d}$.

To determine whether an equation is a proportion, we can check to see whether the product of the extremes is equal to the product of the means.

Self Check 2

Determine whether the equation is a proportion: $\frac{6}{13} = \frac{24}{53}$

Now Try Problems 33 and 35

EXAMPLE 2 Determine whether each equation is a proportion:

a. $\frac{3}{7} = \frac{9}{21}$ **b.** $\frac{8}{3} = \frac{13}{5}$

Strategy We will check to see whether the product of the extremes is equal to the product of the means.

WHY If the product of the extremes equals the product of the means, the equation is a proportion. If the cross products are not equal, the equation is not a proportion.

Solution

In each case, we check to see whether the product of the extremes is equal to the product of the means.

a. The product of the extremes is $3 \cdot 21 = 63$. The product of the means is $7 \cdot 9 = 63$. Since the products are equal, the equation is a proportion: $\frac{3}{7} = \frac{9}{21}$.

$$3 \cdot 21 = 63 \qquad 7 \cdot 9 = 63$$

$$\frac{3}{7} = \frac{9}{21} \qquad \text{The product of the extremes and the product of the means are also known as cross products.}$$

b. The product of the extremes is $8 \cdot 5 = 40$. The product of the means is $3 \cdot 13 = 39$. Since the cross products are not equal, the equation is not a proportion: $\frac{8}{3} \neq \frac{13}{5}$.

$$8 \cdot 5 = 40 \qquad 3 \cdot 13 = 39$$

$$\frac{8}{3} \times \frac{13}{5} \qquad \text{One cross product is 40 and the other is 39.}$$

Suppose that we know three terms in the proportion

$$\frac{x}{5} = \frac{24}{20}$$

To find the unknown term, we can multiply both sides of the equation by 20 to clear it of fractions, and then solve for x. However, with proportions, it is often easier to simply compute the cross products, set them equal, and solve for the variable.

$$\frac{x}{5} = \frac{24}{20}$$

$20 \cdot x = 5 \cdot 24$ In a proportion, the product of the extremes equals the product of the means.

$20x = 120$ Do the multiplication: $5 \cdot 24 = 120$.

$\dfrac{20x}{20} = \dfrac{120}{20}$ To undo the multiplication by 20, divide both sides by 20.

$x = 6$ Do the divisions.

Thus, x is 6. To check this result, we substitute 6 for x in $\dfrac{x}{5} = \dfrac{24}{20}$ and find the cross products.

$$\frac{6}{5} \stackrel{?}{=} \frac{24}{20} \qquad \begin{array}{l} 6 \cdot 20 = 120 \\ 5 \cdot 24 = 120 \end{array}$$

Since the cross products are equal, this is a proportion. The result, 6, is correct.

EXAMPLE 3   Solve: $\dfrac{12}{18} = \dfrac{3}{x}$

Strategy To solve for x, we will set the cross products equal.

WHY Since the equation is a proportion, the product of the means equals the product of the extremes.

Solution

$\dfrac{12}{18} = \dfrac{3}{x}$ This is the given proportion.

$12 \cdot x = 18 \cdot 3$ In a proportion, the product of the extremes equals the product of the means.

$12x = 54$ Multiply: $18 \cdot 3 = 54$.

$\dfrac{12x}{12} = \dfrac{54}{12}$ To undo the multiplication by 12, divide both sides by 12.

$x = \dfrac{9}{2}$ Simplify: $\dfrac{54}{12} = \dfrac{9 \cdot \overset{1}{6}}{\underset{1}{6} \cdot 2} = \dfrac{9}{2}$.

Thus, x is $\dfrac{9}{2}$. Check the result.

Self Check 3

Solve: $\dfrac{15}{x} = \dfrac{25}{40}$

Now Try Problem 39

Success Tip Since proportions are rational equations, they can also be solved by multiplying both sides by the LCD. Here an alternate approach is to multiply both sides by $18x$.

Chapter 7 Rational Expressions and Equations

> **Caution!** Remember that a cross product is the product of the means or extremes of a *proportion*. For example, it would be incorrect to try to compute cross products to solve the rational equation $\frac{12}{18} = \frac{3}{x} + \frac{1}{2}$. The right side is not a ratio, so the equation is *not* a proportion.

Using Your CALCULATOR Solving Proportions with a Calculator

To solve the proportion $\frac{3.5}{7.2} = \frac{x}{15.84}$ with a calculator, we can proceed as follows.

$$\frac{3.5}{7.2} = \frac{x}{15.84}$$

$$\frac{3.5(15.84)}{7.2} = x \quad \text{To undo the division by 15.84 and isolate } x,$$
$$\text{multiply both sides of the equation by 15.84.}$$

We can find x by entering these numbers into a scientific calculator.

3.5 $\boxed{\times}$ 15.84 $\boxed{\div}$ 7.2 $\boxed{=}$ $\boxed{7.7}$

Using a graphing calculator, we enter these numbers and press these keys.

3.5 $\boxed{\times}$ 15.84 $\boxed{\div}$ 7.2 $\boxed{\text{ENTER}}$ $\boxed{\begin{array}{r}3.5*15.84/7.2\\7.7\end{array}}$

Thus, x is 7.7.

Self Check 4

Solve: $\dfrac{3x - 1}{2} = \dfrac{12.5}{5}$

Now Try Problem 49

EXAMPLE 4 Solve: $\dfrac{2a + 1}{4} = \dfrac{10}{8}$

Strategy To solve for a, we will set the cross products equal.

WHY Since the equation is a proportion, the product of the means equals the product of the extremes.

Solution

$\dfrac{2a + 1}{4} = \dfrac{10}{8}$ This is the given proportion.

$8(2a + 1) = 40$ In a proportion, the product of the extremes equals the product of the means.

$16a + 8 = 40$ Distribute the multiplication by 8.

$16a + 8 - 8 = 40 - 8$ To undo the addition of 8, subtract 8 from both sides.

$16a = 32$ Combine like terms.

$\dfrac{16a}{16} = \dfrac{32}{16}$ To undo the multiplication by 16, divide both sides by 16.

$a = 2$ Do the divisions.

Thus, a is 2. Check the result.

Self Check 5

Solve: $\dfrac{6}{c} = \dfrac{c - 1}{5}$

Now Try Problem 53

EXAMPLE 5 Solve: $\dfrac{a}{2} = \dfrac{4}{a - 2}$

Strategy To solve for a, we will set the cross products equal.

WHY Since this equation is a proportion, the product of the means equals the product of the extremes.

Solution

$$\frac{a}{2} = \frac{4}{a-2}$$ This is the given proportion.

$a(a-2) = 2 \cdot 4$ Find each cross product and set them equal. Don't forget to write the parentheses.

$a^2 - 2a = 8$ On the left side, distribute the multiplication by a. This is a quadratic equation.

$a^2 - 2a - 8 = 0$ To get 0 on the right side of the equation, subtract 8 from both sides.

$(a + 2)(a - 4) = 0$ Factor $a^2 - 2a - 8$.

$a + 2 = 0$ or $a - 4 = 0$ Set each factor equal to 0.

$a = -2$ | $a = 4$ Solve each equation.

The solutions are -2 and 4, and the solution set is $\{-2, 4\}$. Verify this using a check.

3 Use proportions to solve problems.

We can use proportions to solve many real-world problems. If we are given a ratio (or rate) comparing two quantities, the words of the problem can be translated to a proportion, and we can solve it to find the unknown.

EXAMPLE 6   **Grocery Shopping**   If 6 apples cost $1.38, how much will 16 apples cost?

Analyze   We know the cost of 6 apples; we are to find the cost of 16 apples.

Assign   Let c = the cost of 16 apples.

Form   If we compare the number of apples to their cost, we know that the two rates are equal.

6 apples is to $1.38 as 16 apples is to c.

Number of apples → $\dfrac{6}{1.38} = \dfrac{16}{c}$ ← Number of apples
Cost of the apples →   ← Cost of the apples

Solve

$6 \cdot c = 1.38(16)$ In a proportion, the product of the extremes equals the product of the means.

$6c = 22.08$ Do the multiplication: $1.38(16) = 22.08$.

$\dfrac{6c}{6} = \dfrac{22.08}{6}$ To undo the multiplication by 6, divide both sides by 6.

$c = 3.68$ Divide: $\dfrac{22.08}{6} = 3.68$.

State   Sixteen apples will cost $3.68.

Check   If 16 apples are bought, this is about 3 times as many as 6 apples, which cost $1.38. If we multiply $1.38 by 3, we get an estimate of the cost of 16 apples: $1.38 · 3 = $4.14. The result, $3.68, seems reasonable.

In Example 6, we could have compared the cost of the apples to the number of apples: $1.38 is to 6 apples as c is to 16 apples. This would have led to the proportion

Cost of the apples → $\dfrac{1.38}{6} = \dfrac{c}{16}$ ← Cost of the apples
Number of apples →   ← Number of apples

If we solve this proportion for c, we will obtain the same result: $c = 3.68$.

Self Check 6
If 9 tickets to a concert cost $112.50, how much will 15 tickets cost?

Now Try   Problem 77

> **Caution!** When solving problems using proportions, we must make sure that the units of both numerators are the same and the units of both denominators are the same. In Example 6, it would be incorrect to write
>
> Cost of the apples → $\dfrac{1.38}{6} = \dfrac{16}{c}$ ← Number of apples
> Number of apples → ← Cost of the apples

Self Check 7

The scale for a blueprint indicates $\frac{1}{4}$ inch on the print is equivalent to 1 foot for the actual building. If the width of the building on the print is 7.5 inches, what is the width of the actual building?

Now Try Problem 90

EXAMPLE 7 Scale Models

A **scale** is a ratio (or rate) that compares the size of a model, drawing, or map to the size of an actual object. The scale shown in the figure indicates that 1 inch on the model carousel is equivalent to 160 inches on the actual carousel. How wide should the model be if the actual carousel is 35 feet wide?

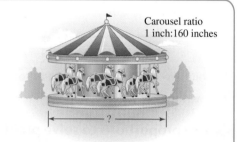

Carousel ratio
1 inch:160 inches

Analyze We are asked to determine the width of the miniature carousel, if a ratio of 1 inch to 160 inches is used. We would like the width of the model to be given in inches, not feet, so we will express the 35-foot width of the actual carousel as $35 \cdot 12 = 420$ inches.

Assign Let w = the width of the model.

Form The ratios of the dimensions of the model to the corresponding dimensions of the actual carousel are equal.

1 inch is to 160 inches as w inches is to 420 inches.

$$\text{model} \rightarrow \dfrac{1}{160} = \dfrac{w}{420} \leftarrow \text{model}$$
$$\text{actual} \rightarrow \qquad\qquad \leftarrow \text{actual}$$

Solve

$420 = 160w$ In a proportion, the product of the extremes is equal to the product of the means.

$\dfrac{420}{160} = \dfrac{160w}{160}$ To undo the multiplication by 160, divide both sides by 160.

$2.625 = w$ Do the division: $\dfrac{420}{160} = 2.625$.

State The width of the miniature carousel should be 2.625 in., or $2\dfrac{5}{8}$ in.

Check A width of $2\dfrac{5}{8}$ in. is approximately 3 in. When we write the ratio of the model's approximate width to the width of the actual carousel, we get $\dfrac{3}{420} = \dfrac{1}{140}$, which is about $\dfrac{1}{160}$. The answer seems reasonable. ∎

Self Check 8

How many cups of sugar will be needed to make several cakes that will require a total of 25 cups of flour?

Now Try Problem 83

EXAMPLE 8 Baking

A recipe for rhubarb cake calls for $1\frac{1}{4}$ cups of sugar for every $2\frac{1}{2}$ cups of flour. How many cups of flour are needed if the baker intends to use 3 cups of sugar?

Analyze The baker needs to maintain the same ratio between the amounts of sugar and flour as is called for in the original recipe.

Assign Let f = the number of cups of flour to be mixed with the 3 cups of sugar.

Form The ratios of the cups of sugar to the cups of flour are equal.

$1\frac{1}{4}$ cups sugar is to $2\frac{1}{2}$ cups flour as 3 cups sugar is to f cups flour.

Cups sugar → $\dfrac{1\frac{1}{4}}{2\frac{1}{2}} = \dfrac{3}{f}$ ← Cups sugar
Cups flour →   ← Cups flour

Solve

$\dfrac{1.25}{2.5} = \dfrac{3}{f}$ Change the fractions to decimals.

$1.25f = 2.5 \cdot 3$ In a proportion, the product of the extremes equals the product of the means.

$1.25f = 7.5$ Do the multiplication: $2.5 \cdot 3 = 7.5$.

$\dfrac{1.25f}{\mathbf{1.25}} = \dfrac{7.5}{\mathbf{1.25}}$ To undo the multiplication by 1.25, divide both sides by 1.25.

$f = 6$ Divide: $\frac{7.5}{1.25} = 6$.

State The baker should use 6 cups of flour.

Check The recipe calls for about 2 cups of flour for about 1 cup of sugar. If 3 cups of sugar are used, 6 cups of flour seems reasonable.

4 Use proportions to solve problems that involve similar triangles.

If two angles of one triangle have the same measures as two angles of a second triangle, the triangles have the same shape. Triangles with the same shape are called **similar triangles**. In the figure, $\triangle ABC \sim \triangle DEF$. (Read the symbol $\sim$ as "is similar to.")

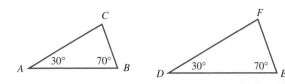

> **Property of Similar Triangles**
>
> If two triangles are **similar**, all pairs of corresponding sides are in proportion.

In the similar triangles shown in the figure above, the following proportions are true.

$\dfrac{AB}{DE} = \dfrac{BC}{EF}$, $\dfrac{BC}{EF} = \dfrac{CA}{FD}$,   and $\dfrac{CA}{FD} = \dfrac{AB}{DE}$ Read AB as "the length of segment AB."

EXAMPLE 9 **Finding the Height of a Tree** A tree casts a shadow 18 feet long at the same time as a woman 5 feet tall casts a shadow 1.5 feet long. Find the height of the tree.

Analyze The figure shows the similar triangles determined by the tree and its shadow and the woman and her shadow. Since the triangles are similar, the lengths

Self Check 9

Find the height of the tree in Example 9 if the woman is 5 feet 6 inches tall and her shadow is 1.5 feet long.

Now Try Problem 97

of their corresponding sides are in proportion. We can use this fact to find the height of the tree.

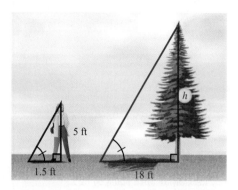

Each triangle has a right angle. Since the sun's rays strike the ground at the same angle, the angles highlighted with a tick mark have the same measure. Therefore, two angles of the smaller triangle have the same measures as two angles of the larger triangle; the triangles are similar.

Assign We let $h =$ the height of the tree.

Form We can find h by solving the following proportion.

$$\frac{h}{5} = \frac{18}{1.5} \qquad \frac{\text{Height of the tree}}{\text{Height of the woman}} = \frac{\text{Length of shadow of the tree}}{\text{Length of shadow of the woman}}$$

Solve

$1.5h = 5(18)$ In a proportion, the product of the extremes equals the product of the means.

$1.5h = 90$ Do the multiplication.

$\dfrac{1.5h}{1.5} = \dfrac{90}{1.5}$ To undo the multiplication by 1.5, divide both sides by 1.5.

$h = 60$ Divide: $\frac{90}{1.5} = 60$.

State The tree is 60 feet tall.

Check $\dfrac{18}{1.5} = 12$ and $\dfrac{60}{5} = 12$. The ratios are the same. The result checks.

ANSWERS TO SELF CHECKS

1. a. $\dfrac{15}{2}$ **b.** $\dfrac{1}{4}$ **2.** no **3.** 24 **4.** 2 **5.** $-5, 6$ **6.** $187.50 **7.** 30 ft **8.** $12\dfrac{1}{2}$ cups **9.** 66 ft

---

SECTION 7.8   STUDY SET

VOCABULARY

Fill in the blanks.

1. A _____ of two numbers is the quotient of two quantities with the same units.

2. A _____ is a quotient of two quantities that have different units.

3. A _____ is a mathematical statement that two ratios or two rates are equal.

4. In the proportion $\dfrac{a}{b} = \dfrac{c}{d}$, a and d are called the _____ of the proportion.

5. In the proportion $\dfrac{a}{b} = \dfrac{c}{d}$, b and c are called the _____ of the proportion.

6. The product of the extremes and the product of the means of a proportion are also known as _____ products.

7. If two triangles have the same _____, they are said to be *similar*.

8. If two triangles are _____, their corresponding sides are in proportion.

CONCEPTS

Fill in the blanks.

9. **WEST AFRICA** Write the ratio (in fractional form) of the number of red stripes to the number of white stripes on the flag of Liberia.

10. The equation $\dfrac{a}{b} = \dfrac{c}{d}$ is a proportion if the cross product ___ is equal to the cross product ___.

11. Is 45 a solution of $\dfrac{5}{3} = \dfrac{75}{x}$?

12. Consider: $\dfrac{2}{3} = \dfrac{x}{15}$

    a. Solve the proportion by multiplying both sides by the LCD.

    b. Solve the proportion by setting the cross products equal.

13. **MINIATURES** A high-wheeler bicycle is shown below. A model of it is to be made using a scale of 2 inches to 15 inches. The following proportion was set up to determine the height of the front wheel of the model. Explain the error.

    $\dfrac{2}{15} \cancel{=} \dfrac{48}{h}$

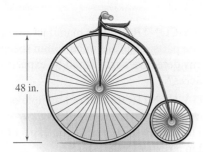

14. Two similar triangles are shown below. Fill in the blanks to make the proportions true.

    $\dfrac{AB}{DE} = \dfrac{}{EF}$ $\dfrac{BC}{} = \dfrac{CA}{FD}$ $\dfrac{CA}{FD} = \dfrac{AB}{}$

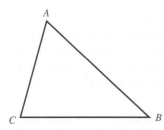

NOTATION

Complete each step.

15. Solve for x: $\dfrac{12}{18} = \dfrac{x}{24}$

    $12 \cdot 24 = 18 \cdot $

    $ = 18x$

    $\dfrac{288}{} = \dfrac{18x}{}$

    $16 = x$

16. Solve for x: $\dfrac{14}{x} = \dfrac{49}{17.5}$

    $14 \cdot = 49x$

    $ = 49x$

    $\dfrac{245}{} = \dfrac{49x}{}$

    $5 = x$

17. We read "$\triangle ABC$" as "_____ ABC."

18. The symbol $\sim$ is read as "_____."

GUIDED PRACTICE

Translate each ratio into a ratio in simplest form. See Example 1.

19. 4 boxes to 15 boxes
20. 2 miles to 9 miles
21. 18 watts to 24 watts
22. 11 cans to 121 cans
23. 30 days to 24 days
24. 45 people to 30 people
25. 90 minutes to 3 hours
26. 20 inches to 2 feet
27. 8 quarts to 4 gallons
28. 6 feet to 12 yards
29. 6,000 feet to 1 mile (*Hint:* 1 mi = 5,280 ft)
30. 5 tons to 4,000 pounds (*Hint:* 1 ton = 2,000 lb)

Determine whether each equation is a proportion. See Example 2.

31. $\dfrac{9}{7} = \dfrac{81}{70}$
32. $\dfrac{5}{2} = \dfrac{20}{8}$
33. $\dfrac{7}{3} = \dfrac{14}{6}$
34. $\dfrac{13}{19} = \dfrac{65}{95}$
35. $\dfrac{9}{19} = \dfrac{38}{80}$
36. $\dfrac{40}{29} = \dfrac{29}{22}$
37. $\dfrac{10.4}{3.6} = \dfrac{41.6}{14.4}$
38. $\dfrac{13.23}{3.45} = \dfrac{39.96}{11.35}$

Solve each proportion. See Example 3.

39. $\dfrac{2}{3} = \dfrac{x}{6}$
40. $\dfrac{3}{6} = \dfrac{x}{8}$
41. $\dfrac{5}{10} = \dfrac{3}{c}$
42. $\dfrac{7}{14} = \dfrac{2}{x}$

Solve each proportion. See Example 4.

43. $\dfrac{x+1}{5} = \dfrac{3}{15}$

44. $\dfrac{x-1}{7} = \dfrac{2}{21}$

45. $\dfrac{x+3}{12} = \dfrac{-7}{6}$

46. $\dfrac{x+7}{-4} = \dfrac{1}{4}$

47. $\dfrac{13}{4-x} = \dfrac{26}{11}$

48. $\dfrac{17}{5-x} = \dfrac{34}{13}$

49. $\dfrac{14}{3} = \dfrac{2x+1}{18}$

50. $\dfrac{9}{54} = \dfrac{2x-1}{18}$

Solve each proportion. See Example 5.

51. $\dfrac{y}{4} = \dfrac{4}{y}$

52. $\dfrac{2}{3x} = \dfrac{6x}{36}$

53. $\dfrac{2}{c} = \dfrac{c-3}{2}$

54. $\dfrac{b-5}{3} = \dfrac{2}{b}$

55. $\dfrac{a-4}{a} = \dfrac{15}{a+4}$

56. $\dfrac{s}{s-5} = \dfrac{s+5}{24}$

57. $\dfrac{t+3}{t+5} = \dfrac{-1}{2t}$

58. $\dfrac{5h}{14h+3} = \dfrac{1}{h}$

TRY IT YOURSELF

Solve each proportion.

59. $\dfrac{6}{x} = \dfrac{8}{4}$

60. $\dfrac{4}{x} = \dfrac{2}{8}$

61. $\dfrac{x}{3} = \dfrac{9}{3}$

62. $\dfrac{x}{2} = \dfrac{18}{6}$

63. $\dfrac{2}{x+6} = \dfrac{-2x}{5}$

64. $\dfrac{x-1}{x+1} = \dfrac{2}{3x}$

65. $\dfrac{x+1}{4} = \dfrac{3x}{8}$

66. $\dfrac{x-1}{9} = \dfrac{2x}{3}$

67. $\dfrac{3}{4x} = \dfrac{x-4}{x+\frac{5}{3}}$

68. $\dfrac{3}{x-1} = \dfrac{x}{4}$

69. $\dfrac{y-4}{y+1} = \dfrac{y+3}{y+6}$

70. $\dfrac{r-6}{r-8} = \dfrac{r+1}{r-4}$

71. $\dfrac{c}{10} = \dfrac{10}{c}$

72. $\dfrac{-6}{r} = \dfrac{r}{-6}$

73. $\dfrac{m}{3} = \dfrac{4}{m+1}$

74. $\dfrac{n}{2} = \dfrac{5}{n+3}$

75. $\dfrac{3}{3b+4} = \dfrac{2}{5b-6}$

76. $\dfrac{2}{4d-1} = \dfrac{3}{2d+1}$

APPLICATIONS

Set up and solve a proportion. Use a calculator if it is helpful. See Examples 6–8.

77. GROCERY SHOPPING If 3 pints of yogurt cost $1, how much will 51 pints cost?

78. SHOPPING FOR CLOTHES If shirts are on sale at two for $25, how much will five shirts cost?

79. ADVERTISING In 2013, a 30-second TV ad during the Super Bowl telecast cost $4 million. At this rate, what was the cost of a 45-second ad?

80. COOKING A recipe for spaghetti sauce requires four 16-ounce bottles of ketchup to make 2 gallons of sauce. How many bottles of ketchup are needed to make 10 gallons of sauce?

81. MIXING PERFUME A perfume is to be mixed in the ratio of 3 drops of pure essence to 7 drops of alcohol. How many drops of pure essence should be mixed with 56 drops of alcohol?

82. CPR A first aid handbook states that when performing cardiopulmonary resuscitation on an adult, the ratio of chest compressions to breaths should be 15:1. If 210 compressions were administered to an adult patient, how many breaths should have been given?

83. COOKING A recipe for wild rice soup is shown. Find the amounts of chicken broth, rice, and flour needed to make 15 servings.

Wild Rice Soup	
A sumptuous side dish with a nutty flavor	
3 cups chicken broth	1 cup light cream
$\frac{2}{3}$ cup uncooked rice	2 tablespoons flour
$\frac{1}{4}$ cup sliced onions	$\frac{1}{8}$ teaspoon pepper
$\frac{1}{2}$ cup shredded carrots	Serves: 6

84. QUALITY CONTROL In a manufacturing process, 95% of the parts made are to be within specifications. How many defective parts would be expected in a run of 940 pieces?

85. QUALITY CONTROL Out of a sample of 500 men's shirts, 17 were rejected because of crooked collars. How many shirts would you expect to be rejected in a run of 15,000 shirts?

86. GAS CONSUMPTION If a car can travel 42 miles on 1 gallon of gas, how much gas is needed to travel 315 miles?

87. HIP-HOP According to the *Guinness Book of World Records,* Rebel X.D. of Chicago rapped 674 syllables in 54.9 seconds. At this rate, how many syllables could he rap in 1 minute? Round to the nearest syllable.

88. BANKRUPTCY After filing for bankruptcy, a company was able to pay its creditors only 15 cents on the dollar. If the company owed a lumberyard $9,712, how much could the lumberyard expect to be paid?

89. COMPUTING A PAYCHECK Billie earns $412 for a 40-hour week. If she missed 10 hours of work last week, how much did she get paid?

90. **MODEL RAILROADS** A model railroad engine is 9 inches long. If the scale is 87 feet to 1 foot, how long is a real engine?

91. **MODEL RAILROADS** A model railroad caboose is 3.5 inches long. If the scale is 169 feet to 1 foot, how long is a real caboose?

92. **NUTRITION** The following table shows the nutritional facts about a 10-oz chocolate milkshake sold by a fast-food restaurant. Use the information to complete the table for the 16-oz shake. Round to the nearest unit when an answer is not exact.

	Calories	Fat (gm)	Protein (gm)
10-oz chocolate milkshake	355	8	9
16-oz chocolate milkshake			

93. **DRIVER'S LICENSES** Of the 50 states, Alabama has one of the largest ratios of licensed drivers to residents. If the ratio is 796:1,000 and Alabama's population is about 4,800,000, how many residents of that state have a driver's license?

94. **MIXING FUEL** The instructions on a can of oil intended to be added to lawnmower gasoline read as follows:

Recommended	Gasoline	Oil
50 to 1	6 gal	16 oz

Are the instructions correct? (*Hint:* There are 128 ounces in 1 gallon.)

95. **PHOTO ENLARGEMENT** In the illustration, the 3-by-5 photograph is to be blown up to the larger size. Find x.

96. **BLUEPRINTS** The scale for the blueprint shown in the next column tells the reader that a $\frac{1}{4}$-inch length $\left(\frac{1}{4}''\right)$ on the drawing corresponds to an actual size of 1 foot (1′0″). Suppose the length of the kitchen is $2\frac{1}{2}$ inches on the drawing. How long is the actual kitchen?

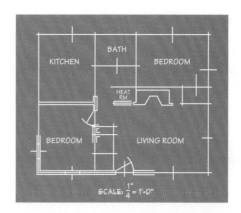

Use similar triangles to solve each problem. See Example 9.

97. **HEIGHT OF A TREE** A tree casts a shadow of 26 feet at the same time as a 6-foot man casts a shadow of 4 feet. Find the height of the tree.

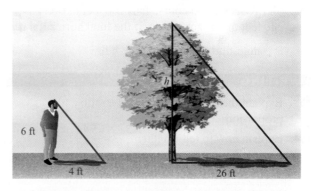

98. **HEIGHT OF A BUILDING** A man places a mirror on the ground and sees the reflection of the top of a building, as shown. The two triangles in the illustration are similar. Find the height, h, of the building.

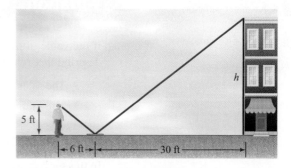

99. **WIDTH OF A RIVER** Use the dimensions in the illustration to find w, the width of the river. (The two triangles in the illustration are similar.)

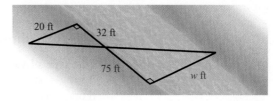

100. FLIGHT PATHS The airplane shown below ascends 100 feet as it flies a horizontal distance of 1,000 feet. How much altitude will it gain as it flies a horizontal distance of 1 mile? (*Hint:* 5,280 feet = 1 mile.)

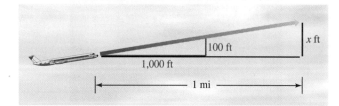

101. FLIGHT PATHS An airplane descends 1,350 feet as it flies a horizontal distance of 1 mile. How much altitude is lost as it flies a horizontal distance of 5 miles?

102. SKI RUNS A ski course falls 100 feet in every 300 feet of horizontal run. If the total horizontal run is $\frac{1}{2}$ mile, find the height of the hill.

WRITING

103. Explain the difference between a ratio and a proportion.

104. Explain how to tell whether $\frac{3.2}{3.7} = \frac{5.44}{6.29}$ is a proportion.

105. Explain why the concept of cross products cannot be used to solve this equation:

$$\frac{x}{3} - \frac{3x}{4} = \frac{1}{12}$$

106. Write a problem about a situation you encounter in your daily life that could be solved by using a proportion.

REVIEW

107. Change $\frac{9}{10}$ to a percent.

108. Change $33\frac{1}{3}\%$ to a fraction.

109. Find 30% of 1,600.

110. SHOPPING Maria bought a dress for 25% off the original price of $98. How much did the dress cost?

111. Find the slope of the line passing through $(-2, -2)$ and $(-12, -8)$.

112. What are the slope and the *y*-intercept of the graph of $y = 2x - 3$?

Objectives

1. Solve direct variation problems.
2. Solve inverse variation problems.

SECTION 7.9
Variation

If the value of one quantity depends on the value of another quantity, we can often describe that relationship using the language of variation:

- The sales tax on an item varies as the price.
- The intensity of light varies as the distance from its source.
- The pressure exerted by water on an object varies as the depth of the object beneath the surface.

In this section, we will discuss two types of variation, and we will see how to represent them algebraically using equations.

1 Solve direct variation problems.

One type of variation, called **direct variation**, is represented by an equation of the form $y = kx$, where k is a constant (a number). Two variables are said to *vary directly* if one is a constant multiple of the other.

Direct Variation

The words *y varies directly as x* mean that

$$y = kx$$

for some constant k, called the **constant of variation**.

EXAMPLE 1 Suppose y varies directly as x. If $y = 12$ when $x = 4$, find y when $x = 6$.

Strategy We will use the equation $y = kx$ to solve this problem.

WHY The words *varies directly* indicate that we should use the direct variation equation $y = kx$.

Solution
We can use the given pair of values of x and y to determine the constant of variation k.

$\quad y = kx \quad$ This is the equation that models direct variation.
$\quad 12 = k(4) \quad$ Substitute 4 for x and 12 for y.
$\quad 3 = k \quad$ To isolate k on the right side, divide both sides by 4. This is the constant of variation.

Since $y = kx$ and $k = 3$, we have $y = 3x$.

We can use $y = 3x$ to find other pairs of values of x and y. When $x = 6$, we see that

$\quad y = 3(6) = 18 \quad$ Substitute 6 for x and do the multiplication.

Thus, when $x = 6$, the value of y is 18.

Self Check 1
Suppose y varies directly as x. If $y = 24$ when $x = 3$, find y when $x = 5$.

Now Try Problem 29

Success Tip If we divide both sides of $y = kx$ by x, we obtain $\frac{y}{x} = k$. Thus, for the direct variation model, k is simply the quotient of one pair of values of x and y.

Scientists have found that the distance a spring will stretch varies directly as the force applied to it. The more force applied to the spring, the more it will stretch. If d represents the distance stretched and f represents the force applied, this relationship can be expressed by the equation

$\quad d = kf \quad$ where k is the constant of variation

Suppose that a 150-pound weight stretches a spring 18 inches. (See the figure to the right.) We can find the constant of variation for the spring by substituting 150 for f and 18 for d in the equation $d = kf$ and solving for k:

$\quad d = kf$
$\quad 18 = k(150)$
$\quad \frac{18}{150} = k \quad$ Divide both sides by 150 to isolate k.
$\quad \frac{3}{25} = k \quad$ Simplify the fraction: $\frac{18}{150} = \frac{\overset{1}{\cancel{6}} \cdot 3}{\underset{1}{\cancel{6}} \cdot 25} = \frac{3}{25}$.

Therefore, the equation describing the relationship between the distance the spring will stretch and the amount of force applied to it is $d = \frac{3}{25}f$. To find the distance that the same spring will stretch when a 50-pound weight is attached, we proceed as follows:

$$d = \frac{3}{25}f \qquad \text{This is the equation describing the direct variation.}$$

$$d = \frac{3}{25}(50) \qquad \text{Substitute 50 for } f.$$

$$d = 6 \qquad \text{Do the multiplication.}$$

The spring will stretch 6 inches when a 50-pound weight is attached.

The table shows some other possible values for f and d as determined by the equation $d = \frac{3}{25}f$. When these ordered pairs are graphed and a straight line is drawn through them, it is apparent that as the force f applied to a spring increases, the distance d it stretches increases. Furthermore, the slope of the graph is $\frac{3}{25}$, the constant of variation.

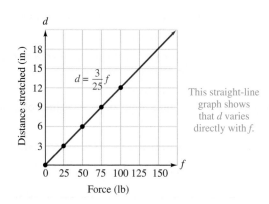

This straight-line graph shows that d varies directly with f.

Success Tip The value of k is for this specific example. Another spring made out of a different type of steel will more than likely have a different value of k.

We can use the following steps to solve variation problems.

Strategy for Solving Variation Problems

1. Translate the verbal model into an equation.
2. Substitute the first set of values into the equation from step 1 to determine the value of k.
3. Substitute the value of k into the equation from step 1.
4. Substitute the remaining set of values into the equation from step 3 and solve for the unknown variable.

Self Check 2

The cost of a bus ticket varies directly as the number of miles traveled. If a ticket for a 180-mile trip cost $45, what would a ticket for a 1,500-mile trip cost?

EXAMPLE 2 *Geology* The weight of an object on Earth varies directly as its weight on the moon. If a rock weighs 5 pounds on the moon and 30 pounds on Earth, what would be the weight on Earth of a larger rock weighing 26 pounds on the moon?

Strategy We will follow the strategy for solving a direct variation problem.

WHY In the words of the problem, the phrase *varies directly* indicates that a direct variation model should be used.

Solution

Step 1 We let *e* represent the weight of the object on Earth and *m* the weight of the object on the moon. Translating the words *weight on Earth varies directly with weight on the moon*, we get the equation

$$e = km$$

Step 2 To find the constant of variation, *k*, we substitute 30 for *e* and 5 for *m*.

$$e = km$$
$$30 = k(5)$$
$$6 = k \quad \text{To isolate k, divide both sides by 5.}$$

Step 3 The equation describing the relationship between the weight of an object on Earth and on the moon is

$$e = 6m$$

Step 4 We can find the weight of the larger rock on Earth by substituting 26 for *m* in the equation from step 3.

$$e = 6m$$
$$e = 6(26)$$
$$e = 156 \quad \text{Do the multiplication.}$$

The rock would weigh 156 pounds on Earth.

Now Try Problem 45

A. T. Willett/Alamy

2 Solve inverse variation problems.

Another type of variation, called **inverse variation**, is represented by an equation of the form $y = \dfrac{k}{x}$, where *k* is a constant. Two variables are said to *vary inversely* if one is a constant multiple of the reciprocal of the other.

Inverse Variation

The words *y varies inversely as x* mean that

$$y = \frac{k}{x}$$

for some constant *k*, called the **constant of variation**.

EXAMPLE 3 Suppose *y* varies inversely as *x*. If $y = 5$ when $x = 20$, find *y* when $x = 50$.

Strategy We will use the equation $y = \dfrac{k}{x}$ to solve this problem.

WHY The words *varies inversely* indicate that we should use the inverse variation equation $y = \dfrac{k}{x}$.

Solution

We can use the given pair of values of *x* and *y* to determine the constant of variation, *k*, in $y = \dfrac{k}{x}$.

Self Check 3

Suppose *y* varies inversely as *x*. If $y = 25$ when $x = 3$, find *y* when $x = 15$.

Now Try Problem 37

$$y = \frac{k}{x} \quad \text{This is the equation that models inverse variation.}$$

$$5 = \frac{k}{20} \quad \text{Substitute 20 for } x \text{ and 5 for } y.$$

$$20 \cdot 5 = k \quad \text{To isolate } k \text{ on the right side, multiply both sides by 20.}$$

$$100 = k \quad \text{This is the constant of variation.}$$

Since $y = \frac{k}{x}$ and $k = 100$, we have

$$y = \frac{100}{x}$$

We can use $y = \frac{100}{x}$ to find other pairs of values of x and y. When $x = 50$, we see that

$$y = \frac{100}{50} = 2 \quad \text{Substitute 50 for } x \text{ and do the division.}$$

Thus, when $x = 50$, the value of y is 2.

Suppose that the time (in hours) it takes to paint a house varies inversely as the size of the painting crew. As the number of painters increases, the time that it takes to paint the house decreases. If n represents the number of painters and t represents the time it takes to paint the house, this relationship can be expressed by the equation

$$t = \frac{k}{n} \quad \text{where } k \text{ is the constant of variation}$$

If we know that a crew of 8 can paint the house in 12 hours, we can find the constant of variation by substituting 8 for n and 12 for t in the equation $t = \frac{k}{n}$ and solving for k:

$$t = \frac{k}{n} \quad \text{This is the equation that models inverse variation.}$$

$$12 = \frac{k}{8} \quad \text{Substitute 8 for } n \text{ and 12 for } t.$$

$$12 \cdot 8 = k \quad \text{Multiply both sides by 8 to isolate } k.$$

$$96 = k \quad \text{Do the multiplication.}$$

The equation describing the relationship between the size of the painting crew and the time it takes to paint the house is $t = \frac{96}{n}$. We can use this equation to find the time it will take a crew of any size to paint the house. For example, to find the time it would take a 4-person crew, we substitute 4 for n in the equation $t = \frac{96}{n}$.

$$t = \frac{96}{n} \quad \text{This is the equation describing the inverse variation.}$$

$$t = \frac{96}{4} \quad \text{Substitute 4 for } n.$$

$$t = 24 \quad \text{Do the division.}$$

It would take a 4-person crew 24 hours to paint the house.

The table shows some possible values for n and t as determined by the equation $t = \frac{96}{n}$. When these ordered pairs are graphed and a smooth curve is drawn through them, it is clear that as the number of painters n increases, the time t decreases.

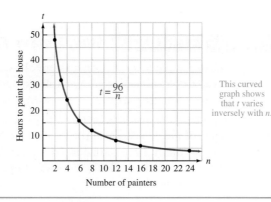

This curved graph shows that t varies inversely with n.

EXAMPLE 4 Chemistry
The volume occupied by a fixed weight of gas (held at a constant temperature) varies inversely as the pressure placed on it. If hydrogen gas occupies a volume of 22.5 cubic inches when placed under 3 pounds per square inch (psi) of pressure, find the volume occupied by the hydrogen gas when the pressure is 7.5 psi.

The volume occupied by gas decreases as pressure increases.

Strategy We will follow the strategy for solving an inverse variation problem.

WHY In the words of the problem, the phrase *varies inversely* indicates that an inverse variation model should be used.

Self Check 4
Find the volume occupied by the hydrogen gas in Example 4 when the pressure placed on it is 5 psi.

Now Try Problem 49

Solution
Step 1 We let V represent the volume occupied by the gas and p represent the pressure. Translating the words *volume occupied by a gas varies inversely as the pressure*, we get the equation

$$V = \frac{k}{p} \quad \text{The equation models inverse variation.}$$

Step 2 To find the constant of variation, k, we substitute 22.5 for V and 3 for p.

$$V = \frac{k}{p}$$

$$22.5 = \frac{k}{3} \quad \text{Substitute 3 for } p \text{ and 22.5 for } V.$$

$$67.5 = k \quad \text{To isolate } k, \text{ multiply both sides by 3. This is the constant of variation.}$$

Step 3 The equation describing the relationship between the volume occupied by the gas and the pressure placed on it is

$$V = \frac{67.5}{p} \quad \text{Substitute 67.5 for } k \text{ in } V = \frac{k}{p}.$$

Step 4 We can find the volume occupied by the gas when a pressure of 7.5 psi is placed on it by substituting 7.5 for p in the equation and evaluating the right side.

$$V = \frac{67.5}{p}$$

$$V = \frac{67.5}{7.5} \quad \text{Substitute 7.5 for } p.$$

$$V = 9 \quad \text{Do the division.}$$

The hydrogen gas will occupy a volume of 9 cubic inches when the pressure placed on it is 7.5 psi.

THINK IT THROUGH Study Time vs. Effectiveness

"Above all, review regularly and plan to study ahead, so that the night before an exam, all you do is review material. Avoid all-nighters!"

Improve Your Studying Skills, Counseling Service at The University of North Carolina, Chapel Hill

Each graph below shows a direct or inverse relationship between two components of the educational process as found by researchers in *An Analysis of the Study Time* (Orlando J. Olivares, Department of Psychology, Bridgewater State College, 2002). For each graph, explain why you agree or disagree with the findings.

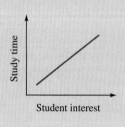
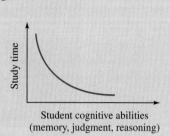
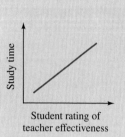

ANSWERS TO SELF CHECKS
1. 40 **2.** $375 **3.** 5 **4.** 13.5 in.^{3}

SECTION 7.9 STUDY SET

VOCABULARY

Fill in the blanks.

1. The equation $y = kx$ defines _____ variation.
2. The equation $y = \dfrac{k}{x}$ defines _____ variation.
3. In $y = kx$, the _____ of variation is k.
4. In $y = \dfrac{k}{x}$, the constant of variation is _____.

CONCEPTS

Determine whether each graph represents direct variation or inverse variation.

5.

6.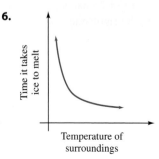

7.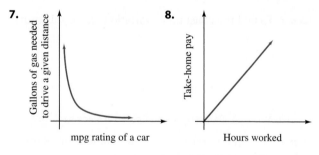

8.

9. Translate the following sentence into mathematical symbols:

   A farmer's harvest, h varies directly   as the number of acres planted, a.

10. If the constant of variation for Problem 9 is $k = 10{,}000$, what will happen to the size of the harvest as the number of acres planted increases?

11. Express the following relationship with an equation:

The number of gallons g of paint needed to paint a room varies directly as the number of square feet f to be painted.

12. Express the following relationship with an equation:

The amount of sales tax t varies directly as the purchase price p of a new car.

13. Translate the following sentence into mathematical symbols:

The time t in hours it takes a commuter to drive from her home to her office	varies inversely as	her average speed s (in mph).

14. In Problem 13, what will happen to the time her commute takes her as her average speed increases?

15. Express this relationship using an equation: The number of hot dogs n that a street vendor sells varies inversely as the price p that he charges.

16. a. If y varies directly as x and $k > 0$, what happens to y as x increases?

b. If y varies inversely as x and $k > 0$, what happens to y as x increases?

17. Suppose $y = \dfrac{k}{x}$. If $y = 15$ when $x = 10$, find k.

18. Suppose $c = \dfrac{k}{d}$. If $c = 9$ when $d = 5$, find k.

NOTATION

Determine whether the equation defines direct variation.

19. $y = kx$ **20.** $y = k + x$

21. $y = \dfrac{k}{x}$ **22.** $m = kc$

Determine whether each equation defines inverse variation.

23. $y = kx$ **24.** $y = \dfrac{k}{x}$

25. $d = \dfrac{k}{g}$ **26.** $y = \dfrac{x}{k}$

Complete each step.

27. Find f if $d = 21$ and $k = \tfrac{7}{5}$.

$$d = kf$$
$$21 = \boxed{} f$$
$$\boxed{} \cdot 21 = \boxed{} \cdot \tfrac{7}{5} f$$
$$15 = f$$

28. Find f if $d = 20$ and $k = 0.75$.

$$d = \dfrac{k}{f}$$
$$\boxed{} = \dfrac{0.75}{f}$$
$$\boxed{} \cdot 20 = \boxed{} \cdot \dfrac{0.75}{f}$$
$$20f = \boxed{}$$
$$f = \dfrac{0.75}{\boxed{}}$$
$$f = 0.0375$$

GUIDED PRACTICE

Solve each direct variation problem. See Examples 1–2.

29. y varies directly as x. If $y = 10$ when $x = 2$, find y when $x = 7$.

30. r varies directly as s. If $r = 21$ when $s = 6$, find r when $s = 12$.

31. l varies directly as m. If $l = 50$ when $m = 200$, find l when $m = 25$.

32. x and y vary directly. If $x = 30$ when $y = 2$, find y when $x = 45$.

33. n_1 and n_2 vary directly. If $n_1 = 315$ when $n_2 = 3$, find n_2 when $n_1 = 10.5$.

34. d is directly proportional to t. If $d = 21$ when $t = 6$, find d when $t = 4$.

35. t varies directly as s. If $t = 21$ when $s = 6$, find k.

36. y varies directly as x. If $y = 10$ when $x = 2$, find k.

Solve each inverse variation problem. See Examples 3–4.

37. y varies inversely as x. If $y = 8$ when $x = 1$, find y when $x = 8$.

38. r varies inversely as s. If $r = 40$ when $s = 10$, find r when $s = 15$.

39. a varies inversely as t. If $a = 600$ when $t = 300$, find a when $t = 15$.

40. t_1 and t_2 vary inversely. If $t_1 = 4$ when $t_2 = 5$, find t_2 when $t_1 = 3\tfrac{1}{3}$.

41. a and r vary inversely. If $a = 9$ when $r = 7$, find r when $a = \tfrac{1}{9}$.

42. q is inversely proportional to s. If $q = 6$ when $s = 9$, find q when $s = 24$.

43. a varies inversely as b^3. If $a = 4$ when $b = 2$, find a when $b = 3$.

44. s varies inversely as t^2. If $s = 180$ when $t = 3$, find s when $t = 9$.

APPLICATIONS

Solve each direct variation problem. **See Example 2.**

45. DRIVING The distance that a car can travel without refueling varies directly as the number of gallons of gasoline in the tank. If a car can go 360 miles on a full tank of gas (15 gallons), how far can it go on 7 gallons?

46. GRAVITY The force of gravity acting on an object varies directly as the mass of the object. The force on a mass of 5 kilograms is 49 newtons. What is the force acting on a mass of 12 kilograms?

47. DOSAGES The recommended dose (in milligrams) of Demerol, a preoperative medication given to children, varies directly as the child's weight (in pounds). If the proper dosage for a child weighing 30 pounds is 18 milligrams, find the proper dosage for a child weighing 45 pounds.

48. MEDICATIONS To fight ear infections in children, doctors often prescribe Ceclor. The recommended dose in milligrams varies directly as the child's body weight in pounds. If the proper dosage for a 20-pound child is 124 milligrams, find the proper dosage for a 28-pound child.

Solve each inverse variation problem. **See Example 4.**

49. TRAVELING The time it takes a car to travel a certain distance varies inversely as its rate of speed. If a certain trip takes 3 hours at 50 mph, how long will the trip take at 60 mph?

50. GEOMETRY For a fixed area, the length of a rectangle is inversely proportional to its width. A rectangle has a width of 12 feet and a length of 20 feet. If its width is increased to 12.5 feet, find the length that will maintain the same area.

51. ELECTRICITY The current in an electric circuit varies inversely as the resistance. If the current in the circuit shown is 30 amps when the resistance is 4 ohms, what will the current be for a resistance of 15 ohms?

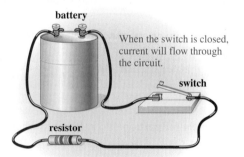

52. FARMING The length of time a given number of bushels of corn will last when feeding cattle varies inversely as the number of animals. If a certain number of bushels will feed 25 cows for 10 days, how long will the feed last for 10 cows?

TRY IT YOURSELF

Use a calculator to help solve each variation problem.

53. g varies directly as t. If $g = 3,616$ when $t = 8,000$, find g when $t = 2,405$.

54. b varies inversely as c. If $b = 0.45$ when $c = 1.6$, find b when $c = 80$.

55. p varies inversely as q. If $p = 55$ when $q = 2.5$, find p when $q = 100$.

56. m varies directly as n. If $m = 4,560$ when $n = 950$, find m when $n = 725$.

57. PULLEYS The speeds, in revolutions per minute (rpm), of two pulleys connected by a belt are inversely proportional to their diameters. If a pulley 24 inches in diameter, making 120 revolutions per minute, is belted to a second pulley 16 inches in diameter, how many rpm does the smaller pulley make?

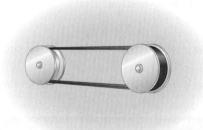

58. CIDER For the following, the number of inches of stick cinnamon to use varies directly as the number of servings of spiced cider to be made. How many inches of stick cinnamon are needed to make 36 servings?

Hot Spiced Cider
8 cups apple cider or apple juice
$\frac{1}{4}$ to $\frac{1}{2}$ cup packed brown sugar
6 inches stick cinnamon
1 teaspoon whole allspice
1 teaspoon whole cloves
8 thin orange wedges or slices (optional)
8 whole cloves (optional)   Makes 8 servings

59. LUNAR GRAVITY Refer to the illustration on the next page. The weight of an object on the moon varies directly as its weight on Earth. If 6 pounds on Earth weighs 1 pound on the moon, what would the scale register if the astronaut were weighed on the moon?

60. SEESAWS When a seesaw is balanced, the distance (in feet) each person is from the fulcrum is inversely proportional to that person's weight. Use the information in the illustration to determine how far away from the fulcrum Brandon is sitting.

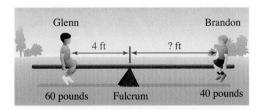

61. HOOKE'S LAW The distance that a spring will stretch varies directly as the force applied to it. Suppose that a 15-kilogram weight stretches a spring 24 centimeters. Find the distance that the same spring will stretch when a 25-kilogram weight is used.

62. ARCHITECTURE The total numbers of windows needed in the construction of an apartment building varies directly as the number of floors. If a 4-story building requires 176 windows, how many windows will an 11-story building require?

63. COMPUTING PRESSURES If the temperature of a gas is constant, the volume occupied varies inversely as the pressure. If a gas occupies a volume of 40 cubic meters under a pressure of 8 atmospheres, find the volume when the pressure is changed to 6 atmospheres.

64. ARCHITECTURE If an office building is to have a fixed total floor space, the number of square feet of ground space it must occupy varies inversely as the number of floors. Suppose a 5-story building will occupy 163,000 square feet of ground space. How many square feet will a 25-story building occupy?

65. HARD DRIVES The number of revolutions made by a computer hard drive varies directly as time. If it makes 16,200 revolutions in 3 minutes, how many revolutions will it make in 45 minutes?

66. DEPRECIATION Assume that the value of a machine varies inversely as its age. If a drill press is worth $300 when it is 2 years old, find its value when it is 6 years old. How much has the machine depreciated over that 4-year period?

WRITING

67. Give two examples of quantities that vary directly and two that do not.

68. What is the difference between direct variation and inverse variation?

69. What is a constant of variation?

Is there a direct variation or an inverse variation between each pair of quantities? Explain why.

70. The time it takes to type a term paper and the speed at which you type.

71. The time it takes to type a term paper (working at a constant rate) and the length of the term paper.

72. The time it takes to type a term paper and the amount of time you did research.

REVIEW

Solve each equation.

73. $x^2 - 5x - 6 = 0$
74. $x^2 - 25 = 0$
75. $(t + 2)(t^2 + 7t + 12) = 0$
76. $2(y - 4) = -y^2$
77. $y^3 - y^2 = 0$
78. $5a^3 - 125a = 0$
79. $(x^2 - 1)(x^2 - 4) = 0$
80. $6t^3 + 35t^2 = 6t$

STUDY SKILLS CHECKLIST

Preparing for the Chapter 7 Test

There are several common mistakes that students make when working with the topics of Chapter 7. To make sure you are prepared for your test, study this checklist of common mistakes below.

☐ When simplifying fractions that have more than one term in the numerator and/or the denominator, factor the numerator and denominator using the factoring strategies from Chapter 6. When you have completely factored the numerator and denominator, remove all common factors.

$$\frac{a^2 - 3a - 10}{a^2 - 25} = \frac{(a-5)(a+2)}{(a-5)(a+5)} = \frac{a+2}{a+5}$$

☐ To multiply fractions, you do not need a common denominator. Simply factor across the numerators and factor across the denominators and remove common factors.

$$\frac{x^2 - x}{x^2 - 1} \cdot \frac{x^2 - x - 2}{x - 2} = \frac{x(x-1)(x-2)(x+1)}{(x+1)(x-1)(x-2)} = x$$

☐ To divide fractions, make sure to multiply the first fraction by the reciprocal of the *second* fraction.

$$\frac{4a^2}{3b^2} \div \frac{8a^2}{3b} = \frac{4a^2}{3b^2} \cdot \frac{3b}{8a^2} = \frac{4 \cdot a^2 \cdot 3 \cdot b}{3 \cdot b^2 \cdot 2 \cdot 4 \cdot a^2} = \frac{1}{2b}$$

☐ When adding or subtracting fractions, you must have a common denominator. Then you add or subtract across the numerator and *keep the common denominator.*

$$\frac{2}{x+5} + \frac{6}{x} = \frac{2x}{x(x+5)} + \frac{6(x+5)}{x(x+5)}$$
$$= \frac{2x + 6x + 30}{x(x+5)} = \frac{8x + 30}{x(x+5)}$$

☐ When subtracting a fraction with more than one term in the numerator, make sure you distribute the negative through the entire numerator being subtracted.

$$\frac{5x+7}{x(x-2)} - \frac{3x-5}{x(x-2)} = \frac{5x + 7 - (3x - 5)}{x(x-2)}$$
$$= \frac{5x + 7 - 3x + 5}{x(x-2)} = \frac{2x + 12}{x(x-2)}$$

☐ When *solving equations* that involve rational expressions, you can remove the fractions by multiplying both sides by the LCD of the entire equation. When working with *rational expressions,* you must work with the fractions that are in the expression.

CHAPTER 7 SUMMARY AND REVIEW

SECTION 7.1 Simplifying Rational Expressions

DEFINITIONS AND CONCEPTS	EXAMPLES
A **rational expression** is an expression of the form $\frac{A}{B}$, where A and B are polynomials and $B \neq 0$.	Rational expressions: $\frac{8}{7t}$, $\frac{a}{a-3}$, and $\frac{4x^2 - 16x}{x^2 - 6x + 8}$

Chapter 7 Summary and Review

To **evaluate a rational expression**, substitute the values of its variables and simplify.	Evaluate $\dfrac{3x+1}{x-2}$ for $x=5$. $\dfrac{3x+1}{x-2} = \dfrac{3(5)+1}{5-2} = \dfrac{16}{3}$ Substitute 5 for x.
To find the real numbers for which a **rational expression is undefined**, find the values of the variable that make the denominator 0.	For which real numbers is $\dfrac{11}{2x-3}$ undefined? $2x-3=0$ Set the denominator equal to 0 and solve for x. $2x=3$ $x=\dfrac{3}{2}$ The expression is undefined for $x=\dfrac{3}{2}$.
Fundamental property of fractions: If A, B, and C are polynomials and B and C are not 0, $\dfrac{AC}{BC} = \dfrac{A}{B}$	$\dfrac{3a}{ab} = \dfrac{3\cdot\cancel{a}}{b\cdot\cancel{a}} = \dfrac{3}{b}$ $\dfrac{4t^3}{bst^2} = \dfrac{4t\cdot\cancel{t^2}}{bs\cdot\cancel{t^2}} = \dfrac{4t}{bs}$
To simplify a rational expression: 1. Factor the numerator and denominator completely. 2. Remove all factors equal to 1. 3. Multiply the remaining factors in the numerator and denominator.	$\dfrac{x^2-4}{x^2-7x+10} = \dfrac{(x+2)\cancel{(x-2)}}{(x-5)\cancel{(x-2)}}$ Factor and simplify. $= \dfrac{x+2}{x-5}$
The quotient of any nonzero expression and its opposite is -1.	$\dfrac{2t-3}{3-2t} = -1$ Because $2t-3$ and $3-2t$ are opposites.

REVIEW EXERCISES

1. Evaluate $\dfrac{x^2-1}{x-5}$ for $x=-2$.

2. Find the values of x for which the rational expression $\dfrac{x-1}{x^2-16}$ is undefined.

Write each fraction in simplest form. If it is already in simplest form, so indicate.

3. $\dfrac{10}{25}$

4. $-\dfrac{12}{18}$

Simplify each rational expression, if possible. Assume that no denominators are 0.

5. $\dfrac{3x^2}{6x^3}$

6. $\dfrac{5xy^2}{2x^2y^2}$

7. $\dfrac{x^2}{x^2+x}$

8. $\dfrac{a^2-4}{a+2}$

9. $\dfrac{3p-2}{2-3p}$

10. $\dfrac{8-x}{x^2-5x-24}$

11. $\dfrac{2x^2-16x}{2x^2-18x+16}$

12. $\dfrac{x^2+x-2}{x^2-x-2}$

13. Explain the error in the following work:

$\dfrac{x+1}{x} = \dfrac{\cancel{x}+1}{\cancel{x}} = \dfrac{2}{1}$

14. Simplify: $\dfrac{4(t+3)+8}{3(t+3)+6}$

SECTION 7.2 Multiplying and Dividing Rational Expressions

DEFINITIONS AND CONCEPTS	EXAMPLES
To **multiply rational expressions**, multiply their numerators and multiply their denominators. For any two rational expressions $\dfrac{A}{B}$ and $\dfrac{C}{D}$, $$\dfrac{A}{B} \cdot \dfrac{C}{D} = \dfrac{AC}{BD}$$	$\dfrac{4b}{b+2} \cdot \dfrac{7}{b} = \dfrac{4b \cdot 7}{(b+2)b}$ Write the product of the numerators. Write the product of the denominators. $= \dfrac{4\overset{1}{b} \cdot 7}{(b+2)\underset{1}{b}}$ Simplify by removing the common factors. $= \dfrac{28}{b+2}$
To write the **reciprocal** of a rational expression, invert its numerator and denominator.	The reciprocal of $\dfrac{c}{c-7}$ is $\dfrac{c-7}{c}$.
To **divide rational expressions**, multiply the first expression by the reciprocal of the second. For any two rational expressions $\dfrac{A}{B}$ and $\dfrac{C}{D}$, where $\dfrac{C}{D} \neq 0$, $$\dfrac{A}{B} \div \dfrac{C}{D} = \dfrac{A}{B} \cdot \dfrac{D}{C}$$	$\dfrac{t}{t+1} \div \dfrac{8}{t^2+t} = \dfrac{t}{t+1} \cdot \dfrac{t^2+t}{8}$ $= \dfrac{t \cdot t(\overset{1}{t+1})}{(\underset{1}{t+1}) \cdot 8}$ Factor and simplify. $= \dfrac{t^2}{8}$
A **unit conversion factor** is a fraction containing units that has a value of 1.	$\dfrac{1 \text{ yd}^2}{9 \text{ ft}^2} = 1$ and $\dfrac{1 \text{ mi}}{5{,}280 \text{ ft}} = 1$

REVIEW EXERCISES

Perform each multiplication and simplify.

15. $\dfrac{3xy}{2x} \cdot \dfrac{4x}{2y^2}$

16. $56x \left(\dfrac{12}{7x} \right)$

17. $\dfrac{x^2 - 1}{x^2 + 2x} \cdot \dfrac{x}{x+1}$

18. $\dfrac{x^2 + x}{3x - 15} \cdot \dfrac{6x - 30}{x^2 + 2x + 1}$

Perform each division and simplify.

19. $\dfrac{3x^2}{5x^2 y} \div \dfrac{6x}{15xy^2}$

20. $\dfrac{x^2 + 5x}{x^2 + 4x - 5} \div \dfrac{x^2}{x - 1}$

21. $\dfrac{x^2 - x - 6}{2x - 1} \div \dfrac{x^2 - 2x - 3}{2x^2 + x - 1}$

22. Simplify: $(b + 2) \div \dfrac{b^2 - 4}{b + 2}$

Determine whether each fraction is a unit conversion factor.

23. $\dfrac{1 \text{ ft}}{12 \text{ in.}}$

24. $\dfrac{60 \text{ min}}{1 \text{ day}}$

25. $\dfrac{1 \text{ gal}}{4 \text{ qt}}$

26. TRAFFIC SIGNS Convert the speed limit on the sign to miles per minute.

SECTION 7.3 Adding and Subtracting with Like Denominators; Least Common Denominators

DEFINITIONS AND CONCEPTS

To **add (or subtract) rational expressions** that have the same denominator, add (or subtract) their numerators and write the sum (or difference) over their common denominator.

For any two rational expressions $\frac{A}{D}$ and $\frac{B}{D}$,

$$\frac{A}{D} + \frac{B}{D} = \frac{A+B}{D}$$

$$\frac{A}{D} - \frac{B}{D} = \frac{A-B}{D}$$

EXAMPLES

Add: $\dfrac{2b}{3b-9} + \dfrac{b}{3b-9} = \dfrac{2b+b}{3b-9}$

$$= \dfrac{\overset{1}{3b}}{\underset{1}{3}(b-3)} \quad \text{Factor and simplify.}$$

$$= \dfrac{b}{b-3}$$

Subtract: $\dfrac{x+1}{x} - \dfrac{x-1}{x} = \dfrac{x+1-(x-1)}{x}$ Don't forget the parentheses.

$$= \dfrac{x+1-x+1}{x}$$

$$= \dfrac{2}{x}$$

To find the **LCD** of several rational expressions, factor each denominator. Then form a product using each different factor the greatest number of times it appears in any one factorization.

Find the LCD of $\dfrac{3}{x^3-x^2}$ and $\dfrac{x}{x^2-1}$.

$$\left. \begin{array}{l} x^3-x^2 = x \cdot x \cdot (x-1) \\ x^2-1 = (x+1)(x-1) \end{array} \right\} \text{LCD} = x \cdot x \cdot (x-1)(x+1)$$

To **build an equivalent rational expression**, multiply the given expression by 1 written in the form $\dfrac{c}{c}$, where $c \neq 0$.

$\dfrac{7}{4t} = \dfrac{7}{4t} \cdot \dfrac{3t}{3t}$

$= \dfrac{21t}{12t^2}$

$\dfrac{x+1}{x-7} = \dfrac{x+1}{x-7} \cdot \dfrac{x-1}{x-1}$

$= \dfrac{(x+1)(x-1)}{(x-7)(x-1)}$

$= \dfrac{x^2-1}{x^2-8x+7}$

REVIEW EXERCISES

Perform each operation. Simplify all answers.

27. $\dfrac{13}{15d} - \dfrac{8}{15d}$

28. $\dfrac{x}{x+y} + \dfrac{y}{x+y}$

29. $\dfrac{3x}{x-7} - \dfrac{x-2}{x-7}$

30. $\dfrac{a}{a^2-2a-8} + \dfrac{2}{a^2-2a-8}$

Find the LCD of each pair of rational expressions.

31. $\dfrac{12}{x}, \dfrac{1}{9}$

32. $\dfrac{1}{2x^3}, \dfrac{5}{8x}$

33. $\dfrac{7}{m}, \dfrac{m+2}{m-8}$

34. $\dfrac{x}{5x+1}, \dfrac{5x}{5x-1}$

35. $\dfrac{6-a}{a^2-25}, \dfrac{a^2}{a-5}$

36. $\dfrac{4t+25}{t^2+10t+25}, \dfrac{t^2-7}{2t^2+17t+35}$

Build each rational expression into an equivalent rational expression having the denominator shown in red.

37. $\dfrac{9}{a}, 7a$

38. $\dfrac{2y+1}{x-9}, x(x-9)$

39. $\dfrac{b+7}{3b-15}, 6(b-5)$

40. $\dfrac{9r}{r^2+6r+5}, (r+1)(r-4)(r+5)$

SECTION 7.4 Adding and Subtracting with Unlike Denominators

DEFINITIONS AND CONCEPTS

To **add (or subtract) rational expressions** with unlike denominators:

1. Find the LCD.
2. Write each rational expression as an equivalent expression whose denominator is the LCD.
3. Add (or subtract) the numerators and write the sum (or difference) over the LCD.
4. Simplify the resulting rational expression, if possible.

EXAMPLES

Add: $\dfrac{4x}{x} + \dfrac{2}{x-1} = \dfrac{4x}{x} \cdot \dfrac{x-1}{x-1} + \dfrac{2}{x-1} \cdot \dfrac{x}{x}$ The LCD is $x(x-1)$.

$= \dfrac{4x(x-1)}{x(x-1)} + \dfrac{2x}{x(x-1)}$

$= \dfrac{4x^2 - 4x + 2x}{x(x-1)}$ Distribute the multiplication by $4x$.

$= \dfrac{4x^2 - 2x}{x(x-1)}$ Combine like terms in the numerator.

$= \dfrac{\overset{1}{2x}(2x-1)}{\underset{1}{x}(x-1)}$ Factor the numerator and simplify.

$= \dfrac{2(2x-1)}{x-1}$

When a polynomial is multiplied by -1, the result is its opposite. This fact is used when adding (or subtracting) rational expressions whose **denominators are opposites**.

Subtract:

$\dfrac{c}{c-4} - \dfrac{1}{4-c} = \dfrac{c}{c-4} - \dfrac{1}{4-c} \cdot \dfrac{-1}{-1}$

$= \dfrac{c}{c-4} - \dfrac{-1}{c-4}$ In the denominator: $-1(4-c) = c - 4$.

$= \dfrac{c-(-1)}{c-4}$

$= \dfrac{c+1}{c-4}$

REVIEW EXERCISES

Perform each operation. Simplify all answers.

41. $\dfrac{1}{7} - \dfrac{1}{c}$

42. $\dfrac{x}{x-1} + \dfrac{1}{x}$

43. $\dfrac{2t+2}{t^2+2t+1} - \dfrac{1}{t+1}$

44. $\dfrac{x+2}{2x} - \dfrac{2-x}{x^2}$

45. $\dfrac{x}{x+2} + \dfrac{3}{x} - \dfrac{4}{x^2+2x}$

46. $\dfrac{6}{b-1} - \dfrac{b}{1-b}$

47. A student added two rational expressions and obtained $\dfrac{-5n^3 - 7}{3n(n+6)}$. Another student obtained $-\dfrac{5n^3 + 7}{3n(n+6)}$. Are the answers equivalent?

48. **VIDEO CAMERAS** Find the perimeter and the area of the LED screen of the camera. The dimensions are measured in cm.

SECTION 7.5 Simplifying Complex Fractions

DEFINITIONS AND CONCEPTS

Complex fractions contain fractions in their numerators and/or their denominators.

EXAMPLES

Complex fractions: $\dfrac{\dfrac{2}{t}}{\dfrac{5}{4t}}$ and $\dfrac{\dfrac{3}{m}+\dfrac{m}{4}}{\dfrac{m}{2}}$

To simplify a complex fraction:

Method 1

Write the numerator and denominator of the complex fraction each as a single rational expression. Then perform the indicated division and simplify.

Simplify: $\dfrac{\dfrac{3}{m}+\dfrac{m}{2}}{\dfrac{m}{4}} = \dfrac{\dfrac{3}{m}\cdot\dfrac{2}{2}+\dfrac{m}{2}\cdot\dfrac{m}{m}}{\dfrac{m}{4}}$ In the numerator, build to have an LCD of $2m$.

$= \dfrac{\dfrac{6}{2m}+\dfrac{m^2}{2m}}{\dfrac{m}{4}}$ The main fraction bar indicates division.

$= \dfrac{\dfrac{6+m^2}{2m}}{\dfrac{m}{4}}$ Add the fractions in the numerator.

$= \dfrac{6+m^2}{2m}\cdot\dfrac{4}{m}$ Multiply by the reciprocal of $\dfrac{m}{4}$.

$= \dfrac{(6+m^2)\cdot\overset{1}{2}\cdot 2}{\underset{1}{2}\cdot m\cdot m}$ Factor and simplify.

$= \dfrac{12+2m^2}{m^2}$ Distribute the multiplication by 2.

Method 2

Determine the LCD of the rational expressions in the complex fraction and multiply the complex fraction by 1, written in the form $\dfrac{\text{LCD}}{\text{LCD}}$. Then simplify, if possible.

Simplify: $\dfrac{\dfrac{3}{m}+\dfrac{m}{2}}{\dfrac{m}{4}} = \dfrac{\dfrac{3}{m}+\dfrac{m}{2}}{\dfrac{m}{4}}\cdot\dfrac{4m}{4m}$ The LCD for all the rational expressions is $4m$.

$= \dfrac{\dfrac{3}{m}\cdot 4m+\dfrac{m}{2}\cdot 4m}{\dfrac{m}{4}\cdot 4m}$ In the numerator, distribute the multiplication by $4m$.

$= \dfrac{12+2m^2}{m^2}$ Do each multiplication by $4m$.

REVIEW EXERCISES

Simplify each complex fraction.

49. $\dfrac{\dfrac{3}{2}}{\dfrac{2}{3}}$

50. $\dfrac{\dfrac{3}{2}+1}{\dfrac{2}{3}+1}$

51. $\dfrac{\dfrac{n^4}{30}}{\dfrac{7n}{15}}$

52. $\dfrac{\dfrac{r^2-81}{18s^2}}{\dfrac{4r-36}{9s}}$

53. $\dfrac{\dfrac{1}{y}+1}{\dfrac{1}{y}-1}$

54. $\dfrac{1+\dfrac{3}{x}}{2-\dfrac{1}{x^2}}$

55. $\dfrac{\dfrac{2}{x-1}+\dfrac{x-1}{x+1}}{\dfrac{1}{x^2-1}}$

56. $\dfrac{\dfrac{1}{x^2y}-\dfrac{5}{xy}}{\dfrac{3}{xy}-\dfrac{7}{xy^2}}$

SECTION 7.6 Solving Rational Equations

DEFINITIONS AND CONCEPTS

To **solve a rational equation**, use these steps:

1. Determine which numbers cannot be solutions.
2. Clear the equation of fractions by multiplying both sides of the equation by the LCD of the rational expressions contained in the equation.
3. Solve the resulting equation.
4. Check all possible solutions in the *original* equation. An apparent solution that does not satisfy the original equation is called an **extraneous solution**.

EXAMPLES

Solve:

$$\frac{y}{y-2} - 1 = \frac{1}{y}$$ Since no denominator can be 0, $y \neq 2$ and $y \neq 0$.

$$y(y-2)\left[\frac{y}{y-2} - 1\right] = y(y-2)\left(\frac{1}{y}\right)$$ The LCD is $y(y-2)$.

$$y(y-2)\left(\frac{y}{y-2}\right) - y(y-2)(1) = y(y-2)\left(\frac{1}{y}\right)$$ Distribute and simplify.

$$y \cdot y - y(y-2) = (y-2) \cdot 1$$

$$y^2 - y^2 + 2y = y - 2$$ Remove parentheses.

$$2y = y - 2$$ Combine like terms.

$$y = -2$$

REVIEW EXERCISES

Solve each equation and check all answers.

57. $\dfrac{3}{x} = \dfrac{2}{x-1}$

58. $\dfrac{a}{a-5} = 3 + \dfrac{5}{a-5}$

59. $\dfrac{2}{3t} + \dfrac{1}{t} = \dfrac{5}{9}$

60. $a = \dfrac{3a-50}{4a-24} - \dfrac{3}{4}$

61. $\dfrac{4}{x+2} - \dfrac{3}{x+3} = \dfrac{6}{x^2+5x+6}$

62. $\dfrac{3}{x+1} - \dfrac{x-2}{2} = \dfrac{x-2}{x+1}$

63. Solve for y: $\dfrac{1}{x} = \dfrac{1}{y} + \dfrac{1}{z}$

64. ENGINEERING The efficiency E of a Carnot engine is given by the following formula. Solve it for T_1.

$$E = 1 - \frac{T_2}{T_1}$$

SECTION 7.7 Problem Solving Using Rational Equations

DEFINITIONS AND CONCEPTS

To solve applications problems, follow these steps:

1. Analyze the problem.
2. Assign a variable.
3. Form an equation.
4. Solve the equation.
5. State the conclusion.
6. Check the result.

EXAMPLES

WASHING CARS Working alone, Carlos can wash the family car in 30 minutes. Victor, his brother, can wash the same car in 20 minutes. How long will it take them if they work together?

Analyze It takes Carlos 30 minutes and it takes Victor 20 minutes. How long will it take working together?

Assign Let $x =$ the number of minutes it will take Carlos and Victor, working together, to wash the SUV.

Form

What Carlos can do in one minute	plus	what Victor can do in one minute	equals	what they can do together in one minute.
$\dfrac{1}{30}$	+	$\dfrac{1}{20}$	=	$\dfrac{1}{x}$

Solve

$$\frac{1}{30} + \frac{1}{20} = \frac{1}{x}$$

$$60x\left(\frac{1}{30} + \frac{1}{20}\right) = 60x\left(\frac{1}{x}\right) \quad \text{Multiply both sides by the LCD, } 60x.$$

$$2x + 3x = 60 \quad \text{Do the multiplications.}$$

$$5x = 60 \quad \text{Combine like terms.}$$

$$x = 12 \quad \text{Divide both sides by 5.}$$

State Working together, it will take Carlos and Victor 12 minutes to wash the family SUV.

Check In 12 minutes, Carlos will do $\frac{12}{30} = \frac{24}{60}$ of the job and Victor will do $\frac{12}{20} = \frac{36}{60}$ of the job. Together they will do $\frac{24}{60} + \frac{36}{60} = \frac{60}{60}$ or 1 whole job. The result checks.

Uniform motion problems:
Distance = rate · time

See Example 3 in Section 7.7.

Investment problems:
Interest = principal · rate · time

See Example 4 in Section 7.7.

REVIEW EXERCISES

65. NUMBER PROBLEM If a number is subtracted from the denominator of $\frac{4}{5}$ and twice as much is added to the numerator, the result is 5. Find the number.

66. HOUSE CLEANING If a maid can clean a house in 4 hours, how much of the house does she clean in 1 hour?

67. PAINTING HOUSES If a homeowner can paint a house in 14 days and a professional painter can paint it in 10 days, how long will it take if they work together?

68. EXERCISE A woman can bicycle 30 miles in the same time that it takes her to jog 10 miles. If she can ride 10 mph faster than she can jog, how fast can she jog?

69. WIND SPEED A plane flies 400 miles downwind in the same amount of time as it takes to travel 320 miles upwind. If the plane can fly at 360 mph in still air, find the velocity of the wind.

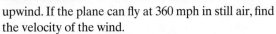

70. INVESTMENTS In one year, a student earned $100 interest on money she deposited at a savings and loan. She later learned that the money would have earned $120 if she had deposited it at a credit union, because the credit union paid 1% more interest at the time. Find the rate she received from the savings and loan.

SECTION 7.8 Proportions and Similar Triangles

DEFINITIONS AND CONCEPTS

A **ratio** is the quotient of two numbers or two quantities with the same units.

A **rate** is the quotient of two quantities with different units.

EXAMPLES

Ratios: $\dfrac{2}{3}$, $\dfrac{1}{50}$, and 2:3

Rates: $\dfrac{4 \text{ oz}}{6 \text{ lb}}$, $\dfrac{525 \text{ mi}}{15 \text{ hr}}$, $\dfrac{\$1.95}{2 \text{ lb}}$

A **proportion** is a statement that two ratios or two rates are equal. In the proportion $\dfrac{a}{b} = \dfrac{c}{d}$, a and d are the **extremes**, and b and c are the **means**.

Proportions: $\dfrac{4}{9} = \dfrac{28}{63}$ Extremes: 4 and 63; $4 \cdot 63 = 252$.
Means: 9 and 28; $9 \cdot 28 = 252$.

In any proportion, the product of the extremes is equal to the product of the means.

To solve a proportion, set the product of the extremes equal to the product of the means and solve the resulting equation.

Solve the proportion: $\dfrac{3}{2} = \dfrac{x}{10}$

$\dfrac{3}{2} = \dfrac{x}{10}$

$3 \cdot 10 = 2 \cdot x$ The product of the extremes is equal to the product of the means.

$30 = 2x$

$15 = x$ Solve for x.

Thus, x is 15.

The measures of corresponding sides of **similar triangles** are in proportion.

In these similar triangles: $\dfrac{a}{d} = \dfrac{b}{e} = \dfrac{c}{f}$

REVIEW EXERCISES

71. Find the ratio of the number of teeth of the larger gear to the number of teeth of the smaller gear.

72. Determine whether $\dfrac{4}{7} = \dfrac{20}{34}$ is a proportion.

Solve each proportion.

73. $\dfrac{3}{x} = \dfrac{6}{9}$

74. $\dfrac{x}{3} = \dfrac{x}{5}$

75. $\dfrac{x-2}{5} = \dfrac{x}{7}$

76. $\dfrac{2x}{x+4} = \dfrac{3}{x-1}$

77. DENTISTRY The diagram in the illustration was displayed in a dentist's office. According to the diagram, if the dentist has 340 adult patients, how many will develop gum disease?

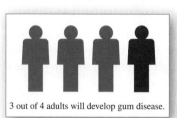

3 out of 4 adults will develop gum disease.

78. A telephone pole casts a shadow 12 feet long at the same time that a man 6 feet tall casts a shadow 3.6 feet long. How tall is the pole?

SECTION 7.9 Variation

DEFINITIONS AND CONCEPTS

Direct variation: As one variable gets larger, the other gets larger as described by the equation $y = kx$, where k is the **constant of variation**.

EXAMPLES

The time t it takes to order at a fast-food drive-through *varies directly* as the number n of cars in line: $t = kn$.

Inverse variation: As one variable gets larger, the other gets smaller as described by the equation $y = \dfrac{k}{x}$, where k is a constant.

The time t it takes to read a book *varies inversely* as the reader's reading rate r: $t = \dfrac{k}{r}$.

Strategy for Solving Variation Problems

1. Translate the verbal model into an equation.
2. Substitute a pair of values to find k.
3. Substitute the value of k into the variation equation.
4. Substitute the remaining given value into the equation from step 3 and answer the question.

Suppose d varies inversely with h. If $d = 5$ when $h = 4$, find d when $h = 10$.

1. The words d *varies inversely as* h translate into $d = \dfrac{k}{h}$.
2. If we substitute 5 for d and 4 for h, we have
$$5 = \dfrac{k}{4} \quad \text{or} \quad k = 20$$
3. Since $k = 20$, the inverse variation equation is $d = \dfrac{20}{h}$.
4. To answer the final question, substitute 10 for h.
$$d = \dfrac{20}{10} = 2$$

REVIEW EXERCISES

Change each verbal model into a variation equation.

79. PHYSICAL FITNESS The number of calories c burned while jogging varies directly as the time t spent jogging.

80. GUITARS The frequency f of a vibrating string varies inversely as the length of the string.

Solve each variation problem.

81. PROFIT The profit made by a strawberry farm varies directly as the number of baskets of strawberries sold. If a profit of $500 was made from the sale of 750 baskets, find the profit if 1,250 baskets are sold.

82. ELECTRICITY For a fixed voltage, the current in an electrical circuit varies inversely as the resistance in the circuit. If a certain circuit has a current of $2\frac{1}{2}$ amps when the resistance is 150 ohms, find the current in the circuit when the resistance is doubled.

83. Give an example of two quantities that vary directly.

84. Does the graph illustrate direct or inverse variation?

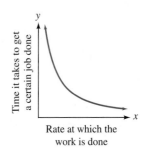

CHAPTER 7 TEST

1. Fill in the blanks.
   a. A quotient of two polynomials, such as $\dfrac{x+7}{x^2+2x}$, is called a _____ expression.
   b. Two triangles with the same shape, but not necessarily the same size, are called _____ triangles.
   c. A _____ is a mathematical statement that two ratios or two rates are equal.
   d. To _____ a rational expression, we multiply it by a form of 1. For example, $\dfrac{2}{5x} \cdot \dfrac{8}{8} = \dfrac{16}{40x}$.
   e. To simplify $\dfrac{x-3}{(x+3)(x-3)}$, we remove common _____ of the numerator and denominator.

2. Translate the ratio 9 letters to 24 letters to a ratio in simplest form.

3. Translate the ratio 2 feet to 12 yards to a ratio in simplest form.

4. Complete the unit conversion:
   $$\dfrac{57 \text{ meters}}{2 \text{ months}} \cdot \dfrac{12 \text{ months}}{1 \text{ year}}$$

5. Find the values of x for which $\dfrac{x}{x^2+x-6}$ is undefined.

6. Simplify: $\dfrac{48x^2y}{54xy^2}$

7. Simplify: $\dfrac{2x^2-x-3}{4x^2-9}$

8. Simplify: $\dfrac{3(x+2)-3}{2x+3-(x+2)}$

9. Multiply and simplify: $-\dfrac{12x^2y}{15xy} \cdot \dfrac{25y^2}{16x}$

10. Multiply and simplify: $\dfrac{x^2+3x+2}{3x+9} \cdot \dfrac{x+3}{x^2-4}$

11. Divide and simplify: $\dfrac{8x^2}{25x} \div \dfrac{16x^2}{30x}$

12. Divide and simplify: $\dfrac{x-x^2}{3x^2+6x} \div \dfrac{3x-3}{3x^3+6x^2}$

13. Simplify: $\dfrac{x^2+x}{x-1} \cdot \dfrac{x^2-1}{x^2-2x} \div \dfrac{x^2+2x+1}{x^2-4}$

14. Add: $\dfrac{5x-4}{x-1} + \dfrac{5x+3}{x-1}$

15. Subtract: $\dfrac{3y+7}{2y+3} - \dfrac{3(y-2)}{2y+3}$

16. Add: $\dfrac{x+1}{x} + \dfrac{x-1}{x+1}$

17. Subtract: $\dfrac{a+3}{a-1} - \dfrac{a+4}{1-a}$

18. Subtract: $\dfrac{2n}{5m} - \dfrac{n}{2}$

19. Simplify: $\dfrac{1+\dfrac{y}{x}}{\dfrac{y}{x}-1}$

20. Solve: $\dfrac{1}{3} + \dfrac{4}{3y} = \dfrac{5}{y}$

21. Solve: $\dfrac{7}{q^2-q-2} + \dfrac{1}{q+1} = \dfrac{3}{q-2}$

22. Solve: $\dfrac{2}{3} = \dfrac{2c-12}{3c-9} - c$

23. Solve: $\dfrac{9n}{n-6} = 3 + \dfrac{54}{n-6}$

24. Solve for B: $H = \dfrac{RB}{R+B}$

25. Is $\dfrac{3}{5} = \dfrac{51}{85}$ a proportion?

26. Solve the proportion: $\dfrac{y}{y-1} = \dfrac{y-2}{y}$

27. HEALTH RISKS A medical newsletter states that a healthy waist-to-hip ratio for men is 19:20 or less. Does the patient shown below fall within the healthy range?

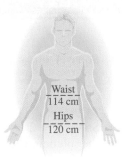

28. CURRENCY EXCHANGE RATES Preparing for a visit to London, a New York resident exchanged 3,500 U.S. dollars for British pounds. (A pound is the basic monetary unit of Great Britain.) If the exchange rate was 100 U.S. dollars for 51 British pounds, how many British pounds did the traveler receive?

29. TV TOWERS A television tower casts a shadow 114 feet long at the same time that a 6-foot-tall television reporter casts a shadow 4 feet long. Find the height of the tower.

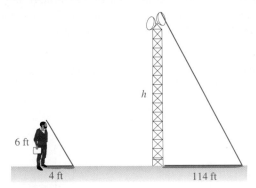

30. FLIGHT PATHS A plane drops 575 feet as it flies a horizontal distance of $\frac{1}{2}$ mile, as shown below. How much altitude will it lose as it flies a horizontal distance of 7 miles?

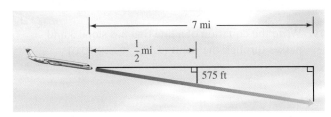

31. POGO STICKS The force required to compress a spring varies directly with the change in the length of the spring. If a force of 130 pounds compresses the spring on the pogo stick 6.5 inches, how much force is required to compress the spring 5 inches?

32. If a varies inversely with d, find the constant of variation if $a = 100$ when $d = 2$.

33. CLEANING HIGHWAYS One highway worker can pick up all the trash on a strip of highway in 7 hours, and his helper can pick up the trash in 9 hours. How long will it take them if they work together?

34. BOATING A boat can motor 28 miles downstream in the same amount of time as it can motor 18 miles upstream. Find the speed of the current if the boat can motor at 23 mph in still water.

35. Explain why we can remove the 5's in $\frac{5x}{5}$ and why we can't remove them in $\frac{5 + x}{5}$.

36. Explain what it means to clear the following equation of fractions.

$$\frac{u}{u-1} + \frac{1}{u} = \frac{u^2 + 1}{u^2 - u}$$

Why is this a helpful first step in solving the equation?

CHAPTERS 1–7 CUMULATIVE REVIEW

1. Evaluate: $9^2 - 3[45 - 3(6 - 4)]$ [Section 1.7]

2. GRAND KING SIZE BEDS Because Americans are taller compared to 100 years ago, bed manufacturers are making larger models. Find the percent of increase in sleeping area of the new grand king size bed compared to the standard king size. [Section 2.3]

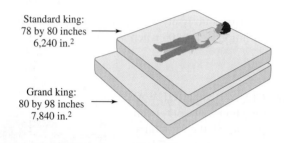

3. Find the average (mean) test score of a student in a history class with scores of 80, 73, 61, 73, and 98. [Section 1.7]

4. What is the value in cents of x 43¢ stamps? [Section 1.8]

5. Solve: $\dfrac{3}{4} = \dfrac{1}{2} + \dfrac{x}{5}$ [Section 2.2]

6. Change 40°C to degrees Fahrenheit. [Section 2.4]

7. Find the volume of a pyramid that has a square base measuring 6 feet on a side and a height of 20 feet. [Section 2.4]

8. Determine whether each statement is true or false.
   a. Every integer is a whole number. [Section 1.3]
   b. 0 is not a rational number. [Section 1.3]
   c. π is an irrational number. [Section 1.3]
   d. The set of integers is the set of whole numbers and their opposites. [Section 1.3]

9. Solve: $2 - 3(x - 5) = 4(x - 1)$ [Section 2.2]

10. Simplify: $8(c + 7) - 2(c - 3)$ [Section 2.2]

11. Solve $A - c = 2B + r$ for B. [Section 2.4]

12. Solve $7x + 2 \geq 4x - 1$ and graph the solution. Then describe the graph using interval notation. [Section 2.7]

13. Solve: $\dfrac{4}{5}d = -4$ [Section 2.2]

14. BLENDING TEA One grade of tea (worth $3.20 per pound) is to be mixed with another grade (worth $2 per pound) to make 20 pounds of a mixture that will be worth $2.72 per pound. How much of each grade of tea must be used? [Section 2.6]

15. SPEED OF A PLANE Two planes are 6,000 miles apart, and their speeds differ by 200 mph. If they travel toward each other and meet in 5 hours, find the speed of the slower plane. [Section 2.6]

16. Graph: $y = 2x - 3$ [Section 3.3]

17. Graph: $y = (x + 2)^3$ [Section 3.2]

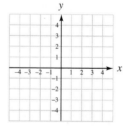 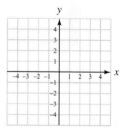

18. Find the slope of the line passing through $(-1, 3)$ and $(3, -1)$. [Section 3.4]

19. Write the equation of a line that has slope 3 and passes through the point $(1, 5)$. [Section 3.6]

20. Graph: $3x - 2y = 6$ [Section 3.3]

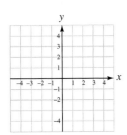

21. Graph: $y = \dfrac{5}{2}$ [Section 3.3]

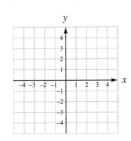

22. What is the slope of a line perpendicular to the line $y = -\dfrac{7}{8}x - 6$? [Section 3.5]

23. a. Is this the graph of a function? [Section 3.8]

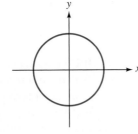

b. Find $f(-4)$ if $f(x) = \dfrac{x^2 - 2x}{2}$. [Section 3.8]

24. CUTTING STEEL The following graph shows the amount of wear (in mm) on a cutting blade for a given length of a cut (in m). Find the rate of change in the length of the cutting blade. [Section 3.4]

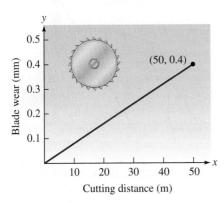

25. Graph: $3x - 2y \leq 6$ [Section 3.7]

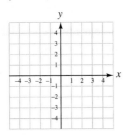

26. Solve the system $\begin{cases} x + y = 1 \\ y = x + 5 \end{cases}$ by graphing. [Section 4.1]

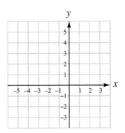

27. Solve the system: $\begin{cases} x = 3y - 1 \\ 2x - 3y = 4 \end{cases}$ [Section 4.2]

28. Solve the system: $\begin{cases} 2x + 3y = -1 \\ 3x + 5y = -2 \end{cases}$ [Section 4.3]

29. POKER After a night of cards, a poker player finished with some red chips (worth $5 each) and some blue chips (worth $10 each). He received $190 when he cashed in the 23 chips. How many of each colored chip did he finish with? [Section 4.4]

30. Evaluate: -5^2 [Section 5.1]

Simplify each expression. Write each answer without using negative exponents.

31. $x^4 x^3$ [Section 5.1] **32.** $(x^2 x^3)^5$ [Section 5.1]

33. $\left(\dfrac{y^3 y}{2yy^2}\right)^3$ [Section 5.1] **34.** $\left(\dfrac{-2a}{b}\right)^5$ [Section 5.1]

35. $(a^{-2}b^3)^{-4}$ [Section 5.2] **36.** $\dfrac{9b^0 b^3}{3b^{-3} b^4}$ [Section 5.2]

37. Write 290,000 in scientific notation. [Section 5.3]

38. What is the degree of the polynomial $5x^3 - 4x + 16$? [Section 5.4]

Perform the operations.

39. $(3x^2 - 3x - 2) + (3x^2 + 4x - 3)$ [Section 5.5]

40. $(2x^2 y^3)(3x^3 y^2)$ [Section 5.6]

41. $(2y - 5)(3y + 7)$ [Section 5.6]

42. $-4x^2 z(3x^2 - z)$ [Section 5.6]

43. $\dfrac{6x + 9}{3}$ [Section 5.7] **44.** $\dfrac{15(r^2 s^3)^2}{-5(rs^5)^3}$ [Section 5.7]

Chapters 1–7 Cumulative Review

45. LICENSE PLATES The number of different license plates of the form three digits followed by three letters, as shown, is $10 \cdot 10 \cdot 10 \cdot 26 \cdot 26 \cdot 26$. Write this expression using exponents. Then evaluate it. [Section 5.1]

46. CONCENTRIC CIRCLES The area of the ring between the two concentric circles of radius r and R is given by the formula

$$A = \pi(R + r)(R - r)$$

Perform the multiplication on the right side of the equation. [Section 5.6]

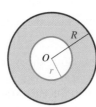

Factor each polynomial completely, if possible.

47. $k^3 t - 3k^2 t$ [Section 6.1]
48. $ax + bx + az + bz$ [Section 6.1]
49. $2a^2 - 200b^2$ [Section 6.4]
50. $b^3 + 125$ [Section 6.5]
51. $u^2 - 18u + 81$ [Section 6.4]
52. $6x^2 - 63 - 13x$ [Section 6.3]
53. $-r^2 + 2 + r$ [Section 6.3]
54. $u^2 + 10u + 15$ [Section 6.2]

Solve each equation by factoring.

55. $5x^2 + x = 0$ [Section 6.7]
56. $6x^2 - 5x = -1$ [Section 6.7]

57. COOKING The electric griddle shown has a cooking surface of 160 square inches. Find the length and the width of the griddle. [Section 6.8]

58. For what values of x is the rational expression $\dfrac{3x^2}{x^2 - 25}$ undefined? [Section 7.1]

Perform the operations and simplify, if possible.

59. $\dfrac{x^2 - 16}{x - 4} \div \dfrac{3x + 12}{x}$ [Section 7.2]

60. $\dfrac{4}{x - 3} + \dfrac{5}{3 - x}$ [Section 7.4]

61. $\dfrac{2 - \dfrac{2}{x+1}}{2 + \dfrac{2}{x}}$ [Section 7.5]

62. $\dfrac{4a}{a - 2} - \dfrac{3a}{a - 3} + \dfrac{4a}{a^2 - 5a + 6}$ [Section 7.4]

Solve each equation.

63. $\dfrac{7}{5x} - \dfrac{1}{2} = \dfrac{5}{6x} + \dfrac{1}{3}$ [Section 7.6]

64. $\dfrac{3}{5} + \dfrac{7}{x + 2} = 2$ [Section 7.6]

65. COMPUTING INTEREST For a fixed rate and principal, the interest earned in a bank account paying simple interest varies directly with the length of time the principal is left on deposit. If an investment earns $700 in 2 years, how much will it earn in 7 years? [Section 7.7]

66. HEIGHT OF A TREE A tree casts a shadow of 29 feet at the same time as a vertical yardstick casts a shadow of 2.5 feet. Find the height of the tree. [Section 7.8]

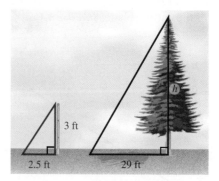

67. DRAINING A TANK If one outlet pipe can drain a tank in 24 hours, and another pipe can drain the tank in 36 hours, how long will it take for both pipes to drain the tank? [Section 7.7]

68. Explain what it means for two variables to vary inversely. [Section 7.9]

8

Roots and Radicals

8.1 Square Roots
8.2 Higher-Order Roots; Radicands That Contain Variables
8.3 Simplifying Radical Expressions
8.4 Adding and Subtracting Radical Expressions
8.5 Multiplying and Dividing Radical Expressions
8.6 Solving Radical Equations; the Distance Formula
8.7 Rational Exponents
  Chapter Summary and Review
  Chapter Test
  Cumulative Review

Dan Porges/Peter Arnold/Getty Images

from Campus to Careers
Archaeologist

Archaeologists examine and recover material evidence, including the ruins of buildings, tools, pottery, and other objects, remaining from past human cultures. They use these items to determine the history, customs, and living habits of earlier civilizations. Many archaeologists specialize in a particular region of the world. They may work under rugged conditions, and their work may involve strenuous physical exertion.

In **Problem 81** of **Study Set 8.1**, you will see how archaeologists can use the Pythagorean theorem to help them determine how far they are from base camp.

In **Problem 103** of **Study Set 8.3**, you will express the distance between pieces of pottery found at an excavation site as a simplified radical expression.

JOB TITLE: Archaeologist
EDUCATION: Graduates with a master's degree in archaeology usually are qualified for positions outside of colleges and universities. A Ph.D. degree may be required for higher-level positions. Training in statistics and mathematics is essential for many archaeologists.
JOB OUTLOOK: Excellent. Jobs are expected to grow 21% from 2010 to 2020.
ANNUAL EARNINGS: In 2012, the median salary for archaeologists working in various capacities was $54,230.
FOR MORE INFORMATION:
www.bls.gov/oes/current/oes193091.htm

637

Objectives

1. Find square roots of perfect squares.
2. Approximate irrational square roots.
3. Graph the square root function.
4. Use the Pythagorean theorem to solve problems.

SECTION 8.1
Square Roots

Addition and subtraction are reverse operations, and so are multiplication and division. In this section, we will discuss another pair of reverse operations: raising a number to a power and finding the root of a number.

1 Find square roots of perfect squares.

When we raise a number to the second power, we are squaring it, or finding its **square**.

- The square of 5 is 25 because $5^2 = 25$.
- The square of -5 is 25, because $(-5)^2 = 25$.

We can reverse the squaring process to find **square roots** of numbers. For example, to find the square roots of 25, we ask ourselves "What number, when squared, is equal to 25?" There are two possible answers.

- 5 is a square root of 25, because $5^2 = 25$.
- -5 is also a square root of 25, because $(-5)^2 = 25$.

In general, we have the following definition.

> **Square Root**
>
> The number b is a **square root** of a if $b^2 = a$.

Every positive number has two square roots, one positive and one negative. For example, the two square roots of 9 are 3 and -3, and the two square roots of 144 are 12 and -12. The number 0 is the only real number with exactly one square root. In fact, it is its own square root, because $0^2 = 0$.

A **radical symbol** $\sqrt{\ }$ represents the **positive** or **principal square root** of a positive number. When reading this symbol, we usually drop the word *positive* (or *principal*) and simply say *square root*. Since 3 is the positive square root of 9, we can write

$\sqrt{9} = 3$ $\sqrt{9}$ represents the positive number whose square is 9.
Read as "the square root of 9 is 3."

> **The Language of Algebra**   The word **radical** comes from the Latin word *radix*, meaning root. The radical symbol has evolved over the years from ℞ in the 1300s, to $\sqrt{\ }$ in the 1500s, to the familiar $\sqrt{\ }$ in the 1600s.

The symbol $-\sqrt{\ }$ is used to represent the **negative square root** of a positive number. It is the opposite of the principal square root. Since -12 is the negative square root of 144, we can write

$-\sqrt{144} = -12$ Read as "the negative square root of 144 is -12" or "the opposite of the square root of 144 is -12." The notation $-\sqrt{144}$ represents the negative number whose square is 144.

If the number under the radical symbol is 0, we have $\sqrt{0} = 0$.

Square Root Notation

If a is a positive real number,

$\sqrt{a}$ represents the **positive** or **principal square root** of a. It is the positive number we square to get a.

$-\sqrt{a}$ represents the **negative square root** of a. It is the opposite of the principal square root of a: $-\sqrt{a} = -1 \cdot \sqrt{a}$.

The principal square root of 0 is 0: $\sqrt{0} = 0$.

The number or variable expression within (under) a radical symbol is called the **radicand**, and the radical symbol and radicand together are called a **radical**.

Radical symbol ⟶ $\sqrt{16}$ ← Radicand
        Radical

An algebraic expression containing a radical is called a **radical expression**. Some examples of radical expressions are

$$\sqrt{100}, \quad \sqrt{2} + 3, \quad \sqrt{x^2}, \quad \text{and} \quad \sqrt{\frac{a-1}{49}}$$

To evaluate (find the value of) square roots, you need to quickly recognize each of the following natural-number **perfect squares** shown in red:

$1 = 1^2$	$25 = 5^2$	$81 = 9^2$	$169 = 13^2$	$289 = 17^2$
$4 = 2^2$	$36 = 6^2$	$100 = 10^2$	$196 = 14^2$	$324 = 18^2$
$9 = 3^2$	$49 = 7^2$	$121 = 11^2$	$225 = 15^2$	$361 = 19^2$
$16 = 4^2$	$64 = 8^2$	$144 = 12^2$	$256 = 16^2$	$400 = 20^2$

EXAMPLE 1 Find each square root:

a. $\sqrt{16}$ **b.** $\sqrt{1}$ **c.** $\sqrt{0.36}$ **d.** $\sqrt{\frac{4}{9}}$ **e.** $-\sqrt{225}$

Strategy In each case, we will determine the positive number, when squared, that produces the radicand.

WHY The radical symbol $\sqrt{}$ indicates that the positive square root (principal square root) of the number under it should be found.

Solution

a. $\sqrt{16} = 4$ Ask: What positive number, when squared, is 16? The answer is 4.

b. $\sqrt{1} = 1$ Ask: What positive number, when squared, is 1? The answer is 1.

c. $\sqrt{0.36} = 0.6$ Ask: What positive number, when squared, is 0.36? The answer is 0.6.

d. $\sqrt{\frac{4}{9}} = \frac{2}{3}$ Ask: What positive number, when squared, is $\frac{4}{9}$? The answer is $\frac{2}{3}$.

e. $-\sqrt{225}$ is the opposite of the square root of 225. Since $\sqrt{225} = 15$, we have $-\sqrt{225} = -15$ $-\sqrt{225} = -1 \cdot \sqrt{225} = -1 \cdot 15 = -15$.

Self Check 1
Find each square root:
a. $\sqrt{121}$ **b.** $-\sqrt{49}$
c. $\sqrt{0.64}$ **d.** $\sqrt{256}$
e. $\sqrt{\frac{1}{25}}$ **f.** $\sqrt{\frac{9}{49}}$

Now Try Problems 25, 29, and 31

Square roots of certain numbers, such as 7, are hard to compute by hand. However, we can approximate $\sqrt{7}$ with a calculator.

2 Approximate irrational square roots.

To find the principal square root of 7, we can enter 7 into a scientific calculator and press the $\sqrt{x}$ key. The approximate value of $\sqrt{7}$ will appear on the display.

$\sqrt{7} \approx 2.6457513$ Read $\approx$ as "is approximately equal to."

Since $\sqrt{7}$ represents the number that, when squared, gives 7, we would expect squares of approximations of $\sqrt{7}$ to be close to 7.

- Rounded to one decimal place, $\sqrt{7} \approx 2.6$ and $(2.6)^2 = 6.76$.
- Rounded to two decimal places, $\sqrt{7} \approx 2.65$ and $(2.65)^2 = 7.0225$.
- Rounded to three decimal places, $\sqrt{7} \approx 2.646$ and $(2.646)^2 = 7.001316$.

Using Your CALCULATOR   Freeway Road Sign

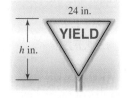

The sign shown in the figure is in the shape of an equilateral triangle, and we can find its height h using the formula

$$h = \frac{\sqrt{3}s}{2}$$

where s is the length of a side of the triangle. In this case, $s = 24$ inches, so we have

$$h = \frac{\sqrt{3}(24)}{2} \quad \sqrt{3}(24) \text{ means } \sqrt{3} \cdot 24.$$

To evaluate this expression with a reverse-entry scientific calculator, we enter these numbers and press these keys.

3 $\boxed{\sqrt{x}}$ $\boxed{\times}$ 24 $\boxed{\div}$ 2 $\boxed{=}$ $\boxed{20.784609}$

To evaluate this expression using a direct-entry or graphing calculator, we press these keys.

$\boxed{\text{2nd}}$ $\boxed{\sqrt{}}$ 3 $\boxed{)}$ $\boxed{\times}$ 24 $\boxed{\div}$ 2 $\boxed{\text{ENTER}}$ $\boxed{\begin{array}{l}\sqrt{}(3)*24/2\\20.78460969\end{array}}$

The height of the sign is approximately 21 inches.

THINK IT THROUGH   Traffic Accidents

"Motor vehicle crashes are the leading cause of death for persons aged 5–34, claiming the lives of 30,000 Americans each year."

The American Association for the Surgery of Trauma

Accident investigators often determine a vehicle's speed prior to braking from the length of the skid marks that it leaves on the street. To do this, they use the formula $s = \sqrt{30Df}$, where s is the speed of the vehicle in mph, D is the skid distance in feet, and f is the drag factor for the road surface. Estimate the speed of each vehicle prior to braking given the following conditions. Round to the nearest mile per hour.

1. Length of skid marks: 71 ft
   Road surface: asphalt, $f = 0.75$
2. Length of skid marks: 133 ft
   Road surface: concrete, $f = 0.90$

EXAMPLE 2   Period of a Pendulum

The *period of a pendulum* is the time required for the pendulum to swing back and forth to complete one cycle. (See the figure.) The period (in seconds) of a pendulum having length L (in feet) is approximated by the function

$$f(L) = 1.11\sqrt{L} \quad \text{Read } 1.11\sqrt{L} \text{ as "1.11 times } \sqrt{L}\text{."}$$

Find the period of a pendulum that is 5 feet long.

Strategy We will substitute 5 for L in the formula, use a calculator to approximate $\sqrt{5}$ and multiply that value by 1.11.

WHY $1.11\sqrt{L}$ means $1.11 \cdot \sqrt{L}$.

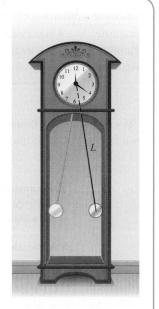

Solution
We substitute 5 for L in the formula and multiply using a calculator.

$f(L) = 1.11\sqrt{L}$
$f(5) = 1.11\sqrt{5}$ $1.11\sqrt{5}$ means $1.11 \cdot \sqrt{5}$.
$f(5) \approx 2.482035455$

The period is approximately 2.5 seconds.

Self Check 2
Find the period of a pendulum that is 3 feet long.
Now Try Problem 37

Whole numbers such as 4, 9, 16, and 49 are called **integer squares**, because each one is the square of an integer. The square root of any integer square is an integer and therefore a rational number:

$$\sqrt{4} = 2, \quad \sqrt{9} = 3, \quad \sqrt{16} = 4, \quad \text{and} \quad \sqrt{49} = 7$$

The square root of any whole number that is not an integer square is an **irrational number**. For example, $\sqrt{7}$ is an irrational number. Recall that the set of rational numbers and the set of irrational numbers together make up the set of real numbers.

It is important to note that the square root of any negative number is *not* a real number. Square roots of negative numbers are studied in more detail in Chapter 9.

EXAMPLE 3

Classify each square root as rational, irrational, or not a real number: **a.** $\sqrt{55}$ **b.** $\sqrt{-81}$ **c.** $-\sqrt{400}$

Strategy We need to determine whether the radicand is positive or negative and whether it is a perfect square.

WHY If a positive number is a perfect square, its square root is rational. If a positive number is not a perfect square, its square root is irrational. The square root of a negative number is not a real number.

Solution
a. Since 55 is positive, but not a perfect square, $\sqrt{55}$ is an irrational number. If we use a calculator and round to two decimal places, we find that $\sqrt{55} \approx 7.42$.
b. $\sqrt{-81}$ is not a real number because it is the square root of a negative number.
c. Since $400 = (20)^2$, it is a perfect square and $-\sqrt{400}$ is rational: $-\sqrt{400} = -20$.

Self Check 3
Classify each square root as rational, irrational, or not a real number:
a. $\sqrt{-6}$ **b.** $-\sqrt{37}$ **c.** $\sqrt{\dfrac{16}{9}}$
Now Try Problem 43

Caution! Square roots of negative numbers are not real numbers. For example, $\sqrt{-4}$ is nonreal, because the square of no real number is -4. The number $\sqrt{-4}$ is an example from a set of numbers called **imaginary numbers**. Remember: *The square root of a negative number is not a real number.*

If we attempt to evaluate $\sqrt{-4}$ using a calculator, an error message like the ones shown below will be displayed.

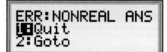

Scientific calculator   Graphing calculator

In this chapter, we will assume that *all radicands under the square root symbols are either positive or zero*. Thus, all square roots will be real numbers.

3 Graph the square root function.

Since there is one principal square root for every nonnegative real number x, the equation $f(x) = \sqrt{x}$ determines a square root function. For example, the value that is determined by $f(x) = \sqrt{x}$ when $x = 4$ is denoted by $f(4)$, and we have $f(4) = \sqrt{4} = 2$.

To graph this function, we make a table of values and plot each ordered pair. Then, from the origin, we draw a smooth curve that passes through the points. In the table, we chose five values for x: 0, 1, 4, 9, and 16, that are integer squares. This made computing $f(x)$ quite simple. The graph appears in the figure below.

$$f(x) = \sqrt{x}$$

x	$f(x)$	$(x, f(x))$
0	0	(0, 0)
1	1	(1, 1)
4	2	(4, 2)
9	3	(9, 3)
16	4	(16, 4)

↑ Values to be input into $\sqrt{x}$
↑ Output values
↑ Ordered pairs to plot

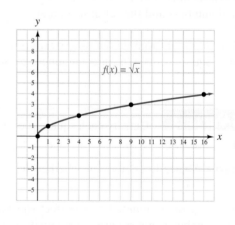

4 Use the Pythagorean theorem to solve problems.

The longest side of a right triangle is the **hypotenuse**, which is the side opposite the right angle. The remaining two sides are the **legs** of the triangle. See the figure to the right. Recall that the **Pythagorean theorem** provides a formula relating the lengths of the three sides of a right triangle.

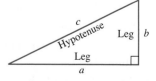

The Pythagorean Theorem

If the length of the hypotenuse of a right triangle is c and the lengths of the two legs are a and b,

$$c^2 = a^2 + b^2$$

Since the lengths of the sides of a triangle are positive numbers, we can use the **square root property of equality** and the Pythagorean theorem to find the length of the third side of any right triangle when the measures of two sides are given.

Square Root Property of Equality

If a and b represent positive numbers, and if $a = b$,

$$\sqrt{a} = \sqrt{b}$$

EXAMPLE 4 *Picture Frames* After gluing together two pieces of picture frame, the maker checks her work by making a diagonal measurement. (See the figure below.) If the sides of the frame form a right angle, what measurement should the maker read on the yardstick?

Analyze The 15- and 20-inch sides of the frame in the figure are the legs of a right triangle, and the diagonal measurement is the hypotenuse. We need to find the length of the diagonal using the Pythagorean theorem.

Assign Let a and b be the two sides of the frame; the diagonal is c.

Form We can use the Pythagorean theorem to form an equation. We substitute 15 for a, 20 for b, and let c represent the length of the hypotenuse.

Solve

$c^2 = a^2 + b^2$	The Pythagorean theorem
$c^2 = \mathbf{15}^2 + \mathbf{20}^2$	Substitute 15 for a and 20 for b.
$c^2 = 225 + 400$	$15^2 = 225$ and $20^2 = 400$.
$c^2 = 625$	Add: $225 + 400 = 625$.

To find c, we must find a number that, when squared, is 625. Since c represents the length of the hypotenuse, c cannot be negative. Thus, we need only determine the positive square root of 625.

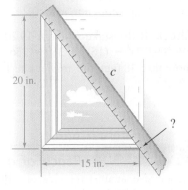

$c^2 = 625$	This is the equation to solve.
$\sqrt{c^2} = \sqrt{625}$	To find c, we undo the operation performed on it by taking the positive square root of both sides.
$c = 25$	$\sqrt{c^2} = c$ because $(c)^2 = c^2$, and $\sqrt{625} = 25$ because $25^2 = 625$.

State The diagonal distance should measure 25 inches. If it does not, the sides of the frame do not form a right angle.

Check If the diagonal is 25 inches, we have $15^2 + 20^2 = 225 + 400 = 625$, which is 25^2. The answer, 25, checks.

Self Check 4

To make certain that the tether ball pole is vertical, a measurement is taken 4 feet up the pole and 3 feet along the ground. What measurement between the point on the pole and the point on the ground would force the angle between the ground and pole to be a right angle?

Now Try Problem 79

> **Success Tip** When using the Pythagorean theorem $c^2 = a^2 + b^2$, we can let a represent the length of either leg of the right triangle in question. We then let b represent the length of the other leg. The variable c must always represent the length of the hypotenuse.

Self Check 5

A support line for a 6-foot badminton pole is 10 feet long. How far from the base of the pole should the line be anchored so that the pole and the ground form a right angle?

Now Try Problem 71

EXAMPLE 5 Building a High Ropes Adventure Course

The builder of a high ropes course wants to use a 25-foot cable to stabilize the vertical pole shown in the figure. To be safe, the ground anchor stake must be farther than 18 feet from the base of the pole. Is the cable long enough to use?

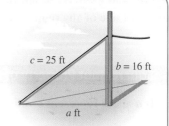

Analyze The pole should make a right angle with the ground, with the cable being the hypotenuse.

Assign We let a represent the distance from the pole to the stake; b is the height of the attachment point and c is the length of the cable.

Form We can use the Pythagorean theorem, with $b = 16$ and $c = 25$, to find a.

Solve

$$c^2 = a^2 + b^2 \quad \text{This is the Pythagorean theorem.}$$
$$25^2 = a^2 + 16^2 \quad \text{Substitute 25 for } c \text{ and 16 for } b.$$
$$625 = a^2 + 256 \quad 25^2 = 625 \text{ and } 16^2 = 256.$$
$$369 = a^2 \quad \text{To isolate } a^2, \text{ subtract 256 from both sides.}$$
$$\sqrt{369} = \sqrt{a^2} \quad \text{To find } a, \text{ undo the operation that is performed on it (squaring) by taking the positive square root of both sides.}$$
$$19.209373 \approx a \quad \text{Use a calculator to approximate } \sqrt{369}.$$

State Since the anchor stake will be more than 18 feet from the base, the 25-foot cable is long enough to use.

Check If the stake is about 19.209373 feet from the base of the pole, we have $(19.209373)^2 + (16)^2 \approx 369 + 256 = 625$, which is 25^2. Since 19.209373 feet is more than 18 feet, the answer checks.

Self Check 6

A 13-foot ladder rests against the side of a building. If the ladder reaches 12 feet up the wall, how far from the building is the base of the ladder?

Now Try Problem 75

EXAMPLE 6 Reach of a Ladder

A 26-foot ladder rests against the side of a building. If the base of the ladder is 10 feet from the wall, how far up the building will the ladder reach?

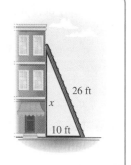

Analyze The wall, the ground, and the ladder form a right triangle, as shown in the figure. In this triangle, the hypotenuse is 26 feet, and one of the legs is the base-to-wall distance of 10 feet.

Assign We can let x = the length of the other leg, which is the distance that the ladder will reach up the wall.

Form We use the Pythagorean theorem to form the equation.

The hypotenuse squared	is	one leg squared	plus	the other leg squared.
26^2	$=$	10^2	$+$	x^2

Solve

$$26^2 = 10^2 + x^2 \quad \text{This is the equation to solve.}$$
$$676 = 100 + x^2 \quad 26^2 = 676 \text{ and } 10^2 = 100.$$
$$676 - 100 = x^2 \quad \text{To isolate } x^2, \text{ subtract 100 from both sides.}$$
$$576 = x^2 \quad 676 - 100 = 576.$$
$$\sqrt{576} = \sqrt{x^2} \quad \text{Take the positive square root of both sides.}$$
$$24 = x \quad \sqrt{576} = 24 \text{ because } 24^2 = 576.$$

State The ladder will reach 24 feet up the side of the building.

Check If the ladder reaches 24 feet up the side of the building, we have $10^2 + 24^2 = 100 + 576 = 676$, which is 26^2. The answer, 24, checks.

EXAMPLE 7 *Roof Design* The gable end of the roof shown in the figure is an isosceles right triangle with a span of 48 feet. Find the distance from the eaves to the peak.

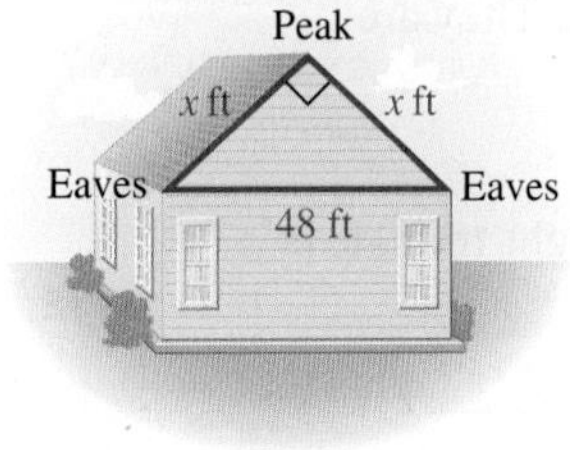

Analyze The two equal sides of the isosceles triangle are the two legs of the right triangle, and the span of 48 feet is the length of the hypotenuse.

Assign We can let x = the length of each leg, which is the distance from eaves to peak.

Form We use the Pythagorean theorem to form the equation.

The hypotenuse squared	is	one leg squared	plus	the other leg squared.
48^2	$=$	x^2	$+$	x^2

Solve

$$48^2 = x^2 + x^2 \quad \text{This is the equation to solve.}$$
$$2{,}304 = 2x^2 \quad 48^2 = 2{,}304 \text{ and } x^2 + x^2 = 2x^2.$$
$$1{,}152 = x^2 \quad \text{To isolate } x^2, \text{ divide both sides by 2.}$$
$$\sqrt{1{,}152} = \sqrt{x^2} \quad \text{Take the positive square root of both sides.}$$
$$33.9411255 \approx x \quad \text{Use a calculator to approximate } \sqrt{1{,}152}.$$

State The eaves-to-peak distance of the roof is approximately 34 feet.

Check If the eaves-to-peak distance is approximately 34 feet, we have $34^2 + 34^2 = 1{,}156 + 1{,}156 = 2{,}312$, which is approximately 48^2. The answer, 34, seems reasonable.

Self Check 7

To find the area of an isosceles right triangle, Sarah must first find its base. If the sides of the triangle are 13 cm, what is the length of the base?

Now Try Problem 87

ANSWERS TO SELF CHECKS

1. a. 11 **b.** -7 **c.** 0.8 **d.** 16 **e.** $\frac{1}{5}$ **f.** $\frac{3}{7}$ **2.** about 1.9 sec **3. a.** not a real number **b.** irrational **c.** rational **4.** 5 ft **5.** 8 ft **6.** 5 ft **7.** about 18.4 cm

SECTION 8.1 STUDY SET

VOCABULARY
Fill in the blanks.

1. b is a _____ root of a if $b^2 = a$.
2. The symbol $\sqrt{}$ is called a _____ symbol.
3. The principal square root of a positive number is a _____ number.
4. The number under the radical sign is called the _____.
5. If a triangle has a right angle, it is called a _____ triangle.
6. The longest side of a right triangle is called the _____, and the other two sides are called _____.

CONCEPTS
Fill in the blanks.

7. The number 25 has _____ square roots. They are _____ and _____.
8. $\sqrt{-11}$ is not a _____ number.
9. If the length of the hypotenuse of a right triangle is c and the legs are a and b, then $c^2 =$ _____.
10. The hypotenuse squared is one leg _____ plus the other leg _____.
11. If a and b are positive numbers and $a = b$, then $\sqrt{a} =$ _____.
12. 2 is a square _____ of 4, because $2^2 = 4$.
13. To isolate x, what step should be used to undo the operation performed on it? (Assume that x is a positive number.)
    a. $2x = 16$
    b. $x^2 = 16$
14. Graph each number on the number line.
    $$\left\{\sqrt{16},\ -\sqrt{\frac{9}{4}},\ \sqrt{1.8},\ \sqrt{6},\ -\sqrt{23}\right\}$$

    ← -5 -4 -3 -2 -1 0 1 2 3 4 5 →

15. Complete the table. Do not use a calculator.

x	$\sqrt{x}$
0	
$\frac{1}{81}$	
0.16	
36	
400	

16. If $f(x) = \sqrt{x}$, find each value. **Do not use a calculator.**
    a. $f(81)$ b. $f(1)$ c. $f(0.25)$
    d. $f\left(\dfrac{1}{121}\right)$ e. $f(900)$

17. a. Use the dashed lines in the following graph to approximate $\sqrt{5}$.
    b. Use the graph to approximate $\sqrt{3}$ and $\sqrt{8}$.

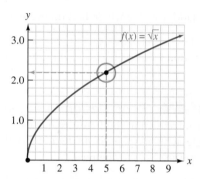

18. A calculator was used to find $\sqrt{-16}$. Explain the message shown on the calculator display.

NOTATION
Complete each step.

19. If the legs of a right triangle measure 5 and 12 centimeters, find the length of the hypotenuse.
$$c^2 = a^2 + b^2$$
$$c^2 = \boxed{}^2 + \boxed{}^2$$
$$c^2 = 25 + \boxed{}$$
$$c^2 = \boxed{}$$
$$\boxed{} = \sqrt{169}$$
$$c = \boxed{}$$

20. If the hypotenuse of a right triangle measures 25 centimeters and one leg measures 24 centimeters, find the length of the other leg.

$$c^2 = a^2 + b^2$$
$$\boxed{}^2 = \boxed{}^2 + b^2$$
$$625 = \boxed{} + b^2$$
$$\boxed{} = b^2$$
$$\sqrt{49} = \boxed{}$$
$$\boxed{} = b$$

21. Is the statement $-\sqrt{9} = \sqrt{-9}$ true or false? Explain your answer.

22. Consider the statement $\sqrt{26} \approx 5.1$. Explain why an $\approx$ symbol is used instead of an $=$ symbol.

GUIDED PRACTICE

Find each square root without using a calculator. See Example 1.

23. $\sqrt{25}$
24. $\sqrt{49}$
25. $\sqrt{196}$
26. $\sqrt{169}$
27. $-\sqrt{81}$
28. $-\sqrt{36}$
29. $\sqrt{1.21}$
30. $\sqrt{1.69}$
31. $\sqrt{\dfrac{9}{256}}$
32. $\sqrt{\dfrac{4}{225}}$
33. $-\sqrt{289}$
34. $-\sqrt{324}$

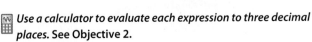
Use a calculator to evaluate each expression to three decimal places. See Objective 2.

35. $\sqrt{2}$
36. $\sqrt{3}$
37. $\sqrt{11}$
38. $\sqrt{53}$
39. $\sqrt{95}$
40. $\sqrt{99}$
41. $\sqrt{428}$
42. $\sqrt{844}$

Determine whether each number in each set is rational, irrational, or imaginary. See Example 3.

43. $\{\sqrt{9}, \sqrt{17}\}$
44. $\{-\sqrt{5}, \sqrt{0}\}$

45. $\{\sqrt{49}, \sqrt{-49}\}$
46. $\{\sqrt{-100}, -\sqrt{225}\}$

Complete the table and graph the function. See Objective 3.

47. $f(x) = 1 + \sqrt{x}$

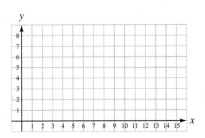

48. $f(x) = -1 + \sqrt{x}$

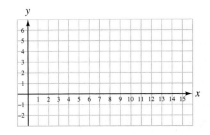

49. $f(x) = -\sqrt{x}$

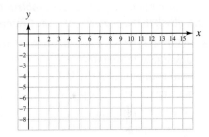

50. $f(x) = 1 - \sqrt{x}$

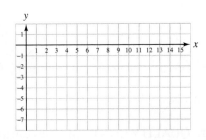

Refer to the following right triangle and find the length of the unknown side. See Example 4.

51. Find c if $a = 4$ and $b = 3$.
52. Find c if $a = 5$ and $b = 12$.
53. Find b if $a = 15$ and $c = 17$.
54. Find b if $a = 21$ and $c = 29$.

TRY IT YOURSELF

 Find each square root. You may use a calculator.

55. $-\sqrt{2{,}500}$
56. $-\sqrt{625}$
57. $\sqrt{3{,}600}$
58. $\sqrt{1{,}600}$
59. $-\sqrt{9{,}876}$
60. $-\sqrt{3{,}619}$
61. $\sqrt{21.35}$
62. $\sqrt{13.78}$
63. $\sqrt{0.3588}$
64. $\sqrt{0.9999}$
65. $-\sqrt{0.8372}$
66. $-\sqrt{0.4279}$
67. $2\sqrt{3}$
68. $3\sqrt{2}$
69. $\dfrac{2+\sqrt{3}}{2}$
70. $\dfrac{2-\sqrt{3}}{2}$

 Refer to the right triangle on the previous page for Problems 71–74 and find the length of the unknown side.

71. Find a if $b = 16$ and $c = 34$.
72. Find a if $b = 45$ and $c = 53$.
73. Find b if $c = 125$ and $a = 44$.
74. Find c if $a = 176$ and $b = 57$.

APPLICATIONS

Use a calculator to help solve each problem. If an answer is not exact, give it to the nearest tenth. See Examples 2–7.

75. **ADJUSTING A LADDER** A 20-foot ladder reaches a window 16 feet above the ground. How far from the wall is the base of the ladder?

76. **PAPER AIRPLANES** The illustration gives the directions for making a paper airplane from a square piece of paper with sides 8 inches long. Find the length l of the plane when it is completed. Give the exact answer and an approximation to two decimal places.

Step 1: Fold up.   Step 2: Fold to make wing.   Step 3: Fold up tip of wing.

77. **QUALITY CONTROL** How can a tool manufacturer use the Pythagorean theorem to verify that the two sides of the carpenter's square meet to form a 90° angle?

78. **GARDENING** A rectangular garden has sides of 28 and 45 feet. Find the length of a path that extends from one corner to the opposite corner.

79. **BASEBALL** A baseball diamond is a square, with each side 90 feet long, as shown. How far is it from home plate to second base?

80. **TELEVISION** The size of a television screen is the diagonal distance from the upper left to the lower right corner. What is the size of the screen shown?

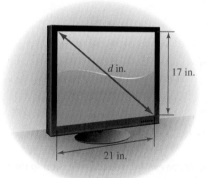

81. **FINDING LOCATION** *from Campus to Careers* — **Archaeologist**
A team of archaeologists travels 4.2 miles east and then 4.0 miles north of their base camp to explore some ancient ruins. "As the crow flies," how far from their base camp are they?

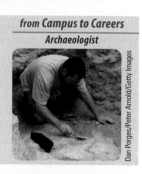

82. **SHORTCUTS** Instead of walking on the sidewalk, students take a diagonal shortcut across the rectangular vacant lot shown. How much distance do they save?

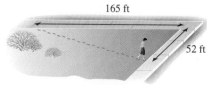

83. THE WIZARD OF OZ In the 1939 classic movie, the Scarecrow was in search of a brain. Once he received an honorary degree in "Thinkology" from the Wizard, he tried to impress his friends by stating: *"The sum of the square roots of any two sides of an isosceles triangle is equal to the square root of the remaining side."* (You can watch this scene on YouTube.) What well-known mathematical fact was the Scarecrow attempting to recite? Explain the errors that he made. (More than 50 years later, in an episode of *The Simpsons,* Homer quotes the Scarecrow word for word after finding a pair of eyeglasses in a public restroom.).

84. GEOMETRY The legs of a right triangle are equal, and the hypotenuse is 2.82843 units long. Find the length of each leg.

85. WRESTLING The sides of a square wrestling ring are 18 feet long. Find the distance from one corner to the opposite corner.

86. PERIMETER OF A SQUARE The diagonal of a square is 3 feet long. Find its perimeter.

87. HEIGHT OF A TRIANGLE Find the area of the isosceles triangle shown.

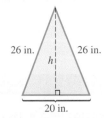

88. INTERIOR DECORATING The following square table is covered by a circular tablecloth. If the sides of the table are 2 feet long, find the area of the tablecloth.

89. DRAFTING Refer to the illustration in the next column. Among the tools used in drafting are 30–60–90 and 45–45–90 triangles.

  a. Find the length of the hypotenuse of the 45–45–90 triangle if it is $\sqrt{2}$ times as long as a leg.

  b. Find the length of the side opposite the 60° angle of the other triangle if it is $\dfrac{\sqrt{3}}{2}$ times as long as the hypotenuse.

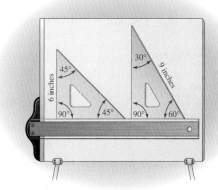

90. ORGAN PIPES The design for a set of brass pipes for a church organ is shown. Find the length of each pipe (to the nearest tenth of a foot), and then find the total length of pipe needed to construct this set.

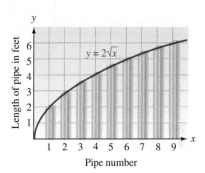

WRITING

91. Explain why the square root of a negative number cannot be a real number.

92. Explain the Pythagorean theorem.

93. Suppose you are told that $\sqrt{10} \approx 3.16$. Explain how another key on your calculator (besides the square root key $\sqrt{}$) could be used to see whether this is a reasonable approximation.

94. Explain the difference between the *square* of a number and the *square root* of a number.

REVIEW

95. Add: $(3s^2 - 3s - 2) + (3s^2 + 4s - 3)$

96. Subtract: $(3c^2 - 2c + 4) - (c^2 - 3c + 7)$

97. Multiply: $(3x - 2)(x + 4)$

98. Divide: $(x^2 + 13x + 12) \div (x + 1)$

Objectives

1. Find the cube roots of perfect cubes.
2. Approximate irrational cube roots.
3. Graph the cube root function.
4. Find higher-order roots.
5. Simplify roots of variable expressions.

SECTION 8.2

Higher-Order Roots; Radicands That Contain Variables

In this section, we will consider higher-order roots such as cube roots and fourth roots.

1 Find the cube roots of perfect cubes.

When we raise a number to the third power, we are cubing it, or finding its **cube**. We can reverse the cubing process to find **cube roots** of numbers. To find the cube root of 8, we ask, "What number, when cubed, is equal to 8?" It follows that 2 is a cube root of 8, because $2^3 = 8$.

In general, we have the following definition.

> **The Definition of Cube Root**
>
> The number b is a **cube root** of the number a if $b^3 = a$.

All real numbers have one real-number cube root. A positive number has a positive cube root, a negative number has a negative cube root, and the cube root of 0 is 0.

> **Cube Root Notation**
>
> The **cube root of a** is denoted by $\sqrt[3]{a}$. By definition,
>
> $\sqrt[3]{a} = b$ if $b^3 = a$

Earlier, we determined that the cube root of 8 is 2. In symbols, we can write: $\sqrt[3]{8} = 2$. The number 3 is called the **index**, 8 is called the **radicand**, and the entire expression is called a **radical**.

Index $\searrow$
$\underbrace{\sqrt[3]{8}}_{\text{Radical}} \leftarrow$ Radicand   Read as "the cube root of 8."

To evaluate cube roots, you need to quickly recognize each of the following **perfect cubes** shown in blue.

$$1 = 1^3 \quad 27 = 3^3 \quad 125 = 5^3 \quad 343 = 7^3 \quad 729 = 9^3$$
$$8 = 2^3 \quad 64 = 4^3 \quad 216 = 6^3 \quad 512 = 8^3 \quad 1{,}000 = 10^3$$

EXAMPLE 1   Evaluate each cube root:
a. $\sqrt[3]{27}$ **b.** $\sqrt[3]{-64}$ **c.** $-\sqrt[3]{125}$

Strategy In each case, we will determine what number, when cubed, produces the radicand.

WHY The symbol $\sqrt[3]{}$ indicates that the cube root of the number written under it should be found.

Solution

a. $\sqrt[3]{27} = 3$ Ask: What number, when cubed, is 27? The answer is 3 because $3^3 = 27$.

b. $\sqrt[3]{-64} = -4$ Read as "the cube root of -64." Ask: What number, when cubed, is -64? The answer is -4 because $(-4)^3 = -64$.

We have seen that the *square root* of a negative number is not a real number. In this example, we see that the *cube root* of a negative number is a real number.

c. $-\sqrt[3]{125}$ is read as "the opposite of the cube root of 125." Since $-\sqrt[3]{125} = -5$, we have

$$-\sqrt[3]{125} = -5 \quad \text{Because} -\sqrt[3]{125} = -1 \cdot \sqrt[3]{125} = -1 \cdot 5 = -5.$$

> **Caution!**
> $\sqrt{-64}$ *is not* a real number.
> $\sqrt[3]{-64}$ *is* a real number; it is -4.

Self Check 1
Find each cube root:
a. $\sqrt[3]{64}$
b. $\sqrt[3]{-64}$
c. $\sqrt[3]{216}$
d. $-\sqrt[3]{1,000}$

Now Try Problems 17 and 21

EXAMPLE 2   Find each cube root: **a.** $\sqrt[3]{\dfrac{1}{8}}$ **b.** $\sqrt[3]{-\dfrac{125}{27}}$

Strategy We will determine what number, when cubed, produces the radicand.

WHY The symbol $\sqrt[3]{}$ indicates that the cube root of the number written under it should be found.

Solution

a. $\sqrt[3]{\dfrac{1}{8}} = \dfrac{1}{2}$, because $\left(\dfrac{1}{2}\right)^3 = \dfrac{1}{2} \cdot \dfrac{1}{2} \cdot \dfrac{1}{2} = \dfrac{1}{8}$.

b. $\sqrt[3]{-\dfrac{125}{27}} = -\dfrac{5}{3}$, because $\left(-\dfrac{5}{3}\right)^3 = \left(-\dfrac{5}{3}\right)\left(-\dfrac{5}{3}\right)\left(-\dfrac{5}{3}\right) = -\dfrac{125}{27}$.

Self Check 2
Find each cube root:
a. $\sqrt[3]{\dfrac{1}{27}}$
b. $\sqrt[3]{-\dfrac{8}{125}}$

Now Try Problem 25

Cube roots of numbers such as 7 are hard to compute by hand. However, we can approximate $\sqrt[3]{7}$ with a calculator.

2 Approximate irrational cube roots.

To find $\sqrt[3]{7}$, we can enter 7 into a reverse-entry scientific calculator, press the root key $\boxed{\sqrt[x]{y}}$, enter 3, and press the $\boxed{=}$ key. The approximate value of $\sqrt[3]{7}$ will appear on the calculator's display.

$$\sqrt[3]{7} \approx 1.912931183$$

If your calculator doesn't have a $\sqrt[x]{y}$ key, you can use the y^x key. We will see later that $\sqrt[3]{7} = 7^{1/3}$. To find the value of $7^{1/3}$, we enter 7 into the calculator and press these keys:

7 y^x (1 ÷ 3) =

The display will read 1.912931183.

Since $\sqrt[3]{7}$ represents the number that, when cubed, gives 7, we would expect cubes of approximations of $\sqrt[3]{7}$ to be close to 7.

- Rounded to one decimal place, $\sqrt[3]{7} \approx 1.9$, and $(1.9)^3 = 6.859$.
- Rounded to two decimal places, $\sqrt[3]{7} \approx 1.91$, and $(1.91)^3 = 6.967871$.
- Rounded to three decimal places, $\sqrt[3]{7} \approx 1.913$, and $(1.913)^3 = 7.000755497$.

Numbers such as 8, −27, −64, and 125 are called **integer cubes**, because each one is the cube of an integer. The cube root of any integer cube is an integer and therefore a rational number:

$$\sqrt[3]{8} = 2, \quad \sqrt[3]{-27} = -3, \quad \sqrt[3]{-64} = -4, \quad \text{and} \quad \sqrt[3]{125} = 5$$

Cube roots of integers such as 7 and −10, which are not integer cubes, are irrational numbers. For example, $\sqrt[3]{7}$ and $\sqrt[3]{-10}$ are irrational numbers.

> **Caution!** Recall that the square root of a negative number (for example, $\sqrt{-27}$) is not a real number, because no real number squared is equal to a negative number. However, the cube root of a negative number is a real number. For example, $\sqrt[3]{-27} = -3$, because $(-3)^3 = (-3)(-3)(-3) = -27$.

3 Graph the cube root function.

Since every real number has one real-number cube root, there is a cube root function $f(x) = \sqrt[3]{x}$. For example, the value that is determined by $f(x) = \sqrt[3]{x}$ when $x = -8$ is denoted as $f(-8)$, and we have $f(-8) = \sqrt[3]{-8} = -2$.

To graph this function, we substitute numbers for x, compute $f(x)$, plot the resulting ordered pairs, and draw a smooth curve through the points as shown in the figure. In the table, we chose five values for x: −8, −1, 0, 1, and 8, which are integer cubes. This made computing $f(x)$ quite simple.

$f(x) = \sqrt[3]{x}$

x	$f(x)$	$(x, f(x))$
−8	−2	(−8, −2)
−1	−1	(−1, −1)
0	0	(0, 0)
1	1	(1, 1)
8	2	(8, 2)

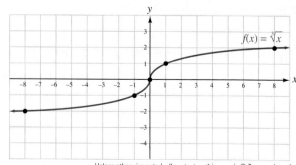

Unless otherwise noted, all content on this page is © Cengage Learning.

Using Your CALCULATOR Radius of a Water Tank

Engineers want to design a spherical tank that will hold 33,500 cubic feet of water, as shown in the figure. They know that the formula for the radius r of a sphere with volume V is given by the formula

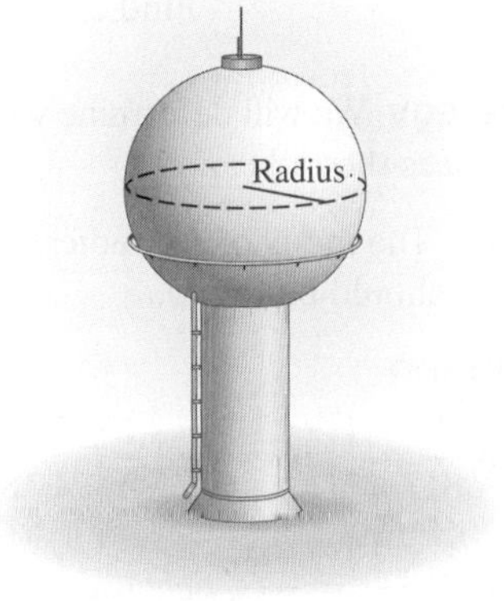

$$r = \sqrt[3]{\frac{3V}{4\pi}} \quad \text{Where } \pi = 3.14159\ldots$$

To use a reverse-entry scientific calculator to find the radius r, we substitute 33,500 for V and enter these numbers and press these keys.

3 × 33500 ÷ (4 × π) = $\sqrt[x]{y}$ 3 = $\boxed{19.99794636}$

To evaluate this expression using a direct-entry or graphing calculator, we press the MATH key. In this mode, arrow down ▼ to highlight the option $\sqrt[3]{\ }$ and ENTER. Then we press the following keys.

3 × 33500 ÷ (4 × 2nd π)) ENTER

$\boxed{\begin{array}{l}\sqrt[3]{\ }(3*33500/(4*\pi)\\ )\quad\quad\quad 19.99794636\end{array}}$

Since the result is 19.99794636, the engineers should design a tank with a radius of 20 feet.

4 Find higher-order roots.

Just as there are square roots and cube roots, there are also fourth roots, fifth roots, sixth roots, and so on. In general, we have the following definition.

> The **nth root of a** is denoted by $\sqrt[n]{a}$, and
>
> $\sqrt[n]{a} = b \quad$ if $\quad b^n = a$
>
> The number n is called the **index** of the radical. If n is an even natural number, a must be positive or zero, and b must be positive.

In the square root symbol $\sqrt{\ }$, the unwritten index is understood to be 2.

$$\sqrt{a} = \sqrt[2]{a}$$

EXAMPLE 3 Find each root: **a.** $\sqrt[4]{81}$ **b.** $\sqrt[5]{32}$ **c.** $\sqrt[5]{-32}$ **d.** $\sqrt[4]{-81}$

Strategy We will determine what number, when raised to the power of the index, produces the radicand.

WHY The index of the radical sign indicates the root of the number written under it that should be found.

Solution
a. $\sqrt[4]{81} = 3$, because $3^4 = 81$.   **b.** $\sqrt[5]{32} = 2$, because $2^5 = 32$.
c. $\sqrt[5]{-32} = -2$, because $(-2)^5 = -32$.
d. $\sqrt[4]{-81}$ is not a real number, because no real number raised to the fourth power is -81.

Self Check 3
Find each root:
a. $\sqrt[4]{16}$
b. $\sqrt[5]{243}$
c. $\sqrt[5]{-1{,}024}$

Now Try Problems 37 and 43

Unless otherwise noted, all content on this page is © Cengage Learning.

Chapter 8 Roots and Radicals

Self Check 4

Find each root:

a. $\sqrt[4]{\dfrac{1}{16}}$

b. $\sqrt[5]{-\dfrac{243}{32}}$

Now Try Problem 45

EXAMPLE 4
Find each root: a. $\sqrt[4]{\dfrac{1}{81}}$ b. $\sqrt[5]{-\dfrac{32}{243}}$

Strategy We will determine what number, when raised to the power of the index, produces the radicand.

WHY The index of the radical sign indicates the root of the number written under it that should be found.

Solution

a. $\sqrt[4]{\dfrac{1}{81}} = \dfrac{1}{3}$, because $\left(\dfrac{1}{3}\right)^4 = \dfrac{1}{81}$.

b. $\sqrt[5]{-\dfrac{32}{243}} = -\dfrac{2}{3}$, because $\left(-\dfrac{2}{3}\right)^5 = -\dfrac{32}{243}$.

5 Simplify roots of variable expressions.

When n is even and $x \geq 0$, we say that the radical $\sqrt[n]{x}$ represents an **even root**. We can find even roots of many quantities that contain variables, provided that these variables represent positive numbers or zero.

Self Check 5

Find each root. Assume that each variable represents a positive number.

a. $\sqrt{a^4}$

b. $\sqrt{m^6 n^8}$

Now Try Problems 51 and 57

EXAMPLE 5
Find each root. Assume that each variable represents a positive number. a. $\sqrt{x^2}$ b. $\sqrt{x^4}$ c. $\sqrt{x^4 y^2}$

Strategy In each case, we will determine what variable expression, when raised to the second power, produces the radicand.

WHY The radical symbol $\sqrt{\ }$ indicates that the positive square root (principal square root) of the expression written under it should be found.

Solution

a. $\sqrt{x^2} = x$, because $(x)^2 = x^2$.

b. $\sqrt{x^4} = x^2$, because $(x^2)^2 = x^4$.

c. $\sqrt{x^4 y^2} = x^2 y$, because $(x^2 y)^2 = x^4 y^2$.

When n is odd, we say that the radical expression $\sqrt[n]{x}$ represents an **odd root**.

Self Check 6

Find each root:

a. $\sqrt[3]{64p^6}$

b. $\sqrt[3]{-27p^9}$

c. $\sqrt[5]{\dfrac{1}{32}n^{15}}$

Now Try Problems 65 and 69

EXAMPLE 6
Find each root: a. $\sqrt[3]{8y^3}$ b. $\sqrt[3]{64x^6}$ c. $\sqrt[5]{32x^{10}}$

Strategy In each case, we will determine what expression, when raised to the power of the index, produces the radicand.

WHY The index of the radical sign indicates the root of the number written under it that should be found.

Solution

a. $\sqrt[3]{8y^3} = 2y$, because $(2y)^3 = 8y^3$.

b. $\sqrt[3]{64x^6} = 4x^2$, because $(4x^2)^3 = 64x^6$.

c. $\sqrt[5]{32x^{10}} = 2x^2$, because $(2x^2)^5 = 32x^{10}$.

ANSWERS TO SELF CHECKS

1. a. 4 b. −4 c. 6 d. −10 2. a. $\dfrac{1}{3}$ b. $-\dfrac{2}{5}$ 3. a. 2 b. 3 c. −4 4. a. $\dfrac{1}{2}$ b. $-\dfrac{3}{2}$ 5. a. a^2 b. $m^3 n^4$ 6. a. $4p^2$ b. $-3p^3$ c. $\dfrac{1}{2}n^3$

SECTION 8.2 STUDY SET

VOCABULARY

Fill in the blanks.

1. If $p^3 = q$, p is called a _____ root of q.
2. If $p^4 = q$, p is called a _____ root of q.
3. We denote the cube root _____ with the notation $f(x) = \sqrt[3]{x}$.
4. If the index of a radical is an even number, the root is called an _____ root.

CONCEPTS

Fill in the blanks.

5. -3 is a cube _____ of -27, because $(-3)^3 = -27$.
6. $\sqrt[3]{a} = b$ if ___ $= a$.
7. $\sqrt[3]{-216} = -6$, because $()^3 = -216$.
8. $\sqrt[5]{32x^5} = 2x$, because $(2x)^{} = 32x^5$.
9. Find each value, if possible.

    a. $\sqrt{-125}$ b. $\sqrt[3]{-125}$

10. Graph each number on the number line.

    $\{\sqrt[3]{16}, -\sqrt[4]{100}, \sqrt[3]{-1.8}, \sqrt[4]{0.6}\}$

    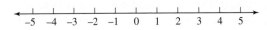

11. If $f(x) = \sqrt[3]{x}$, find each value. **Do not use a calculator.**

    a. $f(1)$ b. $f\left(-\dfrac{1}{27}\right)$

    c. $f(125)$ d. $f(0.008)$

    e. $f(1,000)$

12. a. Use the dashed lines in the following graph to approximate $\sqrt[3]{5}$.

    b. Use the graph to approximate $\sqrt[3]{4}$ and $\sqrt[3]{-6}$.

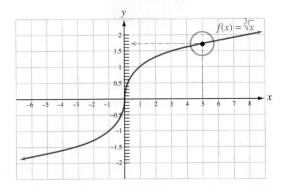

NOTATION

Fill in the blanks.

13. In the notation $\sqrt[3]{x^6}$, 3 is called the _____ and x^6 is called the _____.
14. $\sqrt{}$ is called a _____ symbol.
15. The "understood" index of the radical expression $\sqrt{55}$ is ___.
16. In reading $f(x) = \sqrt[3]{x}$, we say "f ___ x equals the cube root ___ x."

GUIDED PRACTICE

Find each value without using a calculator. See Examples 1–2.

17. $\sqrt[3]{125}$ 18. $\sqrt[3]{27}$

19. $\sqrt[3]{0}$ 20. $\sqrt[3]{1}$

21. $\sqrt[3]{-8}$ 22. $-\sqrt[3]{1}$

23. $-\sqrt[3]{27}$ 24. $\sqrt[3]{-27}$

25. $\sqrt[3]{\dfrac{1}{125}}$ 26. $\sqrt[3]{-\dfrac{1}{1,000}}$

27. $-\sqrt[3]{-1}$ 28. $-\sqrt[3]{-27}$

Use a calculator to find each cube root to the nearest hundredth. See Objective 2.

29. $\sqrt[3]{32,100}$ 30. $\sqrt[3]{-25,713}$

31. $\sqrt[3]{-0.11324}$ 32. $\sqrt[3]{0.875}$

Complete the table and graph the function. See Objective 3.

33. $f(x) = \sqrt[3]{x} + 1$

x	$f(x)$
-8	
-1	
0	
1	
8	

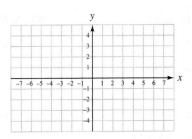

34. $f(x) = \sqrt[4]{x}$

x	$f(x)$
0	
1	
16	

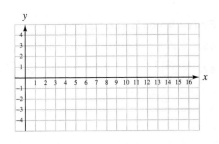

35. $f(x) = -\sqrt[3]{x}$

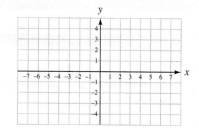

36. $f(x) = \sqrt[4]{x} - 1$

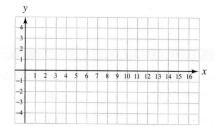

Find each value without using a calculator. See Examples 3–4.

37. $\sqrt[4]{625}$
38. $\sqrt[4]{81}$
39. $-\sqrt[5]{32}$
40. $-\sqrt[5]{243}$
41. $\sqrt[6]{1}$
42. $\sqrt[6]{0}$
43. $\sqrt[5]{-243}$
44. $\sqrt[7]{-1}$
45. $\sqrt[4]{\dfrac{1}{256}}$
46. $\sqrt[4]{\dfrac{16}{81}}$
47. $\sqrt[5]{-\dfrac{1}{32}}$
48. $\sqrt[5]{\dfrac{243}{100,000}}$

Find each root. All variables represent positive numbers. See Example 5.

49. $\sqrt{y^2}$
50. $\sqrt{y^4}$
51. $\sqrt{x^6}$
52. $\sqrt{b^8}$
53. $\sqrt{x^{10}}$
54. $\sqrt{y^{12}}$
55. $\sqrt{4z^2}$
56. $\sqrt{9t^6}$
57. $-\sqrt{x^4 y^2}$
58. $-\sqrt{x^2 y^4}$
59. $\sqrt{36z^{36}}$
60. $\sqrt{64y^{64}}$
61. $-\sqrt{625z^2}$
62. $-\sqrt{729x^8}$
63. $-\sqrt{144x^6}$
64. $-\sqrt{49x^8 y^{12}}$

Find each root. See Example 6.

65. $\sqrt[3]{y^6}$
66. $\sqrt[3]{c^3}$
67. $\sqrt[3]{27y^3}$
68. $\sqrt[3]{-p^6 q^3}$

69. $\sqrt[5]{f^5}$
70. $\sqrt[5]{y^{20}}$
71. $\sqrt[5]{-32t^{10}}$
72. $\sqrt[5]{\dfrac{z^{15}}{32}}$

TRY IT YOURSELF

Find each root.

73. $-\sqrt[3]{64}$
74. $-\sqrt[3]{343}$
75. $\sqrt[3]{729}$
76. $\sqrt[3]{512}$
77. $\sqrt[3]{1,000}$
78. $-\sqrt{0.04y^2}$
79. $-\sqrt{0.81b^6}$
80. $-\sqrt{25x^4 z^{12}}$
81. $-\sqrt{100a^6 b^4}$
82. $\sqrt[3]{64y^6}$
83. $\sqrt[3]{-r^{12} t^6}$
84. $\sqrt[4]{x^4}$
85. $\sqrt[4]{x^8}$
86. $\sqrt[3]{125}$
87. $\sqrt[5]{-\dfrac{x^{10}}{32}}$
88. $\sqrt[5]{\dfrac{x^{15} y^{10}}{100,000}}$

Use a calculator to find each root to the nearest hundredth.

89. $\sqrt[4]{125}$
90. $\sqrt[5]{12,450}$
91. $\sqrt[5]{-6,000}$
92. $\sqrt[6]{0.5}$

APPLICATIONS

Use a calculator to help solve each problem. Give your answers to the nearest hundredth.

93. **PACKAGING** A cubical box has a volume of 2 cubic feet. Substitute 2 for V in the formula $V = s^3$ and solve for s to find the length of each side of the box.

94. **HOT-AIR BALLOONS** If the hot-air balloon shown is in the shape of a sphere, what is its radius? (*Hint:* See the Using Your Calculator feature in this section.)

$V = 15,000$ cubic ft

95. **WINDMILLS** The power generated by a windmill is related to the speed of the wind by the formula

$$S = \sqrt[3]{\dfrac{P}{0.02}}$$

where S is the speed of the wind (in mph) and P is the power (in watts). Find the speed of the wind when the windmill is producing 400 watts of power.

96. ASTRONOMY In the early 17th century, Johannes Kepler, a German astronomer, discovered that a planet's mean distance R from the sun (in millions of miles) is related to the time T (in years) it takes the planet to orbit the sun by the formula

$$R = 93\sqrt[3]{\frac{T^2}{1.002}}$$

Use the information in the illustration to find R for Mercury, Earth, and Jupiter.

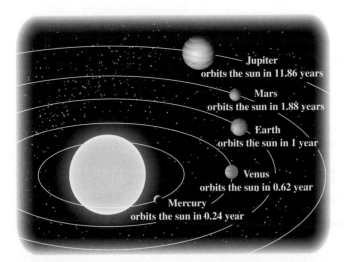

Jupiter orbits the sun in 11.86 years
Mars orbits the sun in 1.88 years
Earth orbits the sun in 1 year
Venus orbits the sun in 0.62 year
Mercury orbits the sun in 0.24 year

97. DEPRECIATION The formula

$$r = 1 - \sqrt[n]{\frac{S}{C}}$$

gives the annual depreciation rate r (in percent) of an item that had an original cost of C dollars and has a useful life of n years and a salvage value of S dollars. Use the information in the illustration in the next column to find the annual depreciation rate for the new piece of sound equipment. (*Hint:* $81K means $81,000.)

OFFICE MEMO

To: Purchasing Dept.
From: Bob Kinsell, Engineering Dept. BK
Re: New sound board

We recommend you purchase the new Sony sound board @ $81K. This equipment does become obsolete quickly but we figure we can use it for 4 yrs. A college would probably buy it from us then. I bet we could get around $16K for it.

98. SAVINGS ACCOUNTS The interest rate r (in percent) earned by a savings account after n compoundings is given by the formula

$$\sqrt[n]{\frac{V}{P}} - 1 = r$$

where V is the current value and P is the original principal. What interest rate r was paid on an account in which a deposit of $1,000 grew to $1,338.23 after five compoundings?

WRITING

99. Explain why a negative number can have a real number for its cube root yet cannot have a real number for its fourth root.

100. To find $\sqrt[3]{15}$, we can use the $\sqrt[n]{x}$ key on a calculator to obtain 2.466212074. Explain how a key other than $\sqrt[n]{x}$ can be used to check the validity of this result.

REVIEW

Simplify each expression.

101. $m^5 m^2$ **102.** $(-5x^3)(-5x)$
103. $(3^2)^4$ **104.** $r^3 r r^5$
105. $(x^2 x^3)^5$ **106.** $(3aa^2a^3)^5$
107. $4x^3(6x^5)$ **108.** $-2x(5x^2)$

SECTION 8.3
Simplifying Radical Expressions

Objectives

1. Use the product rule to simplify square roots.
2. Use prime factorization to simplify square roots.
3. Simplify square roots of variable expressions.
4. Use the quotient rule to simplify square roots.
5. Simplify cube roots.

Square dancing is an American folk dance in which four couples, arranged in a square, perform various moves. The figure on the next page shows a group as they move around a square.

If the square shown in the figure has an area of 12 square yards, the length of a side is $\sqrt{12}$ yards. We can use the formula for the area of a square and the concept of square root to show that this is so.

$A = s^2$ *s is the length of a side of the square.*

$12 = s^2$ *Substitute 12 for A, the area of the square.*

$\sqrt{12} = \sqrt{s^2}$ *Take the positive square root of both sides.*

$\sqrt{12} = s$ *The length of a side of the square is $\sqrt{12}$ yards.*

The form in which we express the length of a side of the square depends on the situation. If an approximation is acceptable, we can use a calculator to find that $\sqrt{12} \approx 3.464101615$, and we can then round to a specified degree of accuracy. For example, to the nearest tenth, each side is 3.5 yards long.

If the situation calls for the *exact* length, we must use a radical expression. As you will see in this section, it is common practice to write a radical expression such as $\sqrt{12}$ in *simplified form*. To simplify radicals, we will use product and quotient rules for radicals.

1 Use the product rule to simplify square roots.

We introduce the first of two properties of square roots with the following examples:

Square root of a product   **Product of square roots**

$\sqrt{4 \cdot 25} = \sqrt{100}$ and $\sqrt{4}\sqrt{25} = 2 \cdot 5$ Read as "the square root of 4 times the square root of 25."

$= 10$ $= 10$

In each case, the answer is 10. Thus, $\sqrt{4 \cdot 25} = \sqrt{4}\sqrt{25}$. Likewise,

$\sqrt{9 \cdot 16} = \sqrt{144}$ and $\sqrt{9}\sqrt{16} = 3 \cdot 4$

$= 12$ $= 12$

In each case, the answer is 12. Thus, $\sqrt{9 \cdot 16} = \sqrt{9}\sqrt{16}$. These results illustrate the **product rule for square roots**.

> **The Product Rule for Square Roots**
>
> For any nonnegative real numbers a and b,
>
> $\sqrt{a \cdot b} = \sqrt{a}\sqrt{b}$

In words, *the square root of the product of two nonnegative numbers is equal to the product of their square roots.*

A square root radical is in **simplified form** when each of the following statements is true.

> **Simplified Form of a Square Root**
>
> 1. Except for 1, the radicand has no perfect-square factors.
> 2. No fraction appears in a radicand.
> 3. No radical appears in the denominator.

We can use the product rule for square roots to simplify square roots whose radicands have perfect-square factors. For example, we can simplify $\sqrt{12}$ as follows:

$$\sqrt{12} = \sqrt{4 \cdot 3} \quad \text{Factor 12 as } 4 \cdot 3.$$
$$= \sqrt{4}\sqrt{3} \quad \text{The square root of } 4 \cdot 3 \text{ is equal to the square root of 4 times the square root of 3.}$$
$$= 2\sqrt{3} \quad \text{Write } \sqrt{4} \text{ as 2. Read as "2 times the square root of 3" or as "2 radical 3."}$$

The square in the figure, which we considered in the introduction to this section, has a side length of $\sqrt{12}$ yards. We now see that the *exact* length of a side can be expressed in simplified form as $2\sqrt{3}$ yards.

To simplify more difficult square roots, it is helpful to know the **natural-number perfect squares**. For example, 81 is a perfect square, because it is the square of 9: $9^2 = 81$. The first 20 natural-number perfect squares are

1, 4, 9, 16, 25, 36, 49, 64, 81, 100, 121, 144, 169, 196, 225, 256, 289, 324, 361, 400

EXAMPLE 1   Simplify: $\sqrt{27}$

Strategy We will factor 27 as $9 \cdot 3$ and then use the product rule for square roots to simplify the radical expression.

WHY Factoring the radicand in this way leads to a square root of a perfect square that we can easily evaluate.

Solution
$$\sqrt{27} = \sqrt{9 \cdot 3} \quad \text{Factor 27 as } 9 \cdot 3.$$
$$= \sqrt{9}\sqrt{3} \quad \text{The square root of a product is equal to the product of their square roots.}$$
$$= 3\sqrt{3} \quad \text{Find the square root of the perfect-square factor: } \sqrt{9} = 3.$$

As a check, recall that $\sqrt{27}$ is the number that, when squared, gives 27. If $3\sqrt{3} = \sqrt{27}$, then $(3\sqrt{3})^2$ should be equal to 27.

$$(3\sqrt{3})^2 = (3)^2(\sqrt{3})^2 \quad \text{Use the power of a product rule for exponents: Raise each factor of the product } 3\sqrt{3} \text{ to the second power.}$$
$$= 9(3) \quad \sqrt{3}, \text{ when squared, gives 3.}$$
$$= 27$$

Self Check 1
Simplify: $\sqrt{28}$
Now Try Problem 21

> **The Language of Algebra** The instructions *simplify* and *approximate* do not mean the same thing.
>
> Simplify: $\sqrt{12} = 2\sqrt{3}$ (exact)   Approximate: $\sqrt{12} \approx 3.464$ (not exact)

EXAMPLE 2   Simplify: $\sqrt{600}$

Strategy We will factor 600 as $100 \cdot 6$ and then use the product rule for square roots to simplify the radical expression.

WHY We want to choose the factorization of 600 that contains the biggest perfect-square factor, which is 100.

Solution
$$\sqrt{600} = \sqrt{100 \cdot 6} \quad \text{Factor 600 as } 100 \cdot 6.$$
$$= \sqrt{100}\sqrt{6} \quad \text{The square root of a product is equal to the product of their square roots.}$$
$$= 10\sqrt{6} \quad \text{Find the square root of the perfect-square factor: } \sqrt{100} = 10.$$

Check the result.

Self Check 2
Simplify: $\sqrt{500}$
Now Try Problem 23

2 Use prime factorization to simplify square roots.

When simplifying square roots, prime factorization can be useful in finding the greatest perfect-square factor of the radicand.

Self Check 3
Simplify, if possible:
a. $\sqrt{140}$
b. $\sqrt{77}$

Now Try Problem 31

EXAMPLE 3 Simplify, if possible: a. $\sqrt{150}$ b. $\sqrt{95}$

Strategy In each case, the greatest perfect-square factor of the radicand (if there is one) is not obvious. Another approach is to find the prime factorization of the radicand and look for pairs of like factors.

WHY Identifying a pair of like factors of the radicand leads to a square root of a perfect square that we can easily evaluate.

Solution

a. $\sqrt{150} = \sqrt{2 \cdot 3 \cdot 5 \cdot 5}$ Write 150 in prime-factored form.

$\phantom{\sqrt{150}} = \sqrt{2 \cdot 3}\sqrt{5 \cdot 5}$ Group the pair of like factors together and use the product rule for square roots.

$\phantom{\sqrt{150}} = \sqrt{6} \cdot 5$ Evaluate the square root of the perfect square: $\sqrt{5 \cdot 5} = \sqrt{25} = 5$.

$\phantom{\sqrt{150}} = 5\sqrt{6}$ Write the factor 5 first. This way, no misunderstanding can occur about exactly what is under the radical symbol.

b. We prime factor 95 to get $95 = 5 \cdot 19$. Since the factorization does not contain a pair of like factors, 95 does not have a perfect-square factor. It follows that $\sqrt{95}$ cannot be simplified.

> **Caution!** As we see in part b, some radical expressions cannot be simplified because the radicand does not have a perfect-square factor.

3 Simplify square roots of variable expressions.

Variable expressions can also be perfect squares. For example, $x^2, x^4, x^6,$ and x^8 are perfect squares because

$$x^2 = (x)^2, \quad x^4 = (x^2)^2, \quad x^6 = (x^3)^2, \quad \text{and} \quad x^8 = (x^4)^2$$

Perfect squares like these are used to simplify square roots involving variable radicands.

Self Check 4
Simplify: $\sqrt{y^5}$

Now Try Problem 33

EXAMPLE 4 Simplify: $\sqrt{x^3}$

Strategy We will factor x^3 as $x^2 \cdot x$ and then use the product rule for square roots to simplify the radical expression.

WHY Factoring the radicand in this way leads to the square root of a perfect square that can be easily simplified.

Solution

$\sqrt{x^3} = \sqrt{x^2 \cdot x}$ Factor x^3 as $x^2 \cdot x$.

$\phantom{\sqrt{x^3}} = \sqrt{x^2}\sqrt{x}$ The square root of a product is equal to the product of their square roots.

$\phantom{\sqrt{x^3}} = x\sqrt{x}$ Find the square root of the perfect-square factor: $\sqrt{x^2} = x$.

8.3 Simplifying Radical Expressions

As a check, recall that $\sqrt{x^3}$, when squared, gives x^3. If $x\sqrt{x} = \sqrt{x^3}$, then $(x\sqrt{x})^2$ should be equal to x^3.

$(x\sqrt{x})^2 = (x)^2(\sqrt{x})^2$ Raise each factor of the product $x\sqrt{x}$ to the 2nd power.
$\quad\quad\quad = x^2(x)$ $\sqrt{x}$, when squared, gives x.
$\quad\quad\quad = x^3$ Keep the base x and add the exponents.

EXAMPLE 5   Simplify: $-7\sqrt{8m}$

Strategy We will write $\sqrt{8m}$ in simplified form and then multiply the result by -7.

WHY The expression $-7\sqrt{8m}$ means $-7 \cdot \sqrt{8m}$.

Solution
By inspection, we see that the radicand, $8m$, has a perfect-square factor of 4. We can write $8m$ in factored form as $4 \cdot 2m$.

$-7\sqrt{8m} = -7\sqrt{4 \cdot 2m}$ Factor 8m as $4 \cdot 2m$.
$\quad\quad\quad = -7\sqrt{4}\sqrt{2m}$ The square root of a product is equal to the product of their square roots.
$\quad\quad\quad = -7(2)\sqrt{2m}$ Find the square root of the perfect-square factor: $\sqrt{4} = 2$.
$\quad\quad\quad = -14\sqrt{2m}$ Multiply: $-7(2) = -14$.

Self Check 5
Simplify: $-2\sqrt{50c}$
Now Try Problem 41

Caution! When writing radical expressions such as $-14\sqrt{2m}$, be sure to extend the radical symbol completely over $2m$, because the expressions $-14\sqrt{2}m$ and $-14\sqrt{2m}$ are not the same. Similar care should be taken when writing expressions such as $\sqrt{3}x$. To avoid any misinterpretation, $\sqrt{3}x$ can be written as $x\sqrt{3}$.

EXAMPLE 6   Simplify: $\sqrt{72x^3}$

Strategy We will determine the greatest perfect-square factor of the numerical part and the greatest perfect-square factor of the variable part of the radicand separately.

WHY It is easier to determine the greatest perfect-square factor of the entire radicand if we consider the numerical and variable factors separately.

Solution
We factor $72x^3$ into two factors, one of which is the greatest perfect square that divides $72x^3$. Since the greatest perfect square that divides $72x^3$ is $36x^2$, such a factorization is $72x^3 = 36x^2 \cdot 2x$. We can then use the product rule for square roots to simplify the expression.

$\sqrt{72x^3} = \sqrt{36x^2 \cdot 2x}$ Factor $72x^3$ as $36x^2 \cdot 2x$.
$\quad\quad\quad = \sqrt{36x^2}\sqrt{2x}$ The square root of a product is equal to the product of their square roots.
$\quad\quad\quad = 6x\sqrt{2x}$ Find the square root of the perfect-square factor: $\sqrt{36x^2} = 6x$.

Self Check 6
Simplify: $\sqrt{48y^3}$
Now Try Problem 45

Self Check 7

Simplify: $5q\sqrt{63p^5q^4}$

Now Try Problem 49

EXAMPLE 7 Simplify: $3a\sqrt{288a^4b^7}$

Strategy We will write $\sqrt{288a^4b^7}$ in simplified form and then multiply the result by $3a$.

WHY The expression $3a\sqrt{288a^4b^7}$ means $3a \cdot \sqrt{288a^4b^7}$.

Solution
To simplify $\sqrt{288a^4b^7}$, we look for the greatest perfect square that divides $288a^4b^7$. Because

- 144 is the greatest perfect square that divides 288,
- a^4 is the greatest perfect square that divides a^4, and
- b^6 is the greatest perfect square that divides b^7,

the factor $144a^4b^6$ is the greatest perfect square that divides $288a^4b^7$.

We can now use the product rule for square roots to simplify the radical.

$$3a\sqrt{288a^4b^7} = 3a\sqrt{144a^4b^6 \cdot 2b} \quad \text{Factor } 288a^4b^7 \text{ as } 144a^4b^6 \cdot 2b.$$

$$= 3a\sqrt{144a^4b^6}\sqrt{2b} \quad \text{The square root of a product is equal to the product of their square roots.}$$

$$= 3a(12a^2b^3)\sqrt{2b} \quad \sqrt{144a^4b^6} = 12a^2b^3.$$

$$= 36a^3b^3\sqrt{2b} \quad \text{Multiply: } 3a(12a^2b^3) = 36a^3b^3.$$

Caution! The product rule for square roots applies to the square root of the product of two numbers. There is no such property for sums or differences. To illustrate this, we consider these correct simplifications:

$$\sqrt{9 + 16} = \sqrt{25} = 5 \quad \text{and} \quad \sqrt{25 - 16} = \sqrt{9} = 3$$

It is incorrect to write

$$\sqrt{9 + 16} = \sqrt{9} + \sqrt{16} \quad \text{or} \quad \sqrt{25 - 16} = \sqrt{25} - \sqrt{16}$$
$$= 3 + 4 \qquad\qquad\qquad\qquad = 5 - 4$$
$$= 7 \qquad\qquad\qquad\qquad\quad = 1$$

Thus, $\sqrt{a + b} \neq \sqrt{a} + \sqrt{b}$ and $\sqrt{a - b} \neq \sqrt{a} - \sqrt{b}$.

4 Use the quotient rule to simplify square roots.

To introduce a second property of radicals, we consider these examples.

Square root of a quotient

$$\sqrt{\frac{100}{25}} = \sqrt{4}$$
$$= 2$$

and

Quotient of square roots

$$\frac{\sqrt{100}}{\sqrt{25}} = \frac{10}{5}$$
$$= 2$$

Read as "the square root of 100 divided by the square root of 25."

Since the answer is 2 in each case,

$$\sqrt{\frac{100}{25}} = \frac{\sqrt{100}}{\sqrt{25}}$$

Likewise,

$$\sqrt{\frac{36}{4}} = \sqrt{9} \quad \text{and} \quad \frac{\sqrt{36}}{\sqrt{4}} = \frac{6}{2}$$
$$= 3 \qquad\qquad\qquad\qquad = 3$$

Since the answer is 3 in each case,

$$\sqrt{\frac{36}{4}} = \frac{\sqrt{36}}{\sqrt{4}}$$

These results illustrate the **quotient rule for square roots**.

> **The Quotient Rule for Square Roots**
>
> For any positive real numbers a and b,
>
> $$\sqrt{\frac{a}{b}} = \frac{\sqrt{a}}{\sqrt{b}}$$

In words, *the square root of the quotient of two numbers is the quotient of their square roots.*

We can use the quotient rule for square roots to simplify square roots that have fractions in their radicands. For example,

$$\sqrt{\frac{59}{49}} = \frac{\sqrt{59}}{\sqrt{49}} \quad \text{The square root of a quotient is equal to the quotient of their square roots.}$$

$$= \frac{\sqrt{59}}{7} \quad \text{Simplify: } \sqrt{49} = 7.$$

EXAMPLE 8

Simplify: $\sqrt{\dfrac{108}{25}}$

Self Check 8

Simplify: $\sqrt{\dfrac{20}{81}}$

Now Try Problem 53

Strategy The square root is not in simplified form because the radicand contains a fraction. To write the radical expression in simplified form, we use the quotient rule for square roots.

WHY Writing the expression in $\dfrac{\sqrt{a}}{\sqrt{b}}$ form leads to a square root of a perfect square in the denominator that we can easily evaluate.

Solution

$$\sqrt{\frac{108}{25}} = \frac{\sqrt{108}}{\sqrt{25}} \quad \text{The square root of a quotient is equal to the quotient of their square roots.}$$

$$= \frac{\sqrt{36 \cdot 3}}{5} \quad \text{Factor 108 using the largest perfect-square factor of 108, which is 36. Write } \sqrt{25} \text{ as 5.}$$

$$= \frac{\sqrt{36}\sqrt{3}}{5} \quad \text{The square root of a product is equal to the product of their square roots.}$$

$$= \frac{6\sqrt{3}}{5} \quad \text{This result can also be written as } \tfrac{6}{5}\sqrt{3}.$$

EXAMPLE 9

Simplify: $\sqrt{\dfrac{44x^3}{9xy^2}}$

Self Check 9

Simplify: $\sqrt{\dfrac{99b^3}{16a^2b}}$

Now Try Problem 57

Strategy Because the radicand is a rational expression, we will remove common factors before using the quotient rule for square roots.

WHY After simplifying the rational radicand, our hope is that the numerator and/or the denominator of the resulting rational expression is a perfect square.

Solution

$$\sqrt{\frac{44x^3}{9xy^2}} = \sqrt{\frac{44x^2}{9y^2}}$$ Simplify the fraction by removing the common factor of x: $\frac{44x^3}{9xy^2} = \frac{44x^2 \cdot \overset{1}{\cancel{x}}}{9\cancel{x}y^2} = \frac{44x^2}{9y^2}$.

$$= \frac{\sqrt{44x^2}}{\sqrt{9y^2}}$$ The square root of a quotient is equal to the quotient of their square roots.

$$= \frac{\sqrt{4x^2}\sqrt{11}}{\sqrt{9y^2}}$$ Factor $44x^2$ as $4x^2 \cdot 11$. The square root of a product is equal to the product of their square roots.

$$= \frac{2x\sqrt{11}}{3y}$$ $\sqrt{4x^2} = 2x$ and $\sqrt{9y^2} = 3y$.

5 Simplify cube roots.

The product and quotient rules for square roots can be generalized to apply to higher-order roots such as cube roots. To simplify cube roots, we must know the following natural-number **perfect cubes**:

$$8,\ 27,\ 64,\ 125,\ 216,\ 343,\ 512,\ 729,\ 1{,}000$$

The Product and Quotient Rules for Cube Roots

For any real numbers a and b,

$$\sqrt[3]{ab} = \sqrt[3]{a}\sqrt[3]{b} \qquad \sqrt[3]{\frac{a}{b}} = \frac{\sqrt[3]{a}}{\sqrt[3]{b}}, \quad \text{provided } b \neq 0.$$

Self Check 10

Simplify: $\sqrt[3]{250}$

Now Try Problems 61 and 63

EXAMPLE 10   Simplify: $\sqrt[3]{54}$

Strategy We will factor 54 as $27 \cdot 2$ and then use the product rule for cube roots to simplify the radical expression.

WHY Factoring the radicand in this way leads to a cube root of a perfect cube that we can easily evaluate.

Solution
The greatest perfect cube that divides 54 is 27.

$$\sqrt[3]{54} = \sqrt[3]{27 \cdot 2} \quad \text{Factor 54 as } 27 \cdot 2.$$

$$= \sqrt[3]{27}\sqrt[3]{2} \quad \text{The cube root of a product is equal to the product of the cube roots.}$$

$$= 3\sqrt[3]{2} \quad \text{Find the cube root of the perfect-cube factor: } \sqrt[3]{27} = 3.$$

As a check, we note that $\sqrt[3]{54}$ is the number that, when cubed, gives 54. If $3\sqrt[3]{2} = \sqrt[3]{54}$, then $\left(3\sqrt[3]{2}\right)^3$ will be equal to 54.

$$\left(3\sqrt[3]{2}\right)^3 = (3)^3\left(\sqrt[3]{2}\right)^3 \quad \text{Raise each factor of the product } 3\sqrt[3]{2} \text{ to the third power.}$$

$$= 27(2) \quad \sqrt[3]{2} \text{ when cubed, gives 2.}$$

$$= 54$$

Variable expressions can also be perfect cubes. For example, $x^3, x^6, x^9,$ and x^{12} are perfect cubes because

$$x^3 = (x)^3, \qquad x^6 = (x^2)^3, \qquad x^9 = (x^3)^3, \qquad \text{and} \qquad x^{12} = (x^4)^3$$

Perfect cubes like these are used to simplify cube roots involving variable radicands.

EXAMPLE 11

Simplify: **a.** $\sqrt[3]{16x^3y^4}$ **b.** $\sqrt[3]{\dfrac{64n^4}{27m^3}}$

Strategy In the first case, we will factor the radicand so one factor is the greatest possible perfect cube, use the product rule for cube roots, and then simplify. For the second case, we will use the quotient rule for cube roots.

WHY Factoring the radicand in this way leads to a cube root of a perfect cube that we can easily evaluate.

Solution

a. We factor $16x^3y^4$ into two factors, one of which is the greatest perfect cube that divides $16x^3y^4$. Since $8x^3y^3$ is the greatest perfect cube that divides $16x^3y^4$, the factorization is $16x^3y^4 = 8x^3y^3 \cdot 2y$.

$$\sqrt[3]{16x^3y^4} = \sqrt[3]{8x^3y^3 \cdot 2y} \qquad \text{Factor } 16x^3y^4 \text{ as } 8x^3y^3 \cdot 2y.$$

$$= \sqrt[3]{8x^3y^3}\sqrt[3]{2y} \qquad \text{The cube root of a product is equal to the product of the cube roots.}$$

$$= 2xy\sqrt[3]{2y} \qquad \text{Find the cube root of the perfect-cube factor: } \sqrt[3]{8x^3y^3} = 2xy$$

b. $\sqrt[3]{\dfrac{64n^4}{27m^3}} = \dfrac{\sqrt[3]{64n^4}}{\sqrt[3]{27m^3}}$ The cube root of a quotient is equal to the quotient of the cube roots.

$$= \dfrac{\sqrt[3]{64n^3}\sqrt[3]{n}}{3m} \qquad \text{In the numerator, use the product rule for square roots. In the denominator, } \sqrt[3]{27m^3} = 3m.$$

$$= \dfrac{4n\sqrt[3]{n}}{3m} \qquad \text{Simplify: } \sqrt[3]{64n^3} = 4n.$$

Self Check 11

Simplify:

a. $\sqrt[3]{54a^3b^5}$

b. $\sqrt[3]{\dfrac{27q^5}{64p^3}}$

Now Try Problems 67 and 69

ANSWERS TO SELF CHECKS

1. $2\sqrt{7}$ **2.** $10\sqrt{5}$ **3. a.** $2\sqrt{35}$ **b.** cannot be simplified **4.** $y^2\sqrt{y}$
5. $-10\sqrt{2c}$ **6.** $4y\sqrt{3y}$ **7.** $15p^2q^3\sqrt{7p}$ **8.** $\dfrac{2\sqrt{5}}{9}$ **9.** $\dfrac{3b\sqrt{11}}{4a}$
10. $5\sqrt[3]{2}$ **11. a.** $3ab\sqrt[3]{2b^2}$ **b.** $\dfrac{3q\sqrt[3]{q^2}}{4p}$

SECTION 8.3 STUDY SET

VOCABULARY

Fill in the blanks.

1. Squares of integers such as 4, 9, and 16 are called _____ squares.
2. Cubes of integers such as 8, 27, and 64 are called perfect _____.
3. "To _____ $\sqrt{8}$" means to write it as $2\sqrt{2}$.
4. The word *product* is associated with the operation of multiplication and the word *quotient* with _____.

CONCEPTS

5. Fill in the blanks.

   a. The square root of the product of two positive numbers is equal to the _____ of their square roots. In symbols,
   $$\sqrt{ab} = \underline{\qquad}$$

   b. The square root of the quotient of two positive numbers is equal to the _____ of their square roots. In symbols
   $$\sqrt{\dfrac{a}{b}} = \underline{\qquad}$$

6. Which of the perfect squares 1, 4, 9, 16, 25, 36, 49, 64, 81, and 100 is the *largest* factor of the given number?
  a. 20 b. 45 c. 72 d. 98

What is wrong with each simplification?

7. $\sqrt{20} = \sqrt{16+4}$
  $= \sqrt{16} + \sqrt{4}$
  $= 4 + 2$
  $= 6$

8. $\sqrt{27} = \sqrt{36-9}$
  $= \sqrt{36} - \sqrt{9}$
  $= 6 - 3$
  $= 3$

9. a. To simplify $\sqrt{40}$, which one of the following factorizations should be used?
  $\sqrt{20 \cdot 2}$ $\sqrt{40 \cdot 1}$ $\sqrt{8 \cdot 5}$ $\sqrt{4 \cdot 10}$

 b. To simplify $\sqrt{n^5}$, which one of the following factorizations should be used?
  $\sqrt{n^2 \cdot n^3}$ $\sqrt{n^3 \cdot n^2}$ $\sqrt{n^4 \cdot n}$

10. a. To simplify $\sqrt[3]{24}$, which of the following expressions should be used?
  $\sqrt[3]{12 \cdot 2}$ $\sqrt[3]{4 \cdot 6}$ $\sqrt[3]{8 \cdot 3}$ $\sqrt[3]{24 \cdot 1}$

 b. To simplify $\sqrt[3]{n^5}$, which of the following expressions should be used?
  $\sqrt[3]{n^4 \cdot n}$ $\sqrt[3]{n^3 \cdot n^2}$ $\sqrt[3]{n^2 \cdot n^2 \cdot n}$

Evaluate the expression $\sqrt{b^2 - 4ac}$ for the given values. Perform the operations within the radical and simplify the radical.

11. $a = 5, b = 10, c = 3$ **12.** $a = 2, b = 6, c = 1$

13. $a = -1, b = 6, c = 9$ **14.** $a = 1, b = -2, c = -11$

NOTATION

Complete each step.

15. $\sqrt{80a^3b^2} = \sqrt{16 \cdot \cdot a^2 \cdot a \cdot b^2}$
  $= \sqrt{16a^2b^2 \cdot }$
  $= \sqrt{} \sqrt{5a}$
  $= \sqrt{5a}$

16. $\sqrt[3]{\dfrac{27a^4b^2}{64}} = \dfrac{\sqrt[3]{27a^4b^2}}{\sqrt[3]{64}}$
  $= \dfrac{\sqrt[3]{27a^3 \cdot }}{\sqrt[3]{64}}$
  $= \dfrac{\sqrt[3]{}\sqrt[3]{ab^2}}{\sqrt[3]{64}}$
  $= \dfrac{3a\sqrt[3]{ab^2}}{}$

17. What operation is indicated between the two radicals in the expression $\sqrt{4}\sqrt{3}$?

18. Fill in each blank to make a true statement.
  a. $16x^2 = \bigl(\bigr)^2$ **b.** $27a^3b^6 = \bigl(\bigr)^3$

19. Write each expression in a better form.
  a. $\sqrt{5} \cdot 2$ **b.** $\sqrt{7}a$
  c. $9\sqrt{x^2}\sqrt{6}$ **d.** $\sqrt{y} \cdot \sqrt{25z^4}$

20. a. Explain the difference between $\sqrt{5x}$ and $\sqrt{5}x$.
  b. Why is it better to write $\sqrt{5}x$ as $x\sqrt{5}$?

GUIDED PRACTICE

Simplify each square root. See Examples 1–2.

21. $\sqrt{20}$ **22.** $\sqrt{18}$
23. $\sqrt{50}$ **24.** $\sqrt{75}$

Use prime factorization to simplify each square root. See Example 3.

25. $\sqrt{45}$ **26.** $\sqrt{54}$
27. $\sqrt{98}$ **28.** $\sqrt{147}$
29. $\sqrt{48}$ **30.** $\sqrt{128}$
31. $\sqrt{200}$ **32.** $\sqrt{300}$

Simplify each square root. Assume that all variables represent positive numbers. See Examples 4–5.

33. $\sqrt{n^3}$ **34.** $\sqrt{x^5}$
35. $\sqrt{5a^5}$ **36.** $\sqrt{7b^7}$
37. $\sqrt{4k}$ **38.** $\sqrt{9p}$
39. $\sqrt{12x}$ **40.** $\sqrt{20y}$
41. $-6\sqrt{75t}$ **42.** $-2\sqrt{24s}$
43. $128\sqrt{8c}$ **44.** $11\sqrt{242d}$

Simplify each square root. Assume that all variables represent positive numbers. See Examples 6–7.

45. $\sqrt{25x^3}$ **46.** $\sqrt{36y^3}$
47. $\sqrt{192a^3}$ **48.** $\sqrt{88t^5}$
49. $2x\sqrt{9x^4y^3}$ **50.** $3y\sqrt{32xy^2}$
51. $12x\sqrt{16x^2y^3}$ **52.** $-4y^3\sqrt{72x^3y^3}$

Write each quotient as the quotient of two radicals and simplify. See Examples 8–9.

53. $\sqrt{\dfrac{25}{9}}$ **54.** $\sqrt{\dfrac{36}{49}}$
55. $\sqrt{\dfrac{81}{64}}$ **56.** $\sqrt{\dfrac{121}{144}}$

8.3 Simplifying Radical Expressions

57. $\sqrt{\dfrac{72x^3}{y^2}}$
58. $\sqrt{\dfrac{108b^2}{d^4}}$
59. $\sqrt{\dfrac{125n^5}{64n}}$
60. $\sqrt{\dfrac{72q^7}{25q^3}}$

Simplify each cube root. See Examples 10–11.

61. $\sqrt[3]{24}$
62. $\sqrt[3]{32}$
63. $\sqrt[3]{-128}$
64. $\sqrt[3]{-250}$
65. $\sqrt[3]{8x^3}$
66. $\sqrt[3]{-64x^5}$
67. $\sqrt[3]{54x^3z^6}$
68. $\sqrt[3]{-24x^3y^5}$
69. $\sqrt[3]{\dfrac{27m^2}{8n^6}}$
70. $\sqrt[3]{\dfrac{125t^9}{27s^6}}$
71. $\sqrt[3]{\dfrac{r^4s^5}{1{,}000t^3}}$
72. $\sqrt[3]{\dfrac{54m^4n^3}{r^3s^6}}$

TRY IT YOURSELF

Simplify each radical. Assume that all variables represent positive numbers.

73. $\sqrt{250}$
74. $\sqrt{1{,}000}$
75. $2\sqrt{24}$
76. $3\sqrt{32}$
77. $-2\sqrt{28}$
78. $-3\sqrt{72}$
79. $\sqrt{a^2b}$
80. $\sqrt{rs^4}$
81. $\dfrac{1}{5}x^2y\sqrt{50x^2y^2}$
82. $\dfrac{1}{5}x^5y\sqrt{75x^3y^2}$
83. $-\dfrac{2}{5}\sqrt{80mn^4}$
84. $\dfrac{5}{6}\sqrt{180ab^6}$
85. $\sqrt{\dfrac{26}{25}}$
86. $\sqrt{\dfrac{17}{169}}$
87. $-\sqrt{\dfrac{20}{49}}$
88. $-\sqrt{\dfrac{50}{9}}$
89. $\sqrt{\dfrac{48}{81}}$
90. $\sqrt{\dfrac{27}{64}}$
91. $\sqrt{\dfrac{32}{25}}$
92. $\sqrt{\dfrac{75}{16}}$
93. $\sqrt[3]{27x^3}$
94. $\sqrt[3]{-16x^4}$
95. $\sqrt[3]{-81x^2y^3}$
96. $\sqrt[3]{81y^2z^3}$
97. $\sqrt{\dfrac{128m^3n^5}{81mn^7}}$
98. $\sqrt{\dfrac{75p^3q^2}{p^5q^4}}$
99. $\sqrt{\dfrac{12r^7s^7}{r^5s^2}}$
100. $\sqrt{\dfrac{m^2n^9}{100mn^3}}$

APPLICATIONS

Use a calculator to help solve each problem.

101. **AMUSEMENT PARK RIDES** The illustration in the next column shows a pirate ship ride. The time (in seconds) it takes to swing from one extreme to the other is given by

$$t = \pi\sqrt{\dfrac{L}{32}}$$

a. Find t and express it in simplified radical form. Leave π in your answer.

b. Express your answer to part a as a decimal. Round to the nearest tenth of a second.

$L = 54$ ft

102. **HERB GARDENS** Refer to the illustration below. The perimeter of the herb garden is given by

$$p = 2\pi\sqrt{\dfrac{a^2+b^2}{2}}$$

a. Find the length of fencing (in meters) needed to enclose the garden. Express the result in simplified radical form. Leave π in your answer.

b. Express the result from part a as a decimal. Round to the nearest tenth of a meter.

$b = 6$ m
$a = 8$ m

103. **ARCHAEOLOGY** *from Campus to Careers* — Archaeologist

Framed grids, made up of 20 cm × 20 cm squares, are often used to record the location of artifacts found during an excavation. See the illustration on the next page.

a. Use the Pythagorean theorem to determine the *exact* distance between a piece of pottery found at point A and a cooking utensil found at point B.

b. Approximate the distance to the nearest tenth of a centimeter.

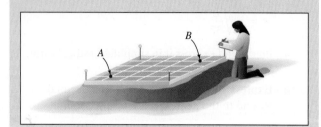

104. ENVIRONMENTAL PROTECTION Refer to the illustration below. A new campground is to be constructed 2 miles from a major highway. The proposed entrance, although longer than the direct route, bypasses a grove of old-growth redwood trees.

  a. Use the Pythagorean theorem to find the length of the proposed entrance road. Express the result as a radical in simplified form.

  b. Express the result from part a as a decimal. Round to the nearest hundredth of a mile.

  c. How much longer is the proposed entrance as compared to the direct route into the campground?

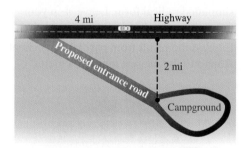

105. A crossword puzzle in a newspaper occupies an area of 28 square inches. See the illustration at right.

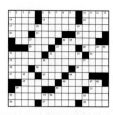

  a. Express the exact length of a side of the square-shaped puzzle in simplified radical form.

  b. What is the length of a side to the nearest tenth of an inch?

106. See the illustration.

  a. Find the exact length of a side of the cube written in simplified radical form.

Volume = 40 ft³

  b. What is the length of a side to the nearest tenth of a foot?

WRITING

107. State the product rule for square roots.

108. When comparing $\sqrt{8}$ and $2\sqrt{2}$, why is $2\sqrt{2}$ called simplified radical form?

REVIEW

109. Multiply: $(-2a^3)(3a^2)$

110. Find the slope of the line passing through $(-6, 0)$ and $(0, -4)$.

111. Write the equation of the line passing through $(0, 3)$ with slope -2.

112. Solve: $-x = -5$

113. Solve: $-x > -5$

114. Find the slope of a line perpendicular to a line with a slope of 2.

SECTION 8.4
Adding and Subtracting Radical Expressions

Objectives

1. Add or subtract square roots.
2. Simplify square roots in a sum or difference.
3. Simplify cube roots in a sum or difference.

We have discussed how to add and subtract like terms. In this section, we will discuss how to add and subtract expressions that contain like radicals.

1 Add or subtract square roots.

When adding monomials, we can often combine *like terms*. For example,

$$3x + 5x = (3 + 5)x \quad \text{Use the distributive property.}$$
$$= 8x \quad \text{Do the addition.}$$

Caution! The expression $3x + 5y$ cannot be simplified, because $3x$ and $5y$ are not like terms.

It is often possible to combine terms that contain *like radicals*.

Like Radicals

Radicals are called **like radicals** when they have the same index and the same radicand.

Like radicals	Unlike radicals
$3\sqrt{2}$ and $5\sqrt{2}$	$3\sqrt{2}$ and $5\sqrt{3}$
The same index and the same radicand	*The same index but different radicands*
$5x\sqrt{3y}$ and $-2x\sqrt{3y}$	$5x\sqrt[3]{3y}$ and $-2x\sqrt{3y}$
The same index and the same radicand	*The same radicands but a different index*

Expressions that contain like radicals can be combined by addition and subtraction. For example, we have

$3\sqrt{2} + 5\sqrt{2} = (3 + 5)\sqrt{2}$ Use the distributive property.
$\qquad\qquad\quad = 8\sqrt{2}$ Do the addition.

Likewise, we can simplify the expression $5x\sqrt{3y} - 2x\sqrt{3y}$.

$5x\sqrt{3y} - 2x\sqrt{3y} = (5x - 2x)\sqrt{3y}$ Use the distributive property.
$\qquad\qquad\qquad\quad = 3x\sqrt{3y}$ Do the subtraction: $5x - 2x = 3x$.

Caution! The expression $3\sqrt{2} + 5\sqrt{3}$ cannot be simplified, because the radicals are unlike. For the same reason, we cannot simplify $5x\sqrt[3]{3y} - 2x\sqrt{3y}$.

EXAMPLE 1   Simplify: **a.** $\sqrt{6} + 6 + 5\sqrt{6}$ **b.** $-2\sqrt{m} - 3\sqrt{m}$

Strategy First, we will see if the radicands of the square roots are the same. If they are, we can use the distributive property in reverse to add (or subtract) like radicals.

WHY We must check the radicands first because only square roots with the same radicand can be added or subtracted.

Solution

a. The expression contains three terms: $\sqrt{6}$, 6, and $5\sqrt{6}$. The first and third terms have like radicals, and they can be combined.

$\sqrt{6} + 6 + 5\sqrt{6} = 6 + \left(1\sqrt{6} + 5\sqrt{6}\right)$ Group the expressions with like radicals. Write $\sqrt{6}$ as $1\sqrt{6}$.
$\qquad\qquad\qquad = 6 + (1 + 5)\sqrt{6}$ Use the distributive property.
$\qquad\qquad\qquad = 6 + 6\sqrt{6}$ Do the addition.

Note that 6 and $6\sqrt{6}$ do not contain like radicals and cannot be combined.

b. The expressions $-2\sqrt{m}$ and $-3\sqrt{m}$ contain like radicals. We can combine them.

$-2\sqrt{m} - 3\sqrt{m} = (-2 - 3)\sqrt{m}$ Use the distributive property.
$\qquad\qquad\qquad = -5\sqrt{m}$ Do the subtraction: $-2 - 3 = -5$.

Self Check 1

Simplify:
a. $\sqrt{7} + 7 + 7\sqrt{7}$
b. $24\sqrt{m} - 25\sqrt{m}$

Now Try Problem 19

Chapter 8 Roots and Radicals

2 Simplify square roots in a sum or difference.

If a sum or difference involves radicals that are unlike, make sure each one is written in simplified form. After doing so, like radicals may result that can be combined.

Self Check 2
Simplify: $2\sqrt{50} + \sqrt{32}$
Now Try Problems 25 and 29

EXAMPLE 2 Simplify: $3\sqrt{18} + 5\sqrt{8}$

Strategy We will simplify the radicals in each term of the expression. Then we will see if there are like radicals that can be combined.

WHY Only like radicals can be combined.

Solution
The radical $\sqrt{18}$ is not in simplified form, because 18 has a perfect-square factor of 9. The radical $\sqrt{8}$ is not in simplified form either, because 8 has a perfect-square factor of 4. To simplify the radicals and add the expressions, we proceed as follows.

$$3\sqrt{18} + 5\sqrt{8} = 3\sqrt{9 \cdot 2} + 5\sqrt{4 \cdot 2}$$ Factor 18 and 8 using perfect-square factors.
$$= 3\sqrt{9}\sqrt{2} + 5\sqrt{4}\sqrt{2}$$ The square root of a product is equal to the product of their square roots.
$$= 3(3)\sqrt{2} + 5(2)\sqrt{2}$$ $\sqrt{9} = 3$ and $\sqrt{4} = 2$.
$$= 9\sqrt{2} + 10\sqrt{2}$$ Multiply: $3(3) = 9$ and $5(2) = 10$.
$$= 19\sqrt{2}$$ To combine like radicals, combine their coefficients: $9 + 10 = 19$.

Self Check 3
Simplify: $\sqrt{12xy^2} + \sqrt{27xy^2}$
Now Try Problem 39

EXAMPLE 3 Simplify: $\sqrt{44x^2y} + x\sqrt{99y}$

Strategy Since the radicals in each part are unlike radicals, we cannot add them in their current form. However, we will simplify the radicals in each term and hope that like radicals result.

WHY Only like radicals can be combined.

Solution
$$\sqrt{44x^2y} + x\sqrt{99y}$$
$$= \sqrt{4x^2 \cdot 11y} + x\sqrt{9 \cdot 11y}$$ Factor $44x^2y$ and $99y$.
$$= \sqrt{4x^2}\sqrt{11y} + x\sqrt{9}\sqrt{11y}$$ The square root of a product is equal to the product of their square roots.
$$= 2x\sqrt{11y} + 3x\sqrt{11y}$$ Simplify: $\sqrt{4x^2} = 2x$ and $\sqrt{9} = 3$.
$$= 5x\sqrt{11y}$$ To combine like radicals, combine their coefficients: $2x + 3x = 5x$.

Self Check 4
Simplify: $\sqrt{20mn^2} - \sqrt{80m^3}$
Now Try Problem 43

EXAMPLE 4 Simplify: $\sqrt{28x^2y} - 2\sqrt{63y^3}$

Strategy Since the radicals are unlike radicals, we cannot subtract them in their current form. However, we will simplify the radicals in each term and hope that like radicals result.

WHY Only like radicals can be combined.

Solution
$$\sqrt{28x^2y} - 2\sqrt{63y^3}$$
$$= \sqrt{4x^2 \cdot 7y} - 2\sqrt{9y^2 \cdot 7y}$$ Factor $28x^2y$ and $63y^3$.

8.4 Adding and Subtracting Radical Expressions

$$= \sqrt{4x^2}\sqrt{7y} - 2\sqrt{9y^2}\sqrt{7y}$$ The square root of a product is equal to the product of their square roots.

$$= 2x\sqrt{7y} - 2(3y)\sqrt{7y}$$ $\sqrt{4x^2} = 2x$ and $\sqrt{9y^2} = 3y$.

$$= 2x\sqrt{7y} - 6y\sqrt{7y}$$

Since $2x$ and $6y$ are not like terms and therefore cannot be subtracted, the expression does not simplify further.

EXAMPLE 5   Simplify: $\sqrt{27xy} + \sqrt{20xy}$

Strategy Since the radicals are unlike radicals, we cannot add them in their current form. However, we will simplify the radicals in each term and hope that like radicals result.

WHY Only like radicals can be combined.

Solution

$$\sqrt{27xy} + \sqrt{20xy} = \sqrt{9 \cdot 3xy} + \sqrt{4 \cdot 5xy}$$ Factor 27xy and 20xy.

$$= \sqrt{9}\sqrt{3xy} + \sqrt{4}\sqrt{5xy}$$ The square root of a product is equal to the product of their square roots.

$$= 3\sqrt{3xy} + 2\sqrt{5xy}$$ $\sqrt{9} = 3$ and $\sqrt{4} = 2$.

Since the terms have unlike radicals, the expression does not simplify further.

Self Check 5
Simplify: $\sqrt{75ab} + \sqrt{72ab}$
Now Try Problem 47

EXAMPLE 6   Simplify: $\sqrt{8x} + \sqrt{3y} - \sqrt{50x} + \sqrt{27y}$

Strategy Since the radicals are unlike radicals, we cannot add or subtract them in their current form. However, we will simplify the radicals in each term and hope that like radicals result.

WHY Only like radicals can be combined.

Solution

$$\sqrt{8x} + \sqrt{3y} - \sqrt{50x} + \sqrt{27y}$$

$$= \sqrt{4 \cdot 2x} + \sqrt{3y} - \sqrt{25 \cdot 2x} + \sqrt{9 \cdot 3y}$$ Factor 8x, 50x, and 27y.

$$= \sqrt{4}\sqrt{2x} + \sqrt{3y} - \sqrt{25}\sqrt{2x} + \sqrt{9}\sqrt{3y}$$

$$= 2\sqrt{2x} + \sqrt{3y} - 5\sqrt{2x} + 3\sqrt{3y}$$

$$= -3\sqrt{2x} + 4\sqrt{3y}$$ Combine like radicals: $2 - 5 = -3$ and $1 + 3 = 4$.

Self Check 6
Simplify: $\sqrt{32x} - \sqrt{5y} - \sqrt{200x} + \sqrt{125y}$
Now Try Problem 51

3 Simplify cube roots in a sum or difference.

We can extend the concepts used to combine square roots to higher-order radicals.

EXAMPLE 7   Simplify: $\sqrt[3]{81x^4} - x\sqrt[3]{24x}$

Strategy Since the radicals are unlike radicals, we cannot add them in their current form. However, we will simplify the radicals in each term and hope that like radicals result.

WHY Only like radicals can be combined.

Self Check 7
Simplify: $\sqrt[3]{24a^4} + a\sqrt[3]{81a}$
Now Try Problem 57

Solution
We simplify each radical and then combine like radicals.

$$\sqrt[3]{81x^4} - x\sqrt[3]{24x} = \sqrt[3]{27x^3 \cdot 3x} - x\sqrt[3]{8 \cdot 3x} \quad \text{Factor } 81x^4 \text{ and } 24x.$$
$$= \sqrt[3]{27x^3}\sqrt[3]{3x} - x\sqrt[3]{8}\sqrt[3]{3x}$$
$$= 3x\sqrt[3]{3x} - 2x\sqrt[3]{3x}$$
$$= x\sqrt[3]{3x} \quad \text{Combine like terms: } 3x - 2x = x.$$

ANSWERS TO SELF CHECKS
1. a. $7 + 8\sqrt{7}$ b. $-\sqrt{m}$ 2. $14\sqrt{2}$ 3. $5y\sqrt{3x}$ 4. $2n\sqrt{5m} - 4m\sqrt{5m}$
5. $5\sqrt{3ab} + 6\sqrt{2ab}$ 6. $-6\sqrt{2x} + 4\sqrt{5y}$ 7. $5a\sqrt[3]{3a}$

SECTION 8.4 STUDY SET

VOCABULARY

Fill in the blanks.

1. Like _____, such as $2\sqrt{3}$ and $5\sqrt{3}$, have the same index and the same radicand.
2. Like _____, such as $2x$ and $5x$, have the same variables with the same exponents.
3. When $\sqrt{8}$ and $\sqrt{18}$ are written in _____ form, the results are the like radicals $2\sqrt{2}$ and $3\sqrt{2}$.
4. The expression $3\sqrt{2} + \sqrt{8} - 2$ contains three _____.

CONCEPTS

Determine whether the expressions contain like radicals.

5. $5\sqrt{2}$ and $2\sqrt{3}$
6. $7\sqrt{3x}$ and $3\sqrt{3x}$
7. $125\sqrt[3]{13a}$ and $-\sqrt[3]{13a}$
8. $-17\sqrt[4]{5x}$ and $25\sqrt[3]{5x}$

What is wrong with the following work?

9. $7\sqrt{5} - 3\sqrt{2} = 4\sqrt{3}$

10. $12\sqrt{7} + 20\sqrt{11} = 32\sqrt{18}$

11. $7 - 3\sqrt{2} = 4\sqrt{2}$

12. $12 + 20\sqrt{11} = 32\sqrt{11}$

Complete each table.

13.

x	$\sqrt{x} + \sqrt{3}$
3	
12	
27	
48	

14.

x	$3\sqrt{x} - \sqrt{2}$
2	
8	
18	
32	

NOTATION

Complete each step.

15. $9\sqrt{5} - 3\sqrt{20} = 9\sqrt{5} - 3\sqrt{ \cdot 5}$
$= 9\sqrt{5} - 3\sqrt{}\sqrt{5}$
$= 9\sqrt{5} - 3 \cdot $
$= 9\sqrt{5} - $
$= $

16. $3\sqrt{80} + 4\sqrt{125} = 3\sqrt{ \cdot 5} + 4\sqrt{ \cdot 5}$
$= 3\sqrt{16} \cdot + 4\sqrt{25} \cdot $
$= 3()\sqrt{5} + 4(5)\sqrt{5}$
$= 12\sqrt{5} + \sqrt{5}$
$= \sqrt{5}$

8.4 Adding and Subtracting Radical Expressions

GUIDED PRACTICE

Simplify each expression. All variables represent positive numbers. See Example 1.

17. $5\sqrt{7} + 4\sqrt{7}$
18. $3\sqrt{10} + 4\sqrt{10}$
19. $5 + 3\sqrt{3} + 3\sqrt{3}$
20. $\sqrt{5} + 2 + 3\sqrt{5}$
21. $\sqrt{x} - 4\sqrt{x}$
22. $\sqrt{t} - 9\sqrt{t}$
23. $-1 + 2\sqrt{r} - 3\sqrt{r}$
24. $-8 - 5\sqrt{c} + 4\sqrt{c}$

Simplify each expression. See Example 2.

25. $\sqrt{12} + \sqrt{27}$
26. $\sqrt{20} + \sqrt{45}$
27. $\sqrt{18} - \sqrt{8}$
28. $\sqrt{32} - \sqrt{18}$
29. $2\sqrt{45} + 2\sqrt{80}$
30. $3\sqrt{80} + 3\sqrt{125}$
31. $\sqrt{20} + \sqrt{45} + \sqrt{80}$
32. $\sqrt{48} + \sqrt{27} + \sqrt{75}$
33. $\sqrt{200} - \sqrt{75} + \sqrt{48}$
34. $\sqrt{20} + \sqrt{80} - \sqrt{125}$
35. $8\sqrt{6} - 5\sqrt{2} - 3\sqrt{6}$
36. $3\sqrt{2} - 3\sqrt{15} - 4\sqrt{15}$

Simplify each expression. All variables represent positive numbers. See Examples 3–5.

37. $\sqrt{2x^2} + \sqrt{8x^2}$
38. $\sqrt{2d^3} + \sqrt{8d^3}$
39. $\sqrt{49xy^3} + y\sqrt{xy}$
40. $\sqrt{20a^2b} + a\sqrt{180b}$
41. $5\sqrt{2ab^2} - b\sqrt{98a}$
42. $3\sqrt{9b^5} - 2b\sqrt{b^3}$
43. $\sqrt{32x^5} - \sqrt{18x^5}$
44. $\sqrt{3a^3} - a\sqrt{12a}$
45. $\sqrt{18x^2y} - \sqrt{27x^2y}$
46. $\sqrt{3xy^2} - \sqrt{12x^2y}$
47. $\sqrt{18ab} + \sqrt{27ab}$
48. $\sqrt{180x} + \sqrt{252y}$

Simplify each expression. All variables represent positive numbers. See Example 6.

49. $\sqrt{48} - \sqrt{8} + \sqrt{27} - \sqrt{32}$
50. $\sqrt{162} + \sqrt{50} - \sqrt{75} - \sqrt{108}$
51. $5\sqrt{x} + 4\sqrt{4y} - 13\sqrt{x} + 2\sqrt{9y}$
52. $\sqrt{4a} + \sqrt{25b} - 2\sqrt{9a} + 5\sqrt{16b}$

Simplify each expression. See Example 7.

53. $\sqrt[3]{3} + \sqrt[3]{3}$
54. $\sqrt[3]{2} + 5\sqrt[3]{2}$
55. $2\sqrt[3]{x} - 3\sqrt[3]{x}$
56. $4\sqrt[3]{s} - 5\sqrt[3]{s}$
57. $\sqrt[3]{8x^5} + \sqrt[3]{27x^8}$
58. $\sqrt[3]{192x^4y^5} - \sqrt[3]{24x^4y^5}$
59. $\sqrt[3]{24a^5b^4} + \sqrt[3]{81a^5b^4}$
60. $\sqrt[3]{135x^7y^4} - \sqrt[3]{40x^7y^4}$

TRY IT YOURSELF

Simplify each expression. All variables represent positive numbers.

61. $2\sqrt{80} - 3\sqrt{125}$
62. $3\sqrt{245} - 2\sqrt{180}$
63. $2\sqrt{28} + 7\sqrt{63}$
64. $\sqrt{12} - \sqrt{48}$
65. $\sqrt{48} - \sqrt{75}$
66. $\sqrt{288} - 3\sqrt{200}$
67. $\sqrt[3]{32} + \sqrt[3]{108}$
68. $\sqrt[3]{40} + \sqrt[3]{125}$
69. $\sqrt[3]{3,000} - \sqrt[3]{192}$
70. $\sqrt[3]{x^4} - \sqrt[3]{x^7}$
71. $\sqrt{80} - \sqrt{245}$
72. $2\sqrt{28} + 2\sqrt{112}$
73. $4\sqrt{63} + 6\sqrt{112}$
74. $\sqrt{24} + \sqrt{150} + \sqrt{240}$
75. $3\sqrt{54b^2} + 5\sqrt{24b^2}$
76. $3\sqrt{24x^4y^3} + 2\sqrt{54x^4y^3}$
77. $\sqrt[3]{24} - \sqrt[3]{81}$
78. $\sqrt[3]{81} - \sqrt[3]{24}$
79. $\sqrt{28} + \sqrt{63} + \sqrt{18}$
80. $\sqrt{27xy^3} - \sqrt{48xy^3}$
81. $\sqrt[3]{16} + \sqrt[3]{54}$
82. $\sqrt[3]{56a^4b^5} + \sqrt[3]{7a^4b^5}$
83. $y\sqrt{490y} - 2\sqrt{360y^3}$
84. $\sqrt{20x^3y} + \sqrt{45x^5y^3} - \sqrt{80x^7y^5}$
85. $x\sqrt{48xy^2} - y\sqrt{27x^3} + \sqrt{75x^3y^2}$
86. $\sqrt{72p^2q} + \sqrt{54p} - p\sqrt{50q}$

APPLICATIONS

87. ANATOMY Find the length of the patient's arm, shown below, when the arm is straightened.

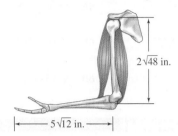

88. PLAYGROUND EQUIPMENT Find the total length of pipe necessary to construct the frame of the swing set shown below.

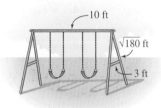

89. BLUEPRINTS Find the length of the motor on the machine shown below.

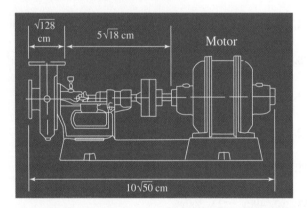

90. TENTS The length of each center support pole for the tents shown in the next column is given by the formula

$$l = 0.5s\sqrt{3}$$

where s is the length of the side of the tent. Find the total length of the four poles (two each) needed for the parents' and children's tents.

91. FENCING Find the number of feet of fencing needed to enclose the swimming pool complex shown below.

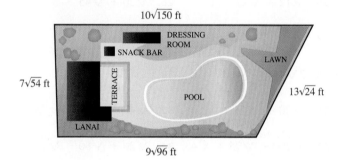

92. HARDWARE Find the difference in the lengths of the arms of the door-closing device shown below.

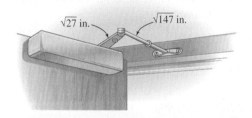

WRITING

93. Explain why $\sqrt{3} + \sqrt{2}$ cannot be combined.

94. Explain why $\sqrt{4x}$ and $\sqrt[3]{4x}$ cannot be combined.

REVIEW

Simplify each expression. Write each answer without using negative exponents.

95. 3^{-2} **96.** $\dfrac{1}{3^{-2}}$

97. -3^2 **98.** -3^{-2}

99. x^{-3} **100.** $\dfrac{1}{x^{-3}}$

101. 3^0 **102.** x^0

SECTION 8.5
Multiplying and Dividing Radical Expressions

Objectives
1. Multiply radical expressions.
2. Find powers of square roots.
3. Multiply radical expressions with more than one term.
4. Divide radical expressions.
5. Rationalize denominators.

In this section, we will discuss the methods used to multiply and divide radical expressions.

1 Multiply radical expressions.

Recall that the *product of the square roots of two nonnegative numbers is equal to the square root of the product of those numbers*. For example,

$$\sqrt{2}\sqrt{8} = \sqrt{2 \cdot 8} \qquad \sqrt{3}\sqrt{27} = \sqrt{3 \cdot 27} \qquad \sqrt{x}\sqrt{x^3} = \sqrt{x \cdot x^3}$$
$$= \sqrt{16} \qquad \qquad = \sqrt{81} \qquad \qquad = \sqrt{x^4}$$
$$= 4 \qquad \qquad = 9 \qquad \qquad = x^2$$

Likewise, the *product of the cube roots of two numbers is equal to the cube root of the product of those numbers*. For example,

$$\sqrt[3]{2}\sqrt[3]{4} = \sqrt[3]{2 \cdot 4} \qquad \sqrt[3]{4}\sqrt[3]{16} = \sqrt[3]{4 \cdot 16} \qquad \sqrt[3]{3x^2}\sqrt[3]{9x} = \sqrt[3]{3x^2 \cdot 9x}$$
$$= \sqrt[3]{8} \qquad \qquad = \sqrt[3]{64} \qquad \qquad = \sqrt[3]{27x^3}$$
$$= 2 \qquad \qquad = 4 \qquad \qquad = 3x$$

These examples illustrate that radical expressions with the same index can be multiplied.

The Product Rules for Square Roots and Cube Roots

For any nonnegative real numbers a and b,

$$\sqrt{a} \cdot \sqrt{b} = \sqrt{a \cdot b} \qquad \text{and} \qquad \sqrt[3]{a} \cdot \sqrt[3]{b} = \sqrt[3]{ab}$$

EXAMPLE 1 Multiply and then simplify if possible:
a. $\sqrt{3}\sqrt{2}$ **b.** $\sqrt{6}\sqrt{8}$ **c.** $\sqrt[3]{4}\sqrt[3]{10}$

Strategy To multiply the radicals, we will multiply their radicands and write the product within the radical. Then we will simplify the result if possible.

WHY The product of the radicals with the same index is the radical of the product of those numbers.

Solution

a. $\sqrt{3}\sqrt{2} = \sqrt{3 \cdot 2}$ The product of the square roots of two numbers is equal to the square root of the product of those numbers.

$\qquad\quad = \sqrt{6}$ Do the multiplication within the radical.

b. $\sqrt{6}\sqrt{8} = \sqrt{6 \cdot 8}$ The product of two square roots is equal to the square root of the product.

$\qquad\quad = \sqrt{48}$ Do the multiplication within the radical. Note that this radical can be simplified.

$\qquad\quad = \sqrt{16}\sqrt{3}$ Factor 48 as $16 \cdot 3$.

$\qquad\quad = 4\sqrt{3}$ Simplify: $\sqrt{16} = 4$.

Self Check 1

Multiply and then simplify if possible:
a. $\sqrt{5}\sqrt{3}$
b. $\sqrt{8}\sqrt{9}$
c. $\sqrt[3]{6}\sqrt[3]{9}$

Now Try Problems 21, 25, and 29

c. $\sqrt[3]{4}\sqrt[3]{10} = \sqrt[3]{4 \cdot 10}$ The product of two cube roots is equal to the cube root of the product.

$\phantom{\sqrt[3]{4}\sqrt[3]{10}} = \sqrt[3]{40}$ Do the multiplication within the radical.

$\phantom{\sqrt[3]{4}\sqrt[3]{10}} = \sqrt[3]{8}\sqrt[3]{5}$ $\sqrt[3]{40} = \sqrt[3]{8 \cdot 5} = \sqrt[3]{8}\sqrt[3]{5}$.

$\phantom{\sqrt[3]{4}\sqrt[3]{10}} = 2\sqrt[3]{5}$ Simplify: $\sqrt[3]{8} = 2$.

To multiply radical expressions having only one term, we multiply the coefficients and multiply the radicals separately and then simplify the result, when possible.

Self Check 2

Multiply and simplify if possible:
a. $(2\sqrt{2x})(-3\sqrt{3x})$
b. $(5\sqrt[3]{2})(2\sqrt[3]{4})$

Now Try Problems 33 and 37

EXAMPLE 2   Multiply and simplify if possible:

a. $3\sqrt{6} \cdot 4\sqrt{3}$ b. $-2\sqrt[3]{7x} \cdot 6\sqrt[3]{49x^2}$

Strategy We will use the commutative and associative properties to multiply the coefficients and the radicals separately. Then we will simplify the result if possible.

WHY This is the rule for multiplying radical expressions that have only one term.

Solution

a. $3\sqrt{6} \cdot 4\sqrt{3} = 3(4)\sqrt{6}\sqrt{3}$ Write the coefficients together and the radicals together.

$\phantom{3\sqrt{6} \cdot 4\sqrt{3}} = 12\sqrt{18}$ Multiply the coefficients and multiply the radicals.

$\phantom{3\sqrt{6} \cdot 4\sqrt{3}} = 12\sqrt{9}\sqrt{2}$ $\sqrt{18} = \sqrt{9 \cdot 2} = \sqrt{9}\sqrt{2}$.

$\phantom{3\sqrt{6} \cdot 4\sqrt{3}} = 12(3)\sqrt{2}$ Simplify: $\sqrt{9} = 3$.

$\phantom{3\sqrt{6} \cdot 4\sqrt{3}} = 36\sqrt{2}$ Do the multiplication: $12(3) = 36$.

b. $-2\sqrt[3]{7x} \cdot 6\sqrt[3]{49x^2} = -2(6)\sqrt[3]{7x}\sqrt[3]{49x^2}$ Write the coefficients together and the radicals together.

$\phantom{-2\sqrt[3]{7x} \cdot 6\sqrt[3]{49x^2}} = -12\sqrt[3]{7x \cdot 49x^2}$ Multiply the coefficients and multiply the radicals.

$\phantom{-2\sqrt[3]{7x} \cdot 6\sqrt[3]{49x^2}} = -12\sqrt[3]{343x^3}$ Do the multiplication within the radical.

$\phantom{-2\sqrt[3]{7x} \cdot 6\sqrt[3]{49x^2}} = -12(7x)$ Simplify: $\sqrt[3]{343x^3} = 7x$.

$\phantom{-2\sqrt[3]{7x} \cdot 6\sqrt[3]{49x^2}} = -84x$ Multiply.

2   Find powers of square roots.

By definition, when the square root of a positive number is squared, the result is that positive number.

Success Tip   Since $(\sqrt{a})^2 = \sqrt{a} \cdot \sqrt{a}$, it also follows that $\sqrt{a} \cdot \sqrt{a} = a$, if a is a positive real number.

Self Check 3

Find: $(6\sqrt{3})^2$

Now Try Problems 43 and 47

EXAMPLE 3   Find: $(2\sqrt{5})^2$

Strategy We will use the rule $(\sqrt{a})^2 = a$.

WHY The expression involves the square of the square root of a positive number.

8.5 Multiplying and Dividing Radical Expressions

4 Divide radical expressions.

To divide radical expressions, we use the division property of radicals. For example, to divide $\sqrt{108}$ by $\sqrt{36}$, we proceed as follows:

$$\frac{\sqrt{108}}{\sqrt{36}} = \sqrt{\frac{108}{36}}$$ The quotient of two square roots is the square root of the quotient.

$$= \sqrt{3}$$ Do the division within the radical: $108 \div 36 = 3$.

EXAMPLE 8

Divide: $\dfrac{\sqrt{22a^2}}{\sqrt{99a^4}}$

Strategy To divide the square roots, we will divide the radicands and write the quotient within a square root symbol. Then we will simplify the result if possible.

WHY Since the denominator contains a square root, the expression is not in simplified form.

Solution

$$\frac{\sqrt{22a^2}}{\sqrt{99a^4}} = \sqrt{\frac{22a^2}{99a^4}}$$

$$= \sqrt{\frac{2}{9a^2}}$$ Simplify the radicand: $\dfrac{22a^2}{99a^4} = \dfrac{\overset{1}{\cancel{11}} \cdot 2 \cdot \overset{1}{\cancel{a^2}}}{\underset{1}{\cancel{11}} \cdot 9 \cdot \underset{1}{\cancel{a^2}} \cdot a^2} = \dfrac{2}{9a^2}$.

$$= \frac{\sqrt{2}}{\sqrt{9a^2}}$$ The square root of a quotient is equal to the quotient of their square roots.

$$= \frac{\sqrt{2}}{3a}$$ Simplify: $\sqrt{9a^2} = 3a$.

Self Check 8

Divide: $\dfrac{\sqrt{30y^9}}{\sqrt{160y^5}}$

Now Try Problems 87 and 91

5 Rationalize denominators.

The length of a diagonal of one of the square tiles shown in the figure is 1 foot. Using the Pythagorean theorem, it can be shown that the length of a side of a tile is $\dfrac{1}{\sqrt{2}}$ feet. Because the expression $\dfrac{1}{\sqrt{2}}$ contains a radical in its denominator, it is not in simplified form. Since it is often easier to work with a radical expression if the denominator does not contain a radical, we now consider how to change the denominator from a radical that represents an irrational number to a rational number. The process is called **rationalizing the denominator**.

To rationalize the denominator of $\dfrac{1}{\sqrt{2}}$, we multiply by 1 in the form $\dfrac{\sqrt{2}}{\sqrt{2}}$.

$$\frac{1}{\sqrt{2}} = \frac{1}{\sqrt{2}} \cdot \frac{\sqrt{2}}{\sqrt{2}}$$ Multiply by 1 in the form $\dfrac{\sqrt{2}}{\sqrt{2}}$.

$$= \frac{\sqrt{2}}{2}$$ In the numerator, $1\sqrt{2} = \sqrt{2}$. In the denominator, $\sqrt{2}\sqrt{2} = (\sqrt{2})^2 = 2$. The denominator is now a rational number.

The length of a side of a patio tile is $\dfrac{1}{\sqrt{2}} = \dfrac{\sqrt{2}}{2}$ feet.

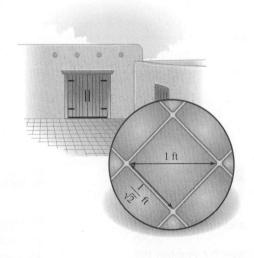

Chapter 8 Roots and Radicals

This example suggests the following procedure for rationalizing denominators.

> **Rationalizing Denominators**
>
> To **rationalize a square root denominator**, multiply the numerator and denominator of the given fraction by the square root that makes a perfect-square radicand in the denominator.
>
> To **rationalize a cube root denominator**, multiply the numerator and denominator of the given fraction by the cube root that makes a perfect-cube radicand in the denominator.

Self Check 9

Rationalize each denominator:

a. $\sqrt{\dfrac{2}{7}}$

b. $\dfrac{5}{\sqrt[3]{5}}$

Now Try Problems 95 and 97

EXAMPLE 9

Rationalize each denominator: a. $\sqrt{\dfrac{5}{3}}$ b. $\dfrac{2}{\sqrt[3]{3}}$

Strategy In each case, we will rationalize the denominator.

WHY The expression will then be in simplest form.

Solution

a. The expression $\sqrt{\dfrac{5}{3}}$ is not in simplified form, because the radicand is a fraction. To write it in simplified form, we use the division property of radicals. Then we use the fundamental property of fractions to rationalize the denominator by multiplying the numerator and the denominator by $\sqrt{3}$.

$$\sqrt{\dfrac{5}{3}} = \dfrac{\sqrt{5}}{\sqrt{3}}$$ The square root of a quotient is the quotient of the square roots. Note that the denominator is the irrational number $\sqrt{3}$.

$$= \dfrac{\sqrt{5}}{\sqrt{3}} \cdot \dfrac{\sqrt{3}}{\sqrt{3}}$$ To build an equivalent fraction, multiply by $\dfrac{\sqrt{3}}{\sqrt{3}} = 1$.

$$= \dfrac{\sqrt{15}}{3}$$ In the numerator, do the multiplication. Simplify in the denominator: $\sqrt{3}\sqrt{3} = (\sqrt{3})^2 = 3$.

b. The denominator contains a cube root. We multiply by the smallest factor that gives an integer cube radicand in the denominator. Since $\sqrt[3]{3}\sqrt[3]{9} = \sqrt[3]{27}$ and 27 is a perfect integer cube, we multiply the numerator and denominator by $\sqrt[3]{9}$ and simplify.

$$\dfrac{2}{\sqrt[3]{3}} = \dfrac{2}{\sqrt[3]{3}} \cdot \dfrac{\sqrt[3]{9}}{\sqrt[3]{9}}$$ To build an equivalent fraction, multiply by $\dfrac{\sqrt[3]{9}}{\sqrt[3]{9}} = 1$.

$$= \dfrac{2\sqrt[3]{9}}{\sqrt[3]{27}}$$ Multiply: $\sqrt[3]{3}\sqrt[3]{9} = \sqrt[3]{27}$.

$$= \dfrac{2\sqrt[3]{9}}{3}$$ $\sqrt[3]{27} = 3$. The denominator is now a rational number.

Self Check 10

Rationalize the denominator and simplify: $\dfrac{6\sqrt{z}}{\sqrt{50y}}$

Now Try Problem 103

EXAMPLE 10

Rationalize the denominator and simplify: $\dfrac{5\sqrt{y}}{\sqrt{20x}}$

Strategy We will start by multiplying the numerator and denominator by the smallest factor that will force a perfect square under the square root in the denominator. Then we will simplify the fraction if possible.

WHY We can then evaluate the square root in the denominator.

8.5 Multiplying and Dividing Radical Expressions

Solution
To rationalize the denominator, we don't need to multiply the numerator and denominator by $\sqrt{20x}$. To keep the numbers small, we can multiply by $\sqrt{5x}$, because $5x \cdot 20x = 100x^2$, which is a perfect square.

$$\frac{5\sqrt{y}}{\sqrt{20x}} = \frac{5\sqrt{y}}{\sqrt{20x}} \cdot \frac{\sqrt{5x}}{\sqrt{5x}}$$
To build an equivalent fraction, multiply by $\frac{\sqrt{5x}}{\sqrt{5x}} = 1$.

$$= \frac{5\sqrt{5xy}}{\sqrt{100x^2}}$$
Multiply: $\sqrt{y}\sqrt{5x} = \sqrt{5xy}$ and $\sqrt{20x}\sqrt{5x} = \sqrt{100x^2}$.

$$= \frac{5\sqrt{5xy}}{10x}$$
Simplify: $\sqrt{100x^2} = 10x$.

$$= \frac{\overset{1}{\cancel{5}}\sqrt{5xy}}{\underset{1}{\cancel{5}} \cdot 2x}$$
Factor $10x$ and remove a common factor of 5.

$$= \frac{\sqrt{5xy}}{2x}$$

We will now discuss a method to rationalize denominators of fractions that have two terms.

One-term denominators
$$\frac{1}{\sqrt{2}}, \quad \frac{\sqrt{5}}{\sqrt{3}}, \quad \frac{4}{\sqrt{24x}}$$

Two-term denominators
$$\frac{5}{\sqrt{6} - 1}, \quad \frac{3}{\sqrt{7} + 2}$$

To rationalize the denominator of $\frac{5}{\sqrt{6}-1}$, we multiply the numerator and denominator by $\sqrt{6} + 1$, because the product $(\sqrt{6} - 1)(\sqrt{6} + 1)$ contains no radicals. Radical expressions such as $\sqrt{6} - 1$ and $\sqrt{6} + 1$ are said to be **conjugates** of each other. Notice that they differ only in the sign separating the terms.

EXAMPLE 11
Rationalize each denominator:

a. $\dfrac{2}{\sqrt{6} - 1}$ **b.** $\dfrac{\sqrt{2} - \sqrt{y}}{\sqrt{3} + \sqrt{y}}$

Strategy To rationalize each two-term denominator, we will multiply the expression by a form of 1 that uses the conjugate of the denominator.

WHY Since the product of a radical expression and its conjugate contains no radicals, this step clears the denominator of radicals.

Solution
a. The conjugate of $\sqrt{6} - 1$ is $\sqrt{6} + 1$.

$$\frac{2}{\sqrt{6}-1} = \frac{2}{\sqrt{6}-1} \cdot \frac{\sqrt{6}+1}{\sqrt{6}+1}$$
To build an equivalent fraction, multiply by $\frac{\sqrt{6}+1}{\sqrt{6}+1} = 1$.

$$= \frac{2(\sqrt{6}+1)}{(\sqrt{6}-1)(\sqrt{6}+1)}$$
Write the product of the numerators and the product of the denominators.

$$= \frac{2(\sqrt{6}+1)}{(\sqrt{6})^2 - 1^2}$$
In the denominator, use a special product rule: $(x - y)(x + y) = x^2 - y^2$.

$$= \frac{2(\sqrt{6}+1)}{6-1}$$
After evaluating $(\sqrt{6})^2$, the denominator no longer contains a radical.

Self Check 11
Rationalize the denominator:

a. $\dfrac{5}{\sqrt{7} + 2}$

b. $\dfrac{\sqrt{x} - \sqrt{7}}{\sqrt{x} - \sqrt{2}}$

Now Try Problems 105 and 109

$$= \frac{2(\sqrt{6}+1)}{5} \quad \text{In the denominator, subtract: } 6-1=5.$$

$$= \frac{2\sqrt{6}+2}{5} \quad \text{In the numerator, distribute the multiplication by 2.}$$

b. The conjugate of the denominator, $\sqrt{3}+\sqrt{y}$, is $\sqrt{3}-\sqrt{y}$.

$$\frac{\sqrt{2}-\sqrt{y}}{\sqrt{3}+\sqrt{y}} = \frac{\sqrt{2}-\sqrt{y}}{\sqrt{3}+\sqrt{y}} \cdot \frac{\sqrt{3}-\sqrt{y}}{\sqrt{3}-\sqrt{y}} \quad \text{To build an equivalent fraction, multiply by } \frac{\sqrt{3}-\sqrt{y}}{\sqrt{3}-\sqrt{y}} = 1.$$

$$= \frac{(\sqrt{2}-\sqrt{y})(\sqrt{3}-\sqrt{y})}{(\sqrt{3}+\sqrt{y})(\sqrt{3}-\sqrt{y})} \quad \text{Write the product of the numerators and the product of the denominators.}$$

$$= \frac{\sqrt{6}-\sqrt{2y}-\sqrt{3y}+y}{(\sqrt{3})^2-(\sqrt{y})^2} \quad \text{In the numerator, use the FOIL method. In the denominator, use a special product rule: } (x+y)(x-y)=x^2-y^2.$$

$$= \frac{\sqrt{6}-\sqrt{2y}-\sqrt{3y}+y}{3-y} \quad \text{After simplifying the denominator, it no longer contains radicals.}$$

ANSWERS TO SELF CHECKS

1. a. $\sqrt{15}$ **b.** $6\sqrt{2}$ **c.** $3\sqrt[3]{2}$ **2. a.** $-6x\sqrt{6}$ **b.** 20   **3.** 108   **4. a.** $9\sqrt{2}-3$
b. $3\sqrt[3]{2x}-2x$ **5.** $5a+\sqrt{5a}-6$ **6. a.** $15-2\sqrt{15n}+n$ **b.** -2
7. $3x-2\sqrt[3]{3x}+\sqrt[3]{9x^2}-2$ **8.** $\frac{y^2\sqrt{3}}{4}$ **9. a.** $\frac{\sqrt{14}}{7}$ **b.** $\sqrt[3]{25}$ **10.** $\frac{3\sqrt{2yz}}{5y}$
11. a. $\frac{5\sqrt{7}-10}{3}$ **b.** $\frac{x+\sqrt{2x}-\sqrt{7x}-\sqrt{14}}{x-2}$

SECTION 8.5 STUDY SET

VOCABULARY

Fill in the blanks.

1. In the radical expression $3\sqrt{7}$, the number 3 is the _____ of the radical.
2. Radical expressions with the same _____ can be multiplied.
3. The method of changing a radical denominator of a fraction into a rational number is called _____ the denominator.
4. $3+\sqrt{2}$ is the _____ of $3-\sqrt{2}$.

CONCEPTS

Fill in the blanks.

5. $\sqrt{a}\cdot\sqrt{b} = \sqrt{}$ 6. $\sqrt{\dfrac{a}{b}} = \dfrac{}{}$

7. To rationalize the denominator of $\dfrac{x}{\sqrt{x}+1}$, multiply the numerator and denominator by _____.
8. To multiply $2\sqrt{x}$ and $6\sqrt{x}$, we first multiply the _____, then multiply the _____, and simplify the result.

Perform each operation, if possible.

9. $\sqrt{2}+\sqrt{3}$ 10. $\sqrt{2}\cdot\sqrt{3}$
11. $\sqrt{2}-\sqrt{3}$ 12. $\dfrac{\sqrt{2}}{\sqrt{3}}$
13. $\sqrt{2}+3\sqrt{2}$ 14. $\sqrt{2}\cdot 3\sqrt{2}$
15. $\sqrt{2}-3\sqrt{2}$ 16. $\dfrac{\sqrt{2}}{3\sqrt{2}}$

Find each special product.

17. $(\sqrt{6}+\sqrt{3})(\sqrt{6}-\sqrt{3})$
18. $(\sqrt{a}+\sqrt{7})(\sqrt{a}-\sqrt{7})$

NOTATION

Complete each step.

19. $(\sqrt{x}+\sqrt{2})(\sqrt{x}-3\sqrt{2})$
$= \sqrt{x} - \sqrt{x}(3\sqrt{2}) + \sqrt{2} - \sqrt{2}(3\sqrt{2})$
$= x - 3 + \sqrt{2x} - 3\sqrt{2}\sqrt{2}$
$= - 2\sqrt{2x} - 3(2)$
$= x - 2\sqrt{2x} - 6$

20. $\dfrac{x}{\sqrt{x}-2} = \dfrac{x(\sqrt{x}+2)}{(\sqrt{x}-2)\boxed{}}$

$= \dfrac{x(\sqrt{x}+2)}{(\boxed{})^2 - 2^2}$

$= \dfrac{x\sqrt{x}+2x}{x-4}$

GUIDED PRACTICE

Perform each multiplication. See Example 1.

21. $\sqrt{7}\sqrt{3}$ 22. $\sqrt{2}\sqrt{11}$
23. $\sqrt{5}\sqrt{7}$ 24. $\sqrt{7}\sqrt{6}$
25. $\sqrt{2}\sqrt{8}$ 26. $\sqrt{27}\sqrt{3}$
27. $\sqrt{8}\sqrt{7}$ 28. $\sqrt{6}\sqrt{8}$
29. $\sqrt[3]{6}\sqrt[3]{4}$ 30. $\sqrt[3]{10}\sqrt[3]{200}$
31. $\sqrt[3]{7}\sqrt[3]{98}$ 32. $\sqrt[3]{16}\sqrt[3]{54}$

Perform each multiplication. See Example 2.

33. $(-5\sqrt{6})(4\sqrt{3})$ 34. $(6\sqrt{3})(-7\sqrt{2})$
35. $(4\sqrt{x})(-2\sqrt{x})$ 36. $(3\sqrt{y})(15\sqrt{y})$
37. $(2\sqrt[3]{4})(3\sqrt[3]{3})$ 38. $(-3\sqrt[3]{3})(\sqrt[3]{5})$
39. $(-3\sqrt[3]{5a})(2\sqrt[3]{25a^2})$ 40. $(2\sqrt[3]{4b^2})(4\sqrt[3]{16b^4})$

Find each power. See Example 3.

41. $(\sqrt{5})^2$ 42. $(\sqrt{11})^2$
43. $(2\sqrt{3})^2$ 44. $(-3\sqrt{5})^2$
45. $(-\sqrt[3]{9})^3$ 46. $(\sqrt[3]{3})^3$
47. $(2\sqrt[3]{9})^3$ 48. $-2(-\sqrt[3]{3})^3$

Perform each multiplication. All variables represent positive numbers. See Example 4.

49. $\sqrt{2}(\sqrt{2}+1)$ 50. $\sqrt{5}(\sqrt{5}+2)$
51. $2\sqrt{2}(\sqrt{8}-1)$ 52. $\sqrt{3}(\sqrt{6}+1)$
53. $\sqrt{x}(\sqrt{3x}-2)$ 54. $\sqrt{y}(\sqrt{y}+5)$
55. $2\sqrt{x}(\sqrt{9x}+3)$ 56. $3\sqrt{z}(\sqrt{4z}-\sqrt{z})$
57. $\sqrt[3]{7}(\sqrt[3]{49}-2)$ 58. $\sqrt[3]{5}(\sqrt[3]{25}+3)$
59. $\sqrt[3]{2x}(\sqrt[3]{27x^2}+3)$ 60. $\sqrt[3]{6a^2}(\sqrt[3]{36a}-5)$

Perform each multiplication. All variables represent positive numbers. See Example 5.

61. $(\sqrt{2}-\sqrt{3})(\sqrt{3}+\sqrt{5})$
62. $(\sqrt{3}+\sqrt{5})(\sqrt{5}-\sqrt{2})$
63. $(\sqrt{2b}+2)(\sqrt{2b}-1)$ 64. $(\sqrt{3p}-5)(\sqrt{3p}-2)$
65. $(2+\sqrt{3t})(3-\sqrt{3t})$ 66. $(5-\sqrt{2q})(2+\sqrt{2q})$
67. $(\sqrt{2x}+3)(\sqrt{8x}-6)$ 68. $(\sqrt{5y}-3)(\sqrt{20y}+6)$

Find each product and simplify if possible. See Example 6.

69. $(5+\sqrt{3})^2$ 70. $(4+\sqrt{2})^2$
71. $(a+\sqrt{7})^2$ 72. $(r+\sqrt{6})^2$
73. $(\sqrt{5}-\sqrt{m})^2$ 74. $(\sqrt{3}-\sqrt{v})^2$
75. $(\sqrt{6}+\sqrt{3})(\sqrt{6}-\sqrt{3})$
76. $(\sqrt{8}+\sqrt{7})(\sqrt{8}-\sqrt{7})$
77. $(\sqrt{11}-y)(\sqrt{11}+y)$ 78. $(\sqrt{2}-s)(\sqrt{2}+s)$
79. $(\sqrt{7c}-3)(\sqrt{7c}+3)$ 80. $(\sqrt{5r}+4)(\sqrt{5r}-4)$

Perform each multiplication. See Example 7.

81. $(\sqrt[3]{2}+1)(\sqrt[3]{2}+3)$ 82. $(\sqrt[3]{5}-2)(\sqrt[3]{5}-1)$
83. $(\sqrt[3]{4y}+3)(\sqrt[3]{2y^2}-1)$ 84. $(\sqrt[3]{3c}-2)(\sqrt[3]{9c^2}+2)$

Perform each division. Assume that all variables represent positive numbers. See Example 8.

85. $\dfrac{\sqrt{12x^3}}{\sqrt{27x}}$ 86. $\dfrac{\sqrt{32}}{\sqrt{98x^2}}$
87. $\dfrac{\sqrt{18x}}{\sqrt{25x}}$ 88. $\dfrac{\sqrt{27x}}{\sqrt{75x}}$
89. $\dfrac{\sqrt{196x}}{\sqrt{49x^3}}$ 90. $\dfrac{\sqrt{50}}{\sqrt{98z^2}}$
91. $\dfrac{\sqrt[3]{16x^6}}{\sqrt[3]{54x^3}}$ 92. $\dfrac{\sqrt[3]{128a^6}}{\sqrt[3]{16a^3}}$

Rationalize each denominator and simplify. All variables represent positive numbers. See Example 9.

93. $\dfrac{1}{\sqrt{3}}$ 94. $\dfrac{1}{\sqrt{5}}$
95. $\sqrt{\dfrac{13}{7}}$ 96. $\sqrt{\dfrac{3}{11}}$

97. $\dfrac{5}{\sqrt[3]{5}}$

98. $\dfrac{7}{\sqrt[3]{7}}$

99. $\dfrac{4}{\sqrt[3]{4}}$

100. $\dfrac{7}{\sqrt[3]{10}}$

Rationalize each denominator and simplify. All variables represent positive numbers. See Example 10.

101. $\dfrac{12}{\sqrt{y}}$

102. $\dfrac{\sqrt{9y}}{\sqrt{2x}}$

103. $\dfrac{4\sqrt{a}}{\sqrt{20b}}$

104. $\dfrac{6\sqrt{m}}{\sqrt{72n}}$

Rationalize each denominator and simplify. Assume that all variables are positive. See Example 11.

105. $\dfrac{3}{\sqrt{3}-1}$

106. $\dfrac{3}{\sqrt{5}-2}$

107. $\dfrac{3}{\sqrt{7}+2}$

108. $\dfrac{5}{\sqrt{8}+3}$

109. $\dfrac{\sqrt{x}+2}{\sqrt{x}-2}$

110. $\dfrac{\sqrt{x}-3}{\sqrt{x}+3}$

111. $\dfrac{\sqrt{2a}+1}{\sqrt{2a}-2}$

112. $\dfrac{\sqrt{3b}+2}{\sqrt{3b}-1}$

TRY IT YOURSELF

Perform the operations and simplify if possible.

113. $(3\sqrt{6})^2$

114. $(-7\sqrt{2})^2$

115. $3\sqrt{2}\sqrt{x}$

116. $4\sqrt{3x}\sqrt{5y}$

117. $\sqrt{8x}\sqrt{2x^3}$

118. $\sqrt{27y}\sqrt{3y^3}$

119. $3\sqrt{3x}(\sqrt{27x}-1)$

120. $\sqrt{2a}(\sqrt{6a}-2)$

121. $(2\sqrt{7}-x)(3\sqrt{2}+x)$

122. $(4\sqrt{2}-\sqrt{x})(\sqrt{x}+2\sqrt{3})$

123. $(\sqrt{6}+1)^2$

124. $(3-\sqrt{3})^2$

125. $\dfrac{9}{\sqrt{27}}$

126. $\dfrac{4}{\sqrt{20}}$

127. $\dfrac{3}{\sqrt{32x}}$

128. $\dfrac{5}{\sqrt{18y}}$

129. $\sqrt{\dfrac{12}{5}}$

130. $\sqrt{\dfrac{24}{7}}$

131. $\sqrt[3]{4}(\sqrt[3]{16}-1)$

132. $\sqrt[3]{5a}(\sqrt[3]{25a}+\sqrt[3]{a^2})$

133. $\dfrac{12}{3-\sqrt{3}}$

134. $\dfrac{10}{5-\sqrt{5}}$

135. $\dfrac{x}{\sqrt{3}+\sqrt{2}}$

136. $\dfrac{a}{\sqrt{3}-\sqrt{2}}$

137. $\dfrac{\sqrt[3]{5}}{\sqrt[3]{2}}$

138. $\dfrac{\sqrt[3]{2}}{\sqrt[3]{5}}$

139. $\dfrac{\sqrt{y}+3}{\sqrt{y}-3}$

140. $\dfrac{\sqrt{t}-1}{\sqrt{t}+4}$

APPLICATIONS

141. **LAWNMOWERS** See the illustration below, which shows the blade of a rotary lawnmower. Use the formula for the area of a circle, $A = \pi r^2$, to find the area of lawn covered by one rotation of the blade. Leave π in the answer.

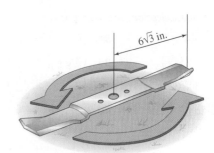

142. **AWARDS PLATFORMS** Find the total number of cubic feet of concrete needed to construct the Olympic Games awards platforms shown below.

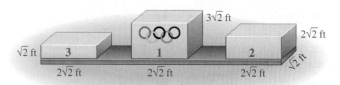

143. **AIR HOCKEY** Find the area of the playing surface of the air hockey game in the illustration.

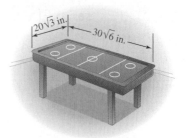

144. **PROJECTOR SCREENS** To find the length l of a rectangle, we can use the formula $l = \dfrac{A}{w}$ where A is the area of the rectangle and w is its width. Find the length of the screen shown below if its area is 54 square feet.

145. **COSTUME DESIGNS** The pattern for one panel of an 1870s English dress is printed on the 1 in. × 1 in. grid shown below. Find the number of square inches of fabric in the trapezoidal panel. (*Hint:* Use the dashed grid lines and the Pythagorean theorem three times to determine the length of the upper base, the length of the lower base, and the height of the trapezoid.)

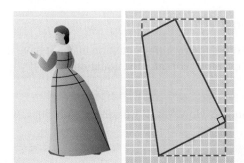

146. **SET DESIGNS** The director of a stage play requested bright down lighting over the portion of the set shown below. Find the area of the rectangle. (*Hint:* Use the dashed grid lines and the Pythagorean theorem twice to determine the length and width of the rectangle.)

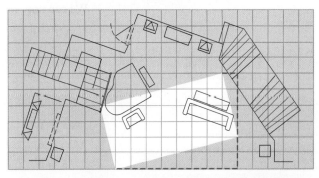

WRITING

147. When rationalizing the denominator of $\dfrac{5}{\sqrt{6}}$, why must we multiply both the numerator and denominator by $\sqrt{6}$?

148. A calculator is used to approximate $\dfrac{2}{\sqrt{6}}$ and $\dfrac{\sqrt{6}}{3}$. In each case, the calculator display reads 0.816496581. Explain why the results are the same.

REVIEW

149. Is -2 a solution of $3x - 7 = 5x + 1$?

150. To evaluate the expression $2 - (-3 + 4)^2$, which operation should be performed first?

151. The graph of a straight line rises from left to right. Is the slope of the line positive or negative?

152. Multiply: $(x - 4)(x + 4)$

SECTION 8.6
Solving Radical Equations; the Distance Formula

Objectives

1. Use the squaring property of equality.
2. Check possible solutions.
3. Solve equations containing one square root.
4. Solve radical equations by squaring a binomial.
5. Solve equations containing two square roots.
6. Solve equations containing cube roots.
7. The distance formula.
8. Solve problems modeled by a radical equation.

Many situations can be modeled mathematically by equations that contain radicals. In this section, we will develop techniques to solve such equations.

1 Use the squaring property of equality.

The equation $\sqrt{x} = 6$ is called a **radical equation**, because it contains a radical expression with a variable radicand. To solve this equation, we isolate x by undoing the operation performed on it. Recall that $\sqrt{x}$ represents the number that, when squared, gives x. Therefore, if we *square* $\sqrt{x}$, we will obtain x.

$$\left(\sqrt{x}\right)^2 = x$$

Using this observation, we can eliminate the radical on the left side of $\sqrt{x} = 6$ by squaring that side. Intuition tells us that we should also square the right side. This is a valid step, because *if two numbers are equal, their squares are equal.*

Unless otherwise noted, all content on this page is © Cengage Learning.

> **Squaring Property of Equality**
>
> For any real numbers a and b, if $a = b$, then $a^2 = b^2$.

We can now solve $\sqrt{x} = 6$ by applying the squaring property of equality.

$\sqrt{x} = 6$ *This is the equation to solve.*

$(\sqrt{x})^2 = (6)^2$ *Square both sides of the equation to eliminate the radical.*

$x = 36$ *Simplify each side: $(\sqrt{x})^2 = x$ and $(6)^2 = 36$.*

Checking this result, we have

$\sqrt{x} = 6$

$\sqrt{36} \stackrel{?}{=} 6$ *Substitute 36 for x.*

$6 = 6$ *Simplify the left side: $\sqrt{36} = 6$.*

Since we obtain a true statement, 36 is the solution, and the solution set is $\{36\}$.

2 Check possible solutions.

If we square both sides of an equation, the resulting equation may not have the same solutions as the original one. For example, consider the equation

$x = 2$

The only solution of this equation is 2. However, if we square both sides, we obtain $(x)^2 = 2^2$, which simplifies to

$x^2 = 4$

This new equation has solutions 2 and -2, because $2^2 = 4$ and $(-2)^2 = 4$.

The equations $x = 2$ and $x^2 = 4$ are not equivalent equations because they do not have the same solutions. The solution -2 satisfies $x^2 = 4$, but it does not satisfy $x = 2$. We see that squaring both sides of an equation can produce an equation with solutions that don't satisfy the original one. We call such numbers *extraneous* solutions. Therefore, we must check each possible solution in the original equation.

3 Solve equations containing one square root.

To solve an equation containing square root radicals, we follow these steps.

> **Solving Radical Equations**
>
> 1. Isolate a radical term on one side of the equation.
> 2. Square both sides of the equation, and solve the resulting equation.
> 3. Check the possible solutions in the original equation. Discard any extraneous solutions. This step is required.

Self Check 1

Solve: $\sqrt{x-4} = 9$

EXAMPLE 1   Solve: $\sqrt{x+2} = 3$

Strategy We will square both sides of the equation.

8.6 Solving Radical Equations; the Distance Formula

WHY The radical is already on a side by itself. Squaring both sides will clear the equation of the square root.

Now Try Problem 27

Solution
Since this might produce an equation with more solutions than the original one, we must check each solution.

$$\sqrt{x+2} = 3 \quad \text{This is the equation to solve.}$$
$$(\sqrt{x+2})^2 = (3)^2 \quad \text{Square both sides.}$$
$$x + 2 = 9 \quad (\sqrt{x+2})^2 = x+2 \text{ and } 3^2 = 9.$$
$$x = 7 \quad \text{Subtract 2 from both sides.}$$

We check by substituting 7 for x in the original equation.

$$\sqrt{x+2} = 3$$
$$\sqrt{7+2} \stackrel{?}{=} 3 \quad \text{Substitute 7 for x.}$$
$$\sqrt{9} \stackrel{?}{=} 3 \quad \text{Perform the addition within the radical symbol.}$$
$$3 = 3$$

Since a true statement results, 7 is the solution. The solution set is $\{7\}$.

EXAMPLE 2 Solve: $\sqrt{5x+1} + 7 = 3$

Self Check 2
Solve: $\sqrt{3x-2} + 6 = 1$
Now Try Problem 31

Strategy Since 7 is a term outside the square root symbol, there are two terms on the left side of the equation. To isolate the radical, we will subtract 7 from both sides.

WHY This will put the equation in a form where we can square both sides to clear the radical.

Solution
We isolate the radical on one side and proceed as follows:

$$\sqrt{5x+1} + 7 = 3 \quad \text{This is the equation to solve.}$$
$$\sqrt{5x+1} = -4 \quad \text{Subtract 7 from both sides.}$$
$$(\sqrt{5x+1})^2 = (-4)^2 \quad \text{Square both sides to eliminate the radical.}$$
$$5x + 1 = 16 \quad (\sqrt{5x+1})^2 = 5x+1 \text{ and } (-4)^2 = 16.$$
$$5x = 15 \quad \text{Subtract 1 from both sides.}$$
$$x = 3 \quad \text{Divide both sides by 5.}$$

We check by substituting 3 for x in the original equation.

$$\sqrt{5x+1} + 7 = 3$$
$$\sqrt{5(3)+1} + 7 \stackrel{?}{=} 3 \quad \text{Substitute 3 for x.}$$
$$\sqrt{16} + 7 \stackrel{?}{=} 3 \quad \text{Evaluate the expression within the radical symbol.}$$
$$4 + 7 \stackrel{?}{=} 3$$
$$11 \neq 3$$

Since $11 \neq 3$, 3 is not a solution. In fact, the equation has no solution. This was apparent in step 2 of the solution. There is no real number x that could make the nonnegative number $\sqrt{5x+1}$ equal to -4. The solution set is $\{\ \}$.

Example 2 shows that squaring both sides of an equation can lead to possible solutions that do not satisfy the original equation. We call such numbers **extraneous solutions**. In Example 2, the number 3 is an extraneous solution of $\sqrt{5x+1} + 7 = 3$.

4 Solve radical equations by squaring a binomial.

Self Check 3
Solve: $b + 4 = \sqrt{b^2 + 6b + 12}$

Now Try Problem 35

EXAMPLE 3 Solve: $a + 2 = \sqrt{a^2 + 3a + 3}$

Strategy We will square both sides to clear the equation of the radical.

WHY The radical is by itself on one side of the equation.

Solution
The radical is isolated on the right side, so we proceed by squaring both sides to eliminate it.

$a + 2 = \sqrt{a^2 + 3a + 3}$ This is the equation to solve.

$(a + 2)^2 = \left(\sqrt{a^2 + 3a + 3}\right)^2$ Square both sides.

$a^2 + 4a + 4 = a^2 + 3a + 3$ Use a special product rule: $(a + 2)^2 = a^2 + 4a + 4$. $\left(\sqrt{a^2 + 3a + 3}\right)^2 = a^2 + 3a + 3$.

$a^2 + 4a + 4 - a^2 = a^2 + 3a + 3 - a^2$ To eliminate a^2, subtract a^2 from both sides.

$4a + 4 = 3a + 3$ Combine like terms: $a^2 - a^2 = 0$.

$a + 4 = 3$ Subtract $3a$ from both sides.

$a = -1$ Subtract 4 from both sides.

We check by substituting -1 for a in the original equation.

$a + 2 = \sqrt{a^2 + 3a + 3}$

$-1 + 2 \stackrel{?}{=} \sqrt{(-1)^2 + 3(-1) + 3}$ Substitute -1 for a.

$1 \stackrel{?}{=} \sqrt{1 - 3 + 3}$ Within the radical symbol, first find the power, then perform the multiplication.

$1 \stackrel{?}{=} \sqrt{1}$ Simplify within the radical symbol.

$1 = 1$

The solution is -1, and the solution set is $\{-1\}$.

Sometimes, after clearing an equation of a radical, the result is a quadratic equation.

Self Check 4
Solve: $\sqrt{x + 4} - x = -2$

Now Try Problem 45

EXAMPLE 4 Solve: $\sqrt{3 - x} - x = -3$

Strategy We will add x to both sides of the equation to get the radical alone on one side of the equation.

WHY This will put the equation in a form where we can square both sides to clear the radical.

Solution

$\sqrt{3 - x} - x = -3$ This is the equation to solve.

$\sqrt{3 - x} = x - 3$ Add x to both sides.

We then square both sides to clear the equation of the radical.

$\left(\sqrt{3 - x}\right)^2 = (x - 3)^2$ Square both sides.

$3 - x = (x - 3)(x - 3)$ The second power indicates two factors of $x - 3$.

$3 - x = x^2 - 6x + 9$ Multiply the binomials.

To solve the resulting quadratic equation, we write it in standard form so that the left side is 0.

$$3 = x^2 - 5x + 9 \quad \text{Add } x \text{ to both sides.}$$
$$0 = x^2 - 5x + 6 \quad \text{Subtract 3 from both sides.}$$
$$0 = (x - 3)(x - 2) \quad \text{Factor } x^2 - 5x + 6.$$
$$x - 3 = 0 \quad \text{or} \quad x - 2 = 0 \quad \text{Set each factor equal to 0.}$$
$$x = 3 \quad | \quad x = 2 \quad \text{Solve each equation.}$$

There are two possible solutions to check in the original equation.

For $x = 3$
$$\sqrt{3 - x} - x = -3$$
$$\sqrt{3 - 3} - 3 \stackrel{?}{=} -3$$
$$\sqrt{0} - 3 \stackrel{?}{=} -3$$
$$0 - 3 \stackrel{?}{=} -3$$
$$-3 = -3$$

For $x = 2$
$$\sqrt{3 - x} - x = -3$$
$$\sqrt{3 - 2} - 2 \stackrel{?}{=} -3$$
$$\sqrt{1} - 2 \stackrel{?}{=} -3$$
$$1 - 2 \stackrel{?}{=} -3$$
$$-1 \neq -3$$

Since a true statement results when 3 is substituted for x, 3 is a solution. Since a false statement results when 2 is substituted for x, 2 is an extraneous solution. The solution set is $\{3\}$.

5 Solve equations containing two square roots.

In the next example, the equation contains two square roots.

EXAMPLE 5 Solve: $\sqrt{x + 12} = 3\sqrt{x + 4}$

Strategy We will square both sides to clear the equation of both radicals.

WHY We can immediately square both sides since each square root term is by itself on one side of the equation. This step will clear the equation of both radicals.

Solution
$$\sqrt{x + 12} = 3\sqrt{x + 4} \quad \text{This is the equation to solve.}$$
$$\left(\sqrt{x + 12}\right)^2 = \left(3\sqrt{x + 4}\right)^2 \quad \text{Square both sides.}$$
$$x + 12 = 9(x + 4) \quad \begin{array}{l}\left(\sqrt{x + 12}\right)^2 = x + 12. \\ \left(3\sqrt{x + 4}\right)^2 = 3^2\left(\sqrt{x + 4}\right)^2 = 9(x + 4).\end{array}$$
$$x + 12 = 9x + 36 \quad \text{Distribute the multiplication by 9.}$$
$$-8x = 24 \quad \text{Subtract } 9x \text{ and } 12 \text{ from both sides.}$$
$$x = -3 \quad \text{Divide both sides by } -8.$$

We check the solution by substituting -3 for x in the original equation.
$$\sqrt{x + 12} = 3\sqrt{x + 4}$$
$$\sqrt{-3 + 12} \stackrel{?}{=} 3\sqrt{-3 + 4} \quad \text{Substitute } -3 \text{ for } x.$$
$$\sqrt{9} \stackrel{?}{=} 3\sqrt{1} \quad \text{Simplify within the radical symbols.}$$
$$3 = 3$$

The solution is -3, and the solution set is $\{-3\}$.

Self Check 5
Solve: $\sqrt{x - 4} = 2\sqrt{x - 16}$
Now Try Problem 47

6 Solve equations containing cube roots.

In the next example, we cube both sides of an equation to eliminate a cube root.

Self Check 6
Solve: $\sqrt[3]{3x - 3} = 3$

Now Try Problem 53

EXAMPLE 6
Solve: $\sqrt[3]{2x + 10} = 2$

Strategy We will cube both sides of the equation.

WHY This will clear the equation of the cube root.

Solution

$$\sqrt[3]{2x + 10} = 2 \quad \text{This is the equation to solve.}$$
$$\left(\sqrt[3]{2x + 10}\right)^3 = (2)^3 \quad \text{Cube both sides.}$$
$$2x + 10 = 8 \quad \left(\sqrt[3]{2x + 10}\right)^3 = 2x + 10 \text{ and } (2)^3 = 8.$$
$$2x = -2 \quad \text{Subtract 10 from both sides.}$$
$$x = -1 \quad \text{Divide both sides by 2.}$$

The solution is -1, and the solution set is $\{-1\}$. Check the result.

7 The distance formula.

We can use the Pythagorean theorem to derive a formula for finding the distance between two points $P(x_1, y_1)$ and $Q(x_2, y_2)$ on a rectangular coordinate system. The distance d between points P and Q is the length of the hypotenuse of the triangle. The two legs have lengths $x_2 - x_1$ and $y_2 - y_1$.

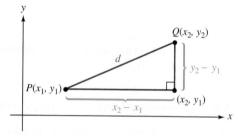

By the Pythagorean theorem, we have

$$d^2 = (x_2 - x_1)^2 + (y_2 - y_1)^2$$

We can take the positive square root of both sides of this equation to get the **distance formula**.

$$d = \sqrt{(x_2 - x_1)^2 + (y_2 - y_1)^2}$$

The Distance Formula

The distance d between the points with coordinates (x_1, y_1) and (x_2, y_2) is given by

$$d = \sqrt{(x_2 - x_1)^2 + (y_2 - y_1)^2}$$

Self Check 7
Find the distance between the points $(-2, 1)$ and $(4, 9)$.

Now Try Problem 59

EXAMPLE 7
Find the distance between the points $(1, 2)$ and $(4, 6)$.

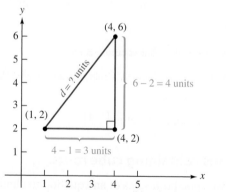

Strategy We will substitute the coordinates into the distance formula and solve for d.

WHY In the formula, d represents the distance between the points.

Solution
We use the distance formula and substitute 1 for x_1, 2 for y_1, 4 for x_2, and 6 for y_2. Then we evaluate the expression under the radical symbol.

$d = \sqrt{(x_2 - x_1)^2 + (y_2 - y_1)^2}$
$d = \sqrt{(4 - 1)^2 + (6 - 2)^2}$ Substitute.
$d = \sqrt{3^2 + 4^2}$ Do the subtractions within the parentheses first.
$d = \sqrt{9 + 16}$ Evaluate the powers.
$d = \sqrt{25}$ Do the addition.
$d = 5$ Find the square root.

The distance between the points is 5 units.

EXAMPLE 8 Find the distance between the points $(-4, 5)$ and $(3, -1)$.

Self Check 8
Find the distance between the points $(-1, 2)$ and $(-6, 4)$.
Now Try Problem 61

Strategy We will substitute the coordinates into the distance formula and solve for d.

WHY In the formula, d represents the distance between the points.

Solution
We use the distance formula and substitute -4 for x_1, 5 for y_1, 3 for x_2, and -1 for y_2.

$d = \sqrt{(x_2 - x_1)^2 + (y_2 - y_1)^2}$
$d = \sqrt{[3 - (-4)]^2 + (-1 - 5)^2}$ Substitute.
$d = \sqrt{7^2 + (-6)^2}$ Do the subtractions.
$d = \sqrt{49 + 36}$ Evaluate the powers.
$d = \sqrt{85}$ Do the addition.
$d \approx 9.219544457$ Use a calculator.

The distance between the points is exactly $\sqrt{85}$ or approximately 9.22 units.

---

8 Solve problems modeled by a radical equation.

Radical equations can be used to model many real-life equations.

EXAMPLE 9 **Height of a Bridge** The distance d (in feet) that an object will fall in t seconds is given by the formula

$$t = \sqrt{\frac{d}{16}}$$

To find the height of the bridge shown in the figure on the next page, a man drops a stone into the water. If it takes the stone 3 seconds to hit the water, how high is the bridge?

Self Check 9
If it takes 4 seconds for the stone in Example 9 to hit the water, how high is the bridge?
Now Try Problem 91

Strategy We will substitute 3 for t in the formula and solve for d.

WHY Since the coin fell for 3 seconds, $t = 3$. The height of the bridge above the water is the same as the distance d that the coin fell.

Solution
We substitute 3 for t in the formula and solve for d.

$$t = \sqrt{\frac{d}{16}}$$

$$3 = \sqrt{\frac{d}{16}} \quad \text{Substitute 3 for } t.$$

$$(3)^2 = \left(\sqrt{\frac{d}{16}}\right)^2 \quad \text{Square both sides to eliminate the radical.}$$

$$9 = \frac{d}{16} \quad 3^2 = 9 \text{ and } \left(\sqrt{\frac{d}{16}}\right)^2 = \frac{d}{16}.$$

$$144 = d \quad \text{Multiply both sides by 16.}$$

The bridge is 144 feet above the water. Check this result in the original equation.

ANSWERS TO SELF CHECKS
1. 85 **2.** no solution **3.** -2 **4.** 5, 0 is extraneous **5.** 20 **6.** 10 **7.** 10
8. $\sqrt{29} \approx 5.39$ **9.** 256 ft

SECTION 8.6 STUDY SET

VOCABULARY
Fill in the blanks.

1. A _____ equation contains one or more radical expressions with a variable radicand.
2. To _____ the radical expression in $\sqrt{x} + 1 = 10$ means to get $\sqrt{x}$ by itself on one side of the equation.
3. A false solution that occurs when you square both sides of an equation is called an _____ solution.
4. The squaring property of equality states that if two numbers are equal, their _____ are equal.

CONCEPTS
Fill in the blanks.

5. The squaring property of equality states that
   If $a = b$, then $a^2 = $ ____.
6. The distance formula states that
   $d = $ ____

To isolate x, what step should be used to undo the operation performed on it? (Assume that x is a positive number.)

7. $x^2 = 4$
8. $\sqrt{x} = 4$

Simplify each expression.

9. $\left(\sqrt{x}\right)^2$
10. $\left(\sqrt{x-1}\right)^2$
11. $\left(2\sqrt{x}\right)^2$
12. $\left(2\sqrt{x} - 1\right)^2$
13. $\left(\sqrt{2x}\right)^2$
14. $\left(\sqrt[3]{x}\right)^3$

What is wrong with each solution?

15. $\sqrt{x - 2} = 3$
    $\left(\sqrt{x - 2}\right)^2 = 3$
    $x - 2 = 3$
    $x = 5$

8.6 Solving Radical Equations; the Distance Formula   693

16. ~~$2 = \sqrt{x-9}$~~
     ~~$(2)^2 = (\sqrt{x-9})^2$~~
     ~~$4 = x - 9$~~
     ~~$-5 = x$~~
     ~~$x = -5$~~

17. ~~$\sqrt{a+2} - 5 = 4$~~
     ~~$(\sqrt{a+2} - 5)^2 = 4^2$~~
     ~~$a + 2 - 25 = 16$~~
     ~~$a - 23 = 16$~~
     ~~$a = 39$~~

18. ~~$\sqrt[3]{x+1} = -2$~~
     ~~$(\sqrt[3]{x+1})^2 = (-2)^2$~~
     ~~$x + 1 = 4$~~
     ~~$x = 3$~~

19. **a.** On the graph below, plot the points $A(-4, 6)$, $B(4, 0)$, $C(1, -4)$, and $D(-7, 2)$.
    b. Draw figure $ABCD$. What type of geometric figure is it?
    c. Find the length of each side of the figure.
    d. Find the perimeter of the figure.

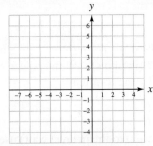

20. **a.** What type of geometric figure is figure $ABCD$ shown below?
    b. Give the coordinates of points A, B, C, and D.
    c. Find the length of each side of the figure.
    d. Find the area of the figure.

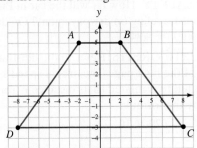

NOTATION

Complete each step.

21. Solve: $\sqrt{x-3} = 5$
              $()^2 = ^2$
              $x - 3 = $
              $x = 28$

22. Solve: $\sqrt{2x - 18} = \sqrt{x-1}$
              $()^2 = (\sqrt{x-1})^2$
              $ = x - 1$
              $ - 18 = -1$
              $x = $

GUIDED PRACTICE

Solve each equation and check all solutions. See Example 1.

23. $\sqrt{x} = 3$ 24. $\sqrt{x} = 5$
25. $\sqrt{2a} = 4$ 26. $\sqrt{3a} = 9$
27. $\sqrt{x+3} = 2$ 28. $\sqrt{x-2} = 3$
29. $\sqrt{6+2x} = 4$ 30. $\sqrt{5-T} = 10$

Solve each equation and check all solutions. See Example 2.

31. $\sqrt{3-T} + 2 = 0$ 32. $\sqrt{7+2x} + 4 = 0$
33. $\sqrt{r} + 6 = 2$ 34. $\sqrt{r} + 5 = 4$

Solve each equation and check all solutions. See Example 3.

35. $x - 3 = \sqrt{x^2 - 15}$ 36. $v - 2 = \sqrt{v^2 - 16}$
37. $x - 1 = \sqrt{x^2 - 4x + 9}$ 38. $\sqrt{15 - 3t} = t - 5$
39. $3d = \sqrt{9d^2 - 2d + 8}$ 40. $\sqrt{4m^2 + 6m + 6} = -2m$
41. $y - 9 = \sqrt{y-3}$ 42. $m - 9 = \sqrt{m-7}$

Solve each equation and check all solutions. See Example 4.

43. $b = \sqrt{2b-2} + 1$ 44. $c = \sqrt{5c+1} - 1$
45. $\sqrt{24 + 10n} - n = 4$ 46. $\sqrt{7 + 6y} - y = 2$

Solve each equation and check all solutions. See Example 5.

47. $\sqrt{3t-9} = \sqrt{t+1}$ 48. $\sqrt{a-3} = \sqrt{2a-8}$
49. $\sqrt{10-3x} = \sqrt{2x+20}$ 50. $\sqrt{1-2x} = \sqrt{x+10}$

Solve each equation and check all solutions. See Example 6.

51. $\sqrt[3]{x} = 7$
52. $\sqrt[3]{x} = -9$
53. $\sqrt[3]{x - 1} = 4$
54. $\sqrt[3]{2x + 5} = 3$
55. $\sqrt[3]{\frac{1}{2}x - 3} = 2$
56. $\sqrt[3]{x + 4} = 1$
57. $\sqrt[3]{7n - 1} + 1 = 4$
58. $\sqrt[3]{12m + 4} + 2 = 6$

Find the distance between the two points. If an answer contains a radical, give an exact answer and an approximate answer to two decimal places. See Examples 7–8.

59. $(3, -4)$ and $(0, 0)$
60. $(0, 0)$ and $(-6, 8)$
61. $(2, 4)$ and $(5, 9)$
62. $(5, 9)$ and $(9, 13)$
63. $(-2, -8)$ and $(3, 4)$
64. $(-5, -2)$ and $(7, 3)$
65. $(6, 8)$ and $(12, 16)$
66. $(4, 6)$ and $(1, 2)$

TRY IT YOURSELF

Solve each equation. Assume that all variables are positive.

67. $-\sqrt{x} = -5$
68. $-\sqrt{x} = -12$
69. $10 - \sqrt{s} = 7$
70. $-4 = 6 - \sqrt{s}$
71. $\sqrt{5x - 5} - 5 = 0$
72. $\sqrt{6x + 19} - 7 = 0$
73. $\sqrt{x + 3} + 5 = 12$
74. $\sqrt{x - 5} - 3 = 4$
75. $\sqrt{3c - 8} - \sqrt{c} = 0$
76. $\sqrt{2x} - \sqrt{x + 8} = 0$
77. $\sqrt{9t^2 + 4t + 20} = -3t$
78. $\sqrt{1 - 8s} = s + 4$
79. $\sqrt{3x + 3} = 3\sqrt{x - 1}$
80. $2\sqrt{4x + 5} = 5\sqrt{x + 4}$
81. $2\sqrt{3x + 4} = \sqrt{5x + 9}$
82. $\sqrt{3x + 6} = 2\sqrt{2x - 11}$
83. $\sqrt{y + 3} = y - 3$
84. $\sqrt{p - 4} + 2 = \sqrt{p}$
85. $\sqrt{3x + 1} + 1 = x$
86. $\sqrt{4x + 1} + 1 = x$
87. $\sqrt[3]{7a - 1} + 1 = 4$
88. $\sqrt[3]{12x + 4} - 2 = 2$

Find the distance between the two points.

89. $(-2, 3)$ and $(4, -5)$
90. $(-2, -8)$ and $(3, 4)$

APPLICATIONS

91. **NIAGARA FALLS** The distance s (in feet) that an object will fall in t seconds is given by the formula

$$t = \frac{\sqrt{s}}{4}$$

The time it took a stuntman to go over Niagara Falls in a barrel was 3.25 seconds. Substitute 3.25 for t and solve the equation for s to find the height of the waterfall.

92. **THE WASHINGTON MONUMENT** Gabby Street, a baseball player of the 1920s, was known for once catching a ball dropped from the top of the Washington Monument. If the ball fell for slightly less than 6 seconds before it was caught, find the approximate height of the monument. (*Hint:* See Problem 91.)

93. **PENDULUMS** The time t (in seconds) required for a pendulum of length L feet to swing through one back-and-forth cycle, called its **period**, is given by the formula

$$t = 1.11\sqrt{L}$$

The Foucault pendulum in Chicago's Museum of Science and Industry is used to demonstrate the rotation of the Earth. The pendulum completes one cycle in 8.91 seconds. To the nearest tenth of a foot, how long is the pendulum?

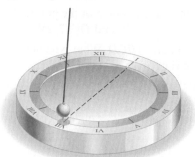

94. **POWER USAGE** The current I (in amperes), the resistance R (in ohms), and the power P (in watts) are related by the formula

$$I = \sqrt{\frac{P}{R}}$$

Find the power (to the nearest watt) used by a space heater that draws 7 amps when the resistance is 10.2 ohms.

95. **ROAD SAFETY** The formula $s = k\sqrt{d}$ relates the speed s (in mph) of a car and the distance d of the skid when a driver hits the brakes. On wet pavement, $k = 3.24$. How far will a car skid if it is going 55 mph?

96. **ROAD SAFETY** How far will the car in Problem 95 skid if it is traveling on dry pavement? On dry pavement, $k = 5.34$.

97. **SATELLITE ORBITS** Refer to the illustration at the top of the next page. The orbital speed s of an Earth satellite is related to its distance r from Earth's center by the formula

$$\sqrt{r} = \frac{2.029 \times 10^7}{s}$$

If the satellite's orbital speed is 7×10^3 meters per second, find its altitude a (in meters) above Earth's surface.

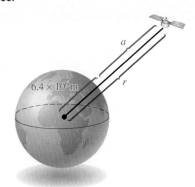

98. HIGHWAY DESIGNS A highway curve banked at 8° will accommodate traffic traveling at speed s (in mph) if the radius of the curve is r (feet), according to the equation $s = 1.45\sqrt{r}$. If highway engineers expect traffic to travel at 65 mph, to the nearest foot, what radius should they specify?

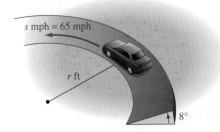

99. GEOMETRY The radius of a cone with volume V and height h is given by the formula

$$r = \sqrt{\frac{3V}{\pi h}}$$

Solve the equation for V.

100. WINDMILLS The power produced by a certain windmill is related to the speed of the wind by the formula

$$s = \sqrt[3]{\frac{P}{0.02}}$$

where P is the power (in watts) and s is the speed of the wind (in mph). How much power will the windmill produce if the wind is blowing at 30 mph?

101. NAVIGATION An oil tanker is to travel from Tunisia to Italy, as shown in the next column. The captain wants to travel a course that is always the same distance from a point on the coast of Sardinia as it is from a point on the coast of Sicily. How far will the tanker be from these points when it reaches

a. position 1?
b. position 2?

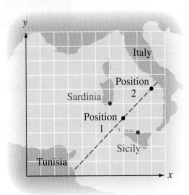

102. DECK DESIGN The plans for a patio deck shown below call for three redwood support braces directly under the hot tub. Find the length of each support. Round to the nearest tenth of a foot.

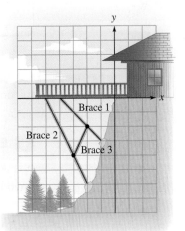

WRITING

103. Explain why a check is necessary when solving radical equations.

104. How would you know, without solving it, that the equation $\sqrt{x + 2} = -4$ has no solution?

REVIEW

Perform the operations.

105. $(3x^2 + 2x) + (5x^2 - 8x)$
106. $(7a^2 + 2a - 5) - (3a^2 - 2a + 1)$
107. $(x + 3)(x + 3)$
108. $x - 1 \overline{)x^2 - 6x + 5}$
109. $(5x - 2)^2$
110. $(3y - 7)^2$

Objectives

1. Evaluate expressions of the form $a^{1/n}$.
2. Evaluate expressions of the form $a^{m/n}$.
3. Apply rules of exponents.

SECTION 8.7
Rational Exponents

We have seen that a positive integer exponent indicates the number of times that a base is to be used as a factor in a product. For example, x^4 means that x is to be used as a factor four times.

$$x^4 = \overbrace{x \cdot x \cdot x \cdot x}^{\text{4 factors of } x}$$

Also, recall the following rules for exponents.

Rules for Exponents

If m and n represent natural numbers and there are no divisions by zero, then

$$x^m x^n = x^{m+n} \qquad (x^m)^n = x^{mn} \qquad (xy)^n = x^n y^n \qquad \left(\frac{x}{y}\right)^n = \frac{x^n}{y^n}$$

$$x^0 = 1 \qquad x^{-n} = \frac{1}{x^n} \qquad \frac{x^m}{x^n} = x^{m-n}$$

In this section, we will extend the definition and rules for exponents to cover fractional exponents.

1 Evaluate expressions of the form $a^{1/n}$.

It is possible to raise numbers to fractional powers. To give meaning to **rational** (or **fractional**) **exponents**, we consider $\sqrt{7}$. Because $\sqrt{7}$ is the positive number whose square is 7, we have

$$\left(\sqrt{7}\right)^2 = 7$$

We now consider the symbol $7^{1/2}$. If fractional exponents are to follow the same rules as integer exponents, the square of $7^{1/2}$ must be 7, because

$$(7^{1/2})^2 = 7^{(1/2)2} \qquad \text{Keep the base 7 and multiply the exponents.}$$
$$= 7^1 \qquad \text{Multiply: } \tfrac{1}{2} \cdot 2 = 1.$$
$$= 7$$

> **The Language of Algebra** Rational exponents are also called *fractional exponents*.

Since $(7^{1/2})^2$ and $\left(\sqrt{7}\right)^2$ are both equal to 7, we define $7^{1/2}$ to be $\sqrt{7}$. Similarly, we make these definitions.

$$7^{1/3} = \sqrt[3]{7}$$
$$7^{1/7} = \sqrt[7]{7}$$

and so on. In general, we have the following definition.

Definition of $a^{1/n}$

If n represents a positive integer greater than 1 and $\sqrt[n]{a}$ represents a real number, then

$$a^{1/n} = \sqrt[n]{a}$$

EXAMPLE 1 Simplify: **a.** $64^{1/2}$ **b.** $64^{1/3}$ **c.** $(-64)^{1/3}$

Strategy We will identify the base and the exponent of the exponential expression so that we can write the exponential expression in equivalent radical form.

WHY We know how to evaluate radicals.

Solution

a. $64^{1/2} = \sqrt{64} = 8$ *The denominator of the fractional exponent is 2. Therefore, we find the square root of the base, 64.*

b. $64^{1/3} = \sqrt[3]{64} = 4$ *The denominator of the fractional exponent is 3. Therefore, we find the cube root of the base, 64.*

c. $(-64)^{1/3} = \sqrt[3]{-64} = -4$ *The denominator of the fractional exponent is 3. Therefore, we find the cube root of the base, −64.*

Self Check 1

Simplify:
a. $81^{1/2}$
b. $125^{1/3}$
c. $(-27)^{1/3}$

Now Try Problems 21, 27, and 29

2 Evaluate expressions of the form $a^{m/n}$.

We can extend the definition of $a^{1/n}$ to cover fractional exponents for which the numerator is not 1. For example, because $4^{3/2}$ can be written as $(4^{1/2})^3$, we have

$$4^{3/2} = (4^{1/2})^3 = (\sqrt{4})^3 = 2^3 = 8$$

Because $4^{3/2}$ can also be written as $(4^3)^{1/2}$, we have

$$4^{3/2} = (4^3)^{1/2} = 64^{1/2} = \sqrt{64} = 8$$

In general, $a^{m/n}$ can be written as $(a^{1/n})^m$ or as $(a^m)^{1/n}$. Since $(a^{1/n})^m = (\sqrt[n]{a})^m$, and $(a^m)^{1/n} = \sqrt[n]{a^m}$, we make the following definition.

The Definition of $a^{m/n}$

If m and n represent positive integers ($n \neq 1$) and $\sqrt[n]{a}$ represents a real number, then

$$a^{m/n} = \sqrt[n]{a^m} = (\sqrt[n]{a})^m$$

EXAMPLE 2 Simplify: **a.** $8^{2/3}$ **b.** $27^{4/3}$

Strategy We will identify the base and the exponent of the exponential expression so that we can write the exponential expression in equivalent radical form.

WHY We know how to evaluate radicals.

Solution

a. $8^{2/3} = (\sqrt[3]{8})^2$
$= 2^2$
$= 4$

b. $27^{4/3} = (\sqrt[3]{27})^4$
$= (3)^4$
$= 81$

Self Check 2

Simplify:
a. $16^{3/2}$
b. $8^{4/3}$

Now Try Problems 33 and 39

Chapter 8 Roots and Radicals

> **Success Tip** The work in Example 2 suggests that in order to avoid large numbers, it is usually easier to take the root of the base first and then find the power.

Self Check 3

Simplify:
a. $-100^{3/2}$
b. $(-8)^{2/3}$

Now Try Problem 47

EXAMPLE 3 Simplify: **a.** $(-125)^{4/3}$ **b.** $-9^{5/2}$ **c.** $-25^{3/2}$

Strategy We will identify the base and the exponent of the exponential expression so that we can write the exponential expression in equivalent radical form.

WHY We know how to evaluate radicals.

Solution

a. $(-125)^{4/3} = \left(\sqrt[3]{-125}\right)^4$
$= (-5)^4$
$= 625$

b. $-9^{5/2} = -\left(\sqrt{9}\right)^5$
$= -(3)^5$
$= -243$

c. $-25^{3/2} = -\left(\sqrt{25}\right)^3$
$= -(5)^3$
$= -125$

Using Your CALCULATOR Fractional Exponents

To use a reverse-entry scientific calculator to evaluate an exponential expression containing a fractional exponent, we can use the $\boxed{y^x}$ key. For example, to evaluate $6^{-2/3}$, we enter these numbers and press these keys.

6 $\boxed{y^x}$ $\boxed{(}$ 2 $\boxed{+/-}$ $\boxed{\div}$ 3 $\boxed{)}$ $\boxed{=}$ $\boxed{0.302853432}$

So $6^{-2/3} \approx 0.302853432$.

To use a direct-entry or graphing calculator to evaluate $6^{-2/3}$, we press the following keys.

6 $\boxed{\wedge}$ $\boxed{(}$ $\boxed{(-)}$ 2 $\boxed{\div}$ 3 $\boxed{)}$ ENTER $\boxed{\begin{array}{l}6\wedge(-2/3)\\ .3028534321\end{array}}$

---

3 Apply rules of exponents.

Because of the way in which $a^{1/n}$ and $a^{m/n}$ are defined, the familiar rules for exponents are valid for rational exponents. The following example illustrates the use of each rule.

Self Check 4

Simplify:
a. $5^{1/3}5^{1/3}$ b. $(5^{1/3})^4$
c. $(3x)^{1/5}$ d. $\dfrac{5^{3/7}}{5^{2/7}}$
e. $\left(\dfrac{2}{3}\right)^{2/3}$ f. $5^{-2/7}$
g. $(12^{1/2})^0$

Now Try Problems 49, 57, and 61

EXAMPLE 4 Simplify: **a.** $4^{2/5}\,4^{1/5}$ **b.** $(5^{2/3})^{1/2}$ **c.** $(3x)^{2/3}$

d. $\dfrac{4^{3/5}}{4^{2/5}}$ **e.** $\left(\dfrac{3}{2}\right)^{2/5}$ **f.** $4^{-2/3}$ **g.** $(5^{1/3})^0$

Strategy In each case, we will identify the correct rule of exponents to be applied.

WHY The rules for exponents are also valid for fractional exponents.

Solution

a. $4^{2/5}4^{1/5} = 4^{2/5\,+\,1/5} = 4^{3/5}$ Use $x^m x^n = x^{m+n}$.

b. $(5^{2/3})^{1/2} = 5^{(2/3)(1/2)} = 5^{1/3}$ Use $(x^m)^n = x^{m \cdot n}$.

c. $(3x)^{2/3} = 3^{2/3}x^{2/3}$ Use $(xy)^m = x^m y^m$.

d. $\dfrac{4^{3/5}}{4^{2/5}} = 4^{3/5 - 2/5} = 4^{1/5}$ Use $\dfrac{x^m}{x^n} = x^{m-n}$.

e. $\left(\dfrac{3}{2}\right)^{2/5} = \dfrac{3^{2/5}}{2^{2/5}}$ Use $\left(\dfrac{x}{y}\right)^n = \dfrac{x^n}{y^n}$.

f. $4^{-2/3} = \dfrac{1}{4^{2/3}}$ Use $x^{-n} = \dfrac{1}{x^n}$.

g. $(5^{1/3})^0 = 1$ Use $x^0 = 1$.

We can use the rules for exponents to simplify expressions containing rational exponents.

EXAMPLE 5
Simplify. All variables represent positive numbers:
a. $64^{-2/3}$ b. $(x^2)^{1/2}$ c. $(x^6 y^4)^{1/2}$ d. $(27x^{12})^{-1/3}$

Strategy In each case, we will identify the correct rule of exponents to be applied.

WHY The rules for exponents are also valid for fractional exponents.

Solution

a. $64^{-2/3} = \dfrac{1}{64^{2/3}}$

$= \dfrac{1}{(64^{1/3})^2}$

$= \dfrac{1}{4^2}$

$= \dfrac{1}{16}$

b. $(x^2)^{1/2} = x^{2(1/2)}$

$= x^1$

$= x$

c. $(x^6 y^4)^{1/2} = x^{6(1/2)} y^{4(1/2)}$

$= x^3 y^2$

d. $(27x^{12})^{-1/3} = \dfrac{1}{(27x^{12})^{1/3}}$

$= \dfrac{1}{27^{1/3} x^{12(1/3)}}$

$= \dfrac{1}{3x^4}$

Self Check 5
Simplify:
a. $25^{-3/2}$
b. $(x^3)^{1/3}$
c. $(x^6 y^9)^{-2/3}$

Now Try Problems 67, 71, and 75

EXAMPLE 6
Simplify: a. $x^{1/3} x^{1/2}$ b. $\dfrac{3x^{2/3}}{6x^{1/5}}$ c. $\dfrac{2x^{-1/2}}{x^{3/4}}$

Strategy In each case, we will identify the correct rule of exponents to be applied.

WHY The rules for exponents are also valid for fractional exponents.

Solution

a. $x^{1/3} x^{1/2} = x^{2/6} x^{3/6}$ Get a common denominator for the fractional exponents.

$= x^{5/6}$ Keep the base x and add the exponents.

b. $\dfrac{3x^{2/3}}{6x^{1/5}} = \dfrac{3x^{10/15}}{6x^{3/15}}$ Get a common denominator for the fractional exponents.

$= \dfrac{1}{2} x^{10/15 - 3/15}$ Simplify $\dfrac{3}{6}$. Keep the base x and subtract the exponents.

$= \dfrac{1}{2} x^{7/15}$ Do the subtraction.

Self Check 6
Simplify:
a. $x^{2/3} x^{1/2}$
b. $\dfrac{x^{2/3}}{2x^{1/4}}$

Now Try Problems 77, 81, and 83

c. $\dfrac{2x^{-1/2}}{x^{3/4}} = \dfrac{2x^{-2/4}}{x^{3/4}}$ Get a common denominator for the fractional exponents.

$\phantom{\dfrac{2x^{-1/2}}{x^{3/4}}} = 2x^{-2/4 - 3/4}$ Keep the base x and subtract the exponents.

$\phantom{\dfrac{2x^{-1/2}}{x^{3/4}}} = 2x^{-5/4}$ Do the subtraction.

$\phantom{\dfrac{2x^{-1/2}}{x^{3/4}}} = \dfrac{2}{x^{5/4}}$ $x^{-5/4} = \dfrac{1}{x^{5/4}}$.

ANSWERS TO SELF CHECKS

1. a. 9 **b.** 5 **c.** -3 **2. a.** 64 **b.** 16 **3. a.** $-1{,}000$ **b.** 4 **4. a.** $5^{2/3}$ **b.** $5^{4/3}$
c. $3^{1/5}x^{1/5}$ **d.** $5^{1/7}$ **e.** $\dfrac{2^{2/3}}{3^{2/3}}$ **f.** $\dfrac{1}{5^{2/7}}$ **g.** 1 **5. a.** $\dfrac{1}{125}$ **b.** x **c.** $\dfrac{1}{x^4 y^6}$
6. a. $x^{7/6}$ **b.** $\dfrac{1}{2}x^{5/12}$

SECTION 8.7 STUDY SET

VOCABULARY

Fill in the blanks.

1. A fractional exponent is also called a _____ exponent.
2. In the expression $27^{1/3}$, 27 is called the _____ and the exponent is ☐.

CONCEPTS

Complete each rule for exponents.

3. $x^m x^n =$
4. $(x^m)^n =$
5. $\left(\dfrac{x}{y}\right)^n =$
6. $x^0 =$
7. $x^{-n} =$
8. $\dfrac{x^m}{x^n} =$
9. $x^{1/n} =$
10. $x^{m/n} =$

11. Write $\sqrt{5}$ using a fractional exponent.
12. Write $5^{1/3}$ using a radical.
13. Write $8^{4/3}$ using a radical.
14. Write $\left(\sqrt{8}\right)^3$ using a fractional exponent.
15. Complete the table.
16. Complete the table.

x	$x^{1/2}$
0	
1	
4	
9	

x	$x^{1/3}$
0	
-1	
-8	
8	

17. Graph each number on the number line.
$\left\{8^{1/3},\ 17^{1/2},\ 2^{3/2},\ -5^{2/3}\right\}$

18. Graph each number on the number line.
$\left\{4^{-1/2},\ 64^{-2/3},\ (-8)^{-1/3}\right\}$

NOTATION

Complete each step.

19. $(-216)^{4/3} = \left(\sqrt[3]{}\right)^4$
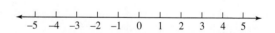
$= \left(\right)^4$
$= 1{,}296$

20. $\dfrac{3x^{-2/3}}{x^{3/4}} = \dfrac{3x^{-8/12}}{\phantom{x^{9/12}}}$
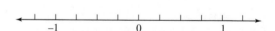
$= 3x^{-8/12 \,-\, }$
$= 3x^{-17/12}$
$= \dfrac{3}{x}$

GUIDED PRACTICE

Simplify each expression. See Example 1.

21. $81^{1/2}$
22. $100^{1/2}$
23. $-144^{1/2}$
24. $-400^{1/2}$
25. $\left(\dfrac{4}{49}\right)^{1/2}$
26. $\left(\dfrac{9}{64}\right)^{1/2}$

27. $27^{1/3}$
28. $8^{1/3}$
29. $-125^{1/3}$
30. $-1{,}000^{1/3}$
31. $\left(\dfrac{27}{64}\right)^{1/3}$
32. $\left(\dfrac{64}{125}\right)^{1/3}$

Simplify each expression. See Example 2.

33. $8^{4/3}$
34. $27^{2/3}$
35. $81^{3/2}$
36. $16^{3/2}$
37. $25^{3/2}$
38. $4^{5/2}$
39. $125^{4/3}$
40. $8^{5/3}$
41. $1{,}000^{2/3}$
42. $27^{4/3}$
43. $\left(\dfrac{8}{27}\right)^{2/3}$
44. $\left(\dfrac{49}{64}\right)^{3/2}$

Simplify each expression. See Example 3.

45. $(-8)^{2/3}$
46. $(-125)^{2/3}$
47. $(-1{,}000)^{2/3}$
48. $(-216)^{4/3}$

Simplify each expression. See Example 4.

49. $6^{3/5}6^{2/5}$
50. $3^{4/7}3^{3/7}$
51. $5^{2/3}5^{4/3}$
52. $2^{7/8}2^{9/8}$
53. $(7^{2/5})^{5/2}$
54. $(8^{1/3})^3$
55. $(5^{2/7})^7$
56. $(3^{3/8})^8$
57. $(2x)^{1/2}$
58. $(5a)^{3/2}$
59. $\dfrac{8^{3/2}}{8^{1/2}}$
60. $\dfrac{11^{9/7}}{11^{2/7}}$
61. $\dfrac{5^{11/3}}{5^{2/3}}$
62. $\dfrac{27^{13/15}}{27^{8/15}}$
63. $\left(\dfrac{5}{2}\right)^{2/3}$
64. $\left(\dfrac{3}{7}\right)^{3/2}$

Simplify each expression. Write your answers without using negative exponents. See Example 5.

65. $4^{-1/2}$
66. $8^{-1/3}$
67. $27^{-2/3}$
68. $36^{-3/2}$
69. $(m^{1/2})^2$
70. $(x^9)^{1/3}$
71. $(x^{12}y^6)^{1/3}$
72. $(x^{18}y^{10})^{1/2}$
73. $16^{-3/2}$
74. $100^{-5/2}$
75. $(-27a^9)^{-4/3}$
76. $(-8b^3)^{-4/3}$

Simplify each expression. All variables represent positive numbers. See Example 6.

77. $x^{5/6}x^{7/6}$
78. $x^{2/3}x^{7/3}$
79. $y^{4/7}y^{10/7}$
80. $y^{5/11}y^{6/11}$

81. $\dfrac{x^{3/5}}{x^{1/5}}$
82. $\dfrac{x^{4/3}}{x^{2/3}}$
83. $\dfrac{2a^{-1/2}}{a^{1/4}}$
84. $\dfrac{5b^{-2/3}}{b^{5/6}}$

TRY IT YOURSELF

85. $\left(\dfrac{1}{4}\right)^{1/2}$
86. $\left(\dfrac{1}{25}\right)^{1/2}$
87. $(-8)^{1/3}$
88. $(-125)^{1/3}$
89. $64^{4/3}$
90. $64^{3/2}$
91. $(-343)^{2/3}$
92. $(-512)^{4/3}$
93. $\dfrac{x^{1/7}x^{3/7}}{x^{2/7}}$
94. $\dfrac{x^{5/6}x^{5/6}}{x^{7/6}}$
95. $x^{2/3}x^{3/4}$
96. $a^{3/5}a^{1/2}$
97. $(b^{1/2})^{3/5}$
98. $(x^{2/5})^{4/7}$
99. $\dfrac{t^{2/3}}{t^{2/5}}$
100. $\dfrac{p^{3/4}}{p^{1/3}}$
101. $\dfrac{4b^{-3/2}}{b^{3/4}}$
102. $\dfrac{p^{-2/3}}{6p^{3/2}}$
103. $\left(\dfrac{x^{4/5}}{x^{2/15}}\right)^3$
104. $\left(\dfrac{y^{2/3}}{y^{1/5}}\right)^{15}$

APPLICATIONS

If an answer is not exact, round to the nearest tenth.

105. **SPEAKERS** The formula $A = V^{2/3}$ can be used to find the area A of one face of a cube if its volume V is known. Find the amount of floor space on the dance floor taken up by the speakers shown below if each speaker is a cube with a volume of 2,744 cubic inches.

106. **MEDICAL TESTS** Before a series of X-rays are taken, a patient is injected with a special contrast mixture that highlights obstructions in his blood vessels. The amount of the original dose of contrast material remaining in the patient's bloodstream h hours after it is injected is given by $h^{-3/2}$. How much of the contrast material remains in the patient's bloodstream 4 hours after the injection?

107. HOLIDAY DECORATING Find the length s of each string of colored lights used to decorate an evergreen tree in the manner shown below if $s = (r^2 + h^2)^{1/2}$.

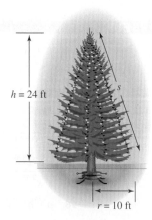

108. VISIBILITY The distance d in miles a person in an airplane can see to the horizon on a clear day is given by the formula $d = 1.22a^{1/2}$, where a is the altitude of the plane in feet. Find d in the illustration.

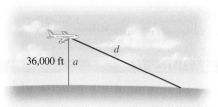

109. TOY DESIGNS Knowing the volume V of a sphere, we can find its radius r using the formula

$$r = \left(\frac{3V}{4\pi}\right)^{1/3}$$

If the volume occupied by a ball is 2π cubic inches, find its radius.

110. EXERCISE EQUIPMENT Find the length l of the incline bench in the illustration, using the formula $l = (a^2 + b^2)^{1/2}$.

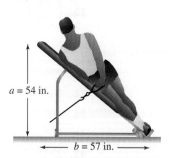

WRITING

111. What is a rational exponent? Give several examples.

112. Explain this statement: *In the expression $16^{3/2}$, the number 3/2 requires that two operations be performed on 16.*

REVIEW

Graph each equation.

113. $x = 3$

114. $y = -3$

115. $-2x + y = 4$

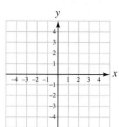

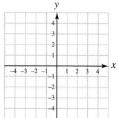

116. $4x - y = 4$

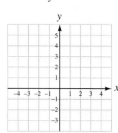

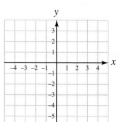

STUDY SKILLS CHECKLIST
Preparing for the Chapter 8 Test

There are several common mistakes that students make when working with the topics of Chapter 8. To make sure you are prepared for the test over this material, read the list below to help you avoid these mistakes.

☐ When adding or subtracting radical expressions, you must have *like radicals*. Like radicals are radical expressions in which the index and the radicand are the same.

$5\sqrt{3} + 2\sqrt{3} = 7\sqrt{3}$ Both have the same index and radicand. Add the coefficients of the terms and keep the same radical.

$5\sqrt{3} + 2\sqrt{7}$ cannot be combined.   The operation is addition but the radicals are not like radicals. The expression is in simplest form.

$$5\sqrt{18} + 2\sqrt{50} = 5\sqrt{9 \cdot 2} + 2\sqrt{25 \cdot 2}$$ Write 18 as $18 = 9 \cdot 2$. Write 50 as $50 = 25 \cdot 2$.
$$= 5\sqrt{9} \cdot \sqrt{2} + 2\sqrt{25} \cdot \sqrt{2}$$ The square root of a product is equal to the product of the square roots.
$$= 5 \cdot 3 \cdot \sqrt{2} + 2 \cdot 5 \cdot \sqrt{2}$$ Evaluate $\sqrt{9} = 3$ and $\sqrt{25} = 5$.
$$= 15\sqrt{2} + 10\sqrt{2}$$ Multiply $5 \cdot 3 = 15$ and $2 \cdot 5 = 10$.
$$= 25\sqrt{2}$$ Both radicals have the same index and the same radicand. To combine them, add the coefficients and keep the radical.

☐ To multiply radicals, only the index has to be the same. Use the product rule for radicals to carry out the multiplication. Be sure to simplify all answers.

$$3\sqrt{15}(2\sqrt{10}) = 3 \cdot 2\sqrt{15} \cdot \sqrt{10}$$ Multiply the integer factors, 3 and 2, and multiply the radicals.
$$= 6\sqrt{150}$$ Use the product rule for radicals.
$$= 6\sqrt{25}\sqrt{6}$$
$$= 6(5)\sqrt{6}$$
$$= 30\sqrt{6}$$

☐ To multiply radical expressions with more than one term, we use the distributive property.

$$3\sqrt{5}(2\sqrt{15} - 6\sqrt{10}) = 3\sqrt{5} \cdot 2\sqrt{15} - 3\sqrt{5} \cdot 6\sqrt{10}$$
$$= 6\sqrt{75} - 18\sqrt{50}$$
$$= 6\sqrt{3 \cdot 5 \cdot 5} - 18\sqrt{2 \cdot 5 \cdot 5}$$
$$= 6 \cdot 5\sqrt{3} - 18 \cdot 5\sqrt{2}$$
$$= 30\sqrt{3} - 90\sqrt{2}$$

☐ When solving radical equations, isolate the radical on one side of the equation before raising both sides to the power that matches the index.

☐ Even if you are certain that no algebraic mistakes were made when solving a radical equation, you must still check your solutions. Raising both sides to a power can introduce extraneous solutions that must be discarded.

CHAPTER 8 SUMMARY AND REVIEW

SECTION 8.1 Square Roots

DEFINITIONS AND CONCEPTS	EXAMPLES
The number b is a **square root** of a if $b^2 = a$. Every positive real number has two square roots.	3 is a square root of 9 because $3^2 = 9$. -3 is also a square root of 9, because $(-3)^2 = 9$. $\dfrac{1}{2}$ is a square root of $\dfrac{1}{4}$ because $\left(\dfrac{1}{2}\right)^2 = \dfrac{1}{4}$. $-\dfrac{1}{2}$ is also a square root of $\dfrac{1}{4}$, because $\left(-\dfrac{1}{2}\right)^2 = \dfrac{1}{4}$.
If a is positive, the expression $\sqrt{a}$ represents the **principal** (or **positive**) **square root** of a. $-\sqrt{a}$ represents the negative square root of a. The principal square root of 0 is 0.	$\sqrt{16} = 4$ 4 is the principal square root of 16. $-\sqrt{\dfrac{9}{25}} = -\dfrac{3}{5}$ $\dfrac{3}{5}$ is thre pincipal square root of $\dfrac{9}{25}$. $\sqrt{0} = 0$ 0 is the principal square root of 0.
The expression within a **radical symbol** $\sqrt{}$ is called the **radicand**.	In the expression $\sqrt{16}$, 16 is the radicand. In the expression $\sqrt{\dfrac{9}{25}}$, $\dfrac{9}{25}$ is the radicand.
If a positive number is not a **perfect square**, its square root is irrational. Square roots of negative numbers are called **imaginary numbers**.	$\sqrt{11}$ is an irrational number. $\sqrt{-5}$ is an imaginary number.
The function $f(x) = \sqrt{x}$ is called the **square root function**.	Evaluate $f(x) = \sqrt{x}$ for $x = 9$ and $x = 16$. $f(9) = \sqrt{9} = 3$ $f(16) = \sqrt{16} = 4$
If a and b are positive numbers, and $a = b$, then $\sqrt{a} = \sqrt{b}$. **The Pythagorean theorem:** If the length of the hypotenuse of a right triangle is c and the lengths of the two legs are a and b, then $c^2 = a^2 + b^2$.	If $h^2 = 169$, then $\sqrt{h^2} = \sqrt{169}$ or $h = 13$. To find the hypotenuse h of a right triangle with sides of 5 and 12 inches, use the Pythagorean theorem: $h^2 = 5^2 + 12^2 = 25 + 144 = 169$ Since $h^2 = 169$, $h = \sqrt{169} = 13$.

REVIEW EXERCISES

Fill in the blank:

1. 4 is a _____ root of 16, because $4^2 = 16$.

2. $\dfrac{1}{3}$ is a square root of $\dfrac{1}{9}$, because ▬ = ▬ .

Find each square root. Do not use a calculator.

3. $\sqrt{25}$

4. $\sqrt{49}$

5. $-\sqrt{144}$

6. $-\sqrt{\dfrac{16}{81}}$

Chapter 8 Summary and Review

7. $\sqrt{900}$
8. $-\sqrt{0.64}$
9. $\sqrt{1}$
10. $\sqrt{0}$

Use a calculator to approximate each expression to three decimal places.

11. $\sqrt{21}$
12. $-\sqrt{15}$
13. $2\sqrt{7}$
14. $\sqrt{751.9}$

15. Determine whether each number is rational, irrational, or imaginary. Which is not a real number? $\{\sqrt{-2},\ \sqrt{68},\ \sqrt{81},\ \sqrt{3}\}$

Complete the table of values for each function and then graph it.

16. $f(x) = \sqrt{x}$

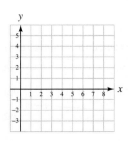

x	f(x)
0	
1	
4	
9	

17. $f(x) = 2 - \sqrt{x}$

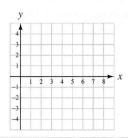

x	f(x)
0	
1	
4	
9	

Refer to the right triangle shown.

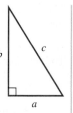

18. Find c where $a = 21$ and $b = 28$.
19. Find b where $a = 1$ and $c = \sqrt{2}$.
20. Find a where $b = 5$ and $c = 6$.

21. **THEATER SEATING** For the theater seats shown, how much higher is the seat at the top of the incline compared to the one at the bottom?

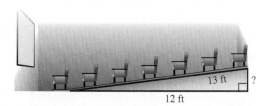

22. **ROAD SIGNS** To find the maximum velocity a car can safely travel around a curve without skidding, we can use the formula $v = \sqrt{2.5r}$, where v is the velocity in mph and r is the radius of the curve in feet. How should the road sign in the illustration be labeled if it is to be posted in front of a curve with a radius of 360 feet?

SECTION 8.2 Higher-Order Roots; Radicands That Contain Variables

DEFINITIONS AND CONCEPTS	EXAMPLES
The number b is a **cube root** of a if $b^3 = a$.	4 is a cube root of 64 because $4^3 = 64$. $\frac{1}{5}$ is a cube root of $\frac{1}{125}$ because $\left(\frac{1}{5}\right)^3 = \frac{1}{125}$.
The cube root of a is denoted by $\sqrt[3]{a}$, and by definition, $\sqrt[3]{a} = b$ if $b^3 = a$.	$\sqrt[3]{343} = 7$ because $7^3 = 343$. $\sqrt[3]{\frac{8}{125}} = \frac{2}{5}$ because $\left(\frac{2}{5}\right)^3 = \frac{8}{125}$.

Chapter 8 Roots and Radicals

The function $f(x) = \sqrt[3]{x}$ is called the **cube root function**.	Evaluate $f(x) = \sqrt[3]{x}$ for $x = 27$ and $x = -64$. $f(27) = \sqrt[3]{27} = 3 \qquad f(-64) = \sqrt[3]{-64} = -4$
The number b is an **nth root of a** if $b^n = a$.	3 is a fourth root of 81 because $3^4 = 81$.
In $\sqrt[n]{a}$, the number n is called the **index** of the radical.	$\dfrac{1}{3}$ is a fifth root of $\dfrac{1}{243}$ because $\left(\dfrac{1}{3}\right)^5 = \dfrac{1}{243}$. The index of $\sqrt[5]{24x}$ is 5. The index of $\sqrt[6]{32ab}$ is 6.
When n is even, we say that the radical $\sqrt[n]{x}$ is an **even root**. When n is odd, $\sqrt[n]{x}$ is an **odd root**. $\sqrt{a} = \sqrt[2]{a}$	$\sqrt{24}$ and $\sqrt[4]{32a^3}$ are even roots. $\sqrt[3]{45mn}$ and $\sqrt[5]{100pq}$ are odd roots. $\sqrt{9}$ means $\sqrt[2]{9}$

REVIEW EXERCISES

Fill in the blanks.

23. $\sqrt[3]{125} = 5$, because ___ $= 125$; 5 is called the ___ root of 125.

24. $\dfrac{1}{3}$ is a cube root of $\dfrac{1}{27}$, because ___ $=$ ___.

Find each root. Do not use a calculator.

25. $\sqrt[3]{-27}$
26. $-\sqrt[3]{125}$
27. $\sqrt[4]{81}$
28. $\sqrt[5]{32}$
29. $\sqrt[3]{0}$
30. $\sqrt[3]{-1}$
31. $\sqrt[3]{\dfrac{1}{64}}$
32. $\sqrt[3]{1}$

Use a calculator to find each root to three decimal places.

33. $\sqrt[3]{16}$
34. $\sqrt[3]{-102.35}$
35. $\sqrt[4]{6}$
36. $\sqrt[5]{34,500}$

Find each root. Each variable represents a positive number.

37. $\sqrt{x^2}$
38. $\sqrt{4b^2}$
39. $\sqrt{x^4y^4}$
40. $-\sqrt{y^{12}}$
41. $\sqrt[3]{x^3}$
42. $\sqrt[3]{y^6}$
43. $\sqrt[3]{27x^3}$
44. $\sqrt[3]{-r^{12}}$

45. **DICE** Find the length of an edge of one of the dice shown if each one has a volume of 1,728 cubic millimeters.

46. If the volume of a cube is 2,744 in.^{3}, find the length of one side.

SECTION 8.3 Simplifying Radical Expressions

DEFINITIONS AND CONCEPTS	EXAMPLES
The product rule for square roots: For any nonnegative real numbers a and b, $\sqrt{ab} = \sqrt{a}\sqrt{b}$	Simplify: $\sqrt{50x^7} = \sqrt{25x^6 \cdot 2x}$ $\qquad = \sqrt{25x^6} \cdot \sqrt{2x}$ $\qquad = 5x^3\sqrt{2x}$

Chapter 8 Summary and Review

Simplified form of a square root: 1. Except for 1, the radicand has no perfect-square factors. 2. No fraction appears in the radicand. 3. No radical appears in the denominator.	The following square roots are not in simplified form: $\sqrt{32}$ because 32 has a perfect-square factor of 16. $\sqrt{\dfrac{1}{2}}$ because the radicand is a fraction. $\dfrac{1}{\sqrt{2}}$ because a radical appears in a denominator.
The quotient rule for square roots: For any positive real numbers a and b, $\sqrt{\dfrac{a}{b}} = \dfrac{\sqrt{a}}{\sqrt{b}}$ where $b \neq 0$	Simplify: $\sqrt{\dfrac{7}{16}} = \dfrac{\sqrt{7}}{\sqrt{16}} = \dfrac{\sqrt{7}}{4}$ Simplify: $\sqrt{\dfrac{22x}{25}} = \dfrac{\sqrt{22x}}{\sqrt{25}} = \dfrac{\sqrt{22x}}{5}$
The product and quotient rules for cube roots: $\sqrt[3]{ab} = \sqrt[3]{a}\sqrt[3]{b}$ $\sqrt[3]{\dfrac{a}{b}} = \dfrac{\sqrt[3]{a}}{\sqrt[3]{b}}$ $(b \neq 0)$	Simplify: $\sqrt[3]{16} = \sqrt[3]{8 \cdot 2}$ $= \sqrt[3]{8} \cdot \sqrt[3]{2}$ $= 2\sqrt[3]{2}$ Simplify: $\sqrt[3]{\dfrac{11}{27}} = \dfrac{\sqrt[3]{11}}{\sqrt[3]{27}}$ $= \dfrac{\sqrt[3]{11}}{3}$

REVIEW EXERCISES

Simplify each expression. All variables represent positive numbers.

47. $\sqrt{32}$
48. $\sqrt{500}$
49. $\sqrt{80x^2}$
50. $-2\sqrt{63}$
51. $-\sqrt{250t^3}$
52. $-\sqrt{700z^5}$
53. $\sqrt{200x^2y}$
54. $\dfrac{1}{5}\sqrt{75y^4}$
55. $\sqrt[3]{8x^2y^3}$
56. $\sqrt[3]{250x^4y^3}$

Simplify each expression. All variables represent positive numbers.

57. $\sqrt{\dfrac{16}{25}}$
58. $\sqrt{\dfrac{60}{49}}$
59. $\sqrt[3]{\dfrac{1{,}000}{27}}$
60. $\sqrt{\dfrac{242x^4}{169x^2}}$

Refer to the sit-up board shown in the illustration.

61. Find its length. Express the answer in simplified radical form.
62. Express your answer to Problem 61 as a decimal approximation rounded to the nearest tenth.

SECTION 8.4   Adding and Subtracting Radical Expressions

DEFINITIONS AND CONCEPTS	EXAMPLES
Square root radicals are called **like radicals** when they have the same radicand.	Like radicals: $4\sqrt{2}$ and $5\sqrt{2}$ Unlike radicals: $3\sqrt{6}$ and $7\sqrt{3}$ The same radicand Different radicands

Radical expressions can be added or subtracted if they contain like radicals. To **combine like radicals** we use the distributive property in reverse.	Add: $3\sqrt{7} + 5\sqrt{7} = (3+5)\sqrt{7}$ $= 8\sqrt{7}$ Subtract: $8\sqrt{2y} - 2\sqrt{2y} = (8-2)\sqrt{2y}$ $= 6\sqrt{2y}$
If a sum or difference involves unlike radicals, make sure that each one is written in simplified form. After doing so, like radicals may result that can be combined.	Add: $\sqrt{12} + \sqrt{75} = \sqrt{4}\sqrt{3} + \sqrt{25}\sqrt{3}$ Simplify $\sqrt{12}$ and $\sqrt{75}$. $= 2\sqrt{3} + 5\sqrt{3}$ $= 7\sqrt{3}$ Combine like radicals.

REVIEW EXERCISES

Perform the operations. All variables represent positive numbers.

63. $\sqrt{2} + \sqrt{8} - \sqrt{18}$
64. $\sqrt{3} + 4 + \sqrt{27} - 7$
65. $5\sqrt{28} - 3\sqrt{63}$
66. $3y\sqrt{5xy^3} - y^2\sqrt{20xy}$
67. $\sqrt[3]{16} + \sqrt[3]{54}$
68. $\sqrt[3]{2,000x^3} - \sqrt[3]{128x^3}$
69. Explain why we cannot add $3\sqrt{5}$ and $5\sqrt{3}$.

70. **GARDENING** Find the difference in the lengths of the two wires used to secure the tree shown.

SECTION 8.5 Multiplying and Dividing Radical Expressions

DEFINITIONS AND CONCEPTS	EXAMPLES
The product rules for square roots and cube roots: For any nonnegative real numbers a and b: $\sqrt{a}\sqrt{b} = \sqrt{ab}$ $\sqrt[3]{a}\sqrt[3]{b} = \sqrt[3]{ab}$	Multiply: $\sqrt{2}\sqrt{8} = \sqrt{2 \cdot 8} = \sqrt{16} = 4$ $\sqrt[3]{3}\sqrt[3]{9} = \sqrt[3]{3 \cdot 9} = \sqrt[3]{27} = 3$
To multiply radical expressions containing only one term, first multiply the coefficients, then multiply the radicals separately, and simplify the result.	Multiply: $3\sqrt{8} \cdot 3\sqrt{2} = 3 \cdot 3\sqrt{8}\sqrt{2}$ $= 9\sqrt{8 \cdot 2}$ $= 9\sqrt{16}$ $= 9(4)$ $= 36$
To multiply two binomials, multiply each term of one binomial by each term of the other binomial and simplify.	Multiply: $(\sqrt{5x} + 2)(\sqrt{5x} - 1)$ $= \sqrt{5x}\sqrt{5x} - (1)\sqrt{5x} + 2\sqrt{5x} - 2$ $= 5x + \sqrt{5x} - 2$

If the denominator of a fraction is a square root, **rationalize** the denominator by multiplying the numerator and denominator by some appropriate square root that makes a perfect-square radicand in the denominator.

If the denominator of a fraction is a cube root, **rationalize** the denominator by multiplying the numerator and denominator by the cube root that makes a perfect cube radicand in the denominator.

If the denominator of a fraction contains two terms with square roots, multiply the numerator and denominator by the **conjugate** of the denominator.

Rationalize the denominator.

$$\frac{x}{\sqrt{5}} = \frac{x}{\sqrt{5}} \cdot \frac{\sqrt{5}}{\sqrt{5}} = \frac{x\sqrt{5}}{\sqrt{5}\sqrt{5}} = \frac{x\sqrt{5}}{\sqrt{25}} = \frac{x\sqrt{5}}{5}$$

$$\frac{2}{\sqrt[3]{25}} = \frac{2}{\sqrt[3]{25}} \cdot \frac{\sqrt[3]{5}}{\sqrt[3]{5}} = \frac{2\sqrt[3]{5}}{\sqrt[3]{125}} = \frac{2\sqrt[3]{5}}{5}$$

$$\frac{x}{\sqrt{x}-2} = \frac{x}{\sqrt{x}-2} \cdot \frac{\sqrt{x}+2}{\sqrt{x}+2} = \frac{x(\sqrt{x}+2)}{(\sqrt{x}-2)(\sqrt{x}+2)} = \frac{x(\sqrt{x}+2)}{x-4}$$

REVIEW EXERCISES

Perform the operations.

71. $\sqrt{2}\sqrt{3}$

72. $(-5\sqrt{5})(-2\sqrt{2})$

73. $(3\sqrt{3x})(4\sqrt{6x})$

74. $(\sqrt{15}+3x)^2$

75. $\sqrt{2}(\sqrt{8}-\sqrt{18})$

76. $(\sqrt{3}+\sqrt{5})(\sqrt{3}-\sqrt{5})$

77. $(\sqrt[3]{4})(2\sqrt[3]{4})$

78. $(\sqrt[3]{3}+2)(\sqrt[3]{3}-1)$

Rationalize each denominator.

79. $\dfrac{1}{\sqrt{7}}$

80. $\sqrt{\dfrac{3}{7}}$

81. $\dfrac{\sqrt{9}}{\sqrt{18}}$

82. $\dfrac{\sqrt{c}-4}{\sqrt{c}+4}$

83. $\dfrac{7}{\sqrt{2}+1}$

84. $\dfrac{8}{\sqrt[3]{16}}$

The illustration shows the amount of surface area of a rug suctioned by a vacuum nozzle attachment.

$5\sqrt{3}$ in. $2\sqrt{6}$ in.

85. Find the perimeter and area of this section of rug. Express the answers in simplified radical form.

86. Express your answer to Problem 85 as decimal approximations to the nearest tenth.

SECTION 8.6 Solving Radical Equations; the Distance Formula

DEFINITIONS AND CONCEPTS	EXAMPLES
The squaring property of equality: If $a = b$, then $a^2 = b^2$.	If $\sqrt{x} = 6$, then $(\sqrt{x})^2 = (6)^2$ or $x = 36$.

To solve equations containing square roots:

1. Isolate the radical term on one side of the equation.
2. Square both sides and solve the resulting equation.
3. Check the solution. Discard any **extraneous solutions**.

Solve:

$\sqrt{x-2} + 1 = 5$

$\sqrt{x-2} = 4$ To isolate the radical, subtract 1 from both sides.

$(\sqrt{x-2})^2 = (4)^2$ Square both sides.

$x - 2 = 16$

$x = 18$ Add 2 to both sides.

Check: $\sqrt{x-2} + 1 = 5$ The original equation.

$\sqrt{18-2} + 1 \stackrel{?}{=} 5$ Substitute 18 for x.

$\sqrt{16} + 1 \stackrel{?}{=} 5$ Evaluate the left side.

$4 + 1 \stackrel{?}{=} 5$

$5 = 5$ True

The solution is 18, and the solution set is $\{18\}$.

The distance formula:

$d = \sqrt{(x_2 - x_1)^2 + (y_2 - y_1)^2}$

To find the distance between the points $(2, -3)$ and $(5, -7)$, substitute into the distance formula:

$d = \sqrt{(x_2 - x_1)^2 + (y_2 - y_1)^2}$

$d = \sqrt{(5-2)^2 + [(-7)-(-3)]^2}$ Substitute.

$d = \sqrt{3^2 + (-4)^2}$ $5 - 2 = 3, -7 - (-3) = -4$

$d = \sqrt{9 + 16}$ Evaluate the powers.

$d = \sqrt{25}$ Do the addition.

$d = 5$ Find the square root.

REVIEW EXERCISES

Simplify each expression. All variables represent positive numbers.

87. $(\sqrt{x})^2$ **88.** $(\sqrt[3]{x})^3$

89. $(2\sqrt{t})^2$ **90.** $(\sqrt{e-1})^2$

Solve each equation and check all solutions.

91. $\sqrt{x} = 9$ **92.** $\sqrt{3x+4} + 5 = 3$

93. $\sqrt{24 + 10y} = y + 4$ **94.** $\sqrt{2(r+4)} = 2\sqrt{r}$

95. $\sqrt{p^2 - 3} = p + 3$ **96.** $\sqrt[3]{x-1} = 3$

Find the distance between the points. If an answer contains a radical, round to the nearest hundredth.

97. $(-7, 12), (-4, 8)$ **98.** $(-15, -3), (-10, -16)$

99. FERRIS WHEELS The distance d in feet that an object will fall in t seconds is given by the formula

$$t = \sqrt{\frac{d}{16}}$$

If a person drops a coin from the top of a Ferris wheel and it takes 2 seconds to hit the ground, how tall is the Ferris wheel?

SECTION 8.7 Rational Exponents

DEFINITIONS AND CONCEPTS

To evaluate exponential expressions involving fractional exponents, use the **rules for rational exponents** to write the expressions in an equivalent radical form.

$$x^{1/n} = \sqrt[n]{x}$$

$$x^{m/n} = \sqrt[n]{x^m} = \left(\sqrt[n]{x}\right)^m$$

$$x^{-m/n} = \frac{1}{x^{m/n}}$$

EXAMPLES

Evaluate: $8^{1/3} = \sqrt[3]{8} = 2$

Simplify: $(16x^2)^{1/2} = \sqrt{16x^2} = 4x$

Simplify: $(8x^6)^{1/3} = \sqrt[3]{8x^6} = 2x^2$

Simplify: $(-64)^{4/3} = \left(\sqrt[3]{-64}\right)^4 = (-4)^4 = 256$

Simplify: $(27x^3)^{2/3} = [(27x^3)^{1/3}]^2 = (3x)^2 = 9x^2$

Simplify: $125^{-2/3} = \frac{1}{125^{2/3}}$

$$= \frac{1}{(125^{1/3})^2}$$

$$= \frac{1}{(5)^2}$$

$$= \frac{1}{25}$$

The rules for exponents can be used to simplify expressions involving rational exponents.

Simplify: $7^{1/9} 7^{4/9} = 7^{1/9 + 4/9} = 7^{5/9}$

REVIEW EXERCISES

Simplify each expression. Write answers without using negative exponents.

100. $49^{1/2}$

101. $(-1{,}000)^{1/3}$

102. $36^{3/2}$

103. $\left(\dfrac{8}{27}\right)^{2/3}$

104. $4^{-3/2}$

105. $8^{2/3} 8^{4/3}$

106. $(3^{2/3})^3$

107. $(a^4 b^8)^{-1/2}$

108. $x^{1/3} x^{2/5}$

109. $\dfrac{t^{3/4}}{t^{2/3}}$

110. $\dfrac{x^{2/5} x^{1/5}}{x^{-2/5}}$

111. $\dfrac{x^{17/7}}{x^{3/7}}$

112. Graph each number on the number line: $\{4^{-1/2},\ 12^{1/2},\ 9^{1/3},\ -2^{2/3}\}$.

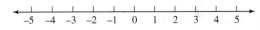

113. DENTISTRY The fractional amount of painkiller remaining in the system of a patient h hours after the original dose was injected into her gums is given by $h^{-3/2}$. How much of the original dose is in the patient's system 16 hours after the injection?

114. Explain why $(-4)^{1/2}$ is not a real number.

CHAPTER 8 TEST

1. Fill in the blanks.
   a. The symbol $\sqrt{}$ is called a _____ symbol.
   b. The _____ of $\sqrt{25a^2}$ is $25a^2$.
   c. The _____ theorem relates the lengths of the sides of a right triangle.
   d. We read $\sqrt[3]{8}$ as "the ___ ___ of 8."
   e. For the expression $\sqrt[5]{32}$, the _____ is 5.

2. Evaluate $\sqrt{b^2 - 4ac}$ for $a = 2$, $b = 10$, and $c = 6$. Round to the nearest tenth.

Simplify each radical.

3. $\sqrt{100}$

4. $-\sqrt{\dfrac{400}{9}}$

5. $\sqrt[3]{-27}$

6. $\sqrt{\dfrac{50}{49}}$

7. A 26-foot ladder reaches a point on a wall 24 feet above the ground. How far from the wall is the ladder's base?

8. a. Suppose a square has an area of 24 square yards. Express the length of a side of the square in simplified radical form.

   b. Round the length of a side of the square to the nearest tenth.

Simplify each expression. Assume that x and y represent positive numbers.

9. $\sqrt{4x^2}$

10. $\sqrt{54x^3}$

11. $\sqrt{\dfrac{18x^2y^3}{2xy}}$

12. $\sqrt[3]{x^6y^3}$

Perform each operation and simplify.

13. $\sqrt{12} + \sqrt{27}$

14. $\sqrt{8x^3} - x\sqrt{18x}$

15. $(-2\sqrt{8x})(3\sqrt{12x})$

16. $\sqrt{3}(\sqrt{8} + \sqrt{6})$

17. $(\sqrt{2} + \sqrt{3})(\sqrt{2} - \sqrt{3})$

18. $(2\sqrt{x} + 2)(\sqrt{x} - 3)$

19. SEWING A corner of fabric is folded over to form a collar and stitched down as shown below. From the dimensions given in the figure, determine the exact number of inches of stitching that must be made. Then give an approximation to one decimal place. (All measurements are in inches.)

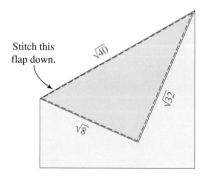

Chapter 8 Test

Rationalize each denominator.

20. $\dfrac{2}{\sqrt{2}}$

21. $\dfrac{\sqrt{3x}}{\sqrt{x}+2}$

31. Explain why we cannot perform the subtraction $4\sqrt{3} - 7\sqrt{2}$.

Solve each equation.

22. $\sqrt{x} = 15$

23. $\sqrt{2-x} - 2 = 6$

24. $\sqrt{3x+9} = 2\sqrt{x+1}$

25. $x - 1 = \sqrt{x-1}$

26. $\sqrt{3a+4} + 2 = 0$

27. $\sqrt[3]{x-2} = 3$

32. CARPENTRY In the illustration below, a carpenter is using a tape measure to see whether the wall he just put up is perfectly square with the floor. Explain what mathematical concept he is applying. If the wall is positioned correctly, what should the measurement on the tape read?

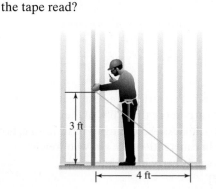

28. Find the distance between points $(-2, -3)$ and $(-8, 5)$.

29. Complete the table and graph the function. Round to the nearest tenth when necessary.

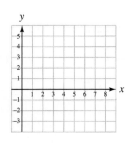

$f(x) = \sqrt{x}$

x	f(x)
0	
1	
2	
4	
6	
9	

33. Explain why $\sqrt{-9}$ is not a real number.

34. Evaluate: $8^{2/3}$

Simplify each expression. All variables represent positive numbers. Write your answers without using negative exponents.

35. $\dfrac{a^{3/4}}{a^{2/3}}$

36. $p^{2/3} p^{4/3}$

30. Is 0 a solution of the radical equation $\sqrt{3x+1} = x - 1$? Explain your answer.

37. $(x^4 y^8)^{-1/2}$

38. $(16n^2)^{1/2}$

CHAPTERS 1–8 CUMULATIVE REVIEW

1. Determine whether each statement is true or false.
   a. All whole numbers are integers. [Section 1.3]
   b. π is a rational number. [Section 1.3]
   c. A real number is either rational or irrational. [Section 1.3]

2. Evaluate: $\dfrac{-3(3+2)^2 - (-5)}{17 - 3|-4|}$ [Section 1.7]

3. Simplify: $3p - 6(p + z) + p$ [Section 2.2]

4. Solve: $2 - (4x + 7) = 3 + 2(x + 2)$ [Section 2.2]

5. BACKPACKS Pediatricians advise that children should not carry more than 20% of their own body weight in backpacks. According to this warning, how much weight can a fifth-grade girl who weighs 85 pounds safely carry in her backpack? [Section 2.3]

6. ENERGY DRINKS The graph below shows the increase in the sales of energy drinks in the United States. Determine the percent change from 2006 to 2011. Round to the nearest tenth of a percent. [Section 2.3]

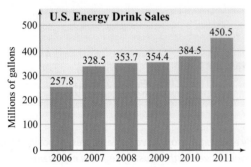

7. Solve $3 - 3x \geq 6 + x$, graph the solution, and use interval notation to describe the solution. [Section 2.7]

8. Solve $0 \leq \dfrac{4 - x}{3} < 2$, graph the solution, and use interval notation to describe the solution. [Section 2.7]

9. SEARCH AND RESCUE Two search and rescue teams leave base at the same time, looking for a lost boy. The first team, on foot, heads north at 2 mph and the other, on horseback, south at 4 mph. How long will it take them to search a distance of 21 miles between them? [Section 2.6]

10. BLENDING COFFEES A store sells regular coffee for $4 a pound and gourmet coffee for $7 a pound. Using 40 pounds of the gourmet coffee, the owner makes a blend to put on sale for $5 a pound. How many pounds of regular coffee should he use? [Section 2.6]

11. SURFACE AREA The total surface area A of a box with dimensions l, w, and h is given by the formula

    $A = 2lw + 2wh + 2lh$

    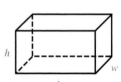

    If $A = 202$ square inches, $l = 9$ inches, and $w = 5$ inches, find h. [Section 2.4]

Graph each equation or inequality.

12. $3x - 4y = 12$ [Section 3.3]

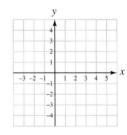

13. $y = \dfrac{1}{2}x$ [Section 3.3]

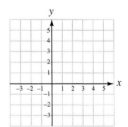

14. $x = 5$ [Section 3.3]

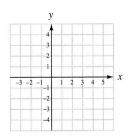

15. $3x + 4y \leq 12$ [Section 3.7]

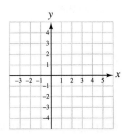

16. What is true about the slopes of two
   a. parallel lines? [Section 3.4]
   b. perpendicular lines? [Section 3.4]

17. ONLINE HOLIDAY SALES On the graph below, the line approximates the growth in U.S. online holiday retail sales for 2004 to 2012. Find the rate of increase in sales for that time span. [Section 3.4]

18. Find the slope of the line defined by each equation.
   a. $y = 3x - 7$ [Section 3.5]
   b. $2x + 3y = -10$ [Section 3.5]

19. Write an equation of the line passing through $(-2, 5)$ and $(4, 8)$. Express the result in slope–intercept form. [Section 3.6]

20. If $f(x) = x^3 - x + 5$, find $f(-2)$. [Section 3.8]

21. Complete the table and graph the function. Then give the domain and range of the function. [Section 3.8]

$f(x) = |1 - x|$

x	$f(x)$
0	
1	
2	
3	
-1	
-2	

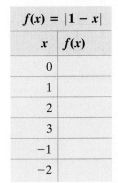
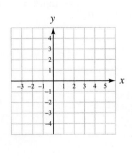

22. BOATING The following graph shows the vertical distance from a point on the tip of a propeller to the centerline as the propeller spins. Is this the graph of a function? [Section 3.8]

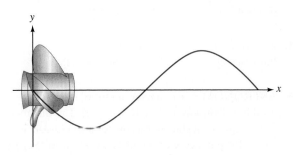

23. Use a check to determine whether $(-2, -3)$ is a solution of the system $\begin{cases} 2x - 3y = 5 \\ 7x + y = -16 \end{cases}$. [Section 4.1]

24. Solve the system $\begin{cases} x + y = 4 \\ y = x + 6 \end{cases}$ by graphing. [Section 4.1]

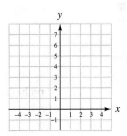

Solve each system of equations.

25. $\begin{cases} x = y + 4 \\ 2x + y = 5 \end{cases}$ [Section 4.2]

26. $\begin{cases} 3s + 4t = 5 \\ 2s - 3t = -8 \end{cases}$ [Section 4.3]

27. FINANCIAL PLANNING In investing $6,000 of a couple's money, a financial planner put some of it into a savings account paying 6% annual interest. The rest was invested in a riskier mini-mall development plan paying 12% annually. The combined interest earned for the first year was $540. How much money was invested at each rate? Use two variables to solve this problem. [Section 4.4]

28. Graph: $\begin{cases} 3x + 2y \geq 6 \\ x + 3y \leq 6 \end{cases}$ [Section 4.5]

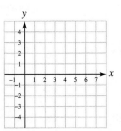

Chapters 1–8 Cumulative Review

Simplify each expression. Write each answer without using parentheses or negative exponents.

29. $(x^5)^2(x^7)^3$ [Section 5.1]

30. $\left(\dfrac{a^3b}{c^4}\right)^5$ [Section 5.1]

31. $4^{-3} \cdot 4^{-2} \cdot 4^5$ [Section 5.2]

32. $(a^{-2}b^3)^{-4}$ [Section 5.2]

33. **ASTRONOMY** The **parsec**, a unit of distance used in astronomy, is 3×10^{16} meters. The distance to Betelgeuse, a star in the constellation Orion, is 1.6×10^2 parsecs. Use scientific notation to express this distance in meters. [Section 5.3]

34. **GEOMETRY** Write a polynomial that represents the area of the figure. [Section 5.6]

a.

b.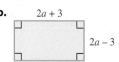

Perform the operations.

35. $(r^4st^2)(2r^2st)(rst)$ [Section 5.6]

36. $(-3t + 2s)(2t - 3s)$ [Section 5.6]

37. $(3a^2 - 2a + 4) - (a^2 - 3a + 7)$ [Section 5.5]

38. $(y - 6)^2$ [Section 5.6]

39. $\dfrac{4x - 3y + 8z}{4xy}$ [Section 5.7]

40. $2 + x \overline{\smash{)}3x + 2x^2 - 2}$ [Section 5.8]

Factor each expression completely.

41. $3x^2y - 6xy^2$ [Section 6.1]

42. $2x^2 + 2xy - 3x - 3y$ [Section 6.1]

43. $25p^4 - 16q^2$ [Section 6.4]

44. $3x^3 - 243x$ [Section 6.4]

45. $x^2 - 11x - 12$ [Section 6.2]

46. $a^3 + 8b^3$ [Section 6.5]

47. $6a^2 - 7a - 20$ [Section 6.3]

48. $16m^2 - 20m - 6$ [Section 6.3]

Solve each equation.

49. $x^2 + 3x + 2 = 0$ [Section 6.7]

50. $5x^2 = 10x$ [Section 6.7]

51. $6x^2 - x - 2 = 0$ [Section 6.7]

52. $2y^2 = 12 - 5y$ [Section 6.7]

53. **CHILDREN'S STICKERS** This rectangular-shaped sticker has an area of 20 cm². The width is 1 cm shorter than the length. Find the length of the sticker. [Section 6.8]

54. For what value of x is $\dfrac{4x}{x - 6}$ undefined? [Section 7.1]

Simplify each expression.

55. $\dfrac{x^2 + 2x + 1}{x^2 - 1}$ [Section 7.1]

56. $-\dfrac{15a^2}{25a^3}$ [Section 7.1]

Perform the operation(s) and simplify when possible.

57. $\dfrac{p^2 - p - 6}{3p - 9} \div \dfrac{p^2 + 6p + 9}{p^2 - 9}$ [Section 7.2]

58. $\dfrac{x^2y^2}{cd} \cdot \dfrac{d^2}{c^2x}$ [Section 7.2]

59. $\dfrac{3x}{x + 2} + \dfrac{5x}{x + 2} - \dfrac{7x - 2}{x + 2}$ [Section 7.3]

60. $\dfrac{x + 2}{x + 5} - \dfrac{x - 3}{x + 7}$ [Section 7.4]

61. $\dfrac{3a}{2b} - \dfrac{2b}{3a}$ [Section 7.4]

62. $\dfrac{\dfrac{1}{x} + \dfrac{1}{y}}{\dfrac{1}{x} - \dfrac{1}{y}}$ [Section 7.5]

Solve each equation.

63. $\dfrac{4}{a} = \dfrac{6}{a} - 1$ [Section 7.6]

64. $\dfrac{a+2}{a+3} - 1 = \dfrac{-1}{a^2 + 2a - 3}$ [Section 7.6]

65. Solve the formula $\dfrac{1}{r} = \dfrac{1}{r_1} + \dfrac{1}{r_2}$ for r.

[Section 7.6]

66. FILLING A POOL An inlet pipe can fill an empty swimming pool in 5 hours, and another inlet pipe can fill the pool in 4 hours. How long will it take both pipes to fill the pool? [Section 7.7]

67. ONLINE SALES A company found that, on average, it made 9 online sales transactions for every 500 hits on its Internet website. If the company's website had 360,000 hits in one year, how many sales transactions did it have that year? [Section 7.8]

68. Assume that y varies inversely with x. If $y = 8$ when $x = 2$, find y when $x = 8$. [Section 7.9]

Simplify each expression. All variables represent positive numbers.

69. $\sqrt{\dfrac{49}{225}}$ [Section 8.1]   **70.** $-\sqrt[3]{-27}$ [Section 8.2]

71. $-12x\sqrt{16x^2 y^3}$ [Section 8.3]

72. $\sqrt{48} - \sqrt{8} + \sqrt{27} - \sqrt{32}$ [Section 8.4]

73. $(\sqrt{y} - 4)(\sqrt{y} - 5)$ [Section 8.5]

74. $(-5\sqrt{6})(4\sqrt{3})$ [Section 8.5]

75. $\dfrac{4}{\sqrt{20}}$ [Section 8.5]

76. $\dfrac{\sqrt{x} - 3}{\sqrt{x} + 3}$ [Section 8.5]

77. Solve: $\sqrt{6x + 19} - 5 = 2$ [Section 8.6]

78. CARGO SPACE How wide a piece of plywood can be carried diagonally in the back of the van shown? [Section 8.1]

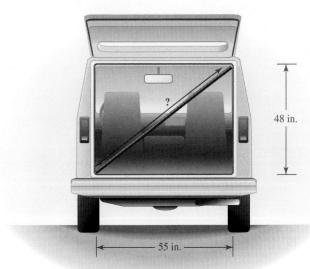

9

Quadratic Equations

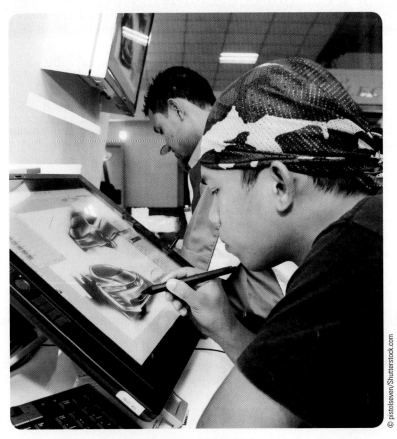

9.1 Solving Quadratic Equations: The Square Root Property

9.2 Solving Quadratic Equations: Completing the Square

9.3 Solving Quadratic Equations: The Quadratic Formula

9.4 Graphing Quadratic Equations

9.5 Complex Numbers

Chapter Summary and Review

Chapter Test

Cumulative Review

from Campus to Careers
Graphic Designer

Graphic designers combine their artistic talents with their mathematical skills to create attractive advertisements, photography, packaging, and websites. They perform arithmetic computations with fractions and decimals to determine proper font sizes and page margins. They apply algebraic concepts such as graphing and equation solving to plan page layouts. They also use formulas to create production schedules and calculate costs.

Graphic designers often begin a project with the final product in mind and work backwards to determine the specifics. In **Problem 70** of **Study Set 9.3**, you will do just that. Given the area of a triangular poster, you are to find the length of its base and its height by writing and then solving a quadratic equation.

JOB TITLE: Graphic Designer
EDUCATION: A bachelor's degree is required for most entry-level positions.
JOB OUTLOOK: Employment is expected to increase about 13% through the year 2020.
ANNUAL EARNINGS: Median annual salary for entry-level designers is $48,730.
FOR MORE INFORMATION:
www.bls.gov/ooh/Arts-and-Design
/Graphic-designers.htm

Objectives

1 Use the square root property to solve equations of the form $x^2 = c$.

2 Use the square root property to solve equations of the form $(ax + b)^2 = c$.

3 Solve problems modeled by quadratic equations.

SECTION 9.1
Solving Quadratic Equations: The Square Root Property

Recall that a **quadratic equation** can be written in the form $ax^2 + bx + c = 0$, where a, b, and c represent real numbers and $a \neq 0$. Some examples are:

$$x^2 - x - 6 = 0, \quad 4x^2 + 4x + 1 = 0, \quad \text{and} \quad x^2 - 16 = 0$$

We have solved quadratic equations like these using factoring in combination with the zero-factor property. To review this method, let's solve $x^2 - 16 = 0$.

$$x^2 - 16 = 0$$
$$(x + 4)(x - 4) = 0 \qquad \text{Factor the difference of two squares.}$$
$$x + 4 = 0 \quad \text{or} \quad x - 4 = 0 \qquad \text{Set each factor equal to 0.}$$
$$x = -4 \qquad \qquad x = 4 \qquad \text{Solve each equation.}$$

The solutions are -4 and 4.

1 Use the square root property to solve equations of the form $x^2 = c$.

We will now solve $x^2 - 16 = 0$ in another way. This time, we will ignore the zero-factor property condition that requires 0 on one side of the equation. Instead, we will add 16 to both sides to isolate x^2.

$$x^2 - 16 = 0$$
$$x^2 = 16 \qquad \text{Add 16 to both sides.}$$

We see that x must be a number whose square is 16. Therefore, x must be a square root of 16. Since every positive number has two square roots, one positive and one negative, we have

$$x = \sqrt{16} \quad \text{or} \quad x = -\sqrt{16}$$
$$x = 4 \qquad \qquad x = -4$$

As with the factor method, we find that the solutions of $x^2 - 16 = 0$ are 4 and -4. This approach illustrates how the following *square root property* can be used to solve certain types of quadratic equations.

The Square Root Property of Equations

For any nonnegative real number c, if $x^2 = c$, then

$$x = \sqrt{c} \quad \text{or} \quad x = -\sqrt{c}$$

We can write the conclusion of the square root property in more compact form, called **double-sign notation**:

$$x = \pm\sqrt{c} \qquad \text{Read as "x is equal to the positive or negative square root of c."}$$

9.1 Solving Quadratic Equations: The Square Root Property

EXAMPLE 1 Solve: **a.** $x^2 - 5 = 0$ **b.** $x^2 = 12$ **c.** $3a^2 + 1 = 11$ **d.** $n^2 = -4$

Strategy We will use properties of equality to isolate the *squared term* on one side of the equation. Then we will use the square root property to isolate the variable itself.

WHY To solve the original equation, we want to find a simpler equivalent equation of the form $x = \pm$ **a number**, whose solutions are obvious.

Solution

a. We can use the addition property of equality to isolate x^2.

$x^2 - 5 = 0$ This is the equation to solve.
$x^2 = 5$ Add 5 to both sides.

Now we use the square root property to isolate x.

$x = \sqrt{5}$ or $x = -\sqrt{5}$
$x = \pm\sqrt{5}$ Use double-sign notation.

The check for $\sqrt{5}$:
$x^2 - 5 = 0$
$(\sqrt{5})^2 - 5 \stackrel{?}{=} 0$
$5 - 5 \stackrel{?}{=} 0$
$0 = 0$ True

The check for $-\sqrt{5}$:
$x^2 - 5 = 0$
$(-\sqrt{5})^2 - 5 \stackrel{?}{=} 0$
$5 - 5 \stackrel{?}{=} 0$
$0 = 0$ True

Since each statement is true, the solutions are $\sqrt{5}$ and $-\sqrt{5}$, or $\pm\sqrt{5}$. The solution set is written as $\{-\sqrt{5}, \sqrt{5}\}$ or $\{\pm\sqrt{5}\}$.

b. Since x^2 is already isolated, we simply apply the square root property.

$x^2 = 12$ This is the equation to solve.
$x = \sqrt{12}$ or $x = -\sqrt{12}$ To isolate x, use the square root property.
$x = \pm\sqrt{12}$ Use double-sign notation.
$x = \pm 2\sqrt{3}$ Simplify the radical: $\sqrt{12} = \sqrt{4 \cdot 3} = \sqrt{4}\sqrt{3} = 2\sqrt{3}$.

Verify that $2\sqrt{3}$ and $-2\sqrt{3}$ are solutions by checking each in the original equation. The solution set is $\{\pm 2\sqrt{3}\}$.

Caution! When using the square root property to solve an equation, always write the $\pm$ symbol, or you will lose one of the solutions.

c.
$3a^2 + 1 = 11$ This is the equation to solve.
$3a^2 = 10$ To isolate the term $3a^2$, subtract 1 from both sides.
$a^2 = \dfrac{10}{3}$ To isolate a^2, divide both sides by 3.
$a = \pm\sqrt{\dfrac{10}{3}}$ To isolate a, use the square root property: $a = \sqrt{\dfrac{10}{3}}$ or $a = -\sqrt{\dfrac{10}{3}}$. Use double-sign notation.

To rationalize the denominator of the result, we proceed as follows:

$x = \pm\sqrt{\dfrac{10}{3}} = \pm\dfrac{\sqrt{10}}{\sqrt{3}} \cdot \dfrac{\sqrt{3}}{\sqrt{3}} = \pm\dfrac{\sqrt{30}}{3}$

Verify that $\dfrac{\sqrt{30}}{3}$ and $-\dfrac{\sqrt{30}}{3}$ are solutions by checking each in the original equation.

Self Check 1
Solve: **a.** $x^2 - 21 = 0$
b. $b^2 = 54$ **c.** $5m^2 - 1 = 6$
d. $x^2 = -25$

Now Try Problems 23, 25, 31, and 33

The Language of Algebra The *exact* solutions are $\pm \frac{\sqrt{30}}{3}$. To the nearest hundredth, the *approximate* solutions are ± 1.83.

d. $n^2 = -4$ This is the equation to solve.

   $n = \pm \sqrt{-4}$ Use the square root property.

Since the square root of -4 is not a real number, the equation has no real-number solutions.

2 Use the square root property to solve equations of the form $(ax + b)^2 = c$.

We can extend the square root property to solve equations that involve the square of a binomial.

Self Check 2

Solve:
a. $(x - 2)^2 = 64$
b. $(x + 3)^2 = 98$
c. $(3r - 1)^2 = 3$

Now Try Problems 35, 41, and 45

EXAMPLE 2   Solve: **a.** $(x - 3)^2 = 36$ **b.** $(x + 1)^2 = 50$
c. $(2s - 4)^2 = 7$

Strategy Instead of a variable squared on the left side of the equation, we have a quantity squared. We still use the square root property to solve each equation.

WHY We want to eliminate the square on the binomial, so that we can eventually isolate the variable on one side of the equation.

Solution

a.
$(x - 3)^2 = 36$ This is the equation to solve.

$x - 3 = \sqrt{36}$ or $x - 3 = -\sqrt{36}$ Use the square root property.

$x - 3 = \pm\sqrt{36}$ Use double-sign notation.

$x - 3 = \pm 6$ Evaluate: $\sqrt{36} = 6$.

$x = 3 \pm 6$ To isolate x, add 3 to both sides. It is standard practice to write the 3 in front of the $\pm$ symbol.

We read 3 ± 6 as "3 plus or minus 6." To find the solutions, we perform the calculation using a plus symbol $+$ and then using a minus symbol $-$.

$x = 3 + 6$ or $x = 3 - 6$
$x = 9$ $x = -3$

The check for 9:
$(x - 3)^2 = 36$
$(9 - 3)^2 \stackrel{?}{=} 36$
$(6)^2 \stackrel{?}{=} 36$
$36 = 36$ True

The check for -3:
$(x - 3)^2 = 36$
$(-3 - 3)^2 \stackrel{?}{=} 36$
$(-6)^2 \stackrel{?}{=} 36$
$36 = 36$ True

The solutions are 9 and -3. The solution set is $\{-3, 9\}$.

Caution! It might be tempting to square the binomial on the left side of $(x - 3)^2 = 36$. However, that causes unnecessary, additional steps to be used to solve the equation in another way.

~~$(x - 3)^2 = 36$~~
~~$x^2 - 6x + 9 = 36$~~

b. $(x + 1)^2 = 50$ This is the equation to solve.

$x + 1 = \pm\sqrt{50}$ By the square root property, $x + 1 = \sqrt{50}$ or $x + 1 = -\sqrt{50}$. Use double-sign notation.

$x + 1 = \pm 5\sqrt{2}$ $\sqrt{50} = \sqrt{25 \cdot 2} = \sqrt{25}\sqrt{2} = 5\sqrt{2}$.

$x = -1 \pm 5\sqrt{2}$ To isolate x, subtract 1 from both sides (or add -1 to both sides). It is standard practice to write the -1 in front of the $\pm$ symbol.

The solutions are $-1 + 5\sqrt{2}$ and $-1 - 5\sqrt{2}$, and the solution set is $\{-1 \pm 5\sqrt{2}\}$.

c. $(2s - 4)^2 = 7$ This is the equation to solve.

$2s - 4 = \pm\sqrt{7}$ By the square root property, $2s - 4 = \sqrt{7}$ or $2s - 4 = -\sqrt{7}$. Use double-sign notation.

$2s = 4 \pm \sqrt{7}$ To isolate the variable term $2s$, add 4 to both sides.

$s = \dfrac{4 \pm \sqrt{7}}{2}$ To isolate s, divide both sides by 2.

Caution! Since 2 is not a common factor of the entire numerator, it would be incorrect to simplify the solutions as shown:

$$s = \dfrac{\cancel{4}^{\,2} \pm \sqrt{7}}{\cancel{2}_{\,1}}$$

Use a check to verify that the solutions are $\dfrac{4 + \sqrt{7}}{2}$ and $\dfrac{4 - \sqrt{7}}{2}$.

> **Caution!** Equations like $(x - 3)^2 = -16$ and $(5y + 6)^2 = -4$ have no real-number solutions because no real number squared is negative.

If one side of the equation factors as the square of a binomial, we can use the methods of Example 2 to solve the equation.

EXAMPLE 3   Solve: $x^2 + 16x + 64 = 2$

Strategy We will attempt to factor the trinomial on the left side of the equation. Our hope is that it factors as the square of a binomial.

WHY If the equation can be written in the form $(ax + b)^2 = c$, we can use the square root property to solve it.

> **The Language of Algebra** Recall that trinomials that are squares of binomials are called **perfect-square trinomials**.

Solution

Since $x^2 + 16x + 64 = x^2 + 2 \cdot x \cdot 8 + 8^2$, it is a perfect-square trinomial and factors as $(x + 8)^2$.

$x^2 + 16x + 64 = 2$ This is the equation to solve.

$(x + 8)^2 = 2$ Factor the perfect-square trinomial $x^2 + 16x + 64$.

$x + 8 = \pm\sqrt{2}$ By the square root property, $x + 8 = \sqrt{2}$ or $x + 8 = -\sqrt{2}$. Use double-sign notation.

$x = -8 \pm \sqrt{2}$ To isolate x, subtract 8 from both sides.

Self Check 3
Solve: $x^2 - 14x + 49 = 11$
Now Try Problem 47

As an informal check, we can use a calculator to approximate $-8 + \sqrt{2}$ and $-8 - \sqrt{2}$. Then we can substitute each approximation into the original equation.

Success Tip To check, we must substitute $-8 + \sqrt{2}$ and then $-8 - \sqrt{2}$ for x in $x^2 + 16x + 64 = 2$. Those calculations would be difficult by hand. Instead, we can perform an informal check using approximations of the possible solutions.

The check for $-8 + \sqrt{2} \approx -6.6$:
$$x^2 + 16x + 64 = 2$$
$$(-6.6)^2 + 16(-6.6) + 64 \stackrel{?}{=} 2$$
$$1.96 \approx 2$$

The check for $-8 - \sqrt{2} \approx -9.4$:
$$x^2 + 16x + 64 = 2$$
$$(-9.4)^2 + 16(-9.4) + 64 \stackrel{?}{=} 2$$
$$1.96 \approx 2$$

In each case, the sides are approximately equal. This suggests that the results, $-8 + \sqrt{2}$ and $-8 - \sqrt{2}$, are reasonable.

3 Solve problems modeled by quadratic equations.

The equation solving methods discussed in this section can be used to solve a variety of real-world applications that are modeled by quadratic equations.

Self Check 4

If a penny is dropped from a bridge that is 48 feet above a river, how long will it take for it to hit the water?

Now Try Problem 76

EXAMPLE 4 *Movie Stunts*

In a scene for an action movie, a stuntwoman falls from the top of a 95-foot-tall building into a 10-foot-tall airbag directly below her on the ground. The formula $d = 16t^2$ gives the distance d in feet that she falls in t seconds. For how many seconds will she fall before making contact with the airbag?

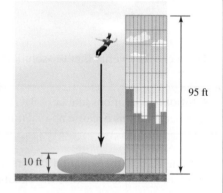

Solution
The woman will fall a distance of $95 - 10 = 85$ feet before making contact with the airbag. To find the number of seconds that the fall will last, we substitute 85 for d in the formula and solve for t, the time.

$d = 16t^2$ This is the formula that models the situation.

$85 = 16t^2$ Substitute 85 for d, the distance in feet, that the stuntwoman falls.

The resulting quadratic equation is easily solved by the square root property.

$\dfrac{85}{16} = t^2$ To isolate t^2, divide both sides by 16.

$\pm\sqrt{\dfrac{85}{16}} = t$ To isolate t, use the square root property. Use double-sign notation.

$\pm\dfrac{\sqrt{85}}{\sqrt{16}} = t$ Use the quotient rule for square roots: The square root of a quotient is the quotient of square roots.

$\pm\dfrac{\sqrt{85}}{4} = t$ Evaluate: $\sqrt{16} = 4$. Since $85 = 5 \cdot 17$, we cannot simplify $\sqrt{85}$.

The stuntwoman will fall for $\dfrac{\sqrt{85}}{4}$ seconds (approximately 2.3 seconds) before making contact with the airbag. We discard the other solution, $-\dfrac{\sqrt{85}}{4}$, because a negative time does not make sense in this example.

ANSWERS TO SELF CHECKS

1. a. $\pm\sqrt{21}$ b. $\pm 3\sqrt{6}$ c. $\pm\dfrac{\sqrt{35}}{5}$ d. no real-number solutions   2. a. $10, -6$
   b. $-3 \pm 7\sqrt{2}$ c. $\dfrac{1 \pm \sqrt{3}}{3}$ 3. $7 \pm \sqrt{11}$ 4. $\sqrt{3}$ sec ≈ 1.7 sec

SECTION 9.1 STUDY SET

VOCABULARY

Fill in the blanks.

1. $x^2 - 15 = 0$ is an example of a _____ equation.
2. $x^2 + 6x + 9$ is a perfect _____ trinomial because $x^2 + 6x + 9 = (x + 3)^2$.

CONCEPTS

Fill in the blanks.

3. The square root property of equations: If $x^2 = c$, then $x = $ ____ or $x = $ ____.
4. a. If $x^2 = 5$, then $x = \pm$ ____.
   b. If $(x - 2)^2 = 7$, then $x - 2 = $ ____ $\sqrt{7}$.
5. Use a property of equality to isolate the variable on the left side of the equation.
   a. $x + 9 = \pm\sqrt{2}$
   b. $6x = 3 \pm \sqrt{2}$
6. Rationalize the denominator: $x = \pm\sqrt{\dfrac{7}{2}}$
7. Is $2\sqrt{5}$ a solution of $x^2 = 20$?
8. Is -4 a solution of $(x - 1)^2 = 25$?

NOTATION

9. Write the statement $x = \sqrt{6}$ or $x = -\sqrt{6}$ using a $\pm$ symbol (double-sign notation).
10. Fill in the blanks: $2 \pm \sqrt{3}$ is read as "Two ____ or ____ the square root of three."

GUIDED PRACTICE

Use the square root property to solve each equation, if possible. See Example 1.

11. $x^2 - 36 = 0$
12. $x^2 - 4 = 0$
13. $b^2 - 9 = 0$
14. $a^2 - 25 = 0$
15. $x^2 = \dfrac{49}{16}$
16. $x^2 = \dfrac{81}{121}$
17. $4x^2 = 400$
18. $3m^2 = 27$
19. $5x^2 = 125$
20. $4x^2 = 16$
21. $x^2 - 6 = 0$
22. $x^2 - 7 = 0$
23. $b^2 - 17 = 0$
24. $a^2 - 26 = 0$
25. $m^2 = 20$
26. $n^2 = 32$
27. $t^2 = 72$
28. $n^2 = 75$
29. $2x^2 + 8 = 23$
30. $3m^2 + 5 = 18$
31. $6r^2 - 3 = 4$
32. $2w^2 - 9 = 12$
33. $x^2 = -81$
34. $y^2 = -100$

Use the square root property to solve each equation. See Example 2.

35. $(x + 1)^2 = 25$
36. $(x - 1)^2 = 49$
37. $(x + 2)^2 = 81$
38. $(x + 3)^2 = 16$
39. $(x - 2)^2 = 8$
40. $(x + 2)^2 = 50$
41. $(s + 9)^2 = 63$
42. $(t - 11)^2 = 45$
43. $(3x + 1)^2 - 18 = 0$
44. $(6y + 5)^2 - 72 = 0$
45. $(5c - 10)^2 - 6 = 0$
46. $(4n + 8)^2 - 17 = 0$

Use the square root property to solve each equation. See Example 3.

47. $x^2 + 2x + 1 = 10$
48. $x^2 + 8x + 16 = 6$
49. $x^2 - 18x + 81 = 7$
50. $x^2 - 14x + 49 = 19$
51. $a^2 - 6a + 9 = 40$
52. $b^2 - 10b + 25 = 90$
53. $m^2 + 4m + 4 = 75$
54. $m^2 + 16m + 64 = 80$

TRY IT YOURSELF

Use the square root property to solve each equation, if possible.

55. $(x + 12)^2 = 27$
56. $(m + 1)^2 = 32$
57. $m^2 = 98$
58. $n^2 = 99$
59. $b^2 - 12b + 36 = 2$
60. $a^2 - 18a + 81 = 5$

61. $(y - 15)^2 - 8 = 0$ **62.** $(y - 5)^2 - 12 = 0$

63. $t^2 = \dfrac{1}{144}$ **64.** $d^2 = \dfrac{1}{9}$

65. $4(t - 7)^2 - 12 = 0$ **66.** $2(t - 6)^2 - 22 = 0$

67. $h^2 + 25 = 0$ **68.** $4r^2 + 16 = 0$

69. $5x^2 + 1 = 18$ **70.** $7x^2 + 3 = 6$

71. $x^2 - 14 = 0$ **72.** $x^2 - 46 = 0$

73. $(8y + 9)^2 = 44$ **74.** $(6y + 13)^2 = 99$

APPLICATIONS

75. LIGHTHOUSES The 144-foot-tall Tybee Island Lighthouse is located near Savannah, Georgia. If an object is dropped from the top of the lighthouse, how long will it take for it to hit the ground? (*Hint:* Refer to Example 4.)

76. SKYSCRAPERS A downtown office building on the north side of the Chicago River is 784 feet tall. If an object is dropped from the top of the building, how long will it take for it to hit the ground? (*Hint:* Refer to Example 4.)

77. SCIENCE HISTORY Legend has it that Galileo Galilei (1564–1642) dropped two objects having different weights from the leaning tower of Pisa in order to prove that they fall at the same rate. If a steel ball is dropped from the lowest side of the tower, and falls 183 feet, how long will it take to hit the ground? Round to the nearest tenth of a second. (*Hint:* Refer to Example 4.)

78. STUDYING MICROGRAVITY NASA's Glenn Research Center in Cleveland, Ohio, has a 435-foot drop tower that begins on the surface and descends into Earth like a mineshaft. How long will it take a sealed container to fall 435 feet? Round to the nearest tenth of a second. (*Hint:* Refer to Example 4.)

79. DAREDEVILS In 1873, Henry Bellini combined a tightrope walk with a leap into the Niagara River below, where he was picked up by a boat. If the rope was 200 feet above the water, for how many seconds did he fall before hitting the water? Round to the nearest tenth of a second. (*Hint:* Refer to Example 4.)

80. ROLLER COASTERS Soaring to a height of 420 feet, the *Top Thrill Dragster* at Cedar Point Amusement Park in Sandusky, Ohio, is one of the tallest and fastest roller coasters in the world. If a rider accidentally dropped a camera as the coaster reached its highest point, how long would it take the camera to hit the ground? Round to the nearest tenth. (*Hint:* Refer to Example 4.)

81. PRO WRESTLING A WWE (World Wrestling Entertainment) ring is square in shape and has an area of 400 ft². Find the length of a side of the ring. (*Hint:* Use the formula for the area of a square, $A = s^2$.)

82. CHESS A tournament chessboard is square in shape and has an area of 441 in². Find the length of a side of the board. (*Hint:* Use the formula for the area of a square, $A = s^2$.)

WRITING

83. Explain why the equation $x^2 + 16 = 0$ has no real-number solutions.

84. Explain why the notation $6 \pm \sqrt{2}$ represents two real numbers.

85. Explain the error in the following work.

Solve: $x^2 = 7$

$x = \sqrt{7}$

86. Explain the error in the following work.

Solve: $x^2 = 28$

$x = \pm\sqrt{28}$

$x = 2 \pm \sqrt{7}$

REVIEW

Solve each equation.

87. $\sqrt{5x - 6} = 2$ **88.** $\sqrt{6x + 1} + 2 = 7$

89. $2\sqrt{x} = \sqrt{5x - 16}$ **90.** $\sqrt{22y + 86} = y + 9$

SECTION 9.2
Solving Quadratic Equations: Completing the Square

Objectives
1. Complete the square to write perfect-square trinomials.
2. Solve quadratic equations with leading coefficients of 1 by completing the square.
3. Solve quadratic equations with leading coefficients other than 1 by completing the square.

In Section 9.1, we used the square root property to solve equations, such as $(x - 3)^2 = 36$ and $(x + 1)^2 = 50$, whose left side is a binomial squared and right side is a constant. We solved equations such as $x^2 + 16x + 64 = 2$ in a similar way by first factoring the perfect-square trinomial on the left side.

In this section, we will discuss a procedure that enables us to solve quadratic equations such as $x^2 + 4x = -3$, whose left side is not a perfect-square trinomial. To make the left side a perfect-square trinomial, we will use a procedure called *completing the square*.

1 Complete the square to write perfect-square trinomials.

Consider the following perfect-square trinomials and their factored forms.

$$x^2 + 2bx + b^2 = (x + b)^2 \quad \text{and} \quad x^2 - 2bx + b^2 = (x - b)^2$$

In each of these perfect-square trinomials, the third term is the square of one-half of the coefficient of x.

- In $x^2 + 2bx + b^2$, the coefficient of x is $2b$. If we find $\frac{1}{2} \cdot 2b$, which is b, and square it, we get the third term, b^2.
- In $x^2 - 2bx + b^2$, the coefficient of x is $-2b$. If we find $\frac{1}{2}(-2b)$, which is $-b$, and square it, we get the third term: $(-b)^2 = b^2$.

We can use these observations to change certain binomials into perfect-square trinomials. For example, to change $x^2 + 12x$ into a perfect-square trinomial, we find one-half of the coefficient of x, square the result, and add the square to $x^2 + 12x$.

$$x^2 + 12x + \square$$

First, find one-half of the coefficient of x.

$$\frac{1}{2} \cdot 12 = 6 \qquad 6^2 = 36$$

Then square the result.

Finally, add the square to the binomial to create a trinomial.

We obtain the perfect-square trinomial $x^2 + 12x + 36$ that factors as $(x + 6)^2$. By adding 36 to $x^2 + 12x$, we **completed the square** on $x^2 + 12x$.

> **Completing the Square**
>
> To complete the square on $x^2 + bx$, add the square of one-half of the coefficient of x:
>
> $$x^2 + bx + \left(\frac{1}{2}b\right)^2$$

EXAMPLE 1 Complete the square and factor the resulting perfect-square trinomial: **a.** $x^2 + 4x$ **b.** $x^2 - 6x$ **c.** $x^2 - 5x$

Strategy In each case, we will add the square of one-half the coefficient of x to the given binomial.

WHY Adding such a term will change the binomial into a perfect-square trinomial that will factor.

Self Check 1

Complete the square and factor the resulting perfect-square trinomial:
a. $y^2 + 6y$
b. $y^2 - 8y$
c. $y^2 + 3y$

Now Try Problems 18, 19, and 21

Solution

a. Since the coefficient of x is 4, we add the square of one-half of 4.

$$x^2 + 4x + \left[\frac{1}{2}(4)\right]^2 = x^2 + 4x + (2)^2 \quad \text{Multiply: } \tfrac{1}{2}(4) = 2.$$

$$= x^2 + 4x + 4 \quad \text{Square 2 to get 4.}$$

This perfect-square trinomial factors as $(x + 2)^2$.

b. Since the coefficient of x is -6, we add the square of one-half of -6.

$$x^2 - 6x + \left[\frac{1}{2}(-6)\right]^2 = x^2 - 6x + (-3)^2 \quad \text{Multiply: } \tfrac{1}{2}(-6) = -3.$$

$$= x^2 - 6x + 9 \quad \text{Square } -3 \text{ to get 9.}$$

This perfect-square trinomial factors as $(x - 3)^2$.

c. Since the coefficient of x is -5, we add the square of one-half of -5.

$$x^2 - 5x + \left[\frac{1}{2}(-5)\right]^2 = x^2 - 5x + \left(-\frac{5}{2}\right)^2 \quad \text{Multiply: } \tfrac{1}{2}(-5) = -\tfrac{5}{2}.$$

$$= x^2 - 5x + \frac{25}{4} \quad \text{Square } -\tfrac{5}{2} \text{ to get } \tfrac{25}{4}.$$

This perfect-square trinomial factors as $\left(x - \frac{5}{2}\right)^2$.

2 Solve quadratic equations with leading coefficients of 1 by completing the square.

If the quadratic equation $ax^2 + bx + c = 0$ has a leading coefficient of 1, it's easy to solve by completing the square.

Self Check 2

Solve $x^2 + 5x = 3$ by completing the square. Approximate the solutions to the nearest hundredth.

Now Try Problem 33

EXAMPLE 2 Solve $x^2 - 7x = 2$ by completing the square.

Strategy We will use the addition property of equality and add the square of one-half of the coefficient of x to both sides of the equation.

WHY This will create a perfect-square trinomial on the left side that will factor as the square of a binomial. Then we can use the square root property to solve for x.

Solution

$$x^2 - 7x = 2 \quad \text{This is the equation to solve.}$$

$$x^2 - 7x + \left[\frac{1}{2}(-7)\right]^2 = 2 + \left[\frac{1}{2}(-7)\right]^2 \quad \text{Since the coefficient of } x \text{ is } -7, \text{ add the square of one-half of } -7 \text{ to both sides.}$$

$$x^2 - 7x + \frac{49}{4} = 2 + \frac{49}{4} \quad \tfrac{1}{2}(-7) = -\tfrac{7}{2}. \text{ Then square } -\tfrac{7}{2} \text{ to get } \tfrac{49}{4}.$$

$$\left(x - \frac{7}{2}\right)^2 = \frac{8}{4} + \frac{49}{4} \quad \text{Factor the left side. Write 2 as } \tfrac{8}{4}.$$

$$\left(x - \frac{7}{2}\right)^2 = \frac{57}{4} \quad \text{The fractions have a common denominator. Add them.}$$

$$x - \frac{7}{2} = \pm\sqrt{\frac{57}{4}} \quad \text{Use the square root property to solve for } x - \tfrac{7}{2}.$$

$$x - \frac{7}{2} = \pm\frac{\sqrt{57}}{2} \quad \text{Simplify the radical:}$$

$$\sqrt{\tfrac{57}{4}} = \tfrac{\sqrt{57}}{\sqrt{4}} = \tfrac{\sqrt{57}}{2}.$$

9.2 Solving Quadratic Equations: Completing the Square

$$x = \frac{7}{2} \pm \frac{\sqrt{57}}{2} \quad \text{Add } \tfrac{7}{2} \text{ to both sides.}$$

$$x = \frac{7 \pm \sqrt{57}}{2} \quad \text{Since the fractions have a common denominator of 2, we can combine them.}$$

The solutions are $\frac{7 \pm \sqrt{57}}{2}$, and the solution set is $\left\{\frac{7 \pm \sqrt{57}}{2}\right\}$. If we approximate the solutions to the nearest hundredth, we have

$$\frac{7 + \sqrt{57}}{2} \approx 7.27 \qquad \frac{7 - \sqrt{57}}{2} \approx -0.27$$

EXAMPLE 3 Solve $x^2 + 4x - 13 = 0$ by completing the square. Give the exact solutions, and then approximate them to the nearest hundredth.

Strategy We will use the addition property of equality and add 13 to both sides. Then we will proceed as in Example 2.

WHY To prepare to complete the square, we need to isolate the variable terms, x^2 and $4x$, on the left side of the equation and the constant term on the right.

Solution
Since the coefficient of x^2 is 1, we can complete the square as follows:

$$x^2 + 4x - 13 = 0 \quad \text{This is the equation to solve.}$$
$$x^2 + 4x = 13 \quad \text{Add 13 to both sides so that the constant term is on the right side.}$$

We then find one-half of the coefficient of x, square it, and add the result to both sides to make the left side a perfect-square trinomial.

$$x^2 + 4x + \left[\tfrac{1}{2}(4)\right]^2 = 13 + \left[\tfrac{1}{2}(4)\right]^2 \quad \text{Since the coefficient of } x \text{ is 4, add the square of one-half of 4 to both sides.}$$
$$x^2 + 4x + 4 = 13 + 4 \quad \tfrac{1}{2}(4) = 2. \text{ Then square 2 to get 4.}$$
$$(x + 2)^2 = 17 \quad \text{Factor } x^2 + 4x + 4 \text{ and simplify.}$$
$$x + 2 = \pm\sqrt{17} \quad \text{Use the square root method to solve for } x + 2.$$
$$x = -2 \pm \sqrt{17} \quad \text{Subtract 2 from both sides to isolate } x. \text{ Write } -2 \text{ in front of the radical.}$$

We can use a calculator to approximate each solution.

$$x = -2 + \sqrt{17} \qquad \text{or} \qquad x = -2 - \sqrt{17}$$
$$x \approx -2 + 4.123105626 \qquad \qquad x \approx -2 - 4.123105626$$
$$x \approx 2.12 \qquad \qquad x \approx -6.12$$

Self Check 3
Solve $x^2 + 10x - 4 = 0$ by completing the square. Give the exact solutions, and then approximate them to the nearest hundredth.

Now Try Problem 41

3 Solve quadratic equations with leading coefficients other than 1 by completing the square.

If the quadratic equation $ax^2 + bx + c = 0$ has a leading coefficient other than 1, we can make the leading coefficient 1 by dividing both sides of the equation by a.

EXAMPLE 4 Solve $4x^2 + 4x - 3 = 0$ by completing the square.

Strategy We will use the addition property of equality to get the variable terms on one side of the equation and the constant term on the other. Then we will use the division property of equality and divide both sides by 4 so that the coefficient of x^2 is 1.

Self Check 4
Solve $2x^2 - 5x - 3 = 0$ by completing the square.

Now Try Problem 46

WHY This will create a leading coefficient of 1 so that we can complete the square to solve the equation.

Solution

$4x^2 + 4x - 3 = 0$	This is the equation to solve.
$x^2 + x - \dfrac{3}{4} = 0$	Divide both sides by 4: $\dfrac{4x^2}{4} + \dfrac{4x}{4} - \dfrac{3}{4} = \dfrac{0}{4}$.
$x^2 + x = \dfrac{3}{4}$	Add $\dfrac{3}{4}$ to both sides so that the constant term is on the right side.
$x^2 + 1x + \left[\dfrac{1}{2}(1)\right]^2 = \dfrac{3}{4} + \left[\dfrac{1}{2}(1)\right]^2$	Since the coefficient of x is 1, add the square of one-half of 1 to both sides.
$x^2 + x + \dfrac{1}{4} = \dfrac{3}{4} + \dfrac{1}{4}$	$\dfrac{1}{2}(1) = \dfrac{1}{2}$. Then square $\dfrac{1}{2}$ to get $\dfrac{1}{4}$.
$\left(x + \dfrac{1}{2}\right)^2 = 1$	Factor the trinomial. Add the fractions.
$x + \dfrac{1}{2} = \pm 1$	Solve for $x + \dfrac{1}{2}$ using the square root method.
$x = -\dfrac{1}{2} \pm 1$	Subtract $\dfrac{1}{2}$ from both sides to isolate x.

$x = -\dfrac{1}{2} + 1 \quad \text{or} \quad x = -\dfrac{1}{2} - 1$

$x = \dfrac{1}{2} \qquad\qquad\qquad x = -\dfrac{3}{2}$

The solutions are $\dfrac{1}{2}$ and $-\dfrac{3}{2}$, and the solution set is $\left\{-\dfrac{3}{2}, \dfrac{1}{2}\right\}$. Check each one.

Success Tip In Example 4, you may have noticed that $4x^2 + 4x - 3$ can be factored. Therefore, we could have solved $4x^2 + 4x - 3 = 0$ by factoring. This example illustrates an important fact: Completing the square can be used to solve any quadratic equation.

The previous examples illustrate that to solve a quadratic equation by completing the square, we follow these steps.

Completing the Square to Solve a Quadratic Equation in x

1. If the coefficient of x^2 is 1, go to step 2. If it is not 1, make it 1 by dividing both sides of the equation by the coefficient of x^2.
2. Get all variable terms on one side of the equation and constants on the other side.
3. Complete the square by finding one-half of the coefficient of x, squaring the result, and adding the square to both sides of the equation.
4. Factor the perfect-square trinomial as the square of a binomial.
5. Solve the resulting equation using the square root property.
6. Check your answers in the original equation.

9.2 Solving Quadratic Equations: Completing the Square

EXAMPLE 5 Solve $2x^2 - 2 = 4x$ by completing the square.

Strategy We will use the addition and subtraction properties of equality to get the variable terms on one side of the equation and the constant term on the other. Then we will use the division property of equality and divide both sides by 2 so that the coefficient of x^2 is 1.

WHY This will create a leading coefficient of 1 so that we can complete the square to solve the equation.

Solution

$2x^2 - 2 = 4x$ This is the equation to solve.

$2x^2 - 4x - 2 = 0$ Subtract 4x from both sides to get 0 on the right side.

$x^2 - 2x - 1 = 0$ Divide both sides by 2: $\frac{2x^2}{2} - \frac{4x}{2} - \frac{2}{2} = \frac{0}{2}$.

Since this equation cannot be solved by factoring, we complete the square.

$x^2 - 2x = 1$ Add 1 to both sides.

$x^2 - 2x + \left[\frac{1}{2}(-2)\right]^2 = 1 + \left[\frac{1}{2}(-2)\right]^2$ Since the coefficient of x is −2, add the square of one-half of −2 to both sides.

$x^2 - 2x + 1 = 1 + 1$ $\frac{1}{2}(-2) = -1$. Then square −1 to get 1.

$(x - 1)^2 = 2$ Factor the trinomial and simplify.

$x - 1 = \pm\sqrt{2}$ Use the square root method to solve for x − 1.

$x = 1 \pm \sqrt{2}$ Add 1 to both sides.

The solutions are $1 \pm \sqrt{2}$, and the solution set is $\{1 \pm \sqrt{2}\}$. Check each one.

Self Check 5

Solve $3x^2 - 18x = -12$ by completing the square.

Now Try Problem 51

> **Caution!** A common error is to add a constant to one side of an equation to complete the square and forget to add it to the other side.

EXAMPLE 6 Solve $4x^2 - 3 = 24x$ by completing the square. Give the exact solutions, and then approximate them to the nearest hundredth.

Strategy We will use the addition and subtraction properties of equality to get the variable terms on one side of the equation and the constant term on the other. Then we will use the division property of equality and divide both sides by 4 so that the coefficient of x^2 is 1.

WHY This will create a leading coefficient of 1 so that we can complete the square to solve the equation.

Solution

$4x^2 - 3 = 24x$ This is the equation to solve.

$4x^2 - 24x = 3$ To have both variable terms on the left side, subtract 24x from both sides. To have the constant term on the right, add 3 to both sides.

$x^2 - 6x = \frac{3}{4}$ To make the coefficient of the x^2 term 1, divide both sides by 4: $\frac{4x^2}{4} - \frac{24x}{4} = \frac{3}{4}$.

$x^2 - 6x + 9 = \frac{3}{4} + 9$ Complete the square: $\frac{1}{2}(-6) = -3$ and $(-3)^2 = 9$. Add 9 to both sides.

$(x - 3)^2 = \frac{39}{4}$ On the left side, factor. On the right side, express 9 as $\frac{36}{4}$ and add to $\frac{3}{4}$ to get $\frac{39}{4}$.

$x - 3 = \pm\sqrt{\frac{39}{4}}$ Use the square root property.

Self Check 6

Solve $4d^2 - 1 = 32d$ by completing the square. Give the exact solutions, and then approximate them to the nearest hundredth.

Now Try Problem 55

$$x - 3 = \pm \frac{\sqrt{39}}{2}$$ Use the quotient rule to simplify: $\sqrt{\frac{39}{4}} = \frac{\sqrt{39}}{\sqrt{4}} = \frac{\sqrt{39}}{2}$.

$$x = 3 \pm \frac{\sqrt{39}}{2}$$ To isolate x, add 3 to both sides.

$$x = \frac{6}{2} \pm \frac{\sqrt{39}}{2}$$ To write the solutions in compact form, express 3 as a fraction with denominator 2: $3 = \frac{3}{1} \cdot \frac{2}{2} = \frac{6}{2}$.

$$x = \frac{6 \pm \sqrt{39}}{2}$$ Write the sum (and difference) over the common denominator 2.

The exact solutions are $\frac{6 \pm \sqrt{39}}{2}$, and the solution set is $\left\{\frac{6 \pm \sqrt{39}}{2}\right\}$. Check each one in the original equation. We can approximate the solutions using a calculator. To the nearest hundredth, we have

$$\frac{6 + \sqrt{39}}{2} \approx 6.12 \qquad \frac{6 - \sqrt{39}}{2} \approx -0.12$$

ANSWERS TO SELF CHECKS

1. a. $y^2 + 6y + 9 = (y + 3)^2$ **b.** $y^2 - 8y + 16 = (y - 4)^2$ **c.** $y^2 + 3y + \frac{9}{4} = \left(y + \frac{3}{2}\right)^2$

2. $\frac{-5 \pm \sqrt{37}}{2}$; 0.54, −5.54   **3.** $-5 \pm \sqrt{29}$; 0.39, −10.39   **4.** $3, -\frac{1}{2}$ **5.** $3 \pm \sqrt{5}$

6. $\frac{8 \pm \sqrt{65}}{2}$; 8.03, −0.03

SECTION 9.2   STUDY SET

VOCABULARY

Fill in the blanks.

1. Since $x^2 + 12x + 36 = (x + 6)^2$, we call the trinomial a perfect-_____ trinomial.
2. When we add 9 to $x^2 + 6x$, we say that we have completed the _____ on $x^2 + 6x$.
3. The _____ coefficient of $5x^2 - 4x + 8 = 0$ is 5 and the _____ term is 8.
4. If the polynomial in the equation $ax^2 + bx + c = 0$ doesn't factor, we can solve the equation by _____ the square.

CONCEPTS

Fill in the blanks.

5. Find one-half of the given number and then square the result.
   - **a.** 6
   - **b.** −12
   - **c.** 3
   - **d.** −5
6. To complete the square on $x^2 + 8x$, we add the _____ of one-half of 8, which is 16.
7. To complete the square on $x^2 - 10x$, we add the square of _____ of −10, which is 25.
8. The solutions of $x^2 = c$, where $c > 0$, are _____ and _____.
9. If $(x - 2)^2 = 7$, then $x - 2 = \pm$ _____.
10. What is the first step if we solve $x^2 - 2x = 35$
    - **a.** by the factoring method?
    - **b.** by completing the square?
11. Why can't $x^2 - 2x - 1 = 0$ be solved by the factoring method?
12. Find the result when both sides of $2x^2 + 4x - 8 = 0$ are divided by 2.

NOTATION

Complete each step to solve the equation.

13. $(y - 1)^2 = 9$

$y - 1 = $ _____   or $y - 1 = -\sqrt{9}$

_____ $= 3$ $y - 1 = $ _____

$y = 4$ $y = -2$

14. $y^2 + 2y - 3 = 0$
$y^2 + 2y = \square$
$y^2 + 2y + 1 = 3 + \square$
$(y + 1)^2 = \square$
$\square = \sqrt{4}$ or $y + 1 = -\square$
$y + 1 = \square \quad | \quad \square = -2$
$y = 1 \quad | \quad y = -3$

15. a. In solving a quadratic equation, a student obtains $x = \pm\sqrt{10}$. How many solutions are represented by this notation? List them.

b. In solving a quadratic equation, a student obtains $x = 8 \pm \sqrt{3}$. List each solution separately. Round each one to the nearest hundredth.

16. Solve $x + 1 = \pm\sqrt{2}$ for x.

GUIDED PRACTICE

Complete the square and factor the perfect-square trinomial. See Example 1.

17. $x^2 + 2x$ **18.** $x^2 + 12$

19. $x^2 - 4x$ **20.** $x^2 - 14x$

21. $x^2 + 7x$ **22.** $x^2 + 21x$

23. $x^2 + x$ **24.** $x^2 - x$

25. $a^2 - 3a$ **26.** $b^2 - 13b$

27. $b^2 + \frac{2}{3}b$ **28.** $c^2 - \frac{5}{2}c$

Solve each equation by completing the square. See Example 2.

29. $g^2 + 5g = 6$ **30.** $s^2 + 5s = 14$
31. $x^2 + 6x = -8$ **32.** $x^2 + 8x = -12$
33. $a^2 + 4a = 5$ **34.** $y^2 + 6y = 7$
35. $h^2 - 2h = 15$ **36.** $k^2 - 8k = -12$

Solve each equation by completing the square. Give the exact solutions, and then approximate them to the nearest hundredth. See Example 3.

37. $x^2 + 8x - 6 = 0$ **38.** $x^2 + 6x - 2 = 0$

39. $x^2 + 6x + 4 = 0$ **40.** $x^2 + 8x + 6 = 0$

41. $x^2 + 4x + 1 = 0$ **42.** $x^2 + 6x + 2 = 0$

43. $x^2 - 2x - 4 = 0$ **44.** $x^2 - 4x - 2 = 0$

Solve each equation by completing the square. See Example 4.

45. $3x^2 + 5x - 2 = 0$ **46.** $4x^2 - 4x - 3 = 0$

47. $6x^2 - 5x - 6 = 0$ **48.** $10x^2 + 21x - 10 = 0$

Solve each equation by completing the square. See Example 5.

49. $x^2 = 4x + 3$ **50.** $x^2 = 6x - 3$

51. $4x^2 + 4x + 1 = 20$ **52.** $9x^2 = 8 - 12x$

Solve each equation by completing the square. Give the exact solutions, and then approximate them to the nearest hundredth. See Example 6.

53. $t^2 - 2t - 4 = 0$ **54.** $m^2 - 4m - 2 = 0$

55. $2t^2 = -6t - 1$ **56.** $3p^2 = -4p + 3$

57. $4x^2 + 4x + 1 = 20$ **58.** $9x^2 = 8 - 12x$

59. $3q^2 - 4 = -2q$ **60.** $2y^2 + 3 = 10y$

TRY IT YOURSELF

Solve each equation by completing the square.

61. $2x^2 = 4 - 2x$ **62.** $3q^2 = 3q + 6$

63. $k^2 - 8k + 12 = 0$ **64.** $p^2 - 4p + 3 = 0$

65. $x^2 - 2x = 15$ **66.** $x^2 - 2x = 8$

67. $3x^2 + 9x + 6 = 0$ **68.** $3d^2 + 48 = -24d$

69. $2x^2 = 3x + 2$ **70.** $3x^2 = 2 - 5x$

71. $4x^2 = 2 - 7x$ **72.** $2x^2 = 5x + 3$

73. $4x^2 - 24x - 13 = 0$ **74.** $4d^2 - 16d - 9 = 0$

75. $4a^2 - 9a + 1 = 0$ **76.** $4a^2 - 11a + 1 = 0$

77. $2x^2 + 6x = 8$ **78.** $3x^2 - 6x = 9$

79. $6x^2 - 6 = 5x$ **80.** $6x^2 - 4 = 5x$

81. $x^2 + 3x - \frac{1}{2} = -2$ **82.** $x^2 + x = 3x - \frac{2}{3}$

APPLICATIONS

83. CAROUSELS In 1999, the city of Lancaster, Pennsylvania, considered installing a classic Dentzel carousel in an abandoned downtown building. After learning that the circular carousel would occupy 2,376 square feet of floor space and that it was 26 feet high, the proposal was determined to be impractical because of the large remodeling costs. Find the diameter of the carousel to the nearest foot.

84. ESCAPE VELOCITY The speed at which a rocket must be fired for it to leave the Earth's gravitational attraction is called the **escape velocity**. If the escape velocity v_e, in mph, is given by

$$\frac{v_e^2}{2g} = R$$

where $g = 78,545$ and $R = 3,960$, find v_e. Round to the nearest mi/hr.

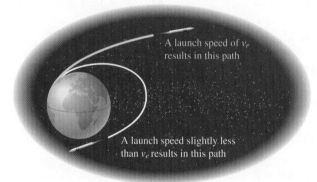

85. BICYCLE SAFETY See the illustration in the next column. A bicycle training program for children uses a figure-8 course to help them improve their balance and steering. The course is laid out over a paved area covering 800 square feet. Find its dimensions.

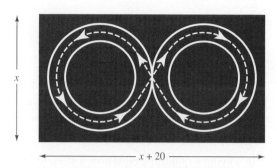

86. BADMINTON The badminton court shown below occupies 880 square feet of the floor space of a gymnasium. If its length is 4 feet more than twice its width, find its dimensions.

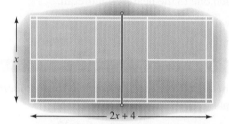

WRITING

87. Explain how to complete the square on $x^2 - 5x$.

88. Explain the error in the following work.

Solve:
$$x^2 = 28$$
$$x = \pm\sqrt{28}$$
$$x = 2 \pm \sqrt{7}$$

89. Rounded to the nearest hundredth, one solution of the equation $x^2 + 4x + 1 = 0$ is -0.27. Use your calculator to check it. How could it be a solution if it doesn't make the left side zero? Explain.

90. Give an example of a perfect-square trinomial. Why do you think the word *perfect* is used to describe it?

REVIEW

Perform each operation.

91. $(y - 1)^2$ **92.** $(z + 2)^2$

93. $(x + y)^2$ **94.** $(a - b)^2$

95. $(2z)^2$ **96.** $(xy)^2$

SECTION 9.3
Solving Quadratic Equations: The Quadratic Formula

Objectives
1. Use the quadratic formula to solve quadratic equations.
2. Identify quadratic equations with no real-number solutions.
3. Determine the most efficient method to solve a quadratic equation.
4. Solve problems modeled by quadratic equations.

We can solve any quadratic equation by completing the square, but the work is often lengthy and involved. In this section, we will develop a formula, called the *quadratic formula*, that will enable us to solve quadratic equations with much less effort.

1 Use the quadratic formula to solve quadratic equations.

We can solve the **general quadratic equation** $ax^2 + bx + c = 0$, where $a \neq 0$, by completing the square.

$$ax^2 + bx + c = 0$$

$$\frac{ax^2}{a} + \frac{bx}{a} + \frac{c}{a} = \frac{0}{a} \quad \text{Divide both sides by } a \text{ so that the coefficient of } x^2 \text{ is 1.}$$

$$x^2 + \frac{b}{a}x + \frac{c}{a} = 0 \quad \text{Simplify } \frac{ax^2}{a} = x^2. \text{ Write } \frac{bx}{a} \text{ as } \frac{b}{a}x.$$

$$x^2 + \frac{b}{a}x = -\frac{c}{a} \quad \text{Subtract } \frac{c}{a} \text{ from both sides.}$$

Since the coefficient of x is $\frac{b}{a}$, we can complete the square on $x^2 + \frac{b}{a}x$ by adding $\left(\frac{1}{2} \cdot \frac{b}{a}\right)^2$, which is $\frac{b^2}{4a^2}$ to both sides:

$$x^2 + \frac{b}{a}x + \frac{b^2}{4a^2} = \frac{b^2}{4a^2} - \frac{c}{a}$$

After factoring the perfect-square trinomial on the left side, we have

$$\left(x + \frac{b}{2a}\right)\left(x + \frac{b}{2a}\right) = \frac{b^2}{4a^2} - \frac{4ac}{4aa} \quad \text{The lowest common denominator on the right side is } 4a^2. \text{ Build the second fraction.}$$

$$\left(x + \frac{b}{2a}\right)^2 = \frac{b^2 - 4ac}{4a^2} \quad \text{Subtract the numerators and write the difference over the common denominator.}$$

The resulting equation can be solved by the square root method to obtain

$$x + \frac{b}{2a} = \sqrt{\frac{b^2 - 4ac}{4a^2}} \quad \text{or} \quad x + \frac{b}{2a} = -\sqrt{\frac{b^2 - 4ac}{4a^2}}$$

$$x + \frac{b}{2a} = \frac{\sqrt{b^2 - 4ac}}{\sqrt{4a^2}} \quad \quad x + \frac{b}{2a} = -\frac{\sqrt{b^2 - 4ac}}{\sqrt{4a^2}}$$

$$x = -\frac{b}{2a} + \frac{\sqrt{b^2 - 4ac}}{2a} \quad \quad x = -\frac{b}{2a} - \frac{\sqrt{b^2 - 4ac}}{2a}$$

$$x = \frac{-b + \sqrt{b^2 - 4ac}}{2a} \quad \quad x = \frac{-b - \sqrt{b^2 - 4ac}}{2a}$$

These solutions are usually written in one formula called the **quadratic formula**.

The Quadratic Formula

The solutions of the quadratic equation $ax^2 + bx + c = 0$ are

$$x = \frac{-b \pm \sqrt{b^2 - 4ac}}{2a} \quad \text{where } a \neq 0$$

Caution! When you write the quadratic formula, draw the fraction bar so that it includes the complete numerator. Do not write

$$x = -b \pm \frac{\sqrt{b^2 - 4ac}}{2a}$$

Self Check 1
Solve: $x^2 + 6x + 5 = 0$
Now Try Problem 29

EXAMPLE 1 Solve: $x^2 + 5x + 6 = 0$

Strategy We will compare the given equation to the general quadratic equation $ax^2 + bx + c = 0$ to identify a, b, and c.

WHY To use the quadratic formula, we need to know what numbers to substitute for a, b, and c in $x = \dfrac{-b \pm \sqrt{b^2 - 4ac}}{2a}$.

Solution
The equation is written in $ax^2 + bx + c = 0$ form with $a = 1$, $b = 5$, and $c = 6$.

$$\mathbf{1}x^2 + \mathbf{5}x + \mathbf{6} = 0$$
$$ax^2 + bx + c = 0$$

To find the solutions, we substitute these values into the quadratic formula and evaluate the right side.

$x = \dfrac{-b \pm \sqrt{b^2 - 4ac}}{2a}$ This is the quadratic formula.

$x = \dfrac{-5 \pm \sqrt{5^2 - 4(1)(6)}}{2(1)}$ Substitute 1 for a, 5 for b, and 6 for c.

$x = \dfrac{-5 \pm \sqrt{25 - 24}}{2}$ Evaluate the power and multiply within the radical symbol.

$x = \dfrac{-5 \pm \sqrt{1}}{2}$ Do the subtraction within the radical symbol.

$x = \dfrac{-5 \pm 1}{2}$ Simplify: $\sqrt{1} = 1$.

This notation represents two solutions. We simplify them separately, first using the $+$ sign and then using the $-$ sign.

$$x = \frac{-5 + 1}{2} \quad \text{or} \quad x = \frac{-5 - 1}{2}$$
$$x = \frac{-4}{2} \quad\quad\quad\quad x = \frac{-6}{2}$$
$$x = -2 \quad\quad\quad\quad\quad x = -3$$

The solutions are -2 and -3, and the solution set is $\{-3, -2\}$.

9.3 Solving Quadratic Equations: The Quadratic Formula

Success Tip In Example 1, you may have noticed that we could have solved $x^2 + 5x + 6 = 0$ by factoring.

EXAMPLE 2 Solve: $2x^2 = 5x + 3$

Strategy We will use the subtraction property of equality to get 0 on the right side of the equation. Then we will compare the resulting equation to the general quadratic equation $ax^2 + bx + c = 0$ to identify a, b, and c.

WHY To use the quadratic formula, we need to know what numbers to substitute for a, b, and c in $x = \dfrac{-b \pm \sqrt{b^2 - 4ac}}{2a}$.

Solution

$2x^2 = 5x + 3$ This is the equation to solve.

$2x^2 - 5x - 3 = 0$ Subtract 5x and 3 from both sides.

In this equation, $a = 2$, $b = -5$, and $c = -3$. To find the solutions, we substitute these values into the quadratic formula and evaluate the right side.

$x = \dfrac{-b \pm \sqrt{b^2 - 4ac}}{2a}$ This is the quadratic formula.

$x = \dfrac{-(-5) \pm \sqrt{(-5)^2 - 4(2)(-3)}}{2(2)}$ Substitute 2 for a, −5 for b, and −3 for c.

$x = \dfrac{5 \pm \sqrt{25 - (-24)}}{4}$ −(−5) = 5. Evaluate the power and multiply within the radical symbol.

$x = \dfrac{5 \pm \sqrt{49}}{4}$ Do the subtraction within the radical symbol: 25 − (−24) = 25 + 24 = 49.

$x = \dfrac{5 \pm 7}{4}$ Simplify: $\sqrt{49} = 7$.

Thus,

$x = \dfrac{5 + 7}{4}$ or $x = \dfrac{5 - 7}{4}$

$x = \dfrac{12}{4}$ $x = \dfrac{-2}{4}$

$x = 3$ $x = -\dfrac{1}{2}$

The solutions are 3 and $-\dfrac{1}{2}$. Check each one in the original equation.

Self Check 2
Solve: $4x^2 - 11x = 3$
Now Try Problem 33

EXAMPLE 3 Solve $3x^2 = 2x + 4$. Round each solution to the nearest hundredth.

Strategy We will use the subtraction property of equality to get 0 on the right side of the equation. Then we will compare the resulting equation to the general quadratic equation $ax^2 + bx + c = 0$ to identify a, b, and c.

WHY To use the quadratic formula, we need to know what numbers to substitute for a, b, and c in $x = \dfrac{-b \pm \sqrt{b^2 - 4ac}}{2a}$.

Solution
We begin by writing the given equation in $ax^2 + bx + c = 0$ form.

Self Check 3
Solve $2x^2 - 1 = 2x$. Round each solution to the nearest hundredth.
Now Try Problem 37

$$3x^2 = 2x + 4 \quad \text{This is the equation to solve.}$$
$$3x^2 - 2x - 4 = 0 \quad \text{Subtract } 2x \text{ and } 4 \text{ from both sides.}$$

In this equation, $a = 3$, $b = -2$, and $c = -4$. To find the solutions, we substitute these values into the quadratic formula and evaluate the right-hand side.

$$x = \frac{-b \pm \sqrt{b^2 - 4ac}}{2a} \quad \text{This is the quadratic formula.}$$

$$x = \frac{-(-2) \pm \sqrt{(-2)^2 - 4(3)(-4)}}{2(3)} \quad \text{Substitute 3 for } a, -2 \text{ for } b, \text{ and } -4 \text{ for } c.$$

$$x = \frac{2 \pm \sqrt{4 + 48}}{6} \quad -(-2) = 2. \text{ Simplify within the radical symbol.}$$

$$x = \frac{2 \pm \sqrt{52}}{6} \quad \text{Do the addition within the radical symbol.}$$

$$x = \frac{2 \pm 2\sqrt{13}}{6} \quad \sqrt{52} = \sqrt{4 \cdot 13} = 2\sqrt{13}.$$

$$x = \frac{\overset{1}{\cancel{2}}(1 \pm \sqrt{13})}{\underset{1}{\cancel{2}} \cdot 3} \quad \text{In the numerator, factor out 2: } 2 \pm 2\sqrt{13} = 2(1 \pm \sqrt{13}). \text{ Write 6 as } 2 \cdot 3. \text{ Then remove the common factor of 2.}$$

$$x = \frac{1 \pm \sqrt{13}}{3} \quad \text{Simplify.}$$

The solutions are $\frac{1 \pm \sqrt{13}}{3}$, and the solution set is $\left\{\frac{1 \pm \sqrt{13}}{3}\right\}$. We can use a calculator to approximate each of them. To the nearest hundredth,

$$\frac{1 + \sqrt{13}}{3} \approx 1.54 \qquad \frac{1 - \sqrt{13}}{3} \approx -0.87$$

2 Identify quadratic equations with no real-number solutions.

The next example shows that some quadratic equations have no real-number solutions.

Self Check 4

Does the equation

$$2x^2 + x + 1 = 0$$

have any real-number solutions?

Now Try Problem 41

EXAMPLE 4 Does the equation $x^2 + 2x + 5 = 0$ have any real-number solutions?

Strategy We will compare the given equation to the general quadratic equation $ax^2 + bx + c = 0$ to identify a, b, and c.

WHY To use the quadratic formula, we need to know what numbers to substitute for a, b, and c in $x = \frac{-b \pm \sqrt{b^2 - 4ac}}{2a}$.

Solution
In this equation $a = 1$, $b = 2$, and $c = 5$. We substitute these values into the quadratic formula.

$$x = \frac{-b \pm \sqrt{b^2 - 4ac}}{2a} \quad \text{This is the quadratic formula.}$$

$$x = \frac{-2 \pm \sqrt{2^2 - 4(1)(5)}}{2(1)} \quad \text{Substitute 1 for } a, 2 \text{ for } b, \text{ and 5 for } c.$$

$$x = \frac{-2 \pm \sqrt{4 - 20}}{2} \quad \text{Evaluate the power and multiply within the radical symbol.}$$

$$x = \frac{-2 \pm \sqrt{-16}}{2} \quad \text{Do the subtraction within the radical symbol. The result is a negative number, } -16.$$

Since $\sqrt{-16}$ is not a real number, there are no real-number solutions.

3 Determine the most efficient method to solve a quadratic equation.

We have discussed four methods that are used to solve quadratic equations. The following table shows some advantages and disadvantages of each method.

Method	Advantages	Disadvantages	Examples
Factoring and the zero-factor property	It can be very fast. When each factor is set equal to 0, the resulting equations are usually easy to solve.	Some polynomials may be difficult to factor and others impossible.	$x^2 - 2x - 24 = 0$ $4a^2 + a = 0$
Square root property	It is the fastest way to solve equations of the form $ax^2 = n$ or $(ax + b)^2 = n$, where n is a number.	It only applies to equations that are in these forms.	$x^2 = 27$ $(2y + 3)^2 = 25$
Completing the square*	It can be used to solve any quadratic equation. It works well with equations of the form $x^2 + bx = n$, where b is even.	It involves more steps than the other methods. The algebra can be cumbersome if the leading coefficient is not 1.	$t^2 - 14t = 9$ $x^2 + 4x + 1 = 0$
Quadratic formula	It can be used to solve any quadratic equation.	It involves several computations where sign errors can be made. Often the result must be simplified.	$x^2 + 3x - 33 = 0$ $4s^2 - 10s + 5 = 0$

*The quadratic formula is just a condensed version of completing the square and is usually easier to use. However, you need to know how to complete the square because it is used in more advanced mathematics courses.

To determine the most efficient method for a given equation, we can use the following strategy.

Strategy for Solving Quadratic Equations

1. See whether the equation is in a form such that the **square root method** is easily applied.
2. See whether the equation is in a form such that the **completing the square method** is easily applied.
3. If neither step 1 nor step 2 is reasonable, write the equation in $ax^2 + bx + c = 0$ form.
4. See whether the equation can be solved using the **factoring method**.
5. If you can't factor, solve the equation by the **quadratic formula**.

4 Solve problems modeled by quadratic equations.

The equation solving methods discussed in this section can be used to solve a variety of real-world applications that are modeled by quadratic equations.

Self Check 5

A rectangular garden is 4 feet longer than it is wide. If the garden has an area of 96 square feet, find the garden's length and width.

Now Try Problem 69

EXAMPLE 5

Sailing The height of a triangular sail is 4 feet more than the length of the base. If the sail has an area of 30 square feet, find the length of its base and the height.

Analyze
- The height of the sail is 4 feet more than the length of the base.
- The area of the sail is 30 ft².
- Find the length of the base and height of the sail.

Assign If we let $b =$ the length of the base in feet of the triangular sail, then $b + 4 =$ the height in feet.

Form We can use the formula for the area of a triangle, $A = \frac{1}{2}bh$, to form an equation.

$\frac{1}{2}$	times	the length of the base	times	the height	equals	the area of the triangle.
$\frac{1}{2}$	$\cdot$	b	$\cdot$	$(b + 4)$	$=$	30

Success Tip It is usually easier to clear quadratic equations of fractions before attempting to solve them.

Solve

$\frac{1}{2}b(b + 4) = 30$ *This is the equation to solve.*

$b(b + 4) = 60$ *To clear the equation of the fraction, multiply both sides by 2.*

$b^2 + 4b = 60$ *Distribute the multiplication by b.*

Since the coefficient of the b-term is the even number 4, this equation can be solved quickly by completing the square.

$b^2 + 4b = 60$

$b^2 + 4b + 4 = 60 + 4$ *Complete the square: $\frac{1}{2}(4) = 2$ and $(2)^2 = 4$. Add 4 to both sides.*

$(b + 2)^2 = 64$ *On the left side, factor the perfect-square trinomial. On the right, add.*

$b + 2 = \pm\sqrt{64}$ *Use the square root property.*

$b = -2 \pm 8$ *To isolate b, subtract 2 from both sides. Evaluate: $\sqrt{64} = 8$.*

$b = -2 + 8$ or $b = -2 - 8$ *To find the solutions, perform the calculation using a + symbol and then using a − symbol.*

$b = 6$ ~~$b = -10$~~   *Discard the solution −10. The length of the base cannot be negative.*

State The length of the base of the sail is 6 feet. Since the height is given by $b + 4$, the height of the sail is $6 + 4 = 10$ feet.

Check A height of 10 feet is 4 feet more than the length of the base, which is 6 feet. Also, the area of the triangle is $\frac{1}{2}(6)(10) = 30$ ft². The results check.

EXAMPLE 6

Televisions A television's screen size is measured diagonally. For the 42-inch plasma television shown in the illustration, the screen's height is 16 inches less than its width. To the nearest tenth of an inch, what are the height and width of the screen?

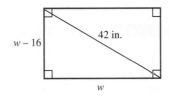

Self Check 6
Find the width and the height of the screen of a laptop if the diagonal measurement is 15.4 inches and its width is 4.5 inches more than its height. Round to the nearest hundredth.

Now Try Problem 72

Analyze A sketch of the screen shows that two adjacent sides and the diagonal form a right triangle. The length of the hypotenuse is 42 inches.

Assign If we let w = the width of the screen in inches, then $w - 16$ represents the height of the screen in inches.

Form We can use the Pythagorean theorem to form an equation.

$$a^2 + b^2 = c^2 \quad \text{This is the Pythagorean theorem.}$$
$$w^2 + (w - 16)^2 = 42^2 \quad \text{Substitute } w \text{ for } a, w - 16 \text{ for } b, \text{ and } 42 \text{ for } c.$$
$$w^2 + w^2 - 32w + 256 = 1{,}764 \quad \text{Find } (w - 16)^2 \text{ and } 42^2.$$
$$2w^2 - 32w - 1{,}508 = 0 \quad \text{To get 0 on the right side of the equation, subtract 1,764 from both sides.}$$
$$w^2 - 16w - 754 = 0 \quad \text{Divide both sides of the equation by 2: } \tfrac{2w^2}{2} - \tfrac{32w}{2} - \tfrac{1{,}508}{2} = \tfrac{0}{2}.$$

Solve Because of the large constant term, -754, we will not attempt to solve this quadratic equation by factoring. Instead, we will use the quadratic formula, with $a = 1$, $b = -16$, and $c = -754$.

$$w = \frac{-b \pm \sqrt{b^2 - 4ac}}{2a} \quad \text{In the quadratic formula, replace } x \text{ with } w.$$

$$w = \frac{-(-16) \pm \sqrt{(-16)^2 - 4(1)(-754)}}{2(1)} \quad \text{Substitute 1 for } a, -16 \text{ for } b, \text{ and } -754 \text{ for } c.$$

$$w = \frac{16 \pm \sqrt{256 - (-3{,}016)}}{2} \quad \text{Evaluate the power and multiply within the radical. Multiply in the denominator.}$$

$$w = \frac{16 \pm \sqrt{3{,}272}}{2} \quad \text{Subtract within the radical.}$$

We can use a calculator to approximate each one to the nearest tenth. The negative solution is discarded because the width of the screen cannot be negative.

$$\frac{16 + \sqrt{3{,}272}}{2} \approx 36.6 \quad \text{or} \quad \cancel{\frac{16 - \sqrt{3{,}272}}{2} \approx -20.6}$$

State The width of the television screen is approximately 36.6 inches. Since the height is $w - 16$, the height is approximately $36.6 - 16$ or 20.6 inches.

Check The sum of the squares of the lengths of the sides is $(36.6)^2 + (20.6)^2 = 1{,}763.92$. The square of the length of the hypotenuse is $42^2 = 1{,}764$. Since these are approximately equal, the results seem reasonable.

ANSWERS TO SELF CHECKS

1. $-1, -5$ 2. $3, -\frac{1}{4}$ 3. $\frac{1+\sqrt{3}}{2} \approx 1.37, \frac{1-\sqrt{3}}{2} \approx -0.37$ 4. no
5. width: 8 ft, length: 12 ft   6. width: 12.90 in., height: 8.40 in.

SECTION 9.3 STUDY SET

VOCABULARY

Fill in the blanks.

1. To _____ a quadratic equation means to find all the values of the variable that make the equation true.
2. $\sqrt{-16}$ is not a _____ number.
3. The general _____ equation is $ax^2 + bx + c = 0$.
4. The formula $x = \dfrac{-b \pm \sqrt{b^2 - 4ac}}{2a}$ is called the _____ formula.

CONCEPTS

Fill in the blanks.

5. In the quadratic equation $ax^2 + bx + c = 0$, $a \neq$ ___.
6. Before we can determine a, b, and c for $x = 3x^2 - 1$, we must write the equation in _____ (general) form.
7. In the quadratic equation $3x^2 - 5 = 0$, $a =$ ___, $b =$ ___, and $c =$ ___.
8. In the quadratic equation $-4x^2 + 8x = 0$, $a =$ ___, $b =$ ___, and $c =$ ___.
9. The formula for the area of a rectangle is $A =$ ___, and the formula for the area of a triangle is $A =$ ___.
10. If a, b, and c are three sides of a right triangle and c is the hypotenuse, then $c^2 =$ ___.
11. In evaluating the numerator of $\dfrac{-5 \pm \sqrt{5^2 - 4(2)(1)}}{2(2)}$ what operation should be performed first?
12. Consider the expression $\dfrac{3 \pm 6\sqrt{2}}{3}$
    a. How many terms does the numerator contain?
    b. What common factor do the terms have?
    c. Simplify the expression.
13. A student used the quadratic formula to solve an equation and obtained $x = \dfrac{-3 \pm \sqrt{15}}{2}$
    a. How many solutions does the equation have?
    b. What are they exactly?
    c. Approximate them to the nearest hundredth.

14. The solutions of a quadratic equation are $x = 2 \pm \sqrt{3}$. Graph them on the number line.

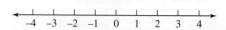

NOTATION

Complete each step.

15. Solve: $x^2 - 5x - 6 = 0$

$$x = \dfrac{-b \pm \sqrt{b^2 - 4ac}}{2a}$$

$$x = \dfrac{-(\;\;) \pm \sqrt{(-5)^2 - 4(1)(-6)}}{2(1)}$$

$$x = \dfrac{\pm \sqrt{25 + \;\;}}{2}$$

$$x = \dfrac{5 \pm \sqrt{\;\;}}{2}$$

$$x = \dfrac{\pm 7}{2}$$

$$x = \dfrac{5 \;\; 7}{2} = \;\; \text{ or } x = \dfrac{5 \;\; 7}{2} = \;\;$$

16. Solve: $3x^2 + 2x - 2 = 0$

$$x = \dfrac{-b \pm \sqrt{b^2 - 4ac}}{2a}$$

$$x = \dfrac{-2 \pm \sqrt{\;\;^2 - 4(3)(\;\;)}}{2(\;\;)}$$

$$x = \dfrac{-2 \pm \sqrt{4 \;\; 24}}{6}$$

$$x = \dfrac{-2 \pm \sqrt{\;\;}}{6}$$

$$x = \dfrac{-2 \pm \;\;\sqrt{7}}{6}$$

$$x = \dfrac{\left(-1 \pm \sqrt{7}\right)}{\;\; \cdot 3}$$

$$x = \dfrac{-1 \pm \sqrt{7}}{\;\;}$$

17. What is wrong with the following work?

Solve: $x^2 + 4x - 5 = 0$

$$x = -4 \pm \frac{\sqrt{16 - 4(1)(-5)}}{2}$$

18. In reading $\dfrac{-b \pm \sqrt{b^2 - 4ac}}{2a}$ we say, "the _____ of b, plus or minus the _____ root of b squared minus 4 _____ a times c, all _____ $2a$."

GUIDED PRACTICE

Change each equation into quadratic form, if necessary, and find the values of a, b, and c. Do not solve the equation. See Example 1.

19. $x^2 + 4x + 3 = 0$
20. $x^2 - x - 4 = 0$
21. $3x^2 - 2x + 7 = 0$
22. $4x^2 + 7x - 3 = 0$
23. $4y^2 = 2y - 1$
24. $2x = 3x^2 + 4$
25. $x(3x - 5) = 2$
26. $y(5y + 10) = 8$

Use the quadratic formula to find all real solutions. See Example 1.

27. $x^2 - 5x + 6 = 0$
28. $x^2 + 5x + 4 = 0$
29. $x^2 + 7x + 12 = 0$
30. $x^2 - x - 12 = 0$

Use the quadratic formula to find all real solutions. See Example 2.

31. $2x^2 - x = 1$
32. $2x^2 + 3x = 2$
33. $3x^2 + 2 = -5x$
34. $3x^2 + 1 = 4x$

Use the quadratic formula to find all real solutions. Round each solution to the nearest hundredth. See Example 3.

35. $x^2 - 2x - 1 = 0$
36. $b^2 = 18$
37. $2x^2 + x = 5$
38. $3x^2 - x = 1$

Does each equation have any real-number solutions? If so, find them. See Example 4.

39. $3m^2 - 2m + 5 = 0$
40. $4n^2 + 12n - 3 = 0$
41. $2d^2 + 8d + 5 = 0$
42. $9c^2 - 2c + 4 = 0$

Use the most convenient method to find all real solutions. If a solution contains a radical, give the exact solution and then approximate it to the nearest hundredth. See Objective 3.

43. $(2y - 1)^2 = 25$
44. $m^2 + 14m + 49 = 0$
45. $2x^2 - x + 2 = 0$
46. $x^2 + 2x + 7 = 0$
47. $x^2 - 2x - 35 = 0$
48. $x^2 + 5x + 3 = 0$
49. $4c^2 + 16c = 0$
50. $t^2 - 1 = 0$
51. $18 = 3y^2$
52. $25x - 50x^2 = 0$

TRY IT YOURSELF

Write each equation in $ax^2 + bx + c = 0$ form. Then identify a, b, and c.

53. $7(x^2 + 3) = -14x$
54. $(2a + 3)(a - 2) = (a + 1)(a - 1)$

Use the quadratic formula to find all real solutions.

55. $4x^2 + 4x - 3 = 0$
56. $4x^2 + 3x - 1 = 0$
57. $x^2 + 3x + 1 = 0$
58. $x^2 + 3x - 2 = 0$
59. $3x^2 - x = 3$
60. $5x^2 = 3x + 1$
61. $x^2 + 5 = 2x$
62. $2x^2 + 3x = -3$
63. $x^2 = 1 - 2x$
64. $x^2 = 4 + 2x$
65. $3x^2 = 6x + 2$
66. $3x^2 = -8x - 2$

Solve each equation. Round each solution to the nearest tenth.

67. $2.4x^2 - 9.5x + 6.2 = 0$
68. $-1.7x^2 + 0.5x + 0.9 = 0$

APPLICATIONS

69. **HEIGHT OF A TRIANGLE** The triangle shown has an area of 30 square inches. Find its height.

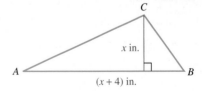

70. A poster that shows former UCLA basketball coach John Wooden's Pyramid of Success has an area of 80 square inches. The base of the triangular poster is 6 inches longer than the height. Find the length of the base and the height of the poster.

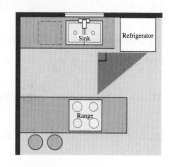

71. KITCHEN FLOOR PLANS To minimize the number of steps that a cook must take when preparing meals, designers carefully plan the *kitchen work triangle* (the area between the sink, refrigerator, and range). One leg of the work triangle shown is 2 feet longer than the other leg, and the area covered is 24 ft². Find the length of each leg of the triangle.

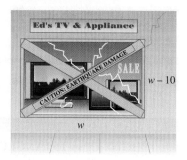

72. EARTHQUAKES After an earthquake, a store owner nailed a 50-inch-long board across a broken window. Find the height and width of the window if the height is 10 inches less than the width.

73. FLAGS According to the *Guinness Book of World Records*, the largest flag flown from a flagpole was a Brazilian national flag, a rectangle having an area of 3,102 ft². If the flag is 19 feet longer than it is wide, find its width and length.

74. COMICS See the illustration. A comic strip occupies 96 square centimeters of space in a newspaper. The length of the rectangular space is 4 centimeters more than twice its width. Find its dimensions.

75. COMMUNITY GARDENS See the illustration. Residents of a community can work their own 16 ft × 24 ft plot of city-owned land if they agree to the following stipulations:

- The area of the garden cannot exceed 180 square feet.
- A path of uniform width must be maintained around the garden.

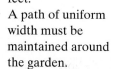

Find the dimensions of the largest possible garden.

76. DECKING The owner of the pool shown below wants to surround it with a concrete deck of uniform width (shown in gray). If he can afford 368 square feet of decking, how wide can he make the deck?

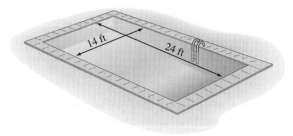

77. FALLING OBJECTS A tourist drops a penny from the observation deck of a skyscraper 1,377 feet above the ground. How long will it take for the penny to hit the ground? (*Hint:* Refer to Example 4 in Section 9.1.)

78. ABACUS The Chinese abacus shown consists of a frame, parallel wires, and beads that are moved to perform arithmetic computations. If the frame is 21 centimeters wider than it is high, find its dimensions.

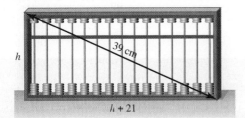

79. SIDEWALKS A 170-meter-long sidewalk from the mathematics building M to the student center C is shown in red in the illustration. However, students prefer to walk directly from M to C. How long are the two segments of the existing sidewalk?

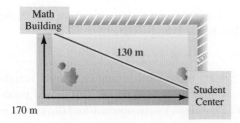

80. NAVIGATION Two boats leave port at the same time, one sailing east and one sailing south. If one boat sails 10 nautical miles more than the other and they are then 50 nautical miles apart, how far does each boat sail?

81. NAVIGATION One plane heads west from an airport, flying at 200 mph. One hour later, a second plane heads north from the same airport, flying at the same speed. When will the planes be 1,000 miles apart?

82. INVESTING We can use the formula $A = P(1 + r)^2$ to find the amount $A that $P will become when invested at an annual rate of r% for 2 years. What interest rate is needed to make $5,000 grow to $5,724.50 in 2 years?

83. INVESTING What interest rate is needed to make $7,000 grow to $8,470 in 2 years? (See Problem 82.)

84. MANUFACTURING A firm has found that its revenue for manufacturing and selling x television sets is given by the formula $R = -\frac{1}{6}x^2 + 450x$. How much revenue will be earned by manufacturing 600 television sets?

85. RETAILING When a wholesaler sells n CD players, his revenue R is given by the formula $R = 150n - \frac{1}{2}n^2$. How many players would he have to sell to receive $11,250? (*Hint:* Multiply both sides of the equation by -2.)

86. METAL FABRICATION A square piece of tin, 12 inches on a side, is to have four equal squares cut from its corners, as shown. If the edges are then to be folded up to make a box with a floor area of 64 square inches, find the depth of the box.

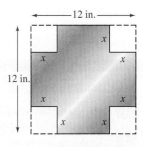

87. MAKING GUTTERS A piece of sheet metal, 18 inches wide, is bent to form a gutter, as shown. If the cross-sectional area is 36 square inches, find the depth of the gutter.

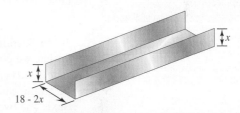

WRITING

88. Do you agree with the following statement? Explain your answer.

    The quadratic formula is the easiest method to use to solve quadratic equations.

89. Explain the meaning of the $\pm$ symbol.

90. Use the quadratic formula to solve $x^2 - 2x - 4 = 0$. What is an exact solution, and what is an approximate solution of this equation? Explain the difference.

91. Rewrite in words:

$$x = \frac{-b \pm \sqrt{b^2 - 4ac}}{2a}$$

REVIEW

Solve each equation for the indicated variable.

92. $A = p + prt$ for r

93. $F = \dfrac{GMm}{d^2}$ for M

Write an equation in slope–intercept form for the line that has the given properties.

94. Slope of $\frac{3}{5}$ and passes through $(0, 12)$

95. Passes through $(6, 8)$ and the origin

Simplify each expression.

96. $\sqrt{80}$

97. $2\sqrt{x^3 y^2}$

Rationalize each denominator and simplify.

98. $\dfrac{x}{\sqrt{7x}}$

99. $\dfrac{\sqrt{x}+2}{\sqrt{x}-2}$

Objectives

1. Understand the vocabulary used to describe parabolas.
2. Find the intercepts of a parabola.
3. Determine the vertex of a parabola.
4. Graph equations of the form $y = ax^2 + bx + c$.
5. Solve quadratic equations graphically.

SECTION 9.4
Graphing Quadratic Equations

In this section, we will combine our graphing skills with our equation-solving skills to graph *quadratic equations in two variables*.

1 Understand the vocabulary used to describe parabolas.

Equations that can be written in the form $y = ax^2 + bx + c$, where $a \neq 0$, are called **quadratic equations in two variables**. Some examples are

$$y = x^2 - 2x - 3 \qquad y = -2x^2 - 8x - 8 \qquad y = x^2 + x$$

In Section 3.2, we graphed $y = x^2$, a quadratic equation in two variables. To do this, we constructed a table of solutions, plotted points, and joined them with a smooth curve, called a **parabola**. The parabola opens upward, and the lowest point on the graph, called the **vertex**, is the point $(0, 0)$. If we fold the graph paper along the y-axis, the two sides of the parabola match. We say that the graph is *symmetric about the y-axis* and we call the y-axis the **axis of symmetry**.

> **The Language of Algebra** An *axis of symmetry* (or *line of symmetry*) divides a parabola into two matching sides. The sides are said to be *mirror images* of each other.

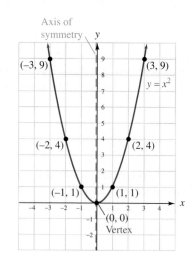

Using the same steps, we can also graph $y = -x^2 + 2$, another quadratic equation in two variables. The resulting parabola opens downward, and the **vertex** (in this case, the highest point on the graph) is the point $(0, 2)$. The axis of symmetry is the y-axis.

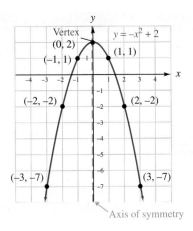

$y = -x^2 + 2$		
x	y	(x, y)
-3	-7	$(-3, -7)$
-2	-2	$(-2, -2)$
-1	1	$(-1, 1)$
0	2	$(0, 2)$
1	1	$(1, 1)$
2	-2	$(2, -2)$
3	-7	$(3, -7)$

For the equation $y = x^2$, the coefficient of the x^2 term is the positive number 1. For $y = -x^2 + 2$, the coefficient of the x^2 term is the negative number -1. These observations suggest the following fact.

> **Graphs of Quadratic Equations**
>
> The graph of $y = ax^2 + bx + c$, where $a \neq 0$, is a parabola. It opens upward when $a > 0$ and downward when $a < 0$.

> **The Language of Algebra** In the equation $y = ax^2 + bx + c$, each value of x determines exactly one value of y. Therefore, the equation defines y to be a function of x and we could write $f(x) = ax^2 + bx + c$. Your instructor may ask you to use the vocabulary and notation of functions throughout this section.

Parabolic shapes can be seen in a wide variety of real-world settings. These shapes can be modeled by quadratic equations in two variables.

The path of a stream of water

Parabola

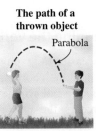

The shape of a satellite antenna dish

Parabola

The path of a thrown object

Parabola

> **The Language of Algebra** The word *parabolic* (pronounced par · a · BOL · ic) means having the form of a parabola. For example, the light bulb in most flashlights is surrounded by a *parabolic* reflecting mirror.

2 Find the intercepts of a parabola.

When graphing quadratic equations, it is helpful to know the x- and y-intercepts of the parabola. To find the intercepts of a parabola, we use the same steps that we used to find the intercepts of the graphs of linear equations.

Finding Intercepts

To find the *y*-intercept, substitute 0 for *x* in the given equation and solve for *y*.

To find the *x*-intercepts, substitute 0 for *y* in the given equation and solve for *x*.

Self Check 1

Find the *y*- and *x*-intercepts of the graph of $y = x^2 + 6x + 8$.

Now Try Problem 17

EXAMPLE 1 Find the *y*-intercept and the *x*-intercepts of the graph of $y = x^2 - 2x - 3$.

Strategy To find the *y*-intercept of the graph, we will let $x = 0$ and find *y*. To find the *x*-intercepts of the graph, we will let $y = 0$ and solve the resulting equation for *x*.

WHY A point on the *y*-axis has an *x*-coordinate of 0. A point on the *x*-axis has a *y*-coordinate of 0.

Solution
We let $x = 0$ and evaluate the right side to find the *y*-intercept.

$y = x^2 - 2x - 3$ This is the given equation in two variables.
$y = 0^2 - 2(0) - 3$ Substitute 0 for x.
$y = -3$ Evaluate the right side.

The *y*-intercept is $(0, -3)$. We note that the *y*-coordinate of the *y*-intercept is the same as the value of the constant term *c* on the right side of $y = x^2 - 2x - 3$.

Next, we let $y = 0$ and solve the resulting quadratic equation to find the *x*-intercepts.

$y = x^2 - 2x - 3$ This is the given equation.
$0 = x^2 - 2x - 3$ Substitute 0 for y.
$0 = (x - 3)(x + 1)$ Factor the trinomial.
$x - 3 = 0$ or $x + 1 = 0$ Set each factor equal to 0.
$x = 3$ $x = -1$

Since there are two solutions, the graph has two *x*-intercepts: $(3, 0)$ and $(-1, 0)$.

3 Determine the vertex of a parabola.

It is usually easier to graph a quadratic equation when we know the coordinates of the vertex of its graph. Because of symmetry, if a parabola has two *x*-intercepts, the *x*-coordinate of the vertex is exactly midway between them. We can use this fact to derive a formula to find the vertex of a parabola.

In general, if a parabola has two *x*-intercepts, they can be found by solving $0 = ax^2 + bx + c$ for *x*. We can use the quadratic formula to find the solutions. They are

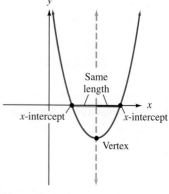

$$x = \frac{-b - \sqrt{b^2 - 4ac}}{2a} \quad \text{and} \quad x = \frac{-b + \sqrt{b^2 - 4ac}}{2a}$$

Thus, the parabola's *x*-intercepts are $\left(\frac{-b - \sqrt{b^2 - 4ac}}{2a}, 0\right)$ and $\left(\frac{-b + \sqrt{b^2 - 4ac}}{2a}, 0\right)$.

Since the *x*-value of the vertex of a parabola is halfway between the two *x*-intercepts, we can find this value by finding the average, or $\frac{1}{2}$ of the sum of the *x*-coordinates of the *x*-intercepts.

$$x = \frac{1}{2}\left(\frac{-b - \sqrt{b^2 - 4ac}}{2a} + \frac{-b + \sqrt{b^2 - 4ac}}{2a}\right)$$

$$x = \frac{1}{2}\left(\frac{-b - \sqrt{b^2 - 4ac} + (-b) + \sqrt{b^2 - 4ac}}{2a}\right) \quad \text{Add the numerators and keep the common denominator.}$$

$$x = \frac{1}{2}\left(\frac{-2b}{2a}\right) \quad \text{In the numerator combine like terms: } -b + (-b) = -2b \text{ and } -\sqrt{b^2 - 4ac} + \sqrt{b^2 - 4ac} = 0.$$

$$x = \frac{-b}{2a} \quad \text{Simplify.}$$

This result is true even if the graph has no *x*-intercepts.

Finding the Vertex of a Parabola

The graph of the quadratic equation $y = ax^2 + bx + c$ is a parabola whose vertex has an *x*-coordinate of $\frac{-b}{2a}$. To find the *y*-coordinate of the vertex, substitute $\frac{-b}{2a}$ for *x* into the equation and find *y*.

EXAMPLE 2 Find the vertex of the graph of $y = x^2 - 2x - 3$.

Strategy We will compare the given equation to the general form $y = ax^2 + bx + c$ to identify *a* and *b*.

WHY To use the vertex formula, we need to know *a* and *b*.

Solution
From the following diagram, we see that $a = 1$, $b = -2$, and $c = -3$.

$$y = \mathbf{1}x^2 - \mathbf{2}x - \mathbf{3}$$
$$y = ax^2 + bx + c$$

To find the *x*-coordinate of the vertex, we substitute the values for *a* and *b* into the formula $x = \frac{-b}{2a}$.

$$x = \frac{-b}{2a} = \frac{-(-2)}{2(1)} = 1$$

The *x*-coordinate of the vertex is 1. To find the *y*-coordinate, we substitute 1 for *x* in the original equation.

$y = x^2 - 2x - 3$ This is the given equation in two variables.
$y = 1^2 - 2(1) - 3$ Substitute 1 for x.
$y = -4$ Evaluate the right side.

The vertex of the parabola is $(1, -4)$.

Self Check 2
Find the vertex of the graph of $y = x^2 + 6x + 8$.

Now Try Problem 21

Success Tip An easy way to remember the vertex formula is to note that $x = \frac{-b}{2a}$ is part of the quadratic formula:

$$x = \frac{-b \pm \sqrt{b^2 - 4ac}}{2a}$$

750 Chapter 9 Quadratic Equations

4 Graph equations of the form $y = ax^2 + bx + c$.

Much can be determined about the graph of $y = ax^2 + bx + c$ from the coefficients a, b, and c. We can use this information to help graph the equation.

Graphing a Quadratic Equation $y = ax^2 + bx + c$

1. **Test for opening upward/downward.** Determine whether the parabola opens upward or downward. If $a > 0$, the graph opens upward. If $a < 0$, the graph opens downward.

2. **Find the vertex/axis of symmetry.** The x-coordinate of the vertex of the parabola is $x = \dfrac{-b}{2a}$. To find the y-coordinate, substitute $\dfrac{-b}{2a}$ for x into the equation and find y. The axis of symmetry is the vertical line passing through the vertex.

3. **Find the intercepts.** To find the y-intercept, substitute 0 for x in the given equation and solve for y. The result will be c. Thus, the y-intercept is $(0, c)$.

   To find the x-intercepts (if any), substitute 0 for y and solve the resulting quadratic equation $ax^2 + bx + c = 0$. If no real-number solutions exist, the graph has no x-intercepts.

4. **Plot points, using symmetry.** To find two more points on the graph, select a convenient value for x and find the corresponding value of y. Plot that point and its mirror image on the opposite side of the axis of symmetry.

5. **Draw the parabola.** Draw a smooth curve through the located points.

Self Check 3

Use your results from Self Checks 1 and 2 to help graph:
$y = x^2 + 6x + 8$

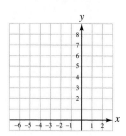

Now Try Problem 25

EXAMPLE 3 Graph: $y = x^2 - 2x - 3$

Strategy We will use the five-step procedure described above to sketch the graph of the equation.

WHY This strategy is usually faster than making a table of solutions and plotting points.

Solution

Upward/downward: The equation is in the form $y = ax^2 + bx + c$, with $a = 1$, $b = -2$, and $c = -3$. Since $a > 0$, the parabola opens upward.

Vertex/axis of symmetry: In Example 2, we found that the vertex of the graph is $(1, -4)$. The axis of symmetry will be the vertical line passing through $(1, -4)$. See figure (a) on the next page.

Intercepts: In Example 1, we found that the y-intercept is $(0, -3)$ and x-intercepts are $(3, 0)$ and $(-1, 0)$.

If the point $(0, -3)$, which is 1 unit to the left of the axis of symmetry, is on the graph, the point $(2, -3)$, which is 1 unit to the right of the axis of symmetry, is also on the graph. See figure (a) on the next page.

Plotting points/using symmetry: It would be helpful to locate two more points on the graph. To find a solution of $y = x^2 - 2x - 3$, we select a convenient value for x, say -2, and find the corresponding value of y.

$y = x^2 - 2x - 3$ This is the equation to graph.
$y = (-2)^2 - 2(-2) - 3$ Substitute -2 for x.
$y = 5$ Evaluate the right side.

Thus, the point $(-2, 5)$ lies on the parabola. If the point $(-2, 5)$, which is 3 units to the left of the axis of symmetry, is on the graph, the point $(4, 5)$, which is 3 units to the right of the axis of symmetry, is also on the graph. See figure (a).

$y = x^2 - 2x - 3$		
x	y	(x, y)
-2	5	$(-2, 5)$

Drawing the curve: Draw a smooth curve through the points. The completed graph of $y = x^2 - 2x - 3$ is shown in figure (b).

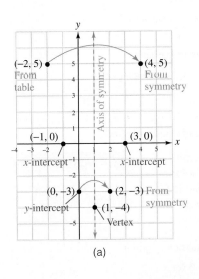
(a)

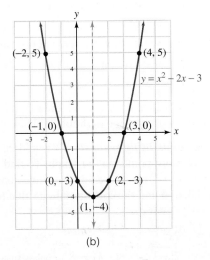
(b)

Success Tip The most important point to find when graphing a quadratic equation in two variables is the vertex.

EXAMPLE 4 Graph: $y = -2x^2 - 8x - 8$

Strategy We will follow the five-step procedure to sketch the graph of the equation.

WHY This strategy is usually faster than making a table of solutions and plotting points.

Solution
Upward/downward: The equation is in the form $y = ax^2 + bx + c$, with $a = -2$, $b = -8$, and $c = -8$. Since $a < 0$, the parabola opens downward.

Vertex/axis of symmetry: To find the x-coordinate of the vertex, we substitute -2 for a and -8 for b into the formula $x = \dfrac{-b}{2a}$.

$$x = \frac{-b}{2a} = \frac{-(-8)}{2(-2)} = -2$$

The x-coordinate of the vertex is -2. To find the y-coordinate, we substitute -2 for x in the original equation and find y.

$y = -2x^2 - 8x - 8$ This is the equation to graph.
$y = -2(-2)^2 - 8(-2) - 8$ Substitute -2 for x.
$y = 0$ Evaluate the right side.

The vertex of the parabola is the point $(-2, 0)$. The axis of symmetry is the vertical line passing through $(-2, 0)$. See figure (a) on the next page.

Self Check 4
Graph: $y = -2x^2 + 4x - 2$

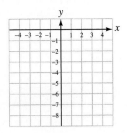

Now Try Problem 29

Intercepts: Since $c = -8$, the y-intercept of the parabola is $(0, -8)$. The point $(-4, -8)$, which is 2 units to the left of the axis of symmetry, must also be on the graph. See figure (a).

To find the x-intercepts, we let $y = 0$ and solve the resulting quadratic equation.

$$y = -2x^2 - 8x - 8 \quad \text{This is the equation to graph.}$$
$$0 = -2x^2 - 8x - 8 \quad \text{Substitute 0 for } y.$$
$$0 = x^2 + 4x + 4 \quad \text{Divide both sides by } -2\colon \frac{0}{-2} = \frac{-2x^2}{-2} - \frac{8x}{-2} - \frac{8}{-2}.$$
$$0 = (x + 2)(x + 2) \quad \text{Factor the trinomial.}$$
$$x + 2 = 0 \quad \text{or} \quad x + 2 = 0 \quad \text{Set each factor equal to 0.}$$
$$x = -2 \qquad\qquad x = -2$$

Since the solutions are the same, the graph has only one x-intercept: $(-2, 0)$. This point is the vertex of the parabola and has already been plotted.

Plotting points/using symmetry: It would be helpful to know two more points on the graph. To find a solution of $y = -2x^2 - 8x - 8$, we select a convenient value for x, say -3, and find the corresponding value for y.

$y = -2x^2 - 8x - 8$		
x	y	(x, y)
-3	-2	$(-3, -2)$

$$y = -2x^2 - 8x - 8 \quad \text{This is the equation to graph.}$$
$$y = -2(-3)^2 - 8(-3) - 8 \quad \text{Substitute } -3 \text{ for } x.$$
$$y = -2 \quad \text{Evaluate the right side.}$$

Thus, the point $(-3, -2)$ lies on the parabola. We plot $(-3, -2)$ and then use symmetry to determine that $(-1, -2)$ is also on the graph. See figure (a).

Drawing the curve: Draw a smooth curve through the points. The completed graph of $y = -2x^2 - 8x - 8$ is shown in figure (b) below.

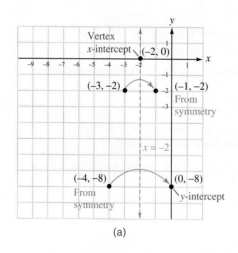
(a)

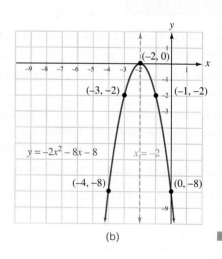
(b)

EXAMPLE 5   Graph: $y = x^2 + 2x - 2$

Strategy We will follow the five-step procedure to sketch the graph of the equation.

WHY This strategy is usually faster than making a table of solutions and plotting points.

Solution

Upward/downward: The equation is in the form $y = ax^2 + bx + c$, with $a = 1$, $b = 2$, and $c = -2$. Since $a > 0$, the parabola opens upward.

Vertex/axis of symmetry: To find the x-coordinate of the vertex, we substitute 1 for a and 2 for b into the formula $x = \frac{-b}{2a}$.

$$x = \frac{-b}{2a} = \frac{-2}{2(1)} = -1$$

The x-coordinate of the vertex is -1. To find the y-coordinate, we substitute -1 for x in the original equation and find y.

$y = x^2 + 2x - 2$	This is the equation to graph.
$y = (-1)^2 + 2(-1) - 2$	Substitute -1 for x.
$y = -3$	Evaluate the right side.

The vertex of the parabola is the point $(-1, -3)$. The axis of symmetry is the vertical line passing through $(-1, -3)$. See figure (a) on the next page.

Intercepts: Since $c = -2$, the y-intercept of the parabola is $(0, -2)$. The point $(-2, -2)$, which is one unit to the left of the axis of symmetry, must also be on the graph. See figure (a) on the next page.

To find the x-intercepts, we let $y = 0$ and solve the resulting quadratic equation.

$y = x^2 + 2x - 2$	This is the equation to graph.
$0 = x^2 + 2x - 2$	Substitute 0 for y.

Since $x^2 + 2x - 2$ does not factor, we will use the quadratic formula to solve for x.

$x = \dfrac{-2 \pm \sqrt{2^2 - 4(1)(-2)}}{2(1)}$ In the quadratic formula, substitute 1 for a, 2 for b, and -2 for c.

$x = \dfrac{-2 \pm \sqrt{12}}{2}$ Evaluate the right side.

$x = \dfrac{-2 \pm 2\sqrt{3}}{2}$ Simplify the radical: $\sqrt{12} = \sqrt{4 \cdot 3} = 2\sqrt{3}$.

$x = \dfrac{\overset{1}{\cancel{2}}(-1 \pm \sqrt{3})}{\underset{1}{\cancel{2}}}$ Factor out the GCF, 2, from the terms in the numerator. Then simplify by removing the common factor of 2 in the numerator and denominator.

$x = -1 \pm \sqrt{3}$

The x-intercepts of the graph are $\left(-1 + \sqrt{3}, 0\right)$ and $\left(-1 - \sqrt{3}, 0\right)$. To help to locate their position on the graph, we can use a calculator to approximate the two irrational numbers. See figure (a) on the next page.

$$-1 + \sqrt{3} \approx 0.7 \quad \text{and} \quad -1 - \sqrt{3} \approx -2.7$$

Plotting points/using symmetry: To find two more points on the graph, we let $x = 2$, substitute 2 for x in $y = x^2 + 2x - 2$, and find that y is 6. We plot $(2, 6)$ and then use symmetry to determine that $(-4, 6)$ is also on the graph. See figure (a) on the next page.

$y = x^2 + 2x - 2$		
x	y	(x, y)
2	6	$(2, 6)$

Drawing the curve: Draw a smooth curve through the points. The completed graph of $y = x^2 + 2x - 2$ is shown in figure (b) on the next page.

Self Check 5

Graph: $y = x^2 - 4x - 3$

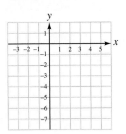

Now Try Problem 43

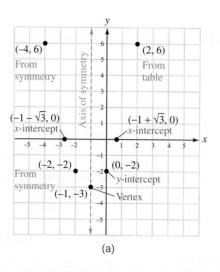
(a)

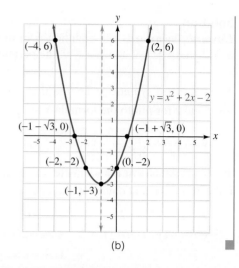
(b)

Success Tip To plot the x-intercepts, we use the approximations $(0.7, 0)$ and $(-2.7, 0)$ to locate their position on the x-axis. However, we label the points on the graph using the exact values: $\left(-1 + \sqrt{3}, 0\right)$ and $\left(-1 - \sqrt{3}, 0\right)$.

5 Solve quadratic equations graphically.

The number of distinct x-intercepts of the graph of $y = ax^2 + bx + c$ is the same as the number of distinct real-number solutions of $ax^2 + bx + c = 0$. For example, the graph of $y = x^2 + x - 2$ in figure (a) below has two x-intercepts, and $x^2 + x - 2 = 0$ has two real-number solutions. In figure (b), the graph has one x-intercept, and the corresponding equation has one real-number solution. In figure (c), the graph does not have an x-intercept, and the corresponding equation does not have any real-number solutions. Note that the solutions of each equation are given by the x-coordinates of the x-intercepts of each respective graph.

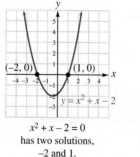
$x^2 + x - 2 = 0$
has two solutions,
-2 and 1.
(a)

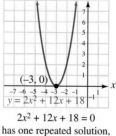
$2x^2 + 12x + 18 = 0$
has one repeated solution,
-3.
(b)

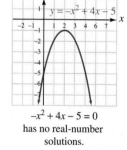
$-x^2 + 4x - 5 = 0$
has no real-number
solutions.
(c)

ANSWERS TO SELF CHECKS

1. y-intercept: $(0, 8)$; x-intercepts: $(-2, 0), (-4, 0)$ 2. $(-3, -1)$
3. 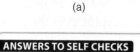 4. 5.

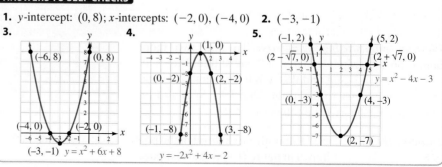

SECTION 9.4 STUDY SET

VOCABULARY

Fill in the blanks.

1. $y = 3x^2 + 5x - 1$ is a _____ equation in two variables. Its graph is a cup-shaped figure called a _____.

2. The lowest point on a parabola that opens upward, and the highest point on a parabola that opens downward, is called the _____ of the parabola.

3. Points where a parabola intersects the x-axis are called the x-_____ of the graph and the point where a parabola intersects the y-axis is called the y-_____ of the graph.

4. The vertical line that splits the graph of a parabola into two identical parts is called the axis of _____.

CONCEPTS

Fill in the blanks.

5. The graph of $y = ax^2 + bx + c$ opens downward when a ___ 0 and upward when $a > $ ___.

6. The graph of $y = ax^2 + bx + c$ is a parabola whose vertex has an x-coordinate given by ___.

7. **a.** To find the y-intercepts of a graph, substitute ___ for x in the given equation and solve for y.
   b. To find the x-intercepts of a graph, substitute 0 for y in the given equation and solve for ___.

8.
$y = x^2 - 3x - 1$

x	y	(x, y)
3		(3,)

9.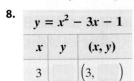
   a. What do we call the curve shown in the graph?
   b. What are the x-intercepts of the graph?
   c. What is the y-intercept of the graph?
   d. What is the vertex?
   e. Draw the axis of symmetry on the graph.

10. Does the graph of each quadratic equation open upward or downward?
    a. $y = 2x^2 + 5x - 1$ **b.** $y = -6x^2 - 3x + 5$

11. The vertex of a parabola is $(1, -3)$, its y-intercept is $(0, -2)$, and it passes through the point $(3, 1)$. Draw the axis of symmetry and use it to help determine two other points on the parabola.

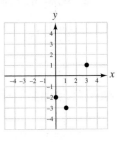

12. Sketch the graph of a quadratic equation using the given facts about its graph.
    - Opens upward
    - Vertex: $(3, -1)$
    - y-intercept: $(0, 8)$
    - x-intercepts: $(2, 0), (4, 0)$

x	y	(x, y)
1	3	(1, 3)

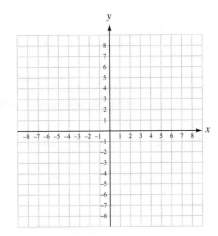

13. Examine the graph of $y = x^2 + 2x - 3$. How many real-number solutions does the equation $x^2 + 2x - 3 = 0$ have? Find them.

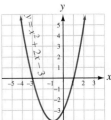

14. Examine the graph of $y = x^2 + 4x + 4$. How many real-number solutions does the equation $x^2 + 4x + 4 = 0$ have? Find them.

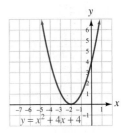

NOTATION

15. Consider the equation $y = 2x^2 + 4x - 8$.
    a. What are a, b, and c?
    b. Find $\dfrac{-b}{2a}$.

16. Evaluate: $\dfrac{-(-12)}{2(-3)}$

GUIDED PRACTICE

Find the y- and x-intercepts of the graph of the quadratic equation. See Example 1.

17. $y = x^2 - 6x + 8$
18. $y = 2x^2 - 4x$
19. $y = -x^2 - 10x - 21$
20. $y = 3x^2 + 6x - 9$

Find the vertex of the graph of each quadratic equation. See Example 2.

21. $y = 2x^2 - 4x + 1$
22. $y = 2x^2 + 8x - 4$
23. $y = -x^2 + 6x - 8$
24. $y = -x^2 - 2x - 1$

Graph each quadratic equation by finding the vertex, the x- and y-intercepts, and the axis of symmetry of its graph. See Examples 3 and 4.

25. $y = x^2 + 2x - 3$

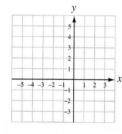

26. $y = x^2 + 6x + 5$

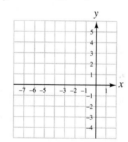

27. $y = 2x^2 + 8x + 6$

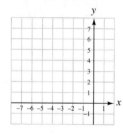

28. $y = 3x^2 - 12x + 9$

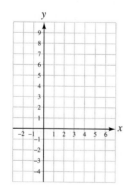

29. $y = -x^2 + 2x + 3$

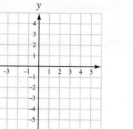

30. $y = -2x^2 + 4x$

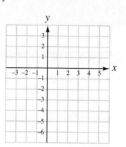

31. $y = -x^2 + 5x - 4$

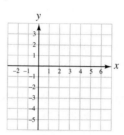

32. $y = -x^2 + 2x - 1$

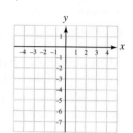

33. $y = x^2 - 2x$

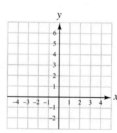

34. $y = x^2 + x$

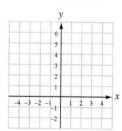

35. $y = x^2 + 4x + 4$

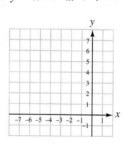

36. $y = x^2 - 6x + 9$

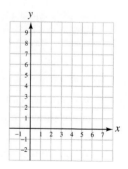

37. $y = -x^2 - 4x$

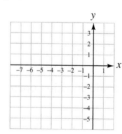

38. $y = -x^2 + 2x$

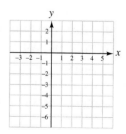

39. $y = 2x^2 + 3x - 2$

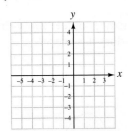

40. $y = 3x^2 - 7x + 2$

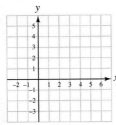

41. $y = 4x^2 - 12x + 9$

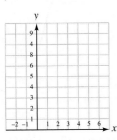

42. $y = -x^2 - 2x - 1$

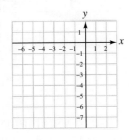

Graph each quadratic equation by finding the vertex, the x- and y-intercepts, and the axis of symmetry of its graph. See Example 5.

43. $y = x^2 - 4x - 1$

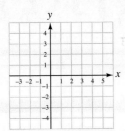

44. $y = x^2 + 2x - 5$

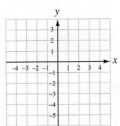

45. $y = -x^2 - 2x + 2$

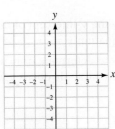

46. $y = -x^2 - 4x + 3$

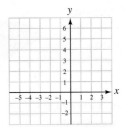

47. $y = -x^2 - 6x - 4$

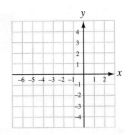

48. $y = x^2 - 6x + 4$

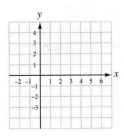

49. $y = x^2 - 6x + 10$

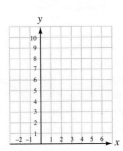

50. $y = x^2 - 2x + 4$

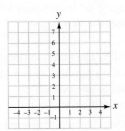

APPLICATIONS

51. BIOLOGY Draw an axis of symmetry over the sketch of the butterfly.

52. CROSSWORD PUZZLES Darken the appropriate squares to the right of the dashed blue line so that the puzzle has symmetry with respect to that line.

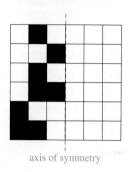

axis of symmetry

53. COST ANALYSIS A company has found that when it assembles x carburetors in a production run, the manufacturing cost of $\$y$ per carburetor is given by the graph below. What important piece of information does the vertex give?

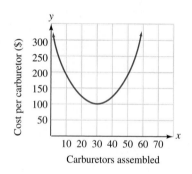

54. HEALTH DEPARTMENT The number of cases of flu seen by doctors at a county health clinic each week during a 10-week period is described by the graph. Write a brief summary report about the flu outbreak. What important piece of information does the vertex give?

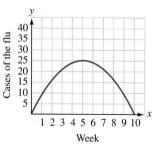

55. TRAMPOLINES The graph in the next column shows how far a trampolinist is from the ground (in relation to time) as she bounds into the air and then falls back down to the trampoline.

  a. How many feet above the ground is she $\frac{1}{2}$ second after bounding upward?

  b. When is she 9 feet above the ground?

  c. What is the maximum number of feet above the ground she gets? When does this occur?

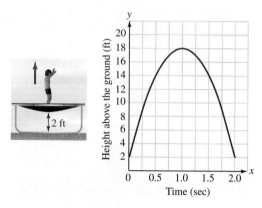

56. TRACK AND FIELD Sketch the parabolic path traveled by the long-jumper's center of gravity from the take-off board to the landing. Let the x-axis represent the ground.

WRITING

57. A mirror is held against the y-axis of the graph of a quadratic equation. What fact about parabolas does this illustrate?

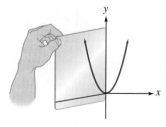

58. Use the example of a stream of water from a drinking fountain to explain the concept of the vertex of a parabola.

59. Explain why the y-intercept of the graph of the quadratic equation $y = ax^2 + bx + c$ is $(0, c)$.

60. Explain how to determine from its equation whether the graph of a parabola opens upward or downward.

61. Is it possible for the graph of a parabola with equation of the form $y = ax^2 + bx + c$ not to have a y-intercept? Explain.

62. Sketch the graphs of parabolas with zero, one, and two x-intercepts. Can the graph of a quadratic equation of the form $y = ax^2 + bx + c$ have more than two x-intercepts? Explain why or why not.

REVIEW

Simplify each expression.

63. $\sqrt{8} - \sqrt{50} + \sqrt{72}$

64. $(4\sqrt{x})(-2\sqrt{x})$

65. $3\sqrt{z}(\sqrt{4z} - \sqrt{z})$

66. $\sqrt[3]{27y^3z^6}$

SECTION 9.5
Complex Numbers

Objectives

1. Express square roots of negative numbers in terms of i.
2. Write complex numbers in $a + bi$ form.
3. Add and subtract complex numbers.
4. Multiply and divide complex numbers.
5. Solve quadratic equations with complex-number solutions.

Earlier in this chapter, we saw that some quadratic equations do not have real-number solutions. When we used the square root property or the quadratic formula to solve them, we encountered the square root of a negative number.

Section 9.1 Example 1d
Solve: $n^2 = -4$
$n = \pm\sqrt{-4}$
No real-number solutions

Section 9.3 Example 4
Solve: $x^2 + 2x + 5 = 0$
$x = \dfrac{-2 \pm \sqrt{-16}}{2}$
No real-number solutions

To solve such equations, we must define the square root of a negative number.

1 Express square roots of negative numbers in terms of *i*.

Recall that the square root of a negative number is not a real number. However, an expanded number system, called the *complex number system*, has been devised to give meaning to expressions such as $\sqrt{-4}$ and $\sqrt{-16}$. To define complex numbers, we use a new type of number that is denoted by the letter i.

> **The Number *i***
>
> The **imaginary number** i is defined as
> $i = \sqrt{-1}$
> From the definition, it follows that $i^2 = -1$.

We can use extensions of the product and quotient rules for radicals to write the square root of a negative number as the product of a real number and i.

EXAMPLE 1 Write each expression in terms of i:

a. $\sqrt{-4}$ **b.** $\sqrt{-16}$ **c.** $-\sqrt{-48}$ **d.** $\sqrt{-\dfrac{54}{25}}$

Strategy We will write each radicand as the product of -1 and a positive number. Then we will apply the appropriate rules for radicals.

WHY We want our work to produce a factor of $\sqrt{-1}$ so that we can replace it with i.

Self Check 1

Write each expression in terms of i:

a. $\sqrt{-81}$ **b.** $-\sqrt{-11}$

c. $\sqrt{-28}$ **d.** $\sqrt{-\dfrac{27}{100}}$

Now Try Problems 13, 17, and 23

Solution
After factoring the radicand, use the product rule for radicals.

a. $\sqrt{-4} = \sqrt{-1 \cdot 4} = \sqrt{-1}\sqrt{4} = i \cdot 2 = 2i$ Replace $\sqrt{-1}$ with i.

b. $\sqrt{-16} = \sqrt{-1 \cdot 16} = \sqrt{-1}\sqrt{16} = i\sqrt{16} = 4i$ Replace $\sqrt{-1}$ with i.

c. $-\sqrt{-48} = -\sqrt{-1 \cdot 16 \cdot 3} = -\sqrt{-1}\sqrt{16}\sqrt{3} = -i \cdot 4 \cdot \sqrt{3} = -4i\sqrt{3}$ or $-4\sqrt{3}i$

d. After factoring the radicand, use the product and quotient rules for radicals.

$$\sqrt{-\frac{54}{25}} = \sqrt{-1 \cdot \frac{54}{25}} = \frac{\sqrt{-1 \cdot 54}}{\sqrt{25}} = \frac{\sqrt{-1}\sqrt{9}\sqrt{6}}{\sqrt{25}} = \frac{3i\sqrt{6}}{5} \text{ or } \frac{3\sqrt{6}}{5}i$$

Success Tip Since it is easy to confuse $\sqrt{23}i$ with $\sqrt{23i}$, we write i first so that it is clear that the i is not within the radical symbol. However, both $\sqrt{23}i$ and $i\sqrt{23}$ are correct.

2 Write complex numbers in $a + bi$ form.

The imaginary number i is used to define *complex numbers*.

Complex Numbers

A **complex number** is any number that can be written in the form $a + bi$, where a and b are real numbers and $i = \sqrt{-1}$.

Complex numbers of the form $a + bi$, where $b \neq 0$ are also called **imaginary numbers**. Complex numbers of the form bi are also called **pure imaginary numbers**.

For a complex number written in the standard form $a + bi$, we call the real number a the **real part** and the real number b the **imaginary part**.

Self Check 2
Write each number in the form $a + bi$:
a. -18
b. $\sqrt{-36}$
c. $1 + \sqrt{-24}$

Now Try Problems 25 and 29

EXAMPLE 2 Write each number in the form $a + bi$:
a. 6 b. $\sqrt{-16}$ c. $-8 + \sqrt{-45}$

Strategy To write each number in the form $a + bi$, we will determine a and bi.

WHY The standard form $a + bi$ of a complex number is composed of two parts, the real part a and the imaginary part b.

Solution
a. $6 = 6 + 0i$ The real part is 6 and the imaginary part is 0.

b. To simplify $\sqrt{-16}$, we need to write $\sqrt{-16}$ in terms of i:
$$\sqrt{-16} = \sqrt{-1}\sqrt{16} = 4i$$
Thus, $\sqrt{-16} = 0 + 4i$. The real part is 0 and the imaginary part is 4.

c. To simplify $-8 + \sqrt{-45}$, we need to write $\sqrt{-45}$ in terms of i:
$$\sqrt{-45} = \sqrt{-1}\sqrt{45} = \sqrt{-1}\sqrt{9}\sqrt{5} = 3i\sqrt{5}$$
Thus, $-8 + \sqrt{-45} = -8 + 3i\sqrt{5}$. The real part is -8 and the imaginary part is $3\sqrt{5}$.

The following illustration shows the relationship between the real numbers, the imaginary numbers, and the complex numbers.

Complex numbers

Real numbers				Imaginary numbers		
−6	$\frac{5}{16}$	−1.75	π	$9 + 7i$	$-2i$	$\frac{1}{4} - \frac{3}{4}i$
48	0	$-\sqrt{10}$	$-\frac{7}{2}$	$0.56i$		$6 + i\sqrt{3}$

3 Add and subtract complex numbers.

Adding and subtracting complex numbers is similar to adding and subtracting polynomials.

Addition and Subtraction of Complex Numbers

1. To add complex numbers, add their real parts and add their imaginary parts.
2. To subtract complex numbers, add the opposite of the complex number being subtracted.

EXAMPLE 3 Find each sum or difference:
a. $(5 + 2i) + (1 + 8i)$ **b.** $(-6 - 5i) - (3 - 4i)$ **c.** $11i + (-2 + 6i)$

Strategy To add the complex numbers, we will add their real parts and add their imaginary parts. To subtract the complex numbers, we will add the opposite of the complex number to be subtracted.

WHY We perform the indicated operations as if the complex numbers were polynomials with i as a variable.

Solution
a. $(5 + 2i) + (1 + 8i) = (5 + 1) + (2 + 8)i$

— The sum of the imaginary parts.
— The sum of the real parts.

$= 6 + 10i$

b. $(-6 - 5i) - (3 - 4i) = (-6 - 5i) + (-3 + 4i)$ To find the opposite, change the sign of each term of $3 - 4i$.

the opposite

$= [-6 + (-3)] + (-5 + 4)i$ Add the real parts. Add the imaginary parts.

$= -9 - i$

c. $11i + (-2 + 6i) = -2 + (11 + 6)i$ Add the imaginary parts.
$= -2 + 17i$

Self Check 3
Find the sum or difference. Write each result in the form $a + bi$:
a. $(3 - 5i) + (-2 + 6i)$
b. $(-4 - i) - (-1 - 6i)$
c. $9 + (16 - 4i)$

Now Try Problems 33 and 35

Success Tip i is not a variable, but it is helpful to think of it as one when adding, subtracting, and multiplying. For example:

$4i + 3i = 7i$
$8i - 6i = 2i$
$i \cdot i = i^2$

4 Multiply and divide complex numbers.

Multiplying complex numbers is similar to multiplying polynomials.

Self Check 4

Find the product. Write each result in the form $a + bi$:
a. $4i(3 - 9i)$
b. $(3 - 2i)(5 - 4i)$
c. $(2 + 6i)(2 - 6i)$

Now Try Problems 43 and 45

EXAMPLE 4 Find each product:
a. $5i(4 - i)$
b. $(2 + 5i)(6 + 4i)$
c. $(3 + 5i)(3 - 5i)$

Strategy We will use the distributive property or the FOIL method to find the products.

WHY We perform the indicated operations as if the complex numbers were polynomials with i as a variable.

Solution

a. $5i(4 - i) = 5i \cdot 4 - 5i \cdot i$ Distribute the multiplication by 5i.
$= 20i - 5i^2$
$= 20i - 5(-1)$ Replace i^2 with -1.
$= 20i + 5$
$= 5 + 20i$ Write the result in the form $a + bi$.

b. $(2 + 5i)(6 + 4i) = 12 + 8i + 30i + 20i^2$ Use the FOIL method.
$= 12 + 38i + 20(-1)$ $8i + 30i = 38i$. Replace i^2 with -1.
$= 12 + 38i - 20$
$= -8 + 38i$ Subtract: $12 - 20 = -8$.

c. $(3 + 5i)(3 - 5i) = 9 - 15i + 15i - 25i^2$ Use the FOIL method.
$= 9 - 25(-1)$ Add: $-15i + 15i = 0$. Replace i^2 with -1.
$= 9 + 25$
$= 34$

Written in the form $a + bi$, the product is $34 + 0i$.

In Example 4c, we saw that the product of the imaginary numbers $3 + 5i$ and $3 - 5i$ is the real number 34. We call $3 + 5i$ and $3 - 5i$ *complex conjugates* of each other.

Complex Conjugates

The complex numbers $a + bi$ and $a - bi$ are called **complex conjugates**.

For example,

- $7 + 4i$ and $7 - 4i$ are complex conjugates.
- $5 - i$ and $5 + i$ are complex conjugates.
- $-6i$ and $6i$ are complex conjugates, because $-6i = 0 - 6i$ and $6i = 0 + 6i$.

Success Tip To find the conjugate of a complex number, write it in $a + bi$ form and change the sign between the real and imaginary parts from $+$ to $-$ or from $-$ to $+$.

In general, the product of the complex number $a + bi$ and its complex conjugate $a - bi$ is the real number $a^2 + b^2$, as the following work shows:

$$(a + bi)(a - bi) = a^2 - abi + abi - b^2i^2 \quad \text{Use the FOIL method.}$$
$$= a^2 - b^2(-1) \quad \text{Add: } -abi + abi = 0. \text{ Replace } i^2 \text{ with } -1.$$
$$= a^2 + b^2$$

We can use this fact when dividing by a complex number. The process that we use is similar to rationalizing two-term denominators.

EXAMPLE 5

Write each quotient in the form $a + bi$: **a.** $\dfrac{6}{7 - 4i}$ **b.** $\dfrac{3 - i}{2 + i}$

Strategy We will build each fraction by multiplying it by a form of 1 that uses the conjugate of the denominator.

WHY This step produces a real number in the denominator so that the result can then be written in the form $a + bi$.

Solution

a. We want to build a fraction equivalent to $\dfrac{6}{7 - 4i}$ that does not have i in the denominator. To make the denominator, $7 - 4i$, a real number, we need to multiply it by its complex conjugate, $7 + 4i$. It follows that $\dfrac{7 + 4i}{7 + 4i}$ should be the form of 1 that is used to build $\dfrac{6}{7 - 4i}$.

$\dfrac{6}{7 - 4i} = \dfrac{6}{7 - 4i} \cdot \dfrac{7 + 4i}{7 + 4i}$ To build an equivalent fraction, multiply by $\dfrac{7 + 4i}{7 + 4i} = 1$.

$= \dfrac{42 + 24i}{49 - 16i^2}$ To multiply the numerators, distribute the multiplication by 6. To multiply the denominators, find $(7 - 4i)(7 + 4i)$.

$= \dfrac{42 + 24i}{49 - 16(-1)}$ Replace i^2 with -1. The denominator no longer contains i.

$= \dfrac{42 + 24i}{49 + 16}$ Simplify the denominator.

$= \dfrac{42 + 24i}{65}$ This notation represents the sum of two fractions that have the common denominator 65: $\dfrac{42}{65}$ and $\dfrac{24i}{65}$.

$= \dfrac{42}{65} + \dfrac{24}{65}i$ Write the result in the form $a + bi$.

b. We can make the denominator of $\dfrac{3 - i}{2 + i}$ a real number by multiplying it by the complex conjugate of $2 + i$, which is $2 - i$.

$\dfrac{3 - i}{2 + i} = \dfrac{3 - i}{2 + i} \cdot \dfrac{2 - i}{2 - i}$ To build an equivalent fraction, multiply by $\dfrac{2 - i}{2 - i} = 1$.

$= \dfrac{6 - 3i - 2i + i^2}{4 - i^2}$ To multiply the numerators, find $(3 - i)(2 - i)$. To multiply the denominators, find $(2 + i)(2 - i)$.

$= \dfrac{6 - 5i + (-1)}{4 - (-1)}$ Replace i^2 with -1. The denominator no longer contains i.

$= \dfrac{5 - 5i}{5}$ Simplify the numerator and the denominator.

$= \dfrac{5}{5} - \dfrac{5i}{5}$ Write each term of the numerator over the denominator, 5.

$= 1 - i$ Simplify each fraction.

Self Check 5

Write each quotient in the form $a + bi$:

a. $\dfrac{5}{4 - i}$

b. $\dfrac{5 - 3i}{4 + 2i}$

Now Try Problems 49 and 53

Caution! A common mistake is to replace i with -1. Remember, $i \neq -1$. By definition, $i = \sqrt{-1}$ and $i^2 = -1$.

As the results of Example 5 show, we use the following rule to divide complex numbers.

Division of Complex Numbers

To divide complex numbers, multiply the numerator and denominator by the complex conjugate of the denominator.

5 Solve quadratic equations with complex-number solutions.

We have seen that certain quadratic equations do not have real-number solutions. In the complex number system, all quadratic equations have solutions. We can write their solutions in the form $a + bi$.

Self Check 6

Solve each equation. Express the solutions in the form $a + bi$:
a. $x^2 + 121 = 0$
b. $(d - 3)^2 = -48$

Now Try Problems 57 and 63

EXAMPLE 6 Solve each equation. Express the solutions in the form $a + bi$:
a. $x^2 + 25 = 0$ b. $(y - 3)^2 = -54$

Strategy We will use the square root property to solve each equation.

WHY It is the fastest way to solve equations of this form.

Solution

a. $x^2 + 25 = 0$ This is the equation to solve.

$x^2 = -25$ To isolate x^2, subtract 25 from both sides.

$x = \pm\sqrt{-25}$ Use the square root property. Note that the radicand is negative.

$x = \pm 5i$ Write $\sqrt{-25}$ in terms of i: $\sqrt{-25} = \sqrt{-1}\sqrt{25} = 5i$.

To express the solutions in the form $a + bi$, we must write 0 for the real part, a. Thus, the solutions are $0 + 5i$ and $0 - 5i$, or more simply, $0 \pm 5i$.

b. $(y - 3)^2 = -54$ This is the equation to solve.

$y - 3 = \pm\sqrt{-54}$ Use the square root property. Note that the radicand is negative.

$y - 3 = \pm 3i\sqrt{6}$ Write $\sqrt{-54}$ in terms of i: $\sqrt{-54} = \sqrt{-1}\sqrt{9}\sqrt{6} = 3i\sqrt{6}$.

$y = 3 \pm 3i\sqrt{6}$ To isolate y, add 3 to both sides.

The solutions are $3 + 3i\sqrt{6}$ and $3 - 3i\sqrt{6}$, or more simply, $3 \pm 3i\sqrt{6}$. The solution set is $\{3 \pm 3i\sqrt{6}\}$.

Self Check 7

Solve $a^2 + 2a + 3 = 0$. Express the solutions in the form $a + bi$.

Now Try Problem 65

EXAMPLE 7 Solve $4t^2 - 6t + 3 = 0$. Express the solutions in the form $a + bi$.

Strategy We use the quadratic formula to solve the equation.

WHY $4t^2 - 6t + 3$ does not factor.

Solution

$t = \dfrac{-b \pm \sqrt{b^2 - 4ac}}{2a}$ In the quadratic formula, replace x with t.

$t = \dfrac{-(-6) \pm \sqrt{(-6)^2 - 4(4)(3)}}{2(4)}$ Substitute 4 for a, -6 for b, and 3 for c.

$$t = \frac{6 \pm \sqrt{36 - 48}}{8}$$ Evaluate the power and multiply within the radical. Multiply in the denominator.

$$t = \frac{6 \pm \sqrt{-12}}{8}$$ Subtract within the radical: $36 - 48$. Note that the radicand is negative.

$$t = \frac{6 \pm 2i\sqrt{3}}{8}$$ Write $\sqrt{-12}$ in terms of i: $\sqrt{-12} = \sqrt{-1}\sqrt{12} = \sqrt{-1}\sqrt{4}\sqrt{3} = 2i\sqrt{3}$

$$t = \frac{\overset{1}{\cancel{2}}(3 \pm i\sqrt{3})}{\underset{1}{\cancel{2}} \cdot 4}$$ Factor out the GCF 2 from $6 \pm 2i\sqrt{3}$ and factor 8 as $2 \cdot 4$. Then remove the common factor 2: $\frac{2}{2} = 1$.

$$t = \frac{3 \pm i\sqrt{3}}{4}$$ This notation represents the sum and difference of two fractions that have the common denominator 4: $\frac{3}{4}$ and $\frac{i\sqrt{3}}{4}$.

Writing each result as a complex number in the form $a + bi$, the solutions are

$$\frac{3}{4} + \frac{\sqrt{3}}{4}i \quad \text{and} \quad \frac{3}{4} - \frac{\sqrt{3}}{4}i \quad \text{or more simply} \quad \frac{3}{4} \pm \frac{\sqrt{3}}{4}i$$

ANSWERS TO SELF CHECKS

1. a. $9i$ **b.** $-i\sqrt{11}$ **c.** $2i\sqrt{7}$ **d.** $\frac{3i\sqrt{3}}{10}$ **2. a.** $-18 + 0i$ **b.** $0 + 6i$ **c.** $1 + 2i\sqrt{6}$
3. a. $1 + i$ **b.** $-3 + 5i$ **c.** $25 - 4i$ **4. a.** $36 + 12i$ **b.** $7 - 22i$ **c.** $40 + 0i$
5. a. $\frac{20}{17} + \frac{5}{17}i$ **b.** $\frac{7}{10} - \frac{11}{10}i$ **6. a.** $0 \pm 11i$ **b.** $3 \pm 4i\sqrt{3}$ **7.** $-1 \pm i\sqrt{2}$

SECTION 9.5 STUDY SET

VOCABULARY

Fill in the blanks.

1. $9 + 2i$ is an example of a _____ number. The _____ part is 9 and the _____ part is 2.
2. The _____ number i is used to define complex numbers.

CONCEPTS

Fill in the blanks.

3. **a.** $i = $ _____   **b.** $i \cdot i = i^{__} = $ _____
4. $\sqrt{-25} = \sqrt{__} \cdot \sqrt{25} = $ _____ $\sqrt{25} = 5$ _____
5. **a.** $5i + 3i = $ _____   **b.** $5i - 3i = $ _____
6. The product of any complex number and its complex conjugate is a _____ number.
7. To write the quotient $\dfrac{2 - 3i}{6 - i}$ as a complex number in standard form, we multiply it by _____.

8. $\dfrac{3 \pm \sqrt{-4}}{5} = \dfrac{3 \pm \_\_}{5}$

9. Determine whether each statement is true or false.
   a. Every real number is a complex number.
   b. $2 + 7i$ is an imaginary number.
   c. $\sqrt{-16}$ is a real number.
   d. In the complex number system, all quadratic equations have solutions.

10. Give the complex conjugate of each number.
    a. $2 - 9i$ **b.** $-8 + i$
    c. $4i$ **d.** $-11i$

NOTATION

11. Write each expression so it is clear that i is not within the radical symbol.
    a. $\sqrt{7}i$ **b.** $2\sqrt{3}i$

12. Write $\dfrac{3 - 4i}{5}$ in the form $a + bi$.

GUIDED PRACTICE

Write each expression in terms of i. See Example 1.

13. $\sqrt{-9}$
14. $\sqrt{-4}$
15. $\sqrt{-7}$
16. $\sqrt{-11}$
17. $\sqrt{-24}$
18. $\sqrt{-28}$
19. $-\sqrt{-32}$
20. $-\sqrt{-72}$
21. $5\sqrt{-81}$
22. $6\sqrt{-49}$
23. $\sqrt{-\dfrac{25}{9}}$
24. $\sqrt{-\dfrac{121}{144}}$

Write each number in the form a + bi. See Example 2.

25. 12
26. -27
27. $\sqrt{-100}$
28. $\sqrt{-64}$
29. $6 + \sqrt{-16}$
30. $14 + \sqrt{-25}$
31. $-9 - \sqrt{-49}$
32. $-45 - \sqrt{-36}$

Perform the operations. Write all answers in the form a + bi. See Example 3.

33. $(3 + 4i) + (5 - 6i)$
34. $(5 + 3i) - (6 - 9i)$
35. $(7 - 3i) - (4 + 2i)$
36. $(8 + 3i) + (-7 - 2i)$
37. $(14 - 4i) - 9i$
38. $(20 - 5i) - 17i$
39. $15 + (-3 - 9i)$
40. $-25 + (18 - 9i)$

Perform the operations. Write all answers in the form a + bi. See Example 4.

41. $3(2 - i)$
42. $9(-4 - 4i)$
43. $-5i(5 - 5i)$
44. $2i(7 + 2i)$
45. $(3 - 2i)(2 + 3i)$
46. $(3 - i)(2 + 3i)$
47. $(4 + i)(3 - i)$
48. $(1 - 5i)(1 - 4i)$

Write each quotient in the form a + bi. See Example 5.

49. $\dfrac{5}{2 - i}$
50. $\dfrac{26}{3 - 2i}$
51. $\dfrac{-4i}{7 - 2i}$
52. $\dfrac{5i}{6 + 2i}$

53. $\dfrac{2 + 3i}{2 - 3i}$
54. $\dfrac{2 - 5i}{2 + 5i}$
55. $\dfrac{4 - 3i}{7 - i}$
56. $\dfrac{4 + i}{4 - i}$

Solve each equation. Write all solutions in the form a + bi. See Example 6.

57. $x^2 + 9 = 0$
58. $x^2 + 100 = 0$
59. $d^2 + 8 = 0$
60. $a^2 + 27 = 0$
61. $(x + 3)^2 = -1$
62. $(x + 2)^2 = -25$
63. $(x - 11)^2 = -75$
64. $(x - 22)^2 = -18$

Solve each equation. Write all solutions in the form a + bi. See Example 7.

65. $x^2 - 3x + 4 = 0$
66. $y^2 + y + 3 = 0$
67. $2x^2 + x + 1 = 0$
68. $2x^2 + 3x + 3 = 0$

TRY IT YOURSELF

Perform the operations. Write all answers in the form a + bi.

69. $\dfrac{3}{5 + i}$
70. $\dfrac{-4}{7 - 2i}$
71. $(6 - i) + (9 + 3i)$
72. $(5 - 4i) + (3 + 2i)$
73. $(-3 - 8i) - (-3 - 9i)$
74. $(-1 - 8i) - (-1 - 7i)$
75. $(2 + i)(2 + 3i)$
76. $2i(7 + 2i)$
77. $\dfrac{3 - 2i}{3 + 2i}$
78. $\dfrac{3 + 2i}{3 + i}$
79. $(10 - 9i) + (-1 + i)$
80. $(32 - 3i) + (-44 + 15i)$
81. $-4(3 + 4i)$
82. $-7(5 - 3i)$
83. $(2 + i)^2$
84. $(3 - 2i)^2$

Solve each equation. Write all solutions in the form a + bi.

85. $2x^2 + x = -5$
86. $4x^2 = -7x - 4$
87. $(x - 4)^2 = -45$
88. $(x - 9)^2 = -80$
89. $b^2 + 2b + 2 = 0$
90. $t^2 - 2t + 6 = 0$

91. $x^2 = -36$ **92.** $x^2 = -49$

93. $x^2 = -\dfrac{16}{9}$ **94.** $x^2 = -\dfrac{25}{4}$

95. $3x^2 + 2x + 1 = 0$ **96.** $3x^2 - 4x + 2 = 0$

APPLICATIONS

97. ELECTRICITY In an AC (alternating current) circuit, if two sections are connected in series and have the same current in each section, the voltage is given by $V = V_1 + V_2$. Find the total voltage in a given circuit if the voltages in the individual sections are $V_1 = 10.31 - 5.97i$ and $V_2 = 8.14 + 3.79i$.

98. ELECTRONICS The impedance Z in an AC (alternating current) circuit is a measure of how much the circuit impedes (hinders) the flow of current through it. The impedance is related to the voltage V and the current I by the formula

$V = IZ$

If a circuit has a current of $(0.5 + 2.0i)$ amps and an impedance of $(0.4 - 3.0i)$ ohms, find the voltage.

WRITING

99. What unusual situation discussed at the beginning of this section illustrated the need to define the square root of a negative number?

100. Explain the difference between the opposite of a complex number and its conjugate.

101. What is an imaginary number?

102. In this section, we have seen that $i^2 = -1$. From your previous experience in this course, what is unusual about that fact?

REVIEW

Rationalize the denominator.

103. $\dfrac{1}{\sqrt{7}}$ **104.** $\dfrac{\sqrt{3}}{\sqrt{10}}$

105. $\dfrac{8}{\sqrt{x} - 2}$ **106.** $\dfrac{\sqrt{3}}{5 + \sqrt{x}}$

STUDY SKILLS CHECKLIST

Preparing for the Chapter 9 Test

The material in Chapter 9 focused on quadratic equations, graphing quadratic equations, and complex numbers. Be sure to review the following checklist in addition to your other studying as you prepare for the exam on this material.

☐ When solving a quadratic equation using the factoring method, one side of the equation must be 0. Example:

Solve:

$x^2 = 9x$	This is the equation to solve.
$x^2 - 9x = 0$	Subtract 9x from both sides to get 0 on the right side.
$x(x - 9) = 0$	Factor the left side.
$x = 0 \text{ or } x - 9 = 0$	Set each factor equal to 0.
$x = 9$	Solve each linear equation.

☐ When using the square root property to solve a quadratic equation, always write the $\pm$ sign. Example:

Solve:

$x^2 - 7 = 0$	This is the equation to solve.
$x^2 = 7$	Add 7 from both sides to isolate the x^2 term.
$\sqrt{x^2} = \pm\sqrt{7}$	Use the square root property and remember the $\pm$.
$x = \pm\sqrt{7}$	Simplify the left side of the equation.

☐ We can find the vertex of the graph of a quadratic equation in two variables by completing the square or by using the fact that the x-coordinate of the vertex is $x = \dfrac{-b}{2a}$. To find the y-coordinate of the vertex, substitute $x = \dfrac{-b}{2a}$ for x into the equation and find y.

☐ When solving a quadratic equation using the completing the square method, make sure the coefficient of the x^2 term is 1 before you complete the square. If it is not 1, you must divide both sides of the equation by the coefficient of the x^2 to make it 1. Example:

Solve by completing the square.

$2x^2 - 12x = 6$	This is the equation to solve.
$x^2 - 6x = 3$	Divide both sides by 2 to make the coefficient of the x^2 term 1.
$x^2 - 6x + 9 = 3 + 9$	Complete the square on the left side by $\frac{1}{2}(-6) = -3, (-3)^2 = 9$; add 9 to both sides.
$(x - 3)^2 = 12$	Factor the left side and simplify the right side of the equation.
$\sqrt{(x-3)^2} = \pm\sqrt{12}$	Use the square root property and remember the $\pm$.
$x - 3 = \pm 2\sqrt{3}$	Simplify the left and right sides of the equation.
$x = 3 \pm 2\sqrt{3}$	Isolate x by adding 3 to both sides of the equation.

☐ Adding, subtracting, and multiplying complex numbers is similar to these operations with polynomials. Remember to replace i^2 with -1 when simplifying the expression.

CHAPTER 9 SUMMARY AND REVIEW

SECTION 9.1 Solving Quadratic Equations: The Square Root Property

DEFINITIONS AND CONCEPTS

We can use the **square root property** to solve equations of the form $x^2 = c$, where $c > 0$. The two solutions are

$$x = \sqrt{c} \quad \text{or} \quad x = -\sqrt{c}$$

We can write $x = \sqrt{c}$ or $x = -\sqrt{c}$ in more compact form using **double-sign notation**:

$$x = \pm\sqrt{c}$$

EXAMPLES

Solve: $x^2 = 27$

$x = \sqrt{27} \quad \text{or} \quad x = -\sqrt{27}$ Use the square root property.
$x = \pm\sqrt{27}$ Use double-sign notation.
$x = \pm 3\sqrt{3}$ Simplify: $\sqrt{27} = \sqrt{9 \cdot 3} = \sqrt{9} \cdot \sqrt{3} = 3\sqrt{3}$.

The solutions are $3\sqrt{3}$ and $-3\sqrt{3}$. The solution set is $\{\pm 3\sqrt{3}\}$.

Solve: $(x - 3)^2 = 5$

$x - 3 = \pm\sqrt{5}$ Use the square root property and double-sign notation.
$x = 3 \pm \sqrt{5}$ To isolate x, add 3 to both sides.

The exact solutions are $3 + \sqrt{5}$ and $3 - \sqrt{5}$. The solution set is $\{3 \pm \sqrt{5}\}$. If we approximate the solutions to the nearest hundredth, we have

$3 + \sqrt{5} \approx 3 + 2.236067978 \approx 5.24$ Use a calculator.
$3 - \sqrt{5} \approx 3 - 2.236067978 \approx 0.76$

Solve: $x^2 - 22x + 121 = 25$

$(x - 11)^2 = 25$ Factor the perfect-square trinomial.
$x - 11 = \pm\sqrt{25}$ Use the square root property.
$x = 11 \pm 5$ Add 11 to both sides and simplify $\sqrt{25}$.
$x = 11 + 5 \quad \text{or} \quad x = 11 - 5$
$x = 16 \qquad\qquad\qquad x = 6$

The solutions are 16 and 6. The solution set is $\{6, 16\}$.

REVIEW EXERCISES

Use the square root property to solve each equation.

1. $x^2 = 64$
2. $t^2 - 8 = 0$
3. $2x^2 - 1 = 149$
4. $(x - 1)^2 = 25$
5. $(9x - 8)^2 = 40$
6. $4(x - 2)^2 - 9 = 0$
7. $p^2 - 20p + 100 = 9$
8. $9m^2 + 6m + 1 = 6$

Use the square root property to find all real-number solutions of each equation. Round each solution to the nearest hundredth.

9. $x^2 = 12$
10. $(x - 1)^2 = 55$
11. $m^2 + 36 = 0$
12. $(2x - 3)^2 = -8$

13. **CLIFF DIVERS** The La Quebrada Cliff Divers of Acapulco, Mexico, perform daily for the public by diving 148 feet from ocean-side cliffs into the sea below. Find the length of time of a dive. Round to the nearest one tenth of one second. (*Hint:* Use the formula $d = 16t^2$.)

14. **RUBIK'S CUBE** The area of one of the square faces of a Rubik's cube is $\frac{81}{16}$ in.^{2}. Find the length of one side of a Rubik's cube. Express your answer as a mixed number. (*Hint:* Use the formula for the area of a square, $A = s^2$.)

SECTION 9.2 Solving Quadratic Equations: Completing the Square

DEFINITIONS AND CONCEPTS

To **complete the square** on $x^2 + bx$, add the square of one-half of the coefficient of x.

$$x^2 + bx + \left(\frac{1}{2}b\right)^2$$

To solve a quadratic equation in x by completing the square:

1. If necessary, divide both sides of the equation by the coefficient of x^2 to make its coefficient 1.
2. Get all variable terms on one side of the equation and all constants on the other side.
3. Complete the square by finding one-half of the coefficient of x, squaring the result, and adding the square to both sides of the equation.
4. Factor the perfect-square trinomial.
5. Solve the resulting equation by using the square root property.
6. Check your answers in the original equation.

EXAMPLES

Complete the square on $x^2 + 12x$ and factor the resulting perfect-square trinomial.

$x^2 + 12x + 36$ The coefficient of x is 12. To complete the square: $\frac{1}{2} \cdot 12 = 6$ and $6^2 = 36$. Add 36 to the binomial.

This trinomial factors as $(x + 6)^2$. We can check using multiplication.

Solve: $3x^2 - 12x + 6 = 0$

$\frac{3x^2}{3} - \frac{12}{3}x + \frac{6}{3} = 0$ To make the leading coefficient 1, divide both sides by 3.

$x^2 - 4x + 2 = 0$ Do the divisions.

$x^2 - 4x = -2$ Subtract 2 from both sides so that the constant term, -2, is on the right side.

$x^2 - 4x + 4 = -2 + 4$ The coefficient of x is -4. To complete the square: $\frac{1}{2}(-4) = -2$ and $(-2)^2 = 4$. Add 4 to both sides.

$(x - 2)^2 = 2$ Factor the perfect-square trinomial on the left side. Add on the right side.

$x - 2 = \pm\sqrt{2}$ Use the square root property.

$x = 2 \pm \sqrt{2}$ To isolate x, add 2 to both sides.

The solutions are $2 + \sqrt{2}$ and $2 - \sqrt{2}$. The solution set is $\{2 \pm \sqrt{2}\}$. We can approximate each solution. To the nearest one hundredth

$2 + \sqrt{2} \approx 3.41 \qquad 2 - \sqrt{2} \approx 0.59$

REVIEW EXERCISES

Complete the square to make each expression a perfect-square trinomial. Then factor.

15. $x^2 + 4x$
16. $t^2 - 5t$

Solve each quadratic equation by completing the square.

17. $x^2 - 8x + 15 = 0$
18. $x^2 = -5x + 14$

19. $x^2 + 2x = 5$
20. $4x^2 - 16x = 7$
21. $2x^2 - 2x - 1 = 0$
22. $3x^2 + 5x + 2 = 0$

Solve each quadratic equation by completing the square. Round each solution to the nearest hundredth.

23. $x^2 + 4x + 1 = 0$
24. $x^2 - 7x = 5$

SECTION 9.3 Solving Quadratic Equations: The Quadratic Formula

DEFINITIONS AND CONCEPTS

To solve a quadratic equation in x using the quadratic formula:

1. Write the equation in standard form:
   $$ax^2 + bx + c = 0$$

2. Identify a, b, and c.

3. Substitute the values for a, b, and c in the quadratic formula
   $$x = \frac{-b \pm \sqrt{b^2 - 4ac}}{2a}$$
   and evaluate the right side to obtain the solutions.

EXAMPLES

Use the quadratic formula to solve: $3x^2 - 2x = 2$

$3x^2 - 2x - 2 = 0$ To get 0 on the right side, subtract 2 from both sides.

Here, $a = 3$, $b = -2$, and $c = -2$.

$x = \dfrac{-b \pm \sqrt{b^2 - 4ac}}{2a}$ This is the quadratic formula.

$x = \dfrac{-(-2) \pm \sqrt{(-2)^2 - 4(3)(-2)}}{2(3)}$ Substitute 3 for a, -2 for b, and -2 for c.

$x = \dfrac{2 \pm \sqrt{4 - (-24)}}{6}$ Evaluate within the radical. Multiply in the denominator.

$x = \dfrac{2 \pm \sqrt{28}}{6}$ Add the opposite: $4 - (-24) = 4 + 24 = 28$.

$x = \dfrac{2 \pm 2\sqrt{7}}{6}$ Simplify: $\sqrt{28} = \sqrt{4}\sqrt{7} = 2\sqrt{7}$.

$x = \dfrac{\overset{1}{\cancel{2}}(1 \pm \sqrt{7})}{\underset{1}{\cancel{2}} \cdot 3}$ Factor out the GCF, 2, in the numerator. In the denominator, factor 6.

$x = \dfrac{1 \pm \sqrt{7}}{3}$ Remove the common factor, 2.

The solutions are $\dfrac{1 \pm \sqrt{7}}{3}$, and the solution set is $\left\{\dfrac{1 \pm \sqrt{7}}{3}\right\}$.

A **strategy for solving quadratic equations** is given on page 739.

A suggested method for solving each quadratic equation is given.

Factor method
$x^2 - 3x - 18 = 0$

Square root property
$(x - 1)^2 = 18$

Complete the square
$x^2 + 6x - 11 = 0$
↑
even coefficient

Quadratic formula
$3x^2 - 9x + 1 = 0$

REVIEW EXERCISES

Write each equation in $ax^2 + bx + c = 0$ form and find a, b, and c.

25. $x^2 + 2x = -5$
26. $6x^2 = 2x + 1$

Use the quadratic formula to find all real-number solutions of each equation.

27. $x^2 - 2x - 15 = 0$
28. $6x^2 = 7x + 3$
29. $p^2 - 4 = 2p$
30. $x^2 + 7 = 6x$
31. $3x^2 + 3x = 1$
32. $5x^2 + x = 1$
33. $7x^2 - x + 2 = 0$
34. $2x^2 + 6x = 5$

Use the most efficient method to find all real-number solutions of each equation.

35. $4x^2 + 16x = 0$
36. $(y + 3)^2 = 16$
37. $3g^2 - 81 = 0$
38. $3x^2 - 6x = -1$

39. $2x^2 + 2x - 5 = 0$ **40.** $a^2 = 4a - 4$

41. $(2x - 5)^2 = 64$ **42.** $a^2 - 2a + 5 = 0$

43. Use the quadratic formula to solve $3x^2 + 2x - 2 = 0$. Give the solutions in exact form and then rounded to the nearest hundredth.

44. SECURITY GATES The length of the frame for an iron gate is 14 feet more than the width. A diagonal cross brace is 26 feet long. Find the width and length of the gate frame.

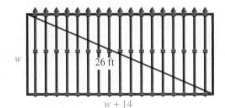

45. THE GRAND CANYON The depth of the Grand Canyon at the South Rim is almost one mile. Suppose a visitor standing on the rim tosses a rock upward over the canyon. The time t (in seconds) that it takes for the rock to hit the bottom of the canyon can be found by solving the quadratic equation $0 = -16t^2 + 8t + 5{,}040$. Find t.

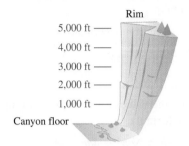

46. GEOMETRY The triangle shown has an area of 24 square inches. Find its height.

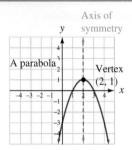

SECTION 9.4 Graphing Quadratic Equations

DEFINITIONS AND CONCEPTS	EXAMPLES
The **vertex** of a parabola is the lowest (or highest) point on the parabola. A vertical line through the vertex of a parabola that opens upward or downward is called its **axis of symmetry**.	
Equations that can be written in the form $y = ax^2 + bx + c$, where $a \neq 0$, are called **quadratic equations in two variables**. The graph of $y = ax^2 + bx + c$ is a **parabola**. Much can be determined about the graph from the coefficients a, b, and c. The parabola opens **upward** when $a > 0$ and downward when $a < 0$.	Graph: $y = x^2 + 6x + 8$ *Upward/downward:* The equation is in the form $y = ax^2 + bx + c$, with $a = 1$, $b = 6$, and $c = 8$. Since $a > 0$, the parabola opens upward. *Vertex/axis of symmetry:* The x-coordinate of the vertex is $$\frac{-b}{2a} = \frac{-6}{2(1)} = -3$$

The x-coordinate of the **vertex** of the parabola is $x = \frac{-b}{2a}$. To find the y-coordinate of the vertex, substitute $\frac{-b}{2a}$ for x in the equation of the parabola and find y.

To find the **y-intercept**, substitute 0 for x in the given equation and solve for y. To find the **x-intercepts**, substitute 0 for y in the given equation and solve for x.

The number of distinct x-intercepts of the graph of a quadratic equation $y = ax^2 + bx + c$ is the same as the number of distinct **real-number solutions** of $ax^2 + bx + c = 0$.

The y-coordinate of the vertex is:

$y = x^2 + 6x + 8$ The equation to graph.
$y = (-3)^2 + 6(-3) + 8$
$y = 9 - 18 + 8$
$y = -1$

The vertex of the parabola is $(-3, -1)$.

Intercepts: Since $c = 8$, the y-intercept of the parabola is $(0, 8)$. Since this point is 3 units to the right of the axis of symmetry, the point $(-6, 8)$, which is 3 units to the left of the axis of symmetry, must also be on the graph.

To find the x-intercepts of the graph of $y = x^2 + 6x + 8$, we set $y = 0$ and solve for x.

$$0 = x^2 + 6x + 8$$
$$0 = (x + 4)(x + 2)$$
$$x + 4 = 0 \quad \text{or} \quad x + 2 = 0$$
$$x = -4 \quad | \quad x = -2$$

The x-intercepts of the graph are $(-4, 0)$ and $(-2, 0)$.

Plotting points/using symmetry: To locate two more points on the graph, we let $x = -1$ and find the corresponding value of y.

$$y = (-1)^2 + 6(-1) + 8 = 3$$

Thus, the point $(-1, 3)$ lies on the parabola. We use symmetry to determine that $(-5, 3)$ is also on the graph.

Drawing the curve: Draw a smooth curve through the points. The completed graph of $y = x^2 + 6x + 8$ is shown here.

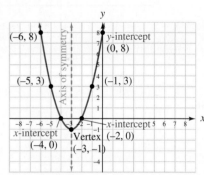

Since the graph of $y = x^2 + 6x + 8$ (shown above) has two x-intercepts, $(-4, 0)$ and $(-2, 0)$, the equation $x^2 + 6x + 8 = 0$ has two distinct real-number solutions, -4 and -2.

REVIEW EXERCISES

47. Refer to the figure.

a. What are the x-intercepts of the parabola?

b. What is the y-intercept of the parabola?

c. What is the vertex of the parabola?

d. Draw the axis of symmetry of the parabola on the graph.

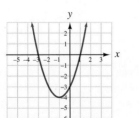

48. The point $(0, -3)$ lies on the parabola in Problem 47. Use symmetry to determine the coordinates of another point that lies on the parabola.

Find the vertex of the graph of each quadratic equation and tell in which direction the parabola opens. Do not draw the graph.

49. $y = 2x^2 - 4x + 7$ **50.** $y = -3x^2 + 18x - 11$

Find the x- and y-intercepts of the graph of each quadratic equation.

51. $y = x^2 + 6x + 5$ **52.** $y = x^2 + 2x + 3$

Graph each quadratic equation by finding the vertex, the x- and y-intercepts, and the axis of symmetry of its graph.

53. $y = x^2 + 2x - 3$ **54.** $y = -2x^2 + 4x - 2$

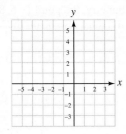 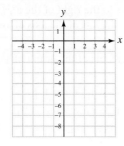

55. $y = -x^2 - 2x + 5$

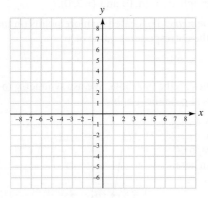

56. $y = x^2 + 4x - 1$

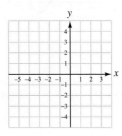

57. The graphs of three quadratic equations in two variables are shown. Fill in the blanks.

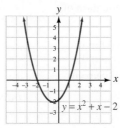

$x^2 + x - 2 = 0$ has ___ real-number solution(s). Give the solution(s): ___

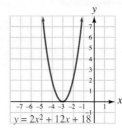

$2x^2 + 12x + 18 = 0$ has ___ repeated real-number solution. Give the solution(s): ___

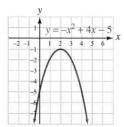

$-x^2 + 4x - 5 = 0$ has ___ real-number solution(s).

58. MANUFACTURING What important information can be obtained from the vertex of the parabola in the graph below?

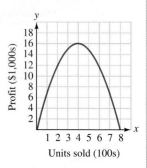

SECTION 9.5 Complex Numbers

DEFINITIONS AND CONCEPTS	EXAMPLES
The **imaginary number** i is defined as $$i = \sqrt{-1}$$ From the definition, it follows that $i^2 = -1$.	Write each expression in terms of i: $$\sqrt{-100} = \sqrt{-1 \cdot 100} \qquad \sqrt{-75} = \sqrt{-1 \cdot 75}$$ $$= \sqrt{-1}\sqrt{100} \qquad\quad = \sqrt{-1}\sqrt{75}$$ $$= i \cdot 10 \qquad\qquad\quad = i\sqrt{25}\sqrt{3}$$ $$= 10i \qquad\qquad\qquad = 5i\sqrt{3} \text{ or } 5\sqrt{3}i$$
A **complex number** is any number that can be written in the form $a + bi$, where a and b are real numbers and $i = \sqrt{-1}$. We call a the **real part** and b the **imaginary part**. If $b \neq 0$, $a + bi$ is also an **imaginary number**.	Show that each number is a complex number by writing it in the form $a + bi$. $$10 = 10 + 0i \quad \text{10 is the real part and 0 is the imaginary part.}$$ $$5i = 0 + 5i \quad \text{0 is the real part and 5 is the imaginary part.}$$ $$-2 - \sqrt{-16} = -2 - 4i \quad \text{-2 is the real part and -4 is the imaginary part.}$$
Adding and subtracting complex numbers is similar to adding and subtracting polynomials. To **add two complex numbers**, add their real parts and add their imaginary parts. To **subtract two complex numbers**, add the opposite of the complex number being subtracted.	Add: $$(3 - 4i) + (5 + 7i) = (3 + 5) + (-4 + 7)i \quad \text{Add the real parts. Add the imaginary parts.}$$ $$= 8 + 3i$$ Subtract: $$(3 - 4i) - (5 + 7i) = (3 - 4i) + (-5 - 7i) \quad \text{Add the opposite of $5 + 7i$.}$$ $$= (3 - 5) + [-4 + (-7)]i \quad \text{Add the real parts. Add the imaginary parts.}$$ $$= -2 - 11i$$
Multiplying complex numbers is similar to multiplying polynomials.	Find each product: $$7i(4 - 9i) = 28i - 63i^2 \qquad (2 + 3i)(4 - 3i) = 8 - 6i + 12i - 9i^2$$ $$= 28i - 63(-1) \qquad\qquad\qquad\quad = 8 + 6i - 9(-1)$$ $$= 28i + 63 \qquad\qquad\qquad\qquad\quad = 8 + 6i + 9$$ $$= 63 + 28i \qquad\qquad\qquad\qquad\quad = 17 + 6i$$
The complex numbers $a + bi$ and $a - bi$ are called **complex conjugates**.	The complex numbers $4 - 3i$ and $4 + 3i$ are complex conjugates.

To write the **quotient of two complex numbers** in the form $a + bi$, multiply the numerator and denominator by the complex conjugate of the denominator. The process is similar to rationalizing two-term denominators.

Write each quotient in the form $a + bi$:

$$\frac{6}{4 - 3i} = \frac{6}{4 - 3i} \cdot \frac{4 + 3i}{4 + 3i}$$

$$= \frac{24 + 18i}{16 - 9i^2}$$

$$= \frac{24 + 18i}{16 - 9(-1)}$$

$$= \frac{24 + 18i}{16 + 9}$$

$$= \frac{24 + 18i}{25}$$

$$= \frac{24}{25} + \frac{18}{25}i$$

$$\frac{2 + i}{1 + i} = \frac{2 + i}{1 + i} \cdot \frac{1 - i}{1 - i}$$

$$= \frac{2 - 2i + i - i^2}{1 - i^2}$$

$$= \frac{2 - i - (-1)}{1 - (-1)}$$

$$= \frac{3 - i}{2}$$

$$= \frac{3}{2} - \frac{1}{2}i$$

Some quadratic equations have **complex solutions that are imaginary numbers**.

Use the quadratic formula to solve: $p^2 + p + 3 = 0$

Here, $a = 1$, $b = 1$, and $c = 3$.

$$p = \frac{-b \pm \sqrt{b^2 - 4ac}}{2a}$$ In the quadratic formula, replace x with p.

$$p = \frac{-1 \pm \sqrt{1^2 - 4(1)(3)}}{2(1)}$$ Substitute 1 for a, 1 for b, and 3 for c.

$$p = \frac{-1 \pm \sqrt{-11}}{2}$$ Evaluate the power and multiply within the radical. Multiply in the denominator.

$$p = \frac{-1 \pm i\sqrt{11}}{2}$$ Write $\sqrt{-11}$ in terms of i.

$$p = -\frac{1}{2} \pm \frac{\sqrt{11}}{2}i$$ Write the solutions in the form $a + bi$.

The solutions are $-\frac{1}{2} \pm \frac{\sqrt{11}}{2}i$. The solution set is $\left\{ -\frac{1}{2} \pm \frac{\sqrt{11}}{2}i \right\}$.

REVIEW EXERCISES

Write each expression in terms of i.

59. $\sqrt{-25}$ **60.** $\sqrt{-18}$

61. $-\sqrt{-49}$ **62.** $\sqrt{-\dfrac{9}{64}}$

63. Complete the diagram.

Complex numbers	
numbers	numbers

64. Determine whether each statement is true or false.
  a. Every real number is a complex number.
  b. $3 - 4i$ is an imaginary number.
  c. $\sqrt{-4}$ is a real number.
  d. i is a real number.

Give the complex conjugate of each number.

65. $3 + 6i$ **66.** $-1 - 7i$

67. $19i$ **68.** $-i$

Perform the operations. Write all answers in the form $a + bi$.

69. $(3 + 4i) + (5 - 6i)$ **70.** $(7 - 3i) - (4 + 2i)$

71. $3i(2 - i)$ **72.** $(2 + 3i)(3 - i)$

73. $\dfrac{2 + 3i}{2 - 3i}$ **74.** $\dfrac{3}{5 + i}$

Solve each equation. Write all solutions in the form $a + bi$.

75. $x^2 + 9 = 0$ **76.** $3x^2 = -16$

77. $(p - 2)^2 = -24$ **78.** $(q + 3)^2 = -54$

79. $x^2 + 2x = -2$ **80.** $2x^2 - 3x + 2 = 0$

CHAPTER 9 TEST

1. Fill in the blanks.
   a. A _____ number is any number that can be written in the form $a + bi$, where a and b are real numbers and $i = \sqrt{-1}$.
   b. The graph of the equation $y = x^2$ is a cup-shaped figure called a _____.
   c. A _____ equation can be written in the form $ax^2 + bx + c = 0$, where a, b, and c represent real numbers and $a \neq 0$.
   d. For $7 + 8i$, the real part is 7 and the _____ part is 8.
   e. The _____ coefficient of $3x^2 + 8x - 9$ is 3.

2. Write the statement $x = \sqrt{5}$ or $x = -\sqrt{5}$ using double-sign notation.

Solve each equation by the square root method.

3. $x^2 = 17$ 4. $(x - 2)^2 = 3$

5. $4y^2 - 25 = 0$ 6. $x^2 + 16x + 64 = 24$

7. Explain why the equation $m^2 + 49 = 0$ has no real-number solutions.

8. ARCHERY The area of a circular archery target is 5,026 cm². What is the radius of the target? Round to the nearest centimeter.

Complete the square and factor the resulting perfect-square trinomial.

9. $x^2 - 14x$

10. $a^2 - \dfrac{5}{3}a$

11. Complete the square to solve $a^2 + 2a - 4 = 0$. Give the exact solutions and then round them to the nearest hundredth.

12. Complete the square to solve: $2x^2 = 3x + 2$

Use the quadratic formula to solve each equation.

13. $2x^2 - 5x - 12 = 0$ 14. $5x^2 + 11x = -3$

15. Solve $3x^2 - 2x - 2 = 0$ using the quadratic formula. Give the exact solutions, and then approximate them to the nearest hundredth.

16. NEW YORK CITY The rectangular Samsung sign in Times Square is a full-color LED screen that has an area of 2,665 ft². Its length is 17 feet less than twice its width. Find the width and length of the sign.

17. Use a check to determine whether $1 + \sqrt{5}$ is a solution of $x^2 - 2x - 4 = 0$.

18. ST. LOUIS On October 28, 1965, workers "topped out" the final section of the Gateway Arch in St. Louis, Missouri. It is the tallest national monument in the United States at 630 feet. If a worker dropped a tool from that height, how long would it take to reach the ground? Round to the nearest tenth. (*Hint:* Use the formula $d = 16t^2$.)

Use the most efficient method to solve each equation.

19. $x^2 - 4x = -2$ **20.** $(3b + 1)^2 = 16$

21. $u^2 - 24 = 0$ **22.** $3x^2 + 2x + 1 = 0$

23. ADVERTISING When a business runs x advertisements per week on television, the number y of air conditioners it sells is given by the graph. What important information can be obtained from the vertex?

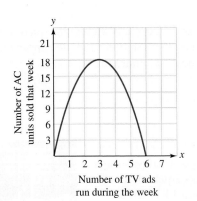

24. Fill in the blanks: The graph of $y = ax^2 + bx + c$ opens downward when a ___ 0 and upward when a ___ 0.

Graph each quadratic equation by finding the vertex, the x- and y-intercepts, and the axis of symmetry of its graph.

25. $y = x^2 + 6x + 5$ **26.** $y = -x^2 + 6x - 7$

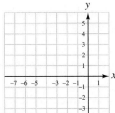 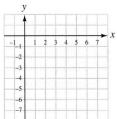

Write each expression in terms of i.

27. $\sqrt{-100}$ **28.** $-\sqrt{-18}$

Perform the operations. Write all answers in the form a + bi.

29. $(8 + 3i) + (-7 - 2i)$ **30.** $(5 + 3i) - (6 - 9i)$

31. $(2 - 4i)(3 + 2i)$ **32.** $\dfrac{3 - 2i}{3 + 2i}$

CHAPTERS 1–9 CUMULATIVE REVIEW

1. Determine whether each statement is true or false. [Section 1.3]
   a. Every rational number can be written as a ratio of two integers.
   b. The set of real numbers corresponds to all points on the number line.
   c. The whole numbers and their opposites form the set of integers.

2. DRIVING SAFETY In cold climates, salt is spread on roads to keep snow and ice from bonding to the pavement. This allows snowplows to remove accumulated snow quickly. According to the graph, when is the accident rate the highest? [Section 1.4]

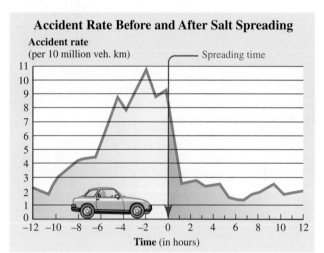

Based on data from the Salt Institute

3. Evaluate: $-4 + 2[-7 - 3(-9)]$ [Section 1.7]

4. Evaluate: $\left| \dfrac{4}{5} \cdot 10 - 12 \right|$ [Section 1.7]

5. Evaluate $(x - a)^2 + (y - b)^2$ for $x = -2$, $y = 1$, $a = 5$, and $b = -3$. [Section 1.8]

6. Simplify: $3p - 6(p - 9) + p$ [Section 1.9]

7. Solve $\dfrac{5}{6}k = 10$ and check the result. [Section 2.1]

8. Solve $-(3a + 1) + a = 2$ and check the result. [Section 2.2]

9. LOOSE CHANGE The Coinstar machines that are in many grocery stores count unsorted coins and print out a voucher that can be exchanged for cash at the checkout stand. However, to use this service, a processing fee is charged. If a boy turned in a jar of coins worth $50 and received a voucher for $45.55, what was the processing fee (expressed as a percent) charged by Coinstar? [Section 2.3]

10. Solve $T = 2r + 2t$ for r. [Section 2.4]

11. SELLING A HOME At what price should a home be listed if the owner wants to make $330,000 on its sale after paying a 4% real estate commission? [Section 2.5]

12. BUSINESS LOANS Last year, a women's professional organization made two small business loans totaling $28,000 to young women beginning their own businesses. The money was lent at 7% and 10% simple interest rates. If the annual income the organization received from these loans was $2,560, what was each loan amount? [Section 2.6]

13. Solve $5x + 7 < 2x + 1$ and graph the solution set. Then use interval notation to describe the solution. [Section 2.7]

14. Use a check to determine whether $(-5, -3)$ is a solution of $2x - 3y = -1$. [Section 3.1]

Graph each equation or inequality.

15. $y = -x + 2$ [Section 3.2]

16. $2y - 2x = 6$ [Section 3.3]

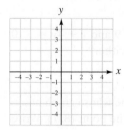
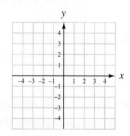

17. $y = -3$
[Section 3.3]

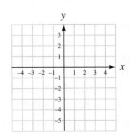

18. $y < 3x$
[Section 3.7]

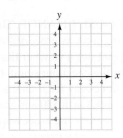

19. Find the slope of the line passing through $(-2, -2)$ and $(-12, -8)$. [Section 3.4]

20. TV NEWS The graph in red approximates the evening news viewership on all networks for the years 1995–2012. Find the rate of decrease over this period of time. [Section 3.4]

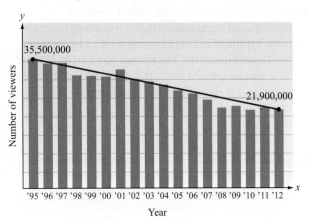

Source: The State of the News Media, 2013

21. What is the slope of the line defined by $4x + 5y = 6$? [Section 3.5]

22. Write the equation of the line whose graph has slope -2 and y-intercept $(0, 1)$. [Section 3.5]

23. Are the graphs of $y = 4x + 9$ and $x + 4y = -10$ parallel, perpendicular, or neither? [Section 3.5]

24. Write the equation of the line whose graph has slope $\frac{1}{4}$ and passes through the point $(8, 1)$. Write the equation in slope–intercept form. [Section 3.6]

25. Graph the line passing through $(-2, -1)$ and having slope $\frac{4}{3}$. [Section 3.6]

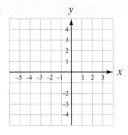

26. If $f(x) = 3x^2 + 3x - 8$, find $f(-1)$. [Section 3.8]

27. Find the domain and range of the relation: $\{(1, 8), (4, -3), (-4, 2), (5, 8)\}$ [Section 3.8]

28. Is this the graph of a function? Explain why or why not. [Section 3.8]

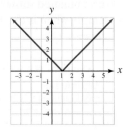

29. Solve using the graphing method. [Section 4.1]

$$\begin{cases} x + y = 1 \\ y = x + 5 \end{cases}$$

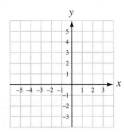

30. Solve using the substitution method.

$$\begin{cases} y = 2x + 5 \\ x + 2y = -5 \end{cases}$$ [Section 4.2]

31. Solve using the elimination (addition) method.

$$\begin{cases} \frac{3}{5}s + \frac{4}{5}t = 1 \\ -\frac{1}{4}s + \frac{3}{8}t = 1 \end{cases}$$ [Section 4.3]

32. AVIATION With the wind, a plane can fly 3,000 miles in 5 hours. Against the wind, the trip takes 6 hours. Find the airspeed of the plane (the speed in still air). Use two variables to solve this problem. [Section 4.4]

33. **MIXING CANDY** How many pounds of each candy must be mixed to obtain 48 pounds of candy worth $4.50 per pound? Use two variables to solve this problem. [Section 4.4]

34. Solve the system of linear inequalities.

$$\begin{cases} 3x + 4y > -7 \\ 2x - 3y \geq 1 \end{cases}$$

[Section 4.5]

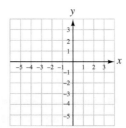

Simplify each expression. Write each answer without using parentheses or negative exponents.

35. $y^3(y^2y^4)$ [Section 5.1]
36. $\left(\dfrac{b^2}{3a}\right)^3$ [Section 5.1]

37. $\dfrac{10a^4 a^{-2}}{5a^2 a^0}$ [Section 5.2]
38. $\left(\dfrac{21x^{-2}y^2z^{-2}}{7x^3 y^{-1}}\right)^{-2}$ [Section 5.2]

39. **FIVE-CARD POKER** The odds against being dealt the hand shown are about 2.6×10^6 to 1. Express 2.6×10^6 using standard notation. [Section 5.3]

40. Write 0.00073 in scientific notation. [Section 5.3]

41. Graph: $y = x^3 - 2$ [Section 5.4]

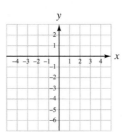

42. Write a polynomial that represents the perimeter of the rectangle. [Section 5.5]

(rectangle with sides $2x^3 - x$ and $x^3 + 3x$)

Perform the operations.

43. $4(4x^3 + 2x^2 - 3x - 8) - 5(2x^3 - 3x + 8)$ [Section 5.5]

44. $(-2a^3)(3a^2)$ [Section 5.6]

45. $(2b - 1)(3b + 4)$ [Section 5.6]

46. $(3x + y)(2x^2 - 3xy + y^2)$ [Section 5.6]

47. $(2x + 5y)^2$ [Section 5.7]

48. $(9m^2 - 1)(9m^2 + 1)$ [Section 5.7]

49. $\dfrac{12a^3b - 9a^2b^2 + 3ab}{6a^2b}$ [Section 5.8]

50. $x - 3 \overline{)2x^2 - 3 - 5x}$ [Section 5.8]

Factor each expression completely.

51. $6a^2 - 12a^3b + 36ab$ [Section 6.1]
52. $2x + 2y + ax + ay$ [Section 6.1]

53. $x^2 - 6x - 16$ [Section 6.2]
54. $30y^5 + 63y^4 - 30y^3$ [Section 6.3]

55. $t^4 - 16$ [Section 6.4]
56. $b^3 + 125$ [Section 6.5]

Solve each equation by factoring.

57. $3x^2 + 8x = 0$ [Section 6.7]
58. $15x^2 - 2 = 7x$ [Section 6.7]

59. **HEIGHT OF A TRIANGLE** The triangle shown has an area of 22.5 square inches. Find its height. [Section 6.8]

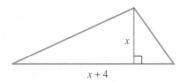

60. For what value is $\dfrac{x}{x+8}$ undefined? [Section 7.1]

Simplify each expression.

61. $\dfrac{3x^2 - 27}{x^2 + 3x - 18}$ [Section 7.1]

62. $\dfrac{a - 15}{15 - a}$ [Section 7.1]

Perform the operations and simplify when possible.

63. $\dfrac{x^2 - x - 6}{2x^2 + 9x + 10} \div \dfrac{x^2 - 25}{2x^2 + 15x + 25}$ [Section 7.2]

64. $\dfrac{1}{s^2 - 4s - 5} + \dfrac{s}{s^2 - 4s - 5}$ [Section 7.3]

65. $\dfrac{x + 5}{xy} - \dfrac{x - 1}{x^2 y}$ [Section 7.4]

66. $\dfrac{x}{x - 2} + \dfrac{3x}{x^2 - 4}$ [Section 7.4]

Simplify each complex fraction.

67. $\dfrac{\dfrac{9m - 27}{m^6}}{\dfrac{2m - 6}{m^8}}$ [Section 7.5]

68. $\dfrac{\dfrac{5}{y} + \dfrac{4}{y + 1}}{\dfrac{4}{y} - \dfrac{5}{y + 1}}$ [Section 7.5]

Solve each equation.

69. $\dfrac{2p}{3} - \dfrac{1}{p} = \dfrac{2p - 1}{3}$ [Section 7.6]

70. $\dfrac{7}{q^2 - q - 2} + \dfrac{1}{q + 1} = \dfrac{3}{q - 2}$ [Section 7.6]

71. Solve the formula $\dfrac{1}{a} + \dfrac{1}{b} = 1$ for a. [Section 7.6]

72. **ROOFING** A homeowner estimates that it will take him 7 days to roof his house. A professional roofer estimates that he can roof the house in 4 days. How long will it take if the homeowner helps the roofer? [Section 7.7]

73. **LOSING WEIGHT** If a person cuts his or her daily calorie intake by 100, it will take 350 days for that person to lose 10 pounds. How long will it take for the person to lose 25 pounds? [Section 7.8]

74. $\triangle ABC$ and $\triangle DEC$ are similar triangles. Find x. [Section 7.8]

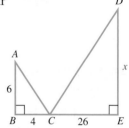

75. Suppose w varies directly as x. If $w = 1.2$ when $x = 4$, find w when $x = 30$. [Section 7.9]

76. **GEARS** The speed of a gear varies inversely with the number of teeth. If a gear with 10 teeth makes 3 revolutions per second, how many revolutions per second will a gear with 25 teeth make? [Section 7.9]

Simplify each radical expression. All variables represent positive numbers.

77. $\sqrt{100x^2}$ [Section 8.1]

78. $-\sqrt{18b^3}$ [Section 8.2]

Perform the indicated operation.

79. $3\sqrt{24} + \sqrt{54}$ [Section 8.3]

80. $(\sqrt{2} + 1)(\sqrt{2} - 3)$ [Section 8.4]

Rationalize the denominator.

81. $\dfrac{8}{\sqrt{10}}$ [Section 8.4]

82. $\dfrac{\sqrt{2}}{3 - \sqrt{a}}$ [Section 8.4]

Solve each equation.

83. $\sqrt{6x+1} + 2 = 7$ [Section 8.5]

84. $\sqrt{3t+7} = t + 3$ [Section 8.5]

Simplify each radical expression. All variables represent positive numbers.

85. $\sqrt[3]{\dfrac{27m^3}{8n^6}}$ [Section 8.6]

86. $\sqrt[4]{16}$ [Section 8.6]

Evaluate each expression.

87. $25^{3/2}$ [Section 8.6]

88. $(-8)^{-4/3}$ [Section 8.6]

Solve each equation.

89. $t^2 = 75$ [Section 9.1]

90. $(6y+5)^2 - 72 = 0$ [Section 9.1]

91. STORAGE CUBES The diagonal distance across the face of each of the stacking cubes is 15 inches. What is the height of the entire storage arrangement? Round to the nearest tenth of an inch. [Section 9.1]

92. Solve $x^2 + 8x + 12 = 0$ by completing the square. [Section 9.2]

93. Solve $4x^2 - x - 2 = 0$ using the quadratic formula. Give the exact solutions, and then approximate each to the nearest hundredth. [Section 9.3]

94. QUILTS According to the *Guinness Book of World Records*, the world's largest quilt was made by the Seniors' Association of Saskatchewan, Canada, in 1994. If the length of the rectangular quilt is 11 feet less than twice its width and it has an area of 12,865 ft², find its width and length. [Section 9.3]

95. Graph the quadratic equation $y = 2x^2 + 8x + 6$. Find the vertex, the *x*- and *y*-intercepts, and the axis of symmetry of the graph. [Section 9.4]

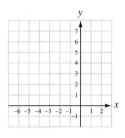

96. POWER OUTPUT The graph shows the power output (in horsepower, hp) of a certain engine at various engine speeds (in revolutions per minute, rpm). At what engine speed does the power output reach a maximum? [Section 9.4]

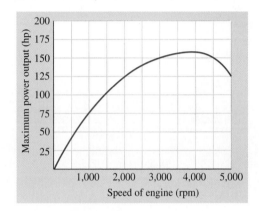

Write each expression in terms of i.

97. $\sqrt{-49}$ [Section 9.5]

98. $\sqrt{-54}$ [Section 9.5]

Perform the operations. Express each answer in the form a + bi.

99. $(2 + 3i) - (1 - 2i)$ [Section 9.5]

100. $(7 - 4i) + (9 + 2i)$ [Section 9.5]

101. $(3 - 2i)(4 - 3i)$ [Section 9.5]

102. $\dfrac{3 - i}{2 + i}$ [Section 9.5]

Solve each equation. Express the solutions in the form a + bi.

103. $x^2 + 16 = 0$ [Section 9.5]

104. $x^2 - 4x = -5$ [Section 9.5]

APPENDIX 1

Statistics

Statistics is a branch of mathematics that deals with the analysis of numerical data. In statistics, three types of averages are commonly used as **measures of central tendency** of a distribution of numbers: the *mean*, the *median*, and the *mode*.

In this appendix, you will learn about:

1. The mean
2. The median
3. The mode

1 The mean

We have previously discussed the mean of a distribution.

The Mean

The **mean**, or the **average**, of several values is the sum of those values divided by the number of values.

$$\text{Mean} = \frac{\text{sum of the values}}{\text{number of values}}$$

EXAMPLE 1

Physiology As part of a class project, a student measured 10 people's reaction time to a visual stimulus. Their reaction times (in seconds) were

0.36	0.24	0.23	0.41	0.28
0.25	0.20	0.28	0.39	0.26

Find the mean reaction time.

Strategy We will add the values and divide by the number of values.

WHY This is the formula for finding the mean of several values.

Solution

$$\text{Mean} = \frac{0.36 + 0.24 + 0.23 + 0.41 + 0.28 + 0.25 + 0.20 + 0.28 + 0.39 + 0.26}{10}$$

$$\text{Mean} = \frac{2.9}{10}$$

$$\text{Mean} = 0.29$$

The mean reaction time is 0.29 second.

Self Check 1

Find the mean of 2.3, 4.1, 5.2, 6.3, 3.7, 5.1, 4.6, 5.3

Now Try Problem 6

> **Success Tip** The mean (average) is a single value that is "typical" of a set of values. It can be, but is not necessarily, one of the values in the set. In the previous example, note that the mean reaction time was 0.29; however, none of the 10 reaction times was 0.29 second.

EXAMPLE 2

Banking When the mean (average) daily balance of a checking account falls below $500 in any week, the customer must pay a $20 service charge. What minimum balance must a customer have on Friday to avoid a service charge? See the table on the next page.

A-1

Self Check 2

GRADING To receive a grade of A in math, Lindsey must have a mean (average) of at least 90 on five exams. On the first four exams, she received scores of 94, 97, 80, and 87. What is the lowest score she can get on her final exam and still earn an A?

Now Try Problem 1

Security Savings Bank		
Day	Date	Daily balance
Mon	5/09	$670.70
Tues	5/10	$540.19
Wed	5/11	−$60.39
Thurs	5/12	$475.65
Fri	5/13	

Analyze We can find the mean (average) daily balance for the week by adding the daily balances and dividing by 5. If the mean is $500 or more, there will be no service charge.

Assign We can let x = the minimum balance needed on Friday and translate the words into mathematical symbols.

Form

| The sum of the five daily balances | divided by | 5 | is | $500. |

$$\frac{670.70 + 540.19 + (-60.39) + 475.65 + x}{5} = 500$$

Solve

$$\frac{670.70 + 540.19 + (-60.39) + 475.65 + x}{5} = 500$$

$$\frac{1{,}626.15 + x}{5} = 500 \qquad \text{Simplify the numerator.}$$

$$5\left(\frac{1{,}626.15 + x}{5}\right) = 5(500) \qquad \text{Multiply both sides by 5.}$$

$$1{,}626.15 + x = 2{,}500$$

$$x = 873.85 \qquad \text{Subtract 1,626.15 from both sides.}$$

State On Friday, the account balance must be at least $873.85 to avoid a service charge.

Check Check the result by adding the five daily balances and dividing by 5.

2 The median

The mean is not always the best measure of central tendency. It can be affected by very high or very low values. Another measure of central tendency that determines a "middle" value is the **median**.

The Median

The **median** of several values is the middle value. To find the median of several values:

1. Arrange the values in increasing order.
2. If there is an odd number of values, choose the middle value.
3. If there is an even number of values, add the middle two values and divide by 2.

EXAMPLE 3   Finding the Median
In Example 1, the following values were the reaction times of 10 people to a visual stimulus.

| 0.36 | 0.24 | 0.23 | 0.41 | 0.28 |
| 0.25 | 0.20 | 0.28 | 0.39 | 0.26 |

Find the median of these values.

Strategy We will put the values in increasing order. Then, since there is an even number of values, we will add the middle two values and divide by 2.

WHY These are the steps to find the median of an even number of values.

Solution
To find the median, we first arrange the values in increasing order:

| 0.20 | 0.23 | 0.24 | 0.25 | 0.26 |
| 0.28 | 0.28 | 0.36 | 0.39 | 0.41 |

Because there is an even number of values, the median will be the sum of the middle two values, 0.26 and 0.28, divided by 2. Thus, the median is

$$\text{Median} = \frac{0.26 + 0.28}{2} = 0.27$$

The median reaction time is 0.27 second.

Self Check 3
For the values in Self Check 1, find the median.
Now Try Problem 4

3   The mode

The mean and median are not always the best measures of central tendency. Another measure of central tendency is the value that occurs most often, called the **mode**.

> **The Mode**
>
> The **mode** of several values is the value that occurs most often.

EXAMPLE 4   Finding the Mode
Find the mode of the following values.

| 0.36 | 0.24 | 0.23 | 0.41 | 0.28 |
| 0.25 | 0.20 | 0.28 | 0.39 | 0.26 |

Strategy We will count the number of times each value occurs.

WHY The mode is the value that occurs most often.

Solution
Since the value 0.28 occurs most often, it is the mode.

If two different numbers in a distribution tie for occurring most often, there are two modes, and the distribution is called **bimodal**.

Although the mean is probably the most common measure of average, the median and the mode are frequently used. For example, workers' salaries are usually compared to the median (average) salary. To say that the modal (average) shoe size is 10 means that a shoe size of 10 occurs more often than any other shoe size.

Self Check 4
Find the mode of 12, 16, 13, 13, 12, 15, 14, 13
Now Try Problem 5

ANSWERS TO SELF CHECKS
1. 4.575 **2.** 92 **3.** 4.85 **4.** 13

APPENDIX 1 STUDY SET

PRACTICE

In Exercises 1–3, use the following distribution of values:

7 5 9 10 8 6
6 7 9 12 9

1. Find the mean.
2. Find the median.
3. Find the mode.

In Exercises 4–6, use the following distribution of values:

8 12 23 12 10 16
26 12 14 8 16 23

4. Find the median.
5. Find the mode.
6. Find the mean.
7. Find the mean, median, and mode of the following values:

   24 27 30 27
   31 30 27

8. Find the mean, median, and mode of the following golf scores:

   85 87 88 82
   85 91 88 88

APPLICATIONS

9. **FOOTBALL** The gains and losses made by a running back on seven plays were:

   −8 yd   2 yd   −6 yd   6 yd
   4 yd   −7 yd   −5 yd

   Find his average (mean) yards per carry.

10. **SALES** If a clerk had the sales shown at right for one week, find the mean of her daily sales.

Monday	$1,525
Tuesday	$ 785
Wednesday	$1,628
Thursday	$1,214
Friday	$ 917
Saturday	$1,197

11. **VIRUSES** The table at right gives the approximate lengths (in centimicrons) of the viruses that cause five common diseases. Find the mean length of the viruses.

Polio	2.5
Influenza	105.1
Pharyngitis	74.9
Chicken pox	137.4
Yellow fever	52.6

12. **SALARIES** Ten workers in a small business have monthly salaries of

    $2,500 $1,750 $2,415 $3,240 $2,790
    $3,240 $2,650 $2,415 $2,415 $2,650

    Find the average (mean) salary.

13. **JOB TESTING** To be accepted into a police training program, a recruit must have an average (mean) score of 85 on a battery of four tests. If a candidate scored 78 on the oral test, 91 on the physical test, and 87 on the psychological test, what is the lowest score she can obtain on the written test and be accepted into the program?

14. **GAS MILEAGE** Mileage estimates for four cars owned by a small business are shown in the table. If the business buys a fifth car, what must its mileage average be so that the five-car fleet averages 30.0 mpg?

Model	City mileage (mpg)
Chevrolet Malibu	29.9
Jeep Cherokee	23.1
Ford Fusion	34.0
Dodge Caravan	25.2

15. **SPORT FISHING** The weights (in pounds) of the trophy fish caught one week in Catfish Lake were

    4 7 4 3 3
    5 6 9 4
    5 8 13 4
    5 4 6 9

    Find the median and modal averages of the fish caught.

16. **SALARIES** Find the median and mode of the 10 salaries given in Exercise 12.

17. FUEL EFFICIENCY The 10 most fuel-efficient cars in 2012, based on manufacturer's estimated city and highway average miles per gallon (mpg), are shown in the table below.

  a. Find the mean, median, and mode of the city mileage.

  b. Find the mean, median, and mode of the highway mileage.

Model	mpg city/hwy
Mitsubishi i	126/99
Nissan Leaf	106/92
Ford Transit	62/62
Chevy Volt	58/62
Toyota Prius	51/48
Honda Civic Hybrid	44/44
Toyota Prius V	44/40
Lexus CT200h	43/40
Honda Insight	41/44
Toyota Camry Hybrid	43/39

Source: autoguide.com

18. NUTRITION Refer to the table below.

  a. Find the mean number of calories in one serving of the meats shown.

  b. Find the median.

  c. Find the mode.

NUTRITIONAL COMPARISONS Per 3.5 oz. serving of cooked meat	
Species	Calories
Bison	143
Beef (Choice)	283
Beef (Select)	201
Pork	212
Chicken (Skinless)	190
Sockeye Salmon	216

Source: The National Bison Association

WRITING

19. Explain why the mean of two numbers is halfway between the numbers.

20. Can the mean, median, and mode of a distribution be the same number? Explain.

21. Must the mean, median, and mode of a distribution be the same number? Explain.

22. Can the mode of a distribution be greater than the mean? Explain.

APPENDIX 2

Roots and Powers

n	n^2	$\sqrt{n}$	n^3	$\sqrt[3]{n}$	n	n^2	$\sqrt{n}$	n^3	$\sqrt[3]{n}$
1	1	1.000	1	1.000	51	2,601	7.141	132,651	3.708
2	4	1.414	8	1.260	52	2,704	7.211	140,608	3.733
3	9	1.732	27	1.442	53	2,809	7.280	148,877	3.756
4	16	2.000	64	1.587	54	2,916	7.348	157,464	3.780
5	25	2.236	125	1.710	55	3,025	7.416	166,375	3.803
6	36	2.449	216	1.817	56	3,136	7.483	175,616	3.826
7	49	2.646	343	1.913	57	3,249	7.550	185,193	3.849
8	64	2.828	512	2.000	58	3,364	7.616	195,112	3.871
9	81	3.000	729	2.080	59	3,481	7.681	205,379	3.893
10	100	3.162	1,000	2.154	60	3,600	7.746	216,000	3.915
11	121	3.317	1,331	2.224	61	3,721	7.810	226,981	3.936
12	144	3.464	1,728	2.289	62	3,844	7.874	238,328	3.958
13	169	3.606	2,197	2.351	63	3,969	7.937	250,047	3.979
14	196	3.742	2,744	2.410	64	4,096	8.000	262,144	4.000
15	225	3.873	3,375	2.466	65	4,225	8.062	274,625	4.021
16	256	4.000	4,096	2.520	66	4,356	8.124	287,496	4.041
17	289	4.123	4,913	2.571	67	4,489	8.185	300,763	4.062
18	324	4.243	5,832	2.621	68	4,624	8.246	314,432	4.082
19	361	4.359	6,859	2.668	69	4,761	8.307	328,509	4.102
20	400	4.472	8,000	2.714	70	4,900	8.367	343,000	4.121
21	441	4.583	9,261	2.759	71	5,041	8.426	357,911	4.141
22	484	4.690	10,648	2.802	72	5,184	8.485	373,248	4.160
23	529	4.796	12,167	2.844	73	5,329	8.544	389,017	4.179
24	576	4.899	13,824	2.884	74	5,476	8.602	405,224	4.198
25	625	5.000	15,625	2.924	75	5,625	8.660	421,875	4.217
26	676	5.099	17,576	2.962	76	5,776	8.718	438,976	4.236
27	729	5.196	19,683	3.000	77	5,929	8.775	456,533	4.254
28	784	5.292	21,952	3.037	78	6,084	8.832	474,552	4.273
29	841	5.385	24,389	3.072	79	6,241	8.888	493,039	4.291
30	900	5.477	27,000	3.107	80	6,400	8.944	512,000	4.309
31	961	5.568	29,791	3.141	81	6,561	9.000	531,441	4.327
32	1,024	5.657	32,768	3.175	82	6,724	9.055	551,368	4.344
33	1,089	5.745	35,937	3.208	83	6,889	9.110	571,787	4.362
34	1,156	5.831	39,304	3.240	84	7,056	9.165	592,704	4.380
35	1,225	5.916	42,875	3.271	85	7,225	9.220	614,125	4.397
36	1,296	6.000	46,656	3.302	86	7,396	9.274	636,056	4.414
37	1,369	6.083	50,653	3.332	87	7,569	9.327	658,503	4.431
38	1,444	6.164	54,872	3.362	88	7,744	9.381	681,472	4.448
39	1,521	6.245	59,319	3.391	89	7,921	9.434	704,969	4.465
40	1,600	6.325	64,000	3.420	90	8,100	9.487	729,000	4.481
41	1,681	6.403	68,921	3.448	91	8,281	9.539	753,571	4.498
42	1,764	6.481	74,088	3.476	92	8,464	9.592	778,688	4.514
43	1,849	6.557	79,507	3.503	93	8,649	9.644	804,357	4.531
44	1,936	6.633	85,184	3.530	94	8,836	9.695	830,584	4.547
45	2,025	6.708	91,125	3.557	95	9,025	9.747	857,375	4.563
46	2,116	6.782	97,336	3.583	96	9,216	9.798	884,736	4.579
47	2,209	6.856	103,823	3.609	97	9,409	9.849	912,673	4.595
48	2,304	6.928	110,592	3.634	98	9,604	9.899	941,192	4.610
49	2,401	7.000	117,649	3.659	99	9,801	9.950	970,299	4.626
50	2,500	7.071	125,000	3.684	100	10,000	10.000	1,000,000	4.642

APPENDIX 3

Answers to Selected Exercises

Study Set Section 1.1 (page 8)

1. sum, difference **3.** Variables **5.** equation **7.** line
9. equation **11.** algebraic expression **13.** algebraic expression **15.** equation **17. a.** multiplication, subtraction **b.** x **19. a.** addition, subtraction **b.** m
21. a. They help us determine that 15-year-old machinery is worth $35,000. **b.** It decreases. **23.** $5 \cdot 6, 5(6)$
25. $34 \cdot 75, 34(75)$ **27.** $4x$ **29.** $3rt$ **31.** lw **33.** Prt
35. $2w$ **37.** xy **39.** $\dfrac{32}{x}$ **41.** $\dfrac{90}{30}$ **43.** 250 **45.** 200
47. The product of 18 and 24 is equal to 432.
49. The difference of 11 and 9 is equal to 2.
51. The product of 2 and x is equal to 1.
53. The quotient of 66 and 11 is equal to 6.
55. $p = 100 - d$ **57.** $7d = h$ **59.** 390, 400, 405 **61.** 1,300, 1,200, 1,100 **63.** $s = 3c$ **65.** $w = e + 1,200$
67. $p = r - 600$ **69.** $\dfrac{l}{4} = m$ **71.** 12 **73.** 2
75. $l = 4c, b = c, a = 2c, p = 2c, S = c, s = 20c$
77.

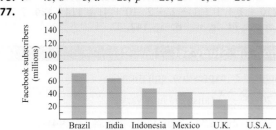

79. a. $90°, 60°, 45°, 30°, 0°$
b.

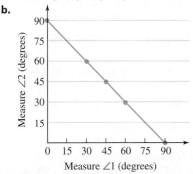

Think It Through (page 24)

Housing: $630; food: $315; savings, utilities, transportation, debt: $210; entertainment, car insurance, clothing: $105

Study Set Section 1.2 (page 26)

1. multiplied **3.** prime-factored **5.** equivalent
7. least or lowest **9. a.** 1 **b.** a **c.** $\dfrac{a \cdot c}{b \cdot d}$ **d.** $\dfrac{a \cdot d}{b \cdot c}$
e. $\dfrac{a+b}{d}$ **f.** $\dfrac{a-b}{d}$ **11. a.** 1 **b.** 1 **13. a.** $\dfrac{5}{6} \cdot \dfrac{5}{5} = \dfrac{25}{30}$
b. $\dfrac{2 \cdot 2 \cdot 3}{2 \cdot 3 \cdot 7} = \dfrac{2}{7}$ **15.** $3 \cdot 5 \cdot 5$ **17.** $2 \cdot 2 \cdot 7$ **19.** $3 \cdot 3 \cdot 3 \cdot 3$
21. $3 \cdot 3 \cdot 13$ **23.** $2 \cdot 2 \cdot 5 \cdot 11$ **25.** $2 \cdot 3 \cdot 11 \cdot 19$
27. $\dfrac{5}{48}$ **29.** $\dfrac{21}{55}$ **31.** $\dfrac{15}{8}$ **33.** $\dfrac{42}{25}$ **35.** $\dfrac{3}{9}$ **37.** $\dfrac{24}{54}$ **39.** $\dfrac{35}{5}$
41. $\dfrac{35}{7}$ **43.** $\dfrac{1}{3}$ **45.** $\dfrac{6}{7}$ **47.** $\dfrac{3}{8}$ **49.** lowest terms **51.** $\dfrac{2}{3}$
53. $\dfrac{4}{25}$ **55.** $\dfrac{6}{5}$ **57.** $\dfrac{4}{7}$ **59.** $\dfrac{5}{24}$ **61.** $\dfrac{22}{35}$ **63.** $\dfrac{41}{45}$ **65.** $\dfrac{3}{20}$
67. 24 **69.** 4 **71.** $\dfrac{7}{9}$ **73.** $\dfrac{7}{20}$ **75.** $32\dfrac{2}{3}$ **77.** $2\dfrac{1}{2}$ **79.** $\dfrac{5}{9}$
81. $5\dfrac{19}{48}$ **83.** $\dfrac{19}{15}$ **85.** 70 **87.** $13\dfrac{3}{4}$ **89.** $\dfrac{1}{2}$ **91.** $\dfrac{14}{5}$ **93.** $\dfrac{8}{5}$
95. $\dfrac{3}{35}$ **97.** $\dfrac{9}{4}$ **99.** $\dfrac{3}{10}$ **101.** $1\dfrac{9}{11}$ **103.** $\dfrac{1}{7}$ **105. a.** $\dfrac{31}{32}$ in.
b. $\dfrac{11}{32}$ in. **107.** $5\dfrac{3}{4}$ lb **109.** $40\dfrac{1}{2}$ in. **115.** 150, 180

Section 1.3 Study Set (page 37)

1. natural **3.** integers **5.** inequality **7.** set-builder
9. irrational **11.** opposites **13.** $\dfrac{6}{1}, \dfrac{-9}{1}, \dfrac{-7}{8}, \dfrac{7}{2}, \dfrac{-3}{10}, \dfrac{283}{100}$
15. 13 and -3 **17. a.** $<$ **b.** $>$ **c.** $>$ and $<$ **19.** $\sqrt{2}$ in.
21. a. true **b.** false **c.** false **d.** true **23.** -4 **25.** 2.236
27. 9.950 **29.** 3.142 **31.** 1.571 **33.** square root
35. is approximately equal to **37. a.** $0.\overline{6}$, nonterminating
b. 0.8, terminating **39.** inequalities **41.** -5 **43.** $\dfrac{7}{8}$
45. 10 **47.** 2.3 **49.** -3
51. ⟵─┼──┼──┼──┼──┼──┼──┼──┼──┼──⟶ $-3\ -2\ -1\ 0\ 1\ 2\ 3\ 4\ 5\ 6$
53. ⟵─┼──┼──┼──┼──┼──┼──┼──┼──┼──┼──┼──┼──┼──┼──⟶ $-6\ -5\ -4\ -3\ -2\ -1\ 0\ 1\ 2\ 3\ 4\ 5\ 6\ 7\ 8$
55. $>$ **57.** $>$ **59.** $>$ **61.** $<$
63. natural: none; whole: 0; integers: 0, -50; rational: $-\dfrac{5}{6}$, 35.99, 0, $4\dfrac{3}{8}$, -50, $\dfrac{17}{5}$; irrational: $\sqrt{2}$; real: all
65. natural: 2, 8; whole: 0, 2, 8; integers: 0, 8, -3, 2; rational: 0, $1\dfrac{3}{5}$, 8, -3, 2.6, 2; irrational: $-\pi, \pi, \sqrt{11}$; real: all
67. ⟵─┼──┼──┼──┼──┼──┼──┼──┼──┼──┼──⟶
$-\dfrac{35}{8}\quad -\pi\quad -1\dfrac{1}{2}\ -0.333...\ \sqrt{2}\quad 3\quad 4.25$
$-5\ -4\ -3\ -2\ -1\ 0\ 1\ 2\ 3\ 4\ 5$
69. ⟵─┼──┼──┼──┼──┼──┼──┼──┼──┼──┼──⟶
$-3\dfrac{5}{8}\qquad 0.333...\ \sqrt{3}\ 2\qquad \dfrac{17}{5}$
$-4\ -3\ -2\ -1\ 0\ 1\ 2\ 3\ 4\ 5\ 6$
71. 17 **73.** -2.5 **75.** $=$ **77.** $>$ **79.** $>$ **81.** $=$ **83.** $=$
85. $>$ **87.** $>$ **89.** $<$ **91.** natural, whole, integers: 750, 5,000; rational: all; irrational: none; real: all
93. a. 2006, $-\$90$ billion **b.** 2009, $-\$43$ billion
95. 1 in., $\sqrt{3}$ in., 2 in. **97.** 81.7 in.

99. Arrow 1, $|-6| > |5|$ **105.** $\frac{4}{9}$ **107.** $\frac{6}{5}$ **109.** $\frac{13}{30}$
111. 0.375

Section 1.4 Study Set (page 48)

1. positive, negative   **3.** commutative   **5.** associative   **7.** 5
9. 1   **11.** add   **13.** $1 + (-5)$ **15.** $(20 + 4)$
17. $-6 + (2 + 8)$ **19.** $(-96 + 4) + 200$ **21.** -7 **23.** $\frac{1}{2}$
25. $-13, 10$ **27.** -77 **29.** -9.8 **31.** -2 **33.** 4   **35.** 16
37. -15 **39.** -21 **41.** 59   **43.** 215   **45.** -4 **47.** 0
49. -6 **51.** 0   **53.** 2   **55.** -35 **57.** -8.2 **59.** -20.1
61. $-\frac{1}{8}$ **63.** -26 **65.** $\frac{5}{12}$ **67.** -3.216 **69.** 273
71. $-22,470$ **73.** 67   **75.** $-1,632.51$ **77.** 2,150 m
79. a. Woods: -18, Kite: -6, Tolles: -5, Watson: -4
b. 12 strokes   **81.** 2,772   **83.** It fell $0.38.
85. $-$26,605,538   **87.** $-$120 (overdrawn $120)
89. $-$167 million   **93.** true   **95.** -9 and 3

Section 1.5 Study Set (page 56)

1. Subtraction   **3.** opposite   **5.** a **7.** negative   **9.** 15
11. -7 **13.** 11   **15.** -21 **17.** -2 **19.** -2.3 **21.** $-\frac{1}{2}$
23. $-\frac{5}{16}$ **25.** 22   **27.** -2.5 **29.** -11 **31.** -50 **33.** -7
35. -1 **37.** 4.6   **39.** -88 **41.** 12   **43.** 0   **45.** -4
47. 12,466   **49.** $-44,571.251$ **51.** 3   **53.** -47.5 **55.** -12
57. 149   **59.** 2   **61.** $-\frac{19}{12}$ **63.** $-\frac{11}{15}$ **65.** 160°F
67. 1,030 ft   **69.** 6.85   **71.** $-5.75°$ **73.** 5,189 ft   **75. a.** iii
b. $-$116.1 billion   **77.** $-40, -55, -68$ **83.** $2 \cdot 3 \cdot 5$
85. 89

Section 1.6 Study Set (page 66)

1. product, quotient   **3.** commutative   **5.** undefined
7. $4(-5)$ **9.** positive   **11.** 1   **13. a.** One of the numbers
is 0.   **b.** They are reciprocals (multiplicative inverses).
15. a. associative property of multiplication
b. multiplicative inverse   **c.** commutative property of
multiplication   **d.** multiplication property of 1
17. a. NEG   **b.** not possible to tell   **c.** POS   **d.** NEG
19. If we are multiplying or dividing, this is true. If we are
adding, the sum of a negative number and a positive number
could be positive. For example, $-6 + 7 = 1$.
21. Since division by 0 is undefined, the calculator was
unable to perform the division.   **23.** $5, 10$ **25.** -60
27. -24 **29.** -800 **31.** -54 **33.** 36   **35.** $\frac{3}{8}$ **37.** 60
39. 0.084   **41.** 3   **43.** -2 **45.** -4 **47.** -1 **49.** 0
51. undefined   **53.** -520 **55.** 0   **57.** 0   **59.** 120
61. $-1,109.2$ **63.** -417.3624 **65.** 2.4   **67.** -0.48
69. $\frac{15}{16}$ **71.** -5.5 **73.** 1   **75.** -4 **77.** -20 **79.** -0.005
81. $-\frac{5}{12}$ **83.** $\frac{9}{32}$ **85.** 10   **87.** -6 **89.** -670 **91.** -280

93. $-72°$ **95.** $-70°$ **97.** $-193°$F   **99.** $-$15.3 billion
101. $-$614,516   **103. a.** $5, -10$ **b.** $2.5, -5$ **c.** $7.5, -15$
d. $10, -20$ **107.** yes   **109.** $1.08\overline{3}$ **111.** 475

Section 1.7 Study Set (page 78)

1. base, exponent   **3.** power   **5.** order   **7. a.** addition and
multiplication   **b.** $54, 34$ **c.** 34, multiplication is to be done
before addition   **9. a.** Divide the sum of the scores by the
number of scores.   **b.** 64   **11. a.** addition, power,
multiplication   **b.** power, multiplication, addition
13. multiplication, subtraction   **15. a.** power   **b.** addition
c. subtraction   **17.** 12^6 **19.** $9, 54$ **21.** $3, 9, 18$ **23.** 3^4
25. $10^2 \cdot 12^3$ **27.** $8\pi r^3$ **29.** $6x^2y^3$ **31.** 49   **33.** 36
35. -8 **37.** $-\frac{8}{125}$ **39.** 0.16   **41.** -36 **43.** 187
45. -77 **47.** 38   **49.** 80   **51.** 23   **53.** -28 **55.** 194
57. 58   **59.** 49   **61.** 1,000   **63.** 13   **65.** 201   **67.** -396
69. 50   **71.** -27 **73.** 69   **75.** -10 **77.** 10   **79.** 12
81. 1   **83.** -8 **85.** 31   **87.** 1   **89.** 11   **91.** $-1,296$
93. -44 **95.** 12   **97.** 8   **99.** -24 **101.** 20   **103.** 343
105. 360   **107.** 12   **109.** 59   **111.** 28   **113.** -24 **115.** -2
117. $\frac{1}{8}$ **119.** -39 **121.** -8 **123.** $-\frac{8}{9}$ **125.** -1
127. 427   **129.** 3,337,154.31   **131.** 2^2 square units, 3^2 square
units, 4^2 square units   **133.** $2,106   **135. a.** $11,875   **b.** $95
137. $3(10 + 1 + 1 + 1 + 1 + 2 \cdot 10 + 4) = 114$ **139.** $2, 7, 5$
145. a. ii   **b.** iii   **c.** iv   **d.** i   **147.** 1

Section 1.8 Study Set (page 91)

1. evaluate   **3.** expression, equation
5. $6 + 20x$, $\frac{6 - x}{20}$ (answers may vary)
7. We would obtain $34 - 6$; it looks like 34, not $3(4)$.
9. a. $2x - 500$ **b.** 3,500 lb   **11.** Value in cents: $5, 10, 50$;
Total value in cents: $30, 10d, 50(x + 5)$ **13.** $5, 5, 25, 45$
15. $4, -5, 7$ **17.** $9, -4$ **19.** term   **21.** factor   **23.** $l + 15$
25. $50x$ **27.** $\frac{w}{l}$ **29.** $P + p$ **31.** $k^2 - 2,005$ **33.** $J - 500$
35. $\frac{1,000}{n}$ **37.** $p + 90$ **39.** $(x + 2)$ in.   **41.** $x + 150$
43. $\frac{c}{6}$ **45.** $\frac{x}{2}$ **47.** $5b$ **49.** $2w$ **51.** x ft, $(12 - x)$ ft
53. $2x + 25$ **55. a.** x = age of Apple,
$x + 80$ = age of IBM, $x - 9$ = age of Dell
b. IBM: 112 years, Dell: 23 years
57. Illinois: x, Florida: $x + 27$, California: $x + 32$ **59. a.** 300
b. $60h$ **61. a.** $3y$ **b.** $\frac{f}{3}$ **63.** -12 **65.** 20   **67.** 156
69. -5 **71.** $-\frac{1}{5}$ **73.** -2 **75.** 17   **77.** 230
79. $-1, -2, -28$ **81.** $41, 11, 2$ **83.** $150, -450$ **85.** $0, 0, 5$
87. $35 + h + 300$ **89.** $p - 680$ **91.** $4d - 15$
93. $2(200 + t)$ **95.** $|a - 2|$ **97.** $2x + 4$ **99.** $0, 48, 64, 48, 0$
101. $-37°$C, $-64°$C   **103.** $1\frac{23}{64}$ in.^{2} **105.** 235 ft^2 **111.** 0
113. $\frac{2}{3}$ **115.** c^4 **117.** 83

Section 1.9 Study Set (page 103)

1. simplify **3.** distributive **5.** distributive property
7. $ab + ac + ad$ **9. a.** + **b.** − **11. a.** − **b.** +
13. $x + 20 - x = 20$, 20 ft **15.** yes **17.** $10x$, can't be simplified **19.** $18x$, $3x + 5$ **21.** 7, 7, 14 **23. a.** no **b.** yes
25. a. $5x + 1$ **b.** $16t - 6$ **27.** $63m$ **29.** $-35q$ **31.** $5x$
33. $20bp$ **35.** $40r^2$ **37.** $48q^2$ **39.** $5x + 15$ **41.** $-2b + 2$
43. $24t - 16$ **45.** $-15t - 12$ **47.** $-r + 10$ **49.** $-x + 7$
51. $-2w + 4$ **53.** $6x - 2y$ **55.** $34x - 17y + 34$
57. $1.4 - 0.3p + 0.1t$ **59.** $8p, -5p$ **61.** no like terms
63. $20x$ **65.** $3x^2$ **67.** 0 **69.** 0 **71.** $1.1h - 2p$
73. $2a + b$ **75.** $x + 3y$ **77.** $-11x^2 - 3y^2$ **79.** $b + 2$
81. $-2x^2 + 3x$ **83.** $-3c - 1$ **85.** $2c + 9$ **87.** $0.4x - 1.6$
89. $7X - 2x$ **91.** $-3x$ **93.** t **95.** $\frac{4}{5}t$ **97.** $12y - 6$
99. $7y - 5$ **101.** $6y$ **103.** $0.4r$ **105.** $7z - 15$ **107.** $12x$
109. $(4x + 8)$ ft **111.** $(8x - 8)$ ft **117.** 0 **119.** 2

Chapter 1 Review Exercises (page 106)

1. 100 **2.** 7 P.M. **3.** The difference of 15 and 3 is 12.
4. The sum of 15 and 3 is 18. **5.** The quotient of 15 and 3 is 5. **6.** The product of 15 and 3 is 45. **7.** equation
8. algebraic expression **9.** algebraic expression
10. equation **11.** 10, 15, 25 **12.** $f = 50c$, the total fees are the product of 50 and the number of children. (Answers may vary depending on the variables chosen.) **13.** $2 \cdot 12$, $3 \cdot 8$ (answers may vary) **14.** $2 \cdot 2 \cdot 6$ (answers may vary)
15. $2 \cdot 3^3$ **16.** $3 \cdot 7^2$ **17.** $5 \cdot 7 \cdot 11$ **18.** prime **19.** $\frac{4}{7}$
20. $\frac{4}{3}$ **21.** $\frac{40}{64}$ **22.** $\frac{36}{3}$ **23.** $\frac{5}{21}$ **24.** 10 **25.** $\frac{16}{45}$ **26.** $3\frac{1}{4}$
27. $\frac{2}{5}$ **28.** 2 **29.** $\frac{5}{22}$ **30.** $\frac{11}{12}$ **31.** $81\frac{13}{24}$ **32.** $20\frac{5}{18}$
33. $\frac{17}{96}$ in. **34.** no **35.** −\$65 billion **36.** -206 ft **37.** <
38. > **39.** $\frac{5}{1}$ **40.** $\frac{-12}{1}$ **41.** $\frac{7}{10}$ **42.** $\frac{14}{3}$
43. number line with $-\frac{17}{4}$, -2, $0.333...$, $\frac{7}{8}$, π, 3.75
44. $\sqrt{2} \approx 1.41$ **45.** false **46.** false **47.** true **48.** true
49. true **50.** natural: 8; whole: 0, 8; integers: $0, -12, 8$; rational: $-\frac{4}{5}, 99.99, 0, -12, 4\frac{1}{2}, 0.666\ldots, 8$; irrational: $\sqrt{2}$; real: all **51.** $\frac{9}{16}$ **52.** 0 **53.** > **54.** > **55.** 45 **56.** -82
57. 22 **58.** 12 **59.** -11 **60.** -4 **61.** -12.3 **62.** $-\frac{3}{16}$
63. 11 **64.** -45 **65.** commutative property of addition
66. associative property of addition **67.** -19 **68.** -49
69. -15 **70.** 5.7 **71. a.** 14 **b.** 13.78 **72.** 65,233 ft
73. -56 **74.** 54 **75.** 12 **76.** -24 **77.** -24 **78.** 36
79. 6.36 **80.** -2 **81.** $-\frac{2}{15}$ **82.** $-\frac{3}{4}$
83. associative property of multiplication
84. commutative property of multiplication **85.** 3
86. $-\frac{1}{3}$ **87.** 2 **88.** -4 **89.** 3 **90.** 0 **91.** $-\frac{6}{5}$
92. undefined **93.** -4.5 **94.** 1 **95.** At least one of the numbers is 0. **96. a.** high: 2, low: -3 **b.** high: 4, low: -6
97. 8^5 **98.** $5^3 \cdot 9^2$ **99.** a^4 **100.** $9\pi r^2$ **101.** x^3y^4
102. 1^6 **103.** 81 **104.** 8 **105.** 32 **106.** 50 **107.** -48
108. -9 **109.** 44 **110.** -420 **111.** $-\frac{14}{19}$
112. 113 **113.** -7 **114.** 0 **115.** \$20 **116.** parentheses, brackets, absolute value, fraction bar **117.** $h + 25$
118. $s - 15$ **119.** $\frac{1}{2}t$ **120.** $6x$ **121. a.** $(n + 4)$ in.
b. $(b - 4)$ in. **122.** nickel: 5, 30; dime: 10, 10d
123. $0, 19, -16$ **124.** 17.7 in.^{3} **125.** 110 **126.** 40
127. 432 **128.** -36 **129.** $-28w$ **130.** $24x$ **131.** $2.08f$
132. r **133.** $5x + 15$ **134.** $-2x - 3 + y$ **135.** $3c - 6$
136. $12.6c + 29.4$ **137.** $9p$ **138.** $-7m$ **139.** $4n$
140. $-p - 18$ **141.** $0.1k^2$ **142.** $8a^2 - 1$ **143.** w **144.** 0
145. x **146.** $-x$ **147.** $4x + 1$ **148.** $4x - 1$
149. $(16x + 6)$ in. **150.** $4x + 1$

Chapter 1 Test (page 116)

1. a. equivalent **b.** sum **c.** opposite **2.** \$24 **3.** 5 hr
4. 3, 20, 70 **5.** $2 \cdot 2 \cdot 3 \cdot 3 \cdot 5 = 2^2 \cdot 3^2 \cdot 5$ **6.** $\frac{2}{5}$ **7.** $\frac{30}{42}$
8. \$3.57 **9.** $\frac{3}{2} = 1\frac{1}{2}$ **10.** $\frac{25}{36}$ **11.** $20\frac{1}{15}$ **12.** 49
13. A: 0.73, 0.83; B: 0.593, 0.693 **14.** $0.8\overline{3}$
15. number line with $-3.75, -3, -1\frac{1}{4}, 0.5, \sqrt{2}, \frac{7}{2}$
16. a. true **b.** false **c.** true **d.** true **17.** iii
18. a. > **b.** < **c.** < **d.** > **19.** 0.6
20. -2 **21.** $\frac{3}{8}$ **22.** -6 **23.** -30 **24.** 2
25. a. 0 **b.** 0 **c.** -3 **d.** 0 **26. a.** associative property of addition **b.** commutative property of multiplication
27. a. 9^5 **b.** $3x^2z^3$ **28.** 170 **29.** 36 **30.** -12
31. 36 **32.** $4, 17, -59$ **33.** $x - 2$ **34.** $25q$ cents
35. a. An equation is a mathematical sentence that contains an = sign. An expression does not contain an = sign.
b. Subtraction is the same as addition of the opposite.
c. The set of real numbers corresponds to all points on a number line. A real number is any number that is either a rational number or an irrational number.
36. \$290 million **37.** $-20x$ **38.** $224t$ **39.** $-4a + 4$
40. $-5.9d^2$ **41.** $14x + 3$ **42.** $(18x + 6)$ ft
43. $(23x + 85.8)$ mi **44.** $34x$ in.

Study Set Section 2.1 (page 128)

1. equation **3.** solve **5.** equivalent **7. a.** $x + 6$
b. neither **c.** no **d.** yes **9. a.** c, c **b.** c, c **11. a.** x
b. y **c.** t **d.** h **13.** $5, 5, 50, 50, \stackrel{?}{=}, 45, 50$
15. a. is possibly equal to **b.** yes **17.** no **19.** no **21.** no
23. no **25.** yes **27.** no **29.** no **31.** yes **33.** yes
35. yes **37.** 71 **39.** 18 **41.** -0.9 **43.** 3 **45.** $\frac{8}{9}$ **47.** 3
49. $-\frac{1}{25}$ **51.** -2.3 **53.** 45 **55.** 0 **57.** 21 **59.** -2.64
61. 20 **63.** 15 **65.** -6 **67.** 4 **69.** 4 **71.** 7 **73.** 1
75. -6 **77.** 20 **79.** 0.5 **81.** -18 **83.** $-\frac{4}{21}$ **85.** 13

87. 2.5 **89.** $-\dfrac{8}{3}$ **91.** $\dfrac{13}{20}$ **93.** 4 **95.** -5 **97.** -200
99. 95 **101.** $65°$ **103.** $\$1,000,000$ **109.** 0 **111.** $45 - x$

Study Set Section 2.2 (page 137)

1. equation **3.** identity **5.** subtraction, multiplication
7. a. $-2x - 8 = -24$ **b.** $-20 = 3x - 16$ **9. a.** $12x$
b. $2x$ **11.** 10 **13.** 7, 7, 2, 2, 14, $\stackrel{?}{=}$, 28, 21, 14 **15.** 6
17. 5 **19.** -7 **21.** 18 **23.** 16 **25.** 12 **27.** $\dfrac{10}{3}$ **29.** $-\dfrac{5}{2}$
31. 5 **33.** -0.25 **35.** 2.9 **37.** -4 **39.** $\dfrac{11}{5}$
41. -1 **43.** -6 **45.** 0.04 **47.** -6 **49.** -11 **51.** 7
53. -11 **55.** 1 **57.** $\dfrac{9}{2}$ **59.** 3 **61.** -20 **63.** 6 **65.** $\dfrac{2}{15}$
67. $-\dfrac{12}{5}$ **69.** $\dfrac{27}{5}$ **71.** 5 **73.** 200 **75.** 1,000 **77.** 200
79. $\dfrac{5}{4}$ **81.** -1 **83.** 1 **85.** 80 **87.** all real numbers
89. no solution **91.** no solution **93.** all real numbers
95. $\dfrac{1}{4}$ **97.** 30 **99.** -11 **101.** no solution **103.** 1
105. $\dfrac{52}{9}$ **107.** -6 **109.** -5
115. commutative property of multiplication
117. associative property of addition

Think It Through (page 145)

24%; 27%; 13%

Study Set Section 2.3 (page 145)

1. Percent **3.** multiplication, is **5.** $\dfrac{51}{100}, 0.51, 51\%$
7. a. 3,957 **b.** 3,957, what, 14,792 **9. a.** $12 = 0.40 \cdot x$
b. $99 = x \cdot 200$ **c.** $x = 0.66 \cdot 3$ **11. a.** 0.35 **b.** 0.085
c. 1.5 **d.** 0.0275 **13.** 312 **15.** 26% **17.** 300 **19.** 46.2
21. 2.5% **23.** 1,464 **25.** 0.48 oz **27. a.** about $1,333
billion **b.** about $865 billion **29.** $10.45 **31.** $24.20
33. 60%, 40% **35.** 19% **37.** no (66%) **39.** 120
41. a. 5 g, 25% **b.** 20 g **43.** 2008–2009, about 15%
45. 12% **47.** 17% **49.** $75 **51.** $300 **53.** $95,000
55. $25,600 **61.** $\dfrac{12}{5} = 2\dfrac{2}{5}$ **63.** no

Study Set Section 2.4 (page 157)

1. formula **3.** volume **5. a.** $d = rt$ **b.** $r = c + m$
c. $p = r - c$ **d.** $I = Prt$ **7.** 11,176,920 mi, 65,280 ft
9. Ax, Ax, B, B, B **11.** $110 million **13.** $931 **15.** 3.5%
17. $6,000 **19.** 2.5 mph **21.** 4.5 hours **23.** $185°C$
25. $-454°F$ **27.** 20 in. **29.** 1,885 mm^3 **31.** $c = r - m$
33. $b = P - a - c$ **35.** $R = \dfrac{E}{I}$ **37.** $l = \dfrac{V}{wh}$ **39.** $r = \dfrac{C}{2\pi}$
41. $h = \dfrac{3V}{B}$ **43.** $f = \dfrac{s}{w}$ **45.** $r = \dfrac{T - 2t}{2}$ **47.** $x = \dfrac{C - By}{A}$
49. $m = \dfrac{2K}{v^2}$ **51.** $c = 3A - a - b$ **53.** $t = T - 18E$
55. $r^2 = \dfrac{s}{4\pi}$ **57.** $v^2 = \dfrac{2Kg}{w}$ **59.** $r^3 = \dfrac{3V}{4\pi}$
61. $M = 4.2B + 19.8$ **63.** $h = \dfrac{S - 2\pi r^2}{2\pi r}$
65. $y = -3x + 9$ **67.** $y = \dfrac{1}{3}x + 3$ **69.** $y = -\dfrac{3}{4}x - 4$
71. $b = \dfrac{2A}{h} - d$ or $b = \dfrac{2A - hd}{h}$ **73.** $c = \dfrac{72 - 8w}{7}$
75. 87, 89, 91 **77.** 7,154.2, 6,552.3 **79.** 14 in. **81.** 50 in.
83. 5 ft, 3 in.; 10 ft, 6 in. **85.** 1,870 in.^{2} **87.** 7,958 mi
89. about 132 in. **91.** about 85 in.^{3} **93.** 17 in. **95.** 589 ft^3
97. 85,503,750 ft^3 **99.** 25.1 ft **101.** $p = \dfrac{G - U + TS}{V}$
107. 137.76 **109.** 15%

Study Set Section 2.5 (page 168)

1. consecutive **3.** vertex, base
5.

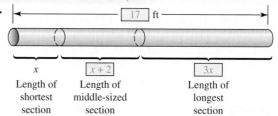

7. $\$0.03x$ **9.** $180°$ **11. a.** $x + 1$ **b.** $x + 2$ **13.** 4 ft, 8 ft
15. 102 mi, 108 mi, 114 mi, 120 mi **17.** 7.3 ft, 10.7 ft
19. 250 calories in ice cream, 600 calories in pie **21.** 7
23. 580 **25.** 20 **27.** $50,000 **29.** $5,250 **31.** Ronaldo: 15, Mueller: 14 **33.** *Friends*: 236 episodes, *Leave It to Beaver*: 234 episodes **35.** Gomez: July 22; Lopez, July 24; Bullock, July 26 **37.** width: 27 ft, length: 78 ft **39.** 21 in. by 30.25 in. **41.** 7 ft, 7 ft, 11 ft **43.** $20°$ **45.** $42.5°, 70°, 67.5°$
47. $22°, 68°$ **53.** -24 **55.** $-\dfrac{40}{37}$ **57.** 1

Study Set Section 2.6 (page 179)

1. investment, motion **3.** $30,000 - x$
5. $r - 150$ **7.** $35t + 45t = 80, 35t, t, 45t, 80$
9. a. $0.50(6) + 0.25x = 0.03(6 + x), 0.50(6), 0.25x, 6 + x, 0.30(6 + x)$ **b.** $0.06x + 0.03(10 - x) = 0.05(10), 0.06x, 10 - x, 0.03(10 - x), 0.05(10)$ **11.** 0.06, 0.152 **13.** 4
15. 6,000 **17.** $15,000 at 4%, $10,000 at 7% **19.** silver: $1,500, gold: $2,000 **21.** $26,000 **23.** 822: $9,000, 721: $6,000 **25.** $4,900 **27.** 2 hr **29.** $\dfrac{1}{4}$ hr = 15 min **31.** 1 hr
33. 4 hr **35.** 55 mph **37.** 50 gal **39.** 4%: 5 gal, 1%: 10 gal
41. 32 ounces of 8%, 32 ounces of 22% **43.** 6 gal
45. 50 lb **47.** 20 scoops **49.** 15 **51.** $4.25 **53.** 17
55. 90 **57.** 40 pennies, 20 dimes, 60 nickels
59. 2-pointers: 50, 3-pointers: 4 **63.** $-50x + 125$
65. $-3x + 3$ **67.** $16y - 16$

Study Set Section 2.7 (page 193)

1. inequality **3.** interval **5. a.** same **b.** positive
c. negative **7.** $x > 32$ **9. a.** $\leq$ **b.** ∞ **c.** [or] **d.** $>$

11. 5, 5, 12, 4, 4, 3 **13. a.** yes **b.** no **15. a.** no **b.** yes
17. $(-\infty, 5)$
19. $(-3, 1]$
21. $x < -1, (-\infty, -1)$ **23.** $-7 < x \le 2, (-7, 2]$
25. $(3, \infty)$
27. $[10, \infty)$
29. $(-\infty, 6)$
31. $(-\infty, 48]$
33. $[2, \infty)$
35. $[3, \infty)$
37. $(7, \infty)$
39. $(-\infty, 0.4]$
41. $[16, \infty)$,
43. $(-\infty, 0)$
45. $[-10, \infty)$
47. $(-\infty, -2)$
49. $(-5, \infty)$
51. $(-\infty, 1.5]$
53. $(-\infty, 20]$
55. $(0, \infty)$
57. $\left(\dfrac{5}{4}, \infty\right)$
59. $\left(-\infty, \dfrac{3}{2}\right]$
61. $(7, 10)$
63. $[-10, 0]$
65. $[-6, 10]$
67. $[2, 3)$
69. $(-3, 6]$
71. $(-5, -2)$
73. $\left[\dfrac{9}{4}, \infty\right)$
75. $(-\infty, -40]$
77. $(-2, 1]$
79. $(-\infty, 2]$
81. $(-\infty, -27)$
83. $\left(-\infty, \dfrac{1}{8}\right]$
85. $[-13, \infty)$
87. $(6, \infty)$
89. $[-32, 48]$
91. $\left(-\infty, -\dfrac{11}{4}\right)$
93. $\left[-\dfrac{3}{8}, \infty\right)$
95. $(-\infty, -1]$
97. $\left(\dfrac{6}{7}, \infty\right)$

99. 98% or better **101.** more than 27 mpg **103.** 19 ft or less **105.** more than 5 ft **107.** 40 or less **109.** 12.5 in. or less **111. a.** $100 \le b \le 118$ **b.** $136 \le b \le 160$ **c.** $134 \le b \le 171$ **115.** $1, -3, 6$

Chapter 2 Review Exercises (page 197)

1. yes **2.** no **3.** no **4.** no **5.** yes **6.** yes **7.** equation
8. true **9.** 21 **10.** 32 **11.** -20.6 **12.** 107 **13.** 24
14. 2 **15.** -9 **16.** -7.8 **17.** 0 **18.** $-\dfrac{16}{5}$ **19.** 2
20. -30.6 **21.** 30 **22.** -19 **23.** 4 **24.** 1 **25.** $\dfrac{5}{4}$
26. $\dfrac{47}{13}$ **27.** 6 **28.** $-\dfrac{22}{75}$ **29.** 5 **30.** 1
31. identity, all real numbers **32.** contradiction, no solution
33. a. Percent **b.** discount **c.** commission **34.** 192.4
35. 142.5 **36.** 12% **37. a.** April 2012 **b.** about 22,000,000
38. $26.74 **39.** no **40.** $450 **41.** $150 **42.** 1,567%
43. $176 **44.** $11,800 **45.** 8 min **46.** 4.5% **47.** 1,949°F
48. a. 168 in. **b.** 1,440 in.^{2} **c.** 4,320 in.^{3} **49.** 76.5 m^2
50. 144 in.^{2} **51. a.** 50.27 cm **b.** 201 cm^2
52. 9.4 ft^3 (Answers may vary, depending on which approximation of π is used.) **53.** 120 ft^3 **54.** 381.70 in.^{3}
55. $h = \dfrac{A}{2\pi r}$ **56.** $G = 3A - 3BC + K$ **57.** $t = \dfrac{4C}{s} + d$
58. $y = \dfrac{3}{4}x + 4$ **59.** 8 ft **60.** 200 **61.** $2,500,000

62. Menard: 27; Harvick: 29
63. 24.875 in. × 29.875 in. $\left(24\frac{7}{8}\text{ in.} \times 29\frac{7}{8}\text{ in.}\right)$
64. 76.5°, 76.5° **65.** $16,000 at 7%, $11,000 at 9%
66. 20 min **67.** $1\frac{2}{3}$ hr = 1 hr 40 min
68. TV celebrities: 12, movie stars: 4 **69.** 10 lb of each
70. 2 gal
71. $(-\infty, 1)$
72. $(-\infty, 12]$
73. $\left(\frac{5}{4}, \infty\right)$
74. $[3, \infty)$
75. $(-\infty, 40]$
76. $(9, \infty)$
77. $(6, 11)$
78. $\left(-\frac{7}{2}, \frac{3}{2}\right]$
79. 2.40 g ≤ w ≤ 2.53 g **80.** 0 in. < l ≤ 48 in., 48 in. or less

Chapter 2 Test (page 206)

1. a. solve **b.** Percent **c.** circumference **d.** inequality
e. multiplication, equality **2.** no **3.** 2 **4.** −5 **5.** 22
6. $-\frac{1}{4}$ **7.** 160.32 **8.** all real numbers (an identity)
9. $\frac{7}{4}$ **10.** −4 **11.** no solution (a contradiction)
12. 0 **13.** 12.16 **14.** $176,000 **15.** 9.3% **16.** $30
17. $295 **18.** −10°C **19.** 393 in.³ **20.** $r = \frac{A - P}{Pt}$
21. 20 in.² **22.** 22 min, 8 min **23.** $120,000
24. 380 mi, 280 mi **25.** green: 16 lb, herbal: 4 lb
26. 412, 413 **27. a.** $5x, 6x, 10(x − 2), 5x$
b. $5x + 6x + 10(x − 2) + 5x = 110$
28. 106° **29.** $\frac{3}{5}$ hr or 36 min **30.** 10 liters **31.** 68°
32. $5,250 **33.** $[-3, \infty)$
34. $(-\infty, 6.4)$
35. $[-7, 4)$
36. 180 words

Cumulative Review Chapters 1-2 (page 208)

1. a. expression **b.** equation **2.** 3, 4, 5 **3. a.** $w + 12$
b. $n − 4$ **4.** $2 \cdot 2 \cdot 5 \cdot 5 = 2^2 \cdot 5^2$ **5.** $\frac{2}{3}$ **6.** $-\frac{2}{9}$ **7.** 6
8. $\frac{22}{15} = 1\frac{7}{15}$ **9.** $12\frac{11}{24}$ **10.** 0.9375 **11. a.** 65 **b.** −12

12. a. true **b.** true **c.** false **13. a.** −10 **b.** −14
c. −64 **d.** 0 **14. a.** 4^3 **b.** $\pi r^2 h$ **15.** 4 **16.** 1, −3, 6
17. 0 **18.** −2 **19.** 16 **20.** 0 **21.** −32d
22. $10x − 15y + 5$ **23.** $5x$ **24.** $−8a$ **25.** $8q^2 − 5q$
26. $27t − 20$ **27.** 9 **28.** 20 **29.** −0.6 **30.** 19 **31.** −20
32. −2 **33.** 1 **34.** $\frac{5}{4}$ **35.** 300 **36.** 87.5 **37.** 25%
38. $2 **39.** 65 m² **40.** 376.99 cm³ **41.** $t = \frac{A - P}{Pr}$
42. 37.5 ft-lb **43.** 9.45 lb **44.** 55° **45.** $4,000
46. 10 oz **47.** $x > -2, (-2, \infty)$
48. $x \leq 2, (-\infty, 2]$
49. $x \geq -1, [-1, \infty)$
50. $x < -14, (-\infty, -14)$
51. $-1 \leq x < 2, [-1, 2)$
52. The score must be greater than 5.2.

Study Set Section 3.1 (page 218)

1. rectangular **3.** origin **5.** quadrants **7.** x-coordinate,
y-coordinate **9.** origin, left, up **11.** no **13.** quadrant II
15. a. 3 **b.** −4 **c.** 0 **d.** −4 **e.** 0 **f.** 5 **17.** (3, 5) is an
ordered pair, 3(5) indicates multiplication, and 5(3 + 5) is an
expression containing grouping symbols. **19.** yes
21. **23.**
25. (2, 3) **27.** (−3, −4) **29.** (0, 0) **31.** (−4, 0)
33. 10 min before the workout, her heart rate was
60 beats/min. **35.** 150 beats/min **37.** approximately 5 min
and 50 min after starting **39.** 10 beats/min faster after cool-
down **41.** $20 **43.** $70 **45.** rivets: (−6, 0), (−2, 0), (2, 0),
(6, 0); welds: (−4, 3), (0, 3), (4, 3); anchors: (−6, −3), (6, −3)
47. (G, 2), (G, 3), (G, 4) **49. a.** 60°, 4 ft **b.** 30°, 4 ft
51. (28°, −89°) **53. a.** 8 **b.** It represents the patient's left side.
55. 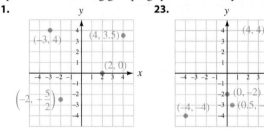 **a.** A 3-year-old car is worth $7,000. **b.** $1,000 **c.** 6 **61.** 12 **63.** 8 **65.** 7 **67.** −49

Study Set Section 3.2 (page 231)

1. one, two 3. table, output 5. a. 2 b. yes c. yes
d. infinitely many 7. solution, point 9. 0, −1, −8, 1, 8
11. He should have checked his computations. At least one of his "solutions" is wrong. 13. A smooth curve should be drawn through the points. 15. 6, −2, 2 17. yes 19. no
21. no 23. yes 25. (1, −1) 27. (−3, 18)
29. −3, −2, −5 31. −3, 1, 1

33. 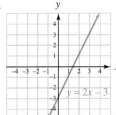 35.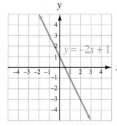

37. 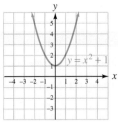 39.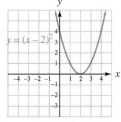

1 unit higher   2 units to the right

41. 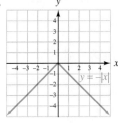 43.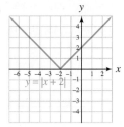

It is turned upside down.   2 units to the left

45. 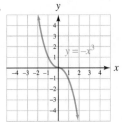 47.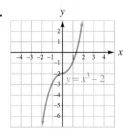

It is turned upside down.   2 units lower

49. −2, 0, 4; 4, 0, −4   51. 0, 1, 4, 1, 4
53. a. It costs 8¢ to make a 2-in. bolt. b. 12¢ c. a 4-in. bolt
d. It decreases as the length increases to 4 in., then increases as the length increases to 7 in. 55. a. $90,000
b. the 3rd yr after being bought c. after the 6th yr
d. It decreased in value for 3 yr to a low of $60,000, then increased in value for 5 yr to a high of $110,000.
63. −96 65. an expression 67. 1.25 69. 0.1

Study Set Section 3.3 (page 244)

1. linear 3. y-intercept 5. x, y 7. vertical 9. the y-axis
11. y: 1st power, x: 1st power 13. y: 1st power, x: 3rd power
15. because A is on the line 17. He made a mistake. The points should lie on a straight line. 19. a. $(-3, 0)$ b. $(0, -1)$
21. and 23.

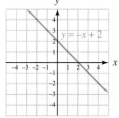

25. $4x - y = 6$
27. $x + 3y = 9$
29. $y = 8 - x$
31. $y = -6x + 8$
33. nonlinear 35. nonlinear
37. nonlinear 39. linear
41. 6, −5, 4   43. $-2, 4, -\dfrac{3}{2}$

Answers may vary.

45.   47.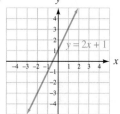

49. 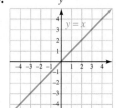 51.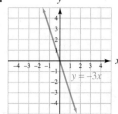

53. 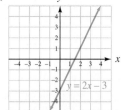 55.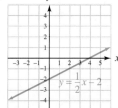

57. 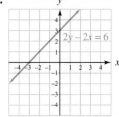 59.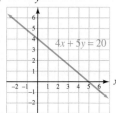

61. 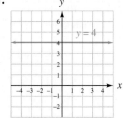 63.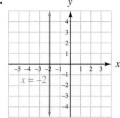

65. 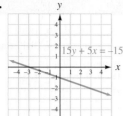   **67.**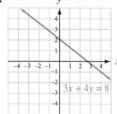

85. a. 56.2, 62.1, 64.0

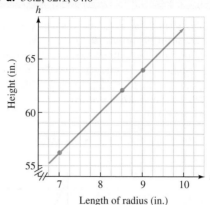

b. taller the woman is.
c. 58 in.
91. $5 + 4c$
93. -4

69. 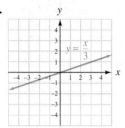   **71.**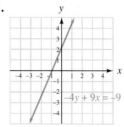

73. 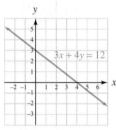   **75.**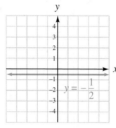

95. profit = revenue − costs   **97.** 491

Think It Through (page 254)

$80 per year

Study Set Section 3.4 (page 258)

1. ratio   **3.** slope   **5.** run   **7.** parallel   **9.** Horizontal
11. rises   **13.** l_2 **15.** l_4 **17.** $\frac{1}{2}$ **19.** $\frac{1}{2}$ **21.** -1
23. $m = \frac{y_2 - y_1}{x_2 - x_1}$ **25.** $m = \frac{2}{3}$ **27.** $m = \frac{4}{3}$ **29.** $m = -\frac{7}{8}$
31. $m = -\frac{1}{5}$ **33.** 1   **35.** -3 **37.** $\frac{5}{4}$ **39.** $-\frac{1}{2}$ **41.** 0
43. undefined   **45.** 0   **47.** undefined

77. 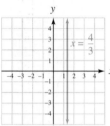   **79.**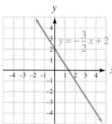

49. 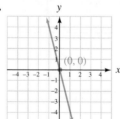   **51.**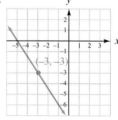

81.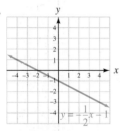

53. 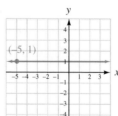   **55.**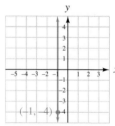

57. parallel   **59.** perpendicular   **61.** neither   **63.** neither
65. $\frac{5}{9}$ **67.** $-\frac{2}{3}$ **69.** $-\frac{2}{5}$ **71.** $\frac{1}{20}$, 5%   **73. a.** $\frac{1}{8}$ **b.** $\frac{1}{12}$
c. 1: less expensive, steeper; 2: not as steep, more expensive
75. $-$$2,500/year   **77.** 3 hp/40 rpm   **79.** 329 lb/yr
85. quadrant II   **87.** no   **89.** linear

83. a. $c = 50 + 25u$
b. 150, 250, 400
c. $850
d. The service fee is $50.

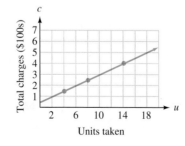

Think It Through (page 264)

The rate of increase in beginning teacher salary is $1,150 per year. In 1990, the average beginning teacher salary was $32,850. In 2020, it will be $67,350.

Study Set Section 3.5 (page 267)

1. slope–intercept 3. y-intercept, slope 5. no 7. yes
9. $-\dfrac{2}{3}$ 11. -8 13. 1 15. $-\dfrac{1}{2}$ 17. $y = -\dfrac{1}{2}x - 4$
19. $(0, 0)$ 21. $-\dfrac{1}{2}$ 23. $-\dfrac{1}{2}$ 25. a. $(0, 0)$
b. same slope, different y-intercepts 27. parallel
29. neither 31. $6x, -6x, -2, -2, -2, 3x, 3, (0, -5)$
33. $4, (0, 2)$ 35. $\dfrac{1}{4}, \left(0, -\dfrac{1}{2}\right)$ 37. $4, (0, -2)$ 39. $\dfrac{1}{6}, (0, -1)$
41. $\dfrac{4}{3}, (0, -4)$ 43. $2, \left(0, -\dfrac{12}{5}\right)$ 45. parallel
47. perpendicular 49. neither 51. parallel 53. $-1, (0, 8)$
55. $-\dfrac{2}{3}, (0, 2)$ 57. $0, \left(0, \dfrac{13}{3}\right)$ 59. $-5, (0, 0)$

61. 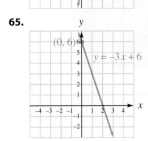 63.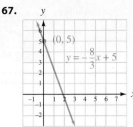

65. 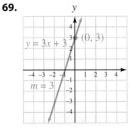 67.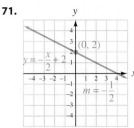

69. 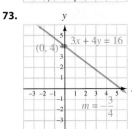 71.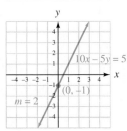

73. 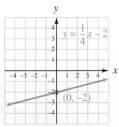 75.

77. a. $y = 2{,}000x + 5{,}000$ b. $21{,}000$ 79. $y = 5x - 10$

81. a. $y = 0.25x + 2.00$
b.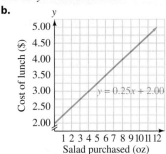
c. same slope, different y-intercept d. same y-intercept, steeper slope
83. a. $y = 2.50x + 20$
b. and c.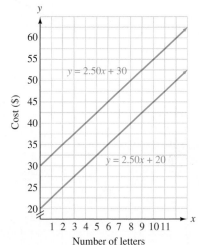

85. $y = -20x + 500$ 91. $-\dfrac{1}{4}$ 93. 0 95. subtraction
97. 25%

Study Set Section 3.6 (page 276)

1. point–slope 3. The slope is 2. The y-intercept is $(0, -3)$.
5. point–slope form 7. $(-4, -2), (3, 2)$
9. $y + 2 = \dfrac{4}{7}(x + 4)$ or $y - 2 = \dfrac{4}{7}(x - 3)$ 11. no 13. yes
15. yes 17. $-3x, 12, 2, 2$ 19. $5, -1, -2x$
21. $y - 1 = 3(x - 2)$ 23. $y + 1 = -\dfrac{4}{5}(x + 5)$
25. $y = \dfrac{1}{5}x - 1$ 27. $y = -5x - 37$ 29. $y = -\dfrac{4}{3}x + 4$
31. $y = -\dfrac{2}{3}x + 2$ 33. $y = 2x + 5$ 35. $y = \dfrac{1}{10}x + \dfrac{1}{2}$
37. $y = 5$ 39. $y = -\dfrac{1}{2}x + 1$ 41. $x = 4$ 43. $y = 5$
45. $y = 8x + 4$ 47. $y = -3x$ 49. $x = -8$ 51. $x = 12$
53. $y = 3x + 4$ 55. $y = -5x - 9$ 57. $y = \dfrac{7}{3}x + 7$
59. $y = \dfrac{9}{8}x$ 61. $x = \dfrac{2}{5}$ 63. $y = -32$
65. a. $y = -40x + 920$ b. 440 yd^3
67. $y = -\dfrac{2}{5}x + 4, y = -7x + 70, x = 10$
69. $c = 30t + 45$ 71. a. $(0, 32), (100, 212)$
b. $F = \dfrac{9}{5}C + 32$ 73. a. $(6, 10{,}808), (15, 14{,}570)$
b. $E = 418a + 8{,}300$ c. $15{,}406$ 79. $-\dfrac{1}{2}$
81. 113.1 ft^2 83. -1 85. 6

Study Set Section 3.7 (page 286)

1. inequality **3.** boundary **5.** no **7.** yes **9.** no **11.** no
13. no, dashed **15.** no **17.** yes **19.** yes **21.** no
23. The test point must be on one side of the boundary.
25. is less than or equal to **27.** is less than **29.** yes
31. no **33.** no **35.** yes

37. 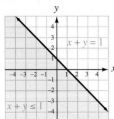   **39.**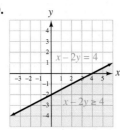

41. 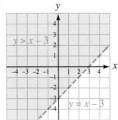   **43.**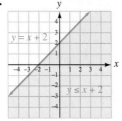

45. 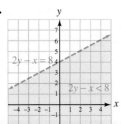   **47.**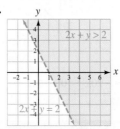

49. 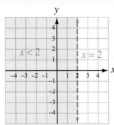   **51.**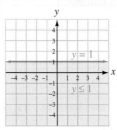

53. 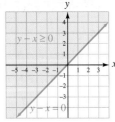   **55.**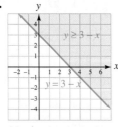

57. 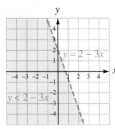 **59.**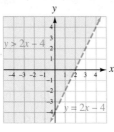

61. 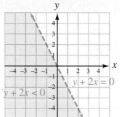 **63.**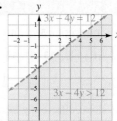

65. 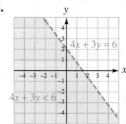 **67.**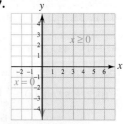

69.

Based on data from *Los Angeles Times* (March 24, 1999)

71. 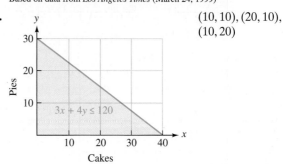   (10, 10), (20, 10), (10, 20)

73. 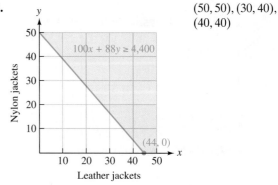   (50, 50), (30, 40), (40, 40)

Appendix 3 Answers to Selected Exercises A-17

75.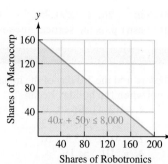

(80, 40), (80, 80), (120, 40)
77. ii **81.** 15%
83. $d = \dfrac{c}{\pi}$
85. $d = rt$
87. $t = \dfrac{A - P}{Pr}$

Study Set Section 3.8 (page 300)

1. relation **3.** independent **5.** domain, range **7.** positive numbers **9.** 0 **11.** all real numbers, domain **13.** $f(-1)$
15. a. $(-2, 4), (-2, -4)$ **b.** No, the x-value -2 is assigned to more than one y-value (4 and -4). **17.** x, $-5, 4, -5$
19. of, is **21.** domain: $\{-6, -1, 6, 8\}$, range: $\{-10, -5, -1, 2\}$
23. domain: $\{-8, 0, 6\}$, range: $\{9, 50\}$ **25.** yes **27.** yes
29. no, $(4, 2), (4, -2)$ **31.** yes **33.** yes **35.** no, $(4, 2)$, $(4, 4), (4, 6)$ (answers may vary) **37.** D: all reals, R: all reals
39. D: all reals, R: real numbers greater than or equal to 0
41. a. 3 **b.** -9 **c.** 0 **d.** 199
43. a. 0.32 **b.** 18 **c.** 2,000,000 **d.** $\dfrac{1}{32}$
45. $-2, -5, 1, 4$ **47.** $-3, -2, -1$

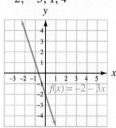 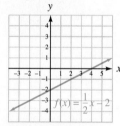

49. yes **51.** no, $(3, 4), (3, -1)$ (answers may vary)
53. yes **55.** no, $(3, 1), (3, 2)$ **57.** no, $(-1, 0), (-1, 2)$
59. no, $(3, 4), (3, -4)$ or $(4, 3), (4, -3)$ **61.** no, $(0, 2)$, $(0, -4)$ (answers may vary) **63.** yes **65.** $f(x) = |x|$
67. a. all real numbers from 0 through 24 **b.** 0.5
c. 1.5 **d.** -1.4 **e.** The low tide mark was -2.5 m.
69. 78.5 ft², 314.2 ft², 1,256.6 ft² **71.** 32, 36, 20, 0
75. $y = 6$ **77.** profit = revenue − costs **79.** $-6x + 12$

Chapter 3 Review Exercises (page 305)

1. a.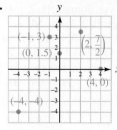

b. quadrant III **2.** $-1, 0, 1$
3. (158, 21.5) **4. a.** 2 ft
b. 2 ft **c.** 6 ft **5. a.** 2,500, week 2 **b.** 1,000 **c.** 1st week and 5th week **6.** not a solution

7. a.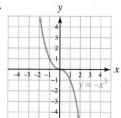

x	y	(x, y)
-2	8	$(-2, 8)$
-1	1	$(-1, 1)$
0	0	$(0, 0)$
1	-1	$(1, -1)$
2	-8	$(2, -8)$

b. It would be 2 units higher **8. a.** 9,000 **b.** It tells us that 40 trees on an acre give the highest yield, 18,000 oranges.
9. nonlinear **10.** linear **11.** linear **12.** nonlinear
13. $A = 5, B = 2, C = 10$ **14.** $-6, -6, -8, -8$
15. 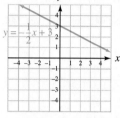 **16.** x-intercept: $(-2, 0)$, y-intercept: $(0, 4)$

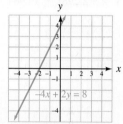

17. 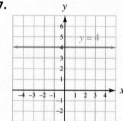 **18.**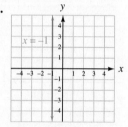

19. If the three points do not lie on a line, then at least one of them is in error. **20. a.** True **b.** False
21. $\dfrac{1}{4}$ **22.** $-\dfrac{7}{8}$ **23.** -7 **24.** 0 **25.** $-\dfrac{3}{2}$ **26.** $\dfrac{3}{4}$
27.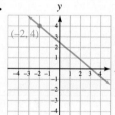

28. a. -4.5 million people per yr
b. 3.45 million people per yr
29. $m = \dfrac{3}{4}$, y-intercept: $(0, -2)$
30. $m = -4$, y-intercept: $(0, 0)$

31. $m = 3$, y-intercept: $(0, -5)$

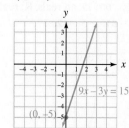

32. $m = -\dfrac{2}{5}$, $(0, 3)$;
$y = -\dfrac{2}{5}x + 3$
33. parallel
34. perpendicular
35. a. $c = 300w + 75,000$
b. 90,600 **36.** $c = 7.8m + 220$

Unless otherwise noted, all content on this page is © Cengage Learning.

37. $y - 5 = 3(x - 1)$, $y = 3x + 2$

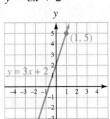

38. $y + 1 = -\frac{1}{2}(x + 4)$, $y = -\frac{1}{2}x - 3$

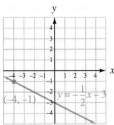

39. $y = \frac{2}{3}x + 5$ **40.** $y = -8$ **41.** $f(x) = -35x + 450$
42. $y = -2{,}298x + 19{,}984$ **43.** yes **44.** yes **45.** yes **46.** no

47. 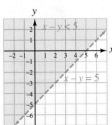 **48.**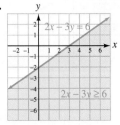

49. 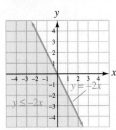 **50.**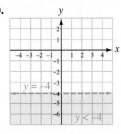

51. a. true **b.** false **c.** false
52. $(2, 4), (5, 3), (6, 2)$; answers will vary

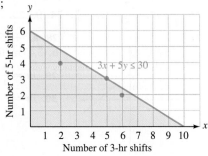

53. domain: $\{-5, 0, 4, 7\}$, range $\{-11, -3, 4, 9\}$ **54.** domain: $\{-6, 1, 2, 15\}$, range: $\{-8, -2, 9\}$ **55.** yes **56.** no **57.** no
58. yes **59.** D: all reals, R: all reals **60.** D: all reals, R: real numbers greater than or equal to 0 **61.** -5 **62.** 37
63. -2 **64.** -8
65. $1, 0, -1, 0, -1, -2$ **66.** $2, 3, 6, 3, 6$

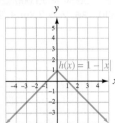 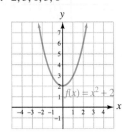

67. no **68.** yes **69.** 1,004.8 in.³ **70. a.** $(200, 25)$, $(200, 90)$, $(200, 105)$ **b.** It doesn't pass the vertical line test.

Chapter 3 Test (page 315)

1. a. 10 **b.** 60 **c.** 1 day before and the 3rd day of the holiday **d.** 50 dogs were in the kennel when the holiday began.
2. **3.** $A(2, 4), B(-3, 3), C(-2, -3), D(4, -3), E(-4, 0), F(3.5, 1.5)$
4. a. III **b.** IV

5. **6.**

7. yes **8.**

x	y	(x, y)
2	1	$(2, 1)$
-6	3	$(-6, 3)$

9. no **10.** x-intercept: $(3, 0)$, y-intercept: $(0, -2)$
11. $m = -\frac{1}{2}, (0, 4)$

12. **13.**

14. -1 **15.** undefined **16.** $\frac{8}{7}$ **17.** $\frac{3}{2}$ **18.** parallel
19. -15 ft per mi **20.** 25 ft per mi
21. $v = -1{,}500x + 15{,}000$ **22.** $y = 7x + 19$
23. $y = -\frac{1}{5}T + 41$ **24. a.** yes **b.** no **c.** no **25.** yes
26. **27.** domain: $\{-2, 0, 6, 7\}$, range: $\{1, 4, 5, 6\}$ **28.** no
29. yes, domain: $\{1, 2, 3, 4\}$, range: $\{1, 2, 3, 4\}$ **30.** no, $(-3, 9), (-3, -7)$ **31.** yes
32. D: all real numbers, R: real numbers less than or equal to 0 **33.** -13 **34.** 822
35. $C(45) = 28.50$; it cost $28.50 to make 45 calls. **36.**

Cumulative Review Chapters 1-3 (page 317)

1. $2^2 \cdot 3^3$ 2. 0.004 3. a. true b. true c. true 4. -18
5. 6 6. -2 7. 1 8. 28 9. 32 10. $500 - x$ 11. 3
12. -2 13. $2x + 8$ 14. $2x - 8$ 15. $-2x - 8$
16. $-2x + 8$ 17. $4a + 10$ 18. $4b^2$ 19. 4 20. $-3y$
21. 6 22. 2.9 23. 9 24. -19 25. $\dfrac{1}{7}$ 26. 1 27. $-\dfrac{55}{6}$
28. no solution 29. 1, 100 30. $x = \dfrac{y - b}{m}$
31. $3\dfrac{1}{8}$ in., $\dfrac{39}{64}$ in.^{2} 32. $79°$
33. $0.05x$, $0.25(13 - x)$, $0.30(13)$ 34. $1,900 at 3%, $600 at 8% 35. 7.5 hr 36. 80 lb candy corn, 120 lb gumdrops
37. $x \le 48$, ———]——, $(-\infty, 48]$
                    48
38. $x > 0$, ——(——→, $(0, \infty)$
              0
39. a. 1 tsp b. 3 tsp 40. no
41. [graph $y = |x - 2|$] 42. [graph $4y + 2x = -8$]
43. $\dfrac{7}{12}$ 44. 0 45. $-\dfrac{10}{7}$ 46. $\dfrac{2}{3}$, $(0, 2)$ 47. $y = -2x + 1$
48. $y + 9 = -\dfrac{7}{8}(x - 2)$
49. 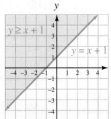 50. [graph $x < 4$]
51. 10 52. no; $(1, 2)$, $(1, -2)$

Think It Through (page 325)

(1978, 50). In 1978, 50% of the associate degrees that were awarded went to men and 50% went to women. Since then, the percent awarded to women has increased, while the percent awarded to men has decreased.

Study Set Section 4.1 (page 329)

1. system 3. independent 5. inconsistent 7. true
9. false 11. true 13. $(-4, -1)$ 15. 1 solution, consistent
17. The method is not accurate enough to find a solution such as $\left(\dfrac{2}{5}, -\dfrac{1}{3}\right)$. 19. $6, 6, 6, \dfrac{1}{3}y$ 21. yes 23. yes 25. no
27. no

29. $(3, 2)$
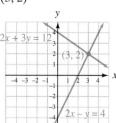

31. $(-2, -1)$
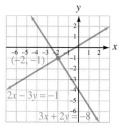

33. $(3, -1)$
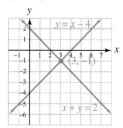

35. $(4, -2)$
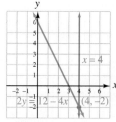

37. $(4, -4)$
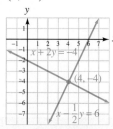

39. $(-4, 0)$
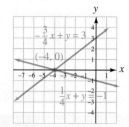

41.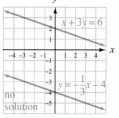
Inconsistent system

43.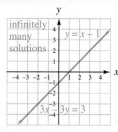
Dependent equations

45. 1 solution 47. same line, infinitely many solutions
49. no solution 51. 1 solution 53. no 55. yes
57. $(-3, 4)$ 59. $(2, 1)$

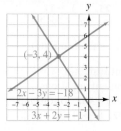
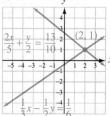

61.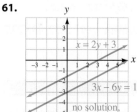
no solution, inconsistent system

63.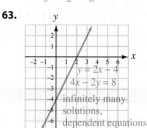
infinitely many solutions, dependent equations

65. (1, 3)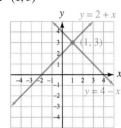
67. 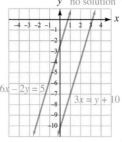 Inconsistent system
69. (SW 149th St, S May Ave) **71. a.** Houston, New Orleans, St. Augustine **b.** St. Louis, Memphis, New Orleans **c.** New Orleans
73. yes, (2, 2)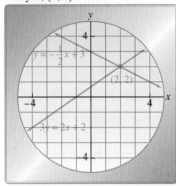
75. (2025, 38) **79.** parallel **81.** −21 **83.** $x = 0$ **85.** (5, 2)

Study Set Section 4.2 (page 340)

1. y, terms **3.** distributive **5.** infinitely **7. a.** 2 **b.** 1
9. a. $x = 2y - 10$ **b.** $y = \dfrac{x}{2} + 5$
c. x, it involved only one step. **11. a.** Parentheses must be written around $x - 4$ in line 2. **b.** $\dfrac{13}{3}$
13. a. The coordinates of the intersection point are not integers. **b.** $\left(\dfrac{4}{5}, \dfrac{2}{5}\right)$
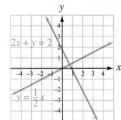
15. $3x, 4, -2, (-2, -6)$
17. (2, 4) **19.** (3, 0) **21.** $\left(\dfrac{2}{3}, -\dfrac{1}{3}\right)$
23. (3, 2) **25.** (−1, −1) **27.** (−3, −1) **29.** (−2, 3)
31. $\left(\dfrac{1}{5}, 4\right)$ **33.** no solution, inconsistent system
35. infinitely many solutions, dependent equations
37. (3, −2) **39.** (4, −2) **41.** (1, 4) **43.** no solution, inconsistent system **45.** (4, 2) **47.** (−6, 4)
49. $\left(\dfrac{1}{2}, \dfrac{1}{3}\right)$ **51.** (5, 5) **53.** 22.5°, 67.5° **55.** melon, because it's the same price as hash browns **57.** 2032; 433,235
63. $-\dfrac{5}{8}$ **65.** (0, −6) **67.** no **69.** x-axis

Study Set Section 4.3 (page 349)

1. coefficient **3.** standard **5.** The second equation should be written in standard form: $3x - 2y = 10$ **7.** Multiply both sides by 15.

9. a. (2, −3)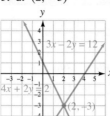
b. (2, −3) **11.** $2x, 1, 1, (1, 4)$
13. (−2, 3) **15.** (−1, 1)
17. (−3, 4) **19.** (0, 8)
21. (2, 3) **23.** (3, −2)
25. (2, 7) **27.** $\left(1, -\dfrac{5}{2}\right)$
29. (−1, 2) **31.** (0, 1)
35. (4, 0) **33.** (3, 0)
37. no solution, inconsistent system
39. infinitely many solutions, dependent equations
41. (8, −3) **43.** $\left(\dfrac{10}{3}, \dfrac{10}{3}\right)$ **45.** (5, −6) **47.** (−1, 2)
49. (4, −2) **51.** infinitely many solutions **53.** $\left(\dfrac{10}{3}, \dfrac{10}{3}\right)$
55. (1, −1) **57.** 1991 **63.** 4 **65.** 0 **67.** 7.5 ft^2 **69.** $x - 10$

Think It Through (page 355)

12,000 hr, 18,000 hr

Study Set Section 4.4 (page 360)

1. variable **3.** system **5.** $x + c, x - c$ **7. a.** $y **b.** 5%
c. $0.05x, 0.11y$ **9. a.** two **b.** graphing, substitution, addition **11.** $A = lw$ **13.** $d = rt$ **15.** $2l + 2w = 90$
17. 32, 64 **19.** 8, 5 **21.** 22 ft, 29 ft **23.** Alaska: 201 mi; Canada: 1,221 mi **25.** 120,000 from accidents, 600,000 from heart disease **27.** $29.50, $21 **29.** 85 cones **31.** 72°
33. length 70 in., width 48 in. **35. a.** 400 tires
b.   **c.** the second mold
37. nursing: $2,000, business: $3,000
39. 5 mph
41. 50 mph
43. 4 L 6% salt water, 12 L 2% salt water
45. 32 lb peanuts, 16 lb cashews **47.** $640
51.   **53.**   **55.** 30
57. all real numbers

Study Set Section 4.5 (page 370)

1. inequalities **3.** shaded **5.** true **7.** false **9.** true **11.** ii
13. iv **15.** yes **17.** no **19.** $\begin{cases} x \le -2 \\ y > 2 \end{cases}$ **21.** ≤
25.   **27.**

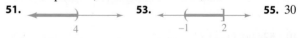

Appendix 3 Answers to Selected Exercises A-21

29.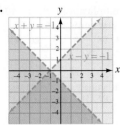
31.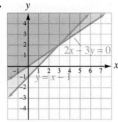
33.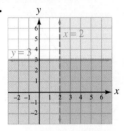
35.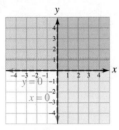
37.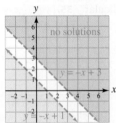
39.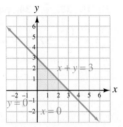
41.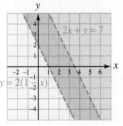
43.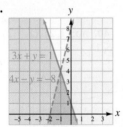
45.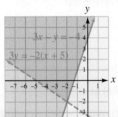
47.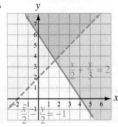
49.
51. 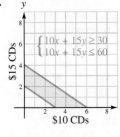 1 $10 CD and 2 $15 CDs, 4 $10 CDs and 1 $15 CD, answers vary

53. 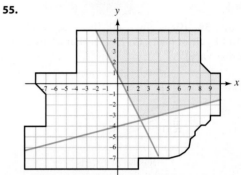 2 desk chairs and 4 side chairs, 1 desk chair and 5 side chairs, answers vary

55.

61. 128, 8
63. 4, −23

Chapter 4 Review Exercises (page 374)

1. yes **2.** (1980, 480,000). In 1980, the number of men and women awarded bachelor's degrees in the United States was the same, 480,000.

3.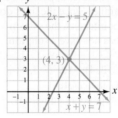
4.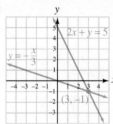
5.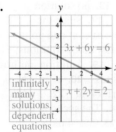
6.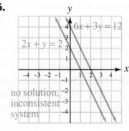

7. (3, 3) **8.** (5, 0) **9.** $\left(-\frac{1}{2}, \frac{7}{2}\right)$ **10.** (−2, 1)
11. infinitely many solutions, dependent equations
12. (12, 10) **13. a.** no solutions **b.** two parallel lines **c.** inconsistent system **14.** one **15.** (3, −5) **16.** $\left(3, \frac{1}{2}\right)$
17. (−1, 7) **18.** (0, 9) **19.** infinitely many solutions, dependent equations **20.** (0, 0) **21.** (−5, 2) **22.** no solution, inconsistent system **23.** Addition, no variables have a coefficient of 1 or −1. **24.** Substitution, equation 1 is solved for x. **25.** Las Vegas: 2,000 ft; Baltimore: 100 ft
26. base: 21 ft, extension: 14 ft **27.** 10,800 yd²
28. a. $y = 5x + 30,000$, $y = 10x + 20,000$ **b.** 2,000 **c.** the athlete **29. a.** $0.02x$, $0.09y$, $0.08(100)$
b. $5(s + w)$, $7(s - w)$ **c.** $0.11x$, $0.06y$ **d.** $4x$, $8y$, $10(5)$
30. 12 lb gummy worms, 18 lb gummy bears **31.** 3 mph

32. $16.40, $10.20 **33.** $750 **34.** $13\frac{1}{3}$ gal 40%, $6\frac{2}{3}$ gal 70%
35. 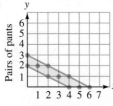 **36.**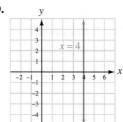

7. $0.\overline{6}$ **8.** associative property of multiplication **9.** 0 **10.** 1
11. 4 **12.** -2 **13.** $250 - x$ **14.** $10d$ cents **15.** r
16. $18x$ **17.** $3d - 11$ **18.** $-78c + 18$ **19.** 13 **20.** 41
21. $\frac{10}{9}$ **22.** -24 **23.** 140 **24.** $h = \frac{2A}{b + B}$
25. 20 lb of $3.80 candy, 10 lb of $4.40 candy
26. $(-\infty, -14)$ **27.** II **28.** no

29. 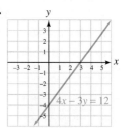 **30.**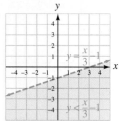

37. a. yes **b.** no **38.** $10x + 20y \geq 40$, $10x + 20y \leq 60$; (3, 1): 3 shirts and 1 pair of pants; (1, 2): 1 shirt and 2 pairs of pants (answers may vary)

31. $\frac{1}{2}$ **32.** 0 **33.** $\frac{2}{3}$ **34.** perpendicular **35.** $y = \frac{2}{3}x + 5$
36. $y = \frac{3}{4}x + \frac{11}{2}$ **37.** $y = 4$ **38.**

Chapter 4 Test (page 381)

1. a. solution **b.** consistent **c.** inconsistent **d.** independent **e.** dependent **2. a.** (30, 3); if 30 items are sold, the salesperson is paid the same amount under both plans, $3,000. **b.** Plan 2 **3.** yes **4.** no
5. (2, 1) 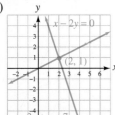 **6.** The lines appear to be parallel. Since the lines do not intersect, the system does not have a solution. **7.** $(-2, -3)$ **8.** $(-1, -1)$ **9.** (2, 4) **10.** $(-3, 3)$ **11.** inconsistent, independent equations

39. 1 **40.** no; (1, 2), (1, -2) (answers may vary)
41. $(-6, 1)$ **42.**

12. consistent, dependent equations **13.** no solution
14. $(-1, -1)$ **15.** (5, 14) **16.** (0, 0) **17.** Since the lines have different slopes, they will intersect at one point. The system has 1 solution. **18.** infinitely many solutions; (0, 2), (3, 1), (6, 0) (answers may vary) **19.** Addition method; the coefficients of y are already opposites. **20. a.** iii **b.** ii **c.** i **d.** iv **21.** 1st part: 8 mi; 2nd part: 14 mi **22.** 3 adult tickets, 4 child tickets **23.** $4,000 **24.** The speed in calm air: 165 mph; speed of wind: 15 mph **25.** 5%: 4 pints; 20%: 8 pints **26.** $1.50 sunscreen: 3 oz; $0.80 sunscreen: 7 oz
27. 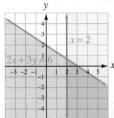 **28.** (1, 2), (2, 2), (3, 1) (answers may vary)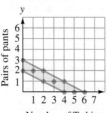

43. (4, 1) **44.** $\left(\frac{2}{3}, \frac{3}{2}\right)$ **45.** noodles: 2 servings, rice: 3 servings **46.** $3,000 at 10%, $5,000 at 12%

Think It Through (page 388)

$10^4 = 10,000$

Study Set Section 5.1 (page 394)

1. base, exponent **3.** power **5.** $3x, 3x, 3x, 3x$ **7.** x^{m+n}
9. $\frac{a^n}{b^n}$ **11.** x^{m-n} **13.** x^{mn} **15.** $(3x^2)^6$ (answers may vary)
17. 3 **19.** x^6 **21.** 16 **23.** -16 **25.** base 4, exponent 3
27. base x, exponent 5 **29.** base $-3x$, exponent 2
31. base y, exponent 6 **33.** $\left(\frac{a}{3}\right)^3$ **35.** $-\left(\frac{b}{6}\right)^5$ **37.** $6x^5$
39. $(4t)^4$ **41.** $-4t^3$ **43.** $(x + y)^2$ **45.** 12^7 **47.** 2^5 **49.** a^6
51. x^7 **53.** a^9 **55.** y^9 **57.** a^5b^6 **59.** $(m + n)^5$ **61.** 8^8
63. 12^3 **65.** x^{12} **67.** c **69.** xy^4 **71.** c^2d^6 **73.** m^{11}
75. a **77.** 3^8 **79.** 2^{15} **81.** y^{15} **83.** m^{500} **85.** x^{25}
87. b^{14} **89.** y^{14} **91.** p^{15} **93.** $64p^9$ **95.** a^9b^6 **97.** p^4q^8

Cumulative Review Chapters 1-4 (page 383)

1. sum, difference, product, quotient **2.** $2^2 \cdot 5^2$ **3.** $\frac{5}{8}$ **4.** $\frac{22}{35}$
5. irrational **6.**

99. $-8r^6s^9$ **101.** x^3y^4 **103.** p^5q^2 **105.** $\dfrac{a^3}{b^3}$ **107.** $\dfrac{32x^{10}}{y^{15}}$
109. $27z^3$ **111.** $216k^3$ **113.** x^3y^3 **115.** $243z^{30}$
117. x^{31} **119.** u^4v^4 **121.** $-\dfrac{32a^5}{b^5}$ **123.** $a^{12}b^6$ **125.** ab^4
127. $r^{13}s^3$ **129.** $\dfrac{y^3}{8}$ **131.** $\dfrac{27t^{12}}{64}$ **133.** $\dfrac{t^6}{8}$ **135.** c^2d^5
137. y^4 **139.** a^{10} mi² **141.** x^9 m³ **143. a.** $25x^2$ ft²
b. $9a^2\pi$ ft² **145. b.** 16 ft, 8 ft, 4 ft, 2 ft
147. a. 11000001, 11010001, 11001101, 11000011 (answers may vary) **b.** 2^8 **153.** c **155.** d

Study Set Section 5.2 (page 404)

1. base, exponent **3.** reciprocal **5. a.** 4 − 4, 0
b. 6, 6, 6, 6, 1 **c.** 1, 1 **7.** 9, 3, 1, $\dfrac{1}{3}, \dfrac{1}{9}$ **9.** 81, −9, 1, $-\dfrac{1}{9}, \dfrac{1}{81}$

11.

x	y	y as a power of 2
2	4	2^2
1	2	2^1
0	1	2^0
−1	$\tfrac{1}{2}$	2^{-1}
−2	$\tfrac{1}{4}$	2^{-2}

13. y^8, −40 **15.** base x, exponent −2
17. a. 4, 2, −16
b. 4, −2, $\dfrac{1}{16}$
c. 4, −2, $-\dfrac{1}{16}$ **19.** 1
21. 1 **23.** 2 **25.** 1
27. $\dfrac{1}{144}$ **29.** $-\dfrac{1}{4}$
31. 125 **33.** $\dfrac{3}{16}$
35. $\dfrac{1}{x^2}$ **37.** $\dfrac{b^3}{a^2}$ **39.** $\dfrac{1}{16y^2}$ **41.** $-\dfrac{2}{b^5}$ **43.** $\dfrac{8}{7}$ **45.** $\dfrac{16}{9}$
47. $\dfrac{1}{a^{12}}$ **49.** $\dfrac{1}{a^5}$ **51.** $\dfrac{1}{a^3b^6}$ **53.** y^3 **55.** $\dfrac{3q}{p}$ **57.** $\dfrac{1}{b^{14}}$
59. 1 **61.** x^{3m} **63.** $\dfrac{1}{u^m}$ **65.** y^m **67.** 1 **69.** $\dfrac{5}{2}$
71. $-\dfrac{1}{64}$ **73.** $\dfrac{1}{64}$ **75.** y^5 **77.** c^5 **79.** $\dfrac{8}{9}$ **81.** $\dfrac{1}{16y^4}$ **83.** 8
85. 1 **87.** 1 **89.** $\dfrac{1}{y}$ **91.** $\dfrac{1}{r^6}$ **93.** 2 **95.** $\dfrac{1}{a^2b^4}$ **97.** $\dfrac{1}{x^6y^3}$
99. $\dfrac{1}{x^3}$ **101.** $\dfrac{a^8}{b^{12}}$ **103.** $-\dfrac{y^{10}}{32x^{15}}$ **105.** a^{14} **107.** $\dfrac{256x^{28}}{81}$
109. $\dfrac{16y^{14}}{z^{10}}$ **111.** $10^2, 10^1, 10^0, 10^{-1}, 10^{-2}, 10^{-3}, 10^{-4}$
113. approximately $4,603 **115.** $\dfrac{7}{3}$ **119.** 13.5 yr
121. $y = \dfrac{3}{4}x - 5$

Think It Through (page 411)

3.24×10^7 mi

Study Set Section 5.3 (page 412)

1. scientific **3.** 250 **5.** 0.000025 **7.** 10^5 **9.** 10^{-3}
11. 5, right **13.** 10^{-4} **15.** positive **17.** 6.37×10^1, 1
19. 2.3×10^4 **21.** 6.25×10^5 **23.** 6.2×10^{-2}
25. 7.3×10^{-4} **27.** 5.43×10^5 **29.** 8.75×10^{-6}
31. 1.7×10^6 **33.** 9.09×10^8 **35.** 5.02×10^{-3}

37. 5.1×10^{-6} **39.** 4.25×10^3 **41.** 2.5×10^{-3}
43. 2.018×10^{17} **45.** 7.3×10^{-5} **47.** 230 **49.** 812,000
51. 0.00115 **53.** 0.000976 **55.** 714,000 **57.** 30,000
59. 25,000,000 **61.** 0.00051 **63.** 6.17 **65.** 699
67. 0.0043 **69.** 200,000 **71.** 9,038,030,748,000,000
73. 1,881,676,423,000 **75.** 847,288,609,400
77. 2.57×10^{13} mi **79.** 63,800,000 mi² **81.** 6.22×10^{-3} mi
83. g, x, u, v, i, m, r **85.** 1.7×10^{-18} g
87. 3.099363×10^{16} ft **89.** 2.148×10^{11} dollars
91. a. 3.5×10^6; 3,500,000 **b.** 1.4×10^6; 1,400,000
97. 5 **99.** commutative property of addition **101.** 6

Study Set Section 5.4 (page 419)

1. polynomial **3.** degree **5.** monomial **7.** trinomial
9. of **11.** decreasing or descending **13.** $3x^2 + x - 9$
15. 2, 2, 4, 6, 1, −2 **17.** yes **19.** no **21.** yes **23.** yes
25. binomial **27.** trinomial **29.** monomial **31.** binomial
33. trinomial **35.** none of these **37.** 4 **39.** 2
41. 1 **43.** 4 **45.** 12 **47.** 0 **49.** 7 **51.** −8
53. −2 **55.** −7.5 **57.** −4 **59.** −5 **61.** −5.69
63. −188.96 **65.** 3 **67.** −1 **69.** 1.929 **71.** −512.241
73. −1 **75.** −7 **77.** 72 in.³ **79.** 22 ft, 2 ft
81. about 404 ft **83.** 18.75 ft, 20 ft, 15 ft
87. $y \geq -3$, $[-3, \infty)$ **89.** x^{18} **91.** y^9

Study Set Section 5.5 (page 427)

1. polynomials **3.** Like **5.** terms **7.** like **9.** changing
11. $2x^2 + 3x - 4$ **13.** $-7x$ **15.** $3x, 2x, 7x^2$ **17.** $9y$
19. $12t^2$ **21.** $14r$ **23.** $9ab$ **25.** $3a^3$ **27.** $-48u^3$
29. $-7m - 6$ **31.** $-1.9p^2 - 4.5p$ **33.** $7x + 4$
35. $6x^2 + x - 5$ **37.** $5x^2 + x + 11$
39. $-7x^3 - 7x^2 - x - 1$ **41.** $2a + 7$ **43.** $2a^2 + a - 3$
45. $5x^2 + 6x - 8$ **47.** $-x^3 + 6x^2 + x + 14$
49. $t^3 + 3t^2 + 6t - 5$ **51.** $-3x^2 + 5x - 7$ **53.** $6x - 2$
55. $-5x^2 - 8x - 4$ **57.** $-4y^2 + 8y + 4$
59. $a^2b + 2ab^2 - 6ab - 4a - b$ **61.** $-0.1x$ **63.** $2st$
65. $15x^2$ **67.** $7x - 7y$ **69.** $3x - 4y$ **71.** $3x + 1$
73. $13.54x^2 - 6.55x + 6.15$ **75.** $20a - 14b - 22c$
77. $2x^2 + 4x + 13$ **79.** $-12x^2y^2 - 9xy + 36y^2$
81. $(3x^2 + 6x - 2)$ yd **83.** $(x^2 - 8x + 12)$ ft
85. $(3x^2 + 11x + 4.5\pi)$ in. **87.** $314,000
89. $f(x) = 1,900x + 525,000$ **91.** $f(x) = -1,100x + 6,600$
93. $f(x) = -2,800x + 15,800$ **95. a.** $22t + 20$ **b.** 108 ft
97. $(11x - 12)$ ft **103.** 180° **105.** $x \leq 2$, $(-\infty, 2]$

Study Set Section 5.6 (page 438)

1. binomials **3.** first, outer, inner, last **5.** $6x^2$ **7.** $15x$
9.

$(2x + 5)(3x + 4) = 6x^2 + 8x + 15x + 20$

First Outer Inner Last

11. $7x, 7x, 7x$ **13.** $12x^5$ **15.** $-5x^5y^8$ **17.** $-3x^4y^5z^8$
19. $\frac{1}{3}x^5y^5$ **21.** $-4t^2 - 28t$ **23.** $6s^4 - 18s^3$
25. $3xy + 12y^2$ **27.** $6x^4 + 8x^3 - 14x^2$
29. $4a^2b^2 + 6ab^3 - 4a^3b^2$ **31.** $0.62p^4 - 0.8p^2q$
33. $a^2 + 9a + 20$ **35.** $3x^2 + 10x - 8$ **37.** $6a^2 + 2a - 20$
39. $6x^2 - 7x - 5$ **41.** $6t^2 + 7st - 3s^2$
43. $-\frac{1}{8}t^2 + \frac{3}{4}tu - u^2$ **45.** $8x^2 - 12x - 8$ **47.** $3a^3 - 3ab^2$
49. $5t^2 - 11t$ **51.** $2x^2 + xy - y^2$ **53.** $x^2 + 8x + 16$
55. $t^2 - 6t + 9$ **57.** $16x^2 - 25$ **59.** $x^2 - 4xy + 4y^2$
61. $x^3 - 3x + 2$ **63.** $12x^3 + 17x^2 - 6x - 8$
65. $-6x^4 - 24x^3$ **67.** $-4x^3 - 6x^2 + 18x$ **69.** -3 **71.** -8
73. -1 **75.** 0 **77.** $3x^2 - 6x$ **79.** $-6x^4 + 2x^3$
81. $3x^2y + 3xy^2$ **83.** $r^2 - 16$ **85.** $4s^2 + 4s + 1$
87. $x^2 + xz + xy + yz$ **89.** $x^3 - x + 6$
91. $4t^3 + 11t^2 + 18t + 9$
93. $-3x^3 + 25x^2y - 56xy^2 + 16y^3$ **95.** $x^2 + 10x + 25$
97. $4a^2 - 12ab + 9b^2$ **99.** $16x^2 + 40xy + 25y^2$
101. $5a^2 - 11a$ **103.** $(4x^2 - 6x + 2)$ cm^2
105. $(x^2 + 6x + 9)\pi$ in.^{2}
107. $(24x + 14)$ cm, $(35x^2 + 43x + 12)$ cm^2
109. $(6x^2 + x - 1)$ cm^2
111. a. x^2 ft^2, $6x$ ft^2, $5x$ ft^2, 30 ft^2; $(x^2 + 11x + 30)$ ft^2
b. $(x + 6)$ ft, $(x + 5)$ ft; $(x^2 + 11x + 30)$ ft^2
c. They are the same. **113.** 5 and 6 **115.** 4 m **117.** 90 ft
119. $\left(\frac{1}{2}h^2 - 2h\right)$ in.^{2} **123.** 1 **125.** $-\frac{2}{3}$ **127.** $(0, 2)$

Study Set Section 5.7 (page 446)

1. polynomial **3.** two **5.** quotient **7.** $9, 9$ **9.** $m - n$
11. a. $t = \frac{d}{r}$ **b.** $3x^2$ **13.** $2x + 7$ **15.** b, b, b, a, a, a, b, a
17. $\frac{1}{3}$ **19.** $-\frac{5}{3}$ **21.** $\frac{3}{4}$ **23.** 1 **25.** x^3 **27.** $\frac{r^2}{s}$ **29.** $\frac{2x^2}{y}$
31. $-\frac{3u^3}{v^2}$ **33.** a^8b^8 **35.** $-\frac{3r}{s^9}$ **37.** $-\frac{x^3}{4y^3}$ **39.** $-\frac{16}{y^6}$
41. $2x + 3$ **43.** $\frac{1}{5y} - \frac{2}{5x}$ **45.** $\frac{1}{y^2} + \frac{2y}{x^2}$ **47.** $3a - 2b$
49. $\frac{1}{y} - \frac{1}{2x} + \frac{2z}{xy}$ **51.** $3x^2y - 2x - \frac{1}{y}$ **53.** $\frac{10x^2}{y} - 5x$
55. -2 **57.** $\frac{4r}{y^2}$ **59.** $-\frac{13}{3rs}$ **61.** $\frac{x^4}{y^6}$ **63.** a^8
65. $5x - 6y + 1$ **67.** $-\frac{4x}{3} + \frac{3x^2}{2}$ **69.** $xy - 1$ **71.** 2
73. $(2x^2 - x + 3)$ in. **75.** $(3x - 2)$ ft **77.** yes **79.** no
83. binomial **85.** none of these **87.** 2

Study Set Section 5.8 (page 452)

1. divisor, dividend **3.** remainder **5.** $4x^3 - 2x^2 + 7x + 6$
7. $6x^4 - x^3 + 2x^2 + 9x$ **9.** $0x^3$ and $0x$ **11. a.** $r = \frac{d}{t}$
b. $x - 3$ **13.** It is correct. **15.** $x, 2x, 2x$ **17.** $x + 6$
19. $y + 12$ **21.** $3a - 2$ **23.** $b + 3$ **25.** $x + 4 + \frac{14}{2x - 3}$
27. $2x + 2 + \frac{-3}{2x + 1}$ **29.** $2x + 1$ **31.** $x - 7$ **33.** $x + 1$
35. $2x - 3$ **37.** $3x + 2$ **39.** $2x - 1$ **41.** $x^2 + 2x - 1$
43. $2x^2 + 2x + 1$ **45.** $x^2 + x + 1$ **47.** $x^2 + 2x + 1$
49. $x^2 + 2x - 1 + \frac{6}{2x + 3}$ **51.** $2x^2 + 8x + 14 + \frac{31}{x - 2}$
53. $x^2 - x + 1$ **55.** $a^2 - 3a + 10 + \frac{-30}{a + 3}$
57. $5x^2 - x + 4 + \frac{16}{3x - 4}$ **59. a.** $(x - 6)$ in.
b. $(4x - 4)$ in. **61.** $(4x^2 + 3x + 7) + 1 = (4x^2 + 3x + 8)$
poles **65.** x^{22} **67.** $8x^2 - 6x + 1$ **69.** They are the same.

Chapter 5 Review Exercises (page 455)

1. $-3 \cdot x \cdot x \cdot x \cdot x$ **2.** $\left(\frac{1}{2}pq\right)\left(\frac{1}{2}pq\right)\left(\frac{1}{2}pq\right)$ **3.** 125 **4.** 64
5. -64 **6.** 4 **7.** x^5 **8.** $-3y^6$ **9.** y^{21} **10.** $81x^4$ **11.** b^{12}
12. $-y^2z^5$ **13.** $256s^3$ **14.** $4x^4y^2$ **15.** x^{15} **16.** $\frac{x^2}{y^2}$ **17.** x^4
18. $125yz^4$ **19.** $64x^{12}$ in.^{3} **20.** y^4 m^2 **21.** 1 **22.** 1 **23.** 9
24. $\frac{1}{1,000}$ **25.** $\frac{4}{3}$ **26.** $-\frac{1}{25}$ **27.** $\frac{1}{x^5}$ **28.** $-\frac{6}{y}$ **29.** $\frac{1}{x^{10}}$
30. x^{14} **31.** $\frac{1}{x^5}$ **32.** $\frac{1}{27z^3}$ **33.** y^{7n} **34.** $\frac{1}{z^{2c}}$
35. 7.28×10^2 **36.** 9.37×10^6 **37.** 1.36×10^{-2}
38. 9.42×10^{-3} **39.** 1.8×10^{-4} **40.** 7.53×10^5
41. $726,000$ **42.** 0.000391 **43.** 2.68 **44.** 57.6 **45.** 0.03
46. 160 **47.** $7,098,000,000$, 7.098×10^9
48. $1.0 \times 10^5 = 100,000$ **49.** yes **50.** no **51.** no
52. yes **53.** 4 **54.** $3x^3$ **55.** -1 **56.** 10
57. 7, monomial **58.** 3, monomial **59.** 2, binomial
60. 5, trinomial **61.** 6, binomial **62.** 4, none of these
63. 34 **64.** 1 **65.** 9 **66.** 0.72 **67.** 8 in. **68.** $2x^6 + 5x^5$
69. $-2x^2y^2$ **70.** $8x^2 - 6x$ **71.** $5x^2 + 19x + 3$
72. $3x^2 + 2x + 8$ **73.** $8x^3 - 7x^2 + 19x$ **74.** $10x^3$
75. $-6x^{10}z^5$ **76.** $-6r^3s^4t^5$ **77.** $120b^{11}$ **78.** $5x + 15$
79. $3x^4 - 5x^2$ **80.** $x^2y^3 - x^3y^2$ **81.** $-2y^4 + 10y^3$
82. $6x^6 + 12x^5$ **83.** $-3x^3 + 3x^2 - 6x$ **84.** $x^2 + 5x + 6$
85. $2x^2 - x - 1$ **86.** $6a^2 - 6$ **87.** $6a^2 - 6$
88. $2a^2 - ab - b^2$ **89.** $-6x^2 - 5xy - y^2$ **90.** $x^2 + 6x + 9$
91. $x^2 - 25$ **92.** $a^2 - 6a + 9$ **93.** $x^2 + 8x + 16$
94. $4y^2 - 4y + 1$ **95.** $y^4 - 1$ **96.** $3x^3 + 7x^2 + 5x + 1$
97. $8a^3 - 27$ **98.** 1 **99.** -1 **100.** 7 **101.** 0
102. $(6x + 10)$ in., $(2x^2 + 11x - 6)$ in.^{2},
$(6x^3 + 33x^2 - 18x)$ in.^{3}
103. $\frac{2x}{3y^2}$ **104.** $\frac{1}{x}$ **105.** $4x + 3$ **106.** $2 - \frac{3}{y}$
107. $3a + 4b - 5$ **108.** $-\frac{x}{y} - \frac{y}{x}$ **109.** $x + 5$
110. $x + 1 + \frac{3}{x + 2}$ **111.** $x - 5$ **112.** $2x + 1$
113. $x + 5 + \frac{3}{3x - 1}$ **114.** $3x^2 + 2x + 1 + \frac{2}{2x - 1}$
115. $3x^2 - x - 4$ **117.** $(4x + 3)$ in./min

Chapter 5 Test (page 461)

1. a. base, exponent b. monomial, binomial, trinomial c. degree d. special 2. $2x^3y^4$ 3. y^6
4. $32x^{21}$ 5. 3 6. $\dfrac{2}{y^3}$ 7. y^3 8. $\dfrac{64a^3}{b^3}$ 9. $\dfrac{m^{12}}{64}$
10. $-6ab^9$ 11. $1{,}000y^{12}$ in.^{3} 12. $\dfrac{1}{4^2}, \dfrac{1}{16}$
13. a. 6.25×10^{18} b. 0.00025 m
14. 1.116×10^7 mi 15. binomial
16. trinomial

Term	Coefficient	Degree
x^4	1	4
$8x^2$	8	2
-12	-12	0

Degree of the polynomial: 4

17. 5
18. 0 ft; the rock hits the canyon floor 18 seconds after being dropped. 19. 0
20. $-3x^2y^2$
21. $-7x + 2y$ 22. $-x^2 - 5x + 4$ 23. $(10a^2 + 8a - 20)$ in.
24. a. $(8x^2 - 2x)$ in.^{2} b. $(16x^3 + 12x^2 - 4x)$ in.^{3}
25. $-4x^5y$ 26. $3y^4 - 6y^3 + 9y^2$ 27. $x^2 - 81$
28. $9y^2 - 24y + 16$ 29. $6x^2 - 7x - 20$
30. $2x^3 - 7x^2 + 14x - 12$ 31. $\dfrac{1}{2}$ 32. $\dfrac{y}{2x}$ 33. $\dfrac{a}{4b} - \dfrac{b}{2a}$
34. $x - 2$ 35. $3x^2 + 2x + 1 + \dfrac{2}{2x - 1}$ 36. $x - 2$
38. $(x - 5)$ ft

Cumulative Review Chapters 1-5 (page 463)

1. $2 \cdot 3^3 \cdot 5$ 2. a. $a + b = b + a$ b. $(xy)z = x(yz)$
3. -37 4. 28 5. $18x$ 6. 0 7. -2 8. 15
9. $2.414 billion 10. 1.2 ft^3 11. 30 12. $6,250
13. mutual fund: $25,000, bonds: $20,000
14. $\left(-\infty, -\dfrac{11}{4}\right)$,

15. 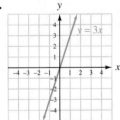 16.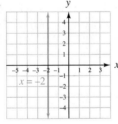

17. $-\dfrac{4}{9}$ 18. $m = 3$, $(0, -2)$; $y = 3x - 2$ 19. perpendicular
20. $y = -4x + 2$ 21. no 22. 26 23. no
24. $(4, 1)$ 25. $(-4, 3)$
26. $(-2, 4)$
27. adult: $84, child: $76

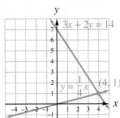

28.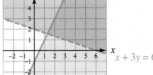

29. $9x^4y^8$ 30. v^{22}
31. $a^2b^7c^6$ 32. $\dfrac{64t^{12}}{27}$
33. $\dfrac{1}{16y^4}$ 34. a^7
35. $-\dfrac{1}{25}$ 36. $\dfrac{x^{10}}{a^{10}}$
37. 6.15×10^5
38. 1.3×10^{-6}

39.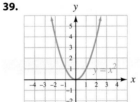

40. 1.5 in. 41. $7c^2 + 7c$
42. $-6x^4 - 17x^2 - 68x + 11$
43. $6t^2 + 7st - 3s^2$
44. $12x^3 + 36x^2 + 27x$
45. $\dfrac{1}{8} - \dfrac{2}{x}$
46. $2x + 1$

Study Set Section 6.1 (page 472)

1. factoring 3. prime 5. factored out 7. The 0 in the first line should be 1. 9. The answer should be $(a + 1)(b + 1)$. 11. $3x$ 13. 2 15. 4 17. 4, x 19. b^2, $(b - 6)$, $(b^2 + 2)$ 21. 1 23. 2 25. 15 27. $5y$ 29. $20pq$
31. $6t$ 33. $15ab$ 35. $3(x + 2)$ 37. $6(6 - x)$
39. $t^2(t + 2)$ 41. $a^2(a - 1)$ 43. $8xy^2(3xy + 1)$
45. $6uv(2 - 3v)$ 47. $6(2x^2 - 1 - 4a)$ 49. $3(1 + y - 2z)$
51. $a(b + c - 1)$ 53. $3r(4r - 1 + 3rs^2)$
55. $(x + 2)(3 - x)$ 57. $(14 + r)(h^2 + 2)$ 59. $-(a + b)$
61. $-(3m + 4n - 1)$ 63. $-3x(x + 2)$
65. $-4a^2b^2(b - 3a - 1)$ 67. $(x + y)(2 + a)$
69. $(r + s)(7 - k)$ 71. $(b + c)(a + 1)$
73. $(m - n)(p - q)$ 75. $x^2(a + b)(x + 2y)$
77. $4a(b + 3)(a - 2)$ 79. $\pi(R^2 - ab)$ 81. $-(2x - 5y)$
83. $-(3ab + 5ac - 9bc)$ 85. $-2ab^2c(2ac - 7a + 5c)$
87. $(b + c)(2a + 3)$ 89. $(3x - 1)(2x - 5)$
91. $(3p + q)(3m - n)$ 93. $(2x + y)(y - 1)$
95. $(2z^3 + 3)(4z^2 - 5)$ 97. $y(x^2 - y)(x - 1)$
99. a. $12x^3$ in.^{2} b. $20x^2$ in.^{2} c. $4x^2(3x - 5)$ in.^{2}
101. a. $4r$ in., $16r^2$ in.^{2} b. $4\pi r^2$ in.^{2}
c. $16r^2 - 4\pi r^2 = 4r^2(4 - \pi)$ in.^{2} 107. $\dfrac{y^3}{8}$ 109. yes

Study Set Section 6.2 (page 483)

1. trinomial 3. factors 5. prime
7. $4, 1, 2, 2, -4, -1, -2, -2$ 9. common 11. $-6, -3$
13. $1 + (-9) = -8, 3 + (-3) = 0, -1 + 9 = 8$
15. a. 1 b. 15, 8 c. 5, 3 17. $+, +$ 19. $-, -$ 21. $+, -$
23. $5x, 2$ 25. $(z + 11)(z + 1)$ 27. $(p + 2)(p + 7)$
29. $(m - 3)(m - 2)$ 31. $(y - 3)(y - 10)$
33. $(b + 7)(b - 1)$ 35. $(a + 8)(a - 2)$
37. $(a - 5)(a + 1)$ 39. $(z - 6)(z + 3)$
41. $-(x + 5)(x + 2)$ 43. $-(t + 17)(t - 2)$
45. $(x + 2y)(x + 2y)$ 47. $(m + 5n)(m - 2n)$
49. $2(x + 3)(x + 2)$ 51. $5p(p + 7)(p - 2)$
53. $(r + 3x)(r + x)$ 55. $a(a - 2b)(a - b)$ 57. prime
59. prime 61. $(p + 6)(p - 5)$ 63. $(m - 4n)(m + n)$
65. $3x(x - 4)(x - 5)$ 67. $4y(y - 5)(y - 2)$
69. $(a - 13)(a + 3)$ 71. $(s + 13)(s - 2)$ 73. prime

75. $(a - 6b)(a + 2b)$ **77.** $-(r - 10)(r - 4)$
79. $-(a + 3b)(a + b)$ **81.** $(x - 4)(x - 1)$
83. $(y + 9)(y + 1)$ **85.** $-(r - 2)(r + 1)$
87. $-5(a - 3)(a - 2)$ **89.** $-4x(x + 3y)(x - 2y)$
91. $(m - 4)(m + 3)$ **93.** $3ab(a + b)$ **95.** $-4a(a + 2)$
97. $-(x - 7y)(x + y)$ **99.** $4y(x + 6)(x - 3)$
101. $(3a + 2)(p + q)$ **103.** $3(z - 4)(z - 1)$
105. $(x + 9)$ in., $(x + 3)$ in., x in.
113.
115.

Study Set Section 6.3 (page 492)

1. leading, last **3.** outer, inner **5.** middle **7.** GCF
9. $5, -8$ **11.** positive, negative, negative **13.** negative, different **15.** $+, +$ **17.** $-, -$ **19.** $-, +$ **21.** $24, -11$
23. $9, 3t, 2, 4t + 3$ **25. a.** $x^2 + 2x + 3$ (answers may vary)
b. $2x^2 + 2x + 3$ (answers may vary) **27.** $(3a + 1)(a + 4)$
29. $(z + 3)(4z + 1)$ **31.** $(2t - 1)(2t - 1)$
33. $(2x - 1)(x - 1)$ **35.** $(2u - 3)(4u + 5)$
37. $(4y + 1)(3y - 1)$ **39.** $(3r + 2s)(2r - s)$
41. $(2b - c)(b - 2c)$ **43.** $2(2x - 1)(x + 3)$
45. $-y(y + 12)(y + 1)$ **47.** $3x(2x + 1)(x - 3)$
49. $3r^3(5r - 2)(2r + 5)$ **51.** $(5y + 1)(2y - 1)$
53. $(3y - 2)(4y + 1)$ **55.** $y^2(3y + 1)(4y - 1)$
57. $-2mn(4m + 3n)(2m + n)$ **59.** $(2x + 3)(2x + 1)$
61. $(7x - 2)(x - 1)$ **63.** prime **65.** $-(8y + 1)(2y + 1)$
67. $-(4x + 1)(4x + 3)$ **69.** $(4b - c)(b + 4c)$
71. $(4x + 3y)(3x - y)$ **73.** $(3a - 7)(a + 2)$
75. $(5a - 4)(5a - 4)$ **77.** $(3u + 2v)(u + 3v)$
79. $(4y - 3)(3y - 4)$ **81.** $(3x + y)(2x - 5)$
83. $(3a + 2)(2a - 5)$ **85.** $(3p + 2q)(4p - q)$
87. $-3a(a + 3)(a - 1)$ **89.** $-2uv^3(7u - 3v)(2u - v)$
91. $-2x^2y^3(8x + y)(x - 2y)$ **93.** $(3m - 2n)(4m - n)$
95. $(2x + 11)$ in., $(2x - 1)$ in.; 12 in. **101.** x^{14} **103.** 8

Study Set Section 6.4 (page 499)

1. perfect **3. a.** $5x$ **b.** 3 **c.** $5x, 3$ **5.** twice, sum
7. $1, 4, 9, 16, 25, 36, 49, 64, 81, 100$ **9. a.** $6x$ **b.** $10x^2$ **c.** $2x, 3$
11. 3 **13.** $+$ **15.** $7, 7$ **17.** t, w **19.** $9a^2 - 30ab + 25b^2$
21. $36x^2 - 25y^2$ **23.** $(x + 8)^2$ **25.** yes **27.** no **29.** no
31. yes **33.** $(x + 3)^2$ **35.** $(t - 10)^2$ **37.** $(a + b)^2$
39. $(4x - y)^2$ **41.** $y(y - 4)^2$ **43.** $2x(2x + 3)^2$
45. $(x + 4)(x - 4)$ **47.** $(t + 7)(t - 7)$ **49.** $(7 + c)(7 - c)$
51. $(12 + 5a)(12 - 5a)$ **53.** prime **55.** prime
57. $(2y + 1)(2y - 1)$ **59.** $(7a + 13)(7a - 13)$
61. $(3x + y)(3x - y)$ **63.** $(4a + 5b)(4a - 5b)$
65. $8(a + 2)(a - 2)$ **67.** $7(1 + a)(1 - a)$ **69.** $(z - 1)^2$
71. $(2x - 1)^2$ **73.** $(a^2 + 12b)(a^2 - 12b)$
75. $(tz + 8)(tz - 8)$ **77.** $6x^2(x + y)(x - y)$
79. $(ab + 12)(ab - 12)$ **81.** $(5z - 4)^2$ or $(4 - 5z)^2$
83. $2xy(2ax + b)(2ax - b)$ **85.** $8(x + 2y)(x - 2y)$
87. $8mn^3(m - 3n)$ **89.** prime **91.** $-(3xy - 1)^2$
93. $(c + 2d)(2a + b)$ **95.** prime **97.** $a(6a - 1)(a + 6)$
99. $7p^4q^2(10q - 5 + 7p)$ **101.** $(c + d^2)(a^2 + b)$
103. $(p + q)^2$ **105.** $0.5g(t_1 + t_2)(t_1 - t_2)$
107. $16(3 + t)(3 - t)$ **111.** $\dfrac{x}{y} + \dfrac{2y}{x} - 3$ **113.** $3a + 2$

Study Set Section 6.5 (page 504)

1. cubes **3.** F, L **5.** $x^3 - 8$ **7.** $1, 8, 27, 64, 125$ **9.** $3m$
11. $2x, 3$ **13.** $2a$ **15.** $b + 3$ **17.** $(y + 1)(y^2 - y + 1)$
19. $(y + 7)(y^2 - 7y + 49)$ **21.** $(a + 4)(a^2 - 4a + 16)$
23. $(m + 8)(m^2 - 8m + 64)$ **25.** $(x - 2)(x^2 + 2x + 4)$
27. $(z - 7)(z^2 + 7z + 49)$ **29.** $(s - 2t)(s^2 + 2st + 4t^2)$
31. $(s - 4t)(s^2 + 4st + 16t^2)$ **33.** $-(x - 6)(x^2 + 6x + 36)$
35. $2(x + 3)(x^2 - 3x + 9)$
37. $8x(2m - n)(4m^2 + 2mn + n^2)$
39. $xy(x + 6y)(x^2 - 6xy + 36y^2)$ **41.** $(2 + x)(4 - 2x + x^2)$
43. $(2u + w)(4u^2 - 2uw + w^2)$
45. $(4x - 3)(16x^2 + 12x + 9)$ **47.** $(a^2 - b)(a^4 + a^2b + b^2)$
49. $(x^3 + y^2)(x^6 - x^3y^2 + y^4)$
51. $3rs^2(3r - 2s)(9r^2 + 6rs + 4s^2)$
53. $3(a + 2b)(a^2 - 2ab + 4b^2)$
55. $(2p^2 - 3q^2)(4p^4 + 6p^2q^2 + 9q^4)$ **57.** $(x + 9)(x - 9)$
59. $(a + 4)(a - 4)$ **61.** $(9r + 16s)(9r - 16s)$
63. $(x - a)(a + b)(a - b)$ **65.** $(y^2 - 2)(x + 1)(x - 1)$
67. $(9p^2 + 4q^2)(3p + 2q)(3p - 2q)$
69. $2(3x + 5y^2)(9x^2 - 15xy^2 + 25y^4)$ **71.** $(1,000 - x^3)$ in.^{3};
$(10 - x)(100 + 10x + x^2)$ **75.** $\dfrac{21}{5}$ **77.** no

Study Set Section 6.6 (page 509)

1. product **3.** factor out the GCF **5.** perfect-square trinomial **7.** sum of two cubes **9.** trinomial factoring
11. Is there a common factor? **13.** $14m, m$
15. $2(b + 6)(b - 2)$ **17.** $4p^2q^3(2pq^4 + 1)$
19. $2(2y + 1)(10y + 1)$ **21.** $8(x^2 + 1)(x + 1)(x - 1)$
23. $(c + 21)(c - 7)$ **25.** prime
27. $-2x^2(x - 4)(x^2 + 4x + 16)$ **29.** $(c + d^2)(a^2 + b)$
31. $-(3xy - 1)^2$ **33.** $-5m(2m + 5)^2$
35. $(2c + d)(c - 3d)$ **37.** $(p - 2)^2(p^2 + 2p + 4)$
39. $(x - a)(a + b)(a - b)$ **41.** $7p^4q^2(10q - 5 + 7p)$
43. $(x + 5)(2x^2 + 1)$ **45.** $v^2(v^2 - 14v + 8)$
47. $(x + 2)(x - 2)(x + 3)(x - 3)$
49. $2x(2ax + b)(2ax - b)$ **51.** $2(3x - 4)(x - 1)$
53. $y^2(2x + 1)^2$ **55.** $4m^2(m + 5)(m^2 - 5m + 25)$
57. $(x + 2)(x - 2)(x^2 + 2)$ **59.** prime
61. $2a^2(2a - 3)(4a^2 + 6a + 9)$ **63.** $27(x - y - z)$
65. $(x - t)(y + s)$ **67.** $x^6(7x + 1)(5x - 1)$
69. $5(x - 2)(1 + 2y)$ **71.** $(7p + 2q)^2$ **73.** $4(t^2 + 9)$
75. $(n + 3)(n - 3)(m^2 + 3)$
81.

83.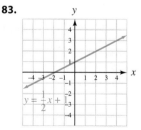

Study Set Section 6.7 (page 516)

1. quadratic 3. $0, 0, 0$ 5. zero 7. a. yes b. no
9. $3, -4, 5$ 11. $7y, y + 2$ 13. $2, -3$ 15. $\frac{5}{2}, -6$ 17. $0, 3$
19. $1, -2, 3$ 21. $0, 7$ 23. $0, -\frac{8}{3}$ 25. $-10, 10$ 27. $-\frac{9}{2}, \frac{9}{2}$
29. $-3, 7$ 31. $12, 1$ 33. $-\frac{1}{2}, -\frac{1}{2}$ 35. $\frac{2}{3}, -\frac{1}{5}$ 37. $-\frac{5}{2}, 4$
39. $\frac{1}{8}, 1$ 41. $0, -1, -2$ 43. $0, 9, -3$ 45. $0, \frac{5}{2}$ 47. $0, 2$
49. $-5, 5$ 51. $-\frac{1}{2}, \frac{1}{2}$ 53. $-\frac{2}{3}, \frac{2}{3}$ 55. $8, 1$ 57. $-3, -5$
59. $-4, 2$ 61. $\frac{1}{2}, 2$ 63. $\frac{1}{5}, 1$ 65. $1, -2, -3$ 67. $0, -3, -\frac{1}{3}$
69. $0, -3, -3$ 73. $12b^2 - 25b$ 75. $4b^2 - 16b + 15$
77. a 79. $2z + 3$

Study Set Section 6.8 (page 522)

1. consecutive 3. hypotenuse 5. lw 7. $-16, 1, 0, 0, 3, -1$
9. 9 sec 11. $\frac{3}{2} = 1.5$ sec 13. 2 sec 15. 8
17. Ambrose: 9, Patrick: 10 19. 12 21. $(11, 13)$
23. 4 m by 9 m 25. 3 ft by 7 ft 27. 3 ft by 5 ft
29. 5 m, 12 m, 13 m 31. 3 mm, 4 mm, 5 mm 33. $h = 5$ ft, $b = 12$ ft 35. 5 in. 37. 4 m 39. 5 sec
43. $25b^2 - 20b + 4$ 45. $s^4 + 8s^2 + 16$ 47. $81x^2 - 36$

Chapter 6 Review Exercises (page 527)

1. 5 2. 9 3. 6 4. 15 5. $18pq$ 6. $7p^2q$ 7. $3(x + 3y)$
8. $5a(x^2 + 3)$ 9. $7s(s + 2)$ 10. $\pi a(b - c)$
11. $2x(x^2 + 2x - 4)$ 12. $xyz(x + y + 1)$
13. $-5ab(b - 2a + 3)$ 14. $(x - 2)(4 - x)$ 15. $-(a + 7)$
16. $-(4t^2 - 3t + 1)$ 17. $(c + d)(2 + a)$
18. $(y + 3)(3x - 2)$ 19. $(2a^2 - 1)(a + 1)$
20. $4m(n + 3)(m - 2)$ 21. $(x + 6)(x - 4)$
22. $(x - 6)(x + 2)$ 23. $(n - 5)(n - 2)$ 24. prime
25. $-(y - 5)(y - 4)$ 26. $(y + 9)(y + 1)$
27. $(c + 5d)(c - 2d)$ 28. $(m - 2n)(m - n)$
29. $5(a + 10)(a - 1)$ 30. $-4x(x + 3y)(x - 2y)$
31. $(p + 5)(p - 4)$ 32. $-4q(q + 2)(q - 3)$
33. $(2x + 1)(x - 3)$ 34. $(2y + 5)(5y - 2)$
35. $-(3x + 1)(x - 5)$ 36. $3p(2p + 1)(p - 2)$
37. $(4b - c)(b - 4c)$ 38. prime 39. $(3p - 2)(4p + 1)$
40. $q(4q - 3)(3q + 2)$ 41. $-2(4p + q)(2p + 3)$
42. $(4x + 1)$ in., $(3x - 1)$ in. 43. $(x + 5)^2$ 44. $(3y - 4)^2$
45. $-(z - 1)^2$ 46. $(5a + 2b)^2$ 47. $(x + 3)(x - 3)$
48. $(7t + 5y)(7t - 5y)$ 49. $(xy + 20)(xy - 20)$
50. $8a(t + 2)(t - 2)$ 51. $4c(c + 4)(c - 4)$ 52. prime
53. $(h + 1)(h^2 - h + 1)$ 54. $(5p + q)(25p^2 - 5pq + q^2)$
55. $(x - 3)(x^2 + 3x + 9)$
56. $2x^2(2x - 3y)(4x^2 + 6xy + 9y^2)$ 57. $2y^2(3y - 5)(y + 4)$
58. $(t + u^2)(s^2 + v)$ 59. $(j^2 + 4)(j + 2)(j - 2)$
60. $-3(j + 2k)(j^2 - 2jk + 4k^2)$ 61. $3(2w - 3)^2$
62. prime 63. $5xy(3x - y + 1)$ 64. $2(x^3 + 6)$ 65. $0, -2$
66. $0, 6$ 67. $-3, 3$ 68. $3, 4$ 69. $-2, -2$ 70. $6, -4$
71. $\frac{1}{5}, 1$ 72. $0, -1, 2$ 73. 10 sec 74. 6 and 7 75. 15 m
76. 3 ft by 9 ft 77. 5 m

Unless otherwise noted, all content on this page is © Cengage Learning.

Chapter 6 Test (page 534)

1. a. greatest common factor b. product c. Pythagorean d. difference e. binomials 2. a. $2^2 \cdot 7^2$ b. $3 \cdot 37$
3. $4(x + 4)$ 4. $5ab(6ab^2 - 4a^2b + c)$ 5. $(q + 9)(q - 9)$
6. prime 7. $(4x^2 + 9)(2x + 3)(2x - 3)$ 8. $(x + 3)(x + 1)$
9. $-(x - 11)(x + 2)$ 10. $(a - b)(9 + x)$
11. $(2a - 3)(a + 4)$ 12. $2(3x - 5y)^2$
13. $(x + 2)(x^2 - 2x + 4)$ 14. $2(a - 3)(a^2 + 3a + 9)$
15. $r^2(4 - \pi)$ 16. $4x - 3$ 17. $2ab^2$
18. $(x - 9)(x + 6)$; to check, multiply the binomials:
$(x - 9)(x + 6) = x^2 + 6x - 9x - 54 = x^2 - 3x - 54$
19. $-3, 2$ 20. $-5, 5$ 21. $0, \frac{1}{6}$ 22. $-3, -3$ 23. $\frac{1}{3}, -\frac{1}{2}$
24. $-1, -6$ 25. $0, -1, -6$ 26. 6 ft by 9 ft
27. 5 sec 28. base: 6 in.; height: 11 in. 29. 12, 13
30. 10 31. A quadratic equation is an equation that can be written in the form $ax^2 + bx + c = 0$; $x^2 - 2x + 1 = 0$. (answers may vary) 32. At least one of them is 0.

Cumulative Review Chapters 1–6 (page 536)

1. about 35 beats/min difference 2. $2 \cdot 5^3$ 3. $\frac{24}{25}$
4. 0.992 5. a. false b. true c. true 6. $\frac{5}{0}$ 7. -39
8. -5 9. -27 10. $\$20x$ 11. 3 12. $8, -1, 9$
13. $-13y^2 + 6$ 14. $3y + z$ 15. $-\frac{3}{2}$ 16. -2 17. 9
18. $-\frac{55}{6}$ 19. 248 lb 20. 330 mi 21. $I = Prt$
22. 12.6 in.² 23. $t = \frac{A - P}{Pr}$
24. 22nd president, 24th president 25. 4 L
26. $(-\infty, -2)$,  27. yes

28. 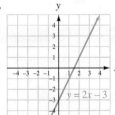 29. vertical line
30. They are the same.
31. -1.5 degree/hr
32. $1, (0, -2)$

33. 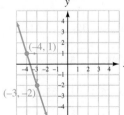 34. $y = 7x + 19$

35. 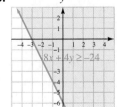 36. 17 37. yes

38. (4, 3)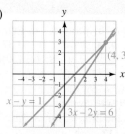
39. $\left(\dfrac{3}{4}, -2\right)$
40. $\left(-2, \dfrac{2}{3}\right)$
41. newspaper: 8 tons, cardboard: 6 tons
42.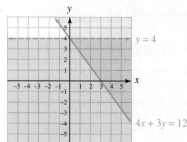
43. $-4y^5$
44. $x^4 y^{23}$
45. b^7
46. 2
47. 9.011×10^{-5} **48.** 1.7×10^6 **49.** 3
50.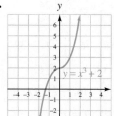
51. $-2x^2 - 4x + 5$
52. $8b^5 - 8b^4$
53. $3x^2 + 10x - 8$
54. $y^2 - 12y + 36$
55. $3ab - 2a - 1$
56. $2x + 1$
57. a. $(4x + 8)$ in.
 b. $(x^2 + 4x + 3)$ in.^{2}
 c. $(x^3 + 4x^2 + 3x)$ in.^{3}
58. $6x^5 y$ **59.** $9b^2(b - 3)$ **60.** $(x + y)(a + b)$
61. $(u + 3)(u - 1)$ **62.** $(2x + 1)(5x - 2)$ **63.** $(2a - 3)^2$
64. $(3z + 1)(3z - 1)$ **65.** $(t - 2)(t^2 + 2t + 4)$
66. $3(b^2 - 2)(a + 1)(a - 1)$ **67.** $0, \dfrac{4}{3}$ **68.** $\dfrac{1}{2}, 2$

Study Set Section 7.1 (page 547)

1. numerator, denominator **3.** undefined **5.** simplify
7. a. 0 **b.** 6 **c.** -6 **9. a.** 1 **b.** -1 **c.** $x - 8$
11. $6, 1, x + 1$ **13.** 4 **15.** $-\dfrac{3}{7}$ **17.** 2 **19.** none
21. $-6, 6$ **23.** $-2, 1$ **25.** $\dfrac{4}{5}$ **27.** $\dfrac{1}{3}$ **29.** $-\dfrac{3}{4}$ **31.** $-\dfrac{7}{5}$
33. $\dfrac{5}{a}$ **35.** $\dfrac{2}{3}$ **37.** in simplified form **39.** $\dfrac{3x}{y}$ **41.** $\dfrac{2x + 1}{y}$
43. $\dfrac{a}{4}$ **45.** $\dfrac{3x - 2}{x + 4}$ **47.** $\dfrac{x + 1}{x - 1}$ **49.** x^3 **51.** $\dfrac{x + 2}{x^2}$
53. $\dfrac{4}{3}$ **55.** $\dfrac{2}{3}$ **57.** $\dfrac{1}{2}$ **59.** $x + 3$ **61.** -1 **63.** -6
65. $-\dfrac{1}{a + 1}$ **67.** -1 **69.** $\dfrac{1}{x + 5}$ **71.** $\dfrac{2}{5xy^2}$ **73.** $\dfrac{2}{z}$ **75.** $\dfrac{a}{3}$
77. $\dfrac{3}{2}$ **79.** $\dfrac{1}{3}$ **81.** $\dfrac{2x}{x - 2}$ **83.** $\dfrac{2x - 3}{x + 1}$
85. $\dfrac{3(x + 3)}{2x + 1}$ or $\dfrac{3x + 9}{2x + 1}$ **87.** $\dfrac{m - n}{2(m + n)}$ or $\dfrac{m - n}{2m + 2n}$
89. in simplified form **91.** $\dfrac{4 - x}{4 + x}$ or $-\dfrac{x - 4}{x + 4}$
93. in simplified form **95.** $\dfrac{2}{3}$ **97.** $\dfrac{3x}{5y}$ **99.** $\dfrac{-3(x + 2)}{x + 1}$
101. $\dfrac{x - 5}{x + 2}$ **103.** $85\dfrac{1}{3}$ times **105.** $\dfrac{x + 2}{x - 2}$
111. $(a + b) + c = a + (b + c)$ **113.** One of them is zero.
115. $\dfrac{5}{3}$

Study Set Section 7.2 (page 556)

1. reciprocal **3.** numerators, denominators **5.** reciprocal
7. 1 **9.** $(x - 2), (3x - 6), (x + 1), (x - 2), 3$ **11.** $\dfrac{3}{2}$
13. $-\dfrac{14}{9}$ **15.** $x^2 y^2$ **17.** $2xy^2$ **19.** $-3y^2$
21. $\dfrac{4z + 8}{7z}$ **23.** x **25.** $\dfrac{x}{5}$ **27.** $x + 2$ **29.** $\dfrac{3}{2x}$
31. $\dfrac{x^2 - 4x + 4}{x}$ **33.** $\dfrac{m^2 - 5m + 6}{2(m + 2)}$ **35.** 5 **37.** 6
39. $x + 1$ **41.** $2y + 16$ **43.** $x + 5$ **45.** $10h - 30$ **47.** $\dfrac{3}{2y}$
49. 3 **51.** $\dfrac{6}{y}$ **53.** 6 **55.** $-\dfrac{2x}{3}$ **57.** $\dfrac{2}{y}$ **59.** $\dfrac{x + 2}{3}$ **61.** 1
63. $\dfrac{2x - 14}{x + 9}$ **65.** $d + 5$ **67.** $\dfrac{1}{3 - x}$ **69.** 1 **71.** 450 ft
73. $\dfrac{3}{4}$ gal **75.** $\dfrac{1}{2}$ mi per min **77.** 1,800 m per min **79.** y
81. $x - 2$ **83.** x **85.** $\dfrac{b - 3}{b}$ **87.** $\dfrac{2}{3x}$ **89.** $\dfrac{2(z - 2)}{z}$
91. $\dfrac{5z^2 - 35z}{z + 2}$ **93.** $\dfrac{1}{12(2r - 3s)}$ **95.** $\dfrac{12x^2 + 12x + 3}{2}$ in.^{2}
101. $-6x^5 y^6$ **103.** $\dfrac{1}{81y^4}$ **105.** $4y^3 + 16y^2 - 8y + 8$

Study Set Section 7.3 (page 565)

1. denominator **3.** build **5.** numerators, common denominator **7.** $3y$ **9.** $2a + 3, 2, (4a + 1), 2$
11. $\dfrac{x}{3}$ **13.** $\dfrac{4x}{y}$ **15.** $\dfrac{2}{y}$ **17.** $\dfrac{y + 3}{5z}$ **19.** 9 **21.** $\dfrac{1}{b - 2}$
23. $-\dfrac{y}{8}$ **25.** $\dfrac{x}{y}$ **27.** $\dfrac{x}{y}$ **29.** $\dfrac{2}{3}$ **31.** 0 **33.** $\dfrac{4x - 2y}{y + 2}$
35. $\dfrac{3}{a^2 + 2a + 1}$ **37.** $\dfrac{4p}{p^2 - 3p + 2}$ **39.** $6x$ **41.** $30a^3$
43. $c(c + 2)$ **45.** $2(3x + 1)(3x - 1)$ **47.** $12(b + 2)$
49. $(b + 3)^2 (b - 3)$ **51.** $\dfrac{125x}{20x}$ **53.** $\dfrac{8xy}{x^2 y}$ **55.** $\dfrac{2xy + 2y}{x(x + 1)}$
57. $\dfrac{3x^2 + 3x}{(x + 1)^2}$ **59.** $\dfrac{z^2 + 3z + 2}{z(z - 6)(z + 1)}$ **61.** $\dfrac{3x + 3}{2(x + 1)(x + 2)}$
63. $\dfrac{y}{x}$ **65.** $\dfrac{1}{y}$ **67.** $\dfrac{4x}{3}$ **69.** $\dfrac{2x + 10}{x - 2}$ **71.** $\dfrac{3}{t - 1}$
73. $\dfrac{t - 5}{(t + 1)(t - 1)}$ **75.** $\dfrac{2}{a}$ **77.** $3b - 2$ **79.** $\dfrac{20x + 9}{6x^2}$ cm
85. 7^2 **87.** $2^3 \cdot 17$

Study Set Section 7.4 (page 573)

1. like, unlike 3. $2 \cdot 2 \cdot 5 \cdot x \cdot x$ 5. $36a^2$
7. $(x+6)(x+3)$ 9. $3x, 5, 15x, 15x, 35$ 11. $\dfrac{5y}{9}$ 13. $\dfrac{13x}{21}$
15. $\dfrac{5a}{2}$ 17. $\dfrac{73x}{42}$ 19. $\dfrac{43}{60a}$ 21. $\dfrac{53}{75p}$ 23. $\dfrac{y^2+7y+6}{15y^2}$
25. $\dfrac{2xy+x-y}{xy}$ 27. $\dfrac{53x}{42}$ 29. $\dfrac{7t-32}{8t}$ 31. $\dfrac{a^2-2a-4}{a(a+2)}$
33. $\dfrac{-d^2+5d+5}{d(d+1)}$ 35. $\dfrac{y-2}{y-1}$ 37. $\dfrac{d^2+d-15}{(d+5)(d+1)(d+4)}$
39. $\dfrac{s^2+8s+4}{(s+4)(s+1)^2}$ 41. $\dfrac{5b-1}{2(b+1)^2}$ 43. $\dfrac{1}{b+4}$
45. $\dfrac{b-1}{2(b+1)}$ 47. $\dfrac{6x+8}{x}$ 49. $\dfrac{x^2-4x+9}{x-4}$ 51. $\dfrac{a^2b-3}{a^2}$
53. $\dfrac{-4x-3}{x+1}$ or $-\dfrac{4x+3}{x+1}$ 55. $\dfrac{a+2}{a-b}$ 57. $-\dfrac{2}{a-4}$
59. $\dfrac{4x+2}{x-2}$ 61. $\dfrac{1}{r+2}$ 63. $\dfrac{3b+4ab-2a}{a^2b^2}$
65. $\dfrac{5x-2}{(x+1)(x-1)}$ 67. $\dfrac{x}{x-2}$ 69. $\dfrac{5x+3}{x+1}$ 71. $\dfrac{4y-5xy}{10x}$
73. $\dfrac{7+2m}{m^2}$ 75. $\dfrac{4xy+6x}{3y}$ 77. $\dfrac{2x^2-1}{x(x+1)}$
79. $\dfrac{x^2+4x+1}{x^2y}$ 81. $\dfrac{x+2}{x-2}$ 83. $\dfrac{b+4}{b-6}$ 85. $\dfrac{8x-11}{4(2x-3)}$
87. $\dfrac{1}{a+1}$ 89. $-\dfrac{1}{2(x-2)}$ 91. $\dfrac{10a-14}{3a(a-2)}$ 93. $\dfrac{x+3}{x+2}$
97. $8, (0, 2)$ 99. 0

Study Set Section 7.5 (page 581)

1. complex fraction 3. single, division, reciprocal
5. $\div, ab, ab, (b+2a), 2b-a$ 7. $\dfrac{8}{9}$ 9. $\dfrac{3}{8}$ 11. $\dfrac{5x}{12}$
13. $\dfrac{x^2}{y}$ 15. $\dfrac{5t^2}{27}$ 17. $\dfrac{t^3}{2}$ 19. $\dfrac{5}{7}$ 21. $\dfrac{9y+15}{5y^2-6y}$ 23. $\dfrac{2-5y}{6+y}$
25. $\dfrac{24-c^2}{12}$ 27. $\dfrac{s^2-s}{2s+2}$ 29. $\dfrac{2b+a}{b-3a}$ 31. $\dfrac{5}{4}$
33. $\dfrac{1-3x}{5+2x}$ 35. $\dfrac{1+x}{2+x}$ 37. $\dfrac{3-x}{x-1}$ 39. $\dfrac{32h-1}{96h+6}$
41. $\dfrac{d-3}{2d}$ 43. $\dfrac{x-12}{x+6}$ 45. $a-7$ 47. $\dfrac{m^2+n^2}{m^2-n^2}$
49. $\dfrac{3y+12}{4y^2-6y}$ 51. $\dfrac{4c+5c^2}{8}$ 53. $\dfrac{y}{x-2y}$ 55. $\dfrac{5}{x+2}$
57. $\dfrac{1}{x-3}$ 59. $\dfrac{x}{x-2}$ 61. $2m-1$ 63. $\dfrac{3-x}{x-1}$ 65. $\dfrac{3}{5x+6}$
67. $\dfrac{1}{x+1}$ 69. $\dfrac{1+3y}{3-2y}$ 71. $\dfrac{xy}{y+x}$ 73. $\dfrac{y}{x-2y}$
75. $\dfrac{x^2}{(x-1)^2}$ 77. $\dfrac{7x+3}{-x-3}$ 79. $\dfrac{x-2}{x+3}$ or $\dfrac{-x+2}{-x-3}$
81. -1 83. $\dfrac{3}{14}$ 85. $\dfrac{R_1R_2}{R_2+R_1}$ 89. t^9 91. $-2r^7$
93. $\dfrac{81}{256r^8}$ 95. $\dfrac{r^{10}}{9}$

Study Set Section 7.6 (page 590)

1. rational equations 3. check 5. yes 7. x
9. a. $\dfrac{5}{x}+\dfrac{2}{3}=\dfrac{7}{4x}$ b. $12x$ 11. a. 3 b. $3x+18$
13. 6 15. 12 17. y 19. $10x$
21. $2a, 2a, 2a, 2a, 2a, 4, 4, 4$ 23. $\dfrac{3}{5}+\dfrac{7}{x-1}=2$
25. 4 27. 0 29. 3 31. $\dfrac{11}{4}$ 33. 2 35. -1
37. 1 39. 1 41. $-4, 4$ 43. $0, 3$ 45. $4, -\dfrac{3}{2}$
47. $-2, 1$ 49. No solution; 2 is extraneous.
51. No solution; 5 is extraneous. 53. $a=\dfrac{b}{b-1}$
55. $r=\dfrac{E-IR}{I}$ 57. $1, 1$ 59. 6 61. 1 63. 4
65. No solution; -2 is extraneous. 67. 1 69. 2 71. 0
73. $2, -5$ 75. $-4, 3$ 77. 5; 3 is extraneous. 79. $3, -4$
81. $d=\dfrac{bc}{a}$ 83. $f=\dfrac{d_1d_2}{d_1+d_2}$ 85. $R=\dfrac{HB}{B-H}$
89. $x(x+4)$ 91. $(2x+3)(x-1)$
93. $(x^2+4)(x+2)(x-2)$ 95. $a(ab^2-1)$

Study Set Section 7.7 (page 596)

1. work 3. investment 5. $d=rt$ 7. $6\dfrac{1}{9}$ days 9. 0.09
11. 4 13. 2 15. 5 17. $\dfrac{2}{3}, \dfrac{3}{2}$ 19. 8
21. No; after the pipes are opened, the swimming is scheduled to take place in 6 hr. It takes 7.2 hr (7 hr 12 min) to fill the pool.
23. $4\dfrac{4}{9}$ hr 25. 8 hr 27. 20 min
29. Garin: 16 mph, Froome: 26 mph
31. first: $1\dfrac{1}{2}$ ft per sec, second: $\dfrac{1}{2}$ ft per sec
33. Canada goose: 30 mph, great blue heron: 20 mph
35. 340 mph, 170 mph 37. $\dfrac{300}{255+x}, \dfrac{210}{255-x}$, 45 mph
39. 25 mph 41. credit union: 4%; bonds: 6% 43. 7%, 8%
45. 30 radios 51. $(1, 3)$ 53. yes 55. 0

Think It Through (page 601)

$\dfrac{1}{4}, \dfrac{1}{2}, \dfrac{2}{5}$, yes, no, yes

Study Set Section 7.8 (page 608)

1. ratio 3. proportion 5. means 7. shape 9. $\dfrac{6}{5}$
11. yes 13. The ratio on the right side should be $\dfrac{h}{48}$.
15. $x, 288, 18, 18$ 17. triangle 19. $\dfrac{4}{15}$ 21. $\dfrac{3}{4}$ 23. $\dfrac{5}{4}$

25. $\frac{1}{2}$ **27.** $\frac{1}{2}$ **29.** $\frac{25}{22}$ **31.** no **33.** yes **35.** no **37.** yes
39. 4 **41.** 6 **43.** 0 **45.** -17 **47.** $-\frac{3}{2}$ **49.** $\frac{83}{2}$
51. $4, -4$ **53.** $4, -1$ **55.** $-1, 16$ **57.** $-\frac{5}{2}, -1$ **59.** 3
61. 9 **63.** $-5, -1$ **65.** 2 **67.** $-\frac{1}{4}, 5$ **69.** $-\frac{27}{2}$
71. $-10, 10$ **73.** $-4, 3$ **75.** $\frac{26}{9}$ **77.** \$17 **79.** \$6 million
81. 24 drops **83.** $7\frac{1}{2}$ cups, $1\frac{2}{3}$ cups, 5 tablespoons
85. 510 shirts **87.** 737 syllables **89.** \$309 **91.** 49 ft, $3\frac{1}{2}$ in.
93. 3,820,800 residents **95.** $3\frac{3}{4}$ in. **97.** 39 ft
99. $46\frac{7}{8}$ ft **101.** 6,750 ft **107.** 90% **109.** 480 **111.** $\frac{3}{5}$

Study Set Section 7.9 (page 618)

1. direct **3.** constant **5.** direct variation
7. inverse variation **9.** $h = ka$ **11.** $g = kf$ **13.** $t = \frac{k}{s}$
15. $n = \frac{k}{p}$ **17.** 150 **19.** yes **21.** no **23.** no **25.** yes
27. $\frac{7}{5}, \frac{5}{7}, \frac{5}{7}$ **29.** 35 **31.** 6.25 **33.** 0.1 **35.** $\frac{7}{2}$ **37.** 1
39. 12,000 **41.** 567 **43.** $\frac{32}{27}$ **45.** 168 mi **47.** 27 mg
49. $2\frac{1}{2}$ hr **51.** 8 amps **53.** 1,087.06 **55.** 1.375
57. 180 rpm **59.** 55 lb **61.** 40 cm **63.** $53\frac{1}{3}$ m^3
65. 243,000 revolutions **73.** $-1, 6$ **75.** $-2, -3, -4$
77. $0, 0, 1$ **79.** $1, -1, 2, -2$

Chapter 7 Review Exercises (page 623)

1. $-\frac{3}{7}$ **2.** $4, -4$ **3.** $\frac{2}{5}$ **4.** $-\frac{2}{3}$ **5.** $\frac{1}{2x}$ **6.** $\frac{5}{2x}$ **7.** $\frac{x}{x+1}$
8. $a - 2$ **9.** -1 **10.** $-\frac{1}{x+3}$ **11.** $\frac{x}{x-1}$
12. in simplest form **13.** x is not a common factor of the numerator and the denominator.
14. $\frac{4}{3}$ **15.** $\frac{3x}{y}$ **16.** 96 **17.** $\frac{x-1}{x+2}$ **18.** $\frac{2x}{x+1}$
19. $\frac{3y}{2}$ **20.** $\frac{1}{x}$ **21.** $x + 2$ **22.** $\frac{b+2}{b-2}$ **23.** yes
24. no **25.** yes **26.** $\frac{3}{4}$ mi/min $= 0.75$ mi/min
27. $\frac{1}{3d}$ **28.** 1 **29.** $\frac{2x+2}{x-7}$ **30.** $\frac{1}{a-4}$ **31.** $9x$
32. $8x^3$ **33.** $m(m-8)$ **34.** $(5x+1)(5x-1)$
35. $(a+5)(a-5)$ **36.** $(2t+7)(t+5)^2$ **37.** $\frac{63}{7a}$

38. $\frac{2xy + x}{x(x-9)}$ **39.** $\frac{2b+14}{6(b-5)}$ **40.** $\frac{9r^2 - 36r}{(r+1)(r-4)(r+5)}$
41. $\frac{c-7}{7c}$ **42.** $\frac{x^2 + x - 1}{x(x-1)}$ **43.** $\frac{1}{t+1}$ **44.** $\frac{x^2 + 4x - 4}{2x^2}$
45. $\frac{x+1}{x}$ **46.** $\frac{b+6}{b-1}$ **47.** yes
48. $\frac{14x+28}{(x+6)(x-1)}$ cm, $\frac{12}{(x+6)(x-1)}$ cm^2
49. $\frac{9}{4}$ **50.** $\frac{3}{2}$ **51.** $\frac{n^3}{14}$ **52.** $\frac{r+9}{8s}$ **53.** $\frac{1+y}{1-y}$
54. $\frac{x(x+3)}{2x^2 - 1}$ **55.** $x^2 + 3$ **56.** $\frac{y - 5xy}{3xy - 7x}$ **57.** 3
58. no solution, 5 is extraneous **59.** 3 **60.** $2, 4$ **61.** 0
62. $-4, 3$ **63.** $y = \frac{xz}{z-x}$ **64.** $T_1 = \frac{T_2}{1-E}$ **65.** 3 **66.** $\frac{1}{4}$
67. $5\frac{5}{6}$ days **68.** 5 mph **69.** 40 mph **70.** 5% **71.** $\frac{3}{2}$
72. no **73.** $\frac{9}{2}$ **74.** 0 **75.** 7 **76.** $4, -\frac{3}{2}$ **77.** 255 **78.** 20 ft
79. $c = kt$ **80.** $f = \frac{k}{L}$ **81.** \$833.33 **82.** $1\frac{1}{4}$ amps or 1.25 amps **83.** answers vary **84.** inverse variation

Chapter 7 Test (page 632)

1. a. rational **b.** similar **c.** proportion **d.** build
e. factors **2.** $\frac{3}{8}$ **3.** $\frac{1}{18}$ **4.** 342 m per year **5.** $-3, 2$
6. $\frac{8x}{9y}$ **7.** $\frac{x+1}{2x+3}$ **8.** 3 **9.** $-\frac{5y^2}{4}$ **10.** $\frac{x+1}{3(x-2)}$
11. $\frac{3}{5}$ **12.** $-\frac{x^2}{3}$ **13.** $x + 2$ **14.** $\frac{10x-1}{x-1}$ **15.** $\frac{13}{2y+3}$
16. $\frac{2x^2 + x + 1}{x(x+1)}$ **17.** $\frac{2a+7}{a-1}$ **18.** $\frac{4n - 5mn}{10m}$ **19.** $\frac{x+y}{y-x}$
20. 11 **21.** 1 **22.** $1, 2$ **23.** no solution, 6 is extraneous
24. $B = \frac{HR}{R-H}$ **25.** yes **26.** $\frac{2}{3}$ **27.** yes
28. 1,785 British pounds **29.** 171 ft **30.** 8,050 ft
31. 100 lb **32.** 200 **33.** $3\frac{15}{16}$ hr **34.** 5 mph
35. We can remove only common factors, as in the first expression. We can't remove common terms, as in the second expression. **36.** We multiply both sides of the equation by the LCD of the rational expressions appearing in the equation. The resulting equation is easier to solve.

Cumulative Review Chapters 1-7 (page 634)

1. -36 **2.** about 26% **3.** 77 **4.** 43x¢ **5.** $\frac{5}{4}$ **6.** 104°F
7. 240 ft^3 **8. a.** false **b.** false **c.** true **d.** true **9.** 3
10. $6c + 62$ **11.** $B = \frac{A - c - r}{2}$
12. $x \geq -1$, $[-1, \infty)$, ⟵─●──────⟶
                                    -1
13. -5
14. 12 lb of the \$3.20 tea and 8 lb of the \$2 tea
15. 500 mph

16. 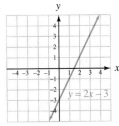 **17.**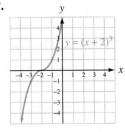

17. a. $\sqrt{5} \approx 2.2$ **b.** $\sqrt{3} \approx 1.7$, $\sqrt{8} \approx 2.8$
19. $5, 12, 144, 169, \sqrt{c^2}, 13$ **21.** False; $-\sqrt{9} = -3, \sqrt{-9}$ is not a real number. **23.** 5 **25.** 14 **27.** -9 **29.** 1.1
31. $\dfrac{3}{16}$ **33.** -17 **35.** 1.414 **37.** 3.317 **39.** 9.747
41. 20.688 **43.** $\sqrt{9}$ rational, $\sqrt{17}$ irrational
45. $\sqrt{49}$ rational, $\sqrt{-49}$ imaginary
47. 1, 2, 3, 4, 5

18. -1 **19.** $y = 3x + 2$
20. 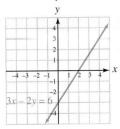 **21.**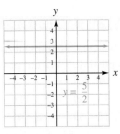

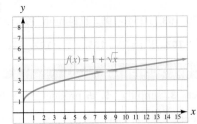

49. $0, -1, -2, -3, -4$

22. $\dfrac{8}{7}$ **23. a.** no **b.** 12 **24.** 0.008 mm/m

25. 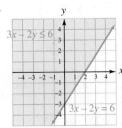 **26.**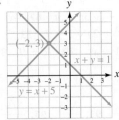

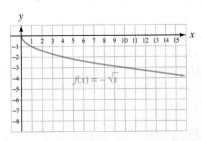

51. 5 **53.** 8 **55.** -50 **57.** 60
59. -99.378 **61.** 4.621 **63.** 0.599 **65.** -0.915 **67.** 3.464
69. 1.866 **71.** 30 **73.** 117 **75.** 12 ft **77.** The diagonal measurement should be $\sqrt{16^2 + 30^2} = 34$ in. **79.** 127.3 ft
81. 5.8 mi **85.** 25.5 ft **87.** 240 in.² **89. a.** 8.5 in. **b.** 7.8 in.
95. $6s^2 + s - 5$ **97.** $3x^2 + 10x - 8$

27. $(5, 2)$ **28.** $(1, -1)$ **29.** red: 8; blue: 15 **30.** -25
31. x^7 **32.** x^{25} **33.** $\dfrac{y^3}{8}$ **34.** $-\dfrac{32a^5}{b^5}$ **35.** $\dfrac{a^8}{b^{12}}$ **36.** $3b^2$
37. 2.9×10^5 **38.** 3 **39.** $6x^2 + x - 5$ **40.** $6x^5y^5$
41. $6y^2 - y - 35$ **42.** $-12x^4z + 4x^2z^2$ **43.** $2x + 3$
44. $-\dfrac{3r}{s^9}$ **45.** $10^3 \cdot 26^3$; 17,576,000 **46.** $A = \pi R^2 - \pi r^2$
47. $k^2t(k - 3)$ **48.** $(a + b)(x + z)$
49. $2(a + 10b)(a - 10b)$ **50.** $(b + 5)(b^2 - 5b + 25)$
51. $(u - 9)^2$ **52.** $(2x - 9)(3x + 7)$ **53.** $-(r - 2)(r + 1)$
54. prime **55.** $0, -\dfrac{1}{5}$ **56.** $\dfrac{1}{3}, \dfrac{1}{2}$ **57.** 10 in., 16 in.
58. $5, -5$ **59.** $\dfrac{x}{3}$ **60.** $-\dfrac{1}{x - 3}$ **61.** $\dfrac{x^2}{(x + 1)^2}$ **62.** $\dfrac{a}{a - 3}$
63. $\dfrac{17}{25}$ **64.** 3 **65.** $2,450 **66.** 34.8 ft **67.** $14\dfrac{2}{5}$ hr
68. One variable is a constant multiple of the reciprocal of the other; $y = \dfrac{k}{x}$.

Study Set Section 8.2 (page 655)

1. cube **3.** function **5.** root **7.** -6 **9. a.** not a real number **b.** -5 **11. a.** 1 **b.** $-\dfrac{1}{3}$ **c.** 5 **d.** 0.2 **e.** 10
13. index, radicand **15.** 2 **17.** 5 **19.** 0 **21.** -2 **23.** -3
25. $\dfrac{1}{5}$ **27.** 1 **29.** 31.78 **31.** -0.48

33. $-1, 0, 1, 2, 3$

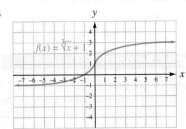

Think It Through (page 640)

1. 40 mph **2.** 60 mph

Study Set Section 8.1 (page 646)

1. square **3.** positive **5.** right **7.** two, 5, -5 **9.** $a^2 + b^2$
11. $\sqrt{b}$ **13. a.** Divide both sides by 2. **b.** Take the positive square root of both sides. **15.** $0, \dfrac{1}{9}, 0.4, 6, 20$

35. $2, 1, 0, -1, -2$

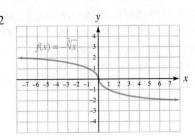

37. 5 **39.** -2 **41.** 1 **43.** -3 **45.** $\dfrac{1}{4}$
47. $-\dfrac{1}{2}$ **49.** y **51.** x^3 **53.** x^5 **55.** $2z$ **57.** $-x^2y$
59. $6z^{18}$ **61.** $-25z$ **63.** $-12x^3$ **65.** y^2 **67.** $3y$ **69.** f
71. $-2t^2$ **73.** -4 **75.** 9 **77.** 10 **79.** $-0.9b^3$
81. $-10a^3b^2$ **83.** $-r^4t^2$ **85.** x^2 **87.** $-\dfrac{x^2}{2}$ **89.** 3.34
91. -5.70 **93.** 1.26 ft **95.** 27.14 mph **97.** $33\dfrac{1}{3}$%
101. m^7 **103.** 3^8 or 9^4 **105.** x^{25} **107.** $24x^8$

Study Set Section 8.3 (page 665)

1. perfect **3.** simplify **5. a.** product, $\sqrt{a}\sqrt{b}$
b. quotient, $\dfrac{\sqrt{a}}{\sqrt{b}}$ **7.** Line 2 is not true. There is no addition property of radicals. **9. a.** $\sqrt{4\cdot 10}$ **b.** $\sqrt{n^4\cdot n}$
11. $2\sqrt{10}$ **13.** $6\sqrt{2}$ **15.** $5, 5a, 16a^2b^2, 4ab$
17. multiplication **19. a.** $2\sqrt{5}$ **b.** $a\sqrt{7}$ **c.** $9x\sqrt{6}$
d. $5z^2\sqrt{y}$ **21.** $2\sqrt{5}$ **23.** $5\sqrt{2}$ **25.** $3\sqrt{5}$
27. $7\sqrt{2}$ **29.** $4\sqrt{3}$ **31.** $10\sqrt{2}$ **33.** $n\sqrt{n}$
35. $a^2\sqrt{5a}$ **37.** $2\sqrt{k}$ **39.** $2\sqrt{3x}$ **41.** $-30\sqrt{3t}$
43. $256\sqrt{2c}$ **45.** $5x\sqrt{x}$ **47.** $8a\sqrt{3a}$ **49.** $6x^3y\sqrt{y}$
51. $48x^2y\sqrt{y}$ **53.** $\dfrac{5}{3}$ **55.** $\dfrac{9}{8}$ **57.** $\dfrac{6x\sqrt{2x}}{y}$
59. $\dfrac{5n^2\sqrt{5}}{8}$ **61.** $2\sqrt[3]{3}$ **63.** $-4\sqrt[3]{2}$ **65.** $2x$
67. $3xz^2\sqrt[3]{2}$ **69.** $\dfrac{3\sqrt[3]{m^2}}{2n^2}$ **71.** $\dfrac{rs\sqrt[3]{rs^2}}{10t}$ **73.** $5\sqrt{10}$
75. $4\sqrt{6}$ **77.** $-4\sqrt{7}$ **79.** $a\sqrt{b}$ **81.** $x^3y^2\sqrt{2}$
83. $-\dfrac{8n^2\sqrt{5m}}{5}$ **85.** $\dfrac{\sqrt{26}}{5}$ **87.** $-\dfrac{2\sqrt{5}}{7}$ **89.** $\dfrac{4\sqrt{3}}{9}$
91. $\dfrac{4\sqrt{2}}{5}$ **93.** $3x$ **95.** $-3y\sqrt[3]{3x^2}$ **97.** $\dfrac{8m\sqrt{2}}{9n}$
99. $2rs^2\sqrt{3s}$ **101. a.** $\dfrac{3\pi\sqrt{3}}{4}$ sec **b.** 4.1 sec
103. a. $60\sqrt{2}$ cm **b.** 84.9 cm **105. a.** $2\sqrt{7}$ in.
b. 5.3 in. **109.** $-6a^5$ **111.** $y=-2x+3$
113. $x<5$, ⟵————, $(-\infty, 5)$

Study Set Section 8.4 (page 672)

1. radicals **3.** simplified **5.** no **7.** yes **9.** Because the radicals don't have the same radicand, they can't be combined. **11.** Because the terms are not like terms, they cannot be combined. **13.** $2\sqrt{3}, 3\sqrt{3}, 4\sqrt{3}, 5\sqrt{3}$
15. $4, 4, 2\sqrt{5}, 6\sqrt{5}, 3\sqrt{5}$ **17.** $9\sqrt{7}$ **19.** $5+6\sqrt{3}$
21. $-3\sqrt{x}$ **23.** $-1-\sqrt{r}$ **25.** $5\sqrt{3}$ **27.** $\sqrt{2}$ **29.** $14\sqrt{5}$
31. $9\sqrt{5}$ **33.** $10\sqrt{2}-\sqrt{3}$ **35.** $5\sqrt{6}-5\sqrt{2}$ **37.** $3x\sqrt{2}$
39. $8y\sqrt{xy}$ **41.** $-2b\sqrt{2a}$ **43.** $x^2\sqrt{2x}$
45. $3x\sqrt{2y}-3x\sqrt{3y}$ **47.** $3\sqrt{2ab}+3\sqrt{3ab}$

49. $7\sqrt{3}-6\sqrt{2}$ **51.** $-8\sqrt{x}+14\sqrt{y}$ **53.** $2\sqrt[3]{3}$
55. $-\sqrt[3]{x}$ **57.** $2x\sqrt[3]{x^2}+3x^2\sqrt[3]{x^2}$ **59.** $5ab\sqrt[3]{3a^2b}$
61. $-7\sqrt{5}$ **63.** $25\sqrt{7}$ **65.** $-\sqrt{3}$ **67.** $5\sqrt[3]{4}$ **69.** $6\sqrt[3]{3}$
71. $-3\sqrt{5}$ **73.** $36\sqrt{7}$ **75.** $19b\sqrt{6}$ **77.** $-\sqrt[3]{3}$
79. $5\sqrt{7}+3\sqrt{2}$ **81.** $5\sqrt[3]{2}$ **83.** $-5y\sqrt{10y}$ **85.** $6xy\sqrt{3x}$
87. $18\sqrt{3}$ in. **89.** $27\sqrt{2}$ cm **91.** $133\sqrt{6}$ ft **95.** $\dfrac{1}{9}$
97. -9 **99.** $\dfrac{1}{x^3}$ **101.** 1

Study Set Section 8.5 (page 682)

1. coefficient **3.** rationalizing **5.** $a\cdot b$ **7.** $\sqrt{x}-1$
9. not possible **11.** not possible **13.** $4\sqrt{2}$ **15.** $-2\sqrt{2}$
17. 3 **19.** $\sqrt{x}, \sqrt{x}, \sqrt{2x}, x$ **21.** $\sqrt{21}$ **23.** $\sqrt{35}$ **25.** 4
27. $2\sqrt{14}$ **29.** $2\sqrt[3]{3}$ **31.** $7\sqrt[3]{2}$ **33.** $-60\sqrt{2}$ **35.** $-8x$
37. $6\sqrt[3]{12}$ **39.** $-30a$ **41.** 5 **43.** 12 **45.** -9 **47.** 72
49. $2+\sqrt{2}$ **51.** $8-2\sqrt{2}$ **53.** $x\sqrt{3}-2\sqrt{x}$
55. $6x+6\sqrt{x}$ **57.** $7-2\sqrt[3]{7}$ **59.** $3x\sqrt[3]{2}+3\sqrt[3]{2x}$
61. $\sqrt{6}+\sqrt{10}-3-\sqrt{15}$ **63.** $2b+\sqrt{2b}-2$
65. $6+\sqrt{3t}-3t$ **67.** $4x-18$ **69.** $28+10\sqrt{3}$
71. $a^2+2a\sqrt{7}+7$ **73.** $5-2\sqrt{5m}+m$ **75.** 3
77. $11-y^2$ **79.** $7c-9$ **81.** $\sqrt[3]{4}+4\sqrt[3]{2}+3$
83. $2y-\sqrt[3]{4y}+3\sqrt[3]{2y^2}-3$ **85.** $\dfrac{2x}{3}$ **87.** $\dfrac{3\sqrt{2}}{5}$ **89.** $\dfrac{2}{x}$
91. $\dfrac{2x}{3}$ **93.** $\dfrac{\sqrt{3}}{3}$ **95.** $\dfrac{\sqrt{91}}{7}$ **97.** $\sqrt[3]{25}$ **99.** $2\sqrt[3]{2}$
101. $\dfrac{12\sqrt{y}}{y}$ **103.** $\dfrac{2\sqrt{5ab}}{5b}$ **105.** $\dfrac{3\sqrt{3}+3}{2}$ **107.** $\sqrt{7}-2$
109. $\dfrac{x+4\sqrt{x}+4}{x-4}$ **111.** $\dfrac{2a+3\sqrt{2a}+2}{2a-4}$ **113.** 54
115. $3\sqrt{2x}$ **117.** $4x^2$ **119.** $27x-3\sqrt{3x}$
121. $6\sqrt{14}+2x\sqrt{7}-3x\sqrt{2}-x^2$ **123.** $7+2\sqrt{6}$
125. $\sqrt{3}$ **127.** $\dfrac{3\sqrt{2x}}{8x}$ **129.** $\dfrac{2\sqrt{15}}{5}$ **131.** $4-\sqrt[3]{4}$
133. $6+2\sqrt{3}$ **135.** $x\sqrt{3}-x\sqrt{2}$ **137.** $\dfrac{\sqrt[3]{20}}{2}$
139. $\dfrac{y+6\sqrt{y}+9}{y-9}$ **141.** 108π in.^{2} **143.** $1{,}800\sqrt{2}$ in.^{2}
145. 90 in.^{2} **149.** no **151.** positive

Study Set Section 8.6 (page 692)

1. radical **3.** extraneous **5.** b^2 **7.** Take the positive square root of both sides. **9.** x **11.** $4x$ **13.** $2x$
15. On the second line, both sides of the equation were not squared—only the left side. **17.** On the second line, $\sqrt{a+2}$ wasn't isolated before squaring both sides. Also, $(\sqrt{a+2}-5)^2 \neq a+2-25$.

19. a.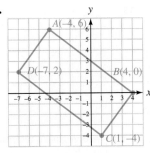
b. rectangle **c.** AB: 10, BC: 5, CD:10, DA: 5
d. 30 units
21. $\sqrt{x-3}, 5, 25$
23. 9
25. 8
27. 1
29. 5
31. no solution, -1 is extraneous
33. no solution, 16 is extraneous
35. 4 **37.** 4 **39.** 4 **41.** 12, 7 is extraneous **43.** 1, 3
45. $-2, 4$ **47.** 5 **49.** -2 **51.** 343 **53.** 65 **55.** 22
57. 4 **59.** 5 **61.** $\sqrt{34}, 5.83$ **63.** 13 **65.** 10 **67.** 25
69. 9 **71.** 6 **73.** 46 **75.** 4 **77.** -5 **79.** 2 **81.** -1
83. 6, 1 is extraneous **85.** 5, 0 is extraneous **87.** 4
89. 10 **91.** 169 ft **93.** 64.4 ft **95.** about 288 ft
97. about 2×10^6 m **99.** $V = \dfrac{\pi r^2 h}{3}$
101. a. $\sqrt{2} \approx 1.4$ units **b.** $\sqrt{10} \approx 3.2$ units
105. $8x^2 - 6x$ **107.** $x^2 + 6x + 9$ **109.** $25x^2 - 20x + 4$

Study Set Section 8.7 (page 700)

1. rational **3.** x^{m+n} **5.** $\dfrac{x^n}{y^n}$ **7.** $\dfrac{1}{x^n}$ **9.** $\sqrt[n]{x}$ **11.** $5^{1/2}$
13. $(\sqrt[3]{8})^4$ or $\sqrt[3]{8^4}$ **15.** 0, 1, 2, 3
17.
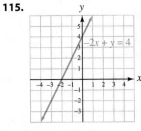
19. $-216, -6$ **21.** 9 **23.** -12 **25.** $\dfrac{2}{7}$ **27.** 3 **29.** -5
31. $\dfrac{3}{4}$ **33.** 16 **35.** 729 **37.** 125 **39.** 625 **41.** 100
43. $\dfrac{4}{9}$ **45.** 4 **47.** 100 **49.** 6 **51.** 25 **53.** 7 **55.** 25
57. $2^{1/2}x^{1/2}$ **59.** 8 **61.** 125 **63.** $\dfrac{5^{2/3}}{2^{2/3}}$ **65.** $\dfrac{1}{2}$ **67.** $\dfrac{1}{9}$
69. m **71.** x^4y^2 **73.** $\dfrac{1}{64}$ **75.** $\dfrac{1}{81a^{12}}$ **77.** x^2 **79.** y^2
81. $x^{2/5}$ **83.** $\dfrac{2}{a^{3/4}}$ **85.** $\dfrac{1}{2}$ **87.** -2 **89.** 256 **91.** 49
93. $x^{2/7}$ **95.** $x^{17/12}$ **97.** $b^{3/10}$ **99.** $t^{4/15}$ **101.** $\dfrac{4}{b^{9/4}}$
103. x^2 **105.** 392 in.² **107.** 26 ft **109.** 1.1 in.
113. **115.**

Chapter 8 Review Exercises (page 704)

1. square **2.** $\left(\dfrac{1}{3}\right)^2 = \dfrac{1}{9}$ **3.** 5 **4.** 7 **5.** -12 **6.** $-\dfrac{4}{9}$
7. 30 **8.** -0.8 **9.** 1 **10.** 0 **11.** 4.583 **12.** -3.873
13. 5.292 **14.** 27.421 **15.** $\sqrt{-2}$: imaginary, $\sqrt{68}$: irrational, $\sqrt{81}$: rational, $\sqrt{3}$: irrational, $\sqrt{-2}$ is not real
16. 0, 1, 2, 3

17. 2, 1, 0, -1
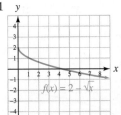

18. 35 **19.** 1
20. $\sqrt{11}$ **21.** 5 ft
22. 30 mph
23. 5^3, cube
24. $\left(\dfrac{1}{3}\right)^3 = \dfrac{1}{27}$
25. -3 **26.** -5
27. 3 **28.** 2 **29.** 0
30. -1 **31.** $\dfrac{1}{4}$ **32.** 1 **33.** 2.520 **34.** -4.678
35. 1.565 **36.** 8.083 **37.** x **38.** $2b$ **39.** x^2y^2
40. $-y^6$ **41.** x **42.** y^2 **43.** $3x$ **44.** $-r^4$ **45.** 12 mm
46. 14 in. **47.** $4\sqrt{2}$ **48.** $10\sqrt{5}$ **49.** $4x\sqrt{5}$ **50.** $-6\sqrt{7}$
51. $-5t\sqrt{10t}$ **52.** $-10z^2\sqrt{7z}$ **53.** $10x\sqrt{2y}$ **54.** $y^2\sqrt{3}$
55. $2y\sqrt[3]{x^2}$ **56.** $5xy\sqrt[3]{2x}$ **57.** $\dfrac{4}{5}$ **58.** $\dfrac{2\sqrt{15}}{7}$ **59.** $\dfrac{10}{3}$
60. $\dfrac{11x\sqrt{2}}{13}$ **61.** $2\sqrt{10}$ ft **62.** 6.3 ft **63.** 0
64. $-3 + 4\sqrt{3}$ **65.** $\sqrt{7}$ **66.** $y^2\sqrt{5xy}$ **67.** $5\sqrt[3]{2}$
68. $6x\sqrt[3]{2}$ **69.** They do not contain like radicals; the radicands are different. **70.** $13\sqrt{5}$ in. **71.** $\sqrt{6}$
72. $10\sqrt{10}$ **73.** $36x\sqrt{2}$ **74.** $15 + 6x\sqrt{15} + 9x^2$ **75.** -2
76. -2 **77.** $4\sqrt[3]{2}$ **78.** $\sqrt[3]{9} + \sqrt[3]{3} - 2$ **79.** $\dfrac{\sqrt{7}}{7}$
80. $\dfrac{\sqrt{21}}{7}$ **81.** $\dfrac{\sqrt{2}}{2}$ **82.** $\dfrac{c - 8\sqrt{c} + 16}{c - 16}$ **83.** $7\sqrt{2} - 7$
84. $2\sqrt[3]{4}$ **85.** $(4\sqrt{6} + 10\sqrt{3})$ in., $30\sqrt{2}$ in.²
86. 27.1 in., 42.4 in.² **87.** x **88.** x **89.** $4t$ **90.** $e - 1$
91. 81 **92.** no solution, 0 is extraneous **93.** $-2, 4$
94. 4 **95.** -2 **96.** 28 **97.** 5 **98.** 13.93
99. 64 ft **100.** 7 **101.** -10 **102.** 216 **103.** $\dfrac{4}{9}$
104. $\dfrac{1}{8}$ **105.** 64 **106.** 9 **107.** $\dfrac{1}{a^2b^4}$ **108.** $x^{11/15}$
109. $t^{1/12}$ **110.** x **111.** x^2
112.

113. $\dfrac{1}{64}$ of the original dose
114. $(-4)^{1/2} = \sqrt{-4}$; there is no real number that, when squared, gives -4

Chapter 8 Test (page 712)

1. a. radical **b.** radicand **c.** Pythagorean **d.** cube root
e. index   **2.** 7.2   **3.** 10   **4.** $-\dfrac{20}{3}$ **5.** -3 **6.** $\dfrac{5\sqrt{2}}{7}$
7. 10 ft   **8. a.** $2\sqrt{6}$ yd   **b.** 4.9 yd   **9.** $2x$ **10.** $3x\sqrt{6x}$
11. $3y\sqrt{x}$ **12.** $x^2 y$ **13.** $5\sqrt{3}$ **14.** $-x\sqrt{2x}$
15. $-24x\sqrt{6}$ **16.** $2\sqrt{6} + 3\sqrt{2}$ **17.** -1
18. $2x - 4\sqrt{x} - 6$ **19.** $\left(6\sqrt{2} + 2\sqrt{10}\right)$ in., 14.8 in.
20. $\sqrt{2}$ **21.** $\dfrac{x\sqrt{3} - 2\sqrt{3x}}{x - 4}$ **22.** 225   **23.** -62 **24.** 5
25. 2, 1   **26.** no solution, 0 is extraneous   **27.** 29   **28.** 10
29. 0, 1, 1.4, 2, 2.4, 3

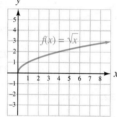

30. No; when 0 is substituted for x, the result is not a true statement: $1 \neq -1$.   **31.** The terms do not contain like radicals — the radicands are different.
32. the Pythagorean theorem; 5 ft   **33.** There is no real number that, when squared, gives -9.   **34.** 4   **35.** $a^{1/12}$
36. p^2 **37.** $\dfrac{1}{x^2 y^4}$ **38.** $4x$

Cumulative Review Chapters 1–8 (page 714)

1. a. true   **b.** false   **c.** true   **2.** -14 **3.** $-2p - 6z$
4. -2 **5.** 17 lb   **6.** 75.0% increase
7. $x \leq -\dfrac{3}{4}, \left(-\infty, -\dfrac{3}{4}\right]$

8. $-2 < x \leq 4, (-2, 4]$

9. 3.5 hr   **10.** 80   **11.** 4 in.
12. 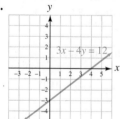   **13.**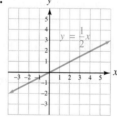

14. 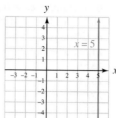   **15.**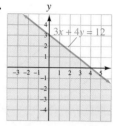

16. a. They are the same.   **b.** They are negative reciprocals.
17. \$5.8 billion per yr   **18. a.** 3   **b.** $-\dfrac{2}{3}$
19. $y = \dfrac{x}{2} + 6$ **20.** -1
21. 1, 0, 1, 2, 2, 3

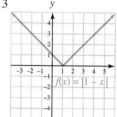

D: all reals
R: all real numbers greater than or equal to 0
22. yes
23. Not a solution

24. $(-1, 5)$

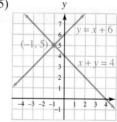

25. $(3, -1)$
26. $(-1, 2)$
27. 6%: \$3,000; 12%: \$3,000

28.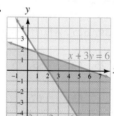

29. x^{31} **30.** $\dfrac{a^{15} b^5}{c^{20}}$
31. 1   **32.** $\dfrac{a^8}{b^{12}}$
33. 4.8×10^{18} m
34. a. $16a^2 + 24a + 9$
b. $4a^2 - 9$ **35.** $2r^7 s^3 t^4$
36. $-6t^2 + 13st - 6s^2$
37. $2a^2 + a - 3$
38. $y^2 - 12y + 36$ **39.** $\dfrac{1}{y} - \dfrac{3}{4x} + \dfrac{2z}{xy}$ **40.** $2x - 1$
41. $3xy(x - 2y)$ **42.** $(x + y)(2x - 3)$
43. $(5p^2 + 4q)(5p^2 - 4q)$ **44.** $3x(x + 9)(x - 9)$
45. $(x - 12)(x + 1)$ **46.** $(a + 2b)(a^2 - 2ab + 4b^2)$
47. $(3a + 4)(2a - 5)$ **48.** $2(4m + 1)(2m - 3)$ **49.** $-1, -2$
50. 0, 2   **51.** $\dfrac{2}{3}, -\dfrac{1}{2}$ **52.** $\dfrac{3}{2}, -4$ **53.** 5 cm   **54.** 6
55. $\dfrac{x+1}{x-1}$ **56.** $-\dfrac{3}{5a}$ **57.** $\dfrac{(p+2)(p-3)}{3(p+3)}$ **58.** $\dfrac{xy^2 d}{c^3}$
59. 1   **60.** $\dfrac{7x + 29}{(x+5)(x+7)}$ **61.** $\dfrac{9a^2 - 4b^2}{6ab}$ **62.** $\dfrac{y+x}{y-x}$
63. 2   **64.** 2   **65.** $r = \dfrac{r_1 r_2}{r_2 + r_1}$ **66.** $2\dfrac{2}{9}$ hr   **67.** 6,480
68. 2   **69.** $\dfrac{7}{15}$ **70.** 3   **71.** $-48x^2 y \sqrt{y}$ **72.** $7\sqrt{3} - 6\sqrt{2}$

73. $y - 9\sqrt{y} + 20$ **74.** $-60\sqrt{2}$ **75.** $\dfrac{2\sqrt{5}}{5}$
76. $\dfrac{x - 6\sqrt{x} + 9}{x - 9}$ **77.** 5 **78.** 73 in.

Study Set Section 9.1 (page 725)

1. quadratic **3.** $\sqrt{c}, -\sqrt{c}$ **5. a.** $x = -9 \pm \sqrt{2}$
b. $x = \dfrac{3 \pm \sqrt{2}}{6}$ **7.** yes **9.** $x = \pm\sqrt{6}$ **11.** ± 6 **13.** ± 3
15. $\pm\dfrac{7}{4}$ **17.** ± 10 **19.** ± 5 **21.** $\pm\sqrt{6}$ **23.** $\pm\sqrt{17}$
25. $\pm 2\sqrt{5}$ **27.** $\pm 6\sqrt{2}$ **29.** $\pm\dfrac{\sqrt{30}}{2}$ **31.** $\pm\dfrac{\sqrt{42}}{6}$
33. no real-number solutions **35.** $-6, 4$ **37.** $7, -11$
39. $2 \pm 2\sqrt{2}$ **41.** $-9 \pm 3\sqrt{7}$ **43.** $\dfrac{-1 \pm 3\sqrt{2}}{3}$
45. $\dfrac{10 \pm \sqrt{6}}{5}$ **47.** $-1 \pm \sqrt{10}$ **49.** $9 \pm \sqrt{7}$
51. $3 \pm 2\sqrt{10}$ **53.** $-2 \pm 5\sqrt{3}$ **55.** $-12 \pm 3\sqrt{3}$
57. $\pm 7\sqrt{2}$ **59.** $6 \pm \sqrt{2}$ **61.** $15 \pm 2\sqrt{2}$ **63.** $\pm\dfrac{1}{12}$
65. $7 \pm \sqrt{3}$ **67.** no real-number solutions **69.** $\pm\dfrac{\sqrt{85}}{5}$
71. $\pm\sqrt{14}$ **73.** $\dfrac{-9 \pm 2\sqrt{11}}{8}$ **75.** 3 sec **77.** 3.4 sec
79. 3.5 sec **81.** 20 ft **87.** 2 **89.** 16

Study Set Section 9.2 (page 732)

1. square **3.** leading, constant **5. a.** 9 **b.** 36 **c.** $\dfrac{9}{4}$ **d.** $\dfrac{25}{4}$
7. one-half **9.** $\sqrt{7}$ **11.** $x^2 - 2x - 1$ doesn't factor
13. $\sqrt{9}, y - 1, -3$ **15. a.** two; $\sqrt{10}, -\sqrt{10}$
b. $8 + \sqrt{3}, 8 - \sqrt{3}; 9.73, 6.27$ **17.** $x^2 + 2x + 1 = (x + 1)^2$
19. $x^2 - 4x + 4 = (x - 2)^2$
21. $x^2 + 7x + \dfrac{49}{4} = \left(x + \dfrac{7}{2}\right)^2$ **23.** $x^2 + x + \dfrac{1}{4} = \left(x + \dfrac{1}{2}\right)^2$
25. $a^2 - 3a + \dfrac{9}{4} = \left(a - \dfrac{3}{2}\right)^2$
27. $b^2 + \dfrac{2}{3}b + \dfrac{1}{9} = \left(b + \dfrac{1}{3}\right)^2$ **29.** $1, -6$ **31.** $-2, -4$
33. $-5, 1$ **35.** $-3, 5$ **37.** $-4 \pm \sqrt{22}; 0.69, -8.69$
39. $-3 \pm \sqrt{5}; -0.76, -5.24$ **41.** $-2 \pm \sqrt{3}; -0.27, -3.73$
43. $1 \pm \sqrt{5}; 3.24, -1.24$ **45.** $\dfrac{1}{3}, -2$ **47.** $\dfrac{3}{2}, -\dfrac{2}{3}$
49. $2 \pm \sqrt{7}$ **51.** $\dfrac{-1 \pm 2\sqrt{5}}{2}$ **53.** $1 \pm \sqrt{5}; 3.24, -1.24$
55. $\dfrac{-3 \pm \sqrt{7}}{2}; -0.18, -2.82$ **57.** $\dfrac{-1 \pm 2\sqrt{5}}{2}; -2.74, 1.74$
59. $\dfrac{-1 \pm \sqrt{13}}{3}; 0.87, -1.54$

61. $1, -2$ **63.** $2, 6$ **65.** $5, -3$ **67.** $-1, -2$ **69.** $2, -\dfrac{1}{2}$
71. $-2, \dfrac{1}{4}$ **73.** $\dfrac{13}{2}, -\dfrac{1}{2}$ **75.** $\dfrac{9 \pm \sqrt{65}}{8}$ **77.** $1, -4$
79. $\dfrac{3}{2}, -\dfrac{2}{3}$ **81.** $\dfrac{-3 \pm \sqrt{3}}{2}$ **83.** 55 ft **85.** 20 ft by 40 ft
91. $y^2 - 2y + 1$ **93.** $x^2 + 2xy + y^2$ **95.** $4z^2$

Study Set Section 9.3 (page 742)

1. solve **3.** quadratic **5.** 0 **7.** $3, 0, -5$ **9.** $lw, \dfrac{1}{2}bh$
11. Evaluate 5^2. **13. a.** 2 **b.** $\dfrac{-3 + \sqrt{15}}{2}, \dfrac{-3 - \sqrt{15}}{2}$
c. $0.44, -3.44$ **15.** $-5, 5, 24, 49, 5, +, 6, -, -1$
17. The fraction bar is not extended to underline the complete numerator. **19.** $a = 1, b = 4, c = 3$
21. $a = 3, b = -2, c = 7$ **23.** $a = 4, b = -2, c = 1$
25. $a = 3, b = -5, c = -2$ **27.** $2, 3$ **29.** $-3, -4$
31. $1, -\dfrac{1}{2}$ **33.** $-1, -\dfrac{2}{3}$ **35.** $1 \pm \sqrt{2}; -0.41, 2.41$
37. $\dfrac{-1 \pm \sqrt{41}}{4}; -1.85, 1.35$ **39.** no **41.** yes, $\dfrac{-4 \pm \sqrt{6}}{2}$
43. $-2, 3$ **45.** no real solutions **47.** $-5, 7$ **49.** $-4, 0$
51. $\pm\sqrt{6}, \pm 2.45$ **53.** $a = 7, b = 14, c = 21$ **55.** $\dfrac{1}{2}, -\dfrac{3}{2}$
57. $\dfrac{-3 \pm \sqrt{5}}{2}$ **59.** $\dfrac{1 \pm \sqrt{37}}{6}$ **61.** no real solutions
63. $-1 \pm \sqrt{2}$ **65.** $\dfrac{3 \pm \sqrt{15}}{3}$ **67.** $0.8, 3.1$ **69.** 6 in.
71. 6 ft, 8 ft **73.** width: 47 ft; length: 66 ft **75.** 10 ft by 18 ft
77. about 9.3 sec **79.** 50 m and 120 m **81.** 3 hr after the second plane takes off **83.** 10% **85.** 150 players
87. 3 in. or 6 in. **93.** $M = \dfrac{Fd^2}{Gm}$ **95.** $y = \dfrac{4}{3}x$ **97.** $2xy\sqrt{x}$
99. $\dfrac{x + 4\sqrt{x} + 4}{x - 4}$

Study Set Section 9.4 (page 755)

1. quadratic, parabola **3.** intercepts, intercept **5.** $<, 0$
7. a. 0 **b.** x **9. a.** parabola **b.** $(1, 0), (3, 0)$ **c.** $(0, -3)$
d. $(2, 1)$ **e.** It is a vertical line through $(2, 1)$.
11. $(2, -2), (-1, 1)$

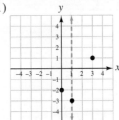

13. two; $-3, 1$ **15. a.** $2, 4, -8$ **b.** -1
17. $(0, 8); (4, 0), (2, 0)$ **19.** $(0, -21); (-3, 0), (-7, 0)$
21. $(1, -1)$ **23.** $(3, 1)$

25. 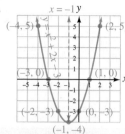 **27.** 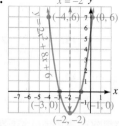 **47.**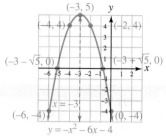

29. 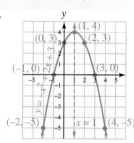 **31.** 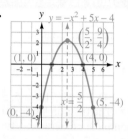 **49.**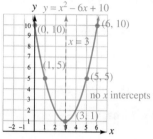

33. 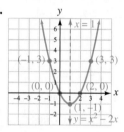 **35.**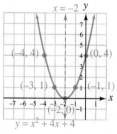

51. It is a vertical line through the body of the butterfly.
53. The cost to manufacture a carburetor is lowest ($100) for a production run of 30 units. **55. a.** 14 ft **b.** 0.25 sec and 1.75 sec **c.** 18 ft; 1.0 sec **63.** $3\sqrt{2}$ **65.** $3z$

Study Set Section 9.5 (page 765)

1. complex, real, imaginary **3. a.** $\sqrt{-1}$ **b.** 2, −1
5. a. $8i$ **b.** $2i$ **7.** $\dfrac{6+i}{6+i}$ **9. a.** true **b.** true **c.** false
d. true **11. a.** $i\sqrt{7}$ **b.** $2i\sqrt{3}$ **13.** $3i$ **15.** $i\sqrt{7}$ or $\sqrt{7}i$
17. $2i\sqrt{6}$ or $2\sqrt{6}i$ **19.** $-4i\sqrt{2}$ or $-4\sqrt{2}i$ **21.** $45i$
23. $\dfrac{5}{3}i$ **25.** $12+0i$ **27.** $0+10i$ **29.** $6+4i$ **31.** $-9-7i$
33. $8-2i$ **35.** $3-5i$ **37.** $14-13i$ **39.** $12-9i$
41. $6-3i$ **43.** $-25-25i$ **45.** $12+5i$ **47.** $13-i$
49. $2+i$ **51.** $\dfrac{8}{53}-\dfrac{28}{53}i$ **53.** $-\dfrac{5}{13}+\dfrac{12}{13}i$ **55.** $\dfrac{31}{50}-\dfrac{17}{50}i$
57. $0\pm 3i$ **59.** $0\pm 2i\sqrt{2}$ **61.** $-3\pm i$ **63.** $11\pm 5i\sqrt{3}$
65. $\dfrac{3}{2}\pm\dfrac{\sqrt{7}}{2}i$ **67.** $-\dfrac{1}{4}\pm\dfrac{\sqrt{7}}{4}i$ **69.** $\dfrac{15}{26}-\dfrac{3}{26}i$ **71.** $15+2i$
73. $0+i$ **75.** $1+8i$ **77.** $\dfrac{5}{13}-\dfrac{12}{13}i$ **79.** $9-8i$
81. $-12-16i$ **83.** $3+4i$ **85.** $-\dfrac{1}{4}\pm\dfrac{\sqrt{39}}{4}i$
87. $4\pm 3i\sqrt{5}$ **89.** $-1\pm i$ **91.** $0\pm 6i$ **93.** $0\pm\dfrac{4}{3}i$
95. $-\dfrac{1}{3}\pm\dfrac{\sqrt{2}}{3}i$ **97.** $18.45-2.18i$ **103.** $\dfrac{\sqrt{7}}{7}$
105. $\dfrac{8\sqrt{x}+16}{x-4}$

37. 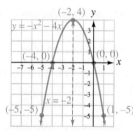 **39.**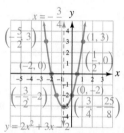

41. 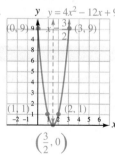 **43.**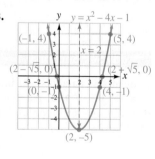

45.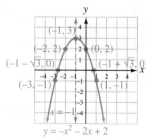

Chapter 9 Review (page 769)

1. ± 8 **2.** $\pm 2\sqrt{2}$ **3.** $\pm 5\sqrt{3}$ **4.** $-4, 6$ **5.** $\dfrac{8\pm 2\sqrt{10}}{9}$
6. $\dfrac{7}{2},\dfrac{1}{2}$ **7.** $13, 7$ **8.** $\dfrac{-1\pm\sqrt{6}}{3}$ **9.** ± 3.46 **10.** $-6.42, 8.42$

11. no real-number solutions **12.** no real-number solutions
13. 3.0 sec **14.** $2\frac{1}{4}$ in. **15.** $x^2 + 4x + 4 = (x + 2)^2$
16. $t^2 - 5t + \frac{25}{4} = \left(t - \frac{5}{2}\right)^2$ **17.** 3, 5 **18.** 2, −7
19. $-1 \pm \sqrt{6}$ **20.** $\frac{4 \pm \sqrt{23}}{2}$ **21.** $\frac{1 \pm \sqrt{3}}{2}$ **22.** $-1, -\frac{2}{3}$
23. −0.27, −3.73 **24.** −0.65, 7.65 **25.** $x^2 + 2x + 5 = 0$; 1, 2, 5 **26.** $6x^2 - 2x - 1 = 0$; 6, −2, −1 **27.** 5, −3
28. $\frac{3}{2}, -\frac{1}{3}$ **29.** $1 \pm \sqrt{5}$ **30.** $3 \pm \sqrt{2}$ **31.** $\frac{-3 \pm \sqrt{21}}{6}$
32. $\frac{-1 \pm \sqrt{21}}{10}$ **33.** no real-number solutions
34. $\frac{-3 \pm \sqrt{19}}{2}$ **35.** 0, −4 **36.** 1, −7 **37.** $\pm 3\sqrt{3}$
38. $\frac{3 \pm \sqrt{6}}{3}$ **39.** $\frac{-1 \pm \sqrt{11}}{2}$ **40.** 2 **41.** $\frac{13}{2}, -\frac{3}{2}$
42. no real-number solutions **43.** $\frac{-1 \pm \sqrt{7}}{3}$; −1.22, 0.55
44. width: 10 ft; length: 24 ft **45.** 18 sec **46.** 6 in.
47. a. $(-3, 0), (1, 0)$ **b.** $(0, -3)$ **c.** $(-1, -4)$
d. a vertical line through $(-1, -4)$ **48.** $(-2, -3)$
49. $(1, 5)$, upward **50.** $(3, 16)$, downward
51. $(-5, 0), (-1, 0); (0, 5)$ **52.** no x-intercepts; $(0, 3)$
53. 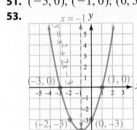 **54.**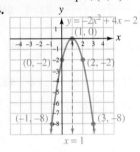

55.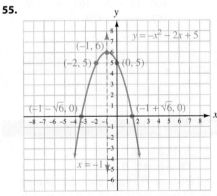

56.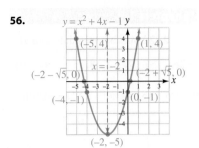

57. 2; −2, 1; 1, −3; no **58.** The maximum profit of $16,000 is obtained from the sale of 400 units. **59.** $5i$ **60.** $3i\sqrt{2}$
61. $-7i$ **62.** $\frac{3}{8}i$ **63.** Real, Imaginary
64. a. true **b.** true **c.** false **d.** false **65.** $3 - 6i$
66. $-1 + 7i$ **67.** $0 - 19i$ **68.** $0 + i$ **69.** $8 - 2i$
70. $3 - 5i$ **71.** $3 + 6i$ **72.** $9 + 7i$ **73.** $-\frac{5}{13} + \frac{12}{13}i$
74. $\frac{15}{26} - \frac{3}{26}i$ **75.** $0 \pm 3i$ **76.** $0 \pm \frac{4\sqrt{3}}{3}i$ **77.** $2 \pm 2i\sqrt{6}$
78. $-3 \pm 3i\sqrt{6}$ **79.** $-1 \pm i$ **80.** $\frac{3}{4} \pm \frac{\sqrt{7}}{4}i$

Chapter 9 Test (page 777)

1. a. complex **b.** parabola **c.** quadratic **d.** imaginary
e. leading **2.** $x = \pm\sqrt{5}$ **3.** $\pm\sqrt{17}$ **4.** $2 \pm \sqrt{3}$
5. $\pm\frac{5}{2}$ **6.** $-8 \pm 2\sqrt{6}$ **7.** $m = \pm\sqrt{-49}$, and $\sqrt{-49}$ is not a real number. **8.** 40 cm **9.** $x^2 - 14x + 49 = (x - 7)^2$
10. $a^2 - \frac{5}{3}a + \frac{25}{36} = \left(a - \frac{5}{6}\right)^2$ **11.** $-1 \pm \sqrt{5}$; −3.24, 1.24
12. $2, -\frac{1}{2}$ **13.** $-\frac{3}{2}, 4$ **14.** $\frac{-11 \pm \sqrt{61}}{10}$ **15.** $\frac{1 \pm \sqrt{7}}{3}$; −0.55, 1.22 **16.** 41 ft, 65 ft **17.** It is a solution.
18. 6.3 sec **19.** $2 \pm \sqrt{2}$ **20.** $1, -\frac{5}{3}$ **21.** $\pm 2\sqrt{6}$
22. $-\frac{1}{3} \pm \frac{\sqrt{2}}{3}i$ or $\frac{-1 \pm i\sqrt{2}}{3}$
23. The most air conditioners sold in a week (18) occurred when 3 ads were run. **24.** $<, >$
25. 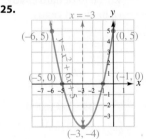 **26.**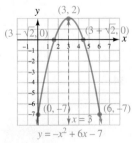

27. $10i$ **28.** $-3i\sqrt{2}$ **29.** $1 + i$ **30.** $-1 + 12i$
31. $14 - 8i$ **32.** $\frac{5}{13} - \frac{12}{13}i$

Cumulative Review Chapters 1-9 (page 779)

1. a. true b. true c. true 2. 2 hours before salt is spread
3. 36 4. 4 5. 65 6. $-2p + 54$ 7. 12 8. $-\dfrac{3}{2}$ 9. 8.9%
10. $r = \dfrac{T - 2t}{2}$ 11. $343,750 12. $8,000 at 7%, $20,000 at 10% 13. $(-\infty, -2)$,   14. yes

15. 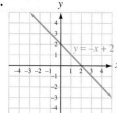 16.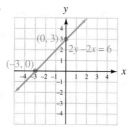

17. 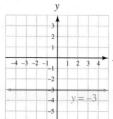 18.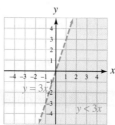

19. $\dfrac{3}{5}$ 20. a decrease of 800,000 viewers per year 21. $-\dfrac{4}{5}$
22. $y = -2x + 1$ 23. perpendicular 24. $y = \dfrac{1}{4}x - 1$

25. 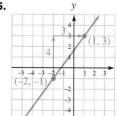 26. -8
27. domain: $\{-4, 1, 4, 5\}$, range: $\{-3, 2, 8\}$ 28. Yes, it passes the vertical line test.

29. $(-2, 3)$ 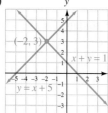 30. $(-3, -1)$
31. $(-1, 2)$
32. 550 mph
33. 36 lb of hard candy, 12 lb of soft candy

34. 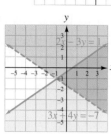 35. y^9 36. $\dfrac{b^6}{27a^3}$ 37. 2
38. $\dfrac{x^{10}z^4}{9y^6}$ 39. 2,600,000
40. 7.3×10^{-4}

41.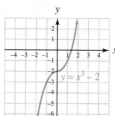

42. $6x^3 + 4x$
43. $6x^3 + 8x^2 + 3x - 72$
44. $-6a^5$
45. $6b^2 + 5b - 4$
46. $6x^3 - 7x^2y + y^3$
47. $4x^2 + 20xy + 25y^2$
48. $81m^4 - 1$
49. $2a - \dfrac{3}{2}b + \dfrac{1}{2a}$
50. $2x + 1$ 51. $6a(a - 2a^2b + 6b)$ 52. $(x + y)(2 + a)$
53. $(x + 2)(x - 8)$ 54. $3y^3(5y - 2)(2y + 5)$
55. $(t^2 + 4)(t + 2)(t - 2)$ 56. $(b + 5)(b^2 - 5b + 25)$
57. $0, -\dfrac{8}{3}$ 58. $\dfrac{2}{3}, -\dfrac{1}{5}$ 59. 5 in. 60. -8 61. $\dfrac{3(x+3)}{x+6}$
62. -1 63. $\dfrac{x-3}{x-5}$ 64. $\dfrac{1}{s-5}$ 65. $\dfrac{x^2 + 4x + 1}{x^2 y}$
66. $\dfrac{x^2 + 5x}{x^2 - 4}$ 67. $\dfrac{9m^2}{2}$ 68. $\dfrac{9y + 5}{4 - y}$ 69. 3 70. 1
71. $a = \dfrac{b}{b-1}$ 72. $2\dfrac{6}{11}$ days 73. 875 days 74. 39
75. 9 76. 1.2 revolutions per second 77. $10x$
78. $-3b\sqrt{2b}$ 79. $9\sqrt{6}$ 80. $-1 - 2\sqrt{2}$ 81. $\dfrac{4\sqrt{10}}{5}$
82. $\dfrac{3\sqrt{2} + \sqrt{2a}}{9 - a}$ 83. 4 84. $-2, -1$ 85. $\dfrac{3m}{2n^2}$
86. 2 87. 125 88. $\dfrac{1}{16}$ 89. $\pm 5\sqrt{3}$ 90. $\dfrac{-5 \pm 6\sqrt{2}}{6}$
91. 21.2 in. 92. $-2, -6$ 93. $\dfrac{1 \pm \sqrt{33}}{8}$; $-0.59, 0.84$
94. 83 ft × 155 ft

95.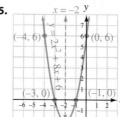

96. 4,000 rpm 97. $7i$
98. $3i\sqrt{6}$ 99. $1 + 5i$
100. $16 - 2i$ 101. $6 - 17i$
102. $1 - i$ 103. $0 \pm 4i$
104. $2 \pm i$

Study Set Appendix I (page A-4)

1. 8 3. 9 5. 12 7. 28, 27, 27 9. -2 yd
11. 74.5 centimicrons 13. 84 15. 5 lb, 4 lb
17. a. mean: 61.8 mpg; median: 47.5 mpg; bimodal: 44 mpg and 43 mpg b. mean: 57 mpg; median: 46 mpg; trimodal: 62 mpg, 44 mpg, 40 mpg

INDEX

Absolute value, 35, 109
Absolute value function, 298, 313
Absolute value symbols, 74, 77
Addition
   associative property of, 45, 46, 110
   commutative property of, 45, 110
   identity property of, 46, 47
   inverse property of, 47
   of monomials, 422–423
   of polynomials, 422–430, 424, 458
   properties of, 45–47, 110
   solving systems of equations with, 343–345, 349, 377–378
Addition property
   of equality, 121, 198
   of inequality, 186
   of opposites (inverse property of addition), 47
   of zero (identity property of addition), 46, 47
Additive inverse, 47
Algebraic expressions, 5, 82–96, 113–114
   definition of, 106
   evaluating, 88–91, 113
   simplifying, 97
Angles, 104
Area
   definition of, 153, 201
   formulas for, 153
Arithmetic mean or average, 77, 112, A-1
Arrow diagrams, 293
Associative property
   of addition, 45, 46, 110
   of multiplication, 61, 111
Average, arithmetic, 77, 112, A-1
Axis of symmetry, 746, 772

Bar graphs, 2, 106
Base
   in exponential expressions, 386–387, 455
   exponents and, 69, 112
Bimodal distributions, A-3
Binomials
   definition of, 416
   difference of two squares, 494–502
   multiplying, 432, 459
   multiplying (FOIL method), 433, 459
   squaring, 688–689
Boundary lines, 282
Braces { }, 29
Brackets [ ], 74, 75, 186, 204
Break point, 356
Business formulas, 149–151

Calculators. *See also* Graphing calculators
   approximating irrational numbers with, 33
   entering numbers in scientific notation on, 412
   evaluating algebraic expressions with, 90
   finding powers on, 72
   fractional exponents on, 698
   order of operations on, 76
   scientific notation on, 409
   sign change key, 44
   solving proportions with, 603

Cartesian coordinate system. *See* Rectangular coordinate system (Cartesian)
Circle graphs, 141
Circumference, 153
Coefficients
   definition of, 82, 113
   examples, 415
   implied, 83
   leading, 417, 528, 529
Commission problems, 143–145
Common denominators, 19, 20. *See also* Least or lowest common denominator (LCD)
Common factors, 19. *See also* Greatest common factor (GCF)
Commutative property
   of addition, 45, 110
   of multiplication, 60, 61, 111
Complementary angles, 104
Completing the square, 727–734, 770
   solving quadratic equations by, 739, 771
   solving quadratic equations in x by, 770
Complex conjugates, 762, 775
Complex fractions. *See* Fractions, complex
Complex numbers, 759–767, 775–776
   adding, 761, 775
   definition of, 775
   dividing, 762–764
   imaginary part, 760, 761, 775
   multiplying, 762–764, 775
   quotient of, 776
   real part, 760, 761, 775
   subtracting, 761, 775
Complex rational expressions, 575
Composite numbers, 13, 107
Compound inequalities, 190–191
Conjugates
   complex, 762, 775
   multiplying with, 709
Consecutive integers, 166, 518, 532
Consistent systems, 322, 327, 375
Constant(s), 5, 415
Constant of variation, 613, 614, 615, 631
Constant terms
   definition of, 83, 415
   in expressions, 101
Contradictions, 136, 137, 199
Converting temperature, 151
Converting units of measurement, 555–556
Coordinate plane, 212
Cube root function, 652–653, 706
Cube roots
   definition of, 650, 705
   irrational, 651–652
   notation, 650
   simplifying, 664–665, 671–672
Cubes
   difference of two, 502, 504, 531
   integer, 652
   perfect, 650, 664
   sum of two, 502, 504, 531

Decimal equivalents, 32
Decimals
   changing to percents, 140
   clearing equations of, 134–136, 199
   nonrepeating, 33

nonterminating, 33
repeating, 109
terminal, 109
Degrees
   of a monomial, 417, 457
   of a polynomial, 417, 457
Denominators
   common, 19, 20. *See also* Least or lowest common denominator (LCD)
   conjugate of, 709
   cube root, 680
   definition of, 14
   opposites, 626
   rationalizing, 680, 709
   square root, 680
Dependent equations
   definition of, 327, 375
   identifying by addition (elimination), 348–349
   identifying by graphing, 326–327
   identifying by substitution, 339–340
Dependent variables, 225, 306
Descartes, René, 212
Descending powers, 528
Difference
   definition of, 4, 25, 52, 106
   of two cubes, 504, 531
   of two squares
      definition of, 435, 497
      factoring, 497–499, 530
Direct variation, 612, 613, 631
Discount, 143
Distance formula, 690, 710
Distributive property
   definition of, 97, 98
   extended, 100
   simplifying algebraic expressions with, 97–101, 114
Division
   by 1, 543
   properties of, 64–66
   of real numbers, 58–69, 63, 111–112
   simplifying complex fractions by, 576–578
   with zero, 64, 65, 111
Division ladders, 14
Division property
   of equality, 126, 198
   of inequality, 188
Domain
   definition of, 291, 294, 313
   of a function, 294, 313
   of a relation, 291, 313
Double-sign notation, 720, 769
Dry mixture problems, 177–178

Elements, 29
Elimination (addition) method. *See* Addition, solving systems of equations with
Empty set, 136
Equal sign, 4, 6
Equality
   addition property of, 121, 198
   division property of, 126, 198
   multiplication property of, 124, 198
   square root property of, 643
   squaring property of, 686, 709
   subtraction property of, 123–124, 198

Equations, 119–207, 292. *See also* Linear equations; Quadratic equations
   clearing fractions and decimals from, 134–136, 199
   definition of, 5, 106, 119, 197
   dependent, 327, 375
   equivalent, 121, 197
   of horizontal lines, 242, 274
   independent, 322, 327, 375
   linear, 236, 307, 511
   nonlinear, 236
   in one variable, 222
   radical, 685–695, 709–710
   solving, 120, 130–139, 198
      by multiplying polynomials, 437–438
      with properties of equality, 119–130, 197–198
      by simplifying expressions, 132–134
   square root property of, 720
   substitution, 334, 335, 336–337
   systems of
      consistent, 322, 327, 375
      definition of, 321
      inconsistent, 326, 327, 375
      third-degree, 515
   in two variables, 222–236, 306–307
   of vertical lines, 242, 274
Equivalent equations, 121, 197
Equivalent expressions, 96, 563–565, 625
Equivalent fractions, 16–17
Equivalent inequalities, 186
Equivalent rational expressions, 563–565, 625
Evaluation, 69. *See also* Simplification
   of algebraic expressions, 88–91, 113
   of exponential expressions, 69–72
   of expressions with or without grouping symbols, 73–77
   of rational expressions, 540, 623
Exponent rules, 394, 399–401, 401–403, 455, 456, 696
   applying, 698–700
   for dividing monomials by monomials, 442–443
   negative integer exponent rule, 398–399
   for variable exponents, 403
   zero exponent rule, 397–398
Exponential expressions
   dividing, 388–390
   equivalent, 399–401
   evaluating, 69–72
   multiplying, 387–388
   parts of, 69, 112
   raising to a power, 390–392
   simplifying, 401–403
Exponents, 69–82, 112–113, 385–462
   of 0, 402
   of 1, 394, 402
   definition of, 69, 112, 455
   in descending order, 460
   fractional, 696, 698
   identifying, 386–387
   natural-number, 386–387, 455
   negative, 399, 402, 456
      changing to positive, 400
      in fractions, 402
      and reciprocals, 401

I-1

Exponents (*continued*)
  negative integer exponent rule, 398–399
  power rule for, 391, 394, 402
  product rule for, 387, 394, 402
  quotient rule for, 389, 394, 402
  rational, 696–702, 711
  variable, 403
  zero exponent rule, 397–398
Expressions. *See also* Algebraic expressions; Exponential expressions
  equivalent, 96
  with grouping symbols, 74–77
  with no grouping symbols, 73–74
  rational, 539–633
  simplifying, 132–134
  variable, 654, 660–662
Extended distributive property, 100
Extraneous solutions, 589, 710
Extremes (in a proportion), 602, 630

Factor trees, 14
Factoring, 465–535
  definition of, 13, 466
  by grouping, 470
  solving quadratic equations by, 512–515, 532, 739, 771
  solving third-degree equations by, 515
Factors
  common, 19
  definition of, 13
  greatest common factor (GCF), 527–528
    definition of, 466, 527
    factoring out, 468–470
    finding, 466–468, 527
FOIL method, 433, 459
Formulas, 149–162, 200–202
  for area, 153
  from business, 149–151
  for circle, 153
  for converting temperature, 151
  definition of, 7, 106, 149, 200
  for distance, 690, 710
  for distance traveled, 151, 173
  equivalent forms of, 445
  from geometry, 152–155
  for perimeter, 152
  for profit, 149
  for retail price, 149
  from science, 151–152
  for simple interest, 150, 172
  for special products, 435
  for specified variables, 155–157, 201, 589–590
  for volume, 155
Fraction bar, 4, 74, 76
Fractions, 13–28, 107–108
  adding, 20–23, 108
  building, 17, 108, 563
  clearing equations of, 134–136, 199
  complex
    definition of, 575, 627
    simplifying, 575–584, 627
  decimal equivalents of, 32
  definition of, 14
  dividing, 16, 107
  equivalent, 16
  form of 1, 14
  fundamental property of, 17, 20, 541, 623
  improper, 25
  in lowest terms, 18, 108
  multiplication property of 1, 16
  multiplying, 15, 107
  negative exponents in, 402

simplest form, 18, 23–24, 108, 442, 443
simplified form, 541
simplifying, 18–20, 23–24, 108, 442–443
special forms, 14–15
subtracting, 20–23, 108, 443, 559
Function notation, 294–296
Functions, 291–303, 313–314
  absolute value, 298, 313
  definition of, 293, 313
  domain of, 294, 313
  graphing, 296–298
  identifying, 292–294
  identity, 298, 313
  linear
    definition of, 297
    graph of, 298, 313
  range of, 294, 313
  square root, 642, 704
  squaring, 298, 313
  vertical line test of, 298–299, 313
Fundamental property
  of fractions, 17, 20, 541, 623
  of proportions, 602

GCF. *See* Greatest common factor
Generation time, 407
Geometric figures, 519–520
Geometric formulas, 152–155
Graphing
  definition of, 30, 305
  functions, 296–298
  identifying inconsistent systems and dependent equations by, 326–327
  inequalities, 185–186, 280–291, 311–312
  linear equations, 238, 307
    by intercept method, 240–241
    by plotting points, 225–229, 238–240
    with slope and *y*-intercept, 265
  in two variables, 306
  lines, 241–242, 256–257
  on number lines, 30–31
  paired data, 215
  points, 213–214
  procedure, 322
  quadratic equations, 750–754, 772–774
  with rectangular coordinate system, 212–222, 305–306
  solution sets, 184–186
  solving systems of equations by, 374–376
  solving systems of inequalities by, 367–368
  solving systems of linear equations by, 322–325, 349, 380
  solving systems of linear inequalities by, 364–367
Graphing calculators, 642
  graphing equations with, 228–229
  solving systems of equations with, 325
Graphs and tables
  bar graphs, 2, 106
  circle graphs, 141
  finding slope from, 251–252
  line graphs, 2–3, 106, 216
  of linear equations, 212, 243–244, 292
  of quadratic equations, 747
  rectangular coordinate graphs, 212
  step graphs, 217–218
  tables, 106
  two-column tables, 2

Greatest common factor (GCF), 466–474, 527–528
  definition of, 466, 527
  factoring out, 468–470
  finding, 466–468, 527
Grouping
  factoring by, 470
  factoring polynomials by, 527
  factoring trinomials by, 490–492
  factoring trinomials with leading coefficient of 1 by, 481, 528
  factoring trinomials with leading coefficient other than 1 by, 491, 529
Grouping symbols, 45, 74–77

Half-plane, 281
Higher-order roots, 653–654, 705–706
Horizontal axis, 2
Horizontal lines
  equations of, 242, 274
  identifying and graphing, 241–242
  slope of, 255–256, 308
Hypotenuse, 520, 642

Identity, 136–137, 199
Identity function, 298, 313
Identity property of addition, 46, 47
Imaginary numbers
  complex numbers as, 760–761
  definition of, 642, 704, 759, 775
  square roots of negative numbers in terms of, 759–760
Implied coefficients, 83
Improper fractions, 25
Inconsistent systems, 327
  definition of, 326, 375
  identifying by addition (elimination), 348–349
  identifying by graphing, 326–327
  identifying by substitution, 339–340
Independent equations, 322, 327, 375
Independent variables, 225, 306
Index of a radical, 650, 653, 706
Inequality, 119–207, 183
  addition property of, 186
  compound, 190–191
  definition of, 183, 204, 280
  division property of, 188
  equivalent, 186
  graphing, 185
  multiplication property of, 187, 188
  ordered-pair solutions, 281
  solutions of, 184, 204
  solving, 183–197, 204–205
  subtraction property of, 186
Inequality symbols, 31, 184
Input value, 224, 306
Integers, 29–30, 34
  consecutive, 166, 518, 532
  definition of, 29, 109
Intercept, 747–748, 773
Intercept method, 240–241
Interest, 150, 172
Interval notation, 185, 204
Inverse property
  of addition, 47
  of multiplication, 62
Inverse variation, 615, 631
Investment problems, 172–173, 595–596, 629
Irrational numbers, 32–33, 34
  approximating, 33, 640–642
  cube roots, 651–652
  definition of, 33, 109, 641
  square roots, 640–642

Key number, 480, 490

LCD. *See* Least or lowest common denominator
Lead terms, 417
Leading coefficients
  definition of, 417
  positive, 528
Least or lowest common denominator (LCD)
  definition of, 21, 108
  finding
    with prime factorization, 22, 108
    in rational expressions, 562, 625
  multiplying by, 578
Like radicals
  combining, 669, 708
  definition of, 669, 707
Like terms
  combining, 101–103, 114, 423, 668
  definition of, 101, 114, 422
  identifying, 101
Line graphs
  algebraic use of, 2–3, 106
  reading, 216
Line of symmetry, 746, 772
Linear equations, 211–316
  definition of, 236, 307, 511
  dependent, 327
  with different variables, 230–231
  graphing, 236–248, 238, 307–308
    with graphing calculators, 228–229
    by intercept method, 240–241
    by plotting points, 225–229, 238–240
    with slope and *y*-intercept, 265
  in one variable, 130, 135
  ordered-pair solutions of, 223, 237–238
  point-slope form of, 272, 311
  slope-intercept form of, 263
  standard or general form of, 236, 307
  strategy for solving, 198
  systems of
    solving, 349, 380
    solving by addition (elimination), 343–345, 349
    solving by graphing, 322–325, 349
    solving by substitution, 334–336, 349
    solving without graphing, 328
  in *x* and *y*, 236
Linear functions
  definition of, 297
  graph of, 298, 313
Linear inequalities
  graphing, 280–291, 311–312
  in one variable, 184
  ordered-pair solutions, 311
  solving, 186–190
  systems of
    with no solutions, 367–368
    solving, 364–374
    solving application problems that involve, 369
  in two variables
    examples, 281
    graphing, 280–291
    solving application problems that involve, 285–286
Lines
  boundary, 282
  graphing, 256–257
  horizontal
    equations of, 242, 274
    identifying and graphing, 241–242
    slope of, 255–256, 308

nonvertical, 251, 253, 308
parallel, 310
  definition of, 257
  identifying, 257–258
  recognizing, 265–266
  slope of, 257
perpendicular, 310
  definition of, 257
  recognizing, 265–266
  slope of, 257
slope of, 249–263, 308–309
of symmetry, 746, 772
vertical
  equations of, 242, 274
  identifying and graphing, 241–242
  slope of, 255–256, 308
Liquid mixture problems, 176–177
Long division symbol, 4
Lowest common denominator. *See* Least or lowest common denominator (LCD)

Mapping diagrams, 293
Markup, retail, 149
Mean
  arithmetic, 77, 112, A-1
  in proportions, 602, 630
Minuend, 425
Mixed numbers, 24–25
Mixture problems, 176–178, 359–360
Models
  algebraic, 2
  linear equations, 266–267, 274–275
  verbal, 5, 165
Monomials
  adding, 422–423
  definition of, 416
  degree of, 416, 417, 457
  dividing, 459
  dividing by monomials, 442–443
  dividing polynomials by, 443–445, 459–460
  multiplying, 431
  multiplying by polynomials, 431–432, 458
  of sixth degree or degree 6, 416
  subtracting, 423–424
Motion problems. *See* Uniform motion problems
Multiplication
  associative property of, 61, 111
  of binomials, 432–434, 459
  commutative property of, 60, 61, 111
  of complex numbers, 762–764, 775
  eliminating variables with, 345–348
  FOIL method for, 433, 459
  of fractions, 15, 107
  inverse property of, 62
  by LCD, 578
  with mixed numbers, 24–25
  of monomials, 431–432
  of numbers with different (unlike) signs, 59
  of numbers with same (like) signs, 60
  of polynomials, 431–432, 436–437, 458–459
    FOIL method for, 433, 459
  to solve equations, 437–438
  properties of, 60–62
  of radical expressions, 675–676, 708–709
  of radical expressions with more than one term, 677–678
  of rational expressions, 550–553, 624
  vertical form, 436

Multiplication property
  of 0, 62
  of 1, 16, 62, 107
  of equality, 124, 198
  of inequality, 187, 188
Multiplicative inverses, 62

Natural numbers
  definition of, 13, 29, 34, 109
  exponents, 386–387, 455
  perfect squares, 659
Negative exponents, 399, 456
  changing to positive, 400
  in fractions, 402
Negative numbers
  definition of, 30
  even and odd powers of, 71
  multiplication by, 572
  real numbers, 41, 42
  square roots of, 759–760
Negative reciprocals, 257, 456
Negative slope, 255, 309
Nonlinear equations, 236
Nonvertical line slope, 251, 253, 308
Notation
  of algebra, 4–5
  cube root, 650
  double-sign, 720, 769
  function, 294–296
  interval, 184–186, 185, 204
  overbar, 32
  scientific, 407–415, 456–457
  set-builder, 32, 185
  square root, 639
  standard, 410, 456
  subscript, 252
nth root of a, 653, 706
Null set, 136
Number lines, 30
  graphing inequalities on, 185
  graphing sets of numbers on, 30–31
  intervals on, 185
Number problems, 593
Number-value problems, 178–179
Numbers. *See also* Complex numbers; Irrational numbers; Negative numbers; Rational numbers; Real numbers
  composite, 13, 107
  graphing, 30–31
  imaginary
    definition of, 642, 704, 759, 775
    square roots of negative numbers in terms of, 759–760
  irrational, 32–33, 34, 109, 640–642, 651–652
  mixed, 24–25
  natural
    definition of, 13, 29, 34, 109
    exponents, 386–387, 455
    perfect squares, 659
  negative, definition of, 30
  opposites, 35, 47, 109
  positive, 30, 41
  prime, 13, 107
  rational, 31–32, 34
  in scientific notation, 408–409
    entering on calculators, 412
  signed, 41
    dividing, 62–64
    multiplying, 58–60
  that cause rational expressions to be undefined, 540–541
  whole, 29, 30, 34, 109
Numerators, 14
Numerical coefficients, 415

Odd powers, 71
Odd roots, 654, 706

One (1)
  division by, 543
  exponent of 1, 394, 402
  fraction form of, 14
  leading coefficients of
    quadratic equations with, 728–729
    trinomials with, 474–485, 528
  multiplication property of, 16, 62, 107
Operations
  hidden, 87–88, 113
  order of, 69–82, 112
    on calculators, 76
    rules for, 72–73, 97
Opposite numbers, 35, 47, 109
Opposites of opposites, 51
Order of operations, 69–82, 112
  on calculators, 76
  rules for, 72–73, 97
Ordered pairs
  definition of, 212, 213
  plotting, 213–214
  as solutions of inequalities, 311
  as solutions of linear equations
    completing, 223, 237–238
    definition of, 281, 306
  as solutions of linear equations in two variables, 222–223
  as solutions of systems of equations, 320–321
Origin (zero point), 41, 212, 305
Output value, 224
Overbar notation, 32

Paired data, 215
Parabolas
  definition of, 226, 746, 772
  graph of, 772–773
Parallel lines
  definition of, 257, 310
  identifying, 257–258
  slope of, 257
Parentheses ( ), 4, 45, 74, 75, 76, 185, 204
Percent
  changing decimals to, 140
  changing to decimals, 140
  of decrease, 142–143, 199–200
  definition of, 140
  of increase, 142–143, 199–200
Percent problems, 140–149
  solving, 140–141, 141–142, 199
  types of, 140
Percent sentences, 141
Perfect cubes, 650, 664
Perfect integer cube, 503
Perfect squares, 704
  natural-number, 659
  square roots of, 638–639
Perfect-square trinomials. *See* Trinomials, perfect-square
Perimeter, 152, 201
Perpendicular lines
  definition of, 257, 310
  identifying, 257–258
  slope of, 257
Pi (π), 33
Pie charts, 141
Plotting, 213–214, 773
Point-slope form, 272, 311
Points
  graphing or plotting, 213–214, 773
  test, 283
Polynomial functions, 418
Polynomials, 415–422, 457–458
  adding, 422–430
    method for, 458
    process of, 424

  adding multiples of, 426–427
  definition of, 415, 457
  degree of, 417, 457
  dividing by monomials, 443–445, 459–460
  dividing by polynomials, 448–450, 460
  evaluating, 418–419
  factoring
    definition of, 466, 527
    by grouping, 527
    steps for, 506
    strategy for, 506–509
  four-term, 415
  with missing terms, 451, 460
  multiplying, 436–437, 458–459
    FOIL method for, 433, 459
    by monomials, 431–432, 458
    by polynomials, 436, 459
    rational expressions by, 552–553
    to solve equations, 437–438
  in one variable, 415
  opposites, 545, 571
  prime, 479, 528
  random, 506–509, 531
  second-degree, 511
  subtracting, 425–426, 458
  subtracting multiples of, 426–427
  in two variables, 415
  vocabulary for, 415–417
Positive infinity symbol (number line), 185
Positive leading coefficients, 528
Positive numbers, 30
Positive slope, 255, 309
Positive square roots, 638, 639, 704
Powers, 69, A-6
  of a, 69
  on calculators, 72
  descending, 450–451, 528
  even, 71
  odd, 71
  of products, 392–394, 402, 455
  of quotients, 392–394, 402, 455
  raising exponential expressions to, 390–392
  rule for exponents, 391, 394, 402, 455
  of square roots, 676–677
Prime factorization
  definition of, 13
  finding LCD with, 22
  simplifying square roots with, 660
Prime numbers, 13, 107
Prime polynomials, 479, 528
Principal (or positive) square roots, 638, 639, 704
Problem solving, 119–207
  with Pythagorean theorem, 642–645
  with rational equations, 593, 628–629
  six-step strategy for, 163, 202, 203, 352, 378, 532
  with systems of equations, 352–364, 378–379
  with two unknowns, 352–355
Product rule
  for cube roots, 664, 675, 707, 708
  for exponents, 387, 394, 402, 455
  simplifying square roots with, 658–659
  for square roots, 658, 675, 706, 708
Products
  definition of, 4, 106
  simplifying, 96–97
  of square roots, 658
Profit, 149

Properties
  of addition
    associative, 45, 46, 110
    commutative, 45, 110
    of equality, 121, 198
    of inequality, 186
    of opposites (inverse property of addition), 47
    of zero (identity property of addition), 46, 47
  distributive property
    definition of, 97, 98
    extended, 100
    simplifying algebraic expressions with, 97–101, 114
  of division, 64–66
    of equality, 126, 198
    of inequality, 188
  of equality, 119–130, 130–132, 197–198
  fundamental
    of fractions, 17, 20, 541, 623
    of proportions, 602
  of multiplication, 60–62
    of 0, 62
    of 1, 16, 62, 107
    of equality, 124, 198
    of inequality, 187, 188
    of real numbers, 96–106
  of subtraction
    of equality, 123, 198
    of inequality, 186
Proportions, 599–612
  definition of, 599, 601, 630
  fundamental property of, 602
  solving, 601–608, 630
Pythagorean theorem, 520, 533, 642
  definition of, 521, 643, 704
  solving problems with, 520–521, 642–645

Quadrants, 212, 305
Quadratic equations, 465–535, 719–778
  applications of, 517–525, 532–533
  with complex-number solutions, 764–765, 776
  definition of, 511, 532
  of form $(ax + b)^2 = c$, 722–724
  of form $x^2 = c$, 720–722
  of form $y = ax^2 + bx + c$, 750–754
  graphing, 746–759, 772–774
  with imaginary-number solutions, 776
  with leading coefficients of 1, 728–729
  with no real-number solutions, 738
  with real-number solutions, 773
  solving, 720–726, 727–734, 769–770, 771–772
    by completing the square, 728–729, 731, 739, 771
    by factoring, 512–515, 532, 739, 771
    by graphing, 750–754
    methods for, 739, 771
    with quadratic formula, 735–738, 739, 771
    with square root property, 720–722, 722–724, 739, 771
    strategy for, 739, 771
    tips for success, 730
    with zero-factor property, 512, 739
  standard form, 511
  in two variables, 746, 772
  in x, 731

Quadratic formula, 735–746, 771
  caution for writing, 736
  definition of, 736
  general, 735
  solving quadratic equations with, 735–738, 739, 771
Quotient rule, 455
  for exponents, 389, 394, 402
  simplifying square roots with, 662–664
  for square roots, 663, 707
Quotients
  definition of, 4, 106
  of opposites, 546
  of square roots, 662
  of two complex numbers, 776

Radical equations, 685
  solving, 685–695, 709–710
  solving by squaring a binomial, 688–689
  solving problems modeled by, 691–692
Radical expressions
  adding, 668–674, 707–708
  definition of, 639
  dividing, 679, 708–709
  with more than one term, 677–678
  multiplying, 675–676, 677–678, 708–709
  simplifying, 657–668, 706–707
  subtracting, 668–674, 707–708
Radical symbol, 638, 704
Radicals, 650
  definition of, 639
  index of, 650, 653, 706
  like
    combining, 708
    definition of, 669, 707
  unlike, 669
Radicands, 639, 650, 704
Raised dot, 4
Range
  of a function, 294, 313
  of a relation, 291, 313
Rates
  of change, 249–250, 308–309
  of commission, 144
  definition of, 249, 600, 630
  of discount, 143
  in simplest form, 599–600
Rational equations, 539–633
  definition of, 584
  problem solving with, 593, 628–629
  solving, 584–589, 628
    for specified variables in formulas, 589–590
    strategy for, 584
    tips for success, 588
Rational exponents, 696–702, 711
Rational expressions, 539–633
  building, 564
  building into equivalent expressions, 563–565, 625
  complex, 575
  definition of, 540, 622
  dividing, 553–555, 624
  equivalent, 563–565, 625
  evaluating, 540, 623
  least common denominator (LCD) of, 562, 625
  multiplying, 550–552, 624
  multiplying by polynomials, 552–553
  with opposite denominators, 571–573
  with opposite factors, 545–547

  reciprocal of, 624
  with same (like) denominators, 559–561, 625
  simplifying, 542, 622–623
  undefined, 540–541, 623
  with unlike denominators, 568–571, 626
Rational numbers, 31–32, 34
  definition of, 31, 109
  set of, 32
Rationalizing denominators, 679–682, 709
Ratios
  definition of, 249, 600, 630
  examples, 599
  in simplest form, 599–600
Reading line graphs, 216
Reading step graphs, 217–218
Real numbers, 29–40
  absolute value of, 35–36
  adding, 41–51, 110
  defining the set of, 34–35
  definition of, 29, 34–35, 109
  with different (unlike) signs
    adding, 43–45, 110
    dividing, 63, 111
    multiplying, 59, 111
  dividing, 58–69, 111–112
  multiplying, 58–69, 111–112
  negative, 41
  positive, 41
  properties of, 96–106, 114–115
  with same (like) signs
    adding, 41–43, 110
    dividing, 62–63, 111
    multiplying, 60, 111
  solutions for quadratic equations, 773
  subtracting, 51–58, 110–111
    definition of, 52
    rule for, 53
Reciprocals
  definition of, 15, 62, 107
  negative (or opposite), 257, 401, 456
  of rational expressions, 624
Rectangles
  area of, 153
  perimeter of, 152, 519–520
Rectangular coordinate systems (Cartesian)
  constructing, 212 213
  definition of, 212, 305
  graphing with, 212–222, 305–306
  x-coordinate, 213
  y-coordinate, 213
Relations
  definition of, 291, 313
  domain of, 291, 313
  range of, 291, 313
Repeating decimals, 109
Retail price, 149
Revenue, 149
Right triangles
  definition of, 520, 521
  legs of, 642
Rise, of a graphed line, 251
Roots, 650, 651–652, 664, 675, 707–708, A-6
Run, 251

Science formulas, 151–152
Scientific calculators, 642
Scientific notation, 407–415, 456–457
  on calculators, 409
  computations with, 410–411
  conversion to standard notation, 410
  definition of, 408, 456

  entering numbers in, 412
  writing numbers in, 408–409
Second-degree polynomials, 511
Set-builder notation, 32, 185
Sets
  definition of, 29
  elements of, 29
  empty or null, 136
  graphing solution sets, 184–186
  of integers, 29–30
  of rational numbers, 32
  subsets, 29
Shared-work problems, 593–594
Signed numbers, 41
  dividing, 62–64
  multiplying, 58–60
Similar terms, 101. See also Like terms
Similar triangles, 599–612, 630
  definition of, 607
  property of, 607
  solving problems that involve, 607–608
Simplification. See also Evaluation
  of algebraic expressions, 96–106, 114–115
    with distributive property, 97–101
    to solve equations, 132–134
  of complex fractions, 575–584, 627
    definition of, 576
    by division, 576–578
    with LCD, 578–581
    tips for success, 577, 579, 581
  of cube roots, 664–665
  of cube roots in sum or difference, 671–672
  definition of, 659
  of fractions, 18–20, 23–24, 108
  of products, 96–97
  of radical expressions, 657–668, 706–707
  of rational expressions, 541, 542, 622–623
  of rational expressions that have factors that are opposites, 545–547
  of roots of variable expressions, 654
  of square roots
    with prime factorization, 660
    with product rule, 658–659
    with quotient rule, 662–664
  of square roots in sum or difference, 670–671
  of square roots of variable expressions, 660–662
Six-step strategy for problem solving
  steps of, 163, 202, 203, 352, 532
  with systems of equations, 378
Slope, 249–263, 308–309
  finding
    given two points, 252–254
    from graph, 251–252
    with slope-intercept form, 263–264
  graphing linear equations with slope and y-intercept, 265
  graphing lines with, 256–257
  of horizontal lines, 255–256
  negative, 255, 309
  of nonvertical lines, 251, 253, 308
  of parallel lines, 257
  of perpendicular lines, 257
  point-slope form, 271–280
  positive, 255, 309
  undefined, 309
  of vertical lines, 255–256
  zero, 309

Slope-intercept form
  definition of, 263, 310
  identifying slope and y-intercept with, 263–264
  writing linear equations in, 265, 266–267
Solution sets
  definition of, 120, 184
  graphing, 184–186
Solutions
  definition of, 120, 197
  extraneous, 589, 710
  of inequalities, 184
  for systems of equations, 375
Special product formulas, 434–435
Specified variables, 155–157
Square root function, 642, 704
Square root notation, 639
Square root property, 720–726, 769–770
  definition of, 769
  of equality, 643
  of equations, 720
  solving quadratic equations with, 720–722, 722–724, 739, 771
Square roots, 638–649, 704–705
  adding, 668–669
  definition of, 638, 704
  denominators, 680
  equations containing, 686–687, 689, 710
  irrational, 640–642
  negative, 638, 639
  of negative numbers, 759–760
  of perfect squares, 638–639
  positive or principal, 638, 639, 704
  powers of, 676–677
  product of, 658
  product rule for, 658, 675, 706, 708
  quotient rule for, 663, 707
  quotients of, 662
  simplified form of, 658, 707
  simplifying
    with prime factorization, 660
    with product rule, 658–659
    with quotient rule, 662–664
    in sum or difference, 670–671
  subtracting, 668–669
    in sum or difference, 670–671
  of variable expressions, 660–662
Squares
  completing, 727–734, 770, 771
  of a difference, 434, 497
  difference of two squares, 494–502, 530
    definition of, 497
    factoring, 497–499
  integer, 641
  perfect, 704
    natural-number, 659
    square roots of, 638–639
Squaring function, 298, 313
Squaring property of equality, 685–686, 709
Standard notation, 410
Statistics, A-1–A-5
Step graphs, 217–218
Subscript notation, 252
Subsets, 29
Substitution, 7
  identifying inconsistent systems and dependent equations by, 339–340
  solving systems of equations by, 334–343, 349, 376–377
  for variables, 88

Substitution equations, 334, 335, 336–337
Subtraction
  of complex numbers, 761, 775
  definition of, 51–54
  of fractions, 20–23
  of fractions with different (unlike) denominators, 23, 108
  of fractions with same (like) denominators, 20, 108, 443, 559
  of monomials, 423–424
  of multiples of polynomials, 426–427
  of polynomials, 422–430, 425–426, 458
  of radical expressions, 668–674, 707–708
  of rational expressions with opposite denominators, 571–573
  of rational expressions with same (like) denominators, 559–561, 625
  of rational expressions with unlike denominators, 568–571, 626
  of real numbers, 51–58, 110–111
  solving application problems with, 54–55
  of square roots, 668–669
  of two cubes, 502
  vertical form, 425
Subtraction property
  of equality, 123, 198
  of inequality, 186
Subtrahend, 425
Sum and a difference, 435
Sum of two cubes, factoring, 502, 504, 531
Sums, 4, 106
Supplementary angles, 104
Symbols
  absolute value, 74, 77
  division, 4
  grouping, 45, 74–77
  inequality, 31, 184
  infinity, 185, 186
  minus, 4, 51
  multiplication, 4
  plus symbol, 4
  radical, 638, 704
Symmetry
  axis or line of, 746, 772
  in parabolas, 773
Systems of equations
  consistent, 322, 327, 375
  definition of, 321, 374
  inconsistent, 327, 375
    definition of, 326, 375
    identifying, 326–327, 339–340, 348–349
  ordered-pair solutions, 320–321
  problem solving with, 352–364, 378–379
  solving
    by addition (elimination), 343–345, 349, 377–378
    checking your solution, 324
    by graphing, 320–333, 322–325, 349, 374–376, 375, 380
    with graphing calculator, 325
    selecting most efficient method for, 349
    by substitution, 334–336, 349, 376–377
Systems of inequalities, 319–384
  application problems that involve, 369

  with no solutions, 367–368
  solving, 364–374
Tables, 106, 215, 292. See also Graphs and tables
  of data, 7
  of function values, 297
  of solutions, 224, 306
  of values, 224
Tax rates, 142
Terminal decimals, 109
Terms, 113, 415
  coefficient of, 82–83
  constant, 83, 415
  examples, 415
  identifying, 82–83
  lead, 417
  like or similar, 101, 114, 422
    combining, 101–103, 423
    identifying, 101
  unlike, 422
Test point, 283
Third-degree equations, 515
Translation, direct, 140
Trial-and-check method
  factoring trinomials by, 486–489
  factoring trinomials with leading coefficient other than 1 by, 529
Triangles
  area of, 153
  legs of, 520
  perimeter of, 152
  right, 520, 521
  similar, 599–612, 630
    definition of, 607
    property of, 607
    solving problems that involve, 607–608
  slope triangle, 251
Trinomials
  definition of, 416
  factoring, 474–485, 485–494
    after factoring out GCF, 490
    by grouping, 490–492
    tips for success, 475, 488
    by trial-and-check method, 486–489
  factoring completely, 479, 528
  with leading coefficient of 1, 474–485, 528
  with leading coefficient other than 1, 485–494, 529
  perfect-square, 723
    completing the square for, 727–728
    examples, 495
    factoring, 496–497, 530
    recognizing, 495–496
  prime, 479
  as products of two binomials, 489

Undefined rational expressions, 540–541, 623
Undefined slope, 309
Uniform motion problems, 173–175, 358–359, 594–595, 629
Unit conversion factors, 555, 624
Unknowns. See Variables

Variable exponents, 403
Variable expressions, 654, 660–662
Variables, 86
  definition of, 5, 106, 119
  dependent, 225, 306
  eliminating, with multiplication, 345–348
  equations in one variable, 222

  equations in two variables, 222–236, 306–307
  equations in x and y, 225
  equations that use different variables, 230–231
  independent, 225, 306
  isolated, 122
  isolating, 130
  linear inequalities in one variable, 184
  linear inequalities in two variables, 211–316
  radicands that contain, 650–657, 705–706
  specific, 201
  specified, 155–157, 589–590
  substituting for, 88
Variation, 612–621, 631
  constant of, 613, 614, 615, 631
  direct, 612, 613, 631
  inverse, 615, 631
Variation problems, 614, 631
Verbal models, 5, 6, 165
Vertex, 748–749, 772, 773
Vertical axis, 2
Vertical form addition, 424
Vertical form multiplication, 436
Vertical form subtraction, 425
Vertical line test, 298–299, 313
Vertical lines
  equations of, 242, 274
  identifying and graphing, 241–242
  slope of, 255–256, 308
Vocabulary
  basic, 4–5
  to describe parabolas, 746
  language of algebra, 2–12, 106–107
  for polynomials, 415–417
Volume
  definition of, 154, 201
  formulas for, 155

Whole numbers, 29, 30, 34, 109

x-axis, 212, 305
  change in x, 253
  y as function of x, 293
x-coordinate, 213
x-intercept
  definition of, 240
  finding, 307, 773

y as function of x, 293
y-axis, 212, 305
  change in y, 253
y-coordinate, 213
y-intercept
  definition of, 240
  finding, 307, 773
  graphing linear equations with slope and y-intercept, 265
  identifying, 263–264

Zero (0), 15, 41
  addition property of, 46, 47
  denominator of, 540–541
  division with, 64, 65, 111
  exponent of 0, 402
  multiplication property of, 62
Zero exponent rule, 397–398
Zero-factor property
  definition of, 512, 532
  solving quadratic equations with, 512, 739
Zero point (origin), 41
Zero slope, 309